Neutrons, Nuclei and Matter

An Exploration of the Physics of Slow Neutrons

James Byrne

Dover Publications, Inc.
Mineola, New York

Bibliographical Note

This Dover edition, first published in 2011, is an unabridged, corrected republication of the work originally published in 1994 by the Institute of Physics Publishing, Philadelphia. The author has provided a new Preface for this Dover edition.

Library of Congress Cataloging-in-Publication Data

Byrne, J.
Neutrons, nuclei, and matter: an exploration of the physics of slow neutrons/ James Byrne.—Dover ed.
p. cm.
Originally published: Bristol; Philadelphia: Institute of Physics Pub., ©1994.
Includes bibliographical references and index.
ISBN-13: 978-0-486-48238-5 (pbk.)
ISBN-10: 0-486-48238-3 (pbk.)
1. Slow neutrons. 2. Neutrons. I. Title.

qc793.5.s62b9 2011
539.7'213—dc23

2011017574

Manufactured in the United States by Courier Corporation
48238301
www.doverpublications.com

To Astrid

Contents

Preface to the Dover Edition

Almost alone among simple physical systems, the neutron participates in all four fundamental interactions, strong, electromagnetic, weak and gravitational, and, since the free neutron lives on average for about fifteen minutes, this makes it an ideal candidate for the study of both scattering and decay processes. From an experimental viewpoint, the neutron possesses certain additional advantages. In the first place neutrons are available in vast numbers with fluxes up to $10^{10}\,cm^{-2}\,sec^{-1}$. Secondly they come in a wide range of energy from thermal (~ 0.025 eV), through cold (< 5. 10^{-5} eV) to ultra-cold (<2. 10^{-7} eV). Ultra-cold neutrons can be stored in bottles made from materials such as quartz which have positive scattering lengths, thereby generating a Fermi pseudo-potential which is repulsive to the neutron. Thirdly, neutrons, although uncharged, have magnetic moments allowing easy manipulation of the spin and the production of close to 100% polarization in both longitudinal and transverse modes.

There is an additional feature in the case of the parity-violating weak interaction responsible for neutron β-decay, in that the neutron-proton mass difference is so small that the momentum transfer dependence of all form factors may be neglected. In the Standard Model maximal parity violation is a built-in feature in that the weak currents which participate in the interactions, are *assumed* to be exclusively left-handed. Why the weak interaction is so special in this regard is unknown and suggests a question which has been attributed to Pauli: *Why does the Lord still appear to be left-right symmetric when He expresses himself strongly?*

The late Valentine Telegdi once observed that all experiments should be feasible, exciting and interpretable, although most rarely enjoyed more than two of these three properties, frequently failing the interpretability test. However, following electro-weak unification and the genesis of the Standard Model, many experiments involving the weak and electromagnetic interactions of the neutron lead to results with clear interpretations for particle physics at the fundamental level. This explains why weak interaction phenomena in particular have been subjected to increasingly precise and detailed experimental investigation over the seventeen years which have passed since the appearance of the first edition this book.

The electromagnetic interaction enters this scheme of things directly in several places. One example is the Foldy term identified in the scattering of neutrons by electrons bound in spinless atoms which implies that the mean square charge radius of the neutron is negative. A second example is the phenomenon of weak magnetism, which is a direct consequence of unification, but which has so far resisted detection in neutron decay, since it contributes to the hadronic matrix element of the vector current in recoil order only. However of far greater significance are the radiative corrections which influence all aspects of neutron β-decay at the level of about one per cent, and whose detailed study has advanced in unison with the experimental work.

Following the discovery of parity violation in 1956 the radiative corrections were studied in the context of local (V-A) theory which, however, showed up several difficulties. In the first place these corrections appeared to go in the wrong direction, a weakness which was only removed with the identification of the Cabibbo angle θ_c leading to an understanding as to why ΔS=1 weak decays were suppressed relative to ΔS=0 decays. The second problem arose from the fact that in this theory the radiative corrections were logarithmically divergent requiring the introduction of an energy cut-off Λ conventionally identified with the nucleon mass. Since the spontaneously broken local gauge theory which underpins the Standard Model is renormalizable, this problem has now been resolved with the result that, at least to a high degree of approximation, the conclusions of local (V-A) theory continue to be valid when Λ is identified with the mass of the Z-boson.

A precise and reliable measure of the neutron lifetime continues in its traditional role as one of the two major challenges facing experimentalists, the other being the search for the neutron's electric dipole moment *d*, whose presence would signal the simultaneous violation of both parity and time – reversal invariance in the same part of the weak Hamiltonian. The search for an electric dipole moment in the neutron using the Ramsey separated oscillatory coil resonance technique applied to ultra-cold neutrons stored in a bottle has continued almost without interruption since the mid seventies setting an upper limit $d < 10^{-26}$ e cm which falls well above the Standard Model predication $d < 10^{-31}$ e cm. Thus these experiments are motivated by the search for new physics, e.g. supersymmetry, beyond the Standard Model. Several new studies are currently in progress based on the use of a superthermal source of ultra-cold neutrons produced by down-scattering of 8.9A° cold neutrons in superfluid helium, which has the potential to increase the number of stored neutrons by several orders of magnitude. However measurements of the neutron lifetime using the same technique have turned up some unexpected problems and further progress may not prove easy.

Although the neutron lifetime is itself of direct importance in big-bang cosmology, since it determines the rate at which hydrogen is converted into helium in the early Universe, currently the emphasis has switched to its significance in relation to the question of the unitarity of the Cabibbo-Kobayashi-Maskawa quark mixing matrix and obtaining a precise value for the magnitude $|V_{ud}|$ of its largest element. Over the half-century which has elapsed since the pioneering work of Snell, Robson and Spivak, two quite distinct techniques with different systematic errors have evolved. These are (a) bottle methods which rely on observing the decay rate of ultra-cold neutrons trapped in material or magnetic bottles and (b) Penning trap methods which measure the number of decay protons stored in a given time in a variable length of cold neutron beam of measured neutron density. Up to quite recently both techniques converged to a value of $\tau_n = 885.8 \pm 0.9$ sec. However, with the recent appearence of new data which diverge from the standard value the situation becomes fluid once again.

In addition to the main three-body decay mode of the neutron, two weaker decay branches also exist: (a) the inner bremsstrahlung process which involves the simultaneous emission of a γ-ray in company with the leptons and has a branching ratio of about 10^{-3} and (b) two-body decay into an antineutrino and a hydrogen atom which has a branching ratio of about 4.10^{-6}. The radiative process has recently been observed at NIST and found to be in good agreement with theory while the two-body decay branch is currently the subject of an extremely difficult and exciting project at TUM Garching. By recording separately the relative populations of the decoupled hyperfine levels of the $2S_{1/2}$ state in hydrogen, it may prove possible to isolate the contribution of a weak right-handed current, which is forbidden in the Standard Model

Of course the neutron lifetime alone does not suffice to determine a value for $| V_{ud} |$ and it is necessary also to determine the magnitude and sign of the axial vector/ polar vector coupling constant ratio λ. In the past this has been derived from an extensive series of measurements of the electron momentum-neutron spin correlation coefficient *A*, carried out by the Heidelberg group at the ILL. However a new series of projects has been initiated to measure the parity-conserving electron-antineutrino angular correlation coefficient *a* which does not require that the neutrons be polarized and which shares with the coefficient *A* the property that it determines the anomaly $\lambda|-1|$ rather than $| \lambda |$ itself. Concerning the question of the time reversal invariance of the weak interaction, this can be tested through observation of the parity conserving *D*-coefficient of the triple correlation between the neutron spin and electron and antineutrino momenta. This has recently been measured independently by groups at the ILL and at NIST which agree in the conclusion $| D | < 10^{-3}$.

There remains the question of the weak nucleon-nucleon interaction, an area in which the Standard Model has less predictive power and one must therefore have recourse to an effective theory of light meson exchange involving the π, ρ and ω mesons. This theory is characterised by five amplitudes of which the iso-vector f_π and the iso-scalar h^0_ρ have been measured, the former by the observation of the parity-violating neutron spin-γ-ray angular correlation in the process p (n, γ) ^{2}H, and the latter through the observation of the weak spin rotation of polarized neutrons in helium which is sensitive to both amplitudes.

Coming now to the gravitational interaction, we know that, like all massive bodies, neutrons undergo free fall in a gravitational field, and the effect was observed long ago. This is, however, a purely classical phenomenon, and the influence of gravity in quantum systems is not that easy to demonstrate. However a highly versatile approach to the problem was initiated in the Collela-Overhauser-Werner series of gravitationally induced quantum interference experiments, in which a monochromatic beam of neutrons enters a perfect crystal interferometer which is tilted in such a way that the emergent beams traverse regions of differing gravitational potential. These are then made to interfere thereby recording a phase shift proportional to both wavelength and the sine of the angle of tilt, in a formula which depends explicitly on Planck's constant *h* and the acceleration *g* due to gravity

However the earliest observation of a quantum phenomenon in which gravity plays a central role goes back to Koester's neutron gravity refractometer which measures nuclear scattering lengths by determining the critical heights reached by neutrons reflected from the surface of a liquid mirror. More recently it has been confirmed that the gravitational potential can be viewed just as any other potential when inserted into the Schrodinger equation, through the observation of quantum states in the earth's gravitational field formed when ultra-cold neutrons fall on a horizontal mirror. These states are separated in energy by a few peV and by about 10μm in the vertical, with a probability density in the nth quantum state proportional to $|\Psi_n|^2$.

It is not possible to do more than register a brief comment on progress in the field of neutron scattering in matter This is an independent discipline in its own right, having come to maturity in 1994 when C.G. Shull and B.N. Brockhouse were jointly awarded the Nobel prize in Physics; the former for his pioneering work on neutron diffraction in single crystals, applicable to the study of elastic scattering, and the latter for his invention of the triple axis spectrometer which incorporates a second diffracting crystal to record the energies of the scattered neutron. Along with time of flight spectroscopy this has proved by far the most successful method for the study of inelastic processes in physics, chemistry, biology and materials science. There is also in intermediate range of instrument, which includes the spin echo spectrometer, matched to the study of quasi-elastic processes, which require the highest energy resolution.

The most interesting innovations during the past decade relate to the development of a range of hybrid instruments which can operate in both monochromatic and time-of- flight modes. Specific examples are the reflectometer D17 at the ILL which is particularly suitable for the study of solid and solid/liquid interfaces, and the instrument Vivaldi which is a neutron-sensitive image-plate Laue diffractometer ideal for the study of pressure- and temperature-dependent phase transitions. The time of flight option has also been combined with the spin echo principle in the spectrometer IN15 at the ILL, which has the advantage that the energy resolution is independent of the degree of monochromaticity of the incident neutron beam.

To conclude this Preface to the Dover edition of my book, I should like to thank those of my colleagues with whom I have been privileged to work at various stages over the past decade. These include Ferenc Gluck, Nathal Severijns and Oliver Zimmer in Europe and Tim Chupp, Scott Dewey, Jeff Nico, Fred Wietfeldt and Boris Yerozolimsky in the United States. A special note of thanks goes to my former student David Worcester from the University of Missouri who first introduced me to Dover Books. Finally there is a long-delayed recognition of the supportive role played at several critical points in my professional life by Sir Denys Wilkinson FRS, formerly Chairman of the Nuclear Physics Board at the Science Research Council in the United Kingdom, and subsequently Vice Chancellor of the University of Sussex.

James Byrne
July, 2011

Preface

The late George Bernard Shaw was famous for his extended prefaces in which he contrived simultaneously to delight and infuriate his readers by parading his views on all manner of topics, not necessarily central to the theme of the drama which followed. However, Shaw was concerned, not with natural science, but with art and society; it was not a Shavian hero who was given to musing on the nature of moon and stars, and on the roles of atoms and molecules in determining human destiny. Dramatists, whose business it is to mould and manipulate words, can expect a greater degree of indulgence from their audiences than can scientists, whose colleagues quite properly expect them to write succinctly and to the point. Nevertheless I hope I may be forgiven if, in the course of introducing one more book on slow neutron physics, I do not confine my remarks entirely to matters of a technical nature. In particular I wish to emphasize my view that the benefits which 60 years of neutron research have brought to applied science cannot themselves provide the main justification for expanding and refining that research. The proper objective of all science is the description of nature and, in an era when science tends to be viewed as little more than a technical resource of industrial society, this is a truth which cannot be restated too often.

In planning the contents of this book it has been my intention, from the beginning, to fashion a text aimed principally at those experimental physicists whose interests, like my own, are directed towards the interactions and structure of the neutron itself, rather than to its function as a tool in nuclear and condensed matter research. While I naturally hope that potential readers in these disciplines may find some new insights within its pages, it is not to this audience that the book is primarily addressed. Indeed it is my firm conviction that those who wish to pursue fundamental studies on the neutron and its interactions can hope to advance their subject only by fully exploiting those techniques which have hitherto been developed in an applied context. In arriving at this view I have been strongly influenced by my exposure to many aspects of applied neutron research at the Institut Laue Langevin in Grenoble, where I made many extended visits between 1975 and 1990.

At Grenoble two main lines of research have been pursued. These are, first, the investigation of nuclear processes, particularly nuclear fission, initiated by slow neutron capture, and second, the study of the structure and dynamics of condensed materials, physical, chemical and biological, by elastic and inelastic neutron scattering. These activities have been supported by an instrumentation base, constructed to the highest possible technological standards, which has generated quite astonishing amounts of superb scientific data. However, the directorate of the Institute has also provided substantial support for experimental work on the strong, electromagnetic, weak and gravitational interactions of the neutron, and for novel technologies with potential applications in these fields. This has proved a wise policy, resulting in advances in our understanding of the neutron quite unmatched anywhere else in the world. The reason is to be found, I believe, in the enormous range of ideas and skills generated in a continuously floating population of nuclear and condensed matter specialists, each one of them anxious to exert an influence in the wider scientific world outside their own particular areas of expertise.

My first objective, therefore, was to collect together as much information about the neutron and its interactions as could be conveniently assembled in one place, or at least as much as could be discussed at a fairly elementary level. My second objective was to condense this knowledge into a comprehensive review of the role of the neutron in the physical world as a whole. This role encompasses a vast range of physical phenomena involving consideration of the ultimate constitution of matter, the structures of nuclei and materials, and the influence of the neutron in shaping the astrophysical environment we see about us today. In retrospect I fear that this programme may have proved somewhat over-ambitious as evidenced by the rather uneven levels of treatment of the various topics which make up its subject matter.

To take an example: although more than 10% of the identifiable hadronic matter in the Universe is composed from neutrons, the free neutron is itself unstable, living, on average, for about a quarter of an hour. Thus neutrons exist in finite numbers only because most of them pass uneventful lives locked safely away inside nuclei. One cannot therefore hope to understand the neutron without first gaining a detailed understanding of nuclei. In contrast, the interactions of neutrons with condensed materials, made up from many nuclei, are quite easily characterized by the relevant scattering lengths and magnetic parameters. Thus, in any investigation of the neutron itself, it is inevitable that the nuclear physics element be analysed in greater depth than the condensed matter element. By the same reasoning a third level of sophistication must be scaled to make any progress in understanding the neutron as a composite structure whose constituents are quarks and gluons.

In order to smooth out these difficulties of presentation I have therefore adopted a scheme in which each well-defined division of the subject matter is introduced at a fairly elementary level, and is subsequently developed in more and more advanced stages. The intention is that the reader may break off at any point and move on to a new topic without any serious disadvantage. Within this programme chapters 1 and 2 are devoted to the elementary properties of the neutron viewed as a quantum mechanical particle, whose free and bound states are defined by appropriate solutions of the Schrödinger–Pauli wave equation. These introductory chapters are then followed by three further chapters which describe the interactions of neutrons and nuclei from various viewpoints. Thus chapter 3 is primarily concerned with the nature of the nucleon–nucleon force, chapter 4 with nuclear structure, and chapter 5 with the application of neutron technologies in the development of nuclear energy sources. There are difficult areas in this treatment which many readers might wish to avoid, for example the use of group theory in the treatment of certain nuclear models in chapter 4, and the application of complex variable techniques to obtain the solutions of the Boltzmann transport equation which are described in chapter 5. These topics could well be omitted without hindering understanding of later material.

With the completion of the relevant nuclear physics content in chapter 5 the half-way stage has been reached, and at this point there is a sharp change in direction. Thus the next three chapters are devoted entirely to phenomena associated with the propagation of beams of slow neutrons in condensed matter systems, in which the nuclei are characterized only by their mechanical, nuclear and electromagnetic properties. Nevertheless the approach to these phenomena parallels to a large degree the previous discussion of the nuclear dimension. For example, chapter 6 develops the theory of neutron propagation in homogeneous media in the 'neutron optics' approximation, while chapter 7 describes in some detail the various wave phenomena which are encountered with neutrons. It would not be a wholly inaccurate description of the subject matter of these chapters to summarize their content as the 'geometrical' and the 'physical' optics, respectively, of the neutron. This trio of chapters also ends on an applied note since chapter 8 is concerned, throughout, with practical applications. In keeping with the general aim of laying the emphasis on the behaviour of the neutron itself, rather than on the structure of the bulk material with which it interacts, such solid state physics as is introduced is kept to a minimum, and treated at a fairly elementary level.

In the final two chapters of the book I return to the discussion of those matters which were introduced in the opening two chapters, namely the properties and interactions of the neutron, but this time from a much more advanced viewpoint, and taking into account those discoveries of high energy particle physics which in recent years have transformed our understanding of the nature of protons and neutrons, and have revealed many new features of their substructure. With the obvious exception of the high energy phenomena demonstrated by electron scattering experiments, most of the information on the neutron has been derived from very precise experiments carried out on slow neutron beams, and this work is described in detail in chapter 9. How the results of these and other experiments are interpreted in terms of the Standard Model of particle physics is discussed in chapter 10, which also includes a description of several current experiments on slow neutrons whose results could ultimately lead to the discovery of new physics beyond the Standard Model. The book concludes with a summary of the various ways in which the interactions of the neutron, established by terrestrial experiments, influence very directly the observed behaviour of astrophysical systems such as the sun, and even more exotic objects such as pulsars. Indeed they provide a very simple link between the Universe we observe in our immediate neighbourhood, and that other Universe which modern cosmological theories tell us emerged from the big bang some fifteen thousand million years ago.

Physicists of my generation who received their first introduction to nuclear physics from printed notes of Fermi's lectures first delivered in 1949, will recall the sense of excitement and expectation which that little volume stimulated in its readers. With this example before me, I have made a determined effort to go, wherever possible, to the original source of a theoretical idea or an experimental technique. Not least of the advantages which such a policy reveals, is the recognition that many of the early papers in a subject are often easy to read and understand. This contrasts with the difficulty, frequently encountered by research students, of coming to grips with a new field, whose current research papers tend to assume, in the reader, a familiarity with the background material which is often absent. A second advantage is that it enhances appreciation of the work of the great thinkers in nuclear physics, of Bethe, Weisskopf and Wigner, and above all of Niels Bohr who has dominated nuclear theory thoughout much of this century, in the same degree that nuclear experiment was driven by Rutherford and his associates in the pre-war era, and by Fermi and his associates in its aftermath.

Fermi also founded the new science of neutron optics, although in this context there is a long history of relevant investigation, preceding the discovery of the neutron itself, and stretching backwards in time through the work of Rayleigh and Hamilton to Huygens, who thought and wrote more than three centuries ago. To anyone dipping into his works for the first time, Huygens comes as a revelation: for here is a true modern whose views place him closer in spirit to the scientific New Testament of Heisenberg and Schrödinger, rather than to the Old Testament of Newton and his followers. The flavour of Huygens' writing may be judged from the following extract from his *Traité de la Lumière* (1690)

> There will be seen in it demonstrations of those kinds which do not produce as great a certitude as those of Geometry, and which even differ much therefrom, since whereas the Geometers prove their Propositions by fixed and incontestable Principles, here the Principles are verified by the conclusions to be drawn from them; the nature of these things not allowing of this being done otherwise. It is always possible to attain thereby to a degree of probability which very often is scarcely less than complete proof. To wit, when things which have been demonstrated by the Principles that have been assumed correspond perfectly to the phenomena which experiment has brought under observation; especially when there are a great number of them, and further, principally, when

> one can imagine and foresee new phenomena which ought to follow from the hypotheses which one employs, and when one finds that therein the fact corresponds to our prevision.

It is difficult to formulate a clearer or more concise summary of the scientific method whose spectacular successes have created the modern world in which we live.

According to the romantic and visionary poet Shelley, whose bicentenary we have recently celebrated, it is the poet who 'beholds the future in the present and his thoughts are the germ of the flower and the fruit of latest time'. While this may have been true in Shelley's century it certainly has not been true in ours; at least up to a few decades ago it was the scientist rather than the poet who 'comprised and united the characters of legislator and prophet'. Unfortunately this is no longer the case in our society, where the standing of the scientist appears to be in a state of irreversible decline. The nuclear scientist, in particular, has become something of a social outcast who, like the hero of Paradist Lost, has 'witnessed huge affliction and dismay'.

Of course art and science have always existed in a state of tension and mutual mistrust. In Hazlitt's view 'the progress of knowledge and refinement has a tendency to circumscribe the limits of the imagination and to clip the wings of poetry'. Lamb and Keats were united in their opinion of Newton as a 'fellow who believed nothing unless it was as clear as three sides of a triangle . . . who had destroyed the poetry of the rainbow by reducing it to the prismatic colours'. What would they have had to say about Berry's 'falling neutron rainbow' just supposing that, transported to this century, either would have thought this natural phenomenon of the nuclear world worthy of attention? Shelley himself, we may be sure, would have appreciated that modern science has revealed many beautiful phenomena, and might have written as enthusiastically about the 'solar wind', as about the 'west wind'. Indeed in his poem *The Cloud*:

> I pass through the pores of the ocean and shores, I change but I cannot die

he comes as close to a statement of the law of baryon conservation as was possible given the state of natural knowledge in his day. At least he would have understood what many of our legislators and educators appear to have forgotten: that science is concerned first and foremost with revealing the secrets of nature, and scientists have more in common with artists than they have with accountants, lawyers or politicians.

To conclude this Preface I must thank those of my colleagues who have been kind enough to read various chapters of my book in manuscript form, and who have enlightened me with their criticisms and suggestions. In this respect my particular thanks go to Peter Dawber, Mike Pendlebury, Ken Smith and Brian Smith from the Physics and Astronomy Division here at

Sussex, to Roger Scott from the Scottish Universities Research and Reactor Centre at the University of Glasgow and to Keith Green in the Particle Physics Department at the Rutherford–Appleton Laboratory. Looking further afield there are a great many others to whom I am indebted in various ways, and of these I must mention at least a few: Peter Farago, who introduced me to the field of electron spin polarization, resulting in a fruitful collaboration extending over many years; John Robson for the help, support and encouragement I and my close colleagues Peter Dawber and Roger Scott received from him in our efforts to carry forward the work on the beta-decay of the neutron which he himself initiated more than 40 years ago, and Roger Tayler for stimulating my interest in the part played by the neutron in astrophysics and cosmology, an area of physics in which he has himself carried out much original and distinguished research. Finally I wish to acknowledge my debt to Norman Ramsey, and to thank him for the enthusiasm, encouragement and generosity of spirit he always displays when faced with any scientific endeavour, great or small, and for other kindnesses too numerous to mention.

The manuscript was typed here at the Physics and Astronomy Division by Sally Church, who saw it through countless restructurings, revisions and changes of direction, and to her I offer my sincerest thanks; also to Pauline Cherry and Janet Robertson in the Design Office of the School of Engineering and Applied Sciences who, between them, drew the illustrations.

J Byrne

Foreword

During the sixty years since its discovery, man's knowledge of the properties and interactions of the neutron has progressed at an ever increasing pace and has produced all kinds of applications of neutron-related technology in science, medicine, engineering and industrial activity. The discovery and understanding of neutron induced fission led to the development of the controlled nuclear reactor which, despite controversies regarding its safety and environmental impact, has provided, and will continue to provide, mankind with a reliable and economically competitive supply of electrical power. It also made possible intense sources of neutrons which have enabled the ongoing experimental investigations into the properties of the neutron and its interactions with nuclei and matter to be not only possible but extremely fascinating to those fortunate enough to have been able to take advantage of their availability. The constant and invigorating interplay between experimental data and theoretical implications and predictions has enhanced the progress and vitality of these investigations, and stimulated their use in ever widening areas of human endeavour.

Many people have contributed to the astonishing increase in man's knowledge of nature which has been a consequence of these studies and, with a few exceptions, they have tended to concentrate their efforts and make their contributions in rather restricted and well defined areas. This is not to say that they have been uninterested in the rest, rather that they have not had the time to delve into the ever expanding mass of literature necessary to fully appreciate these developments. Attendance at seminars and review talks has often stimulated a real interest in another area and occasionally a dramatic shift in the line of an individual's research. What has been lacking, however, is a concise, easily approachable summary which presents a basic elementary introduction and a gradual presentation of the fundamental theoretical basis of each subject.

This book provides such a resource. It covers all the major areas of neutron physics and gives the reader a solid overview of their theoretical background as well as the present experimental situation. The extensive references not only enable him/her to delve further into a particular subject but allow him/her to appreciate how it developed. Many of us who worked in this field in its earlier days would have deeply welcomed such a resource. To present day researchers it will be available as an often used and very valuable reference book as well as an entertaining and informative one in which to browse.

While reading it I have been fascinated by the astonishing progress that this book describes, but I have also been aware of the way it subtly indicates the incompleteness of our understanding of many areas of neutron physics. It may well help others to develop them further.

John M Robson
Emeritus Professor of Physics
McGill University
Montréal, Quebec

1 The neutron as an elementary particle

1.1 The nuclear atom

1.1.1 The classical era: Democritus to Mendeleev

The modern atomic theory of matter owes its origins to a view of nature propounded in the fifth century BC, by the Greek philosopher Leucippus and his pupil Democritus, who taught that matter was not infinitely divisible, but was ultimately composed of fundamental entities to which Democritus gave the name atoms. Perhaps because Aristotle took an opposing view, these ideas made little headway over the next two millenia, until they were revived during the time of the Renaissance. As Russell (1946) has observed 'Throughout modern times, practically every advance in science, in logic, or in philosophy has had to be made in the teeth of opposition from Aristotle's disciples'. Galileo, Boyle, Newton and Bernoulli were each atomists in their various ways. To Boyle we owe the credit for recognizing the distinction between chemical mixtures, compounds and elements. That air is a mixture of oxygen and nitrogen was demonstrated by the French chemist Lavoisier in 1774 whereas the chemical composition of water as a compound of oxygen and hydrogen was established in 1781 through the researches of Priestley. However, it was Dalton who, in his *New System of Chemical Philosophy*, published in 1808, developed these ideas on a firm quantitative basis, and from this time onwards the science of chemistry has progressed within a totally atomistic theoretical framework.

Perhaps the most important initiative undertaken by Dalton was his systematic attack on the problem of determining atomic weights, for which he took the weight of the lightest atom, hydrogen, as the basic unit. This was a natural choice but, from a practical viewpoint, somewhat unfortunate, since not many elements form stable compounds with hydrogen. In the modern era this system of units has been replaced, first by a system based on oxygen, and more recently by a scale based on a single isotope of carbon. The importance of distinguishing between atoms and molecules was first emphasized in 1811 by Avogadro, whose simplifying hypothesis that, under standard conditions, equal volumes of gases contain equal numbers of molecules, allowed the determination of atomic weights, or atomic masses as they are more properly called, with some precision. In the hands of Joule, Maxwell and Boltzmann, Avogadro's ideas were fashioned into the kinetic theory of gases, yet almost a century was to elapse before Perrin (1908, 1908a) succeeded in deriving a value for Avogadro's famous number from the results of his studies of the Brownian motion.

Subsequent to the publication of Avogadro's ideas it was noted by Prout (1815) that many atomic weights had integral or near integral values, from which observation be deduced that the hydrogen atom was a fundamental building block of matter to which he attached the name 'protyle'. Since many elements violate Prout's empirical rule, in particular chlorine which has atomic weight 35.46, the protyle hypothesis fell into disfavour although the more perceptive, notably Rayleigh (1882), remarked that the appearance of integral atomic weights was too frequent an occurrence to be accounted for entirely on the basis of chance. However, in more recent times Prout's insights have achieved a somewhat belated confirmation, since the integer he identified in the system of atomic weights is now recognized as the mass number A, representing the number of nucleons in the atomic nucleus of the given chemical element.

In comparing the chemical properties of elements and their compounds certain periodicities become evident, allowing elements to be classified in families having a common property, e.g. valency. These regularities, although pointed out by Chancourtois as early as 1862, and subsequently by Newlands (1864) and Meyer (1869), were first systematized by Mendeleev in his paper *The Periodic Properties of the Chemical Elements* published in 1871, in which he ordered the known elements of similar properties in columns with atomic weight increasing from top to bottom and from left to right. It was observed that the ordinal number Z of an element arranged in this scheme had a value approximately equal to one-half the atomic weight although there were pairs of elements, e.g. cobalt and nickel, whose measured atomic weights seem to place them in the wrong order. One important consequence of the Mendeleev classification was the recognition that some elements were missing, and in this way the elements gallium, scandium and germanium were predicted and subsequently identified.

During the course of an investigation of the solar eclipse observed from India in 1868, the rare gas helium was discovered by P J C Janssen, who recorded a hitherto unknown yellow line in the spectrum of the solar chromosphere. The existence of the new element was confirmed by Ramsey (1895) who detected it in thorium and uranium ores. Eventually it was found necessary to add a whole new column to the periodic table to take account of the rare gases which are present in minute quantities in the earth's atmosphere or generated as products of radioactive decay of heavy elements (Ramsey 1907).

1.1.2 The modern era: electrons and nuclei

The demonstration that the physical atom was not the indivisible and immutable atom of Democritus, but had an internal structure of its own, was inspired, not from chemistry, but from the great discoveries at the turn of the century concerning the nature of X-rays, radioactive decay and the conduction of electricity in gases. The central point in these developments was Thomson's discovery of the electron which he identified during the course of his researches on cathode rays (Thomson 1897). These particles he showed to be carriers of negative electricity of specific charge, i.e. charge to mass ratio, some 1836 times larger than the specific charge on the hydrogen ion in solution. Identical results were obtained with 'photo-electrons' and with 'thermionic electrons' for which the absolute charge *e* was found to be comparable with that on the hydrogen ion. From these results Thomson was forced to the conclusion that the electron, a term introduced in 1874 by Stoney to describe the electronic charge, but henceforth adopted as the name of the particle iself, was approximately 1836 times less massive than the lightest atom and had therefore to be viewed itself as an ultimate constituent of matter.

Since measurements of the specific charges of positive rays in electrical discharges carried out by Wien (1898, 1898a), had shown that these were massive particles, probably ionized atoms or molecules, the conclusion was inevitable that the mass of an atom was intimately associated with positive charge and, on this basis, Thomson advanced the idea that the neutral atom was a stable structure consisting of light negative electrons moving throughout a sphere of positive electricity where the mass of the atom was concentrated (Thomson 1904). To see how Thomson's atomic model was put to the test, a test which it ultimately failed, we must consider very briefly the phenomenon of radioactivity, so called by Marie Curie (1898) following its discovery by Becquerel (1896, 1896a), which in the hands of Rutherford and his associates led directly to the discovery of the atomic nucleus, and ultimately to the modern science of nuclear physics.

Radioactive atoms originally attracted attention because of their photochemical properties and their ability to ionize gases. The most prominent radiations are of three types: (1) γ-rays, which are electromagnetic radiations of the same nature as X-rays but of much shorter wavelength (Villard 1900); (2) β-rays which are identical with the electrons discovered by Thomson in the cathode rays (Becquerel 1900); and (3) α-rays which are doubly ionized atoms of helium (Rutherford and Geiger 1908, Rutherford and Royds 1909). By 1911 some 40 radioactive species had been isolated and yet only 12 free spaces were available in the periodic table in which to accommodate them. This difficulty was resolved by recognizing that there could be more than one radioactive species of a given chemical element, and indeed three radioactive gases were shown to be chemically identical with rare gases previously known only in stable form. Different radioactive species of the same Z were called isotopes by Soddy (1913), who demonstrated that radioactive decay and chemical change were associated in a group displacement law which states that an α-transition decreases Z by two units, a β-transition increases Z by one unit, while a γ-transition leaves Z unchanged.

The discovery of isotopes suggested an immediate explanation for the occurrence of non-integral atomic weights which had so discredited Prout's hypothesis and this was confirmed by Aston (1919, 1920) who showed that, with the significant exception of hydrogen which had an atomic weight of 1.008 on the oxygen scale, all isotopes had integral atomic weights to within an accuracy of 0.1%. However later improvements in Aston's mass spectrograph leading to an accuracy of 0.01%, showed that, at this level, small but significant deviations from the integral mass rule were observed in all isotopes.

The availability of energetic α-particles with which to bombard atoms opened a new avenue of attack on the problem of atomic structure culminating in the discovery by Geiger (1908) and by Geiger and Marsden (1909, 1910), that approximately one α-particle in 10^3 scattered from a heavy gold atom, was deflected through an angle of more than 90°. Since the Thomson model did not permit an accumulation of electric charge sufficient to account for such large deflections, Rutherford (1911) advanced the hypothesis that the positive charge and associated mass of the atom were concentrated in a central nucleus, of radius about 10^{-14} m, with the electrons circulating in orbits of about 10^{-10} m in radius, comparable with the known dimensions of atoms. Similar suggestions had indeed been made by Nagaoka (1904) but without any supporting experimental evidence. Assuming that the charged nucleus exerts an electrostatic force, varying as the inverse square, on the incoming α-particle, Rutherford deduced a law of scattering which, expressed in terms of a differential scattering cross-section, is given by

$$\frac{d\sigma}{d\Omega}=\frac{r_c^2}{16}\operatorname{cosec}^4(\theta/2) \tag{1.1}$$

where θ is the angle of scatter and r_c is the collision diameter

$$r_c=2e^2Q/E_\alpha \tag{1.2}$$

In these formulae E_α is the kinetic energy of the α-particle and Q is the nuclear charge expressed in units of e.

The predictions of Rutherford's scattering law were tested by Geiger and Marsden (1913) and shown to account for the observed angular distribution of scattering for a range of angles between 5° and 150° to an accuracy of about 20%, provided Q was assigned a value equal to approximately half the atomic weight of the scattering atom. At this stage it was pointed out by Van den Broek (1913) that, since the ordinal number Z in the periodic table was also approximately equal to half the atomic weight, as was also the number of electrons per atom derived from Barkla's X-ray scattering experiments (Barkla 1911), interpreted according to a formula of Thomson (1906), the number Z should be understood as the number of positive charges on the atomic nucleus. Such an interpretation is consistent with Soddy's group displacement law (1913) and received further direct support from Chadwick's detailed studies of α-particle scattering from a range of elements (Chadwick 1920).

A particularly important contribution to the process of uncovering the precise significance of Z in atomic theory was made by Moseley (1913, 1914) who used the newly discovered technique of X-ray diffraction to make accurate measurements of the frequencies of characteristic K X-rays emitted from a variety of substances which he showed could be represented by the formula

$$\nu_K = 0.248 \times 10^{16}(Z-1)^2 \text{ s}^{-1} \tag{1.3}$$

Mosely also introduced the nomenclature 'atomic number' for Z which henceforth is identified both as the number of orbital electrons in the atom, and as the number of elementary positive charges on the nucleus.

It is of course well known that Rutherford's model of a planetary atom conflicted totally with established electrodynamic theory and was quite powerless to account for the observed optical spectrum of hydrogen. Thus some radically new ideas were required and these were supplied by Bohr (1913) who combined his notion of the stationary state with the old quantum theory of Planck and Einstein to effect an accommodation between the apparently irreconcilable results of classical optics and the new concept of a nuclear atom. Ultimately the Rutherford–Bohr atom was placed on firm theoretical foundations following the development of the new quantum mechanics over the period 1924–27 by Heisenberg, Born, Schrödinger and Dirac.

1.2 The constituents of nuclei

1.2.1 The proton–electron nuclear model

A decade after his pioneering studies of the scattering of α-particles by heavy atoms Rutherford initiated a new series of experiments on the interaction of α-particles with light atoms, nitrogen in particular, in which he obtained convincing evidence for the actual disintegration of atomic nuclei. These observations, and the very fundamental and wide-ranging conclusions he drew from them, are described in great detail in his famous Bakerian Lecture to the Royal Society (1920). Concerning the experimental findings themselves he writes:

> It is thus clear that some of the nitrogen atoms are disintegrated by their collision with swift α-particles and that swift atoms of positively charged hydrogen are expelled. It is to be inferred that the charged atom of hydrogen is one of the components of which the nucleus of nitrogen is built up .... This is the first time that evidence has been obtained that hydrogen is one of the components of the nitrogen nucleus.

The essential truth of Rutherford's conclusion, that a nuclear transmutation had occurred with the ejection of a hydrogen nucleus, was confirmed by Blackett (1925) who observed the nuclear reaction

$$^{14}N + ^{4}He \rightarrow ^{17}O + ^{1}H \tag{1.4}$$

in a cloud chamber. However, it still remained to account for the existence of isotopes, which could not be explained on the assumption that atomic nuclei were composed solely of hydrogen nuclei. Rutherford's solution to this problem is also expressed clearly:

> In considering the possible constitution of the elements, it is natural to suppose that they are built up ultimately of hydrogen nuclei and electrons. On this view the helium nucleus is composed of four hydrogen nuclei and two negative electrons with a resultant charge of two.

At this time the evidence in favour of the electron–proton nuclear model was compelling, since electron emission from nuclei was well known and positron emission had not yet been observed (Curie and Joliot 1934). As a consequence the model became universally adopted and survived in popular expositions of atomic phenomena long after Chadwick's discovery of the neutron (1932) had forced its abandonment. Indeed, in the most famous passage of his Bakerian lecture Rutherford had used the electron–proton model to predict the existence both of the deuteron, subsquently isolated by Urey and associates (1932), and of the neutron:

> . . . it seems very likely that one electron can bind two H nuclei and possibly also one H nucleus. In the one case, this entails the possible existence of an atom of mass 2 carrying one charge, which is to be regarded as an isotope of hydrogen. In the other case, it involves the idea of the possible existence of an atom of mass 1 which has zero nuclear charge.

It is clear in retrospect that the remarkable properties which Rutherford ascribes to his hypothetical neutral nucleus are precisely those which make the neutron the unique probe of matter which it is:

> Its external field would be practically zero, except very close to the nucleus, and in consequence it should be able to move freely through matter . . . it should readily enter the structure of atoms, and may either unite with the nucleus or be disintegrated by its intense field, resulting possibly in the escape of a charged H atom or an electron or both.

At about the same time similar ideas had occurred both to Masson (1921), and to Harkins (1921), who introduced the name 'neutron' for this hypothetical particle, although Sutherland (1899) had earlier used the same name to describe the neutral particle which we now call positronium, and Pauli (1930) was later to apply it to the particle now known as the neutrino. In a footnote to Masson's paper Rutherford describes how the British Association at its Cardiff meeting (cf. Eddington 1920) accepted his suggestion that, recalling Prout's hypothetical 'protyle', the hydrogen nucleus should be called a 'proton'. Although Rutherford (1919) was the first to detect protons emitted in nuclear processes the word does not occur in the Bakerian lecture but 'protons' and 'neutrons' as descriptions of the constituents of nuclei eventually received unanimous endorsement (Feather 1960). It may be of some historical interest that Masson proposed the name 'baron' for the proton, anticipating by several decades the description 'baryon', which is today applied to protons, neutrons and hyperons (cf. section 9.2.1.1).

Rutherford also drew attention to the role that neutrons might play in the synthesis of heavy nuclei:

> The existence of such atoms seems almost necessary to explain the building up of the nuclei of heavy elements; for unless we suppose the production of charged particles of very high velocities it is difficult to see how many positively charged particles can reach the nucleus of a heavy atom against its intense repulsive field.

This viewpoint was further elaborated by Glasson (1921) who, reporting on his failure to detect neutrons in hydrogen discharges, commented on the significance the neutron might have for astrophysics: 'This building up process is apparently at work in the evolution of stellar systems from the nebular state'.

Subsequent developments in cosmological theory have provided confirmation for these ideas and, according to current views, most of the helium in existence at the present epoch, was formed in the early Universe in a complex of reactions involving neutrons and protons (Peebles 1966).

$$p+n\leftrightarrow {}^2H+\gamma;\ {}^2H+{}^2H\leftrightarrow {}^3He+n\leftrightarrow {}^3H+p$$

$$^3H+{}^2H\leftrightarrow {}^4He+n \qquad (1.5)$$

Writing at about the same time, but from the viewpoint of an astronomer, Eddington (1920) drew attention to the fact that, because the mass of four protons exceeds that of the nucleus ^{4}He by about 0.8%, direct conversion of hydrogen into helium represented a possible solution to the outstanding problem of the sun's energy source.

> But is it possible to admit that such a transformation is occurring? It is difficult to assert, but perhaps more difficult to deny, that this is going on. Sir Ernest Rutherford has recently been breaking down the atoms of oxygen and nitrogen driving out an isotope of helium from them, and what is possible in the Cavendish Laboratory may not be too difficult in the sun. I think that the suspicion has been greatly entertained that the stars are the crucibles in which the lighter atoms which abound in the nebulae are compounded into more complex atoms.

The difficulty was that the Kelvin–Helmholtz theory of solar energy generation, which postulated the conversion of the sun's gravitational energy into heat as it contracted (Kelvin 1861), was unable to explain a solar age $>10^8$ years, a conclusion which, incidentally, caused consternation among the disciples of Charles Darwin whose revolutionary treatise on *The Origin of Species* had appeared a couple of years earlier. Ranged against this result was the geological evidence which, particularly that derived at a somewhat later stage from the radioactive dating of terrestrial rocks, required a solar age $>4\times10^9$ years (Rutherford 1929). The energy source suggested by Eddington did not require the intervention of neutrons but relied on the high temperatures at the centre of the sun to overcome the Coulomb forces between colliding protons. He was able to show that, by this means, a solar life-span of the order of 10^{11} years was possible. An important link in the chain was completed with the demonstration (Atkinson and Houtermans 1929) that quantum mechanical tunnelling through the Coulomb barrier would permit the exothermic fusion of charged particles in stellar interiors at temperatures in excess of 10^6 K (cf. section 5.6). Along with the explanation of the Geiger–Nutall rule governing α-decay as a manifestation of quantum mechanical tunnelling (Geiger and Nutall 1911, Gamow 1928, Gurney and Condon 1928), this was one of the first applications of quantum mechanics in nuclear physics.

We know now of course that Eddington's speculation was essentially correct and that thermonuclear fusion of hydrogen into helium in the proton–proton cycle of reactions is the main power source in the sun. A typical solar proton has a lifetime of about 10^9 years before it undergoes the weak capture reaction

$$\mathrm{p}+\mathrm{p}\rightarrow {}^2\mathrm{H}+e^+ +\nu_e \tag{1.6}$$

Within about 4 s the deuteron ^{2}H is converted into ^{3}He via the electromagnetic interaction

$$\mathrm{p}+{}^2\mathrm{H}\rightarrow {}^3\mathrm{He}+\gamma \tag{1.7}$$

The ^{3}He nucleus lives for about 4×10^2 years before it reacts strongly with another ^{3}He nucleus.

$${}^3\mathrm{He}+{}^3\mathrm{He}\rightarrow {}^4\mathrm{He}+\mathrm{p}+\mathrm{p} \tag{1.8}$$

leading to normal ^{4}He plus two further protons which continue the cycle. The total energy release is about 27 MeV of which electron–positron annihilation contributes about 2 MeV. About 2% of this energy is carried off by the neutrinos, the remainder being converted into radiant heat.

1.2.2 The Pauli neutrino hypothesis

In spite of its successes the proton–electron nuclear model soon encountered great, and eventually insurmountable, problems following the rapid developments in quantum mechanics in the years from 1924 onwards. The most obvious objection arose through the application of the uncertainty relation to an electron confined in a nucleus of linear dimension of order 10^{-12} m. For such an electron the uncertainty in momentum should be of order 20 MeV/c, a prediction which contrasts rather sharply with the observed momenta of β-particles ejected from the nucleus which are typically of order 1 MeV/c. In addition nuclei do not have magnetic moments of electronic magnitude as might have been expected according to this model. However, much more powerful arguments could be deployed against the model based on observed spins and statistics of nuclei derived from the molecular spectra of diatomic homonuclear molecules.

It is observed that alternate lines are reduced in intensity, or in some cases vanish altogether, in some molecular band spectra, and the fact that this happens only when the two nuclei are identical was first appreciated by Mecke (1925). The explanation for the phenomenon was put on a firm quantum mechanical foundation by Hund (1927). The nuclear spin of a diatomic molecule formed from identical nuclei with spin I may be assigned a quantum number K, which can take any value in the range 0, 1, . . ., $2I$. The number of possible spin states is $(2I+1)^2$ of which $(2I+1)(I+1)$ are symmetric, and $(2I+1)I$ antisymmetric, with respect to nuclear exchange. In a diatomic molecule the nuclear spin state is coupled to a rotational state characterized by an integral quantum number J, which is symmetric for even J, and anti-symmetric for odd J. Assuming the electronic wavefunction is symmetric (which it usually is), it follows that even values of J are associated with even (odd) values of K if the nucleus is a boson (fermion) and conversely for odd values of J. If the electronic state is anti-symmetric the rule is reversed. In either case there is an alternation in the statistical weights of successive values of J in the ratio $(I+1)/I$ which in certain circumstances can be reflected in the observed molecular band spectrum.

Because transitions between states of opposite symmetry are forbidden for homonuclear diatomic molecules, in this case there is a separation into an ortho modification with the greater statistical weight and a para modification with lesser statistical weight (Dennison 1927). Because of the very long relaxation times between ortho and para modifications these will not in general be in thermal equilibrium, except at high temperatures where the population ratio will be $(I+1)/I$. Furthermore rotation, or rotation–vibration bands, with dipole selection rule $\Delta J = \pm 1$, do not occur. On the other hand Raman transitions, with selection rule $\Delta J = 0$, ± 2, do occur, since in this case initial and final states have the same symmetry.

These expectations are confirmed for the case of the hydrogen molecule for which the lines in the Raman spectrum characterized by odd (even) values of J are observed to be strong (weak), with an intensity ratio 3/2, thus proving the proton to be an $I=\frac{1}{2}$ fermion (Hund 1927, Rasetti 1929). The same conclusion is required to account for observed specific heat anomalies in hydrogen at low temperatures associated with the ortho–para separation (Dennison 1927). However, in the case of the molecule $^{14}N^{14}N$ the even-valued lines were observed to be the strong lines with a relative intensity 2, showing conclusively that ^{14}N is a boson with $I=1$. The same conclusion was reached in the case of the nucleus ^{6}Li (Rasetti 1930). These results utterly contradicted the predictions of the proton electron model, which required these nuclei to be fermions, since each was thought to be composed of an odd number of fermions (Ehrenfest and Oppenheimer 1931).

An ingenious solution to these difficulties was advanced by Pauli (1930), who linked them to the observation that, not only was energy conservation apparently violated in β-decay as shown by calorimetric measurements (Ellis and Wooster 1927, Meitner and Orthmann 1930) but angular momentum could not be conserved either if the electron was the only spin $\frac{1}{2}$ fermion emitted in the decay process. He therefore proposed that, in addition to protons and electrons, the nucleus contained very light neutral spin $\frac{1}{2}$ fermions which he called 'neutrons'. These are supposed to be ejected along with electrons in the process of nuclear β-decay thereby offering simultaneous solutions to the problems of nuclear spin and statistics, and the apparent violation of energy and angular momentum conservation laws. This solution does not however explain why nuclei do not have magnetic moments comparable with those of the electron.

Although Pauli's use of the word 'neutron' for his hypothetical neutral particle caused a certain amount of confusion in the literature of the period (Wigner 1933), following the Rome conference of 1931 the word 'neutrino' was adopted for this particle at Fermi's suggestion, and in future the word 'neutron' was to be reserved exclusively for the heavy neutral particle discovered by Chadwick in 1932. In 1933 at the Solvay Congress Pauli returned to his explanation for the apparent violation of the mechanical conservation laws in nuclear β-decay and suggested that the hypothetical neutrinos should be spin $\frac{1}{2}$ fermions. In all this he was absolutely correct except that neutrinos do not exist in the nucleus but are created along with the electron at the moment of nuclear disintegration. This is the essence of Fermi's theory of β-decay (1934) which likens the creation of the electron–neutrino pair in β-decay to the creation of a γ-quantum in electromagnetic decay. Pauli's neutrino was first detected through the observation of inverse neutron β-decay (cf. section 1.5.1.1) (Cowan *et al* 1956). With some major modifications to take account of parity violation in weak interactions and the existence of heavy intermediate vector bosons which mediate the decay, the main pillars of the Fermi theory survive intact today.

1.3 Discovery of the neutron

1.3.1 First evidence for the neutron

Nuclear disintegration of nitrogen by α-particles followed by proton emission had been observed first by Rutherford in the reaction $^{14}N(\alpha, p)^{17}O$ (cf. section 1.2), and the detection of fine structure in the proton spectrum from the reaction $^{10}B(\alpha, p)^{13}C$ accompanied by γ-radiation had been reported by Bothe and Fränz (Bothe and Fränz 1928, Fränz 1930, Bothe 1930, Bothe and Becker 1930). Subsequently Bothe and Becker initiated a systematic search for γ-radiation emitted in (α, p) reactions with light nuclei, which indeed they observed in targets of B, Al and Mg (Bothe and Becker 1930). In the cases of Li and Be they detected the presence of an intense highly penetrating radiation, unaccompanied by protons and particularly prominent in the case of Be, which they identified as being electromagnetic in origin. These conclusions were supported in most important respects by the researches of Curie (1931).

Subsequently Webster reported (1932) that the radiation emitted from Be was absorbed by matter to a degree which would imply a quantum energy of about 7 MeV, more than twice as energetic as the most energetic known γ-ray, the 2.62 MeV γ-ray emitted from the nucleus ^{208}Pb in the thorium active deposit. He also made the observation, highly significant in retrospect, that the radiation emitted in the forward direction with respect to the incident α-momentum, appeared to be more energetic than that emitted in the backward direction, and speculated that the observed effects could be caused by 'high speed corpuscles consisting e.g. of a proton and an electron in close combination'. This was a possibility which he considered 'interesting in view of the ... usefulness of the conception of neutrons in accounting for astrophysical and nuclear phenomena'.

Unfortunately the failure to observe any 'neutron' tracks in the cloud chamber led him to abandon this line of enquiry and revert to the electromagnetic explanation for want of anything better. The cloud chamber results were confirmed by Dee (1932) at about the same time that Chadwick was announcing the discovery of the neutron, much to the displeasure of Rutherford and Chadwick, so firmly were they wedded to the view of the neutron as an intimate bound state of proton and electron.

Concurrently with Webster's studies Curie and Joliot (1932) noted that the radiations from beryllium ejected protons from hydrogenic substances such as paraffin. These events they interpreted as Compton scattering of 50 MeV γ-rays by protons, an explanation which, bizarre as it may seem today, was in line with that offered to account for certain cosmic ray events which were ultimately shown to be examples of electron–positron pair production. Almost immediately Perrin (1932) suggested that the mystery rays were neutrons and Majorana observed that Joliot and Curie had found the 'neutral proton' and had not recognized it.

1.3.2 Chadwick's experiment: the proton–neutron nuclear model

At this stage the problem was taken up by Chadwick, who, using the apparatus shown in figure 1.1 with a range of converters, H, Li, Be, B and N in place of the paraffin converter, and a variety of ionization chamber gases H_2, Ne, N_2, O_2 and A, was able to demonstrate conclusively that it was impossible to account for the energy spectra of all the recoil atoms observed, in terms of a Compton process for a fixed incoming γ-ray energy. For example, to explain the detection of a maximum recoil proton energy of 5.7 MeV it was necessary to postulate an incident quantum energy of 54.6 MeV. In similar circumstances a recoiling nitrogen atom could have a maximum energy of 0.45 MeV as compared with the 1.6 MeV actually observed. However, the observations could be readily explained on the assumption that the incident radiation was a neutral particle of mass comparable with that of the proton, i.e. a neutron, and kinetic energy 5.7 MeV. In particular this hypothesis was quite adequate to account for the forward–backward asymmetry in energy detected by Webster. A detailed account of his researches leading up to the discovery of the neutron was given by Chadwick in his Bakerian lecture to the Royal Society (Chadwick 1933).

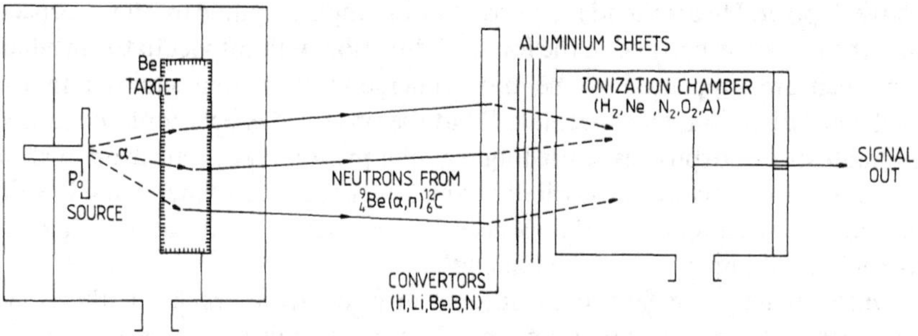

Figure 1.1. Apparatus used by Chadwick (1932) in his discovery of the neutron. Neutrons in the energy range 0–13 MeV from the reaction $^9Be(\alpha, n)^{12}C$ undergo elastic collisions with nuclei of H, Li, Be, B and N in convertor targets, whose energies after transmission through aluminium sheets are measured in an ionization chamber.

We know now that the generation of neutrons in beryllium targets bombarded with α-particles arises from two competing reactions

$$\begin{aligned} \alpha + {}^9Be &\rightarrow {}^{12}C + n \\ &\rightarrow 3\,{}^4He + n \end{aligned} \tag{1.9}$$

but this was not known at the time of Chadwick's discovery nor indeed had the mass of the ^9Be nucleus been measured. To determine the neutron mass Chadwick therefore studied the reaction $^{11}B(\alpha, n)^{14}N$ and deduced that the masses of neutron and proton were equal to within about 0.3%. Final verification of Chadwick's conclusions was provided by Feather (1932) who observed the inverse reaction $^{14}N(n, \alpha)^{11}B$ in the cloud chamber.

The suggestion that the neutron is not a bound state of electron and proton but is itself a spin $\frac{1}{2}$ fermion which is a fundamental constituent of the atomic nucleus was advanced by Heisenberg (1932), and also by Iwanenko (1932), who proposed that the nucleus ^{14}N be considered as a bound state of three α-particles, a proton and a neutron. Heisenberg considered the properties of a nucleus composed entirely of neutrons and protons and, by postulating attractive forces between proton and neutron and neutron and neutron was able to account, in a qualitative sense, for the range of values of the neutron-to-proton ratio observed in stable nuclei. These ideas were further developed by Majorana (1933) who, as noted earlier, had been among the first to suggest that the radiations produced in reactions of α-particles in beryllium were in fact neutrons, and had independently developed a nuclear model based on neutrons and protons alone.

1.4 Neutron capture

1.4.1 Artificial radioactivity

In 1933–4 Joliot and Curie observed that positrons continued to be emitted from targets of B, Mg and Al irradiated with α-particles, even when the α-source had been removed, and deduced that they had observed (α, n) reactions leading to a radioactive product, e.g.

$$^{27}Al + \alpha \rightarrow {}^{30}P + n \tag{1.10}$$

Since ^{31}P is the only naturally occurring isotope of phosphorous, the conclusion was that the neutron-deficient isotope ^{30}P decays by positron emission

$$^{30}P \rightarrow {}^{30}Si + e^{+} + \nu_e \tag{1.11}$$

This was the first observation of artificial radioactivity and ^{30}P the first of at least six radio-isotopes of phosphorous to be produced by artificial means. To confirm that phosphorous was indeed the active product, the irradiated aluminium target was dissolved in acid and small amounts of sodium phosphate and zirconium salts were added to the solution. The positron activity was always observed to be concentrated with the precipitated zirconium phosphate.

Immediately following this discovery, Fermi and his collaborators initiated a study of the effects of neutron bombardment, arguing that the relative weakness of available neutron sources should be compensated by the uncharged neutron's ability to move easily through matter and approach the atomic nucleus without having to encounter a repulsive field, exactly as Rutherford had suggested in his 1920 Bakerian lecture. A systematic investigation followed in the period 1934–35 using a radium–beryllium source of 7–8 MeV neutrons with a range of target elements, and chemical separation with appropriate carriers to concentrate the active products. These experiments revealed that the great majority of elements when bombarded with neutrons produce β^--active isotopes, in many cases with several different half-lives, and that, in general, three types of reaction occur. These are (1) (n, α), (2) (n, p) and (3) (n, γ) reactions, all of which increase the neutron-to-proton ratio in the daughter nucleus as compared with the (α, n) reaction which decreases this ratio. Reactions (1) and (2) are more likely in light nuclei and reaction (3) in heavy nuclei, although in some cases all three reactions occur giving three different activities, e.g.

$$^{27}\mathrm{Al}+\mathrm{n}\rightarrow{}^{24}\mathrm{Ne}+\alpha \quad \text{or} \quad \rightarrow{}^{27}\mathrm{Mg}+\mathrm{p} \quad \text{or} \quad \rightarrow{}^{28}\mathrm{Al}+\gamma \qquad (1.12)$$

All three product nuclei are neutron rich and therefore β^--active; they are identified by adding small amounts of sodium or magnesium carrier to the dissolved target and observing which activities go with which residues. Conversely the radio-isotope ^{24}Na can be produced three ways using targets of ^{27}Al, ^{24}Mg and ^{23}Na.

1.4.2 Neutron capture resonances

One very important result which emerged during the course of this work was that activities generated in various targets could be enhanced by factors of as much as 100 by enveloping the neutron source in paraffin or similar hydrogenous material. This effect was traced to the 'moderating' properties of the paraffin which slowed down the neutrons by elastic collisions with protons, about 20 collisions being required to slow a 1 MeV neutron down to thermal energies, depending on the temperature of the moderator. Evidently the capture cross-section increased at lower neutron energies and it could be shown that, on the assumption that the neutron energy was much less than the separation between neighbouring levels in the 'compound' nucleus formed when the incident neutron enters the target nucleus, the capture cross-section varies in inverse proportion to the relative velocity of neutron and target nucleus. In general it was found that one or more absorption bands were superimposed on this monotone cross-section, usually at neutron energies <100 eV (Moon and Tillman 1935, Szilard 1935, Bjerge and Westcott 1935). This effect was explained by Bohr (1936), and by Breit and Wigner (1936) as being due to resonance with a virtual level of the many-body compound nucleus. The great frequency with which such resonances occur is explained by the fact that, at an excitation energy of about 10 MeV, which corresponds to the binding energy of a neutron in the compound nucleus, the energy level spacings for a medium mass nucleus are on average <1 eV.

1.4.3 Nuclear fission

The observations of neutron capture in uranium targets were anomalous in that, although four β-activities were observed, with half-lives of 10 s, 40 s, 13 min and 90 min, respectively, neither of the two longer lived radio-isotopes was isotopic with any of the chemical elements in the range $86 \leqslant Z \leqslant 92$. It was deduced therefore that these activities were associated with transuranic elements, i.e. $Z > 92$ (Fermi *et al* 1934). This conclusion was quite mistaken as subsequent investigations by Hahn and Strassmann (1938, 1939) were to show. They found that the activity in uranium targets was concentrated by a barium carrier and at first deduced that the active element was radium which, like barium, is an alkaline earth. Further work showed however that the active product concentrated with barium ($Z = 56$) in preference to radium ($Z = 88$), and they were reluctantly driven to the conclusion that the active product was indeed barium. The implications of this finding were realized almost immediately by Meitner and Frisch who advanced the true explanation, namely that the uranium nucleus is only barely stable and, following neutron capture, undergoes fission into two nuclei of approximately equal size (Meitner and Frisch 1939, Frisch 1939). As a matter of historical record similar ideas had been advanced by Noddack (1934), the discoverer of the element rhenium, during a critique of the claim of Fermi's group to have discovered the transuranic elements, but the suggestion appears not to have been taken seriously.

As it turns out there is also a 23 min activity in neutron-activated ^{238}U which Hahn *et al* (1938) considered could be due to a transuranic element. This conclusion was later confirmed by McMillan and Abelson (1940) who identified 23 min and 2.3 day β-active products, non-recoiling and hence not fission products, in thin ^{238}U targets bombarded with neutrons. The events which they detected were correctly interpreted as the sequence of processes

$$\begin{aligned} {}^{238}\mathrm{U} + \mathrm{n} \rightarrow {}^{239}\mathrm{U} + \gamma \rightarrow {}^{239}\mathrm{Np} + e^- + \bar{\nu}_e \ (23.5\ \mathrm{min}) \\ \rightarrow {}^{239}\mathrm{Pu} + e^- + \bar{\nu}_e \ (2.34\ \mathrm{day}) \end{aligned} \tag{1.13}$$

where ^{239}Np and ^{239}Pu are the transuranic elements neptunium and plutonium.

1.5 Mechanical properties of the neutron

1.5.1 Inertial mass

1.5.1.1 Mass spectrometry: exothermic and endothermic reactions

Nuclear masses are determined experimentally by combining mass spectroscopic data with the results of nuclear reaction studies. There are several versions of the mass spectrograph but all of them employ some arrangement of electric and magnetic fields to measure the energy and momentum of charged particles. These are usually ionized atoms rather than bare nuclei and it is customary therefore for nuclear mass tables to list the masses $M(A, Z)$ of neutral atoms of mass number A and atomic number Z. The mass of the atomic nucleus is then derived from $M(A, Z)$ by subtracting the masses of Z electrons and adding on the electronic binding energy. Common exceptions to the rule that the bombarding particles are ionic species rather than bare nuclei are the proton p(^{1}H), the deuteron d(^{2}H), the triton t(^{3}H) and the α-particle (^{4}He).

Up to 1961 two mass scales were in use; a physical scale in which the mass of the neutral atom of ^{16}O was arbitrarily assigned the value 16 atomic mass units (amu), and a chemical scale in which the average mass of natural oxygen (99.76% ^{16}O, 0.037% ^{17}O and 0.20% ^{18}O) was given the value 16. However in that year the International Union of Physics and Chemistry adopted a new scale based on the neutral atom of ^{12}C (98.9% of natural carbon) (Kohman *et al* 1958, 1958a) which is assigned the value of 12 mass units (u), and this scale is now in universal use.

In a nuclear reaction two atoms of masses $M(A_1, Z_1)$ and $M(A_2, Z_2)$ collide to produce one or more atoms of mass $M(A_j, Z_j)$ and the energy released, i.e. the Q-value, is given by

$$Q = \left[M(A_1, Z_1) + M(A_2, Z_2) - \sum_j M(A_j, Z_j) \right] c^2 \tag{1.14}$$

When $Q > 0$ the reaction can take place even when the target particle and the bombarding particle are relatively at rest, and is said to be exothermic. Such a reaction is the ^{11}B(α, n)^{14}N reaction, i.e.

$$\alpha + {}^{11}\mathrm{B} \rightarrow {}^{14}\mathrm{N} + \mathrm{n} \tag{1.15}$$

which Chadwick first used to determine the mass of the neutron, by measuring the Q-value for the reaction and using the known masses of ^{4}He, ^{11}B and ^{14}N. One should note that the Q-value is a relativistic invariant and can also describe reactions in which the incident particle has zero mass, e.g. photonuclear reactions, or where there is no incident particle at all as in a pure decay process.

When $Q < 0$ the reaction is said to be endothermic and in this case there is a minimum or threshold energy for the bombarding particle (M_1) in the rest-frame of the target particle, in order for the reaction to be energetically possible. The relation between Q-value and kinetic energy T_t at threshold is

$$T_t = |Q| \left\{ 1 + \frac{M_1 c^2}{M_2 c^2} + \frac{|Q|}{2 M_2 c^2} \right\} \tag{1.16}$$

and in the non-relativistic region this reduces to

$$|Q| = T_t \frac{M_2}{M_1 + M_2} = T_t \frac{\mu_1}{M_1} \tag{1.17}$$

where μ_1 is the reduced mass of the bombarding particle.

In the description of nuclear reactions the natural unit of energy is the electron volt (eV) and fractions or multiples thereof (keV, MeV, GeV, etc.) and to convert from mass units (u) to MeV we only require a knowledge of the velocity of light c, and the Faraday constant of electrolysis $F=N_A e/c$ where N_A is Avogadro's number. The required relation is

$$1\,\text{u}=10^{-9}(c^2/F)\,\text{MeV}=931.494\,32\pm0.0028\,\text{MeV} \tag{1.18}$$

where F is expressed in traditional units of C/g molecule. To convert mass units into kg we require to know N_A and the conversion formula is (Review of Particle Properties 1992)

$$1\,\text{u}=10^{-3}N_A^{-1}\,\text{kg}=(1.660\,5402\pm0.000\,0010)\times10^{-27}\,\text{kg} \tag{1.19}$$

It should be observed that the nuclear mass as determined by the methods outlined above is the 'inertial mass' since it is this quantity which enters into the equations of motion of ions in the mass spectrograph and into the relativistic energy–momentum relation which underpins the quantitative analysis of nuclear reactions. When it becomes important to distinguish between 'inertial' and 'gravitational' mass, one needs to enquire more deeply into how conversion factors such as F and N_A are determined experimentally, for indeed the methods employed may assume the equality of 'inertial' and 'gravitational' mass which is precisely the question at issue.

1.5.1.2 Neutron β-decay and other routes to the neutron mass

When Chadwick and Goldhaber (1934, 1935) first detected the photo-disintegration of the deuteron

$$\gamma+{}^2\text{H}\rightarrow{}^1\text{H}+\text{n} \tag{1.20}$$

they were able to determine the Q-value ($|Q|=2.23$ MeV) from the energies of the incident γ-rays and the photo-protons. This result allowed them to make an accurate estimate of the neutron mass, which turned out to be substantially greater than the mass of the hydrogen atom. They were therefore led to conclude that the neutron would be β-active

$$\text{n}\rightarrow\text{p}+e^-+\bar{\nu}_e \tag{1.21}$$

with a Q-value of about 0.78 MeV, a prediction which was verified in detail some considerable time later (Snell and Miller 1948, Snell *et al* 1950, Robson 1950, 1950a). The lifetime of the free neutron is now known to have a value slightly less than 15 min (cf. section 9.3.7) but, prior to the successful determination of the Q-value, the existing best evidence pointed to a lifetime of about $3\frac{1}{2}$ days (Bethe 1935, Konopinski and Uhlenbeck 1935, Motz and Schwinger 1935).

The spectrum of electrons from β-decay is normally measured using a magnetic spectrometer and provides a measure of the end-point energy E_0, which is the maximum relativistic energy of electrons emitted in the decay.

The kinetic energy end point T_0 is defined by

$$T_0 = E_0 - m_e c^2 \tag{1.22}$$

and this is just the Q-value for the decay (assuming the neutrino rest-mass $m_\nu c^2 = 0$). In the case of neutron β-decay the Q-value is just the energy corresponding to the mass difference

$$\Delta_1 m_\mathrm{n} = m_\mathrm{n} - M(1, 1) \tag{1.23}$$

whose determination clearly provides a new route for the measurement of the neutron mass. However the study of neutron β-decay is by no means the only way, and certainly not the easiest way, of finding $\Delta_1 m_\mathrm{n}$. In fact there are altogether eight combinations or cycles of nuclear reactions which lead directly to $\Delta_1 m_\mathrm{n}$ (Li *et al* 1951) and this result is important since it allows the possibility of establishing an internal consistency check for all such cycles. We may mention three of these cycles in addition to neutron β-decay $\mathrm{n}(\beta^-)\mathrm{p}$; these are: (1) $^3\mathrm{H}(\mathrm{p,n})^3\mathrm{He}$, $^3\mathrm{H}(\beta^-)^3\mathrm{He}$; (2) $^{13}\mathrm{C}(\mathrm{p,n})^{13}\mathrm{N}$, $^{13}\mathrm{N}(\beta^+)^{13}\mathrm{C}$; (3) $^{14}\mathrm{C}(\mathrm{p,n})^{14}\mathrm{N}$, $^{14}\mathrm{C}(\beta^-)^{14}\mathrm{N}$, of which (1) is the most thoroughly studied. We write it out explicitly

$$\begin{aligned} &\mathrm{p} + {}^3\mathrm{H} \rightarrow {}^3\mathrm{He} + \mathrm{n} \qquad && Q = -0.7637 \pm 0.001 \text{ MeV} \\ &{}^3\mathrm{H} \rightarrow {}^3\mathrm{He} + e^- + \bar{\nu}_e \qquad && Q = 0.0185 \pm 0.0002 \text{ MeV} \end{aligned} \tag{1.24}$$

and from the arithmetic sum of these Q-values we find

$$\Delta_1 m_\mathrm{n} = 0.7822 \pm 0.001 \text{ MeV} \tag{1.25}$$

The weakness of this technique is that at least two absolute energies are required, i.e. the threshold of the $^3\mathrm{H}(\mathrm{p,n})^3\mathrm{He}$ reaction and the end-point of the $^3\mathrm{H}$ β-spectrum, neither of which can be determined easily to high precision.

A much more suitable reaction cycle is one which leads to a determination of the mass difference

$$\Delta_2 m_\mathrm{n} = m_\mathrm{n} + M(1, 1) - M(2, 1) \tag{1.26}$$

and there are six nuclear reaction cycles of this type. Again we list three of these: (1) $^1\mathrm{H}(\mathrm{n},\gamma)^2\mathrm{H}$; (2) $^2\mathrm{H}(\mathrm{d,p})^3\mathrm{H}$, $^2\mathrm{H}(\mathrm{n},\gamma)^3\mathrm{H}$ and (3) $^9\mathrm{Be}(\mathrm{p,d})^8\mathrm{Be}$, $^9\mathrm{Be}(\gamma,\mathrm{n})^8\mathrm{Be}$, although all six cycles are available for purposes of checking internal consistency. However, the first cycle, which is just the radiative capture of neutrons on protons

$$\mathrm{n} + {}^1\mathrm{H} \rightarrow {}^2\mathrm{H} + \gamma \tag{1.27}$$

and is the reverse reaction to the photodisintegration of the deuteron discussed above, is by far the most important. This is because only one reaction is involved and because, being exothermic, this reaction can take place with zero-energy neutrons. In this case the Q-value, which is just $\Delta_2 m_\mathrm{n}$, is the sum of the γ-ray energy E_γ and the energy $E_\gamma^2/2m_\mathrm{d}c^2$ of the recoiling deuteron, the last term contributing a small correction of order 0.06% of the total. The quantity E_γ may be determined to very high precision by measuring the γ-ray wavelength using X-ray diffraction techniques (Bell and Elliott 1948).

The most accurate measurement of $\Delta_2 m_n$ obtained by this method to date is

$$\Delta_2 m_n = 2.224\,575 \pm 0.000\,009 \text{ MeV} \tag{1.28}$$

Combined with mass measurements for the proton and the deuteron this yields a result for the neutron mass (Greene *et al* 1986, Review of Particle Properties 1992)

$$m_n = 1.008\,664\,904 \pm 0.000\,000\,014 \text{ u} \tag{1.29}$$

The value for the neutron mass expressed in MeV, i.e. 939.565 63 ± 0.000 28 MeV, is much less accurately known because of uncertainties in the conversion from atomic mass units to MeV given in equation (1.18) (Cohen and Taylor 1987).

1.5.2 Gravitational mass

1.5.2.1 The principle of equivalence

The inertial or kinetic mass of a body is that mass which appears in any expression of the laws of motion of the body, both in classical and quantum mechanics, whereas the gravitational mass is defined as that mass which determines the force acting on the body, or the potential it experiences, when placed in a gravitational field. It has been recognized since the time of Galileo, and subjected to direct experimental test by Newton (1686) and by Bessel (1832), that all bodies fall equally fast in a uniform gravitational field. This result implies that the inertial and gravitational masses are always in a fixed proportion, irrespective of chemical composition, and, with a suitable choice of units, may be regarded as equal. The most famous demonstration of the equality of inertial mass m_i and gravitational mass m_g derives from the static torsion balance experiments of Eötvös (1922) who showed that, for a range of substances including light and heavy elements, the ratio m_i/m_g is unity to within an accuracy of 3×10^{-9}. Because the presumed law of nature expressed in this equality lies at the foundations of all gravity theories, Dicke carried out a more precise version of the Eötvös experiment, using the sun rather than the earth as the source of gravitational field and succeeded in establishing that the accelerations for gold and aluminium are equal to within an accuracy of 10^{-11} (Roll *et al* 1967). More recent findings based on a re-analysis of the Eötvös data suggesting the presence of a feeble macroscopic ‘fifth force’ mediated by a low mass boson (Fischbach *et al* 1986) have not been confirmed by subsequent investigations (Adelberger *et al* 1991).

In constructing his theory of general relativity Einstein (1916) took as point of departure the assumed equality of m_i and m_g and was thereby led to the principle of equivalence. This states in effect that the interaction of a body with a gravitational field can be eliminated locally by employing an appropriately chosen accelerated coordinate system. The principle is applied locally in the sense that it holds at a single point in space-time, or more precisely to a spatial neighbourhood of a point on a time-like world line. Ocean tides, for example, represent non-local phenomena, and cannot be transformed away by a suitable choice of coordinate system.

The principle enunciated above is the 'strong' equivalence principle which requires that, in a freely falling coordinate system, the laws of physics including their number content, e.g. mass ratios, fine structure constant, etc., should be the same as in gravity-free space. Taken together with the general principle of covariance, the strong principle of equivalence is almost sufficient to determine the equations of general relativity uniquely, provided an 'appropriate' coordinate system is selected. This is understood as one in which the metric tensor reduces to the Minkowski metric of special relativity. The 'weak' equivalence principle on the other hand is essentially just the statement of the universal equality of m_i and m_g. It is implied by the strong principle but the converse is not true (Dicke 1959).

The experiments of Eötvös and Dicke provide direct evidence only for the weak principle of equivalence and their support for the strong principle is indirect and less precise. However, it is still significant; thus detailed analysis of the results shows that the ratio m_i/m_g for the neutron is equal to that of the hydrogen atom to an accuracy of 2×10^{-6} and equal to that of nuclear binding energy to an accuracy of 10^{-4} (Wapstra and Nijgh 1955). The verification to an accuracy of 10% of the predicted gravitational red shift for Mossbaüer γ-rays also lends indirect support to the strong principle (Pound and Rebka 1960, Pound and Snider 1965).

1.5.2.2 Neutron gravity experiments

We now consider briefly the nature and significance of the evidence relating to the equality of inertial and gravitational mass which is available from gravity experiments carried out with beams of free neutrons. The very first experimental study of this type, carried out at Brookhaven National Laboratory, confirmed that neutrons fall in the earth's gravitational field with an acceleration equal to the local value of g, which is determined by conventional observations on the motions of macroscopic bodies, to within an accuracy of about 7% (McReynolds 1951). Later developments of technique within the same general method resulted in an improvement in accuracy by an order of magnitude and showed also that the gravitational acceleration of a free neutron was independent of the sign of its vertical spin component (Dabbs *et al* 1965). More recently a neutron gravity refractometer has been employed to measure the scattering lengths in a variety of substances, for slow neutrons critically reflected from the surface of a liquid mirror after falling freely under gravity through a measured height. The principle of operation of this device is illustrated in figure 1.2. Comparison of the results obtained with those determined by more conventional methods confirmed that (m_i/m_g) was equal to unity to an accuracy of 3×10^{-4} (Koester 1967, 1975, 1976, Sears 1982). A similar result is obtained, in this case to the somewhat lower accuracy of 3.6×10^{-3}, from measurements of gravitationally induced phase shifts in the neutron interferometer (Staudenmann *et al* 1980).

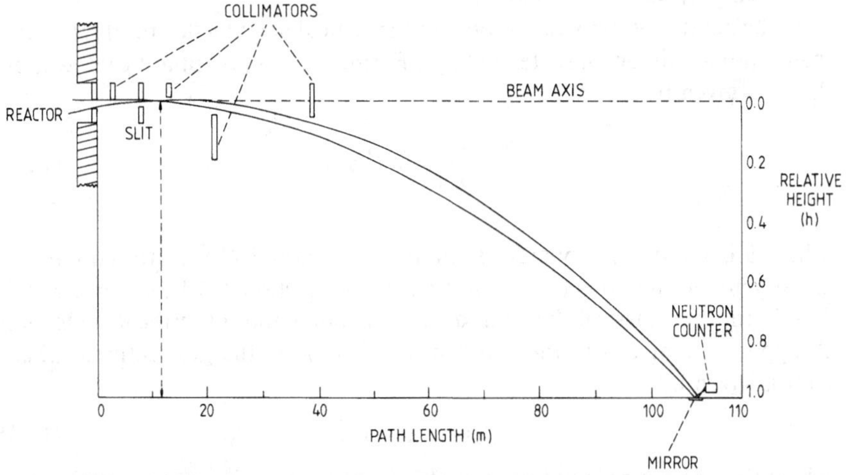

Figure 1.2. The neutron gravity refractometer designed by Koester (1965) to measure scattering lengths by determining the critical heights for reflection of neutrons from the surface of a liquid mirror. By operating with a known scattering length the instrument has been used to demonstrate the equality of the inertial and gravitational masses of the neutron.

The accuracy of all these measurements is several orders of magnitude below that which can be obtained by experiments on macroscopic bodies and the question might reasonably be asked as to what is the justification for persisting with such experiments. The proper answer would appear to be that, whereas the experiments of Eötvös and Dicke belong solely to the domain of classical physics, some at least of the free neutron studies provide evidence in support of the weak principle of equivalence which is valid in the quantum limit. In the case of simple free-fall experiments, application of the uncertainty relation shows that if the vertical width of the horizontal neutron beam is limited by an aperture of spread Δz, free fall under gravity is detectable over a horizontal distance Δx provided only that the inequality is satisfied

$$\Delta x \, \Delta z \geqslant l_c^2 \qquad (1.30)$$

where l_c is a macroscopic critical length defined by

$$l_c = \hbar/(\lambda m_i m_g g)^{1/2} \tag{1.31}$$

and λ is the de Broglie wavelength (cf. section 1.5.3.3). For thermal neutrons $l_c = 4$ mm. Thus quantum mechanics establishes a figure of merit $\Delta x\, \Delta z/l_c^2$ for the accuracy with which these experiments can be performed. For the two experiments which have been carried out this figure improved from 54 (McReynolds 1951) to 225 (Dabbs *et al* 1965). In the classical limit $\hbar \to 0$, $l_c \to 0$ and this limitation on the accuracy disappears. Hence these experiments are essentially classical in nature.

In the neutron gravity refractometer the critical height h_c for total reflection from a liquid mirror characterized by a Fermi pseudo-potential V_F (cf. section 2.5.5) is given by

$$\frac{h}{c} = \frac{V_F}{m_g g} = 2\pi l_c^2 N b \lambda \tag{1.32}$$

where b is the bound coherent scattering length and N the atomic number density in the liquid (cf. section 1.6.5.1 and section 6.3.1.2). Since $b \propto \lambda$, $h_c \to 0$ in the classical limit and the phenomenon of critical reflection disappears. Similarly in the neutron interferometer the gravitational phase shift is given by

$$\beta_g = A' \sin \varphi / l_c^2 \tag{1.33}$$

where the angle φ specifies the orientation of the interferometer with respect to the horizontal and A' is an area determined by the interferometer Bragg angle and the area enclosed by the beam paths. In the classical limit $\beta_g \to \infty$ and the interference pattern ceases to be observable. Thus in both cases the equality of inertial and gravitational mass is established from observations on phenomena which are purely quantum in nature.

1.5.3 Neutron momentum analysis

1.5.3.1 Mechanical velocity selectors

For non-relativistic neutrons the determination of neutron momentum (or velocity) is a problem in neutron optics, and there is a clear distinction between methods based on geometrical optics and those based on wave optics. The former category includes mechanical velocity selectors and choppers, while the latter features crystal monochromators and momentum filters, both of which rely on Bragg diffraction in crystals (cf. section 7.2.3). Although it is difficult to draw a sharp dividing line between mechanical velocity selectors and choppers the former may be described as devices which transmit a narrow band of neutron velocities with minimal time structure, whereas the latter transmit a broad band of energies in the form of pulses sharply separated in time and space (Brugger 1965).

The first mechanical velocity selector was based directly on Fizeau's apparatus for measuring the velocity of light and was used in conjunction with very weak sources (Dunning *et al* 1935). Nevertheless the device successfully demonstrated two very important properties of slow neutrons: (i) neutrons slowing down in a paraffin wax moderator eventually arrive at a Maxwellian distribution of velocities (cf. section 1.6.5.3); and (ii) the $1/v$ law for neutron absorption is valid for slow neutrons lying outside the resonance region (cf. section 1.7.1.1). The instrument consisted of two thin circular discs each marked out in the form of identical patterns of alternating aluminium and cadmium sectors, mounted a fixed distance L apart, coaxially with the direction of the neutron beam. When corresponding sectors are misaligned by a phase angle $\Delta\varphi$ a paraxial neutron transmitted by an aluminium sector on the first disc is absorbed by a cadmium sector on the second disc. If the second disc is now rotated about an axis through its centre, and normal to its plane, with angular velocity ω, then neutrons of velocity v arrive at the second disc just in time to be transmitted by an aluminium sector provided that

$$v = \omega L/\Delta\varphi \tag{1.34}$$

To take a realistic example, with $L = 54$ cm and $\Delta\varphi = 10.6°$, the velocity selector transmits neutrons of energy 0.0005 eV when the revolving disc turns at a rate of 10^3 rpm. However the energy resolution is poor, and the device is useful only for applications with very slow neutrons.

Velocity selectors have also been constructed with suitably shaped slits in place of the two-disc arrangement, and in this form are used in combination with more versatile dispersing systems to provide some initial energy selection. For example a helical velocity selector is used in combination with a conventional chopper in studies of small-angle scattering as shown in figure 8.12. Similarly the same selector acts as a band-pass filter to provide a range of operating energies with a resolution of 12–25% when used with the neutron spin echo spectrometer shown in figure 8.19.

1.5.3.2 Neutron choppers

In essence a neutron chopper is a rotating wheel with one or more holes in it, which periodically interrupts a neutron beam and transmits only for a very short time Δt. It therefore has a duty cycle $2\Delta t/T$, where T is the period of repetition. Choppers may rotate about an axis normal to the neutron beam (vertical configuration), or about an axis parallel to the neutron beam, (horizontal configuration); in either case the transmitting aperture is always cut normal to the axis of rotation. The first neutron chopper was built by Fermi in 1943 (Fermi *et al* 1947), and was constructed, in the form of a cylinder, as a sandwich of alternating copolanar layers of aluminium and cadmium as shown in figure 1.3. This rotates in the vertical configuration with angular velocity ω, and neutrons can be transmitted only during a time Δt given by

$$\Delta t = d/\omega R \qquad (1.35)$$

where d is the thickness of a transmitting layer or slit, and R is the cylinder radius.

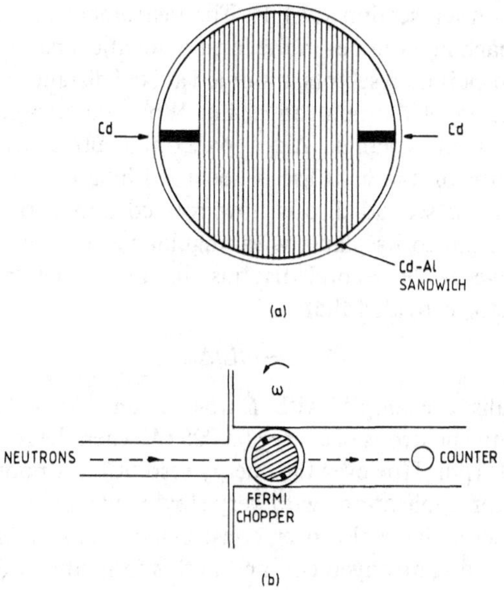

Figure 1.3. (a) Cross-section normal to the neutron beam of the Fermi chopper designed to produce pulses of slow neutrons for use in a time-of-flight spectrometer. Depending on the application and the time-resolution of the pulse, the collimating slits may be straight or curved (Egelstaff 1954, Marseguerra and Pauli 1959, Maliszewskii 1960, Chrien and Reich 1967). (b) Orientation of the chopper with respect to the neutron beam with the axis of rotation normal to the beam axis. The Fermi chopper operates with a duty cycle typically $\leqslant 1\%$. An example of its use in a time-focusing time-of-flight array is shown in figure 8.15.

Since there is insufficient time to penetrate the transparent layer for neutrons of velocity $v < v_c$, where v_c is a cut-off velocity given by

$$v_c = 2\omega R^2/d \qquad (1.36)$$

there is a minimum energy $E_c = \frac{1}{2}m_n v_c^2$ of neutrons in the transmitted pulse. Thus the device does provide some energy selection. There is of course no upper limit to the energy of a transmitted neutron and the pulse therefore spreads out in time and space as it travels towards the detector. To avoid 'frame overlap' between slow neutrons in one pulse and fast neutrons in the succeeding pulse it is necessary to preset a minimum velocity v_c. By eliminating d between equations (1.35) and (1.36) this fixes a linear relation between R and Δt

$$R = v_c\,\Delta t/2 \tag{1.37}$$

For a given flux distribution $\varphi(E_n)$ from the source the differential neutron number density reaching the detector is (Stanford 1961)

$$\frac{dn(E_n)}{dE_n} = \varphi(E_n)[n_s A_i A_o/l^2]\left(\frac{\Delta t}{T}\right) f_c(E_n, E_c) \tag{1.38}$$

where n_s is the number of slits, and the entrance and exit apertures A_i, A_o are separated by a distance l. The cut-off function $f_c(E_n, E_c)$ is defined as the ratio of the time averaged probability for transmission of a neutron of energy E_n, to the same probability for a neutron of infinite energy. For plane slits and paraxial neutrons $f(E_n, E_c)$ is a function of $\beta = v_c/v$ only, and is given by (Stone and Slovacek 1959)

$$\begin{aligned} f_c(\beta) &= 1-(8/3)\beta^2 & 0 \leqslant \beta \leqslant \tfrac{1}{4} \\ &= (8/3)\beta^2 - 8\beta + (16/3)\beta^{1/2} & \tfrac{1}{4} \leqslant \beta \leqslant 1 \\ &= 1 & \beta \geqslant 1 \end{aligned} \tag{1.39}$$

Since all neutrons travel the same distance $L = vt(v)$ the energy resolution obtained when the time of arrival $t(v)$ is measured with respect to a start pulse generated at the chopper is given by

$$\frac{\Delta E_n}{E_n} = \frac{2\Delta t}{t(v)} = 2(2E_n/m_n)^{1/2}\left(\frac{\Delta t}{L}\right) \tag{1.40}$$

Thus the parameter $\Delta t/L$ provides a figure of merit for the joint system considered as a spectrometer. Typical experimental values are $L = 10-120$ m and $t = 1-10$ μs. The time of arrival may be determined to better than 0.1 μs.

The use of a pair of Fermi choppers in a time focusing time-of-flight spectrometer is shown in figure 8.15. For slow neutrons the horizontal rotor configuration is equally useful and examples of horizontally mounted single and multichopper combinations are shown in figure 8.12 and in figure 8.14. For fast neutrons with energy $E > 0.25$ cadmium is not a suitable absorber and is replaced by borated plastics or Cu–Ni alloys. Although fast neutron choppers have been operated successfully in the horizontal configuration (Selove 1952, Pawlicki and Smith 1952), the advantages of the vertical configuration with respect to superior mechanical performance at high rotation speeds have proved decisive (Seidl 1954, Chrien and Reich 1967). In any case fast choppers have largely been abandoned in favour of pulsed accelerator sources which are far easier to control.

1.5.3.3 Crystal monochromators

Crystal monochromators provide the best possible energy resolution where the question is posed of how to select a specific neutron energy from a heterogeneous beam of neutrons. The technique exploits the fact that a neutron of energy $E_n = p_n^2/2m_n$ has associated with it a de Broglie wavelength (cf. section 6.1.2)

$$\lambda = h/p_n = h/(2m_n E_n)^{1/2} \tag{1.41}$$

Since slow neutrons have wavelengths λ typically in the range 1–10 Å they may be made to undergo diffraction in the *n*th order from a set of crystal planes of lattice spacing *d*, provided they satisfy the Bragg condition (cf. figure 7.7)

$$2d \sin\theta = n\lambda \tag{1.42}$$

Since the theory of Bragg diffraction is treated in detail in section 7.2.3, further discussion of the details of momentum analysis by these techniques will be deferred to section 8.5.1.2.

There is, however a further experimental application of equation (1.42) of some importance, which follows from the fact that for $\lambda > \lambda_B = 2d$ there exists no angle θ for which this equation can be satisfied. It follows that low energy neutrons satisfying the inequality

$$E_n < E_B = h^2/8m_n d^2 \tag{1.43}$$

propagate freely through the crystal. Conversely for neutrons with energy above the Bragg cut-off, the Bragg condition can always be satisfied. Thus a polycrystalline sample of the material containing a random distribution of values of $\sin\theta$ acts as an effective filter for fast neutrons which are scattered out of the beam. Beryllium, graphite and bismuth (useful also for its γ-absorbing properties) are the most widely used filters of this type. We shall return to this topic again in section 8.5.1.2 and section 8.5.2.3 where specific applications are discussed.

1.5.4 Spin and statistics

1.5.4.1 Neutron scattering in ortho and para hydrogen

In the band spectrum of molecular deuterium the even-valued lines are the more intense, rather than the odd-valued lines as in hydrogen, and the intensity ratio is 2. This shows that the deuteron is an $I = 1$ boson, which implies in turn that the neutron is either an $I = \frac{1}{2}$ or an $I = \frac{3}{2}$ fermion. The question as to which is the correct value of the neutron spin was settled almost as a byproduct of the solution to a related problem concerning the structure of the deuteron and the nature of the neutron–proton force. We shall outline the arguments involved at this stage since we will wish to make use of some of the conclusions at a later stage.

If the neutron spin is assumed to be $\frac{1}{2}$, then the deuteron is a triplet bound state of the neutron–proton system, and from its measured binding energy of 2.23 MeV, we can calculate the value $\sigma_3 \simeq 3.5$ b for the triplet scattering cross-section of zero-energy neutrons on protons (where $1\ \mathrm{b} = 1$ barn $= 10^{-24}\ \mathrm{cm}^2$). However, the measured cross-section σ_0 is a statistically weighted average of triplet and singlet scattering cross-sections

$$\sigma_0 = \tfrac{3}{4}\sigma_3 + \tfrac{1}{4}\sigma_1 \tag{1.44}$$

and it has a measured value of about 20.4 b. From this result we deduce that the neutron–proton force is highly spin-dependent with a large singlet scattering cross-section $\sigma_1 \simeq 71$ b. Using the relation between zero-energy scattering cross-section σ_s, and scattering length a_s in a scattering channel with multiplicity $(2s+1)$

$$\sigma_s = 4\pi |a_s|^2 \tag{1.45}$$

we find that $|a_3| = 5.3$ fm and $|a_1| = 23.8$ fm (where 1 fm = 1 fermi $= 10^{-13}$ cm).

We know that $a_3 > 0$ because the triplet state of the deuteron is bound, but this analysis does not tell us anything about the sign of a_1 which is negative if the singlet state of the deuteron is virtual, i.e. unbound. To determine the relative sign of a_1 and a_3 an interference experiment is required, and it was pointed out by Schwinger and Teller (1937) that such interference effects could be detected in the scattering of neutrons from ortho and para hydrogen provided the characteristic neutron wavelength λ was so large that $\lambda > 0.768$ Å which is the mean internuclear separation in the hydrogen molecule. In these circumstances the neutrons scatter coherently from the two nuclei in the molecule, corresponding to an addition of amplitudes rather than an addition of cross-sections. Ignoring common factors associated with the reduced mass of the neutron and the structure of the hydrogen molecule, the elastic scattering cross-sections for scattering from para and ortho hydrogen molecules are given respectively by

$$\begin{aligned} \sigma_o^P &= 16\pi\left(\frac{3a_3}{4} + \frac{1}{4}a_1\right)^2 \\ \sigma_o^O &= 16\pi\left[\left(\frac{3a_3}{4} + \frac{1}{4}a_1\right)^2 + 2\left(\frac{a_3 - a_1}{4}\right)^2\right] \end{aligned} \tag{1.46}$$

The ratio of the cross-sections is therefore

$$\sigma_o^O / \sigma_o^P = 1 + 2\frac{(a_3 - a_1)^2}{(3a_3 + a_1)^2} \tag{1.47}$$

The contributions of inelastic scattering processes have been ignored in the above expressions because it is assumed that the neutron energy

$E_n < 0.023$ eV which is the energy gap between the two lowest rotational levels with $J=0, 1$. Thus para to ortho scattering transitions $J=0 \rightarrow J=1$ cannot occur. Ortho to para scattering transitions $J=1 \rightarrow J=0$ can occur but their effect is relatively unimportant; at $E_n = 0.001\,46$ eV ($\lambda = 12$ Å) and $T = 20$ K the result is to alter the factor 2 in equation (1.47) to 2.26. We have already noted that at ordinary temperatures the ratio of ortho to para populations in hydrogen is 3:1 and, because ortho$\rightarrow$para spontaneous transitions are forbidden, the ratio persists at low temperature even when only the $J=1$ and $J=0$ levels are populated. However, the transition $J=1 \rightarrow J=0$ can be brought about by adsorption in the presence of a catalyst, e.g. active charcoal, so that at temperatures below 100 K 99.9% pure para hydrogen can be prepared. It is therefore possible to measure σ_o^O and σ_o^P separately, and the results show that $\sigma_o^O/\sigma_o^P \simeq 30$. Application of equation (1.47) which corresponds to the case $I_n = \frac{1}{2}$ shows that for $a_1/a_3 = +4.5$, $\sigma_o^O/\sigma_o^P \simeq 1.4$, whereas for $a_1/a_3 = -4.5$, $\sigma_o^O/\sigma_o^P \simeq 27.9$, very close to the measured ratio. The conclusion is that $a_1 < 0$ and the singlet state of the deuteron is virtual (Halpern *et al* 1937).

Aside from the precise result the essential deduction from the measurement is that $\sigma_o^O/\sigma_o^P \gg 1$ and, when this conclusion was first established, Schwinger (1937) posed the question whether it was possible to account for this observation with a neutron spin of $\frac{3}{2}$. Under this assumption the theoretical ratio, including elastic processes only, is

$$\sigma_o^O/\sigma_o^P = 1 + 10\left(\frac{a_3 - a_5}{5a_5 + 3a_3}\right)^2 \qquad (1.48)$$

where the triplet and singlet scattering lengths a_3, a_1, are replaced by the corresponding quintet and triplet scattering lengths a_5, a_3. Inclusion of inelastic ortho$\rightarrow$para scattering processes alters the factor 10 to 13.9. Application of equation (1.48) shows that, for $a_3/a_5 = -4.5$, $\sigma_o^O/\sigma_o^P \simeq 6.8$, a value which is too small by a factor of about 4. The conclusion is that a neutron spin $I_n = \frac{3}{2}$ cannot account for the data and the neutron spin must be assigned a value $I_n = \frac{1}{2}$. In these circumstances the observed large value for the cross-section ratio is a consequence of the almost complete destructive interference in the singlet and triplet scattering amplitudes in the scattering from para hydrogen.

1.5.4.2 Neutron magnetic dipole moment: the Stern–Gerlach experiment

The deuteron has a magnetic dipole moment equal to 0.857, measured in units of nuclear magnetons (n.m.) $\mu_N = e\hbar/2m_pc$. This is much less than the proton magnetic moment $\mu_p = 2.793$ n.m. and implies that, since the deuteron is predominantly a $^3S_1$ bound state of proton and neutron, the neutron must have a magnetic moment which is large and negative. Experimentally $\mu_n = -1.913$ n.m. (cf. section 7.5.1). The existence of a finite neutron magnetic moment affords several different means for confirming that the neutron spin has indeed the value $\frac{1}{2}$.

A pronounced weakness of the method described in section 1.5.4.1 above is that it relies for its conclusions on a detailed interpretation of the theory of the neutron–proton force. It was therefore pointed out (Hamermesh and Eisner 1950) that the angular distribution of neutrons reflected from magnetized mirrors through the interaction with the magnetic dipole moment is a much simpler phenomenon to interpret, and is equally conclusive when applied to the question of determining the neutron spin. For $I_n = \frac{1}{2}$ there should be two critical angles in the reflection pattern as is observed (Hughes and Burgy 1949) whereas for $I_n = \frac{3}{2}$ there should be three, not four as might be expected since the spin component $I_{nz} = -\frac{3}{2}$ does not give rise to total reflection. Thus the observed pattern of reflection maxima rules out the value $I_n = \frac{3}{2}$ and this result relies only on an understanding of the interaction of a magnetic moment with a static magnetic field. A similar advantage holds for magnetic resonance experiments of the type first used to determine the gyromagnetic ratio of the neutron (Alverez and Bloch 1940). In these experiments also the neutron spin transition probability, measured as a function of the amplitude of an oscillating magnetic field at resonance, has a value which agrees with the theoretical prediction for $I_n = \frac{1}{2}$ but is quite inconsistent with $I_n = \frac{3}{2}$, or indeed with any other value of the neutron spin (Stanford *et al* 1954).

The original determination of the neutron spin using the method proposed by Schwinger was an impressive *tour de force* and, in combination with the results of subsequent experiments involving non-resonant and resonant magnetic manipulation of the neutron spin, allows no escape from the conclusion that the neutron spin has the value $\frac{1}{2}$. Nevertheless in more recent times a much more direct demonstration of this result has become possible by means of the Stern–Gerlach experiment, which brings about a complete spatial separation of the two spin components in an initially unpolarized neutron beam (Sherwood *et al* 1954, Barkan *et al* 1968, Jones and Williams 1980).

In a Stern–Gerlach apparatus, whose mode of separation is illustrated in figure 1.4, a neutron with magnetic moment $\boldsymbol{\mu}_n$ enters a region of inhomogeneous magnetic field $\boldsymbol{B}$ where its magnetic potential energy is given by

$$V_{mag} = -\boldsymbol{\mu}_n \cdot \boldsymbol{B} \qquad (1.49)$$

It is necessary to assume that the neutron motion is adiabatic in this field, i.e. the magnetic field does not change appreciably during one period of oscillation of the neutron spin so that the axis of quantization remains well defined and the populations of the magnetic substates remain constant. Taking the dominant component of the magnetic field to be in the z-direction, and the neutrons to enter it in the x-direction, one need take account only

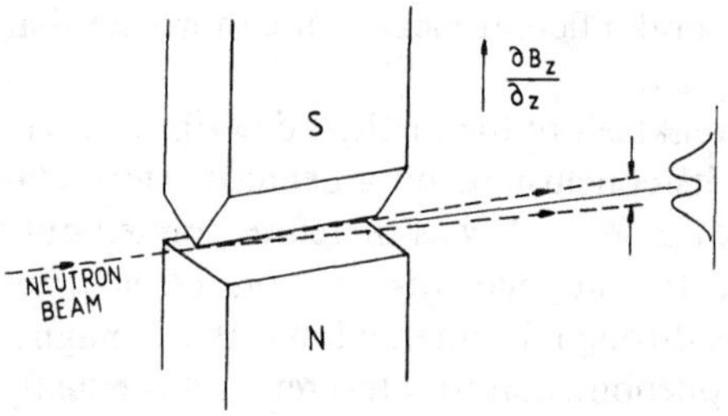

Figure 1.4. Operating principle of a Stern–Gerlach apparatus used to split an unpolarized neutron beam into two spatially separated beams of opposite polarization. The system makes use of the fact that the sign of the force exerted on a neutron magnetic moment, moving transverse to an inhomogeneous magnetic field, depends on the sign of the spin component measured with respect to the magnetic field.

of the z-component of the field gradient. In this case a neutron in the spin state $I_{nz} = \pm\frac{1}{2}$ experiences a force in the z-direction

$$F^{\pm} = \mu_z \frac{\partial \boldsymbol{B}_z}{\partial z} = \pm\tfrac{1}{2}\hbar\gamma_n \frac{\partial \boldsymbol{B}_z}{\partial z} \tag{1.50}$$

where $\gamma_n = -1.832 \times 10^7\ \mathrm{kg^{-1}\,s^{-1}}$ is the neutron gyromagnetic ratio. Since the magnetic field divergence vanishes there must be other magnetic field components but in adiabatic conditions these components may be ignored.

Suppose now an unpolarized beam of neutrons of mean momentum p_x, passes through the region of magnetic field inhomogeneity for an interval of time Δt, then it is split into two beams, corresponding to the two values of the spin component, with an angular separation

$$\Delta\theta = \tfrac{1}{2}\hbar\,\Delta t \left| \gamma_n \frac{\partial \boldsymbol{B}_z}{\partial z} \bigg/ p_x \right| \tag{1.51}$$

and we see that, for a given path length Δx through the apparatus, $\Delta\theta \propto \lambda^2$ and the separation of spin states will be a maximum for very slow neutrons. However, the beam has to be confined within the pole-pieces of the magnet, and if we suppose that this is effected by a slit of width Δz, then quantum mechanics requires that the z-component of momentum has an uncertainty $\hbar/\Delta z$ giving the beam a minimum divergence $\Delta\theta_{min} = \hbar/p_x\,\Delta z$. Thus, for a real physical separation of the two emergent beams to be detectable we must have $\Delta\theta \gg \Delta\theta_{min}$, i.e.

$$\left| \gamma_n \frac{\partial \boldsymbol{B}_z}{\partial z} \right| \Delta z\,\Delta t \gg 1 \tag{1.52}$$

This condition sets limits on minimum field inhomogeneities and passage times which must be met for state-separation to be observed, and may be interpreted as a statement to the effect that, if the apparatus is to separate two spin states of the neutron which differ in energy by an amount $\Delta E = \Delta(\hbar\gamma_n \boldsymbol{B})$, it can do so only if the neutron is acted on for a time Δt such that it satisfies the uncertainty relation

$$\Delta E\,\Delta t \geqslant \hbar \tag{1.53}$$

Since neutron beams which are available in practice have some degree of intrinsic divergence caused by departures from monochromaticity, and the necessity of using finite collimating apertures, values of $\Delta\theta/\Delta\theta_{\min}$ of order $(1\text{–}2)\times 10^3$ are required to achieve good separation. This is illustrated in figure 1.4 where an initially unpolarized beam is completely separated into two polarized beams when the inhomogeneous magnetic field is applied, thereby confirming that the neutron spin has indeed the value $\frac{1}{2}$. It should be noted that the experiment works only because the neutron is a neutral particle; it does not work with protons or electrons because in either case the action of the Lorentz force generated by other magnetic field components ignored above, and acting on the electric charge, always introduces an additional angular divergence in the z-direction much greater than that due to the forces exerted on the spin.

1.6 Neutron sources

1.6.1 Natural and artificial neutron sources

Neutrons have been detected at the surface of the earth coming from both extra-terrestrial and terrestrial sources. The presence of small numbers of neutrons in the cosmic radiation was first confirmed by observing proton recoils in photographic emulsions carried to high altitudes in balloons (Rambaugh and Rocher 1936). Subsequent investigation with BF_3 counters, with or without cadmium shielding and using various degrees of enrichment with ^{10}B (cf. section 1.7.1.2) confirmed these findings. In particular it was shown that the neutron flux reached a maximum at a pressure of about 120 mb, decreasing by some two orders of magnitude at sea level (Funfer 1937, Simpson 1948, Yuan 1949). Such findings are consistent with the view that high energy neutrons lose energy by scattering as they travel down through the atmosphere, and are eventually absorbed via the reaction $^{14}N(n, p)^{14}C$ (cf. section 1.7.1.2).

Since neutrons are β-active with a mean life-time of about 900 s, it follows that even a 100 MeV neutron has a better than 50% chance of decaying in a distance equal to the distance between the sun and the earth. One may conclude therefore that cosmic ray neutrons originate inside the solar system. Indeed, since the flux is strongly latitude dependent (Staker 1950), and varies inversely with solar activity (Soberman 1956, Hayms 1959) in the same way as does the charged component of the cosmic radiation (Neher 1956), the implication is that neutrons are not present to any significant degree in the primary cosmic radiation but are secondary particles generated by high-energy collisions between protons and nuclei at the top of the atmosphere.

The principal terrestrial source of neutrons derives from spontaneous fission of heavy elements (Petrzhak and Flerov 1940) of which the most important source is $^{238}_{92}U$ (Segré 1952). Indeed up to the time of the discovery of the Oklo natural reactor (Neuilly *et al* 1972) this was the only confirmed terrestrial neutron source, with not more than a 10% contribution from slow neutron induced fission in $^{235}_{92}U$.

Five types of artificial neutron source are in general laboratory use: (1) radioisotope sources; (2) photoneutron sources; (3) accelerator sources; (4) nuclear reactors; (5) spallation sources. Of these the first two are small and easily accommodated in a standard laboratory. The remaining three are large expensive facilities each with its own particular features. For carrying out precise measurements on the properties of the neutron itself, and for producing radioactive isotopes, the nuclear reactor has in the past been by far the most important source. For condensed matter studies spallation sources are likely to be of increasing significance in the future.

1.6.2 Radioisotope sources

These sources are mainly based on the (α, n) reaction in beryllium whose observation led to the discovery of the neutron in the first place. Beryllium is a mononuclidic element in the sense that ^9Be is its only naturally occurring isotope, and the dominant reaction is $^9Be(\alpha, n)^{12}C$ which is exothermic with $Q = 5.71$ MeV. These sources produce neutrons of energy in the range 0–13 MeV, with a mean energy of about 5 MeV, which can be thermalized by enclosing the source in paraffin or other hydrogeneous material. A useful source may be constructed using mixtures of metallic beryllium with

Table 1.1. Properties of the most commonly used (α, n) sources

Source	E_α (MeV)	$T_{1/2}$	Yield (n/α)
$^{210}Po+Be$	5.30	138.4 days	0.7×10^{-4}
$^{226}Ra+Be$	4.78, 4.59	1.622×10^3 years	1.8×10^{-4}
$^{239}Pu+Be$	5.15	2.44×10^4 years	0.6×10^{-4}
$^{241}Am+Be$	5.48	4.60×10^2 years	0.7×10^{-4}
$^{242}Cm+Be$	6.11	163 days	1.1×10^{-4}
$^{241}Am+B$	5.48	2.44×10^4 years	1.0×10^{-5}
$^{241}Am+F$	5.48	2.44×10^4 years	4.0×10^{-6}

compounds of polonium, radium or transuranic elements, although the intense background of γ-radiation can be a disadvantage. Sources with substantially lower yields of neutrons per α-particle can be based on (α, n) reactions in natural boron (81.3%, ^{11}B, 18.7% ^{10}B) or fluorine (100% ^{19}F). The characteristics of a range of (α, n) sources are summarized in Table 1.1.

The single isotope ^{252}Cf, which has a half-life for spontaneous fission of about 10^2 years, is also a useful source of neutrons of energy $E_n \simeq 2.3$ MeV. Sources of this type emitting up to 10^9 neutrons/s have been constructed.

1.6.3 Photoneutron sources

The only photonuclear reactions with neutron yields sufficiently high to serve as useful sources of neutrons are $^2H(\gamma, n)^1H$ $(Q = -2.23$ MeV) and $^9Be(\gamma, n)^8Be$ $(Q = -1.67$ MeV). These produce homogeneous beams of neutrons of energy <1 MeV when excited by energetic γ-rays from radio-isotopes which are sufficiently long lived. The most commonly used isotopes are the β^- emitters ^{24}Na, ^{72}Ga, ^{124}Sb and ^{140}La which are enclosed in targets of deuterium oxide (D_2O) or metallic beryllium. Since the stable isotope ^{123}Sb forms 42.8% of natural antimony, the ^{124}Sb/Be source is particularly convenient because it can be repeatedly reconstructed by neutron irradiation in a nuclear reactor.

The main features of practical photoneutron sources are summarized in Table 1.2. The yields of neutron/β-decay are estimated on the assumption that the target is a spherical shell of given surface mass density measured in $g\,cm^{-2}$.

Table 1.2. Properties of the most commonly used photoneutron sources

Source	E_γ (MeV)	$T_{1/2}$	Yield ($n/\beta\ cm^2\ gm^{-1}$)
$^{24}Na + Be$	2.757	15.1 h	3.8×10^{-6}
$^{24}Na + D_2O$	2.757	15.1 h	7.3×10^{-6}
$^{72}Ga + D_2O$	2.51	14.3 h	1.9×10^{-6}
$^{124}Sb + Be$	1.691	60.9 days	5.1×10^{-6}
$^{140}La + Be$	2.51	40.2 h	0.08×10^{-6}

1.6.4 Accelerator sources

The simplest way to use an accelerator to generate neutrons is to use the bremsstrahlung from electron accelerators such as the synchrotron or the betatron to create photoneutrons. However, these neutrons have a continuous distribution in energy and a time of flight technique must be used to resolve the neutron velocity spectrum. Pulsed electron linacs, which may be regarded as sources of virtual photons fulfill the same general purpose.

For example the (e, n) process in uranium can generate 10^{-2} neutrons per electron at an energy of 30 MeV with a total yield of 2×10^{13} neutrons/s. By raising the electron energy this yield may be increased by at least an order of magnitude.

An alternative technique, first introduced by Crane *et al* (1933), is to use accelerators of nuclear particles and in general this approach yields monoenergetic neutrons. The two most important neutron-producing reactions which can be obtained with small electrostatic accelerators in the laboratory are the (d, n) reactions in deuterium and tritium. Since these are exothermic reactions, i.e.

$$\mathrm{d} + {}^{2}\mathrm{H} \rightarrow {}^{3}\mathrm{He} + \mathrm{n} \qquad Q = 3.26\ \mathrm{MeV}$$

$$\mathrm{d} + {}^{3}\mathrm{H} \rightarrow {}^{4}\mathrm{He} + \mathrm{n} \qquad Q = 17.6\ \mathrm{MeV} \tag{1.54}$$

they can produce very high neutron yields, of order $10^{10}\ \mathrm{s}^{-1}$, with deuterons in the low energy range 100–300 keV. The relation between deuteron energy E_{d} and neutron energy E_{n}, for neutrons emitted at an angle θ to the incident deuteron momentum, is given by

$$4E_{\mathrm{n}} = E_{\mathrm{d}} + 2\{2E_{\mathrm{d}}E_{\mathrm{n}}\}^{1/2} \cos\theta + 3Q \tag{1.55}$$

Thus, for $E_{\mathrm{d}} = 300$ keV, monoenergetic neutrons are emitted at right angles to the incident beam with energies of 2.5 MeV for the ${}^{2}\mathrm{H}(\mathrm{d, n}){}^{3}\mathrm{He}$ reaction, and 14 MeV for the ${}^{3}\mathrm{H}(\mathrm{d, n}){}^{4}\mathrm{He}$ reaction.

From the point of view of increasing the neutron energy no advantage is gained by raising the energy of the bombarding deuteron above 4.5 MeV for deuterium targets, or above 4.0 MeV for tritium targets, because of the large background of low energy neutrons generated by the disintegration of the deuteron.

To produce neutrons of energy >30 MeV large accelerators such as the cyclotron or the Van de Graaff electrostatic generator are required. Used with fast deuterons, (d, n) stripping reactions such as ${}^{2}\mathrm{H}(\mathrm{d, n}){}^{3}\mathrm{He}$, ${}^{3}\mathrm{H}(\mathrm{d, n}){}^{4}\mathrm{He}$ and ${}^{9}\mathrm{Be}(\mathrm{d, n}){}^{10}\mathrm{B}$ are very suitable (Batty *et al* 1969). At energies >300 MeV charge exchange reactions with protons such as ${}^{3}\mathrm{H}(\mathrm{p, n}){}^{3}\mathrm{He}$ or ${}^{7}\mathrm{Li}(\mathrm{p, n}){}^{7}\mathrm{Be}$ provide useful sources.

1.6.5 The nuclear reactor

1.6.5.1 Fission chain reaction

In the phenomenon of neutron-induced nuclear fission discovered by Hahn and Strassmann (cf. section 1.4) a nucleus captures a neutron and subsequently breaks up into two nuclei of approximately equal mass. The process is accompanied by a huge release of energy, typically 200 MeV/fission, plus two or more fast neutrons which are available in principle to keep the process going. A controlled self-sustaining or critical assembly of this type is called a chain-reacting pile or nuclear reactor. The first reactor was constructed by Fermi in 1942 (Fermi 1947) and the modern high-flux research reactor operating on the same principle is a prolific source of neutrons.

Some nuclei, e.g. ^{233}U, ^{235}U and ^{239}Pu undergo fission under slow neutron bombardment whereas others, e.g. ^{232}Th, ^{238}U and ^{237}Np, require neutrons of at least 1 MeV energy. In natural uranium (99.3% ^{238}U, 0.7% ^{235}U) a chain reaction is possible only with slow neutrons, for in this case the slow neutron fission cross-section in ^{235}U is sufficient to compensate for all other neutron loss mechanisms not resulting in fission. It is therefore necessary to mix the uranium with a moderator whose role is to reduce the 2.5 fast neutrons per fission, to thermal energies, by elastic scattering with atoms of the moderator as first demonstrated by Fermi (cf. section 1.4). Common moderators are heavy water (D_2O) or graphite; ordinary water also makes a good moderator provided the loss of neutrons through the capture process $^1H(n, \gamma)^2H$ is compensated by a suitable enrichment of the fuel in ^{235}U. The 57 MW high-flux beam reactor at the Institut Laue Langevin (ILL), Grenoble is a heavy water moderated reactor with a maximum thermal neutron flux (in the reflector) of 1.2×10^{15} cm^{-2} s^{-1}.

1.6.5.2 Materials for neutron technology

Since uranium is the prime source material for the generation of high neutron fluxes, considerable effort has been expended over the past 50 years in locating and developing commercially exploitable deposits. Although the only naturally occurring isotope of thorium $^{232}_{90}Th$ (1.39×10^{10} years) is non-fissionable, thorium is also classed as a source material. This is because it may be transformed into the fissile isotope $^{233}_{92}U$ (1.62×10^5 years) by the sequence of reactions

$$^{232}_{90}Th(n, \gamma)^{233}_{90}Th \text{ (22.1 min)} \xrightarrow{\beta^-} {}^{233}_{91}Pa \text{ (27 days)} \xrightarrow{\beta^-} {}^{233}_{92}U$$

Thus $^{232}_{90}Th$ is a potential fuel for the thermal breeder reactor (cf. section 5.5.1).

There are more than 100 uranium-bearing minerals, but only three are of commercial importance. These are uranitite, which is a cubic or octahedral crystal made up from UO_2, UO_3 and U_3O_8 which can contain up to 85% uranium content, pitchblende, a non-crystalline form of uranitite which is the principal source of high grade uranium ore, and davidite which is a rare earth–iron–titanium oxide containing up to 10% uranium. Although the minerals thorianite and thorite can each contain up to 80–90% thorium in the form of ThO_2, these are highly refractory materials which are not classified as ores due to the difficulty of extracting the metal. Thus the only important source of thorium is the rare-earth phosphate monazite which can contain up to 15% ThO_2.

Apart from the fuels, important roles in neutron technology are assigned to the light elements boron, lithium and beryllium, the first two because of their high cross-sections for neutron capture (cf. section 1.7.1.2) and beryllium because it has vanishing absorption cross-section, and because of its mechanical, thermal and moderating properties. Boron is a fairly widespread material which is present in seawater, and notably in steam issuing from the Tuscany marshes, in the form of boric acids, all of which are derived from BO_3 by adding various amounts of water. However, the principal commercial source of boron is borax ($Na_2B_4O_7$) which is present in several minerals, particularly trincal and kernite. Lithium is a rare element found for example in lepidalite (lithium aluminium fluosilicate) or in spodumene (lithium aluminium silicate) each containing up to 6% lithium. The only commercial source of beryllium is the mineral beryl, which is formed from hexagonal crystals of beryllium aluminium silicate containing 14% BeO, and includes the gemstones emerald and aquamarine.

1.6.5.3 Energy distribution of reactor neutrons

To describe the various fluxes which exist inside a nuclear reactor we shall adopt a simple and convenient scheme according to which neutrons are defined as fast ($E_n > 1$ MeV), of intermediate energy (1 MeV $\geqslant E_n > 1$ eV) or slow ($E_n < 1$ eV). The energy 1 eV selected to define the boundary of the slow neutron region is taken as being typical of the energy of the lowest resonance in the moderating material. Although perhaps 10% of the neutron flux at the surface of a fuel element is fast, the remaining flux being evenly divided between intermediate and slow components, most neutrons emerging from a reactor are slow. In spite of the fact that neutrons can transfer across energy boundaries by scattering and by the action of gravitational or magnetic forces, the density of neutrons in phase space remains constant as required by Liouville's theorem in classical mechanics.

For nuclear reaction studies the resonance region 100 eV $> E_n >$ 1 eV is important but, for studying the basic properties of the neutron itself, or for studying neutron interactions with materials, slow neutrons are far more useful. Within the slow range there are various subgroups classified as epithermal (1 eV $\geqslant E_n >$ 0.025 eV), thermal ($\simeq$0.025 eV), cold (0.025 eV $\geqslant E_n \geqslant 5\times10^{-5}$ eV), very cold (5×10^{-5} eV $> E_n \geqslant 2\times10^{-7}$ eV) and ultra cold ($<2\times10^{-7}$ eV). By convention a thermal neutron is taken to have an energy of 0.025 eV, which is the mean energy of a neutron in thermal equilibrium with its surroundings at 20 °C. The velocity of a thermal neutron is 2.2 km s^{-1} and its wavelength $\lambda_T = 1.8$ Å.

Experimentally thermal fluxes are distributed in energy according to a law which is closely Maxwellian

$$\varphi_T(E_n)\,dE_n = \varphi_{T0}\exp[-(E_n/kT)](E_n/kT)\,d(E_n/kT) \qquad (1.56)$$

characterized by a temperature T which is slightly higher than the moderator temperature. Epithermal neutrons, on the other hand, have a distribution which varies inversely with energy i.e.

$$\varphi_{\mathrm{e}}(E_{\mathrm{n}})\,\mathrm{d}E_{\mathrm{n}}=\varphi_{\mathrm{e}0}\,\mathrm{d}E_{\mathrm{n}}/E_{\mathrm{n}} \tag{1.57}$$

as determined through slowing down by elastic collisions with the nuclei of the moderator (cf. section 5.4.2.1). The ratio of thermal to epithermal flux depends on the nature of the moderator and the enrichment of the fuel. Typical figures of $\varphi_{\mathrm{T}}/\varphi_{\mathrm{e}}$ are 10 (light water/enriched fuel), 30 (heavy water/natural fuel) and 100 (heavy water/enriched fuel).

To obtain neutrons of a specified energy the velocity may be selected using a rotating collimator or chopper as described in section 1.5.4, and this method provides an energy resolution of order 10^{-4}. To obtain a narrow energy band use is made of the wave properties of the neutron as expressed by de Broglie's relation between wavelength and momentum given in equation (6.16) combined with Bragg's law of diffraction from a crystal lattice stated in equation (7.59). For example in NaI the lattice spacing is $d=2.8$ Å and thermal neutrons can be obtained by selecting the first-order diffracted spectrum at $\theta=18.8°$. Other useful crystal lattices are Be ($d=0.732$ Å) Cu ($d=2.08$ Å) and LiF ($d=2.32$ Å). Energy selection by Bragg scattering is discussed in more detail in sections 7.2.3 and 8.5.1, from the theoretical and experimental viewpoints, respectively.

It seems that Fermi was the first to recognize that cold neutrons propagate in condensed matter in a matter similar to the propagation of light waves, with a refractive index determined by the real part of the coherent scattering amplitude (cf. sections 2.4, 2.5 and 6.3.1.1). Since most nuclei have positive scattering lengths, and hence neutron refractive indices are less than unity, it follows that neutrons are totally reflected when incident on the corresponding material surfaces at glancing angles θ which satisfy the inequality

$$\theta\leqslant\sin^{-1}[(V_{\mathrm{F}}/E_{\mathrm{n}})^{1/2}] \qquad E_{\mathrm{n}}\geqslant V_{\mathrm{F}} \tag{1.58}$$

where

$$V_{\mathrm{F}}=\frac{2\pi\hbar^2}{m_{\mathrm{n}}}Nb \tag{1.59}$$

is the Fermi pseudo-potential (cf. sections 2.5.5 and 6.3.1.1) and b is the bound coherent scattering length (Fermi and Zinn 1946, Fermi and Marshall 1947). This phenomenon provides the principle of operation of the gravity refractometer (cf. section 1.5.2.2).

Total internal reflection may also be used to select neutrons of energy less than a predetermined energy (Hughes and Burgy 1951) and this technique has been applied to the development of high transmission neutron guides which provide a very important advantage to the reactor at the ILL Grenoble (Maier-Leibnitz and Springer 1963, Alefeld *et al* 1965). Neutron guides are usually constructed from nickel-coated boron glass and have rectangular cross-sections typically 10 cm × 5 cm in area. Neutron energy selection is governed by the length and radius of curvature of the guide; a thermal neutron guide may be 80 m long with a radius of curvature 2×10^4 m transmitting only those neutrons with wavelength greater than a critical wavelength $\lambda_0 = 1$ Å. For cold neutrons the guide length will be 10 m, the radius of curvature 25 m and the cut-off wavelength $\lambda_0 = 30$ Å. A very important function of the neutron guide is that not only does it eliminate the faster neutrons from the beam, but it also attenuates the γ-radiation in proportion to the square of the guide length.

1.6.5.4 Pulsed reactors

One of the limitations of the time-of-flight technique (cf. section 1.5.3) is that it provides a very inefficient use of the available neutron flux. Repetitively pulsed fast reactors in which the neutrons are confined to a high density pulse over a short time interval exhibit significantly increased levels of efficiency for the same mean power, a feature they share with pulsed spallation sources (cf. section 1.6.7). An example of a pulsed reactor is the IBR-2 reactor at the Joint Institute for Nuclear Research (JINR), Dubna, near Moscow, which operates at a pulsing frequency of 5 Hz delivering peak fluxes of 10^{16} cm^{-2} s^{-1} at an average power of 4 MW.

Pulsed reactors are driven to supercritically by inserting a sudden increase of reactivity, e.g. by passing a pellet of fissile material rapidly through a barely sub-critical core. Neutron multiplication is initiated by the prompt component of fission neutrons with power excursion limited through a negative temperature coefficient of reactivity (cf. section 5.5.3.2). Pulsed reactors have been successfully applied to the study of activation analysis and radiochemistry of short-lived radio-nuclei (Pillay and Thomas 1971, Miller and Guinn 1976, Trautmann and Hermann 1976) in neutron radiography (cf. section 8.1.1) (Robinson and Barton 1972) and nuclear pumping of lasers, e.g. in CO_2 gas (McArthur and Tollefsrud 1975). Underground nuclear explosions, which may be viewed as single pulse fast reactors (cf. section 5.5.1) have also been used on occasion to measure neutron-nucleus cross-sections and to create new elements (cf. section 5.2.5).

1.6.6 Ultra-cold neutrons

When the neutron energy is less than the Fermi pseudo-potential neutrons are reflected at all angles of incidence, and the possibility exists for storing such neutrons in material containers. This energy range is described as 'ultra-cold'. It has been suggested (Golub and Pendlebury 1979) that Fermi had himself considered the possibility of neutron storage in this fashion, and indeed this seems likely. However the first serious proposal in this respect was put forward by Zel'dovich (1959), and the practical problems encountered in the actual construction of a 'neutron bottle' were subsequently analysed in great detail (Forward 1963, Foldy 1966). The first experimental observations of ultra-cold neutrons were reported from JINR, Dubna (Luschikov *et al* 1969), and from Garching (Steyerl 1969). Since that time ultra-cold neutron sources have been constructed at a number of centres including LNPI Leningrad, ILL Grenoble, and Chalk River, Ontario, Canada (Robson 1976, Ageron *et al* 1977, Altarev *et al* 1980, Steyerl 1989, Golub *et al* 1991).

The principal difficulty in the production of ultra-cold neutrons is that the fraction f of these neutrons in the quasi-Maxwellian spectrum of reactor neutrons is approximately $(V_F/kT)^2$, i.e. $f \simeq 10^{-11}$ at 30 K assuming $V_F \simeq 10^{-7}$ eV. Since most modern research reactors are equipped with D_2O or H_2 cold sources, operating at temperatures in the range 20–30 K (Butterworth *et al* 1957, Webb 1963), it is possible to increase f to about 10^{-9} but no higher. Assuming a capture flux $\varphi_c \simeq 10^{15}$ cm^{-2} s^{-1} (cf. section 1.7.1.1), it follows that the maximum density of ultra-cold neutrons at the reactor centre is about 5 cm^{-3}. Even assuming a 20% efficiency of extraction this gives an expectation of 1 cm^{-3} which, at first glance, is not encouraging.

Ultra-cold neutrons cannot be extracted from the cold source of a nuclear reactor by the conventional method of inserting a beam tube, because they will be reflected from the walls of the tube or absorbed by it on their way through. This problem is generally solved by first accepting neutrons of energy above the ultra-cold range, which are subsequently cooled to ultra-cold temperatures by some other means. The most popular method is to insert a converter in the form of a thin sheet of moderating material at low temperature, whose principal function is to regenerate the low energy tail of the Maxwellian spectrum which is removed by the beam tube. Since it is essential that the neutrons should not gain energy from the convertor, it is important that the Fermi potential of the moderator should be much less than that of the walls. In practice this means that the moderator must be made from a material for which the scattering length is negative and hence $V_F < 0$. The obvious choice is hydrogen (cf. Table 6.1); unfortunately most hydrogenous materials are unstable in the presence of the intense γ-ray fluxes which are present at the centre of a reactor and the practical choices are limited. However polyethylene (Golikov *et al* 1972, 1973) and zirconium hydride (Groshev *et al* 1973) have been found suitable.

A number of methods have been used to extract the cold neutrons from the converter, and both vertical guide tubes (Steyerl 1969, 1972, Lobashev *et al* 1973), and inclined guide tubes (Ageron *et al* 1977), which make use of gravity to slow down the emergent neutrons, have been successfully employed. A particularly effective source of ultra-cold neutrons is the neutron turbine (Steyerl 1975) where neutrons of velocity ≈ 50 m s^{-1} are slowed to about 5 m s^{-1} by collisions with the moving blades of the turbine. The turbine source currently operating at the Institut Laue Langevin, Grenoble is shown

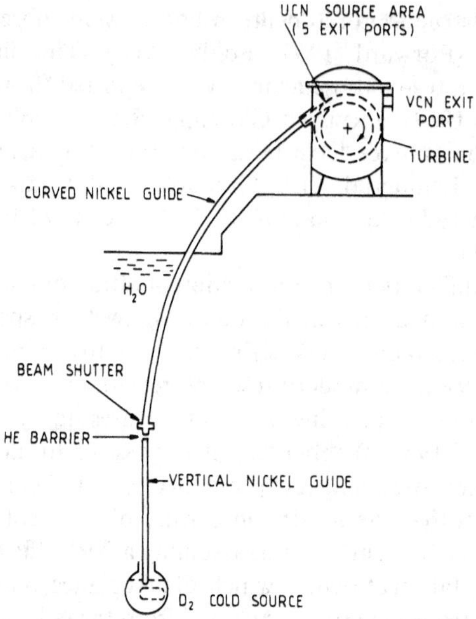

Figure 1.5. The ultra-cold–very cold neutron turbine source constructed by Steyerl (1989) at the Institut Laue Langevin, Grenoble. Neutrons from the 30 K liquid deuterium cold source, conducted vertically upwards by a nickel guide, decelerate under gravity and arrive at a rotating nickel turbine with a wavelength of $\simeq 80$ Å. About 50% of these neutrons are transmitted as very cold neutrons through an exit port, while the remainder are decelerated into the ultra-cold range ($\lambda \simeq 700$ Å) by reflection from a 690-blade receding nickel turbine mirror.

in figure 1.5. An alternative source of ultra-cold neutrons which involves conversion in superfluid helium (Golub and Pendlebury 1977, Kilvington *et al* 1987, Yoshiki *et al* 1992, Golub and Lamoreaux 1993) has attracted considerable attention, although its practicality is, as yet, unproven.

To discuss the confinement problem in a little more detail, we consider the case of a slow neutron of energy $E = \hbar^2 k_0^2 / 2m_n$, incident on a one-dimensional potential barrier

$$V(x) = 0,\ x < 0 \qquad V(x) = V_F,\ x > 0 \tag{1.60}$$

The corresponding Schrödinger equation is elementary, and the solution for $E > E_F$ is

$$\begin{aligned} \Psi(x) &= e^{ik_0 x} + [(k_0 - k)/(k_0 + k)]\, e^{-ik_0 x} & x \leqslant 0 \\ &= [2k_0/(k_0 + k)]\, e^{ikx} & x \geqslant 0 \end{aligned} \tag{1.61}$$

where

$$\hbar k = (2m_n(E - V_F))^{1/2} \tag{1.62}$$

Thus the wave incident from the half-space $x<0$ is reflected or transmitted with coefficients of reflection or transmission given, respectively, by

$$R=[(k_0-k)/(k_0+k)]^2 \qquad T=1-R=[2k_0/(k_0+k)]^2 \tag{1.63}$$

For ultra-cold neutrons where $E<V_F$ the solution is

$$\begin{aligned}\Psi(x)&=e^{ik_0x}-e^{2i\delta-ik_0x} & x\leqslant 0\\ &=(1-e^{2i\delta})e^{-\kappa x} & x\geqslant 0\end{aligned} \tag{1.64}$$

where

$$\hbar\kappa=(2m_n(V_F-E))^{1/2}>0 \tag{1.65}$$

and

$$\cos\delta=\kappa/\sqrt{k_0^2+\kappa^2} \tag{1.66}$$

Hence, although the wave penetrates a distance of order κ^{-1} into the interior of the medium, there is no transmitted wave and the neutrons are totally reflected. Quite generally, neutrons of velocity v will be reflected at all angles of incidence provided $v<v_c$, where v_c is a critical velocity defined by

$$v_c=\sqrt{\frac{2V_F}{m_n}}=\frac{2\pi\hbar}{m_n}\sqrt{\frac{Nb}{\pi}}$$

The values of the Fermi pseudo-potentials and the corresponding critical velocities are listed in Table 1.3 for a range of useful materials. We note that

Table 1.3. Values of the Fermi pseudo-potential (V_F) and critical velocity (v_c) for materials used in the storage of ultra cold neutrons. The values of V_F given in brackets for the magnetic materials (Fe, Co, Ni) refer to the magnetic contribution to V_F under conditions of saturation magnetization at 300 K (Golub and Pendlebury 1979)

Material	V_F (10^{-7} eV)	v_c ($m\,s^{-1}$)
D_2O	1.7	5.6
Be (BeO)	2.5	6.9
C	1.8	5.8
Mg	0.6	3.4
Al	0.5	3.2
SiO_2 (quartz)	1.1	4.6
Cu	1.7	5.6
Fe	2.2 (1.3)	6.5
Co	0.7 (1.1)	3.7
Ni	2.3 (0.4)	6.8

the maximum value $V_F = 2.5 \times 10^{-7}$ eV is achieved for beryllium (conducting) or beryllium oxide (insulating). This energy is approximately equal to the gravitational energy of a neutron raised to a height of 2.5 m, or its magnetic energy in a magnetic field 40 kG. It is partly on account of this rather convenient coincidence of three energies that ultra-cold neutrons may be manipulated by an essentially endless variety of combined strong, magnetic and gravitational forces. In particular, since the Fermi potential (equation (1.59)) for a neutron in a magnetic field $\boldsymbol{B}$ must be replaced by

$$V_F = \frac{2\pi\hbar^2 Nb}{m_n} \pm \boldsymbol{\mu}_n \cdot \boldsymbol{B} \tag{1.68}$$

where the choice in sign reflects the two independent orientations of the spin with respect to the magnetic field, it follows that, for selected ferromagnetic media (e.g. Ni, Co, Fe), it is possible to generate a transmitted beam with $E > V_F$, and a reflected beam with $E < V_F$, which are spin-polarized in opposite senses.

The analysis presented above is contingent on the assumption that there is no loss at the boundary between vacuum and medium due to capture or inelastic scattering, where the latter process can take the incoming neutron out of the ultra-cold domain, so that it is no longer reflected backwards. If we suppose that the medium is weakly absorbing with total cross-section σ_t, we may estimate the probability μ that the neutron is lost at a collision with the boundary from the formula

$$\mu = \int_0^\infty N\sigma_t |\Psi(x)|^2 \, dx = \frac{2N\sigma_t}{\kappa} \sin^2 \delta \tag{1.69}$$

when $\Psi(x)$ is the solution (equation (1.62)) of the Schrödinger equation without absorption, and b is replaced by its real part, Re(b), in equation (1.57). We now express σ_t in equation (1.67) in terms of its value given by the optical theorem (cf. section 2.4.2), which incorporates the $1/v$ law,

$$\sigma_t = \frac{4\pi \, \mathrm{Im} \, f(0)}{k} = -\frac{4\pi}{k_0} \mathrm{Im}\,(b) \tag{1.70}$$

We then find that

$$\mu = -\frac{\mathrm{Im}\,(b)}{\mathrm{Re}\,(b)} \frac{2v}{(v_c^2 - v^2)^{1/2}} \tag{1.71}$$

In the result (1.69), v represents the velocity component normal to the wall, and, if the neutron impacts at an angle of incidence θ_i, v is replaced by $v \cos \theta_i$. When all directions of incidence are equally likely as may be supposed to be the case for neutrons stored in a bottle, then we must average over

$\mu(\theta_i)$ giving the result (Shapiro 1973)

$$\bar{\mu}=2\int\mu(\theta_i)\cos\theta_i\,d(\cos\theta_i)/4\pi=\frac{2\,\mathrm{Im}\,(b)}{\mathrm{Re}\,(b)}[(\sin^{-1}y-y(1-y^2)^{1/2}]/y^2$$

$$y=v/v_c \tag{1.72}$$

Excluding the velocity dependent factor, $\bar{\mu}$ is determined by the phase of the (complex) bound coherent scattering length b, the experimental value for which generally lies within the range 10^{-5}–10^{-4} for those elements listed in Table 1.3. This result implies that a neutron ought to bounce between 10^4 and 10^5 times on average, before it is captured at the wall of the container, whereas in practice these numbers are observed to be smaller by several orders of magnitude. The precise reasons for these findings are still not very clear although surface contaminants (e.g. hydrogen) and non-uniformities evidently play a role. These problems have been subjected to a great deal of analysis and debate (Golub and Pendlebury 1979).

1.6.7 Spallation sources

Since the binding energy per nucleon lies in the range 8 ± 1 MeV over most of the periodic table, it is not unexpected that, for nuclear reactions generated by protons, deuterons or α-particles at energies of 100 MeV and upwards, many, and not just a few, nucleons should be ejected from the struck nucleus. Such reactions are called spallation reactions, and they are very broadly interpreted in terms of a mechanism which takes place in two stages. In the initial phase the reaction is described in terms of collisions between quasi-free nucleons which leads to a prompt cascade of nucleons emitted from the nucleus. This is followed by a slower evaporation phase with the release of further nucleons similar to the de-excitation of the compound nucleus at lower energy.

The spallation process with high energy protons and heavy nuclei can provide a very intense source of neutrons. For example, at a proton energy of 1 GeV the yield in neutrons per proton ranges from about 2 in beryllium to about 40 in uranium (Lone *et al* 1982). In practice the main limitation on available neutron intensity is the resultant heat generation and radiation damage in the target. Spallation sources produce mainly epithermal neutrons, which are useful for neutron scattering experiments in solid state research, along with many other particles, e.g. pions, muons and neutrinos. They have the unique feature of being pulsed, a characteristic which has potential applications in the reduction of background.

An example of a large spallation neutron source is the facility ISIS at the Rutherford Appleton Laboratory which is shown in figure 1.6 (Fender *et al* 1980). This operates with 800 MeV protons from a synchrotron which are fired at a ^{238}U target at a pulse frequency of 50 Hz. Because ^{238}U is fissionable

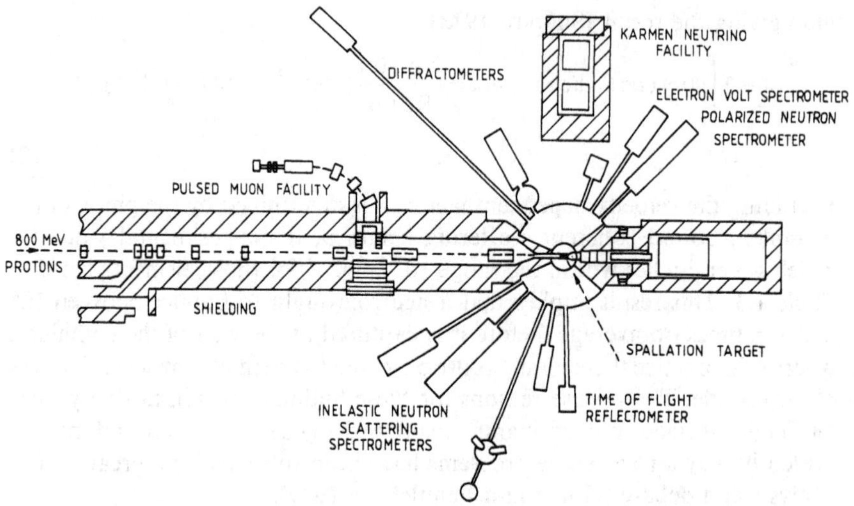

Figure 1.6. A schematic representation of the ISIS spallation neutron source at the Rutherford Appleton Laboratory, Chilton. An 800 MeV proton synchroton with $\approx 100\ \mu$A circulating current operating at 50 Hz, delivers proton pulses containing $\approx 1.5 \times 10^{13}$ protons per pulse onto a uranium spallation target. The source produces about 30 neutrons per proton, mainly epithermal neutrons which can be extracted from 18 beam ports. ISIS is also used as a pulsed source of muons and neutrinos, derived from the decay of pions emitted from the spallation reactions, and from β-radioactivity of fission products produced in the target.

under bombardment with fast neutrons, this results in a doubling of the neutron yield. The beam intensity of 2.5×10^{13} protons per $0.22\ \mu$s pulse gives an average neutron intensity of $4 \times 10^{16}\ s^{-1}$ which is the highest available at the present time. The fast neutrons emitted in the spallation reaction are reduced in energy to the epithermal region by a few cm of hydrogenous moderator. The resultant spread of neutron energies may be separated into selected energy bands by a system of fast choppers and monochromators. A spallation source of comparable intensity also exists at the Los Alamos Meson Physics Facility (LAMPF). This uses pulses of 800 MeV protons from a storage ring coupled to a proton linac.

1.7 Neutron detection and measurement

1.7.1 Slow neutron detectors

1.7.1.1 The '1/v' law

Since neutrons are uncharged they can be counted either by scattering from charged or uncharged particles and recording the recoils, or by generating nuclear reactions whose products are detected. For slow neutrons with $E_n < 0.1$ keV there is insufficient energy in the recoiling particle for it to be easily detected in conventional counters, and only the latter process has any value as a neutron detector. For the nuclear reaction A(n, b)B the cross-section is of the form

$$\sigma(v) = \frac{1}{\pi\hbar^4}|M|^2 \frac{p_b^2}{v_b v}(2I_b+1) \tag{1.73}$$

where the kinetic energy of the emergent particle 'b' is typically of order 1 MeV. In these circumstances the density of final states factor p_b^2/v_b is independent of neutron velocity v, and, provided that the nuclear matrix element M is also independent of v, i.e. there are no nearby resonances, the cross-section is exactly proportional to $1/v$. Thus if $\sigma_0(v_0)$ is the measured cross-section at thermal velocity $v_0 = 2.2$ km s^{-1}, $\sigma(v)$ may be written

$$\sigma(v) = \sigma_0 v_0 / v \tag{1.74}$$

Detectors which satisfy the above relation are called '$1/v$' detectors and they have the following important property. Suppose a beam of slow neutrons traverses a thin target, by which we mean that the neutron beam suffers negligible attenuation on passage through the target, then the total capture activity A in disintegrations/s generated in the target is

$$A = N_t \int n(v) v \sigma(v)\, dv = N_t \sigma_0 v_0 n \tag{1.75}$$

where $n = \int n(v)\, dv$ is the integrated neutron density in the beam and N_t is the total number of target nuclei. Thus, provided N_t is known, i.e. the isotopic content of the target is accurately assayed, a measurement of A provides a value for the spatial average of the neutron density n, rather than for the integrated flux $\varphi = \int n(v) v\, dv$. This result has some very special applications and in general provides a method for standardizing slow neutron beams of varying spectral composition, in terms of their capture flux φ_c, where

$$\varphi_c = n v_0 \tag{1.76}$$

There are many cases where, due to the presence of nearby resonances, the cross-section $\sigma(v)$ does not vary exactly as $1/v$. In such instances we may introduce an effective cross-section $\hat{\sigma}$ (Westcott 1955)

$$\hat{\sigma} = \left(\int_0^\infty n(v) v \sigma(v)\, dv\right) \Big/ \left(v_0 \int_0^\infty n(v)\, dv\right) \tag{1.77}$$

which is defined such that the total reaction rate is just equal to $\hat{\sigma}\varphi_c$. Since $\hat{\sigma}$ reduces to σ_0 in the case of pure $1/v$ behaviour, $\hat{\sigma}$ may be interpreted as the 2.2 km s^{-1} value of the equivalent $1/v$ cross-section which produces the same absolute reaction rate. In a pure Maxwellian spectrum at temperature

T_n, since $n(v)$ varies as $v^2 \exp[-\frac{1}{2}m_n v^2/kT_n]$, we may establish the relationship

$$\hat{\sigma} = \sigma_0 \int_0^\infty \sigma(v) v^3 \exp[-\tfrac{1}{2}m_n v^2/kT_n]\, dv \Big/ v_0\sigma_0 \int_0^\infty v^2 \exp[-\tfrac{1}{2}m_n v^2/kT_n]\, dv$$

$$\equiv g(T_n)\sigma_0 \qquad (1.78)$$

where $g(T_n)$, the Westcott g-factor, has been tabulated over a large temperature range for all processes, and for all nuclei, of interest (Westcott 1960).

In practice the spectra of reactor neutrons are never purely Maxwellian and there is an epithermal component at energies $\geqslant (4-5)kT_n$ which varies as E^{-1}. Hence a more general form of the result (1.78) is

$$\hat{\sigma} = (g(T_n) + r \cdot s(T_n))\sigma_0 \qquad (1.79)$$

where the epithermal index r respresents the proportion of the neutron spectrum in the epithermal range, and $s(T_n)$ is a factor which describes the excess reaction rate which is accounted for by the epithermal component. A useful formula for r is

$$r = (1.97/R_{Cd}(0.5))[T_n/T_0]^{1/2} \qquad (1.80)$$

where $R_{Cd}(0.5)$ is the cadmium ratio for a cut-off at 0.5 eV (cf. section 1.7.1.3). The function $s(T_n)$ may be calculated from the formula

$$s(T_n) = \left(\frac{4T_n}{\pi T_0}\right)^{1/2} \sigma_0 \int_0^\infty \left\{\sigma(E) - g(T_n)\sigma_0 \left(\frac{kT_0}{E}\right)^{1/2}\right\} \frac{\Delta(E)}{E}\, dE \qquad (1.81)$$

where $\Delta(E)$ is the step-function

$$\Delta(E) = 0,\ E < \mu kT \qquad \Delta(E) = 1,\ E \geqslant \mu kT \qquad (1.82)$$

The number μ has a value of about 5 for heavy-water moderated reactors and somewhat less for graphite-moderated reactors. At normal temperatures the epithermal correction is positive and falls in the range 5–15%.

1.7.1.2 Charged particle reactions

There are three exothermic neutron capture processes with charged particles in the final state which have large slow neutron cross-sections and which obey the $1/v$ law. These are $^{10}B(n, \alpha)^7Li$, $^6Li(n, \alpha)^3H$ and $^3He(n, p)^3H$. The characteristics of each reaction are listed in Table 1.4. These reactions are all widely used for neutron detection. The reactions $^{14}N(n, p)^{14}C$ and $^{35}Cl(n, p)^{35}S$ have much smaller cross-sections and are not generally used in this way.

The (n, α) reaction in ^{10}B is the most commonly used reaction for purposes of slow neutron detection, mainly because boron is readily available and because its compounds exist in both solid and gaseous forms. Boron trifluoride BF_3 in particular makes a very suitable gas for use in a proportional counter. This is a very efficient detector because the charged particles are detected inside the active volume and therefore do not degrade in energy as they would do in emerging from a solid target. Boron loaded liquid scintillators or boron glass scintillators used with ZnS(Ag) activators, using boron enriched to 90% in ^{10}B, also make very suitable slow neutron detectors with efficiencies in the range of 20–30%.

A particularly elegant method for determining the capture flux is to use as detector a very thin layer of boron, deposited on a quartz or silicon substrate, in conjunction with a silicon surface barrier detector to record the charged particles emitted from this layer. A typical charged particle spectrum observed in such an arrangement is shown in figure 1.7. Only 6.5% of the cross-section leads to the ground state in ^{7}Li and 93.5% produces an excited state, 0.48 MeV above the ground state, which decays by γ-emission. The two α-branches account for the two upper peaks in the spectrum while the two lower peaks are produced by the ^{7}Li recoils.

The (n, α) reaction in ^{6}Li has a very high Q-value which means that it is very suitable for discriminating against γ-rays in the background. A further advantage is that there is only one α-branch in the spectrum and no γ-rays. Unfortunately, no suitable gaseous compounds of lithium exist which could be used as operating gases for a proportional counter and the ^{6}Li has either to be deposited as a thin layer on a suitable substrate or loaded into a scintillator. Neutron sensitive scintillators can be made with efficiencies of about 50% using lithium enriched to 95% in ^{6}Li, dispersed in a ZnS(Ag) matrix or loaded in a cerium activated silicated glass. LiI(Eu) is also a useful neutron detector similar to the NaI(Tl) scintillator used for γ-detection. From the point of view of efficient discrimination against γ-rays, scintillation counters are never as good as proportional counters because the light output for electrons, and hence for γ-rays, is much higher than for protons or α-particles of the same energy.

The third reaction listed in Table 1.4, ^{3}He(n, p)^{3}H, has the obvious advantage that helium always exists in gaseous form, and makes a natural filling for a proportional counter, and this type of counter can be operated at pressures as high as 12 atmospheres. Further advantages are that the slow

Table 1.4. Pure '$1/v$' (n, α) and (n, p) reactions used for absolute neutron density measurements (Holden 1981)

Nucleus	Abundance (%)	Reaction	Q (MeV)	σ_0(b)	$\delta\sigma/\sigma$ (%)
^{10}B	$19.1 \rightarrow 20.3$	^{10}B(n, α) ^{7}Li	2.792	3838 ± 6	0.16
^{6}Li	$7.34 \rightarrow 7.68$	^{6}Li(n, α) ^{3}H	4.78	941 ± 3	0.32
^{3}He	1.3×10^{-4}	^{3}He(n, p) ^{3}H	0.765	5327 ± 10	0.19

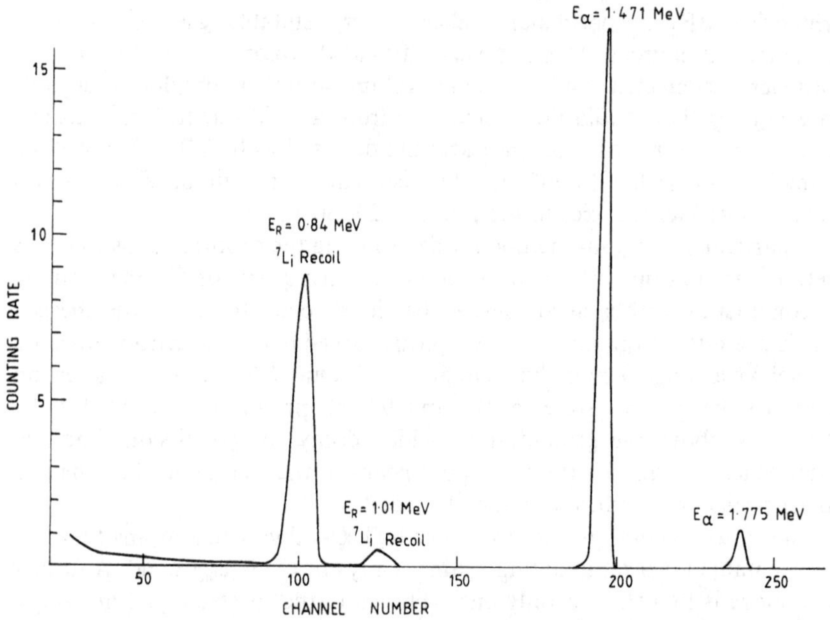

Figure 1.7. Charged particle spectrum from the reaction $^{10}B(n, \alpha)^{7}Li$ observed in a silicon surface barrier detector. The boron targets are prepared by electron bombardment of metallic boron, enriched to 94.5% in ^{10}B, and are vacuum deposited in $\approx 10\ \mu g\,cm^{-2}$ layers on 0.4 mm thick silicon crystal wafers (Pauwels *et al* 1991).

neutron cross-section is large and proportional to $1/v$ over the whole range of energy from 0.01 eV to 0.1 keV, and that the reaction always goes directly to the ground state of ^{3}H. The main disadvantage is one of availability; ^{3}He comprises only about 10^{-4}% of natural helium and must be produced artificially from the β^- decay of 12.3 y ^{3}H.

1.7.1.3 Neutron activation

Neutron detectors based on (n, α) or (n, p) reactions as described above have an instantaneous output, and respond immediately to fluctuations in intensity of the incident neutron beam. Neutron activation on the other hand relies on determining the radioactivity induced by (n, γ) reactions in a thin target of material exposed to a neutron beam. Such a device is therefore an integrating detector and only gives information about the time-averaged neutron flux over the period of irradiation. The method has the advantage, however, that the activated targets can be made so thin as not to introduce any significant perturbation in the flux to be measured, and, after some preselected time, can be removed from the beam and the induced activity measured in an external counter.

Table 1.5. Pure '$1/v$' (n, γ) reactions used for absolute neutron density measurements (Holden 1981)

Nucleus	Abundance (%)	Reaction	$T_{1/2}$	σ(b)	$\delta\sigma/\sigma$ (%)
^{55}Mn	100	^{55}Mn(n, γ)^{56}Mn	2.58 hours	13.3 ± 0.2	1.5
^{197}Au	100	^{197}Au(n, γ)^{198}Au	2.695 days	98.65 ± 0.09	0.09
^{59}Co	100	^{59}Co(n, γ)^{60m}Co	10.4 min	37.18 ± 0.06	0.16

Because the radioisotope produced in the target has a finite half-life it is necessary to keep very accurate records of the absolute times of activation and counting, and a very precise procedure must be followed in order to deduce the flux from the measured activity. One particular advantage of the method is that it can be applied to yield separate estimates of the fluxes in the thermal and resonance energy regions by measuring the so-called cadmium ratio of the activated foil. This technique makes use of the fact that ^{113}Cd has a broad resonance with cross-section $\sigma = 5.7 \times 10^4$ b at $E_n = 0.176$ eV, which effectively removes all neutrons with energies <0.4 eV while transmitting all neutrons at higher energy. The cadmium ratio, which is defined as the ratio of the activity produced under standard conditions in the cadmium covered target foil, to that in the uncovered foil, therefore provide a measure of the relative flux in the resonance region.

Although very many different materials have been used for neutron activation, three elements are particularly useful for this purpose since they are mononuclidic and their $1/v$ cross-sections have all been determined to high accuracy. These are the isotopes ^{55}Mn, ^{197}Au and ^{59}Co and their properties are listed in Table 1.5. Two of the resultant radioisotopes are β^--emitters, i.e. ^{56}Mn and ^{198}Au. The third isotope ^{60m}Co is an isomeric state of ^{60}Co which decays by γ-emission to its ground state. The (n, γ) cross-section in ^{197}Au departs very slightly from the $1/v$ law and is characterized by a value of $g = 1.0053$ at 293 K, with $dg/dT \simeq 6 \times 10^{-5}$ K^{-1}.

Because they have strong (n, γ) resonances at 337 eV and 4.9 eV, respectively, the isotopes ^{55}Mn and ^{197}Au make very good detectors of epithermal neutrons. The isotopes 54 min ^{115}In (1.44 eV), 2.4 h ^{186}W (18.8 eV) and 40 h ^{139}La (73.5 eV) have also been widely used for the same purpose.

1.7.1.4 Fission detectors

As noted already in section 1.6.4 the fissile isotopes ^{233}U, ^{235}U and ^{239}Pu have large fission cross-sections for slow neutrons with positive Q-values of the order of 200 MeV. Since 80% of the energy released per fission appears

Table 1.6. Westcott g-factors and their temperature coefficients for capture (g_c) and fission (g_f) in ^{233}U, ^{235}U and ^{239}Pu (Westcott 1960)

Nucleus	g_c (293 K)	dg_c/dT (K^{-1})	g_f (293 K)	dg_f/dT (K^{-1})
^{233}U	0.9983	3.0×10^{-5}	1.0003	3.3×10^{-5}
^{235}U	0.9780	-6.2×10^{-5}	0.9757	-7.4×10^{-5}
^{239}Pu	1.0723	2.23×10^{-3}	1.0487	1.78×10^{-3}

as kinetic energy of the fission fragments, a detector which records these nuclei makes an excellent neutron detector with almost ideal discrimination against γ-rays. The technique usually employed is to allow the neutrons to enter an ionization chamber lined with fissile material and the fission fragment output is then recorded in pulse mode or in current mode. Because of the very high Q-value, fission chambers provide no information about the energy of the neutrons which they detect.

None of the three fissile isotopes named above has a pure $1/v$ cross-section either for capture or fission, and the corresponding g-factors and their temperature coefficients are listed in Table 1.6.

1.7.2 Fast neutron detectors

1.7.2.1 Fast neutron moderation

The simplest method for detecting a fast neutron is first to thermalize it by elastic scattering, and then to record it in a conventional slow neutron counter. A convenient detector of this type is the so-called Bonner sphere which employs a 20–25 cm spherical moderator of paraffin or polyethylene in conjunction with a LiI(Eu) scintillator or ^3He proportional counter located at the centre of the sphere. The main disadvantages of this type of counter are that all information relating to the energy of the incoming neutron is lost, the efficiency ranging from $10^{-2}\%$ to 1% is low, and the response time is very long ($\simeq 10^{-5}$ s).

1.7.2.2 Recoil detectors

When fast neutrons are elastically scattered from nuclei, the energy carried off by the recoiling nucleus is uniquely determined by the angle of scatter, and only in the case of scattering from protons can the incoming neutron transfer all its energy in a single collision. Thus almost all recoil detectors make use of hydrogenous material. The scattering from hydrogen is isotropic in the centre of mass reference frame and this means that the spectrum of proton recoils is essentially rectangular varying from almost zero for scattering at right angles, to a value equal to E_n for forward scattering. Thus very poor energy resolution is available and matters are made worse when multiple collisions begin to play a role. Detectors of this type have a relatively fast response ($\simeq 0.2\ \mu s$), and efficiencies as high as 5% can be achieved by detecting proton recoils in a hydrogenous plastic scintillator using a ZnS(Ag) phosphor.

To improve the energy resolution of this type of detector a proton recoil telescope may be used. This selects protons ejected at a fixed angle from a hydrogenous converter and records them in two plastic scintillators whose outputs are recorded in coincidence. The first scintillator is of transmission type and removes a small fraction ΔE_p of the energy of the transmitted proton. The second scintillator records the full energy of the proton reaching it and the energy outputs of both counters are then summed.

Gas proportional counters based on ^2H or ^4He gas can also be used in the recoil mode but these are less efficient than scintillators, never achieving efficiencies above 1%.

1.7.2.3 Charged particle reactions

The best energy resolution with fast neutrons is obtained using the reactions $^6Li(n,\alpha)^3H$ and $^3He(n,p)^3H$ where the outgoing charged particle has an energy equal to the sum of the incoming neutron energy and the Q-value for the reaction. The reactions may be observed in the usual way using various types of scintillator or proportional counter. The reaction $^{10}B(n,\alpha)^7Li$ is not suitable for fast neutron spectroscopy because there are two α-branches in the spectrum whose relative intensity is a rapidly varying function of neutron energy up to 2 MeV. In addition to the full-energy peak both ^6Li and ^3He based detectors exhibit a continuous background due to elastic scattering of fast neutrons, plus an epithermal peak at the Q-value produced by neutrons which have been moderated in the detector or its surroundings.

2 Neutron scattering and two-particle interactions

2.1 Fundamental interactions of the neutron

2.1.1 Strong, weak and electromagnetic interactions

An elementary quantum system may be defined as a system whose states transform according to an irreducible representation of the inhomogeneous Lorentz group, labelled by the mass m and the spin I. A particle may be described as 'elementary' provided it continues to exist as a distinguishable entity for a length of time which is large in comparison with its natural period $\hbar/mc^2$. For a neutron this intrinsic time is $\simeq 10^{-24}$ s, a time-scale which is much less than the β-decay lifetime $\tau_n \simeq 10^3$ s. Thus, by this yardstick, it is perfectly legitimate to describe the neutron as 'elementary' (Aleksandrov 1992). The same criterion would also include the deuteron but this particle, like all other atoms and nuclei, can be eliminated from the elementary category if it is agreed to exclude all systems which can be decomposed into simpler constituents by the absorption of electromagnetic radiation. On this view stable matter may be regarded as composed from four elementary spin $\frac{1}{2}$ fermions; n, p, e^- and $\bar{\nu}_e$. Two of these basic fermions, n and $\bar{\nu}_e$, are uncharged and two, p and e^- are charged. The electromagnetic interactions of the charged particles, plus the neutron by virtue of its magnetic moment, are transmitted via the exchange of uncharged vector bosons, called photons (γ), whose vanishing mass is intimately linked with the conservation of electric charge, in a symmetry principle known as gauge invariance.

Electromagnetic interactions are responsible for the existence of stable atoms characterized by definite values of mass and spin and, because nucleons can also bind themselves into stable nuclei similarly characterized, there must exist short-range attractive forces between pairs of nucleons which are an order of magnitude stronger than the Coulomb repulsion between pairs of protons at the same radial separation. The presence of these forces distinguishes in a clear cut way (which electromagnetic forces do not) between the heavy fermions or baryons (n, p), which experience strong interactions, and the light fermions or leptons (e^-, $\bar{\nu}_e$) which do not interact in this way, indicating that, for low energy phenomena at least, a primitive form of baryon–lepton symmetry applies in the physical world.

In addition to their strong and electromagnetic interactions all four fermions participate in the weak interaction responsible for nuclear β-decay, whose simplest manifestation is the disintegration of a neutron into a proton, electron and anti-neutrino, i.e.

$$n \rightarrow p + e^- + \bar{\nu}_e \tag{2.1}$$

Protons bound in nuclei may also be β-active leading to the emission of positrons and neutrinos

$$p \rightarrow n + e^{+} + \nu_{e} \tag{2.2}$$

but such a mode of decay is not available for protons on the 'mass-shell', i.e. for protons satisfying the usual energy–momentum relation of free particles.

The convention which describes a process as strong, electromagnetic or weak is based on the measured relative magnitudes of the matrix elements of the corresponding transition operators evaluated between specified physical states. These matrix elements are observed to fall in the approximate ratio $1:10^{-2}:10^{-7}$ and correspond in the general case to well-defined selection rules. On this scale two-body gravitational matrix elements are of order 10^{-42} and may be disregarded in their entirety.

2.1.2 Strong interactions and potential scattering

In considering the best means of investigating the nature of the strong interactions of nucleons we are naturally led to enquire how the equivalent problems have been tackled in atomic physics. It turns out that the most sensitive tests of quantum electrodynamics (QED), the theory which encompasses all atomic phenomena, rely very heavily on precision measurements of a few selected transition frequencies. Most notable in this respect are the determination of frequency splitting between the cyclotron and Larmor spin precession frequencies of electrons in a magnetic field (electron g-factor anomaly) and the $2S_{1/2} - 2P_{1/2}$ splitting in atomic hydrogen (Lamb shift). Unfortunately systems of comparable simplicity do not exist in strong interaction physics. In particular, although there is an infinity of two-body bound states in atomic physics, not only in baryon–lepton systems such as neutral hydrogen, ionized helium, etc., but also in purely leptonic systems such as positronium or muonium, in the whole realm of nuclear physics there is but a single two-body bound state, namely the triplet ground-state of the deuteron. We, therefore, have this difference that, whereas in atomic physics attention has been directed primarily to the negative energy bound-state solutions of the Schrödinger equation, in nuclear physics the emphasis is placed on the positive energy scattering solutions which behave asymptotically like incoming or outgoing waves. The connection between the two types of solution is provided by the Jost function (cf. section 2.3) and the zeros of this function (poles in the scattering matrix) represent the bound state solutions. Other singularities correspond to virtual and resonance states which, so far as their influence on the scattering parameters is concerned, rank equally in importance with bound states.

In order to establish the main characteristics of the nucleon–nucleon force the basic p–p, n–p and n–n scattering processes are clearly of prime importance, and it is therefore particularly unfortunate that the last named process cannot be studied in the laboratory, because there exist no dense targets composed entirely from neutrons. The best that can be achieved is to study n–d or p–d scattering, but these are three-body reactions, and the experimental data are inherently difficult to interpret since one must take account of the influence of stationary or quasi-stationary states of ^3_1H or ^3_2He, plus the contributions of various inelastic channels which involve the exchange of incident and target particles.

In the general case, since there exists no guiding theory, it is necessary to introduce some simplifying assumptions about the scattering mechanism and it is normally taken for granted, at least in the case of low-energy elastic scattering, that incident and target particles interact through a conservative force. This assumption is expressed by postulating some simple form of scalar potential and the resultant scattering is therefore described as potential scattering. Arranged in order of increasing complexity the most commonly employed two-parameter potential functions are

$$\begin{aligned}
&\text{(i)} \quad \text{Square well} && V(r) = V_0 \quad r \leqslant b \\
& && = 0 \quad r > b \\
&\text{(ii)} \quad \text{Exponential} && V(r) = -V_0 e^{-r/b} \\
&\text{(iii)} \quad \text{Yukawa} && V(r) = -V_0 \frac{b}{r} e^{-r/b}
\end{aligned} \tag{2.3}$$

All of these potential functions give attractive forces for $V_0 > 0$.

The square well potential (i) is the easiest to treat analytically since the elements of the corresponding S-matrix (cf. section 2.3) are regular everywhere in the complex (momentum) k-plane, excluding the bound state poles and an essential singularity at infinity associated with the strictly finite range of the potential. The solution of the scattering problem using this potential is applied in section 2.10 to describe the ground state of the deuteron, and to derive the main features of low-energy neutron–proton scattering.

The exponential potential (ii) also results in a Schrödinger equation which is soluble in terms of standard transcendental functions, and which provides in particular a very simple formula for the bound state energies $-E_b$ ($E_b > 0$). These energies correspond to those values ν which provide solutions to the equations

$$J_\nu(z) = 0 \qquad \nu = 2b\left(\frac{2mE_b}{\hbar^2}\right)^{1/2} \qquad z = 2b\left(\frac{2mV_0}{\hbar^2}\right)^{1/2} \tag{2.4}$$

where m is the reduced mass of the two-body system defined in equation (2.30) and $J_\nu(z)$ is the Bessel function of order ν (non-integral in general). Since the smallest Bessel function zero z_ν occurs at $\nu = 0$, corresponding to a bound state at zero energy, it follows that in general $z_\nu \geqslant z_0 = 2.404$, and, for a bound state to exist, V_0 must satisfy the inequality

$$\frac{2mV_0}{\hbar^2}\cdot b^2 \geqslant 1.445 \tag{2.5}$$

A very useful variant of the exponential potential is the three-parameter Morse potential (Morse 1929):

$$V(r) = V_0\{e^{-2(r-r_m)/b} - 2e^{-(r-r_m)/b}\} \qquad r_m = r_c + b\ln(2) \tag{2.6}$$

which is attractive for $r > r_c$ and repulsive for $r < r_c$. It therefore provides a simple way of incorporating into the scattering potential the short-range repulsive core which is required to account for the phenomenon of saturation in nuclei (cf. sections 3.1 and 3.3), and which is also observed as a component in higher energy nucleon–nucleon scattering (cf. section 3.4).

Finally we consider the Yukawa potential (iii) which is also called the 'screened Coulomb potential', since it was first encountered in the theory of electrolytes where the range 'b' is referred to as the 'Debye screening length'. It is of particular interest in nuclear physics since it arises naturally in meson theories of the nuclear force first proposed by Yukawa (cf. section 3.5) where the range 'b' is identified according to equation (3.95) with the reduced Compton wavelength $\hbar/m_\pi c$ of the exchanged meson. The theory of scattering in a Yukawa potential is not elementary since the corresponding Jost function has a branch point in the finite part of the complex k-plane, a restriction which sets an upper limit on the energy of any bound state

$$E_b \leqslant \hbar^2/8mb^2 = (m_\pi c^2)^2/8(mc^2) \tag{2.7}$$

independent of the strength V_0 of the potential. If $m_\pi = 0$ the Yukawa potential reduces to the Coulomb potential which has an infinite number of bound states and, in this context, describes the exchange of massless scalar bosons. However, no such strongly interacting scalar particles are known to exist, and the Coulomb potential is quite unsuitable as a candidate to describe strong interaction scattering.

The one-pion exchange potential (3.109), also contains derivatives of the Yukawa potential, which behave at the origin like r^{-n} ($n \geqslant 2$) and therefore ought to be excluded from consideration by the regularity condition (2.58). However, this restriction causes no real difficulty in practical calculations, because the immediate neighbourhood of the origin is excluded by introducing the required hard core potential with radius $r_c \simeq 0.5$ fm.

2.1.3 Electromagnetic interactions

2.1.3.1 The quantized radiation field

Quantum electrodynamics is a formal mathematical theory erected on the basis of Einstein's hypothesis (1905), that plane electromagnetic waves in empty space propagate as a beam of free particles of energy $\hbar\omega$ and momentum $\hbar\boldsymbol{k}$, where the wave vector $\boldsymbol{k}$ has magnitude $k=\omega/c$, and is oriented along the energy flux vector $=\boldsymbol{E}\times\boldsymbol{B}/c$. That a quantized theory of the electromagnetic field is a logical necessity follows from the work of Bohr and Rosenfeld (1933) and others, who showed that the field vectors $\boldsymbol{E}$ and $\boldsymbol{B}$ must be treated as quantum mechanical operators rather than as ordinary vector functions, if Heisenberg's uncertainty principle $\Delta x\,\Delta p\simeq\hbar$ is to retain its validity for charged particles. In the following we summarize very briefly how the quantization of the theory is brought about, using a non-covariant formalism based on the Fourier decomposition of the radiation field (Heitler 1954).

It is well known that Maxwell's classical theory, which relates the vectors $\boldsymbol{E}$ and $\boldsymbol{B}$ to their sources through a set of first-order linear differential equations in space and time, can be expressed most economically in terms of a reduced number of second-order equations satisfied by potential functions $\boldsymbol{A}$ and φ, where

$$\boldsymbol{E}=-\frac{1}{c}\frac{\partial\boldsymbol{A}}{\partial t}-\nabla\varphi \qquad \boldsymbol{B}=\nabla\times\boldsymbol{A} \tag{2.8}$$

Equations (2.8) do not define $\boldsymbol{A}$ and φ uniquely, and these functions may be subjected to a range of gauge transformations

$$\boldsymbol{A}\rightarrow\boldsymbol{A}-\nabla X \qquad \varphi\rightarrow\varphi+\frac{1}{c}\frac{\partial X}{\partial t} \tag{2.9}$$

which leave $\boldsymbol{E}$ and $\boldsymbol{B}$ invariant. The symmetry transformation (2.9) is directly related to the conservation of electric charge which is identically satisfied in the Maxwell theory. In the absence of sources it is always possible to select a gauge function X, the so-called radiation gauge, such that

$$\varphi=0 \text{ (i)} \qquad \nabla\cdot\boldsymbol{A}=0 \text{ (ii)} \tag{2.10}$$

in which case the vector potential $\boldsymbol{A}$ satisfies the vector Helmholtz equation

$$\nabla^2\boldsymbol{A}-\frac{1}{c}\frac{\partial^2\boldsymbol{A}}{\partial t^2}=0 \tag{2.11}$$

In the classical theory the solutions of (2.11) describe vector waves propagating with velocity c, and the Lorentz condition (2.10(ii)) expresses the fact that these wave solutions are purely transverse. From a quantum viewpoint equation (2.11) is immediately recognizable as the wave equation (analogous to the Klein–Gordon equation) of a massless vector particle which is restricted by the transversality condition to have two, rather than three, independent states of polarization. Specifically, the photon has available to it two states of longitudinal spin angular momentum, referred to its own momentum, with quantum numbers $m=\pm1$, but the state of transverse spin polarization with $m=0$ does not exist.

The procedure adopted now is to expand the real potential $\boldsymbol{A}(\boldsymbol{r}, t)$ in terms of eigensolutions of (2.11) confined in a cube of volume L^3

$$\boldsymbol{A}(\boldsymbol{r}, t) = \sum_{j=1,2} \sum_{\boldsymbol{k}} (N_k q_{jk}(t)\boldsymbol{a}_j \mathrm{e}^{\mathrm{i}\boldsymbol{k}\cdot\boldsymbol{r}} + N_k^* q_{jk}^*(t)\boldsymbol{a}_j^* \mathrm{e}^{-\mathrm{i}\boldsymbol{k}\cdot\boldsymbol{r}})$$

$$\boldsymbol{k} = 2\pi\boldsymbol{v}/L \qquad \boldsymbol{a}_j \cdot \boldsymbol{k} = 0 \tag{2.12}$$

where $\boldsymbol{v}$ is a triad of positive or negative integers, N_k is a normalization constant, and the amplitudes $q_{jk}(t)$ satisfy the one-dimensional harmonic oscillator equation

$$\ddot{q}_{jk}(t) + \omega_k^2 q_{jk}(t) = 0 \qquad \omega_k = ck \tag{2.13}$$

To write the field equations in Hamiltonian form it is necessary to calculate the total energy U, which after some considerable effort, is found to be

$$\begin{aligned} U &= \frac{1}{8\pi} \int_{L^3} (\boldsymbol{E}^2 + \boldsymbol{B}^2)\,\mathrm{d}^3\boldsymbol{r} \\ &= \frac{1}{8\pi} \int_{L^3} \left\{ \left(-\frac{1}{c}\frac{\partial \boldsymbol{A}}{\partial t} \right)^2 + (\nabla \times \boldsymbol{A})^2 \right\} \mathrm{d}^3\boldsymbol{r} \\ &= \sum_{jk} \frac{4L^3 (N_k)^2}{8\pi} \left(\frac{\omega_k}{c} \right)^2 q_{jk}^* q_{jk} \end{aligned} \tag{2.14}$$

The normalization constant N_k is now chosen to have the value $(2\pi\hbar c^2/L^3\omega_k)^{1/2}$ so that the quantization volume L^3 disappears from the equation and may be chosen indefinitely large, and each characteristic mode $(j, \boldsymbol{k})$ is represented by a term in the Hamiltonian, of a form consistent with Einstein's hypothesis

$$H_{jk} = q_{jk}^* q_{jk} \hbar\omega_k \tag{2.15}$$

The eigenfunction equation for H_{jk} may now be solved by identifying the real and imaginary parts of the time-dependent complex amplitude $q_{jk}(t)$ with time-independent Hermitian operators which are then made to satisfy canonical commutation rules in the usual way, but this is achieved more directly by replacing the complex amplitudes q_{jk}, q_{jk}^* by non-Hermitian operators q_{jk}, $q_{jk}^\dagger$ which are now subjected to the boson quantization rules

$$q_{jk} q_{j'k'}^\dagger - q_{j'k'}^\dagger q_{jk} = \delta_{jj'}\delta_{kk'} \tag{2.16}$$

The resultant normalized eigenfunctions $|n_{jk}\rangle$ are labelled by the photon number n_{jk}, and the energy eigenvalues are given by the familiar formula

$$E_{jk} = n_{jk}\hbar\omega_k \tag{2.17}$$

The operators q_{jk}, $q_{jk}^\dagger$ have non-vanishing matrix elements only between states $|n_{jk}\rangle$ with occupation numbers differing by no more than one photon, i.e.

$$\langle n_{jk}-1|q_{jk}|n_{jk}\rangle=\sqrt{(n_{jk})} \qquad \langle n_{jk}+1|q_{jk}^{\dagger}|n_{jk}\rangle=\sqrt{(n_{jk}+1)} \tag{2.18}$$

Thus the operator q_{jk} is a photon annihilation operator which annihilates the vacuum state $|0\rangle$, while $q_{jk}^{\dagger}$ is a photon creation operator which, acting on the vacuum state $|0\rangle$, creates the one photon state $|1\rangle$. The presence of the term n_{jk} in the amplitude $\sqrt{(n_{jk}+1)}$, shows that photons can be created by induced emission as well as by spontaneous emission, as described by the '1' in the same amplitude. According to (2.18) the operator $q_{jk}^{\dagger}q_{jk}$ has eigenvalue n_{jk} and must be identified with the photon number operator $\hat{n}_{jk}$, thus leading to a simple representation of the single mode Hamiltonian (2.15).

$$H_{jk}=\hat{n}_{jk}\hbar\omega_k \tag{2.19}$$

2.1.3.2 Interaction of field and particle

We consider a class of problem in which an initial state $|\mathrm{i}\rangle$, which specifies the states of all charged particles, and the numbers and types of all photons present, and has total energy E_{i}, undergoes a transition to a final state $|\mathrm{f}\rangle$ of total energy E_{f}. Then, since the coupling between field and particles is characterized by the fine structure constant $\alpha=e^2/\hbar c\simeq 1/137\ll 1$, it is legitimate to apply perturbation theory and the transition probability per unit time is given by

$$w=\frac{2\pi}{\hbar^2}|\langle\mathrm{f}|H_{\mathrm{int}}|\mathrm{i}\rangle|^2\rho_{\mathrm{f}}\delta(E_{\mathrm{i}}-E_{\mathrm{f}}) \tag{2.20}$$

where $\langle\mathrm{f}|H_{\mathrm{int}}|\mathrm{i}\rangle$ is the matrix element of the interaction Hamiltonian connecting initial and final states, and $\rho_{\mathrm{f}}(E_{\mathrm{f}})$ is the density of final states per unit energy interval. The total Hamiltonian of field and particle in the non-relativistic limit may be written in the form (cf. section 9.4.2)

$$H=e\varphi+\frac{1}{2m}\left(\boldsymbol{p}-e\frac{\boldsymbol{A}}{c}\right)^2-\boldsymbol{\mu}\cdot\boldsymbol{B}+\frac{1}{8\pi}\int(\boldsymbol{E}^2+\boldsymbol{B}^2)\,\mathrm{d}^3r \tag{2.21}$$

where $\boldsymbol{\mu}$ represents any intrinsic magnetic moment the particle may carry.

In problems which involve the annihilation or creation of real (transverse) photons, it is customary to separate off the Coulomb potential energy $e\varphi$, together with any part of the magnetic energy $-\boldsymbol{\mu}\cdot\boldsymbol{B}$ which arises from an interaction with a static magnetic field, and treat both these terms as interactions with classical unquantized fields which contribute to the unperturbed Hamiltonian. In this case the interaction energy is given by

$$H_{\mathrm{int}}=-\frac{e}{2mc}(\boldsymbol{p}\cdot\boldsymbol{A}+\boldsymbol{A}\cdot\boldsymbol{p})-\boldsymbol{\mu}\cdot(\nabla\times\boldsymbol{A})+\frac{e^2}{2mc^2}A^2 \tag{2.22}$$

where $\boldsymbol{A}$ is the quantized radiation field given in (2.12). Of the three terms in H_{int}, two are linear in $\boldsymbol{A}$, and hence in q and $q^{\dagger}$, corresponding to the absorption or emission of a single quantum. These two terms contain both field and particle variables and are scalars formed from the contraction of the two polar vectors $\boldsymbol{p}$ and $\boldsymbol{A}$, or the two axial vectors $\boldsymbol{\mu}$ and $\nabla \times \boldsymbol{A}$ (cf. section 3.3.1). The term quadratic in $\boldsymbol{A}$ contains no particle variables and corresponds to a two quantum process involving the field variables only, and, since it is of order $\sqrt{\alpha}$ in comparison with the leading terms, may be neglected in first order. In second order its contribution must be taken along with second-order contributions of the single quantum terms. The presence of the two quantum term is a feature of the non-relativistic theory which does not persist when the fully relativistic Hamiltonian is employed. Its existence nevertheless signals the fact that to each first-order single quantum process there corresponds a weaker coherent scattering process in second order. For example in the case of the photodisintegration of the deuteron discussed in section 2.10, the corresponding coherent scattering process is referred to as nuclear Thomson scattering and represents the scattering of the incoming γ-radiation from the total charge on the deuteron.

One should take note of the fact that, because of the presence of the normalization constant N_k in $\boldsymbol{A}$, the radiation matrix elements contribute a factor proportional to L^{-3} in w, the transition probability per unit time given by (2.20), corresponding either to one incoming or one outgoing photon. This is automatically cancelled by the volume factor L^3 which appears in ρ_f and no difficulty arises. In the case of a scattering process there is a difficulty, because in this case there is an additional incoming or outgoing particle, whose normalization in the quantization volume produces an additional factor L^{-3}, expressing the fact that the larger the quantization volume, the smaller the probability of interaction. This difficulty is avoided by defining a scattering cross-section σ_s through the relation

$$w = \sigma_s v_r / L^3 \tag{2.23}$$

where v_r is the relative velocity of incident and target particles. In general, therefore, it is always permissible to assume that L^3 is so large that sums may be replaced by integrals, and L can be set equal to unity without loss of generality.

The electromagnetic interaction of two particles, expressed as a sum of terms in a perturbation expansion in H_{int}, can be most conveniently visualized by means of Feynman diagrams which associate incoming and outgoing external lines with each initial or final state particle, including both material particles and photons, and the number of interaction vertices is equal to the order of the corresponding term in the expansion. This is illustrated in figure 2.1 which shows the two lowest-order Feynman diagrams which contribute to electron–positron scattering. To each such topologically distinct diagram corresponds a definite matrix element which can be evaluated according to the Feynman rules (Jauch and Rohrlich 1955), and the totality of such

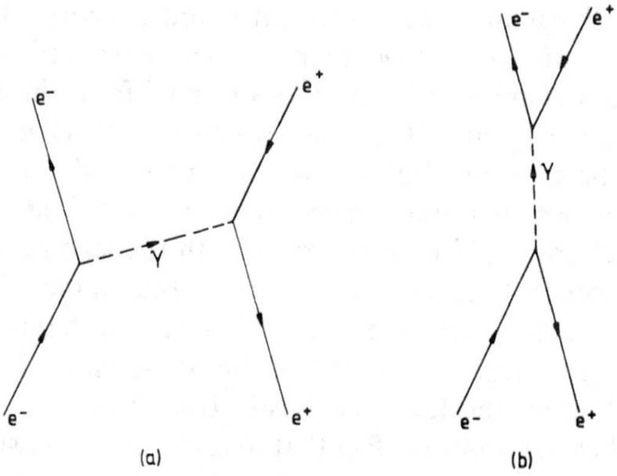

Figure 2.1. Feynman diagrams representing electron–positron (Bhabha) scattering in lowest order. According to the formalism a positron is treated as an electron with its four-momentum reversed, and with the roles of initial and final states exchanged. The process is represented as the exchange of (a) spacelike and (b) time-like virtual photons.

matrix elements is used to evaluate the transition probability for the process in question. In practical cases, only the lowest-order diagrams give rise to finite matrix elements and the higher orders always lead to divergent integrals. Nevertheless, the rigorous enforcement of Lorentz and gauge invariance, leading to the elimination of those features of the theory which are inherently unobservable, has led to the development of methods, based on the renormalization of charge and mass, whereby these infinities may be dealt with in a consistent manner. This renormalization programme has been highly successful and has permitted the extraction of finite results from the theory which are in the closest accord with the most precise measurements as described above. However, these are concerns which cannot be pursued here and Feynman diagrams will be employed, where appropriate, for purposes of illustration only.

2.2 Two-particle collisions

2.2.1 Laboratory and centre of mass reference frames

We consider the system composed of two particles in non-relativistic motion, of masses m_1 and m_2 respectively, located at positions $\boldsymbol{r}_1$ and $\boldsymbol{r}_2$ in an arbitrary reference frame. For such a system we can define a relative coordinate $\boldsymbol{r}$, and a centre of mass coordinate $\boldsymbol{R}$, through the relations

$$(m_1+m_2)\boldsymbol{R}=m_1\boldsymbol{r}_1+m_2\boldsymbol{r}_2 \qquad \boldsymbol{r}=\boldsymbol{r}_1-\boldsymbol{r}_2 \tag{2.24}$$

An obvious frame of reference in which to desribe the interaction of the two particles is the laboratory frame in which one of the particles, say that of mass m_2, is initially at rest, and in a collision process this particle is identified as the target particle. There are of course an infinite number of laboratory frames but it is usually convenient to choose that one for which the initial coordinate $\boldsymbol{r}_2(0)=0$.

Another important class of reference frames is composed of those frames in which the total momentum is zero. These are the centre of mass frames for which

$$m_1\dot{\boldsymbol{r}}_1+m_2\dot{\boldsymbol{r}}_2=0 \qquad \dot{\boldsymbol{R}}=0 \tag{2.25}$$

and in these frames the centre of mass is at rest.

For an arbitrary reference frame the two-particle Schrödinger equation may be written in an obvious notation

$$\left[-\frac{\hbar^2}{2m_1}\nabla_1^2-\frac{\hbar^2}{2m_2}\nabla_2^2+V(\boldsymbol{r}_1,\boldsymbol{r}_2)\right]\Psi=E\Psi \tag{2.26}$$

where $V(\boldsymbol{r}_1,\boldsymbol{r}_2)$ is a real potential function which, in order to satisfy translational invariance may be assumed to depend on the relative position, but not on the absolute positions of the two particles, i.e.

$$V(\boldsymbol{r}_1,\boldsymbol{r}_2)=V(\boldsymbol{r}_1-\boldsymbol{r}_2)=V(\boldsymbol{r}) \tag{2.27}$$

In these circumstances $\Psi(\boldsymbol{r}_1,\boldsymbol{r}_2)$ can be represented as a product function

$$\Psi(\boldsymbol{r}_1,\boldsymbol{r}_2)=\Psi(\boldsymbol{r})\varphi(\boldsymbol{R}) \tag{2.28}$$

whose separate factors satisfy the equations

$$\text{(i)}\quad \nabla_r^2\Psi+\frac{2m}{\hbar^2}(E_{\text{I}}-V(\boldsymbol{r}))\Psi=0$$

$$\text{(ii)}\quad \nabla_R^2\varphi+\frac{2M}{\hbar^2}(E_{\text{F}})\varphi=0 \tag{2.29}$$

where M is the total mass, and m the reduced mass of the system defined by

$$M=m_1+m_2 \qquad m=m_1m_2/(m_1+m_2) \tag{2.30}$$

The total energy E is now expressed as the sum of two terms

$$E=E_{\text{F}}+E_{\text{I}} \tag{2.31}$$

where E_{F} is the energy associated with the free motion of a particle of mass M located at the centre of mass and E_{I} is the energy associated with the relative motion of the two particles.

The solutions of (2.29)(ii) are eigenfunctions, not of $\boldsymbol{R}$ but of the centre of mass momentum $\boldsymbol{P}=\boldsymbol{p}_1+\boldsymbol{p}_2$, which is canonically conjugate to $\boldsymbol{R}$. The energy eigenvalues E_{F} are therefore equal to $\hbar^2K^2/2M$ where $\hbar\boldsymbol{K}$ are the eigenvalues of $\boldsymbol{P}$. The details of the division of the total energy into its two components depend of course on the choice of reference frame, but a centre of mass frame represents an obvious choice for in such a frame E_F is zero and φ is constant over all space. The physical situation is therefore completely described by the Schrödinger equation

$$\nabla^2\Psi + \frac{2m}{\hbar^2}(E - V(\boldsymbol{r}))\Psi = 0 \tag{2.32}$$

which represents the motion of a free particle of mass equal to the reduced mass, moving in the central potential $V(\boldsymbol{r})$. An alternative view is to regard the functions $\Psi(\boldsymbol{r}_1, \boldsymbol{r}_2)$ defined in (2.28) as degenerate eigenfunctions of a Hamiltonian obtained by subtracting the centre of mass energy $\boldsymbol{P}^2/2M$ from the Hamiltonian defined in (2.26).

The relations between the energy (E), momentum (p) and angle of scatter (θ) in the centre of mass frame, and the corresponding quantities (E_L, p_L and θ_L) in the laboratory frame are

$$E_L = [(m_1 + m_2)/m_2]E \qquad p_L = [(m_1 + m_2)/m_2]p$$
$$\cos\theta_L = (m_1 + m_2\cos\theta)/(m_1^2 + 2m_1m_2\cos\theta + m_2^2)^{1/2} \tag{2.33}$$

For a neutron moving in the field of a heavy nucleus, $m_1 = m_n$, $m_2 = M(A, Z) \gg m_n$ and there is no significant difference between the laboratory and centre of mass scattering parameters. However, for the scattering of neutrons on protons the situation is very different, for in this case $m_1 = m_n \simeq m_2 = m_p$ and we have

$$m = m_p/2 = m_n/2 \qquad E_L = 2E \qquad p_L = 2p \qquad \theta_L = \theta/2 \tag{2.34}$$

In particular, the momentum relation $p_L = 2p$ implies that the wavelength λ, which is appropriate to a theoretical discussion of scattering in the centre of mass frame, is twice the wavelength λ_L selected by a neutron crystal monochromator at rest in the laboratory.

2.2.2 An integral equation for scattering

To discuss the scattering of two particles of given spin, interacting through a scalar potential $V(\boldsymbol{r})$ in the centre of mass, it is convenient to rewrite the Schrödinger equation (2.32) in the form of a free-particle equation with a source term, i.e.

$$\nabla^2\Psi + \frac{2m}{\hbar^2}E\Psi = \frac{2m}{\hbar^2}V(\boldsymbol{r})\Psi \tag{2.35}$$

whose solution for positive energy E may be expressed as an integral equation for $\Psi(\boldsymbol{r})$

$$\Psi(\boldsymbol{r}) = \Psi_0(\boldsymbol{r}) + \int \mathrm{d}^3r'\, G(\boldsymbol{r} - \boldsymbol{r}')\Psi(\boldsymbol{r}')V(\boldsymbol{r}') \tag{2.36}$$

Here $\Psi_0(\boldsymbol{r}) = \mathrm{e}^{i\boldsymbol{k}\cdot\boldsymbol{r}}$ is a solution of the homogeneous equation

$$\nabla^2\Psi_0(\boldsymbol{r}) + \frac{2m}{\hbar^2}E\Psi_0(\boldsymbol{r}) = 0 \tag{2.37}$$

having the same energy E as $\Psi(\boldsymbol{r})$, and $G_0(\boldsymbol{r}-\boldsymbol{r}')$ is the free particle Green function defined by

$$G_0(\boldsymbol{r}) = \frac{-m}{2\pi\hbar^2}\frac{\mathrm{e}^{ikr}}{r} \tag{2.38}$$

where k is the positive square root

$$k = \sqrt{(2mE)}/\hbar \tag{2.39}$$

$G_0(\boldsymbol{r})$ is a solution of the free-particle inhomogeneous equation

$$\nabla^2 G_0(\boldsymbol{r}) + \frac{2m}{\hbar^2}EG_0(\boldsymbol{r}) = \delta^3(\boldsymbol{r}) \tag{2.40}$$

which represents spherical waves travelling outwards from a singular source located at the origin.

Since the potential function $V(\boldsymbol{r})$ is real the probability current density given by

$$\boldsymbol{j}(\boldsymbol{r}) = \frac{\hbar}{2m\mathrm{i}}\{\Psi^*\nabla\Psi - \nabla\Psi^*\Psi\} \tag{2.41}$$

is conserved, i.e.

$$\nabla\cdot\boldsymbol{j} + \frac{\partial}{\partial t}|\Psi(\boldsymbol{r})|^2 = 0 \tag{2.42}$$

For convenience it is customary to select the axes so that the incident particle is described by the plane wave e^{ikz}, multiplied by a spin function $\chi(\mathrm{s})$ specifying whether, for example, the collision of two spin $\frac{1}{2}$ particles takes places in a singlet or in a triplet state. The scattered particles are recorded at very large distances from the scattering centre thus we may write in (2.36)

$$k|\boldsymbol{r}-\boldsymbol{r}'| \simeq kr - kr'\cdot\hat{\boldsymbol{r}} \equiv kr - \boldsymbol{k}'\cdot\boldsymbol{r}' \tag{2.43}$$

where $\hat{\boldsymbol{r}} = \boldsymbol{r}/r$, $\boldsymbol{k}' = k\hat{\boldsymbol{r}}$, and the asymptotic solutions of (2.36) are of the form

$$\Psi(r) \underset{r\to\infty}{=} \mathrm{e}^{ikz} + f(\theta,\varphi)\frac{\mathrm{e}^{ikr}}{r} \tag{2.44}$$

where

$$f(\theta,\varphi) = \frac{-m}{2\pi\hbar^2}\int \mathrm{d}^3r' V(\boldsymbol{r}')\mathrm{e}^{-i\boldsymbol{k}'\cdot\boldsymbol{r}'}\Psi(\boldsymbol{r}') \tag{2.45}$$

has the dimensions of length and is called the scattering amplitude. When $V(\mathbf{r}')$ is a central potential $f(\theta, \varphi)$ is a function of θ only.

To relate $f(\theta, \varphi)$ to the differential scattering cross-section $\mathrm{d}\sigma(\theta, \varphi)/\mathrm{d}\Omega$, where $\mathrm{d}\sigma(\theta, \varphi)$ may be defined as the effective area presented by the target nucleus to the incident particle for scattering into the element of solid angle $\mathrm{d}\Omega$ about $\mathbf{k}'(\theta, \varphi)$, it is necessary to evaluate the flux $\mathrm{d}\Phi_s$ of the scattered current $\mathbf{j}_s(\mathbf{r})$ across an element of area $R^2\,\mathrm{d}\Omega$ located a large distance R from the target. The vector $\mathbf{j}_s(\mathbf{r})$ is derived from (2.41) and (2.44) by retaining the scattered wave only and the resultant flux is given by

$$\mathrm{d}\Phi_s = j_{sr}(R, \theta, \varphi)R^2\,\mathrm{d}\Omega = \frac{\hbar k}{mR^2}\,|f(\theta, \varphi)|^2 R^2\,\mathrm{d}\Omega \tag{2.46}$$

To determine the differential cross-section, $\mathrm{d}\Phi_s$ must be divided by the flux, $(\hbar k/m)\,\mathrm{d}\sigma(\theta, \varphi)$ of the incident wave across the target nucleus leading to the result

$$\frac{\mathrm{d}\sigma_s}{\mathrm{d}\Omega}(\theta, \varphi) = \left(\frac{\hbar k}{m}\right)^{-1}\frac{\mathrm{d}\Phi_s}{\mathrm{d}\Omega} = |f(\theta, \varphi)|^2 \tag{2.47}$$

Alternatively (2.46) may be arrived at by multiplying the scalar flux (cf. section 5.3.1.2) of scattered particles $|(\mathrm{e}^{ikR}/R)f(\theta, \varphi)|^2\,(\hbar k/m)$ by the element of area $R^2\,\mathrm{d}\Omega$. To avoid confusion between the term 'scalar flux' and the flux of the current $\mathbf{j}(\mathbf{r})$ over a specified surface, the latter quantity is frequently referred to as the 'fluence'.

2.3 Scattering in a central potential

2.3.1 The Born approximation

Although the scattering amplitude $f(\theta, \varphi)$ defined according to (2.45) has the form of a matrix element of the operator V, this is not a strict interpretation because, although the eigenfunctions $\Psi(\mathbf{r}')$ and $\mathrm{e}^{i\mathbf{k}'\cdot\mathbf{r}} = \langle \mathbf{r}' | k' \rangle$ represent states of the same energy, they satisfy different Schrödinger equations and hence correspond to basis states in different representations. We therefore introduce a transition operator T, corresponding to V, whose matrix elements between free particle states of the same energy E are defined by

$$\langle \mathbf{k}' | T | \mathbf{k} \rangle = \langle k' | V | \Psi(k) \rangle \tag{2.48}$$

where $\Psi(r) = \langle \mathbf{r} | \Psi(k) \rangle$. Applying the integral equation (2.36) to calculate these matrix elements we find that

$$\langle k' | T | k \rangle = \int \mathrm{d}^3 r\, \mathrm{e}^{i(\mathbf{k}-\mathbf{k}')\cdot\mathbf{r}}\, T(\mathbf{r})$$

$$= \int d^3r\, e^{i(\boldsymbol{k}-\boldsymbol{k}')\cdot\boldsymbol{r}}\, V(\boldsymbol{r}) + \int d^3r\, d^3r'\, e^{i\boldsymbol{k}\cdot\boldsymbol{r}}\, V(\boldsymbol{r}) G_0(\boldsymbol{r}-\boldsymbol{r}') T(\boldsymbol{r}')\, e^{-i\boldsymbol{k}'\cdot\boldsymbol{r}'} \tag{2.49}$$

The integral equation (2.49) for $T(\boldsymbol{r}) = \langle \boldsymbol{r}|T|\boldsymbol{r}\rangle$ can be expressed as a symbolic relationship between operators

$$T = V + VG_0T \tag{2.50}$$

where G_0 is interpreted, not as the kernal of the integral equation, but as the operator (cf. section 8.2.2)

$$G_0 = \lim_{\varepsilon \to 0+} [E - H_0 + i\varepsilon]^{-1} \tag{2.51}$$

whose matrix elements are given by

$$\langle \boldsymbol{r}|G_0|\boldsymbol{r}'\rangle = G_0(\boldsymbol{r}-\boldsymbol{r}') \tag{2.52}$$

and the operator H_0 appearing in (2.51) is the free particle Hamiltonian. Formally the solution of (2.50) may be written in the form

$$T = (1 - VG_0)^{-1} V \tag{2.53}$$

which can then be expanded in the Born series of ascending powers of VG_0

$$T = \left(\sum_{j=0}^{\infty} (VG_0)^j \right) V \tag{2.54}$$

The Born approximation consists in neglecting all but the leading term in (2.54) so that $T \simeq V$. Equivalently it consists in replacing $\Psi(\boldsymbol{r}')$ in (2.45) by the unperturbed wavefunction $\Psi_0(\boldsymbol{r}') = e^{i\boldsymbol{k}\cdot\boldsymbol{r}'}$, so that the scattering amplitude is given approximately by the Fourier transform of the potential

$$f(\theta, \varphi) \simeq -\frac{m}{2\pi\hbar^2} \int e^{i\boldsymbol{q}\cdot\boldsymbol{r}'}\, V(\boldsymbol{r}')\, d^3r' \qquad \boldsymbol{q} = \boldsymbol{k} - \boldsymbol{k}' \tag{2.55}$$

The Born approximation is valid only when the potential $V(\boldsymbol{r}')$ is a small perturbation on the total energy so that first order perturbation theory can be applied. This is never true for the scattering of slow neutrons by nuclei since, within the range of the nuclear potential, the interaction energy far exceeds the kinetic energy of the neutron. Nevertheless, as first appreciated by Fermi (1936), in the scattering of neutrons in condensed media, the overall influence of the nuclear scattering is much reduced since the total volume occupied by the nuclei is negligible in comparison with the volume occupied by the atomic system as a whole. Indeed near zero energy, the only effect of the nuclear scattering is to introduce a spatial shift in the wavefunction by an amount equal to the scattering length (cf. section 2.5.1) which is several orders of magnitude less than the mean interatomic separation. Fermi therefore chose to represent the neutron–nucleus interaction in bulk matter in the form of a contact potential, parameterized in such a way that the experimental value of the scattering length agrees with the value calculated on the basis of the Born approximation. Fermi's result is derived in section 2.5.5 and applied to the derivation of the 'neutron refractive index' in section 6.3.1.

2.3.2 Partial wave analysis

The precise nature of the forces experienced by a neutron in the field of a nucleus is difficult to represent in exact mathematical form but, in many cases of practical interest, including the case that there is a significant spin–spin component in this force (cf. section 3.3.2.3), it is sufficient to assume that the scattering potential is a spherically symmetric function $V(r)$. In these circumstances, since the z-axis is a symmetry axis for the scattering problem the solution $\Psi(\boldsymbol{r})$ for the scattered wave may be expanded as a sum of eigenwaves of the orbital angular momentum

$$\Psi(\boldsymbol{r})=\frac{1}{kr}\sum_{l=0}^{\infty} \mathrm{i}^l(2l+1)P_l(\cos\theta)u_l(r) \tag{2.56}$$

where $P_l(\cos\theta)$ are the usual Legendre polynomials and the radial functions $u_l(r)$ satisfy the one-dimensional Schrödinger equation

$$\left[\frac{-\hbar^2}{2m}\frac{\mathrm{d}^2}{\mathrm{d}r^2}+\frac{\hbar^2}{2m}\frac{l(l+1)}{r^2}+V(r)-E\right]u_l(r)=0 \qquad \text{for } 0\leqslant r<\infty \tag{2.57}$$

We note that the scattering potential $V(r)$ is supplemented by a repulsive centrifugal barrier 'potential' $(\hbar^2/2m)(l(l+1)/r^2)$ which acts to repel particles with non-zero angular momentum from the neighbourhood of the origin. To obtain a solution of (2.57) which is regular at the origin we must assume that the centrifugal barrier dominates there and this means that $V(r)$ must be restricted to satisfy

$$\lim_{r\to 0} r^2V(r)=0 \tag{2.58}$$

We now rewrite the equation for the radial function $u_l(r)$ in the form

$$\left[\frac{\mathrm{d}^2}{\mathrm{d}r^2}\frac{-l(l+1)}{r^2}-U(r)+k^2\right]u_l(r)=0 \tag{2.59}$$

where

$$U(r)=\frac{2mV(r)}{\hbar^2} \tag{2.60}$$

In the limit of vanishing scattering potential $U(r)\to 0$, a solution of (2.60) is

$$\Psi^0(\boldsymbol{r})=e^{ikz}=\frac{i}{kr}\sum_{l=0}^{\infty} i^l(2l+1)P_l(\cos\theta)[krj_l(kr)] \tag{2.61}$$

where $j_l(kr)$ is the spherical Bessel function of order l. The corresponding regular solution of (2.59) is

$$u_l^0(r)=krj_l(kr)\to(kr)^{l+1}/(2l+1)!! \qquad \text{as } r\to 0 \tag{2.62}$$

We therefore see that the function $\varphi_l^0(r)=u_l^0(r)/k^{l+1}$ is a real even function of k for real k. This will be true for solutions $\varphi_l(k,r)$ of (2.41) for $V(r)$ restricted by (2.58) which satisfy the boundary condition

$$\lim_{r\to 0}\,(2l+1)!!\,r^{-(l+1)}\varphi_l(k,r)=1 \tag{2.63}$$

We now turn attention to those solutions of (2.59) which behave asymptotically like incoming or outgoing waves. For $U(r)=0$ these solutions are expressible in terms of the spherical Hankel functions $h_l^{(1)}(kr)$ and $h_l^{(2)}(kr)$ and may be written as

$$\chi_l^0(k,r)=-ikrh_l^{(2)}(kr)\to i^l e^{-ikr}\,(r\to\infty) \tag{2.64}$$

which represents incoming waves and

$$\chi_l^{0*}(k,r)=ikrh_l^{(1)}(kr)=(-1)^l\chi_l^0(-k,r)\to(-i)^l e^{ikr}\,(r\to\infty) \tag{2.65}$$

which represents outgoing waves. The corresponding independent solutions of (2.59) which obey the crossing symmetry

$$\chi_l^*(k,r)=(-1)^l\chi_l(-k,r) \tag{2.66}$$

are determined by the boundary conditions at infinity

$$\lim_{r\to\infty} e^{ikr}\chi_l(k,r)=i^l \tag{2.67}$$

where, in the manner of (2.64–2.65) they represent incoming and outgoing waves, respectively. The existence of such asymptotic solutions requires that $V(r)$ fall off sufficiently fast at infinity that the incoming wave is not distorted by the potential in the asymptotic region. This requirement is met if $\mu>0$ exists such that

$$\lim_{r\to\infty} e^{2\mu r}\,V(r)=0 \tag{2.68}$$

a condition which we note excludes the Coulomb potential which does not have solutions representing free partial waves at infinity.

We can now express the regular solutions $\varphi_l(kr)$ in terms of the independent asymptotic solutions $\chi_l(k, r)$ and $(-1)^l \chi_l(-k, r)$

$$\varphi_l(k, r) = \frac{i}{2k^{l+1}} \{g_l(-k)\chi_l(k, r) - g_l(k)(-1)^l \chi_l(-k, r)\} \tag{2.69}$$

where the symmetry of the expansion coefficient

$$g_l(k) = g_l^*(-k) = |g_l(k)| \, e^{i\delta_l(k)} \tag{2.70}$$

follows from the crossing symmetry (2.66), and the fact that $\varphi_l(k, r)$ is an even function of k. Applying the representation (2.69) to the unperturbed regular functions $\varphi_l^0(k, r)$ we have the result

$$\varphi_l^0(k, r) = [krj_l(kr)]/k^{l+1} = \frac{i}{2k^{l+1}} \{-ikrh_l^{(2)}(kr) - ikrh_l^{(1)}(kr)\}$$

$$= \frac{i}{2k^{l+1}} \{\chi_l^0(k, r) - (-1)^l \chi_l^0(-k, r)\} \tag{2.71}$$

which shows that for the unperturbed solution $g_l^0(k) = g_l^0(-k) = 1$.

We can now fix the relationship between the physical radial functions $u_l(k, r)$ and the standard regular solutions $\varphi_l(k, r)$ by the requirement that the scattering potential $U(r)$ does not alter the incoming waves at infinity, which means that the asymptotic terms in the expansions of $u_l(k, r)$ and $u_l^0(k, r)$ which represent incoming waves should coincide. This will be true if

$$u_l(k, r) = k^{l+1} \varphi_l(k, r)/g_l(-k) \tag{2.72}$$

We find then that at infinity

$$u_l(k, r) \to u_l^0(k, r) + (-i)^l \left[\frac{S_l(k) - 1}{2i}\right] e^{ikr} \tag{2.73}$$

where

$$S_l(k) = S_l^{-1}(-k) = g_l(k)/g_l(-k) = e^{2i\delta_l(k)} \tag{2.74}$$

is the diagonal element of the collision matrix introduced to scattering theory by Wheeler (1937) and termed the S-matrix by Heisenberg (1943).

Although the S-matrix only determines the asymptotic behaviour of the wavefunction it is sufficient to allow a calculation of all those features of the scattering process which are observable. Because we so restricted the potential $V(r)$ as to decouple the radial equations for the various partial waves which contribute, the S-matrix case is purely diagonal. In general however the S-matrix has off-diagonal elements, e.g. in the case of scattering from a potential field with a tensor component. In this case for two spin $\frac{1}{2}$ nucleons scattering in the channel with $S = 1$, $I = 1$, the partial waves with $l = 0$ and $l = 1$ satisfy coupled differential equations and the single element S_l becomes a 2×2 sub-matrix. The S-matrix is completely determined by the Jost function $g_l(k)$ (Jost 1947) and, because of the relations (2.70) and (2.64) is unitary for real k, i.e.

$$S_l^{-1}(k) = S_l^*(k) \qquad |S(k)|^2 = 1 \tag{2.75}$$

The asymptotic form of the physical wavefunction $u_l(k, r)$ can now be written in the form

$$u_l(k, r) \rightarrow u_{l,\infty}(k, r) = e^{i\delta_l(k)} \sin\left(kr \frac{-l\pi}{2} + \delta_l(k) \right) \tag{2.76}$$

where the phase shift $\delta_l(k)$ is an odd function of k from (2.70). The functions $u_{l,\infty}(k, r)$ satisfy the normalization condition

$$\int_0^\infty u_{l,\infty}(k, r) u_{l,\infty}(k', r)\, dr = \frac{\pi}{2} \delta(k - k') \tag{2.77}$$

The Jost function may be found explicitly by computing the Wronskian of the functions $\chi_l(k, r)$ and $\varphi_l(k, r)$ which, being solutions of a linear second-order differential equation without a first-order derivative, have a Wronskian which is independent of r. Thus we find

$$g_l(k) = k^l W(\varphi_l(k, r), \varphi_l(k, r)) \tag{2.78}$$

and $g_l(k)$ may be determined by evaluating the Wronskian $W(\chi_l, \varphi_l)$ at some suitable radius. In the case of a potential which vanishes identically for $r > b$, the value $r = b$ represents a convenient choice. If we evaluate $W(\chi_l, \varphi_l)$ at $r = 0$ and use the boundary condition (2.63) we obtain the result

$$g_l(k) = \lim_{r \to 0} (kr)^l \chi_l(k, r)[(l+1)/(2l+1)!!] \tag{2.79}$$

We are now in a position to complete the calculation of the asymptotic form of the wave function for, substituting (2.73) into (2.76) we find

$$\Psi(r) \xrightarrow[r \to \infty]{} e^{ikz} + \sum_{l=0}^{\infty} (2l+1) P_l(\cos\theta)[(e^{2i\delta_l(k)} - 1)/2ik]\, e^{ikr}/r \tag{2.80}$$

and this gives us the result for the scattering amplitude $f(\theta)$ from (2.54)

$$f(\theta) = \sum_{l=0}^{\infty} (2l+1) P_l(\cos\theta) f_l(k) \tag{2.81}$$

where

$$f_l(k) = (e^{2i\delta_l(k)} - 1)/2ik = (S_l(k) - 1)/2ik = e^{i\delta_l(k)} \sin(\delta_l(k)/k \tag{2.82}$$

is called the partial wave scattering amplitude.

2.3.3 Elastic scattering of identical particles

If we ignore any spin–orbit coupling and represent the wavefunction $\Psi(1, 2)$ of two identical particles in product form

$$\Psi(1, 2)=\psi(\boldsymbol{r}_1, \boldsymbol{r}_2)\chi(\boldsymbol{s}_1, \boldsymbol{s}_2) \tag{2.83}$$

where $\psi(\boldsymbol{r}_1, \boldsymbol{r}_2)$ is the orbital, and $\chi(\boldsymbol{s}_1, \boldsymbol{s}_2)$ the spin wavefunction, then Pauli's exclusion principle states that $\Psi(1, 2)$ must be either symmetric (bosons) or anti-symmetric (fermions) with respect to exchange of the two particles, i.e.

$$\Psi(2, 1)=\pm\Psi(1, 2) \tag{2.84}$$

where the upper sign refers to bosons and the lower sign refers to fermions, e.g. neutrons and protons.

The mechanics of exchange is dealt with more fully in section 3.3 below where it is shown that the singlet ($s=0$) spin wavefunction of two nucleons is anti-symmetric, and the triplet ($s=1$) spin wavefunction symmetric, with respect to exchange of the two spins. It is further shown that, if the collision of two nucleons is treated in their centre of mass frame, the coordinate exchange $\boldsymbol{r}_1 \rightleftarrows \boldsymbol{r}_2$ is equivalent to the space reflection transformation on the relative coordinate $\boldsymbol{r}_1-\boldsymbol{r}_2=\boldsymbol{r}$, i.e. $\boldsymbol{r}\to-\boldsymbol{r}$, $r\to r$, $\theta=\pi-\theta$, $\varphi\to\pi+\varphi$. The conclusion follows therefore that, for a pair of identical nucleons, the asymptotic orbital wavefunction is given, not by (2.44) which only applies to non-identical particles, but by the symmetric and anti-symmetric combinations

$$\psi(\boldsymbol{r})=\mathrm{e}^{ikz}\pm\mathrm{e}^{-ikz}+\frac{\mathrm{e}^{ikr}}{r}[f(\theta)\pm f(\pi-\theta)] \tag{2.85}$$

where the upper (lower) sign refers to the singlet (triplet) spin-states. The differential scattering cross-section for unpolarized particles becomes therefore

$$\frac{\mathrm{d}\sigma}{\mathrm{d}\Omega}(\theta)=\frac{1}{4}\,|{}^1f(\theta)+{}^1f(\pi-\theta)|^2+\frac{3}{4}\,|{}^3f(\theta)-{}^3f(\pi-\theta)|^2 \tag{2.86}$$

where ${}^1f(\theta)({}^3f(\theta))$ represents the singlet (triplet) scattering amplitude.

Recalling that the Legendre polynomials $P_l(\cos\theta)$ satisfy the symmetry relations

$$P_l(\cos(\pi-\theta))=P_l(-\cos\theta)=(-1)^l P_l(\cos\theta) \tag{2.87}$$

it may be shown directly that the singlet (triplet) scattered component contains only even (odd)-valued partial waves and we find

$$\frac{\mathrm{d}\sigma}{\mathrm{d}\Omega}(\theta)=\frac{4\pi}{k^2}\left\{\frac{1}{4}\sum_{\substack{l,l'\\ \text{even}}}\left[(2l+1)(2l'+1)\sin{}^1\delta_l\sin{}^1\delta_{l'}\cos({}^1\delta_l-{}^1\delta_{l'})P_l(\cos\theta)P_{l'}(\cos\theta)\right]\right.$$

$$+\frac{3}{4}\left[\sum_{\substack{l,l'\\ \text{odd}}}(2l+1)(2l'+1)\sin{}^3\delta_l\sin{}^3\delta_{l'}\cos({}^3\delta_l-{}^2\delta_{l'})P_l(\cos\theta)P_{l'}(\cos\theta)\right]\Bigg\}$$

(2.88)

where ${}^1\delta_l({}^3\delta_l)$ is the singlet (triplet) phase shift of the lth partial wave.

It is important to note that, in both sums in (2.88), the direct terms with $l=l'$ always make positive contributions to the differential scattering cross-section whereas the cross-product terms with $l\neq l'$ may be positive or negative leading to either constructive or destructive interference between partial waves whose l-values differ by an even number of units. In particular since $|P_l(\pm1)|=1$, every partial wave gives a maximum which is symmetric between forward and backward directions. However, since $P_l(0)$ vanishes for odd l and alternates in sign for successive even values of l, partial waves with neighbouring even values of l always interfere destructively for scattering at right angles, provided that the corresponding phase shifts have the same sign.

2.4 Scattering parameters

2.4.1 The scattering cross-section

We can now compute the total flux of scattered particles by integrating the radial component j_r of the probability current density (2.41) over the surface of a large sphere of radius R in the asymptotic zone surrounding the scattering centre. Corresponding to an incoming flux Φ_{pi} associated with the plane wave e^{ikz} the outgoing flux is made up from three components.

$$\Phi_{\text{out}}=\Phi_{\text{po}}+\Phi_{\text{s}}+\Phi_{\text{int}} \tag{2.89}$$

where Φ_{po} is the outgoing plane wave flux, Φ_{s} is the scattered flux characterized by $f(\theta)$ and Φ_{int} is a flux arising from the interference of the scattered wave and the outgoing plane wave. Clearly the plane wave carries the same flux out of the sphere as it carries into it and we have the result

$$\Phi_{\text{in}}=\Phi_{\text{pi}}=\Phi_{\text{po}}=\left(\frac{\hbar k}{m}\right)\pi R^2 \tag{2.90}$$

The calculation of the scattered flux follows from (2.26) and we find

$$\begin{aligned}\Phi_{\text{s}}&=\frac{\hbar k}{mR^2}\int|f(\theta)|^2R^2\,\mathrm{d}\Omega=\frac{\hbar k}{m}\cdot2\pi\int_0^\pi|f(\theta)|^2\sin\theta\,\mathrm{d}\theta\\&=\frac{\hbar k}{m}\cdot4\pi\sum_{l=0}^{\infty}(2l+1)|f_l(k)|^2 \qquad (2.91)\\&=\frac{\hbar k}{m}\cdot\frac{4\pi}{k^2}\sum_{l=0}^{\infty}(2l+1)\sin{}^2\delta_l(k)\end{aligned}$$

where we have used the expressions for $f(\theta)$ and $f_l(k)$ in (2.81) and (2.82) together with the orthogonality condition on the Legendre polynomials

$$\int_{-1}^{1} P_l(\mu)P_{l'}(\mu)\,d\mu = 2\delta_{l,l'}/(2l+1) \tag{2.92}$$

The scattering cross-section σ_s now follows from (2.27) and (2.91) with the result

$$\sigma_s = \frac{\pi}{k^2}\sum_{l=0}^{\infty}(2l+1)|1-S_l(k)|^2 = \frac{4\pi}{k^2}\sum_{l=0}^{\infty}(2l+1)\sin^2\delta_l(k) \tag{2.93}$$

We note that σ_s is just the sum of the scattering cross-sections σ_{sl} for each of the partial waves and that each partial cross-section $\sigma_{sl} \leqslant 4\pi(2l+1)/k^2$ is uniquely determined by the phase shift $\delta_l(k)$ at the specified energy.

2.4.2 The optical theorem

To determine the interference flux Φ_{int} we need the asymptotic form of the plane wave e^{ikz} which according to (2.61) is given by

$$e^{ikz} = \sum_{l=0}^{\infty} i^l(2l+1)P_l(\cos\theta)u_l(kr) \xrightarrow[r\to\infty]{} \sum_{l=0}^{\infty} i^l(2l+1)P_l(\cos\theta)\sin(kr - l\pi/2)/kr \tag{2.94}$$

Thus the asymptotic plane wave may be decomposed into an incoming component

$$-\sum_{l=0}^{\infty}(2l+1)(-1)^l P_l(\cos\theta)\frac{e^{-ikr}}{2ikr} = \frac{-e^{-ikr}}{2ikr}\delta(1+\cos\theta) \tag{2.95}$$

together with an outgoing component

$$\sum_{l=0}^{\infty}(2l+1)P_l(\cos\theta)\frac{e^{ikr}}{2ikr} = \frac{e^{ikr}}{2ikr}\delta(1-\cos\theta) \tag{2.96}$$

We find therefore for the interference flux

$$\begin{aligned}\Phi_{int} &= \frac{\hbar k}{m}\left\{\frac{2}{k}\int d\Omega\left[\frac{f^*(\theta)-f(\theta)}{2i}\right]\delta(1-\cos\theta)\right\} \\ &= \frac{\hbar k}{m}\left\{\frac{-4\pi}{k}\,\mathrm{Im}\,f(0)\right\}\end{aligned} \tag{2.97}$$

where $\mathrm{Im}\,f(\theta)$ is the imaginary part of the scattering amplitude. The total outgoing flux defined in (2.89) may now be written

$$\Phi_{out} = \frac{\hbar k}{m}\left\{\pi R^2 + \sigma_s - \frac{4\pi}{k}\,\mathrm{Im}\,f(0)\right\} \tag{2.98}$$

The capture cross-section σ_c is defined by

$$\sigma_c = (\Phi_{in} - \Phi_{out}) \Big/ \frac{\hbar k}{m} = -\sigma_s + \frac{4\pi}{k} \operatorname{Im} f(0) \tag{2.99}$$

a relation which, using (2.81) and (2.82) can be rewritten as an expression for the total cross-section

$$\sigma_t = \sigma_c + \sigma_s = \frac{4\pi}{k} \operatorname{Im} f(0) = \frac{2\pi}{k} \sum_{l=0}^{\infty} (2l+1)(1 - \operatorname{Re}(S_l(k))) \tag{2.100}$$

The result (2.100) is known as the 'optical theorem' and states in effect that the flux removed from the forward beam is either scattered (σ_s) or captured (σ_c). It is the direct analogue of a theorem in classical optics which makes a connection between the imaginary part of the complex refractive index describing the propagation of light in a medium and the forward scattering amplitude $f(\theta = 0)$ for scattering from the individual scattering centres which make up the medium (Feenberg 1932, Bohr *et al* 1939).

We can now combine (2.93) and (2.100) to express the capture cross-section in the form

$$\sigma_c = \frac{\pi}{k^2} \sum_{l=0}^{\infty} (2l+1)(1 - |S_l(k)|^2) \tag{2.101}$$

and it follows from the unitarity condition (2.75) that σ_c must vanish identically for purely elastic scattering. Thus to incorporate the capture process, which includes inelastic scattering, (n, n'), we must extend the definition of the S-matrix by the addition of off-diagonal elements, having the same value of l, in order to maintain overall unitarity. If therefore in place of S_l we suppose that $S_{l,nn}$ represents the elastic scattering matrix element, and $S_{l,nm}$ represents an inelastic process e.g. (n, n'), (n, α), (n, γ), etc., then the unitarity relation for the extended S-matrix is written in place of (2.75).

$$|S_{l,nn}|^2 + \sum_{m \neq n} |S_{l,nm}|^2 = 1 \tag{2.102}$$

Thus, substituting (2.102) into (2.101) the total capture cross-section σ_c can be written

$$\sigma_c = \frac{\pi}{k^2} \sum_{l=0}^{\infty} (2l+1) \sum_{m \neq n} |S_{l,nm}|^2 = \sum_{m \neq n} \sigma_{nm} \tag{2.103}$$

where σ_{nm} represents the cross-section for the capture of the neutron (n), followed by the emission of a particle 'm' (e.g. n', α, γ, etc.).

2.4.3 Complex potentials

We know from experiment that many nuclei have very large cross-sections for the capture of slow neutrons, and this effect may be accommodated within the present model of potential scattering using the concept of a complex potential as originally suggested by Bethe (1940). Within this model all inelastic processes are lumped together as 'capture' and we are not able to differentiate between (n, α), (n, γ) processes, etc. This is not an important restriction if our concern is principally with the fate of the interacting neutron rather than the nature of the particular absorptive process.

We assume that the probability current $\boldsymbol{j}(\boldsymbol{r})$ of neutrons is no longer conserved but satisfies a continuity relation of the form

$$\nabla \cdot \boldsymbol{j} + \frac{\partial}{\partial t} |\Psi(\boldsymbol{r}, t|^2 = \frac{-2W(\boldsymbol{r})}{\hbar} |\Psi(\boldsymbol{r}, t)|^2 \tag{2.104}$$

where the function $W(\boldsymbol{r}) > 0$ represents a sink for neutrons entering the nucleus and is defined to have dimensions of energy. By constructing the current density $\boldsymbol{j}(\boldsymbol{r}, t)$ in the usual way (2.51) one can show directly that to give rise to (2.104) the wavefunction $\Psi(\boldsymbol{r}, t)$ must satisfy a new Schrödinger equation.

$$\nabla^2 \Psi + \frac{2m}{\hbar^2} (E - V(\boldsymbol{r}) + \mathrm{i}W(\boldsymbol{r})\Psi) = 0 \tag{2.105}$$

in which the real potential $V(\boldsymbol{r})$ is replaced by a complex potential $V(\boldsymbol{r}) - \mathrm{i}W(\boldsymbol{r})$. Of course the new Hamiltonian is no longer Hermitian nor indeed should it be since it incorporates a term specifically designed to account in an empirical way for processes which lead to the annihilation of neutrons of the specified energy. The absorption potential $W(\boldsymbol{r})$ is not a function of energy which means in effect that the corresponding absorption cross-section obeys the $1/v$ law (cf. section 1.7.1.1) and the only requirement we place on the complex potential is that it should vary not too rapidly at the nuclear boundary so that neutrons can enter the nucleus and be absorbed there rather than be scattered at surface discontinuities. In general the capture probability will be a monotonic increasing function of d, where d is a length characterizing the depth of the boundary layer at the nuclear surface.

The introduction of a complex potential to describe the interactions of neutrons and nuclei has a direct optical analogue in the use of a complex dielectric constant to describe the propagation of light in a conducting medium. Equation (2.105) is also the point of departure in constructing the optical model of the nucleus (Fernbach *et al* 1949, Feshbach *et al* 1954, Watson 1953).

2.4.4 Dispersion relations

When an atom interacts with a radiation field, the centre of the electronic charge distribution is displaced with respect to the positive nucleus and the atom acquires an induced electric dipole moment which oscillates with the frequency of the exciting radiation. In the classical view it is the radiation from this induced dipole which comprises the scattered radiation. In the absence of a radiation field we may regard the electrons as existing in stationary states of free oscillation with angular frequency ω_0 without radiating. Radiation from the electrons only occurs in the forced motion, and we can take account of the reaction of this radiation on the radiating electrons by including among the forces a damping force of such magnitude that it correctly accounts for the energy lost. The induced electric dipole moment $\boldsymbol{P}(\omega, t)$ is then given in terms of the exciting electric field $\boldsymbol{E}(\omega, t)$ by

$$\boldsymbol{P}(\omega, t) = (c/\omega)^2 f(\omega)\boldsymbol{E}(\omega, t) \tag{2.106}$$

where $f(\omega)$ is the coherent forward scattering amplitude defined by

$$f(\omega) = \frac{r_e\omega^2}{(\omega_0^2 - \omega^2) - i\omega\gamma} \tag{2.107}$$

$r_e = e^2/m_e c^2$ is the classical electron radius, and $\gamma = 2r_e\omega^2/3c$ is the damping constant which characterizes the radiation reaction. The function $f(\omega)$, expressed in terms of the variable $\omega = ck$, is the optical analogue of the neutron scattering amplitude $f(\theta)$, defined in (2.81), evaluated in the forward direction $\theta = 0$ (cf. 2.97).

Equation (2.106) is usually adopted as the point of departure for a classical theory of optical dispersion, and it leads directly to an expression for the refractive index

$$n^2(\omega) = 1 + \frac{4\pi Nc^2}{\omega^2} f(\omega) \tag{2.108}$$

where N is the number of dispersion electrons per unit volume. For transparent materials as a whole, n does not depart much from unity, and hence satisfies the approximate relationship

$$n(\omega) = 1 + \frac{2\pi Nc^2 f(\omega)}{\omega^2} \tag{2.109}$$

The existence of an imaginary part of the refractive index is connected by its very origin with a damping effect, and is clearly associated with the total probability of re-emission of radiation. The direct connection is

$$\operatorname{Im}(n(\omega)) = \frac{c}{2\pi} N\sigma_t(\omega) \tag{2.110}$$

where $\sigma_t(\omega)$ is the total absorption cross-section. In view of the relation between $n(\omega)$ and the coherent forward scattering amplitude $f(\omega)$, the result (2.110) may equally well be expressed as

$$\operatorname{Im} f(\omega)=\frac{k}{4\pi}\sigma_t \tag{2.111}$$

and in this form is known as the optical theorem (cf. 2.100).

It was observed by Kronig (1926) and Kramers (1927) that the dispersive and absorptive parts of the complex refractive index in (2.109) are connected through the relation

$$\operatorname{Re}(n(\omega))=1+\frac{2}{\pi}P\int_0^\infty \frac{\omega' \operatorname{Im}(n(\omega'))\,d\omega'}{\omega'^2-\omega^2} \tag{2.112}$$

where the symbol P indicates that the Cauchy principal value of the integral is to be taken. Expressed in terms of the coherent forward scattering amplitude, (2.112) may be written

$$\operatorname{Re} f(\omega)=\frac{2\omega^2}{\pi}P\int_0^\infty \frac{\operatorname{Im} f(\omega')\,d\omega'}{\omega'(\omega'^2-\omega^2)} \tag{2.113}$$

and this result is known as the Kramers–Kronig dispersion relation.

In the classical theory of dispersion, the expression for $f(\omega)$ in (2.107) is derived under the assumption that the incident light wave may be represented by a stationary electric field $E_0\,e^{-i\omega t}$ of complex exponential form. Since such a field possesses infinite extension in space and time, this assumption inevitably focuses attention on the stationary solutions of the scattering problem. A description of scattering in these terms not only obscures the absorption re-emission nature of the process, but disregards altogether the causal relationship which exists between incident wave, induced dipole moment and scattered wave. If we remove the implication of stationarity, and suppose that the light wave arrives at the scattering centre at $t=0$, $f(\omega)$ may be regarded as the Fourier transform of a function which vanishes for negative t. This restriction on the possible forms of the function $f(\omega)$ expresses the fact that signals cannot propagate in the medium with speeds in excess of the speed of light, which is the basic principle underlying the relation (2.113) (Kronig 1946).

The function $f(\omega)$ has a physical interpretation in the optical problem for real positive values of ω only. In certain circumstances $f(\omega)$ may be continued analytically into the whole of the complex ω-plane, in which case (2.113) may be shown to follow as a consequence of a theorem relating to Fourier transformation.

In particular if we assume that

(1) all singularities of $f(\omega)$, defined for complex ω, are confined to the lower half-plane,

(2) $f(\omega)$ satisfies the crossing relation $f(-\omega)=f^*(\omega)$, (cf. 2.66),

(3) $f(\omega)$ is of a form consistent with causality, i.e. it is the Fourier transform of a function which vanishes for negative t,

(4) $f(\omega)$ and its real and imaginary parts obey some suitable integrability conditions,

then $\operatorname{Re} f(\omega)$ and $\operatorname{Im} f(\omega)$ satisfy the 'dispersion relations'

$$\text{(i)}\quad \operatorname{Re} f(\omega)=\frac{1}{\pi} P \int_{-\infty}^{\infty} \frac{\operatorname{Im} f(\omega')\,d\omega'}{\omega'-\omega}$$

$$\text{(ii)}\quad \operatorname{Im} f(\omega)=-\frac{1}{\pi} P \int_{-\infty}^{\infty} \frac{\operatorname{Re} f(\omega')\,d\omega'}{\omega'-\omega} \tag{2.114}$$

In view of assumption (2) above, equations (2.114) may be rewritten in the form

$$\text{(i)}\quad \operatorname{Re} f(\omega)=\frac{2}{\pi} \int_{-\infty}^{\infty} \frac{\omega' \operatorname{Im} f(\omega')\,d\omega'}{\omega'^2-\omega^2}$$

$$\text{(ii)}\quad \operatorname{Im} f(\omega)=-\frac{2\omega}{\pi} P \int_{-\infty}^{\infty} \frac{Re\, f(\omega')\,d\omega'}{\omega'^2-\omega^2} \tag{2.115}$$

Evaluating these relations at $\omega=0$, we find $\operatorname{Im} f(0)=0$, and

$$\operatorname{Re} f(0)=-f(0)=\frac{2}{\pi} P \int_{0}^{\infty} \frac{\operatorname{Im} f(\omega')\,d\omega'}{\omega'} \tag{2.116}$$

Subtracting (2.116) from (2.115(i)) leads to the result

$$\operatorname{Re} f(\omega)=f(0)+\frac{2\omega^2}{\pi} P \int_{0}^{\infty} \frac{\operatorname{Im} f(\omega')\,d\omega'}{\omega'(\omega'^2-\omega^2)} \tag{2.117}$$

Since the classical amplitude for scattering from bound electrons vanishes in the limit of zero energy, the Kramers–Kronig dispersion relation (2.113) follows from the general result (2.117). For scattering by free electrons $f(0)\neq 0$, and the full form of (2.117) is required. The subtraction procedure adopted to obtain (2.117) serves to improve the convergence of (possibly divergent) integrals, at the expense of requiring additional information, e.g. specifying the value of $f(0)$.

The dispersion relations (2.115(i)) and (2.115(ii)) may be applied directly to the propagation of slow neutrons of kinetic energy $E=\hbar\omega$ in condensed media, provided allowance is made for the possible contribution of bound-state poles on the negative energy axis (Landau and Lifshitz 1965). By combining these relations with each other and with the optical theorem new relations are produced which have been extensively used in strong interaction theory. In particular, by assuming analyticity in the energy and in the momentum transfer, they may be made to yield the Mandelstam representation of scattering amplitudes. However, for our purposes only two consequences of these relations are of importance.

The first important result follows from combining (2.115(i)) with the optical theorem (2.111). Substituting in the integral for $\operatorname{Im} f(\omega)$ we find

$$\operatorname{Re} f(\omega) = \frac{1}{2\pi^2 c} P \int_0^\infty \frac{\omega'^2 \sigma_t(\omega')\, d\omega'}{\omega'^2 - \omega^2} \tag{2.118}$$

From a measured or calculable total cross-section, we may infer the absorptive part of the forward scattering amplitude from the optical theorem; by means of (2.118) we may also find the dispersive part and hence the complete amplitude. The second important result follows from (2.118) on letting ω tend to infinity. This gives us an expression for the energy integral of the total cross-section into the level specified by $f(\omega)$, i.e.

$$\int_0^\infty \sigma_t(E')\, \mathrm{d}E' = 2\pi\hbar c \operatorname{Re} f(\infty) \qquad E' = \hbar\omega' \tag{2.119}$$

Applying (2.119) to the classical amplitude $f(\omega)$, for scattering by the dispersion electron of natural frequency ω_0, we obtain the result

$$\int_0^\infty \sigma_t(\omega_0, E')\, \mathrm{d}E' = \frac{2\pi^2 e^2 \hbar}{m_e c} \tag{2.120}$$

When we sum over all frequencies ω_0, i.e. over all Z electrons in the scattering atom, we obtain finally

$$\int_0^\infty \sigma_t(E')\, \mathrm{d}E' = \frac{2\pi^2 e^2 \hbar}{m_e c} \tag{2.121}$$

This result is the classical electric dipole sum rule; with some modifications it is also applicable to the theory of photonuclear reactions (Gell-Mann *et al* 1954, Goldberger 1955) (cf. section 4.7.5).

2.5 Scattering length and effective range

2.5.1 Scattering length

Since our principal concern will be with the interactions of slow neutrons it is convenient at this stage to apply the formalism developed so far to the case of non-resonant zero-energy scattering. What we mean by zero energy in this context is that the incident particles are so slow that their minimum positional uncertainty is much greater than the range of the scattering potential. This we suppose may be characterized by some length b and we assume therefore that $\lambda\!\!\!^{-} \gg b$. Since this condition may be expressed as a limitation on the magnitude of the orbital angular momentum $mvb \ll \hbar$, we are essentially restricted to consider s-wave scattering only.

We can now rewrite the expression for the partial wave scattering amplitude $f_l(k)$ introduced in (2.82) in the form

$$1/f_l(k) = k\cot(\delta_l(k)) - ik \tag{2.122}$$

where the function

$$\mathrm{Re}(f_l^{-1}(k)) = k\cot(\delta_l(k)) = k\,\mathrm{Re}(g_l(k))/\mathrm{Im}(g_l(k)) \tag{2.123}$$

is an even function of k. This follows from the crossing symmetry relation (2.70) which impies that $\mathrm{Re}(g_l(k))$ is even, and $\mathrm{Im}(g_l(k))$ odd in k. According to (2.79) $g_l(k)$ is determined by the limiting value of $\chi_l(k, r)$ as $r \to 0$, and, since $kb \ll 1$, $\mathrm{Re}(g_l(k))$ will behave like the irregular part of $\chi_l(k, b)$ in the neighbourhood of the origin and vary as $(kb)^{-(l+1)}$, while $\mathrm{Im}(g_l(k))$ will behave like the regular part and vary as $(kb)^l$. It follows that, for $k \to 0$

$$k\cot(\delta_l(k)) \simeq (kb)^{-2l} \tag{2.124}$$

and, for $l \neq 0$, $k\cot\delta_l(k)$ becomes infinite in the limit of zero energy and the phase shift $\delta_l(k)$ tends to zero on the order of k^{2l+1}. For $l=0$ the function $k\cot(\delta_0(k))$ tends to a finite limit and $\delta_0(k)$ tends to zero linearly with k as indeed it must do to preserve the finiteness of $\chi_0(k, r)$ at infinity. We therefore have the result

$$\lim_{k\to 0} k\cot\delta_0(k) = -\frac{1}{a} \tag{2.125}$$

where the parameter $a = -f_0(0)$ is termed the 'scattering length'. It completely determines the scattering cross-section in the zero energy range for, according to (2.93), we now have the result

$$\begin{aligned}\sigma_s &= \frac{4\pi}{k^2}\sin^2\delta_0(k) = \frac{4\pi}{k^2\cot^2(\delta_0(k)) + k^2} \\ &\simeq 4\pi a^2/(1 + a^2k^2)\end{aligned} \tag{2.126}$$

For potentials which have a finite range, i.e. which vanish for $r > b$, we may evaluate the function $k\cot(\delta_0(k))$ rather easily, since, for $r \geqslant b$ we have the explicit result $\chi_0(k, r) = \mathrm{e}^{-ikr}$ from (2.64). We may therefore determine the Jost function $g_0(k)$ directly from (2.78) by evaluating the Wronskian at the boundary $r = b$ and we find

$$\begin{aligned}k\cot(\delta_0(k)) &= -k\left\{\frac{\varphi_0'(k, b)\cos(kb) + k\varphi_0(k, b)\sin(kb)}{\varphi_0'(k, b)\sin(kb) - k\varphi_0(k, b)\cos(kb)}\right\} \\ &\approx \frac{\beta(k) + k^2 b}{1 - \beta(k)b}\end{aligned} \tag{2.127}$$

where

$$\beta(k) = \varphi_0'(k, b)/\varphi_0(k, b) \tag{2.128}$$

is the logarithmic derivative of the interior wavefunction evaluated at the boundary of the potential. In particular letting $k \to 0$ in (2.127) we obtain an expression for the scattering length

$$a = b - 1/\beta(0) \tag{2.129}$$

The various cases which can arise in matching the interior and exterior wavefunctions at the boundary of the potential are illustrated in figure 2.2 which shows that the scattering length a can be interpreted as the intercept

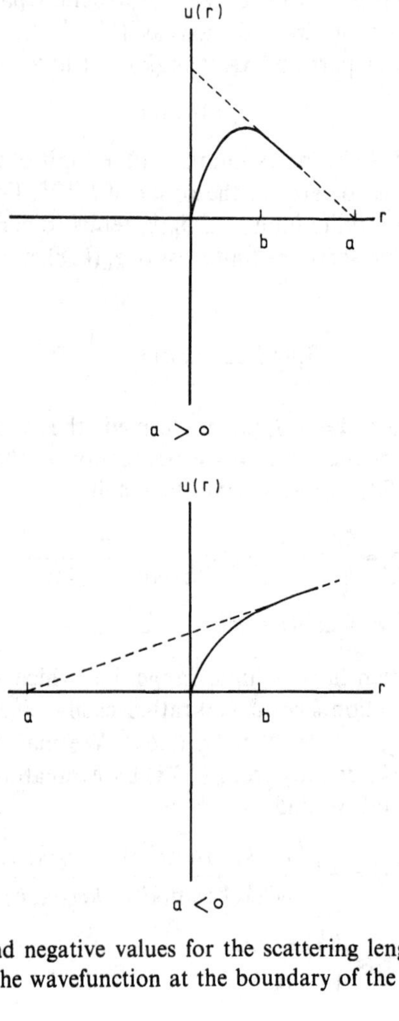

Figure 2.2. Positive and negative values for the scattering length 'a' interpreted in terms of the shape of the wavefunction at the boundary of the potential.

on the r-axis of the tangent to the wavefunction at the matching point. A plot of a against β is shown in figure 2.3; the discontinuity at $\beta(0)=0$ corresponds to a phase shift $\delta_0(0)=\frac{\pi}{2}$ which is the characteristic of a zero energy resonance. We note also that the scattering length is positive unless $\beta(0)$ falls in the range $0<\beta(0)<1/b$ in which case the scattering length is negative. To see how this exceptional case can arise we consider the case of scattering by a finite rectangular well. This is given by the potential distribution

$$U(r)=-U_0 \qquad r\leqslant b \qquad U(r)=0 \qquad r>b \tag{2.130}$$

for which the corresponding regular solution for $r\geqslant b$ is

$$\varphi_0(r,k)=\sin(K(k)r)/K(k) \tag{2.131}$$

where $K(k)$ is defined as the positive square root

$$K(k)=\sqrt{(k^2+U_0)} \tag{2.132}$$

The logarithmic derivative $\beta(0)$ is now given from (2.131) as

$$\beta(0)=K(0)\cot(K(0)b) \tag{2.133}$$

a result which shows, using (2.128), that the scattering length becomes negative only if the angle $K(0)b$ lies in the first or third quadrants and if

$$\tan(K(0)b)>K(0)b \tag{2.134}$$

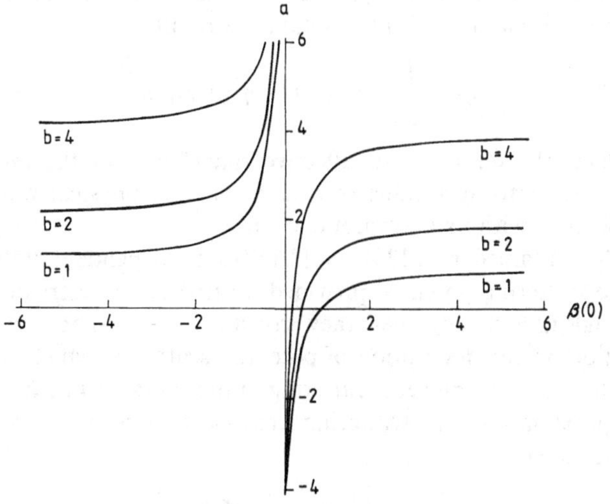

Figure 2.3. The scattering length 'a' displayed as a function of the zero-energy logarithmic derivative $\beta(0)$ of the wavefunction at the boundary of the potential. The discontinuity in 'a' at a zero of $\beta(0)$ is characteristic of a zero-energy resonance.

We know from the study of nuclear systematics that, for a given spin state, the nuclear potential well does not vary very much from one nucleus to the next whereas the characteristic radius of the nuclear potential being proportional to $A^{1/3}$, can vary by a factor of six or more. Thus in general $K(0)b$ may fall anywhere in the range $\pi < K(0)b < 6\pi$ whereas $\tan(K(0)b)$ only becomes large and positive when $K(0)b \leqslant (2n+1)\frac{\pi}{2}$. As we shall see in section 2.7 this is precisely what happens when the potential generates a 'virtual state' near zero energy.

2.5.2 Effective range

Since $k\cot(\delta_0(k))$ is an even function of k it may be expanded as a Taylor series to include terms quadratic in k

$$k\cot(\delta_0(k)) = -\frac{1}{a} + \frac{1}{2}r_0k^2 + \ldots \tag{2.135}$$

where, for an attractive potential, r_0 is a positive quantity having the dimension of length which is called the 'effective range' of the potential (Schwinger 1947). In fact it may be shown that an exact expression for $k\cot(\delta_0(k))$ is

$$k\cot\delta_0(k) = -\frac{1}{a} + k^2\int_0^\infty [\eta_\infty(k,r)\eta_\infty(o,r) - \eta(k,r)\eta(o,r)]\,\mathrm{d}r \tag{2.136}$$

where $\eta(k, r)$ is a regular solution of the Schrödinger equation normalized so that its asymptotic form $\eta_\infty(k, o) = 1$ (Bethe 1949, Salpeter 1951). Comparing (2.135) and (2.136) we obtain the result

$$r_0 = 2\int_0^\infty \{\eta_\infty^2(o,r) - \eta^2(o,r)\}\,\mathrm{d}r \tag{2.137}$$

which explains the terminology 'effective range' in that the integrand in (2.137) becomes zero at a value of r such that the physical wavefunction becomes identical with its asymptotic form.

The results contained in (2.135) and (2.137) are independent of the detailed shape of the scattering potential provided it satisfies the restrictions set out in (2.58) and (2.68). Together they constitute the 'shape independent approximation' in the description of potential scattering which is adequate to account for neutron–nucleus scattering in the energy range $E_n < 10$ MeV. In this approximation the scattering cross-section is given by the two parameter formula

$$\sigma_s(k) = 4\pi \bigg/ \left\{k^2 + \left(\frac{1}{a} - \frac{r_0k^2}{2}\right)^2\right\} \tag{2.138}$$

Since any well-behaved hypothetical potential, of specified range and strength, may be used to calculate values for the parameters a and r_0, which are the quantities determined from experiment, it follows that low-energy neutron–nucleus scattering can determine values for the parameters which characterize any given potential, but cannot enable a choice to be made between alternative candidates.

2.5.3 Absorption at zero energy

Finally we consider briefly the case of s-wave scattering from a complex potential of the type discussed at the end of section 2.4. In this case the S-matrix is no longer unitary and the phase shift $\delta_0(k)$ acquires an imaginary part, i.e.

$$\delta_0(k) \to \delta_0(k) + \mathrm{i}\varepsilon_0(k) \qquad \varepsilon_0(k) > 0 \tag{2.139}$$

and the s-wave scattering cross-section is given by

$$\sigma_s = 4\pi|\delta_0(k)|^2 = \frac{4\pi}{k^2}\left\{\left(\frac{1-\mathrm{e}^{-2\varepsilon_0(k)}}{2}\right)^2 + \mathrm{e}^{-2\varepsilon_0(k)}\sin^2\delta_0(k)\right\} \tag{2.140}$$

Since for s-waves the scattering is isotropic the optical theorem (2.100) can be applied directly to the scattering amplitude $f_0(k)$ to give

$$\sigma_t = \sigma_s + \sigma_c = \frac{4\pi}{k}\,\mathrm{Im}\, f_0(k) = \frac{4\pi}{k^2}\left\{\left(\frac{1-\mathrm{e}^{-2\varepsilon_0(k)}}{2}\right)\right\} + \mathrm{e}^{-2\varepsilon_0(k)}\sin^2\delta_0(k)\Big\} \tag{2.141}$$

The capture cross-section σ_c follows simply from 2.140 and 2.141 but can also be obtained from the following useful result

$$\frac{\sigma_c}{\sigma_s} = \frac{4\pi}{k}\left(\mathrm{Im}\, f_0(k) - \frac{4\pi}{k^2}|f_0(k)|^2\right)\Bigg/\left[\frac{4\pi}{k^2}(|f_0(k)|^2) = \frac{\mathrm{Im}\, f_0(k)}{k|f_0(k)|^2} - 1\right]$$

$$= -\left\{\frac{\mathrm{Im}\,(1/f(k)) + 1}{k}\right\} = -\mathrm{Im}\cot[\delta_0(k) + \mathrm{i}\,\varepsilon_0(k)] \tag{2.142}$$

By either means we obtain the result

$$\sigma_c = \frac{4\pi}{k^2}\left\{\frac{1-\mathrm{e}^{-4\varepsilon_0(k)}}{4}\right\} \simeq \frac{4\pi\varepsilon_0(k)}{k^2} \qquad |\varepsilon_0(k)| \ll 1 \tag{2.143}$$

2.5.4 Shadow scattering

Examination of (2.140) yields the remarkable result that, for $\varepsilon_0(k) \neq 0$, the scattering cross-section does not vanish even for the case $\delta_0(k) = 0$ and capture is accompanied by scattering in all circumstances. Comparison of (2.140) and (2.143) shows that, for $\delta_0(k) = 0$, $\sigma_s/\sigma_a = \tanh(\varepsilon_0(k)) \leqslant 1$, but for very strong absorption, $\varepsilon_0(k) \gg 1$ and σ_c and σ_s become equal. The reason for this anomalous scattering is that, in addition to the unscattered component e^{ikr}, the outgoing wave has an out-of-phase component $e^{i(kr+\pi)}(1-e^{-2\varepsilon_0(k)})$ which owes its origin to the effect of diffraction by the partially opaque nucleus. This scattering is therefore called diffraction or shadow scattering (Placzek and Bethe 1940). At energies where the wavelength is comparable with nuclear dimensions and the contribution from states of higher and higher angular momentum becomes significant, the diffracted wave has maximum amplitude in the shadow of the nucleus and this scattering is characterized by an angular distribution similar to the Fraunhofer diffraction pattern of a circular disc (cf. section 6.4.4).

The shadow scattering is much reduced in comparison with absorption in the zero energy limit $\varepsilon_0(k) \ll 1$. Since in this energy region $\varepsilon_0(k)/k$ must tend to a finite limit of the order of d, the depth of the transition layer at the nuclear boundary, we therefore have the result $\sigma_c \simeq 4\pi\varepsilon_0(k)/k^2 \simeq 4\pi d/k$, which expresses the empirically observed $1/v$ law of absorption for very slow neutrons.

2.5.5 The Fermi pseudo-potential

According to the discussion in section 2.5.1, the scattering amplitude, which is the negative of the scattering length in the low-energy region, tends to a constant value independent of k under these conditions. Since, according to (2.55), in Born approximation $f(\theta)$ and $V(r)$ constitute a Fourier transform pair; following Fermi (1936), we may represent $V(\boldsymbol{r})$ in the form of a contact potential

$$V(\boldsymbol{r}) = B\delta^3(\boldsymbol{r}) \tag{2.144}$$

where B is determined from (2.55) giving

$$f(\theta) = -a = \frac{-mB}{2\pi\hbar^2} \tag{2.145}$$

Thus $B = 2\pi a\hbar^2/m$ and we find using (2.60)

$$U(\boldsymbol{r}) = 4\pi a\delta^3(\boldsymbol{r}) \tag{2.146}$$

This expression for $U(\boldsymbol{r})$ is known as Fermi's pseudo-potential (Cohen 1984), and using it we deduce that a medium containing nuclei characterized by a positive scattering length looks like a repulsive potential for slow neutrons while a negative scattering length looks like an attractive potential, i.e. a slow neutron is repelled by the bound state but attracted by the virtual state. These results are of great importance to the practical problem of confining ultra-cold neutrons (cf. section 1.6.6).

The conditions under which Fermi's approximation is valid have been analysed by Summerfield (1964).

2.6 Discrete states of the scattering potential

2.6.1 Bound states

In addition to the continuum of positive energy states labelled by their asymptotic momenta k, for an attractive potential $V(r)$ the Hamiltonian may possess a discrete spectrum of eigenstates which occur at negative values of the energy

$$E = E_n = -|E_n| \tag{2.147}$$

These bound-state solutions exist in the absence of an incoming wave, are regular at the origin, and decrease exponentially at infinity. Since their asymptotic probability current density vanishes, there is no net flux of particles associated with these states which in some sense are analogues of the evanescent wave solutions encountered in optics. In some simple cases the bound state and scattering data may suffice to determine the potential (Roberts 1991).

The route by which the bound states may be accommodated within the S-matrix formalism was pointed out by Kramers and Heisenberg (Heisenberg 1943, Dresden 1987) who noted that the wavefunction $\Psi(E)$ belonging to the continuous spectrum of energy eigenstates is an analytic function of E considered as a complex variable, and may be defined uniquely by the method of analytic continuation in some domain of the complex E-plane connected to the positive real axis. The function $\Psi(E)$ so defined has a physical interpretation for $E = -|E_n|$ on the negative real axis where it coincides with the corresponding bound state solutions of Schrödinger's equation (Møller 1946, Heitler and Hu 1947, Hu 1948).

With this extension to complex E, the function k defined for real E in (2.39) must now be understood as defined in terms of that branch of the square root function $E^{\frac{1}{2}}$ which is positive on the upper edge of the real axis $\arg(E) = 0$. With this definition the values of k corresponding to the bound states (2.147) will be given by

$$k = k_n = \mathrm{i}\left(\frac{2m|E_n|}{\hbar^2}\right)^{\frac{1}{2}} = \mathrm{i}\kappa_n \qquad \kappa_n > 0 \tag{2.148}$$

These values of k_n therefore lie on the imaginary k-axis in the upper k-plane. Since there is a discontinuity in $E^{\frac{1}{2}}$ going from $\sqrt{|E|}$ at $\arg(E) = 0$ to $-\sqrt{|E|}$ at $\arg(E) = 2\pi$, it is necessary to insert a branch line in the complex E-plane along the real axis from 0 to ∞ in order to make k single valued. This cut plane constitutes the first Riemann surface or physical sheet on which $\Psi(E)$ is defined and equation (2.39) may be regarded as a conformal transformation which maps the physical sheet onto the upper half-plane in the complex k-plane. The second Riemann surface or unphysical sheet $2\pi \leqslant \arg(E) < 4\pi$ is then mapped onto the lower half of the complex k-plane. The relationship between the complex E and complex k-planes is illustrated in figure 2.4.

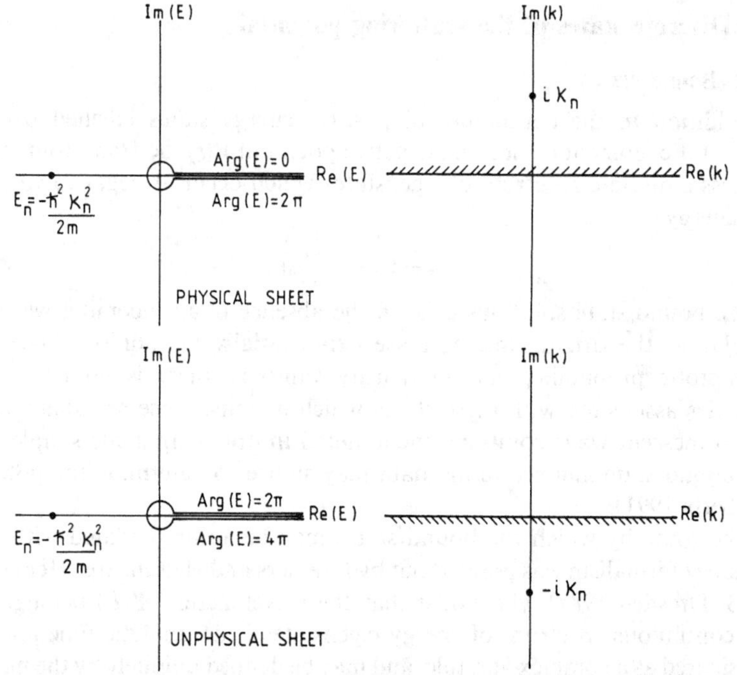

Figure 2.4. Mapping of the complex E-plane onto the complex k-plane. The physical sheet $0 \leqslant \arg(E) < 2\pi$ in the E-plane is mapped onto the upper k-plane, while the unphysical sheet $2\pi \leqslant \arg(E) < 4\pi$ is mapped onto the lower k-plane.

Since a principal characteristic of a bound state is that this solution exists in the absence of an incoming wave, then, for $k = k_n$, according to (2.69) we must have

$$g_l(-k_n) = 0 \tag{2.149}$$

This means, on application of (2.78) that $W(\varphi_l(-k_n, r), \varphi_l(-k_n, r)) = 0$, which in turn implies that $\varphi_l(k, r)$ and $\varphi_l(-k, r)$ coincide to within a multiplicative factor. Thus $\varphi_l(-k_n, r)$ is regular at the origin and vanishes exponentially at infinity and may therefore be used to construct a normalized wavefunction (cf. (2.148))

$$u_{l,n}(k_n, r) = (-i)^l A_n \varphi_l(-k_n, r) \xrightarrow[r \to \infty]{} A_n e^{ik_n r} \tag{2.150}$$

The connection between the bound states and the S-matrix is given by (2.149) and (2.74) which together show that a zero in $g_l(-k)$ on the positive imaginary k-axis (which corresponds to a zero in $g_l(k)$ at the mirror image position on the negative imaginary k-axis) gives a pole in $S(k)$ on the positive imaginary k-axis. Expressed in terms of complex E we see that $S(k)$ will be analytic in some domain on the physical sheet except for simple poles at $E=-|E_n|$ which corresponds to bound states. Of course poles in $g_l(k)$ on the positive imaginary k-axis would also give rise to poles in $S_l(k)$ which would not represent bound states since they do not correspond to the physical situation that there is no incoming wave. Rather the existence of such redundant poles points to the breakdown of the asymptotic approximation in their neighbourhood.

In considering this point it is important to recognize that an asymptotic expansion is divergent in general and always contains an intrinsic error. The asymptotic wavefunction on the physical sheet (Im $k\geqslant 0$) has two terms; a dominant term $\sim \exp(r(\operatorname{Im} k))$ and a subdominant term $\sim \exp(-r(\operatorname{Im} k))$. At the transition point $k=0$, i.e. at the branch point $E=0$ in the complex E-plane, there is no valid asymptotic expansion. Only when Im $k=0$ ($k=0$) is neither exponential dominant, and from this viewpoint the positive energy physical solutions are neutral. Since the asymptotic zone is defined as that region where the potential energy $V(r)$ may be neglected in comparison with the total energy E, to have a valid asymptotic expansion all that is required is that the ratio of the subdominant to the dominant exponential be greater than $V(r)/E$. According to (2.68) this means that a true asymptotic expansion exists only if

$$\mathrm{e}^{-2r(\operatorname{Im} k)} > \mathrm{e}^{-2\mu r} \tag{2.151}$$

Thus $S_l(k)$ is an analytic function of k on the physical sheet only for Im $k\leqslant\mu$, excluding of course the simple poles at the bound states $k_n=\mathrm{i}\kappa_n$ which now must satisfy $\kappa_n \ll \mu$ in order to fall within the region of analyticity of $S_l(k)$. For a bound state the dominant exponential vanishes of course which is precisely the condition for the appearance of an optical evanescent wave.

The net conclusion is that poles in $S(k)$ which occur at values of Im $k>\mu$ cannot be associated with bound states. All these problems are avoided of course if the potential falls off faster than any exponential; in particular if $V(r)$ has finite range, $V(r)=0$, $r>b$, then $S(k)$ is analytic everywhere in Im $k>0$ except for its bound state poles and a possible singularity at infinity.

In this case the residues of $S(k)$ at its bound-state poles may be found by substituting the asymptotic solutions $u_{l,\infty}(k,r)$ (2.76) and $A_n \mathrm{e}^{\mathrm{i}k_n r}$ (2.150) in the completeness condition for the set of eigenfunctions

$$\frac{2}{\pi}\int_0^\infty u_l(k,r)u_l^*(k,r')\,\mathrm{d}k + \sum_{n=0}^{\infty} u_{l,n}(r)u_{l,n}^*(r') = \delta(r-r') \tag{2.152}$$

by which operation we deduce the results (Hu 1948, Corinaldesi 1956, Newton 1960)

$$\frac{(-1)^l}{2\pi}\int_{-\infty}^{\infty} S_l(k)\,\mathrm{e}^{\mathrm{i}k(r-r')} + \sum_{n=0}^{\infty} |A_n|2\,\mathrm{e}^{-\kappa_n(r+r')} = 0 \tag{2.153}$$

In arriving at (2.153) we have used the property that $\delta_l(-k)=-\delta_l(k)$ (cf. 2.70). Noting that $S_l(k)$ is analytic in the finite upper k-plane except for simple poles at the points $k_n=i\kappa_n$ (cf. 2.150) the line integral in (2.153) can be converted into a contour integral in the upper k-plane leading to the results

$$\text{(i)} \qquad \oint_{k_n} S_l(k)\,dk = 2\pi(-1)^l|A_n|^2$$

$$\text{(ii)} \qquad \int_\Gamma S_l(k)\,e^{ik(r+r')}\,dk = 0 \tag{2.154}$$

where Γ is a large semi-circle in the upper half-plane centred on the origin.

Equation (2.153) gives the residue of $S_l(k)$ at $k=k_n$ as $-i(-1)^l|A_n|^2$ while equation (2.154) makes a statement about the behaviour of $S_l(k)$ at infinity. In particular if $S_l(k)$ is regular at infinity (2.154)(ii) will hold for all $(r+r')>0$, which cannot be true since the asymptotic bound-state wavefunction $A_n\,e^{ik_nr}$ is only valid for $r>b$. We conclude therefore that (2.154)(ii) can be valid for all $(r+r')>0$ only if $S_l(k)$ contains a factor $\exp(-2ibk)$ giving rise to an essential singularity at infinity.

We can now apply (2.154)(i) to the most interesting case of low energy s-wave scattering of neutrons by using (2.122) to write $S_0(k)$ in the form of the identity

$$S_0(k) = -\left(\frac{k - ik\cot(\delta_0(k))}{k + ik\cot(\delta_0(k))}\right) \tag{2.155}$$

where $k\cot(\delta_0(k)) \simeq -1/a$ (2.125) provided the energy is sufficiently low. Then, provided $a>0$, $S_0(k)$ has a pole at $k=i\kappa=1/a$, with residue $-iA_0^2=-2i\kappa=-2i/a$ and we have the result

$$a = \frac{1}{\kappa} = 2/A_0^2 \tag{2.156}$$

Thus the scattering length is essentially given by the characteristic dimension κ^{-1} of the bound state. The result (2.156) also implies that the asymptotic wavefunction $A_0\,e^{-\kappa r}$ is valid up to $r=0$ which means that the finite range of the scattering potential has been neglected. If instead of (2.125) we use the effective range approximation (2.135) then we find applying (2.155) that

$$\frac{r_0}{2} = \frac{1}{2\kappa} - \frac{1}{A_0^2} \qquad a = \frac{1}{\kappa\left(1 - \dfrac{r_0\kappa}{2}\right)} \simeq \frac{r_0}{2} + \frac{1}{\kappa} \tag{2.157}$$

and the low-energy scattering amplitude is determined entirely by the properties of the near-by bound state, expressed in terms of its energy κ and its normalization constant A_0.

To conclude this discussion of bound states we take brief note of a remarkable result due to Levinson, which relates the number n_l of bound states of angular momentum l to the corresponding zero-energy phase shift $\delta_l(0)$ (Levinson 1949, Newton 1977). The theorem is valid for a range of central potentials, including those which simultaneously satisfy (2.58) and (2.68), and is based on the regularity of the coefficient $g_l(k)$ (cf. 2.80) which is finite at infinity where its phase can be set arbitrarily equal to zero. Since $g_l(k)$ is analytic in the lower k-plane, the number n_l of its zeros, each corresponding to a bound state, may be evaluated by applying the principle of the argument (Copson 1935) to a semi-circular contour, indented at the origin to allow for a possible zero-energy bound state when $g_l(0)=0$. When $g_l(0)\neq 0$ this procedure immediately leads to the result

$$\delta_l(0)=\pi n_l \qquad g_l(0)\neq 0 \tag{2.158}$$

When $g_l(0)=0$ $(l\neq 0)$, the contour integral over the identation makes a further contribution of π to the phase so that (2.158) still remains valid when n_l is taken to include the zero-energy bound state. However, when $g_0(0)=0$, the indentation makes a contribution of $\pi/2$ to $\delta_0(0)$, a situation what is characteristic of a resonance rather than of a bound state (cf. section 2.8.2), since under these circumstances the zero-energy scattering amplitude $f_0(0)\rightarrow\infty$ and the scattering length becomes infinite. Such a 'zero-energy resonance' is sometimes referred to as a 'half-bound state' in reference to the fact that even the smallest increase in the strength of the potential will create a new bound state of zero angular momentum.

2.6.2 Virtual states

A simple argument based on the uncertainty relation shows that a bound state can exist for an attractive potential of characteristic strength U_0 and range b provided that the inequality

$$U_0 b^2 > \eta \tag{2.159}$$

is satisfied, where $\eta>0$ is a number of order unity. Thus, as U_0 is progressively decreased, the bound state energy approaches zero and the corresponding pole in the S-matrix moves down the imaginary k-axis towards the origin. There even exists a potential which has an s-wave bound state of zero energy although it decreases at infinity on the order of r^{-2} and therefore does not satisfy the condition (2.68) (Moses and Tuan 1959). Now the condition (2.150) which permits a valid analytic continuation into the upper half-plane will also apply in the lower half-plane provided $\operatorname{Im} k > -\mu$ and, for a weak potential U_0, the bound-state pole will move into the lower half-plane. However, since the dominant and subdominant solutions exchange places in this circumstance, the pole $k=-\kappa$ on the negative imaginary axis corresponds to a vanishing subdominant solution and the surviving dominant term represents an unbound state of negative energy $E=-\hbar\kappa^2/2m$ which grows exponentially at infinity and therefore does not yield a physically realizable state. Such unphysical states are therefore called 'virtual states' and the second Riemann surface of the complex E-plane which corresponds to the lower half plane Im $k<0$ is called the unphysical sheet. In fact there exists a more general definition of a virtual state as one characterized by a pole in the lower k-plane at $k=\pm\kappa_1-i\kappa_2$ where $\kappa_2>\kappa_1\geqslant 0$. Such poles also represent unbound states of negative energy and finite width

$$\text{(i) } E=-\frac{\hbar^2}{2m}(\kappa_2^2-\kappa_1^2) \qquad \text{(ii) } \Gamma=\frac{\hbar^2}{2m}4\kappa_1\kappa_2 \tag{2.160}$$

Although unbound states characterized by poles on the negative imaginary k-axis cannot exist as stationary states and do not contribute to the complete set of states contained in (2.152), they can influence the behaviour of the scattering amplitude for positive real values of k and in particular the analysis of low-energy scattering based on (2.155) will go through with only one change, namely that the bound state pole at $k=i\kappa$ now becomes a virtual state pole at $k=-i\kappa$. Thus for zero-energy scattering, the scattering length a becomes negative and (2.156) goes over to

$$a=-1/\kappa \tag{2.161}$$

Thus the observation of a large negative scattering length is associated with the existence of a virtual state near zero energy, a result whose significance in the case of scattering of slow neutrons in ortho and para hydrogen has been discussed in some detail in section 1.5.3.

We illustrate the important role played by virtual states by considering a simple concrete example, namely s-wave scattering described by the Jost function $g_0(k)=(k-i\mu)/(k+i\mu)$ for which the corresponding attractive potential falls off at infinity like $e^{-2\mu r}$ and therefore satisfies the condition (2.68) (Bargmann 1949). The corresponding S-matrix element is

$$S_0(k)=\left(\frac{k-i\kappa}{k+i\kappa}\right)\left(\frac{k+i\mu}{k-i\mu}\right) \tag{2.162}$$

Thus the zero in $g_0(-k)$ at $k=-i\kappa$ gives rise to a pole in $S_0(k)$ at the same point, which represents a virtual state of energy $-\hbar^2k^2/2m$. We can now deduce the value of Re $(1/f_0(k))$ from (2.162) to give the result

$$k\cot(\delta_0(k))=\frac{\kappa\mu}{\mu-\kappa}+\frac{k^2}{\mu-\kappa} \tag{2.163}$$

Thus in the very special case given by (2.161) the effective range approximation (2.135) is exact and the corresponding parameters are

$$a=\frac{-1}{\kappa}+\frac{1}{\mu}<0 \qquad \frac{r_0}{2}=\frac{1}{\mu-\kappa}>0 \tag{2.164}$$

In the limit $\kappa \to 0$, the scattering length $a \to -\infty$ and we have a virtual state at zero energy. The corresponding potential is

$$U(r) = -2\mu^2/\cosh^2 \mu r \tag{2.165}$$

The example given above is interesting for another reason in that it illustrates the role of the redundant role at $k = i\mu$ on the boundary of the strip of analyticity in the upper half-plane. Assuming $\mu \gg \kappa$ the position of this pole does not influence the value of the scattering length to any significant degree but it does determine the effective range of the potential $r_0 \simeq 2/\mu$.

2.6.3 Resonances

2.6.3.1 Resonance reactions with slow neutrons

The earliest attempts at theories of nuclear reactions with slow neutrons were based on a two-body approximation for the neutron–nucleus collision with a square well potential of depth $V_0 \simeq 40$ MeV and range $b = r_0 A^{1/3}$ with $r_0 \simeq 1.5 \times 10^{-13}$ cm (Bethe 1937). All of these theories predicted that elastic scattering would predominate with resonance maxima in the scattering cross-section corresponding to excited nuclear levels with separations of a few MeV and resonance widths of order 1 MeV. The predicted capture rates for neutrons making transitions into bound states with emission of γ-rays were small and only weakly dependent on energy.

As we have already remarked in section 1.4, strong neutron capture is centred at well-defined resonance energies grouped quite closely together on the energy scale, and having resonance widths $\leqslant 1$ eV, with large partial cross-sections for reactions and inelastic scattering. These posed great problems for potential scattering models, and for this reason Bohr introduced his strong coupling or compound nucleus model which treated the neutron–nucleus interaction as a many-body problem (Bohr 1936, Bohr and Kalckar 1937). In this picture the reaction is viewed as taking place in two independent stages. In the first stage the neutron is absorbed by the target nucleus of mass A and a quasi-stationary state of the compound nucleus of mass $(A+1)$ is formed. In the second stage the compound nucleus decays by one of several channels, e.g. elastic or inelastic scattering, γ-decay or charged particle emission. If the incident energy is close to a positive energy quasi-stationary state of the compound nucleus then a resonance is observed in the cross-section with a width inversely proportional to the lifetime of the compound state. Figure 2.5 shows a plot of the total cross-section in ^{115}In which shows a strong s-wave resonance at 1.456 eV and a weaker resonance at 3.86 eV. There are also two unresolved resonances at 9.1 eV and 12.1 eV.

The detailed theory of the compound nucleus resonances was worked out by Breit and Wigner (1936) for a single isolated resonance, and by Bethe

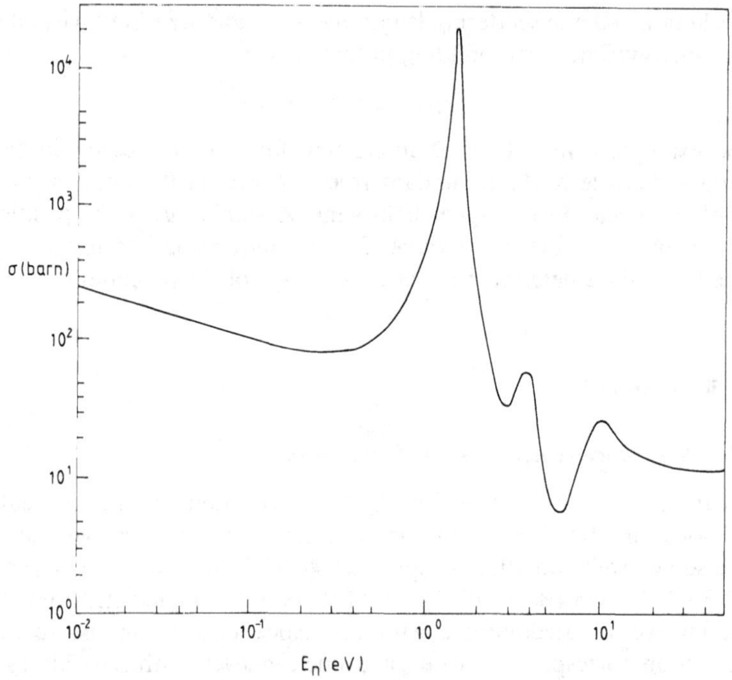

Figure 2.5. Thermal and epithermal resonances observed in the total neutron cross-section in ^{115}In. The narrow widths observed indicate that these are compound nucleus resonances and not single particle resonances.

and Placzek (1937) for overlapping resonances, using essentially the same form of second-order perturbation theory which had been successfully applied to the analysis of optical dispersion (Weisskopf and Wigner 1930, Weisskopf 1931). However the treatment of the neutron–nucleus interaction as a perturbation is hardly a valid procedure and in any case a perturbative approach, although it arrives at the correct Lorenzian shape of the resonance peak, does not suggest any general rules governing specific reaction rates which are consistent with observation. This contrasts sharply with the case of optical dispersion where the perturbation theory correctly predicts that the inelastic process (photoelectric effect) will predominate over the elastic process (Rayleigh scattering) when both are energetically possible. Thus a more general approach to the analysis of resonance phenomena is required when, as in nuclear physics, the energy of interaction of the colliding systems is larger than, or comparable with, their relative kinetic energy.

2.6.3.2 The Breit–Wigner formulae

Although originally arrived at on the basis of a perturbation theory the

Breit–Wigner formula for elastic resonance scattering can be derived quite easily in potential scattering models and the results generalized to include inelastic processes by the introduction of a complex potential as described in section 2.4.3. For $l \neq 0$ quasi-stationary states arise quite naturally in models with an attractive potential, for in this case the centrifugal barrier term $l(l+1)\hbar^2/mr^2$ dominates the effective potential both for very small and very large values of the radius, which means that quasi-stationary states of positive energy can exist in the potential well in the region $0 < r \leqslant b$ as shown in figure 2.6. These are of course not true positive energy bound states such as harmonic oscillator states, since particles trapped in them eventually tunnel through the potential barrier and escape to infinity. This is exactly what happens in radioactive decay by α-particle emission and, by analogy, resonance states are sometimes referred to as 'radioactive states'.

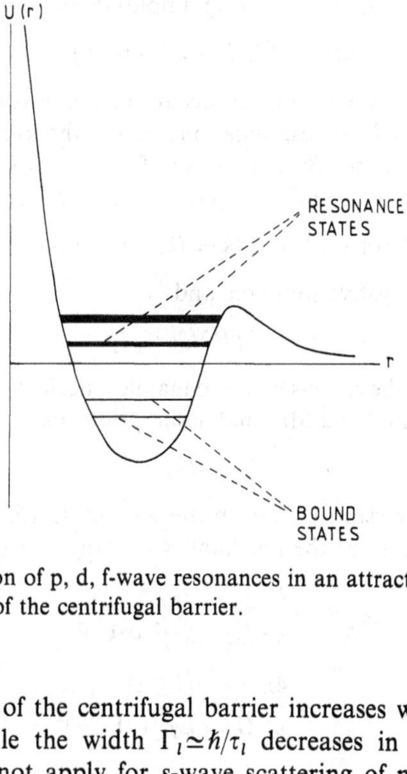

Figure 2.6. Generation of p, d, f-wave resonances in an attractive potential brought about by the action of the centrifugal barrier.

Since the height of the centrifugal barrier increases with l, so also must the lifetime τ_l while the width $\Gamma_l \simeq \hbar/\tau_l$ decreases in proportion. These considerations do not apply for *s*-wave scattering of neutral particles for which there is no centrifugal barrier although for charged particles the Coulomb barrier plays an equivalent role. Thus for neutron scattering there is no maximum in the cross-section unless $\partial^2\beta(k)/\partial k^2 > (2k)^{-\frac{1}{2}}$ at the resonance energy and we should not be surprised that the *s*-wave resonances predicted in potential scattering theory tend to have resonance widths related to lifetimes on the order of the time required for a neutron to travel across the nucleus as already noted in section 2.6.3.1 above.

In deriving the Breit–Wigner formula for the scattering of neutrons at an s-wave resonance it is simplest to assume that the potential has a finite range i.e. $V(r)=0$ for $r>b$. In this case we can rewrite (2.127) in the form

$$k\cot(\delta_0(k)+bk)=\beta(k) \tag{2.166}$$

where the additional phase bk is associated with the factor e^{-2ibk} in $S(k)$ (cf. (2.153) *et seq.*). This factor represents non-resonant potential scattering from an impenetrable sphere of radius b, and it is convenient to separate off the source of possible resonant effects in $S(k)$ by introducing a resonant phase shift $\delta_{or}(k)$ and writing

$$S_r(k)=S(k)\,e^{2ibk}=e^{2i(\delta_0(k)+kb)}=e^{2i\delta_{or}(k)} \tag{2.167}$$

The corresponding resonant scattering amplitude is given by

$$f_r(k)=e^{2ibk}f(k)+e^{ibk}\sin(kb)/k \tag{2.168}$$

According to (2.167) a resonance occurs in the scattering cross-section when $\delta_{or}(k)=\pi/2$, which corresponds to a zero of the logarithmic derivative $\beta(k)$ given in (2.166). Since $\beta(k)$ is an even function of k we may expand it in a Taylor series in the neighbourhood of this zero, i.e.

$$k\cot(\delta_{or}(k))=\beta(k)=(k_r^2-k^2)/\gamma+\ldots \tag{2.169}$$

where k_r^2 and γ are positive numbers and

$$\gamma^{-1}=\partial\beta(k)/\partial k^2|_{k=k_r} \tag{2.170}$$

For the maximum in the cross-section to manifest itself as a narrow resonance we shall have to make the additional assumption that

$$0<\gamma<k_r \tag{2.171}$$

Substituting the expression (2.169) in the identify (2.155) it is now easy to calculate an expression for the resonant scattering cross-section

$$\begin{aligned}\sigma_{sr}(k)&=\frac{4\pi}{k^2}\frac{(k\gamma)^2}{(k_r^2-k^2)^2+(k\gamma)^2}\\&=\frac{4\pi}{k^2}\frac{(\Gamma_s/2)^2}{(E_r-E)^2+(\Gamma_s/2)^2}\end{aligned} \tag{2.172}$$

where

$$\frac{\Gamma_s}{2}=\frac{\hbar^2}{2m}(\gamma k) \tag{2.173}$$

The result (2.172) is the Breit–Wigner formula for scattering at an isolated resonance. The maximum in $\sigma_{sr}(k)$ occurs at the resonance energy $E=E_r$ and, where $\delta_{or}(k)=\pi/2\pm\pi/4$, $\sigma_{sr}(k)$ falls to half its maximum value and this happens when $E=E_r\pm\Gamma_s/2$. For this reason Γ_s is called the 'full width at half maximum (FWHM)' of the resonance or, more concisely, the 'width' of the resonance. Using the relation (2.168) between $f(k)$ and $f_r(k)$ the total elastic scattering cross-section can be written

$$\sigma_s(k)=\sigma_{sr}(k)+\frac{4\pi\sin^2(bk)}{k^2}-2\,\mathrm{Re}\left(f_r(k)\,\mathrm{e}^{-2ibk}\frac{\sin(bk)}{k}\right) \tag{2.174}$$

a result which shows that a resonance in $\sigma_{sr}(k)$ is also a resonance in $\sigma_s(k)$ provided $kb\ll\pi/2$. In general for the maximum in $\sigma_{sr}(k)$ to be observable as a true resonance we also require that kb should not change significantly as k sweeps through its resonance value $k=k_r$.

At this point we may remark that (2.170) may be regarded as the definition of the even function $\gamma(k)$ rather than as the first term in a Taylor expansion, and in principle we can expand $\gamma(k)$ to terms of order k^2 and higher even powers. However, if these terms are significant then the denominator of $f_r(k)$ expressed as a rational function becomes a quartic in k and in these circumstances we are entering the regime of overlapping resonances.

We can now extend the result (2.172) to take account of inelastic processes by introducing an imaginary part to the scattering potential as described in section 2.4.3. Adopting the notation (2.139) we now write

$$\begin{aligned} k\cot(\delta_{or}(k)) &\to k\cot(\delta_{or}(k)+\mathrm{i}\varepsilon_{or}(k)) \\ &=\frac{k\cot(\delta_{or}(k))\,\mathrm{sech}^2(\varepsilon_{or}(k))-ik\,\mathrm{cosec}^2(\delta_{or}(k))\tanh(\varepsilon_{or}(k))}{1+\cot^2(\delta_{or}(k))\tanh^2(\varepsilon_{or}(k))} \end{aligned} \tag{2.175}$$

where $\delta_{or}(k)$ now represents the real part of the complex phase-shift. In this case the condition for a resonance is given by

$$\mathrm{Re}(k\cot(\delta_{or}(k)+\mathrm{i}\varepsilon_{or}(k)))=\mathrm{Re}\,\beta(k)=0 \tag{2.176}$$

and it follows from (2.175) that, as before, $\delta_{or}(k)=\pi/2$. However taking (2.169) to apply to $\mathrm{Re}(\beta(k))$ the total width of the resonance is determined in this case, not by γ alone, but also by the value of

$$\eta(k)=-\mathrm{Im}\,\beta(k) \tag{2.177}$$

evaluated at $k=k_r$. According to (2.175)

$$\eta(k)>0 \quad \text{with} \lim_{\delta_{or}\to 0}\eta(k)=\coth(\varepsilon_{or}(k))$$

and

$$\lim_{\delta_{or}\to\pi/2}\eta(k)=\tanh(\varepsilon_{or}(k))$$

Again we shall have to assume that, like $\gamma(k)$, $\eta(k)$ does not vary much on passing through the resonance. What this means is that the product $\operatorname{cosec}^2(\delta_{or}(k))\tanh(\delta_{or}(k))$ remains constant so that when $\operatorname{cosec}^2(\delta_{or}(k))$ decreases and then increases again on passing the resonance point, $\tanh(\varepsilon_{or}(k))$ increases and decreases correspondingly. Thus the scattering resonance $\delta_{or}(k)\to\pi/2$ is always accompanied by a capture resonance $\varepsilon_{or}(k)\to\infty$ at exactly the same resonance energy as one might expect. The assumption of constant $\eta(k)$ is merely a demand to the effect that Im $\beta(k)$ should not become singular near a zero of Re $(\beta(k))$, and when this condition is relaxed we are leaving the regime of isolated resonances.

We can now complete the computation of σ_{sr} using in place of the expression for $f_r(k)$, derived from (2.168), the modified expression

$$f_r(k)=\frac{1}{(1+\eta)k+i(k_r^2-k^2)/\gamma} \tag{2.178}$$

The resultant resonant scattering cross-section is

$$\sigma_{sr}(k)=\frac{4\pi}{k^2}\frac{(\Gamma_s/2)^2}{(E-E_r)^2+(\Gamma_t/2)^2} \tag{2.179}$$

where Γ_t is the total width of the resonance given by

$$\frac{\Gamma_t}{2}=\frac{\hbar^2}{2m}\gamma(1+\eta) \tag{2.180}$$

The resonant capture cross-section σ_{rc} can now be expressed in terms of σ_{rs} using (2.142), i.e.

$$\sigma_{rc}/\sigma_{rs}=-\text{Im}\,[k\cot(\delta_{or}(k)+i\varepsilon_{or}(k))]=\eta k \tag{2.181}$$

which allows us to express σ_{rc} in the form

$$\sigma_{rc}(k)=\frac{4\pi}{k^2}\frac{(\Gamma_s/2)(\Gamma_c/2)}{(E-E_r)^2+(\Gamma_t/2)^2} \tag{2.182}$$

where

$$\frac{\Gamma_c}{2}=\frac{\hbar}{2m}\gamma\eta k=\frac{\Gamma_t-\Gamma_s}{2} \tag{2.183}$$

is the capture width. The results (2.179) and (2.182) together constitute the Breit–Wigner formulae for scattering and capture at a single isolated resonance.

2.6.3.3 The Kapur–Peierls formalism

The weakness of the preceding analysis of the resonance scattering problem is its lack of generality. In the first place it is limited by various approximations such as the quadratic behaviour of $\beta(k)$ in the neighbourhood of the resonance and the assumed constancy of the width functions $\gamma(k)$ and $\eta(k)$, but, more importantly, it applies only to the case of scattering by a potential and is not generalizable in any direct way to describe scattering in a many-body system. These limitations do not apply to the dispersion theory of Kapur and Peierls (1938) or to equivalent theories such as the R-matrix formalism of Wigner and Eisenbud (1947) which is discussed in some detail in section 4.7.2.

The central feature of the Kapur–Peierls method is the introduction of a complete set of states to describe all the particles in the nucleus, defined by imposing energy-dependent boundary conditions on the wavefunction on the surface of a sphere of radius r_0 completely enclosing the nucleus. Although the formalism is most conveniently presented in the context of a two-body problem with r_0 set equal to b, the range of the potential $V(r)$, it is readily adapted to a many-body system where the states of the complete set are identified with the quasi-stationary states of the compound nucleus (Brown *et al* 1959).

To illustrate the method we consider the solution $\varphi_0(k, r)$ of the radial equation (2.59) for $l=0$, which satisfies the boundary condition (2.63) and which may be represented in the interior region as

$$u_0(r)=\sum_n a_\lambda v_\lambda(r) \qquad 0\leqslant r\leqslant b \tag{2.184}$$

where the Kapur–Peierls eigenfunctions $v_n(r)$ are defined as solutions of the radial equation

$$\frac{\mathrm{d}^2 v_\lambda(r)}{\mathrm{d}r^2}+\frac{2m}{\hbar^2}(E_\lambda - V(r))v_\lambda=0 \tag{2.185}$$

which are finite at the origin and which satisfy the boundary conditions

$$\frac{\mathrm{d}v_\lambda(r)}{\mathrm{d}r}\bigg/ v_\lambda(r)=\beta(k)=ik \tag{2.186}$$

at $r=b$. Although the boundary conditions (2.186) cannot be satisfied by a physical wavefunction, which must match to an incoming wave as well as to an outgoing wave on the boundary, the expansion (2.184) is nonetheless legitimate because the eigenfunctions $v_\lambda(r)$ constitute a complete orthonormal set (Peierls 1947). What is unusual or surprising about these eigenfunctions is that there is a distinct discrete set for each value of the incident momentum k, and corresponding to each eigenfunction of the set there appears a resonant term in the S-matrix which may be written as

$$S(k)=\mathrm{e}^{-2ikb}\left\{1-\sum_\lambda \Gamma_\lambda/(E-E_\lambda)\right\} \tag{2.187}$$

where E_λ is a complex energy eigenvalue and Γ_λ a complex width.

The principal advantage of using (2.187), which is a representation of $S(k)$ in terms of the resonances, is that it is exact, and it is this property of the formalism which allows its generalization to many-body systems. In the neighbourhood of zero energy the width function Γ_λ has a large real part which is approximately equal to twice the imaginary part of the energy eigenvalue and, under these conditions, the Breit–Wigner formula (2.172) is reproduced with the resonance energy E_r identified with the real part Re (E_λ) of the corresponding complex energy eigenvalue.

We have noted above, in the context of the two-body problem (cf. section 2.3.2), that the choice $r_0 = b$ was convenient rather than necessary, which means that, not only do the Kapur–Peierls states which contribute to the S-matrix vary with energy, but the division into resonant and non-resonant contributions is not invariant under changes of r_0. It was first pointed out by Siegert (1939) that both these weaknesses could be overcome if the boundary condition (2.186) was replaced by

$$\frac{dv_\lambda}{dr}(r)/v_\lambda(r) = ik_\lambda \qquad (2.188)$$

which indeed is just the condition that the Jost function $g_0(-k_\lambda)$ should vanish (cf. 2.149) expressing the fact that there is no incoming wave.

For a real potential the hermiticity of the Hamiltonian forbids the presence of complex zeros of $g_l(-k)$ in the upper k-plane, except on the imaginary axis where they represent true bound states (cf. section 2.6). However no such restriction applies to the lower k-plane which corresponds to the unphysical energy sheet. The only requirement which must be fulfilled is that if $k_\lambda = \kappa_1 - i\kappa_2$ is a complex zero of $g_l(-k)$, so also is $-k_\lambda^* = -\kappa_1 - i\kappa_2$; this follows from the crossing symmetry relation (2.70) for real k, which implies that the analytic continuation of $g_l(k)$ off the real axis in the k-plane must satisfy

$$g_l(k) = g_l^*(-k^*) \qquad (2.189)$$

This result implies incidentally that the S-matrix in (2.74) is no longer unitary for complex k.

If we now write the complex energy of the resonance state corresponding to k_n as $E_\lambda - iW_\lambda$ then we have the results

$$E_\lambda = \frac{\hbar^2}{2m}(\kappa_1^2 - \kappa_2^2) \qquad W_\lambda = \frac{\hbar^2}{2m}2\kappa_1\kappa_2 \qquad (2.190)$$

where the second relation in (2.190) can be rewritten as

$$\frac{\hbar\kappa_2}{W_\lambda} = \frac{m}{\hbar\kappa_1} = v_\lambda \qquad (2.191)$$

and v_λ is the velocity of the particle emitted from the decaying resonance state. The asymptotic form of the time-dependent wavefunction can now be written as

$$u_{l,\infty}(k, r, t) \simeq e^{i[(\kappa_1 - i\kappa_2)r - E_\lambda t/\hbar]} = e^{i(\kappa_1 r - E_\lambda t/\hbar)}\, e^{-W_\lambda(t - r/v_\lambda)/\hbar} \tag{2.192}$$

which represents an outgoing wave whose amplitude $\exp[-W_\lambda(t-r/v_\lambda)/\hbar]$ is determined by the amplitude of the decaying resonance at the retarded time $(t-r/v_\lambda)$. The second pole at $-k_\lambda^* = -\kappa_1 - i\kappa_2$ has an asymptotic solution describing an incoming wave whose amplitude increases in time. Such states have therefore been called 'capture states' since there is an incoming wave but no outgoing wave. However, the two states characterized by the poles at $k_\lambda = \pm\kappa_1 - i\kappa_2$ are time-reversed transforms of each other, and correspond to the same resonance state on the unphysical sheet of the complex E-plane. Thus for every pair of complex poles in the lower k-plane we have a single resonance state, and if, for example, the denominator of $f_r(k)$ is a quartic in k, a possibility canvassed in section 2.6.3.2 above, then $f_r(k)$ has four complex poles and receives contributions from two nearby resonance states.

We have already remarked in section 2.6.2 that, when $\kappa_2 > \kappa_1$ the poles at $k_\lambda = \pm\kappa_1 - i\kappa_2$ represent virtual states of negative energy rather than resonances and the location of the poles associated with bound, virtual and resonance states is shown in figure 2.7. With a potential satisfying (2.68), which includes finite range potentials, the numbers of bound and virtual states are both finite, but in general there is an infinite number of complex poles in the lower k-plane of which only a finite number represent true resonances at which the phase shift is an odd multiple of $\pi/2$. This is because, at sufficiently large energy, the phase bk arising from the potential scattering contribution always changes so rapidly as k passes through the prospective resonance, that the resonance is wiped out by destructive interference. Of course for a complex potential the Schrödinger equation has complex energy eigenvalues represented by complex poles in $S(k)$ in the upper k-plane. However, unlike the resonance states, these states have normalizable wavefunctions and must be included in the complete set of states (2.152).

2.6.4 Regge poles, trajectories and recurrences

Although the theory of potential scattering as developed above is valid for non-negative integral values of the orbital angular momentum l, interesting results follow from continuing the theory into the complex l-plane as first demonstrated by Regge (1959, 1960). Such an extension is possible because both the radial wave equation (2.59) and the boundary condition (2.63) imposed on the solution $\varphi_l(k, r)$ are formulated in terms of regular functions of l. Thus $\varphi_l(k, r)$ is an analytic function of l as indeed are the expansion coefficients $g_l(k)$ and $g_l(-k)$. However, for complex l, these coefficients no

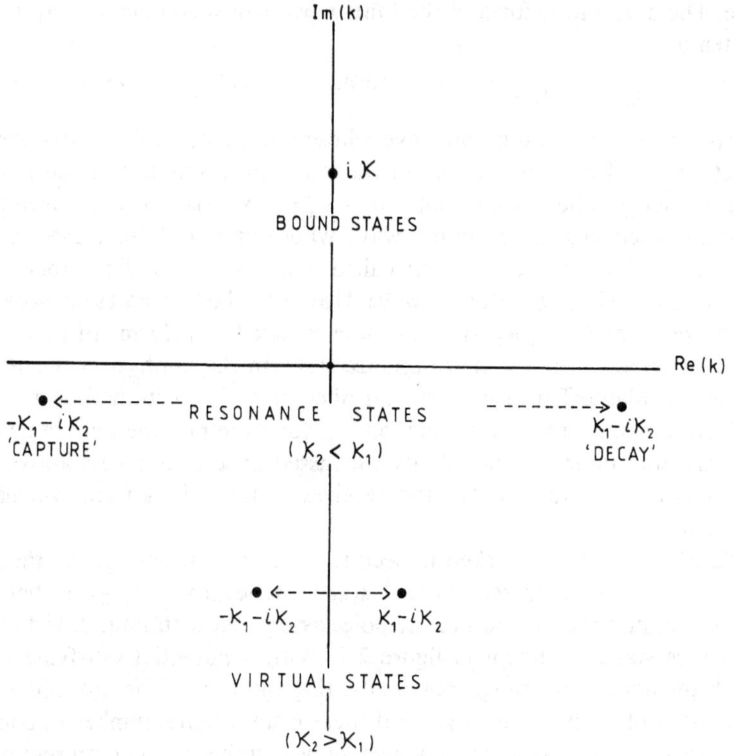

Figure 2.7. Location of poles of the matrix element $S(k)$ in the complex k-plane associated with bound, virtual and resonance states.

longer satisfy the symmetry relation (2.80) and, in consequence, the phase shift $\delta_l(k)$ defined in (2.74) becomes complex.

The S-matrix element $S_l(k)$ remains an even function of k which, in analogy with (2.149), has poles in the complex l-plane wherever $g_l(k)$ has zeros. Quite generally each Regge pole $\lambda_j(E)$ $(j=1, 2, \ldots)$ is a complex number, which, as the (real) energy E varies, moves along a curve in the l-plane

$$l=\lambda_j(E) \tag{2.193}$$

known as a Regge trajectory. One may then easily reformulate the analysis of section 2.6.1 to show that, whenever a Regge trajectory crosses the real l-axis at a non-negative integral value of l, there is a bound state at the corresponding (negative) energy.

The states which are generated at positive energy when $\mathrm{Re}\,\lambda_j=n$ (a non-negative integer) and $\mathrm{Im}\,\lambda_j\neq 0$ have been described by Regge as 'shadow states' because they may be associated with rays which are diffracted into the geometrical shadow according to the Levy–Keller formulation of geometrical optics (Regge 1960). When Im $\lambda_j \ll$ Re λ, they correspond to resonance states as may readily be shown.

The starting point is Regge's representation of the partial wave expansion (2.81) as a Watson–Sommerfeld contour integral

$$f(\theta)=\frac{\mathrm{i}}{2}\int_C (2l+1)f_l(k)[P_l(-\mu)/\cos(\pi l+\tfrac{1}{2}))]\,\mathrm{d}l \qquad (2.194)$$

where $\mu=\cos\theta$ and, provided that $V(r)$ satisfies the condition (2.68), $f_l(k)$ is regular for Re $l>-\frac{1}{2}$, except for a finite number N of poles which are confined to the quadrant Im $l\geqslant 0$. The contour C, which is shown in figure 2.8, loops

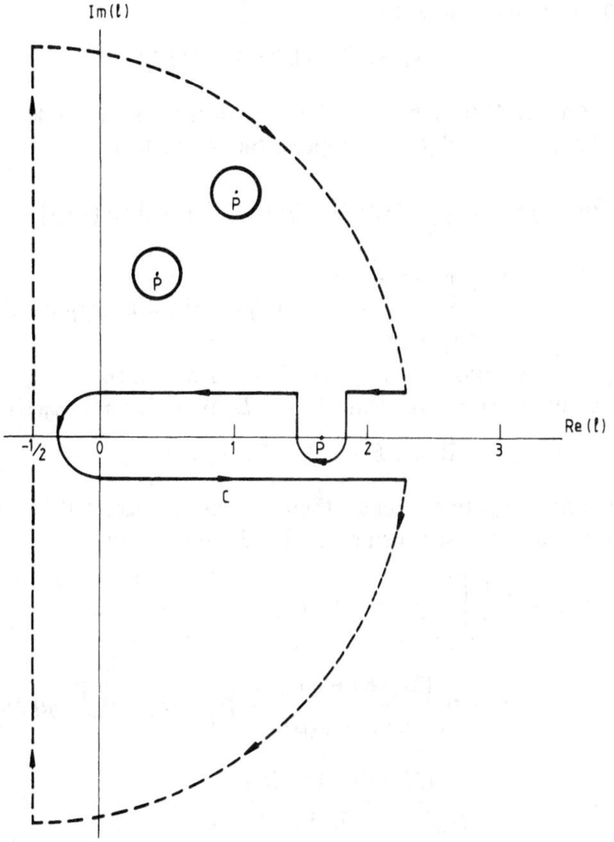

Figure 2.8. The path of integration in the complex l-plane for the Watson–Sommerfeld representation of the scattering amplitude. Completing the path C by a large semi-circle at infinity allows the contour integral to be expressed in terms of a line integral along the line Re $l=-\frac{1}{2}$ plus the residues at the poles of $(S_l(k)-1)$ which are confined to the quadrant Re $l>-\frac{1}{2}$, Im $l\geqslant 0$.

around the zeros of cos $(\pi(l+\frac{1}{2})$, avoiding the poles P of $f_l(k)$ including any which may fall on the real axis.

If, in addition $V(r)$ is assumed to have the form of a superposition of Yukawa potentials (cf. 2.3(iii))

$$rV(r) = -V_0 b \int_{1/b}^{\infty} e^{-\eta r} p(\eta)\, d\eta \tag{2.195}$$

where $p(\eta)$ is a probability density, an interaction which provides a good representation of many processes of physical interest (cf. section 3.5.1), then for Re $l > -\frac{1}{2}$ and $|l| \to \infty$, $f_l(k)$ has the asymptotic behaviour

$$2ikf_l(k) = O\,[\exp(i\varepsilon l - \alpha l)] \qquad 0 < \varepsilon \ll 1 \tag{2.196}$$

where (for a reduced mass m)

$$\cosh \alpha = 1 + (1/2b^2)(\hbar^2/2mE) \tag{2.197}$$

Thus the contour C may be completed by a large semi-circle at infinity as shown in figure 2.8, and $f(\theta)$ is expressible in the form

$$2ikf(\theta) = -\pi \sum_{j=1}^{N} (2\lambda_j(E)+1)\rho_j(E)\,[P_{\lambda_j}(-\mu)/\sin(\pi\lambda_j)]$$

$$+\frac{i}{2}\int_{-\frac{1}{2}-i\infty}^{-\frac{1}{2}+i\infty} (2l'+1)[P_{l'}(-\mu)/\sin(\pi l')]f_{l'}(k)\, dl' \tag{2.198}$$

where $\rho_j(E)$ is the residue of $[S_l(E)-1]$ at its pole $\lambda_j(E)$.

If we now suppose that, at an energy $E = E_r$, there is a pole $\lambda_0(E)$ such that

$$\text{Re}\ \lambda_0(E_r) = l \qquad \text{Im}\ \lambda_0(E_r) \ll l \tag{2.199}$$

where l is a non-negative integer, then we may project out its contribution to the lth partial wave scattering amplitude by writing

$$2ikf_l(k) = \frac{1}{2}\int_{-1}^{1} f(\theta)P_l(\mu)\, d\mu$$

$$\simeq -\pi\left(\frac{(2\lambda_0(E)+1)\rho_0(E)}{\sin(\pi\lambda_0(E))}\right)\int_{-1}^{1} P_l(\mu)P_{\lambda_0}(-\mu)\, d\mu$$

$$= -\frac{(2\lambda_0(E)+1)\rho_0(E)}{(\lambda_0(E)-l)(\lambda_0(E)+l+1)} \tag{2.200}$$

In deriving this result we have made use of the orthogonality relation valid for integral l (Abramowitz and Segun 1965)

$$\int_{-1}^{1} P_l(\mu)P_{\lambda_0}(-\mu)\, d\mu = \frac{2\sin(\pi\lambda_0)}{\pi(\lambda_0 - l)(\lambda_0 + l + 1)} \tag{2.201}$$

It remains only to expand $\lambda_0(E)$ to first order in $(E-E_r)$ using (2.199)

$$\lambda_0(E) \simeq l + (E-E_r)(\mathrm{d}\,\mathrm{Re}\,(\lambda_0(E))/\mathrm{d}E)_{E=E_r} + \mathrm{i}\,\mathrm{Im}\,\lambda_0(E_r) \qquad (2.202)$$

and $f_l(k)$ reduces to

$$2\mathrm{i}f_l(k) \simeq \frac{-\rho_0(E)/(\mathrm{d}\,\mathrm{Re}\,(E_r)/\mathrm{d}E)}{(E-E_r) + \mathrm{i}\,\mathrm{Im}\,\lambda_0(E_r)/(\mathrm{d}\,\mathrm{Re}\,\lambda(E_r)/\mathrm{d}E)} \qquad (2.203)$$

which is immediately recognizable as a Breit–Wigner resonance of half-width

$$\Gamma/2 = \mathrm{Im}\,\lambda_0(E_r)/(\mathrm{d}\,\mathrm{Re}\,\lambda_0(E_r)/\mathrm{d}E) \qquad (2.204)$$

According to the optical theorem (2.100) the residue $\rho_0(E_r)$ must satisfy

$$\rho_0(E_r) \leqslant 2\,Im\,\lambda_0(E_r) \qquad (2.205)$$

the equality holding when there is only one pole $\lambda_0(E)$, and the background integral in (2.198) may be neglected.

It is possible of course, for a given pole $\lambda_j(E)$ to satisfy (2.199) with many different pairs of values l and E_r. Such states, all lying on a single Regge trajectory, are described as Regge recurrences since, apart from their angular momentum and energy, they all have the same quantum numbers. For example, the single-particle bound states of the nuclear shell model shown in figure 4.1, with the same principal quantum number and $j=l+\frac{1}{2}$, i.e. $1s_{1/2}$, $1p_{3/2}$, $1d_{5/2}$, etc. lie on a single Regge trajectory, while the states with $j=l-\frac{1}{2}$ lie on an adjacent trajectory. More generally, when an exchange contribution is added to the potential, which changes sign when l alternates between even and odd, states with l even or odd lie on different trajectories, all states on the one trajectory have the same parity, and consecutive states differ by two units in l (cf. sections 3.3.1 and 3.3.2.2).

Although Regge theory provides new insights into the nature of bound states, resonances, etc. in non-relativistic two-particle scattering, it has found its widest application in relativistic strong interaction physics, where many (not entirely successful) attempts have been made to assign elementary particles, which are stable with respect to strong interactions, and unstable resonances, to Regge trajectories. For example in the pion–nucleon excitation function shown in figure 9.1, the nucleon N(940), which is interpreted as a $P_{1/2}$ bound state of the pion–nucleon system, lies on a common trajectory with the resonances $N^0(1688)$ $(F_{5/2})$ and $N^0(2200)$ $(H_{9/2})$. These states are therefore interpreted as Regge recurrences of the nucleon. In the same way the Δ-particles, $\Delta^0(1236)$, $(P_{3/2})$, $\Delta^{++}(1950)$ $(F_{7/2})$ and $\Delta^{++}(2420)$ $(H_{11/2})$ lie on a neighbouring trajectory and differ from the nucleon in terms of quantum numbers other than the mass and the spin.

2.7 The neutron–proton system

2.7.1 The deuteron

To conclude this discussion of two-body scattering with neutrons we present a simple model of the neutron–proton system, laying particular emphasis on the role of the deuteron which historically played such an important part in the early attempts to unravel the complexities of the nucleon–nucleon force. Of course these simple quantum mechanical theories have long since been superseded by more sophisticated techniques, founded in quantum field theory and aided by the rapid development of modern computing methods. Nevertheless the low-energy experimental results still retain their absolute validity and serve to provide calibrations for the various effective interactions used in nuclear physics, and in the analysis of high energy nuclear scattering. The description of these experimental results in terms of simple potential models is also of great value in demonstrating the power of scattering theory, and in exhibiting the links which connect the scattering parameters with the discrete states of the two-body system.

We therefore represent the neutron–proton system as a pair of equal mass non-identical particles in a relative s-state, interacting through the attractive square well potential defined in (2.3(i)). The corresponding Schrödinger equation (2.59) for the radial wavefunction u_0 is therefore given as

$$\left(\frac{\mathrm{d}^2}{\mathrm{d}r^2} - U(r) + k^2\right)u_0 = 0 \tag{2.206}$$

where

$$U(r) = -U_0 \quad r<b \qquad U(r)=0 \quad r>b \tag{2.207}$$

In all cases the solution of (2.206) for the internal wavefunction ($r<b$) is

$$u_0 = A\sin(Kr) \qquad K(k) = \sqrt{(U_0+k^2)} \qquad r<b \tag{2.208}$$

but the form of the external solution ($r>b$) depends on whether the total energy is negative or positive. In particular for a bound state the energy must be negative, and the external wavefunction is the decaying exponential

$$u_0 = B\,\mathrm{e}^{-\gamma r} \qquad r>b \qquad \gamma = \left(\frac{2mE_b}{\hbar^2}\right)^{\frac{1}{2}} > 0 \tag{2.209}$$

where $m \simeq m_p/2 \simeq m_n/2$ is the reduced mass, and $-E_b$ is the energy of the bound state. The values of the normalization constants A and B must be chosen consistent with the overall normalization condition.

$$\int_0^\infty u_0^2(r)\,\mathrm{d}r = 1 \tag{2.210}$$

while their ratio is eliminated by matching the values of the logarithmic derivatives at the boundary of the potential with the result

$$K_0 \cot(K_0 b) = -\gamma \qquad K_0 = \sqrt{(U_0 - \gamma^2)} \tag{2.211}$$

Assuming that (2.211) describes the ground state it follow that U_0 and b must be assigned values restricted to the range $K_0 b > \pi/2$, which can at the same time account for the measured value of the deuteron binding energy $E_b = 2.226$ MeV, but neither the well depth nor the range of the potential are uniquely determined by these requirements.

To enquire in a little more detail into the conditions which must apply for the presence of other discrete states of the neutron–proton system, an expression is needed for the Jost function (2.70). The relevant result is (Nussenzweig 1959)

$$g_0(k) = e^{-ikb}\{\cos(Kb) + i(k/K)\sin(Kb)\} \tag{2.212}$$

and it follows that discrete states corresponding to the zeros of the function $g_0(-k)$ are given for each solution of the equation

$$Kb \cot(Kb) = ikb \tag{2.213}$$

Clearly the ground state solution (2.211) corresponds to the zero of $g_0(-k)$ at $k = i\gamma$, and bound states exist in general at the zeros $k = i\gamma_n$, $\gamma_n > 0$ only if $U_0 > \gamma_n^2$ and $K(i\gamma_n)b > \pi/2$. If $U_0 > \gamma_n^2$ and at the same time $K(i\gamma_n)b < \pi/2$, then $\gamma_n < 0$ and the corresponding negative energy state is virtual (cf. section 2.7).

The condition (2.213) can be represented in simple graphical form by setting x equal to the angle Kb appearing in (2.213) and writing $y = ikb$, so that y is real for a negative energy state, bound or virtual, and identifying the discrete states of the potential which correspond to the points of intersection of the curves

$$y = x \cot x \qquad x^2 + y^2 = b^2 U_0 > 0 \tag{2.214}$$

As may be verified from figure 2.9 the two curves intersect in the upper half-plane $y > 0$ when $b\sqrt{U_0} < \pi/2$, and the corresponding state is virtual, whereas points of intersection in the lower half-plane $y < 0$ are associated with stronger potentials $b\sqrt{U_0} > \pi/2$ representing bound states. If the parameter $b\sqrt{U_0}$ is progressively enlarged so that the angle x advances into the third and fourth quadrants, a new virtual state and subsequently a new bound state of the potential is generated. Experimentally it is known that the $^3S_1$ state of the deuteron is bound whereas the $^1S_0$ state at an energy at $E \simeq -0.066$ MeV is virtual (cf. section 1.5.3). The deuteron has no other discrete states, bound or virtual.

In carrying out calculations involving deuteron wavefunctions it is convenient in many cases to replace the true wavefunction $u_0(r)$ by its asymptotic form $N e^{-\gamma r}$, where N is a normalization constant which may be determined in terms of U_0 and b, but which it is preferable to express in

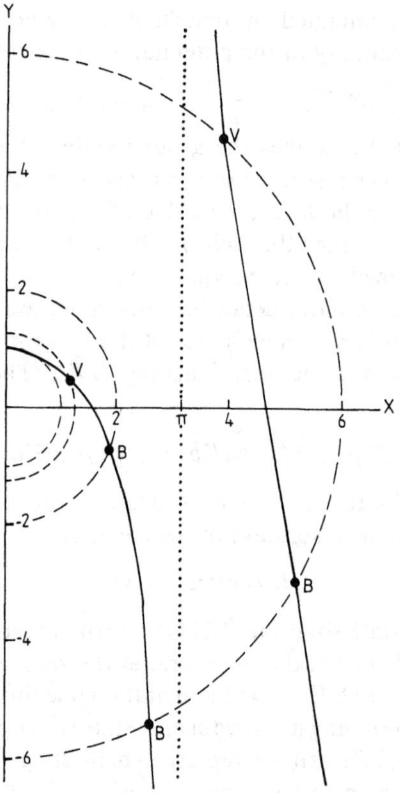

Figure 2.9. Graphical representation in the x-y plane, with $x=Kb$, $y=ikb$, of the condition that negative energy states, virtual (V) or bound (B), should exist in the deuteron, assuming neutron and proton interact through a square well potential. States exist for given well depth V_0 where the full curves $y=x\cot x$ intersect the hatched circles of constant potential $x^2+y^2=b^2U_0$.

terms of directly measurable quantities. Specifically we introduce a function $\eta_0(r)=u_0(r)/N$, which is a regular solution of the bound-state Schrödinger equation normalized such that its asymptotic form $\eta_\infty(r)$ satisfies the condition (cf. section 2.5.2)

$$\eta_\infty(r)=e^{-\gamma r} \qquad \eta_\infty(0)=1 \tag{2.215}$$

We can now express $\eta_0^2(r)$ in the form

$$\eta_0^2(r)=\eta_\infty^2(r)-\{\eta_\infty^2(r)-\eta_0^2(r)\} \tag{2.216}$$

and application of the normalization condition (2.210) then leads to the result

$$\frac{1}{N^2}=\frac{1}{2\gamma}-\int_0^\infty (\eta_\infty^2(r))\,dr=\frac{1}{2\gamma}-\frac{r_0}{2} \tag{2.217}$$

where r_0 is the effective range defined in (2.137), with this difference that the integral in (2.217) is evaluated at an energy $E=-E_b$, rather than at $E=0$. This modification has no appreciable effect on the integral, since the integrand in (2.217) vanishes in the external region $r>b$, which is the only part of its range in which $\eta_\infty(r)$ depends significantly on energy. The net result is that $u_0(r)$ is replaced by its finite-range approximation

$$u_0(r) \simeq \frac{2\gamma}{1-\gamma r_0} e^{-\gamma r} \tag{2.218}$$

whose parameters are the experimentally determined quantities γ and r_0 (cf. (2.138) and (2.209)) (Austern 1958). The replacement of the theoretical wavefunction $u_0(r)$ by its asymptotic form $N e^{-\gamma r}$ is an example of the 'shape independent approximation' introduced in section 2.5.2.

2.7.2 Neutron–proton scattering

Apart from the binding energy of the deuteron the only other parameters of the low-energy neutron–proton system which can be determined experimentally to any precision, are the triplet and singlet scattering lengths, and these are derived from measurements of coherent and incoherent scattering of low-energy neutrons in hydrogen, properly corrected for the molecular binding of the two protons in a molecule. To calculate the scattering length the Schrödinger equation (2.206) must be solved for positive energy $k^2>0$ and this means that the exponentially decaying external solution (2.209) must be replaced by the phase-shifted wave solutions

$$u_0(r) = B \sin(kr+\delta_0) \qquad r>b \tag{2.219}$$

The phase shift δ_0 is determined in the usual way by matching the logarithmic derivatives of the wavefunction at the boundary of the potential giving the result

$$\beta(k) = k \cot(kb+\delta_0) = K \cot(Kb) \tag{2.220}$$

By evaluating $\beta(0)$, an expression for the scattering length in terms of U_0 and b is immediately obtained by application of (2.129).

In the case of scattering in the triplet state for which the binding energy E_b is known, the well-depth U_0 can be eliminated between (2.211) and (2.220), and this permits us to write the scattering length in the form

$$^3a_{np} = \frac{1}{\gamma} + \frac{b}{2} + O(\gamma b^2) \tag{2.221}$$

where the dominant term γ^{-1} is known from (2.209) to have the value $\gamma^{-1}=4.31$ fm. Estimating $b \simeq 2$ fm, we find $a_3 \simeq 5.3$ fm, and the zero-energy triplet scattering cross-section has the approximate value $\sigma_3 \simeq 3.5b$. The significance of this result for the spin-dependence of the neutron–proton force, and the demonstration that the singlet scattering length ${}^1a_{np} \simeq -23.7b$ is large and negative, showing that the singlet state of the deuteron is virtual, have already been discussed in some detail in section 1.5.4.1.

It is difficult to make a direct comparison of results for neutron–proton and proton–proton scattering, and this is so for two reasons. In the first place in proton–proton scattering the two particles are identical, and this means that the ${}^3S_1$ scattering channel does not exist, and that the ${}^1S_0$ scattering is complicated by interference effects as described in section 2.3.3. A second, and much more important difference, arises from the presence of the Coulomb interaction which contributes an essential singularity to the scattering amplitude at $k^2=0$, and hence prevents the direct application of the effective range expansion (2.135). Thus the measured scattering length ${}^1a_{pp} = -7.82 \pm 0.01$ fm has to be subjected to a large correction (cf. section 3.2.1) before the nuclear contribution can be assessed. The corresponding neutron–neutron scattering length ${}^1a_{nn}$ can only be measured from the study of many-body interactions involving at least two neutrons in the final state, and unfortunately the achievable accuracy is no better. However, the essential result is clear; the singlet scattering lengths for p–p and n–n are both negative showing that the corresponding singlet states are unbound. The absolute values of these scattering lengths are approximately equal, and somewhat less than the absolute magnitude of the singlet scattering length in the n–p system, and the significance of these findings in relation to fundamental theory is discussed in section 3.4.1. These results apply of course to free nucleons; within the nucleus the $I=0$ state of two identical nucleons is strongly bound and is responsible for the pairing gap observed in heavy even–even nuclei (cf. section 4.5.2). The $I=2$ state is also bound, while the $I=4$ state is only weakly bound or possible unbound. The pairing force in states of higher angular momentum is repulsive.

The measured values of the scattering lengths and effective ranges in the lowest-energy S-states of the two-nucleon system are assembled in Table 2.1, together with estimates of the corresponding square well parameters. However these data are not sufficient to determine either the potential or the deuteron wavefunction uniquely (Newton 1957).

Table 2.1. Measured values of low energy nucleon–nucleon scattering lengths and effective ranges, and estimates of corresponding square well potential parameters

State	a (fm)	r_0 (fm)	V_0 (MeV)	b (fm)
(n–p) ${}^3S_1$	5.423 ± 0.005	1.75	36.2	2.0
(n–p) ${}^1S_0$	-23.748 ± 0.015	2.73	14.0	2.7
(p–p) ${}^1S_0$	-17.9 ± 0.45	2.97	12.9	2.7
(n–n) ${}^1S_0$	-18.45 ± 0.45	2.84	—	—

2.7.3 Photodisintegration of the deuteron

The photodisintegration of the deuteron (cf. section 1.5) is an endothermic process in which a deuteron in the $^3S_1$ ground state makes a transition to the continuum, following the absorption of a photon of energy $E_\gamma = \hbar\omega$. For the transition to proceed, the quantum energy must exceed the deuteron binding energy $E_b = 2.226$ MeV and, to simplify the calculation, it is convenient to assume that $E_\gamma \gg E_b$, in which case it is legitimate to apply the Born approximation (cf. section 2.3) and replace the unbound wavefunction by its asymptotic form (2.86). Adopting the notation of section 2.2, and assigning spin-spatial coordinates $(\boldsymbol{\sigma}_1, \boldsymbol{r}_1)$ and $(\boldsymbol{\sigma}_2, \boldsymbol{r}_2)$ to the proton and neutron respectively, the electromagnetic interaction energy in the radiation gauge (2.10(ii)) is given, according to (2.22) by the expression

$$H_{\mathrm{EM}} = -\left\{\frac{e}{m_p c}(\boldsymbol{p}_1 \cdot \boldsymbol{A}(\boldsymbol{r}_1)) + \boldsymbol{\mu}_p \cdot (\nabla_1 \times \boldsymbol{A}(\boldsymbol{r}_1)) + \boldsymbol{\mu}_n \cdot (\nabla_2 \times \boldsymbol{A}(\boldsymbol{r}_2))\right\} \quad (2.222)$$

where $\boldsymbol{A}(\boldsymbol{r})$ is the quantized radiation field defined in (2.12). The intrinsic magnetic moments of proton and neutron which contribute to the magnetic interaction in (2.222) may then be expressed in the form

$$\boldsymbol{\mu}_p = \frac{g_p}{2}\mu_N \boldsymbol{\sigma}_1 \qquad \boldsymbol{\mu}_n = \frac{g_n}{2}\mu_N \boldsymbol{\sigma}_2 \quad (2.223)$$

where $\mu_N = e\hbar/2m_p c$ is the nuclear magneton and g_p $(=5.59)$ and $g_n = (-3.83)$ are the corresponding gyromagnetic Landé g-factors (cf. section 4.3.1).

Since the photodisintegration process involves the destruction of a single quantum, only the absorptive part of A contributes to the transition and the matrix element describing the absorption of a photon of momentum $\boldsymbol{k}$ and polarization $\boldsymbol{a}$ is given by

$$\langle 0|H_{\mathrm{EM}}|1\rangle = -\sqrt{\left(\frac{\hbar c}{4\pi^2 k}\right)}\left\{\frac{e}{m_p c}\,\boldsymbol{a}\cdot\boldsymbol{p}_1\,\mathrm{e}^{\mathrm{i}\boldsymbol{k}\cdot\boldsymbol{r}_1} + \mathrm{i}\boldsymbol{a}\cdot(\boldsymbol{\mu}_p \times \boldsymbol{k}\,\mathrm{e}^{\mathrm{i}\boldsymbol{k}\cdot\boldsymbol{r}_1} + \boldsymbol{\mu}_n \times \boldsymbol{k}\,\mathrm{e}^{\mathrm{i}\boldsymbol{k}\cdot\boldsymbol{r}_2})\right\} \quad (2.224)$$

To complete the evaluation of $\langle \mathrm{f}|H_{\mathrm{EM}}|\mathrm{i}\rangle$ between initial and final neutron–proton states we need to transform from a description in terms of laboratory coordinates $(\boldsymbol{r}_1, \boldsymbol{r}_2)$ to centre of mass and relative coordinates $(\boldsymbol{R}, \boldsymbol{r})$ defined in (2.24) where $m_1 = m_p \simeq m_2 \equiv m_n \simeq 2m$, and m is the reduced mass. The initial state of the neutron–proton system is then represented in the finite range approximation (2.218) by the asymptotic $^3S_1$ wavefunction

$$\psi_i(\boldsymbol{r}, \boldsymbol{\sigma}_1, \boldsymbol{\sigma}_2) = \sqrt{\left(\frac{2\gamma}{1-\gamma r_{ot}}\right)}\frac{\mathrm{e}^{-\gamma r}}{r} X_1^{m_i} \qquad m_i = 0 \pm 1 \quad (2.225)$$

where $X_1^{m_i}$ is the spin-function, and r_{ot} is the effective range of the neutron proton potential in the triplet state. Following the absorption of a photon the final state neutron–proton wavefunction is

$$\psi_f(\boldsymbol{r}, \boldsymbol{R}, \boldsymbol{\sigma}_1, \boldsymbol{\sigma}_2) = e^{i\boldsymbol{k}\cdot\boldsymbol{R}} \psi_\infty(\boldsymbol{r}, \boldsymbol{\sigma}_1, \boldsymbol{\sigma}_2) \tag{2.226}$$

where the plane wave factor $e^{i\boldsymbol{k}\cdot\boldsymbol{R}}$ represents the recoil of the centre of mass and $\psi_\infty(\boldsymbol{r}, \boldsymbol{\sigma}_1, \boldsymbol{\sigma}_2)$ describes an outgoing wave in the system of relative coordinates.

Because of the factor $e^{-\gamma r}$ in $\psi_i(\boldsymbol{r})$, where $\gamma^{-1} = 4.31$ fm characterizes the spatial extent of the deuteron bound state, the wavelength $2\pi/k$ of the radiation is much larger than the dimension of the effective domain of integration, and therefore only the leading terms in the expansions of the exponential factors $e^{i\boldsymbol{k}\cdot\boldsymbol{r}_{1,2}}$ appearing in (2.224) need be taken into account. Thus the matrix element (2.224) contains an electric dipole interaction $(e/m_p c)\boldsymbol{a}\cdot\boldsymbol{p}_1$ associated with the motion of the proton, together with two magnetic dipole interactions $\boldsymbol{a}\cdot(\boldsymbol{\mu}_{1,2} \times \boldsymbol{k})$ associated with the nucleon spins.

The operator $\boldsymbol{p}_1$ does not act on the proton spin, so each triplet substate $X_1^{m_i}$ remains unchanged during the transition. Also because $\boldsymbol{p}_1$ is a polar vector (cf. section 3.3) it has non-vanishing matrix elements between states of opposite parity only, and the photoelectric distintegration of the deuteron corresponds to a transition from the $^3S_1$ bound state to the $^3P_{0,1,2}$ states of the continuum (cf. section 10.1.1.3). The required orbital angular momentum is obtained by absorption of the dipole quantum (total angular momentum $L = 1$), whose momentum $\boldsymbol{k}$ and polarization $\boldsymbol{a}$ provide a convenient set of coordinate axes. Since the photon has no state of zero angular momentum component referred to its own momentum, there is no proton emission in the forward direction, i.e. along $\boldsymbol{k}$, and the asymptotic wavefunction is

$$\psi_\infty(\boldsymbol{r}_1, \boldsymbol{\sigma}_1, \boldsymbol{\sigma}_2) \simeq i \frac{\sin(\kappa r - \pi/2)}{\kappa r} P_1(\cos\theta) e^{\pm i\varphi} X_1^{m_i} \tag{2.227}$$

where $\boldsymbol{\kappa}$ is the wavevector of the ejected proton and any small p-wave phase shift has been neglected.

In the $(\boldsymbol{R}, \boldsymbol{r})$ coordinate system the operator $\boldsymbol{p}_1$ may be expressed in the form

$$\boldsymbol{p}_1 = \frac{d}{dt}(m\boldsymbol{r} + 2m\boldsymbol{R}) = \frac{-i}{\hbar}[m\boldsymbol{r} + 2m\boldsymbol{R}, H_{np}] = \frac{-im}{\hbar}[\boldsymbol{r}, H_{np}] \tag{2.228}$$

where H_{np} is the unperturbed neutron–proton Hamiltonian which, in this approximation, is momentum independent, aside from the kinetic energy term. The matrix element of $\boldsymbol{p}_1$ becomes then

$$\int \psi_f^* \boldsymbol{p}_1 \psi_i \, d^3r = -im \int \psi_f^* (\boldsymbol{r} H_{np} - H_{np} \boldsymbol{r}) \psi_i \, d^3r$$

$$= im \Delta E_{np} \int \psi_f^* \boldsymbol{r} \psi_i \, d^3r \tag{2.229}$$

where

$$\Delta E_{\text{np}} = E_{\text{b}} + \frac{\hbar^2\kappa^2}{2m} = \hbar ck - \frac{\hbar^2 k^2}{4m} \tag{2.230}$$

represents the energy absorbed by the neutron–proton system, $\hbar\boldsymbol{\kappa}$ is the momentum in the relative motion and $\hbar^2k^2/4m$ is the energy carried off by the recoiling centre of mass. Even at a photon energy of 20 MeV this energy amounts to no more than 1% of the total and may be neglected.

The differential cross-section for photoelectric disintegration is now obtained from (2.229) using (2.20) and (2.24) where the density of states factor ρ_{f} is defined by

$$\rho_{\text{f}}(E_{\text{f}})\,\text{d}E_{\text{f}} = \frac{p^2\,\text{d}p\,\text{d}\Omega}{(2\pi\hbar)^3} \tag{2.231}$$

with

$$p = \hbar\kappa \qquad E_{\text{f}} = \hbar^2\boldsymbol{\kappa}^2/2m \tag{2.232}$$

A final integration over the angles then gives an expression for the photoelectric cross-section

$$\begin{aligned}\sigma_{\text{e}} &= \left(\frac{8\pi\alpha}{3}\right)\frac{\gamma}{(1-\gamma r_{\text{ot}})}\left(\frac{\kappa}{\kappa^2+\gamma^2}\right)^3 \\ &= \left(\frac{8\pi\alpha}{3}\right)\frac{1}{\gamma^2(1-\gamma r_{\text{ot}})}\frac{(\eta-1)^{3/2}}{\eta^3}\end{aligned} \tag{2.233}$$

where

$$\eta = \frac{E_{\text{r}}}{E_{\text{b}}} = 1 + \frac{\kappa^2}{\gamma^2} \tag{2.234}$$

The result (2.233) departs from the original Bethe–Peierls formula (Bethe and Peierls 1935) only by the inclusion of the effective range factor $(1-\gamma r_{\text{ot}})^{-1}$. It shows that σ_{e} increases in proportion to $(E_\gamma - E_{\text{b}})^{3/2}$ above threshold, and falls off at high energy on the order of $E_\gamma^{-3/2}$. The maximum in the cross-section is reached at $E_\gamma \simeq 2E_{\text{b}} = 4.5$ MeV where the theoretical value $\sigma_{\text{e}} = 2.4$ mb is in good agreement with experiment, a result which confirms the necessity for including the finite range factor which has a value of about 1.76 fm. On the other hand integration over the whole energy range yields the result

$$\int \sigma_{\text{e}}(E_\gamma)\,\text{d}E_\gamma = E_{\text{b}}\int_1^\infty \sigma_{\text{e}}(\eta)\,\text{d}\eta = \frac{2\pi^2 e^2\hbar}{m_{\text{p}}c}\,\frac{1}{2(1-\gamma r_{\text{ot}})} \tag{2.235}$$

which, apart from the finite range factor, differs by a factor of $\frac{1}{2}$ from the dispersion theory predictions (2.121) extended to nuclei. This is a consequence of the recoil of the nucleus, which directs a fraction Z/A of the total available oscillator strength associated with the Z protons in the nucleus into Thomson scattering by the nucleus as a whole, the remaining fraction N/A being accounted for by absorption in the internal degrees of freedom of the nucleus. This effect may be described most elegantly by applying the concept of 'effective charges' on neutrons and protons (cf. (4.3.1)) to the absorption of dipole radiation ($L=1$) by nuclei.

Since the dipole sum rule is model independent for nucleons interacting through a scalar potential, which commutes with the relative coordinate $\boldsymbol{r}$, it follows that the substantial overestimate of the dipole sum (2.231) produced by the finite range factor must have a different origin, and indeed this overestimate is due to the fact that the finite range wavefunction (2.218) does not vanish at the origin as required. This defect may be remedied by using the Hulthén wavefunction (Levinger 1949).

$$\psi_H(r)=\frac{2\gamma}{1-\gamma r_{ot}}(e^{-\gamma r}-e^{-\beta r})/r \tag{2.236}$$

to represent the ground state of the deuteron, where the parameter $\beta\simeq 6\gamma$ is selected to reproduce the measured finite range. The result is that the high energy cross-section, which reflects the short-range structure of the deuteron, is reduced as required. Experimentally, the measured value of the integrated cross-section, integrated up to the threshold for photoproduction of π-mesons at $T_t\simeq 145$ MeV, is about 38 mb MeV, as compared with the dipole sum prediction of 30 mb MeV. These results may be accounted for by taking account of exchange and momentum-dependent contributions to the nuclear force (cf. section 3.3), which do not commute with the relative coordinate, and which tend on the whole to enhance the dipole sum.

The calculation of the photomagnetic cross-section is relatively straightforward, for in this case, the relevant particle operators $\boldsymbol{\mu}_p$ and $\boldsymbol{\mu}_n$ in (2.224) act only on the spins and, to conserve angular momentum and parity, the transition must be of the form $^3S_1\rightarrow{}^1S_0$. The interaction between the nucleon spins and the magnetic field associated with the photon may then be written in the form.

$$\begin{aligned}\langle 0|H_{EM}|1\rangle_{mag}&=i\sqrt{\left(\frac{\hbar c}{4\pi^2 k}\right)}(\boldsymbol{a}\times\boldsymbol{k})\cdot(\boldsymbol{\mu}_p+\boldsymbol{\mu}_n)\\&=i\sqrt{\left(\frac{\hbar c}{4\pi^2 k}\right)}\boldsymbol{a}\times\boldsymbol{k}\cdot\boldsymbol{\mu}_{nm}\left\{\left(\frac{g_p+g_n}{2}\right)(\boldsymbol{\sigma}_1+\boldsymbol{\sigma}_2)\right.\\&\qquad\left.+\left(\frac{g_p-g_n}{2}\right)(\boldsymbol{\sigma}_1-\boldsymbol{\sigma}_2)\right\}\end{aligned} \tag{2.237}$$

Since the triplet spin state $X_1^{m_i}$ is an eigenstate of $\boldsymbol{\sigma}_1+\boldsymbol{\sigma}_2$, the first term in

(2.236) cannot cause a transition, so the only non-vanishing matrix element is

$$\langle X_0^0 | \sigma_{1z} - \sigma_{2z} | X_1^0 \rangle = 1 \tag{2.238}$$

The final state asymptotic wavefunction (2.226) then becomes

$$\psi_\infty(\boldsymbol{r}, \boldsymbol{\sigma}_1, \boldsymbol{\sigma}_2) \simeq e^{i^1\delta_0} \frac{\sin(\kappa r + {}^1\delta_0)}{\kappa r} X_0^0 \tag{2.239}$$

where the singlet s-wave phase shift ${}^1\delta_0$ is related via (2.135) to the large singlet scattering length ${}^1a_{\mathrm{np}} \simeq -24$ fm associated with the virtual state of the deuteron. The corresponding cross-section in the zero-range approximation is then given by (Bethe and Longmere 1950)

$$\sigma_{\mathrm{m}} = \frac{2\pi\alpha}{3}\left(\frac{\hbar}{m_{\mathrm{p}}c}\right)^2 (\mu_{\mathrm{p}} - \mu_{\mathrm{n}})^2 \left(\frac{\kappa\gamma}{\kappa^2 + \gamma^2}\right) \frac{(1 - \gamma^1 a_{\mathrm{np}})^2}{(1 + \kappa^2\, {}^1a_{\mathrm{np}}^2)}$$
$$\propto (\eta - 1)^{1/2} \eta^{-1} [1 + (\gamma^2)\, {}^1a_{\mathrm{np}}^2 (\eta - 1)]^{-1} \tag{2.240}$$

Since the coefficient $(\gamma^2)\, {}^1a_{\mathrm{np}}^2$ has a value of about 30, the factor $[1 + (\gamma^2)\, {}^1a_{\mathrm{np}}^2(\eta - 1)]^{-1}$ varies rapidly in the neighbourhood of threshold $\eta \simeq 1$, and σ_{m} grows to a maximum value $\sigma_{\mathrm{m}} \simeq 0.5$ mb at $\eta \approx 1 + ((\gamma^2)\, {}^1a_{\mathrm{np}}^2)^{-1}$ falling rapidly thereafter in proportion to $(\eta - 1)^{+1/2}$. Thus, immediately above threshold, the photomagnetic cross-section dominates the total cross-section but, in the region of maximum photoelectric absorption near $E_\gamma \simeq 4.5$ MeV, the photomagnetic effect contributes $<5\%$ of the total. Finite range effects do not affect σ_{m} significantly, essentially because both initial and final states are S-states but corrections of this type do emerge when the tensor force is taken into account (Feshbach and Schwinger 1951).

2.7.4 Neutron–proton capture: detailed balance

The capture of a neutron by a proton with the simultaneous creation of a deuteron and a γ-ray

$$\mathrm{n} + \mathrm{p} \rightarrow \mathrm{d} + \gamma \tag{2.241}$$

is an exothermic reaction which is the reverse process to the photodisintegration of the deuteron. It is a very important effect from a practical viewpoint, since it sets a limit to the lifetime of a freely propagating neutron in a water-moderated nuclear reactor. In the thermal region, where only s-wave capture need be considered, it is clear from the discussion in section 2.7.3 that the magnetic dipole capture process dominates, and the cross-section for n–p capture can be obtained directly from the photomagnetic cross-section (2.240) by exploiting the reciprocity relation

$$\langle \mathrm{d}, \gamma | H_{\mathrm{EM}} | \mathrm{n}, \mathrm{p} \rangle = \langle \mathrm{n}, \mathrm{p} | H_{\mathrm{EM}} | \mathrm{d}, \gamma \rangle^* \tag{2.242}$$

The relation (2.242) is a direct consequence of the invariance of the electromagnetic interaction with respect to the symmetry operation of time reversal (cf. section 3.3.1). When applied to the general nuclear reaction

$$A+a \rightleftarrows B+b \tag{2.243}$$

time reversal symmetry leads to the principle of detailed balance (Durbin *et al* 1951) which, in its simplest form, establishes the relation between inverse cross-sections

$$\frac{\sigma(A+a\rightarrow B+b)}{\sigma(B+b\rightarrow A+a)}=\frac{(2I_{\rm B}+1)(2I_{\rm b}+1)\,p_{\rm b}^2}{(2I_{\rm A}+1)(2I_{\rm a}+1)\,p_{\rm a}^2} \tag{2.244}$$

when both reactions are viewed in their respective centres of mass reference frames, and when both incident and target particles are unpolarized. When (2.240) is applied to (2.241) with $I_{\rm n}=I_{\rm p}=\frac{1}{2}$, $I_{\rm d}=1$ and $2I_{\gamma}+1=2$ (allowing for the fact that the massless photon has only two independent states of polarization) we find that

$$\frac{\sigma_{\rm n+p}\rightarrow{\rm d}+\gamma}{\sigma_{\rm d+\gamma}\rightarrow{\rm n+p}}=\frac{3\times 2(\hbar k)^2}{2\times 2(\hbar\kappa)^2}=\frac{3}{2}\left(\frac{E_{\rm b}+E_{\rm n}}{\hbar\kappa c}\right)^2 \tag{2.245}$$

where

$$E_{\rm n}=\frac{\hbar^2\kappa^2}{2m}=\frac{\hbar^2\kappa^2}{m_{\rm p}} \tag{2.246}$$

is the relative kinetic energy of neutron and proton in their centre of mass frame.

When (2.240) and (2.245) are combined, an expression is obtained for the capture cross-section of a thermal neutron

$$\sigma_{\rm c}=\pi\alpha\left(\frac{\hbar}{m_{\rm p}c}\right)^2(\mu_{\rm p}-\mu_{\rm n})^2\left(\frac{E_{\rm n}+E_{\rm b}}{m_{\rm p}c^2}\right)\left(\left(\frac{E_{\rm b}}{E_{\rm n}}\right)\frac{(1-\gamma^1a_{\rm np})^2}{(1+(\kappa^2)^1a_{\rm np}^2)}\right)^{1/2} \tag{2.247}$$

which, in the long-wave limit $E_{\rm n}\ll E_{\rm b}$, $(\kappa^2)^1a_{\rm np}^2\ll 1$, is proportional to $(E_{\rm n})^{-1/2}$. This is, of course, just the '$1/v$' law (cf. section 1.7.1). The result (2.247) predicts that $\sigma_{\rm c}\simeq 300$ mb, on the assumption that the singlet scattering length ${}^1a_{\rm np}$ is negative, a conclusion which agrees closely with experiment. If ${}^1a_{\rm np}$ is positive then $\sigma_{\rm c}$ is reduced by about 50%, and the measured capture rate for neutrons on protons therefore provides an additional strong support in favour of the proposition that the singlet state of the deuteron is virtual (cf. section 1.5.3).

We may conclude this rather brief discussion of the neutron–proton system with one final comment, which is to note that, even when calculated with wavefunctions incorporating all the refinements associated with tensor, exchange, hard-core and momentum-dependent contributions to the nuclear force, theories of the neutron–proton capture process and the photo-disintegration of the deuteron still exhibit discrepancies at the level of a few per cent when compared with the most accurate experimental results. These observations are of the greatest significance since they provide the most convincing evidence for the existence of meson exchange currents in nuclei (cf. section 4.1).

3 Neutrons and nuclei

3.1 Nuclear stability and binding energy

3.1.1 Nuclei in nature

A nucleus which is not observed to be radioactive is said to be stable and though it would seem unprofitable to attempt an absolute definition of stability, nuclear disintegration with a half-life $T_{1/2} > 10^{18}$ years is not in practice detectable by conventional techniques, and thus provides a suitable yard-stick to give a meaning to the notion of relative stability. By this criterion 287 stable nuclei occur naturally on earth together with 17 unstable parent species and a large number of unstable nuclei in the range of mass number $A > 209$ which are themselves the products of radioactive decay. There are no known stable nuclei with $Z > 83$ or $N > 126$ and of the more than 10^3 nuclei which have been produced artifically none is a stable nucleus not occurring naturally.

A classification of all stable nuclei according to whether the numbers of protons (Z) or neutrons ($N = A - Z$) are even or odd, discloses the fact that there are 156 even–even, 52 odd–even, 55 even–odd and only four odd–odd, stable species. Among the stable even–even nuclei are 54 isobaric doublets e.g. ${}^{36}_{16}S$, and ${}^{36}_{18}A$ and four isobaric triplets e.g. ${}^{124}_{56}Sn$, ${}^{124}_{54}Xe$ and ${}^{124}_{52}Te$. The four odd–odd stable nuclei are all self-conjugate light nuclei with $Z = N$ i.e. ${}^{2}_{1}H$, ${}^{6}_{3}Li$, ${}^{10}_{5}B$ and ${}^{14}_{7}N$. All stable even–even nuclei have $I = 0$ whereas all the stable odd–odd nuclei have $I = 1$ except ${}^{10}B$ which has $I = 3$. These results strongly suggest that like nucleons tend to pair off in spin states with $I = 0$ whereas no such rule applies to unlike nucleons.

Of the 17 naturally occurring unstable nuclei six are β-emitters of which one, ${}^{187}_{75}Re$ ($T_{1/2} = 5 \times 10^{10}$ years) provides an example of the rare instance where the bare nucleus is stable although the neutral atom is β-active. Another example where the lifetime is affected by the environment occurs with the unstable nucleus ${}^{7}_{4}Be$. In terrestrial conditions this decays exclusively by K-capture with a half-life of 53 days. However, in solar conditions the atom is partially ionized with the result that the half-life is increased to 83 days. The shortest lived natural β-emitter ${}^{40}_{19}K$ ($T_{1/2} \simeq 10^{11}$ years), which is an important source of background in low-level radiation detectors, also has a weak ($\approx 11\%$) electron capture branch to ${}^{40}_{18}A$, which constitutes 99.6% of natural argon. This decay is the basis of the potassium–argon system for dating minerals, particularly the micas. The long-lived odd–odd nucleus ${}^{50}_{23}V$ ($T_{1/2} = 8 \times 10^{15}$ years) decays solely by electron capture. An extremely rare mode of radioactive decay is the second-order weak process called double β-decay, which involves the simultaneous emission of two electrons, and for this to be energetically possible within an even–even isobaric multiplet the mass difference must exceed 1.1 MeV. Although this condition is fulfilled in

58 instances, up to quite recently double β-decay had been detected by indirect means in only two nuclei, i.e. ${}^{130}_{52}Te \rightarrow {}^{130}_{54}Xe$ ($T_{1/2} \simeq 10^{21}$ years) and ${}^{82}_{34}Se \rightarrow {}^{82}_{36}Kr$ ($T_{1/2} \simeq 10^{20}$ years) (Bernatowicz *et al* 1992). In each case the observation had been made possible by the identification of anomalous fractions of ${}^{130}_{54}Xe$ or ${}^{82}_{36}Kr$ in ancient ores of tellurium or selenium. In similar circumstances excess amounts of ${}^{40}_{20}A$ are observed in potassium ores from the β-decay of ${}^{40}_{19}K$. However it has lately proved possible to observe the process directly using germanium radiation detectors highly enriched in the double β-emitting isotope ${}^{76}Ge$ (Balysh *et al* 1992, Reusser *et al* 1992).

The remaining 10 unstable species are all α-emitters and include five rare earths, e.g. ${}^{144}_{60}Nd$ ($T_{1/2} \simeq 10^{16}$ years), one isotope each of platinum (${}^{190}_{78}Pt$) and lead (${}^{204}_{82}Pb$) and three actinides, ${}^{232}_{90}Th$ ($T_{1/2} = 1.39 \times 10^{10}$ years), ${}^{235}_{92}U$ ($T_{1/2} = 7.13 \times 10^{8}$ years) and ${}^{238}_{92}U$ ($T_{1/2} = 4.51 \times 10^{9}$ years), all of which coexist, on laboratory time scales, in secular equilibrium with their daughter products.

We now know that the elements with atomic numbers $Z = 43$ (technetium), 61 (promethium), 85 (astatine) and 87 (francium) do not occur in significant quantities in nature, nor do the transuranic elements with $Z \geqslant 93$ (cf. section 5.2.5). From this group of missing elements technetium is a rather special case since it has three long-lived isotopes, ${}^{97}_{43}Tc$ (2.6×10^{6} years), ${}^{98}_{43}Tc$ (10^{6} years), and ${}^{99}_{43}Tc$ (2.12×10^{5} years). Technetium was first created artifically through the reaction ${}^{96}_{42}Mo(d, n){}^{97}_{43}Tc$ (Perrier and Segré 1937) and the fission product ${}^{99}_{43}Tc$ has been detected in small concentrations in uranium ores (Kenna and Kuroda 1964). There is also some evidence for the presence of ${}^{98}_{43}Tc$ in molybdenite ores (Herr *et al* 1954, Anders *et al* 1956) although the identification is uncertain (Boyd and Larson 1956). However technetium does exist in S-type stars (Merrell 1952), and californium ($Z = 99$) is detectable in Type I supernovae, indicating that, in these astrophysical sites, nucleosynthesis is still actively proceeding (Baade *et al* 1956, Burbidge *et al* 1956). The discovery of abnormal abundances of ${}^{129}Xe$ in meteorites, produced by the decay of ${}^{129}I$ (1.7×10^{10} years), which must also be continuously synthesized, supports this conclusion.

The first isotopes of promethium to be identified were ${}^{147}_{61}Pm$ (2.6 years) and ${}^{149}_{61}Pm$ (53 h) which were detected in mixtures of fission products containing praseodymium ($Z = 59$) and neodymium ($Z = 60$) (Marinsky *et al* 1947). Since the longest-lived promethium isotope is ${}^{145}_{61}Pm$ which has a half-life of 18 years, there is no prospect of finding significant quantities of natural promethium. However ${}^{147}_{61}Pm$ occurs at ultra-low levels of concentration in uranium ores (Attrep and Kuroda 1968); that it existed at one time in nature at much higher abundances may be inferred from the observation of anomalous amounts of its daughter product ${}^{147}_{62}Sm$ in the Oklo fossil reactor (cf. section 5.5.5).

The longest-lived isotopes of astatine ($Z = 85$) and francium ($Z = 87$) are ${}^{220}_{85}As$ (8.3 h) and ${}^{223}_{87}Fr$ (22 min), respectively. The former does not occur in nature because its potential parent ${}^{210}_{84}Po$ is α-active rather than β-active, but ${}^{223}_{87}Fr$ is produced in the actinium series of naturally-occurring radioactive elements with a branching ratio of 1.2%. The astatine isotope ${}^{219}_{85}At$ (54 s) also occurs in the same series with a branching ratio of 4×10^{-3}%. Of the transuranic elements only ${}^{239}_{84}Pu$ (2.44×10^4 years) has been identified with certainty in old uranium ores, where it is produced by fast neutron capture following spontaneous fission of ${}^{238}_{92}U$ (Seaborg and Perlman 1948). The details of the process are shown in equation (1.13). The presence in nature of the long-lived isotope ${}^{244}_{94}Pu$ (7.5×10^7 years) is uncertain (Hoffman *et al* 1971, Fleischer and Naeser 1972).

The detection by neutron activation techniques of anomalous amounts of ${}^{191}Ir$ and ${}^{193}Ir$ in 65-million-year-old clay layers in Italy, Denmark and New Zealand, marking the boundary between the Cretaceous and Tertiary geological periods and the disappearance of the dinosaurs, has been interpreted as evidence that this cataclysmic event is to be associated with the impact of a 10 ± 4 km diameter asteroid originating within the solar system (Alvarez *et al* 1980). The isotopic constitution of the iridium sample, providing a measure of the relative importance of r- and s-process element formation, together with the failure to detect any ${}^{244}Pu$, appear to rule out a supernova explosion as the primary cause (cf. sections 10.5.5–10.5.7).

3.1.2 Nuclear binding energy

A plot of N against Z, known as the Segré chart, is shown in figure 3.1 for the stable nuclei, which cluster about a 'stability line' which coincides with the line $N = Z$ for $A \leqslant 40$ but moves into the region of neutron excess for higher values of A. This deviation from linear behaviour is a manifestation of the Coulomb repulsion between protons whose contribution to the total energy is a quadratic function of Z. The β^--unstable and β^+-unstable (including electron capture) nuclei are distributed respectively above and below the stability line; exceptionally some nuclei, e.g. ${}^{64}_{29}Cu$ are simultaneously β^- and β^+-unstable. Within this broad pattern there are regions of exceptional stability when either Z or N is equal to one of the 'magic numbers' 2, 8, 20, (28), (40), 50, 82 or 126. For example at $Z = 50$ there are 10 stable isotopes, i.e. ${}^{A}_{50}Sn$, where $A = 112$ (1%), 114 (0.65%), 115 (0.35%), 116 (14.2%), 117 (7.6%), 118 (23.9%), 119 (8.6%), 120 (32.9%), 122 (4.8%) and 124 (6.0%). In addition the tin isotopes ${}^{113}_{50}Sn$ and ${}^{123}_{50}Sn$ have half-lives in excess of 100 days. At $N = 50$ there are four stable isotopes, i.e. ${}^{86}_{36}Kr$, ${}^{88}_{38}Sr$, ${}^{89}_{39}Y$ and ${}^{90}_{40}Zr$, and one long-lived isotope ${}^{87}_{37}Rb$ (4.7×10^{10} years) which is an important agent in Rb–Sr dating methods (Symbalisty and Schramm 1981). This general picture is confirmed by the study of the absolute isotopic abundance which, although it decreases by eight orders of magnitude between $A = 1$ and $A = 100$, shows pronounced local maxima near

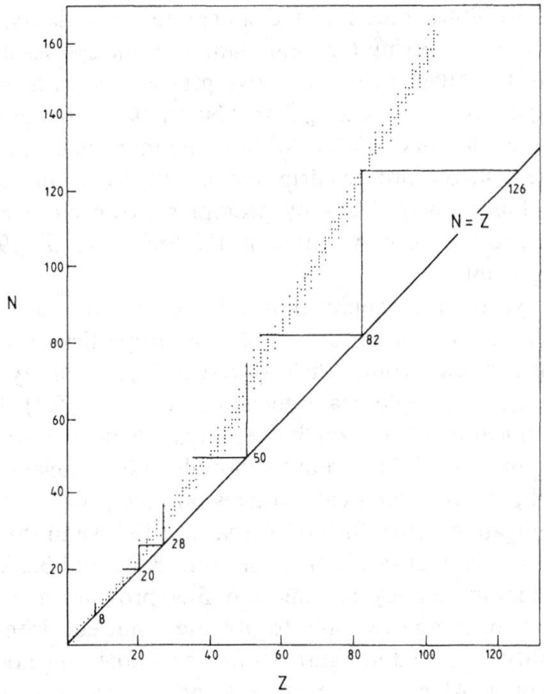

Figure 3.1. Segré chart of the stable nuclei indicating regions of special stability associated with the magic numbers. For $Z \leqslant 20$ the stability line coincides with the line $N=Z$, and the increasing deviation of the stable nuclei from this line as Z increases is associated with the effects of Coulomb repulsion. Nuclei with values of N above the stability line tend to be β^--active while nuclei with values of N below the line are β^+-active.

$^{90}_{40}$Zr ($N=50$), $^{120}_{50}$Sn ($Z=50$), $^{138}_{50}$Sn ($Z=50$), $^{138}_{56}$Ba ($N=82$) and $^{208}_{82}$Pb ($Z=82$, $N=126$).

In characterizing the overall stability of an atom a useful quantity to consider is the binding energy, which is defined as the total amount of energy which is required to break the atom up into Z neutral hydrogen atoms and N neutrons, i.e.

$$B(N, Z) = (Zm_{\mathrm{H}} + Nm_{\mathrm{n}} - M(A, Z))c^2 \tag{3.1}$$

$B(N, Z)$ is maximal for nuclei lying close to the stability line and beyond this valley of stability there is a broad region of the Segré chart which is bounded by the neutron and proton drip-lines. These are curves defined by

(i) $B(N, Z) = B(N-1, Z)$ neutron drip-line
(ii) $B(N, Z) = B(N, Z-1)$ proton drip-line (3.2)

Between the drip-lines nuclei can disintegrate by β-decay, which is an elementary process involving the annihilation of one species of nucleon and the creation of the other, or by collective processes such as α-decay or the emission of α-particle nuclei, e.g. ${}^{12}_{6}C$ or ${}^{20}_{10}Ne$ (Aleksandrov *et al* 1985, Gates *et al* 1984, Rose and Jones 1984). Adding one more neutron (proton) to a nucleus on the neutron (proton) drip-line takes the nucleus into a region of maximal instability where decay by prompt neutron emission (cf. section 5.1.3) or by prompt proton emission (Hofmann *et al* 1982) becomes energetically possible.

In recent years the application of heavy ion accelerators and electromagnetic mass separators to study the properties of nuclei far from stability has generated a fruitful field of research particularly in relation to nuclear reactions of astrophysical interest (cf. section 10.5.7). This work has led to the identification of neutron 'haloes' in highly neutron-rich nuclei such as ${}^{11}_{3}Li$ (Thibault *et al* 1975), and of the doubly magic nucleus ${}^{132}_{50}Sn$ (Kerek *et al* 1972). Of perhaps even greater interest are the proton-rich nuclei such as the self conjugate nucleus ${}^{74}_{37}Rb$ (D'Auria *et al* 1977) and the heavy mirror nucleus ${}^{97}_{48}Cd$ (Elmroth *et al* 1978) (cf. sections 3.4.2 and 3.2.2).

The mean binding energy per nucleon B/A provides a measure of the average amount of energy required to liberate a nucleon from the nucleus, and this quantity is plotted in figure 3.2 for the whole range of stable nuclei as a function of A. After rising through a series of maxima corresponding to mass numbers which are simple multiples of 4, B/A reaches a broad maximum of 8.8 MeV/nucleon near ${}^{58,60}_{28}Ni$ and then falls gradually to about 7.4 MeV/nucleon at $A=238$. The most remarkable feature of the binding energy curve is the essential constancy of B/A, a result which would not be obtained were a nucleon equally attracted to each of the remaining $(A-1)$ nucleons in the nucleus, since in this case the number of force bonds is $\frac{1}{2}A(A-1)$ and the binding energy per nucleon increases in direct proportion with A for large A. Instead it appears that a given nucleon interacts only with those nucleons in its immediate neighbourhood and this is what is understood by the statement that the nuclear force saturates. Examples of forces with this property are the so-called exchange forces which are familiar in chemistry as a sharing of electrons giving rise to chemical bonding in molecules like hydrogen H_2 (two electrons shared) or methane CH_4 (four electrons shared). The suggestion that nuclear forces have an exchange component was made by Heisenberg (1932).

In order to facilitate identification of the saturated unit the binding energy details are listed in Table 3.1 for the low-mass nuclei with $A=2$, 3 or 4, and from these data we deduce that the deuteron ${}^{2}_{1}H$, whose binding energy is just the negative of the Q-value in the photodisintegration process ${}^{2}_{1}H(\gamma, n){}^{1}_{1}H$, is the least tightly bound few-nucleon system and cannot therefore be the saturated unit. This position is evidently filled by the α-particle ${}^{4}_{2}He$, a conclusion which is confirmed by the non-existence of bound states of ${}^{5}_{2}He$

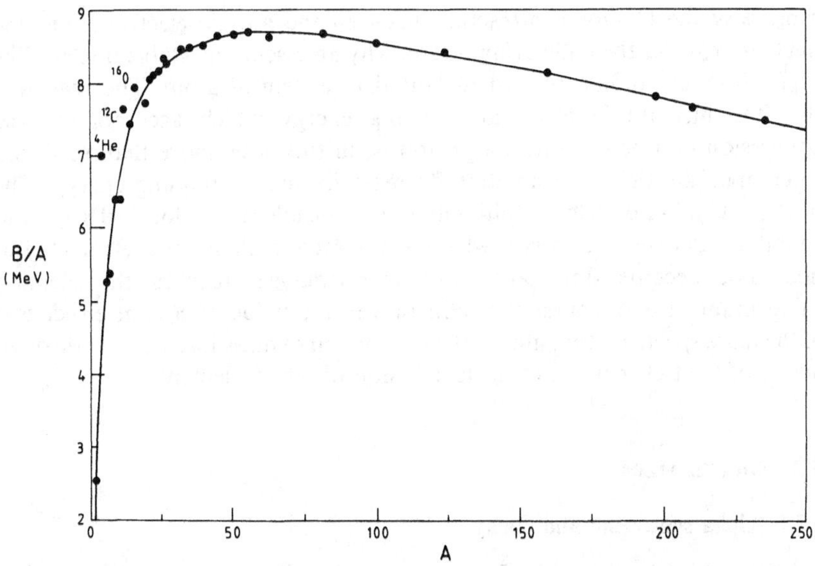

Figure 3.2. The measured binding energy per nucleon for the stable nuclei shown as a function of mass number A. The broad maximum of about 8.8 MeV per nucleon falls in the iron–nickel region of the periodic table and, in principle, energy is released by fusion of low mass nuclei below this maximum. Conversely energy is released by α-decay or fission of heavier mass nuclei falling above the maximum.

Table 3.1. Binding energy characteristics of nuclei with $A = 2$, 3 and 4

Nucleus	B (MeV)	B/A	$\frac{1}{2}A(A-1)$	$2B/[A(A-1)]$
$^{2}_{1}$H	2.20	1.10	1	2.20
$^{3}_{1}$H	8.33	2.78	3	2.77
$^{3}_{2}$He	7.60	2.53	3	2.53
$^{4}_{2}$He	28.11	7.03	6	4.68

and $^{5}_{3}$Li. Furthermore the nucleus $^{8}_{4}$Be, even though it contributes a pronounced maximum in the binding energy curve, is itself unstable and spontaneously disintegrates into two α-particles. The first stable 'α-particle' nucleus is $^{12}_{6}$C and all the low mass peaks in the binding energy up to $^{24}_{12}$Mg are for nuclei of this type and this sequence of stable nuclei continues all the way up to $^{40}_{20}$Ca (Wheeler 1937, Hafstad and Teller 1938, Dennison 1940).

The monotone decline in B/A for high mass nuclei is an indication of the growing importance of the unsaturated Coulomb forces of repulsion between protons which make a negative contribution to the binding energy. In the same way the Coulomb attraction between the atomic electrons and the nucleus provides the underlying reason why an essentially stable nucleus like $^{187}_{75}Re$ becomes radioactive when bound in a neutral atom. The reason is that the increase in electronic binding energy which accompanies the conversion of a neutron into a proton is, in this case, more than sufficient to compensate the corresponding decrease in nuclear binding energy. The nuclear Coulomb field is ultimately responsible both for α-decay and spontaneous fission, a process which is observed in all actinide elements but does not become dominant until the sequence reaches the element californium. It is of course the reduction in the value of B/A at either end of the mass spectrum that allows the conversion of mass into energy, through fusion of light elements and nuclear fission of heavy elements.

3.2 Nuclear sizes

3.2.1 Alpha scattering and decay

The formula (1.1) derived by Rutherford to describe the Coulomb scattering of slow α-particles by heavy nuclei is based on the assumption that the nuclear charge and mass are concentrated at a point. If instead the nucleus is taken to be a uniformly charged sphere of radius R then the nuclear Coulomb potential is

$$\begin{aligned}\varphi(r) &= (Ze^2/R)\left[\frac{3}{2}-\frac{1}{2}\left(\frac{r}{R}\right)^2\right] \qquad r \leqslant R \\ &= (Ze^2/r) \qquad r > R \qquad (3.3)\end{aligned}$$

and deviations from the Rutherford law must be observed when the collision diameter $r_c < R$. Thus even as early as 1911 Rutherford could claim that the observed scattering pattern using 7.68 MeV α-particles from ^{214}Po scattered from gold nuclei was consistent with a nuclear radius $R < 10^{-12}$ cm, but by 1919 clear departures from the point nucleus predictions had been detected, results which were subsequently verified for a range of α-particle energies (Chadwick and Bieler 1927). However these observations were not readily interpretable in terms of classical concepts, a conclusion which, 65 years on, is not at all surprising given that the α-particle experiences the strong interaction with both neutrons and protons in the nucleus, in addition to the Coulomb interaction with the protons alone.

At this point we must glance back to the detailed discussion of the nuclear scattering problem in section 2.2 and emphasize that this analysis rests squarely on the assumption that the nuclear scattering potential is short range in the sense that, at worst, it falls off exponentially at infinity. This restriction means that the asymptotic wavefunction at great distances from the scattering centre describes a free particle and that, at low bombarding energies, only *s*-wave scattering need be taken into account. In the case of Coulomb scattering neither of these assumptions holds good and, in place of the asymptotic form (2.36), for Coulomb scattering the asymptotic wavefunction is

$$\psi_c(r) \to e^{ikz} + e^{i(kr - \gamma \ln[2kr])} f_c(\theta)/r \tag{3.4}$$

where the scattered component has an additional phase shift $-\gamma \ln[2kr]$ which increases indefinitely as $r \to \infty$. This effect is a direct result of the infinite range of the Coulomb potential. For the case of Coulomb scattering of α-particles by nuclei of atomic number Z, γ is a real number given by

$$\gamma = 2m_\alpha Ze^2/\hbar k = 2(\alpha Z)(m_\alpha c/\hbar k) \tag{3.5}$$

where $\alpha = e^2/\hbar c$ is the fine structure constant. The Coulomb scattering amplitude $f_c(\theta)$ is given by the expression

$$f_c(\theta) = -\frac{\gamma}{2k} e^{2i\delta_c(k)} \left[\sin^2 \frac{\theta}{2} \right]^{-1-i\gamma} \tag{3.6}$$

where $\delta_c(k)$ is a real phase shift defined by

$$e^{2i\delta_c(k)} = \Gamma(1+i\gamma)/\Gamma(1-i\gamma) \tag{3.7}$$

The differential scattering cross-section $d\sigma_c/d\Omega = |f_c(\theta)|^2$ derived from (3.6) reduces to the Rutherford scattering law (1.1).

We can now carry out a phase shift analysis of $f_c(\theta)$ in the manner of (2.81) and (2.82) where the Coulomb phase shift $\delta_{cl}(k)$ and the nuclear phase shift $\delta_l(k)$ add coherently so that the total scattering amplitude is

$$\begin{aligned} f(\theta) &= \frac{1}{2ik} \sum_{l=0}^{\infty} (2l+1)[e^{2i(\delta_l + \delta_{cl})} - 1] P_l(\cos\theta) \\ &= f_c(\theta) + \frac{1}{2ik} \sum_{l=0}^{\infty} (2l+1)\, e^{2i\delta_{cl}} [e^{2i\delta_l} - 1] P_l(\cos\theta) \end{aligned} \tag{3.8}$$

and the second term in (3.8) represents a deviation from pure Coulomb scattering including the resonance and shadow-scattering effects discussed in section 2.2. The nuclear phase shifts $\delta_l(k)$ can be complex in general since the incoming α-particle may be captured by the nucleus (cf. section 2.2.4.3). The number of partial waves contributing to the deviation from pure Coulomb scattering is related to the range b of the nuclear potential and $\delta_l(k) = 0$ for $l \geqslant l_{max} \simeq kb$. Thus for high angular momentum states, i.e. distant collisions and small angles of scatter, the nuclear potential makes no contribution. For low angular momentum states, i.e. close encounters and large angles of scatter, the nuclear potential does influence the scattering and its contribution must be estimated in some way. In the sharp cut-off model for example (Blair 1954) it is assumed that the nuclear potential absorbs all partial waves with $l<l_{max}$ and transmits all partial waves with $l \geqslant l_{max}$, i.e.

$$e^{i\delta_l(k)}=0 \qquad l<l_{max}, \qquad e^{i\delta_l(k)}=1 \qquad l \geqslant l_{max} \tag{3.9}$$

The conclusion must be that anomalous α-particle scattering, although it provides clear evidence that nuclei have finite radii, does not give a simple method for measuring these radii. Indeed a simple and more successful approach to the problem may be based on a study of the systematics of α-decay interpreted as a process involving quantum mechanical tunnelling through the Coulomb barrier from the inside rather than from the outside (Condon and Gurney 1928, 1929, Gamow 1928). Although there are severe theoretical difficulties involved since it is necessary to assess, firstly the probability than an α-particle is formed within the nucleus, and secondly the probability that it should leak through the Coulomb barrier, nevertheless consistent, if not very precise results, were obtained by this means quite a long time ago. For example an analysis of the half-lives and energies of zero spin change α-decays for 15 isotopes with $211 \leqslant A \leqslant 238$ (Bethe 1937) showed that nuclear radii R satisfied the empirical formula

$$R=r_0 A^{1/3} \qquad r_0 \simeq 1.5\times 10^{-13}\ \text{cm} \tag{3.10}$$

This result implies that nuclear matter has a constant density $\simeq 10^{14}$ g cm^{-3}, a conclusion which is in complete agreement with the observed constancy of the mean binding energy per nucleon as noted in section 3.1.

3.2.2 Mirror nuclei

The potential advantages offered by the existence of pairs of 'mirror' nuclei for the precise determination of nuclear radii was pointed out by Fowler *et al* (1936) in respect of the isobaric doublet ($^{15}_{7}$N, $^{15}_{8}$O), and by Bethe (1938) for the doublet ($^{11}_{5}$B, $^{11}_{6}$C). The characteristic feature of mirror nuclei is that they each have equal numbers of neutrons and protons in the core plus one additional nucleon. This nucleon is a neutron for one member of the pair, e.g. $^{15}_{7}$N, but it is a proton for the other member, e.g. $^{15}_{8}$O. Since the number of (n−p) force bonds is the same for each member of the mirror pair, while the number of (n−n) bonds in the neutron rich member is equal to the number of (p−p) bonds in the proton-rich member, the mass splitting within the pair should be equal to the difference in Coulomb energy making due allowance for the slightly greater intrinsic mass of the neutron. This conclusion depends crtically on the assumption that the (n−n) nuclear force is equal to the (p−p) nuclear force, an equality which certainly holds good approximately, since otherwise stable nuclei would tend to be either highly neutron rich or highly proton rich. Experimentally nuclei tend to be neutron rich as we have seen in section 3.1, but this discrepancy is readily accounted for in terms of the Coulomb energy of the protons.

The lowest mass mirror pair is the (n, p) system itself, with a negative mass difference $(m_p - m_n)c^2 = -1.29$ MeV which manifests itself as the total (kinetic plus rest) energy of the electron emitted in the β-decay of the neutron. The next pair in the sequence is ($^3_1$H, $^3_2$He) and, although $^3_1$H is an electron emitter, the low value of 18.6 keV for the end-point kinetic energy of the β-spectrum shows that the additional Coulomb energy in $^3_2$He is only barely compensated by the extra mass of the additional neutron in $^3_1$H. The doublet ($^5_3$Li, $^5_4$Be) does not exist in nature (cf. section 2.1.1) and nor does the nucleus $^9_5$B so the mirror pair ($^9_4$Be, $^9_5$B) is incomplete. Thus the doublet ($^7_3$Li, $^7_4$Be) is the first example in which the proton-rich partner is also the more massive, but in this case the mass-splitting is difficult to measure with accuracy since $^7_4$Be decays by electron capture rather than by positron emission. However, the remaining 16 pairs of mirror nuclei, beginning with ($^{11}_4$B, $^{11}_5$C) and ending with ($^{43}_{21}$Sc, $^{43}_{22}$Ti) each contain a positron-emitting proton-rich partner and the end point energy of the β-spectrum yields a value of the mass-splitting within the mirror doublet (cf. section 1.5.1).

The nuclear radius is derived from the mass difference according to the following simple argument. A nucleus containing Z protons distributed with uniform charge density $\rho(r) = Ze/\frac{4}{3}\pi R^3$ throughout a spherical volume of radius R has a Coulomb energy given from (3.3) by

$$E_c(Z) = \frac{1}{2}\int_0^R \rho(r)\varphi(r)4\pi r^2\,dr = \frac{3}{5}(Ze)^2/R \tag{3.11}$$

Thus, if this nucleus decays by positron emission to the daughter nucleus with atomic number $(Z-1)$ the change in Coulomb energy is

$$\Delta E_c = \frac{3}{5}\left(\frac{e^2}{R}\right)(2Z-1) \tag{3.12}$$

and the total mass difference is

$$M(A, Z) - M(A, Z-1) = \Delta E_c/c^2 - (m_n - m_p) \tag{3.13}$$

Systematic application of (3.12) and (3.13) to the whole range of mirror nuclei in the mass range $11 \leqslant A \leqslant 43$ allows us to verify (3.10) for light nuclei, a result which appears to hold quite accurately with a characteristic radius $r_0 \simeq 1.45 \times 10^{-13}$ cm.

There are, however, a number of weaknesses in the above calculation which in the first place is purely classical and it is not quite clear how the localization of the charge on individual protons should be taken into account (Peaslee 1954). Secondly the use of correctly anti-symmetrized proton wavefunctions in a proper quantum mechanical calculation introduces a Coulomb exchange or electrostatic self-energy term which results in a reduction in the Coulomb energy. Finally one must allow for the fact that, in a radioactive nucleus, the decaying proton usually comes from the region of the nuclear surface rather than from the core, and lastly it may be necessary to take account of the departure from sphericity in the nuclear shape. The net action of all these factors is to reduce r_0 giving a best value of $r_0 = 1.28 \pm 0.05 \times 10^{-13}$ cm (Kofoed-Hansen 1948).

Because of all the uncertainties involved it is probably not worth further refining the calculation and in any case the significance of the result is not the precision it gives for the determination of r_0 but rather that this result agrees with the value obtained by more accurate methods. Thus the mass splitting for mirror nuclei provides strong support for the assumption that (n–n) and (p–p) forces are equal. This assumption is known as the charge symmetry hypothesis.

3.2.3 Isotope shift

The difficulties associated with the use of strongly interacting charged particles such as protons or α-particles to probe the nuclear structure can be avoided in their entirety by using charged leptons such as electrons or muons. The earliest electromagnetic effect of this type to be observed in the laboratory is the isotope shift which was detected by Merton (1919) using a Fabry–Perot etalon to study the hyperfine spectrum of lead samples of variable isotopic composition. A possible explanation for these observations was advanced by Ehrenfest (1922) in terms of a correction to the Rydberg constant derived from the variation in the reduced mass for different isotopes, an effect which is important for light atoms and ultimately led to the discovery of deuterium (Urey *et al* 1932). Bohr thought this explanation unlikely since the observed isotopic shift was too large and, in an addendum to Ehrenfest's paper, commented that

> it cannot be excluded that the discrepancies in question are due to a slight difference in the field of force surrounding the nucleus arising from the difference in the internal nuclear structures of the lead isotopes

Bohr was, as ever, absolutely right, and the theoretical studies of Bartlett (1931) and of Rosenthal and Breit (1932) showed that the isotope shift was readily accounted for in terms of a departure from the point source Coulomb field as expressed by (3.3). The essential point is that the nuclear radius R is determined by A rather than by Z and for this reason isotopes of the same element have slightly different electronic energies and these differences are enhanced for s-state electrons which have non-vanishing wavefunctions within the nuclear volume.

To observe the isotope effect cleanly one must study isotopes of even–even type so that each nucleus has spin $I = 0$ and there is no hyperfine splitting. Apart from lead which has three even–even isotopes, a very suitable candidate is mercury which has five even–even isotopes of mass values $A = 196$, 198, 200, 202 and 204 together making up 70% of natural mercury. The effect has been studied for some 20 elements between rubidium ($Z=44$) and lead ($Z=82$) and, though difficult to analyse quantitatively, the results tend to confirm the picture of a spherical nucleus with radius R given by (3.10) and a somewhat smaller value of $r_0 \simeq 1.1 \times 10^{-13}$ cm. The rare earth elements Sm ($Z=62$) and Nd ($Z=60$) are exceptions to this general rule and show anomalously large shifts which have been interpreted as evidence of nuclear deformation away from the spherical shape (Brix and Kopfermann 1949, 1958). These conclusions have since been convincingly confirmed by nuclear spectroscopic studies of nuclei far from the closed shell configurations associated with the magic numbers.

3.2.4 Muonic atoms

The main difficulty with using electronic transition energies as a source of information on nuclear radii is that the effects involved are so small. This is because the first Bohr radius, although inversely proportional to Z, even for a heavy atom like lead, has a value $a_e = 6.45 \times 10^{-11}$ cm which is 100 times greater than the nuclear radius. Thus even an s-state electron spends only a tiny fraction of its time within the nucleus and the resultant perturbation of its energy is infinitesimal. As was first realized by Wheeler (1949) this situation changes dramatically when the electron is replaced by a muon which is 207 times more massive. Since the Bohr orbit radius is further reduced in the ratio (m_e/m_μ), for muonic lead it will have a value $a_\mu = 3.1 \times 10^{-13}$ cm which is less than the corresponding nuclear radius and the energy shift will be very great indeed.

When a μ^- is captured by an atom, it first loses energy by Auger interactions with atomic electrons but, by the time its principal quantum number has fallen to a value $n \leqslant (m_\mu/m_e)^{1/2} \simeq 14$, the wavefunction reaches a maximum amplitude inside the first electronic Bohr orbit where the electron density is small and subsequently it loses energy mainly by the emission of X-rays. The time elapsing between the first capture in a high angular momentum state and the muon's arrival in the $n=2$ level is $\leqslant 10^{-13}$ s and the lifetime of the final 2p–1s X-ray transition, which is most sensitive to the finite nuclear size, is $\simeq 10^{-18}$ s.

The free μ^- is of course unstable and decays via the weak interaction

$$\mu^- \rightarrow e^- + \nu_\mu + \bar{\nu}_e \qquad (3.14)$$

with a lifetime of about 2.2 μs. However, for heavy muonic atoms a far more important decay process is weak muon capture

$$\mu^- + \mathrm{p} \rightarrow \nu_\mu + \mathrm{n} \qquad (3.15)$$

which increases in probability relative to free decay in proportion to $(\alpha Z)^4$ and has a lifetime of about 0.2 ns in muonic lead. However, by the time either of these weak processes become effective the sequence of electromagnetic transitions will already have been completed.

The energy levels of a muon of reduced mass m_μ, moving in the field of a point nucleus of charge Ze, are given by the expression

$$E_{n,j}=m_\mu c^2\{1+(\alpha Z)^2/[(n-j-\tfrac{1}{2})-\sqrt{((j+\tfrac{1}{2})^2-(\alpha Z)^2)}]\}^{-1/2} \quad (3.16)$$

where $j=l\pm\frac{1}{2}$ and screening effects due to atomic electrons are ignored. In any case these are quite negligible for $n\leqslant 3$. For light atoms the change in the muon energy due to the finite nuclear size may be estimated by first order perturbation theory, i.e.

$$\Delta E_{n,j}=e\int \psi^*_{n,j}[\varphi(r)-Ze^2/r]\psi_{n,j}\,d^3r \quad (3.17)$$

where $\varphi(r)$ is the electrical potential produced by the finite nucleus. For a uniformly charged spherical nucleus $\varphi(r)$ is given by (3.3). Carrying through the computation assuming $R\ll a_\mu$ we find for the energy shift in the 1s ground state

$$\Delta E_0=\frac{-(\alpha Z)^6}{3\pi}(m_\mu c^2)\frac{\langle r_c^2\rangle}{(\lambda\!\!\!^{-}_{c\mu})^2} \quad (3.18)$$

where $m_\mu c^2=105.6$ MeV, $\lambda\!\!\!^{-}_{c\mu}=\hbar/m_\mu c=1.87\times10^{-13}$ cm is the reduced Compton wavelength of the muon, and $\langle r_c^2\rangle$ is the mean square nuclear radius.

The result (3.18) shows that at low Z the muonic energy shifts determine only one parameter $\langle r_c^2\rangle$ of the nuclear charge distribution, and from this viewpoint this approach may be compared with the method based on the mass splitting within mirror doublets, although it is not limited by any assumptions about nuclear structure and is potentially much more accurate. For high Z elements the perturbative approach fails and in this regime it is necessary to adopt some suitable form for the charge distribution within the nucleus, solve the corresponding Dirac equation numerically, and determine the parameters of the distribution by comparison with experiment. This is the approach followed in electron scattering experiments which are reviewed immediately below.

The first experimental studies of the nuclear size effect in energy levels of muonic atoms were carried out by Fitch and Rainwater (1953) who measured the energy of the 2p–1s transition in a range of elements and detected energy shifts varying from 2% in aluminium ($Z=13$) to >300% in bismuth ($Z=83$). These measurements have been greatly refined and extended in the intervening period and yield an effective nuclear radius $R=r_0A^{1/3}$ with $r_0\simeq1.17\times10^{-13}$ cm (Anderson *et al* 1963, Backenstoss *et al* 1965).

3.2.5 Electron scattering

Electron scattering from nuclei is a process similar to α-particle scattering except of course that the electron is a much less massive particle and has no strong interactions. Since in order to observe nuclear size effects one requires electrons of reduced de Broglie wavelength $\lambda\!\!\!^{-}_e \leqslant 10^{-12}$ cm and this means electrons of energy >100 MeV, a full relativistic treatment of the scattering problem must be carried out and the simple Rutherford cross-section (1.1) is no longer applicable. In its place we must use the full quantum mechanical formula (Mott 1929, 1932) for electron scattering from a point-charge of magnitude Ze. This is given in second Born approximation by (McKinley and Feshbach 1948)

$$\left(\frac{d\sigma}{d\Omega}\right)_M = \frac{Z^2 r_e^2}{4}\left(\frac{1}{(\beta^2\gamma)^2 \sin^4(\theta/2)}\right)\left\{1-\beta^2\sin^2(\theta/2)+\alpha Z\beta\pi\sin\frac{\theta}{2}\left(1-\sin\frac{\theta}{2}\right)\right\} \tag{3.19}$$

where $\beta = v/c$, $\gamma = (1-\beta^2)^{-1/2}$ and $r_e = e^2/m_e c^2$ is the classical electron radius.

Deviations from the Mott scattering formula are observed when $\lambda\!\!\!^{-}_e$ becomes so small that the target nucleus no longer appears as a point object, and in these circumstances the cross-section becomes

$$\frac{d\sigma}{d\Omega} = \left(\frac{d\sigma}{d\Omega}\right)_M |F_c(q)|^2 \tag{3.20}$$

where $q = 2\sin(\theta/2)/\lambda\!\!\!^{-}_e$ is the momentum transferred to the target nucleus and $F_c(q)$ is the charge form factor defined as the three-dimensional Fourier transform of the normalized charge density function $\rho_c(\boldsymbol{r})$, i.e.

$$F_c(q) = \int e^{i\boldsymbol{q}\cdot\boldsymbol{r}} \rho_c(\boldsymbol{r})\, d^3r = 1 - \frac{q^2}{3!}\langle r_c^2\rangle + \frac{q^4}{5!}\langle r_c^4\rangle + \ldots \tag{3.21}$$

Experimentally the form factor $F_c(q)$ is determined over a range of values of $\lambda\!\!\!^{-}_e$ and θ and the results used to fix the parameters of some assumed density function. A function which has been widely used for this purpose is the Fermi distribution (Yennie *et al* 1954)

$$\rho_c(r) = \frac{\rho_c(0)}{1+\exp[(r-r_{1/2})/d]} \qquad r_{1/2} \gg d \tag{3.22}$$

where $r_{1/2}$ is that radius at which the charge density has decreased to half its value $\rho_c(0)$ at the nuclear centre, and d is a diffuseness parameter which characterizes the thickness of the boundary layer at the nuclear surface. Just as with the muonic X-ray method, at relatively low energy the electron scattering data only yields a value for $\langle r_c^2\rangle$ which for the Fermi density is given by

$$\langle r_c^2\rangle = \tfrac{3}{5} r_{1/2}^2 + \tfrac{7}{5}\pi^2 d^2 \tag{3.23}$$

and a useful way of analysing the data is to express $\langle r_c^2\rangle$ in terms of an equivalent uniform charge radius R given by

$$\langle r_c^2 \rangle = \tfrac{3}{5} R^2 \tag{3.24}$$

At higher energies further parameters can be determined, e.g. the diffuseness parameter d which vanishes for the case of a uniformly charged sphere.

Electron scattering experiments have been extensive and have been carried out at energies up to several GeV and analysed under a great variety of theoretical assumptions (Hofstadter 1956). In general it is found that for $A > 20$ the Fermi distribution gives a satisfactory representation of measured nuclear charge distributions with a constant diffuseness parameter $d = 0.55 \times 10^{-12}$ cm. However, for light nuclei there is no central core of uniform charge density and other hypothethical density functions, e.g. exponential, are found to be more suitable. It is also observed that the equivalent charge radius R obeys the law (3.10) very closely allowing for nuclear deformation in for example rare earth nuclei, except that the intrinsic radius r_0 varies slowly over the mass range. For $A \leqslant 10$, $r_0 \simeq 1.9 \times 10^{-13}$ cm excluding the tightly bound α-particle for which $r_0 = 1.3 \times 10^{-13}$ cm. For $A > 10$, r_0 decreases slowly from about 1.33×10^{-13} cm at ${}^{12}_{6}\mathrm{C}$ to 1.18×10^{-13} cm at ${}^{208}_{82}\mathrm{Pb}$. In general one may say that these results, which in principle are the most accurate, are in very satisfactory agreement with results obtained by other methods, and this is particularly true when comparison is made with the most precise experiments on muonic atoms.

3.2.6 Neutron scattering

All the methods for measuring nuclear radii discussed up to this point rely on the electromagnetic interaction, and therefore determine the parameters of the electric charge density $\rho_c(r)$. This is not in principle identical with the nuclear mass density $\rho_m(r)$ defined by

$$A\rho_m(r) = Z\rho_p(r) + N\rho_n(r) \tag{3.25}$$

where $\rho_p(r)$ and $\rho_n(r)$ are the mass densities of protons and neutrons respectively. Neither are $\rho_c(r)$ and $\rho_p(r)$ identical, and this is because scattering experiments with electrons of energy > 1 GeV ($\lambda\!\!\!^{-}_e \simeq 2 \times 10^{-14}$ cm) impinging on hydrogen targets, show that the proton is a composite object with a mean square charge radius $\langle r_{pc}^2 \rangle = 0.64 \times 10^{-26}$ cm^2. The mean square radius of the nuclear charge distribution is therefore given by

$$\langle r_c^2 \rangle = \langle r_p^2 \rangle + \langle r_{pc}^2 \rangle \tag{3.26}$$

At first glance one might suppose that, because of their Coulomb repulsion, the distribution of protons would be more spread out in space than the distribution of neutrons but in fact the reverse is the case. There are two reasons for supposing that the protons will be concentrated near to the central region of the nucleus where the nuclear potential is lowest. The first reason is that the nuclear force is momentum dependent in general and is less attractive at higher momenta. Therefore, because of their positive Coulomb potential energy, protons will have a lower kinetic energy, and experience a stronger attractive nuclear force than neutrons of the same total energy. The second reason concerns the known properties of the two-nucleon potential which is most strongly attractive in the triplet s-state, a state which is forbidden to like nucleons by the Pauli principle. Since there are more neutrons than protons in the nucleus, a nuclear proton can form more triplet bonds than can a nuclear neutron and therefore experiences an attractive force which is stronger on average. This effect is the origin of the (positive) asymmetry term proportional to $(A/2-Z)^2/A$ in the semi-empirical mass formula derived from the liquid drop model of the nucleus.

Since the Fermi levels of the degenerate proton and neutron distributions must coincide in order to maintain nuclear stability against β-decay (Johnson and Teller 1958) there is a tendency for neutron and proton energies to equalize, with the result that the protons concentrate in a region of lower nuclear potential energy with a mean square radius given by

$$\langle r_{\mathrm{p}}^2\rangle \simeq \langle r_{\mathrm{n}}^2\rangle \simeq 1.0\times 10^{-26}\ \mathrm{cm}^2 \tag{3.27}$$

We can now combine the results (3.25–3.27) to give a relation between the distributions of mass and charge

$$\langle r_{\mathrm{m}}^2\rangle \simeq \langle r_{\mathrm{c}}^2\rangle + [0.4N/A - 0.6Z/A]\times 10^{-26}\ \mathrm{cm}^2 \tag{3.28}$$

Since $N>Z$ we see that the nuclear and Coulomb forces between nucleons conspire to yield a balance within the nucleus so that $\langle r_{\mathrm{m}}^2\rangle \simeq \langle r_{\mathrm{c}}^2\rangle$.

These conclusions may be tested directly by studying neutron–nucleus scattering since neutrons, in perfect contrast with electrons, are uncharged but do interact strongly. Selecting as a target an intermediate mass nucleus (e.g. copper) with $R\simeq 5\times 10^{-13}$ cm, we need to use neutrons with reduced de Broglie wavelength $\lambda\!\!\!^{-}_{\mathrm{n}} \leqslant R$, i.e. with energy $E_{\mathrm{n}} \geqslant 10$ MeV. The 14 MeV neutrons from the reaction ${}^3_1\mathrm{H}(\mathrm{d},\mathrm{n}){}^4_2\mathrm{He}$ are therefore ideally suited for this purpose. At these energies the nucleus may be regarded as a perfectly 'black' target for neutrons and the scattering process may be analysed within the sharp cut-off approximation (cf. section 3.2.1) with a maximum angular momentum quantum number $l_{\max}=kR$. We then find for the total cross-section (2.100).

$$\begin{aligned}\sigma_{\mathrm{t}} &= \frac{2\pi}{k^2}\sum_{l=0}^{\infty}(2l+1)[1-Re(S_l(k))]\\ &= \frac{2\pi}{k^2}\sum_{l=0}^{kR}(2l+1) = 2\pi\left(R+\frac{1}{k}\right)^2 \simeq 2\pi R^2\end{aligned} \tag{3.29}$$

The factor of 2 in the total cross-section, giving it a value twice as large as the geometrical cross-section πR^2, is associated with the fact that the capture process is always accompanied by shadow scattering, the two processes making equal contributions to the cross-section in the limit of a perfectly black nucleus.

At higher neutron energies $E_n \geqslant 90$ MeV the capture cross-section begins to fall below the predicted value and this is because the neutron interacts with the individual neutrons and protons rather than with the nucleus as a whole (Serber 1947). In this situation the nucleus is no longer perfectly black but becomes partially transparent, an effect first discussed by Bethe (1940) in terms of concepts subsequently developed into the highly successful optical model of the nucleus (section 2.2). Like the nuclear shell model the optical model considers the nucleons as moving independently in a central potential which, in addition, contains an imaginary term. In its original version the optical potential was assumed to have the form of a square well (Feshbach *et al* 1954) but a far more realistic choice is the Woods–Saxon potential (Woods and Saxon 1954).

$$V(r) = \frac{V \times iW}{1 + e^{(r - r_{1/2})/d}} \qquad r_{1/2} \gg d \tag{3.30}$$

This potential has an analytical form which is identical to that of the Fermi charge density (3.22) and its parameters may be similarly interpreted. The measurement of nuclear radii via neutron scattering shows very satisfactory agreement with results obtained by electromagnetic methods giving a value for the radial parameter $r_0 = 1.25 \pm 0.05 \times 10^{-13}$ cm, and for the diffuseness parameter $d = 0.65 \times 10^{-13}$ cm.

The optical model has also been applied successfully in the low energy region 0–3 MeV where the measured neutron cross-sections fluctuate very rapidly with energy with a structure of narrow resonances which varies in an unpredictable way from one nucleus to the next. However, when these cross-sections are averaged over an energy interval containing very many narrow resonances, they are found to exhibit broad maxima, with widths on the order of 1 MeV, and these so-called 'giant resonances' cannot be interpreted in terms of a compound nucleus model which predicts that the cross-section should decrease monotonically with energy according to the $1/v$ law. It is found however that the sizes and positions of these maxima vary in a regular way with mass number in a pattern which can be correlated with the predictions of the optical model. This topic is discussed in much greater detail in section 4.7.4.

3.3 Nuclear forces

3.3.1 Invariance principles

The atomic nucleus is an extremely complex system which is made up from two types of spin $\frac{1}{2}$ fermion, coupled together by forces about which we know for certain only that they are strong, of very short range and are spin and velocity dependent. Reflecting this profound ignorance concerning the precise nature of the forces holding the nucleus together, nuclear theory has developed along two quite dissimilar, not to say incompatible, lines of investigation. In the first class of nuclear model which received its stimulus from ideas propounded by Bohr, the nucleus is conceived of as a strongly-coupled condensed structure, somewhat akin to a 'liquid drop', which totally absorbs all strongly interacting nuclear particles which collide with it, leading to the formation of an unstable compound nucleus which can then decay through a number of channels, eventually returning to a stable state. In the second class of model the nucleus is viewed as a system of essentially free particles composed from two 'Fermi' gases, whose interactions establish a uniform potential field with a set of permitted energy levels, which then fill up in sequence as demanded by the Pauli principle, exactly as electrons arrange themselves in atoms. This latter picture is the genesis of the nuclear shell model which has been modified and extended in various directions, e.g. in the collective model only those nucleons in unclosed shells are regarded as moving independently, while the core of closed shells possesses modes of excitation appropriate to a totally condensed system.

Although the shell model has been highly successful in predicting and correlating the static properties of nuclear levels, e.g. spins, parities and electromagnetic moments, it has not fared so well in the analysis of scattering and reaction phenomena where the observational evidence on the whole lends support to Bohr's concept of a compound nucleus. Although the gap between these extreme viewpoints is bridged to some extent by the optical model, which treats the nucleus as a semi-transparent medium characterized by a complex refractive index, there exists in general no all-embracing theory of the nucleus which presents us with a uniquely defined nuclear potential, or even allows us to infer that nuclear forces can be derived from such a potential. We are therefore obliged to make some simplifying assumptions and the simplest hypothesis to adopt is that nuclear forces are two-body forces, derivable from an inter-nucleon potential, whose parameters are restricted first of all by the experimental data, but also by a number of invariance principles, some exact and others only approximate, which are expressed in the laws of conservation of energy, momentum, angular momentum, parity, charge, isospin, etc.

In setting up the centre of mass coordinate system in section 2.2 we have already exploited two of these principles, namely the principle of translational invariance which allows us to assert that a two-particle potential $V(\boldsymbol{r}_1, \boldsymbol{r}_2)$ can depend only on the relative coordinate $\boldsymbol{r}=\boldsymbol{r}_1-\boldsymbol{r}_2$, and the principle of Galilean invariance (or Lorentz invariance in the fully relativistic context) which establishes the existence of a centre of mass whose motion is uniform. There is, however, a further invariance, namely charge conjugation or C-invariance, which is a direct consequence of Lorentz invariance. This follows from the fact that the negative energy solutions of relativistic wave equations cannot be rejected without sacrificing the principle that the eigenstates of the Hamiltonian should form a complete set. Dirac's solution to this difficulty was to propose the existence of antiparticles, which have the same mass as the corresponding particles, but which differ in the sign of all additive quantum numbers, e.g. the charge. On one interpretation the antiparticle may be viewed as a particle propagating backwards in time.

That it is impossible to dispense with the concept of antiparticles follows from the fact that the temporal order of two events with a space-like separation is not a Lorentz invariant. A classical particle may not propagate between such space–time points but, because of the uncertainty principle, this restriction does not apply to a quantum particle. However, a sequence of events which may be viewed in one reference frame as a particle moving in one direction may be viewed from another frame as an antiparticle moving in the opposite direction. For example, if the neutron–proton pair represented in figure 3.6 have a space-like separation, the conclusion as to whether a π^+ or a π^- is exchanged depends on the viewpoint of the observer. We shall now go on to discuss the principles of rotational, reflection and time-reversal invariance as applied to the description of an isolated system of interacting particles.

Rotational invariance is an exact symmetry principle which finds its physical expression in the law of angular momentum conservation. The strong and electromagnetic interactions, although not the weak interaction, also satisfy reflection invariance, which means they contribute terms to the Hamiltonian which are invariant under the parity operator P which reverses the sign of the spatial coordinates of each of the interacting particles of the system. With the exception of a single phenomenon, namely the weak decay of the neutral K-meson, which through the application of the so-called CPT theorem (Lüders 1957) appears to indicate the presence of a small T-violating component in the weak interaction, all nuclear interactions are invariant under the operation of time reversal. The transformations of rotation and reflection are unitary, which means that they preserve the norms of all state vectors; also the angular momentum operators are Hermitian and represent observables. The parity operation is not Hermitian and therefore does not strictly correspond to an observable. However, with suitable arbitrary assignments of a finite number of intrinsic parities, e.g. proton parity, electron parity, the principle of reflection invariance does permit a prediction of relative parity, a result which is frequently expressed as 'the law of conservation of parity'.

The first deduction we make from the principle of rotational invariance is that the nuclear potential V must be a scalar, i.e. spherically symmetric, and it can be a function only of scalars, e.g. $\boldsymbol{r}\cdot\boldsymbol{r}=r^2$, $\boldsymbol{p}\cdot\boldsymbol{p}=p^2$, $\boldsymbol{r}\cdot\boldsymbol{p}$, etc., where the position and momentum vectors $\boldsymbol{r}$, $\boldsymbol{p}$ are Cartesian vectors. We do not at this stage distinguish between proper scalars and pseudo-scalars which behave identically under rotations. Neither does it matter whether a rotation is interpreted in the active sense (rotation of the physical system under investigation) or in the passive sense (rotation of the coordinate system). A Cartesian system of coordinates is of course quite unsuitable for treating a finite system like a nucleus which is spherically symmetric or almost so, although the geometry of the coordinate system does not matter much when we are dealing with single component scalar operators (spin 0), or even with three-component vector operators (spin 1). This is because a Cartesian vector with three components is also a spherical vector in the sense that it transforms under rotations in the same way as the $(2l+1)$ components of the spherical harmonic $Y_l^m(\theta, \varphi)$ with $l=1$. For example the position vector $\boldsymbol{r}$, which is a first rank Cartesian tensor with components $x\ y\ z$, has spherical components.

$$+(x-iy)/\sqrt{2}=r\sqrt{(4\pi/3)}Y_1^{-1}(\theta,\varphi) \qquad z=r\sqrt{(4\pi/3)}Y_1^0(\theta,\varphi)$$

$$-(x+iy)/\sqrt{2}=r\sqrt{(4\pi/3)}Y_1^1(\theta,\varphi) \tag{3.31}$$

and therefore transforms under rotations like the spherical harmonics $Y_1^{0,\pm1}(\theta, \varphi)$ which characterize the state of a system with angular momentum 1. This is why we can refer to vector operators like $\boldsymbol{r}$ as 'spin 1' tensor operators.

The situation is very different for the second-rank Cartesian tensor $\mathbf{T}_{\alpha\beta}$ which by definition has two indices $\alpha, \beta(=x, y, z)$ and nine components in general. These do not transform under rotations like linear combinations of the components of a spherical tensor of spin 4. Rather they can be organized into sub-units each of which transforms like a spherical tensor. In this analysis we use the summation convention which requires that a repeated index be summed over the three possible values of that index. For example the trace T of the tensor $\mathbf{T}_{\alpha\beta}$, which is defined as the sum of the diagonal elements $T_{xx}+T_{yy}+T_{zz}$ can be written as

$$\mathrm{Tr}(\mathbf{T}_{\alpha\beta})=\mathbf{T}_{\alpha\alpha}=\mathrm{T} \tag{3.32}$$

The process of summing over a repeated index is known as contraction and, since it allows us to reduce the rank of the tensor by 2, permits the construction of scalars by contraction of the indices of second-rank tensors. Similarly we can construct a fourth-rank tensor by direct multiplication of two second-rank tensors and produce further scalars by contraction of both pairs of indices.

In general the nine components of the second-rank Cartesian tensor $\mathbf{T}_{\alpha\beta}$ can be organized into three spherical tensors according to the relation

$$\mathbf{T}_{\alpha\beta}=\mathbf{A}_{\alpha\beta}+\mathbf{S}_{\alpha\beta}+\frac{1}{3}\mathrm{T}\delta_{\alpha\beta} \tag{3.33}$$

where the properties of the spherical tensors T, $\mathbf{A}_{\alpha\beta}$ and $\mathbf{S}_{\alpha\beta}$ are listed in Table 3.2. We note that, where antisymmetric tensors $\mathbf{A}_{\alpha\beta}$ and $\mathbf{B}_{\alpha\beta}$ are

Table 3.2. Reduction of a second-rank cartesian tensor

Linear combinations of components	Number of components	Spherical tensor	Spin
$\mathsf{T}=\mathsf{T}_{\alpha\alpha}$	1	scalar	0
$\mathsf{A}_{\alpha\beta}=\frac{1}{2}(\mathsf{T}_{\alpha\beta}-\mathsf{T}_{\beta\alpha})$	3	anti-symmetric tensor (or axial vector)	1
$\mathsf{S}_{\alpha\beta}=\frac{1}{2}[(\mathsf{T}_{\alpha\beta}+\mathsf{T}_{\beta\alpha})-(2/3)\mathrm{T}\,\delta_{\alpha\beta}]$	5	traceless symmetric tensor	2

represented as axial vectors $\boldsymbol{a}$ and $\boldsymbol{b}$, with $a_x=A_{yz}$, $b_y=B_{zx}$, etc., their contraction $\mathsf{A}_{\alpha\beta}\mathsf{B}_{\alpha\beta}$ is equal to $2\boldsymbol{a}\cdot\boldsymbol{b}$.

We illustrate the above ideas by a simple example. The tensor $\mathsf{T}_{\alpha\beta}=r_\alpha p_\beta$ is constructed by direct multiplication of the components of $\boldsymbol{r}$ and $\boldsymbol{p}$ and according to the decomposition given in Table 3.2 reduces to the scalar $\mathsf{T}_{\alpha\alpha}=\boldsymbol{r}\cdot\boldsymbol{p}$, the anti-symmetric tensor $\mathsf{A}_{\alpha\beta}=\frac{1}{2}(\boldsymbol{r}\times\boldsymbol{p})$ and the traceless symmetric tensor $\mathsf{S}_{\alpha\beta}$ where

$$\mathsf{S}_{\alpha\beta}=\begin{pmatrix} xp_x-\frac{1}{3}(\boldsymbol{r}\cdot\boldsymbol{p}) & xp_y+yp_x & xp_z+zp_x \\ yp_x+xp_y & yp_y-\frac{1}{3}(\boldsymbol{r}\cdot\boldsymbol{p}) & yp_z+zp_y \\ zp_x+xp_z & zp_y+yp_z & zp_z-\frac{1}{3}(\boldsymbol{r}\cdot\boldsymbol{p}) \end{pmatrix} \tag{3.34}$$

It is clear that, since $\mathsf{A}_{\alpha\beta}$ and $\mathsf{S}_{\alpha\beta}$ are both traceless tensors, no new invariants can be constructed by contractions from either tensor, a result which merely reinforces the fact that $\boldsymbol{r}\cdot\boldsymbol{p}$ is the only invariant that can be constructed from the vectors $\boldsymbol{r}$ and $\boldsymbol{p}$.

This analysis must be generalized to some extent when reflections as well as rotations are included among the acceptable symmetry transformations for in this case we must distinguish between proper (polar) vectors such as the momentum $\boldsymbol{p}$, which change sign under inversion of the coordinate system, and pseudo (axial) vectors such as the orbital angular momentum $\boldsymbol{l}=\boldsymbol{r}\times\boldsymbol{p}$ or the spin $\boldsymbol{s}(=\frac{1}{2}\hbar\boldsymbol{\sigma}$ for a spin $\frac{1}{2}$ particle) which do not change sign. In the same way we need to distinguish between proper scalars such as $\boldsymbol{r}\cdot\boldsymbol{r}$, $\boldsymbol{p}\cdot\boldsymbol{p}$ or $\boldsymbol{r}\cdot\boldsymbol{p}$ which are invariant under the parity operation and pseudo-scalars such as the helicity $\langle\boldsymbol{\sigma}\cdot\boldsymbol{p}\rangle$ which change sign. The general rule is that the orbital parity of a spherical tensor of order l is $(-1)^l$ and this may be multiplied by an additional factor of (-1) giving rise to pseudo-scalars, pseudo-vectors, etc. For example the Cartesian tensor $\mathsf{T}_{\alpha\beta}=r_\alpha p_\beta$ defined above reduces to the scalar $T_{\alpha\alpha}$, the pseudo-vector $A_{\alpha\beta}$ and the proper symmetric tensor $\mathsf{S}_{\alpha\beta}$. An example of the second type of tensor is the Cartesian tensor constructed by direct multiplication of the components $\boldsymbol{\sigma}$ and $\boldsymbol{p}$. This process gives rise to a pseudo-scalar $\boldsymbol{\sigma}\cdot\boldsymbol{p}$, a proper vector $\boldsymbol{\sigma}\times\boldsymbol{p}$, and a symmetric pseudo-tensor.

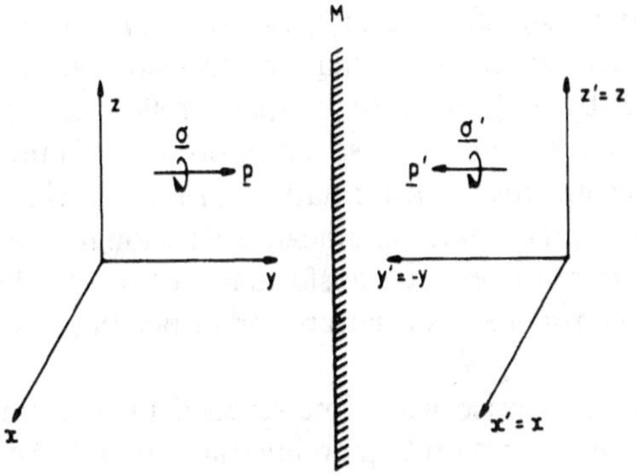

Figure 3.3. Diagram illustrating how the parity transformation P converts the right-handed coordinate system (xyz) into its left-handed mirror image $(x'y'z')$. Vectors such as momentum $\boldsymbol{p}$ whose mirror images are reversed in direction are described as polar vector. Vectors such as spin angular momentum $\boldsymbol{\sigma}$ which define a screw sense, which is preserved in the mirror image, are described as axial vectors.

In strong interactions we say that the law of parity conservation applies and what this means is that strong interactions are symmetric between right and left and there exists no strong interaction phenomenon which allows an absolute criterion to be established which distinguishes between right and left. This implies that the nuclear potential can contain no pseudo-scalar terms since such terms change sign under parity and therefore appear with different signs in right-handed and left-handed coordinate systems. We can illustrate this idea by considering the case of nuclear β-decay which is a weak interaction and does not conserve parity. Experimentally it is found that more electrons are emitted with their spins anti-parallel to their momenta than are emitted with their spins parallel to their momenta which means that the helicity $\langle \boldsymbol{\sigma} \cdot \boldsymbol{p} \rangle$ is negative. Now since $\boldsymbol{p}$ is a proper vector and $\boldsymbol{\sigma}$ a pseudo-vector, if we transform from the conventional right-handed coordinate system to a left-handed coordinate system as shown in figure 3.3, $\boldsymbol{p}$ changes sign while $\boldsymbol{\sigma}$ remains invariant. Thus if $\boldsymbol{p}$ and $\boldsymbol{\sigma}$ are parallel in a right-handed coordinate system they are anti-parallel in a left-handed coordinate system. Thus the statement that the helicity of the electron emitted in β-decay is negative is true only in a right-handed coordinate system and the observation of a non-vanishing value of $|\langle \boldsymbol{\sigma} \cdot \boldsymbol{p} \rangle|$ permits us to establish a meaning to 'right-handedness' which is absolute in the sense that it is not determined by some historical accident, for example that the sun rises to the right and sets to the left of an observer facing the polar star.

The final symmetry we must touch on briefly is that of time reversal (T) which, like the parity transformation (P) is a discrete symmetry. Another property it shares with parity is that it is always a passive rather than an active transformation. Although the parity transformation tends to be thought of as having significance only for quantum phenomena, in fact it has a direct classical analogue in that the transformation from a state of motion, specified by $\boldsymbol{r}(t)$, $\boldsymbol{p}(t)$, to a space reflected state $\boldsymbol{r}_{\mathrm{p}}(t) = -\boldsymbol{r}(t)$, $\boldsymbol{p}_{\mathrm{p}}(t) = -\boldsymbol{p}(t)$, is canonical in the sense that it preserves Hamilton's equations of motion. However, the transformation (T) to a time reversed state $\boldsymbol{r}_{\mathrm{T}}(t) = \boldsymbol{r}(t)$, $\boldsymbol{p}_{\mathrm{T}}(t) = -\boldsymbol{p}(t)$ is not canonical, even though the laws of motion are evidently invariant under the transformation $t \to -t$. This is because the time is normally considered as a parameter rather than as a coordinate in classical mechanics.

Equivalent problems arise with time-reversal in quantum mechanics in that the Schrödinger equation is not invariant under the transformation $t \to -t$ unless $\psi(\boldsymbol{r}, t)$ is replaced by $\psi^*(\boldsymbol{r}, -t)$. This is a non-linear or anti-unitary transformation and has the consequence that there is no 'time parity' quantum number analogous to the parity quantum number discussed above. Nevertheless the assumption of T-invariance does forbid the presence of scalars in the Hamiltonian which change sign under T provided they arise in the first order of a perturbation. This restriction to first order terms may have important experimental consequences in the search for T-violating first order terms in the weak interaction which in practice may be mimicked by second-order T-conserving terms in the electromagnetic interaction (Callan and Treiman 1967).

3.3.2 The nucleon–nucleon potential

3.3.2.1 Static potentials

Let us begin by introducing the simplest possible hypothesis, that the nucleon–nucleon potential is independent of momentum and, to construct this potential, we have at our disposal only three vectors namely the relative coordinate $\boldsymbol{r} = \boldsymbol{r}_1 - \boldsymbol{r}_2$ and the spin angular momenta of the two nucleons $\boldsymbol{s}_1 = \frac{1}{2}\hbar\boldsymbol{\sigma}_1$ and $\boldsymbol{s}_2 = \frac{1}{2}\hbar\boldsymbol{\sigma}_2$. Conservation of parity tells us that only even powers of $\boldsymbol{r}$ can enter the interaction and the observation that neutrons are scattered differently in para hydrogen (anti-parallel spins) than in ortho hydrogen (parallel spins) implies that the spins of the two particles should enter the interaction in a symmetric combination only. Thus the interaction must be constructed from the second-rank Cartesian tensors $r_\alpha r_\beta$, $\sigma_{1\alpha}\sigma_{2\beta}$ and, by evaluating the traces as described in section 3.3.1 above, we can pick out the spin 0 components r^2 and $\boldsymbol{\sigma}_1 \cdot \boldsymbol{\sigma}_2$. Since the spin 1 component $\boldsymbol{r} \times \boldsymbol{r}$ of $r_\alpha r_\beta$ vanishes identically the only other scalar possible is constructed by contracting the spin 2 component $\mathbf{S}_{\alpha\beta}$ of $r_\alpha r_\beta$ with the spin 2 component $\Sigma_{\alpha\beta}$ of $\sigma_{1\alpha}\sigma_{2\beta}$ giving

$$S_{\alpha\beta}\,\Sigma_{\alpha\beta} = (\boldsymbol{\sigma}_1 \cdot \boldsymbol{r})(\boldsymbol{\sigma}_2 \cdot \boldsymbol{r}) - (1/3)r^2\boldsymbol{\sigma}_1 \cdot \boldsymbol{\sigma}_2 = (1/3)r^2 S_{12}(\boldsymbol{\sigma}_1 \cdot \boldsymbol{\sigma}_2, \hat{\boldsymbol{r}}) \qquad (3.35)$$

where

$$S_{12}(\boldsymbol{\sigma}_1, \boldsymbol{\sigma}_2, \hat{\boldsymbol{r}}) = 3(\boldsymbol{\sigma}_1 \cdot \hat{\boldsymbol{r}})(\boldsymbol{\sigma}_2 \cdot \hat{\boldsymbol{r}}) - \boldsymbol{\sigma}_1 \cdot \boldsymbol{\sigma}_2 \tag{3.36}$$

and $\hat{\boldsymbol{r}}$ is a unit vector in the direction of $\boldsymbol{r}$. Because it is constructed from the contraction of two spin 2 spherical tensors, S_{12} is called the tensor force operator between the two nucleons. Also since ∇S_{12} has non-vanishing non-radial components, the tensor force is not a central force.

Tensor forces also play an important role in other branches of physics and the classic example is that of the energy of interaction between two magnetic dipoles $\boldsymbol{\mu}_1$ and $\boldsymbol{\mu}_2$. Since the magnetic field $\boldsymbol{B}$ in a region of space which contains no currents can be represented as the gradient of a pseudo-scalar magnetostatic potential ψ ($\boldsymbol{B}$ is a pseudo-vector) where

$$\boldsymbol{B} = -\nabla\psi \tag{3.37}$$

and a magnetic moment $\boldsymbol{\mu}_1$ at the origin produces a magnetostatic potential

$$\psi = -\boldsymbol{\mu}_1 \cdot \nabla\left(\frac{1}{r}\right) \tag{3.38}$$

we can write the energy of interaction with a second magnetic dipole $\boldsymbol{\mu}_2$ at $\boldsymbol{r}$ in the form

$$U = -\boldsymbol{\mu}_2 \cdot \boldsymbol{B} = \boldsymbol{\mu}_2 \cdot \nabla\psi = -(\boldsymbol{\mu}_2 \cdot \nabla)(\boldsymbol{\mu}_1 \cdot \nabla)\left(\frac{1}{r}\right) \tag{3.39}$$

Therefore, since $\nabla(1/r) \equiv \hat{\boldsymbol{r}}(\partial/\partial r)(1/r)$ we can write U in the form

$$\begin{aligned} U &= -\boldsymbol{\mu}_2 \cdot \nabla\left[\boldsymbol{\mu}_1 \cdot \nabla\left(\frac{1}{r}\right)\right] = -\boldsymbol{\mu}_2 \cdot \nabla\left[\frac{1}{r}\frac{\partial}{\partial r}\left(\frac{1}{r}\right)\boldsymbol{\mu}_1 \cdot \boldsymbol{r}\right] \\ &= -3(\boldsymbol{\mu}_2 \cdot \hat{\boldsymbol{r}})(\boldsymbol{\mu}_1 \cdot \hat{\boldsymbol{r}})\frac{1}{3}\left[\frac{\partial^2}{\partial r^2}\left(\frac{1}{r}\right) - \frac{1}{r}\frac{\partial}{\partial r}\left(\frac{1}{r}\right)\right] - \boldsymbol{\mu}_2 \cdot \boldsymbol{\mu}_1 \frac{1}{r}\frac{\partial}{\partial r}\left(\frac{1}{r}\right) \\ &= -[3(\boldsymbol{\mu}_1 \cdot \hat{\boldsymbol{r}})(\boldsymbol{\mu}_2 \cdot \hat{\boldsymbol{r}}) - \boldsymbol{\mu}_1 \cdot \boldsymbol{\mu}_2]\left(\frac{1}{r^3}\right) - \boldsymbol{\mu}_1 \cdot \boldsymbol{\mu}_2\left[\frac{2}{3r}\frac{\partial}{\partial r}\left(\frac{1}{r}\right) + \frac{1}{3}\frac{\partial^2}{\partial r^2}\left(\frac{1}{r}\right)\right] \end{aligned} \tag{3.40}$$

We now use the result that $(1/r)$ is the solution of Poisson's equation for a unit point source at the origin, i.e.

$$\frac{2}{3r}\frac{\partial}{\partial r}\left(\frac{1}{r}\right) + \frac{1}{3}\frac{\partial^2}{\partial r^2}\left(\frac{1}{r}\right) = \frac{1}{3}(-4\pi\delta^3(r)) \tag{3.41}$$

which leads us to the conclusion that

$$U = -S_{12}(\boldsymbol{\mu}_1, \boldsymbol{\mu}_2, \hat{\boldsymbol{r}}) + \frac{4\pi}{3}\boldsymbol{\mu}_1 \cdot \boldsymbol{\mu}_2\,\delta^3(r) \tag{3.42}$$

In practice it is usual to ignore the δ-function part of the interaction since the existence of mechanical constraints keeps the two magnetic dipoles well apart. The same thing happens in nuclei because of the short range repulsive core which is known from high energy scattering data to operate between nucleons (cf. section 3.4). Another example of a tensor force is the Van der Waals' force between two hydrogen atoms with relative coordinate $\boldsymbol{R}$ which may be derived from a potential of the form

$$V(\boldsymbol{R}, \boldsymbol{r}_1, \boldsymbol{r}_2) = -\left(\frac{e^2}{R^3}\right) S_{12}(\boldsymbol{r}_1, \boldsymbol{r}_2, \hat{\boldsymbol{R}}) \tag{3.43}$$

where $\boldsymbol{r}_1$, $\boldsymbol{r}_2$ are the position vectors of the two electrons relative to the mass centres of their respective atoms. Clearly the Van der Waals' force can be regarded as originating in the interaction between two electric dipoles $e\boldsymbol{r}_1$ and $e\boldsymbol{r}_2$ a distance R apart.

We conclude that the most general momentum-independent two-nucleon potential must have the form

$$V = V_c(r) + V_s(r)\boldsymbol{\sigma}_1 \cdot \boldsymbol{\sigma}_2 + V_T(r) S_{12}(\boldsymbol{\sigma}_1, \boldsymbol{\sigma}_2, \hat{\boldsymbol{r}}) \tag{3.44}$$

where $V_c(r)$ is a conventional central potential, the so-called Wigner potential which, when entered in the two body Schrödinger equation (2.32), yields energy eigenfunctions which are simultaneous eigenfunctions of the angular momentum operators l^2, l_z, s^2 and s_z. Furthermore since the spin–spin interaction operator $\boldsymbol{\sigma}_1 \cdot \boldsymbol{\sigma}_2$ commutes with all of these operators, having the eigenvalues -3 (1) for the singlet (triplet) two-nucleon spin-state, the term $V_s(r)\boldsymbol{\sigma}_1 \cdot \boldsymbol{\sigma}_2$ introduces no additional difficulty and requires only that different central potentials be used for the singlet and triplet states, i.e.

$$\begin{aligned} V_1(r) &= V_c(r) - 3V_s(r) \qquad \text{singlet } s=0 \\ V_3(r) &= V_c(r) + V_s(r) \qquad \text{triplet } s=1 \end{aligned} \tag{3.45}$$

Expressing this result in the language of matrices we can state that the potential has no off-diagonal elements in the basis of orbital angular momentum eigenstates provided that we use separate representations for singlet and triplet states.

These conclusions are no longer true when we take account of the tensor potential $V_T(r)S_{12}$ because, although S_{12} commutes with s^2, j^2 and j_z, it does not commute with l^2, l_z and s_z. Thus the Schrödinger equation with a tensor potential does not have energy eigenfunctions which are simultaneous eigenfunctions of l^2, l_z and s_z. For the singlet state this causes no problems since this is an eigenstate of S_{12} with eigenvalue zero and there is no tensor force. For a triplet state with spin $S=1$ and total angular momentum j the solutions of the Schrödinger equation with a tensor force are represented as admixtures of three eigenstates of l^2 with eigenvalues $l=j$, $j \pm 1$. In the case of the most important example, the deuteron, which has a measured spin

$I=1$ in its ground state, this state must be represented as a superposition of $s(l=0)$ and $d(l=2)$ states, the p-state ($l=1$) being excluded on the grounds that it has the wrong orbital parity.

It is known from experiment (Kellogg *et al* 1939, 1939a, 1940, Glendenning and Kramer 1962) that the deuteron has a small positive electric quadrupole moment $Q=2.875\times10^{-27}$ cm^2, which shows that the distribution of electric charge is not spherically symmetric but is slightly elongated in the direction of the angular momentum. Since $Q=\langle 3z^2-r^2\rangle$ determines the expectation value of $S_{\alpha\beta}$, the spin 2 component in the reduction of $r_\alpha r_\beta$, it must vanish identically in an s-state. Furthermore, the measured magnetic moment of the deuteron is about $2\frac{1}{2}\%$ less than the arithmetic sum of the neutron and proton magnetic moments and hence must contain some contribution from the orbital motion. Both these observations demonstrate conclusively that there must be some d-state admixed into the ground state of the deuteron, estimated at $\simeq 7\%$, and hence there must exist a tensor component to the nucleon–nucleon potential. However, because the tensor contribution is small and because its inclusion adds to the complexity of the formalism without altering the main conclusions, we shall as a general rule omit it in subsequent discussion of the problem.

3.3.2.2 Exchange forces

If we assume that the nuclear potential is just the sum of the potential energies associated with a short-range two body interaction between nucleons, and if in addition we assume that all nucleons interact with all other nucleons, then we must expect to have a nuclear radius comparable with the range of the two nucleon potential and a binding energy increasing with A approximately as A^2. As we have already emphasized in section 3.1, except for the very lightest nuclei this does not happen and the binding energy per nucleon is essentially constant right across the mass range, and the slight fall off at high mass numbers can be accounted for as a consequence of the increasing importance of the Coulomb repulsion between protons.

All of these observations suggest that the nuclear force saturates, i.e. two nucleons interact strongly only if they are in a relative s-state. Heisenberg (1932) proposed to explain these facts by postulating that there was a component in the nuclear force which arose from the exchange of electric charge between two nucleons, a process which one may visualize as being represented by a compound operator $\mathrm{P_H}$ given by

$$\mathrm{P_H}=\mathrm{P}_\sigma\mathrm{P_r} \tag{3.46}$$

where the Majorana operator $\mathrm{P_r}$ exchanges the positions of the two nucleons (Majorana 1933), i.e.

$$\mathrm{P_r}\psi(\boldsymbol{r}_1,\boldsymbol{r}_2)=\psi(\boldsymbol{r}_2,\boldsymbol{r}_1) \tag{3.47}$$

and the Bartlett operator P_σ exchanges their spins (Bartlett 1936), i.e.

$$P_\sigma \chi(s_1, s_2) = \chi(s_2, s_1) \tag{3.48}$$

Since the exchange of two particles in a coordinate system with origin at their centre of mass has the same meaning as changing their relative coordinate $\boldsymbol{r}$ to $-\boldsymbol{r}$, the exchange operator P_r is identical with the parity operator P. Also since, neglecting tensor forces, the orbital wavefunction $\psi(\boldsymbol{r})$ is an eigenfunction l^2, we have the result that

$$P_r \psi_l(\boldsymbol{r}_1, \boldsymbol{r}_2) = P\psi_l(\boldsymbol{r}) = (-1)^l \psi_l(\boldsymbol{r}) \tag{3.49}$$

If we have a system of two particles of the same spin s, coupled to a total spin S, the spin function χ_S is a spinor of order $2S$ which satisfies the exchange symmetry rule

$$P_\sigma \chi_S = (-1)^{2s+S} \chi_S \tag{3.50}$$

In the case of two nucleons for which $s = \frac{1}{2}$ we can construct either the symmetric triplet state

$$\begin{aligned} \chi_1^{-1} &= \chi_{1/2}^{-1/2}(1)\chi_{1/2}^{-1/2}(2) \\ \chi_1^0 &= \frac{1}{\sqrt{2}}(\chi_{1/2}^{-1/2}(1)\chi_{1/2}^{1/2}(2) + \chi_{1/2}^{1/2}(1)\chi_{1/2}^{-1/2}(2)) \\ \chi_1^1 &= \chi_{1/2}^{1/2}(1)\chi_{1/2}^{1/2}(2) \end{aligned} \tag{3.51}$$

or the anti-symmetric singlet state

$$\chi_0^0 = \frac{1}{\sqrt{2}}(\chi_{1/2}^{-1/2}(1)\chi_{1/2}^{1/2}(2) - \chi_{1/2}^{1/2}(1)\chi_{1/2}^{-1/2}(2)) \tag{3.52}$$

Thus to construct the operator P_σ which exchanges the spins of two-nucleons we require an operator whose eigenvalue is $+1$ for the triplet states $\chi_1^{-1,0,1}$ and -1 for the singlet state χ_0^0. Since the operator $\boldsymbol{\sigma}_1 \cdot \boldsymbol{\sigma}_2$ has eigenvalue -3 (1) for the singlet (triplet) state clearly the operator

$$P_\sigma = \frac{1}{2}(1 + \boldsymbol{\sigma}_1 \cdot \boldsymbol{\sigma}_2) \tag{3.53}$$

has the required property. The Heisenberg operator P_H operating on two nucleon states therefore produces the result

$$P_H \psi_l(\boldsymbol{r})\chi_S = P\psi_l(\boldsymbol{r}) P_\sigma \chi_S = (-1)^{l+S+1} \psi_l(\boldsymbol{r})\chi_S \tag{3.54}$$

We know from experiment that the deuteron has a bound state with $l=0$ and $S=1$ which means that, adding a proton (neutron) in a relative s-state to the one already present, the two protons (neutrons) must be in a singlet spin state to satisfy the requirements of the Pauli principle. Thus a Heisenberg potential $V_H(r)P_H$ which is attractive in the deuteron becomes repulsive for the singlet state interaction between like nucleons and the binding energy per nucleon in either ${}^3_1H$ or ${}^3_2He$ must be less than it is in ${}^2_1H$ which is assuredly not the case (cf. Table 3.1). It follows that the dominant exchange force cannot be of the Heisenberg type although there could indeed be a Heisenberg component associated with the tensor force which does not exist in the singlet state.

It is clear that a Bartlett force, which involves pure spin exchange only, does not discriminate between the relative orbital states of two nucleons and cannot therefore lead to saturation. However, if the exchange force is of Majorana type, independent of spin, then four nucleons all in central and relative s-states can bind together so that the α-particle, rather than the deuteron, becomes the saturated unit in agreement with observation. However, if we attempt to add a fifth nucleon to ${}^4_2He$, then this has to be inserted into a central p-orbital, and into a relative p-orbital with respect to the other four nucleons present. Under these circumstances the Majorana potential $V_m(r)P_r$ changes sign, and the binding energy is reduced, a result which tallies exactly with the observation that no stable nuclei exist with $A=5$. We conclude therefore that the dominant exchange potential must be of Majorana type.

Direct evidence for the presence of exchange forces is derived from the observed peaking in backward as well as forward directions in the scattering of neutrons from protons at energies above 100 MeV (Stahl and Ramsey 1954, Thresher *et al* 1955). Although the forward scattered peak is easily interpreted in terms of elastic and diffraction scattering of incident neutrons, the backward peaking can be explained only on the basis of a Majorana exchange force which interchanges the coordinates of neutrons and protons in the scattered wave. Broadly speaking, the results imply at least approximate equality of the direct Wigner potential $V_W(r)$ (Wigner 1933) and the Majorana potential $V_M(r)$, at least for energies below about 200 MeV, as originally suggested by Serber (1947). These observations however introduce new difficulties of interpretation, for a detailed calculation of the nuclear binding energy shows that it is not possible to ensure saturation and prevent the nucleus from collapsing, unless the space averaged Wigner and Majorana potentials V_W and V_M satisfy the inequality $V_M > 4V_W$ (Blatt and Weisskopf 1952). Since this requirement is evidently not fulfilled, it is necessary to postulate the existence of a repulsive core potential which is sufficiently long range to maintain the nucleus as a dilute system of constant density, and sufficiently short range that nucleons can move freely within the nucleus as is suggested by the overwhelming success of independent particle nuclear models. The effect on the wavefunction of a hard core contribution to the nuclear potential is illustrated in figure 3.4.

In fact the hypothesis of a repulsive hard core component in the nucleon–nucleon potential (Jastrow 1951, Bethe and Goldstone 1957) finds impressive support in the observation that proton–proton scattering in the

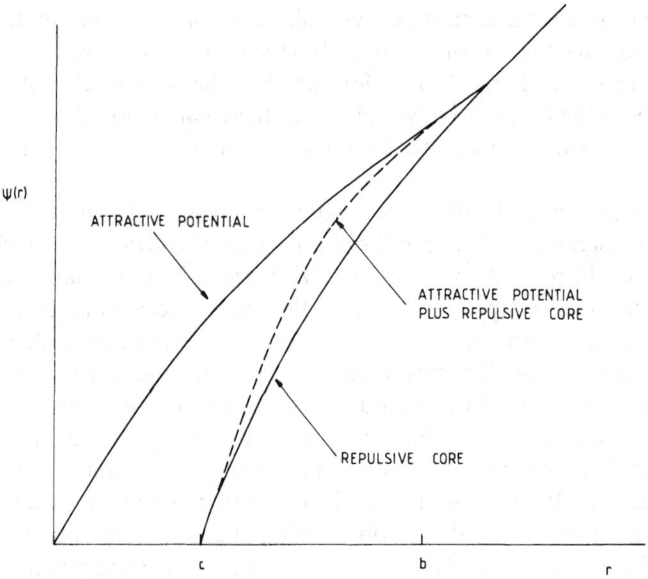

Figure 3.4. An illustration of the effect on the two-nucleon wavefunction $\psi(r)$ when the attractive potential of range b is supplemented by a repulsive hard-core potential of range $c<b$.

energy range 100–300 MeV, below the single pion production threshold at $\simeq 340$ MeV, is isotropic in the centre of mass (Hess 1958). Because in this case incident and target particles are identical, and the even (odd) valued partial waves are associated exclusively with singlet (triplet) spin states, the various partial waves each contribute to the scattering in sequence as the energy is progressively raised, and there is no interference between even and odd valued partial waves. Thus up to 2 MeV only ^{1}s-wave scattering contributes and the ^{3}p-wave component does not become effective until the energy is raised above 10 MeV. The ^{1}d-wave begins to influence the scattering at energies above 100 MeV, and ^{1}s–^{1}d interference terms become important at higher energies. Since these terms are positive at $\theta=0$ and negative at $\theta=\pi/2$ if the scattering potential is attractive throughout, the effect of s–d interference is to enhance the forward scattered component in direct conflict with observation. If however a short-range repulsive core is added to the singlet potential this will effect a change of sign in the s-wave phase shift at a relatively low energy, with the result that the s–d interference terms change sign resulting in an isotropic distribution as observed. The conclusion is that high energy unpolarized proton–proton scattering can be accounted for by a singlet potential with a long-range attractive component and a short-range repulsive core, together with a long-range attractive triplet potential containing central and tensor components. As we shall see presently, studies using polarized protons at energies above 300 MeV show that the triplet potential also has a short-range repulsive spin–orbit potential.

3.3.2.3 Momentum dependence: spin–orbit coupling

The momentum-dependent force with which we are most familiar is of course, the electromagnetic force, and this is obtained by making the 'minimal' substitution

$$p_\mu \rightarrow p_\mu - \frac{e}{c} A_\mu \tag{3.55}$$

in the charged particle Hamiltonian, where p_μ is the four-momentum of the particle and A_μ ($\equiv \boldsymbol{A}, i\varphi/c$) is the four-potential of the field. The electric and magnetic fields in which the particle moves are then given by the usual rules

$$\boldsymbol{E} = -\frac{1}{c}\frac{\partial}{\partial t}\boldsymbol{A} - \nabla\varphi \qquad \boldsymbol{B} = \nabla \times \boldsymbol{A} \tag{3.56}$$

The electromagnetic interaction is of course the most comprehensively studied example of an interaction which is mediated by the exchange of a single gauge boson, i.e. the photon A_μ, and interactions involving other gauge fields, such as the neutral and charged vector bosons which mediate the weak interaction, are introduced in a manner analogous to (3.55). Unfortunately the strong internucleon force cannot be described in this way.

Following the prescription (3.55), the Hamiltonian of a classical non-relativistic charged particle is given by

$$\begin{aligned} H &= e\varphi + \frac{1}{2m}\left(\boldsymbol{p} - e\frac{\boldsymbol{A}}{c}\right)^2 \\ &= e\varphi + \frac{p^2}{2m} + \frac{e^2 A^2}{2mc^2} - \frac{e}{2mc}(\boldsymbol{p}\cdot\boldsymbol{A} + \boldsymbol{A}\cdot\boldsymbol{p}) \end{aligned} \tag{3.57}$$

where the transition to quantum mechanics is obtained in the usual way by replacing $\boldsymbol{p}$ by the operator $-i\hbar\nabla$. Thus, in the presence of a non-vanishing vector potential, the static Hamiltonian is supplemented by terms which are linear in momentum. One should be aware however that the generalized momentum $\boldsymbol{p}$ is no longer identical with the kinetic momentum $m\dot{\boldsymbol{r}}$ and we have instead

$$\boldsymbol{p} = m\dot{\boldsymbol{r}} + e\frac{\boldsymbol{A}}{c} \tag{3.58}$$

Thus if we choose to define the potential energy as the difference between the total energy H and the kinetic energy $(\boldsymbol{p} - e(\boldsymbol{A}/c))^2/2m$, this potential energy is just $e\varphi(\boldsymbol{r})$ and is independent of momentum. These comments serve to emphasize that a momentum-dependent contribution of the form $\boldsymbol{p}\cdot\boldsymbol{A}$, which is the contraction of two Cartesian polar vectors, is a rather special case.

The Hamiltonian (3.57) describes the motion of a non-relativistic spinless particle and, for a spin $\frac{1}{2}$ electron, is supplemented by a spin-dependent contribution

$$H_\sigma = -\boldsymbol{\mu}_e\cdot\left[\boldsymbol{B}+\frac{\boldsymbol{E}\times\boldsymbol{p}}{mc}\right]+\boldsymbol{\mu}_e\cdot\left[\frac{\boldsymbol{E}\times\boldsymbol{p}}{2mc}\right] \tag{3.59}$$

where $\boldsymbol{\mu}_e=(e/mc)\boldsymbol{s}$ is the Dirac magnetic moment of the electron. H_σ is seen to be composed of three terms: (i) the term $-\boldsymbol{\mu}_e\cdot\boldsymbol{B}$ which gives the interaction between the electron magnetic moment and the magnetic field $\boldsymbol{B}$ as viewed in the laboratory frame of reference; (ii) the term $-\boldsymbol{\mu}_e[\boldsymbol{E}\times\boldsymbol{p}]/mc$ which describes the magnetic interaction with the additional magnetic field $\boldsymbol{E}\times\boldsymbol{p}/mc$ which the electron experiences in its own rest-frame; and (iii) the term $\boldsymbol{\mu}_e\cdot[\boldsymbol{E}\times\boldsymbol{p}]/2mc$ which is the so-called Thomas precession term which stems from the fact that the rest frame of an electron, accelerating in an electric field, is non-inertial (Thomas 1926). Nucleons moving in an electromagnetic field will also experience spin-dependent forces similar to (i) and (ii) above except that, in the case of the proton, its Dirac magnetic moment is supplemented by a Pauli moment associated with its internal electromagnetic structure. The neutron, being uncharged, has a Pauli moment only.

It is important to appreciate that the Thomas precession term (iii) has nothing to do with electromagnetism and is purely of kinetic origin. It arises from the fact that the application of successive Lorentz transformations to a frame of reference, i.e. the electron's, which are not along the same spatial axis, does not result in a single Lorentz transformation but includes a net rotation of the coordinate system which is observable as a counter rotation of the spin. It will therefore contribute to the Hamiltonian of any spin $\frac{1}{2}$ particle, moving in an arbitrary central potential $V(r)$, a term

$$H_T=\frac{-\hbar}{(2mc)^2}\boldsymbol{\sigma}\cdot(\nabla V\times\boldsymbol{p})=\frac{-\hbar}{(2mc)^2}\left(\frac{1}{r}\frac{\mathrm{d}V}{\mathrm{d}r}\right)\boldsymbol{\sigma}\cdot\boldsymbol{r}\times\boldsymbol{p} \tag{3.60}$$

which has the form of a spin–orbit coupling obtained by contracting the axial vectors $\boldsymbol{\sigma}$ and $\boldsymbol{l}=\boldsymbol{r}\times\boldsymbol{p}$. Such a term therefore appears when a proton, but not a neutron, moves in a central electric field. Of course both protons and neutrons will experience a magnetic spin–orbit coupling

$$H_M=\frac{g}{2}\left(\frac{e}{m_p^2c^2}\right)\left(\frac{1}{r}\frac{\partial\varphi}{\partial r}\right)\boldsymbol{s}\cdot\boldsymbol{l} \tag{3.61}$$

where the Landé factor g has the value 5.59 for a proton and -3.83 for a neutron.

There is another way of looking at the question of momentum dependence which is suggested by the fact that, when we permit the potential energy of a particle to depend on its momentum, we are stating in effect that its energy depends on the value of its wavefunction not only at its own position $\boldsymbol{r}$, but at other points $\boldsymbol{r}' \neq \boldsymbol{r}$. The Schrödinger equation describing the motion of a particle in such a non-local potential $V(\boldsymbol{r}, \boldsymbol{r}')$ becomes therefore

$$\frac{-\hbar^2}{2m}\nabla^2\psi(\boldsymbol{r}) + \int d^3r'\, V(\boldsymbol{r}, \boldsymbol{r}')\psi(\boldsymbol{r}') = E\psi(\boldsymbol{r}) \tag{3.62}$$

To see how the momentum dependence emerges we may suppose that the particle is moving with momentum $\boldsymbol{p}$ in an infinite Fermi gas where the potential well $-V_0$ is supplemented by a small non-local potential $V_{nl}(\boldsymbol{r}, \boldsymbol{r}')$ $(\boldsymbol{r}' \neq \boldsymbol{r})$, i.e.

$$V(\boldsymbol{r}, \boldsymbol{r}') = -V_0\delta^3(\boldsymbol{r}' - \boldsymbol{r}) + V_{nl}(\boldsymbol{r}, \boldsymbol{r}') \tag{3.63}$$

Thus, inserting the wavefunction $\psi(\boldsymbol{r}) = A\, e^{i\boldsymbol{p}\cdot\boldsymbol{r}/\hbar}$ in (3.62) we find an expression for the total energy

$$E = \frac{p^2}{2m} - V_0 + \int d^3r'\, V_{nl}(\boldsymbol{r}, \boldsymbol{r}')\, e^{i\boldsymbol{p}\cdot(\boldsymbol{r}' - \boldsymbol{r})/\hbar} \tag{3.64}$$

In an infinite Fermi gas we may assume that translational invariance is satisfied, i.e.

$$V_{nl}(\boldsymbol{r}, \boldsymbol{r}') = V_{nl}(|\boldsymbol{r} - \boldsymbol{r}'|) \tag{3.65}$$

in which case the momentum-dependent contribution to the energy is just the Fourier transform of the non-local potential which we may write as

$$\int d^3r'\, V_{nl}(|\boldsymbol{r} - \boldsymbol{r}'|)\, e^{i\boldsymbol{p}\cdot(\boldsymbol{r} - \boldsymbol{r}')/\hbar} \equiv V_{nl}(p) = a_2p^2 + a_4p^4 + \ldots \tag{3.66}$$

Thus to lowest order in p^2 the energy momentum relation is

$$E = \frac{p^2}{2m} - V_0 + a_2p^2 = \frac{p^2}{2m^*} - V_0 \tag{3.67}$$

where m^* is the effective mass defined by

$$\frac{1}{m^*} = \frac{1}{m} + 2a_2 \tag{3.68}$$

We may conclude therefore that the principal effect which arises when a non-local potential is introduced may be expressed as a renormalization of the nucleon mass arising from its interaction with other nucleons. It is of course the action of the Pauli exclusion principle which justifies the use of plane-wave states even though the nucleons experience strong interactions with other nucleons, and this cannot continue to hold good in the immediate neighbourhood of a collision. However, detailed analysis of two-particle scattering in nuclear matter shows that the plane-wave approximation is justified unless the nucleons come within what has been termed the 'healing distance' which is estimated to have a value of about 10^{-13} cm (Gomes *et al* 1958). For a nucleon moving in a finite nucleus one may of course expect a more complicated momentum dependence.

In general if we allow the internucleon potential to depend explicitly on momentum we have at our disposal one new polar vector $\boldsymbol{p}$ which provides us with two new scalar invariants p^2 and $\boldsymbol{r}\cdot\boldsymbol{p}$ which may appear as arguments in the potential function. We must however immediately discard terms which are linear in $\boldsymbol{r}\cdot\boldsymbol{p}$ because the vector $\boldsymbol{r}$, unlike the vectors $\boldsymbol{p}$ and the electromagnetic vector potential $\boldsymbol{A}(\boldsymbol{r})$, does not change sign under the operation of time reversal. Thus terms linear in $\boldsymbol{r}\cdot\boldsymbol{p}$ are not time reversal invariant and are therefore forbidden. In addition, however, we can construct the Cartesian tensor $r_\alpha p_\beta$, whose properties we have already summarized in section 3.3.1 and this tensor provides us with a new axial vector $\boldsymbol{l}=\boldsymbol{r}\times\boldsymbol{p}$ which may be contracted with the symmetric spin axial vector $\boldsymbol{\sigma}_1+\boldsymbol{\sigma}_2$ to give a spin–orbit interaction which is the only surviving term linear in the components of $\boldsymbol{p}$ which is consistent with a charge symmetric interaction (cf. section 3.5.1). There are of course quadratic spin–orbit terms which have this property, e.g. $(\boldsymbol{\sigma}_1\cdot\boldsymbol{\sigma}_2)l^2$, $(\boldsymbol{\sigma}_1\cdot\boldsymbol{l})(\boldsymbol{\sigma}_2\cdot\boldsymbol{l})$ (Okubo and Marshak 1958) and such terms play a role in certain phenomenological nucleon potentials such as the Yale potential (Lassila *et al* 1962) or the Hamada–Johnston potential (Hamada and Johnston 1962).

The spin–orbit coupling operator $\boldsymbol{l}\cdot\boldsymbol{s}$ commutes with $\boldsymbol{l}^2$, $\boldsymbol{s}^2$, $\boldsymbol{j}^2$ and j_z and has the eigenvalue $l/2$, $(-(l+1)/2)$ for the state with $j=l+\frac{1}{2}$, $(l-\frac{1}{2})$. It does not commute with l_z or s_z and clearly its expectation value vanishes in s-states. In atoms the spin–orbit potential is positive which means that the substate with $j=l+\frac{1}{2}$ has the higher energy. It is of course well known that the original authors of the nuclear shell model had proposed to introduce an inverted spin–orbit potential to explain the existence of magic numbers, a possibility which had been subjected to theoretical study more than a decade previously (Breit 1937, 1938). That the observed splitting is indeed due to an inverted spin–orbit coupling was shown by Adair (1952) who studied the angular distribution of neutrons in the energy range 0–20 MeV scattered by ${}^4_2$He and showed that the scattering cross-section was dominated by a strong $p_{3/2}$-wave resonance near 1 MeV with the contribution from the $p_{1/2}$-wave making a significant contribution to the total cross-section at energies of order 5 MeV and upwards. Since ${}^4_2$He is a spinless nucleus the only way in which a p-wave resonant state can be split in an independent particle model is when the scattering potential contains a spin–orbit term.

Confirmation of the existence of a spin–orbit term of the correct sign and magnitude in the nucleon–nucleon potential emerges from studies of polarized proton–proton scattering at high energy where the measured p-wave phase shifts do not agree with predictions based on central, tensor and hard core potentials acting alone (Stapp *et al* 1957). However, the data can be accounted for on the assumption that a long-range attractive tensor potential exists in the $^3P_0$ scattering state and is supplemented by a short-range repulsive spin–orbit potential with the properties required by the shell model (Gammel and Thaler 1957, Signell and Marshak 1957, 1958), and calculations based on the Hamada–Johnston phenomenological nucleon–nucleon potential correctly predict both the sign and the magnitude of the doublet splitting in the doubly magic nuclei $^{16}_{8}O$ and $^{40}_{20}Ca$ (Landé and Svenne 1969).

3.4 Nuclear isospin

3.4.1 Charge symmetry and independence

The charge symmetric property of nuclear forces was first recognized by Heisenberg (1932) and, in its simplest form, requires that the neutron–neutron force be the same as the proton–proton force, provided both pairs of nucleons are in the same spin and orbital states. The main evidence in favour of the principle of charge symmetry as it operates in nuclear physics derives from the observation noted previously (section 3.2), that the values of nuclear radii deduced from measured mass differences between mirror pairs on the basis of charge symmetry, agrees with those obtained from electron scattering distributions, or from muonic atom energy level shifts, neither of which are sensitive to the details of the nuclear force.

The example which has been studied in the greatest detail is the mirror pair (^3_2He, ^3_1H) for which the measured Coulomb energy difference determined from the mass difference according to (3.13) is

$$\begin{aligned}\Delta E_c &= (m_n - m_p - m_e)c^2 - (M(3,1) - M^*(3,2) - m_e)c^2 \\ &= (1294 - 511) - (18.6) = 764.4\ keV \qquad (3.69)\end{aligned}$$

where $M^*(3, 2)$ is the mass of the singly ionized ^3He atom. This value is to be compared with that obtained by a direct calculation of the electromagnetic contribution to the binding energy, carried out using the most accurate nuclear wavefunctions, and including electromagnetic spin–orbit coupling and the neutron–proton mass difference correction to the kinetic energy, which gives the result $\Delta E_c = 638 \pm 29$ keV (Brandenburg *et al* 1978). This conclusion is supported by the results of a more recent calculation which arrives at the value $\Delta E_c = 648 \pm 4$ keV for the Coulomb energy difference (Ishikawa and Sasakawa 1990). For heavier nuclei also there is some evidence that the calculated energy differences are systematically too small (Nolen and Schiffer 1969); one might therefore conclude that the evidence from nuclear physics is that the charge symmetry principle, although very closely obeyed, is not exact.

Since the Pauli principle rules out the s-wave triplet state for two identical nucleons, and we know from experiment that neither $^2_2$He nor the di-neutron exist as bound states (Cohen and Handley 1953), the only way in which we test the charge symmetry hypothesis at the most fundamental level is by studying the scattering of identical nucleons in the singlet state. Since neutron targets do not exist in the laboratory, proton–proton scattering is a unique process whose fundamental importance as a probe of the nuclear force was appreciated a very long time ago (Breit *et al* 1936). The main problem with interpreting the results of low-energy proton–proton scattering is the difficulty of dealing with the Coulomb potential which contributes both directly and by interference with the nuclear scattering (cf. section 3.2). The measured scattering length corrected for Coulomb effect is $a_{pp} = -(17.9 \pm 0.45)$ fm (Ericson and Miller 1983), where the negative sign confirms that, as with the singlet state of the deuteron, the ^1S state of two protons is a virtual state.

Although studies have been undertaken at various times of the conditions necessary for carrying out a successful neutron–neutron scattering experiment (Muehlhause 1963, West 1967, Dickinson *et al* 1970), the neutron–neutron scattering length has not yet been directly determined. This is unfortunate because the interaction takes place entirely in the singlet state and no corrections are required for the long-range Coulomb potential which complicates the interpretation of proton–proton scattering. However the neutron–neutron scattering length can be determined indirectly by studying nuclear reactions leading to the production of two neutrons which then interact in the final state. Examples of such reactions are d(n, p)2n, $^3_1$H($^3_1$H, α)2n and d(π^-, γ)2n, the last named reaction yielding the result $a_{nn} = -(18.45 \pm 0.45)$ fm (Slaus 1982). This result confirms that the di-neutron is unbound in agreement with experiment and shows that there is no direct evidence from the two-nucleon system of any violation of the charge symmetry principle.

The principle of charge independence represents an extension of the charge symmetry idea (Cassen and Condon 1936, Breit and Feenberg 1936) which postulates that all three force pairs, n–n, n–p, p–p are the same in given spin and orbital states. If the principle is valid we may expect to observe in analogy with the mirror pairs, trios of even-A isobars which share a common level corresponding to two nucleons in the combination (n, n), (n, p), (p, p) outside a common even–even core. A good example of such a system is provided by the $A = 12$ triplet $^{12}_5$B, $^{12}_6$C, $^{12}_7$N whose relevant properties are summarized in the energy level diagram in figure 3.5. The nucleus $^{12}_6$C is a tightly bound self-conjugate nucleus of the α-particle variety which contributes a low-mass peak in the binding energy curve in figure 3.2 and, according to shell-model assignments, has a ground state configuration $(1s)^2(^1p_{3/2})^4_p$; $(1s)^2(^1p_{3/2})^4_n$. The nucleus $^{12}_5$B($^{12}_7$N) has instead an additional neutron (proton) in the $^1p_{1/2}$ orbital with an unfilled proton (neutron) hole in the $^1p_{3/2}$ orbital. Thus in both $^{12}_5$B and $^{12}_7$N we have two nucleons, one

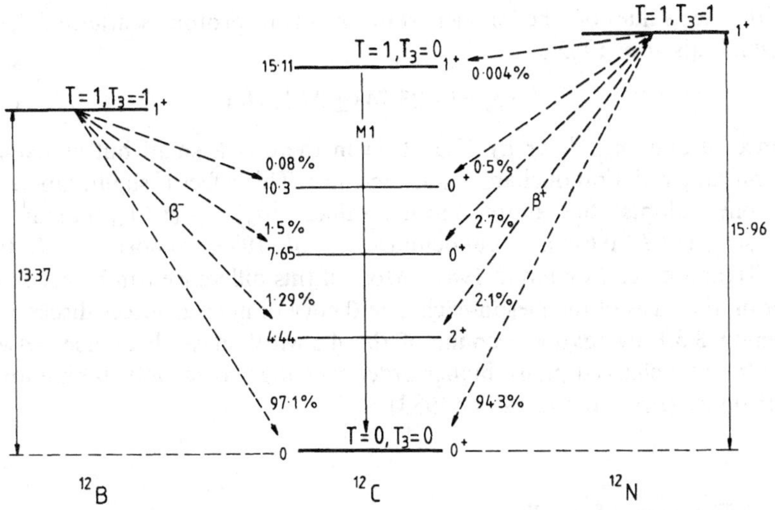

Figure 3.5. The triad of $A=12$ nuclei which constitute an isospin triplet with $T=1$, ^{12}B ($T_3=-1$), ^{12}C* ($T_3=0$) and ^{12}N ($T_3=1$). The ground state of ^{12}C is an isosinglet with $T=0$. The β and γ-transitions within this triad have also been studied in the search for 'weak magnetism' in nuclear β-decay (cf. section 9.4.8) (Calaprice and Holstein 1976).

neutron and one proton in $^1\mathrm{p}_{1/2,3/2}$ orbitals outside a closed core, i.e. in a relative s-state with anti-parallel spins. Clearly the analogue state in $^{12}_{6}$C cannot be the ground state but rather the first excited state at 15.11 MeV which presumably is a mixture of configurations with one neutron (or proton) promoted to the $^1\mathrm{p}_{1/2}$ orbital coupling with the unpaired neutron (or proton) in the $^1\mathrm{p}_{3/2}$ orbital. The energy differences

$$[M^*(12,6)-M(12,5)]c^2=1.7\ \mathrm{MeV}$$
$$[M(12,7)-M^*(12,6)]c^2=2.6\ \mathrm{MeV} \qquad (3.70)$$

are in clear qualitative agreement with the trend suggested by the semi-classical formula (3.12) and may be accounted for in detail as due to the difference in Coulomb energy for the individual pairs (Wilkinson 1956).

More recent analyses of the energy differences between analogue states of nuclei in the $1\mathrm{d}_{3/2}$ and $1\mathrm{f}_{7/2}$ shells (Sherr and Talmi 1975) has revealed that these differences can be accounted for in terms of Coulomb energy only if the nuclear potential is assumed to be more attractive in the n–p than in the n–n or p–p states, and this result is confirmed by comparing the value of the mean scattering length for identical nucleons (Ericson and Millar 1983),

$$^1\bar{a}=\tfrac{1}{2}(^1a_{\mathrm{nn}}+{}^1a_{\mathrm{pp}})=-(18.2\pm0.3)\ \mathrm{fm} \qquad (3.71)$$

with the value of the singlet state neutron–proton scattering length (Dumbrajs *et al* 1983).

$$^1a_{np} = (-23.748 \pm 0.015)\ \text{fm} \tag{3.72}$$

Since $^1\bar{a}$ and $^1a_{np}$ differ by (5.5 ± 0.3) fm there is a small but statistically significant violation of change independence. This effect is slightly increased if one adopts the more recent values $^1a_{pp} = -(17.1 \pm 0.2)$ fm and $^1a_{nn} = -(18.7 \pm 0.6)$ fm (Brandenburg *et al* 1988, Schori *et al* 1987, de Teramond and Gabioud 1987). Most of this difference can be accounted for on the basis of the meson-exchange theory of nuclear forces discussed in section 3.5.1 by taking account of the 4.5 MeV mass difference between neutral and charged pions, higher order exchange corrections taking up the remainder (Ericson and Miller 1983).

3.4.2 The isospin formalism

The observational fact of charge independence of nuclear forces is expressed in the theoretical notion, originally advanced by Heisenberg (1932), that proton and neutron should be viewed collectively as two substates of the one elementary particle, the nucleon. The dichotomic variable associated with these two nucleon substates was originally given the name 'isotopic spin' (Wigner 1937, 1939), in analogy with the two magnetic substates of a spin $\frac{1}{2}$ particle, although the more appropriate description 'isobaric' spin later became popular reflecting the fact that neutrons and protons are isobars rather than isotopes (Inglis 1953). In recent usage both these terms have been abandoned and the more economic terminology 'isospin' is employed. Thus in the same way as a fermion with spin $s = \frac{1}{2}$, has two magnetic substates with $s_z = \pm\frac{1}{2}$, the nucleon is assigned an isospin quantum number $t = \frac{1}{2}$, with components $t_3 = \frac{1}{2}$ representing the proton, and $t_3 = -\frac{1}{2}$ representing the neutron. The corresponding isospin wavefunctions are the two component isospinors

$$\varphi_p = \begin{pmatrix} 1 \\ 0 \end{pmatrix} \qquad \varphi_n = \begin{pmatrix} 0 \\ 1 \end{pmatrix} \tag{3.73}$$

which diagonalize the third component of isospin $t_3 = \frac{1}{2}\tau_3$ where τ_3 is the Hermitian matrix

$$\tau_3 = \begin{pmatrix} 1 & 0 \\ 0 & -1 \end{pmatrix} \tag{3.74}$$

In the representations (3.73–3.74) the nucleon charge operator Q is given by

$$Q = \tfrac{1}{2}(1 + \tau_3) \tag{3.75}$$

and the eigenvalues $Q=1, 0$ give the nucleon charge in the substates φ_p, φ_n respectively, expressed in units of the elementary charge $+e$. Since the charge is absolutely conserved the operator τ_3 is conserved by the strong and electromagnetic interactions. For the same reason no meaning can be attached to isospin states which are coherent superposition states of the basis states φ_p and φ_n, and this is one feature in which isospin is very different from spin. In the weak interactions which play a major role in nuclear physics, i.e. neutron and (bound) proton β-decay, charge is transferred from the nucleon to the lepton sector in units $\Delta Q = \pm 1$, so that τ_3 is no longer conserved. Thus to include these weak interactions within the isospin formalism it is necessary to construct operators which annihilate neutrons (protons) and create protons (neutrons). In the basis (3.73) such operators are given by

$$\tau_+ = \tfrac{1}{2}(\tau_1 + i\tau_2) = \begin{pmatrix} 0 & 1 \\ 0 & 0 \end{pmatrix} \qquad \tau_- = \tfrac{1}{2}(\tau_1 - i\tau_2) = \begin{pmatrix} 0 & 0 \\ 1 & 0 \end{pmatrix} \tag{3.76}$$

which satisfy the relations

$$\tau_+\varphi_n = \varphi_p \qquad \tau_+\varphi_p = 0 \qquad \tau_-\varphi_p = \varphi_n \qquad \tau_-\varphi_n = 0 \tag{3.77}$$

where τ_1 and τ_2 are Hermitian matrices defined by

$$\tau_1 = (\tau_+ + \tau_-) = \begin{pmatrix} 0 & 1 \\ 1 & 0 \end{pmatrix} \qquad \tau_2 = -i(\tau_+ - \tau_-) = \begin{pmatrix} 0 & -i \\ i & 0 \end{pmatrix} \tag{3.78}$$

Since the matrix vector $\boldsymbol{\tau} = (\tau_1, \tau_2, \tau_3)$, which operates in the complex two dimensional space spanned by φ_p and φ_n, is formally identical with the Pauli spin matrix vector $\boldsymbol{\sigma} = (\sigma_x, \sigma_y, \sigma_z)$, and its components satisfy the same commutation relations

$$\boldsymbol{\tau} \times \boldsymbol{\tau} = 2i\boldsymbol{\tau} \tag{3.79}$$

the algebra of isospin is identical with the algebra of spin in every last detail. Thus just as the group SU(2) of unitary unimodular transformations in the complex two-dimensional spin-space of a spin $\frac{1}{2}$ particle is isomorphic on the group R_3 of rotations in three dimensional position space, so also unitary transformations on elementary isospinors may be visualized as rotations in a real three dimensional isospin space.

An important unitary operator is the operator $i\tau_2$ which generates the transformations

$$i\tau_2\varphi_p = -\varphi_n \qquad i\tau_2\varphi_n = +\varphi_p \tag{3.80}$$

and therefore transforms a proton state into a neutron state and conversely. It is therefore called the charge symmetry operator and is represented by the symbol P_{cs}. In isospin space the charge symmetry operator may be viewed as a rotation through an angle π about the 2-axis and its action may be described in the equivalent forms

$$P_{cs} = e^{i\pi\tau_2} = \cos\left(\frac{\pi\tau_2}{2}\right) + i\sin\left(\frac{\pi\tau_2}{2}\right) = \tau_0 \cos\frac{\pi}{2} + i\tau_2 \sin\frac{\pi}{2} = i\tau_2 \quad (3.81)$$

where τ_0 denotes the 2×2 unit matrix. The choice of the 2-axis in isospin space as the axis of rotation is conventional and any other axis in the 1–2 plane would be just as suitable. The essential point is that all such transformations exhibit the relative phase difference of -1 shown in (3.80) reflecting the fact that a double application of the transformation introduces the factor (-1) which is a characteristic feature of spinor rotations.

All of these results can be generalized to many-nucleon systems by adding isospins together according to the usual rules of angular momentum algebra leading to higher dimensional representations of the rotation group in isospin space. Thus in analogy with the scalar, spinor, vector and tensor representations in position space we have isoscalar, isospinor, etc. representations in isospin space. Obviously the most important system we have to consider is the two-nucleon system and the most important problem we have to deal with is how to express the ideas of charge symmetry and charge independence in the language of isospin.

Suppose therefore that we have two nucleons with isospin vectors $\boldsymbol{t}(1)$ and $\boldsymbol{t}(2)$, then the two nucleon system will have total isospin $\boldsymbol{T}$ where

$$\boldsymbol{T} = \boldsymbol{t}(1) + \boldsymbol{t}(2) \quad (3.82)$$

which, according to the rules of angular momentum algebra, means that the eigenvalues $T(T+1)$ and T_3 of the operators $\boldsymbol{T}^2$ and T_3 are restricted by

$$|t(1) - t(2)| \leqslant T \leqslant t(1) + t(2) \qquad T_3 = t_3(1) + t_3(2) \quad (3.83)$$

Thus, since $t(1) = t(2) = \frac{1}{2}$, the two-nucleon states can be resolved into an isoscalar with $T = T_3 = 0$ and an isovector with $T = 1$, $T_3 = 0 \pm 1$. In analogy with (3.51–3.52) the corresponding isosinglet and isotriplet wavefunctions are

$$\left.\begin{array}{ll} \varphi_n(1)\varphi_n(2) & T_3 = -1 \\ \frac{1}{\sqrt{2}}(\varphi_p(1)\varphi_n(2) + \varphi_p(2)\varphi_n(1)) & T_3 = 0 \\ \varphi_p(1)\varphi_p(2) & T_3 = 1 \end{array}\right\} \quad T = 1$$

$$\frac{1}{\sqrt{2}}(\varphi_p(1)\varphi_n(2) - \varphi_p(2)\varphi_n(1)) \qquad T_3 = 0 \qquad T = 0 \quad (3.84)$$

The isotriplet state $(T=1)$ is symmetric and the isosinglet state $(T=0)$ anti-symmetric under the action of the isospin exchange operator P_τ which is defined in analogy with (3.53) by

$$P_\tau = \tfrac{1}{2}(1 + \boldsymbol{\tau}(1)\cdot\boldsymbol{\tau}(2)) \quad (3.85)$$

It is very important to distinguish between the isospin exchange operator P_τ and the charge symmetry operator P_{cs} and to note in particular that, although the isotriplet (isosinglet) substate with $T_3=0$ is an eigenstate of P_τ with eigenvalue $+1(-1)$, it is simultaneously an eigenstate of P_{cs} with eigenvalue $-1(+1)$.

There are corresponding generalizations for a mass A nucleus of the charge and charge symmetry operators introduced in (3.76) and (3.81) and these operators are defined in general by

$$Q=\sum_{i=1}^{A}\frac{1}{2}(1+\tau_3(i))=\frac{A}{2}+T_3 \tag{3.86}$$

and

$$P_{cs}=\exp\left[\frac{i\pi}{2}\sum_{i=1}^{A}\tau_2(i)\right]=e^{i\pi T_2} \tag{3.87}$$

In general P_{cs} does not commute with Q except in the special case of a self-conjugate nucleus for which $T_3=0$. In this case it is possible to construct simultaneous eigenstates of Q and P_{cs} and, for a state of definite T, the eigenvalue of P_{cs} is $(-1)^T$. We have already seen immediately above an example of this rule operating on the states of a two-nucleon system for which $T_3=0$.

In seeking a means of expressing the idea of charge symmetry in the language of isospin we must be aware that there is a weak and a strong version of this principle (Heller 1967). The weak principle is that stated earlier, namely that, in equivalent states, neutron–neutron and proton–proton forces are equal, and clearly this version has no implications for the neutron–proton force. The strong principle (which incorporates the weak principle) postulates that the nucleon–nucleon force is invariant with respect to the simultaneous exchange of all neutrons and protons, i.e. that the nucleon–nucleon interaction commutes with the charge symmetry operator P_{cs}. We shall adopt the strong form of the charge symmetry hypothesis and indeed have already done so implicitly in section 3.3 where we accepted terms in the nuclear potential such as $(\boldsymbol{r}\cdot\boldsymbol{\sigma}_1)(\boldsymbol{r}\cdot\boldsymbol{\sigma}_2)$, $\boldsymbol{l}\cdot(\boldsymbol{\sigma}_1\times\boldsymbol{\sigma}_2)$ which commute with P_{cs}, while rejecting terms such as $\boldsymbol{l}\cdot(\boldsymbol{\sigma}_1-\boldsymbol{\sigma}_2)$ or $\boldsymbol{l}\cdot\boldsymbol{\sigma}_1\times\boldsymbol{\sigma}_2$ which anti-commute, even though they conserve angular momentum and parity.

Hence, if we represent the state of two protons in the isospin formalism by

$$\Psi_{pp}(\boldsymbol{r}_1,\boldsymbol{\sigma}_1;\boldsymbol{r}_2,\boldsymbol{\sigma}_2)=\Psi(1,2)\varphi_p(1)\varphi_p(2) \tag{3.88}$$

the strong charge symmetry principle states that the corresponding two-neutron system is

$$\begin{aligned}\Psi_{nn}(\boldsymbol{r}_1,\boldsymbol{\sigma}_1;\boldsymbol{r}_2,\boldsymbol{\sigma}_2)&=P_{cs}\Psi_{pp}(\boldsymbol{r}_1,\boldsymbol{\sigma}_1;\boldsymbol{r}_2,\boldsymbol{\sigma}_2)\\&=\Psi(1,2)[-\varphi_n(1)][-\varphi_n(2)]=\Psi(1,2)\varphi_n(1)\varphi_n(2)\end{aligned} \tag{3.89}$$

Since two-proton and two-neutron states are both restricted by the Pauli principle, i.e.

$$\Psi(2,1)=P_\sigma P_r \Psi(1,2)=-\Psi(1,2) \tag{3.90}$$

and since both the isospin states $\varphi_p(1)\varphi_p(2)$ and $\varphi_n(1)\varphi_n(2)$ are symmetric with respect to exchange of the isospin coordinates, it follows that the principle of charge symmetry applied to systems of identical nucleons can be expressed in the form of a generalized Pauli principle

$$P_\sigma P_r P_\tau = -1 \tag{3.91}$$

The state $\Psi_{np}(\boldsymbol{r}_1, \boldsymbol{\sigma}_1; \boldsymbol{r}_2, \boldsymbol{\sigma}_2)$ of two non-identical nucleons is not subject to the Pauli principle and need not have any definite symmetry with respect to exchange of space and spin coordinates. However, if we assume charge symmetry, so that the Hamiltonian commutes with P_{cs}, then, since $T_3=0$, $\Psi_{np}(1,2)$ must be an eigenstate of P_{cs} with eigenvalue $(-1)^T$, and a simultaneous eigenstate of P_τ with eigenvalue $(-1)(-1)^T$. Since in the conventional (non-isospin) formulation the mutual conversion of neutrons into protons is accomplished by the Heisenberg operator $P_H=P_\sigma P_r$ (cf. section 3.3), it follows that for the exchange of neutrons and protons the generalized Pauli principle (3.91) also holds. For a system of two nucleons in a state characterized by definite values of l and s, this rule implies that $l+s+T$ is an odd number. We conclude that the ground state of the deuteron is an isosinglet with $T=T_3=0$ while the virtual state is the $T_3=0$ component of an isotriplet with $T=1$, whose companion states with $T_3=\pm 1$, i.e. the di-proton ${}^2_2$He and the di-neutron, are also unbound.

We can therefore summarize the principle of charge symmetry by the statement that the internucleon force is invariant with respect to rotations about the 2-axis in isospin space. This requires that, ignoring electromagnetic interactions, nuclear states should be characterized by even or odd values of T; it does not require T to be defined uniquely although for given $T_3=\frac{1}{2}(Z-N)$, we must have $T \geqslant T_3$. The principle of charge independence removes the special status of the 3-axis in isospin space by stating that the internucleon force is invariant with respect to all rotations in isospin space. Thus, in analogy with angular momentum conservation, all nuclear states are characterized by definite values of T and T_3, and the freedom allowed by charge symmetry for T to be merely even or odd is suppressed. In the limit of exact charge independence, states of the same angular momentum and parity and given T, and approximately of the same energy, can occur in a range of isobars defined by $-T \leqslant \frac{1}{2}(Z-N) \leqslant T$, and these states constitute the members of an isospin multiplet. To completely describe such states, it is necessary to specify, in addition to I and T, both the seniority quantum number v which, in the isospin formalism, represents the number of nucleons in a level after the removal of all saturated pairs, and the reduced isospin T' which is the isospin of the remaining v nucleons. This is because the addition of a saturated pair to a state of the configuration $(nlj)^v$ does not change the spin value I, but may alter the isospin value T' to T, because a saturated pair has $I=0$ but $T=1$.

Since according to the definition of the nuclear charge operator Q (3.87) the Coulomb contribution to the nuclear Hamiltonian is

$$V_Q = \sum_{i<j} \frac{e^2}{r_{ij}} [\tfrac{1}{2}(1+\tau_3(i))\tfrac{1}{2}(1+\tau_3(j))] \tag{3.92}$$

and V_Q does not commute with $\boldsymbol{T}^2$, T can be regarded as a good quantum number only so long as V_Q is a small correction to the Hamiltonian whose contribution to the total energy may be evaluated by first-order perturbation theory (Radicati 1953, 1954). This proviso is in line with the fundamental assumption of the theory, namely that isospin symmetry should be regarded as exact, except that mass and energy differences should be retained. In the neutron–proton doublet which is the simplest isospin system, there is an intrinsic mass difference whose origin has been assumed traditionally to be electromagnetic in origin although, with the arrival of quark theories, it now seems less likely that the total effect is electromagnetic, and it is necessary to take account of the neutron–proton mass difference to maintain energy conservation and to provide the necessary phase space in nuclear β-decay.

As a general rule isospin is a good quantum number for light nuclei with $A<50$, where neutrons and protons fill states within the same (nlj) shell. At higher values of A the Coulomb repulsion between protons becomes so large that not only do neutron and proton states have different energies but they arise from different configurations. In this case T is no longer a good quantum number and the practical advantages of the isospin concept in nuclear physics are less evident.

For very light nuclei circumstances may arise in which both Coulomb and spin-dependent forces may be neglected in first order so that only Wigner and Majorana exchange forces are operative (Wigner 1937, Barkas 1939, Wigner and Feenberg 1942). In this case the nuclear Hamiltonian is invariant with respect to the interchange of spin and isospin variables of all nucleons, with the result that a state specified by its spin S (not its total angular momentum I), and its isospin T, is degenerate to first order with a state specified by spin T and isospin S. In this case the two isospin multiplets involved are partners in a larger isospin super-multiplet. In section 3.1 it was remarked that all four stable odd–odd nuclei which occur naturally are self-conjugate nuclei with $T_3=0$; it is perhaps also significant that all these nuclei are members of supermultiplets containing isobaric stable even–even nuclei. For example the $S=1$, $T=0$ isosinglet ground state of ${}^6_3$Li is a member of a supermultiplet along with the $S=0$, $T=1$ isotriplet ${}^6_2$He ($T_3=1$), ${}^6_3$Li* ($T_3=0$) and ${}^6_4$Be ($T_3=1$). Even in this company the deuteron is an exception for its accompanying isotriplet, the di-neutron, the virtual state of the deuteron and the di-proton are all unbound. We return to this matter again in connection with the discussion of the one-pion exchange potential in section 3.5.4.

When we apply the isospin formalism to the trio of nuclei ${}^{12}_{5}B$, ${}^{12}_{6}C$, ${}^{12}_{6}N$ (cf. figure 3.5) we note that ${}^{12}_{6}C$ is self-conjugate so that all its states have $T_3=0$ and in particular its ground state must have $T=0$, for higher values of T would give rise to companion states in ${}^{12}_{5}B$ to which the ground state of ${}^{12}_{6}C$ would decay by positron emission or electron capture. Because the measured width $\Gamma_2=37.0\pm1.1$ eV (Chertak *et al* 1973) of the 15.11 MeV γ-transition in ${}^{12}_{6}C$ shows that this is an unhindered magnetic dipole transition, we may conclude that the 15.11 MeV level is almost pure $T=1$ (there is in fact a small admixture of $T=0$ (Calaprice *et al* 1976)). By the same reasoning the fact that the β^- and β^+ ground-state transitions in ${}^{12}_{5}B$ and ${}^{12}_{7}N$ are unhindered allowed transitions allows us to classify the ground states of ${}^{12}_{5}B$ and ${}^{12}_{7}N$ as having $T=1$. Thus the trio of states with $T=1$ in these nuclei comprises an isospin triplet which is an ideal system in which to test the charge independence hypothesis. It should be noted that the isosinglet ${}^{12}_{6}C$ and the isotriplet ${}^{12}_{5}B$, ${}^{12}_{6}C^*$ and ${}^{12}_{7}N$ are not members of a common supermultiplet since the isosinglet with $S=0$, $T=0$ and the isotriplet with $S=1$, $T=1$, do not satisfy the required spin–isospin exchange symmetry.

3.5 Pions

3.5.1 The Yukawa hypothesis

Charged particles at rest exert electrostatic forces on each other whose magnitude and direction are given by Coulomb's law, a process which in quantum theory is described as a continuous emission and absorption of virtual photons, the infinite range of the Coulomb potential expressing the fact that free photons have zero rest-mass. The nuclear force is at least partly of exchange type, and has a finite range of order 1–2 fm, and, if the force is mediated by the exchange of some entity, then the nature of that entity is the major question to be settled. Heisenberg originally suggested that the exchange of electric charge could be important (1932) but the most significant advance was made by Yukawa (1935), who pointed out that the exchange of virtual mesons, particles intermediate in mass between electron and proton, would give a nuclear force of the observed range. He therefore postulated the existence of a static meson potential

$$\varphi_y(r)=g\,e^{-\kappa r}/r \qquad (3.93)$$

in the region of space surrounding a nucleon, which satisfies an equation which, for massive field quanta, is the analogue of Poisson's equation in electrostatics

$$\nabla^2\varphi_y(\mathbf{r})-\kappa^2\varphi_y(\mathbf{r})=4\pi g\delta^3(\mathbf{r}) \qquad (3.94)$$

containing a source term representing an infinitely heavy point nucleon at the origin carrying 'mesonic charge' g.

The mass m of the exchanged meson is related to the range κ^{-1} of the nuclear force by application of the uncertainty relation, which states that energy conservation can be violated by an amount $\Delta E = mc^2$, corresponding to the emission of the virtual particle, for a time $\Delta t = \hbar/mc^2$. Setting $c\,\Delta t \simeq \kappa^{-1}$, we find $mc^2 \simeq 165$ MeV. To account for the observed strength of the nuclear force the mesonic charge or strong coupling constant g must have a value such that $\alpha_s = g^2/\hbar c \simeq 15$.

After a prolonged period of confusion during which the muon (a spin $\frac{1}{2}$ charged lepton with no strong interactions discovered by Neddermeyer and Anderson in 1937) was mistakenly identified as the Yukawa meson, the true meson was finally isolated as a component in the cosmic radiation (Lattes *et al* 1948). These pi-mesons, or pions as they are now called, are the lightest members of a whole family of mesons to be discovered subsequently, and come in three charge states with masses.

$$m_\pi^{\pm} c^2 = 139.5679 \pm 0.0007 \text{ MeV} \tag{3.95}$$

$$m_\pi^0 c^2 = 134.9743 \pm 0.0008 \text{ MeV} \tag{3.96}$$

The uncharged pion, the π^0, is its own anti-particle and is unstable against electromagnetic interactions, decaying into two γ-rays with a very short lifetime $\tau_0 = (0.8 \pm 0.1) \times 10^{-16}$ s. The charged pions $\pi^\pm$ constitute a particle–antiparticle pair which are stable against both strong and electromagnetic interactions but decay weakly into leptons according to the scheme

$$\pi^+(\pi^-) \to \mu^+(\mu^-) + \nu_\mu(\bar{\nu}_\mu) \qquad \tau_\pm = (2.602 \pm 0.002) \times 10^{-8} \text{ s} \tag{3.97}$$

They also have several other low intensity weak decay modes of which the most significant, from a theoretical viewpoint, is pion β-decay

$$\pi^+(\pi^-) \to \pi^0 + e^+(e^-) + \nu_e(\bar{\nu}_e) \tag{3.98}$$

which is observed with a branching ratio of $1.02 \pm 0.07 \times 10^{-8}\%$.

Until the pions could be produced artificially in the laboratory it was not known whether the Yukawa mesons were scalar or vector particles. However, it was soon shown that pions have spin $I^\pi = 0^-$; for the π^0 this is an immediate deduction from the observed two photon decay mode, and for the π^+ it follows from the application of detailed balance arguments to the reactions $\pi^+ + \mathrm{d} \leftrightarrows \mathrm{p} + \mathrm{p}$ (Durbin *et al* 1951). A major discovery was the demonstration that the pions are pseudo-scalar rather than scalar particles (Kemmer 1938) and the pion field therefore has more in common with the magnetostatic potential $\psi(r)$ treated in (3.37)–(3.38) than with the electrostatic potential $\varphi(r)$ which provided the model for (3.94).

3.5.2 The one-pion exchange potential

The pseudo-scalar nature of the pion was established from a detailed study of the strong interaction process

$$\pi^- + \mathrm{d} \rightarrow \mathrm{n} + \mathrm{n} \tag{3.99}$$

which occurs when a π^- is captured by a deuteron into the lowest s-orbital of pionic deuterium (Panofsky *et al* 1951). Since the deuteron has spin $I_\mathrm{d} = 1$, and the triplet s-state is excluded for two neutrons, it follows that they must be created in a negative parity p-state to conserve angular momentum, and the π^- must therefore have negative intrinsic parity if parity is to be conserved. Conversely in the strong emission of virtual pions by nucleons

$$\mathrm{N} \rightarrow \mathrm{N} + \pi \tag{3.100}$$

the pions must be emitted in p-states to conserve parity. This property alone is sufficient to determine the form of the pion–nucleon coupling in the static limit, for the pion momentum $\boldsymbol{p}_\pi$ is the only vector operator available to effect the transfer of one unit of angular momentum, and the only pseudo-scalar nucleon source term we can construct must be proportional to $\boldsymbol{\sigma}\cdot\boldsymbol{p}_\pi$. It follows that the static pseudo-scalar pion field in the neighbourhood of an infinitely massive point nucleon must satisfy an equation of the form

$$\nabla^2\pi(r) - \kappa_\pi^2\pi(r) = -\sqrt{(4\pi)} f \boldsymbol{\sigma}\cdot\nabla\delta^3(r) \tag{3.101}$$

The solution of (3.101), i.e.

$$\pi(r) = \frac{f}{\sqrt{(4\pi)}} \boldsymbol{\sigma}\cdot\nabla\left(\frac{\mathrm{e}^{-\kappa_\pi r}}{r}\right) \tag{3.102}$$

is the mesonic analogue of the magnetostatic potential (3.38) in the neighbourhood of a magnetic dipole, and we can follow a similar procedure to that followed in the magnetostatic example, to construct the one-pion exchange potential (OPEP) V_π between point nucleons separated by a distance r, i.e.

$$V_\pi = \frac{f^2}{4\pi}\frac{\kappa_\pi}{3}\{\boldsymbol{\sigma}_1\cdot\boldsymbol{\sigma}_2\varphi_0(\kappa_\pi r) + S_{12}(\boldsymbol{\sigma}_1, \boldsymbol{\sigma}_2, \hat{\boldsymbol{r}})\varphi_2(\kappa_\pi r)\} \tag{3.103}$$

where

$$\varphi_0(\kappa_\pi r) = \frac{\mathrm{e}^{-\kappa_\pi r}}{\kappa_\pi r} - \frac{4\pi}{\kappa_\pi^3}\delta^3(r) \tag{3.104}$$

and

$$\varphi_2(\kappa_\pi r) = \frac{\mathrm{e}^{-\kappa_\pi r}}{\kappa_\pi r}\left\{1 + \frac{3}{\kappa_\pi r} + \frac{3}{(\kappa_\pi r)^2}\right\} \tag{3.105}$$

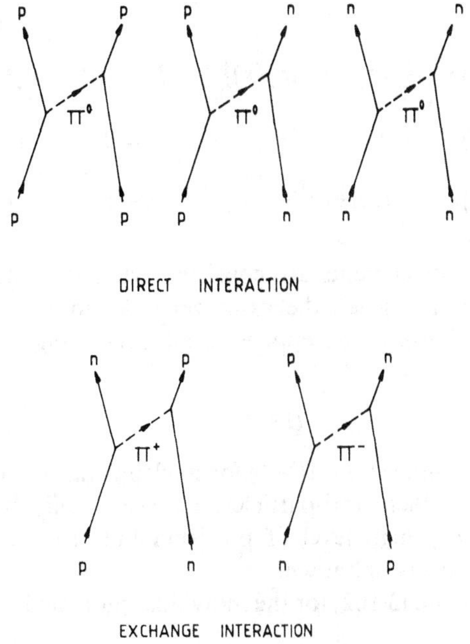

Figure 3.6. Feynman diagrams illustrating the role of single pion exchange in generating direct and exchange forces in thé nucleus.

Because of the short-range repulsive core in the internucleon potential the δ-function term in (3.104) plays no effective role and is normally omitted in theoretical calculations.

It remains to take account of the existence of three pion charge-states and to incorporate the notion of charge independence into the formalism. The one-pion exchange potential contribution to the nuclear force is illustrated by the diagrams in figure 3.6 which draw attention to the fact that, although neutral and charged pions can contribute to the neutron–proton force in lowest order, only neutral pions contribute to the proton–proton and neutron–neutron forces in lowest order. Since the pion field $\pi^+(x)$ describes the annihilation of a π^+ and the simultaneous conversion of a neutron into a proton at the vertex where it is absorbed, it must be proportional to the isospin operator τ_+ acting on the nucleon field, in order to conserve the charge. Similarly the pion fields $\pi^-(x)$ and $\pi^0(x)$ must be proportional to the nucleon isospin operators τ_- and τ_3 respectively. These requirements are accommodated naturally if the pions are assumed to behave like the components of an isovector $(T=1)$ under rotations in isospin space. Formally if we let π_1, π_2, π_3 represent the pion field components along the 1, 2, 3 axes in isospin space then, in analogy with (3.31) we can write the spherical components as

$$\begin{aligned} \pi^-(x) &= \frac{1}{\sqrt{2}}(\pi_1(x) - i\pi_2(x)) \qquad T_3 = -1 \qquad Q = -1 \\ \pi^0(x) &= \pi_3(x) \qquad T_3 = 0 \qquad Q = 0 \\ \pi^+(x) &= \frac{1}{\sqrt{2}}(\pi_1(x) + i\pi_2(x)) \qquad T_3 = 1 \qquad Q = 1 \end{aligned} \tag{3.106}$$

where the charge assignments are consistent with the rule (3.87) provided the mass number A is given the value zero. A more general form of the charge equation (3.86), which applies to all non-strange hadrons (strongly interacting particles) is

$$Q = T_3 + \tfrac{1}{2}B \tag{3.107}$$

where the baryon number $B = 1(-1)$ for nucleons (anti-nucleons) and $B = 0$ for all mesons and their anti-particles. Experimentally baryon number is conserved to a very high level of precision but the exact nature of the underlying symmetry is unknown.

We can now rewrite (3.102) for the individual pion field components in the form

$$\pi_i(x) = -\frac{1}{4\pi}\frac{f}{\kappa_\pi}\tau_i \boldsymbol{\sigma}\cdot\nabla\left(\frac{e^{-\kappa_\pi r}}{r}\right) \qquad i = 1-3 \tag{3.108}$$

and the resulting expression for the one-pion exchange potential

$$V_\pi = \frac{f^2}{4\pi}\left(\frac{\kappa_\pi}{3}\right)\boldsymbol{\tau}(1)\cdot\boldsymbol{\tau}(2)\{\boldsymbol{\sigma}_1\cdot\boldsymbol{\sigma}_2\varphi(\kappa_\pi r) + S_{12}(\boldsymbol{\sigma}_1, \boldsymbol{\sigma}_2, \hat{\boldsymbol{r}})\varphi_2(\kappa_\pi r)\} \tag{3.109}$$

is manifestly charge independent, since the isospin factor $\boldsymbol{\tau}(1)\cdot\boldsymbol{\tau}(2)$ is an isoscalar, and is therefore invariant with respect to rotations in isospin space. It is worth noting that the small difference between the charged pion range $(\kappa\pm)^{-1} = 1.41$ fm, and the neutral pion range $(\kappa_0)^{-1} = 1.45$ fm, has been ignored in (3.108) and (3.109). This effect breaks the charge independence, but not the charge symmetry, of the one-pion exchange theory at the level of a few per cent in agreement with observation (cf. section 3.4.1).

There are several points of immediate interest which arise in connection with the details of V_π, and the first of these concerns the central part of the potential which may be written in the form

$$V_{c\pi}(r) = \frac{-f^2}{4\pi}\kappa_\pi\left[P_r + \frac{1}{3}\boldsymbol{\sigma}_1\cdot\boldsymbol{\sigma}_2(1 - P_r)\right]\varphi_0(\kappa_\pi r) \tag{3.110}$$

where P_r is the Majorana exchange operator, and we have made use of the generalized Pauli principle given in (3.91). The sign of $V_{c\pi}(r)$ we note is negative, and hence attractive, in relative s-states with $P_r = 1$, and positive, i.e. repulsive, in relative p-states with $P_r = -1$. This is exactly the exchange behaviour which is necessary to make the α-particle the saturated unit, although this condition is not in general sufficient to ensure saturation, which is a co-operative effect involving the Pauli principle, exchange forces and a repulsive core. So far as the neutron–proton system is concerned, the s-state attractive central force is the same for singlet and triplet states, an identity which is a consequence of the invariance of $V_{c\pi}$ with respect to interchange of the spin and isospin coordinates according to (3.109–3.110), so that the bound and virtual states of the deuteron are members of the same isospin supermultiplet. This symmetry is certainly not reflected in the observed nuclear structure of the (virtual) singlet and (bound) triplet states of the deuteron.

The ground state of the deuteron is an isosinglet with $T=0$, and there is no spin–orbit term of the kind which is detected in high energy p–p scattering which is only effective in states with $T=1$. It follows that the tensor potential $V_{T\pi}$ must exert a critical influence in maintaining the deuteron as a bound system in the triplet state. Now since $\langle \boldsymbol{\tau}(1)\cdot\boldsymbol{\tau}(2)\rangle = -3$ for the deuteron and the d-wave matrix element of the tensor force $\langle {}^3D_1|S_{12}|{}^3D_1\rangle = -2$ it follows that the tensor force is repulsive in a pure $l=2$ neutron–proton state and it is not immediately obvious how the contribution of the $l=2$ state to the ground state of the deuteron can enhance the binding. The explanation lies in the presence of a large positive mixing matrix element $\langle {}^3D_1|S_{12}|{}^3S_1\rangle = \sqrt{8}$ which couples the Schrödinger equations for the s-wave and d-wave components in the ground state of the deuteron and accounts for the existence of a bound state in a natural way (Rarita and Schwinger 1941, Gartenhaus 1955, de Swart and Marshak 1959, Noyes 1965). Nevertheless, since the admixture of d-state is only of the order of 7% it is still permissible in most situations to describe the states of the deuteron in terms of effective central potentials according to (3.45).

A further remark is directed towards the contribution of the tensor potential $V_{T\pi}$ to proton–proton scattering. Since $\boldsymbol{\tau}(1)\cdot\boldsymbol{\tau}(2)=1$ for this case and the matrix element $\langle {}^3P_0|S_{12}|{}^3P_0\rangle = -4$, there is a long range attractive tensor force in the ${}^3P_0$ state as required to account for the observed features of polarized proton–proton scattering (cf. section 3.3.2.3).

The broad conclusion appears to be that the one pion exchange potential V_π accounts at least qualitatively for the long-range behaviour of the nuclear force but, at distances <2 fm, it must be supplemented by additional components (Moravcsik and Noyes 1961). It was thought at one time that the spin–orbit potential could be derived as a second-order contribution from the tensor potential but attempts to do so ended in failure (Nigam and Sundaresan 1958). Also the calculation of 2π and 3π exchanges in perturbation theory is prohibitive in effort and in any case involves many approximations of doubtful validity (Lomon and Partovi 1973). Furthermore the experimental discovery of a whole family of nucleonic Δ-resonances allowing the possibility of introducing excited nucleons as intermediate states added more complications (cf. section 9.1.2.1).

A new, and ultimately more fruitful line of research was opened up with the discovery of the pionic resonances, the η (549 MeV) ρ (767 MeV) and the ω (782 MeV). The η-meson is a pseudoscalar isoscalar 3π resonance which is stable against strong interactions. It does not couple strongly to nucleons and makes no significant contribution to the nuclear force. However the ρ-meson and the ω-meson are vector mesons with masses close to the 6π threshold and either or both may make important contributions to the nuclear potential at short range. The ρ-meson is a 2π resonance which can be treated as an element in the 2π-exchange contribution but the ω-meson is a 3π resonance with a very narrow width of about 8.7 MeV which can be treated as a stable particle in this context, and represents a possible candidate as the source of the repulsive core.

Recent computations of the $(\pi+2\pi+\omega)$ exchange potential, the so-called 'Paris potential', based on a dispersion theoretic rather than a perturbative approach, and involving a certain degree of non-locality, have pointed to the existence of central, tensor, spin–orbit and quadratic spin–orbit components in the nuclear force, and show very satisfactory agreement with the Yale and Hamada–Johnston phenomenological potentials for distances >0.8 fm (Cottingham *et al* 1973, Vinh Mau 1983). Thus some measure of understanding of the nuclear force has been achieved at a fundamental level at least so far as the long- and medium-range behaviour is concerned.

4 Neutrons and nuclear structure

4.1 Models of the nucleus

4.1.1 Effective forces in the nucleus

At the present state of development, both in basic nuclear theory and in computational techniques, the provision of an exact theoretical description of even the simplest nucleus with $A>2$ is not practicable, and this is true partly because the nucleus is a many-body quantum system, but principally because current descriptions of the internucleon forces are neither complete nor unique. Thus velocity-dependent forces which play an important part in all nuclear phenomena are entirely equivalent to non-local potentials in the two-nucleon system, which is the source of the most precise data, but this is not true in a many-body system whose description requires the specification of two-body matrix elements between single particle states off the mass shell. Phenomenological potentials, no matter how accurate the parameterization, are no substitute for a quantum field theory of the nuclear force and, although various forms of meson theory have been in existence for more than half a century (cf. section 3.5.3) the reliable calculation of strong interaction processes on the basis of these theories has proved to be an intractible problem, and it is doubtful whether any real progress has been made. It remains to be seen whether the new gauge theories which dominate the scene in today's particle physics will achieve any greater success.

The prime concern of any theory of the nucleus is the specification of the states of the nucleons of which it is composed and, although the meson degrees of freedom are important in nuclei, and cannot be ignored, the only practical way of dealing with them is by the introduction of effective interactions between nucleons. This is done by writing down an approximate nuclear Hamiltonian

$$H=\sum_{i=1}^{A}\frac{p_1^2}{2m}+\sum_{i<j}^{A}V_{ij}^{(2)}+\sum_{i<j<k}^{A}V_{ijk}^{(3)}+\dots \tag{4.1}$$

where $V_{ij}^{(2)}$ and $V_{ijk}^{(3)}$ are two- and three-body effective potentials whose choice is dictated by experimental data and certain well-established invariance principles or conservation laws. Then any approximate method for solving the A-body problem with the approximate Hamiltonian (4.1) may be given the title of 'nuclear model' and the resultant theory has some predictive value provided the set of available experimental data is larger than the set of parameters required to characterize the model.

Having fixed the model and determined the spectrum of states, clearly the first problem to be tackled is the prediction of electromagnetic properties, i.e. magnetic dipole moments, electric quadrupole moments, electric and magnetic transition rates, etc., for electromagnetism is the one interaction for which a complete and unambiguous theory exists. However, the attempt to undertake such a programme immediately introduces a new problem, or perhaps a new aspect of the same old problem, for it is not possible to introduce the electromagnetic coupling via the minimal substitution (3.55) where p_μ is the nucleon four-momentum, because the charged mesons which are exchanged between nucleons also contribute to the electromagnetic current. To incorporate these meson exchange currents into the theory the ordinary nucleonic current must therefore be supplemented by two- and three-particle exchange currents, i.e.

$$j_\mu^{\text{em}}(\boldsymbol{r})=\sum_{i=1}^{A} j_\mu^{(1)}(\boldsymbol{r}_i)\,\delta^3(\boldsymbol{r}-\boldsymbol{r}_i)+\sum_{i<j}^{A} j_\mu^{(2)}(\boldsymbol{r}_i,\boldsymbol{r}_j)(\delta^3(\boldsymbol{r}-\boldsymbol{r}_i)+\delta^3(\boldsymbol{r}-\boldsymbol{r}_j)+\ldots) \quad (4.2)$$

Thus the nuclear model is not completely determined by the Hamiltonian (4.1) but requires a prescription for the electromagnetic current (4.2), and also for the weak currents which participate in the weak interactions.

At this stage the exploitation of invariance principles becomes important since the principle of gauge invariance in electromagnetic interactions requires that the electromagnetic current be absolutely conserved no matter what its composition. A very important consequence of this conservation law is the result known as Siegert's theorem which states in effect that meson exchange currents make no contribution to electric multipole transition operators in the long-wave limit, i.e. for γ-ray wavelengths which are much larger than the nuclear radius (Siegert 1937, Sachs and Austern 1951, Foldy 1953). Siegert's theorem also applies to the weak vector current in nuclear β-decay which, on the strong CVC hypothesis (cf. section 9.3.3), is conserved at the same level to which isospin is conserved (cf. section 3.5). It does not apply to magnetic multipole operators (Sachs 1948) (or to the weak axial vector current), so that the effective magnetic moment operator is

$$\boldsymbol{\mu}=\sum_{i=1}^{A}\boldsymbol{\mu}_i+\sum_{i<j}^{A}\boldsymbol{\mu}_{ij}+\ldots \quad (4.3)$$

The significance of exchange magnetic moments was first recognized by Breit and Condon (1936) in the context of the photodisintegration of the deuteron, and indeed some experimental evidence for their existence may be derived from observed discrepancies between measured and computed magnetic moments of the nuclei $^3_1$H and $^3_2$He, which are the only nuclei whose wavefunctions are known from numerical solutions of the Schrödinger equation obtained using phenomenological nucleon–nucleon potentials as a direct input. However, even this procedure is not free from ambiguity, for there is a danger of error due to double counting when the matrix elements of effective operators are evaluated on the basis of wavefunctions which also include the effects of meson exchange at some level.

4.1.2 Independent particle models

Independent particle models of the nucleus are based on the hypothesis that the motion of a nucleon inside the nucleus is not chaotic in the sense that particle motion in a classical gas is chaotic, but is determined by some average potential due to all the other nucleons. The problem therefore reduces to one of determining the average field, and the single particle states in that average field, for a prescribed Hamiltonian of the form (4.1). In atomic physics there is no difficulty in justifying the concept of an average field since the one-body nuclear Coulomb potential makes the dominant contribution to the potential while the interelectron forces may be treated as perturbations. In the case of the internal nuclear potential the Hamiltonian (4.1) contains no equivalent one-body potential term, and it is not at all obvious why the idea of an average field should work so successfully. The reason in broad terms appears to be that the nuclear force is only moderately strong in the sense that the condition (2.206) necessary for the existence of a two-nucleon bound state is satisfied only in exceptional circumstances. Thus two-body scattering cannot generate high momentum components in the two-nucleon wavefunction while low momentum transfers for a nucleon in a degenerate Fermi gas in the nuclear interior are suppressed by the Pauli principle (Weisskopf 1951). Conversely, the high probability for the capture of an incident nucleon into a state of the compound nucleus is associated with the high density of states into which it may be scattered at high levels of excitation.

Formally the concept of an average field may be interpreted as the result of a Hartree–Fock treatment of the many-body problem whose objective it is to derive a set of N single-particle wavefunctions which, when properly anti-symmetrized in the form of a Slater determinant

$$\psi = (N!)^{-1/2} \det(\varphi_1, \varphi_2, \ldots, \varphi_N) \tag{4.4}$$

gives precisely that wave function ψ which minimizes the expectation value of the energy

$$\langle E \rangle = \int \psi^* H \psi \, \mathrm{d}^3 r \bigg/ \int \psi^* \psi \, \mathrm{d}^3 r \tag{4.5}$$

The self-consistent field V_{scf} is then defined in terms of its matrix elements as that potential which yields the Hartree–Fock equations

$$\frac{-\hbar^2}{2m} \nabla^2 \varphi_i(\boldsymbol{r}_i) + V_{\mathrm{scf}}(\boldsymbol{r}_i)\varphi_i(\boldsymbol{r}_i) = \varepsilon_i \varphi_i(\boldsymbol{r}_i) \tag{4.6}$$

Because of the anti-symmetrization the potential in (4.6) contains exchange terms which act as a non-local potential and ultimately contribute to the effective mass of the nucleon as described in section 3.3.2.

In practice the Hartree–Fock programme is not carried through and instead some convenient potential is selected to represent the average field with parameters chosen to reproduce a given set of experimental data. It is therefore implicitly assumed that the potential adopted is a close approximation to the self-consistent field although, since the Hamiltonian is itself only an approximation, both approaches to the solution of the problem are probably equally valid. However, independent particle models also make use of the same average field and the same set of single-particle states to construct the excited states of the nucleus, and this extension of the theory represents a clear departure from the rationale which underpins the Hartree–Fock method.

4.1.3 Motion of the mass centre

The approximate A-body nuclear Hamiltonian H given by (4.1) is defined as a rotational and translational invariant provided only that the potentials are functions of relative positions only. Thus its eigenfunctions are simultaneous eigenfunctions of the angular momentum and the total momentum $\boldsymbol{P}$ defined by

$$\boldsymbol{P}=\sum_{i=1}^{A} \boldsymbol{P}_i=-i\hbar\nabla_{\mathrm{R}} \tag{4.7}$$

where

$$\boldsymbol{R}=\sum_{i=1}^{A} \boldsymbol{r}_i/A \tag{4.8}$$

is the centre of mass, neglecting the neutron–proton mass difference. The eigenfunctions of H are then given by

$$\psi_{K,n}(\boldsymbol{r})=\exp\left[\mathrm{i}\boldsymbol{K}\cdot\boldsymbol{R}\right]\psi_{0,n}(\boldsymbol{r}) \tag{4.9}$$

where $\boldsymbol{r}$ represent all nucleon coordinates $\boldsymbol{r}_i$ ($i=1, 2, \ldots, A$), $\hbar\boldsymbol{K}$ are the eigenvalues of $\boldsymbol{P}$ and n represents all quantum numbers describing the internal motion of the system. The corresponding energy eigenvalues are

$$E_{K,n}=E_n+(\hbar K)^2/2Am \tag{4.10}$$

where E_n is the energy of the zero momentum state $\psi_{0,n}$ which is referred to as the 'intrinsic' state.

Unfortunately the many particle states which arise in nuclear models cannot be defined as eigenfunctions of $\boldsymbol{P}$, a condition which could be imposed only if the single-particle states were plane wave states or if their motions were correlated. The last condition is ruled out by the requirement of independence (Lipkin 1960). Thus the centre of mass carries out a non-uniform or 'shuffling' motion which is not a characteristic of the intrinsic state (Inglis 1937). A number of approximate techniques have been devised to eliminate this irregular motion most of which are variants on the following procedure (Lipkin 1958).

First of all the model states $\Phi_n(\boldsymbol{r})$ are expressed as functions of $\boldsymbol{R}$ and $3(A-1)$ independent relative coordinates $\boldsymbol{r}'$. From the model state $\Phi_n(\boldsymbol{R}, \boldsymbol{r}')$ so defined we project out a zero momentum state $\Psi_{0,n}(\boldsymbol{r}')$ which satisfies the condition

$$\boldsymbol{P}\psi_{0,n}(\boldsymbol{r}') = -i\hbar\nabla_R\Psi_{0,n}(\boldsymbol{r}') \equiv 0 \tag{4.11}$$

This is not a unique procedure but a convenient way to do it is to define $\Psi_{0,n}(\boldsymbol{r}')$ by the equation

$$\Psi_{0,n}(\boldsymbol{r}') = \int_{-\infty}^{\infty} G(\boldsymbol{R})\Phi_n(\boldsymbol{R}, \boldsymbol{r}')\,\mathrm{d}^3R \tag{4.12}$$

where the generating function $G(\boldsymbol{R})$ is also not uniquely defined. Thus $\Psi_{0,n}(\boldsymbol{r}')$ is determined as an average over $\Phi_n(\boldsymbol{R}, \boldsymbol{r}')$ and differs in energy from $\Phi_n(\boldsymbol{R}, \boldsymbol{r}')$ by an amount which approximates to the centre of mass energy, the accuracy of the approximation depending on the choice of $G(\boldsymbol{R})$.

In the case of the two-body problem discussed in section 2.2.1 the result (4.12) is exact and independent of the choice of $G(\boldsymbol{R})$ because the function $\Phi_n(\boldsymbol{R}, \boldsymbol{r}')$ can be factorized into independent functions of $\boldsymbol{R}$ and $\boldsymbol{r}'$. The same convenient decomposition is true also for the oscillator wavefunction in which case the wavefunction factorizes into an intrinsic function $\Psi_{0,n}(\boldsymbol{r}')$ and a zero angular momentum eigenstate describing the motion of the centre of mass (Bethe and Rose 1937). The validity of this result requires that both proton and neutron configurations each contain no more than one partially filled shell with all shells of lower energy filled. When this condition fails spurious states can arise associated with higher angular momentum states of motion of the mass centre. These states may be eliminated by selecting linear combinations of degenerate oscillator states which force the mass centre motion into its lowest state. This is most conveniently done by diagonalizing the operator $\boldsymbol{R}^2$ in oscillator states of the same energy (Elliott and Skyrme 1955).

4.2 The nuclear shell model

4.2.1 Magic numbers and inverted spin–orbit coupling

Although the earliest speculations about possible nuclear shell structure effects (Beck 1928, 1930) predate the discovery of the neutron, the first serious proposal along these lines was advanced by Bartlett (1932) who pointed to the possible occurrence of S, P and D shells in the structure of light nuclei. Since nuclear magnetic moments are very small, and even in heavy atoms spin-orbit splitting is of order $(v/c)^2$ in comparison with the Coulomb energy, it was tacitly assumed that in any non-singular central potential shell closure would occur when the number of identical nucleons in a shell is equal to

$2(2l+1)$. Thus special stability might be expected at N or $Z=2$, 8, 18 or even at N or $Z=20$ in an isotropic harmonic oscillator potential where the 1d and 2s levels are degenerate (Bethe and Bacher 1936). In nuclear physics single-particle states or orbitals are labelled *n*s, *n*p, etc., where the quantum number n is the number of nodes in the radial wave function and is not the same as the principal quantum number used in atomic physics which is more appropriate for the discussion of energy levels in a highly singular potential like the Coulomb potential.

The evidence for exceptional stability or abundance (cf. section 3.1) at values of $Z=50$ or 82, and $N=50$, 82 and 126 was first assembled by Elsasser (1934), Guggenheimer (1934) and Goldschmidt (1938) and supportive evidence from a different field was published by Griffiths (1939), who recorded anomalously low cross-sections for scattering of 0.4 MeV neutrons in $^{89}_{39}Y$ ($N=50$), $^{139}_{57}La$ ($N=82$) and $^{141}_{59}Pm$ ($N=82$). Subsequent work (Harris and Muehlhause 1950) extended these observations to $^{209}_{83}Bi$ ($N=126$) which explains why pure bismuth makes such a suitable γ-ray filter for slow neutron beams. However, none of the early work attracted much interest until the appearance of a review by Mayer (1948) stimulated renewed activity in the field, one outcome of which was the demonstration that the higher magic numbers 50, 82 and 126 could be brought about by a comparatively minor reordering of orbitals for a range of potentials including harmonic oscillator, deep square well and hybrids of these (Feenberg and Hammack 1949, 1951, Nordheim 1949).

The following year witnessed the most spectacular advance in nuclear theory since the appearance of the proton–neutron nuclear model nearly two decades earlier, when it was realized that the whole array of magic numbers, including the numbers 14, 28 and 40 whose semi-magic status is associated with closure of the $1d_{5/2}$, $1f_{7/2}$ and $2p_{5/2}$ sub-shells, could be explained by the addition of a strong spin–orbit term, with inverted level ordering, to the single-particle potential (Mayer 1950, Haxel *et al* 1948, 1949, 1950). Inversion of the level ordering means that, unlike the equivalent situation in atoms, that component of the spin–orbit doublet with the higher j-value has lowest energy. This was an idea which had been canvassed some years earlier (Dancoff and Inglis 1936, Furry 1936, Breit 1937, 1938) although no physical basis for the effect was known, and the suggestion that the tensor force could be responsible (Keilson 1951) turns out to be incorrect. It now seems probable that the spin–orbit coupling originates in a higher order meson-exchange effect in the internucleon potential although it is unlikely that the last word on the subject has been spoken. The single-particle level ordering in a potential intermediate between infinite square well and three-dimensional harmonic oscillator potential plus inverted spin–orbit coupling is illustrated in figure 4.1.

In atomic theory spin- and velocity-dependent forces are neglected in first approximation, so that electrons move in a spherically symmetric average

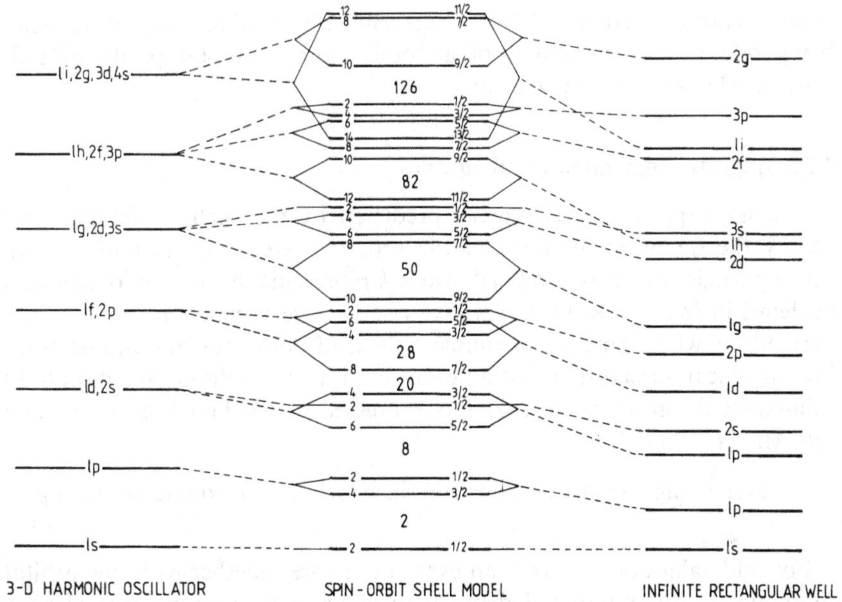

Figure 4.1. Single particle levels in the nuclear shell model (Klinkenberg 1952, Feenberg 1955, Zeldes 1956). The central potential is assumed to have a form intermediate between the three-dimensional simple harmonic oscillator potential which extends to infinity, and the three-dimensional infinite rectangular potential well which has a sharp cut-off at a finite radius. The magic numbers 2, 8, 20, 50, 82, 126 associated with major shell closures are accounted for by introducing a strong inverted spin–orbit interaction into the potential. The existence of a quasi-magic number at $N=152$ can be explained on the assumption that the higher neutron shells fill in the order $2g_{9/2}$, $4s_{1/2}$, $2g_{7/2}$ and $3d_{5/2}$, before the $1i_{11/2}$ and $3d_{3/2}$ shells are completed (cf. section 5.2.5).

potential. The single-particle wavefunctions φ_i can therefore be chosen as eigenfunctions of the commuting angular momentum operators and labelled by eigenvalues $(l, m_l; s, s_z)$. The spin–orbit potential is then included as a small perturbation so that atomic states are labelled by the usual quantum numbers (L, S, J, M). This is the approximation known as Russel–Saunders coupling. In the Mayer–Jensen nuclear shell model the spin–orbit interaction must be included as part of the unperturbed Hamiltonian so that single-particle states are labelled by quantum numbers (l, s, j, m_j). This treatment is essential to the theory and reflects the experimental fact that the energy splitting between subshells labelled by $j=l\pm\frac{1}{2}$ is comparable to the energy splitting between shells labelled by adjacent values of l. Thus the rules of j–j coupling apply and nuclear states are labelled by quantum numbers I^π where π represents the parity. The parity of a many-particle system is not a function of L but is determined by its single-particle content, being odd if it is composed of an odd number of odd parity orbitals ($\pi(l)=(-1)^l$) and otherwise even.

4.2.2 Angular momentum coupling rules

The single-particle level scheme predicted by the shell model does not specify the quantum numbers of the ground state uniquely and a given single-particle configuration $(nlj)^k$, where k represents the number of identical nucleons in the orbital (nlj), can give rise to a very wide variety of states, particularly when there are incomplete shells of both neutrons and protons. The empirical observation that all even–even nuclei without exception have spin-parity 0^+ in the ground state is expressed by the first Mayer coupling rule Mayer (1950).

(i) An even number of identical nucleons in an orbital (nlj) couples to spin $I=0$

$$ \tag{4.13} $$

For odd values of A there is an even–even core together with one orbital containing an odd number of identical nucleons. In the vast majority of cases odd-A nuclei have spin values I equal to the angular momentum j of that orbital containing an odd number of nucleons, although there are a number of exceptions to the rule which have instead $I=(j-1)$; e.g. ${}^{21}_{10}\text{Ne}$ ($j=\frac{5}{2}$, odd N), ${}^{23}_{11}\text{Na}$ ($j=\frac{5}{2}$, odd Z), ${}^{47}_{22}\text{Ti}$ ($j=\frac{7}{2}$, odd N), ${}^{55}_{25}\text{Mn}$ ($j=\frac{5}{2}$ odd N), ${}^{79}_{34}\text{Se}$ ($j=\frac{9}{2}$, odd N). In the normal case there is a single unpaired nucleon and this represents the extreme independent particle picture which is expressed by Mayer's second coupling rule.

(ii) An odd number of identical nucleons in an orbital (nlj) couples to spin $I=j$.

$$ \tag{4.14} $$

We have already noted in the discussion of nuclear stability in section 3.1 that even–even nuclei are energetically favoured in comparison with even–odd, odd–even and odd–odd nuclei. This is an observation which may be interpreted in shell-model terms as a tendency for pairs of identical nucleons in an orbital (nlj) to pair off in magnetic substates with equal but opposite values of m_j, giving rise to a saturated pair with total angular momentum zero. The pair states of two identical nucleons in an (nlj) orbital must be anti-symmetrized according to (4.4) and, since the resultant wavefunction is spherically symmetric, all substates labelled by $\pm m_j$ are equally represented in it. In general if the orbital contains one nucleon in the substate $|jm_j\rangle$, together with p pairs distributed over substates $|j\pm m_j\rangle$ for $p<(j\pm\frac{1}{2})$, all substates $|j\pm m_j'\rangle$ are equally populated by pairs of particles excluding the state with $|m_j'|=|m_j|$. If we successively add two identical nucleons to those already present in an orbital, such that they form a saturated pair, then the pairing energy may be defined approximately as the binding energy of the pair less twice the binding energy of the first additional nucleon. The behaviour of the pairing energy as a function of j is summarized in Mayer's third coupling rule.

(iii) The pairing energy of two nucleons in an (nlj) orbital increases with j. (4.15)

The main consequence of the pairing rule is that saturated pairs may go into the higher of a pair of neighbouring orbitals of different j, leaving the lower energy orbital free to accommodate a single odd nucleon. As an example the odd A nuclei ${}^{203}_{81}T_l$ $(N=122)$ and ${}^{205}_{81}T_l$ $(N=124)$ each have spin-parity $\frac{1}{2}^+$ although on the basis of coupling rule (ii) one might have expected a spin parity assignment $\frac{11}{2}^-$. However, these examples do not represent exceptions to rule (ii) since the explanation is that the state with the saturated pair in the $1h_{11/2}$ orbital, and the unpaired proton in the $3s_{1/2}$ orbital, has a higher binding energy, and therefore a lower total energy, than the state in which the orbitals occupied by the paired and unpaired protons are exchanged.

In the case of odd–odd nuclei, where each odd group occupies an orbital $(n_i l_i j_i)$, $i=1, 2$, the ground state spin I must be in the range

$$|j_n - j_p| \leqslant I \leqslant j_n + j_p \tag{4.16}$$

a result which, given that in any range of N and Z, there are at least two orbitals in competition, is not of much predictive value for determining I. To deal with this situation Nordheim (1950) proposed two rules derived from the systematic study of nuclear β-decay data

(i) strong rule $I=|j_n-j_p|$ $\quad j_n+j_p+l_n+l_p=$ even integer

(ii) weak rule $I>|j_n-j_p|$ $\quad j_n+j_p+l_n+l_p=$ odd integer (4.17)

A modified version of Nordheim's rules (Brennan and Bernstein 1960) states that

(i)′ strong rule $I=|j_n-j_p|$ $\quad N_0=0$

(ii)′ weak rule $I=|j_n-j_p|$ or $|j_n+j_p|$ $\quad N_0=\pm 1$ (4.18)

where N_0 is Nordheim's number defined by

$$N_0=(j_n+j_p)-(l_n+l_p) \tag{4.19}$$

The weak rule does not in general provide a unique prediction for the ground state spin, e.g. all odd–odd stable nuclei listed in section 3.1 have $I>0$ as required by the weak rule (ii) which does not distinguish between $I=1$ (${}^{6}_{3}Li$) and $I=3$ (${}^{10}_{5}B$). The nucleus ${}^{6}_{3}Li$ violates version (ii)′ of the weak rule. On the other hand the most famous example of all, the nucleus ${}^{210}_{83}Bi$, which has one proton and one neutron in excess of the magic numbers 82 and 126, respectively, has a ground state spin-parity 1^- which is consistent with neither of Nordheim's rules in their original formulation. It does however agree with the weak rule (ii)′ assuming the additional proton is in the $2f_{7/2}$ orbital and the additional neutron is in the $2g_{9/2}$ orbital.

These ideas may be formalized by introducing the seniority quantum number v, which represents the number of uncoupled identical nucleons in a state of the configuration $(nlj)^k$, where $p=\frac{1}{2}(k-v)$ gives the number of saturated pairs of identical numbers. Mayer's coupling rules (i) and (ii) may then be expressed by the rule that the ground state of the configuration $(nlj)^k$ has seniority $v=0$ for k even (always), and $v=1$ for k odd (usually).

A state of the configuration $(nlj)^k$ of seniority v has many properties in common with the state of the configuration $(nlj)^v$ from which it was constructed by the addition of p pairs. In particular, since the multipole moments of odd (even) order, although equal in magnitude in the substates $|j, m_j\rangle$, have opposite (equal) signs, a saturated pair has vanishing multipole moments of odd order and corresponding states have equal multipole moments of odd order. This is why the magnetic moment in the ground state of an odd A nucleus is just the magnetic moment of the uncoupled nucleon in its shell model orbital. Multipole moments of even order do not vanish for a saturated pair and are therefore not equal for corresponding states. They are however reduced in a state of seniority v containing p pairs relative to their value in the corresponding state containing no pairs. The contribution made to the quadrupole moment by adding p pairs to the single nucleon in the (nlj) orbital can be calculated very simply. Since the measured single particle quadrupole moment is just the matrix element $\langle jj|Q|jj\rangle$, and since the p pairs are distributed over the magnetic substates, $|m_j|\neq j$, with uniform probability $p/(j-\frac{1}{2})$ per substate, the measured quadrupole moment with p pairs present is given by

$$Q(p)=\langle jj|Q|jj\rangle+\sum_{m_j=-(j-1)}^{(j-1)}\frac{p}{(j-\frac{1}{2})}\langle jm_j|Q|jm_j\rangle$$

$$=\left(1-\frac{2p}{j-\frac{1}{2}}\right)\langle jj|Q|jj\rangle \tag{4.20}$$

where, for a closed shell

$$\sum_{m=-j}^{i}\langle jm_j|Q|jm_j\rangle=0$$

Thus, according to the extreme independent particle model as expressed by Mayer's coupling rules, the net effect of adding pairs successively to the single nucleon outside closed shells, is to convert the negative quadrupole moment at $p=0$ (oblate nucleus) into a positive quadrupole moment at $p=(j-\frac{1}{2})$ (prolate nucleus) where there is a single hole in the closed shell system. Thus the quadrupole moment is predicted to change sign approximately in the middle of the shell.

4.3 Simple shell-model tests

4.3.1 Electromagnetic multiple moments

The nuclear shell model works best when applied to the description of nuclear ground-states in ranges of Z and N not too far removed from closed shells. Apart from the question of testing the angular momentum coupling rules against the observed spins and parities, the most obvious way of bringing about a direct confrontation between theory and experiment is through a comparison between calculated and measured electromagnetic multipole moments. In general the matrix element of the 2^L-pole moment vanishes from angular momentum conservation in the (nlj) orbital unless $j \geqslant \frac{1}{2}L$ and, to conserve parity, these matrix elements are non-vanishing for even-valued electric multipoles and for odd-valued magnetic multipoles only. The vanishing of odd-valued electric multipoles also follows from time-reversal invariance which explains why the attempt to observe a finite neutron electric dipole moment has attracted so much effort (cf. section 9.2.5.4). Thus in practice only magnetic dipole and electric quadrupole moments are of interest and of these only the former is uniquely determined, for a specified angular momentum coupling scheme, by the quantum numbers j and v. For the calculation of electric quadrupole moments in particular (Schuler and Schmidt 1935), and electromagnetic transition matrix elements in general, some additional theoretical input is required.

In the extreme single particle model as defined by Mayer's coupling rules (3.75)–(3.77), the magnetic moment of an odd-A nucleus is given by

$$\boldsymbol{\mu} = g_j \sum_{i=1}^{A} \boldsymbol{j}_i = g_j \boldsymbol{I} \tag{4.21}$$

where j is the angular momentum of the orbital containing an odd number of identical nucleons, g_j is the Landé g-factor

$$\begin{aligned} g_j &= g_l + \tfrac{1}{2}(g_s - g_l)/j & j &= l + \tfrac{1}{2} \\ &= g_l - \tfrac{1}{2}(g_s - g_l)/(j+1) & j &= l - \tfrac{1}{2} \end{aligned} \tag{4.22}$$

and the orbital and spin g-factors are given by

$$\begin{aligned} g_l &= 1 & \tfrac{1}{2}g_s &= 2.72985 & &\text{odd } Z \\ g_l &= 0 & \tfrac{1}{2}g_s &= -1.91304 & &\text{odd } N \end{aligned} \tag{4.23}$$

The measured magnetic moment is therefore given in terms of the nuclear magneton $e\hbar/2m_p c$ by

$$\mu = g_j I \qquad j = l \pm \tfrac{1}{2} \tag{4.24}$$

The rules (4.22)–(4.24) imply that, for odd Z, the orbital with $j = l + \frac{1}{2}$ has the larger magnetic moment in the doublet and $\mu_j > 0$ except for the $p_{1/2}$ orbital where, because of the large anomaly in the proton moment, the nuclear magnetic moment is negative, e.g. in ${}^{15}_{7}$N. For odd N nuclei $\mu_j > 0$ for $j = l - \frac{1}{2}$

and $\mu_j<0$ for $j=l+\frac{1}{2}$. Exceptional cases can arise however, even in odd A nuclei, where both neutrons and protons contribute to the magnetic moment, e.g. in $^{19}_{9}$F. To extend these results to include odd–odd nuclei (and exceptionally odd–even or even–odd nuclei) it is necessary to assume that each odd group maintains the same quantum numbers it had when associated with an even number of the opposite nucleonic species. In this case (4.21) goes over to

$$\boldsymbol{\mu}=g_{j_p}\boldsymbol{I}_p+g_{j_n}\boldsymbol{I}_n \tag{4.25}$$

and, evaluating matrix elements in a nuclear state of spin I, we find in place of (4.24).

$$\mu=\tfrac{1}{2}I\left\{(g_{j_p}+g_{j_n})+(g_{j_p}-g_{j_n})\left(\frac{I_p(I_p+1)-I_n(I_n+1)}{I(I+1)}\right)\right\} \tag{4.26}$$

Either or both of I_p and I_n may be integral or half-integral in (4.26), which reduces to (4.24) when either I_p or I_n is zero. Clearly the next simplest case occurs for a self-conjugate nucleus with $Z=N$ and $I_p=I_n$.

Table 4.1 lists the measured magnetic moments for a range of light nuclei in the s, p and d shells, together with shell-model configurations and theoretical values for the magnetic moments calculated using pure j–j coupling. Experimentally it is found that the measured moments for odd-A nuclei lie either very close to or within the 'Schmidt lines' defined by plotting μ as a function of I for $j=l\pm\frac{1}{2}$ according to (4.22) and (4.24) (Schmidt 1939). For odd–odd nuclei also, the correlation between experiment and theory is so close as to confirm the essential validity of the jj-coupling scheme. To explain the observed deviations between experiment and theory a much more detailed theoretical description is required and this usually hinges on the idea of constructing single particle states by superposition of neighbouring shell-model orbitals. This procedure is called 'configuration mixing' and is necessary to account both for the binding energy, and for the observed multipole moments of the deuteron (cf. section 3.3.2.1 and 3.5.3). To account for the binding energy in $^3_1$H and $^3_2$He, an admixture of about 4% $d_{1/2}$ state is required (Blatt and Weisskopf 1952) but, even after this correction has been applied, there remain residual anomalies in the magnetic moments which have been interpreted as evidence in favour of exchange moments (cf. section 3.4.1) (Villars 1947, Sachs 1948).

The evidence from electric quadrupole moments is not so decisive for, although we should expect from (4.20) that Q be negative (positive) for a closed shell plus (minus) one proton, and this is indeed observed to be the case, the theory breaks down totally between major closed shells where quadrupoles ten times larger than predicted are observed and most frequently with the wrong sign. There is also another very striking feature of the quadrupole moment which is well illustrated by the nucleus $^{17}_{8}$O. According to the data in table 4.1 this nucleus is assumed to have an odd neutron in

Table 4.1. Experimental and theoretical magnetic moments calculated on the basis of pure *j–j* coupling for a range of stable and long-lived light nuclei in the s-, p- and d-shells

odd Z	I^π	p-state	odd Z odd N	I^π	n-state	odd N	I^π	μ_{th}	μ_{exp}
$^1_1$p	$\frac{1}{2}^+$	$s_{1/2}$							2.79277
					$s_{1/2}$	$^1_0$n	$\frac{1}{2}^+$		-1.91314
		$s_{1/2}$	$^2_1$H	1^+	$s_{1/2}$			0.8797	0.857406
$^3_1$H[a]	$\frac{1}{2}^+$	$s_{1/2}$						2.7928	2.97885
					$s_{1/2}$	$^3_2$He	$\frac{1}{2}^+$	-1.9131	-2.12755
		$p_{3/2}$	$^6_3$Li	1^+	$p_{3/2}$			0.6265	0.822010
$^7_3$Li	$\frac{3}{2}^-$	$p_{3/2}$						3.7928	3.25628
					$p_{3/2}$	$^9_4$Be	$\frac{3}{2}^-$	-1.9131	-1.17744
		$(p_{3/2})^{-1}$	$^{10}_5$B	3^+	$(p_{3/2})^{-1}$			1.8797	1.80063
$^{11}_5$B	$\frac{3}{2}^-$	$(p_{3/2})^{-1}$						3.7928	2.68857
					$(p\frac{1}{2})^{-1}$	$^{13}_6$C	$\frac{1}{2}^-$	0.6369	0.702381
		$(p_{1/2})^{-1}$	$^{14}_7$N	1^+	$(p_{1/2})^{-1}$			0.3667	0.40361
$^{15}_7$N	$\frac{1}{2}^-$	$(p_{1/2})^{-1}$						-0.2643	-0.28309
					$d_{5/2}$	$^{17}_8$O	$\frac{5}{2}^+$	-1.9131	-1.89370
$^{19}_9$F	$\frac{1}{2}^+$	[c]$(d_{5/2})^1$			[c]$(d_{5/2})^2$			2.7470	2.6289
					[d]$(d_{5/2})^3$	$^{21}_{10}$Ne	$\frac{3}{2}^+$	-1.1479	-0.66176
		[d]$(d_{5/2})^3$	$^{22}_{11}$N[b]	3^+	[d]$(d_{5/2})^3$			1.7278	1.746
$^{23}_{11}$Na	$\frac{3}{2}^+$	[d]$(d_{5/2})^3$						2.8757	2.21751
					$d_{5/2}$	$^{25}_{12}$Mg	$\frac{5}{2}^+$	-1.9131	-0.85512

[a] $T_{1/2} = 12.26$ y.
[b] $T_{1/2} = 2.6$ y.
[c] $I_p = \frac{5}{2}$, $I_n = 2$.
[d] $I = (j-1) = \frac{3}{2}$.

the $d_{5/2}$ shell and yet the measured quadrupole moment $Q = -2.6\ \text{fm}^2$, is directly comparable with the result $Q = -3.5\ \text{fm}^2$ one might expect from (4.20) for an odd proton in the $d_{5/2}$ shell and a value of $\langle r^2 \rangle$ given by (3.10) and (3.24) with $r_0 = 1.25$ fm. This is an effect due to the recoil of the even–even core which, for a neutron at r_n, corresponds to a displacement of the nuclear charge Ze from the centre of mass by an amount r_n/A. Thus the induced quadrupole moment is of order $Z\langle r_n^2 \rangle / A^2$ as observed. In general nuclear recoil effects can be described by attaching effective charges to neutrons and protons which, for an electric multipole of order L are given by

$$q_{Lp} = [(A-1)/A]^L + (-1)^L[(Z-1)/A^L]$$
$$q_{Ln} = (-1)^L Z/A^L \qquad (4.27)$$

For a quadrupole ($L=2$), q_p and q_n always have the same sign whereas for a dipole ($L=1$) the signs are opposite. For a self-conjugate nucleus $q_{1p}+q_{1n}=0$, a result which gives rise to an important selection rule for $E1$ transitions (Morpurgo 1954).

The physical phenomenon which is responsible for the observation of very large quadrupole moments in nuclei lying between closed shells cannot be described within the confines of the single particle shell model and to encompass these effects some radical alteration to the model must be invoked.

4.3.2 Nuclear isomerism

One outstanding success of the nuclear shell model with inverted spin–orbit coupling was that it provided a natural explanation for the observation that odd A nuclear isomers are confined to well defined 'islands' in the range of values of the odd member of the (Z, N) pair. An isomer is a long-lived excited nuclear state whose anomalous lifetime, typically on the scale of seconds to years, is associated with a large multipolarity L, i.e. a large nuclear spin change, in the de-exciting γ-transition (Weizsacker 1936). Thus the 'island' of isomers in the range 39–49 is explained by the fact that the lowest 1g orbital with $j=\frac{9}{2}$ and even parity, is only slightly higher in energy than the highest 2p orbital with $j=\frac{1}{2}$ and odd parity.

Examples of isomers are found in the isotopes $^{87}_{39}Y$, $^{89}_{39}Y$ and $^{91}_{39}Y$, and all arise in the same way. Since the 2p shell closes with 40 identical nucleons in the nucleus, and the 1g shell closes with 50 identical nucleons, a γ-transition $1g_{9/2}\rightarrow 2p_{1/2}$ for N or Z in the range 39–49 requires a change in parity and a spin change of 4. It is therefore classified as a magnetic 2^4-pole ($M4$) electromagnetic transition and would be listed as 'fourth forbidden' on a scale of forbiddeness in which electric dipole ($E1$) transitions are regarded as 'allowed'.

4.3.3 Nuclear β-decay

The weak interaction responsible for nuclear β-decay is similar to the electromagnetic interaction in that both interactions are mediated by vector gauge bosons. They differ primarily in two respects; first, whereas the electromagnetic current which couples to the gauge field is purely polar vector, the corresponding weak current is a mixture of polar vector and axial vector components whose interference provides a method for observing the phenomenon of parity violation. Second, the electromagnetic interaction is long range, corresponding to the fact that the photon is massless, whereas the weak interaction is short range and the mediating vector bosons are very massive ($m_w \simeq 10^2 m_p$). The effective weak interaction is therefore a point interaction between the four participating fermions and the corresponding weak Hamiltonian can be expressed as a contraction of hadronic ($n\rightarrow p$) and leptonic ($\nu_e \rightarrow e^-$) weak transition currents, which are charged in the sense that one unit of charge is always exchanged between hadronic and leptonic participants.

For allowed polar vector (Fermi) β-decay the leptons are emitted in an s-wave singlet state and the nuclear spin-parity selection rule is

$$\Delta I = 0 \qquad \Delta\pi = \text{no} \qquad \text{(Fermi)} \tag{4.28}$$

where $\Delta\pi$ represents the nuclear parity change. There is no analogous electromagnetic ($E0$) transition since massless photons are purely transverse. For allowed axial vector (Gamow–Teller) decays the leptons are emitted in an s-wave triplet state and the selection rule is

$$\Delta I = 0 \pm 1 \qquad \text{no } 0 \rightarrow 0 \qquad \Delta\pi = \text{no} \qquad \text{(Gamow–Teller)} \tag{4.29}$$

This is the same selection rule as applies to magnetic dipole ($M1$) electromagnetic transitions.

Experimentally both pure Fermi ($0^+ \rightarrow 0^+$) transitions (e.g. ${}^{14}_{8}\mathrm{O} \rightarrow {}^{14}_{7}\mathrm{N}$) and pure Gamow–Teller ($\Delta I = \pm 1$) transitions (e.g. ${}^{6}_{2}\mathrm{He} \rightarrow {}^{6}_{3}\mathrm{Li}$) are observed, and the absence of Fierz interference excludes the presence in significant amounts of other (scalar, tensor) Fermi or Gamow–Teller interactions (Fierz 1937). In these circumstances the electron spectrum shape in allowed transitions is determined by phase-space alone, and β-transitions are classified as allowed or forbidden according to their '$\log_{10} ft$-values' where t is the half-life ($T_{1/2}$) and f is the Fermi phase-space integral. For normal allowed transitions this number lies in the range

$$3 < \log_{10} ft < 5 \qquad \text{(allowed)} \tag{4.30}$$

although in special circumstances certain hindrance factors can operate and $\log_{10} ft$-values as large as 7 or 8 may be observed (Alaga 1955, 1957). Of much greater importance from the viewpoint of fundamental theory are the 'superallowed' transitions whose significance to the description of the weak interaction properties of the neutron will be discussed in section 9.3.7. These are allowed β-transitions with values of $\log_{10} ft < 3.5$, which include β-transitions between mirror pairs (isospin doublets), and $0^+ \rightarrow 0^+$ pure Fermi transitions within an isospin triplet (cf. figure 3.5).

Transitions arising from p-wave lepton emission, and relativistic corrections to s-wave emission, are termed 'first-forbidden' and are governed by the selection rules

$$\begin{aligned} &\Delta I = 0 \pm 1 \text{ (non-unique)} \qquad \Delta\pi = \text{yes} \qquad 5 < \log_{10} ft < 7 \\ &\Delta I = \pm 2 \text{ (unique)} \qquad \Delta\pi = \text{yes} \qquad 8 < \log_{10} ft < 9 \end{aligned} \tag{4.31}$$

The 'unique' transitions are defined in general by the rule $|\Delta I| = n + 1$, where n is the degree of forbiddeness, and may be recognized experimentally by their characteristic spectrum shapes. The first recorded example of a unique transition is the decay of 58.8d ${}^{91}_{39}Y \to {}^{91}_{40}Zr$ (Langer and Price 1949). At the next level down second forbidden transitions involve spin changes $\Delta I = \pm 2$, ± 3 and $\Delta\pi = $ no, with $\log_{10} ft$-values in the range 11–13. Of particular interest is that group, also considered as second forbidden, for which $\Delta l = \pm 2$, but which otherwise have the same selection rules as allowed Gamow–Teller decays. These are called '*l*-forbidden' transitions and have $\log_{10} ft$-values typically in the range

$$5 < \log_{10} ft < 9 \qquad (l\text{-forbidden}) \tag{4.32}$$

Experimentally the vast majority of β-transitions are allowed or first forbidden and the systematic study of the lifetimes, end-point energies and spectrum shapes of β-transitions has enabled the spins and parities of very many nuclear ground-states and excited states to be assigned in a consistent way. Second forbidden transitions are also fairly common and the element technetium, which does not occur naturally, provides two good examples namely 2.6×10^6 years ${}^{97}_{43}Tc \to {}^{97}_{43}Mo$ and 2.12×10^5 years ${}^{99}_{43}Tc \to {}^{99}_{44}Ru$, both having $\Delta I = 2$, $\Delta\pi = $ no and $\log_{10} ft \simeq 12$. The natural β-emitters 1.26×10^9 years ${}^{40}_{19}K \to {}^{40}_{20}Cu$ ($\Delta I = 4$, $\Delta\pi = $ yes, $\log_{10} ft = 18.1$) and 6×10^{14} year ${}^{115}_{49}In \to {}^{115}_{50}Sb$ ($\Delta I = 4$, $\Delta\pi = $ no, $\log_{10} ft = 22.7$) are third and fourth forbidden, respectively.

The significance of β-decay for the testing of shell-model assignments may be understood by reference to figure 4.2 which illustrates the pattern of β-decays involving phosphorous isotopes in the middle of the 1d–2s shell. An examination of the corresponding $\log_{10} ft$-values shows that, with two notable exceptions, all these transitions have $\log_{10} ft$-values in the range 3.5–5.5, and are therefore allowed. This is consistent with the shell model predictions that the ground states should have even parity and spins in the range $0^+ \to 1^+$ (even A) or $\frac{1}{2}^+ \to \frac{3}{2}^+$ (odd A). The two exceptions are the pure ground-state to ground-state Gamow–Teller decays

$$\begin{aligned} &{}^{32}_{14}Si(0^+) \to {}^{32}_{15}P(1^+) \qquad && \log_{10} ft = 8.7 \\ &{}^{32}_{15}P(1^+) \to {}^{32}_{16}S(0^+) && \log_{10} ft = 7.9 \end{aligned} \tag{4.33}$$

which lead eventually to the stable self-conjugate nucleus ${}^{32}_{16}S$.

The single-particle shell-model interpretation of these two anomalous decays is quite clear. In ${}^{32}_{14}Si$ there is an even–even core composed of two (minor) closed shells with $N = Z = 14$, plus two saturated neutron pairs in the configuration $(d_{3/2})^4$. One of these neutrons decays leading to the odd–odd nucleus ${}^{32}_{15}P$ in a state of the configuration $(s_{1/2})^1_p\,(d_{3/2})^3_n$ with $I = 1$. This transition involves a change in orbital angular momentum of two units $|\Delta l| = 2$ and is therefore classed as *l*-forbidden, an interpretation supported by its large $\log_{10} ft$-value. So also is the subsequent β-transition in which a second $d_{3/2}$ neutron decays to an $s_{1/2}$ proton leaving the nucleus ${}^{32}_{16}S$ with two saturated pairs $(s_{1/2})^2_p(d_{3/2})^2_n$ outside minor closed shells. The conclusion

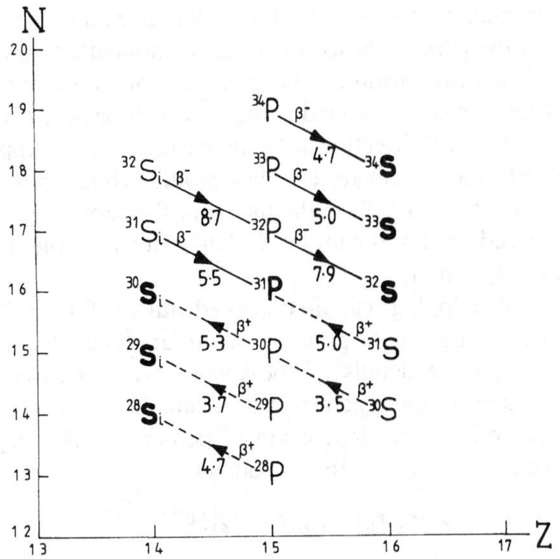

Figure 4.2. The sequence of allowed β-transitions within the $1d_{3/2}-2s_{1/2}$ shell which confirms the nuclear shell model prediction that all nuclear ground states in this region of the periodic table have even parity and spin 0 or 1 (even A), or spin $\frac{1}{2}$ or $\frac{3}{2}$ (odd A). The nuclei indicated in bold-face type are stable.

must be that all three nuclei involved have ground states derived from essentially pure single particle configurations. Conversely other β-decays in the complex, e.g. ${}^{34}_{15}P(1^+) \rightarrow {}^{34}_{16}S(0^+)$ with $\log_{10} ft = 5.2$, are not l-forbidden experimentally, indicating that these nuclei must have a substantial degree of $1d_{3/2} - 2s_{1/2}$ configuration mixing in their ground states.

4.3.4 Deuteron stripping reactions

The deuteron is a charged particle which does not readily penetrate the nuclear Coulomb barrier, and compound nucleus formation is therefore strongly suppressed in nuclear reactions with slow deuterons. However, the deuteron is also a large loosely bound structure, with a mean neutron–proton separation of about 2 fm, and, for this reason, it is possible for the neutron to be captured at the nuclear surface, while the residual proton is repelled by the Coulomb field. This is the process predicted by Oppenheimer and Philips (1935) and subsequently described as 'stripping' by Serber (1947). At energies above the Coulomb barrier, protons as well as neutrons can be stripped from the incident deuterons.

That the stripping reaction $({}^{A}_{Z})(d, p)({}^{A+1}_{Z})$ is a direct reaction which does not proceed through the formation of a compound nucleus is confirmed by the experimental observation that a larger number of partial waves contributes to the process. Since the angular momentum l_n of the captured neutron may in most instances be unambiguously determined from the angular distribution of the ejected protons, which peaks further and further away from the forward direction with increasing l_n, the stripping reaction is uniquely suited as a tool for testing shell model orbital assignments. In any case the measurement of l_n fixes the parity of the capturing state relative to that of the ground state of the target nucleus, which are the same if l_n is even and opposite if l_n is odd.

The theory of stripping was first worked out by Butler (1951) who was also the first to recognize its potential as an analyser of nuclear spins and parities. Although the details of the theory are somewhat complex, the essential mechanics of the reaction can be summarized fairly simply. In the initial state there is an incident beam of deuterons, of energy 5–10 MeV, which may be represented by the wavefunction

$$\psi_d(\boldsymbol{r}_n, \boldsymbol{r}_p) = \varphi_d(|\boldsymbol{r}_n - \boldsymbol{r}_p|)\, e^{i\boldsymbol{k}_d\cdot(\boldsymbol{r}_n+\boldsymbol{r}_p)/2} \tag{4.34}$$

where $\varphi_d(|\boldsymbol{r}_n - \boldsymbol{r}_p|)$ is the $^3S_1$ deuteron ground-state wavefunction and the deuteron mass is taken to be $m_d \simeq 2m_n \simeq 2m_p$. Stripping takes place on the surface of the target nucleus of spin $\boldsymbol{I}_i$, leaving a final nucleus of spin $\boldsymbol{I}_f$ containing an additional neutron, and an outgoing proton represented by the plane-wave $e^{i\boldsymbol{k}_p\cdot\boldsymbol{r}_p}$ neglecting Coulomb effects. The proton energy $E_p = \hbar^2 k_p^2/2m_p$ is fixed in terms of the deuteron energy $E_d = \hbar^2 k_d^2/4m_p$ by the relation

$$E_p = E_d + Q \tag{4.35}$$

where Q is positive when the neutron is captured into a bound state, which is normally the case. Thus $E_p > E_d$ as a general rule and, since the typical separation energy of the last neutron in a nucleus is of order 8 MeV, proton energies up to double the deuteron energy or higher are possible. This result also illustrates one advantage that stripping enjoys over compound nucleus reactions in that it picks out ground states and low excited states in preference to the highly excited states which are populated for example in neutron capture. Also individual nuclear states can be selected by recording only those protons which are emitted within a predetermined energy band.

The final state neutron is in a shell model orbital $u_l^n(\boldsymbol{r}_n)$ of the final state nucleus which recoils with momentum

$$\boldsymbol{q} = \boldsymbol{k}_d - \boldsymbol{k}_p \tag{4.36}$$

and, taking the z-axis along the direction of $\boldsymbol{q}$, the angular distribution of the ejected protons is expressed in terms of the polar angle θ defined from (3.98) by

$$q^2 = k_d^2 + k_p^2 - 2k_p k_d \cos\theta \tag{4.37}$$

The differential cross-section describing the proton emission is given approximately by an expression of the form

$$\frac{d\sigma(\theta)}{d\Omega}=4A_0^2\frac{k_p}{k_d}\frac{2I_f+1}{2I_i+1}\left|\int_R^{\infty} u_{l_n}(r)j_{l_n}(qr)r^2\,dr\right|^2 \tag{4.38}$$

where A_0 is the normalization constant for the asymptotic deuteron bound-state wavefunction (cf. section 2.6.11) and the statistical factor $(2I_f+1)/(2I_i+1)$ comes from averaging over the initial, and summing over the final, nuclear spin states. The relevant S-matrix element is given by the integral which is cut off at the nuclear surface $r=R$ in order to exclude compound nucleus contributions to the reaction process. The angular momentum $\boldsymbol{l}_n$ of the captured neutron must satisfy the conservation rule

$$\boldsymbol{I}_f=\boldsymbol{I}_i+\boldsymbol{l}_n+\boldsymbol{s}_n \tag{4.39}$$

which means that, for capture into a given state, any value of l_n can contribute to (4.38) provided it satisfies the relation

$$||I_i-I_f|-\tfrac{1}{2}|\leqslant l_n\leqslant I_i+I_f+\tfrac{1}{2} \tag{4.40}$$

and is either even or odd depending on the ground-state parity of the target nucleus. Also, from the point of view of reducing the complexity of the scattering pattern, there is a clear advantage in choosing $I_i=0$ since under these circumstances there are at most two values of I_f which can be reached by stripping and only one, i.e. $I_f=\frac{1}{2}$, if $l_n=0$.

The results obtained from a range of stripping reactions, mostly in light elements including several of the self-conjugate variety, are assembled in Table 4.2 where the measured values of l_n may be compared with the l-values predicted on the basis of shell model assignments for the orbital into which the stripped neutron is assumed to be captured. The agreement in every case is seen to be excellent. At higher mass values where there may be many competing shell model orbitals the stripping reaction is particularly effective in picking out which orbital makes the maximum contribution to the ground state configuration, and this is illustrated by the results observed with the heavy nucleus $^{140}_{58}$Ce, which has a magic neutron number $N=82$, showing conclusively that the additional neutron goes into the $2f_{7/2}$ orbital.

4.4 The extended shell model

4.4.1 The binding energy problem

The nuclear shell model is a phenomenological model in the sense that it permits the detailed features of nuclear energy levels, i.e. spins, parities and electromagnetic matrix elements, to be correlated and interpreted in terms of a simple picture of nucleons moving independently in a central potential.

Table 4.2. Determination of the angular momentum l_n of the captured neutron in deuteron stripping reactions leading to the ground state of the residual nucleus. These data may be compared with shell model predictions of the angular momentum l of the nearest available neutron orbital, which are also listed in the table

Stripping reaction	N_i	I_i^π	l_n	N_f	I_f^π	Neutron orbital	l
${}^{11}_{5}B(d, p){}^{12}_{5}B$	6	$\frac{3}{2}^-$	1	7	1^+	$1p_{1/2}$	1
${}^{14}_{7}N(d, p){}^{15}_{7}N$	7	1^+	1	8	$\frac{1}{2}^-$	$1p_{1/2}$	1
${}^{16}_{8}O(d, p){}^{17}_{8}O$	8‡	0^+	2	9	$\frac{5}{2}^+$	$1d_{5/2}$	2
${}^{24}_{12}Mg(d, p){}^{25}_{12}Mg$	12	0^+	2	13	$\frac{5}{2}^+$	$1d_{5/2}$	2
${}^{27}_{13}Al(d, p){}^{28}_{13}Al$	14†	$\frac{5}{2}^+$	0	15	3^+	$2s_{1/2}$	0
${}^{32}_{16}S(d, p){}^{33}_{16}S$	16	0^+	2	17	$\frac{3}{2}^+$	$1d_{3/2}$	2
${}^{40}_{20}Ca(d, p){}^{41}_{20}Ca$	20‡	0^+	3	21	$\frac{7}{2}^-$	$1f_{7/2}$	3
${}^{140}_{58}Ce(d, p){}^{141}_{58}Ce$	82‡	0^+	3	83	$\frac{7}{2}^-$	$1h_{9/2}$	5
						$2f_{7/2}$	3
						$1i_{13/2}$	6

† Minor closed shell.
‡ Major closed shell.

This potential is assumed to approximate, in first order, to the average interaction between pairs of nucleons, supplemented by a very specific spin–orbit term of unknown origin, which is necessary to arrive at the correct level order. In respect of its stated aims to correlate and interpret, the shell model is brilliantly successful and the detailed evidence which may be assembled to support its predictions is impressive. However, if the model is to be accepted as an accurate representation of a real nucleus, then it must also be able to describe the gross properties of nuclei such as their size, mass and density, and this it signally fails to do.

From the viewpoint of calculating total energies, as distinct from relative energies, there are two principal difficulties in the model. The first relates to the saturation properties of nuclei as expressed in the relation (3.10) for the nucleus radius R and the constancy of $\langle B/A \rangle$, the mean binding energy per nucleon (cf. section 3.1). The second difficulty concerns the observation that the average value of $\langle B/A \rangle$ of about 8 MeV is also equal to the average value of the separation energy S, which is defined as the energy required to release the least tightly bound nucleon from the nucleus. These statements must not be taken too literally, for neither B/A nor S are exactly constant but fluctuate from one nucleus to the next, each showing pronounced maxima at closed shells. Nevertheless, they are closely correlated and their approximate equality is clearly incompatible with the existence of a fixed shell model potential which, in any given nucleus, implies that $B/A > S$.

In any discussion of the average properties of nuclei it is permissible to ignore spin–orbit and tensor forces whose effects for a large nucleus tend to average to zero, and the problem may be further simplified by letting $N=Z$ and assuming that all forces are independent of charge and spin. The total kinetic energy may then be estimated by treating the nucleus as a Fermi gas of A nucleons confined in a volume $\omega(R)=\frac{4}{3}\pi R^3$, for which the Fermi energy is

$$E_F=\frac{\hbar^2 k_F^2}{2m}=\frac{h^2}{2m}\left(\frac{3\pi^2 A}{2\omega}\right)^{2/3} \tag{4.41}$$

This gives an average kinetic energy per nucleon $\frac{3}{5}E_F$, and a total nuclear kinetic energy

$$T(A, R)=\frac{3}{5}AE_F=\alpha_T A^{5/3} \tag{4.42}$$

where α_T is a constant. If pairs of nucleons are assumed to interact via an attractive square well potential of depth v_0 and range b, then the nuclear potential energy is

$$V(A, R) = -\tfrac{1}{2}(A)(A-1)v_0 p(R, b) \simeq -A^2 v_0 p/2 \tag{4.43}$$

where $p(R, b)$ is the probability that two nucleons confined in $\omega(R)$ are within the range b of their mutual interaction. The precise value of p will depend in general on the spatial distribution within $\omega(R)$ but, for values of the nuclear radius R, small and large, respectively, in comparison with b, p must have the value

(i) $p(R, b)=1 \qquad R\leqslant\frac{b}{2}$

(ii) $p(R, b)\simeq\omega(b)/\omega(R)=(b/R)^3 \qquad R\gg\frac{b}{2}$ (4.44)

Corresponding to the limits (i) and (ii) given in (4.43) the nuclear potential energy is

(i) $V(A, R)\simeq -A^2 v_0/2 = -\alpha_V^{(1)}A^2 \qquad R\leqslant\frac{b}{2}$

(ii) $V(A, R)\simeq -A^2 v_0 b^3/2R^3 = -\alpha_V^{(2)}A^2/R^3 \qquad R\gg\frac{b}{2}$ (4.45)

The total nuclear energy is now given by

$$E(A, R)=T(A, R)+V(A, R) \tag{4.46}$$

and the ground state of the nucleus is given by that equilibrium value of the radius $R=R_e$ which minimizes the total energy, i.e.

$$\frac{\partial E}{\partial R}(A, R_e)=0 \tag{4.47}$$

Since, according to (4.41) and (4.44), $T(A, R)$ increases in proportion to R^{-2}, while $V(A, R)$ decreases monotonically for $R \geqslant b/2$ and remains constant for $R<b/2$, it follows that $E(A, R)$ reaches its minimum value at $R_e=b/2$, independent of A, and all nuclei should have the same volume $\omega(R_e)$ contrary to observation. This result is essentially model-independent and, as discussed in section 3.3.2.2, cannot be altered except in detail, for any realistic combination of Wigner and Majorana exchange forces.

Evidently what is required is some mechanism which increases the equilibrium radius to a value $R_e \gg b/2$ under which conditions $V(A, T)=-\alpha_V^{(2)}A^2/R^3$ according to (4.45(ii)). In these circumstances the energy reaches a minimum at a value of the nuclear radius

$$R_e=\frac{3\alpha_V^{(2)}}{2\alpha_T}A^{1/3} \tag{4.48}$$

which reproduces the experimental result corresponding to saturation. The necessary conditions may be achieved by postulating a short-range repulsive force between nucleons which acts to increase their average separation and thereby to increase the size of the volume $\omega(R)$ required to contain A nucleons. For example, if nucleons are treated as hard spheres of radius r_c, there is an additional positive contribution to the nuclear potential energy which has the required dependence on A and R.

$$V_c(A, R)=A\frac{k_F^3 r_c}{\pi m}=\alpha_V^{(c)}A^2/R^3 \tag{4.49}$$

It follows that, with the addition of a hard core potential of the kind believed to play a necessary part in high-energy nucleon–nucleon scattering (cf. section 3.3.2.2), the shell-model can account, at least qualitatively, for the observed saturation properties of real nuclei.

The observed equality of S and B/A cannot be explained within the restrictions imposed by the shell-model scheme without introducing the additional assumption that nucleons in different shells move in different potential wells. To show that this is so we may suppose that a nucleon at the Fermi surface moves in a potential well $V_F<0$ so that the separation energy is given by

$$S=-(E_F+V_F) \tag{4.50}$$

If now a nucleon of total energy E_i moves in a potential well V_i then the mean binding energy per nucleon is

$$\langle B/A\rangle=-(\tfrac{3}{5}E_F+\langle V_i\rangle) \tag{4.51}$$

and the postulated equality of S and $\langle B/A \rangle$ leads to the result

$$\langle V_i \rangle = V_F + \tfrac{2}{5}E_F = V_F + \tfrac{2}{3}\langle T_i \rangle \tag{4.52}$$

where $T_i(E_i)$ represents the kinetic energy. Clearly the result (4.51) can be brought about if $V_i(E_i)$ is assumed to have a momentum-dependence of the form

$$V_i = V_F + \tfrac{2}{3}p_i^2/2m \tag{4.53}$$

a result which, following the discussion in section 3.3.2.3, relies on the fact that the energy–momentum relation for a nucleon moving in nuclear matter differs from that for a nucleon moving in free space. The result (3.76) corresponds to a value of the effective mass $m^* = 0.6m$ where m^* is defined in (3.68). Since the consequence of having a reduced effective mass $m^* < m$ is to lower the attractive potential, it also helps to reduce the nuclear density, and detailed calculations show that both a repulsive core and a momentum-dependent potential are necessary to produce a nuclear radius $R = r_0 A^{1/3}$ with $r_0 \simeq 1.2$ fm in agreement with observation (Brueckner 1955, 1955a, Brueckner and Levinson 1955).

4.4.2 Configuration mixing

One of the main weaknesses of the independent particle model is that it takes no account of configuration mixing whose effects have already been encountered in connection with the suppression of the l-forbidden selection rules (4.32) in allowed β-decay, and in deviations of measured magnetic moments from the Schmidt lines (4.24) (Horie and Arima 1954, 1955). Although, in the case of odd A $p_{1/2}$-nuclei, agreement with the single-particle predictions is almost exact, and examples listed in Table 4.1 are $^{13}_{6}C$ (odd N) and $^{15}_{7}N$ (odd Z), in fact almost all nuclear magnetic moments fall within the region bounded by the Schmidt lines. The principal exceptions to this rule are $^{3}_{1}H$ ($\Delta\mu = 0.25$) and $^{3}_{2}He$ ($\Delta\mu = -0.25$) whose moments lie outside this region, results which have been interpreted as evidence in favour of exchange moments (Villars 1947).

Another important observation is that, for odd A nuclei, the even-numbered nucleons appear to make little or no contribution to the deviations although again there are exceptional cases. For example, the nucleus $^{153}_{63}Eu$ ($I^\pi = \tfrac{5}{2}^+$) has a magnetic moment $\mu = 1.52$, whereas the isotope $^{151}_{63}Eu$ ($I^\pi = \tfrac{5}{2}^+$) has a magnetic moment $\mu = 3.42$, much closer to the predicted value $\mu = 4.79$ for an odd proton in a $2d_{5/2}$ orbital. One must conclude that the major deviation originates in an interaction of the odd particle with the odd-numbered closed shells, and a lesser part from the interaction with the even-numbered closed shells. This interaction is expressed as configuration mixing in states which, in lowest approximation, are derived from the lowest unoccupied single-particle orbitals.

Of the large number of possible admixtures clearly the most important are those whose contribution to the magnetic moment are linear in the mixing amplitudes, and the corresponding excited states of the core are those which have non-vanishing magnetic dipole matrix elements with the core ground state. Since the operator $\boldsymbol{\mu}$ is an axial vector ($I^\pi = 1^+$), the significant admixtures are those states involving the excitation of a nucleon from a closed shell, leaving a hole to which it couples to form a state with $I^\pi = 1^+$ and $m_I = 0$. This can only be achieved by mixing the two components in the spin-orbit doublet, which involves the excitation of a nucleon from the $j = l + \frac{1}{2}$ orbital to the $j = l - \frac{1}{2}$ orbital, although the argument fails if there is a single hole in the $j = l = \frac{1}{2}$ orbital, because in this case the Pauli principle forbids admixtures which contribute to the magnetic moment in first order with the required spin value $I = l - \frac{1}{2}$ (Blin-Stoyle 1953, Blin-Stoyle and Perks 1954, Blin-Stoyle 1956). This failure explains why the deviations observed in odd A $p_{1/2}$ nuclei are minimal.

Since two protons (or neutrons) interact most strongly in relative singlet s-states, the odd A nucleon tends to excite an identical nucleon into the opposite spin state, thereby effectively quenching its own intrinsic moment, with the result that the total moment moves in the direction of the opposite Schmidt line. The same result follows if the excited nucleon is of the opposite variety because, although neutrons and protons interact most strongly in a relative triplet s-state, they also have magnetic moments of opposite sign so that the effect of having parallel spin is negated. The net effect therefore still represents a quenching of the intrinsic magnetic moment.

This type of configuration mixing is nicely exemplified by the nucleus ${}^{209}_{83}\mathrm{Bi}$ which has a single proton outside a doubly magic ${}^{208}_{82}\mathrm{Pb}$ core and might therefore be expected to have a magnetic moment close to the appropriate Schmidt line. ${}^{209}_{83}\mathrm{Bi}$ has $I^\pi = \frac{9}{2}^-$ which means that the additional proton goes into the $1h_{9/2}$ orbital for which the predicted magnetic moment is $\mu = 2.6241\mu_N$, which is much less than the measured magnetic moment $\mu = 4.0802\mu_N$. Since in this case $I = l - \frac{1}{2}$, the spin and orbital contributions to the magnetic moment have opposite signs according to (4.22) so that the observation of a positive deviation $\Delta\mu = 1.3561$, implies a quenching of the intrinsic magnetic moment as described above. On the configuration mixing hypothesis this result is explained in terms of the particle–hole pair $[(1h_{9/2})(1h_{11/2})^{-1}]^{m_I=0}_{I=1}$ which generates an admixture in a single particle orbital $(1h^m_{9/2})$ of the form $[1h_{9/2}[(1h_{9/2})(1h_{11/2})^{-1}]]^m_{9/2}$ with amplitude proportional to the ratio of the pairing energy of the two protons to the splitting of the spin–orbit doublet. This of course represents a very simple example of configuration mixing and a much more complex situation arises with open-shell nuclei where the tensor forces produce very significant mixing with configurations at higher levels of excitation in the central field, as indeed happens with crucial effect in the deuteron bound state (cf. section 3.5.2).

4.5 Collective motion

4.5.1 Deformed nuclei

In the Fermi gas model of the nucleus it is assumed that nucleons are confined within the nuclear volume by some rather ill-defined 'nuclear force', and thereafter the internucleon forces are totally neglected. A significant advance on this model is made when the average interaction between nucleons is represented as an effective central potential and, with the addition of spin–orbit coupling and pairing effects, the final result is the shell model. However, as has already been emphasized above, the shell model predictions for electric quadrupole moments are supported by the experimental evidence only for nuclei where both Z and N differ from closed shell values by a few units at most. Away from the region of major shell closure the experimental data does not match shell model predictions even approximately, and it is evident that an element of major importance is missing. Thus both the qualitative and quantitative aspects of the quadrupole moment problem are of the greatest significance.

In the first place there are far more positive than negative quadrupole moments, representing an excess of prolate over oblate shapes. This is quite contrary to shell model ideas, which suggest that both shapes should occur with about equal frequency, corresponding to the expected sign change in mid-shell. Second, the absolute magnitudes of measured quadrupole moments are much larger than can be accounted for in terms of single particle motion plus pairing, suggesting that the effects are co-operative involving many nucleons in closed shells. These anomalous results are specially remarkable in the regions $A \simeq 25$, in the rare-earth region $150 < A < 190$, and for actinide elements with $A > 222$. The strongest departures from shell model theory are observed in the mid-shell range for both protons $(50 < Z < 82)$ and neutrons $(82 < N < 126)$. For example, in the even–odd nucleus $^{167}_{68}\mathrm{Er}$, the quadrupole moment is positive and approximately 16 times larger than its single-particle value, and, in the odd–odd nucleus $^{176}_{71}\mathrm{Lu}$, the quadrupole moment is also positive and approximately 36 times too large. Nuclei also exist with very large negative moments, particularly among the isotopes of antimony $(Z = 51)$ and iodine $(Z = 53)$.

These observations show that, far from closed shells, the assumption of independent particle motion in a spherically symmetric central potential modified by short-range pairing forces breaks down, and the closed shell nucleons co-operate to make a significant contribution to the total quadrupole moment. The suggestion that these phenomena have a collective nature (Rainwater 1950, Bohr 1952, Bohr and Mottelson 1953, 1955) is strongly supported by the observation of associated level structure, with level spacings which are far too small to be accounted for in terms of single-particle excitations, and $E2$ transition rates which are faster than single-particle transition rates by factors of order $10^{3\pm1}$. These low-lying excited states are heavily populated in the α-decays of actinide elements, they can be reached most easily by Coulomb excitation of the nuclear ground states, and indeed have many features in common with the rotational band structures observed in diatomic molecules, including the characteristic $I(I+1)$ level spacing rule.

The very earliest attempt at a theoretical description of collective motion in nuclei interpreted the observed excitations as modes of propagation of irrotational waves on the surface of a liquid drop (Bohr and Mottelson 1955), an idea which represented a return to a nuclear model which had proved very useful in the analysis of nuclear fission, but in other areas of nuclear science was somewhat outmoded. However, the predicted value of the moment of inertia associated with surface wave deformation was too small by at least a factor of three to reproduce the observed level spacings, and the hypothesis of quantized surface waves was soon abandoned. In fact such a radical departure from the basic philosophy of the shell model proves quite unnecessary, and a far more fruitful approach may be based on an extension of this model which assumes that the nuclear shape, whether prolate ellipsoid, oblate ellipsoid or completely lacking in cylindrical symmetry, is determined by single-particle motion in a deformed nuclear potential modified by residual short-range interactions between nucleons (Nilsson 1955).

To see how nuclear deformation arises we return to a many-body description of the nucleus and represent the two-body scalar potential $V(r_{12})$ as a multipole expansion

$$V(r_{12})=\sum_{\lambda=0}^{\infty} f_\lambda(r_1, r_2)P_\lambda(\cos\theta_{12}) \tag{4.54}$$

where odd values of λ are excluded by parity conservation. Since $P_0(\cos\theta_{12})=1$, the monopole potential $f_0(r_1, r_2)$ is spherically symmetric and may be applied, at least in principle, to determine a self-consistent central shell-model field according to the Hartree–Fock procedure. For $\lambda \geqslant 2$, $P_\lambda(\cos\theta_{12})$ is an oscillatory function of θ_{12}, very similar to the optical diffraction pattern of a single slit, with a central lobe of width $\Delta\theta_{12} \simeq \lambda^{-1}$. It follows that the λ-multipole force is negligible outside a range $b_\lambda \simeq R/\lambda$, where R is a characteristic nuclear radius, and the greater the value of λ, the shorter the range of the corresponding force.

It now becomes clear that, aside from the monopole term ($\lambda=0$) which determines the average field in which nucleons move, the most important long-range force is the quadrupole term ($\lambda=2$), and it is this term which is responsible for nuclear shape deformation and the resultant collective modes of excitation. One cannot of course entirely ignore the higher multipoles which describe the short-range forces, and it is convenient to lump all these terms together and represent them as a single pairing force. If this procedure is valid, then it is not at all surprising that the simple shell model gives incorrect results when applied to nuclei with many valence nucleons, since it entirely ignores the most important long-range component of the residual force between nucleons, i.e. the quadrupole force, which has precisely those properties which are required to generate collective motion and to lead to enhanced values of the quadrupole moment.

To investigate the nature of the excited core configuration, which the quadrupole interaction between core and valence nucleons admixes into the single-particle states, we may suppose that this interaction excites a particle from the closed shell orbital $(n_1l_1j_1)$ into the valence orbital $(n_2l_2j_2)$. Then, to repeat an argument already advanced in connection with anomalous magnetic moments in section 3.3.4.2, the resultant particle hole pair must couple to a state $[(n_2l_2j_2)(n_1l_1j_1)^{-1}]_{I=2}^{m_I=0}$ to produce a contribution to the quadrupole moment which is linear in the mixing amplitude. For a given single particle orbital (nlj), with $j>\frac{1}{2}$, there are many values of j_1 and j_2 which can produce an admixture of the required form $[(nlj)(n_2l_2j)(n_1l_1j_1)^{-1}]_j^m$, and therefore very many core nucleons which can contribute to the quadrupole moment. To calculate the resultant configuration mixing it is necessary to have an exact expression for the quadrupole factor $f_2(r_1, r_2)$ in the interaction potential, but, if we adopt an approximation originally introduced by Elliott in a wider context (Elliott 1958, Moszkowski 1958)

$$f_2(r_1, r_2)=\kappa r_1^2 r_2^2 \tag{4.55}$$

then the matrix element of the interaction separates in the radial coordinates r_1, r_2, and the mixing amplitude is proportional to the quadrupole moment of the valence nucleon.

The net result is that the closed shells are aligned by the valence nucleons and the quadrupole force produces an effect which may be described as an interaction between each nucleon and the total quadrupole moment of the nucleus. Because the single-particle quadrupole moment in the state $|jm\rangle$ depends only on $|m|$, the quadrupole interaction produces an alignment rather than a polarization of the closed shells with respect to a given direction in space, since it does not distinguish between the positive and negative senses of that direction. There is no equivalent magnetic dipole 'polarization' of the closed shell since $\lambda=1$ multipoles do not exist. There is however a weak magnetic octupole alignment $(\lambda=4)$ for $j\geqslant\frac{5}{2}$, and indeed there is also some experimental evidence for the existence of static magnetic octupoles (Jaccarino *et al* 1954, Kusch and Eck 1954, Daly and Holloway 1954, Schwartz 1955). The main consequence of the quadrupole alignment is that the equilibrium shape of the nucleus is not spherical even for just one nucleon outside closed shells, and this deformation of the nuclear surface grows steadily as the occupation number of the valence shell increases.

Whether the oblate or the prolate deformation is favoured energetically depends on the detailed form of the central potential and the number k of valence nucleons. The tensor forces tend to produce an effect which favours prolate shapes, i.e. positive quadrupole moments as in the deuteron, and this expectation is in accord with observation. On the other hand the centrifugal forces associated with particles in high angular momentum states tend to produce oblate forms for a few nucleons, and prolate forms for a few holes, outside closed shells, and this prediction also accords with the experimental evidence. Of course the tendency of the quadrupole force to amplify any existing nuclear deformation is always opposed by the pairing forces which act to restore the spherical symmetry, and the net quadrupole moment in any particular nucleus represents the resultant of these opposing tendencies. However, since the pairing energy is proportional to k, and the quadrupole energy to k^2, the quadrupole force must inevitably dominate in mid-shell.

The adoption of a non-spherically symmetric average field implies that the parameters which characterize the shape and orientation of the deformed nucleus must be included as dynamical variables whose variation is observed in states of collective motion. In the region remote from closed shells it is legitimate to assume that intrinsic and collective modes may be separated, and the nuclear wavefunction expressed in product form (Mottelson and Nilsson 1959)

$$\Psi = \psi \chi_{\mathrm{rot}} \varphi_{\mathrm{vib}} \tag{4.56}$$

where ψ describes the intrinsic motion of the valence nucleons in the field of the deformed core, χ_{rot} describes the rotation of the deformed field without change of shape, and φ_{vib} represents the vibrational motion about the equilibrium shape. There are also transition regions in the spectrum when the separation of the wavefunction fails and coupling between intrinsic and rotational motion, and between rotation and vibration is important.

In general the shape of a deformed surface may be represented by Rayleigh's formula

$$R(\theta, \varphi) = R_0 \left\{ 1 \times \sum_{l=0}^{\infty} \sum_{m=-l}^{l} \alpha_{lm} Y_l^m(\theta, \varphi) \right\} \tag{4.57}$$

where, in the nuclear context, R_0 is just the mean charge radius (cf. section 3.2). To maintain the invariant status of $R(\theta, \varphi)$ the coefficients α_{lm} must transform under rotations like Y_l^{m*}. For nuclei the monopole term $l=0$ vanishes because the nuclear density is constant, while the dipole term also vanishes for an isolated nucleus because it implies a displacement of the centre of mass. In a photonuclear reaction however the motion of the centre of mass can be separated off as nuclear Thomson scattering, and there remains a residual dipole oscillation which, in the hydrodynamic model, is described as a vibration of neutrons against protons, and gives rise to the giant dipole resonance at an excitation energy of 10–20 MeV in most nuclei (Goldhaber and Teller 1948). This phenomenon can be described in terms of the effective charge formalism introduced in section 4.3.1. Thus, for an isolated nucleus, the most important deformation is the quadrupole deformation described by the terms with $l=2$ in (4.57), which is characterized by the five generalized collective coordinates $\alpha_{2,m}$, with $-2 \leqslant m \leqslant 2$. To second order in $\alpha_{2,m}$ and $\dot{\alpha}_{2,m}$, the Hamiltonian describing small oscillations about the equilibrium state with $\alpha_{2,m} \equiv 0$, is identical with that of the five-dimensional harmonic oscillator.

In the immediate vicinity of a closed shell it is possible for a nucleus to undergo collective vibrations about the spherical equilibrium shape with a restoring force provided by the short-range pairing interaction. Since these are quadrupole rather than dipole vibrations, and may be regarded as oscillations in the quadrupole moment about a mean value zero, the corresponding phonons have even parity and carry two units of angular momentum. The energy levels with n phonons of excitation are given by the familiar spectrum with constant level spacing.

$$E_n = (n + \tfrac{5}{2})\hbar\omega \tag{4.58}$$

where the $\frac{5}{2}\hbar\omega$ zero point energy signifies that this is a five-dimensional harmonic oscillation. As with the three-dimensional oscillator level scheme shown in figure 4.1, states with $n > 2$ are degenerate, although the degeneracy is lifted when two-phonon interactions are taken into account. This state of collective motion is referred to as the anharmonic vibrator limit. An example of vibrational motion in the nucleus ${}^{110}_{48}Cd$ is shown in figure 4.3.

When there is a permanent deformation, such that the potential energy is a minimum at values of $\alpha_{2,m}$ not all zero, this deformation is most simply described, in the principal axes coordinate system, by the collective coordinates

$$\alpha'_{2,1} = \alpha'_{2,-1} = 0 \qquad \alpha'_{2,0} = \beta \cos \gamma \qquad \alpha'_{2,2} = \alpha'_{2,-2} = \frac{\beta}{\sqrt{2}} \sin \gamma \tag{4.59}$$

Thus, under the action of a quadrupole force, the nucleus assumes an ellipsoidal shape, with semi-principal axes given by the formula

$$R_\kappa = R_0 \left\{ 1 + \left(\frac{5}{4\pi} \right)^{1/2} \beta \cos\left(\gamma - \frac{2\pi\kappa}{3} \right) \right\} \qquad \kappa = 1, 2, 3 \tag{4.60}$$

The transformation from the $\alpha_{2,m}$ to the $\alpha'_{2,m}$ is accomplished by a rotation specified by Euler angles θ_i which describe the orientation of the intrinsic (principal axes) system with respect to the laboratory system. The five collective coordinates $\alpha_{2,m}$ are therefore replaced by β, γ together with the three Euler angles. However the potential energy, being a scalar, is a function of β and γ only. The coordinate γ, which may be restricted to the range $0 \leqslant \gamma \leqslant \pi/3$ without loss of generality, measures the departure from cylindrical symmetry. However, the vast majority of nuclei are

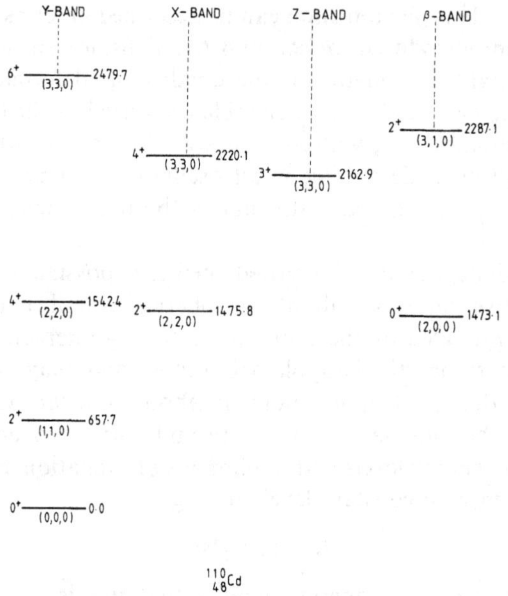

Figure 4.3. Quadrupole vibrational modes in the nucleus $^{110}_{48}$Cd. According to the interacting boson model these states are characterized by SU(5) quantum numbers (n_d, τ, n_Δ) for $N = N_\pi + N_\nu = 1 + 6 = 7$. All four vibration bands have $n_\Delta = 0$ with $\tau = n_d$ (Y, X, Z) or $n_d - 2$ (β). The angular momentum quantum numbers are $L = 2\tau$ (Y, β), $2\tau - 2$ (X) or $2\tau - 4$ (Z) (Arima and Iachello 1976).

cylindrically symmetric corresponding to the values $\gamma = 0$ (prolate nucleus) or $\gamma = \pi/3$ (oblate nucleus). This reflects the fact that the minimum of the potential energy, evaluated to terms of fourth order in $\alpha_{2,m}$, always occurs at $\gamma = 0$ or $\gamma = \pi/3$. If ΔR represents the difference between the polar radius and the equatorial radius in a spheroidal nucleus, then an asymmetry parameter δ may be defined by

$$\delta = \frac{\Delta R}{R_0} = \frac{3}{2}\left(\frac{5}{4\pi}\right)^{1/2} \beta \simeq 0.95\beta \text{ (prolate)}$$

$$= -\frac{3}{2}\left(\frac{5}{4\pi}\right)^{1/2} \beta \simeq -0.95\beta \text{ (oblate)} \qquad (4.61)$$

and the parameter $\beta > 0$ is referred to as the deformation parameter.

The restriction to cylindrical symmetry in a nucleus where the intrinsic and collective degrees of freedom separate according to (4.56) implies that, in addition to the total angular momentum quantum number I, and its component m_I along an arbitrary space-fixed axis, the component of angular momentum K along the symmetry axis is also a constant of the motion. Also, since there is no collective motion about a symmetry axis, $K=\sum_{i=1}^{k} K_i$, where K_i is the angular momentum component of a valence nucleon. Thus K is a characteristic of the intrinsic state and is strictly an adiabatic invariant rather than a true constant of the motion. The adiabatic condition fails if the intrinsic motion decouples from the collective motion, a state of affairs which may be brought about through the action of Coriolis forces in high angular momentum states. Since the Coriolis potential couples states differing in K by ± 1, it has diagonal elements in states with $K=\pm\frac{1}{2}$ and in these states must be taken into account. Also, because of the reflection symmetry in a plane through the nuclear centre normal to the symmetry axis, states with $K=\pm|K|$ are degenerate and it suffices to specify the positive value K. For the same reason collective states built on a given intrinsic state have the parity of that state.

In the Nilsson model of a deformed nucleus (Nilsson 1955, Mottelson and Nilsson 1959), the single-particle states are calculated as a function of δ in the spheroidal field of a three-dimensional axisymmetric harmonic oscillator, with principal angular frequencies given by

$$\omega_1^2=\omega_2^2=\omega_0^2\left\{1+\frac{2}{3}\delta\right\} \qquad \omega_3^2=\omega_0^2\left\{1-\frac{4}{3}\delta\right\} \tag{4.62}$$

The asymmetry parameter δ may be determined from the intrinsic quadrupole Q_0 where

$$Q_0=\frac{4}{5}ZR_0^2\delta\left(1+\frac{\delta}{2}\right) \tag{4.63}$$

is the quadrupole moment of a uniformly charged ellipsoid of revolution. In an angular momentum coupling scheme when K is a good quantum number, Q_0 is related to the observed quadrupole moment Q by

$$Q=Q_0\frac{I(2I-1)}{(I+1)(2I+3)} \tag{4.64}$$

Thus, although a nucleus can have a permanent deformation and a non-vanishing Q_0, this cannot lead to an observable Q if $I=0$ or $\frac{1}{2}$, because there is no axis of alignment, specified by distinct values of $|m_I|$, along which the intrinsic quadrupole moment can be oriented.

In the limit $\delta\rightarrow 0$, the Nilsson model reduces to the usual harmonic oscillator shell-model with its characteristic $(2j+1)$-fold degenerate single particle orbitals (nlj). For $\delta\neq 0$, j is no longer a good quantum number, but, to first approximation, the orbital splits into $\frac{1}{2}(2j+1)$ states labelled by values of K_i in the range $-j\leqslant K_i\leqslant j$ which are degenerate in pairs with $K_i=\pm|K_i|$. The energy levels of a rotational band built on an intrinsic state with

$K = \sum_{i=1}^{k} K_i$ are given by (Alaga *et al* 1955)

$$E(I) = \frac{\hbar^2}{2\theta}\{I(I+1) - K^2 - (-1)^{I+1/2}a(I+\tfrac{1}{2})\delta_{K,1/2}\}$$
$$I = K,\ K+1,\ K+2,\ \ldots \tag{4.65}$$

where the moment of inertia θ falls somewhere within the range

$$\frac{2}{5}M(A, Z)(\Delta R)^2 < \theta < \frac{2}{5}M(A, Z)R_0^2 \tag{4.66}$$

and the lower and upper limits represent irrotational and rigid body moments of inertia respectively. The level formula (4.65) applies to all intrinsic states of nuclei, odd A and even A, for which $K \neq 0$. The special status of states with $K = \frac{1}{2}$ is recognized by the admission of a decoupling coefficient a which, for heavy nuclei, takes values in the range $|a| \leqslant 1$ but which, for light nuclei with $A \simeq 25$, can have values as large as $|a| \simeq 3$. For intrinsic states with $K = 0$, a condition which includes the ground states of all even–even nuclei, there is an additional invariance of the wavefunction with respect to rotation through an angle π about an axis in the symmetry plane through the nuclear centre, which excludes odd values of I.

Experimentally the energy spectrum predicted by (4.65) is closely realized in very many even–even nuclei and the rotational bands have been traced to values of I as high as 30. However, to achieve the best agreement it is necessary to take account of terms in $E(I) \propto I^2(I+1)^2$ which describe coupling between the rotational and the vibrational degrees of freedom. This comes about because the moment of inertia is not strictly a constant, but depends weakly on β and γ.

In addition to the pure vibration or rotation modes described above it is also possible to have rotation–vibration modes in which the rotation band is built on a vibrational state exactly as happens with diatomic molecules. This effect is encountered most frequently in even–even nuclei with prolate deformation and the vibrations are of two types: β-vibrations which represent oscillations of the quadrupole moment about some mean value specified by the deformation parameter β, and γ-vibrations which can be visualized as oscillations of the ellipsoidal cross-section about its equilibrium circular shape characterized by the value $\gamma = 0$. The β-vibrations describe oscillations in the collective coordinate $\alpha_{2,0}$ which preserve axial symmetry; they carry no component of angular momentum along the symmetry axis and therefore have rotation bands with $K = 0$ and $I = 0, 2, 4, \ldots$, etc. The γ-vibrations describe oscillations of the collective coordinate $\alpha_{2,\pm 2}$ and specifically demonstrate departures from axial symmetry. Since these vibrations carry two units of angular momentum along the symmetry axis, the one phonon vibrational state has $K = 2$ and $I = 2, 3, 4, \ldots$, etc. For an even–even nucleus the corresponding spectrum is

$$E(I, K, n_\beta, n_\gamma)=\frac{\hbar^2}{2\theta}[I(I+1)-K^2]+(n_\beta+\tfrac{1}{2})\hbar\omega_\beta+(n_\gamma+1)\hbar\omega_\gamma \quad (4.67)$$

where in general, for given n_γ, K can take values $2n_\gamma$, $2n_\gamma-4$, ..., 2 or 0. An example of a rotation–vibration band spectrum in the nucleus $^{234}_{92}U$ is shown in figure 4.4. This shows a ground state band ($n_\beta=n_\gamma=0$; $K=0$) together with a β-band ($n_\beta=1$, $n_\gamma=0$; $K=0$) and a γ-band ($n_\beta=0$, $n_\gamma=1$; $K=2$). The higher energy bands are termed $\beta\beta$ ($n_\beta=2$, $n_\gamma=0$; $K=0$) and $\beta\gamma$ ($n_\beta=n_\gamma=1$; $K=2$) bands, respectively.

Finally there is an extreme case, of which examples are to be found among even–even nuclei with spin sequences 0^+, 2^+, 2^+, e.g. $^{114}_{48}Cd$, $^{122}_{52}Te$, $^{196}_{78}Pt$ (Scharff-Goldhaber and Wenesser 1955) where the restoring force is

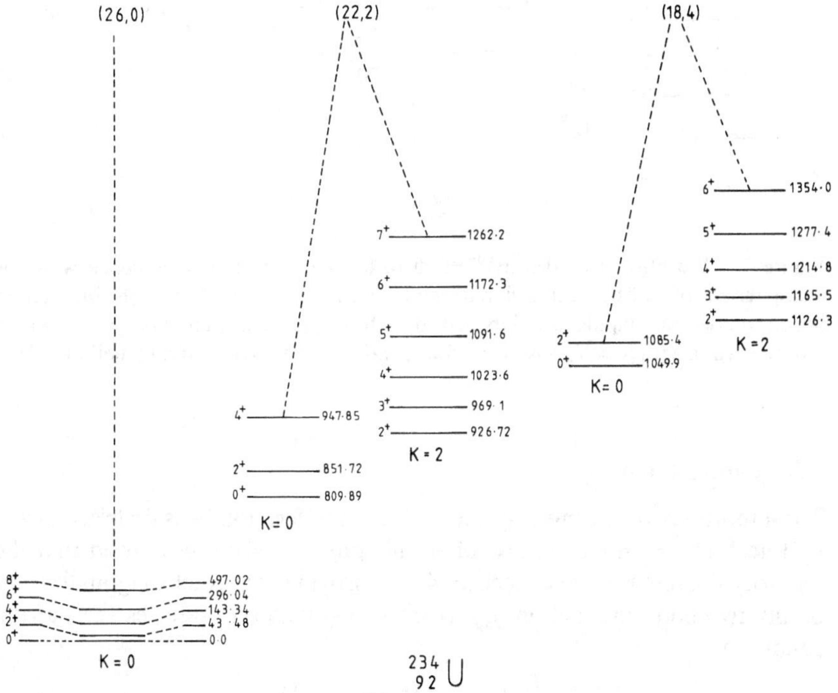

Figure 4.4. Vibration–rotation bands in $^{234}_{92}U$. According to the collective model there is a ground state rotation band ($n_\beta=n_\gamma=K=0$), a β-band ($n_\beta=1$, $n_\gamma=K=0$), a γ-band ($n_\beta=0$, $n_\gamma=1$, $K=2$), together with $\beta\beta$ and $\beta\gamma$-bands with ($n_\beta=2$, $n_\gamma=K=0$) and ($n_\beta=n_\gamma=1$, $K=2$) respectively. In the interacting boson model the bands are labelled by SU(3) quantum numbers (λ, μ) in a representation with $N=N_\pi+N_\nu=5+8=13$ and non-negative integers $l=0$, $k=0$, 1 and 2 (Arima and Iachello 1978).

independent of γ and the resultant excitations are referred to as γ-unstable (Wilets and Jean 1956). An example is shown in figure 4.5. These spectra are similar to vibrational spectra in many respects with the characteristic feature that the 0^+ state is missing from the vibrational triplet with phonon number $n=2$. Other modes of collective motion have also been identified, e.g. octupole vibrations (Lane and Pendlebury 1960), but as a general rule collective modes associated with quadrupole deformation are the most frequent.

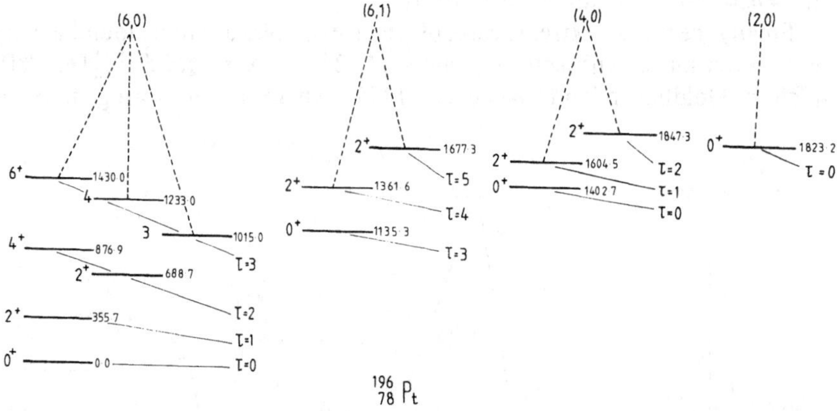

Figure 4.5. Vibrational modes in $^{196}_{78}$Pt showing the 0^+2^+2 spin sequence which is characteristic of γ-unstable nuclei in the collective model. According to the interacting boson model the bands are labelled by O(6) quantum numbers (σ, n_Δ) for a representation with $N=N_\pi+N_\nu=2+4=6$, and $\sigma_{\max}=N$ (Arima and Iachello 1979).

4.5.2 Pairing forces

The importance of pairing in systems of identical fermions was first recognized by Racah (1943) in the context of atomic physics, when he showed that the seniority coupling scheme (section 4.2.2) provides an exact diagonalization of any two-body interaction $q_{\alpha\beta}$ which in j–j coupling satisfies the pairing condition

$$\langle jjJM|q_{\alpha\beta}|jjJM\rangle=(2j+1)\delta_{J,0} \tag{4.68}$$

The idea was later adopted as an integral feature of the shell model by Mayer (1950) who showed that, for a short-range interaction proportional to $\delta^3(\boldsymbol{r}_1-\boldsymbol{r}_2)$, the $L=0$ (singlet) state of a pair of identical nucleons in an (nlj) orbital is highly depressed in energy compared with states with $L=1, 2, \ldots, 2l$ which are approximately degenerate. Assuming the existence of a short-range interaction of this type she was able to establish a theoretical basis for the three coupling rules set out in section 4.2.2. Subsequently it was shown that the pairing property is shared by all scalar two-body interations of the type $\boldsymbol{\sigma}_1 \cdot \boldsymbol{\sigma}_2\ V(r_{12})$ (Racah and Talmi 1952). However, the wider possibilities associated with the action of pairing forces in nuclei were not recognized until parallel research into the phenomenon of superconductivity revealed that even every feeble coherent interactions between pairs of fermions, having many features in common with the pairing force between nucleons, could alter the intrinsic properties of an assembly of identical fermions in a very fundamental way (Cooper 1956, Bardeen *et al* 1957, Bogoliubov 1958).

In a superconductor correlations between pair of electrons in time-reversed states of equal and opposite momentum close to the Fermi surface, which arise through their coherent interaction with the vibrations of the lattice, permit the formation of a quasi-bound state for pairs of electrons which alters the distribution of occupied states over an energy interval Δ in the neighbourhood of the Fermi level E_{F}, depleting states with energies $E_j \leqslant E_{\mathrm{F}}$ and increasing the populations of state with energies $E_j \geqslant E_{\mathrm{F}}$ in proportion. The 'Cooper pairs' of electrons which are the charge carriers in superconductors exist in states of opposite momentum, whereas the paired nucleons which populate unfilled shells in nuclei are in time-reversed states of equal and opposite angular momentum, but nevertheless the analogy is very close, the main difference being that, whereas the electrons are correlated in position the size of a bound pair being of order $\hbar k_{\mathrm{F}}/m\Delta$, the paired nucleons are correlated in angle. Another difference of course is that there are two varieties of nucleon, and it is not easy to construct a theory of n–p pairing. not because the n–p force is not sufficiently strong, but because there is no selective lowering in energy of the $L=0$ state in comparison with all other states.

The significance of these results for nuclear structure was first appreciated by Bohr *et al* (1958), who suggested that the occurrence of a pairing gap in nuclei might be able to account for the observation of a 1 MeV energy difference between the ground and first excited states in heavy even–even nuclei, which is much larger than the single-particle level spacing in that region of A. They further speculated that the pairing effect could account for the fact that the theoretical rigid-body moment of inertia associated with nuclear deformation was too large by a factor of about 2, when compared with the experimental results.

An important feature of the δ-function interaction is that its matrix elements $\langle l'm'|\delta^3|\boldsymbol{r}_1-\boldsymbol{r}_2||l'-m'\rangle$ all have the same phase and it is indeed precisely this coherence property which depresses the $L=0$ state with respect to all other states. Thus the effect of pairing correlations in the configuration $(nlj)^k$ may be studied by constructing a pairing potential V_{12} whose matrix elements are defined by

$$\langle jm', j-m'|V_{12}|jm, j-m\rangle = -G \tag{4.69}$$

where $G>0$ gives an attractive force. The problem of k particles interacting through this pairing force may be solved exactly giving the expected result that the gound state has seniority $v=0$ (k even) or $v=1$ (k odd). For even k the energy gap between the ground and first excited states is given by

$$\Delta E_{10}=E_1-E_0=G\Omega \tag{4.70}$$

where $\Omega=\frac{1}{2}(2j+1)$ is the total number of pair states in the (nlj) orbital. This result provides a theoretical justification for Mayer's coupling rule (iii) (4.15). For odd k the corresponding energy gap is $G(\Omega-1)$, reflecting the fact that, since neither the state occupied by the odd nucleon nor its conjugate is available to accept pairs, the effective degeneracy is reduced by unity.

It does not of course suffice to restrict attention to the Ω degenerate pair states within a single orbital, since, in a real nucleus, the valence nucleons have a choice of several competing orbitals. For example, three orbitals are available to accommodate the 12 particles required to complete the s–d shell and five orbitals are required to complete the next major shell between the magic numbers at 50 and 82. It is therefore necessary to investigate the possibility that the pairing force has non-vanishing matrix elements between states of different j and the pairing effect is then evaluated for k particles distributed over a range of non-degenerate single particle states. This is not a problem for which an exact solution is attainable and the most successful approximate solutions have been found using a technique pioneered in the BCS theory of superconductivity (Bardeen *et al* 1957) in which the chemical potential λ (i.e. the Fermi energy E_F), rather than the total number of particles k is taken to be fixed (Balyaev 1959, Bayman 1960). This approach has a parallel in statistical mechanics where in general it is easier to study the equilibrium states of quantum gases in the grand canonical ensemble (fixed temperature and chemical potential), rather than in the canonical ensemble (fixed temperature and particle number).

The solution to the non-degenerate problem is obtained using the method of Lagrange multipliers to minimize the expectation value of the Hamiltonian

$$H=H_0-\lambda N \tag{4.71}$$

where H_0 is the Hamiltonian of an N-particle system in a central potential with pairing interactions. The undetermined multiplier λ (i.e. the chemical potential) is related to the number k of particles in the real nucleus by making the identification

$$\langle N\rangle=\sum_{\alpha>0}\left\{1-\frac{(E_\alpha-\lambda)}{\sqrt{(E_\alpha-\lambda)^2+\Delta_\alpha^2}}\right\}=k \tag{4.72}$$

where the gap parameters Δ_α satisfy the coupled equations

$$\Delta_\alpha=-\frac{1}{2}\sum_{\alpha'}\frac{\langle\alpha,-\alpha|V_{12}|\alpha',-\alpha'\rangle}{\sqrt{(E_{\alpha'}-\lambda)^2+\Delta_{\alpha'}^2)}}\Delta_{\alpha'} \tag{4.73}$$

The indices $\pm\alpha$ represent the set of quantum numbers $(nlj\pm m_j)$, respectively, and the restriction $\alpha>0$ on the summation index in (4.72) means that the sum is taken over all pair states labelled by j and $|m_j|$. Since there are two particles in each pair state it is implicit that k is even. In a spherical potential the single particle energies E_α are of course independent of $|m_j|$ but in a deformed cylindrically symmetric potential they will in general depend on the absolute value of the angular momentum projection along the symmetry axis.

It follows from (4.73) that, to reach a finite value of Δ_α, the separate terms in the sum must add coherently to avoid cancellations, and this requires that the matrix elements of V_{12} maintain a constant phase over a broad range of states. Within this range it is convenient to replace these matrix elements, and the gap parameters Δ_α, by their average values, i.e.

$$\overline{\langle\alpha, -\alpha|V_{12}|\alpha', -\alpha'\rangle} = -G \qquad \overline{\Delta_\alpha} = \Delta \tag{4.74}$$

when G is real and positive. In this case the result (4.73) reduces to

$$\sum_{\alpha>0} \{E_\alpha - \lambda)^2 + \Delta^2\}^{-1/2} = 2/G \tag{4.75}$$

The relations (4.72) and (4.75) are referred to as the 'gap equations', and may be solved for λ and Δ for a specified set of single-particle energies E_α and a given value of G. For heavy nuclei G may be determined from the systematic variation in the mass differences for even–even, even–odd and odd–odd nuclei and is given approximately by the empirical rule $G=22/A$ MeV.

The gap equation (4.72) describes the modification of the Fermi sea in the energy range $(\lambda\pm\Delta/2)$ introduced by pairing and, in the limit $\Delta\to 0$, reduces to the statement that k is equal to twice the number of pair-states with $E_\alpha<\lambda$, all pair states with $E_\alpha>\lambda$ remaining empty. For $|E_\alpha-\lambda|\gg\Delta$ and $E_\alpha>\lambda$ the modified excitation energy

$$\tilde{E}_\alpha = \{(E_\alpha - \lambda)^2 + \Delta^2\}^{1/2} \tag{4.76}$$

may be identified as the energy required to create a particle in the orbital $|\alpha\rangle$, and, for $|E_\alpha-\lambda|\gg\Delta$ with $E_\alpha<\lambda$, as the energy required to create a hole in the filled orbital $|\alpha\rangle$. Conversely, for $|E_\alpha-\lambda|\leqslant\Delta/2$, the excitation energy $\tilde{E}_\alpha$ cannot be related specifically to either particle creation or hole creation, because in this region the occupation probability of the pair state $|\alpha\rangle$ varies from zero at $E_\alpha\simeq\lambda-\frac{1}{2}\Delta$ to unity at $E_\alpha\simeq\lambda+\frac{1}{2}\Delta$. Reflecting this confusion of identity the excitations $\tilde{E}_\alpha$ are referred to as 'quasi-particles'; they share many characteristics in common with real physical particles including the same angular momentum quantum numbers as the single particle state $|\alpha\rangle$. The parameter Δ may therefore be thought of as a gap in the energy spectrum of quasi-particles.

The ground state of an even-k nucleus is the quasi-particle vacuum (it being understood of course that the numbers of valence neutrons and protons are specified by separate values of k). Thus, to generate an excited state, at least one nucleon pair must be broken, and this requires an excitation energy

$$\tilde{E}(\alpha_1, \alpha_2) = \tilde{E}_{\alpha_1} + \tilde{E}_{\alpha_2} \tag{4.77}$$

where $|\alpha_1\rangle$ and $|\alpha_2\rangle$ are pair states which, in the excited nuclear state, contain no saturated pairs. Thus an intrinsic excited state of an even-k nucleus (i.e. excluding states of collective motion) must contain at least two quasi-particles with total energy $\tilde{E}(\alpha_1, \alpha_2) \geqslant 2\Delta$. This result may be compared with the equivalent result for a single degenerate level given in (4.70) by defining an effective degeneracy Ω_{eff} (Nilsson and Prior 1961) given by

$$G\Omega_{\text{eff}} = 2\Delta \tag{4.78}$$

To construct an odd-k nucleus from an even k nucleus, pairs must be excluded from one pair state $|\alpha'\rangle$, and this result may be achieved by adding one quasi-particle of energy $\tilde{E}_{\alpha'}$. For the resultant nucleus to be in its ground state the excluded pair state must have $E_{\alpha'} \simeq \lambda$. Thus $\tilde{E}_{\alpha'} \simeq \Delta$ and the binding energy of an odd-k nucleus is less than that for the neighbouring even-k nuclei by about Δ. The excited states of an odd k nucleus will then occur at excitation energies $\Delta\tilde{E}_\alpha \simeq \tilde{E}_\alpha - \Delta \simeq \frac{1}{2}(E_\alpha - \lambda)^2/\Delta$ for $\Delta \gg |E_\alpha - \lambda|$. These excited states therefore cluster more closely in energy than they would in the absence of pairing effects.

These ideas are neatly illustrated by the nucleus $^{117}_{50}\text{Sn}$ for which the gap equations have been solved for a value $G = 0.188$ and $k = 17$ (Kisslinger and Sorensen 1960). $^{117}_{50}\text{Sn}$ has a magic number $Z = 50$ and 17 valence neutrons which places it almost mid-way between the magic numbers $N = 50$ and $N = 82$. In this region of N there are four even parity orbitals $2d_{5/2}$, $1g_{7/2}$, $3s_{1/2}$ and $2d_{3/2}$ in competition, plus one odd parity orbital $1h_{11/2}$. With a spin-parity $I^\pi = \frac{1}{2}^+$ and a magnetic moment $\mu = -0.9998$, the shell model interpretation of this nucleus would assume that the $2d_{5/2}$ and $1g_{7/2}$ orbitals were full, and the $1h_{11/2}$ orbital empty. The odd neutron would go into the $3s_{1/2}$ orbital with a saturated pair of neutrons in the $2d_{3/2}$ orbital about 0.3 MeV higher in energy. This is an example of the empirical pairing rule (4.15) similar to the case of the thallium isotopes $^{203}_{81}\text{Tl}$ and $^{205}_{81}\text{Tl}$ discussed earlier.

The solution of the gap equations gives values $\lambda = 1.81$ MeV, measured relative to the $d_{5/2}$ level, and an energy gap $\Delta = 1.0$ MeV. The quasi-particle energies in order of magnitude are $\tilde{E}_{1/2} = 1.004$ MeV, $\tilde{E}_{3/2} = 1.073$ MeV, $\tilde{E}_{11/2} = 1.407$ MeV, $\tilde{E}_{7/2} = 1.878$ MeV and $E_{5/2} = 2.068$ MeV, and these results predict a $\frac{1}{2}^+$ ground state and excitated states at energies 0.069 MeV $(\frac{3}{2}^+)$, 0.403 MeV $(\frac{11}{2}^-)$, 0.874 $(\frac{7}{2})^+$ and 1.064 MeV $(\frac{5}{2})^+$ in very good agreement with the observed level scheme. In contrast the first intrinsic excited states in the neighbouring even–even nuclei ${}^{116}_{50}Sn$ and ${}^{118}_{50}Sn$ occur at 1.724 MeV and 2.05 MeV respectively, in remarkably good agreement with the predicted values $2\Delta \simeq 2$ MeV.

4.5.3 Moments of inertia

The preceding discussion has been concerned in the main with the very general question as to how to arrive at an understanding of the behaviour of atomic nuclei through the study of idealized systems or models, and in particular with the detailed properties of two extreme nuclear models which rely on contrasting coupling schemes. In the pairing coupling scheme the emphasis is placed on independent particle motion in a central field modified by the action of short-range pairing forces, whereas in the aligned coupling scheme the motion is essentially collective in nature, reflecting the dominance of the long-range quadrupole forces. The pairing force tends to cluster the particles into saturated pairs with large intrinsic quadrupole moments which are uncorrelated (cf. section 4.1) and its characteristic feature is the 'pairing gap'. The quadrupole force, on the other hand, acts to align the quadrupole moments of individual particles rather than of pairs and, in doing so, breaks up the saturated pairs (cf. 4.43). Its characteristic feature is the rotational band.

To construct a complete theory of the nucleus it is clearly necessary to bring about some state of co-existence between these opposing modes of description, first of all by establishing procedures whereby the collective parameters can be calculated within the independent particle formation, and second, by providing an understanding of the nature of the transition between the two extreme coupling schemes. This is not a smooth transition as might be expected, but takes place quite abruptly within the space of a few mass units.

The general nature of the unification problem may best be illustrated by considering a particular example, namely how to calculate the moment of inertia in a coupling regime where the system of rotational bands is securely established. In this mode of collective motion the valence nucleons move in a deformed potential field which means that the angular momentum of the deformed shell model state is no longer a good quantum number. It is therefore necessary to project out of the shell-model state an intrinsic state which, for an even–even nucleus, has zero angular momentum. This may be done by following a procedure similar to that applied in section 4.1.3 to project out a state of zero centre-of-mass momentum. More generally a state of orbital angular momentum l may be projected from a deformed state $\Phi(\boldsymbol{r})$ by writing, in analogy with (4.12)

$$\psi_{l,m}(\boldsymbol{r}) = \int \mathrm{d}\theta \, \mathrm{d}\varphi \; Y_{lm}(\theta, \varphi)\Phi(R_{\theta\varphi}\boldsymbol{r}) \tag{4.79}$$

where $R_{\theta\varphi}$ is the operator which takes the axis of symmetry of the deformed well into the direction (θ, φ). In this case the energy of the state $\psi_{l,m}(\boldsymbol{r})$ has an additional contribution proportional to $l(l+1)$ as compared with the state $\Phi(\boldsymbol{r})$ (Peierls and Yoccoz 1957).

A more transparent procedure which may be adopted is to force the deformed field to rotate with some angular velocity ω_z, which is such that, in the rotating frame, states of differing angular momentum are degenerate but overall conservation of the intrinsic angular momentum is restored. The rotation may be thought of as imposed on the nucleus by some external agency, hence the description 'cranking model' (Inglis 1954, Amado and Brueckner 1959). The cranking action is formally incorporated in the theory by adding a term $-\hbar\omega_z I_z$ to the Hamiltonian rather in the same way that separation may be brought about between the intrinsic motion and the centre-of-mass motion in the two-body system (cf. section 2.1). The difference is that the angular velocity ω_z cannot be specified in advance, but must be treated as a Lagrange multiplier which is determined to maintain a specified value of the total angular momentum component I_z. The additional energy required by the valence nucleons to follow the rotating field adiabatically is then calculated as a perturbation to second order in ω_z and the result equated to $\frac{1}{2}\theta\omega_z^2$, where θ is the moment of inertia.

The result of this calculation leads to the rigid body value

$$\theta = 2\hbar^2 \sum_{j \neq 0} \frac{|\langle j | I_z | 0 \rangle|^2}{\varepsilon_j - \varepsilon_0} \tag{4.80}$$

where the index j is summed over all intrinsic states of the system of energy ε_j excluding the ground state. The values of θ deduced from this highly convoluted computation are too large by a factor of about two when compared with experimental values derived from rotational spectra, and the reason is that the calculation does not take account of the self-consistency condition which requires that the potential and density distributions be the same. To bring about this result the short range correlations, i.e. the pairing forces, must be taken into the equation and, when this is done, the moment of inertia is reduced to a value which brings theory and experiment into close accord (Griffin and Rich 1959, 1960, Nilsson and Prior 1961, Prange 1961). This result illustrates the general conclusion that in no case is it possible to dispense with either the short-range or the long-range component of the residual force; it also brings to the fore the very important connection that exists between collective motion and degeneracy of states.

4.5.4 High spin states

The first experiments aimed at the production of high spin states in nuclei were carried out using Coulomb excitation, whereby time-varying electric fields associated with the motion of charged particles of energy below the

Coulomb barrier are coupled to the quadrupole excitations of deformed nuclei (Huss and Zupancic 1953, Stephens *et al* 1959). Subsequently α-particle induced nuclear reactions followed by multiple emission of neutrons were employed (Morinaga and Gugelot 1963, Diamond *et al* 1964), and more recently heavy ion beams used in combination with Compton suppressed germanium γ-ray detectors to disentangle the decay schemes (Twin 1983, Nolan *et al* 1985).

Studies of high spin states have concentrated in the main on the so-called 'yrast' states, namely those states which, for a given value of the energy have the highest angular momentum. The description 'yrast' is a neoteric word derived from the Nordic 'yr' meaning 'dizzy' (Grover 1967). Since for given energy the states of highest angular momentum are the states of a rotational band, the yrast state is reached when all the energy is converted into rotational energy and the nucleus is thermodynamically cold.

For an even–even nucleus the rotational energy is given by (4.65), which means that the moment of inertia θ is related to the energy separation between adjacent levels according to the formula

$$\hbar^2/2\theta = [(E_I - E_{I-2})/(4I-2)] \tag{4.81}$$

Significant departures from the energy level formula (4.65) over and above those centrifugal stretching effects associated with higher powers of $I(I+1)$ were first detected in the nucleus ^{162}Er (Johnson *et al* 1971, 1972), which exhibits the effect known as 'back-bending'. The origin of this descriptive term will become clear from figure 4.6 (Lieder and Ryde 1978). This represents a plot for the nucleus ^{158}Er of the parameter $2\theta/\hbar^2$ against the rotational quantum energy $\hbar\omega_I$ defined by

$$(\hbar\omega_I)^2 = (I^2 - I + 1)[(E_I - E_{I-2})/(2I-1)]^2 \tag{4.82}$$

If the energy level formula (4.65) was valid even for the highest values of I considered as a parameter, one would expect the yrast line to run parallel to the $(\hbar\omega_I)^2$-axis in figure 4.6, or, at some value of $(\hbar\omega_I)^2$, to become two parallel lines when two rotational bands cross in energy without interaction. The simplest explanation of the back-bending effect observed in ^{158}Er is that the yrast line coincides with the levels of the ground state band for $I < 12$, at which point it switches over to the levels of the higher band. This could be explained on the assumption that at $I \simeq 12$, the moment of inertia makes a discrete jump from the lower values observed in oblate nuclei to a value approximating to the rigid body value given by (4.80). Since the larger value of θ is associated with the smaller energy at higher angular momentum, a point must be reached at which the spin I of the higher band exceeds that of the lower band at the same energy, at which point the yrast line switches from the ground state band to the higher band.

The back-bending effect observed in deformed rare earth nuclei was in fact predicted by Mottelson and Valatin (1960) who calculated that, at a critical

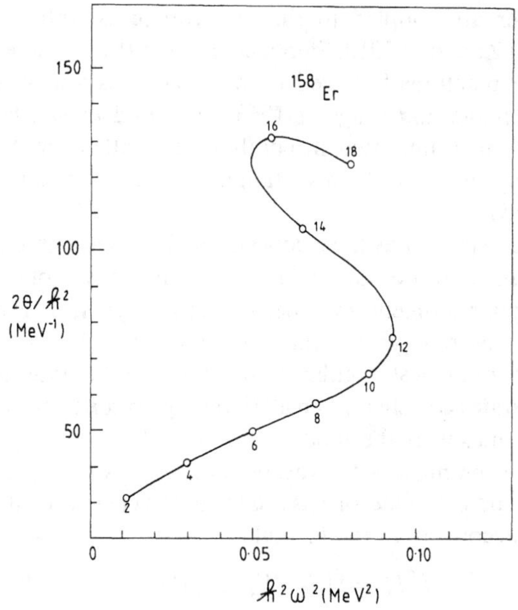

Figure 4.6. Diagram illustrating the back-bending anomaly in the moment of inertia as observed in the nucleus ^{158}Er (Lieder and Ryde 1978). This is explained on the assumption that, at $I \simeq 12$, the ground state rotational band crosses a higher band in which two $i_{11/2}$ quasineutrons are rotationally aligned.

value of I estimated to lie near $I = 12$, the Coriolis force acting on individual nucleons in a rapidly rotating nucleus would be sufficient to break the pairing interaction. This rotational alignment effect induces a sudden transition from superfluid to rigid body motion (Stephens and Simon 1972). The assumption is then that pairing is broken in the upper band and a phase transition from the paired to the unpaired state occurs where the two bands cross (Faessler *et al* 1965). Since the Coriolis force derives from a potential

$$V_c = -(\hbar/\theta) \sum_i \boldsymbol{j}_i \cdot \boldsymbol{I} \tag{4.83}$$

it follows that particles with large values of j_i are most influenced by the Coriolis action and the breaking of a single pair may suffice to bring about the transition. The theory then assigns the first back-bending anomaly in even–even nuclei to a crossing between the ground state rotational band and a higher band in which two $i_{11/2}$ quasi-neutrons are rotationally aligned. In the same way the second anomaly is associated with crossing a band containing two aligned quasi-protons (Lee *et al* 1977, Faessler and Ploszajczak 1978).

At even higher angular momenta the centripetal forces became comparable with the central forces and states of prolate symmetry give way, first to triaxial shapes, and ultimately to states of oblate symmetry. For values of Z in the range 64–66 and N in the range 80–86 and values of I of order 50 to 60 'superdeformed nuclei' with ellipsoidal axes in the ratio 2:1:1 are formed, e.g. $^{156}_{66}Dy$ (Nyako 1984). The nucleus $^{132}_{58}Ce$ which has ellipsoidal axes in the ratio 3:2:2 may be described as 'almost superdeformed' (Kirwan 1987).

4.6 Collective motion and unitary symmetry

4.6.1 The Elliott model

To carry out a calculation of the kind described in section 4.5.3 above in the case of a large nucleus with partially completed shells of protons and neutrons poses an insoluble problem of computation simply because of the enormous number of competing shell model configurations. It has been remarked for example (Otsuka *et al* 1978) that, in the case of the nucleus $^{152}_{62}Sm$, with 10 and 12 valence neutrons and protons respectively, there are more than 4×10^{13} states available with $I=0$, plus many more with $I \neq 0$. Thus in any feasible calculation the number of contributing states must be truncated in some way, and a reliable prescription for achieving this aim can be realized only by exploiting to a maximum the symmetries of the system. Inevitably, the detailed procedures employed lean very heavily on group theoretical methods.

The impressive power of group theory to impose a degree of order, based on an underlying symmetry, on the highly complex system of states available to a nucleus, has been vividly demonstrated by Elliott (1958), in addressing the question as to whether collective properties emerge naturally in conventional shell-model theory with configuration mixing, when the motion is solved in a central potential with a very specific symmetry and the associated degeneracy of states is broken by the addition of residual interactions whose symmetry is also well defined. Ultimately it was shown that the essential features of the rotational band could be derived by standard shell-model techniques by mixing configurations which are degenerate in the three-dimensional isotropic harmonic oscillator potential, the principal difference between this result and the usual collective model result being that the number of states in each rotational 'band' is finite.

We have already remarked in section 4.2.1 on the existence of a degeneracy, e.g. 1d2s or 1f2p, in the energy eigenstates of the three-dimensional isotropic harmonic oscillator potential (cf. figure 4.1). This observation may be traced to the face that the corresponding Schrödinger equation is simultaneously separable in spherical polar and in Cartesian coordinates. From the former property one may construct energy eigenstates with n phonons of excitation which are also eigenstates of $\boldsymbol{L}^2$ with L give by

$$L = n,\ n-2,\ \ldots,\ 1 \text{ or } 0 \tag{4.84}$$

while the latter property reveals the existence of states characterized by phonon numbers n_x, n_y, n_z restricted to satisfy

$$n_x + n_y + n_z = n \tag{4.85}$$

which are also degenerate.

Another way of arriving at these results is to note that the oscillator Hamiltonian

$$H_0 = \frac{\boldsymbol{p}^2}{2M} + \tfrac{1}{2}M\omega^2\boldsymbol{r}^2 \tag{4.86}$$

is invariant with respect to transformations generated by the nine components of the second-rank Cartesian tensor $\mathbf{a}_i^\dagger\mathbf{a}_j$, where

$$\boldsymbol{a} = (\boldsymbol{p} - \mathrm{i}M\omega\boldsymbol{r})/\sqrt{(2M\hbar\omega)} \qquad \boldsymbol{a}^\dagger = (\boldsymbol{p} + \mathrm{i}M\omega\boldsymbol{r})/\sqrt{(2M\hbar\omega)} \tag{4.87}$$

represent phonon annihilation and creation operators, respectively. Thus the degenerate eigenfunctions of H_0 with eigenvalue n provide a representation space for the group U(3) of unitary transformations whose nine generators are linear combinations of the elements $\mathbf{a}_i^\dagger\mathbf{a}_j$. The phonon number operator $\hat{n}$ defined by

$$\hat{n} = (H_0 - \tfrac{3}{2}\hbar\omega)/\hbar\omega = \boldsymbol{a}^\dagger \cdot \boldsymbol{a} = \mathrm{Tr}(\mathbf{a}_i^\dagger\mathbf{a}_j) \tag{4.88}$$

commutes with each of the remaining eight operators of the tensor because each such element describes the simultaneous creation and annihilation of one phonon. It follows from Schur's lemma that, in any representation of U(3), $\hat{n}$ is a multiple of the unit matrix, and may be eliminated from the set of nine generators without loss of generality, since its deletion means only that the phase of the determinant of any transformation of the group is defined to be zero. The remaining eight operators constitute the generators of the special group SU(3) of unitary unimodular transformations, and the result has been to effect the reduction $\mathrm{U(3)} \supset \mathrm{SU(2)} \times \mathrm{U(1)}$ where the representations of the U(1) group are labelled by the quantum number n.

When the tensor $\mathbf{a}_i^\dagger\mathbf{a}_j$ is decomposed into its spherical components according to the rules set out in Table 3.2, the SU(3) generators may be organized into two spherical tensors comprising the three orbital angular momentum operators L_0, $L_\pm$ plus the five components of the quadrupole operator

$$\tilde{\mathrm{Q}}_m^{(2)} = \frac{4\pi}{5}\frac{M^2\omega^2}{2}\{\boldsymbol{r}^2 Y_2^m(\theta, \varphi) + \boldsymbol{p}^2 Y_2^m(\theta_{\mathrm{p}}, \varphi_{\mathrm{p}})\} \qquad m = 0, \pm 1 \pm 2 \tag{4.89}$$

where (θ, φ) and $(\theta_{\mathrm{p}}, \varphi_{\mathrm{p}})$ represent the orientations of the vectors $\boldsymbol{r}$ and $\boldsymbol{p}$

respectively. The operators $\tilde{Q}_m^{(2)}$ describing quadrupole deformations are unconventional in that they are constructed from the components of $\boldsymbol{p}$ as well as from the components of $\boldsymbol{r}$ but, because of the symmetry between $\boldsymbol{r}$ and $\boldsymbol{p}$ in the Hamiltonian, the matrix elements of $\tilde{Q}_m^{(2)}$ are identical with those of the quadrupole moment operator when evaluated between states of the same oscillator shell. However, because $\tilde{Q}_m^{(2)}$ are group generators, their matrix elements vanish when evaluated between states of different shells, a property which makes them very suitable for constructing residual two-body interactions which do not bring about the mixing of configurations in different major shells.

Since SU(3) is a symmetry group of the oscillator Hamiltonian, it follows that single-particle states within the oscillator shell of phonon number n transform according to irreducible representations of SU(3). This is a second-rank group with quadratic and cubic invariant (Casimir) operators whose eigenvalues are expressible in terms of a pair of integers (λ, μ) which suffice to label the corresponding irreducible representations of dimension $\frac{1}{2}(\lambda+1)(\mu+1)(\lambda+\mu+2)$ (Elliott and Dawber 1979). For example the six states in the 1d,2s shell with $n=2$ transform according to the (2, 0) representation of SU(3). Since H_0 is also rotationally invariant, states of a given representation (λ, μ) can be organized into sets of eigenstates of $\boldsymbol{L}^2$ which transform among themselves according to irreducible representations of the subgroup $SU(2)_L$ labelled by the quantum number L, which is related to the eigenvalue $L(L+1)$ of the single quadratic Casimir operator $\hat{\boldsymbol{L}}\cdot\hat{\boldsymbol{L}}$. The group $SU(2)_L$ has three generators L_0, $L_\pm$ and is isomorphic on the group R_3 of rotations in a three-dimensional real space. The subscript 'L' has no group theoretical significance and is used merely to distinguish the subgroup SU(2) (angular momentum) from the subgroup SU(2) (two-dimensional harmonic oscillator) which is identified below as $SU(2)_\Lambda$. The states of the $n=2$ oscillator shell transform according to a reducible representation of $SU(2)_L$ constructed as a direct sum of $L=0$ (2s-state) and $L=2$ (1d-state) representations, a result which shows how the 2s–1d degeneracy arises as a consequence of the higher SU(3) symmetry.

The Hamiltonian (4.86) is not only rotationally invariant; it is also parity conserving, and this means that it is invariant under transformations of the extended group which is obtained by adjoining to $SU(2)_L$ the parity operator P. This extended group is called the orthogonal group in three dimensions and is identified symbolically as O(3). Corresponding to each irreducible tensor representation of $SU(2)_L$ there exist two tensor representations of O(3) which may be described as 'proper' and 'pseudo' respectively (cf. section 3.3.2). In the present context it may be assumed that all representations of O(3) are proper so that the symmetry groups $SU(2)_L$ and O(3) are equivalent (cf. section 4.6.3).

These results for single particle states generalize in a not quite straightforward manner to the case of k particles in an oscillator shell of phonon number n, with the difference that the k-particle states now transform according to reducible representations of SU(3). For example, the 10^2 states of 2 particles in the $n=3$ oscillator shell are classified in Table 4.3 according to their SU(3) and $SU(2)_L$ labels, together with values of a new quantum number K, which associates states of given (λ, μ) into sets which strongly resemble rotational bands, and which is the analogue of the projection of the total angular momentum along the nuclear symmetry axis in the collective model.

Table 4.3. Group classification of 2 particle states in the $n=3$ oscillator shell

SU(3) labels λ	μ	Dimension $d(\lambda\mu)$	$SU(2)_L$ labels L	Intrinsic quantum numbers Λ	K
6	0	28	0, 2, 4, 6	0	0
2	2	27	0, 2	1	0
			2, 3, 4		2
4	1	35	1, 2, 3, 4, 5	$\frac{1}{2}$	1
0	3	10	1, 3	0	0

The degenerate states of an oscillator shell are usually labelled by quantum numbers L and M so that $\boldsymbol{L}^2$ and L_0 are simultaneously diagonalized, and in these states angular momentum is conserved. However, because of the degeneracy discussed above it is possible to construct deformed energy eigenstates which are eigenstates of $\tilde{Q}_0^{(2)}$ and $(\tilde{Q}_2^{(2)}+\tilde{Q}_{-2}^{(2)})$ with eigenvalues proportional to $(2n_z-n_x-n_y)$ and (n_x-n_y), respectively. These states transform according to irreducible representations of a subgroup $SU(2)_\Lambda \times U(1)$ of SU(3) which has generators L_0, $Q_2^{\pm 2}$ $(SU(2)_\Lambda)$ and $Q_0^{(2)}$ (U(1)). The representations of $SU(2)_\Lambda$ are labelled by an integral or half-integral quantum number Λ (analogous to L) and the states of the representation are labelled by the eigenvalue K of L_0 which takes the values

$$K=-2\Lambda,\ -2\Lambda+1,\ \ldots,\ -1, 0, 1, \ldots, 2\Lambda \qquad (4.90)$$

From these states one may select an intrinsic state for which the quantum numbers n_z, and subsequently n_x are assigned maximum values subject to (4.85), and the resultant numbers $\lambda=n_z-n_x$, $\mu=n_x-n_y$ are sufficient to label the corresponding representation of SU(3). It then turns out, rather conveniently, that the remaining states of the respresentation (λ, μ) may be obtained by acting on the intrinsic state with the angular momentum operators $L_\pm$, and each state thereby generated is assigned to an $SU(2)_\Lambda$ multiplet with $\Lambda=(n_x+n_y)/6$.

For single-particle states the representation $(n, 0)$ is the only SU(3) representation in a given shell, and the intrinsic state has $\Lambda=K=0$. This is not true in general, however, and the intrinsic state may be expanded as a superposition of eigenstates of L_0 and L^2, with values of K restricted by (4.90), and L-values associated with a given K which are just those of a K-band cut-off at a maximum value $L_{max} = K + \max(\lambda, \mu)$ (cf. Table 4.3). The final set of states $|KLM\rangle$ can then be generated by operating on the component states $|KL\rangle$ of this expansion with the operators $L_\pm$ in the usual way.

All these states are of course degenerate in a pure oscillator field but the degeneracy may be lifted by appending to H_0 a two-body quadrupole interaction $H_Q \sim \tilde{\boldsymbol{Q}} \cdot \tilde{\boldsymbol{Q}}$ which breaks the SU(3) symmetry within a given oscillator shell, without admixing states of neighbouring shells. Since $SU(2)_L$ is a symmetry group of the modified Hamiltonian $H_0 + H_Q$, L is still a good quantum mumber and the effect of the residual interaction is to contribute a term $\sim L(L+1)$ to the energy, independent of K, which is characteristic of the rotational band. Incidentally the same conclusion follows from the addition of a residual interaction $H_L \sim \boldsymbol{L} \cdot \boldsymbol{L}$, a result which was exploited in the early Nilsson model to depress the energy of the higher angular momentum states (Nilsson 1955) (cf. section 4.6.3).

The quantum number K is not quite a good quantum number when the residual interactions are switched on, since it labels states which transform according to irreducible representations of the group $SU(2)_\Lambda \times U(1)$ which does not contain $SU(2)_L$ as a subgroup. The slight ambiguity which arises thereby is also illustrated in Table 4.3, where the rotational state of the SU(3) representation (2, 2) with $L = 2$ may be associated with intrinsic states with $K = 0$ (β-vibrations) or $K = 2$ (γ-vibrations). However, the general conclusion stands that the presence of rotation–vibration bands may be associated with the symmetry-breaking chain $SU(3) \supset SU(2)_L$.

4.6.2 The interacting boson model

The interacting boson model (IBM) of nuclear structure was originally introduced (Arima and Iachello 1974, 1975, 1976, 1977, 1978, 1979) with the quite specific aim of describing the main observational features of collective motion in even–even nuclei, e.g. the quadrupole nature of the excitations, in terms of a model with a restricted number of degrees of freedom. Thus it provides a prescription for truncating the set of shell-model states which is quite different from that adopted in the collective model, where the equilibrium nuclear shape is assumed *a priori*, and oscillations about this shape are quantized according to the usual rules. In the IBM the basic states are laid down in advance and the various nuclear shapes are then associated with certain dynamical symmetries imposed on the Hamiltonian.

In its simplest form (IBM 1) the model represents the nucleus as a system of N spinless bosons which can occupy levels with $L = 0$ (s-bosons) and $L = 2$ (d-bosons), and which interact via one-boson and two-boson forces. One possible physical interpretation of this scheme is to identify s- and d-bosons with the lowest 0^+ and 2^+ states of a Cooper pair of valence nucleons, a conclusion which would seem to imply that these are the only fermion states which are of significance for the description of low-energy collective excitations in even–even nuclei. Thus, the d-boson is to be associated with the quadrupole nature of these excitations but there are no monopole excitations associated with the s-boson because the total number of bosons is conserved. Monopole vibrations are also forbidden in the collective model by the requirement that the nuclear volume be conserved. In extended versions of the model (IBM 2) (Elliott 1985, Iachello and Talmi 1987) a distinction is drawn between proton (π) and neutron (ν) bosons and, in addition, the introduction of a neutron–proton boson allows the possibility of incorporating odd–odd nuclei into the scheme. To account for the collective properties of odd-A nuclei the Hamiltonian must be further enlarged to include boson, fermion and fermion–boson interactions, but all of these refinements fall quite outside the very limited scope of the present discussion.

Since the model postulates the existence of one s-boson, and five d-bosons denoted by d_m ($m=0, \pm 1, \pm 2$), a total of 36 one-boson operators may be constructed from the 12 creation and annihilation operators ($s^\dagger$, s; $\boldsymbol{d}^\dagger$, $\boldsymbol{d}$), and these may be organized into eight spherical tensor operators

$$\begin{aligned} & s^\dagger s^{(0)}, \; \boldsymbol{d}^\dagger \boldsymbol{d}^{(L)} \qquad L=0, 1, 2, 3, 4 \\ & s^\dagger \boldsymbol{d}^{(2)} \qquad \boldsymbol{d}^\dagger s^{(2)} \end{aligned} \tag{4.91}$$

which are the generators of the group U(6) of unitary transformations in the six-dimensional space of one-boson states. Since the boson number operator $\hat{N}$ defined by

$$\hat{N} = s^\dagger s^{(0)} + \sum_\mu d_\mu{}^\dagger d_\mu = s^\dagger s^{(0)} + \sqrt{5}\, \boldsymbol{d}^\dagger \boldsymbol{d}^{(0)} \tag{4.92}$$

commutes with all the generators of U(6) defined in (4.91), and is therefore a constant multiple of the unit operator, it may be excluded from the set of generators and the states of N bosons may be classified according to the irreducible representations of SU(6). These representations are totally symmetric, and the single number N suffices as a label. When the energy difference $\varepsilon = \varepsilon_d - \varepsilon_s$ between s- and d-boson states vanishes, and there are no two-body interactions between bosons, all states of the representation (N) of SU(6) are degenerate.

In general the Hamiltonian H may be constructed from the scalar operators in (4.91) together with two-body interactions obtained by contraction from the tensor operators with $L \geqslant 1$, but in general the resulting eigenvalue problem can be solved only in the circumstance that H can be expressed in terms of the invariant (Casimir) operators of a chain of subgroups G_i of SU(6).

$$\mathrm{SU}(6) \supset G_1 \supset G_2, \ldots, \supset \mathrm{O}(3) \tag{4.93}$$

ending on the angular momentum subgroup O(3) which must be an exact symmetry group of H. Within such a chain there is no mixing of representations of a given subgroup G, although the states of such a representation may be split by invariant interactions of lower symmetry, thus giving rise to a dynamical or accidental symmetry, which expresses the fact that certain interactions are dominant while others may be neglected, and does not imply the presence of some more fundamental invariance. In the case of the group SU(6) there are three such symmetry breaking chains which may be represented symbolically in the form

$$\mathrm{SU}(6) \supset \mathrm{SU}(5) \supset \mathrm{O}(5) \supset \mathrm{O}(3) \tag{4.94a}$$

$$\mathrm{SU}(6) \supset \mathrm{O}(6) \supset \mathrm{O}(5) \supset \mathrm{O}(3) \tag{4.94b}$$

$$\mathrm{SU}(6) \supset \mathrm{SU}(3) \supset \mathrm{O}(3) \tag{4.94c}$$

The dynamical symmetry (4.94a) arises in the circumstance that the number n_{d} of d-bosons is conserved, and all two-boson interactions which do not conserve n_{d} and n_{s} separately are neglected. In this case the lowest-order Hamiltonian can be written in the form

$$H_0 = \varepsilon_{\mathrm{s}}\hat{n} + \varepsilon_{\mathrm{d}}\hat{n}_{\mathrm{d}} = \varepsilon_{\mathrm{s}}\hat{N} + \varepsilon\hat{n}_{\mathrm{d}} \tag{4.95}$$

to which may be added two-boson interactions constructed from the invariant operators of the subgroup O(5), whose 10 generators are the operators $\boldsymbol{d}^{\dagger}\boldsymbol{d}^{(3)}$ and $\boldsymbol{d}^{\dagger}\boldsymbol{d}^{(1)}$ and O(3), whose three generators are the angular momentum operators $\boldsymbol{d}^{\dagger}\boldsymbol{d}^{(1)}$. The SU(5) invariant Hamiltonian given by (4.95) is of course the Hamiltonian of the five-dimensional harmonic oscillator whose spectrum is given in equation (4.58). Thus the symmetry-breaking chain (4.94a) applies when the energy difference ε is much larger than all surviving n_{d}-conserving two-boson matrix elements, and this broken SU(5) symmetry describes exactly the case of the anharmonic quadrupole vibrator discussed in the context of collective oscillations in spherical nuclei in section 4.4.3. In this limit states are labelled by four quantum numbers, n_{d}, τ, n_{Δ} and L, where τ is the O(5) 'seniority' quantum number defined with respect to states of d-boson pairs coupled to zero spin, and n_{Δ} represents the number of boson triplets coupled to zero spin. The quantum number τ is restricted to the range $n_{\mathrm{d}}, n_{d-2}, \ldots, 1$ or 0 and the rotational quantum number L takes the values $\lambda, \lambda+1, \ldots, 2\lambda-2, 2\lambda$ ($2\lambda-1$ excluded), where $\lambda = \tau - 3n_{\Delta} \geqslant 0$. The corresponding energy spectrum is

$$E(n_{\mathrm{d}}, \tau, L) = \varepsilon n_{\mathrm{d}} + \frac{\alpha}{2} n_{\mathrm{d}}(n_{\mathrm{d}}-1) + \beta(n_{\mathrm{d}}-\tau)(n_{\mathrm{d}}+\tau+3) + \gamma[L(L+1) - 6n_{\mathrm{d}}] \tag{4.96}$$

The first two terms describe a sequence of phonon multiplets which, because of the term quadratic in n_{d}, are not quite equidistant. This is the spectrum of an anharmonic vibrator. The remaining two terms describe the splitting of these phonon multiplets due to O(5)-symmetric and O(3)-symmetric interactions respectively.

A good example of an SU(5) spectrum as observed in the nucleus ${}^{110}_{48}\text{Cd}$ is shown in figure 4.3. All four bands shown have $n_\Delta = 0$. The Y, X and Z-bands have $\tau = n_d$ with $L = 2\tau$, $2\tau - 2$ and $2\tau - 4$ respectively, and the β-band has $\tau = n_d - 2$ with $L = 2\tau$.

The symmetry breaking chain (4.94b) describes the case where n_d is not conserved in two-boson interactions, so that the basic SU(6) symmetry is broken directly to the group O(6) (which is isomorphic on SU(4)), which numbers the five s–$\boldsymbol{d}$ mixing operators $\boldsymbol{d}^\dagger s + s^\dagger \boldsymbol{d}$ among its generators, along with those of its subgroup O(5). In this limit the states are labelled by an additional O(6) 'seniority' quantum number σ, which replaces n_d, and this broken symmetry is associated with the states of deformed nuclei with γ-instability (Meyer-ter-Vehn 1979). The corresponding energy spectrum is

$$E(\sigma_d, \tau, L) = \frac{1}{4} A(N - \sigma)(N + \sigma + 4) + B\tau(\tau + 3) + CL(L + 1) \qquad (4.97)$$

This describes a set of O(6) multiplets labelled by the quantum number σ which varies in the range $\sigma = N, N - 2, \ldots, 0$ or 1 with $\tau = 0, 1, 2, \ldots, \sigma$. The splitting of these multiplets follows the same rule as for the anharmonic vibrator.

A good example of an O(6) spectrum as observed in the nucleus ${}^{196}_{78}\text{Pt}$ is shown in figure 4.5. Each group of bands is labelled by the quantum numbers (σ, n_Δ) and the selection rule for $E2$ transitions within each band is $\Delta\sigma = 0$, $\Delta\tau = \pm 1$. Within the (6, 0) bands the spin sequence $0^+\ 2^+\ 2^+\ 4^+$, which is characteristic of a γ-unstable nucleus, corresponds to $\tau = 0$, $L = 0$; $\tau = 1$, $L = 2$ ($L = 1$ excluded) and $\tau = 2$, $L = 2, 4$ ($L = 3$ excluded). In the (6, 1) group of bands the sequence is repeated for $\tau = 3$, 4 and 5.

The final symmetry chain (4.94c) applies in the case that the energy difference ε in (4.95) tends to zero and, apart from terms linear and quadratic in $\hat{N}$, H is constructed entirely from invariant operators of the subgroup SU(3) whose eight generators are

$$\boldsymbol{L} = \boldsymbol{d}^\dagger \boldsymbol{d}^{(1)} \qquad \mathrm{Q}^{(2)} = (\boldsymbol{d}^\dagger s + s^\dagger \boldsymbol{d})^{(2)} - \sqrt{(7/4)}\boldsymbol{d}^\dagger \boldsymbol{d}^{(2)} \qquad (4.98)$$

Thus the Hamiltonian may be written

$$H = \varepsilon_1 \hat{N} + \varepsilon_2 \hat{N}^2 + \sigma \boldsymbol{Q}^{(2)} \cdot \boldsymbol{Q}^{(2)} + \beta \boldsymbol{L} \cdot \boldsymbol{L}$$

$$= \varepsilon_1 \hat{N} + \varepsilon_2 \hat{N}^2 + \frac{\alpha}{2} \hat{C}_2 + (\beta - \tfrac{3}{8}\alpha) \boldsymbol{L} \cdot \boldsymbol{L} \qquad (4.99)$$

where $\hat{C}_2$ is the quadratic Casimir operator in SU(3) whose eigenvalues in the representation (λ, μ) are

$$C_2(\lambda, \mu) = \lambda^2 + \mu^2 + \lambda\mu + 3(\lambda + \mu) \qquad (4.100)$$

Within a given representation (N) of SU(6) the integers λ and μ are restricted to the ranges

$$\lambda = 2N - 6l - 4k \geqslant 0 \qquad \mu = 2k \geqslant 0 \tag{4.101}$$

where k and l are non-negative integers. Thus the eigenvalues of H are

$$E(N, \lambda, \mu, L) = \varepsilon_1 N + \varepsilon_2 N^2 + \frac{\alpha}{2} C_2(\lambda, \mu) + (\beta - \tfrac{3}{8}\alpha) L(L+1) \tag{4.102}$$

The eigenfunctions of $\hat{C}_2$ are not of course eigenfunctions of H since they provide a reducible representation of the angular momentum subgroup $SU(2)_L$ and do not correspond to a definite value of L. This is a problem which has been discussed already in section 4.6.2 and the same conclusion applies; states of the same L within the representation (λ, μ) must be distinguished by another label which will play the same role in SU(3) as the quantum number n_Δ in SU(5) and O(6). We therefore introduce the number K where

$$K = 0, 2, \ldots, \min(\lambda, \mu) \tag{4.103}$$

and the allowed values of L within the respresentation (λ, μ) are

$$\begin{aligned} &L = 0, 2, 4, \ldots, \max(\lambda, \mu) && K = 0 \\ &L = K, K+1, \ldots, K + \max(\lambda, \mu) && K > 0 \end{aligned} \tag{4.104}$$

States with the same L and different K are of course degenerate since the energy eigenvalues do not depend on K.

Apart from the cut-off in the energy spectrum at $L = K + \max(\lambda, \mu)$ equations (4.102)–(4.104) describe the case of the axi-symmetric rotor with degenerate β and γ-bands (cf. section 4.5.1) where the operator $Q^{(2)}$ defined in (4.98), which is understood to be summed over all N bosons where it appears in the Hamiltonian (4.99), is the analogue of the quadrupole operator $\tilde{Q}^{(2)}$ defined in (4.89) in relation to fermion states in the isotropic oscillator potential which also provide representations of SU(3). In the boson space, however, only totally symmetric representations of SU(3) (e.g. the respresentations (6, 0) and (2, 2) in Table 4.3) are acceptable.

An interpretation of the spectrum of $^{234}_{92}U$ in terms of SU(3) symmetry is shown in figure 4.4. The number N of bosons is taken to be equal to half the number of valence protons (10) and valence neutrons (16) outside the closed shells at 82 and 126 respectively (Arima and Iachello 1978). Here the (26, 0) representation corresponds to the ground state band with $L = 0, 2, \ldots, 8$. Similarly the (22, 2) representation corresponds to the β and γ bands with $K = 0$ and $K = 2$, respectively, while the (18, 4) representation contains the $\beta\beta$ and $\beta\gamma$ bands of the collective model as discussed in section 4.5.1. Also the degeneracy in K is broken which means that there is a preferential coordinate system. This result is consistent with the interpretation of K as the projection of the angular momentum along the nuclear symmetry axis.

It appears therefore that the various unitary groups which play a role in the IBM may be associated with a range of nuclear shapes, exactly in the same way that crystal shapes, monoclinic, triclinic, etc., are classified by means of point groups. Real nuclei, however, do not correspond exactly to any of the three symmetry limits. For example, β-bands always tend to fall somewhat higher in energy than γ-bands, so there is a small admixture of O(6)-symmetric states in the states of SU(3) symmetry. Furthermore, and as noted previously, the transition between symmetry schemes can occur very rapidly. The SU(5) to SU(3) transition is one which has been studied in some detail and a good example is found in the samarium isotopes where the spectrum of $^{146}_{62}Sm$ is SU(5)-symmetric with $N=2$, while the spectrum of $^{156}_{62}Sm$ is SU(3)-symmetric with $N=12$ (Scholten *et al* 1978). Clearly the value of the IBM as a tool in the phenomenological analysis of collective spectra is not in doubt, even if the microscopic foundations of the theory in its present state of development seem somewhat insecure.

4.7 Neutron reactions and nuclear structure

4.7.1 Phenomenology of neutron–nucleus reactions

Over the past 50 years a vast store of information concerning the interactions of neutrons and nuclei has been accumulated, whose range and significance it is neither sensible, nor indeed possible, to review in the limited space available here. Most of these data come in the form of measured excitation functions, i.e. cross-sections for the various reactions of interest which involve neutrons in the initial or final state, measured as a function of incident particle energy, and have been systematically assembled for specific applications in the nuclear power industry (cf. sections 5.5.1–2). For thermal fission reactors the most important and thoroughly studied nuclear parameters are those characterizing the fission process in the fissile nuclei ^{233}U, ^{235}U, ^{239}Pu and ^{241}Pu, and the same data have also been determined for ^{252}Cf which undergoes spontaneous fission and is widely used as a fission standard. Because of its central role in the generation of nuclear power, nuclear fission is excluded from the present summary, and discussed in detail in chapter 5.

Almost all artificially produced radioactive nuclei are β^--active and have been created by multiple (n, γ) reactions in stable nuclei subjected to neutron bombardment in nuclear reactors, and almost all of the available information on low-lying nuclear levels has been obtained by studying the decay schemes of these radioactive nuclei. The energies and lifetimes of the high excited levels have been determined to a large extent by populating these levels through (n, p) and (p, n) reactions. Most neutron induced nuclear reactions of significance to the operation of fast reactors take place at energies $E_n \geqslant 30$ keV, rather than at thermal energies, and the relevant cross-sections may be determined by exposing the sample nuclei to high neutron fluxes for which the energy spectra are accurately known. The required information is then extracted by performing activation analysis, radiochemistry or mass spectroscopy on the irradiated targets. The (γ, n) or photoneutron reaction, which is important in the electromagnetic excitation of collective nuclear modes at energies above the nucleon emission threshold, is the subject of a separate treatment in section 4.7.5.

Empirical neutron reaction data are also used as inputs in astrophysical theory, since the (n, γ), (n, p) and (n, α) capture processes contribute to heavy element nucleosynthesis in stars (cf. section 9.6.7), and the inverse processes $^{13}C\ (\alpha, n)\ ^{16}O$ and $^{22}Ne\ (\alpha, n)\ ^{25}Mg$ are recognized as primary neutron sources at astrophysical sites. At the terrestrial level the inverse processes (p, n), (d, n), (t, n) and (α, n) generated by bombarding light nuclei with charged particles of energy between 2 and 12 MeV, are important for the production of polarized neutrons of a few MeV energy (cf. Table 7.2).

For medium mass and heavy nuclei, elastic scattering (n, n), radiative capture (n, γ), and fission (n, f) $(Z \geqslant 90)$, are the only reactions which are energetically allowed for slow neutrons in the energy range $0 \leqslant E_n \leqslant 0.5$ MeV, and below 1 keV the corresponding excitation functions exhibit a large number of narrow resonances. These are typically spaced ≈ 10 eV apart and have widths of order 0.1 eV. Since this width corresponds to a mean lifetime of about 10^{-14} s, which is perhaps 10^4 times greater than the time it takes for a slow neutron to traverse a nucleus, these resonances are classified as compound nucleus resonances for which the processes of formation and decay are statistically independent (cf. section 2.6.3.1). Neutron resonances tend to be predominantly capture resonances below 10 eV, and scattering resonances above 10 eV. Thus, always allowing for the contribution from shadow scattering (cf. section 2.5.4), the (n, γ) process is dominant in heavy nuclei, where the level density is high and the spacing between levels much reduced. The heavy nuclei ^{113}Cd and ^{135}Xe have resonances near 0.1 eV extending down to zero energy, a property which makes these elements strong absorbers of thermal neutrons in nuclear reactors, the former as an agent in the dynamical control system, and the latter as a poison (cf. section 5.5.2).

In light nuclei where the Coulomb barrier is low, several exothermic reactions occur involving charged particles in the final state, e.g. (n, p) reactions in ^3He, ^{14}N and ^{35}Cl, and (n, α) reactions in ^{10}B and ^6Li, of which the most intense, those with $Z \leqslant 10$, are used as neutron detectors (cf. Table 1.4). Quite generally the (n, p) reactions are exothermic provided the Q-value for the β-decay of the daughter nucleus satisfies the condition $Q < (m_n - m_H)c^2$. In these cases the daughter nuclei, e.g. ^3H, ^{14}C and ^{35}S will have β-spectra with low end-point energies. Since resonances in light nuclei tend to be displaced towards higher energy, e.g. $E_r \simeq 100$ keV in ^{11}B, the cross-sections for all these reactions satisfy the $1/v$ law to high accuracy (cf. section 1.1.1.1). However, deviations do occur which provide significant indicators of nuclear structure in the compound nucleus. An interesting case in point is the reaction ^{3}He(n, p)^{3}H whose cross-section shows departures from the $1/v$ law for $E_n \geqslant 10$ keV. Taken together with data from the inverse (endothermic) reaction ^{3}H(p, n)^{3}He, these observations point to the presence of 0^+ and 1^- resonant states in the α-particle at 19.94 MeV and 21.24 MeV respectively (Bergman *et al* 1957, Parker *et al* 1965).

At neutron energies $E_n > 0.5$ MeV, inelastic scattering (n, n′) is observed in most nuclei, together with (n, p) and (n, α) reactions, although in heavy nuclei the last named process is significant only when $E_n \geqslant 10$ MeV. At these energies (n, 2n), (n, np) and (n, 2p) processes also become prominent and, in contrast with the situation at lower energies, these reactions are all classified as direct reactions.

4.7.2 Theory of neutron–nucleus reactions

4.7.2.1 The Wigner–Eisenbud formalism

We confine attention to the simplest two-body reactions $A(a, b)B$, where 'A' is the target nucleus and the incident particle 'a' is (usually) a neutron. In this case it is convenient to separate neutron–nucleus reactions into 'elastic scattering' and 'absorption', where the latter description is understood to include inelastic scattering, capture and fission. The objective of a theory of neutron–nucleus reactions is to calculate the various partial cross-sections $\sigma_{ab}(E)$ for given 'A', and the total cross-section $\sigma_{at}(E) = \sum_b \sigma_{ab}(E)$, where E is the energy in the centre of mass of the colliding system. The initial state is specified by a set of quantum numbers which we denote collectively by the index 'i'. These quantum numbers characterize the nature and internal states of 'a' and 'A', when these systems are separated by distances large in comparison with the range of their nuclear interaction (although they may interact electromagnetically). In this case 'a' and 'A' have a joint internal wavefunction which we denote by χ_i. In the same way the outgoing particles 'b' and 'B' have a joint asymptotic internal state denoted by χ_f. In the present context we choose to describe particles 'b' or 'B' in various states of excitation 'b^*' or 'B^*' as different particles.

We then have the following angular momentum coupling rules

$$\boldsymbol{I}_i = \boldsymbol{I}_a + \boldsymbol{I}_A \qquad \boldsymbol{I}_f = \boldsymbol{I}_b + \boldsymbol{I}_B$$
$$\boldsymbol{I} = \boldsymbol{I}_i + \boldsymbol{l}_i = \boldsymbol{I}_f + \boldsymbol{l}_f \tag{4.105}$$

where $\boldsymbol{I}_i$ and $\boldsymbol{I}_f$ are channel spins and $\boldsymbol{l}_i$ and $\boldsymbol{l}_f$ are orbital angular momenta in the incident and exit channels respectively. In the case of neutron interactions

$$I_i = I_A \pm \tfrac{1}{2} \qquad I_A \neq 0 \qquad I_i = \tfrac{1}{2} \qquad I_A = 0 \tag{4.106}$$

For given E, I_i and l_i, only certain exit channels are allowed by the conservation laws and these channels are said to be 'open'. Thus the allowed values of the total angular momentum I are those which fall in the range

$$|I_i - l_i| \leqslant I \leqslant I_i + l_i \tag{4.107}$$

and I is just the spin of the resonant state of the compound nucleus. The number of contributing partial waves characterized by $l_i = 0, 1, 2, \ldots$ increases with increasing E.

The simplest description of the initial state is one in terms of the coupled angular momentum states $|I_i l_i I, M\rangle$, whose components defined by different values of I may interfere insofar as the angular distribution of final state particles is concerned. However, these interference terms disappear when integrated over all angles, and each partial cross-section reduces to a sum of terms corresponding to individual values of I. Thus we may write

$$\sigma_{nn}(E) = \frac{\pi}{k^2} \sum_I g_I |1 - S^I_{nn}|^2 \tag{4.108a}$$

$$\sigma_{na}(E) = \frac{\pi}{k^2} \sum_I g_I |S^I_{na}|^2 \tag{4.108b}$$

when S^I_{nn} and S^I_{na} are the relevant S-matrix elements, and the statistical factor g_I is given by

$$g_I = \frac{2I+1}{(2I_a+1)(2I_A+1)} = \frac{2I+1}{2(2I_A+1)} \qquad \sum_I g_I = 1 \tag{4.109}$$

Because of the unitarity of the S-matrix as expressed by (2.102), the total cross-section $\sigma_{nt}(E)$ obtained by summing the partial cross-sections in (4.108) is

$$\sigma_{nt} = \frac{2\pi}{k^2} \sum_I g_I \operatorname{Re}(1 - S^I_{nn}) \tag{4.110}$$

The theoretical problem presented is the calculation of the matrix elements S^I_{if} whose solution requires a knowledge of the stationary states of the joint system throughout the whole of the configuration space. It is this calculation which enables the observations as represented by $\sigma_{nn}(E)$, $\sigma_{at}(E)$ to be related to the nuclear structure. In the Wigner–Eisenbud method the configuration space is divided into two regions; an internal region (the 'compound nucleus') where all the nucleons of the joint system are close together, and can interact in an arbitrary way, and an external region where the joint system is subdivided into two well defined sub-systems, e.g. $A+a$, $B+b$, etc. (Wigner

1946, Wigner and Eisenbud 1947). When these sub-systems are both charged their electrostatic repulsion reduces the cross-section by a penetration factor $p \approx \exp[-G(E)\}$, where $G(E)$ is the Gamow factor defined in (5.265).

In the following simplified version of the Wigner–Eisenbud method we suppress the spin degree of freedom so that $I = l_i = l_f = l$. In this case an expression for the joint wavefunction Ψ_l which is valid in the external region, and on the boundary surface Σ which separates internal and external regions, is given by

$$\Psi_l = (m_i/\hbar)^{1/2} P_l(\mathbf{\Omega}_i)[I_l(r_i)/r_i]\chi_i - \sum_f S_{l,if}(m_f/\hbar)^{1/2} P_l(\mathbf{\Omega}_f)[O_l(r_f)/r_f]\chi_f \quad (4.111)$$

where $\mathbf{\Omega}_i$ is a unit vector aligned along the collision axis, and $\mathbf{\Omega}_f$ is a unit vector specifying the angular distribution of the emitted particles. $I_l(r_i)$ and $O_l(r_f)$ represent radial wavefunctions for incoming and outgoing waves respectively, normalized so that $(m_i/\hbar)^{1/2} I_l(r_i)$ and $(m_f(\hbar)^{1/2} O_l(r_f)$ each carry unit current per steradian, where m_i and m_f are the corresponding reduced masses.

In the R-matrix method, the matrix elements $S_{l,if}$ are determined by matching the logarithmic derivative of I_l on Σ, to its value in the internal region. The wavefunction in the internal region is determined in its turn by an expansion over a complete set of states which are the solutions of the many-body Hermitian eigenvalue problem, valid in the internal region, and specified by a given boundary condition on Σ. The relationship between the R-matrix and the S-matrix is illustrated in figure 4.7. These states act the

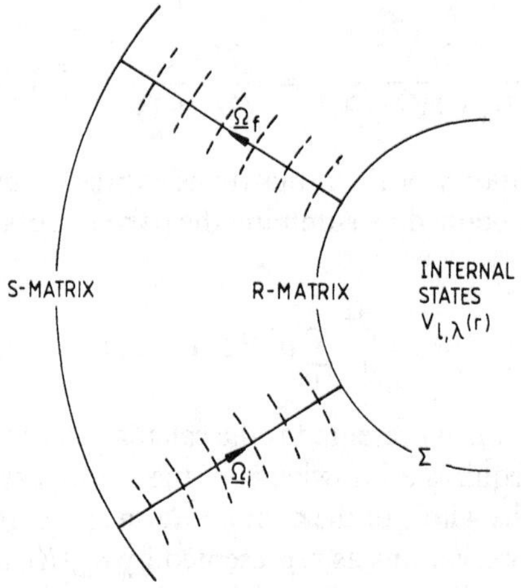

Figure 4.7. Diagram illustrating the relationship of the R-matrix to the S-matrix. In the case of elastic scattering the elements of R are obtained by matching the logarithmic derivatives of the radial wavefunction on the surface Σ separating internal and external regions, whereas the elements of S are determined by comparing the phases of scattered and unscattered waves in the asymptotic region.

part of resonance states which, in a limited sense, characterize the nuclear structure. Since there is a degree of arbitrariness in the selection of Σ, the R-matrix representation is not unique, a weakness which it shares with the Kapur–Peierls formalism to which it is closely related (Lane and Thomas 1958, Adler and Adler 1962, 1964). In theories of nuclear reactions based on dispersion theory with complex energies (cf. section 2.6.3.3), there is a corresponding degree of uncertainty associated with the behaviour of $S(k)$ at infinity as discussed in section 2.6.1 (Humblet and Rosenfeld 1961, Rosenfeld 1961, Humblet 1964).

4.7.2.2 Elastic scattering ignoring spin

In the case of elastic scattering discussed in section 2.3.2 the radial wavefunction corresponding to the expansion (4.111) of Ψ_l, is given according to (2.69), (2.72) and (2.74) by

$$u_l(k, r) = \frac{1}{2} ik^{1/2}\{I_l(k, r) - S_l(k)O_l(k, r)\} \tag{4.112}$$

where, in the notation of (2.69)

$$k^{1/2}I_l(k, r) = \chi_l(k, r)$$

$$k^{1/2}O_l(k, r) = \chi_l^*(k, r) = (-1)^l\chi_l(-k, r) \tag{4.113}$$

If we now assume that the interaction potential vanishes for $r > b$, we may choose the surface Σ at $r = b$. The R-matrix element $R_l(k)$ is then defined in terms of the logarithmic derivative of $u_l(k, r)$ on Σ, i.e.

$$R_l(k) \equiv \left.\frac{u_l(k, r)}{ru_l'(k, r)}\right|_{r=b} = \left.\frac{I_l(k, r) - S_l(k)O_l(k, r)}{r(I_l'(k, r) - S_l(k)O_l'(k, r))}\right|_{r=b} \tag{4.114}$$

which is a generalization of the results obtained in section 2.5.1 for s-wave scattering. $R_l(k)$ is a real function of k so long as the scattering potential is real.

To evaluate $R_l(k)$ we need an expression for $u_l(k, r)$ valid for $r \leqslant b$ and, to this end, we introduce a set of functions $v_{l,\lambda}(r)$, which satisfy the radial Schrödinger equation (2.59) for $0 \leqslant r \leqslant b$, subject to the boundary conditions

$$v_{l,\lambda}(0) = 0 \tag{4.115a}$$

$$v_{l,\lambda}'(b) = 0 \tag{4.115b}$$

The boundary condition (4.115a) is necessary so that $\psi_l(r) \approx v_l(r)/r$ should be finite at the origin. The boundary conditions (4.115) define a Hermitian boundary value problem, analogous to the Kapur–Peierls problem discussed in section 2.6.3.3, except that the real boundary condition (4.115b) is independent of energy. The real eigenfunctions $v_{l,\lambda}(r)$, for which the corresponding real eigenvalues are denoted by $E_\lambda(l)$, satisfy the orthonormal condition

$$\int_0^b v_{l,\lambda}(r)v_{l,\lambda'}(r)\,dr=\delta_{\lambda\lambda'} \tag{4.116}$$

and hence constitute a complete set on [0, b]. We may therefore expand $u_l(k, r)$ in the form

$$u_l(k,r)=\sum_\lambda A_{l,\lambda}v_{l,\lambda}(r) \tag{4.117}$$

Noting that $u_l(k, r)$ and $v_{l,\lambda}(r)$ are each eigensolutions of (2.59) with eigenvalues E and $E_\lambda(l)$, respectively, we may derive the relation

$$\int_0^b [v_{l,\lambda}(r)u_l''(k,r)-v_{l,\lambda}''(r)u_l(k,r)]\,dr+\frac{2m}{\hbar^2}(E-E_\lambda(l))\int_0^b u_l(k,r)v_{l,\lambda}(r)\,dr=0 \tag{4.118}$$

Integration by parts gives the value $v_{l,\lambda}(b)u_l'(k, b)$ for the first term in (4.118), while the value $(2m/\hbar)^2(E-E_\lambda(l))A_{l,\lambda}$ for the second term follows from the orthonormal condition (4.116). We then find, on substituting for $A_{l,\lambda}$ in (4.117) that

$$u_l(k,r)=\frac{\hbar^2}{2mb}\sum_\lambda \frac{v_{l,\lambda}(b)v_{l,\lambda}(r)}{E_\lambda(l)-E}[bu_l'(k,b)] \tag{4.119}$$

Evaluation of $R_l(k)$ using (4.114) and (4.119) then leads to the result

$$R_l(k)\equiv\frac{u_l(k,b)}{bu_l'(k,b)}=\frac{\hbar^2}{2mb}\sum_\lambda\frac{v_{l,\lambda}^2(b)}{E_\lambda(l)-E}=\sum_\lambda\frac{\gamma_\lambda^2(l)}{E_\lambda(l)-E} \tag{4.120}$$

where the reduced width $\gamma_\lambda^2(l)$ is defined by

$$\gamma_\lambda^2(l)=(\hbar^2/2mb)(v_{l,\lambda}(b))^2 \tag{4.121}$$

Since, as we show immediately below for the square well potential, $(v_{l,\lambda}(b))^2$ has a maximum value equal to $2/b$ (cf. 4.128–4.129), $\gamma_\lambda^2(l)$ has a maximum value equal to $(\hbar^2/mb^2)$ which is termed the single-particle, or Wigner, reduced width.

Solving (4.114) for $S_l(k)$ in terms of $R_l(k)$ we find that

$$S_l(k)=\frac{I_l(b)-bI_l'(b)R_l(k)}{O_l(b)-bO_l'(b)R_l(k)}=\frac{I_l(b)}{O_l(b)}\left\{\frac{1-(bO_l'(b)/O_l(b))^*R_l(k)}{1-(bO_l'(b)/O_l(b))R_l(k)}\right\} \tag{4.122}$$

since $I_l(r)$ and $O_l(r)$ are complex conjugate functions (cf. 4.113). Setting $\beta(k)$ equal to the logarithmic derivative of the outgoing wave at the boundary of the potential, we may rewrite this result in the form

$$S_l(k)=e^{-2i\varphi_l}\left\{\frac{1-b\beta^*(k)R_l(k)}{1-b\beta(k)R_l(k)}\right\} \tag{4.123}$$

where φ_l is the phase shift produced by scattering from an impenetrable sphere of radius b and is given by

$$\tan \varphi_l = n_l(kb)/j_l(kb) \tag{4.124}$$

Since $R_l(k)$ is real, $S_l(k) = \exp(2i\delta_l)$ is a number of modulus unity, and the phase shift in the lth partial wave is given by

$$\tan(\delta_l + \varphi_l) = \frac{R_l(k)b \operatorname{Im}(\beta(k))}{1 - R_l(k)b \operatorname{Re}(\beta(k))} \tag{4.125}$$

The function $s_l(k) = b \operatorname{Re}(\beta(k))$ is referred to as the shift factor, while $p_l(k) = b \operatorname{Im}(\beta(k))$ is the penetration factor.

Ignoring the contribution from hard-sphere scattering and assuming that only a single resonant state $v_{l,\lambda}$ plays a role, the result (4.125) reproduces the familiar Breit–Wigner formula for scattering at an isolated resonance

$$\sigma_l(k) = \frac{\pi}{k^2}(2l+1)\frac{\Gamma_\lambda^2(l)}{(E - E_\lambda(l) - \Delta_\lambda(l))^2 + (\Gamma_\lambda(l)/2)^2} \tag{4.126}$$

where

$$\frac{\Gamma_\lambda(l)}{2} = \gamma_\lambda^2(l)p_l(k) \qquad \Delta_\lambda(l) = \gamma_\lambda^2(l)s_l(k) \tag{4.127}$$

As an example we may consider briefly the case of scattering from a potential well of depth V_0. In this case the normalized eigenfunctions are

$$v_{l,\lambda}(r) = a_{l,\lambda}[rj_l(k_{l,\lambda}r)] \tag{4.128}$$

where

$$a_{l,\lambda}^2 = 2b^{-3}\{j_l^2(k_{l,\lambda}b) - j_{l-1}(k_{l,\lambda}b)j_{l+1}(k_{l,\lambda}b)\}^{-1} \tag{4.129}$$

The eigenvalues $E_\lambda(l)$ satisfy

$$E_\lambda(l) + V_0 = \frac{\hbar^2 k_{l,\lambda}^2}{2m} \tag{4.130}$$

where $k_{l,\lambda}$ is the root of the equation $j_{l-1}(k_{l,\lambda}b) = 0$, except in the case of s-waves for which

$$E_{0,\lambda} = \frac{\hbar^2 k_{0,\lambda}^2}{2m} = \frac{\hbar^2}{2mb^2}\pi^2\left(\lambda + \frac{1}{2}\right)^2 \qquad \lambda = 0, 1, \ldots \tag{4.131}$$

According to (4.121) the reduced widths $\gamma_\lambda^2(l)$ which characterize the R-matrix are given by

$$\gamma_\lambda^2(l) = \frac{\hbar^2}{2mb^2} a_{l,\lambda}^2 j_l^2(k_{l,\lambda}b)^2 \tag{4.132}$$

To apply these results to the optical model we replace V_0 by V_0+iW_0 (cf. section 2.4.3), in which case the R-matrix defined in (4.120) is replaced by

$$R_l(k)=\sum_n \frac{\gamma_\lambda^2(l)}{E_\lambda(l)+iW_0-E} \tag{4.133}$$

Thus $R_l(k)$ is no longer real and $S_l(k)$ ceases to be a number of unit modulus, corresponding to the fact that the particle current is not conserved in a complex potential.

4.7.2.3 The general R*-matrix*

Deriving the general R-matrix involving the full apparatus of spin, and determining its precise relationship to the S-matrix, presents far too complex a problem to be analysed in detail here. We shall therefore attempt to provide only an outline of the theory. In the general case the contribution to the total wavefunction in the external region corresponding to the separation of the compound nucleus into a pair of systems identified by the index 'j' is a superposition of states of the form

$$\psi_j=[u_l(r_j)/r_j](i^l Y_{lm}(\Omega_j))\eta(I_jM_j) \tag{4.134}$$

where $\eta(I_jM_j)$ represents the joint spin-state of the pair coupled to total spin I_j and z-component M_j. The wavefunction ψ_j is a solution of the Schrödinger equation for a spherically symmetric potential $V(r_j)$ equal to the energy of interaction of the two systems when their mass centres are separated by a distance r_j. The definition of the states which characterize the compound nucleus in the internal region depends on the boundary conditions which are imposed on the wavefunction. In the Kapur–Peierls formulation of dispersion theory, although the boundary conditions (2.6.40) lead quite naturally to a corresponding expression for the S-matrix in (2.6.41) the resonance parameters are in general both complex and energy dependent.

The Wigner–Eisenbud R-matrix technique avoids these difficulties by imposing energy-independent and real boundary conditions on the internal wavefunction, but at the expense of introducing a description in terms of standing waves rather than progressive waves. The starting point is to express the radial wavefunction $v_l(r_j)$ in terms of linearly independent regular and irregular functions $S_l(r_j)$ and $C_l(r_j)$ respectively, which are the analogues of the functions $(m_j/\hbar)^{1/2}r_jj_l(kr_j)$ and $(m_j/\hbar)^{1/2}(kr_j)n_l(kr_j)$ which enter into the analysis of elastic scattering. Choosing the surface Σ_j as the sphere of radius b_j beyond which the (non-electromagnetic) interaction between the two systems is assumed to vanish, the functions $S_l(r_j)$ and $C_l(r_j)$ are subjected to the boundary conditions

$$S_l(b_j)=0 \qquad C_l'(b_j)=-(l_j/b_j)(m_j/\hbar)^{1/2}$$
$$C_l(b_j)=S_l'(b_j)=(m_j/\hbar)^{112} \tag{4.135}$$

From these radial wavefunctions we construct coupled angular momentum eigenstates $|I_j, l; I, M\rangle$ whose representatives we denote by (cf. section 6.3.1)

$$D^I_{jlM} \equiv D^I_{lM}(r_j\mathbf{\Omega}_jI_j) = \sum_m [S_l(r_j)/r_j](\mathrm{i}^l Y_{lm}(\mathbf{\Omega}_j))\eta(I_jM-m)\langle I_jM-m;\, lm|I_jlIM\rangle \tag{4.136}$$

and

$$V^I_{jlM} \equiv V^I_{lM}(r_j\mathbf{\Omega}_jI_j) = \sum_m [C_l(r_j)/r_j](\mathrm{i}^l Y_{lm}(\mathbf{\Omega}_j))\eta(I_jM-m)\langle I_jM-m;\, lm|I_jlIM\rangle \tag{4.137}$$

where $\langle I_1M_1; I_2M_2|I_1I_2IM\rangle$ are Clebsch–Gordon coefficients describing the coupling of $\boldsymbol{I}_1$ and $\boldsymbol{I}_2$ to form total angular momentum $\boldsymbol{I}$.

The most general solution of the wave-equation in the external region for which I and M are good quantum numbers, may be expressed as a linear combination of the functions D^I_{jlM} and V^I_{jlM} with coefficients which depend on j and l. A solution which accords a particular status to one channel 'i', namely that channel for which the coefficient of the function D^I_{jlM} is non-vanishing, is given by

$$\Psi^I_{il_iM} = D^I_{il_iM} + \sum_{f,l_f} (b_ib_f)^{1/2} R_{il_i;fl_f} V^I_{fl_fM} \tag{4.138}$$

and of course we identify the ith channel with the single channel for which there is an incoming wave. This equation may be taken as the definition of the matrix elements of the R-matrix, which are functions of I and E, but not of M, since all directions in space are equivalent. Because of the boundary condition (4.135) the relation (4.138) for R reduces to (4.114) in the special case of elastic scattering of spinless particles. Similarly it reproduces the external wavefunction (4.112) in the case of hard-sphere scattering with $\varphi_l = -\pi/2$ (cf. 4.124).

The next stage in the development of the formalism is to relate the matrix elements of R to the structure of the compound nucleus as parameterized by the normal derivatives of the internal wavefunction on all the surfaces Σ_j which divide the configuration space into internal and external regions. This wavefunction, which is the continuation of $\Psi^I_{il_iM}$ into the internal region, we shall denote by $\Phi^I_{il_iM}$, and to represent it we introduce a complete orthonormal set of solutions $v^I_{\lambda M}$ of the Schrödinger equation which satisfy the same boundary conditions on every surface Σ_j as do the functions $V^I_{il_jM}$. We then expand $\Phi^I_{il_iM}$ in the form

$$\Phi^I_{il_iM} = \sum_\lambda A_{il_i\lambda} v^I_{\lambda M} \tag{4.139}$$

where the expansion coefficients are defined by

$$A_{il_i\lambda}(E, I) = \int v_{\lambda M}^{I*} \Phi_{il_iM}^{I} \, d\tau \tag{4.140}$$

The integral is computed over the whole volume of the internal region and includes the relevant spin summations.

Exactly as in section 4.7.2.2 we may determine the integral in (4.140) by making use of the facts that $\Phi_{il_iM}^I$ and $v_{\lambda M}^I$ are each solutions of the internal Schrödinger equation with energies E and $E_\lambda(I)$, respectively. Thus following the same procedure that led to (4.118) we find that

$$-\sum_j \frac{\hbar^2}{2m_j} \int \left\{ v_{\lambda M}^{I*} \frac{\partial}{\partial r_j} \Phi_{jl_jM}^I - \Phi_{jl_jM}^I \frac{\partial}{\partial r_j} v_{\lambda M}^I \right\} d\sigma_j = (E - E_\lambda(I)) \int v_{\lambda M}^{I*} \Phi_{il_iM}^I \, d\tau \tag{4.141}$$

When $\Phi_{il_iM}^I$ is matched to $\Psi_{il_iM}^I$ on each Σ_j the first sum in the surface integral in (4.141) vanishes because of the boundary condition on $V_{jl_iM}^I$, while only a single term survives in the second sum, contributed by the surface Σ_i due to the presence of the single function $D_{il_iM}^I$ in $\Psi_{il_iM}^I$. Therefore for $E \neq E_\lambda(I)$ there exists a solution to (4.140)

$$A_{il_i\lambda} = \left(\frac{\hbar}{2b}\right)^{1/2} \frac{\gamma_{\lambda il_i}}{E_\lambda(I) - E} \tag{4.142}$$

where

$$\left(\frac{\hbar}{2b}\right)^{1/2} \gamma_{\lambda il_i} = \frac{\hbar^2}{2m} \int_{\Sigma_i} v_{\lambda M}^{I*} V_{il_iM}^I \, d\sigma_i \tag{4.143}$$

The corresponding expression for the R-matrix is

$$R_{il_i;fl_f} = \frac{\sum_\lambda \gamma_{\lambda il_i} \gamma_{\lambda fl_f}}{E_\lambda(I) - E} \tag{4.144}$$

Note that, in accordance with recent usage, the matrix elements introduced in (4.138) and (4.144) are dimensionless; in the original work of Wigner and Eisenbud they had dimensions of length. The reality of the function $\gamma_{\lambda jl_j}$ is guaranteed from considerations of time reversal invariance and the symmetry properties of the Clelsch–Gordon coefficients (Biedenharn and Rose 1953, Huby 1954).

To determine the elements of the S-matrix as functions of the elements of the R-matrix it is necessary to revert to a description in terms of progressive waves instead of standing waves, starting with an examination of the asymptotic form of the functions $D_{jl_jM}^I$ and $V_{jl_jM}^I$. This is a lengthy procedure involving the construction of linear combinations of the wavefunctions $\Psi_{jl_jM}^I$, constrained in such a way that there is a single incoming wave labelled by the index 'i'. The end result is a very complicated relationship between the two matrices which, given that the number of eigenfunctions and the number of competing channels may be very large, and a significant number of these channels may be closed, may present a major problem of matrix inversion. There are also other ambiguities in the theory associated with the non-uniqueness of the R-matrix; e.g. an R-matrix analysis of the motion of a free particle in the presence of another particle with which it does not interact reveals the existence of purely formal resonances which correspond to no observable phenomenon. As a consequence of these difficulties a number of approximate methods have been introduced, which are valid in certain restricted circumstances and which avoid many of these problems (Teichmann and Wigner 1952, Thomas 1955).

An important example of an approximate theory occurs for the neutron–nucleus interaction, with well-separated resonances in the presence of potential scattering (Lane and Thomas 1958, Vogt 1958, 1960). In this case the S-matrix corresponding to (4.144) is given by

$$S_{if}(I)=\exp\,[i(\varphi_i+\varphi_f)]\left\{\delta_{if}+\sum_{\lambda\mu}[\Gamma_{\lambda i}(I)\Gamma_{\mu f}(I)]^{1/2}A_{\lambda\mu}\right\} \quad (4.145)$$

where δ_{if} denotes the Kroneckr δ-function, φ_i is the hard-sphere scattering phase in channel 'i' and $A_{\lambda\mu}(I)$ is a level matrix defined by

$$A_{\lambda\mu}(I)=\left[(E_\lambda(I)-E)\delta_{\lambda\mu}-\frac{i}{2}\Gamma_{\lambda\mu}(I)\right]^{-1} \quad (4.146)$$

The function $\Gamma_{\lambda\mu}(I)$ is given by

$$\Gamma_{\lambda\mu}(I)=\sum_j(\Gamma_{\lambda j}(I)\Gamma_{\mu j}(I))^{1/2} \quad (4.147)$$

where the partial widths $\Gamma_{\lambda j}(I)$ are those describing the decay of the λth level of spin I into the jth channel. These are related to the reduced widths $\gamma^2_{\lambda j l_j}$ of R-matrix theory

$$\frac{\Gamma_{\lambda j}}{2}=p_{l_j}\gamma^2_{\lambda j l_j} \quad (4.148)$$

where p_{l_j} is the penetration factor determined by the behaviour of the radial wavefunction at $r=b_j$ (cf. 4.147). For s-wave neutrons $p_0=\delta_0=kb_j$. Resonance interference effects become important when the off-diagonal elements of $A_{\lambda\mu}$ are comparable with the diagonal elements.

4.7.3 Analysis of neutron resonance reactions

4.7.3.1 Experimental determination of resonance parameters

In the case of all three versions of nuclear resonance theory introduced above, i.e. the Breit–Wigner, Kapur–Peierls and Wigner–Eisenbud formulations, the total cross-section reduces to the sum of single-level Breit–Wigner terms, provided only that the mean spacing $\langle D \rangle$ between levels is much greater than their average width $\langle \Gamma \rangle$. At the other extreme, in the continuum limit, the total cross-section is a smooth function of energy but the partial cross-sections exhibit structure, the so-called Ericson fluctuations (Ericson 1963). These arise from the circumstance that neither initial nor final states are well-defined, and the conditions for resonance–resonance interference are established. This can happen at low energy for certain fissile nuclei, and at high energy for all nuclei. There is also an intermediate region where $\langle D \rangle$ and $\langle \Gamma \rangle$ are comparable in magnitude; this region is characterized by the appearance of quasi-resonances, i.e. single sharp lines which arise when a sharp resonance is superimposed on a broad resonance.

In addition to the resonance energy E_r itself, the most important parameters characterizing individual resonances are the neutron width Γ_n, the various partial widths Γ_j, the total width Γ and the spin I. In heavy nuclei Γ_γ is the largest width and, in fissile nuclei, the fission width Γ_f is comparable in magnitude. In light nuclei, and quite generally at higher energies ($E_n \geqslant 10$ MeV), Γ_p and Γ_α may also be significant. According to (4.148) a partial width Γ_{jl} is a product of a reduced width γ_{jl}^2 and a penetration factor p_l, which is determined by the height of the centrifugal barrier presented to the incoming neutron. Since, for $l \neq 0$, p_l is very small ($p_1 \simeq (kb)^3$), the vast majority of resonances are s-wave resonances.

By combining pulsed neutron sources such as the electron linac used with a suitable target (cf. section 1.6.4), producing neutron bursts for periods of time 1–5 ns long, with time of flight spectrometers using path length ≈ 200 m, the properties of neutron resonances may be analysed at an energy resolution $\Delta E/E \simeq 0.1\%$, over the whole energy range between 10^{-2} eV and 10 MeV. By these techniques total cross-sections may be determined by transmission through selected targets, and both E_r and Γ determined for whole groups of resonances, by combined shape and area analysis, making due allowance for Doppler broadening. Partial widths Γ_i may be determined by measuring the relevant differential cross-sections. The l-value of a resonance may be derived from its shape according to the Breit–Wigner formulae, or by observing interference effects with potential scattering or with other resonances in its neighbourhood (Merzbacher *et al* 1959).

Although, when $\Gamma_n \simeq \Gamma$, total cross-section measurements may sometimes be used to estimate a value for the statistical factor g_I, leading to a determination of I, the only wholly unambiguous method for measuring I is to study the capture of polarized neutrons in polarized targets (cf. section 7.4.5.3). This technique is of particular importance for the determination of fission resonances. However, in special circumstances indirect methods may be employed successfully; examples are (i) observation of level interference effects in elastic scattering and radiative capture, (ii) measuring the spectra of capture γ-rays or the intensities of low energy γ-rays emitted from heavy nuclei, (iii) exploiting the spin-parity selection rules for (n, α) reactions in light nuclei, and (iv) measuring the angular distribution of scattered neutrons at high energy. The difficulty with indirect methods is that what appears as a single resonance may in fact be two unresolved resonances, and inconsistent results may be obtained.

4.7.3.2 Resonance properties and nuclear structure

At the excitation energies of interest in compound nucleus formation the density of states $\rho(E)$ per unit energy interval is very high, and it is impossible to calculate the detailed properties of all these states on the basis of any single nuclear model. Thus the manner of relating the measured properties of resonances to nuclear structure must be statistical in nature, and the quantities of prime interest in this regard are the statistical distributions of the energy separations D between neighbouring resonances, and of the partial widths Γ_j for the various channels of interest. At a somewhat lower level of significance are the correlations between neighbouring separations and widths, although the former may be measured for s-wave resonances in even–even nuclei and applied to the identification of p-wave resonances in nuclei with non-zero spin.

When absolutely nothing is known about the nuclear Hamiltonian H, the simplest assumption that can be made about the positions of its eigenvalues is that these are distributed at random across the energy scale. In this case the probability of observing n resonances, in the energy interval D above an energy E, is given by the Poisson distribution

$$P(n, D)=\exp[-\rho(E)D](\rho(E)D)^n/n! \qquad (4.149)$$

with mean number $\langle n\rangle=\rho(E)D$. Thus the probability that the next resonance should fall between D and $D+\mathrm{d}D$ above a given state is given by the exponential law

$$p(D)\,\mathrm{d}D=P(0, D)P(1, \mathrm{d}D)=\exp[-\rho(E)D]\,\mathrm{d}(\rho(E)D) \qquad (4.150)$$

The density function given in (4.150) is not confirmed experimentally. In particular, far fewer small values of D are observed for a given $\langle D\rangle$ than the exponential law predicts, an effect which is sometimes expressed in terms of 'repulsion' between neighbouring levels.

If the compound nucleus is described by some version of the shell model plus residual interactions between nucleons then, at high levels of excitation, a large number of closely packed levels will be generated as the high degeneracies of the shell model states are destroyed by the residual interactions. Thus any determination of $p(D)$ must begin with some hypothesis as to how these splittings are distributed. This approach was first adopted by Wigner, who showed that, if the energies E_λ are the eigenvalues of a real symmetric matrix, whose elements are normally distributed random variables, then the distribution of D is described by the density function (Wigner 1956, 1957, Mehta and Gaudin 1960)

$$p(D)\,dD = [\pi D/2\langle D\rangle]\exp[-\pi(D/2\langle D\rangle)^2] \qquad (4.151)$$

Detailed studies of more than 200 resonances in each of the nuclei ^{282}Th and ^{238}U, plus an equal number of resonances observed in 25 other nuclei, have confirmed that the Wigner distribution is obeyed to a high degree of accuracy (Garg *et al* 1964, Garrison 1964, Haq *et al* 1982).

The distribution (4.151) which characterizes Wigner's Gaussian orthogonal ensemble is an indicator of 'quantum chaos', which is a feature of quantum systems in which non-linear effects are dominant. These are the analogues of classical systems which display chaotic behaviour of which the most famous example is the Sinai billiard (Lichtenberg and Liebermann 1983). At lower levels of excitation nuclear chaos has been detected in rotational motion where non-linear effects are introduced by centrifugal and Coriolis forces (Pavlichenkov 1989) and in the transition region between the SU(3) and O(6) limits of the interacting boson model (cf. section 4.6.2) (Bohigas and Weidenmüller 1988, Barani 1990).

It was observed long ago that the neutron widths Γ_n display marked fluctuations in magnitude and, since the penetration factor varies smoothly with energy (cf. 4.127), the conclusion must be that these fluctuations are associated with the reduced neutron widths $\gamma^2_{\lambda nl}$. According to (4.107) the reduced width amplitude $\gamma_{\lambda jl_j}$ is determined by an overlap integral on the nuclear surface between an internal wavefunction $v^I_{\lambda M}(E_\lambda)$ and an external standing wave $V^I_{jl_jM}(E)$. Assuming the compound nucleus is constructed from a large number of single-particle configurations, the argument is advanced that the overlap integral is composed of a sum of independent contributions of either sign. Hence, according to the central limit theorem of statistics, the value of the integral should be normally distributed about a zero mean (Porter and Thomas 1956). Thus, since for elastic neutron scattering there is only one exit channel, the width $\Gamma_n \approx \gamma^2_n$ should be distributed according to a χ^2-law with one degree of freedom

$$p(\Gamma_n)\,d\Gamma_n = (2\pi\langle\Gamma_n\rangle)^{-1/2}(\Gamma_n/\langle\Gamma_n\rangle^{1/2}\exp[-\tfrac{1}{2}(\Gamma_n/\langle\Gamma_n\rangle)] \quad (4.152)$$

The Porter–Thomas distribution given in (4.152) has also been derived by methods similar to those applied by Wigner to the analysis of level spacings (Blumberg and Porter 1958, Rosenzweig 1957, 1958). Measurements of the distribution of widths for more than 100 resonances in ^{238}U and other even–even nuclei, provide strong experimental support for the theoretical predictions (Firk *et al* 1963, Garrison 1964).

For levels of the same spin and parity within a given nucleus the radiation widths Γ_γ do not exhibit large fluctuations, a result which remains true even when a range of nuclei is included, always excluding the large discontinuities which may occur near closed shells. These observations are usually explained on the grounds that, in general, the number of exit channels open to radiative decay is quite large. Experimental results for neutron resonances excited in the reaction $^{195}Pt(n, \gamma)^{196}Pt$ indicate that the radiative widths are distributed according to a χ^2-law with between 1 (Porter–Thomas law) and 2 (exponential law) degrees of freedom (Bollinger *et al* 1963).

Because of the very large number of exit channels which are open to fission one might expect to observe reduced fluctuations in the fission widths but this is found not to be the case. The essential reason for this result is that the saddle point configuration which precedes fission has to be treated as part of the internal region in the Wigner–Eisenbud theory and, because of angular momentum and parity conservation, the number N_t of levels in the transition state which are available for fission tends to be rather small (cf. section 5.2.3). If only one such fission channel is open then the Porter–Thomas distribution is applicable. If there are N_t fission channels for which the Gamow tunnelling factor is equal to unity then the distribution obeys a χ^2-law with N_t degrees of freedom. In general, however, the various barrier transmission factors tend to be unequal, and the distribution follows a more complicated law.

4.7.4 Compound nucleus and direct reactions

4.7.4.1 Statistical model of compound nucleus reactions

By convention, nuclear reactions are organized into two categories described as 'compound nucleus reactions' and 'direct reactions', respectively, although it is now recognized that this division is not so distinct as was once thought (Hodgson 1987). Compound nucleus reactions are those which may be described by Bohr's strong coupling, or uniform model, which was briefly introduced in section 2.6.3. This model assumes, in essence, that all incident particles falling on the nuclear surface are absorbed, and that the resultant compound nucleus survives sufficiently long that the processes of formation and decay are statistically independent. There is some impressive experimental evidence which supports this hypothesis, e.g. the branching ratios for decay of the compound nucleus ^{64}Zn into n, 2n and np channels, are independent of whether this nucleus is formed by the reactions of α-particles with ^{60}Ni, or of protons with ^{63}Cu (Ghoshal 1950).

Compound nucleus reaction types dominate at low energy and gradually give way to direct reactions as the energy is increased. There is however an intermediate energy range where both types of process contribute, and in this regime each process must be evaluated separately and the results summed, before a valid comparison with the experimental data may be made. In practice the compound nucleus contribution to the cross-section is negligible at energies above 15 MeV.

Direct reactions take place on time-scales comparable with the transit time across the nucleus. For example an incoming neutron wave may be refracted rather than absorbed in its passage through the nucleus, producing the shadow scattering effect described in section 2.5.4. Alternatively it may strike a proton which is ejected from the nucleus by the knock-on effect, while the neutron itself is captured into a bound orbit. The stripping reaction described in section 4.3.4 is also a direct reaction. The conditions which favour direct reactions are determined by the mean free path in nuclear matter, which is itself a result of competition between two conflicting factors. These are the nucleon–nucleon collision cross-section which falls with increasing energy, and the restrictions on the number of available final states imposed by the exclusion principle which tend to be less severe at higher energy (cf. section 4.1.1).

Once it is accepted that the reactions which follow the creation of the compound nucleus are determined only by the energy of excitation E, it is only necessary to assume in addition that the density of states in the compound nucleus and the residual nucleus are sufficiently high that the cross-sections for decay into specific channels may be calculated by statistical methods. The nucleus is in a state of thermal equilibrium and particles are emitted by a process of 'evaporation', in analogy with molecules evaporating from the surface of a liquid drop (Weisskopf 1937). The probability for decay into a specific channel is then determined by the principle of detailed balance (cf. section 8.3.2), which relates the decay rate into a specific channel to the corresponding reverse process for formation of the compound nucleus. For example the cross-section for elastic scattering of a neutron of energy E_n is predicted to follow Weisskopf's evaporation formula

$$\sigma(E_n) \approx \sigma_c(E_n, E) E_n \exp[-E_n/kT(E)] \rho(E) \qquad (4.153)$$

where $\rho(E)$ is the density of states of the compound nucleus at excitation energy E, $T(E)$ is a characteristic nuclear temperature, and $\sigma_c(E_n, E)$ is the cross-section for the initial neutron–nucleus reaction.

These ideas formed the basis of the earliest statistical theories of the compound nucleus (Weisskopf and Ewing 1940), subsequently improved to take account of angular momentum and parity conservation rules, whose omission was a serious defect of the original Bohr model (Wolfenstein 1951, Hauser and Feshbach 1952). The question as to whether the formation and decay of the compound nucleus are truly independent may be tested by measuring the angular distribution of the emitted particles. This is predicted to be isotropic, provided the density of states depends on the spin only through the statistical factor $(2I+1)$ (Bethe 1937). Experimentally isotropy appears to hold good for energies up to 20 MeV, but at high energies the distribution is equally peaked in forward and backward directions, indicating the presence of non-linear spin-dependent factors in $\rho(E)$ (Bloch 1954, Lang and Le Couteur 1954). Considerable effort has been expended in calculating

$\rho(E)$ in various nuclear models making due allowance for shell-structure effects (Newton 1956, Rosenzweig 1957, Ericson 1960, 1960a, Lang 1961).

In more recent times attention has moved away from further refinement of the details of specific nuclear reactions towards exploring the more general question as to how the compound nucleus is formed in the first place. This may be viewed as forming in a succession of stages of increasing complexity, where the first stage describes the incident particle moving in the field of the target nucleus. Corresponding to these stages there is a hierarchy of states, beginning with the open channel wavefunction and progressing upwards through 'doorway states' through which the system must pass for the full panoply of resonant states in the compound nucleus to be developed (Block and Feshbach 1963, Feshbach 1974). The essential property of a doorway state is that the ratio of its width to the mean separation between doorway states is small. The square of the amplitude of the doorway state is inversely proportional to the number of compound nucleus resonances contained within its width, which is in general a large number. Thus the doorway state has a small amplitude and its importance lies in its dynamical role which is analogous to that played by states in the second minimum of the fission barrier for heavy nuclei (cf. section 5.2.4).

Assuming that the total Hamiltonian is composed of a sum of some average potential plus residual interactions, then the doorway state is generated by one operation on the open channel wavefunction by the residual interaction. It must therefore be described as some simple excitation of the initial state. If for example the target nucleus is described by the shell model, and the residual interaction is composed of a sum of two-body forces, then the doorway state is a two-particle one-hole state having the same angular momentum and parity as the initial state. The next state in the hierarchy of states is a three-particle two-hole state and so on. Similarly if the target nucleus is described as a vibrational nucleus in the collective model, the doorway state contains one vibrational quantum plus a particle in a single-particle state. The end of the chain is reached when equilibrium is arrived at, which is the classic compound nucleus state. However, pre-equilibrium emission is possible at any stage along the chain giving rise to various subtle experimentally observable effects going under the general title of 'intermediate structure'.

In the most fully developed theory of doorway states (Feshbach *et al* 1980) the assumptions are made that the residual Hamiltonian can only connect stages differing by one unit of complexity (one exciton), and that amplitudes for emission from different stages do not interfere. An important feature is the distinction which is drawn between multistep compound processes where all particles are in bound states, and multistep direct processes where all the particles are in the continuum. Of these two non-interfering contributions to the cross-section the former dominates at low energy, giving an angular distribution which is symmetric between forward and backward directions, and joining on smoothly to the evaporation region at the lowest energies. Conversely the multistep direct processes dominate at high energy approaching the regime of direct reactions, with emission preferentially in the forward direction.

4.7.4.2 The optical model

According to Bohr's strong coupling model, a neutron wave incident on the nuclear surface is absorbed in a skin-depth of order K^{-1}, where K is the wavenumber in the nuclear interior. When this feature of the model is incorporated into the theory, by imposing the boundary condition $\beta(k) = -iK$ on the logarithmic derivative of the incident wave at the surface, a smooth dependence on energy E and mass number A is predicted for the average total cross-section, which is expected to decrease monotonically with energy to an asymptotic value $2\pi(R+\lambda\!\!\!^{-})^2$, where $R = r_0 A^{1/3}$ is the nuclear radius. The average is assumed to be carried out over an energy interval ΔE about E, which is much greater than the mean separation $\langle D \rangle$ between resonances, but sufficiently narrow that non-resonant functions of E (e.g. k^2) may be evaluated at E (Feshbach and Weisskopf 1949).

The first indication of significant departures from these predictions was the observation of major discontinuities or 'size resonances' near $A \approx 55$ and $A \approx 150$ in the measured scattering lengths for slow neutrons (Ford and Bohm 1950, Adair 1954). These are shown in figure 4.8 which is a plot of $\bar{a}/R$ against the nuclear radius R, where $\bar{a} = (\sigma_0/4\pi)^{1/2}$ and σ_0 is the low-energy cross-section averaged over narrow resonances. In this respect it is noteworthy that, with the exception of a few very light or very heavy nuclei, all negative scattering lengths fall within the narrow band $A \approx 55 \pm 9$ (cf. Table 6.2). Subsequently 'giant resonances' in the gross scattering cross-section for slow neutrons in the energy range $0.3\ \text{MeV} < E_n < 3\ \text{MeV}$ were detected near mass values $A \approx 50$–55, 90, 140 and 160. These were correlated with similar resonant effects in the forward scattering cross-section, confirming that the measured elastic component of the scattering cross-section was at variance with the model (Barschall 1952, Walt and Barschall 1954). Since the positions of these resonances, considered as functions of E and A, corresponded approximately with s, p and d resonances in the quantity kR for Bethe's square well model (cf. section 4.7.2.2), it was natural to interpret them as size resonances of the potential well.

These experimental observations led to the formulation of the cloudy-crystal-ball or optical model of the nucleus, according to which the incident neutron moves in a complex potential well as described in section 2.6 (Feshbach *et al* 1954, 1954, Fernbach *et al* 1949). The real part of the optical potential V_0 corresponds to the vestiges of the shell model potential which survive to excitation energies of 8 MeV and above, while the imaginary part W_0 takes into account, in a semi-empirical way, all those processes in

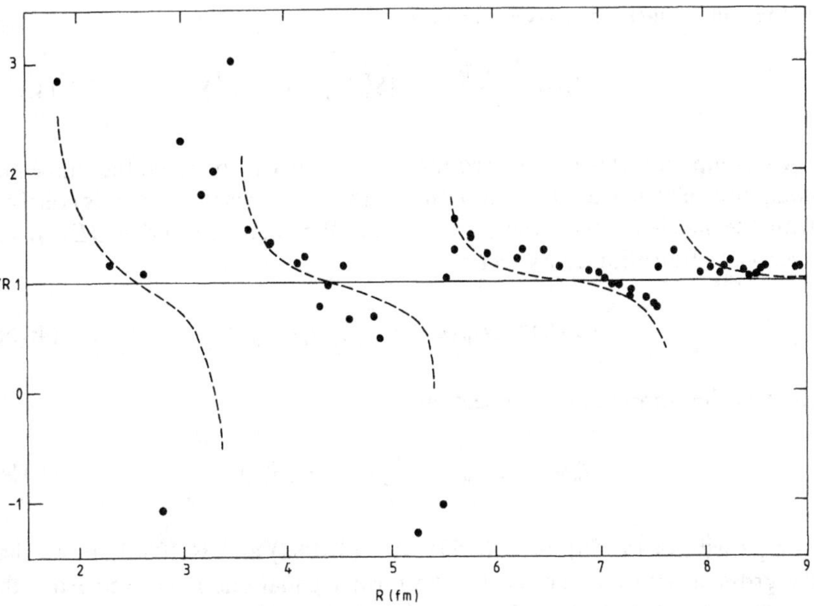

Figure 4.8. Nuclear size resonances indicated by discontinuities in the scattering length at values of the nuclear radius $R \approx 3.5$, 5.3 and 7.4 fm. To obtain smooth curves the scattering lengths are averaged over neutron resonances (Adair 1954).

which the incoming neutron shares its energy with the nuclear constituents. It is related to the mean free path L in the nucleus by the simple equation (Francis and Watson 1953)

$$L = (\hbar c/W_0)\sqrt{((E+V_0)/mc^2)} \tag{4.154}$$

As with the shell model there has been a continuing effort to derive the optical model well-depths from a many-body theory of the nucleus (Brueckner 1958, Brueckner and Goldman 1959).

The essential objective of the optical model has been to account for the gross (i.e. averaged) features of compound nucleus cross-sections in the resonance region $E_n \leqslant 15$ MeV, and the detailed behaviour of direct reaction cross-sections in the continuum region $E_n > 15$ MeV. The point of departure for the theory is to carry out an average of the elastic and absorption cross-sections given by the general formulae in (4.108). The averaged elastic cross-section then splits into two components

$$\langle \sigma_{nn}(E) \rangle = \frac{\pi}{k^2} \sum_I g_I \langle |1 - S^I_{nn}|^2 \rangle = \sigma^s_{nn}(E) + \sigma^c_{nn}(E) \tag{4.155}$$

where

$$\sigma^s_{nn}(E) = \frac{\pi}{k^2} \sum_I g_I |1 - \langle S^I_{nn} \rangle^2 | \tag{4.156}$$

is the shape elastic cross-section and

$$\sigma^{c}_{nn}(E)=\frac{\pi}{k^2}\sum_I g_I(\langle|S^I_{nn}|^2\rangle-\langle S^I_{nn}\rangle^2) \tag{4.157}$$

is the compound elastic cross-section. The distinction between the shape and compound elastic scattering amplitude is that, whereas the former is coherent with the incident wave, the latter is not. We therefore add $\sigma^{c}_{nn}(E)$ to the averaged absorption cross-section

$$\sigma^{c}_{na}(E)\equiv\langle\sigma_{na}(E)\rangle=\frac{\pi}{k^2}\sum_I g_I\langle|S^I_{na}|^2\rangle \tag{4.158}$$

to form the compound cross-section

$$\sigma^{c}_{n}=\sigma^{c}_{nn}+\sigma^{c}_{na}=\frac{\pi}{k^2}\sum_I g_I\{1-\langle S^I_{nn}\rangle^2\} \tag{4.159}$$

Comparison of (4.156) and (4.159) with (4.108a) and (4.108b) shows that the gross structure described by the former equations is derived from the fine structure described by the latter, simply by replacing the matrix element S^I_{nn} by its average value $\langle S^I_{nn}\rangle$.

When the averages indicated in (4.158)–(4.159) are carried out over a set of s-wave resonances described by the Breit–Wigner formulae given in section 2.6.3.2 the results are

$$\sigma^{s}_{nn}(E)=\frac{\pi}{k^2}|(e^{2ika}-1)+\pi\Gamma_s/\langle D\rangle|^2 \tag{4.160a}$$

$$\sigma^{c}_{na}(E)=\frac{2\pi^2}{k^2}\left(\frac{\Gamma_s}{\langle D\rangle}\right)\left(\frac{\Gamma_c}{\Gamma}\right) \tag{4.160b}$$

where a is the scattering length, Γ_c the capture width, Γ the total width and

$$\frac{\Gamma_s}{\langle D\rangle}=\frac{p_0\langle\gamma_s^2\rangle}{\langle D\rangle}=2kb\frac{\langle\gamma_s^2\rangle}{\langle D\rangle}=2kbs_0(E) \tag{4.161}$$

The function $s_0(E)$ is the s-wave strength function defined in (4.168) below. The elastic scattering amplitude $[\exp(2ika)-1]$, corresponding to scattering from a hard sphere of radius a, dominates the shape elastic cross-section, although there is a small contribution ($\simeq 1\%$) due to coherent scattering from the compound nucleus.

The total cross-section leading to the formation of the compound nucleus is given according to (4.159) by

$$\sigma^{c}_{n}=\frac{2\pi}{k^2}\left(\frac{\Gamma_s}{\langle D\rangle}\right)\left(1-\frac{\pi}{2}\frac{\Gamma_s}{\langle D\rangle}\right) \tag{4.162}$$

Finally by subtracting σ_{na}^{c} from σ_{n}^{c} we obtain for the compound elastic (incoherent) cross-section

$$\sigma_{nn}^{c}=\frac{2\pi}{k^2}\left(\frac{\Gamma_s}{\langle D\rangle}\right)\left(\frac{\Gamma_s}{\Gamma}\right) \tag{4.163}$$

This represents the contribution to the elastically scattered flux produced by the processes of absorption and re-emission from the compound nucleus.

We have seen in section 4.7.2.2 that the introduction of an imaginary component into the square well potential produces corresponding damping terms in the R-matrix. We shall now show that this result is exactly what is required by the averaging procedure to maintain the unitarity of the S-matrix. We therefore return to the discussion of spin-independent scattering in section 4.7.2.2, and carry out the average of the matrix element (cf. 4.123)

$$\langle S_l(E)\rangle=(\Delta E)^{-1}\int_{E-1/2\Delta E}^{E+1/2\Delta E} S_l(E)\,\mathrm{d}E \tag{4.164}$$

where the integral is taken along an interval of width ΔE on the positive real axis in the complex E-plane. We know from the discussion in section 2.4.4 and section 2.6.1 that $S_l(E)$ has no singularities in the first quadrant and that singularities produced by resonances fall in the half-plane Re $E<0$. Hence an integral around the rectangle with corners $E\pm\frac{1}{2}\Delta E$, $E\pm\frac{1}{2}\Delta E+\mathrm{i}\delta$ vanishes. Assuming that $\langle\Gamma_\lambda\rangle\ll\delta\leqslant\Delta W$, so that integrals along the two sides of the contour parallel to the imaginary axis cancel, and resonant fluctuations in $S_l(E)$ on the real axis are damped out on the line Im $E=\delta$, then $\langle S_l(E)\rangle$ becomes

$$\langle S_l(E)\rangle=(\Delta E^{-1})\int_{E-1/2\,\Delta E+\mathrm{i}\delta}^{E+1/2\,\Delta E+\mathrm{i}\delta} S_l(E)\,\mathrm{d}E=S_l(E+\mathrm{i}\delta) \tag{4.165}$$

It now follows from (4.123) and (4.120) that $\langle S_l(E)\rangle$ is determined by the matrix element

$$R_l(E+\mathrm{i}\delta)\equiv\sum_\lambda\frac{\gamma_\lambda^2(l)}{E_\lambda(l)-E-\mathrm{i}\delta}=R_l^\infty(E)+\mathrm{i}\pi s_l(E) \tag{4.166}$$

where, assuming a continuous distribution of resonant levels with level density $\rho(E'(l))$

$$R_l^\infty(E)=\int_{-\infty}^{\infty}\frac{\rho(E'(l))\gamma^2(E'(l))[E'(l)-E]\,\mathrm{d}E'}{(E'(l)-E)^2+\delta^2} \tag{4.167}$$

$$s_l(E)=\frac{1}{\pi}\int\frac{\rho(E'(l))\gamma^2(E'(l))\delta\,\mathrm{d}E'}{(E'(l)-E)^2+\delta^2}=\left\langle\frac{\gamma^2(E(l))}{D(E(l))}\right\rangle \tag{4.168}$$

Since the integral in (4.168) contains a Breit–Wigner probability density function which is strongly peaked at $E'(l)=E$, the strength function $s_l(E)$ is a measure of the mean strength at E of all the contributing poles in the R-matrix. Conversely the function $R_l^\infty(E)$ acquires a vanishing contribution from poles inside the interval ΔE, and therefore represents the contribution of distant resonances.

Since it may be shown using arguments similar to those applied in section 2.4.4 that $R_l^\infty(E)$ is connected to $s_l(E)$ by the dispersion relation

$$R_l^\infty(E) = \frac{1}{\pi} P \int_{-\infty}^{\infty} \frac{s_l(E')\,\mathrm{d}E'}{E'-E} \tag{4.169}$$

it follows that $R_l^\infty(E) \simeq 0$ when $s_l(E')$ is slowly varying in the region $E' \simeq E$. On the other hand the existence of a giant resonance in $s_l(E')$ near $E' \simeq E$ will be accompanied by a sign change in $R_l^\infty(E)$, which is reflected in a discontinuity in the scattering length 'a'. As may readily be confirmed by re-evaluating the cross-sections (4.160)–(4.162) in terms of $R_0^\infty(E)$ and $s_0(E)$, a is determined as a function of $R_0^\infty(0)$ and the channel radius b by

$$a = b(1 - R_0^\infty(0)) \tag{4.170}$$

In the original version of the optical model the gross-structure cross-sections were calculated using the complex potential square well described in section 2.4.3, i.e.

$$\begin{aligned} V(r) &= -V_0[1 + \mathrm{i}\zeta] \qquad r \leqslant R \\ &= 0 \qquad r > R \end{aligned} \tag{4.171}$$

with $V_0 = 42$ MeV, $\zeta = 0.03$ and $R = 1.43A^{1/3}$ fm. These calculations were sufficiently accurate to identify an s-wave giant resonance at $A \approx 55$, a p-wave resonance at $A \approx 90$, and d-wave resonances at $A \approx 50$ and $A \approx 140$. It has however been known for a very long time that the discontinuity at $r = R$ in the square well potential produces too high a reflectivity (cf. sections 2.4.2 and 6.2.1), and a much better choice of potential is presented by the Woods–Saxon potential, which has a diffuse surface (cf. section 3.2.6). To maintain the connection with the low-energy shell model it is necessary to include a spin–orbit term which not only broadens and splits giant resonances with $l \neq 0$, but introduces a spin-flipping term which enhances 90° scattering as compared with 180° scattering in accordance with observation. To allow for the non-locality of the true potential the associated well depths are taken to be energy dependent (Hodgson 1984). Finally there is a neutron–proton asymmetry term proportional to $(N-Z)^2/A$ whose role and significance, particularly in relation to fission, are discussed in section 5.1.

4.7.4.3 The intermediate model

If the shell-model description of the nucleus were exact, then an incoming neutron would be captured into a single-particle orbit, assuming the target nucleus was in its ground state, and elastic scattering would be the only allowed process. However, when the residual interactions are introduced, the target nucleus can be raised into an excited state and these states can be split and mixed. In the limit of very strong mixing the compound nucleus loses all memory of how it was formed, and the level system shows no trace of the original single-particle structure of the target nucleus. This is the uniform model which forms the basis of Bohr's theory of nuclear reactions. The success of the optical model in explaining the observation of giant resonances demonstrates that a substantial fraction of the single-particle level structure survives even at the highest excitation energies. However, a much clearer insight into how this comes about is provided by the intermediate model from which both the shell model and the uniform model emerge as limting cases (Thomas 1955, Lane *et al* 1955). The relationship between these three models is represented schematically in figure 4.9.

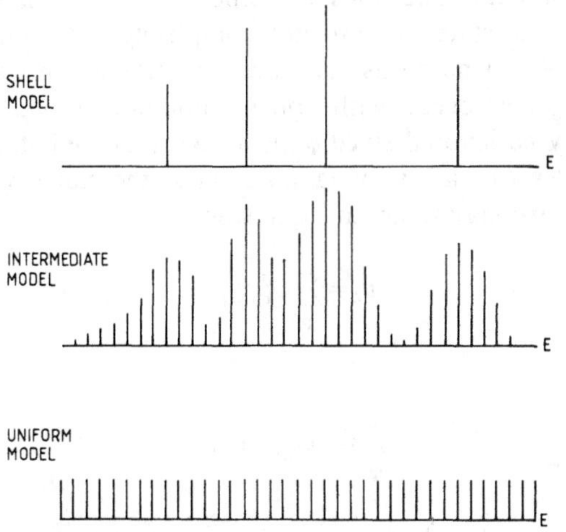

Figure 4.9. Diagram illustrating how the giant resonances of the intermediate model are related to the single-particle states of the shell model and the continuum of states characteristic of the uniform model.

In this model the compound nucleus containing $(A+1)$ nucleons is represented by the set of states v_λ (cf. section 4.7.2.3) which are solutions of the eigenvalue problem

$$Hv_\lambda = E_\lambda v_\lambda \tag{4.172}$$

where H is the exact nuclear Hamiltonian. The states v_λ satisfy the boundary conditions (4.135) and (4.136) on the sphere of radius b_c which defines the limit of the nuclear potential of the target nucleus with A nucleons. At the same time we can construct the set of states

$$\chi_{cp} = \varphi_c \psi_p(\mathbf{r}_n) \tag{4.173}$$

where φ_c are the stationary states of the target nucleus commencing with the ground state given by $c = 0$, and $\psi_p(\mathbf{r}_n)$ denotes a single-particle state for a neutron moving in the average potential $\bar{V}$ of the target nucleus. We may assume that $\bar{V}$ is independent of the state of the target as specified by the index c. The states χ_{cp} obey the wave equation

$$H_0 \chi_{cp} = E_{cp} \chi_{cp} \tag{4.174}$$

where

$$H_0 = H + V_{res} = H_0 + \bar{V} - \sum_{j=1}^{A} V(\mathbf{r}_j, \mathbf{r}_n) \tag{4.175}$$

and the same boundary conditions are imposed on the radial wavefunction $u_p(r_n)$ as on v_λ. The states χ_{cp} are not completely anti-symmetrized with respect to the $(N+1)$ neutrons they describe $(A = Z + N)$ and the single particle state ψ_p can occur with some amplitude in φ_c. However this contribution may be ignored since ψ_p is a highly excited state and φ_c may be assumed to describe a low-lying state. Thus the states χ_{cp} constitute a complete set and we may write an expansion

$$v_\lambda = \sum_{cp} C^\lambda_{cp} \chi_{cp} \tag{4.176}$$

where

$$\sum_{cp} C^\lambda_{cp} C^{\lambda'}_{cp} = \delta_{\lambda\lambda'} \tag{4.177}$$

and, because the v_λ also constitute a complete set,

$$\sum_{\lambda} C^\lambda_{cp} C^\lambda_{c'p'} = \delta_{cc'} \delta_{pp'} \tag{4.178}$$

We may now apply (4.144) and (4.176) to calculate the reduced width of the state v_λ for emission into the channel c, i.e.

$$\gamma^2_{\lambda c} = \left(\frac{\hbar}{2m_c b_c^2}\right) \sum_{pp'} u_p(b_c) u_{p'}(b_c) C^\lambda_{cp} C^\lambda_{cp'} \tag{4.179}$$

If we now assume that the coefficient C^λ_{cp} defined with respect to a given p is equally likely to be positive as negative, then we may ignore the cross-products in (4.179) and write

$$\gamma_{\lambda c}^2 \simeq \frac{h^2}{2m_c b_c^2} \sum_{\text{p}} u_{\text{p}}^2(b_c)(C_{\text{cp}}^\lambda)^2$$

$$= \sum_{\text{p}} (C_{\text{cp}}^\lambda)^2 \gamma_{\text{cp}}^2 \tag{4.180}$$

where γ_{cp}^2 is the single particle reduced width defined in (4.121).

We now introduce the basic assumption of the intermediate model, namely that there exists some energy interval ΔE falling in the range

$$\langle D \rangle \ll \Delta E \ll \langle d \rangle \tag{4.181}$$

where $\langle d \rangle$ is the mean separation between single-particle levels, that only one coefficient C_{cp}^λ differs significantly from zero in (4.180) for all states λ whose energies are restricted to the interval

$$E_{\text{cp}} = \tfrac{1}{2}\Delta E \leqslant E_\lambda \leqslant E_{\text{cp}} + \tfrac{1}{2}\Delta E \tag{4.182}$$

In this case we may write

$$\gamma_{\lambda c}^2 \simeq C_{\text{cp}}^{\lambda 2} \gamma_{\text{cp}}^2 \tag{4.183}$$

Since $C_{\text{cp}}^\lambda \simeq 0$ for all states λ which do not satisfy (4.182) we can sum over all λ according to (4.178) resulting in the sum rule

$$\sum_\lambda \gamma_{\lambda c}^2 = \gamma_{\text{cp}}^2 \tag{4.184}$$

When the residual interaction V is zero there is only one term $\gamma_{\lambda 0}^2$ in (4.184) corresponding to the case where the target nucleus is in its ground state and the neutron is in the single-particle state ψ_{p}. When the residual interaction is switched on, the single-particle reduced width γ_{cp}^2 is spread out over a group of states in the neighbourhood of the unperturbed state which we have denoted by (λ_0). This group of states constitutes the giant resonance.

If we suppose, on the other hand, that it is impossible to establish an energy interval ΔE such that conditions (4.182) and (4.183) are simultaneously satisfied, a conclusion which implies that the matrix elements $\langle \text{p}'|V|\text{p}\rangle$ of the residual interaction connecting single particle states exceed $\langle D \rangle$ in magnitude, then we may visualize a limiting situation where all the coefficients $C_{\text{cp}}^{\lambda 2}$ appearing in (4.180) are comparable in magnitude so that the single-particle structure has disappeared. In this case we may write

$$\frac{\langle d \rangle}{\langle D \rangle} \langle \gamma_{\lambda c}^2 \rangle = \gamma_{\text{cp}}^2 \leqslant \frac{h^2}{m_c b_c^2} \tag{4.185}$$

where $\langle d \rangle / \langle D \rangle$ is the mean number of nuclear states (λ) per single-particle state (p). Using the estimate $\langle d \rangle \approx \pi h^2 K / m_c b_c$ derived from the expression

(4.131) for the s-wave state in a square well potential, we may recast (4.185) into an equation for the s-wave strength function

$$s_0(E) = \frac{\langle \gamma^2_{\lambda c} \rangle}{\langle D \rangle} \leqslant \frac{1}{\pi K b_c} \tag{4.186}$$

This estimate agrees with that derived on the basis of the uniform model (Feshbach *et al* 1954).

4.7.5 Photoneutron production

The giant resonance is the dominant feature in the nuclear absorption cross-section for electromagnetic radiation of energy greater than the nucleon emission threshold, which lies near 8 MeV in medium mass and heavy nuclei. The phenomenon occurs both for real bremsstrahlung photons, and for virtual photons associated with electron bombardment (cf. section 1.6.3). It was first observed for (γ, n) reactions in carbon, copper and tantalum, and for photofission in uranium and thorium (Baldwin and Klaiber 1947, 1948). The giant resonance appears in all nuclei except s-shell nuclei with $A \leqslant 4$. For light nuclei ($A < 30$) the maximum in the resonance occurs near 20 MeV, whereas for medium mass and heavy nuclei it lies between 18 and 14 MeV. For $A \geqslant 120$ the energy of the maximum decreases in proportion to $A^{-1/3}$. The width of the resonance varies between 3 and 8 MeV with narrow resonances appearing near closed shells, and broad resonances at positions in the periodic table where nuclear deformation is greatest. The states of the giant resonance in ^{16}O excited in the photoneutron reaction $^{16}O(\gamma, n)^{15}O$ are shown in figure 4.10.

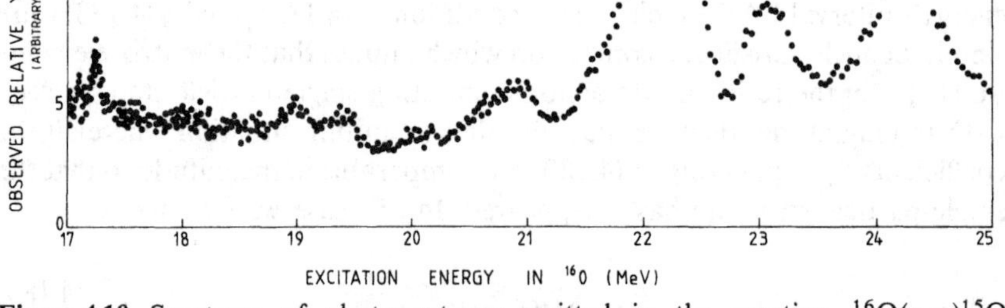

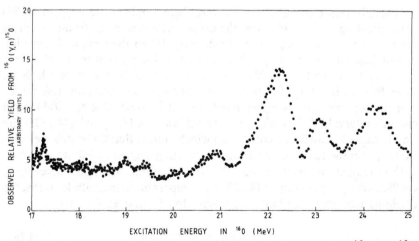

Figure 4.10. Spectrum of photoneutrons emitted in the reaction $^{16}O(\gamma, n)^{15}O$ observed with 25 MeV bremsstrahlung photons from the 33 MeV electron linac at AERE Harwell. The neutron energies were measured using a time of flight spectrometer with a resolution of 0.16 ns m^{-1} (Firk 1964).

For $E_\gamma \leqslant 8$ MeV the excitation function consists mainly of discrete lines. Important exceptions to this general rule are the photonuclear reactions $^9Be(\gamma, n)^8Be$ and $d(\gamma, n)p$ (cf. section 2.7.3) both of which provide useful sources of medium energy neutrons (cf. section 1.6.3). The latter process in particular makes a significant contribution to the delayed neutron component in heavy water moderated reactors (cf. section 5.5.3.3). Above the nucleon emission threshold the fine structure in the absorption spectrum disappears for medium mass and heavy nuclei (Hayward and Fuller 1962) where the shape of the resonance may be represented with fair accuracy by the Breit–Wigner formula given in (2.172). For photons with $E_\gamma = \hbar k$ this takes the form

$$\sigma(E_\gamma) = \frac{\sigma_0 E_\gamma^2 \Gamma_\gamma^2}{(E_{\gamma r}^2 - E_\gamma^2)^2 + E_\gamma^2 \Gamma_\gamma^2} \tag{4.187}$$

For heavy nuclei the giant resonance decays mainly by neutron emission, with some contribution from double neutron emission at higher energies, since for these nuclei the Coulomb barrier B ($\simeq Z^2 A^{-1/3}$ MeV) lies well above the giant resonance and proton emission is strongly suppressed. For light nuclei the fine structure persists through and beyond the giant resonance (Hayward 1963, Danos and Fuller 1965); decay is predominantly by single nucleon emission with $\sigma(\gamma, p) \simeq \sigma(\gamma, n)$, since the reduced Coulomb barrier is compensated to some extent by the slightly higher neutron separation energy S_n (cf. section 5.1.1).

The Hamiltonian describing the interaction of electromagnetic radiation with nucleons in the nucleus may be represented as a sum of single-particle terms as given in (2.22), where the incident field $\boldsymbol{A}(\boldsymbol{r}, t)$ may be expanded as a sum of electric $(A_L^E(\boldsymbol{r}, t))$ and magnetic $(A_L^M(\boldsymbol{r}, t))$ multipole fields, each carrying total angular momentum L. Since the EL transition matrix element (parity $(-1)^L$) varies as $(kR)^L$, where R is the nuclear radius, and the ML transition matrix element (parity $(-1)^{L+1}$) is less by a factor of $(\lambda_c/R)^2$, electric dipole absorption is dominant when $kR < 1$. This condition is well satisfied for the giant resonance since, even for a medium mass nucleus such as copper, $kR \leqslant 0.5$ at 20 MeV.

Since even a high energy photon carries very little momentum, neutrons and protons must move in opposite directions in order to conserve momentum in the struck nucleus. Due to the tight binding between neutrons and protons, photon absorption creates a high frequency dipole which may, broadly speaking, be identified with the giant dipole resonance. To determine the integrated cross-section we may therefore apply the classical dipole sum rule (2.121) modified to take account of nuclear recoil as expressed by the concept of effective charges on neutrons and protons (cf. sections 2.7.3 and 4.3.1). Thus we make the replacement

$$Z \to Z(-N/A)^2 + N(Z/A)^2 = NZ/A \tag{4.188}$$

with the consequence that the relation (2.121) is replaced by (Hayward 1963)

$$\int \sigma(E)\,\mathrm{d}E = \frac{2\pi^2 e^2 \hbar}{m_n c} \cdot \frac{NZ}{A} \tag{4.189}$$

where the integral is extended right up to the photo-pion threshold near 140 MeV. This expression must then be multiplied by an additional enhancement factor to allow for the contribution of Majorana exchange and momentum-dependent forces, which appear in the nuclear force field, but which do not arise in atoms (cf. section 3.3.2) (Feenberg 1936, Levinger and Bethe 1952). Furthermore, by applying the optical theorem (2.100) and the dispersion relations (2.163), the dispersive part of the electromagnetic nuclear scattering amplitude may be determined once the absorption cross-section is known as a function of energy, allowing a confrontation between experiment and theory which is largely model independent.

It is found experimentally that, for light nuclei between ^{6}Li and ^{16}O the giant resonance contributes $\leqslant 50\%$ of the transition strength prediced on the basis of the classical dipole sume rule (4.189) (Hayward and Stovald 1965), which sets a lower limit on the integrated cross-section. A plausible explanation for the discrepancy is that the missing cross-section appears at the highest energies where a significant contribution is made by the (γ, np) process. Here the incident photon couples directly to the electric dipole moment of a quasi-deuteron in the nucleus, there being no equivalent interaction with pairs of like nucleons, which have no dipole moment. The transition rate is amplified in comparison with the photodisintegration rate of a free deuteron, because confinement in the nucleus enhances the high momentum components in the wavefunction (Levinger 1951). Thus, even though a real photon has insufficient momentum to eject high energy particles from the nucleus, this can occur for a photon interacting with a neutron and proton pair. For medium mass nuclei the giant resonance exhausts the sum rule to within about 10%, while for heavy nuclei with $Z \geqslant 50$, the integrated cross-section exceeds the predictions of the sum rule, suggesting an increasing contribution from higher multipoles, first E2+M1, then E3+M2, etc., as the parameter (kR) is progressively raised.

Unlike the conditions which apply for particle-induced nuclear reactions, where the incident nucleons interact strongly with the nucleus as a whole, photons interact only weakly with individual nucleons in the nucleus. Thus two types of nuclear photoeffect are possible: single-particle transitions in which a bound nucleon makes a direct transition into the continuum of states above the nucleon emission threshold, and collective transitions to unbound states, formed from coherent superpositions over many-particle configurations selected from those which are most strongly represented in the nuclear ground state. We may therefore interpret the observed structure of the giant resonance in heavy nuclei as a collective excitation of the target nucleus; the modes of formation and decay of the resonant state are independent and the excitation energy is sufficiently high to justify a statistical average over the phases of all participating states. In this limit the spectrum of emitted neutrons is given by Weisskopf's evaporation formula (4.153), and the angular distribution is isotropic with respect to the momentum of the incident photon.

For light nuclei the persistence of structure may be explained by assuming that neutrons (or protons) are emitted directly from excited resonant states falling within the envelope of the giant resonance. Since the largest electric dipole matrix element is that associated with an increase by one unit in the orbital angular momentum of the ejected nucleon, one may hope to interpret energy spectra and angular distributions in terms of the nuclear shell model exactly as for the deuteron stripping process (cf. section 4.3.4). Thus in p-shell nuclei the transition $1p_{3/2} \rightarrow 1d_{5/2}$ is expected to dominate with an angular distribution varying as $A_2 P_2(\cos\theta)$. Similarly in d-shell nuclei the transition $1d_{5/2} \rightarrow 1f_{7/2}$ should dominate, characterized by an angular distribution varying as $A_3 P_3(\cos\theta)$.

These predictions are confirmed in p-shell nuclei. The angular correlation coefficient $A_2(E_n)$ vanishes for low energy neutrons as expected for an evaporation spectrum and, at higher energies, varies with energy in such a way as to pick out those states in the residual nucleus which can couple with d-wave neutrons to reproduce the original resonant state in the parent nucleus. However for d-shell nuclei the f-wave component is missing, presumably because it is hindered by the centrifugal barrier (cf. section 2.3.2). Experimentally these excited states are most likely to decay by p-wave or s-wave emission, for which there is little or no centrifugal barrier, and the increased lifetime of these states is strongly correlated with an observed reduction in the decay width (Baker and McNeill 1961).

The detailed theory of the giant resonance in heavy nuclei is usually developed in terms of a hydrodynamical model, according to which the nucleus is viewed as a constant density mixture of proton and neutron fluids (Goldhaber and Teller 1948, Steinwedel and Jensen 1950). According to this model, which in many respects resembles the liquid drop model of Bohr and Wheeler (cf. section 5.2.3), the two fluids separate under the impulse communicated by the incident photon, and a local restoring force is established which originates in the asymmetry term which contributes to the Weizsäcker semi-empirical mass formula (cf. section 5.1.1). The coherent mode is damped by the viscous forces between the two fluids leading to a thermalization of the excitation energy and statistical decay (Danos and Greiner 1965). The hydrodynamical model accounts very adequately for the observed $A^{-1/3}$ dependence of the maximum in the resonance and, in the long-range correlation or 'second sound' version (Danos 1958) predicts a correlation between resonance width and nuclear electric quadrupole moment which is confirmed experimentally (Okamato 1958).

For highly deformed rare earth or actinide nuclei, e.g. erbium, holmium, uranium, the giant resonance splits in two, with a mean separation between peaks of about 4 MeV (Hayward and Fuller 1962, Bowman *et al* 1964). These spectra are analysed into two resonances corresponding to dipole oscillations along one long axis ($\Delta K = 0$) and two short axes ($\Delta K = \pm 1$). The measured intensities of the two resonances have a ratio of about 1:2 and, in accordance with the predictions of the collective model for nuclei of positive deformation, the weakest ($\Delta K = 0$) resonance lies at the lower energy (cf. section 4.5.1). These phonomena may be interpreted in much greater detail in a fully unified dynamic theory, which takes account of the coupling between the dipole oscillations of the hydrodynamic model, and the quadrupole modes of the collective model (Danos and Greiner 1964, 1964a, 1965).

5 Neutrons and nuclear energy

5.1 Energetics of nuclear transformation and decay

5.1.1 The liquid drop model and the Weizsäcker mass formula

Industrial production of nuclear energy on macroscopic scales relies on the creation of an environment wherein exothermic encounters between selected nuclear species can be encouraged, under conditions of temperature and pressure and at rates which are always strictly under the control of the operator. To achieve this aim it is necessary that the potential of each and every nucleus as a source or sink of energy be properly assessed. Hence the outstanding role which has been played in the development of nuclear energy systematics by the high resolution mass spectrometer developed by Aston, Bainbridge, Dempster, Nier and their successors (cf. section 1.5.1). Experimentally it is found that the masses $M(A, Z)$ of neutral atoms fall very close to whole numbers, when measured on the ^{16}O scale (cf. section 1.1), and the quantity $\Delta = M(A, Z) - A$ is known as the mass decrement or mass defect. By definition Δ is zero at ^{16}O, and measured values of the packing fraction $P = \Delta/A$ fall in the range $-10^{-3} \leqslant P \leqslant 9 \times 10^{-3}$. P is negative for $16 < A < 209$, indicating that nuclei in this mass range should be regarded as sinks rather than sources of energy. However for $A < 16$ it is possible to convert mass into energy by fusing light elements together, thereby moving the final state nucleus into the region of increased stability. For $A > 209$ the same end is achieved through the mechanism of nuclear fission.

The packing fraction is a relatively crude yardstick of nuclear stability and a much more sensitive measure is the nuclear binding energy $B(A, Z)$ defined in (3.1). Another important parameter is the separation energy S_p (S_n) which may be defined as the binding energy of the last proton (neutron) added to the nucleus. Thus the functions S_p and S_n are given by

$$S_p = B(A, Z) - B(A-1, Z-1) = [M(A-1, Z-1) - M(A, Z) + m_H]c^2 \tag{5.1a}$$

$$S_n = B(A, Z) - B(A-1, Z) = [M(A-1, Z) - M(A, Z) + m_n]c^2 \tag{5.1b}$$

The measured values of S_p and S_n fluctuate from nucleus to nucleus, e.g. for ^{14}N, $S_p = 7.5$ MeV and $S_n = 10.6$ MeV, both numbers falling quite close to the mean binding energy per nucleon $B/A \simeq 7.5$ MeV. Quite generally $\langle S_n \rangle \simeq \langle S_p \rangle \simeq B/A$, an observation which can be accounted for only if it is assumed that internucleon forces are momentum-dependent (cf. section 4.4.1). Even this assumption is insufficient, for the effective mass m^* of a nucleon in the nucleus cannot be determined from the kinetic energy and potential energy alone, but requires some information on the 'rearrangement' energy. This is the energy required to rearrange the remaining nucleons when a nucleon is removed, so that the final nucleus is left in an unexcited state (Brueckner 1958). Although this energy makes the major contribution to the separation energy of a nucleon, it vanishes by definition in a shell model where the average potential is specified *a priori*.

The final energy parameter we need is the pairing energy which is defined for protons (neutrons) by

$$P_p(A, Z) \equiv S_p(A, Z) - S_p(A-1, Z) = [2M(A-1, Z-1) - M(A, Z) - M(A-2, Z-2)]c^2 \quad (5.2a)$$

$$P_n(A, Z) \equiv S_n(A, Z) - S_n(A-1, Z) = [2M(A-1), Z) - M(A, Z) - M(A-2, Z)]c^2 \quad (5.2b)$$

The systematic variation of P_p and P_n shows that the binding energy is increased by about 3 MeV when pairs of like nucleons are coupled in time-reversed substates. The origin of the pairing effect is discussed in section 4.5.2.

The observation that B/A is approximately constant at the value 8.0 ± 0.5 MeV over the mass range $20 < A < 225$ is associated with the saturation property of internucleon forces as discussed in section 3.1.2, and the appearance of pronounced peaks in the binding energy curve, when A is a multiple of 4 for self-conjugate nuclei in the range $4 \leqslant A \leqslant 32$, suggests that the α-particle be identified as the saturated unit. These facts, along with the observation that the nuclear volume is proportional to A, lend support to the hypothesis that, so far as its energy content is concerned, the nucleus may be likened to a charged incompressible liquid drop. This is the physical model underlying the semi-empirical mass formula first proposed by Weizsäcker (Weizsäcker 1935, Bohr and Wheeler 1939), and subsequently elaborated and extended to include additional fine details of nuclear structure (Feenberg 1947, Green 1958, Mozer 1959, Wing and Fong 1964).

In its simplest form the Weiszäcker model provided an analytic expression for the mass surface

$$M(A, Z) = Zm_H + Nm_n - [\alpha A - \beta(N-Z)^2/A - \gamma A^{2/3} - \varepsilon Z(Z-1)/A^{1/3} - \delta(A, Z)] \quad (5.3)$$

where each of the first four binding energy terms bracketed in (5.3) may be simply interpreted on the basis of the liquid drop model. The fifth term, the pairing term $\delta(A, Z)$ is inserted to allow for the effect of short-range correlations for pairs of like nucleons. Unlike the four large terms in the binding energy the pairing term is of microscopic (i.e. quantum mechanical) origin. The coefficients α, β, γ, ε are determined by least squares fitting to the measured mass data subject to the criteria: (i) $M(A, Z)$ must reproduce (a smoothed version of) the measured (A, Z) curve for stable nuclei; (ii) it must reproduce the masses of stable odd-A nuclei for which there are no pairing terms; (iii) considered as a function of Z for fixed A, $M(A, Z)$ must reach its minimum value at the stable isobar.

The dominant term in the binding energy is the saturation energy αA, which is proportional to the nuclear volume, and hence varies linearly with A. It is reduced by the surface tension term $\gamma A^{2/3}$ which allows for the fact that nucleons on the surface (area $\approx A^{2/3}$) do not engage in the full number of interactions. The Coulomb repulsion term $\varepsilon Z(Z-1)/A^{1/3}$ also acts to reduce the binding energy; for a uniformly charged sphere of radius $r_0 A^{1/3}$, ε takes the value $(3/5)(e)^2/r_0$ (cf. section 3.2.2). In a more detailed treatment the Coulomb term has to be modified by exchange corrections and, for the boundary layer effect in the case that the liquid drop has a diffuse surface (Mozer 1959).

The isotopic or asymmetry term $(N-Z)^2/A$ has its origin in two quite distinct effects although both may ultimately be traced to the Pauli principle. This principle implies that, for given A, the populations of the lowest proton and neutron single particle levels are maximized when the isotopic number, $(N-Z)$, takes its minimum value. This is the first effect. The second effect stems from the fact that the $^3S_1$ (symmetric) state of the deuteron is the only bound state of the two-nucleon system (cf. 2.71) and that totally symmetric states (e.g. $^3S_1$, $^1P_1$) are forbidden for pairs of like nucleons. Thus, on average, the contribution of the n–p interactions to the binding energy exceeds the contribution of p–p and n–n interactions. Thus, for given A the binding energy is maximum for minimum isotopic number, when the number of (n–p) pairs is maximized.

The form of the asymmetry term may be adduced from the following simple argument (Cohen 1970). Ignoring spin and degeneracy let the mean level spacing in the nuclear potential well be $\bar{\varepsilon}$. Then $\bar{\varepsilon} \approx A^{-1}$ since it varies inversely as the nuclear volume. When the neutron levels are filled up to $N\bar{\varepsilon}$, and the proton levels are filled up to $Z\bar{\varepsilon}$, the change in kinetic energy compared with the minimum value when $N = Z = \frac{1}{2}A$ is

$$\begin{aligned}\Delta T &= \bar{\varepsilon}N(N+1)/2 + \bar{\varepsilon}Z(Z+1)/2 - 2\bar{\varepsilon}\tfrac{1}{2}A(\tfrac{1}{2}A+1) \\ &= \bar{\varepsilon}(N-Z)^2/4 \approx (N-Z)^2/A \end{aligned} \tag{5.4}$$

To estimate the change in potential energy, we may suppose, for the sake of definiteness, that $N > Z$; then the depth of the effective neutron potential will increase in the ratio $(N-Z)/A$, and the proton potential will decrease by an equivalent amount $(Z-N)/A$. However, the decrease in the energy of Z protons is balanced by a corresponding increase in the energy of Z neutrons, leaving a net increase in energy $\simeq (Z-N)^2/A$ due to the remaining $(N-Z)$ neutrons. Clearly this increase depends only on the magnitude of $(N-Z)$ and not on the sign. Further, the form of the energy shift is model independent, since what we have shown essentially is that there is no energy shift to first order in $|N-Z|/A^{1/2}$.

Finally, we return to a brief consideration of the pairing term $\delta(A, Z)$. This is too small to be determined by a global least-squares fitting algorithm, and is defined to be zero for odd A nuclei. For even A nuclei it is assigned equal and opposite values for even–even nuclei ($\delta(A, Z) < 0$) and odd–odd nuclei ($\delta(A, Z) > 0$). A number of empirical formulae have been suggested from time to time but the relation $\delta(A, Z) = \pm 11A^{-1/2}$ MeV has generally been found adequate for most purposes (Coryell 1953).

The presence of the pairing term means that the mass surface is a surface with three sheets. Treated as a function of Z the curve in which any plane of constant A intercepts the odd-A sheet corresponds to a parabola, and the pair of curves in which this plane intercepts the even-A sheets, corresponds to a pair of identical parabolae, separated in energy by an amount equal to twice the pairing energy at that value of A. The consequence is that, for given odd A, there is at most one stable isobar, whereas for even A there may be more than one stable isobar with values of Z differing by at least two units.

The binding energies and values of three of the four binding energy terms are plotted as a function of A in figure 5.1, for the 18 odd–even naturally occurring nuclei which, for a given value of Z, have a single stable isobar. What this plot clearly demonstrates is that the surface tension term is responsible for reducing the binding energy of low mass nuclei, thereby allowing the possibility of energy release through fusion reactions. The surface energy term reduces as A increases and its role is taken over by the Coulomb energy term. It is this term which is responsible for α-instability, and ultimately for nuclear fission and the huge energy release accompanying that process.

5.1.2 Deformed liquid drop models

The semi-empirical mass formula (5.3) is strictly applicable only to spherical nuclei, and the algebraic form of the relation may be reproduced in a microscopic calculation by setting up the corresponding many-body equations and solving them in some suitable approximation. In one version of this approach (Gomes *et al* 1958, Skyrme 1958) the potential energy is evaluated on the assumption of an internucleon Serber force, i.e. a force derived from an equal admixture of Wigner and Majorana exchange potentials. The central potential is taken to have the form of a square well superimposed on a hard core, which is required to reproduce the observed saturation properties (cf. section 3.1.2). The corresponding single-particle potential is momentum dependent expressing the fact that, at low momentum (large distances) the potential is attractive, whereas, at high momentum (short distances) it is repulsive. The introduction of an effective mass into the kinetic energy takes care of the momentum dependence but, to account for the rearrangement energy which is required to bring about the observed close

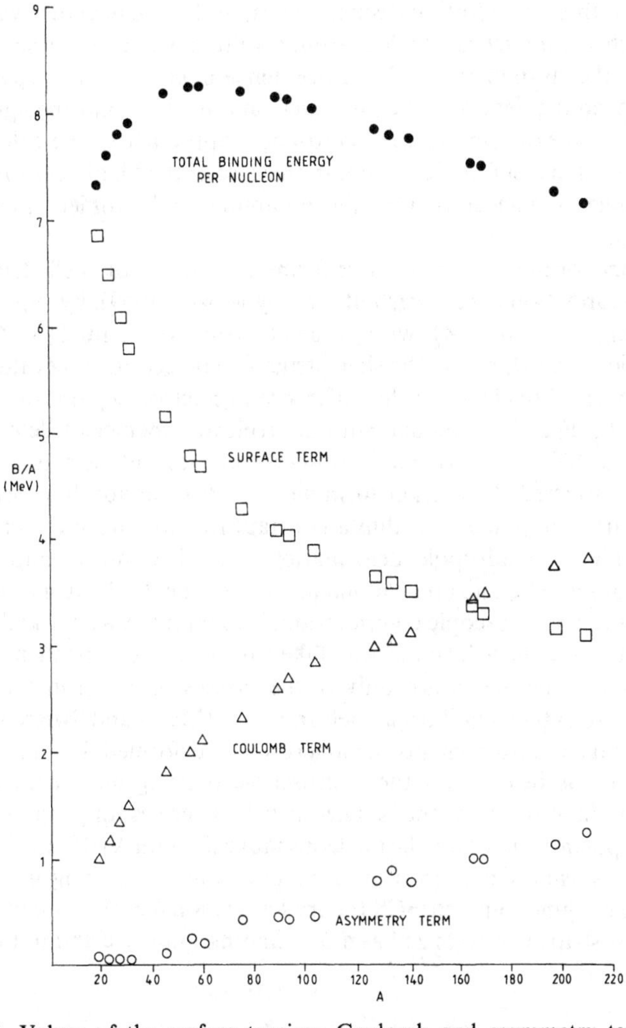

Figure 5.1. Values of the surface tension, Coulomb and asymmetry terms in the Weizsäcker semi-empirical mass formula, computed for 18 odd-A nuclei which, for a given value of Z, have a single stable isobar. These must be subtracted from the saturation exchange energy to arrive at values of B/A, the binding energy per nucleon, which are plotted on the same energy scale. The surface term is responsible for the gain in stability from fusion of light nuclei while the Coulomb term plays the same role in respect of α-decay or fission of heavy nuclei.

equality between nucleon separation energy and mean binding energy, it is necessary to include the many-nucleon effective forces which are generated through the action of the Pauli principle.

This procedure reproduces the essential features of the Weizsäcker formula, together with some additional small terms, and the calculated value of the volume energy matches the least-squares-fitted value quite satisfactorily. However the asymmetry and surface tension coefficients depend rather sensitively on the details of the internucleon potentials and the agreement is not good. The extension of the microscopic approach to the calculation of non-spherical nuclei is difficult and, in the case of the highly deformed nuclei which undergo nuclear fission, the computational problem proves quite intractable.

A scheme for modifying the mass formula to allow for shell-structure and surface deformation effects was initiated by Mozer (1959), by supplementing the pairing term $\delta(A, Z)$ with a shell structure term $S(A, Z)$, and a deformation term $D(A, Z)$. The shell-structure correction was evaluated from the sum over all nucleons of the difference in nucleon separation for a shell model and a Fermi gas model, with an arbitrary coefficient determined by fitting to the difference between the observed masses and the masses predicted by the uncorrected Weizsäcker formula. The deformation function $D(A, Z)$ was similarly computed to within a constant fraction, from the energy gain associated with quadrupole deformation, and determined empirically by fitting to observed quadrupole moments (cf. section 4.5.1). A major advance in these purely macroscopic computational techniques was marked up, when each term in the mass formula was taken to be shape dependent, allowing a prediction to be made, not only of the masses of nuclei in their ground state, but also of their equilibrium deformations (Myers and Swiatecki 1966).

In subsequent developments the macroscopic deformed liquid drop energy is calculated as before, but the microscopic (pairing and shell) terms are computed directly from the single-particle energies appropriate to the deformed potential in which the nucleons move (Strutinsky 1966). The pairing correction is usually described in terms of a constant pairing interaction G applied in conjunction with BCS theory (cf. section 4.5.2). In all of this work the nuclear shape is recognized as a dominating influence and not as a small correction.

5.1.3 Nuclear decay modes

5.1.3.1 Gamma and beta decay

Whether a particular decay channel is open for an excited or quasi-stable nucleus is dictated primarily by energy considerations. The rate at which an energetically allowed process occurs is governed by other factors, e.g. the nature of the interaction, electromagnetic, weak or strong, angular momentum and parity selection rules and, most important in the case of charged particle emission, the height of the Coulomb barrier. Excited nuclei decay predominantly by γ-emission, internal conversion or, less frequently, by internal pair production. Because there is no monopole electromagnetic radiation, γ-transitions between states of zero spin are absolutely forbidden. In very rare instances isomeric states may undergo β-decay or α-decay, and highly excited states may decay by nucleon emission.

For nuclei in their ground-states β-decay is overwhelmingly the most important transformation process. Neglecting the binding energy of the least lightly bound electron in the atom, electron emission

$$(A, Z) \rightarrow (A, Z+1) + e^- + \bar{\nu}_e \qquad (5.5)$$

is energetically allowed if $M(A, Z) > M(A, Z+1)$. Decay by orbital electron capture

$$(A, Z) + e^- \rightarrow (A, Z-1) + \nu_e \qquad (5.6)$$

is allowed if $M(A, Z) > M(A, Z-1)$, but for positron emission

$$(A, Z) \rightarrow (A, Z-1) + e^+ + \nu_e \qquad (5.7)$$

the more stringent condition $M(A, Z) > M(A, Z-1) + 2m_e c^2$ must be fulfilled. When both these processes compete, electron capture is favoured at low energy and high Z, because of the Coulomb well presented by the capturing nucleus. At high energy and low Z, positron emission is more likely, because of the reduced Coulomb barrier. Occasionally electron and positron emission can co-exist as competing processes in the one nucleus, e.g. ^{64}Cu.

The neglect of atomic electron binding energy is almost always justified, but there are exceptional circumstances when it must be taken into account. For example, electron emission is possible even when $M(A, Z) < M(A, Z+1)$ provided the energy deficit is made up by the difference in atomic electron binding energy between parent and daughter nuclei (cf. section 3.1.1). For electron capture to occur the mass difference must be at least equal to the atomic binding energy of the captured electron, so that occasionally LMN-capture is possible even though K-capture is energetically forbidden.

5.1.3.2 Nucleon emission

To determine the conditions under which proton or neutron emission is possible it is necessary to calculate the relevant separation energies defined in (5.1). For proton emission the separation energy is

$$S_p \simeq m_H c^2 - \frac{\partial M}{\partial A}(A, Z)|_N c^2 = \frac{\partial B}{\partial A}(A, Z)|_N = B/A + A\frac{\partial}{\partial A}(B/A)|_N \qquad (5.8)$$

For spontaneous proton emission we must have $S_p < 0$, and this means that $A(\partial/\partial A)(B/A)|_N$ must be negative, and of magnitude $> B/A \simeq 8$ MeV. To identify circumstances in which this may happen for a nucleus in its ground state, we calculate S_p from the semi-empirical mass formula (5.3) with the result

$$S_p = \alpha - \beta(1 - 4N^2/A^2) - \frac{2}{3}\gamma A^{-1/3} + \frac{1}{3}\varepsilon Z(Z-1)A^{-4/3}$$
$$-\varepsilon Z(Z-1)A^{-1/3} + (-1)^N \delta(A, Z) \quad (5.9)$$

It is clear that to arrive at the condition $S_p < 0$ both the pairing term and the asymmetry term must be negative, i.e. Z must be odd and greater than $A/2$.

There are in fact only two cases in nature where the fulfilment of these conditions brings about the required result $S_p < 0$. This happens with the nuclei ^{5}Li and ^{9}B, both of which are unstable against proton emission. There are, however, a number of cases of proton-rich nuclei whose decay leads to proton emission; an example is the decay chain

$$\begin{aligned} {}^{25}\mathrm{Sl} &\to {}^{25}\mathrm{Al}^* + e^+ + \nu_e \\ {}^{25}\mathrm{Al}^* &\to {}^{24}\mathrm{Mg} + \mathrm{p} \end{aligned} \quad (5.10)$$

An expression for the neutron separation energy S_n similar to (5.1.9) is derived by the same procedure and we find

$$S_n = \alpha - \beta(1 - 4Z^2/A^2) - \frac{2}{3}\gamma A^{-1/3} + \frac{1}{3}\varepsilon Z(Z-1) + (-1)^N \delta(A, Z) \quad (5.11)$$

As before a necessary condtition for spontaneous decay via neutron emission to be possible is that N be odd and greater than $A/2$. Once again the mass 5 nucleus ^{5}He is the only nucleus for which $S_n < 0$ in its ground state, so that mass 5 nuclei are not found in nature.

Although neutron emission is observed following β-decay of the light nucleus ^{17}N, i.e.

$$\begin{aligned} {}^{17}\mathrm{N} &\to {}^{17}\mathrm{O}^* + e^- + \bar{\nu}_e \\ {}^{17}\mathrm{O}^* &\to {}^{16}\mathrm{O} + \mathrm{n} \end{aligned} \quad (5.12)$$

by far the most significant cases of neutron emission occur in the β-decay chains of heavy neutron-rich nuclei which are produced in nuclear fission. Since the true neutron-emission lifetime is extremely short, the effective lifetime, as measured from the fission event, is the β-decay lifetime of the last fission product in the chain which is known as the precursor. These 'delayed neutrons', which play a vitally important role in the control of nuclear reactors, are by convention and practice classified in six groups ordered in terms of increasing lifetime. The same groups appear following the fission of the most important fissile isotopes ^{235}U, ^{239}Pu and ^{233}U, but in different proportions. Thus no very precise meaning can be attached to the notion of a group lifetime. The approximate lifetimes and dominant constituents of each of the delayed groups are listed in Table 5.1. Clearly the most important precursors are isotopes of iodine and bromine although rubidium and krypton isotopes also make an appearance.

Table 5.1. Precursors of delayed neutrons following slow neutron fission of ^{235}U, ^{239}Pu and ^{233}U

Group	1	2	3	4	5	6
Lifetime (s)	~0.3	~0.9	~3.3	~9	~33	~80
Precursors	Br, Rb	^{140}I	^{139}I	^{138}I, ^{89}Br	^{137}I, ^{88}Br	^{87}Br

5.1.3.3 Emission of light nuclei

The decay of a nucleus in its ground state, by emission of a particle more massive than the nucleon, is possible only if the sum of the binding energies of the daughter and the emitted particle exceeds the binding energy of the parent. This criterion effectively rules out the emission of weakly bound light nuclei such as the deuteron, the triton and ^{3}He, although these particles may be identified among the decay products of highly excited nuclei. For example the nucleus $^{27}_{13}$Al* can decay with the emission of deuterons or α-particles as well as protons, neutrons or γ-rays. The α-particle, because of its large binding energy 28.3 MeV, is an obvious exception to the general rule. However, the Coulomb barrier has a dominant role in inhibiting α-decay right across the mass range. Thus, if $Q_\alpha \leqslant 3$ MeV the lifetime tends to be so great that the α-transition is not observable. However 1.8 MeV α-particles have been detected in the decay of ^{144}Nd, and ^{8}Be decays with a lifetime of about 10^{-16} s into two α-particles which share the available 90 keV energy. All nuclei with $A > 209$ are α-active.

5.2 Nuclear fission

5.2.1 Neutron induced fission

The circumstances leading up to the discovery of nuclear fission in 1939 have been described in section 1.4. A very early puzzle in the history of fission, which led to at least one very distinguished neutron physicist, namely Placzek, to doubt the very existence of the phenomenon (Wheeler 1989), was the observation that fission in uranium nuclei could be induced by fast neutrons ($E_n \geqslant 1$ MeV), and by thermal neutrons ($E_n \simeq 0.025$ eV) but not by neutrons of intermediate energy. The solution to the puzzle was provided by Bohr who suggested that fission by thermal neutrons is produced only in the rare isotope $^{235}_{92}$U, which is present in natural uranium to the extent of about 0.7%, while fission in the common isotope $^{238}_{92}$U could be brought about only by fast neutrons (Bohr 1939). That this explanation was essentially correct was confirmed when a small sample of pure $^{238}_{92}$U was shown not to be fissionable under bombardment by slow neutrons (Nier *et al* 1940).

Because an incoming slow neutron can form a pair with the unpaired neutron in $^{235}_{92}U$, sufficient energy is released to the compound nucleus $^{236}_{92}U$ for it to become unstable against spontaneous fission, which happens in a time $<5\times10^{-13}$ s. Also, since the fission cross-section obeys the $1/v$ law (cf. section 1.7.1) $^{235}_{92}U$ does not undergo fission to any significant degree when bombarded with fast neutrons. In the case of $^{238}_{92}U$, where no new pair can be formed, there is a fission 'threshold' at $E_n \simeq 1$ MeV. There is, of course, no threshold according to the strict definition (cf. section 1.5); there is an effective threshold in the sense that, at lower energy, the fission rate is so low as to be undetectable. In the same sense the nuclei $^{232}_{91}Pa$ and $^{232}_{90}Th$ have fission 'thresholds' at ≈ 0.5 MeV and ≈ 1 MeV, respectively.

In general fission may be produced in heavy nuclei ($A > 209$) when bombarded by light nuclei, e.g. p, d, t, ^{3}He, α, which are sufficiently energetic to penetrate the Coulomb barrier, as well as by electrons or photons. The cross-sections for fission by light nuclei all approach a maximum of about 2 b at a bombarding energy of ≈ 100 MeV; photofission cross-sections grow from a 'threshold' near 5 MeV, to a maximum of 50 mb near 20 MeV. Fast neutron fission cross-sections are typically of the order of 1 b whereas thermal neutron cross-sections in odd N nuclei are typically two to three orders of magnitude larger. Thus, from the viewpoint of energy production, fission by thermal neutrons is by far the most significant process. The thermal neutron fission cross-sections for the most important fissionable nuclei are listed in Table 5.2, along with the cross-sections for the competing (n, γ) reactions.

Table 5.2. Thermal neutron cross-sections for fission and (n, γ) capture reactions in $^{233}_{92}U$, $^{235}_{92}U$ and $^{239}_{94}Pu$

Nucleus	$^{233}_{92}U$	$^{235}_{92}U$	$^{239}_{94}Pu$
σ_f (b)	524	590	729
σ_γ (b)	69	108	303

5.2.2 Characteristics of the fission process

The most surprising feature of the fission process with thermal neutrons is the marked tendency of the fissioning nucleus to disintegrate into two unequal fragments. For this so-called asymmetric fission mode in $^{285}_{92}U$ the peaks in the mass spectrum occur at $A = 138.5$ and $A = 95$, which, taken together with the average $\nu = 2.5$ (prompt plus delayed) neutrons emitted per fission, accounts for the full mass number of the $^{236}_{92}U$ compound nucleus. The distribution in mass number is shown in figure 5.2. By contrast symmetric binary fission occurs with a frequency of about a part in 10^4, while ternary

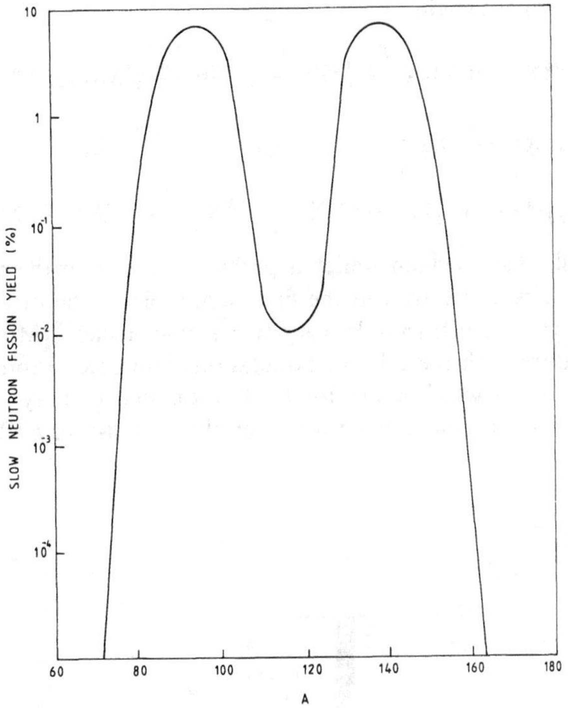

Figure 5.2. The observed distribution in mass number of fission products produced in the slow neutron induced fission of $^{235}_{92}U$. The clear bias in favour of fission into two fragments of unequal mass presented a severe difficulty for the Bohr–Wheeler (1939) theory of fission.

or quaternary fission into three or four roughly equal fragments account for an additional part in 10^5. The relative intensity of symmetric fission increases with incoming energy and dominates at high energy. The explanation for this peculiar state of affairs did not come to light until about 25 years after the discovery of nuclear fission (cf. section 5.2.4).

In the case of $^{235}_{92}U$ the heavy fragment carries away 65 MeV and the light fragment 100 MeV in the most frequent mode. The total energy released is about 200 MeV made up approximately as follows: fission fragments 83.1%, neutrons 3.1%, β-particles 2.6%, neutrinos 5.6%, γ-rays 5.6%. The fission fragments are highly neutron rich and decay by successive β-decays along two chains, in most cases of equal length, with three or more β-decays per chain. An example of the fission process is the following:

(i) Initial fission

$$n + {}^{235}_{92}U \rightarrow {}^{236}_{92}U^* \rightarrow n + {}^{96}_{39}Y^* + {}^{139}_{53}I^*$$

(ii) Light fragment chain (5.13)

$$^{96}_{39}Y^* \rightarrow n + ^{95}_{39}Y \xrightarrow{\beta^-} ^{95}_{40}Zr \xrightarrow{\beta^-} ^{95}_{41}Nb \xrightarrow{\beta^-} ^{95}_{42}Mo \ (15.8\%)$$

(iii) Heavy fragment chain

$$^{139}_{53}I^* \rightarrow n + ^{138}_{53}I \xrightarrow{\beta^-} ^{138}_{54}Xe \xrightarrow{\beta^-} ^{138}_{55}Cs \xrightarrow{\beta^-} ^{138}_{56}Ba \ (71.7\%)$$

This particular fission chain, which is perhaps atypical for the reason that it generates a scission neutron in the first step, replaces the original neutron with three prompt neutrons leaving two stable nuclei $^{95}_{42}Mo$ and $^{136}_{58}Ba$, found in nature with the relative populations indicated. Prompt neutrons are those neutrons which are emitted before the first β-decay.

Delayed neutron emission occurs after the first β-decay (Roberts *et al*

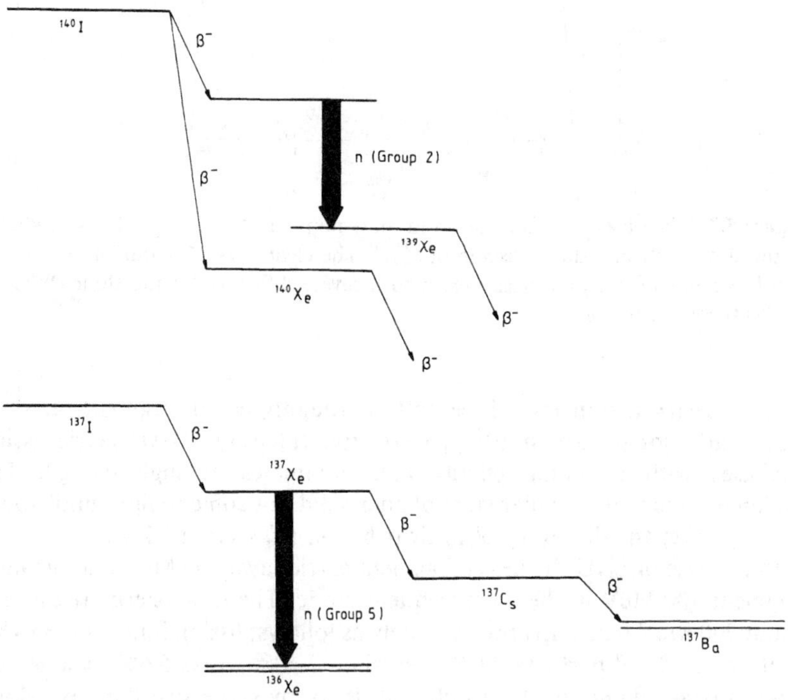

Figure 5.3. Delayed neutron emission in fission, which occurs after the first β-decay, is associated with the heavy fragment chain. This can occur early in the chain because of the large mass difference between an odd–odd parent nucleus (e.g. ^{140}I) and its even–even daughter (e.g. ^{140}Xe). Alternatively it can occur late in the chain when the separation energy S_n is reduced in a nucleus (e.g. ^{137}Xe) whose neutron number differs by unity from a magic number.

1939); it is associated with unequal chain lengths and bifurcation of the long chain. An example is the following:

(i) Initial fission

$$n + {}^{235}_{92}U \rightarrow {}^{236}_{92}U^* \rightarrow {}^{141}_{53}I^* + {}^{95}_{39}Y^*$$

(ii) Light fragment chain (5.14)

$${}^{95}_{39}Y^* \rightarrow n + {}^{94}_{39}Y \xrightarrow{\beta^-} {}^{94}_{40}Zr \ (17.5\%)$$

(iii) Heavy fragment chain

$${}^{141}_{53}I^* \rightarrow n + {}^{140}_{53}I \begin{cases} \xrightarrow{\beta^-} {}^{140}_{54}Xe \rightarrow 4\beta^- \rightarrow {}^{140}_{58}Ce \ (88.45\%) \\ \xrightarrow{\beta^-} {}^{140}_{54}Xe^* \rightarrow n + {}^{139}_{54}Xe \rightarrow 3\beta^- \rightarrow {}^{139}_{57}La \ (99.9\%) \end{cases}$$

In this case there is one prompt neutron and one (group 2) delayed neutron which is emitted with the lifetime of its precursor ${}^{140}_{53}I$. Delayed neutrons can be emitted early on in the chain whenever an odd–odd nucleus (e.g. ${}^{140}_{53}I$) β-decays to an even–even nucleus (e.g. ${}^{140}_{54}Xe$) with an excitation energy greater than the separation energy of a neutron in the same nucleus.

An alternative mode of delayed neutron emission occurs late in the chain when isobars are reached which contain a single (weakly bound) neutron outside closed shells. An example is the following

$${}^{137}_{53}I \xrightarrow{\beta^-} {}^{137}_{54}Xe^* \begin{cases} \xrightarrow{\beta^-} {}^{137}_{55}Cs \xrightarrow{\beta^-} {}^{137}_{56}Ba \ (11.3\%) \\ \rightarrow n + {}^{136}_{54}Xe \ (8.93\%) \end{cases} \qquad (5.15)$$

In this case the (group 5) delayed neutron is emitted with the lifetime of its (relatively long-lived) precursor ${}^{137}_{53}I$. The two alternative modes of delayed neutron emission are shown in figure 5.3.

The spectrum of fission neutrons is highly asymmetric, with a maximum energy of about 17 MeV in the case of fission of ${}^{235}_{92}U$, a mean of 2 MeV and a mode around 0.75 MeV. The energy distribution is approximately Maxwellian in the centre-of-mass frame but is modified by the motion of the fragments in the laboratory frame. A number of empirical formulae have been proposed to represent this spectrum, e.g. (Leachman 1956)

$$p(E_n)\, dE_n = 0.775 E_n^{1/2}\, exp[-0.775 E_n]\, dE_n \qquad (5.16)$$

5.2.3 The Bohr–Wheeler theory of fission

Fission is a violent collective process in which the fissioning nucleus is literally stretched to breaking point, leaving behind two highly deformed fragments of roughly comparable mass. Until the development of heavy ion physics such a phenomenon could not be described by conventional microscopic models and reaction mechanisms. Thus fission theory grew up alongside, but essentially independently of, mainstream nuclear theory. The mechanism of nuclear fission was first described by Bohr and Wheeler, in a famous paper whose substantive conclusions, with one notable exception, retain their validity to this day (Bohr and Wheeler 1939). The missing element in their analysis was, of course, the existence of a double hump in the fission barrier, which owes its existence to single-particle effects, and cannot be accommodated within a purely macroscopic model (Cohen and Swiatecki 1962, 1963).

The Bohr–Wheeler treatment is based on a modified form of the liquid drop model where, for heavy excited nuclei, they postulate a deformed cylindrically symmetric surface defined by (cf. section 4.5.1)

$$R(\theta)=R_0\left[1+\sum_{n\geq 2}\alpha_n P_n(n\theta)\right] \tag{5.17}$$

They then investigate the change in energy which occurs when the nucleus is deformed, under conditions of constant volume, from an initial spherically symmetric state. Under these conditions only the Coulomb energy E_c, and the surface tension energy E_s, contribute to this change which, to second order in the quadrupole deformation parameter α_2 is given by

$$\Delta(E_c+E_s) = -\frac{1}{5}\alpha_2^2\varepsilon Z^2/A^{1/3}+\frac{2}{5}\alpha_2^2\gamma A^{2/3} \tag{5.18}$$

For a heavy nucleus the error incurred in replacing the factor $(Z-1)$ by the factor Z in the Coulomb term is negligible.

When $\Delta(E_c+E_s)>0$ the nucleus retains its stability and will eventually return to its original state of stable equilibrium when its excess energy is released. However, as Z is increased, the long-range Coulomb repulsion will eventually overcome the short-range attractive forces which are responsible for surface tension. According to (5.18) this happens when the Coulomb energy is equal to twice the surface energy, and occurs at a critical value of Z^2/A given by

$$(Z^2/A)_c=2\gamma/\varepsilon\simeq 49.5 \tag{5.19}$$

Thus, defining the fissility parameter x by the relation

$$x=(Z^2/A)/(Z^2/A)_c \tag{5.20}$$

it follows that all nuclei characterized by a value x greater than unity are unstable against spontaneous fission. Experimentally it is indeed confirmed that the systematic variation of spontaneous fission lifetimes can be accounted for on the basis of this analysis (Swiatecki 1956).

In the case of a heavy nucleus for which $x<1$ only a small input of energy may be required for the nucleus to reach the critical state of unstable equilibrium, which can be represented as a saddle point in the multi-dimensional potential energy surface plotted as a function of the parameters α_n characterizing the deformation. The values of the α_n at the saddle point determine the critical shape of the deformed nucleus, and the potential energy at the saddle point defines the fission barrier height E_f. In general E_f can always be expressed in the form

$$E_f=\gamma A^{2/3}f(x) \tag{5.21}$$

where $f(x)$ is a function whose leading term in a power series expansion in $(1-x)$ is negative for $x<1$. For example, when the nucleus is subjected to a parity symmetric distortion specified by α_2 (the fission parameter), and α_4 (the necking parameter), $f(x)$ has the specific form

$$f(x)=0.726(1-x)^3-0.330(1-x)^4 \ldots \tag{5.22}$$

The excited nucleus may therefore be viewed as having available to it two alternative configurations; the normal compound nucleus configuration whose ground state is spherically symmetric (or perhaps slightly deformed), and a stretched saddle configuration or transition nucleus whose ground state is one of unstable equilibrium. These possibilities are illustrated in figure 5.4. Bohr and Wheeler also introduced the concept of a fission exit channel defined in terms of the number $N_t(E)$ of levels in the transition nucleus, which are available for fission at an excitation energy $E>E_t$. $N_t(E)$ is defined by the relation

$$N_t(E)=\int v\,\mathrm{d}p\rho_t(E_t) \tag{5.23}$$

where $\rho_t(E_t)$ is the level density in the transition nucleus at an energy

$$E_t=E-T-E_f \tag{5.24}$$

above the fission barrier, and T is the kinetic energy. The total fission width, expressed in terms of $N_t(E)$, and the level separation $\delta_c(E)=1/\rho_c(E)$ in the compound nucleus, is given by

$$\Gamma_f(E)=N_t(E)\,\delta_c(E)/2\pi \tag{5.25}$$

In the channel theory of nuclear fission developed by A Bohr (1956) it is recognized that the transition nucleus is thermodynamically cold since most of the energy is stored in the deformation. The system of excited states in the transition nucleus will then be primarily collective in nature, and similar

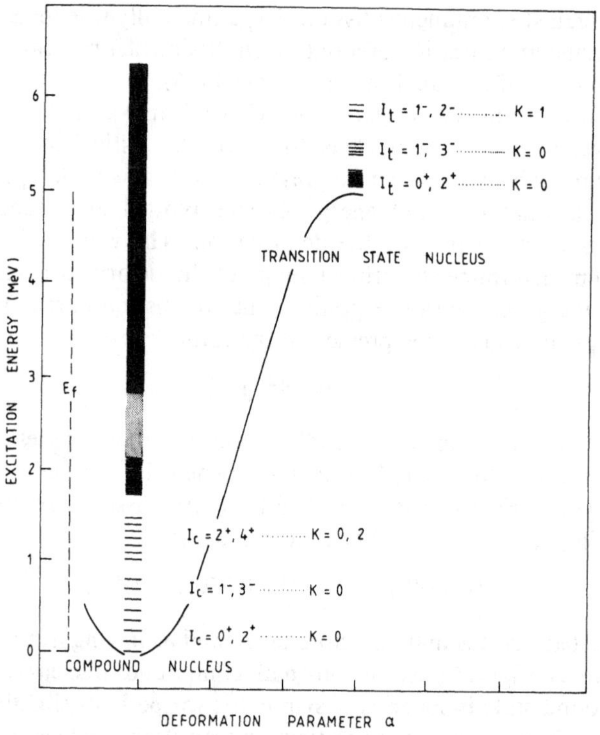

Figure 5.4. Diagram illustrating the Bohr–Wheeler (1939) theory of fission in which the excited nucleus can exist either in the normal compound nucleus configuration with stable ground state, or in a stretched saddle configuration with unstable ground state leading to fission. The collective modes predicted for the transition state all provide symmetric fission channels contrary to observation.

to the collective modes just above the ground state in the compound nucleus. Since the important compound nuclei, i.e. ^{234}U, ^{236}U and ^{240}Pu are all even–even nuclei, the first levels are rotational levels built on the $K=0$ band with $I_c=0^+, 2^+, 4^+$, etc. The same sequence of levels, with $I_t=0^+, 2^+, 4^+$, etc. is expected to appear in the transition nucleus, although with a marked reduction in level spacing due to the much increased moment of inertia in the stretched saddle configuration. These states all correspond to a symmetric division of the fissioning nucleus quite contrary to observation.

The nearest odd parity levels in the transition nucleus, those with $K=0$, $I_t=1^-, 3^-, 5^-$, etc., corresponding to a pear-shaped vibration or 'sloshing mode' leading to asymmetric division, and are expected to be some hundreds of keV higher up the energy level diagram. In principle γ-bands may exist at high excitations (cf. section 4.5.1) but there is no β-band since the corresponding potential energy is negative and this is a fission mode.

In the case of slow neutron induced fission the populated states in the transition nucleus will be those having the same spin I_c, and the same parity π_c, as the resonance states in the compound nucleus which are excited at energy E. Thus $N_t(E)$ in (5.25) will be replaced by

$$N_t(E, I_c, \pi_c) = \sum_i P_i(E, E_{fi})$$
$$= 2\pi\Gamma_f(E, I_c, \pi_c)/\delta_c(I_c, \pi_c) \qquad (5.26)$$

when $P_i(E < E_{fi})$ is the penetrability of the fission barrier corresponding to the ith (non-fission mode) available level in the transition nucleus, with spin-parity $I_c\pi_c$. For an inverted simple harmonic oscillator shaped fission barrier $P_i(E, E_{fi})$ takes the form

$$P_i = \{1 + \exp[-(E - E_i)/\hbar\omega]\}^{-1} \qquad (5.27)$$

where $\hbar\omega$, the transparency parameter or barrier curvature energy, is a function of both barrier shape and inertia tensor (Hill and Wheeler 1953). For odd-N target nuclei $E \gg E_{fi}$ because of the pairing energy and each P_i is unity; for even N target nuclei $E \leqslant E_{fi}$ (cf. section 5.2.1).

5.2.4 The double-humped fission barrier

The Bohr–Wheeler description of fission is essentially a macroscopic one, employing a procedure which is broadly justified on the grounds that the zero-point quantum fluctuations in amplitude of the deformed nucleus are in every case much less than nuclear dimensions. To make further progress it is necessary to apply the macroscopic–microscopic method of calculation whose main features were summarized at the end of section 5.1.2. Considered in the context of fission, the basic assumption of these theories is that, at all points along the fission path from the compound nucleus state, through the transition state, down to the very point of scission, the potential energy can be represented as the sum of a macroscopic (liquid drop) term plus a microscopic (single particle) term. The very first application of these methods to describe the fission of actinide nuclei revealed the presence of the now famous double-humped fission barrier which is shown in figure 5.5 (Strutinsky 1966, 1967).

The existence of a second minimum in the potential energy is associated with fluctuations in the single-particle level density at the Fermi surface, which are correlated with nuclear deformation. Quite generally, normal shell effects in nuclei are associated with degeneracies which are characteristic of spherical symmetry (cf. section 4.2.1), and the shell-structure corrections to the liquid drop model oscillate, with local minima at magic numbers and maxima in between (Mozer 1959, Myers and Swiatecki 1966). Similarly the additional fluctuations are characteristic of the special degeneracies which

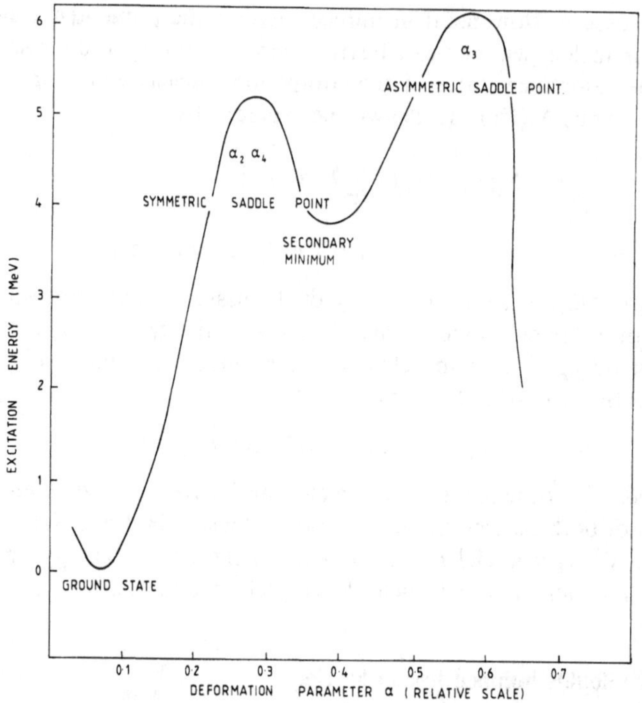

Figure 5.5. The double-humped fission barrier which arises in theories of fission where the liquid-drop potential energy is supplemented by single-particle terms (Strutinski 1966, 1967). For low mass actinides (e.g. U and Pu) the asymmetric saddle point corresponding to odd order deformation lies at the higher energy leading to asymmetric fission as observed. For heavier actinides (e.g. Fm) the symmetric saddle point is highest in energy.

arise in deformed nuclei with special symmetry. For actinide nuclei in particular the energy is substantially reduced when the semi-major axis is approximately twice as large as the semi-minor axis; because the macroscopic energy is close to its saddle point at this deformation, a second minimum is generated in the potential energy.

Detailed calculations confirm the presence of two significant features in this second minimum (Møller and Nilsson 1970): (i) at both minima, and at the first saddle point, the deformation is reflection symmetric, i.e. determined primarily by even order deformation parameters α_2, α_4, etc., whereas the second saddle point is reflection asymmetric, i.e. a pear-shaped deformation associated with the asymmetry parameter α_3; (ii) for low mass actinides the asymmetric saddle point lies at the higher energy, whereas, for high mass actinides, this saddle point falls below the symmetric saddle point. These conclusions agree entirely with the observation that thermal neutron induced fission in isotopes of uranium and plutonium is overwhelmingly asymmetric, whereas for the isotopes of fermium ^{258}Fm and ^{259}Fm, and for the higher transuranic elements, it is primarily symmetric.

The reality of the second minimum is dramatically confirmed by the observation of fission isomers, of which more than 30 have been identified experimentally over the whole range of actinides between uranium and californium. These are interpreted as modes of decay, through the second fission barrier, of states in the second minimum populated by specific nuclear reactions (Specht 1974). The first fission isomer to be discovered was the 14 ms fission activity in ^{243}Am produced by bombarding uranium with neon ions (Polikanov *et al* 1962), and subsequently identified as a product in the reaction ^{243}Am(n, 2n)^{242}Am (Flerov *et al* 1967). Vibrational states in the second minimum have also been associated with the broad resonances observed in sub-threshold fission, and with clusters of weak narrow resonances observed in similar circumstances, which are described as 'intermediate structure'. This latter phenomenon has been interpreted in terms of a resonant coupling between states in the two minima which lie close together, and which have the same spin and parity (Weigmann 1968).

Since the discovery of the double-humped fission barrier, macroscopic–microscopic theories have been extended to cover a wide variety of nuclear shapes, ranging from a single spheroid modified by some combination of deformation parameters α_n $(3 \leqslant n \leqslant 6)$, to a double spheroid joined smoothly at the neck, similarly modified. To determine the microscopic single-particle corrections, various single-particle matching potentials have been constructed, including two-centre potentials required to reproduce the effect of the nascent fragments at the point of scission (Maruhn and Greiner 1972). Much of this work has been carried out using Woods–Saxon potentials (cf. section 3.2.6), or folded Yukawa potentials (cf. section 3.5.1) obtained by folding a Yukawa potential into a predetermined nuclear shape (Nix 1972).

5.2.5 Transuranic and superheavy elements

If the N/Z ratio is taken to have the same value $\simeq 1.5$ which is measured for long-lived nuclei at the upper end of the Segré chart, then all nuclei with $Z \geqslant 124$ will have a fissility parameter $x > 1$ and, according to the simple liquid drop model, should undergo spontaneous fission with lifetimes $\leqslant 10^{-22}$ s. However the observed fission barriers for actinide nuclei are almost constant, implying a dependence on x somewhat weaker than might be expected, and demonstrating once again the importance of shell structure in providing additional stability. The exploitation of this effect has proved crucial in the successful synthesis of new elements much heavier than those found in nature, e.g. the closed neutron subshell at $N = 152$ (Ghiorso *et al* 1954) is of particular importance in maintaining the stability of a range of transuranic isotopes between $Z = 98$ and $Z = 104$ (cf. figure 4.1).

The list of the confirmed transuranic elements is shown in Table 5.3, together with the nuclear transitions in which they were first identified. This list terminates at the nucleus $^{266}(109)$ which has a measured lifetime of a few ms. The identification of element 110 is, as yet, tentative, although efforts are being made to synthesize it by bombarding thorium with calcium. It

Table 5.3. The transuranic elements. With the exception of element 110, whose synthesis has yet to be confirmed, all the transuranic elements have been detected in more than one laboratory. The names of elements 104 (rutherfordium), 105 (hahnium), 107 (nielsbohrium), 108 (hassium) and 109 (meitnerium) have recently (1993) been agreed by the International Union of Pure and Applied Chemistry. A name has not yet been agreed for element 106.

Element	First identified decay or reaction	Reference
$_{93}Np$	$^{238}_{92}U(n,\gamma)^{239}_{92}U \xrightarrow{\beta^-} {}^{239}_{93}Np$	McMillan and Abelson (1940)
$_{94}Pu$	$^{238}_{92}U(d,2n)^{238}_{93}Np \xrightarrow{\beta^-} {}^{238}_{94}Pu$	Seaborg *et al* (1946)
$_{95}Am$	$^{239}_{94}Pu(n,\gamma)^{240}_{94}Pu(n,\gamma)^{241}_{94}Pu \xrightarrow{\beta^-} {}^{241}_{95}Am$	Seaborg *et al* (1949)
$_{96}Cm$	$^{239}_{94}Pu(\alpha,n)^{242}_{96}Cm$	Seaborg *et al* (1949)
$_{97}Bk$	$^{241}_{95}Am(\alpha,2n)^{243}_{97}Bk$	Thompson *et al* (1950)
$_{98}Cf$	$^{242}_{96}Cm(\alpha,n)^{245}_{97}Cf$	Thompson *et al* (1950)
$_{99}E$	$^{253}_{98}Cf \xrightarrow{\beta^-} {}^{253}_{99}E$	Ghiorso *et al* (1955)
$_{100}Fm$	$^{255}_{99}E \xrightarrow{\beta^-} {}^{255}_{100}Fm$	Ghiorso *et al* (1955)
$_{101}Md$	$^{253}_{99}E(\alpha,n)^{256}_{101}Md$	Ghiorso *et al* (1955a)
$_{102}No$	$^{246}_{96}Cm+{}^{12}_{6}C \rightarrow {}^{254}_{102}No+4n$	Ghiorso *et al* (1958)
	$^{241}_{94}Pu+{}^{16}_{8}O \rightarrow {}^{252(253)}_{102}No+5n(4n)$	Flerov *et al* (1958, 1960)
$_{103}Lr$	$^{250,251,252}_{98}Cf+{}^{11}_{5}B({}^{10}_{5}B) \rightarrow {}^{257}_{103}Lr+(3-6)n$	Ghiorso *et al* (1961)
	$^{243}Am+{}^{18}O \rightarrow {}^{256}Lr({}^{257}Lr)+5n(4n)$	Flerov *et al* (1968)
$_{104}Ru$	$^{242}_{94}Pu+{}^{22}_{10}Ne \rightarrow {}^{260(259)}(104)+4n(5n)$	Flerov *et al* (1964)
		Ghiorso *et al* (1970)
$_{105}Hn$	$^{209}_{83}Bi+{}^{50}_{22}Ti \rightarrow {}^{257}(105)+2n$	Oganessian *et al* (1976)
106	$^{249}_{98}Cf+{}^{18}_{8}O \rightarrow {}^{263}(106)+4n$	Ghiorso *et al* (1974)
		Oganessian *et al* (1974)
$_{107}Ns$	$^{209}_{83}Bi+{}^{54}_{24}Cr \rightarrow {}^{262}(107)+n$	Münzenberg *et al* (1981)
$_{108}Hs$	$^{208}_{82}Pb+{}^{58}_{22}Fe \rightarrow {}^{265}(108)+n$	Münzenberg *et al* (1984)
		Oganessian *et al* (1984)
$_{109}Mt$	$^{209}_{83}Bi+{}^{58}_{26}Fe \rightarrow {}^{266}(109)+n$	Münzenberg (1984a)
		Oganessian *et al* (1984, 1984a)
110	$^{238}_{92}U+{}^{40}_{18}A \rightarrow {}^{276-x}(110)+xn$	Oganessian *et al* (1987, 1987a)

would therefore appear that, in the absence of isolated 'islands of stability' containing certain species of (hypothetical) 'super heavy nuclei' as discussed briefly below, the sequence of observable transuranic elements cannot be extended much beyond its present limit.

The role of (n, γ) reactions in ${}^{238}_{92}U$ leading to the first syntheses of neptunium and plutonium has been touched on briefly in section 1.4. Although the long-lived isotope ${}^{239}_{94}Pu$ (2.44×10^4 years) is generated through the β-decay of ${}^{239}_{93}Np$, it is the shorter lived isotope ${}^{238}_{94}Pu$ (90 years) which was the first to be identified unambiguously. The longest lived isotopes of neptunium and uranium, ${}^{237}_{93}Np$ (2.2×10^6 years) and ${}^{244}_{94}Pu$ (7.5×10^7 years) have lifetimes which are too short, by factors of order 10^2–10^3, for any significant fraction of these elements to have survived since the earth's crust was formed more than 4×10^9 years ago. Yet today both are so abundant in the sea and on land, that it is difficult to appreciate that, 50 years ago, these elements did not exist. All the transuranic elements up to mendelevium, ${}_{101}Md$, are described as 'actinides' since they have chemical properties similar to actinium ($Z = 89$), and their existence was confirmed in every case by chemical means. For the 'superactinides' with $Z \geqslant 102$, the identification of a few short-lived atoms of the new elements by chemical methods is not possible, and it is necessary instead to carry out a detailed and complex analysis of all the observed radiations.

Excluding elements 99 (einsteinium) and 100 (fermium) all transuranic elements in the range $96 \leqslant Z \leqslant 101$ were first identified by analysis of the products of (α, n) and (α, 2n) reactions in actinide targets, using α-particles from the 60 inch Berkeley cyclotron. The excluded elements, ${}_{99}E$ and ${}_{100}Fm$, were first detected in the nuclear debris created by the 'Mike' thermonuclear explosion in 1952. The isotope ${}^{253}_{99}E$ was identified as an end-product of five β-decays, beginning with the nucleus ${}^{255}_{92}U$, itself formed by 15 successive neutron captures in ${}^{238}_{92}U$, subjected to the intense flux of 14 MeV neutrons that such an explosion creates (cf. section 5.5.3). In the same event the nucleus ${}^{255}_{100}Fm$ was identified as a product of 17 successive neutron captures in ${}^{238}_{92}U$, followed by a rapid succession of β-decays, terminating with the β-active nucleus ${}^{255}_{99}E$ (18d). These observations provide convincing evidence of the reality of the '*r*-process' for heavy element production (cf. section 10.5.7) by successive neutron capture in nuclei on the neutron-rich slope of the valley of stability in the Segré chart.

For $102 \leqslant Z \leqslant 106$ new elements have been successfully synthesized by means of light ion reactions in actinide targets but, for $Z \geqslant 107$, this technique ceases to be effective. This is not through a lack of suitable target nuclei, e.g. ${}^{249}_{98}Cf$ (1.5×10^9 years) or ${}^{249}_{97}Bk$ (6×10^8 years) but because, when the fission barrier is low, the most probable de-excitation mode for the transition nucleus is through immediate fission rather than through emission of a cascade of neutrons. Fortunately at this stage a brilliant new technique was introduced, involving the cold fusion of medium mass ions of Fe, Cr, etc.

with targets of Pb and Bi, which successfully synthesized several new transuranic elements in addition to new isotopes of those already known. This technique was first developed at JINR, Dubna (Oganessian *et al* 1974) and was subsequently adopted at Berkeley (Nitschke *et al* 1979) and at GSI, Darmstadt (Münzenberg *et al* 1981).

Although there had been some early conjectures about the possible existence of an 'island of stability' at the next doubly magic nucleus above $^{208}_{82}$Pb, separated from the main 'peninsula' of long-lived heavy elements at the head of the Segré chart, by a 'sea' of unstable nuclei (Mayer and Teller 1949, Wheeler 1955, Scharff-Goldhaber 1957), there was a general consensus that this 'island' should emerge in the neighbourhood of the point $Z=126$, $N=184$, which is the next doubly magic nucleus arrived at by following the sequence of major shells in the simple harmonic oscillator with spin orbit coupling (cf. section 4.2.1). Later it was pointed out that the effect of the unsaturated Coulomb interaction would be to lower the next proton magic number to $Z=114$ (Mottelson and Nilsson 1959, Meldner 1965). It was however the suggestion of Myers and Swiatecki (1966) that, because of the effect of shell stabilization, the nucleus $^{298}(114)$ might have a fission barrier of several MeV, that set off the hunt for this, and other superheavy nuclei, e.g. $^{294}(116)$ (Armbruster 1985).

Unfortunately all searches for superheavy nuclei have proved vain (Kratz 1983, Seaborg and Loveland 1987, Münzenberg 1988). This does not imply that semi-stable superheavy nuclei do not exist, it merely reflects the quite extraordinary difficulties which are encountered in attempting their synthesis. It is still a remote possibility that these nuclei may exist in nature having been synthesized in supernovae explosions in the usual way (Meldner 1972) (cf. section 10.5.5). However this is not a view which attracts wide support (Matthews and Viola 1976).

5.3 Neutron transport and diffusion

5.3.1 The one-speed Boltzmann linear transport equation

5.3.1.1 Basic assumptions of transport theory

The transport equation is an integro-differential equation introduced by Boltzmann in 1872 for the express purpose of describing the development in space and time of monatomic classical gases in non-equilibrium conditions. It lies at the foundations of the kinetic theory of gases. In general the Boltzmann equation is non-linear since it contains products of density functions describing the self-scattering of a single atomic species, or the mutual scattering of two atomic species one from the other. However, in the case of a two-component system involving the interaction of n atoms of a dilute species with N atoms of a dense species, such that $n^2 \ll nN \ll N^2$, the

Boltzmann equation describing the behaviour of the dilute species is linear, provided that self-scattering of the dilute system may be neglected in comparison with the mutual scattering of the two systems, which in turn must have a negligible influence on the behaviour of the dense system.

Photons propagating in stellar atmospheres provide one good example of a dilute system as indeed do neutrons moving around in the active volume of a nuclear reactor (Corngold *et al* 1963). In the latter case, not only are there less than 10^{-11} neutrons per core atom, even in very high flux reactors, but the neutron–neutron scattering cross-section is small in comparison with typical neutron–atom cross-sections, and utterly negligible in comparison with atom–atom cross-sections. Thus the conditions necessary for the neutron population to be categorized as a dilute system are easily fulfilled.

The objective of neutron transport theory is to describe the space-time behaviour of the neutrons in terms of the geometry of the particular system of interest, and under certain ideal conditions (Ornstein and Uhlenbeck 1937). The theory can therefore make no generally rigorous statements, and perhaps one of its main applications is to test the accuracy of approximation methods when applied to simple cases where rigorous solutions are known. In setting up the transport equation it is assumed that all neutrons travel with the same speed, so that, in a collision, the magnitude of the velocity is assumed to stay constant, while only the direction is altered. In the theory of thermal neutron diffusion (Amaldi and Fermi 1936) this approximation neglects the Maxwellian distribution of neutron velocities and, applied to fast neutrons, it disregards the slowing down a neutron experiences at every collision. One may generalize this approach to develop a two-group treatment, where neutrons of thermal energy constitute one group, while neutrons emitted from a fission source make up the other. The fast fission neutrons are then assumed to retain their initial energy until they have undergone a specified number of collisions, at which point they are transferred into the thermal group. The slowing down process, known as moderation, plays a pivotal role in the operation of a nuclear reactor and has to be given very careful treatment (cf. section 5.4).

The final idealization is to assume that the neutrons behave as classical point particles moving in a disordered medium. Thus there are no favoured directions of propagation as in a crystalline medium (cf. chapter 7). Also since neutron scattering in the medium is essentially incoherent, quantum interference effects need not be taken into account (Kuscer and Corngold 1965).

5.3.1.2 The continuity relation

The first important quantity to be introduced is the density $n(\boldsymbol{r}, \boldsymbol{\Omega}, t)$, which is defined as the number of neutrons per unit volume per steradian at the space-time point $(\boldsymbol{r}, t)$, moving in the direction of the unit vector $\boldsymbol{\Omega}$. The corresponding neutron density $n(\boldsymbol{r}, t)$ is then given as the integral over all directions

$$n(\boldsymbol{r}, t)=\int n(\boldsymbol{r}, \boldsymbol{\Omega}, t)\, d\Omega \tag{5.28}$$

In practice the most useful function is the scalar flux $\varphi(\boldsymbol{r}, t)$ defined by

$$\varphi(\boldsymbol{r}, t)=n(\boldsymbol{r}, t)v \tag{5.29}$$

which represents the total path length traced out per unit time by all the neutrons $n(\boldsymbol{r}, t)$ in the given unit volume. Thus if σ_0 is the cross-section for a given reaction process, the volume of space sampled per unit time by these neutrons is $\sigma_0\varphi(\boldsymbol{r}, t)$ and the number of reactions per unit volume per unit time is $N\sigma_0\varphi(\boldsymbol{r}, t)=\Sigma_0\varphi(\boldsymbol{r}, t)$, where N is the number of target nuclei per unit volume, and Σ_0 is the corresponding macroscopic cross-section for the particular reaction.

We now define the current density or vector flux

$$\boldsymbol{j}(\boldsymbol{r}, \boldsymbol{\Omega}, t)=v\boldsymbol{\Omega}n(\boldsymbol{r}, \boldsymbol{\Omega}, t) \tag{5.30}$$

which is a vector in the direction $\boldsymbol{\Omega}$ whose magnitude represents the net number of neutrons per unit volume per steradian crossing unit area normal to $\boldsymbol{\Omega}$. Finally we have the total current $\boldsymbol{j}(\boldsymbol{r}, t)$ given by

$$\boldsymbol{j}(\boldsymbol{r}, t)=\int \boldsymbol{j}(\boldsymbol{r}, \boldsymbol{\Omega}, t)\, d\Omega \tag{5.31}$$

The component of $\boldsymbol{j}(\boldsymbol{r}, t)$ in any direction measures the net number of neutrons per unit time crossing unit area normal to that direction, but moving in all possible directions.

If we now define a source function $S(\boldsymbol{r}, \boldsymbol{\Omega}, t)$ which gives the number of neutrons moving in the direction $\boldsymbol{\Omega}$, which are generated per unit volume per unit time per steradian at $(\boldsymbol{r}, t)$, then we may set up a balance relation for neutrons moving in vacuum in a volume V, enclosed by a surface S

$$\frac{\partial}{\partial t}\int_V n(\boldsymbol{r}, \boldsymbol{\Omega}, t)\, d^3r=\int_V S(\boldsymbol{r}, \boldsymbol{\Omega}, t)\, d^3r-\int_S [v\boldsymbol{\Omega}n(\boldsymbol{r}, \boldsymbol{\Omega}, t)]\cdot d\boldsymbol{\sigma} \tag{5.32}$$

This relation states, in essence, that the rate of increase of the number of neutrons within V moving in the direction $\boldsymbol{\Omega}$ is just the number generated by the source, less the number flowing per unit time over the bounding surface. When the surface integral in (5.32) is converted into a volume integral by means of Gauss's theorem, the result is the continuity (or conservation) relation for neutrons in vacuum

$$\frac{\partial n(\boldsymbol{r}, \boldsymbol{\Omega}, t)}{\partial t}+v\boldsymbol{\Omega}\cdot\nabla n(\boldsymbol{r}, \boldsymbol{\Omega}, t)=S(\boldsymbol{r}, \boldsymbol{\Omega}, t) \tag{5.33}$$

Equation (5.33) must be modified substantially to take account of nuclear reactions of neutrons within the volume V, when the vacuum is replaced by a material medium. These reactions are of three kinds: (i) capture, characterized by the macroscopic cross-section $\Sigma_c(\boldsymbol{r})$, in which no neutrons are present in the final state following the reaction; (ii) scattering $(\Sigma_s(\boldsymbol{r}))$ where one neutron is present following the reaction; and (iii) fission $(\Sigma_f(\boldsymbol{r}))$ in which ν neutrons are present following the reaction. In scattering reactions it is necessary to distinguish between elastic scattering in which no internal energy of excitation is transferred to, or accepted from, the scattering nucleus, and inelastic scattering in which there is an exchange of internal energy. Within the one-speed approximation it is therefore necessary to establish criteria to decide whether inelastic scattering should be treated as 'scattering' or 'capture'. A similar difficulty arises in the treatment of fission neutrons, where an inelastically scattered fast neutron cannot induce fast fission but is nevertheless available for producing thermal fission (cf. section 5.5.1). In any case it cannot be regarded as having been captured.

In general the three processes compete and the mean number of neutrons emitted per reaction is given by

$$c(\boldsymbol{r}) = (\Sigma_s(\boldsymbol{r}) + \nu\Sigma_f(\boldsymbol{r}))/\Sigma_t(\boldsymbol{r}) \tag{5.34}$$

where

$$\Sigma_t(\boldsymbol{r}) = \Sigma_c(\boldsymbol{r}) + \Sigma_s(\boldsymbol{r}) + \Sigma_f(\boldsymbol{r}) \tag{5.35}$$

is the total macroscopic reaction cross-section. In addition to their dependence on $\boldsymbol{r}$, all macroscopic cross-sections depend on the neutron velocity v which is treated as a parameter. In a medium which satisfies translational invariance symmetry all the macroscopic cross-sections are independent of $\boldsymbol{r}$. In these circumstances the neutron density $n(\boldsymbol{r}, \boldsymbol{r}_0)$ at $\boldsymbol{r}$ due to a source at $\boldsymbol{r}_0$, is a function of $(\boldsymbol{r}-\boldsymbol{r}_0)$ only.

In a given medium the number of neutrons lost per unit volume per unit time per steradian about $\boldsymbol{\Omega}$ is $\Sigma_c(\boldsymbol{r})vn(\boldsymbol{r}, \boldsymbol{\Omega}, t)$ and the number gained is $\int c(\boldsymbol{r})\Sigma_c(\boldsymbol{r})vn(\boldsymbol{r}, \boldsymbol{\Omega}', t) f(\boldsymbol{\Omega}, \boldsymbol{\Omega}')\, d\Omega'$, where $f(\boldsymbol{\Omega}, \boldsymbol{\Omega}')$ represents the probability per steradian that an incoming neutron moving in the direction $\boldsymbol{\Omega}'$ should be replaced after scattering by an outgoing neutron moving in the direction $\boldsymbol{\Omega}$. The probability density $f(\boldsymbol{\Omega}, \boldsymbol{\Omega}')$ satisfies the normalization condition

$$\int f(\boldsymbol{\Omega}, \boldsymbol{\Omega}')\, d\Omega = \int f(\boldsymbol{\Omega}, \boldsymbol{\Omega}')\, d\Omega' = 1 \tag{5.36}$$

The transport equation is invariant under the operation of time reversal when $f(\boldsymbol{\Omega}, \boldsymbol{\Omega}')$ satisfies the relation

$$f(-\boldsymbol{\Omega}, -\boldsymbol{\Omega}') = f(\boldsymbol{\Omega}', \boldsymbol{\Omega}) \tag{5.37}$$

If the scattering process is rotationally invariant, e.g. spin-independent, then $f(\boldsymbol{\Omega}, \boldsymbol{\Omega}')$ is a function of $\boldsymbol{\Omega}\cdot\boldsymbol{\Omega}'$ only, in which case (5.37) is satisfied identically.

In the special case where the scattering is isotropic in the laboratory $f(\mathbf{\Omega}, \mathbf{\Omega}')$ is given by

$$f(\mathbf{\Omega}, \mathbf{\Omega}') = 1/4\pi \tag{5.38}$$

Allowing for the contribution of scattering and fission the continuity relation (5.33) for neutrons in the one-speed approximation assumes its final form

$$\frac{\partial n}{\partial t}(\mathbf{r}, \mathbf{\Omega}, t) + v\mathbf{\Omega}\cdot\nabla n(\mathbf{r}, \mathbf{\Omega}, t) + \Sigma_t(\mathbf{r})vn(\mathbf{r}, \mathbf{\Omega}, t)$$

$$= S(\mathbf{r}, \mathbf{\Omega}, t) + \int c(\mathbf{r})\Sigma_t(\mathbf{r})vn(\mathbf{r}, \mathbf{\Omega}', t) f(\mathbf{\Omega}, \mathbf{\Omega}')\, \mathrm{d}\Omega' \tag{5.39}$$

We note that, in a non-multiplying medium ($\Sigma_f(\mathbf{r}) \equiv 0$), $c(\mathbf{r})$ can vary between zero (pure capture) and unity (pure scattering). In a multiplying medium $c(\mathbf{r}) > 1$ if $(\nu - 1)\Sigma_f(\mathbf{r}) > \Sigma_c(\mathbf{r})$.

5.3.1.3 Solution for a purely absorbing medium

In a purely absorbing medium $\Sigma_s(\mathbf{r}) = \Sigma_f(\mathbf{r}) = 0$ and $c(\mathbf{r}) = 0$. Thus making the transformation

$$\mathbf{r}' = \mathbf{r} - v\mathbf{\Omega}t \qquad t' = t \tag{5.40}$$

the transport equation (5.39) reduces to the simplified form

$$\frac{\partial}{\partial t'} n(\mathbf{r}' + v\mathbf{\Omega}t', \mathbf{\Omega}, t') + v\Sigma_c n(\mathbf{r}' + v\mathbf{\Omega}t', \mathbf{\Omega}, t') = S(\mathbf{r}' + v\mathbf{\Omega}t', \mathbf{\Omega}, t') \tag{5.41}$$

Evidently the function $\exp\{\int^{t'} v\Sigma_c(\mathbf{r}' + v\mathbf{\Omega}t'')\, \mathrm{d}t''\}$ is an integrating factor for this equation which integrates immediately to

$$n(\mathbf{r}' + v\mathbf{\Omega}t', \mathbf{\Omega}, t') = n(\mathbf{r}', \mathbf{\Omega}, 0) \exp\left\{ -\int_0^{t'} v\Sigma_c(\mathbf{r}' + v\mathbf{\Omega}t'')\, \mathrm{d}t'' \right\}$$

$$+ \int_0^{t'} S(\mathbf{r}' + v\mathbf{\Omega}t'', \mathbf{\Omega}, t'') \exp\left\{ -\int_{t''}^{t'} v\Sigma_c(\mathbf{r} + v\mathbf{\Omega}t''') \right\} \mathrm{d}t''' \tag{5.42}$$

Reverting back to the original variables $(\mathbf{r}, t)$ this solution becomes

$$n(\mathbf{r}, \mathbf{\Omega}, t) = n(\mathbf{r} - v\mathbf{\Omega}t, \mathbf{\Omega}, 0) \exp\left\{ -\int_0^{t} v\Sigma_c(\mathbf{r} - v\mathbf{\Omega}(t - t''))\, \mathrm{d}t'' \right\}$$

$$+ \int_0^{t} S(\mathbf{r} - v\mathbf{\Omega}(t - t''), \mathbf{\Omega}, t'') \exp\left\{ -\int_{t''}^{t} v\Sigma_c(\mathbf{r} - v\mathbf{\Omega}(t - t''')\, \mathrm{d}t''' \right\} \mathrm{d}t'' \tag{5.43}$$

If we now assume that (i) the sources $S(\boldsymbol{r}, \boldsymbol{\Omega}, t) = S(\boldsymbol{r}, \boldsymbol{\Omega})$ are constant in time and (ii) the initial distribution $n(\boldsymbol{r}, \boldsymbol{\Omega}, 0)$ is restricted to a finite region of space, then, when we let $t \to \infty$, the first term in (5.43) makes no contribution since $n(-\infty, \boldsymbol{\Omega}, 0)$ is zero. In this case we approach the stationary limit

$$n(\boldsymbol{r}, \boldsymbol{\Omega}) = \frac{1}{v} \int_0^{\infty} S(\boldsymbol{r} - R\boldsymbol{\Omega}, \boldsymbol{\Omega}) \exp\left\{ -\int_0^R \Sigma_c(\boldsymbol{r} - R'\boldsymbol{\Omega})\, \mathrm{d}R' \right\} \mathrm{d}R \quad (5.44)$$

where we have made the substitutions in (5.43)

$$R' = v(t - t''') \qquad R = v(t - t'') \quad (5.45)$$

The exponential factor in the integral in (5.44) can also be expressed in terms of an optical thickness equivalent for neutrons (Chandrasekhar 1960)

$$\alpha(\boldsymbol{r}, \boldsymbol{r}') = \alpha(\boldsymbol{r}', \boldsymbol{r}) = \int_0^R \Sigma_c(\boldsymbol{r} - s\boldsymbol{R}/R) = \bar{\Sigma}_c(\boldsymbol{r}, \boldsymbol{r}')R \quad (5.46)$$

where

$$\boldsymbol{R} = \boldsymbol{r} - \boldsymbol{r}' = R\boldsymbol{\Omega}' \quad (5.47)$$

$$\bar{\Sigma}_c(\boldsymbol{r}, \boldsymbol{r}') = \bar{\Sigma}_c(\boldsymbol{r}', \boldsymbol{r}) = \int_0^1 \Sigma_c(\boldsymbol{r} - u(\boldsymbol{r} - \boldsymbol{r}'))\, \mathrm{d}u \quad (5.48)$$

is a mean macroscopic capture cross-section averaged along a line joining $\boldsymbol{r}$ to $\boldsymbol{r}'$. With this substitution the angular density (5.44) becomes

$$n(\boldsymbol{r}, \boldsymbol{\Omega}) = \frac{1}{v} \int S(\boldsymbol{r} - R\boldsymbol{\Omega}, \boldsymbol{\Omega}) \exp\{-\bar{\Sigma}_c(\boldsymbol{r}, \boldsymbol{r} - R\boldsymbol{\Omega})R\}\, \mathrm{d}R \quad (5.49)$$

In order to convert the line integral (5.49) into a volume integral we introduce the δ-function $(1/2\pi)\,\delta(\boldsymbol{\Omega}\cdot\boldsymbol{\Omega}' - 1)$ where

$$\frac{1}{2\pi} \int \delta(\boldsymbol{\Omega}\cdot\boldsymbol{\Omega}' - 1)\, \mathrm{d}\Omega = \frac{1}{2\pi} \int \delta(\boldsymbol{\Omega}\cdot\boldsymbol{\Omega}' - 1)\, \mathrm{d}\Omega' = \int_{-1}^{1} \delta(\mu - 1)\, \mathrm{d}\mu = 1 \quad (5.50)$$

Thus we may write, adopting the notation given in (5.47),

$$\begin{aligned} n(\boldsymbol{r}, \boldsymbol{\Omega}) &= \frac{1}{v} \int S(\boldsymbol{r} - R\boldsymbol{\Omega}', \boldsymbol{\Omega}') \frac{1}{2\pi} \delta(\boldsymbol{\Omega}\cdot\boldsymbol{\Omega}' - 1) \exp\{-\bar{\Sigma}_c(\boldsymbol{r}, \boldsymbol{r} - R\boldsymbol{\Omega}')R\}\, \mathrm{d}R\, \mathrm{d}\Omega' \\ &= \frac{1}{v} \int S(\boldsymbol{r} - \boldsymbol{R}, \boldsymbol{R}/R) \frac{1}{2\pi} \delta(\boldsymbol{\Omega}\cdot\boldsymbol{R}/R - 1) \exp\{-\bar{\Sigma}_c(\boldsymbol{r}, \boldsymbol{r} - \boldsymbol{R})R\} \frac{d^3R}{R^2} \\ &= \frac{1}{v} \int S(\boldsymbol{r}', \boldsymbol{R}/R) \frac{1}{2\pi} \delta(\boldsymbol{\Omega}\cdot\boldsymbol{R}/R - 1) \exp\{-\bar{\Sigma}_c(\boldsymbol{r}, \boldsymbol{r}')R\} \frac{\mathrm{d}^3r'}{R^2} \end{aligned} \quad (5.51)$$

Integrating (5.51) over all $\boldsymbol{\Omega}$ then leads to an expression for the neutron density

$$n(\boldsymbol{r})=\frac{1}{v}\int S(\boldsymbol{r}', \boldsymbol{R}/R)\exp\{-\bar{\Sigma}_c(\boldsymbol{r}, \boldsymbol{r}')R\}\frac{d^3r'}{R^2} \tag{5.52}$$

A further useful conclusion may be deduced from (5.51) by applying this relation to the case of a directional source at the point $\boldsymbol{r}_0$, emitting in the direction $\boldsymbol{\Omega}_0$, which is described by the source function

$$S(\boldsymbol{r}'', \boldsymbol{\Omega}'')=\delta^3(\boldsymbol{r}''-\boldsymbol{r}_0)\frac{1}{2\pi}\delta(\boldsymbol{\Omega}''\cdot\boldsymbol{\Omega}_0-1) \tag{5.53}$$

Thus the function $n(\boldsymbol{r}, \boldsymbol{\Omega}; \boldsymbol{r}_0, \boldsymbol{\Omega}_0)$, which represents the angular density at $\boldsymbol{r}$ in the direction $\boldsymbol{\Omega}$, of neutrons arriving from this source, is given by

$$\begin{aligned} n(\boldsymbol{r}, \boldsymbol{\Omega}; \boldsymbol{r}_0, \boldsymbol{\Omega}_0)&=\frac{1}{2\pi}\delta\left(\boldsymbol{\Omega}_0\cdot\frac{(\boldsymbol{r}-\boldsymbol{r}_0)}{|(\boldsymbol{r}-\boldsymbol{r}_0)|}\right)\frac{1}{2\pi}\delta\left(\boldsymbol{\Omega}\frac{(\boldsymbol{r}-\boldsymbol{r}_0)}{|(\boldsymbol{r}-\boldsymbol{r}_0')|}\right)e^{-\bar{\Sigma}_c(\boldsymbol{r},\boldsymbol{r}_0)|(\boldsymbol{r}-\boldsymbol{r}_0)|}/|\boldsymbol{r}-\boldsymbol{r}_0|^2 \\ &\equiv n(\boldsymbol{r}_0, -\boldsymbol{\Omega}_0; \boldsymbol{r}, -\boldsymbol{\Omega}) \end{aligned} \tag{5.54}$$

The result (5.54) depends on the fact that, according to (5.46), the averaged macroscopic cross-section $\bar{\Sigma}_c(\boldsymbol{r}, \boldsymbol{r}_0)$ is invariant with respect to interchange of $\boldsymbol{r}$ and $\boldsymbol{r}_0$. It is a statement of the reciprocity theorem which equates the angular density at $\boldsymbol{r}$ in the direction $\boldsymbol{\Omega}$, due to a point source at $\boldsymbol{r}_0$ radiating in the direction $\boldsymbol{\Omega}_0$, with the angular density at $\boldsymbol{r}_0$ in the direction $-\boldsymbol{\Omega}_0$, due to a point source at $\boldsymbol{r}$ radiating in the direction $-\boldsymbol{\Omega}$. The result holds quite generally, even in a scattering medium, provided those conditions are fulfilled which are necessary for time reversal invariance to apply.

It is evident from (5.42) that the solution of the transport equation is uniquely determined provided the sources are given functions of time, and the initial angular density $n(\boldsymbol{r}, \boldsymbol{\Omega}, 0)$ is known. However as a general rule it is not required to determine $n(\boldsymbol{r}, \boldsymbol{\Omega}, t)$ at all points in space but only within a finite volume V. In this case the contribution of sources outside V can always be represented in terms of effective sources on the boundary surface S. It is then possible to prove the following uniqueness theorem: the angular density $n(\boldsymbol{r}, \boldsymbol{\Omega}, t)$ is uniquely determined within V by the sources $S(\boldsymbol{r}, \boldsymbol{\Omega}, t)$ and the initial angular density $n(\boldsymbol{r}, \boldsymbol{\Omega}, 0)$ within V, together with the angular density incident on S. The theorem holds in scattering media with $c(\boldsymbol{r})<1$, and for finite volumes when $c(\boldsymbol{r})=1$.

5.3.1.4 Integral equations in a scattering medium

We may obtain an expression for the angular density appropriate to a scattering medium in the stationary limit directly from (5.49) by replacing the source term $S(\boldsymbol{r}, \boldsymbol{\Omega})$ with the complete expression for the effective sources including neutron regeneration, which is taken from (5.39). In this case the stationary angular density is given by

$$n(\boldsymbol{r}, \boldsymbol{\Omega}) = \frac{1}{v}\int S(\boldsymbol{r} - R\boldsymbol{\Omega}) \exp\{-\bar{\Sigma}_t(\boldsymbol{r}, \boldsymbol{r} - R\boldsymbol{\Omega})R\}\, dR$$

$$+ \int c(\boldsymbol{r} - R\boldsymbol{\Omega})\Sigma_t(\boldsymbol{r} - R\boldsymbol{\Omega}) \exp\{-\bar{\Sigma}_t(\boldsymbol{r} - R\boldsymbol{\Omega})R\} n(\boldsymbol{r} - R\boldsymbol{\Omega}, \boldsymbol{\Omega}'')$$

$$\times f(\boldsymbol{\Omega}, \boldsymbol{\Omega}'')\, d\Omega''\, dR \qquad (5.55)$$

This result is evidently an integral equation for $n(\boldsymbol{r}, \boldsymbol{\Omega})$, rather than a solution for $n(\boldsymbol{r}, \boldsymbol{\Omega})$; to make further progress it is necessary to make some assumption about the properties of the scattering function $f(\boldsymbol{\Omega}, \boldsymbol{\Omega}')$. It is simplest to assume that scattering is isotropic in the laboratory system in which case (5.37) applies and (5.55) reduces to

$$n(\boldsymbol{r}, \boldsymbol{\Omega}) = \int \left[\frac{S(\boldsymbol{r} - R\boldsymbol{\Omega}, \boldsymbol{\Omega})}{v} + \frac{1}{4\pi} (c(\boldsymbol{r} - R\boldsymbol{\Omega})\Sigma_t(\boldsymbol{r} - R\boldsymbol{\Omega}) n(\boldsymbol{r} - R\boldsymbol{\Omega})) \right.$$

$$\left. \times \exp\{-\bar{\Sigma}_t(\boldsymbol{r}, \boldsymbol{r} - R\boldsymbol{\Omega})R\} \right] dR \qquad (5.56)$$

Integrating over the angles this gives a final integral equation for the neutron density

$$n(\boldsymbol{r}) = \int \left[\frac{S(\boldsymbol{r}', \boldsymbol{R}/R)}{v} + \frac{1}{4\pi} (c(\boldsymbol{r}')\Sigma_t(\boldsymbol{r}') n(\boldsymbol{r}')) \exp\{-\bar{\Sigma}_t(\boldsymbol{r}, \boldsymbol{r}')R\} \right] \frac{d^3 r'}{R^2} \qquad (5.57)$$

It should be observed that obtaining a solution of the integral equation (5.57) for $n(\boldsymbol{r})$ also leads immediately to a solution for $n(\boldsymbol{r}, \boldsymbol{\Omega})$ by application of (5.56).

At this stage it may be remarked that the solution of the time-dependent transport equation (5.39), subject to a specified boundary value $n(\boldsymbol{r}, \boldsymbol{\Omega}, 0)$ at zero time, may be reduced to that of the time-independent equation combined with one additional integration. This may be demonstrated by introducing the Laplace transforms

$$\tilde{n}(\boldsymbol{r}, \boldsymbol{\Omega}, p) = \int_0^\infty e^{-pt} n(\boldsymbol{r}, \boldsymbol{\Omega}, t)\, dt \qquad (5.58a)$$

$$\tilde{S}(\boldsymbol{r}, \boldsymbol{\Omega}, t) = \int_0^\infty e^{-pt} S(\boldsymbol{r}, \boldsymbol{\Omega}, t)\, dt \qquad (5.58b)$$

Thus, multiplying both sides of (5.39) by e^{-pt} and integrating from $t = 0$ to

$t=\infty$ we obtain the relation

$$\boldsymbol{\Omega}\cdot\nabla\tilde{n}(\boldsymbol{r},\boldsymbol{\Omega},p)+(\Sigma_t(\boldsymbol{r})+p/v)\tilde{n}(\boldsymbol{r},\boldsymbol{\Omega},p)$$
$$=\frac{\tilde{S}(\boldsymbol{r},\boldsymbol{\Omega},p)+n(\boldsymbol{r},\boldsymbol{\Omega},0)}{v}+\int\frac{c(\boldsymbol{r})}{1+p/\Sigma_t(\boldsymbol{r})v}\left(\Sigma_t(\boldsymbol{r})+\frac{p}{v}\right)\tilde{n}(\boldsymbol{r},\boldsymbol{\Omega}',p)f(\boldsymbol{\Omega},\boldsymbol{\Omega}')\,\mathrm{d}\Omega' \tag{5.59}$$

Formally this equation for $\tilde{n}(\boldsymbol{r},\boldsymbol{\Omega},p)$ is identical with (5.39) with the time-dependence suppressed, and the following replacements made

$$\Sigma_t(\boldsymbol{r})\to\Sigma_t(\boldsymbol{r})+p/v \tag{5.60a}$$

$$c(\boldsymbol{r})\to c(\boldsymbol{r})/(1+p/\Sigma_t(\boldsymbol{r})v) \tag{5.60b}$$

$$S(\boldsymbol{r},\boldsymbol{\Omega})\to\tilde{S}(\boldsymbol{r},\boldsymbol{\Omega},p)+n(\boldsymbol{r},\boldsymbol{\Omega},0) \tag{5.60c}$$

It follows from (5.60a) that $\Sigma_t(\boldsymbol{r})$ is supplemented by an additional pure absorption term p/v; that this term does indeed correspond to pure absorption follows from (5.60b) which shows that $c(\boldsymbol{r})$ is reduced in exactly the correct ratio according to its definition in (5.34)–(5.35). This absorption effect is also described as 'time absorption' for the reason that, if the solution of (5.59) contains an attenuating factor $\approx\exp\{-(\Sigma_t(\boldsymbol{r})+p/v)R\}$, the corresponding term in the time-dependent problem is evaluated at the retarded time $(t-R/v)$, expressing the fact that the density at a point a distance R from the source may appear attenuated or amplified according as the source strength is increasing or decreasing in time.

When the solution of the time-independent equation (5.59) for $\tilde{n}(\boldsymbol{r},\boldsymbol{\Omega},p)$ has been obtained, the time-dependent angular density $n(\boldsymbol{r},\boldsymbol{\Omega},t)$ may then be recovered by application of the Mellin inversion theorem

$$n(\boldsymbol{r},\boldsymbol{\Omega},t)=\frac{1}{2\pi\mathrm{i}}\int_{a-i\infty}^{a+i\infty}\tilde{n}(\boldsymbol{r},\boldsymbol{\Omega},p)\,\mathrm{e}^{pt}\,\mathrm{d}p \tag{5.61}$$

where the integral is computed along a line parallel to the imaginary axis in the p-plane, passing through any point $(a, 0)$, which lies to the right of all singularities in $\tilde{n}(\boldsymbol{r},\boldsymbol{\Omega},p)$.

5.3.2 Neutron diffusion

5.3.2.1 Fick's law of diffusion

In a uniform infinite medium the general time-dependent transport equation (5.39) for the angular density is

$$\frac{\partial n}{\partial t}(\boldsymbol{r},\boldsymbol{\Omega},t)+v\boldsymbol{\Omega}\cdot\nabla n(\boldsymbol{r},\boldsymbol{\Omega},t)+\Sigma_t v n(\boldsymbol{r},\boldsymbol{\Omega},t)$$
$$=S(\boldsymbol{r},\boldsymbol{\Omega},t)+c\Sigma_t v\int n(\boldsymbol{r},\boldsymbol{\Omega}',t)f(\boldsymbol{\Omega},\boldsymbol{\Omega}')\,\mathrm{d}\Omega' \tag{5.62}$$

where the neutron density $n(\boldsymbol{r}, t)$ satisfies the equation obtained by integrating (5.62) over $\boldsymbol{\Omega}$, i.e.

$$\frac{\partial n}{\partial t}(\boldsymbol{r}, t)+\nabla \cdot \boldsymbol{j}(\boldsymbol{r}, t)+\Sigma_t(1-c)\varphi(\boldsymbol{r}, t)=S(\boldsymbol{r}, t) \tag{5.63}$$

The current divergence $\nabla \cdot \boldsymbol{j}(\boldsymbol{r}, t)$ is evidently the troublesome term in (5.63) and this may be simplified considerably if we assume the validity of Fick's law of diffusion

$$\boldsymbol{j}(\boldsymbol{r}, t)=-D\nabla\varphi(\boldsymbol{r}, t) \tag{5.64}$$

where D is the diffusion coefficient for flux, and has the dimensions of length. The physical basis for Fick's law is the recognition that neutrons flow away from regions of maximum concentration, and it is a reasonable expectation that the current should be proportional to the gradient of the density. In the one-speed approximation density and flux are the same to within the factor v. When Fick's law is valid (5.63) can be written as a linear partial differential equation for $\varphi(\boldsymbol{r}, t)$

$$\frac{1}{v}\frac{\partial\varphi(\boldsymbol{r}, t)}{\partial t}=D\nabla^2\varphi(\boldsymbol{r}, t)-\Sigma_t(1-c)\varphi(\boldsymbol{r}, t)+S(\boldsymbol{r}, t) \tag{5.65}$$

which, apart the inhomogeneous source term $S(\boldsymbol{r}, t)$, is similar in structure to the time-dependent Schrödinger equation.

It is the objective of the following analysis to establish conditions under which Fick's law (5.64) is a good approximation, and to find a suitable expression for D in terms of the parameters Σ_t and c. We do this by comparing some simple solutions of (5.65) for the time-independent case with the corresponding exact solutions of (5.56) and (5.57), and identifying the circumstances in which these solutions display maximal correspondence. The simplest case to consider is the source-free time-independent solution of (5.65) which in this case reduces to the scalar Helmholtz equation

$$\nabla^2\varphi(\boldsymbol{r}, t)+(\Sigma_t(c-1)/D)\varphi(\boldsymbol{r}, t)=0 \tag{5.66}$$

A solution of (5.66) is always given by

$$\varphi(\boldsymbol{r}, t)=\mathrm{e}^{\pm i\boldsymbol{k}\cdot\boldsymbol{r}} \qquad k>0 \qquad c>1 \qquad D(c)=\frac{\Sigma_t(c-1)}{k^2}>0$$

$$\varphi(\boldsymbol{r}, t)=\mathrm{e}^{\pm\boldsymbol{\kappa}\cdot\boldsymbol{r}} \qquad \kappa>0 \qquad c<1 \qquad D(c)=\frac{\Sigma_t(1-c)}{\kappa^2}>0 \tag{5.67}$$

For $c>1$ any linear combination of the oscillatory solutions is a solution and to make the solution unique some additional conditions must be imposed. For $c<1$, the solution $\varphi(\boldsymbol{r}, t)=\exp(\boldsymbol{\kappa}\cdot\boldsymbol{r})$ must be rejected if we set the requirement that the solution should be finite at infinity, i.e. $n(\boldsymbol{r})\exp(-\lambda|\boldsymbol{r}|)\to 0$ as $|\boldsymbol{r}|\to\infty$ for all $\lambda>0$.

5.3.2.2 *Diffusion in a uniform infinite medium*

In a uniform infinite source-free medium the relevant transport equations (5.56) and (5.57) reduce to

$$n(\boldsymbol{r}, \boldsymbol{\Omega}) = \frac{c}{4\pi} \int n(\boldsymbol{r} - R\boldsymbol{\Omega})\, \mathrm{e}^{-\Sigma_t R} \mathrm{d}R \tag{5.68}$$

and

$$n(\boldsymbol{r}) = \frac{c\Sigma_t}{4\pi} \int n(\boldsymbol{r}')\, \mathrm{e}^{-\Sigma_t R} \frac{\mathrm{d}^3 r'}{R^2} \tag{5.69}$$

Since c and Σ_t are independent of position, the translational invariance of (5.68)–(5.69) implies that, if $n(\boldsymbol{r})$ is a solution, so also is $n(\boldsymbol{r} - \boldsymbol{r}_0)$ for all $\boldsymbol{r}_0$. Thus any derivative $(\partial/\partial x_i) n(\boldsymbol{r})$ is a solution, which is either genuinely distinct from $n(\boldsymbol{r})$, or else a multiple of $n(\boldsymbol{r})$. In the former case there is an infinity of distinct solutions which seems unlikely, while in the latter case $n(\boldsymbol{r})$ is an exponential and must satisfy a linear differential equation with constant coefficients. We therefore test this last possibility by making the trial substitution in (5.69)

$$n(\boldsymbol{r}) = \mathrm{e}^{i\boldsymbol{k}\cdot\boldsymbol{r}} \qquad \boldsymbol{k} = k\boldsymbol{\omega} \tag{5.70}$$

where $\boldsymbol{\omega}$ is a unit vector.

We then find that (5.70) does indeed satisfy (5.69) provided k is a root of the equation

$$\Psi(k) = 1 - c \tan^{-1}(k/\Sigma_t)/(k/\Sigma_t) = 0 \tag{5.71}$$

The function $\Psi(k)$ is real only if either (a) k is real or (b) k is pure imaginary. Also since $\Psi(k)$ is an even function of k (an expression of the fact that isotropic scattering satisfies reflection invariance), the roots occur in pairs, symmetrically placed with respect to the origin, either on the real axis for $c > 1$, or on the imaginary axis for $c < 1$. For $c = 1$ there is a double root at the origin. These results are illustrated in figure 5.6(a). If we characterize these solutions by $\pm k_0$ $(c > 1)$ and $\pm i\kappa_0$ $(c \leqslant 1)$ then the general solution of (5.69) may be written as

$$n(\boldsymbol{r}) = \int \theta(\boldsymbol{\omega})\, \mathrm{e}^{ik_0\boldsymbol{\omega}\cdot\boldsymbol{r}}\, \mathrm{d}\boldsymbol{\omega} \tag{5.72}$$

where $\theta(\boldsymbol{\omega})$ is an arbitrary function of the unit vector $\boldsymbol{\omega}$. If, as in section 5.3.2.1, we exclude solutions which grow exponentially at infinity the solution $\mathrm{e}^{\kappa_0\boldsymbol{\omega}\cdot\boldsymbol{r}}$ $(c < 1)$ is omitted.

By calculating $\nabla^2 n(\boldsymbol{r})$ directly from (5.72) and setting $\varphi(\boldsymbol{r}) = v n(\boldsymbol{r})$ we can show immediately that $\varphi(\boldsymbol{r})$ satisfies the scalar Helmholtz equation

$$(\nabla^2 + k_0^2)\varphi(\boldsymbol{r}) = 0 \tag{5.73}$$

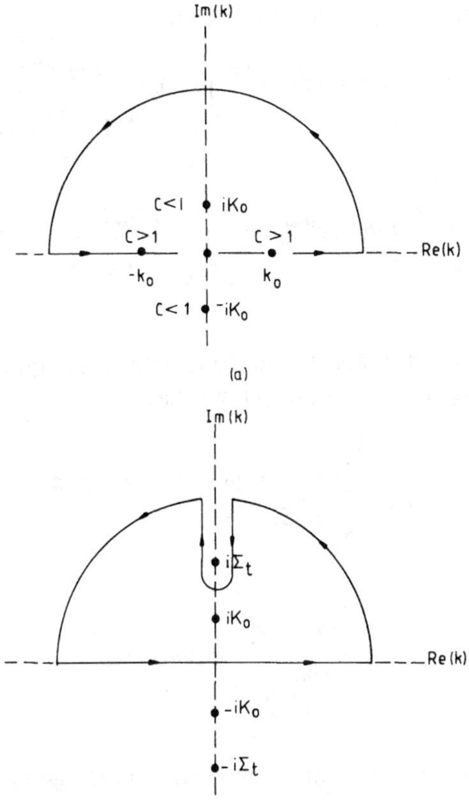

Figure 5.6. (a) Zeros of the function $\Psi(k)$ which determine the number density $n(\boldsymbol{r})$ in a uniform source-free medium (cf. equation (5.71)). The medium is multiplying (non-multiplying) according as $c>1 (c<1)$. (b) Poles and branch points of the integrand determining $n(\boldsymbol{r})$ for a unit isotropic point source at the origin of a uniform medium with $c<1$ (cf. equation (5.83)).

a result which suggests the possibility that, in a uniform infinite source-free medium, Fick's law (5.64) holds exactly, with diffusion coefficient D given by (5.67). We prove this hypothesis directly below.

Substituting (5.72) directly into (5.68) the angular density $n(\boldsymbol{r}, \boldsymbol{\Omega})$ is found to be

$$n(\boldsymbol{r}, \boldsymbol{\Omega}) = \frac{c\Sigma_t}{4\pi} \int \theta(\boldsymbol{\omega})\, e^{ik_0\boldsymbol{\omega}\cdot\boldsymbol{r}} d\omega \int_0^\infty e^{-\Sigma_t R - ik_0\boldsymbol{\omega}\cdot\boldsymbol{R}}\, dR$$

$$= \frac{c}{4\pi} \int \frac{\theta(\boldsymbol{\omega})\, e^{ik_0\boldsymbol{\omega}\cdot\boldsymbol{r}}\, d\omega}{1 + \dfrac{ik_0}{\Sigma_t}\boldsymbol{\omega}\cdot\boldsymbol{\Omega}} \tag{5.74}$$

The total current density is therefore given by

$$j(r)=v\int \boldsymbol{\Omega} n(r,\boldsymbol{\Omega})\,\mathrm{d}\Omega=\frac{cv}{4\pi}\int \boldsymbol{\Omega}\,\mathrm{d}\Omega\int\frac{\theta(\omega)\,\mathrm{e}^{ik_0\omega\cdot r}}{1+\dfrac{ik_0}{\Sigma_t}\,\omega\cdot\boldsymbol{\Omega}}\,\mathrm{d}\omega$$

$$=\frac{cv}{4\pi}\int\theta(\omega)\,\mathrm{e}^{ik_0\omega\cdot r}\,\mathrm{d}\omega\int\frac{\boldsymbol{\Omega}\,\mathrm{d}\Omega}{1+\dfrac{ik_0}{\Sigma_t}\,\omega\cdot\boldsymbol{\Omega}} \tag{5.75}$$

Since ω is the only vector (apart from the variable of integration $\boldsymbol{\Omega}$) appearing in the vector integral in (5.75), it follows that

$$\int\frac{\Omega_i\,\mathrm{d}\Omega}{1+\dfrac{ik_0}{\Sigma_t}\,\omega\cdot\boldsymbol{\Omega}}=\beta\omega_i \qquad i=x,\,y,\,z \tag{5.76}$$

and

$$\int\frac{\boldsymbol{\Omega}\cdot\omega\,\mathrm{d}\Omega}{1+\dfrac{ik_0}{\Sigma_t}\,\omega\cdot\boldsymbol{\Omega}}=\beta\omega\cdot\omega=\beta \tag{5.77}$$

The integral in (5.77) is elementary and immediately gives on evaluation

$$\beta=\frac{c\Sigma_t}{ik_0}\left\{1-\frac{\Sigma_t}{2ik_0}\ln\left(\frac{1+ik_0/\Sigma_t}{1-ik_0/\Sigma_t}\right)\right\}$$

$$\equiv\frac{c\Sigma_t}{ik_0}\left\{1-\frac{\tan^{-1}(k_0/\Sigma_t)}{(k_0/\Sigma_t)}\right\}=\frac{\Sigma_t}{ik_0}(c-1) \tag{5.78}$$

where we have made use of (5.71).

The current is now given from (5.75), (5.76) and (5.78) by

$$j(r)=\frac{v\Sigma_t}{ik}(c-1)\int\theta(\omega)\omega\,\mathrm{e}^{ik_0\omega\cdot r}\,\mathrm{d}\omega$$

$$=-\frac{\Sigma_t}{k^2}(c-1)\nabla\left\{\int v\theta(\omega)\,\mathrm{e}^{ik_0\omega\cdot r}\,\mathrm{d}\omega\right\}$$

$$=-D(c)\nabla\varphi(r) \tag{5.79}$$

where $D(c)$ is given by (5.67). Thus Fick's law (5.64) is proved for transport in a uniform infinite source-free medium. To obtain estimates for $D(c)$ we have to examine the solution of (5.71) in a little more detail and we find

(i) $c \ll 1$ (strong absorption)

$$\kappa_0^2 \simeq \Sigma_t^2 \qquad D(c) \simeq \Sigma_t^{-1} \simeq \Sigma_c^{-1}$$

(ii) $|c-1| \ll 1$ (weak absorption, weak fission) (5.80)

$$k_0^2 = -\kappa_0^2 \simeq 3(c-1)\Sigma_t^2 \qquad D(c) \approx \tfrac{1}{3}\Sigma_t^{-1} \simeq \tfrac{1}{3}\Sigma_s^{-1}$$

(iii) $c \gg 1$ (strong fission)

$$k_0^2 \sim \left(\frac{\pi}{2} c\Sigma_t\right)^2 \qquad D(c) \simeq \frac{4}{\pi^2 c} \Sigma_t^{-1}$$

The conclusion appears to be that, at large distances from sources and boundaries, and assuming isotropic scattering, Fick's law is of general applicability. To see which of the three regimes identified in (5.80) most closely approximates in general to the situation described by diffusion theory, we need to investigate the properties of the transport solutions in the neighbourhood of a source.

In the case of a unit isotropic point source at the origin of a uniform infinite medium the integral equation to be satisfied by $n(\boldsymbol{r})$ is given from (5.57)

$$n(\boldsymbol{r}) = \int \left[\frac{\delta^3(\boldsymbol{r}')}{v} + \frac{c\Sigma_t}{4\pi} n(\boldsymbol{r}') \right] e^{-\Sigma_t R} \frac{d^3 r'}{R^2} \tag{5.81}$$

We can solve (5.81) by taking the Fourier transform of both sides with the result

$$\begin{aligned} \tilde{n}(\boldsymbol{k}) &\equiv \int e^{-i\boldsymbol{k}\cdot\boldsymbol{r}} n(\boldsymbol{r})\, d^3 r = [(v\Sigma_t)^{-1} + c\tilde{n}(\boldsymbol{k})] \frac{\tan^{-1}(k/\Sigma_t)}{k/\Sigma_t} \\ &= \frac{(v\Sigma_t)^{-1} \tan^{-1}(k/\Sigma_t)/(k/\Sigma_t)}{\Psi(k)} \end{aligned} \tag{5.82}$$

where $\Psi(k)$ is defined in (5.71). Noting that $\tilde{n}(\boldsymbol{k})$ is a function of k only we can now invert (5.82) to find $n(r)$, i.e.

$$\begin{aligned} n(r) &\equiv \left(\frac{1}{2\pi}\right)^3 \int e^{-i\boldsymbol{k}\cdot\boldsymbol{r}} \tilde{n}(k)\, d^3 k = \frac{1}{2\pi^2 r} \int_0^\infty \tilde{n}(k) k \sin kr \, dk \\ &= \frac{1}{2\pi r} \times \frac{1}{2\pi i} \int_{-\infty}^{\infty} k\tilde{n}(k) k \sin kr \, dk \end{aligned} \tag{5.83}$$

The integrand in (5.83), considered as a function of complex k, has branch points at $k = \pm i\Sigma_t$ and poles at the zeros of $\Psi(k)$ discussed in section 5.3.2.1 above. If $c < 1$ then the relevant zero of $\Psi(k)$ lies at $k = i\kappa_0$ and the integral is a single-valued function of k in the upper half-plane cut along the imaginary axis from $\mathrm{i}\Sigma_t$ to $\mathrm{i}\infty$. This is shown in figure 5.6(b). There are therefore two contributions to the integral, coming from the residue at the pole, and from the integral around the branch point on the imaginary axis. The result is (Case *et al* 1953)

$$vn(r)=\varphi(r)=\frac{1}{4\pi r\Sigma_t}\left[-\frac{\partial\kappa_0^2}{\partial c}\,\mathrm{e}^{-\kappa_0 r}+\int_0^1 g(c,\mu)\,\mathrm{e}^{-\Sigma_t r/\mu}\frac{\mathrm{d}\mu}{\mu^2}\right] \tag{5.84}$$

where

$$g(c,\mu)=\left[(1-c\mu\tan^{-1}\mu)^2+\left(\frac{\pi}{2}c\mu\right)^2\right]^{-1} \tag{5.85}$$

It is clear that the pole term in (5.84) is dominant at large distances from the source when $\kappa_0\ll\Sigma_t$, i.e. when $|c-1|\ll 1$ (cf. 5.80). In these conditions we may therefore define an asymptotic flux

$$\varphi_{as}(r)=\frac{1}{4\pi r}\left[-\frac{1}{\Sigma_t}\frac{\partial\kappa_0^2}{\partial c}\right]\mathrm{e}^{-\kappa_0 r} \tag{5.86}$$

which is indeed a solution of the time-independent diffusion equation with a point source at the origin. According to Fick's law (5.64) the asymptotic current will be

$$\boldsymbol{j}_{as}(r)=-D\,\nabla\varphi_{as}(r)=\frac{D}{4\pi}\left\{-\frac{1}{\Sigma_t}\frac{\partial\kappa_0^2}{\partial c}\right\}\left\{\frac{\kappa_0 r+1}{r^2}\right\}\mathrm{e}^{-\kappa_0 r}\,\hat{\boldsymbol{r}} \tag{5.87}$$

If we now evaluate the total number of neutrons per second transported from the source by the asymptotic current we find this number to be

$$\lim_{r\to 0} j_r 4\pi r^2=\frac{D(c)}{\Sigma_t}\frac{\partial\kappa_0^2}{\partial c}\simeq\frac{1}{3\Sigma_t^2}3\Sigma_t^2=1 \tag{5.88}$$

Thus the asymptotic current carries away virtually all the neutrons radiated from this source when $|c-1|\ll 1$, a condition which we may take as the criterion that diffusion theory should apply even in the presence of sources. Thus, according to (5.80b), in a weakly capturing medium without regeneration, we have the relation

$$\kappa_0=3(1-c)\Sigma_t^2=3\Sigma_c\Sigma_t \qquad D=1/3\Sigma_t \tag{5.89}$$

The inverse of κ_0 is known as the diffusion length and is usually denoted by L. It may be interpreted in terms of the mean square distance $\langle r^2\rangle_c$ that a neutron diffuses from its point of emission to the point at which it is captured, according to the relation

$$\langle r^2\rangle_c = \int_0^\infty r^2(4\pi r^2\Sigma_c\varphi(r))\,dr \Bigg/ \int_0^\infty (4\pi r^2\Sigma_c\varphi(r))\,dr$$

$$= 6\kappa_0^{-4}/\kappa_0^{-2} = 6\kappa_0^{-2} = 6L^2 \tag{5.90}$$

Combining (5.89) with (5.90) we then have the results

$$L^2 = \kappa_0^{-2} = D/\Sigma_c = \frac{1}{6}\langle r^2\rangle \tag{5.91}$$

The diffusion length L for a thermal neutron in ordinary water is about 3 cm, whereas it is 50 cm in graphite and 100 cm in heavy water ($D_2 0$).

When $c \ll 1$ (strong capture), according to (5.80(i)) $\partial\kappa_0^2/\partial c \simeq 0$ and there is no asymptotic flux. At the other extreme $c \gg 1$ it is impossible to satisfy the source condition (5.88) so the diffusion approximation cannot be applied. However, with $c > 1$ but $(c-1) \ll 1$ diffusion theory does apply although in this case the poles of $\tilde{n}(k)$ move onto the real axis. Thus there is no true stationary solution in an infinite medium because the pole term $\exp(\pm ik_0 r)$ oscillates infinitely rapidly at infinity. However, these static 'solutions' may be used to construct time-dependent solutions as described in section 5.3.1.3. These solutions describe incoming waves, outgoing waves or standing waves depending on which path of integration is prescribed for avoiding the poles on the real k-axis in evaluating the integral (5.83).

5.3.2.3 Time-dependent diffusion in anisotropic conditions

To establish criteria for the validity of the diffusion approximation under the more general conditions of explicit time-dependence with anisotropic source and scattering functions, we retain the assumption that the scattering medium is uniform and adopt equations (5.62)–(5.63) as point of departure. Thus to eliminate the current divergence term from (5.63) we require another relation analogous to Fick's law (5.64), and this is obtained by multiplying (5.62) by $\mathbf{\Omega}$ and integrating over all $\mathbf{\Omega}$. The result of this operation is

$$\frac{1}{v}\frac{\partial}{\partial t}\boldsymbol{j}(\boldsymbol{r},t) + \int \mathbf{\Omega}(\mathbf{\Omega}\cdot\nabla)vn(\boldsymbol{r},\mathbf{\Omega},t)\,d\Omega + \Sigma_t\boldsymbol{j}(\boldsymbol{r},t)$$

$$= \boldsymbol{S}_1(\boldsymbol{r},t) + \Sigma_t c\int \mathbf{\Omega}vn(\boldsymbol{r},\mathbf{\Omega}',t)f(\mathbf{\Omega},\mathbf{\Omega}')\,d\Omega\,d\Omega' \tag{5.92}$$

where

$$\boldsymbol{S}_1(\boldsymbol{r},t) = \int \mathbf{\Omega}S(\boldsymbol{r},\mathbf{\Omega},t)\,d\Omega \tag{5.93}$$

is a vector pointing in the direction of maximum source emissivity.

To make further progress with (5.92) we need to introduce some approximation which permits us to express the integrals in terms of the neutron density and current. We therefore assume that all angular-dependent terms in $n(\boldsymbol{r}, \boldsymbol{\Omega}, t)$ may be neglected, except those which are linear in $\boldsymbol{\Omega}$, i.e.

$$n(\boldsymbol{r}, \boldsymbol{\Omega}, t)=\frac{1}{4\pi}[n(\boldsymbol{r}, t)+3\boldsymbol{\Omega}\cdot\boldsymbol{j}(\boldsymbol{r}, t)/v] \tag{5.94}$$

a relation which is consistent with the definitions (5.28)–(5.31). To maintain consistency with (5.94) we also assume that

$$f(\boldsymbol{\Omega}, \boldsymbol{\Omega}')\equiv f(\boldsymbol{\Omega}, \boldsymbol{\Omega}')=\frac{1}{4\pi}[1+3\bar{\mu}_s(\boldsymbol{\Omega}\cdot\boldsymbol{\Omega}')] \tag{5.95}$$

where $\bar{\mu}_s$ is the mean value of the cosine of the angle of scatter in the laboratory which is defined by

$$\bar{\mu}_s=\int(\boldsymbol{\Omega}\cdot\boldsymbol{\Omega}')f(\boldsymbol{\Omega}\cdot\boldsymbol{\Omega}')\,d\Omega'=\int(\boldsymbol{\Omega}\cdot\boldsymbol{\Omega}')f(\boldsymbol{\Omega}\cdot\boldsymbol{\Omega}')\,d\Omega \tag{5.96}$$

When the scattering is isotropic in the centre of mass system, $\bar{\mu}_s$ take the value $2/3A$, where A is the mass number of the scattering nucleus (cf. section 5.4.1). With the substitution (5.94)–(5.95), equation (5.92) reduces to

$$\frac{1}{v}\frac{\partial}{\partial t}\boldsymbol{j}(\boldsymbol{r}, t)+\frac{1}{3}\nabla\varphi(\boldsymbol{r}, t)+\Sigma_t(1-c\bar{\mu}_s)\boldsymbol{j}(\boldsymbol{r}, t)=\boldsymbol{S}_1(\boldsymbol{r}, t) \tag{5.97}$$

The approximation introduced in (5.94)–(5.95) is described as the 'P_1-approximation' since it amounts to retention of terms $l=0$ and $l=1$ in the expansion of $n(\boldsymbol{r}, \boldsymbol{\Omega}, t)$ and $f(\boldsymbol{\Omega}, \boldsymbol{\Omega}')$ in Legendre polynomials (Marshak *et al* 1949).

By comparing (5.64) with (5.97) we see that the latter equation describes a generalization of Fick's law which introduces two new terms. The term $\boldsymbol{S}_1(\boldsymbol{r}, t)$ may evidently be interpreted as a supplementary current which originates in the source anisotropy; as we shall see below it causes no particular difficulty. However, the time derivative of the current is awkward because its inclusion destroys the simplicity of the diffusion equation. We therefore need to identify conditions under which this term may be neglected. To this end we integrate (5.97) with respect to the time giving the result

$$\boldsymbol{j}(\boldsymbol{r}, t)=\int_{-\infty}^{t}\exp[-v\Sigma_t(1-c\bar{\mu}_s)(t-t')]\left(-\frac{1}{3}\nabla\varphi(\boldsymbol{r}, t')+\boldsymbol{S}_1(\boldsymbol{r}, t')\right)v\,dt' \tag{5.98}$$

Since the main contribution to the integral in (5.98) comes from that part of the path of integration where $t'\simeq t$, it follows that, provided $\varphi(\boldsymbol{r}, t')$ and $\boldsymbol{S}_1(\boldsymbol{r}, t')$ are sensibly constant over the time scale $[v\Sigma_t(1-c\bar{\mu}_s)]^{-1}$, which, apart from the factor $(1-c\bar{\mu}_s)^{-1}$, is just the mean time interval between collisions, then these functions may be evaluated at $t'=t$ and removed from under the integral sign. The result is

$$\boldsymbol{j}(\boldsymbol{r},t)=\left(-\frac{1}{3}\nabla\varphi(\boldsymbol{r},t)+\boldsymbol{S}_1(\boldsymbol{r},t)\right)\Big/[\Sigma_t(1-c\bar{\mu}_s)]^{-1} \tag{5.99}$$

which is equivalent to (5.97) with the term $(1/v)(\partial/\partial t)\boldsymbol{j}(\boldsymbol{r},t)$ omitted.

Substitution of (5.99) into (5.63) then leads to the time-dependent diffusion equation (assuming $\Sigma_t c \simeq \Sigma_s$)

$$\frac{1}{v}\frac{\partial\varphi}{\partial t}(\boldsymbol{r},t)=D\nabla^2\varphi(\boldsymbol{r},t)-\Sigma_c\varphi(\boldsymbol{r},t)+S_d(\boldsymbol{r},t) \tag{5.100}$$

where

$$\begin{aligned}D&=[3\Sigma_t(1-\bar{\mu}_s)(1+\Sigma_c\bar{\mu}_s/\Sigma_t(1-\bar{\mu}_s))^{-1}\\&=\frac{1}{3}\lambda_{tr}(1+\bar{\mu}_s\Sigma_c\lambda_{tr})^{-1}\end{aligned} \tag{5.101}$$

is the diffusion coefficient, and the transport mean free path

$$\lambda_{tr}=(\Sigma_t(1-\bar{\mu}_s))^{-1} \tag{5.102}$$

is just the collision mean free path slightly amplified by the effect of anisotropic scattering, due to the persistence of velocity in the incident direction after the collision. It therefore represents the mean free path of the neutron along the direction of the initial velocity, or the mean free path between random changes in direction (Cohen 1962).

The source function

$$S_d(\boldsymbol{r},t)=S(\boldsymbol{r},t)-3D\nabla\cdot\boldsymbol{S}_1(\boldsymbol{r},t) \tag{5.103}$$

represents the effective source strength in the diffusion approximation. When transport corrections are taken into account as discussed above, D must be divided by $(1-\frac{4}{5}\Sigma_c/\Sigma_t)$ and $S_d(\boldsymbol{r},t)$ multiplied by the same factor. We note that the term $-3D\nabla\cdot\boldsymbol{S}_1(\boldsymbol{r},t)$, which is unchanged by these corrections, stands in relation to the leading term $S(\boldsymbol{r},t)$ as does the bound charge density $-\nabla\cdot\boldsymbol{P}(\boldsymbol{r},t)$ to the free charge density $\rho(\boldsymbol{r},t)$ in a dielectric medium with electric polarization $\boldsymbol{P}(\boldsymbol{r},t)$. It also exhibits many of the same properties; in particular it vanishes in a homogeneous medium and integrates to zero when confined to a finite volume. Thus this term tends to be important only near the boundary between two media.

In the case of the singular source function $S_d(\boldsymbol{r},t)$ defined by

$$S_d(\boldsymbol{r},t)=n\,\delta^3(\boldsymbol{r})\,\delta(t) \tag{5.104}$$

which corresponds to the release of n neutrons at the origin at zero time, the diffusion equation (5.100) may be solved directly using the Laplace transform technique described in section 5.3.2. In this example the transform function $\tilde{\varphi}(\boldsymbol{r}, p)$ is just the Green function for the scalar Helmholtz equation for which we already have the solution (cf. section 3.5.1)

$$\tilde{\varphi}(\boldsymbol{r}, p)=\left(\frac{n}{D}\right)\exp\left[-r\left(\frac{\Sigma_{\mathrm{c}}+p/v}{D}\right)^{1/2}\right]\Big/4\pi r \tag{5.105}$$

Since there is a branch point in $\tilde{\varphi}(\boldsymbol{r}, p)$ at $P=-v\Sigma_{\mathrm{c}}$, this function is regular in the complex p-plane cut along the real axis from $-v\Sigma_{\mathrm{c}}$ to $-\infty$ and the inverse transform $\varphi(\boldsymbol{r}, t)$ may be determined by application of the Mellin theorem in the usual way (cf. section 5.3.1.3). We then find that

$$\varphi(\boldsymbol{r}, t)=(nv)(4\pi Dvt)^{-3/2}\exp[-r^2/4Dvt-v\Sigma_{\mathrm{c}}t] \tag{5.106}$$

Thus $\varphi(\boldsymbol{r}, t)$ is a three-dimensional Gaussian distribution in space which decays exponentially in time proportional to $\exp[-v\Sigma_{\mathrm{c}}t]$, as neutrons emitted at zero time are progressively captured.

The mean square deviation from the origin $\langle r^2(t)\rangle$, of those neutrons which have survived to time t, is then given by

$$\frac{1}{6}\langle r^2(t)\rangle=Dvt \tag{5.107}$$

This quantity must not be confused with $\langle r^2\rangle_{\mathrm{c}}$, which is defined in (5.90) as the mean square distance travelled by a neutron from the point at which it is created to the point where it is ultimately captured. A plot of the distribution (5.106) describing the diffusion of thermal neutrons in graphite ($D\simeq 0.9$ cm) is shown in figure 5.7.

To fix the conditions under which $\varphi(\boldsymbol{r}, t)$ is slowly varying on the time scale $[v\Sigma_{\mathrm{t}}(1-c\bar{\mu}_{\mathrm{s}})]^{-1}$, which are required for the validity of (5.109)–(5.110), we may characterize the time variation of $\varphi(\boldsymbol{r}, t)$ by a time constant $\tau(t)$ defined by

$$\tau(t)^{-1}\equiv\frac{\mathrm{d}}{\mathrm{d}t}[\ln(\varphi(\boldsymbol{r}, t))]=-\frac{3}{2t}+r^2/4Dvt^2-v\Sigma_{\mathrm{c}} \tag{5.108}$$

Since in a weakly capturing medium $v\Sigma_{\mathrm{c}}\ll v\Sigma_{\mathrm{t}}$, it is clear that the necessary condition $\tau(t)^{-1}\ll v\Sigma_{\mathrm{t}}$ is satisfied whenever the following two conditions are separately satisfied,

$$\tfrac{3}{2}t\ll v\Sigma_{\mathrm{t}} \tag{5.109a}$$

$$r^2/4Dvt^2\ll v\Sigma_{\mathrm{t}} \tag{5.109b}$$

Condition (5.109a) clearly implies that (5.105) provides a valid description of affairs only at times t such that $vt\gg\tfrac{3}{2}\Sigma_{\mathrm{t}}^{-1}$, i.e. after sufficient time has elapsed for each neutron to have undergone a large number of collisions.

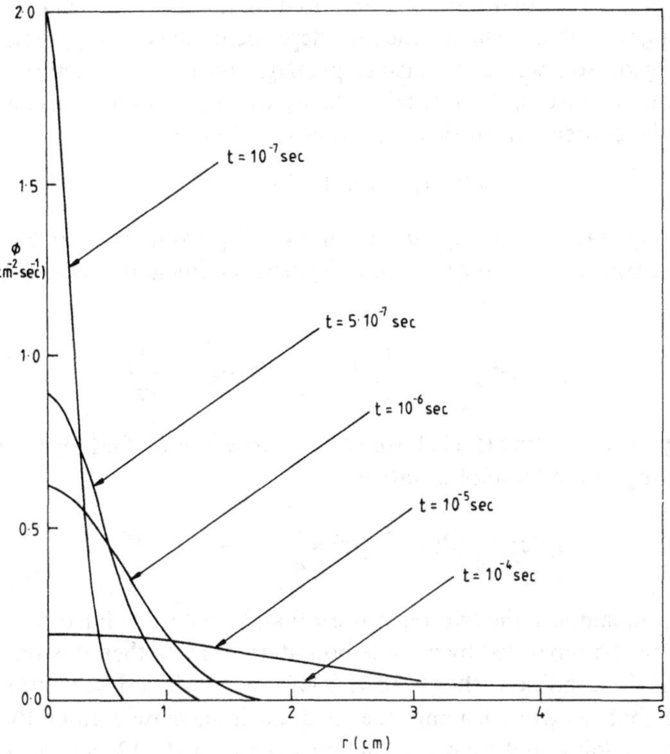

Figure 5.7. Diagram illustrating three-dimensional diffusion of thermal neutrons in graphite starting from an initial concentration at the origin.

When this state has been reached, then (5.109b) is also satisfied provided that

$$r \ll \sqrt{\langle r^2(t) \rangle} \qquad (5.110)$$

The limitation expressed by (5.110) is broadly what one might expect on the basis of the central limit theorem of statistics. For, if we visualize the progress of a neutron leaving the source as a three-dimensional random walk, with positive or negative spatial increments in any direction equally likely at a collision, then this theorem states that the resultant distribution in space may be approximated by a Gaussian, provided a large number of collisions has occurred. However, in the general case this approximation ceases to be reliable beyond one or two standard deviations from the mean, which is the essential content of (5.110).

5.3.2.4 Boundary conditions at a plane surface: Milne's problem

Since formally the time-dependent problem may be reduced to a time-independent problem following the prescription given in section 4.3.1.4, in determining the condition to be imposed on the flux at a plane boundary separating two diffusing media the time dependence may be suppressed. Also, in plane geometry we need consider propagation in one direction only, so that we may replace $n(\boldsymbol{r}, \boldsymbol{\Omega})$ by $(2\pi)^{-1}n(z, \mu)$, where $\mu = \cos\theta$ is the component of $\boldsymbol{\Omega}$ in the z-direction. In this case equation (5.94) becomes

$$n(z, \mu) = \tfrac{1}{2}(n(z) + 3\mu j_z(z)/v) \tag{5.111}$$

Multiplying this relation by μv and integrating from $\theta = 0$ to $\theta = \pi/2$ we obtain an expression for the current $j_z^+(z)$ propagating in the sense of positive z

$$j_z^+(z) = \frac{1}{4}\varphi(z) + \frac{1}{2}j_z(z) = \frac{1}{4}\varphi(z) - \frac{1}{2}D\frac{\partial\varphi_{(z)}}{\partial z} \tag{5.112}$$

Similarly by integrating (5.111) from $\theta = \pi/2$ to $\theta / = \pi$ we find for the current propagating in the sense of negative z

$$j_z^-(z) = \frac{1}{4}\varphi(z) - \frac{1}{2}j_z(z) = \frac{1}{4}\varphi(z) + \frac{1}{2}D\frac{\partial\varphi_{(z)}}{\partial z} \tag{5.113}$$

If we now suppose the two regions $z < 0$ and $z > 0$ to be filled by different diffusing media separated by a plane boundary at $z = 0$, then it is reasonable to impose the conditions that both $j_z^+(z)$ and $j_z^-(z)$ should be continuous at $z = 0$, so that neutrons leaving one medium immediately enter the other. Thus, first adding and then subtracting equations (5.112) and (5.113), we find that both $\varphi(z)$ and $D(\partial\varphi_{(z)}/\partial z)$ are continuous at the boundary, conditions which are closely analogous to those imposed respectively on the electrostatic potential and on the normal component of the electric displacement vector in electrostatic theory.

Evidently these boundary conditions cannot apply when only one part of the space contains a diffusing medium, and there is a source-free vacuum in the other. Suppose therefore that the diffusing medium is confined to the half-space $z > 0$, then, since no neutrons enter from the other half space $z < 0$, the boundary condition must be

$$j_z^+(0) = \frac{1}{4}\varphi(0) - \frac{1}{2}D\frac{\partial\varphi_{(0)}}{\partial z} = 0 \tag{5.114}$$

i.e.

$$\left.\frac{\partial}{\partial z}\ln(\varphi(z))\right|_{z=0} = \frac{1}{2D} = \frac{3}{2\lambda_{\text{tr}}} \tag{5.115}$$

The boundary condition (5.115) is usually imposed by the following neat device. The flux $\varphi(z)$ is linearly extrapolated into the region $z < 0$, so that it vanishes at the point $z = -z_0$, where z_0 is called the 'extrapolated length'.

Thus we have

$$\varphi(z)=\varphi(0)[1+z/z_0] \qquad z<0 \tag{5.116}$$

where z_0 is determined by the matching condition

$$\left.\frac{\partial}{\partial z}(\ln \varphi(z))\right|_{z=0}=\frac{1}{z_0}=\frac{3}{2\lambda_{\mathrm{tr}}} \tag{5.117}$$

The significance of the extrapolated length may be understood by reference to figure 5.8. According to (5.116) the boundary condition to be imposed is that $\varphi(z)$ should vanish at $z=-z_0$, although it must be emphasized that the relation (5.116) is never applied for $z<0$, but only for estimating the behaviour of $\varphi(z)$ for $z\geqslant 0$. The definition of the extrapolated length in (5.117) is strongly reminiscent of the scattering length introduced in section 2.5.1, which may be interpreted in terms of a linear extrapolation of the wavefunction, matched to the logarithmic derivative at the boundary of the potential.

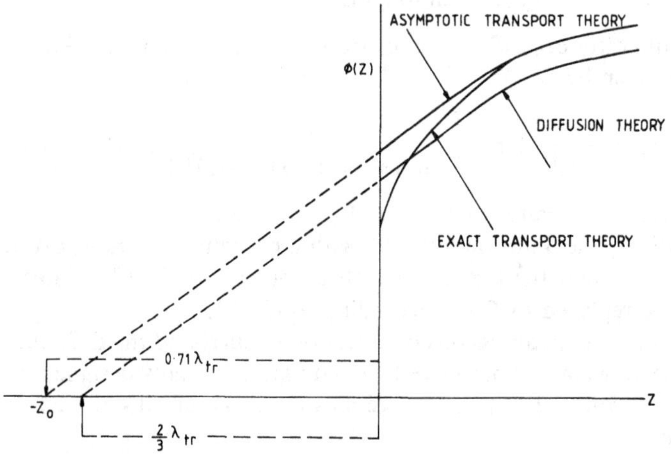

Figure 5.8. Illustrating the concept of 'extrapolated length' applied at a boundary between a diffusing medium and vacuum. The value $z_0=2\lambda_{\mathrm{tr}}/3$ is obtained in the diffusion approximation by imposing the condition that there is no return current at the boundary. An improved result $z_0=0.71\lambda_{\mathrm{tr}}$ is derived from the exact solution of Milne's problem (cf. equation (5.119)).

The value of the numerical coefficient in (5.117) may be improved by an appeal to transport theory where the problem of one-dimensional transport in a uniform semi-infinite half-space with isotropic scattering is also capable of exact solution (Case *et al* 1953, Case 1960, Case and Zweifel 1967). This is known as Milne's problem (LeCaine 1947, Elliott 1955, Kladnik and Kuscer 1961) and is formally equivalent to Kramers' problem in fluid mechanics which is concerned with fluid flow in the x-direction in the semi-infinite half-space $z>0$, with a velocity gradient in the z-direction which tends to a constant at infinity. In Kramers' problem the fluid velocity does not vanish on the plane $z=0$, but is characterized by a 'slip coefficient' which is the analogue of the extrapolation length z_0.

Milne's problem is solved by converting it into a uniform infinite medium problem through the introduction of a hypothetical anisotropic plane source $\mu S(\mu)\delta(z)$. The neutron density is then separated into asymptotic and non-asymptotic components as described in section 5.3.2.3, and the asymptotic component is made to vanish for $z<0$ by adding a suitable linear combination of the solutions of the corresponding homogeneous equation. Imposing the same condition on the non-asymptotic component then leads to an integral equation for $S(\mu)$ which is solved by the Wiener–Hopf technique (Noble 1950). The logarithmic derivative of the asymptotic flux at the boundary is then found to have the value

$$\frac{\partial}{\partial z}\ln(\varphi_{\text{as}}(x))\bigg|_{z=0}=\frac{\kappa_0(c)}{\tanh(\kappa_0(c)z_0(c))}\simeq(z_0(c))^{-1}\qquad |c-1|\ll 1 \quad (5.118)$$

The quantity $cz_0(c)\Sigma_t$ increases from a value of unity at $c=0$ to a value at $c=1$ given by

$$z_0(1)\Sigma_t=-\frac{1}{\pi}\int_0^1 \tan^{-1}\left[\frac{\pi}{(2/s)+\ln[(1-s)/(1+s)]}\right]ds=0.710446 \quad (5.119)$$

rising slowly thereafter to a value of $\frac{3}{4}$ as $c\to\infty$.

Identifying the diffusion flux $\varphi(z)$ with the asymptotic transport flux $\varphi_{\text{as}}(z)$ we then find that the factor $\frac{2}{3}$, which according to (5.117) relates z_0 to λ_{tr}, should be replaced by 0.710 according to (5.119).

It may prove advantageous on occasion to surround the diffusing medium by a reflecting blanket of source-free material specifically designed to enhance the returning flux. For this case we may define an albedo or back-scattering coefficient

$$\beta=j_z^-/j_z^+ \quad (5.120)$$

so that the extrapolated length (5.119) is replaced by

$$z_0=0.710\lambda_{\text{tr}}\left(\frac{1+\beta}{1-\beta}\right) \quad (5.121)$$

For example in an infinite plane reflector, where the flux falls off $\approx\exp(-\kappa_0 z)$ from the boundary, the albedo is given by

$$\beta=\left(\frac{1-2\kappa_0 D}{1+2\kappa_0 D}\right)\approx 1-4(\Sigma_c\lambda_{\text{tr}})^{1/2}\qquad \Sigma_c\ll\Sigma_t \quad (5.122)$$

A reflector may be treated as infinite if its thickness is twice the diffusion length or greater, and in materials commonly used as reflectors, e.g. water or graphite, values of β between 80% and 90% may be achieved with thermal neutrons. It follows from (5.121) that the use of a reflector makes the containing system appear larger than it actually is, by bringing about an effective decrease in the ratio of surface to volume. As a result both neutron leakage and fuel loading are reduced (cf. section 5.5.2).

5.4 Neutron moderation

5.4.1 Kinematics of elastic s-wave scattering

Preliminary to a discussion of the neutron moderation problem, i.e. the slowing down to thermal energies of fast fission neutrons by elastic scattering on nuclei, we need to establish relations connecting $\cos\theta$ and $\cos\theta_L$, the cosines of the scattering angles in the centre of mass (C) system and laboratory (L) system respectively, with the energy change in the scattering process as determined in the L-system, and with each other. We shall assume that the scattering nuclei are at rest in the L-system and that their masses are proportional to mass number A. These are reasonable approximations when applied to neutrons slowing down from ≈ 10 MeV to ≈ 0.025 eV, given also that the neglected packing fractions are in every case less than 1%. The essential basis for the subsequent calculations is that scattering in the C-system is s-wave and therefore isotropic. The analysis is made simple by treating the collision process in the C-system, where neutron and nucleus enter on an equal footing. However the important physical results are obtained on transforming back to the L-system.

If we denote by $\boldsymbol{v}_i$ $(i=n, A)$ velocities measured in the L-system, and by $\boldsymbol{V}_i$ the corresponding velocities measured in the C-system, then the transformation from the L-system to the C-system is given by the relation

$$\boldsymbol{V}_i=\boldsymbol{v}_i-\boldsymbol{v}_{\mathrm{cm}} \qquad v_{\mathrm{cm}}=v_{\mathrm{n}}/(A+1) \tag{5.123}$$

where $\boldsymbol{v}_{\mathrm{cm}}$ is the velocity of the centre of mass as determined in the L-system. The value of v_{cm} given in (5.123) is derived from the condition that the total momentum in the C-system be zero. The component of this momentum along the relative velocity of neutron and nucleus is thus given by

$$V_{\mathrm{n}}+AV_{\mathrm{A}}=(v_{\mathrm{n}}-v_{\mathrm{cm}})+(-Av_{\mathrm{cm}})=0 \tag{5.124}$$

If V'_{n} and V'_{A} denote the corresponding velocity components of neutron and nucleus after the collision, then we may determine these quantities in terms of V_{n} and V_{A} by applying momentum and energy conservation in the C-system

$$V'_{\mathrm{n}}+AV'_{\mathrm{A}}=V_{\mathrm{n}}+AV_{\mathrm{A}}=0 \qquad V'^2_{\mathrm{n}}+AV'^2_{\mathrm{A}}=V^2_{\mathrm{n}}+AV^2_{\mathrm{A}} \tag{5.125}$$

We then arrive at the expected result, i.e.

$$V'_n = V_n = Av_n/(A+1) \qquad V'_A = V_A = -v_n(A+1) \tag{5.126}$$

To derive the required relationships between $\cos\theta$, $\cos\theta_L$ $(=\mu_s)$, $E_n = \frac{1}{2}m_n v_n^2$ and $E'_n = \frac{1}{2}m_n v_n'^2$ we apply (5.123) directly, giving the results

$$V'_n \cos\theta \equiv V_n \cos\theta = v'_n \cos\theta_L - v_{cm} \tag{5.127a}$$

$$V'_n \sin\theta \equiv V_n \sin\theta = v'_n \sin\theta_L \tag{5.127b}$$

Eliminating θ_L between (5.127a and b) we deduce the first of these relationships

$$\begin{aligned} E'_n/E_n &= (v'_n/v_n)^2 = (A^2 + 2A\cos\theta + 1)/(A+1)^2 \\ &= \tfrac{1}{2}[(1+\alpha) + (1-\alpha)\cos\theta] \end{aligned} \tag{5.128}$$

where

$$\alpha = (A-1)^2/(A+1)^2 \tag{5.129}$$

In the same way, by eliminating θ between (5.127a and b) we find that

$$\cos\theta_L = \tfrac{1}{2}[(A+1)(E'_n/E_n)^{1/2} - (A-1)(E_n/E'_n)^{1/2}] \tag{5.130}$$

To derive a relation between $\cos\theta_L$ and $\cos\theta$ (cf. section 2.1) we return to (5.127a) and eliminate V_n, v_n and v'_n using (5.126) and (5.128). The result is

$$\cos\theta = \frac{A\cos\theta + 1}{(A^2 + 2A\cos\theta + 1)^{1/2}} \tag{5.131}$$

We now make use of the condition that scattering is isotropic in the C-system to derive an expression for $\langle\cos\theta_L\rangle = \bar{\mu}_s$ which has been quoted earlier (cf. section 5.3.2)

$$\begin{aligned} \bar{\mu}_s &= \frac{1}{4\pi}\int_0^{2\pi} d\varphi \int_0^1 \frac{(A\cos\theta + 1)\sin\theta\, d\theta}{(A^2 + 2A\cos\theta + 1)^{1/2}} \\ &= \frac{1}{2}\int_{-1}^{1} \frac{(A\mu + 1)\, d\mu}{(A^2 + 2A\mu + 1)^{1/2}} = 2/3A \end{aligned} \tag{5.132}$$

Applied to (5.128), isotropy of scattering in the C-system results in an expression for the probability density for the energy of the scattered neutron

$$\begin{aligned} p(E'_n)\, dE'_n &= d\Omega/4\pi = \tfrac{1}{2}\sin\theta\, d\theta \\ &= -[E_n(1-\alpha)]^{-1}\, dE'_n \qquad \alpha E_n \leqslant E'_n \leqslant E_n \end{aligned} \tag{5.133}$$

There are two very significant features to the result (5.133). The first is that, except in the case $\alpha = 0$ corresponding to scattering from hydrogen nuclei, when all the energy can be transferred in a single collision, final energies in the range $E'_n < \alpha E_n$ cannot be reached following a single collision. This is the reason why functions used to describe the slowing down process, from an initial energy E_n to a final energy E'_n, exhibit discontinuities at $E'_n = \alpha E_n$. The second important feature is that the probability that the final energy should fall in the range E'_n to $E'_n + dE'_n$ is independent of E'_n, but depends only on the fractional energy change in the collision. For this reason it proves convenient to transform to a 'lethargy' variable u, where

$$u = \ln(E_0/E_n) \tag{5.134}$$

and E_0 is the maximum energy in the neutron spectrum (Amaldi *et al* 1935). The lethargy is therefore monotonically increasing in the range $0 \leqslant u < \infty$ and has the property that the increment in the lethargy gained in any number of collisions is just the sum of the lethargies gained at the individual collisions. According to (5.133), the probability density for a gain in lethargy u at a single collision is given by

$$\begin{aligned} f(u)\,du &= (1-\alpha)^{-1}\,e^{-u}\,du \qquad & 0 \leqslant u \leqslant \ln(\alpha^{-1}) \\ &= 0 & u > \ln(\alpha^{-1}) \end{aligned} \tag{5.135}$$

Hence the mean increment in the lethargy per collision is given by

$$\begin{aligned} \xi = \langle u \rangle &= (1-\alpha)^{-1} \int_0^{\ln(1/\alpha)} u\,e^{-u}\,du \\ &= 1 - \alpha(1-\alpha)^{-1} \ln(1/\alpha) \simeq 2/(A + \tfrac{2}{3}) \end{aligned} \tag{5.136}$$

Thus we may estimate that, whereas it requires only about 20 collisions to thermalize a 10 MeV neutron in hydrogen ($A = 1$, $\xi = 1$), the same result requires about 126 collisions in a graphite moderator ($A = 12$, $\xi = 0.158$).

Finally we may apply (5.130) to derive an expression for the scattering function with lethargy increment u in the L-system

$$f(\mathbf{\Omega}, \mathbf{\Omega}', u) = (2\pi)^{-1} f(\mu, u) = (2\pi)^{-1} f(u)\,\delta(\mu - \mu_s(u)) \tag{5.137}$$

where, according to (5.130) and (5.134)

$$\mu_s(u) = \left(\frac{1+\alpha^{1/2}}{1-\alpha}\right) e^{-u/2} - \left(\frac{\alpha+\alpha^{1/2}}{1-\alpha}\right) e^{u/2} \tag{5.138}$$

Integrating over the angles we then confirm that

$$\int f(\mathbf{\Omega}, \mathbf{\Omega}', u)\,d\Omega'\,du = \int f(u)\,du = 1 \tag{5.139}$$

5.4.2 Neutron energy distribution in the steady state

5.4.2.1 An integral equation for the collision density

To determine the distribution in energy (or equivalently the distribution in lethargy) of slowed-down neutrons from a monoenergetic source we return to the neutron transport equation (5.62) appropriate to a uniform infinite medium. However we must now allow for the energy dependence by replacing the angular density $n(\boldsymbol{r}, \boldsymbol{\Omega}, t)$ by $n(\boldsymbol{r}, \Omega, u, t)$ and the scattering function $f(\boldsymbol{\Omega}, \boldsymbol{\Omega}')$ by the new function $f(\boldsymbol{\Omega}, \boldsymbol{\Omega}', u)$ defined in (5.127). Thus the integral over Ω' in the scattering term in (5.62) is replaced by a double integral over Ω' and u'. In addition we express the modified transport equation in terms of the collision density

$$F(\boldsymbol{r}, \boldsymbol{\Omega}, u, t) = v\Sigma_t(u) n(\boldsymbol{r}, \boldsymbol{\Omega}, u, t) \tag{5.140}$$

In section 5.4.4 we shall discuss the spatial dependence of the collision density in the steady state but, for the present, where we are concerned only with the energy spectrum of the whole population of slowed-down neutrons, we integrate over both $\boldsymbol{\Omega}$ and $\boldsymbol{r}$ leading to the integral equation

$$F(u) = S\,\delta(u) + \int_0^u c(u')F(u')f(u-u')\,\mathrm{d}u' \tag{5.141}$$

where

$$F(u) = \int F(\boldsymbol{r}, \boldsymbol{\Omega}, u)\,\mathrm{d}^3r\,\mathrm{d}\Omega \tag{5.142}$$

is the total collision frequency at lethargy u in the steady state, and

$$S\,\delta(u) = \int S(\boldsymbol{r}, \Omega, u)\,\mathrm{d}^3r\,\mathrm{d}\Omega \tag{5.143}$$

is the total source strength.

Because $f(u)$ as defined in (5.135) vanishes for $u > \ln(1/\alpha)$ it is convenient to write the equation for $F(u)$ separately in the two regions $0 \leqslant u \leqslant \ln(1/\alpha)$, and $u > \ln(1/\alpha)$, i.e.

$$F(u) = S\,\delta(u) + \frac{1}{1-\alpha}\int_0^u c(u')F(u')\,\mathrm{e}^{-(u-u')}\,\mathrm{d}u' \qquad 0 \leqslant u < \ln(1/\alpha) \tag{5.144a}$$

$$F(u) = \frac{1}{1-\alpha}\int_{u-\ln(1/\alpha)}^u c(u')F(u')\,\mathrm{e}^{-(u-u')}\,\mathrm{d}u' \qquad u > \ln(1/\alpha) \tag{5.144b}$$

The discontinuity in $F(u)$ and $u = \ln(1/\alpha)$ may be evaluated immediately by introducing the function

$$G(u) \equiv F(u) - S\,\delta(u) \tag{5.145}$$

and calculating the difference of the limits

$$\lim_{u \to \ln(1/\alpha)+} G(u) - \lim_{u \to \ln(1/\alpha)-} G(u) = \frac{S\alpha c(0)}{1-\alpha} \tag{5.146}$$

The origin of the discontinuity (5.146) has already been commented on; it arises from the fact that in a single collision, a neutron of energy E cannot reach an energy $<\alpha E$, and in two collisions it cannot reach an energy $<\alpha^2 E$ and so on. Thus in general the nth derivative of $F(u)$ is discontinuous at $u=(n+1)\ln(1/\alpha)$ as may be shown directly, by repeatedly differentiating (5.144b) under the integral sign. Thus we may expect to observe some singular behaviour in $F(u)$ near all these special points, although in practice these transients are damped out for $u \geqslant 3\ln(1/\alpha)$. This region of u is called the asymptotic region and is by far the most important region since it corresponds to the case where the neutrons are slowed down to near thermal energies. Of course in the case of scattering in hydrogen $\alpha=0$, and there is no discontinuity in $F(u)$; in this case the asymptotic region reaches right down to zero lethargy.

Equation (5.144a) can be solved exactly with the result

$$F(u)=S\delta(u)+Sc(0)\exp\left\{\frac{\alpha u}{1-\alpha}-\int_0^u \frac{[1-c(u')]}{1-\alpha}\,du'\right\} \tag{5.147}$$

but (except in the case of hydrogen) the result is not of much practical value. For hydrogen, however, the result is valid right across the spectrum and it may be applied to mixtures of hydrogen with very heavy elements which contribute to the capture but not to the energy change on scattering. In particular, for hydrogen neglecting capture ($c(u')\equiv 1$), $F(u)\equiv S$, expressing the fact that, in steady-state conditions, the S neutrons produced by the source per unit time are distributed uniformly across the lethargy spectrum. The corresponding energy spectrum varies inversely with the energy.

The integral equations (5.144) can be expressed as a single equation for the function $G(u)$ introduced in (5.145) by making use of the step function $\mu(x)$ and the ramp function $\lambda(x)$ defined by

$$\begin{aligned} &\mu(x)=1 \qquad \lambda(x)=1 \qquad x<1 \\ &\mu(x)=0 \qquad \lambda(x)=x \qquad x>1 \end{aligned} \tag{5.148}$$

Thus we may write

$$e^u G(u)=\frac{1}{1-\alpha}\left\{Sc(0)\mu(\alpha\, e^u)+\int_{\ln\lambda(\alpha e^u)}^u c(u')G(u')\,e^{u'}\,du'\right\} \tag{5.149}$$

This device allows us to derive an alternative form of expression for $G(u)$ which has some practical applications (Placzek 1946). The procedure followed now is to differentiate (5.149) with respect to u, multiply by e^{-u} and integrate between 0 and u allowing for the discontinuity in $G(u)$ given in (5.146). The result is

$$\begin{aligned} G(u)=Sc(0)-&\int_0^u [1-c(u')]G(u')\,du' \\ &+\frac{\alpha}{1-\alpha}\left\{Sc(0)\mu(\alpha\, e^u)+\int_{\ln\lambda(\alpha e^u)}^u c(u')G(u')\,du\right\} \end{aligned} \tag{5.150}$$

For the most important region where $u > \ln(1/\alpha)$ and $G(u) \equiv F(u)$ this result reduces to

$$G(u) = \left\{ Sc(0) - \int_0^u [1 - c(u')]F(u')\,du' + \frac{\alpha}{(1-\alpha)} \int_{u-\ln(1/\alpha)}^u c(u')F(u')\,du' \right\} \tag{5.151}$$

5.4.2.2 *The slowing-down density*

A function which plays a very important part in the description of the mechanism of neutron moderation is the slowing down density $q(E)$ (or $q(u)$), which may be defined as the number of neutrons falling below the energy E (or, equivalently, rising above the lethargy, u) per unit time. Its significance derives from its role as an effective source function in the diffusion equation for thermal neutrons, which are the most lethargic neutrons in the whole population. For neutrons scattering at energy E' it is possible to reach an energy E only if $\alpha E' < E < E'$ and $(1-\alpha)E'$ represents the total range of energies which can conceivably scatter past E. The fraction which can in fact do so is then $(E - \alpha E')/E'(1-\alpha)$ and $q(E)$ is given by

$$q(E) = \int_E^{E/\alpha} c(E')\Phi(E') \left\{ \frac{E - \alpha E'}{E'(1-\alpha)} \right\} dE' \tag{5.152}$$

where

$$\Phi(E') = F(u')\,du'/dE' \tag{5.153}$$

is the collision density per unit energy interval. Expressing (5.152) in lethargy units we then find that (cf. 5.136)

$$\begin{aligned} q(u) &= F(u) - \frac{\alpha}{1-\alpha} \int_{u-\ln(1/\alpha)}^u c(u')F(u')\,du' \\ &\to (F/u)[1 - c(u) + c(u)\xi] \qquad u \to \infty \end{aligned} \tag{5.154}$$

Another useful form of the function $q(u)$ which expresses it in terms of a single integral is obtained by combining (5.154) with (5.151), i.e.

$$q(u) = Sc(0) - \int_0^u [1 - c(u')]F(u')\,du' \tag{5.155}$$

It is clear from the form of (5.154) that, under conditions of weak absorption $q(u) \simeq \xi F(u)$ in the asymptotic range $u \gg \ln(1/\alpha)$ or, expressed in terms of α and ξ, $\alpha u \gg (1-\xi)(1-\alpha)$. This result has a simple interpretation for, since $F(u)\,\Delta u$ is the number of collisions per unit time in the lethargy interval Δu below u, all these neutrons will rise above u in lethargy when $\Delta u = \xi \ll u$, since ξ represents the average gain in lethargy per collision. Also, according to (5.155), if $c(u') = 1$ (i.e. no capture), then the same number of neutrons S

pass every point on the lethargy scale per unit time under steady state conditions. Thus, when $c(u')\neq 1$, the number

$$p(u)\equiv q(u)/S=c(0)-\frac{1}{S}\int_0^u [1-c(u')]F(u')\,du' \tag{5.156}$$

represents the probability that a neutron is not captured in its progress from the source lethargy $u=0$ up to lethargy u. The function $p(u)$ is therefore known as the resonance escape probability since most neutrons are captured as they slow down past individual resonances.

5.4.3 Collision density in the asymptotic regime

5.4.3.1 Solution without capture

When the moderating medium has no capture cross-section we may set $c(u')=1$ in (5.144), and the integral equation for $F(u)$ can be solved exactly over the whole lethargy range $0\leqslant u<\infty$, by application of the Laplace transform technique (Marshak 1947). We therefore introduce the transform functions

$$\tilde{F}(\omega)\equiv\int_0^\infty F(u)\,e^{-\omega u}\,du \tag{5.157}$$

and

$$\tilde{f}(\omega)\equiv\int_0^\infty f(u)\,e^{-\omega u}\,du$$

$$=\int_0^{\ln(1/\alpha)} \exp[-u(1+\omega)]\,du=\frac{(1-\alpha^{1+\omega})}{(1-\alpha)(1+\omega)} \tag{5.158}$$

Since the scattering term in (5.141) has the form of a convolution integral the transformed equation may be written down immediately, i.e.

$$\begin{aligned}\tilde{F}(\omega)&\equiv S+\tilde{F}(\omega)\tilde{f}(\omega)=S/(1-\tilde{f}(\omega))\\&=S+S\tilde{f}(\omega)/(1-\tilde{f}(\omega))\end{aligned} \tag{5.159}$$

To invert $\tilde{F}(\omega)$ we need to identify the poles of the function $\tilde{f}(\omega)/(1-\tilde{f}(\omega))$ in the complex ω-plane, and, according to (5.159) and (5.158), these occur at the roots of the equation

$$(1-\alpha)(1+\omega)=1-\alpha^{1+\omega} \tag{5.160}$$

If we denote these roots by ω_j ($j=0, 1, 2, \ldots$) then we may invert $\tilde{F}(\omega)$ with the result

$$F(u)=S\delta(u)+S\sum_{j=0}^{\infty}\left[\frac{(\omega_j+1)}{1-(1-\xi)\alpha^{\omega_j}}\right]e^{\omega_j u} \tag{5.161}$$

It is evident by inspection that $\omega_0 = -1$ and $\omega_1 = 0$ are both solutions of (5.160) and indeed it may be verified by an appropriate plot that these are the only real solutions. The solution $\omega_0 = -1$ contributes nothing to the sum in (5.161) since we have already extracted the component $S\delta(u)$. Since we are interested in the asymptotic behaviour of $F(u)$ we shall therefore omit this term. The solution $\omega_1 = 0$ contributes the term S/ξ which is indeed the asymptotic value of $F(u)$ as we verify immediately below.

By separating $\omega_j = \eta_j + \mathrm{i}\zeta_j$ into its real and imaginary parts we may rewrite (5.160) as a pair of linked equations

$$(1-\alpha)\zeta_j = \alpha^{\eta_j+1}\sin(\zeta_j\ln(1/\alpha)) \tag{5.162a}$$

$$(1-\alpha)(\eta_j+1) = 1-\alpha^{\eta_j+1}\cos(\zeta_j\ln(1/\alpha)) \tag{5.162b}$$

The purely real solutions described above are obtained from (5.162b) by setting $\zeta_j \equiv 0$. According to (5.162a) the complex solutions ($\zeta_j \neq 0$) must satisfy

$$\alpha^{-\eta_j}(1-\xi)^{-1} = \frac{\sin(\zeta_j\ln(1/\alpha))}{(\zeta_j\ln(1/\alpha))} < 1 \tag{5.163}$$

Since $(1-\xi)^{-1} > 1$ and $\alpha < 1$, in order to satisfy (5.163) we must have $\eta_j < 0$ so that all the complex terms in the sum (5.154) vary as $\exp[(-|\eta_j| \pm \mathrm{i}\zeta_j)u]$ and therefore vanish in the asymptotic limit $u \to \infty$.

If we now retrace our steps to (5.159) and extract the asymptotic part $F_{\mathrm{as}} = \xi^{-1}$ we can rewrite this equation as

$$\tilde{F}(\omega) - \int_0^\infty \mathrm{e}^{-\omega u} F_{\mathrm{as}}(u)\,\mathrm{d}u = S + \frac{S\tilde{f}(\omega)}{1-\tilde{f}(\omega)} - \frac{1}{\omega\xi} \tag{5.164}$$

Since the pole at the origin has now been subtracted away in (5.164), the line integral in the complex ω-plane can now be moved onto the ζ-axis and the inversion of (5.163) may be written in terms of a Fourier cosine integral

$$F(u) = F_{\mathrm{as}} + S\int_0^\infty \cos(\zeta u)\chi(\zeta)\,\mathrm{d}\zeta \qquad u > \ln\left(\frac{1}{\alpha}\right) \tag{5.165}$$

where

$$\chi(\zeta) = \left[\frac{\alpha(1+\alpha)[1-\cos(\zeta\ln(1/\alpha))] - \alpha(1-\alpha)\zeta\sin(\zeta\ln(1/\alpha))}{\zeta^2(1-\alpha)^2 + 2\alpha^2[1-\cos(\zeta\ln(1/\alpha))] - 2\alpha(1-\alpha)\zeta\sin(\zeta\ln(1/\alpha))}\right] \tag{5.166}$$

The collision density in graphite ($\alpha = 0.716$, $\ln(1/\alpha) = 0.334$, $\xi = 0.158$), as evaluated from (5.147) (with $c(u') = 1$), and from (5.158), is plotted as a function of u in figure 5.9. As remarked earlier the asymptotic collision density is reached for values of $u \geqslant 3\ln(1/\alpha)$.

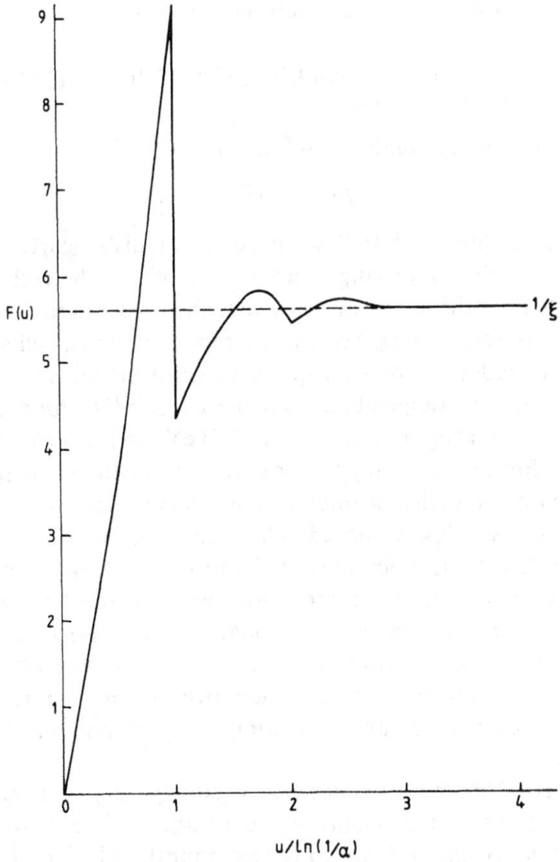

Figure 5.9. Variation of the collision density $F(u)$ with lethargy u in graphite, where $\xi=0.158$ and $\ln(1/\alpha)=0.334$. The discontinuities in $F(u)$ at $u=\ln(1/\alpha)$, $2\ln(1/\alpha)\ldots$ originate in the fact that a neutron cannot increase its lethargy by an amount greater than $\ln(1/\alpha)$ in a single collision.

5.4.3.2 Solution with capture

We cannot, unfortunately, solve equations (5.144) exactly in the case that the macroscopic capture cross-section $\Sigma_t(u)$ is non-zero, and the function $c(u)\leqslant 1$ is an arbitrary function. We may however arrive at approximate solutions of these equations for a number of special cases, which at least provide an indication of the form of the true solution, even though in practical situations such solutions must be obtained through a numerical analysis. The derivation of approximate solutions is much helped by the fact that, in practice, significant capture only sets in at energies much below the initial energy of the fission neutron, where, expressed in terms of the lethargy, the asymptotic regime has already been reached. Hence we are concerned to find a solution to the integral equation (cf. 5.144b)

$$F(u)=\frac{1}{(1-\alpha)}\int_{u-\ln(1/\alpha)}^{u} c(u')F(u')\,e^{-(u-u')}\,du' \qquad c(u')\geqslant 1 \tag{5.167}$$

subject to the boundary condition (cf. (5.164)–(5.165))

$$F(u_0)=S\xi^{-1} \tag{5.168}$$

In seeking solutions to (5.167) we must distinguish clearly between two cases: (i) $c(u)$ is a slowly varying function of u within the collision interval $\ln(1/\alpha)$; this includes the important case that the capture cross-section obeys the $1/v$ law (cf. section 1.7.1); (ii) $c(u)$ varies rapidly within a collision interval; this situation includes resonance capture of neutrons which is a dominant effect in the epithermal range of neutron energy. In ^{238}U, for example, there are important capturing resonances at 274 eV and 6.7 eV. In practical situations it is further necessary to enquire whether the resonance width is comparable with the collision interval and whether individual resonances are widely spaced or closely spaced. The 'width' of interest is the range of energy for which $\Sigma_c(u)$ is much greater than the non-resonant part of $\Sigma_s(n)$, which may be factors of 10 or more greater than the true width of the resonance. The spacing between resonances is important because of complications which arise evaluating the collision density when transients produced by one resonance have not died away before the next resonance is reached. However, these are very complex questions which cannot be pursued here.

To solve (5.167) in the case that $c(u)$ is slowly varying over a collision interval we first suppose that $c(u)$ is a constant $c<1$, and we look for a solution which joins smoothly on to the asymptotic solution of (5.161) with $c=1$. We therefore postulate a trial solution

$$F(u)=S\xi^{-1}\exp[-v(u-u_0)] \tag{5.169}$$

where, on substituting (5.169) into (5.167) it follows that v is a real solution of

$$(1-\alpha)(1-v)=c(1-\alpha^{1-v}) \tag{5.170}$$

Repeating the analysis of the corresponding equation (5.160) for the case $c=1$, it follows that there is a unique solution in the range $0\leqslant v\leqslant 1$ which satisfies the necessary requirements. If we now rewrite (5.170) in the form

$$\frac{1}{c}=\frac{1}{(1-\alpha)}\int_{\alpha}^{1} y^{-v}\,dy=\frac{1}{1-\alpha}\int_{\alpha}^{1}\sum_{j=0}^{\infty}\frac{[-v\ln(y)]^j\,dy}{j!}$$

$$=1+\sum_{j=1}^{\infty} v^j\left[1\times\left(\frac{1}{1-\alpha}\right)\sum_{i=1}^{j}(\ln(\alpha))^i/i!\right]$$

$$\simeq 1+[1-(1-\xi)/\alpha]v+\ldots \tag{5.171}$$

it follows that v can be evaluated as a function of c to any required degree of precision.

When the constant c is replaced by the slowly-varying function $c(u)$, the transformation (5.169) becomes instead

$$\frac{d}{du}[\ln F(u)] = -v(u) \tag{5.172}$$

where $v(u)$ is once again a solution of (5.170) with c replaced by $c(u)$.

The final expression for $F(u)$ is then given by

$$F(u)=S\xi^{-1}\int_{u_0}^{u} \exp[-v(u')u']\, du' \tag{5.173}$$

At the other extreme, if $c(u)\equiv 1$ except for a finite range $u_0<u<\ln(1/\alpha)$, then (5.167) becomes instead

$$\xi F(u)=\frac{1}{(1-\alpha)}\left\{S[e^{-(u-u_0)}-\alpha]+\xi\int_{u_0}^{u} c(u')F(u')\,e^{-(u-u')}\,du'\right\} \tag{5.174}$$

Since only the upper limit in the integral in (5.174) is a function of u this integral equation may be converted into a differential equation. This is achieved by multiplying (5.174) across by e^u and differentiating with respect to u. The resultant equation may then be solved in the usual way using the integrating factor $\exp[\chi(u)]$ where

$$\chi(u)=\frac{1}{(1-\alpha)}\int_{u_0}^{u} [c(u')-1+\alpha]\, du' \tag{5.175}$$

The final result is

$$\xi F(u)=S\exp[\chi(u)]\left\{1-\frac{\alpha}{1-\alpha}\int_{u_0}^{u} \exp[-\chi(u')]\, du'\right\} \tag{5.176}$$

If the capture range is spread out over a number of collision intervals, but the separation between the resonances is sufficiently great that transients may be ignored, then it is evident that the same procedure may be employed for successive intervals. In each case the boundary condition which replaces (5.168) is fixed by the value of $F(u)$ determined in the preceding collision interval.

The slowing down density $q(u)$ may be determined from $F(u)$, for both slowly varying and resonance capture processes by application of (5.155). In this respect it is noteworthy that to fix $q(u)$ requires a knowledge of $F(u)$ only in that range of energy where capture occurs.

5.4.4 Spatial distribution of slowing-down density

5.4.4.1 The Fermi age

When slowing down in light nuclei such as hydrogen or deuterium, a neutron can lose all, or almost all, its energy at a single collision, and there is no close correlation between its energy and its chronological age, i.e. the time which has elapsed since the neutron was created in a fission event. However, when the scattering nuclei are sufficiently heavy that the mean increase in lethargy per collision is small in comparison with the lethargy of a thermal neutron, there is a very close correspondence between lethargy and chronological age. Considered from a more general viewpoint, slowing down in hydrogen and in heavy nuclei are really quite distinct forms of stochastic process. The former may be described as 'Poissonian' in that the energy changes in finite discrete random jumps; the latter is 'Gaussian' with a net change in energy which is the sum of a large number of independent infinitesimal increments. A similar distinction is drawn between 'collision broadening' and 'pressure broadening' in the theory of spectral line shapes.

In the Gaussian limit slowing down may be treated as a continuous process where lethargy and time are related by

$$\frac{\mathrm{d}u}{\mathrm{d}t}(t)=\xi v\Sigma_{\mathrm{t}}(u) \tag{5.177}$$

To simplify matters further we may, for the present, neglect the u-dependence of $\Sigma_{\mathrm{t}}(u)$ so that, for small changes in lethargy, (5.177) integrates to

$$u(t)=2\ \ln\left[1+\frac{1}{2}\xi t v_0\Sigma_{\mathrm{t}}\right]\simeq\xi v_0\Sigma_{\mathrm{t}}t \tag{5.178}$$

The growth of lethargy u with time t given by (5.178) is shown in figure 5.10.

To estimate the average crow's flight distance a neutron of lethargy u has travelled from its point of origin, we start with the familiar result from elementary kinetic theory, that the probability density $p(r)$ that a neutron starting from any point should travel a distance r before colliding with a nucleus, is given by

$$p(r)\,\mathrm{d}r=\mathrm{e}^{-r/\lambda}\,\mathrm{d}r/\lambda \qquad 0\leqslant r<\infty \tag{5.179}$$

where λ is the collision mean free path. The mean square distance travelled between collisions is then

$$\langle r_1^2\rangle=\int_0^\infty p(r)r^2\,\mathrm{d}r=2\lambda^2 \tag{5.180}$$

A neutron which has suffered n collisions follows a zig-zag path from the

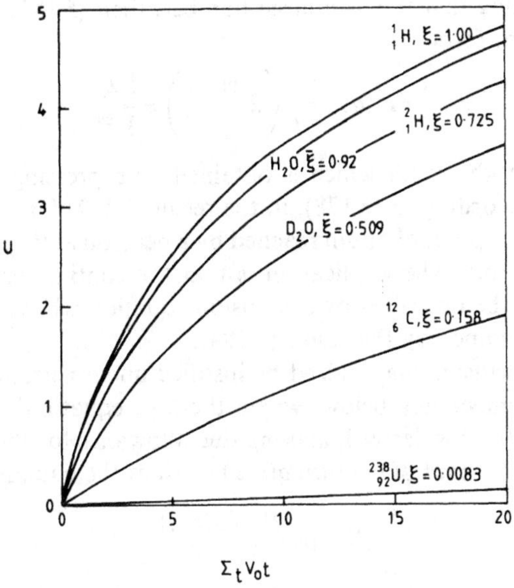

Figure 5.10. Diagram showing the growth of lethargy with chronological age t, measured in units of the initial mean time interval $(v_0 \sum_t)^{-1}$ between collisions, based on the assumption that slowing down is a continuous (Gaussian) process. This assumption is invalid for slowing down in hydrogen where a neutron may, in principle, transfer all its energy in a single collision.

origin to a point $\boldsymbol{r}$ given by

$$\boldsymbol{r} = \sum_{j=1}^{n} \boldsymbol{r}_j \tag{5.181}$$

where its mean square distance of travel is

$$\begin{aligned} \langle \boldsymbol{r}_n^2 \rangle &= \sum_{j=1}^{n} \langle \boldsymbol{r}_j^2 \rangle + \sum_{i<j} \langle \boldsymbol{r}_i \cdot \boldsymbol{r}_j \rangle \\ &= n \langle r_1^2 \rangle = 2n\lambda^2 \end{aligned} \tag{5.182}$$

In arriving at (5.182) we have used the approximate result

$$\langle \boldsymbol{r}_i \cdot \boldsymbol{r}_j \rangle = \langle r_i r_j \cos\theta_{ij} \rangle \simeq 0 \tag{5.183}$$

which would hold exactly were scattering truly isotropic in the laboratory. In fact, since $\bar{\mu}_s = \frac{2}{3}A > 0$ (cf. (5.132)), there is a small bias towards forward scattering; however this bias may be taken into account by replacing the collision mean free path λ by the transport mean free path λ_{tr} which includes the factor $(1-\bar{\mu}_s)^{-1}$ in its definition (cf. 5.102).

Since the mean number of collisions in a time t is just vt/λ_{tr}, provided t

is sufficiently long that this is a large number, then (5.182) may be written in the significant form

$$\frac{1}{6}\langle r^2(u)\rangle = \frac{1}{6}\left(2\frac{vt}{\lambda_{tr}}\lambda_{tr}^2\right) \simeq \frac{1}{3}\frac{\lambda_{tr}u}{\xi\Sigma_t} \tag{5.184}$$

This is precisely the result which is obtained by expressing time t in terms of lethargy u according to (5.178), in the result (5.107) for the mean square deviation from its point of origin reached by a neutron at time t in conditions of diffuse scattering. The implication is that the continuous slowing down model may also be described by a diffusion equation whose parameters are essentially determined by the result (5.184).

That this conclusion may indeed be justified under certain conditions we demonstrate immediately below, where the 'age equation' for the slowing down density $q(\boldsymbol{r}, u)$ is derived, making due allowance for the effect of weak capture. In the absence of capture $q(\boldsymbol{r}, u)$ satisfies the equation

$$\nabla^2 q(\boldsymbol{r}, \tau) = \frac{\partial q(\boldsymbol{r}, \tau)}{\partial \tau} \tag{5.185}$$

where the 'symbolic neutron age' or 'Fermi age' $\tau(u)$ is defined by

$$\tau(u) = \int_0^u \frac{D(u')\,du}{\xi\Sigma_s(u')} \tag{5.186}$$

and $D(u')$ is the diffusion coefficient defined at lethargy u' (cf. 5.101). We note that in a uniform infinite medium, where $\nabla^2 q(\boldsymbol{r}, \tau) \equiv 0$, the slowing down density is independent of $\tau(u)$, a result we have commented on earlier (cf. section 5.4.2.2). In a finite non-capturing medium $\partial q(\boldsymbol{r}, \tau)/\partial z \neq 0$, expressing the fact that neutrons are lost to the system during the slowing down process by leakage across the boundary (cf. section 5.5.2).

The function $\tau(u)$ has the dimensions of area and, expressed in terms of $\tau(u)$, the result (5.184) may be written as

$$\frac{1}{6}\langle r^2(\tau)\rangle = \tau = L_s^2(\tau) \tag{5.187}$$

where, in analogy with (5.91) and (5.107), the quantity $L_s(\tau)$ is termed the slowing down length at age τ. The slowing down time $t_s(\tau)$ is just the chronological age of the neutron at Fermi age τ and is defined by

$$\tau/t_s(\tau) = \int_0^u \frac{vD(u')\,du'}{\xi v\Sigma_s(u')} \bigg/ \int_0^u \frac{du'}{\xi v\Sigma_s(u')} \tag{5.188}$$

This result may also be written in the convenient form

$$t_s(\tau) = \tau/\langle Dv\rangle \tag{5.189}$$

where $\langle Dv\rangle$ is the weighted average of $D(u)v$ over the lethargy range up to

u, with the maximum weight given to that range of v where the collision mean free path $\Sigma_t^{-1}(u)$ is maximum. The quantity $D(u)v$ is frequently referred to as the 'conventional diffusion coefficient' because, in the conventional theory of Brownian motion, Fick's law (5.64) is expressed as a relationship between the current and the gradient of the number density, rather than between the current and the gradient of the scalar flux. In a one-speed theory this distinction does not matter, but in slowing down theory it is evidently the conventional diffusion coefficient which is important.

5.4.4.2 Derivation of the age equation with capture

The age equation (5.185) is derived by a method which parallels exactly that used to obtain the diffusion equation discussed in section 5.4.3.3, and we shall therefore provide only a brief outline of the sequence of steps needed to arrive at it. One additional condition needs to be fulfilled to guarantee the validity of age theory but, aside from that, the theory is subject to all those limitations which restrict the application of diffusion theory in real situations. The point of departure in the derivation is the lethargy-dependent time-independent transport equation which replaces (5.62), i.e.

$$v\mathbf{\Omega}\cdot\nabla n(\mathbf{r}, \mathbf{\Omega}, u)+\Sigma_t n(\mathbf{r}, \mathbf{\Omega}, u)$$
$$=S(\mathbf{r}, \mathbf{\Omega}, u)+\int_0^u \mathrm{d}u' \int \mathrm{d}\Omega' c(u')\Sigma_t(u')vn(\mathbf{r}, \mathbf{\Omega}, u')f(\mathbf{\Omega}, \mathbf{\Omega}', u-u') \quad (5.190)$$

In subsequent developments we shall omit the (arbitrary) source term $S(\mathbf{r}, \mathbf{\Omega}, u)$ which may be replaced by an effective, but equally arbitrary, source term in the final result.

Following the same procedures as before, we postulate a uniform infinite medium and express (5.190) in terms of a lethargy-dependent collision density $F(\mathbf{r}, \mathbf{\Omega}, u)$, which is then expanded to first order in $\mathbf{\Omega}$ (cf. 5.94)

$$F(\mathbf{r}, \mathbf{\Omega}, u)=\frac{1}{4\pi}[F(\mathbf{r}, u)+3\mathbf{\Omega}\cdot\mathbf{F}_1(\mathbf{r}, u)] \quad (5.191)$$

where (cf. 5.93)

$$\mathbf{F}_1(\mathbf{r}, u)=\int \mathbf{\Omega}F(\mathbf{r}, \mathbf{\Omega}, u)\,\mathrm{d}\Omega \qquad |\mathbf{F}_1|\ll|F| \quad (5.192)$$

The approximations defined in (5.191)–(5.192) are an expression of the assumption that the flux is almost isotropic everywhere. A similar expansion is applied to the scattering function (cf. (5.95), (5.137) and (5.138))

$$f(\mu, u)=\frac{1}{4\pi}[f_0(u)+3\mu f_1(u)] \quad (5.193)$$

where

$$f_0(u)=\int d\Omega f(\mu,u)=f(u) \tag{5.194a}$$

$$f_1(u)=\int d\Omega\mu f(\mu,u)=f(u)\lambda_s(u) \tag{5.194b}$$

With these approximations we obtain results analogous to (5.63) and (5.97)

$$(\Sigma_t(u))^{-1}\nabla\cdot\boldsymbol{F}_1(\boldsymbol{r},u)+F(\boldsymbol{r},u)=\frac{1}{(1-\alpha)}\int_{u-\ln(1/\alpha)}^{u} du'c(u')F(\boldsymbol{r},u')\,e^{-(u-u')}\,du'$$

$$(3\Sigma_t(u))^{-1}\nabla F(\boldsymbol{r},u)+\boldsymbol{F}_1(\boldsymbol{r},u)=\frac{1}{(1-\alpha)}\int_{u-\ln(1/\alpha)}^{u} du'c(u')\boldsymbol{F}_1(\boldsymbol{r},u')\,e^{-(u-u')}\,du'$$

(5.195)

We now introduce an important new assumption, namely that the collision density $F(\boldsymbol{r},\boldsymbol{\Omega},u)$ is a slowly varying function of u within a collision interval $\ln(1/\alpha)$. This condition must apply if the slowing down process is to be treated as a continuous one. We therefore write

$$c(u')F(\boldsymbol{r},u')=c(u)F(\boldsymbol{r},u)+(u'-u)\frac{\partial}{\partial u}[c(u)F(\boldsymbol{r},u)]+\ldots \tag{5.196a}$$

$$c(u')\boldsymbol{F}_1(\boldsymbol{r},u')=c(u)\boldsymbol{F}_1(\boldsymbol{r},u)+\ldots \tag{5.196b}$$

The function $c(u')\boldsymbol{F}_1(\boldsymbol{r},u')$ has been evaluated to zero order in (5.196b) since, according to (5.191)–(5.192), the term $\boldsymbol{\Omega}\cdot\boldsymbol{F}_1(\boldsymbol{r},u)$ is itself a small correction to $F(\boldsymbol{r},u)$. When (5.196) is substituted into (5.195) these equations reduce to (cf. (5.63), (5.97))

$$(\Sigma_t(u))^{-1}\nabla\cdot\boldsymbol{F}_1(\boldsymbol{r},u)+(1-c(u))F(\boldsymbol{r},u)+\xi\frac{\partial}{\partial u}[c(u)F(\boldsymbol{r},u)]=0 \tag{5.197a}$$

$$(3\Sigma_t(u))^{-1}\nabla F(\boldsymbol{r},u)+(1-c(u)\bar{\mu}_s)\boldsymbol{F}_1(\boldsymbol{r},u)=0 \tag{5.197b}$$

The result (5.197b) is, of course, a reformulation of Fick's law and, substituting for $\boldsymbol{F}_1(\boldsymbol{r},u)$ in (5.197a) we arrive at the age equation in the form

$$-D(u)\nabla^2F(\boldsymbol{r},u)+\Sigma_t(u)(1-c(u))F(\boldsymbol{r},u)+\xi\Sigma_t(u)\frac{\partial}{\partial u}[c(u)F(\boldsymbol{r},u)]=0$$

(5.198)

To convert (5.198) into a differential equation for $q(\boldsymbol{r},u)$ we need to establish a relation between $F(\boldsymbol{r},u)$ and $q(\boldsymbol{r},u)$ which is valid in the same order of approximation as applies to (5.196)–(5.197). We therefore return to the defining equation (5.154), which, since it applies at all points $\boldsymbol{r}$, may be rewritten as

$$q(\boldsymbol{r}, u) = F(\boldsymbol{r}, u) - \frac{\alpha}{1-\alpha} \int_{u-\ln(1/\alpha)}^{\alpha} c(u')F(\boldsymbol{r}, u')\, du' \tag{5.199}$$

Then, expanding $F(\boldsymbol{r}, u')$ according to (5.196a), we arrive at the approximate relationship

$$\xi\left[c(u)F(\boldsymbol{r}, u) - \gamma \frac{\partial}{\partial u}(c(u)F(\boldsymbol{r}, u) \right] = q(\boldsymbol{r}, u) \tag{5.200}$$

where

$$\gamma = 1 - [\alpha(\ln(1/\alpha))^2/2\,!]/(1-\alpha)(1-\xi) \tag{5.201}$$

Differentiating (5.200) with respect to u, and neglecting second derivatives according to (5.196a), we find that

$$\frac{\partial q}{\partial u}(\boldsymbol{r}, u) = \xi \frac{\partial}{\partial u}[c(u)F(\boldsymbol{r}, u)] \tag{5.202}$$

which allows us to substitute directly for the third term in (5.198). Also by combining (5.202) and (5.200) we may recast the latter equation as an expression for $F(\boldsymbol{r}, u)$, i.e.

$$\xi c(u)F(\boldsymbol{r}, u) = q(\boldsymbol{r}, u) + \gamma \frac{\partial q}{\partial u}(\boldsymbol{r}, u) \tag{5.203}$$

which may then be substituted in the first two terms in (5.199). The net result is

$$\{-D(u)\nabla^2 + \Sigma_t(u)(1-c(u))\}\left[q(\boldsymbol{r}, u) + \gamma \frac{\partial}{\partial u} q(\boldsymbol{r}, u) \right] = \Sigma_t(u)\xi c(u) \frac{\partial q}{\partial u}(\boldsymbol{r}, u) \tag{5.204}$$

The term $-D(u)\nabla^2[\gamma(\partial q/\partial u)(\boldsymbol{r}, u)]$ must be omitted from (5.204) since this term derives from Fick's law (5.197b) where terms proportional to $(\partial/\partial u)\boldsymbol{F}_1(\boldsymbol{r}, u)$, which are of the same order, have already been discarded. We can therefore re-order (5.204) into its final form

$$\nabla^2 q(\boldsymbol{r}, \tau) - \frac{\Sigma_c(\tau)}{D(\tau)} q(\boldsymbol{r}, \tau) = \frac{\partial q}{\partial \tau}(\boldsymbol{r}, \tau) \tag{5.205}$$

where the capture modified Fermi age is given by (Glasstone and Edlund 1952)

$$\tau(u) = \int_0^u \frac{D(u')\, du'}{\xi\Sigma_s(u') + \gamma\Sigma_c(u')} \tag{5.206}$$

It is convenient to remove the dissipative term in (5.205) by introducing the modified slowing down density

$$\tilde{q}(\boldsymbol{r}, \tau)=q(\boldsymbol{r}, \tau) \exp\left\{\int_0^\tau \frac{\Sigma_c(\tau')\,d\tau'}{D(\tau')}\right\}=q(\boldsymbol{r}, \tau) \exp\left\{\int_0^u \frac{\Sigma_c(u')\,du'}{\xi\Sigma_s(u')+\gamma\Sigma_c(u')}\right\} \tag{5.207}$$

which now satisfies an equation similar in form to the age equation without capture (cf. 5.185), i.e.

$$\nabla^2\tilde{q}(\boldsymbol{r}, \tau)=\frac{\partial\tilde{q}}{\partial\tau}(\boldsymbol{r}, \tau) \tag{5.208}$$

Since $(\partial q/\partial\tau)(\boldsymbol{r}, \tau)$ vanishes in a uniform infinite medium, with $\tilde{q}(\boldsymbol{r}, \tau)=\tilde{q}(\boldsymbol{r}, 0)=q(\boldsymbol{r}, 0)$, independent of $\boldsymbol{r}$ in these conditions, it follows that the function

$$p(u)=\exp\left\{-\int_0^u \frac{\Sigma_c(u')\,du'}{\xi\Sigma_s(u')+\gamma\Sigma_c(u')}\right\} \tag{5.209}$$

which describes the attenuation with age τ of $q(\boldsymbol{r}, \tau)$, due to capture, must be identified with the resonance escape probability (cf. 5.156). This conclusion may be confirmed directly by the following argument. First we differentiate (5.155) with respect to u to give the exact result

$$\frac{dq(u)}{du}=-[1-c(u)]F(u) \tag{5.210}$$

Then we integrate (5.203) over $\boldsymbol{r}$, to provide an expression for $F(u)$ to substitute in (5.210). The resulting differential equation for $q(u)$

$$\frac{dq(u)}{du}+\frac{\Sigma_c(u)}{\xi\Sigma_s(u)+\gamma\Sigma_c(u)}q(u)=0 \tag{5.211}$$

immediately integrates to

$$q(u)\equiv q(0)p(u)=q(0)\exp\left\{-\int_0^u \frac{\Sigma_c(u')\,du'}{\xi\Sigma_s(u')+\gamma\Sigma_c(u')}\right\} \tag{5.212}$$

which agrees with (5.209).

The principal advantage of the age equation is that it is formally identical with Fourier's heat conduction equation, whose solutions have been studied in every geometry and under all conceivable boundary conditions. It is one of the most thoroughly investigated equations in the whole of physics. Nevertheless, because it is so limited by the restrictions which govern diffusion theory, as expressed in equations (5.109)–(5.110), with symbolic age replacing neutron age, and because it is not applicable to the analysis of slowing down in light moderators, age theory, although of immense value from a qualitative viewpoint, does not provide reliable quantitative estimates. For this reason, and in spite of the various improvements in age theory which have been proposed from time to time (Verde and Wick 1947, Marshak *et al* 1949, Hurwitz and Zweifel 1955) it is not widely used in modern nuclear engineering.

5.5 The nuclear fission chain reactor

5.5.1 Milestones in nuclear reactor development

The physical principles underlying the operation of a nuclear fission chain reactor, or 'pile' have been summarized very briefly in section 1.6.4. Nuclear reactors are of two types: thermal reactors which make use of a moderator such as graphite or heavy water (deuterium oxide) to slow the fast fission neutrons down to thermal energies, where they induce fission in the nuclear fuel to produce the next generation of fast neutrons; and fast reactors where the neutron multiplication process all takes place at energies $\geqslant 1$ MeV. All reactors used for research purposes or for commercial power generation are of the first type while fast reactors, of which a prototype has been constructed at Dounreay in Scotland, are themselves objects of research. Nuclear explosions may be described as uncontrolled fast reactors. Thermal reactors are fuelled either by odd-N uranium isotopes ^{233}U or ^{235}U, or by the odd-N plutonium isotopes ^{239}Pu or ^{241}Pu, at least in the sense that these isotopes can make positive contributions to the reactivity as the fuel composition changes in time. All of these nuclei can undergo fission following neutron capture at any energy; as discussed in section 5.2.1 this effect is a consequence of the pairing interaction in nuclei.

The even-A isotopes ^{232}Th, ^{234}U, ^{238}U and ^{240}Pu all have fission 'thresholds' and can undergo fission only when bombarded with fast neutrons. Although not 'fissile' these isotopes are described as 'fertile' since they can be made to breed fissile material by neutron capture. The 'breeding ratio' is defined as the ratio of the amount of fissile material produced to the amount consumed in one cycle of operation; if this ratio is greater than unity the reactor is described as a 'breeder reactor'. In a 'fast breeder reactor' ^{238}U is converted to ^{239}Pu by capture of neutrons in the energy range 1 keV to 10 MeV giving a breeding ratio of about 1.4. Thermal breeders are based on the conversion of ^{232}Th to ^{233}U and have a much less favourable breeding ratio.

The first nuclear reactor which successfully supported a self-sustaining chain reaction without the benefit of an external neutron source was the graphite-moderated natural uranium reactor CP-1 constructed by Fermi and his associates in the West Stand of the University of Chicago football stadium (Fermi 1947, 1952). This first went critical on December 2nd 1942 achieving a maximum power of about 200 W. The reactor was constructed from ≈ 50

tons of U_3O_8 made up into $\approx 2 \times 10^4$ fuel elements embedded in ≈ 300 tons of graphite, and was approximately spheroidal in shape with a mean diameter of about 3.5 m. Its successor, CP2, was assembled at the Argonne National Laboratory and reached a power output of ≈ 2 kW. The first reactor specifically designed to breed ^{239}Pu from ^{238}U was built at Clinton, Tennessee, in 1943, yielding a power output of about 4 MW.

The most serious practical problem which arises in constructing a power reactor based on natural uranium is that the fissile isotope ^{235}U is present only to the extent of about 0.7%. Thus a power reactor of this type is inevitably large and costly. Nevertheless several high-flux natural uranium reactors were constructed in the immediate post-war period, two of which used natural uranium in combination with a heavy water moderator. The first of these was the NRX reactor built at Chalk River, Canada in 1947. This used a light water coolant and operated at a power level of 30 MW. The Savannah River reactor which commenced operation in 1954, was of a similar type; this was a breeder reactor as indeed was the high-flux graphite-moderated reactor constructed at Windscale in 1950.

In the USA a reactor contruction policy broadly based on the use of ^{235}U-enriched fuel was adopted almost from the beginning, following the successful completion at Oak Ridge National Laboratory of the first plant to separate ^{235}U from ^{238}U by gaseous diffusion of uranium hexa-fluoride. All modern research reactors use enriched uranium; for example, the 57 MW research reactor at the Institut Laue Langevin uses a single 9 kg fuel element, enriched to 93% in ^{235}U, with heavy water acting as both moderator and coolant. The whole system is immersed in a light water biological shield, whose appearance explains the description 'swimming pool reactor' commonly applied to this type of facility. In the power reactor field a wide range of models has become available; in the USA research was concentrated on the pressurized water reactor (PWR) which uses light water both as moderator and coolant. On the other hand, in the UK and in the USSR graphite was the material most commonly selected as moderator with other materials, e.g. CO_2, used as coolants.

The world's first commercial nuclear power station came into operation at Calder Hall in Cumbria in 17 October 1956. This station generated 150 MW of electrical power, and since that date the production of nuclear power has accelerated all over the world, particularly in France where uranium has largely replaced coal and gas as a fuel. However, these developments have been marred by a number of serious accidents beginning with the accident during start-up at the NRX reactor in 1952, and the Windscale fire in 1957, followed by the near-catastrophe at Three Mile Island in 1979, and culminating in the Chernobyl explosion in 1985, which spread radioactive contaminants over distances measured in thousands of kilometres. The Windscale fire, which was caused by the failure to control the release of the so-called Wigner energy, namely the energy stored in the graphite moderator due to changes in crystal structure induced by neutron bombardment at high flux levels, provides a classic example of the difficulties which may arise in the development of new technologies. These events, plus perceived difficulties in the disposal of nuclear waste, have led to a swift decline in the popularity of the nuclear reactor as a power source and a sharp contraction in the worldwide nuclear power industry.

The loss of confidence in fission based nuclear power has been accompanied to some extent by a growth of interest in the prospects for fusion power, sparked off, it must be admitted, by so far unjustified claims from some quarters, that this could be achieved using palladium cells operating near room temperature (Fleischmann and Pons 1989, Jones *et al* 1989). Unfortunately the evidence to support these findings has failed to be forthcoming (Ziegler *et al* 1989, Leggett and Baym 1989, Alber *et al* 1989). However, there are, as we shall see, certain disadvantages to fusion power, aside from the fact that an operating fusion reactor has yet to be constructed (cf. section 5.6). There is also a school of thought which maintains that the fast breeder fission reactor represents the best hope for the future. Unfortunately the widespread, and quite understandable, association in the public perception, between 'nuclear power' and 'nuclear weapons', and universally-held (and probably justifiable) fears that commercial advantage may be given priority over public health, has led to cutbacks in support for research in this field. This is a pity, for it may take a long time for this option to be properly explored, with the added danger that future programmes will be accelerated too rapidly, leading to the adoption of energy policies even more inept and disastrous than those we have witnessed in the immediate past.

5.5.2 The neutron economy in a thermal reactor

Let us suppose that the population of thermal neutrons in a reactor of effectively infinite size contains n_j members in the jth generation. These neutrons may be produced by an external source or by some other means. Then the number n_{j+1} of thermal neutrons in the $(j+1)$th generation is related to n_j by

$$n_{j+1} = kn_j \tag{5.213}$$

where the number k is called the multiplication factor for the system. This quantity is determined in general by the nature and amounts of fuel and moderator, and by the geometry and spatial concentration of these two components. Real reactors are, of course, finite in size and neutrons may be lost to the chain reaction by leakage across the boundary. Leakage can occur at various stages in the career of a neutron as it progresses down the energy scale from fast to thermal, but this happens most probably during slowing down (fast leakage), with probability $(1-P_s)$, or while diffusing in the moderator (thermal leakage), with probability $(1-P_d)$. The effective multiplication constant is therefore given by

$$k_{eff}=kP_sP_d \tag{5.214}$$

The reactivity ρ of a reactor in a given state is defined by the relation

$$\rho=(k_{eff}-1)/k_{eff} \tag{5.215}$$

If $\rho>0$ the reactor is said to be supercritical, since under these conditions the neutron population increases at each successive generation. When $\rho=0$ the reactor is critical, at which point it will operate at any specified power level chosen by the operator. When the critical condition can be maintained by the prompt neutrons alone, the reactor is said to be prompt critical, and there is a danger that a slight positive perturbation in ρ can cause the reactor to run out of control. If $\rho<0$ the reactor is subcritical and the number of thermal neutrons is attenuated in each succeeding generation. In a laboratory reactor the objective is to maintain conditions at criticality but under control, i.e. in such a way that positive or negative excursions in ρ can be compensated automatically, within a period of time which is short in comparison with the reactor time constant. This compensation mechanism is effected by control rods, constructed from natural boron or cadmium, materials containing significant amounts of ^{10}B or ^{113}Cd, which are highly absorbant to slow neutrons. These control rods can be moved into, or out of, the reactor, as required.

In a sample of pure ^{235}U, because there is no fission threshold, the prompt-critical condition can be reached whenever the production rate of prompt neutrons, which, for a given fuel concentration is proportional to volume, exceeds the leakage rate, which is proportional to surface area. In a uniform spherical reactor this stage is reached at a radius $\geqslant 8.7$ cm, corresponding to a critical mass of approximately 52 kg, and there is danger of a spontaneous nuclear explosion. This critical mass can be reduced to about 11 kg if the uranium is enclosed in a 15 cm thick beryllium reflector. In practice the expansion which follows the onset of such an explosion drives the device sub-critical and the explosion terminates; thus, to initiate a nuclear explosion, it is necessary to drive two sub-critical masses together by a chemical explosion, and to supply an external neutron source to keep the initial neutron population as high as possible. On the other hand, in a fission device based on ^{239}Pu there is danger of premature explosion due to spontaneous fission of ^{240}Pu or ^{241}Pu which are inevitably present as isotopic contaminants (Lynch *et al* 1989).

In an elementary analysis of reactor behaviour it is usually assumed that fuel and moderator are uniformly mixed to form a homogeneous ensemble. This is not the most economical arrangement, neither can it be fulfilled in practice, since the nuclear fuel has to be enclosed in fuel rods manufactured to ensure that the dangerous fission products are securely contained. The fuel rods are arranged in a lattice surrounded by a moderator, the number of moderator atoms in every case greatly exceeding the number of fuel atoms. The assumption of homogeneity does not lead us too far astray provided the mean free path of a neutron, fast or slow, is much greater than a characteristic fuel element dimension. If this condition is satisfied the reactor is said to be quasi-homogeneous; if not the reactor is heterogeneous, but the theoretical analysis of its behaviour becomes immeasurably more complicated.

The neutron cycle in a ^{235}U–^{238}U thermal reactor is illustrated in figure 5.11 starting with *n* first generation thermal neutrons. The fraction of these neutrons which are absorbed in the fuel is given by the thermal utilization factor *f*, where

$$\begin{aligned} f &= (\Sigma_{au}\bar{\varphi}_u V_u)/(\Sigma_{au}\bar{\varphi}_u V_u + \Sigma_{am}\bar{\varphi}_m V_m) \\ &= (1 + \Sigma_{am}\bar{\varphi}_m V_m/\Sigma_{au}\bar{\varphi}_u V_u)^{-1} \end{aligned} \qquad (5.216)$$

Here $\Sigma_{au}(\Sigma_{am})$ represents the macroscopic total absorption (fission plus capture) cross-section in the uranium (moderator), and $V_u(V_m)$ and $\bar{\varphi}_u(\bar{\varphi}_m)$ are the corresponding volumes and mean fluxes. In a homogeneous assembly $V_u = V_m$ and $\bar{\varphi}_u = \bar{\varphi}_m$; in the more general case of a heterogeneous assembly the volume disadvantage factor $\bar{\varphi}_m/\bar{\varphi}_u$ must be calculated using some combination of diffusion theory and transport theory (Amouyal *et al* 1957). Since the prime objective in designing a reactor is to establish conditions most favourable to maintenance of a self-sustaining chain reaction, it is clearly advantageous to make *f* as large as possible. This can be done by increasing the proportion of fuel to moderator, but the amplification in *f* thereby achieved is cancelled to some extent by a decrease in the resonance escape probability *p*. This is because resonance capture in ^{238}U is the dominant reaction mechanism while neutrons are slowing down between 500 eV and 5 eV.

The most important 'parasitic' capture process leading to a reduction in *f* is due to the presence of neutron-absorbing fission fragments or 'poisons' which build up in the reactor over the course of time. The most significant poisons are ^{135}Xe ($\sigma_c \approx 3.5 \times 10^6$ b) and ^{149}Sm ($\sigma_c \approx 5.3 \times 10^4$ b). Xe is an intermediate product in the fission chain

$$^{135}\mathrm{Te} \xrightarrow{\beta^-} {}^{135}\mathrm{I}(6.7\text{ h}) \xrightarrow{\beta^-} {}^{135}\mathrm{Xe}(9.2\text{ h}) \xrightarrow{\beta^-} {}^{135}\mathrm{Cs}(2 \times 10^6\text{ years}) \qquad (5.217)$$

and, because ^{135}Xe is converted by neutron capture to (stable) ^{136}Xe (8.97%) while the reactor is operating, the poisoning effect increases when the reactor is off. During this time the amount of ^{135}Xe grows as the parent nucleus ^{135}I decays, and reaches a maximum concentration about 11 h after shut-down. The other important poison ^{149}Sm is produced in the decay chain

$$^{149}\mathrm{Nd}(1.8\text{ h}) \xrightarrow{\beta^-} {}^{149}\mathrm{Pm}(53\text{ h}) \xrightarrow{\beta^-} {}^{149}\mathrm{Sm}(13.9\%) \qquad (5.218)$$

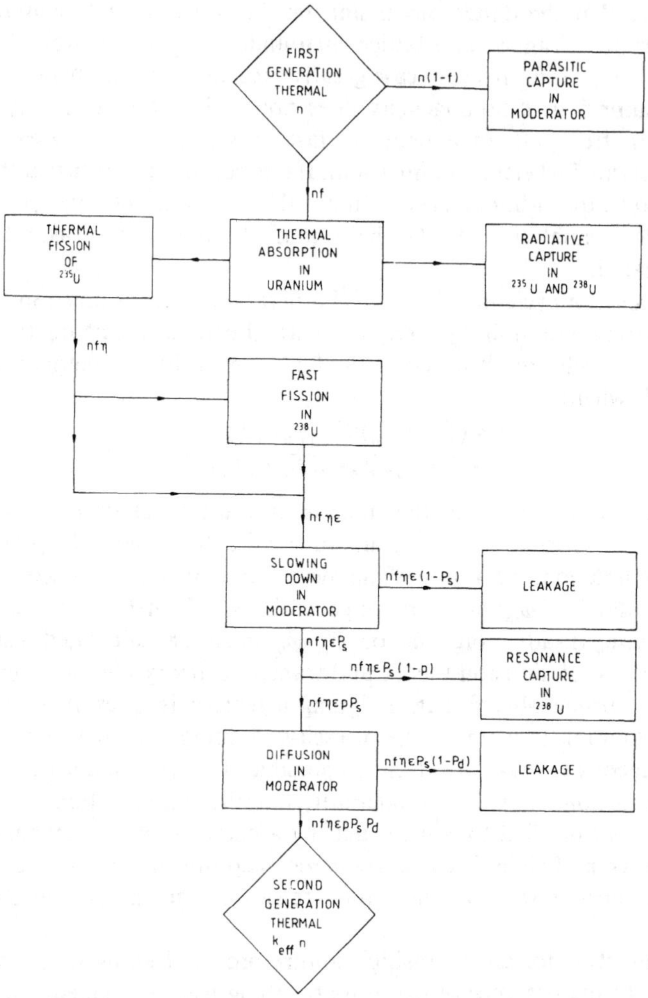

Figure 5.11. The neutron cycle in a ^{235}U–^{238}U thermal nuclear reactor illustrating the significance of the thermal utilization factor (f), the mean number of fast neutrons emitted per thermal neutron absorbed (η), the fast fission factor (ε) and the resonance escape probability (p).

^{149}Sm is a stable isotope which converts to (stable) ^{150}Sm(7.5%) under operating conditions, and increases slowly after shut-down following the decay of 53 h ^{149}Pm.

The next important factor which needs to be taken into account is the mean number η of fast fission neutrons produced per thermal neutron captured in uranium, which is not identical with the mean number ν of fast fission neutrons produced per fission of ^{235}U. Rather it is given by

$$\eta = \nu\Sigma_f^{235}/\Sigma_{au} \tag{5.219}$$

where Σ_f^{235} is the macroscopic thermal fission cross-section in ^{235}U. Since in pure ^{235}U, $\Sigma_f^{235}/\Sigma_{au} \approx 0.85$, whereas, in natural uranium $\Sigma_f^{235}/\Sigma_{au} \approx 0.53$, there is great scope for increasing η by enriching the uranium fuel in its ^{235}U content.

Even in natural uranium η can be further enhanced by the fast fission factor ε, which takes account of the fact that a small proportion of fast neutrons can cause fission in ^{238}U. The process arises in the following way. Let $P_{c'}$ be the probability that a fast neutron makes at least one collision before escaping from the fuel element in which it is born. Then the number of first generation fast neutrons which produce additional fast fission is $(nf\eta)P_{c'}\ \Sigma_{f'}/\Sigma_{t'}$, where $\Sigma_{f'}$ and $\Sigma_{t'}$ are, respectively, the fission and total macroscopic cross-sections for fast neutrons in the given fuel mixture. Thus the number of second generation fast neutrons is $(nf\eta)P_{c'}\nu'\Sigma_{f'}/\Sigma_{t'}$ where ν' is the average number of neutrons produced per fast fission. Fast neutrons also scatter in the fuel element, of which the elastically scattered component $(nf\eta)P_{c'}\Sigma_{s'}$ is available for producing fast fission, and the inelastically scattered component $(nf\eta)P_{c'}\Sigma_{in'}$ cannot produce fast fission, although it may eventually arrive in the thermal region and produce thermal fission. This is because, although the average energy of a fission neutron is 2 MeV, the most probable energy is $\approx$0.75 MeV whereas the fission threshold in ^{238}U is $\approx$1 MeV. Thus an inelastically scattered fast neutron may be assumed to have an energy substantially less than 1 MeV.

Excluding the inelastically scattered component the number of second generation fast neutrons is just $(nf\eta)P_{c'}c'$ where c' is the number of neutrons emitted following each fast collision and is given by (cf. 5.34)

$$c' = (\Sigma_{s'} + \nu'\Sigma_{f'})/\Sigma_{t'} \tag{5.220}$$

If we make a tally after one collision of the number of fast neutrons leaving the fuel element we find $(nf\eta)(1-P_{c'})$ first generation neutrons which did not collide at all, plus $(nf\eta)P_{c'}\Sigma_{in'}/\Sigma_{t'}$ second generation neutrons from inelastic collisions. To these must be added $(nf\eta)P_{c'}c'(1-P_{c'})$ second generation neutrons which escape immediately after they are born. Quite generally the number escaping in the jth generation is $(nf\eta)(1-P_{c'})(P_{c'}\nu')^{j-1} + (nf\eta)(P_{c'})^{j-1}(c')^{j-2}\Sigma_{in'}/\Sigma_{t'}$ where we have assumed that the scattering parameters c', Σ', etc. are independent of generation number. Hence summing up over all generations the number of neutrons leaving the fuel element is

$$\begin{aligned} nf\eta\varepsilon &= nf\eta[1-P_{c'}+P_{c'}\Sigma_{in'}/\Sigma_{t'}][1+(P_{c'}c')+(P_{c'}c')^2+\ldots] \\ &= nf\eta[1-P_{c'}+P_{c'}\Sigma_{in'}/\Sigma_{t'}][1-P_{c'}c']^{-1} \end{aligned} \tag{5.221}$$

It follows from (5.221) that the fast fission factor ε is given by

$$\varepsilon = 1 + P_{c'}[(\nu'-1)\Sigma_{f'} - \Sigma_{c'}]/[\Sigma_{t'} + \Sigma_{in'} - P_{c'}(\Sigma_{s'} + \nu\Sigma_{t'})] \tag{5.222}$$

where $\Sigma_{t'}$ is defined according to (5.35) and does not include inelastic scattering.

In natural uranium $\Sigma_{in'}/\Sigma_{f'} \approx 1.7$ and $\Sigma_{in'}/\Sigma_{c'} \approx 60$ so that inelastic scattering rather than capture is the dominant loss mechanism for fast neutrons which reduces ε, and values of $(\varepsilon-1)$ of order 3–5% may be achieved. The geometry is evidently an important consideration in evaluating ε although in quasi-homogeneous conditions $P_{c'} \simeq \theta a/\lambda$, where a is a typical fuel element dimension, λ is the collision mean free path and θ a number of order unity. Specifically, for an infinite cylinder of radius a, $\theta = \frac{4}{3}$ and, for a sphere of radius a, $\theta = \frac{3}{4}$ (Case *et al* 1953).

To arrive at a final estimate of the number of neutrons which enter the thermal regime after the first cycle of operation we must multiply the number $nf\eta\varepsilon$ of fast neutrons which enter the slowing down phase by the resonance escape probability p. The complete expression for the multiplication factor is then given by the four-factor formula

$$k = f\eta\varepsilon p \tag{5.223}$$

In natural uranium $\eta \approx 1.34$ and $\varepsilon \leqslant 1.05$ which means that, to bring about a self-sustaining chain reaction we require that $fp \geqslant 0.71$. Detailed calculations show that, for a homogeneous mixture of fuel and moderator, $fp \leqslant 0.55$; that there is a maximum achievable value of fp follows from the fact, noted earlier, that the fuel/moderator ratio which favours large value of f results in small values for p. It follows that a homogeneous or quasi-homogeneous light water natural uranium reactor cannot be constructed, although such a system is achievable using heavy water. The usual solution to the problem is to adopt a heterogeneous assembly whereby the flux of resonance neutrons at the centre of the fuel element is depressed due to shielding by the outer layers, while the whole of the fuel element remains relatively accessible to the thermal spectrum. Although there is some reduction in f, the net effect is to provide the required amplification in p.

To obtain an expression for k_{eff} we need to calculate P_s and P_d, and to do this we need to know the spatial distributions of the slowing down neutrons and of the thermal neutrons. This requires us to solve the age equation and the diffusion equation in the particular geometry selected. In the following section we shall outline the solution for a homogeneous cylindrical reactor, and arrive at an expression for the critical condition in such a system.

5.5.3 Elementary theory of the homogeneous cylindrical reactor

5.5.3.1 The moderator

A bare homogeneous thermal reactor is a homogeneous assembly of fuel and moderator placed in a vacuum, where the scalar flux $\varphi(\boldsymbol{r}, t)$ satisifes the condition that it should vanish at the extrapolated boundary (cf. section

5.3.2.4). When the reactor is enclosed by a thin reflector characterized by an albedo β, the main effect is to increase the extrapolated length according to (5.121), but even with a thick reflector, or indeed an infinite reflector with albedo β_∞, no substantial modification of the elementary theory is required. As discussed in sections 5.5.1–2 the behaviour of a fast reactor is primarily determined by the choice of fuel; for a thermal reactor the fuel, at some given degree of enrichment, must be matched to a specified moderator, whose performance is of equal importance.

Since the prime requirement for a good moderator is that ξ, the mean increase in lethargy per collision, should be large, the choice of moderator is effectively limited to a few of the light elements. Excluding helium, which forms no compounds and can therefore be used only in the form of a high pressure gas, and excluding boron and lithium because of their high capture cross-sections for slow neutrons, the choice is further narrowed to hydrogen (or deuterium), beryllium or carbon. Beryllium is expensive, difficult to machine, and highly toxic; this makes it unsuitable for a conventional reactor though it is probably a necessary component in a thermonuclear reactor (cf. section 5.6.4). Since, as noted in section 1.7.5, most hydrogenous substances (other than water) become chemically unstable when subjected to high fluxes of γ-rays, the most suitable moderators are ordinary (or light) water (H_2O), heavy water (D_2O) or graphite (C). To thermalize fast neutrons escaping from reactors through experimental beam ports or by other routes, paraffin wax is a very useful material.

The moderating and diffusion properties of the four principal moderating materials are listed in Table 5.4.

The slowing-down efficiency of a moderator may be quantified by its slowing-down power $\xi\Sigma_s$, which is just the mean increase in lethargy per

Table 5.4. Slowing down and diffusion properties of common moderators. The data, which are taken from Glasstone and Edlund (1952), should be regarded as illustrative rather than definitive

Property	Symbol	H_2O	D_2O	Be	C
Mean lethargy increment	ξ	0.920	0.509	0.209	0.158
Diffusion length (cm)	L	2.88	100	23.6	50.2
Diffusion time (s)	t_d	3.1×10^{-4}	0.15	4.3×10^{-3}	1.2×10^{-3}
Albedo	β_∞	0.821	0.968	0.889	0.930
Slowing-down length (cm)	L_s	5.7	11.0	9.9	18.7
Slowing-down time (s)	t_s	10^{-5}	4.6×10^{-5}	6.7×10^{-5}	1.5×10^{-4}
Migration length (cm)	M	6.4	101	25.6	53.6
Slowing-down power (cm^{-1})	$\xi\Sigma_s$	1.53	0.170	0.176	0.064
Moderating ratio	$\xi\Sigma_s/\Sigma_c$	72	1.2×10^4	159	170

unit path length. Using slowing down power as a measure of efficiency light water clearly comes out best. However, a material is useless as a moderator if it is also a strong absorber of neutrons, and a better figure of merit is provided by the moderating ratio, which is defined as the dimensionless quantity $\xi\Sigma_s/\Sigma_c$. By this criterion light water suffers badly in comparison with heavy water, and this is essentially because the proton has a capture cross-section for slow neutrons which is more than 200 times larger than that of the deuteron. To emphasize this point it may be remarked that light water is the only commonly used moderator for which the slowing down length L_s is greater than the diffusion length L.

Another important point which is clearly brought out in Table 5.4 is that the slowing down time for fast neutrons is between 1 and 3 orders of magnitude less than the diffusion time for slow neutrons, and there are quite different time-scales in the problem. The 'time' introduced in the discussion of the Fermi age in section 5.4.4.2, is the time elapsing from the creation of a neutron of zero lethargy to that epoch at which it reaches lethargy $u(t)$. According to the result (5.178), $u(t)$ depends sensitively on t only for short times of the order of t_s. A more suitable variable with which to plot the varition of u is the age $\tau(u)$ where $\tau(u)^{1/2}$ is just the slowing down length to age $\tau(u)$. On the other hand the time 't' which enters the diffusion equation (5.100) is measured, strictly speaking, from the epoch at which the first generation of fission neutrons reaches thermal energies, and significant variation in the neutron flux $\varphi(\boldsymbol{r}, t)$ may be expected to occur on time-scales of the order of the diffusion time $t_d \gg t_s$. Hence the slowing down density will be a function of time t, as well as of age $\tau(u)$. The average time between successive generations is known as the generation time, and is equal to the sum of the slowing down time and the diffusion time.

Although (with the exception of ordinary water noted above) $L \geqslant L_s$, nevertheless these two lengths are comparable in magnitude. Quite generally the root mean square distance a neutron travels between birth and death is given by the migration length M, where

$$M^2 = L^2 + L_s^2 = (1/6)[\langle r_c^2 \rangle + \langle r_s^2 \rangle] \qquad (5.224)$$

5.5.3.2 The criticality condition

To establish criteria such that a proposed reactor assembly may be made to operate in a critical state with $k_{\text{eff}} = 1$, we need to find simultaneous stationary solutions to the age equation (5.205), and to the diffusion equation (5.100). When these equations are applied to a homogeneous mixture of moderator and fuel we need to replace the macroscopic capture cross-section Σ_c by the mean macroscopic absorption cross-section Σ_a. In addition the source function $S_d(\boldsymbol{r}, t)$ in the diffusion equation must be set equal to the slowing down sensity $q(\boldsymbol{r}, t, \tau_0)$, where τ_0 is the age of a neutron with a most probable thermal velocity $v_0 = 2200\ \mathrm{m\,s^{-1}}$. Under these conditions the diffusion equation assumes the form

$$\nabla^2\varphi(\boldsymbol{r}, t) - L^{-2}\varphi(\boldsymbol{r}, t) + q(\boldsymbol{r}, t, \tau_0)/D(\tau_0) = L^{-2} l_\infty \frac{\partial\varphi}{\partial t}(\boldsymbol{r}, t) \qquad (5.225)$$

where

$$L = \sqrt{(D(\tau_0)/\Sigma_a)}$$

is the corresponding diffusion length (cf. 5.91) and

$$l_\infty = (\Sigma_a v_0)^{-1} \qquad (5.226)$$

is the mean lifetime of a thermal neutron in an infinite reactor. In a finite reactor l_∞ is replaced by

$$l = l_\infty P_s \qquad (5.227)$$

where P_s is defined in section 5.5.2. Typical values of l_∞ range from about 10^{-3} s, in a large natural uranium-graphite reactor, to about 10^{-4} s in a small water-moderated reactor.

To determine $q(\boldsymbol{r}, t, \tau_0)$ we begin with the age equation (5.208) for the modified slowing down density $\tilde{q}(\boldsymbol{r}, t, \tau)$, and assume that this may be solved by separation of variables, i.e.

$$\tilde{q}(\boldsymbol{r}, t, \tau) = \tilde{q}(\boldsymbol{r}, t) f(\tau) \qquad (5.228)$$

Substituting (5.228) into (5.208) we then arrive at the relation

$$\nabla^2\tilde{q}(\boldsymbol{r}, t)/\tilde{q}(\boldsymbol{r}, t) = -\frac{\mathrm{d}}{\mathrm{d}\tau}\ln(f(\tau)) = -B^2 \qquad (5.229)$$

where $B^2 > 0$ so that $f(\tau)$ should decrease with τ. Integrating with respect to τ we then find that

$$f(\tau) = f(0)\exp[-B^2\tau] \qquad (5.230)$$

To determine $f(0)$ we make use of the fact that $q(\boldsymbol{r}, t, 0)$ is just the source function for fast neutrons, and therefore

$$q(\boldsymbol{r}, t, \tau) = p\tilde{q}(\boldsymbol{r}, t, \tau) = k\Sigma_a\varphi(\boldsymbol{r}, t)\exp[-B^2\tau] \qquad (5.231)$$

It follows from (5.231) that $\tilde{q}(\boldsymbol{r}, t)$ is some constant multiple of $\varphi(\boldsymbol{r}, t)$ and therefore according to (5.229), $\varphi(\boldsymbol{r}, t)$ must satisfy the scalar Helmholtz equation

$$\nabla^2\varphi(\boldsymbol{r}, t) + B^2\varphi(\boldsymbol{r}, t) = 0 \qquad (5.232)$$

On the other hand, by substituting (5.231) into (5.225), and assuming steady-state conditions ($\partial\varphi/\partial t \equiv 0$), $\varphi(\boldsymbol{r}, t)$ also satisfies

$$\nabla^2\varphi(\boldsymbol{r}, t) + L^{-2}[k\,\mathrm{e}^{-B^2\tau_0} - 1]\varphi(\boldsymbol{r}) = 0 \qquad (5.233)$$

Thus, to maintain consistency between (5.232) and (5.233) a necessary condition for criticality is

$$B^2L^2 = k\,\mathrm{e}^{-B^2\tau_0} - 1 \tag{5.234}$$

The solution of the transcendental equation (5.234) for given k, L and τ_0, i.e. specified fuel, moderator, and operating temperature is written

$$B = B_{\mathrm{m}} \tag{5.235}$$

where B_{m} is known as the 'material buckling'. The description 'buckling' refers to the fact that, according to (5.229), B is a measure of the degree of curvature of the flux profile at the space-time point $(\boldsymbol{r}, t)$.

There is of course a second condition on B^2, namely that it should coincide with one of the eigenvalues of the operator ∇^2, in the eigenvalue equation (5.232) in the selected geometry. This geometry is shown in figure 5.12, where $(r\theta z)$ represent cylindrical coordinates, and R and $H/2$ represent the extrapolated radius and extrapolated half-height, respectively. Since in the steady state there must be cylindrical symmetry, $\varphi(\boldsymbol{r})$ does not depend on the azimuthal angle θ, and (5.232) is solved by the usual method of separation of variables. Thus we write

$$\varphi(\boldsymbol{r}) = \zeta(z)\rho(r) \tag{5.236}$$

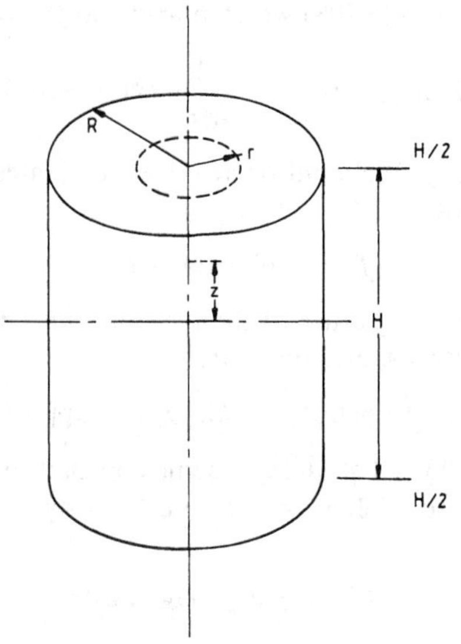

Figure 5.12. Geometry of the bare homogenous cylindrical reactor. The quantities R and $H/2$ represent the extrapolated radius and extrapolated half-height respectively (cf. figure 5.8).

where

$$\frac{d^2\zeta(z)}{dz^2}+B_z^2\zeta(z)=0 \tag{5.237}$$

and

$$r^2\frac{d^2\rho(r)}{dr^2}+r\frac{d\rho(r)}{dr}+r^2B_r^2\rho(r)=0 \tag{5.238}$$

The separation constants B_z and B_r satisfy the condition

$$B_z^2+B_r^2=B^2 \tag{5.239}$$

The only acceptable solutions to (5.237) which satisfy reflection invariance $\zeta(-z)=\zeta(z)$, and which vanish at $z=\pm H/2$, must be in the form

$$\zeta(z)=\zeta(0)\cos[(2n+1)\pi z/H] \qquad n=0, 1, 2, \ldots \tag{5.240}$$

Only the eigenfunction with $n=0$ is non-negative, and is therefore acceptable as a steady-state solution. Thus B_z^2 must be set equal to its lowest eigenvalue

$$B_z^2=(\pi/H)^2 \tag{5.241}$$

Setting $s=rB_r$ in (5.238), this equation reduces to Bessel's equation in s of order zero, for which the solutions are $J_0(s)$ and $Y_0(s)$. The solution $Y_0(s)$ must be rejected, since it tends to (minus) infinity at the origin; thus the only solution which remains finite at the origin and which vanishes at the extrapolated radius is

$$\rho(r)=\rho(0)J_0(2.504\rho/R) \tag{5.242}$$

where $s_0=2.405\ldots$ is the first zero of $J_0(s)$. The corresponding eigenvalue is then

$$B_r^2=[2.405/R]^2 \tag{5.243}$$

The eigensolution of (5.232) for the steady-state flux is then given by

$$\varphi(r)=\varphi(0)J_0(2.405r/R)\cos(\pi z/H) \tag{5.244}$$

where

$$B^2=B_g^2\equiv(2.405/R)^2+(\pi/H)^2 \tag{5.245}$$

is known as the geometric buckling, since it is determined by the shape of the reactor.

Since B^2 must simultaneously satisfy (5.235) and (5.245), the criticality condition may be written as

$$B=B_m=B_g \tag{5.246}$$

where B is defined by (5.234). This relation can now be written in the form

$$k_{\text{eff}}=\frac{k\,\mathrm{e}^{-B^2\tau_0}}{1+L^2B^2}=kP_sP_d=1 \tag{5.247}$$

where

$$1-P_s=1-\exp[-B^2\tau_0] \tag{5.248}$$

is the probability that a neutron should leak from the system during slowing down and

$$1-P_d=\frac{L^2B^2}{1+L^2B^2} \tag{5.249}$$

is the probability of leakage while diffusing as a thermal neutron. For a large reactor where $B^2\tau_0 \ll 1$, the total probability of leakage is given approximately by

$$P_l \simeq M^2B^2/(1+M^2B^2) \tag{5.250}$$

where M^2, the migration area, is defined in (5.224).

To discuss the approach to criticality we may suppose that some arbitrary initial distribution $\varphi(\boldsymbol{r},0)$ is established in the reactor, perhaps by injection of neutrons from an external source. This distribution we assume to be expressible in the form of a Fourier series

$$\varphi(\boldsymbol{r},0)=\sum_{n=0}^{\infty} A_n J_0(2.405r/R)\cos[(2n+1)\pi z/H] \tag{5.251}$$

where the only restriction on the coefficients A_n is that $\varphi(\boldsymbol{r},0)\geqslant 0$. Clearly $\varphi(\boldsymbol{r},0)$ satisfies reflection and rotational invariance and is not the most general form of initial distribution; it is however sufficiently general to illustrate how the critical state is reached. We may now show by direct substitution in (5.225) that the time-dependent flux which reduces to $\varphi(\boldsymbol{r},0)$ at zero time, is given by

$$\varphi(\boldsymbol{r},0)=\sum_{n=0}^{\infty} A_n \exp(\omega_n t)J_0(2.405r/R)\cos[(2n+1)\pi z/H] \tag{5.252}$$

where

$$l_\infty\omega_n=k\,\mathrm{e}^{-B_n^2\tau_0}-1-L^2B_n^2 \tag{5.253}$$

and

$$B_n^2=(2.405/R)^2+[(2n+1)\pi/H]^2 \tag{5.254}$$

We may now rewrite (5.253) in the form

$$\omega_n=(k_n-1)/l_n \tag{5.255}$$

where

$$k_n = k\,e^{-B_n^2\tau_0}/(1+L^2B_n^2) \qquad (5.256)$$

is the effective multiplication constant for neutrons in the nth flux mode, and

$$l_n = l_\infty/(1+L^2B_n^2) \qquad (5.257)$$

is the corresponding lifetime. It follows from (5.255) that, if $k_n<1$ for all n, then every flux-mode in the initial distribution (5.252) decays away in time with a time-constant equal to $l_n/(1-k_n)$. However, if the infinite multiplication constant k is slowly increased until the smallest k_n, namely k_0, reaches unity, then the first term in (5.252) is time-independent and the critical state has been reached. This conclusion agrees with the condition for criticality arrived at in (5.245) above.

5.5.3.3 Reactor control

Let us suppose that the reactor is critical and k_{eff} $(=k_0)$ is increased to a value slightly greater than unity, then the flux grows exponentially at a rate given by

$$\varphi(\mathbf{r},t)=\varphi(\mathbf{r},0)\exp[t/T] \qquad (5.258)$$

where T, the reactor period is defined by

$$T = l_\infty/\rho k_{\text{eff}} \qquad (5.259)$$

and ρ is the reactivity (cf. 5.215). To obtain an estimate of T we assume $l_\infty \simeq 10^{-3}$ s, $k_{\text{eff}}\simeq 1$ and $\rho\simeq 0.01$. It then follows from (5.259) that $T\approx 0.1$ s, the reactor power will grow by a factor $\geqslant 2\times 10^4$ in one second, and a disastrous situation will have developed.

In fact the predicted explosive growth in power does not happen because a small, but significant fraction γ of the fission neutrons are delayed (cf. section 5.1.3.2). In ^{235}U, for example, $\gamma\simeq 0.65\%$, divided in approximately equal proportions among the three main groups 3, 4 and 5 (cf. Table 5.1). If we suppose the ith precursor is produced in a fraction γ_i of fast fissions, and decays by neutron emission with decay constant λ_i, then the precursor density $c_i(\mathbf{r},t)$ satisfies the differential equation

$$\frac{\partial c_i}{\partial t}(\mathbf{r},t) = -\lambda_i c_i(\mathbf{r},t)+\gamma_i(k/p)\Sigma_a\varphi(\mathbf{r},t) \qquad (5.260)$$

and the infinite multiplication constant k is divided between prompt and delayed contributions

$$k=k_p-k_d=(1-\gamma)k+\gamma k \qquad \gamma=\sum_{i=1}^{6}\gamma_i \qquad (5.261)$$

Adding in the effective source due to delayed neutrons the time-dependent diffusion equation becomes

$$\nabla^2\varphi(\boldsymbol{r},t)+L^{-2}(k\,e^{-B^2\tau_0}-1]\varphi(\boldsymbol{r},t)+D^{-1}(\tau_0)\sum_{i=1}^{6}\lambda_i c_i(\boldsymbol{r},t)p\,e^{-B^2\tau_0}$$

$$=L^{-2}l_\infty\frac{\partial\varphi(\boldsymbol{r},t)}{\partial t} \quad (5.262)$$

The solution to the set of seven equations (5.260) and (5.262) is obtained in a straightforward way by the method of Laplace transforms (cf. section 5.3.1.4), yielding a function which is a sum of exponentials

$$\varphi(\boldsymbol{r},t)=\varphi(\boldsymbol{r})\sum_{j=1}^{6}C_j\exp[\Omega_j t] \quad (5.263)$$

where the Ω_j are the seven roots of the characteristic 'Inhour' equation

$$\rho=\frac{l_\infty\Omega}{k_{\text{eff}}}+\sum_{j=1}^{6}\frac{\gamma_j\Omega}{\Omega+\lambda_j} \quad (5.264)$$

For $\rho<0$ (power reduction) all seven roots of (5.264) are negative and the root Ω_0 with the smallest absolute value determines the rate at which the flux decays. For $\rho>0$ (power increase) all but one of the roots are negative, and the positive root Ω_0 determines the rate at which the flux grows. The corresponding period $T_0=\Omega_0^{-1}$ is known as the stable period.

For $\rho\ll\gamma$, T_0 is not very sensitive to the value of l_∞, and in ^{235}U, at $\rho\simeq 0.1\gamma$, $T_0\simeq 100$ s (Keepin and Wimett 1958). This is a stable period which is easily controlled by mechanical means. Also, in heavy-water-moderated reactors there is an additional contribution from photoneutrons produced in the reaction $^2_1$H$(\gamma, n)^1_1$H which tends to increase the stable period (Johns and Sargent 1954). However when $\rho\simeq\gamma$, the prompt critical threshold is reached, and any further increase in reactivity will send the reactor out of control.

5.5.4 The Oklo phenomenon

The Oklo phenomenon is a natural self-sustaining chain reactor in uranium, believed to have occurred some 2×10^9 years ago, at the Oklo uranium deposit in Gabon, West Africa. The discovery was first announced by the French Atomic Energy Commission in September 1972. The possibility that a self-sustaining chain reactor might have existed in uranium ores extracted from old ($>2\times10^9$ years) and well-preserved deposits was first investigated by Kuroda, who carried out detailed calculations for pitchblende ores mined at Johanngeorgenstadt in Saxony (Kuroda 1956). This particular ore was selected because its chemical content was known in detail, and because it was relatively free from neutron absorbing agents such as boron, lithium, thorium and rare earths.

Kuroda was able to demonstrate that a self-sustaining reactor could not be supported in such an ore because, although the factors f and η increased in line with the isotopic ratio $^{235}U/^{238}U$ going backwards in time, the resonance escape probability decreased, indicating a maximum value $k = 0.72$ which was reached about 1.4×10^9 years age. The prospects improved markedly however when the water/uranium ratio was treated as a variable in the problem, the results indicating that criticality could have been reached at some point during the geological age of the earth, for a value of the ratio $H_2O/U > 0.5\%$. However, since many studies of the isotopic ratio $^{235}U/^{238}U$ showed that the ratio is constant at a value $(0.7202 \pm 0.0010)\%$ (Senftle *et al* 1957, Hamer and Robbins 1960), and studies of lunar samples collected on the Apollo 12 mission are in agreement (Rosholt and Tatsumoto 1971) the experimental evidence to support this idea was lacking.

The very first indications of a substantial isotopic anomaly in uranium were reported from the Pierrelate gas-effusion plant in France, where samples were found with isotopic ratios of 0.71%, ten standard deviations away from the mean. Subsequent investigations revealed that the source of the discrepancy was a particular ore, extracted from the Oklo deposit, which was found to have an isotopic ratio of 0.44%, indicating a level of depletion of ^{235}U observed only in spent uranium from natural uranium reactors. The evidence to support the notion that Oklo had indeed been a natural uranium reactor was obtained from a detailed study of neodymium fission products found in the ore (Bodu *et al* 1972, Neuilly *et al* 1972, Baudin *et al* 1972). The choice of neodymium as an indicator was quite deliberate, for neodymium comes close to the heavy fragment maximum of the mass distribution curve in fission (cf. figure 5.4) and has seven stable isotopes. One of these isotopes, namely ^{142}Nd, is shielded from fission by the intervention of a stable isobar in the chain of β-decays, and is therefore available as a control on background.

Experimentally it was found that, with the exception of ^{142}Nd whose relative abundance was normal, all other neodymium isotopes showed substantial excesses ranging from $\simeq 100\%$ (^{150}Nd) to $\simeq 300\%$ (^{143}Nd). Subsequent studies of other fission products, Gd, Sm, Ru and Pd, confirmed that a sustained chain reaction had indeed taken place (Hagemann *et al* 1975). The results of a number of model studies indicate that the reactor operated for a period between 5.4×10^4 and 1.6×10^6 years, at a ^{235}U enrichment of 2–3% and an integrated neutron flux of $\simeq 10^{21}$ cm^{-2} (Naudet 1975, Naudet *et al* 1975). In addition, some 4–10% of all slow neutron induced fissions occurred in ^{239}Pu, generated by neutron capture in ^{238}U.

To date no other fossil reactor has been discovered similar to that at the Oklo deposit which, although not unique, displays a number of highly favourable features. In particular it is very old, and produces pitchblende ore with uranium concentrations $\geqslant 20\%$, which is quite remarkable given that the average concentration is nearer to 0.5%. The high uranium concentration has been explained by Maurette in terms of a series of fractionating processes involving oxygen (Maurette 1976). Since oxygen is believed not to have been present in the earth's atmosphere at times $\gg 2\times 10^9$ years ago, the implication is that, even though the $^{235}U/^{238}U$ ratio would have been much greater at these epochs, the concentration of uranium would have been insufficient to support a self-sustaining chain reaction.

5.6 Controlled nuclear fusion

5.6.1 Laboratory fusion reactions

In a fusion reaction, isotopes of the light elements may fuse together on collision, creating heavier nuclei and releasing kinetic energy equivalent to the difference in binding energy between the colliding species and the emitted fragments. Since the repulsive Coulomb barrier which inhibits the close collision of two charged particles is proportional to the product of their atomic numbers, and even for interactions of hydrogen isotopes has a value of about 1 MeV, fusion is forbidden by the laws of classical mechanics even for the temperatures in excess of 10^6 K which exist at the centre of the sun. Indeed this was the prime argument raised against Eddington's proposal (cf. section 1.2) that thermonuclear fusion was the source of the sun's energy.

According to quantum mechanics, fusion reactions can occur via the process of tunnelling through the Coulomb barrier (Atkinson and Houtermans 1929), which contributes a factor $\exp[-G(E)]$ to the fusion cross-section $\sigma(E)$, where $G(E)$ is the Gamow barrier penetration factor given (for high barriers) by

$$G(E)\simeq 2\pi m^{1/2}e^2Z_1Z_2/\hbar(2E)^{1/2} \tag{5.265}$$

Unfortunately thermonuclear energy cannot be released by bombarding low-Z targets with protons, deuterons or tritons, because the incident energy is rapidly dissipated by ionization, and re-emitted as radiation when the resultant ions and electrons recombine. What is required is an ensemble of completely ionized atoms scattering against each other and occasionally fusing together. However, no material container could sustain the pressures exerted due to Coulomb forces by a charged assembly of this nature (Thonemann 1956). It is this limitation which sets the demand that the ensemble be electrically neutral, i.e. formed from a plasma of electrons and nuclei, maintained at a temperature sufficiently high that fusion via strong interactions occurs at a significant rate, while recombination collision between electrons and nuclei are so infrequent as to be negligible. Such conditions may be established in laboratory plasmas at temperatures in excess of 10^8 K, with particle energies of 5 keV and greater (1 keV $=1.16\times 10^7$ K).

Plasma is of course a very rare state of matter in terrestrial conditions, existing only in artificial devices such as discharge tubes, or in circumstances of severe electrical disturbance such as may arise in thunderstorms. In astrophysical environments plasma is the normal state of matter, and it is the fusion reaction

$$p+p \rightarrow d+e^{+}+\nu_e \tag{5.266}$$

which initiates the p–p cycle of reactions which provide the source of the sun's energy (cf. sections 1.2 and 10.5.4). This is of course a weak interaction, equivalent in many respects to inverse neutron β-decay; it cannot therefore provide the basis for a terrestrial source of thermonuclear energy. For the same reason that they proceed too slowly we leave aside the electromagnetic fusion reactions $d(p, \gamma)^3He$ and $t(p, \gamma)^4He$ and consider as serious candidates only those fusion processes which proceed through strong interactions. The favoured reactions are those listed in Table 5.5; prime among these are the d–d and d–t reactions which produce fast neutrons, whose energy may in principle be extracted to serve as a power source (Post 1956, Teller 1956).

Table 5.5. Laboratory sources of thermonuclear energy. Q_c is the energy carried away by the charged particles emitted in the reaction and T_c is the ignition temperature (cf. section 5.6.2.2)

Reaction	Q (MeV)	Q_c (MeV)	kT_c (keV)
(1) $d+d \rightarrow {}^3He+n$	3.25	0.8	50
(2) $d+d \rightarrow t+p$	4.03	4.0	50
(3) $d+t \rightarrow {}^4He+n$	17.58	3.6	4
(4) $d+{}^3He \rightarrow {}^4He+p$	18.34	18.6	100
(5) $p+{}^6Li \rightarrow {}^3He+{}^4He$	4.0	4.0	900
(6) $p+{}^7Li \rightarrow {}^4He+{}^4He$	17.5	17.5	>900
(7) $p+{}^{11}B \rightarrow {}^4He+{}^4He+{}^4He$	8.7	8.7	300

5.6.2 Characteristics of a self-sustaining fusion device

5.6.2.1 Estimates of fusion power density

Let n_1 and n_2 be the number densities of the two nuclear species colliding in the plasma, and let $\sigma(E)$ be the fusion cross-section at centre-of-mass energy E. Then the fusion rate per collision is $v\sigma(E)$, where v is the relative velocity, and the average power per unit volume transferred to the charged particles emitted in the reaction is (Thompson 1957)

$$\langle P_{\text{fus}} \rangle = n_1 n_2 \langle v\sigma(E) \rangle Q_c \tag{5.267}$$

When there is a single self-interacting nuclear species, as for example in the d–d reaction, the factor $n_1 n_2$ in (5.267) must be replaced by $n_1^2/2$ to avoid double counting. Since the capture cross-section in a two-body collison varies as the square of the de Broglie wavelength (cf. section 2.4.2), the cross-section $\sigma(E)$, including the Coulomb barrier tunnelling factor, may be written in the form

$$\sigma(E) = (A/E)\exp[-G(E)] \tag{5.268}$$

where A is a constant characteristic of the particular fusion transition. For the d–t reaction $A \approx 7.7$ MeV barn, and for each branch of the d–d reaction $A \approx 0.2$ MeV barn.

The procedure normally adopted at this stage is to assume that the collision energy E is distributed according to a Maxwell–Boltzmann density at some effective temperature T, i.e.

$$p(E)\,\mathrm{d}E = 2\pi(\pi kT)^{-3/2}E^{1/2}\exp[-E/kT]\,\mathrm{d}E \tag{5.269}$$

In this case the mean transition rate becomes

$$\langle v\sigma(E)\rangle = 2\pi(\pi kT)^{-3/2}(2/m)^{1/2}A\int_0^\infty \exp[-G(E) - E/kT]\,\mathrm{d}E \tag{5.270}$$

where m is the reduced mass.

The integral in (5.270) may be evaluated to a good approximation using the method of steepest descents. To carry through this procedure we first identify the minimum of the exponent

$$f(E) = G(E) + E/kT \tag{5.271}$$

which occurs at the so-called 'Gamow peak' energy E_g defined by

$$E_g = -kTG'(E_g) = (\pi e^2 m^{1/2} Z_1 Z_2 kT)^{2/3} \tag{5.272}$$

The origin of this peak is shown in figure 5.13, where it represents the product of a monotonic decreasing function $p(E)$ and a monotonic increasing function $\sigma(E)$ (Gamow 1938, Gamow and Teller 1938). The assumption is of course that no resonance of $\sigma(E)$ falls in a region where $p(E)$ is finite; if there is such a resonance then the fusion will be enhanced (cf. section 10.5.2). Expanding $f(E)$ to second order about $E = E_g$, the integral in (5.270) becomes approximately a Gaussian of width

$$\Delta E_g = 2(2E_g kT/3)^{1/2} \tag{5.273}$$

and integration over this Gaussian leads to the result

$$\langle v\sigma(E)\rangle = 2\pi(\pi kT)^{-3/2}(\pi/m)^{1/2}A\,\Delta E_g\exp[-f(E_g)] \tag{5.274}$$

Since $\Delta E_g \approx (kT)^{5/6}$ and $f(E_g) \approx (kT)^{-1/3}$, the behaviour with temperature of $\langle v\sigma(E)\rangle$ may be summarized in the form

$$\langle v\sigma(E)\rangle = \gamma(kT)^{-2/3}\exp[-\delta(kT)^{-1/3}] \tag{5.275}$$

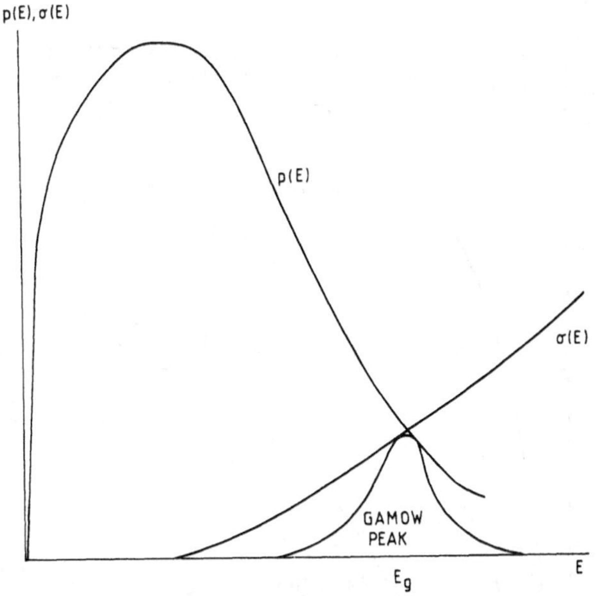

Figure 5.13. Graph showing $p(E)$, the Maxwell–Boltzmann distribution of collision energy E at some effective temperature T, and the cross-section $\sigma(E)$ for fusion of two charged particles. The product of these functions attains a maximum in the region of the Gamow peak $E_g = (\pi e^2 m^{1/2} Z_1 Z_2 kT)^{2/3}$.

where γ and δ are temperature-independent constants. Clearly $\langle v\sigma(E)\rangle$ is limited by the Coulomb barrier at low temperature, rises rapidly with increasing temperature, flattens out at $(kT) \approx (2\delta)^3$ and eventually falls off as $(kT)^{-2/3}$. This behaviour is illustrated in figure 5.14 for the d–d, d–t and d–^{3}He reactions. In the case of the very important d–t reaction, the mean transition rate has reached a value close to its maximum at $(kT) \geqslant 25$ keV.

5.6.2.2 The ignition temperature T_c

When we evaluate the energy loss σT^4 by radiation from a black body at a temperature $kT \approx 25$ keV, we find that this assumes the enormous value of $\approx 2.5 \times 10^{38}$ keV s^{-1} cm^{-2}. Clearly such a huge radiation loss could not be powered by any conceivable laboratory fusion reactor operating under conditions of thermal equilibrium between plasma and radiation field. Fortunately thermal equilibrium can never be reached since the mean free path of a 25 keV photon is vastly greater than the spatial dimension of any practicable plasma machine. Exactly this situation arises in the dilute electron gas which exists in the solar corona where electron energies of order 100 eV may be reached in conditions when the local radiation field has a

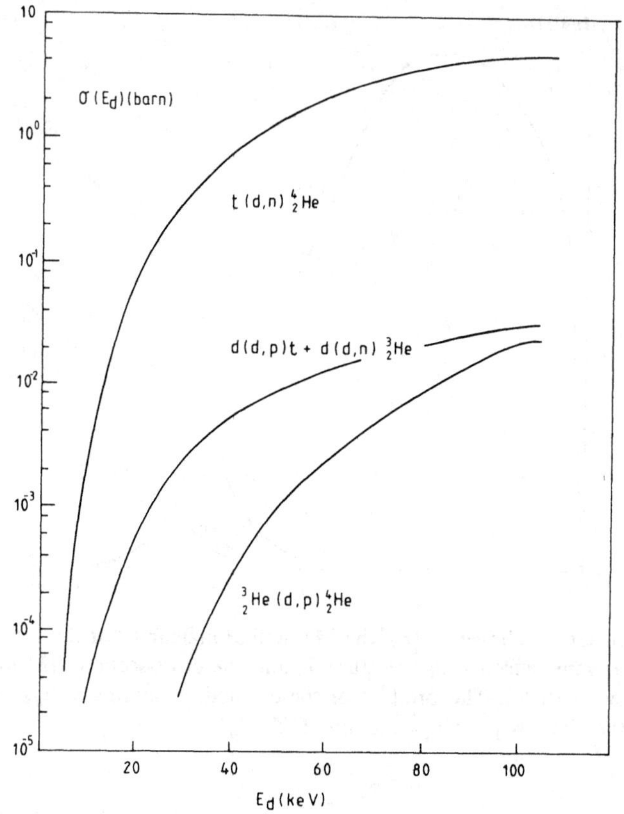

Figure 5.14. The cross-sections $\sigma(E_d)$ for a range of deuteron induced fusion reactions expressed as a function of E_d, the deuteron energy in the laboratory. Only the $t(d, n)$ ${}^4_2$He reaction has a sufficiently large cross-section to allow the possibility of generating fusion energy at temperatures currently achievable in laboratory plasmas.

characteristic temperature $(kT) \approx 0.5$ eV. Thus, in the case of a laboratory plasma, the radiation loss is generated directly by free–free bremsstrahlung collisions between n_e electrons and n_i ions per unit volume of the fully ionized gas.

To determine this energy loss we may suppose, first of all, that ions of only one species are present, and we introduce an energy loss cross-section φ_{rad} defined such that the energy lost by radiation per electron per unit path length is given by (Heitler 1954)

$$\frac{dW}{dx} = -n_i W \varphi_{rad}(W) \tag{5.276}$$

where $W = ((m_e c^2)^2 + c^2 p_e^2)^{1/2}$ is the total (relativistic) energy of the electron.

Since, in the relevant non-relativistic limit $(W - m_e c^2) \ll m_e c^2$, φ_{rad} approaches the constant value

$$\varphi_{rad} \simeq (16/3)\alpha(Zr_e)^2 \qquad W \approx m_e c^2 \tag{5.277}$$

the specific radiation loss per electron is independent of electron energy, and the power generated per unit volume by an electron flux $n_e v_e$ is given by

$$P_{rad} = (n_e v_e)(m_e c^2)(16\alpha/3)Z^2 r_e^2 n_i \tag{5.278}$$

Averaging over a Maxwellian distribution of electron velocity v_e, appropriate to some electron temperature T_e, and including both nuclear species, the average radiated power per unit volume reduces to

$$\langle P_{rad} \rangle = n_e c(8kT_c/\pi m_e c^2)^{1/2}(m_e c^2)(16\alpha/3)(Z_1^2 n_1 + Z_2^2 n_2) r_e^2 \tag{5.279}$$

where $n_e = (Z_1 n_1 + Z_2 n_2)$.

In strong contrast with a black body radiator, a plasma radiates at a rate proportional to $(kT_c)^{1/2}$ and, since the fusion power increases exponentially with temperatures, it follows that, at some ignition temperature T_c defined by (Post 1970)

$$\langle P_{fus}(T_c) \rangle = \langle P_{rad}(T_c) \rangle \tag{5.280}$$

the fusion reaction can sustain itself against radiation, and release net amounts of energy at higher temperatures. The ignition temperature clearly depends on the composition of the plasma but is independent of the absolute ionic number density. In practical systems this number is limited by the heat transfer problem encountered at high number densities and, in magnetically contained plasma, by the magnitude of the magnetic field required to contain the plasma under magnetic forces. At a magnetic field of 100 kG and a temperature $kT \approx 25$ keV the maximum supportable number density is about 10^{16} cm^{-3}, or about 0.4% the number density in an ordinary gas at standard temperature and pressure. Thus the number density in a laboratory plasma is comparable with that in an ordinary discharge tube. In inertially contained plasmas these restrictions on number density do not apply.

Since, according to (5.279) the mean radiated power is proportional to Z^2 it is essential to reduce the flux of plasma near the walls of the containing vessel, and to capture any ionized impurities with $Z > 1$ released there before they can reach the plasma.

5.6.2.3 The Lawson criterion for thermonuclear power production

We have seen that, for a self-sustaining fusion reaction to be possible, a temperature substantially in excess of the ignition temperature must be reached, and, for the favoured d–t reaction, temperatures of the order of 25 keV are required. For the plasma to reach the necessary temperature it must be confined for a minimum time τ, which is defined according to the

Lawson criterion (Lawson 1957), such that the thermonuclear energy released during the time τ, must at least meet the energy required to raise the plasma to its operating temperature. This requirement implies that

$$\langle P_{\text{fus}}\tau\rangle = n_1 n_2 \langle v\sigma(E)\rangle Q_c \geqslant 3/2(n_1+n_2)kT \qquad (5.281)$$

Applying the criterion (5.281) to the d–t reaction, for which $\langle v\sigma(E)\rangle \approx 4.2\times10^{-16}\,\text{cm}^3\,\text{s}^{-1}$ at 25 keV, and assuming that $n_1 = n_2 = n$, we find that

$$n\tau \geqslant (3kT/Q_c)\langle v\sigma(E)\rangle^{-1} \simeq 5\times10^{13}\,\text{cm}^{-3}\,\text{s} \qquad (5.282)$$

Thus even at a maximum number density $n \simeq 10^{16}$ the plasma needs to be confined for periods of at least a millisecond. In such a time a 25 keV hydrogen ion will travel a distance of about 2 km, so that, in the absence of confinement, whether magnetic or inertial or by some other means, the required number density could not be contained, and the energy would be transmitted to the walls of the container rather than to the plasma itself.

An alternative interpretation of the Lawson criterion (5.282) is that it specifies a minimum fraction $f(\tau)$ of the fuel which must be consumed in time τ in order that the reaction be self-sustaining. This fuel is consumed at a rate given by

$$\frac{\mathrm{d}n(t)}{\mathrm{d}t} = -\langle v\sigma(E)\rangle n^2(t) \qquad (5.283)$$

and this simple Ricatti equation integrates immediately to

$$n(t) = \frac{n(0)}{1+\langle v\sigma(E)\rangle n(0)t} \qquad (5.284)$$

Hence rewriting this result in terms of

$$f(\tau) = [n(0) - n(\tau)]/n(0) \qquad (5.285)$$

we can express the Lawson criterion in the form

$$f(\tau)/(1-f(\tau)) = \langle v\sigma(E)\rangle n(0)\tau \geqslant 3kT/Q_c \qquad (5.286)$$

In the case of the d–t reaction at 25 keV we find that $f(\tau) \geqslant 2\frac{1}{2}\%$.

A final question which merits consideration is whether the containment time τ is greater or less than the relaxation time τ_c, due to interparticle Coulomb collisions, which characterizes the rate at which thermal equilibrium is approached. The Rutherford differential scattering cross-section for the long-range Coulomb scattering diverges in the forward direction but, setting some minimum momentum transfer, the integrated cross-section varies inversely as E^2, rather than inversely as E, in the manner of the fusion cross-section given in (5.216), and there is, of course, no barrier penetration factor. Thus to evaluate $\langle v\sigma(E)\rangle$ for Coulomb collisions there is an additional factor E^{-1} in (5.218) and the integral is temperature independent. Thus for Coulomb scattering we may estimate $n\tau_c \approx \langle v\sigma(E)\rangle^{-1} \approx (kT)^{3/2}$. Allowing for shielding by plasma electrons a more rigorous result is (Spitzer 1962)

$$n\tau_c = (2m)^{1/2}(3kT)^{3/2}/5.71\pi(Ze^2)\ln(\Omega) \tag{5.287}$$

where the shielding factor Ω, as given by

$$\Omega = 9(4/3\pi\lambda_d^3)n_e = (3/2Z^2e^3)(k^3T^3)/(\pi n_e)^{1/2} \tag{5.288}$$

is just nine times the number of electrons contained in a sphere surrounding the ionized nucleus, of radius equal to the Debye shielding radius λ_d. Comparing (5.287) with (5.282) and (5.275) we find that

$$(\tau/\tau_c) \approx (kT)^{1/6}\exp[\delta(kT)^{-1/3}] \tag{5.289}$$

We may therefore distinguish two regimes of operation; a low temperature regime with $\tau \gg \tau_c$ and a high temperature regime with $\tau \leqslant \tau_c$. In the low temperature regime a condition approaching thermal equilibrium is established with a scalar pressure p, and well defined electron and ion temperatures.

5.6.3 Confinement of hot plasmas

5.6.3.1 Magnetic confinement

Plasma confinement in the sun and stars is accomplished by the action of gravitational forces, an option which is not available in terrestrial conditions. Most practical systems are based on the application of magnetic or inertial confinement techniques (Kapitza 1979). Magnetic confinement relies on the forces which magnetic fields exert on charged particles to keep them moving in circular orbits about guiding centres which can themselves propagate freely along magnetic field lines (Adams 1966). Confinement geometries may be open or closed, e.g. cylinders or toroids, analogous to the linear and cyclic accelerators of high energy physics. There are also some closed devices, e.g. adiabatic magnetic mirrors, from which trapped particles can escape only through certain escape cones in momentum space. These operate on the same physical principle which is responsible for the Van Allen radiation belts, where charged particles are trapped in the magnetic mirror, formed between the convergent magnetic fields at the polar regions of the earth.

When a volume of hot plasma is confined by a magnetic field, the pressure gradient generates a current $\boldsymbol{J}$, normal to itself and cutting across magnetic field lines. The force exerted on this current by the magnetic field at equilibrium is just sufficient to balance the pressure gradient, i.e.

$$\nabla\cdot\mathbf{P} = \boldsymbol{J}\times\boldsymbol{B}/c \tag{5.290}$$

Here $\mathbf{P}$ is the Cartesian pressure tensor for the plasma, and $\nabla\cdot\mathbf{P}$ is a vector with component $(\nabla \cdot \mathbf{P})_j = \Sigma_i \nabla_i \mathrm{P}_{ij}$. As an example, in a cylindrical tube of plasma aligned parallel to a magnetic field, the pressure gradient at the surface is radial, and the induced surface currents flow in circles about the axis.

In this steady state there is no displacement current and Ampère's law assumes the form

$$\nabla \times \boldsymbol{B} = 4\pi \boldsymbol{J}/c \tag{5.291}$$

Thus we may eliminate $\boldsymbol{J}$ through (5.290) and (5.291) with the result

$$\nabla \cdot \mathbf{P} = (1/4\pi)[(\boldsymbol{B} \cdot \nabla)\boldsymbol{B} - \nabla(\boldsymbol{B}^2/2)] \tag{5.292}$$

The magnetic field factor in (5.292) we recognize as the tensor divergence of the magnetic component of Maxwell's stress tensor, so that this equation may be interpreted as an expression of conservation of momentum in the plasma. If we now assume that we are dealing with a linear plasma with straight magnetic field lines so that $(\boldsymbol{B} \cdot \nabla)\boldsymbol{B} \equiv 0$, equation (5.292) can be rewritten in simple form

$$\nabla_{\mathrm{n}}[P_{\mathrm{n}} + \boldsymbol{B}^2/8\pi] = 0 \tag{5.293}$$

where the subscript 'n' represents the component perpendicular to the magnetic field. A simple integration then yields the relation

$$P_{\mathrm{n}} + \boldsymbol{B}^2/8\pi = \boldsymbol{B}_0^2/8\pi \tag{5.294}$$

where $\boldsymbol{B}$ is the external magnetic field.

The result (5.294) may be interpreted as a pressure balance equation, whereby the total (plasma plus magnetic) pressure in the confinement regime is balanced by the magnetic pressure outside this region. Since the reaction rate is proportional to $n^2 = (P/kT)^2$, this rate can be maximized for a given $\boldsymbol{B}_0$ if conditions can be arranged such that $\boldsymbol{B}$ is zero inside the plasma. We are therefore led to define a thermal utilization factor

$$\beta = P/(B_0^2/8\pi) = 1 - \boldsymbol{B}^2/B_0^2 \qquad 0 \leqslant \beta \leqslant 1 \tag{5.295}$$

such that the fusion rate is proportional to $\beta^2 \boldsymbol{B}_0^2$. Fusion devices may therefore be characterized as low-β or high-β devices depending on whether the pressure of the confined plasma is much less than, or comparable with, the pressure of the confining magnetic field.

One of the earliest examples observed of a magnetic field acting to confine a plasma is the so called θ-pinch (Bennett 1934), where a linear plasma column carrying a high current collapses into a narrow filament along the axis. We can explain this effect as the plasma contracting until the internal pressure P_{n} is balanced by the external pressure $\boldsymbol{B}_0^2/8\pi$, where $\boldsymbol{B}_0$ is the magnetic field generated by the plasma current, and the magnetic field lines form closed loops about the axis. Alternatively we may view the pinch phenomenon as a consequence of the fact that like current elements attract, as opposed to like charge elements which repel. The θ-pinch is an example of a high-β confinement mechanism.

Unfortunately the equilibrium of a linear θ-pinch is unstable and the plasma does not stay in the axis, since even the smallest magnetic gradient immediately drives it to the wall. This is an example of the kink instability (Kruskal and Schwarzschild 1954, Tayler 1957) whereby the magnetic field increases on the concave side of a kink, and decreases on the convex side, so that the plasma always drifts away from the centre of curvature. One consequence of this effect, which is of critical importance in the whole confinement problem, is that it is impossible to establish a plasma in a stable state of pressure equilibrium, in a simple closed torus with axially symmetric magnetic field. Another kind of instability is the tendency for a single filament of plasma to split into two components. This effect is known as the interchange instability, where a state of lower energy is reached when plasma and magnetic field change places. It is related to the Rayleigh–Taylor hydrodynamic instability which occurs when a less dense liquid supports the load of a more dense liquid. In this configuration the interface becomes unstable and the liquids tend to change places.

In addition to these gross (magneto-hydrodynamic) instabilities there also exist many varieties of micro-instability involving the resonant transfer of energy between the particles and the collective excitations (charge density waves, Alvèn waves, etc.) of the plasma. These instabilities are of particular significance in magnetic mirror devices.

Plasmas may be confined for finite times in open (i.e. linear) systems such as high-β thetatrons or low-β adiabatic magnetic mirror devices, in which the interchange instability is suppressed by containing the plasma in a quadrupole magnetic well, where the magnetic field increases in all directions away from the centre. Unfortunately all such systems have inherent plasma leaks at the ends, and are entirely unsuitable for applications in the low-temperature regime, where there is at least approximate thermal equilibrium and hence no pressure gradient along the magnetic field lines. Thus research has concentrated in the main on the development of closed (i.e. toroidal) systems, where the gross instabilities are suppressed, by suitable choice of magnetic field configuration. In the stellarator this is achieved through the application of a rotational transform which breaks the cylindrical symmetry (Spitzer 1962), whereby the closed magnetic field lines within the toroid are twisted by adding external helical field coils inside the main coils wound on the torus. In the tokamak, a name coined from the Russian word meaning 'toroidal chamber with magnetic coils', the rotational transform is provided by a weak poloidal magnetic field, generated by a strong toroidal plasma current. The Joint European Torus (JET) at the Culham Laboratory is a tokamak configured with a D-shaped plasma cross-section designed to enhance the value of β. The action of a tokamak is illustrated schematically in figure 5.15.

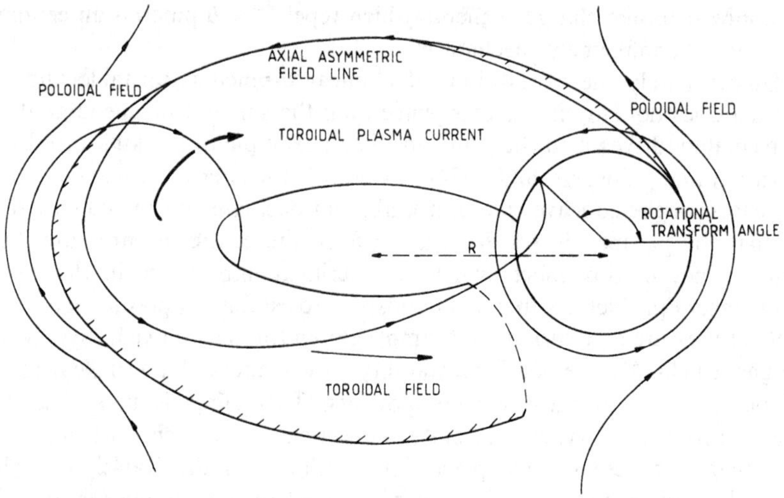

Figure 5.15. The tokamak torroidal configuration for magnetic confinement of a high temperature laboratory plasma. The breaking of the cylindrical symmetry which is necessary for stability is brought about by the poloidal magnetic field associated with the toroidal plasma current.

5.6.3.2 Inertial confinement

The description 'inertial confinement' is applied to systems where the surface layers of a small pellet of plasma are heated so rapidly that they vapourize, and the reaction causes the pellet to contract its radius by factors of the order of 10, leading to extreme compressional heating. Because of its inertia the pellet takes a finite time to explode, and the objective of the device is to obtain a substantial release of thermonuclear energy before the pellet flies apart. This is indeed what happens in an uncontrolled way during an H-bomb explosion. The contraction in the radius reduces the confinement time by a similar amount but this is more than compensated for by the increase in the number density by a factor of $\approx 10^3$ which is necessary to satisfy the Lawson criterion. A further advantage of operating under conditions of high pressure and density is that α-particles produced by the d–t reaction near the centre of the pellet are completely stopped in the surrounding layer leading to a 'thermonuclear burn wave' propagating towards the outer layers of the pellet (Keefe 1982).

Thermonuclear neutrons produced by this means were first observed from a deuterium-loaded lithium pellet irradiated with short-pulse laser light from a neodymium-glass laser (Bosov *et al* 1968). In the intervening period most research on inertially confined fusion plasmas has been conducted using lasers of ever increasing power (Motz 1979). The targets are glass balloons, perhaps 100 μm in diameter and 1 μm thick, filled with deuterium tritium gas and irradiated with an energy flux of 10^{15}–10^{16} W cm^{-2}. To date, central temperatures at the level of 1 keV have been reached with number densities of around 10^{23} cm^{-3} and confinement times of order 10^{-11} s. However, factors of around 10 in temperature and 10^3 in number density are required if break-even conditions are to be achieved. Complementary programmes on electron-beam and ion-beam confinement are also under way (Hirsch 1975).

Although inertial and magnetic confinement systems are superficially very different nevertheless the same problems continue to appear in somewhat different guises. In particular instabilities still play a dominant part, both as regards the efficiency of the initial heating process, and for inhibiting the required high degrees of compression (Brueckner and Jorna 1974).

5.6.4 Design model for a d–t fusion reactor

Until such time that the diagnostic and technical problems associated with the stable confinement and heating of a high temperature plasma have been solved, the detailed design features of a working laboratory fusion reactor must largely remain matters of speculation. Nevertheless sufficient progress has been made on the main objectives of the international fusion programme that a reasonably clear picture of the main characteristics of such a device is already in sight. The fuel cycle in the first fusion reactor will of course be based on the d–t reaction, for which the confinement problem is much reduced in comparison with any of its competitors listed in Table 5.5. Broadly speaking, it is at least necessary to produce temperatures $\geqslant 10$ keV, particle densities $\geqslant 10^{13}$ cm^{-3} and confinement times $\geqslant 5$ ms. Confinement time is the limiting factor for most existing closed geometry fusion devices, since confinement time scales with plasma volume, and the demand for a large volume plasma necessitates a very large financial investment for its fulfillment.

The d–t reaction produces an energy release of 17.6 MeV, of which 14.1 MeV is immediately carried away by neutrons. It is this component of the energy release which must be extracted as heat if the reactor is to function as an energy source. In order to maintain the reaction cycle these very same neutrons must be used to breed tritium with a breeding ratio estimated to lie in the range 1.05 to 1.54 (Alsmiller *et al* 1975). The only practicable sources of neutron-bred tritium are the (n, α) reaction in ^{6}Li and the (n, αn) reaction in ^{7}Li, i.e.

$$\mathrm{n} + {}^6\mathrm{Li} \rightarrow \mathrm{t} + \alpha \tag{5.296a}$$

$$\mathrm{n} + {}^7\mathrm{Li} \rightarrow \mathrm{t} + \alpha + \mathrm{n} \tag{5.296b}$$

The endothermic reaction (5.296b) has a threshold at 2.8 MeV and is far less significant than the exothermic reaction (5.296a). Since the reaction releases an additional 4.6 MeV, the total energy release per d–t reaction is raised to 22.2 MeV with full recovery of the tritium.

In relation to the question of neutron production the direct comparison of fusion and fission reactors is very revealing. In the first place, since each fission event releases about 200 MeV energy, plus 2.5 secondary neutrons, and the same amount of energy is produced by ≈ 10 fusion reactions, there are four times as many neutrons present in a fusion reactor as in a fission reactor at the same power level. Secondly, each fusion neutron carries seven times as much energy as a secondary fission neutron emitted with ≈ 2 MeV energy. This increase in the intensity and hardness of the neutron spectrum is clearly going to create a major shielding problem.

To fulfil the several requirements outlined above, a thermonuclear reactor may be visualized as having the general layout shown schematically in figure 5.16. The fusion zone lies at the centre of the system where the hot plasma is confined by a system of superconducting magnetic field coils which form the outermost layer of the complete apparatus. In tokamak devices the azimuthal currents associated with poloidal fields required to maintain stability are insufficient to heat the plasma to the necessary temperature. Primarily this is because of the excessive time taken to transfer the energy from the electrons to the ions by Coulomb collisions (Kapitza 1975), but also because the plasma conductivity increases with increasing temperature.

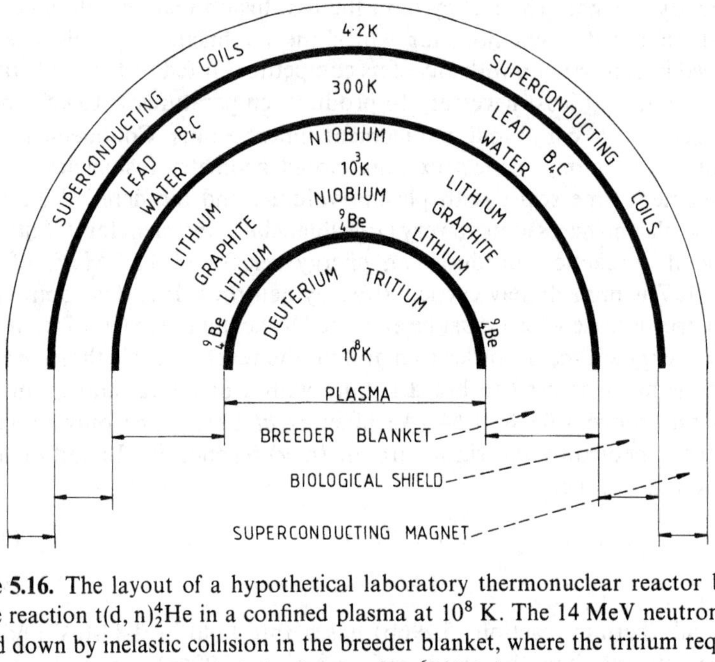

Figure 5.16. The layout of a hypothetical laboratory thermonuclear reactor based on the reaction t(d, n)$^4_2$He in a confined plasma at 10^8 K. The 14 MeV neutrons are slowed down by inelastic collision in the breeder blanket, where the tritium required as a primary fuel is replenished via the reaction ^{6}Li(n, α)t.

This heating is accomplished either by injecting beams of pre-accelerated neutral deuterium atoms (Ribe 1975) or by applying radiofrequency energy resonantly coupled to magnetoacoustic Alvèn waves, where the energy is transferred directly to the ions (Kapitza and Pitaevskii 1975).

The plasma is enclosed in a vacuum wall, which is called on to withstand intense fluxes of neutrons and γ-rays, and presumably must be constructed from some refractory material such as tungsten or molybdenum. This vacuum wall separates the fusion region from the breeder blanket whose principal functions are: (1) to multiply the incident neutron flux by (n, 2n) reactions in suitable nuclei to compensate for those neutrons lost in parasitic (n, γ) reactions; (2) to slow down these 14 MeV neutrons to thermal energies where they are captured in lithium via either of the two reactions given in (5.296). The burnt tritium is thus replaced and returned to the plasma. In the final step the energy given up by the neutron during the slowing down process must be recovered and transferred to an external heat sink by a heat exchanger.

As pointed out above, the slowing down problem is extremely severe for 14 MeV neutrons, but may be solved by inelastic scattering in medium or heavy mass nuclei, e.g. graphite, iron, niobium and molybdenum. Elastic scattering in hydrogenous material is inadequate for this purpose since the relevant cross-sections are much reduced at 14 MeV. The slowing down process will produce an intense γ-ray flux, which will be augmented by (n, γ) reactions. Radiation damage by (n, 2n), (n, p) and (n, α) reactions in the breeder material is also likely to create severe metallurgical problems. Many nuclei suitable for slowing down also have substantial (n, 2n) cross-sections, but the most efficient neutron multiplier is the reaction in beryllium

$$n + {}^{9}Be \rightarrow 2\alpha + 2n \tag{5.297}$$

This endothermic reaction has a threshold at ≈ 3 MeV and the cross-section approaches 0.5 barn at 14 MeV. The lithium required to regenerate the tritium according to (5.296) can be incorporated in the blanket either as a solid, e.g. $LiAl_2O_3$ or as a liquid, e.g. liquid LiF which can also serve as a coolant.

The final element in the system is the biological shield whose purpose is to remove the surviving 14 MeV neutrons and to attenuate the γ-ray flux. For these ends, a heavy element such as lead, which has a large inelastic neutron cross-section, is very desirable. This may be mixed with borated carbon B_4C, to eliminate the residual slow neutrons through the reaction ${}^{10}B(n, \alpha){}^{7}Li$ (cf. Table 1.4). The total radioactivity generated by a thermonuclear reactor operating at $\approx 10^3$ MW, corresponding to the power output of a conventional coal or oil-burning power station, has been estimated to come within the range 10^8–10^9 Ci (Steiner 1975), a not inconsiderable hazard from the viewpoint of health and safety. Thus, although no actinides or long-lived fission products are created in a fusion reactor, and the choice of those radio-nuclides which are produced falls much more under the control of the designer, it is not immediately obvious that the problems of radioactive waste disposal, which have plagued fission power programmes across the world, will be entirely absent in the thermonuclear power plants of the future.

5.6.5 Muon catalysed fusion

Because the muon is 207 times more massive than the electron, and the Bohr radius in a mu-mesic atom $a_\mu = \hbar^2/e^2 m_\mu$ is reduced in proportion, even at low temperatures the two nuclei in a mu-mesic molecular ion, e.g. $(\mathrm{dd}\mu)^+$ or $(\mathrm{dt}\mu^+)$, have a mean separation of order $2a_\mu$, corresponding to a classical distance of closest approach equal to a_μ. This effect, which is a consequence of the shielding of their mutual Coulomb potential by the orbiting muon, implies that their relative energy E which appears in $G(E)$, the Gamow barrier penetration factor (5.265), is effectively replaced by (Jackson 1957)

$$E \to e^2/a_\mu \approx 5.6\ \mathrm{keV} \tag{5.298}$$

Thus the conditions for fusion correspond to those which apply in a d–t plasma with $kT \approx 5.6$ keV; specifically, in the case of the $(\mathrm{dt}\mu)^+$ molecular ion, the lifetime for fusion is about 10^{-10} s, i.e. very much shorter than the 2.2 μs lifetime of the attached muon.

The potential for mu-mesic atom formation as a mechanism for enhancement of p–d fusion was first recognized by Frank (1947), and this very process was observed, and correctly identified, in the course of a study of μ^- interaction in a liquid hydrogen bubble chamber a decade later (Alvarez *et al* 1957). Although muon induced fusion in both d–d and d–t reactions had been subjected to a substantial amount of theoretical study from the beginning (Zel'dovich 1954, Zel'dovich and Sakharov 1957), the possibility of exploiting muon induced fusion in a cold deuterium–tritium gas as an energy source did not attract worldwide attention, until it was predicted that a single muon could catalyse ≈ 100 fusions (Gershstein and Ponomarev 1977). This claim has since been amply verified by a wealth of experimental evidence (Jones *et al* 1983, Jones 1986, 1987). This amplification in efficiency is due to the existence of a weakly bound rotational–vibrational level in $(\mathrm{dt}\mu)^+$, with a binding energy of about 0.66 eV, which may be resonantly created in the scattering of $(\mathrm{t}\mu)$ atoms or deuterium molecules, subsequently to become the nucleus of the mu-mesic molecular complex $[(\mathrm{dt}\mu)(\mathrm{de})]^+$ (Vesman 1967, Vinitsky *et al* 1979). The net result is that, under the best conditions, no more than $\approx 10^{-8}$ s elapses between successive fusions catalysed by the one muon.

The development of a muon catalysed fusion cycle in a deuterium–tritium gas is represented schematically in figure 5.17 (Breunlich *et al* 1989, Ponomarev 1990). Aside from the resonant creation of the $(\mathrm{dt}\mu)^+$ ion, the

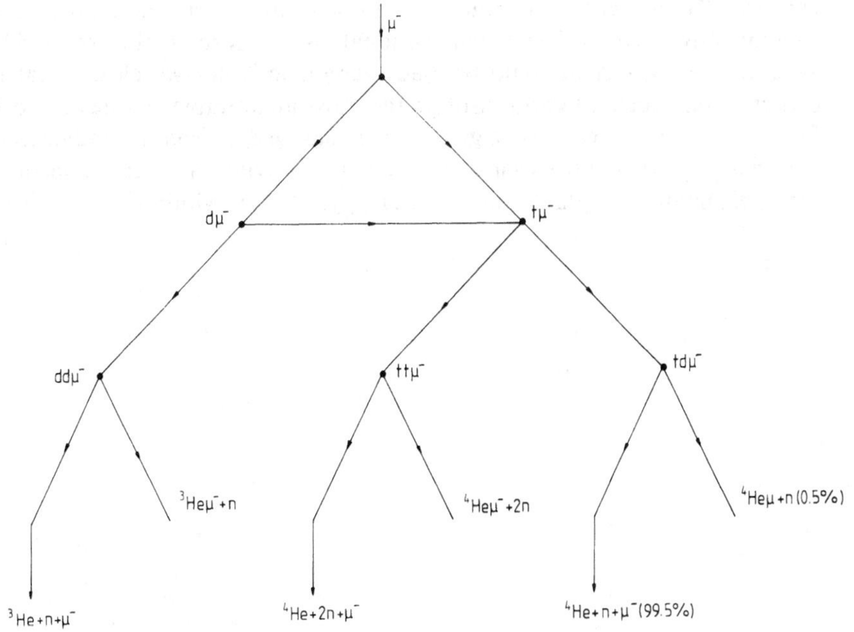

Figure 5.17. Diagram illustrating the reaction chains leading to muon-catalysed fusion in a cold deuterium–tritium gas. The process relies on the resonant creation of the ion $(dt\mu)^+$ where the presence of the muon reduces the Coulomb barrier inhibiting d–t fusion.

most important requirement for the cycle to be efficient, is that the probability ω_s that the muon should stick to the residual nucleus (^{4}He or ^{3}He) should be small (Jackson 1957). Theoretical estimates giving $\omega_s \leqslant 1\%$ (Gershstein *et al* 1981, Bracci and Florentini 1981) have been substantially verified; a recent measurement gives the value $\omega_s = (0.45 \pm 0.05)\%$ (Breunlich *et al* 1987). Taking account of the resonance production of $(dt\mu)^+$ and the small sticking probability, the fusion energy released per initial catalysing muon comes to about 1.8 GeV.

Unfortunately it seems that a minimum 6 GeV energy must be expended to produce the catalysing negative muon, even under the most favourable circumstances. The most promising production technique appears to be one where a 2 GeV deuteron beam interacts with a self-conjugate nucleus producing a negative pion in about one-third of all reactions. This subsequently decays in flight with a lifetime of around 2.10^{-8} s producing a muon which becomes available for catalysis. There remains at least a deficit of a factor of 3 to be made up before muon catalysed fusion can be contemplated as an energy source. To date the favoured means for overcoming this deficit is the proposal to use the 14 MeV neutrons to convert ^{238}U nuclei in a surrounding blanket into ^{239}Pu; these nuclei are subsequently used to fuel a conventional fission reactor (Petrov 1980). Whether such a system could be made economic is not yet clear; what is certain is that such a hybrid displays the most undesirable features of both fusion and fission reactors, e.g. fast neutrons *and* poisonous radioactive materials. It therefore seems unlikely to provide an attractive and environmentally acceptable source of energy, at least within the forseeable future.

6 Foundations of neutron optics

6.1 Wave–particle duality

6.1.1 Matter waves

Matter waves were postulated by de Broglie (1924), and first observed by Davisson and Germer (1927), and by Thomson and Reid (1927), using Bragg diffraction of electrons in crystals as originally proposed by Elsasser (1925). Over the next two decades first Fresnel diffraction, and subsequently interference by division of amplitude, were observed with electrons, so completing the full complement of scalar wave phenomena contained in the Huygens–Young–Fresnel theory of light.

The principal experimental difficulty which arises in demonstrating wave phenomena with electrons is that these are light charged particles whose motion is highly sensitive to the presence of electromagnetic fields in free space or in material media. To reduce the influence of such fields an electron kinetic energy of at least 50 keV is required, and at this energy the de Broglie wavelength is of order 10^5 times less than the wavelength of visible light (Simpson 1956). None of these problems arise with neutrons which are neutral and therefore interact only feebly with electromagnetic fields. In particular, because neutrons are heavy, even at thermal energies ($E_n \simeq 0.025$ eV) they have de Broglie wavelengths $\geqslant 1$ Å, and are therefore ideally matched to the spacings between planes of atoms in crystals (Badurek *et al* 1988).

The superiority of neutrons over electrons is even more clear cut when it comes to the question of demonstrating polarization phenomena with matter waves. The dichotomic variable now associated with electron spin first made its appearance as a fourth quantum number required to establish Pauli's exclusion principle, and was somewhat vaguely associated, in that context, with an internal non-mechanical stress. Its connection with angular momentum was proposed by Uhlenbeck and Goudsmidt (1925, 1926), and the utility of such a notion was confirmed by Thomas (1926) in his explanation of the kinematic effect known today as the Thomas precession (cf. section 3.3.2.3). However, the spin states of free electrons are not easily manipulated in electromagnetic fields; for example they cannot be spatially separated in a Stern–Gerlach apparatus since the resultant separation is always less than the uncertainty in position associated with the action of the Lorentz force on the charge (Mott and Massey 1949). In contrast controlling the neutron spin is rather easy and many efficient methods have been developed for producing intense beams of highly polarized neutrons (cf. section 7.4.3).

The real problem is to understand why optical phenomena observed with slow heavy neutral fermionic particles such as neutrons should resemble so closely the equivalent phenomena observed with relativistic massless bosonic particles such as photons. The problem is even more difficult when it comes to polarization for, although the neutron polarization is an angular momentum, only the circularly polarized component in an electromagnetic wave is directly associated with angular momentum, and this has been demonstrated experimentally (Beth 1936). The linear polarization components are, in fact, not components of an axial vector, but of a second rank tensor (cf. section 3.3.1). The real reason why neutron and electromagnetic polarization effects are so similar is that, because the photon is massless, the longitudinal component is missing. Thus, even though the photon is a spin 1 boson, like the neutron it has only two polarization states. As was recognized long ago (Feynman *et al* 1957) all two-state quantum systems behave in much the same way.

We shall return to the polarization again in section 7.4, but for the moment we shall confine the discussion to scalar waves, and attempt to explore not only those aspects of matter waves and electromagnetic waves in which they closely compare, but also those aspects in which they contrast. It turns out that almost all those properties of electromagnetic waves with which we are familiar, reflection, refraction, interference, etc. can be reproduced with neutrons. The contrasts are few but important (cf. section 1.6.6).

6.1.2 Wave optics

The rectilinear propagation of light and the laws governing its reflection from polished surfaces were well known to the Greek astronomers of the Alexandrian age some two thousand years ago. They also understood the application of these laws to image formation in concave and convex mirrors, and indeed the mechanician Hero (c. 125 BC), whose extensive writings on all manner of scientific topics, including hydraulic organs and steam turbines, have come down to us in Arabic translation, showed that the laws of propagation and reflection of light could be summarized in the proposition that light travels between two points by the shortest of all possible paths. His was the first recorded statement of a set of physical laws in terms of an 'action principle', a procedure subsequently developed and applied by Euler, Lagrange and Hamilton to the whole realm of optics and mechanics, culminating in the 'integral over all paths' formulation of quantum field theory propounded by Feynman (1948) and Schwinger (1951).

The Alexandrian astronomers were also familiar with the phenomenon of optical refraction, and the most renowned of these, Ptolemy (c. AD 140), in order to calculate the deviations produced by atmospheric refraction on the trajectories of light rays from sun and stars, carried out an extensive series of measurements on the refraction of light by water, with the objective of establishing the fundamental law of refraction. He was, however, unsuccessful in this aim and even Kepler (c. 1609) who re-analysed Ptolemy's results, did not succeed in establishing a satisfactory empirical law. Snell's law, which states the law of refraction in terms of the very important concept of refractive index, was not announced until 1621. Subsequently it was shown by Fermat (c. 1657) that Snell's law followed from his own principle of least time, which states that, no matter what sequence of reflections and refractions a light ray pursues in its passage between two points, it always travels by that path which makes the total time of travel a minimum. From this principle Fermat was led to conclude, correctly, that light travels more slowly in the optically denser medium. Fermat's construction for Snell's law of refraction at a plane boundary is shown in figure 6.1.

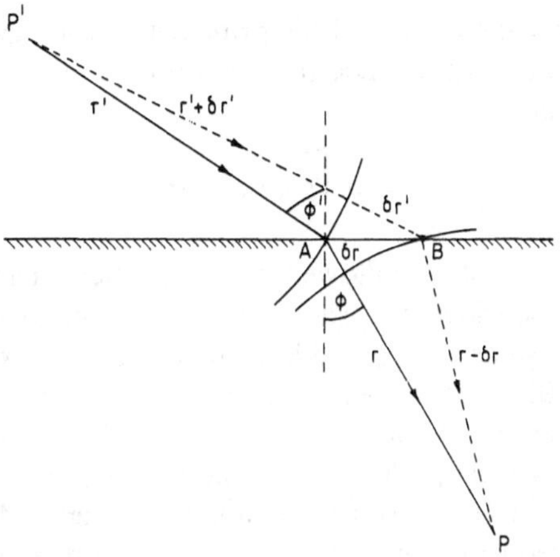

Figure 6.1. Construction based on Fermat's principle of least time leading to Snell's law of refraction at a plane surface. If P'AP is the path actually traversed by the light in its passage from P' to P, and P'BP is a neighbouring path differing infinitesimally from the actual path it is necessary that $\delta t' = \delta r'/v_p' = \delta t = \delta r/v_p$. This is Snell's law since $\delta r' \sin \varphi' = \delta r \sin \varphi$.

That sound is transmitted to the ear from string or pipe through vibrations of the intervening medium is an idea which dates from antiquity, and the discovery of the finite speed of light by Römer in 1676, and the observation during Newtonian times of diffraction effects in optical instruments, were early evidences that light, like sound, is an oscillatory phenomenon. However, the fact that light, unlike sound, propagates in a vacuum, was a severe difficulty for such theories, and Huygens' wave theory of light, proposed in his Traité de la Lumière published in 1690, did not therefore gain immediate acceptance. However, one notable feature of the theory was its ability to attach a simple meaning to the notion of refractive index, which is taken to be the ratio of the velocity of light in vacuum to the velocity of light in the medium. On this interpretation Fermat's principle assumes the form

$$\delta \int_A^B \frac{ds}{v_p} = 0 \tag{6.1}$$

where v_p is the phase velocity of the light, which may be different in different media, and the symbol δ denotes a first order variation from the actual path of the light in its passage from A to B, to a neighbouring path connecting the same end-points. Strictly speaking the expression (6.1) of Fermat's principle implies only that the passage time be stationary rather than minimum, and other forms of variational principle are subject to the same caveat.

In 1744 Maupertuis published his principle of least action in mechanics which, for the motion of a single particle of mass m in a conservative field of force, states that

$$\delta \int_A^B mv\, ds = 0 \tag{6.2}$$

along the actual trajectory followed by the particle. The requirement that all virtual paths considered must have the same energy was a restriction recognized in the same year by Euler. The principle was later extended by Lagrange (c. 1760) to mechanical systems in general, where it leads directly to the Euler–Lagrange equations, and by Hamilton (1835), who removed the restriction to virtual paths of the same total energy.

In applying his principle to the propagation of light Maupertuis adopted the view, advocated with some fervour by the Epicurean poet Lucretius, that light is composed of a stream of minute particles, and he was equally successful with Fermat in deducing Snell's law. However, he reached the diametrically opposite conclusion that light travels faster in the optically denser medium, and this follows from the different roles played by the velocity functions in (6.1) and (6.2). In the Planck–Einstein model of photons there is of course no conflict since a light quantum of energy $E = \hbar\omega$ is assumed to carry a momentum $p = \hbar k = E/v_p$.

The appearance of Hamilton's classic work *A Theory of Systems of Rays* (Hamilton 1828, 1830, 1831) created a revolution in optical theory as profound in its way as that which had reshaped mechanics following the publication of Lagrange's *Méchanique Analytique*, half a century earlier, and the predictive power of the new method was triumphantly confirmed by the observation of conical refraction in biaxial crystals, following an investigation proposed by Hamilton (Lloyd 1833). Without committing himself to any particular view as to the ultimate nature of light, Hamilton defined the action function $\tilde{S}$ through the equation

$$\tilde{S} = \int_A^B n\, ds \tag{6.3}$$

where n is the refractive index. The Euler–Lagrange equations derived from the action principle $\delta\tilde{S}=0$ then gave rise to rectilinear propagation in a homogeneous medium as they must. The novelty of Hamilton's construction consists in recognizing that $\tilde{S}$ is a function of the coordinates of the end points of the path of integration only, when the integral is performed along an actual ray of the system, which varies from one ray to the next according to a law which Hamilton called the principle of varying action. He thereby transformed the mathematical theory of optics from a purely geometric theory to an algebraic theory, whose sole objective is the solution of the first-order and second-degree equation for $\tilde{S}(\boldsymbol{r})$

$$(\nabla\tilde{S})^2=\left(\frac{\partial\tilde{S}}{\partial x}\right)^2+\left(\frac{\partial\tilde{S}}{\partial y}\right)^2+\left(\frac{\partial\tilde{S}}{\partial z}\right)^2=n^2(\boldsymbol{r}) \tag{6.4}$$

In geometric optics the action $\tilde{S}$ is known as the 'eikonal' or image function (Bruns 1895); in mechanics, where the concept was introduced somewhat later (Hamilton 1837, Jacobi 1837), it is known as Hamilton's characteristic function.

In Hamilton's theory of optics the light rays are defined to be the orthogonal trajectories to surfaces of constant $\tilde{S}$ which constitute the wavefront, and their equations are

$$\nabla\tilde{S}=\boldsymbol{n}(\boldsymbol{r}) \tag{6.5}$$

where $\boldsymbol{n}(\boldsymbol{r})$ is a vector of magnitude $n(\boldsymbol{r})$ normal to the wavefront. On the older theory based on Fermat's principle the light rays were defined instead as those curves along which the action integral (6.3) was stationary. It is necessary of course to restrict the range of variation on this integral to be infinitesimal, since in general more than one light ray from a source point can pass through a given field point, and this is the basis of all focusing devices. The locus of all such points of intersection defines a caustic surface with respect to a source point, and is said to be the evolute of the corresponding wavefront surface.

In the Huygens–Fresnel–Maxwell theory of light each component of the optical field satisfies d'Alembert's wave equation

$$\nabla^2\tilde{\Psi}(\boldsymbol{r},t)-\frac{n^2}{c^2}(\boldsymbol{r})\frac{\partial^2\tilde{\Psi}}{\partial t^2}(\boldsymbol{r},t)=0 \qquad n(\boldsymbol{r})=c/v_{\mathrm{p}}(\boldsymbol{r}) \tag{6.6}$$

and, excluding some exceptional but very important cases, the eikonal equation (6.4) may be deduced from the wave equation (6.6) in the limit of very short wavelength. To prove this assertion we assume that $\tilde{\Psi}(\boldsymbol{r},t)$ is purely harmonic, i.e.

$$\tilde{\Psi}(\boldsymbol{r},t)=\tilde{\Psi}(\boldsymbol{r})\,\mathrm{e}^{-\mathrm{i}\omega t} \tag{6.7}$$

where $\tilde{\Psi}(\boldsymbol{r})$ satisfies the scalar Helmholtz equation

$$\nabla^2\tilde{\Psi}+k^2\tilde{\Psi}=0 \qquad k=n(\boldsymbol{r})\omega/c=k_0n(\boldsymbol{r}) \tag{6.8}$$

and $\lambda_0 = 2\pi/k_0$ is the wavelength in free space. We now seek for an oscillatory solution to (6.8)

$$\tilde{\Psi}(\boldsymbol{r}) = \tilde{A}(\boldsymbol{r})\, e^{ik_0\tilde{S}(\boldsymbol{r})} \tag{6.9}$$

where, in comparison with the rapidly oscillating function $e^{ik_0\tilde{S}}$, the amplitude $\tilde{A}(\boldsymbol{r})$ and the 'optical path' $\tilde{S}(\boldsymbol{r})$ are slowly varying real functions of $\boldsymbol{r}$ which, considered as functions of k_0, remain finite in the limit $k_0 \to \infty$. Substituting (6.9) into (6.8) we find, equating real and imaginary parts to zero,

$$2\tilde{A}(\boldsymbol{r})(\tfrac{1}{2}\tilde{A}(\boldsymbol{r})\nabla^2\tilde{S}(\boldsymbol{r}) + \nabla\tilde{A}(\boldsymbol{r})\cdot\nabla\tilde{S}(\boldsymbol{r})) \equiv \nabla\cdot(\tilde{A}^2\nabla\tilde{S}) = 0 \tag{6.10}$$

and

$$(\nabla\tilde{S})^2 = \frac{k^2}{k_0^2} + \frac{\nabla^2\tilde{A}}{k_0^2 A} = n^2(\boldsymbol{r}) + \nabla^2\tilde{A}/k_0^2\tilde{A} \tag{6.11}$$

The eikonal equation (6.4) follows from the dynamical equation (6.11) in the short wavelength limit $k_0 \to \infty$ provided only that $\nabla^2\tilde{A}/k_0^2\tilde{A}$ vanishes in this limit. Thus surfaces of constant action in the eikonal approximation may be identified with the surfaces of constant phase which constitute the wavefront. Further, since the wave intensity is proportional to $\tilde{A}^2(\boldsymbol{r})$, and the ray trajectories point along the vector $\nabla\tilde{S}(\boldsymbol{r})$, the kinematical equation (6.10) expresses the conservation of energy and is the equivalent of Poynting's theorem in Maxwell's theory of electromagnetic waves. Equation (6.10) also fixes the component of $\nabla\tilde{A}(\boldsymbol{r})$ in the direction of $\nabla\tilde{S}(\boldsymbol{r})$ but places no restriction on the components of $\nabla\tilde{A}(\boldsymbol{r})$ in directions normal to the direction of propagation of the wavefront. Thus $A(\boldsymbol{r})$ can vary arbitrarily in directions tangential to a surface of constant $\tilde{S}$, and can even change discontinuously in these directions. Since the diffuse shadow associated with diffraction effects near the boundary of an opaque object in the wave theory of light becomes a sharp shadow in the geometric optics approximation $k_0 \to \infty$, $\nabla\tilde{A}$ varies rapidly in the neighbourhood of the geometric shadow and the approximation $\nabla^2\tilde{A}/k_0^2\tilde{A} \to 0$ may even break down. This failure of the geometric optics approximation can lead to quite dramatic effects in certain circumstances (cf. section 7.1.4).

6.1.3 Wave mechanics

Hamilton's characteristic function S is defined for a classical mechanical system specified by generalized coordinates (q_i, p_i) by the analogue of (6.4); i.e.

$$S = \int_A^B \left(\sum_i p_i\,\delta q_i\right) \tag{6.12}$$

where the variation δS in S is given by

$$\delta S = \left|\sum_i p_i\,\delta q_i - H\,\delta t\right|_A^B \tag{6.13}$$

Thus, for given S, p_i and H may be defined as the coefficients of the differentials in (6.13) or conversely, if H is given as a function of p_i and q_i, application of (6.13) shows that S satisfies the Hamilton–Jacobi equation. For a particle moving in a conservative field $V(\boldsymbol{r})$ this equation is

$$(\nabla S)^2 \equiv \left(\frac{\partial S}{\partial x}\right)^2 + \left(\frac{\partial S}{\partial y}\right)^2 + \left(\frac{\partial S}{\partial z}\right)^2 = p^2 = 2m_0(E-V) \tag{6.14}$$

This result is clearly the mechanical analogue of the eikonal equation (6.4) and shows that something like a wavefront may be associated with the motion of particles even in classical mechanics. It was de Broglie's (1924) rediscovery of this connection that set the scene for the development of wave mechanics. The problem of finding the quantum wave equation which corresponds to (6.14) in the same sense that d'Alembert's classical wave equation (6.6) corresponds to the eikonal equation (6.4) was of course solved by Schrödinger.

De Broglie adopted Planck's quantum hypothesis $E=\hbar\omega$ and assumed that a particle of mass m_0 had associated with it in its rest frame a stationary periodic motion $\exp(-i\omega_0 t)$ where $m_0c^2=\hbar\omega_0$. In a frame of reference in which the particle moves with velocity v along the x-axis, application of relativity theory shows that the time t in the phase of this stationary wave is transformed to $\gamma(t-vx/c^2)$ where $\gamma=(1-(v/c)^2)^{-1/2}$ and the associated periodic motion is a plane progressive wave moving along the x-axis with phase velocity v_p, and wavelength λ, given by

$$v_p = \gamma\hbar\omega_0/\gamma m_0 v = \omega/k \tag{6.15a}$$

$$\lambda = \frac{2\pi}{k} = \frac{2\pi\hbar}{\gamma m_0 v} = \frac{\hbar}{p} \tag{6.15b}$$

These results follow immediately from the dispersion formula

$$\omega = \omega_0\gamma = \omega_0[1+(\hbar k/m_0c)^2]^{1/2} \tag{6.16}$$

which expresses the relation between the energy and momentum of a relativistic particle. The most striking feature of the de Broglie hypothesis is the relation between the group velocity v_g of the wave and the velocity v of the particle

$$v_g = \frac{d\omega}{dk} = v \tag{6.17}$$

a result which not only shows that energy is transported at the same rate by both wave and particle, but also brings about the required accommodation between the variational principles of Fermat (6.1) and Maupertuis (6.2). De Broglie also drew attention to the fact that the Bohr–Sommerfeld quantum conditions which formed the basis of the old quantum theory of the atom could be recovered by postulating that the path of the electron in a stationary orbit always contains an integral number of wavelengths.

In Schrödinger's development of de Broglie's ideas (Schrödinger 1926), he placed the emphasis more on the standing wave rather than on the progressive wave, and it was through addressing the problem of the spatial distribution of these waves that he was led to propose that the amplitude $\Psi_E(\boldsymbol{r})$ should satisfy the partial differential equation

$$-\frac{\hbar^2}{2m}\nabla^2\Psi_E(\boldsymbol{r})+V(\boldsymbol{r})\Psi_E(\boldsymbol{r})=E\Psi_E(\boldsymbol{r}) \tag{6.18}$$

For $E<0$ integral quantum numbers emerge quite naturally in the same way as the node numbers do in the theory of the vibrating string. For $E>0$ equation (6.18) can be rewritten in the form of a scalar Helmholtz equation

$$\nabla^2\Psi_E(\boldsymbol{r})+k_0^2n^2(\boldsymbol{r})\Psi_E(\boldsymbol{r})=0 \tag{6.19}$$

where

$$k_0^2=\left(\frac{2\pi}{\lambda_0}\right)^2=\frac{2mE}{\hbar^2} \tag{6.20}$$

λ_0 is the de Broglie wavelength (6.16) for a free particle of energy E. The 'neutron refractive index' $n(\boldsymbol{r})$ is defined by

$$n^2(\boldsymbol{r})=1-V(\boldsymbol{r})/E \tag{6.21}$$

What this result shows is that a region for which $V(\boldsymbol{r})>0$, is neutron-optically less dense than vacuum and therefore allows the possibility that neutrons can be confined in such a region through the phenomenon of total external reflection.

In wave mechanics the time-dependent wave solution $\Psi_E(\boldsymbol{r}, t)$ which corresponds to the optical solutions (6.7) must have a time-dependence proportional to $\exp(\pm iEt/\hbar)$, but this requirement combined with equation (6.18) is not sufficient to determine uniquely the wave equation satisfied by $\Psi(\boldsymbol{r}, t)$ when the energy E is eliminated. One clear possibility is that both real and imaginary parts of $\Psi(\boldsymbol{r}, t)$ should satisfy d'Alembert's wave equation (6.6), with a phase velocity given according to (6.15)–(6.16) by

$$v_p=\frac{\omega}{k}=\frac{E}{p}=\frac{2E}{\sqrt{2m(E-V)}} \tag{6.22}$$

However, this solution fails for a time-dependent potential $V(t)$, unless the amplitude equation (6.18) is replaced by a fourth-order partial differential equation similar to that satisfied by a vibrating plate.

For these reasons Schrödinger rejected d'Alembert's equation as a suitable candidate for a fundamental wave equation, and adopted in its place the far simpler alternative

$$-\frac{\hbar^2}{2m}\nabla^2\Psi(\boldsymbol{r},t)+V(\boldsymbol{r})\Psi(\boldsymbol{r},t)=i\hbar\frac{\partial\Psi}{\partial t}(\boldsymbol{r},t) \tag{6.23}$$

which can be extended without difficulty to include non-conservative systems. This equation is first order in time and for this reason the interpretation of $|\Psi(\boldsymbol{r},t)|^2$ as a probability density function satisfying the conservation law (2.52) can be upheld.

The real and imaginary parts of $\Psi(\boldsymbol{r},t)$ each satisfy an equation which is second order in time, i.e.

$$\frac{\partial^2\chi}{\partial t^2}(\boldsymbol{r},t)+\omega^2(\boldsymbol{r},\nabla)\chi(\boldsymbol{r},t)=0 \tag{6.24}$$

where $\hbar\omega(\boldsymbol{r},\nabla)$ is the differential operator

$$\hbar\omega(\boldsymbol{r},\nabla)=H(\boldsymbol{r},\boldsymbol{p})=-\frac{\hbar^2}{2m}\nabla^2+V(\boldsymbol{r}) \tag{6.25}$$

Because $\chi(\boldsymbol{r},0)$ and $(\partial\chi/\partial t)(\boldsymbol{r},0)$ can be prescribed arbitrarily when $\chi(\boldsymbol{r},t)$ satisfies an equation which is second order in time, the definition of a conserved density function which is everywhere non-negative is impossible. It follows therefore that the Schrödinger wavefunction $\Psi(\boldsymbol{r},t)$ is inherently complex, since neither equation (6.24) nor d'Alembert's equation (6.6) can be interpreted as a single-particle equation.

6.1.4 Interpretation of the wavefunction

According to Born's (1926) interpretation of quantum mechanics the wavefunction Ψ describes, not a single system, but an ensemble of identical (perhaps many particle) systems prepared in the same quantum state. Knowledge of Ψ then permits a calculation to be made of the probability P_i for observing the eigenvalue a_i when the observable A is measured. When this happens the wave function is said to collapse, i.e. to change discontinuously from its initial form into the corresponding eigenfunction of A. It is not clear how this collapse is supposed to come about or even precisely what is to be understood by a 'measurement' (Everett 1957, Wigner 1963).

There are other conceptual difficulties in the theory, as famously exemplified by 'Schrödinger's cat' which may in principle be prepared in a linear superposition of quantum states describing respectively a 'live cat' and a 'dead cat' (Schrödinger 1935). Thus the cat is neither alive nor dead until the issue is settled by the appropriate measurement, a classic case of the famous complaint 'if looks could kill!'. (For a potential resolution of this paradox cf. section 6.4.2.) However, the great majority of experimental scientists are not too concerned with those problems, and are happy to accept the Copenhagen interpretation of quantum mechanics according to which the wavefunction Ψ is to be regarded only as establishing a prescription for carrying out practical calculations (Bohr 1935). Nevertheless the debate as to whether quantum mechanics is in some sense exact or only a somewhat mystifying approximation to a deeper theory in which all difficulties will be resolved, goes on (Einstein *et al* 1935, Bell 1964, Bohm and Bub 1966).

6.1.5 The Dirac formalism

In order to maintain Born's probabilistic interpretation, the wavefunctions $\Psi(\boldsymbol{r}, t)$ must be square-integrable functions constituting the elements of a Hilbert space, i.e. an infinite-dimensional linear vector space of positive definite metric with a defined scalar product

$$(\chi, \Psi) = \int \chi^*(\boldsymbol{r}, t)\Psi(\boldsymbol{r}, t)\,\mathrm{d}^3r = (\Psi, \chi)^* \tag{6.26}$$

The eigenfunctions of any complete set of commuting observables provide an orthonormal discrete set of basis states in this space. Again, since $\Psi(\boldsymbol{r}, t)$ is square integrable it may be represented in terms of its Fourier transform

$$\Psi(\boldsymbol{r}, t) = (2\pi\hbar)^{-3/2} \int \varphi(\boldsymbol{p}, t)\,\mathrm{e}^{\mathrm{i}\boldsymbol{p}\cdot\boldsymbol{r}/\hbar}\,\mathrm{d}^3p \tag{6.27}$$

where

$$\varphi(\boldsymbol{p}, t) = (2\pi\hbar)^{-3/2} \int \Psi(\boldsymbol{r}, t)\,\mathrm{e}^{-\mathrm{i}\boldsymbol{p}\cdot\boldsymbol{r}/\hbar}\,\mathrm{d}^3r \tag{6.28}$$

The Fourier transform $\varphi(\boldsymbol{p}, t)$ satisfies the Schrödinger equation in momentum space

$$\mathrm{i}\hbar\frac{\partial\varphi}{\partial t}(\boldsymbol{p}, t) = \frac{p^2}{2m}\varphi(\boldsymbol{p}, t) + (2\pi\hbar)^{-3/2}\int \mathrm{d}^3p'\,v(\boldsymbol{p}-\boldsymbol{p}')\varphi(\boldsymbol{p}', t) \tag{6.29}$$

where $\boldsymbol{p}$ is an ordinary vector in configuration space and $v(\boldsymbol{p})$ is the Fourier transform of $V(\boldsymbol{r})$. Equation (6.29) is an integro-differential equation which is somewhat more complicated than the Schrödinger equation in configuration space, but the solution $\varphi(\boldsymbol{p}, t)$ contains exactly the same information as does $\Psi(\boldsymbol{r}, t)$, and $|\varphi(\boldsymbol{p}, t)|^2$ has an equivalent interpretation as a probability density in momentum space.

According to (6.27) the momentum eigenstates $(2\pi\hbar)^{-3/2}\exp(\mathrm{i}\boldsymbol{p}\cdot\boldsymbol{r}/\hbar)$ provide a continuous set of basis states in Hilbert space and, since $\Psi(\boldsymbol{r}, t)$ can always be expressed in the form

$$\Psi(\boldsymbol{r}, t) = \int \Psi(\boldsymbol{r}', t)\,\delta^3(\boldsymbol{r}'-\boldsymbol{r})\,\mathrm{d}^3r' \tag{6.30}$$

the same is true of the functions $\delta^3(\boldsymbol{r}'-\boldsymbol{r})$ labelled by the continuous index $\boldsymbol{r}'$. However, neither of these sets of basis states is composed of square integrable functions and they therefore do not represent physically realizable states. This difficulty may, however, usually be circumvented by confining the system of interest in some large, but finite, volume of phase space. In the most general case the basis states have both a discrete and a continuous component.

Since $\Psi(\boldsymbol{r}, t)$ and $\varphi(\boldsymbol{p}, t)$ are related in a one-to-one correspondence they must be viewed as different representations of the same dynamical state. These ideas are expressed most succinctly in Dirac's formulation of quantum mechanics, which is a presentation of the theory in a form which is independent of specific representations (Dirac 1928, 1928a, 1930, 1947). In the Dirac formalism the dynamical state which corresponds to the coordinate wavefunction $\Psi(\boldsymbol{r}, t)$ is associated with a 'ket' vector $|\Psi\rangle$ in an abstract Hilbert state-space, where the symbol 'Ψ' serves to summarize such information as is available about the state. For example if a linear oscillator is prepared in an eigenstate of the Hamiltonian with $E_{\mathrm{n}}=\hbar\omega(n+\frac{1}{2})$, and is therefore described by the familiar wavefunction

$$\Psi_{\mathrm{n}}(x, t)\approx\exp\left[-\frac{1}{2}\left(x^2\frac{m\omega}{\hbar}\right)\right]H_{\mathrm{n}}\left(x\sqrt{\frac{m\omega}{\hbar}}\right)\mathrm{e}^{-\mathrm{i}E_{\mathrm{n}}t/\hbar} \tag{6.31}$$

the corresponding ket can be written simply as $|n\rangle$, showing that it is an eigenket of the phonon number operator with eigenvalue n. Because the three-dimensional harmonic oscillator Schrödinger equation can be separated both in polar coordinates and Cartesian coordinates it is possible to construct both angular momentum conserving energy eigenkets $|n, l, m\rangle$, and 'deformed' energy eigenkets $|n, 2n_z-n_x-n_y, n_x-n_y\rangle$ whose physical significance is discussed in section 4.6.2.

Since to each pair of ket vectors $|\Psi\rangle$ and $|\chi\rangle$ corresponding to wavefunctions $\Psi(\boldsymbol{r}, t)$ and $\chi(\boldsymbol{r}, t)$ respectively, there corresponds a scalar product defined by (6.26), the set of all such scalar products associated with a given ket $|\Psi\rangle$, may be set in one-to-one correspondence with the elements of a dual linear vector space of 'bra' vectors $\langle\Psi|$, where the bra-vectors $\langle\Psi|$ and $\langle\chi|$ are defined in terms of their scalar products

$$\langle\chi|\Psi\rangle\equiv(\chi, \Psi)=(\Psi, \chi)^*=\langle\Psi|\chi\rangle^* \tag{6.32}$$

With these definitions the improper wavefunctions $(2\pi\hbar)^{-3/2}\exp(-\mathrm{i}\boldsymbol{p}\cdot\boldsymbol{r}/\hbar)$ and $\delta^3(\boldsymbol{r}'-\boldsymbol{r})$ may be associated with generalized kets $|p\rangle$ and $|\boldsymbol{r}\rangle$ respectively. Thus, applying (6.32), the wavefunctions $\Psi(\boldsymbol{r}, t)$ and $\varphi(\boldsymbol{p}, t)$ may be expressed in the convenient form

$$\Psi(\boldsymbol{r}, t)=\langle r|\Psi\rangle \qquad \varphi(\boldsymbol{p}, t)=\langle p|\Psi\rangle \tag{6.33}$$

Thus $\Psi(\boldsymbol{r}, t)$ and $\varphi(\boldsymbol{p}, t)$ are termed respectively the configuration space and momentum space representatives of the dynamical state $|\Psi\rangle$.

In practical calculations in specific systems it is of course necessary to choose a representation, and in most cases the choice is obvious and entails solving an appropriate Schrödinger equation. However, in more formal developments, and especially in those complex problems involving the successive coupling and de-coupling of angular momenta, the power of Dirac's technique is unrivalled. This is particularly true in respect of the ease with which sums and products of linear operators can be manipulated. It is therefore very convenient to employ the techniques of Dirac algebra from time to time, and most frequently when it is required to lay stress on those features of the theory which are independent of representation.

6.2 The eikonal approximation for neutrons

6.2.1 Phase-integral solutions

The phase integral solution (6.9) to the scalar Helmholtz equation (6.8) was first discussed in an optical context by Sommerfeld and Runge (1911), and was subsequently applied by Rayleigh (1912), and by Gans (1915), to the investigation of electromagnetic wave propagation in stratified media. The solution is valid only in the circumstance that k_0^{-1} is small in comparison with any distance over which the refractive index n varies significantly, and violations of this condition can arise in a number of ways. For example $n(\boldsymbol{r})$ can change discontinuously on a scale less than a wavelength near the boundary between two media in which case the assumption $\nabla^2\tilde{A}(\boldsymbol{r})\ll k_0^2\tilde{A}(\boldsymbol{r})$ fails and the eikonal equation (6.4) is no longer valid. The other possibility is that $n(\boldsymbol{r})$ should go through a zero, and in the neighbourhood of such a transition (or turning) point, $n(\boldsymbol{r})$ changes significantly over all length scales, no matter how small, and once again the eikonal approximation breaks down.

Phase integral methods were first introduced into quantum mechanics by Wentzel (1926), Kramers (1926) and Brillouin (1926) (hence the description 'WKB method') and, in analogy with (6.9) the Schrödinger wavefunction $\Psi(\boldsymbol{r})$ is assumed to be expressible in the form

$$\Psi(\boldsymbol{r})=A(\boldsymbol{r})\,\mathrm{e}^{\mathrm{i}S(\boldsymbol{r})/\hbar} \tag{6.34}$$

where the amplitude $A(\boldsymbol{r})$ is real, and $S(\boldsymbol{r})$ has the dimensions of mechanical action. The Bohr–Sommerfeld quantum condition then follows from the requirement that $\Psi(\boldsymbol{r})$ be single-valued even when $S(\boldsymbol{r})$ is permitted to be multi-valued (Keller 1958). The semi-classical, or WKB, approximation applies when $S(\boldsymbol{r})$ is large in comparison with $\hbar$, which means in the limit of large quantum numbers in the sense of Bohr's correspondence principle. An alternative statement of the same limit is that the de Broglie wavelength should be small as measured on any length scale over which $V(\boldsymbol{r})$ varies significantly. It may be of interest to note that, whereas the WKB approximation is normally applied to the solution of barrier penetration problems on the microscopic scale, in the present context it is used to describe the macroscopic motion of neutrons. Quite generally the description of classical charged particle motion in terms of almost conserved adiabatic invariants is an aspect of the WKB approximation in another guise.

The WKB solutions in one dimension may be written in the form

$$\Psi(x)\approx[2m(E-V)]^{-1/4}\{A_+\,e^{(i/\hbar)\int^x\sqrt{(2m(E-V))}dx}+A_-\,e^{-(i/\hbar)\int^x\sqrt{(2m(E-V))}dx}\} \tag{6.35}$$

where A_+ and A_- are arbitrary constants. The correspondence between the optical and quantum mechanical solutions may be established quite easily by solving (6.10–6.11) in one dimension in the limit $k_0\rightarrow\infty$. This gives the results

$$\tilde{A}^2\frac{\partial\tilde{S}}{\partial x}=\text{constant}\qquad \tilde{S}=\pm\int^x n(x)\,dx \tag{6.36}$$

Thus the optical wavefield $\tilde{\Psi}(x)$ becomes

$$\tilde{\Psi}(x)\approx n(x)^{-1/2}\exp\left\{\pm ik_0\int^x n(x)\,dx\right\} \tag{6.37}$$

which corresponds exactly to the WKB solution (6.35).

The WKB approximation in the form presented above is usually applied to obtain an eikonal representation of the wavefunction in configuration space, but it may equally well be employed in momentum space to give a result which is not identical with the Fourier transform of the WKB configuration space wavefunction. In general these two functions differ by terms of the order $\hbar$ except in a spatial neighbourhood where ∇S is multivalued, and $A(x)$ becomes infinite at points on a caustic surface where any two branches of ∇S coincide. In these circumstances the Fourier transform of the configuration space WKB wavefunction becomes infinite while the momentum space WKB approximation remains finite.

When $(E-V(x))>0$ the WKB solutions (6.35) are pure imaginary exponential functions representing progressive waves with phase velocity given by (6.22), and when $(E-V(x))<0$ the exponents are real describing evanescent waves. At a transition point between these two regimes $E=V(x)$, and the phase velocity becomes infinite. According to (6.16) and (6.21) the refractive index vanishes and the de Broglie wavelength $\lambda=\lambda_0/n$ also becomes infinite. This corresponds to a classical turning point where the kinetic energy vanishes. As is true also in optics, the presence of a transition point is invariably associated with strong reflection of an incident wave. Precisely this property is exploited in the production of polarized ultra-cold neutrons by transmission of unpolarized neutrons through magnetized foils, where one spin sense encounters a region of vanishing refractive index at the surface of the film and is totally reflected, while the opposite spin sense is transmitted. Potentials with two transition points, which are often encountered in barrier penetration (progressive wave) and bound state (standing wave) problems, can also be tackled successfully using WKB methods and the same is true for hybrid problems which involve the decay of resonant states. Gamow's (1928) treatment of α-decay as a quantum tunnelling process falls squarely within this category (cf. section 3.2.1).

The first systematic investigation of phase integral solutions valid in the neighbourhood of a transition point was undertaken before the development of wave mechanics by Jeffries (1923), who also succeeded in deriving connection formulae to match the WKB solutions applicable on either side of the transition point. Kramers (1926) also drew attention to the fact that the arbitrary constants $A_\pm$ which appear in (6.35) change discontinuously as the solution proceeds through the transition point. This phenomenon, first discovered and explained by Stokes (1857) as a consequence of the necessity for maintaining the single-valued nature of the solution, is a reminder of the fact that the WKB solutions are valid only in regions remote from transition points, and are truly asymptotic solutions in the sense discussed earlier in connections with the bound state solutions of the Schrödinger equation (cf. section 2.6.1).

Finally there is a class of problem for which $(E-V(x))>0$ everywhere and there is no transition point on the real x-axis. An example of such a potential is shown in figure 6.2, and this is given by the relation

$$V(x)=V_0/(1+e^{-x/d}) \tag{6.38}$$

which describes a potential barrier with a diffuse boundary, of width d, in the neighbourhood of the origin. In the WKB approximation, which is presumably valid for $\lambda \ll d$, that solution constructed to satisfy the boundary condition that there should only be outgoing waves as $x\to\infty$, exhibits no reflected wave as $x\to-\infty$. This indeed is the reason why fast neutrons can be totally absorbed in a nucleus which has a diffuse surface (cf. section 2.4.3). On the other hand the potential (6.38) gives a Schrödinger equation which is exactly soluble (Landau and Lifshitz 1965, Berry and Mount 1972) giving a reflected wave which, for $\lambda \ll d$ and $x\to-\infty$ has an approximate amplitude

$$\psi_r \simeq -\exp\left\{-\frac{2\pi d}{\hbar}2m(E-V_0)\right\} \tag{6.39}$$

which contradicts the WKB result.

This difficulty may be resolved by introducing an improvement to the WKB solutions in (6.35), whereby the constant amplitudes $A_\pm$ are replaced by slowly varying functions $A_\pm(x)$ chosen to satisfy the boundary conditions

$$A_+(-\infty)=1 \qquad A_-(+\infty)=0 \tag{6.40}$$

With this modification (Bremmer 1949, 1951) there is a reflected wave with

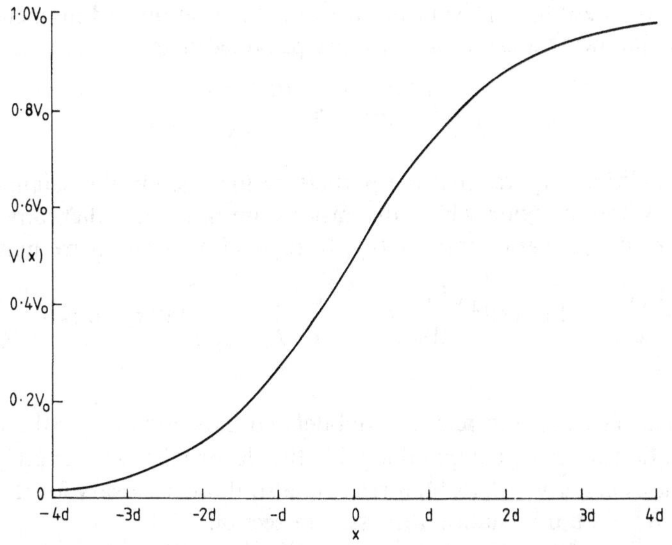

Figure 6.2. A one-dimensional potential barrier with a diffuse boundary layer of width d. The exact solution of the Schrödinger equation for a free particle encountering such a barrier contains a reflected component contrary to the result obtained on the basis of the WKB approximation.

approximate amplitude

$$\psi_r \simeq -A_-(-\infty) \tag{6.41}$$

Not surprisingly it turns out that the resultant value of ψ_r is determined by the position of that complex transition point in the function $(E-V(x))$ which lies closest to the real x-axis.

6.2.2 Wave packets

An alternative form of semi-classical approximation may be constructed by means of Ehrenfest's theorem (1927) using the concept of wave packets (Littlejohn 1986). If we assume that $\Psi(x, t)$ describes the one-dimensional motion of a particle moving in a conservative potential $V(x)$, then Ehrenfest's theorem states that the expectation values $\langle x(t)\rangle$, $\langle p(t)\rangle$ of the position and momentum operators in the state represented by $\Psi(x, t)$ propagate in time according to the equations

$$\frac{\mathrm{d}}{\mathrm{d}t}\langle x(t)\rangle = \left\langle \frac{p(t)}{m} \right\rangle$$

$$\frac{\mathrm{d}}{\mathrm{d}t}\langle p(t)\rangle = -\left\langle \frac{\partial V(x)}{\mathrm{d}x} \right\rangle \tag{6.42}$$

Thus $\langle x(t)\rangle$ and $\langle p(t)\rangle$ vary in time like the position and momentum of a particle following a classical trajectory provided that

$$\left\langle \frac{\partial V(x)}{\partial x} \right\rangle \simeq \frac{\partial V(\langle x\rangle)}{\partial \langle x\rangle} \tag{6.43}$$

This condition requires that the particle be localized in the neighbourhood of $\langle x\rangle$ within a region whose dimensions are much less than any distance over which $V(x)$ varies appreciably. In these circumstances we may write

$$\left\langle \frac{\partial V(x)}{\partial x} \right\rangle = \int |\Psi(x)|^2 \frac{\partial V(x)}{\mathrm{d}x}\,\mathrm{d}x \simeq \left(\frac{\partial V}{\partial x}\right)_{x=\langle x\rangle} \int |\Psi(x)|^2\,\mathrm{d}x = \frac{\partial V(\langle x\rangle)}{\partial \langle x\rangle} \tag{6.44}$$

If the particle is also to retain a well-defined momentum $\langle p\rangle$ the region of localization must be comparable with the de Broglie wavelength, and the conditions under which (6.44) is true are also those necessary for the validity of the WKB approximation discussed in section 6.2.1.

Equation (6.39) is exactly true when $V(x)$ is a constant, i.e. the particle is free, or when $V(x)$ is a quadratic function of x which, by a suitable choice of origin, can be expressed as a harmonic oscillator potential. In general the approximation (6.43) is valid only when the particle is localized in the neighbourhood of a given point in space to such a precision that the potential may be expanded to an acceptable level of accuracy in quadratic order about that point. In this case the wavefunction describing such a localized particle is called a wave packet, and a representation of a one-dimensional Gaussian wave packet is shown in figure 6.3.

To see how a free wave packet propagates in time we may suppose that at $t=0$, $\Psi(x, t)$ is given as the square integrable function $\Psi(x, 0)$, which may therefore be represented as the Fourier integral

$$\Psi(x, 0) = \frac{1}{\sqrt{2\pi\hbar}} \int_{-\infty}^{\infty} \varphi(p, 0)\, \mathrm{e}^{\mathrm{i}px/\hbar}\, \mathrm{d}p \tag{6.45}$$

where

$$\varphi(p, 0) = \frac{1}{\sqrt{2\pi\hbar}} \int_{-\infty}^{\infty} \Psi(x, 0)\, \mathrm{e}^{-\mathrm{i}px/\hbar}\, \mathrm{d}x \tag{6.46}$$

Therefore, at an arbitrary time t in $-\infty < t < \infty$, $\Psi(x, t)$ is given by

$$\Psi(x, t) = \frac{1}{\sqrt{2\pi\hbar}} \int_{-\infty}^{\infty} \varphi(p, 0)\, \mathrm{e}^{\mathrm{i}(kx-\omega(k)t)}\, \mathrm{d}p \tag{6.47}$$

where $p=\hbar k$, and

$$\omega(k) = \frac{\hbar k^2}{2m} \tag{6.48}$$

is the dispersion relation for a free particle.

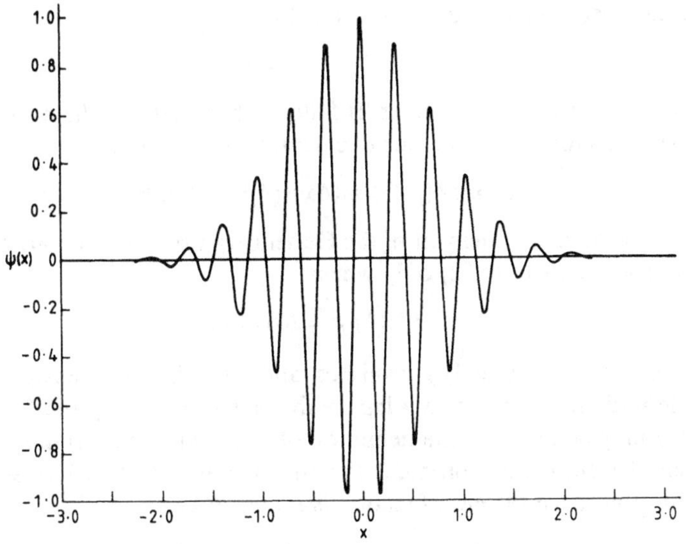

Figure 6.3. The one-dimensional wave packet describing the minimum uncertainty state of a free particle in which the product $\Delta x \Delta p$ takes the minimum value $\frac{1}{2}\hbar$ permitted to it by the Heisenberg uncertainty principle.

The representation (6.45) of the initial wavefunction $\Psi(x, 0)$ as a superposition of plane waves is insufficiently restrictive for present purposes. For example, if $\varphi(p, 0) \approx \delta(p - p(0))$ then the state $|\Psi\rangle$ of the particle occupies a single point in momentum space and is spread over the whole of configuration space. Thus the mean position $\langle x \rangle$ is not defined and $\Psi(x, 0)$ does not describe a wave packet. Similarly if $\varphi(p, 0)$ is independent of p, the mean momentum $\langle p \rangle$ is not defined initially and, as before, no meaning can be attached to the Ehrenfest relations (6.42). Specifically the concept of a wave packet is useful only when the state of the particle is confined within some finite volume about the point $\langle x(t) \rangle$, $\langle p(t) \rangle$ in phase space, whose extent Δx, Δp is specified through position and momentum variances defined in the usual way by

$$(\Delta x)^2 = \langle (x - \langle x \rangle)^2 \rangle \qquad (\Delta p)^2 = \langle (p - \langle p \rangle)^2 \rangle \tag{6.49}$$

In particular, the closest approach to a purely classical system occurs when $\Psi(x, 0)$ is selected to be a 'minimum uncertainty state', that is a state for which the uncertainty product $\Delta x \, \Delta p$ takes the minumum value $\hbar/2$ allowed to it by the Heisenberg uncertainty principle. That this state corresponds to Gaussian probability distributions over eigenstates of either x or p may be demonstrated in a very direct manner.

Let $|\Psi\rangle$ be the state vector represented by the coordinate wavefunction $\Psi(x, 0) = \langle x | \Psi \rangle$ and let $|\chi\rangle$ be another state vector given by

$$|\chi\rangle = (x' + i\alpha p')|\Psi\rangle \tag{6.50}$$

where the operators x' and p' are defined by

$$x' = x - \langle x \rangle \qquad p' = p - \langle p \rangle \tag{6.51}$$

and α is an arbitrary real number. It follows then that, for the norm $\langle \chi | \chi \rangle$ of the state vector $\langle \chi \rangle$ to be positive definite, we must have

$$\langle (x')^2 \rangle \langle (p')^2 \rangle \equiv (\Delta x)^2 (\Delta p)^2 \geqslant (\hbar/2)^2 \tag{6.52}$$

which is the familiar statement of the Heisenberg principle. In particular the equality sign holds in (6.52) only when

$$\alpha = \hbar/2(\Delta p)^2 = 2(\Delta x)^2/\hbar \tag{6.53}$$

in which case the state $|\chi\rangle$ vanishes and $|\Psi\rangle$ is an eigenstate of the (non-Hermitian) operator $x + \mathrm{i}\alpha p$ with eigenvalue $\langle x \rangle + \mathrm{i}\alpha \langle p \rangle$. This property may be taken as the definition of a minimum uncertainty state.

When the eigenvalue problem for the operator $(x + \mathrm{i}\alpha p)$ is solved in configuration space the eigenfunction is found to be

$$\Psi(x, 0) = \frac{1}{(2\pi(\Delta x)^2)^{1/4}} \exp\left\{ \frac{-\mathrm{i}\langle p \rangle x}{\hbar} - \frac{1}{4}\left(\frac{x - \langle x \rangle}{\Delta x} \right)^2 \right\} \tag{6.54}$$

Thus the probability that the particle be found in the range x, $x + \mathrm{d}x$ is determined by the probability density function

$$|\Psi(x, 0)|^2 = \frac{1}{\sqrt{(2\pi(\Delta x)^2)}} \exp\left\{ -\frac{1}{2}\left(\frac{x - \langle x \rangle}{\Delta x} \right)^2 \right\} \tag{6.55}$$

which describes the Gaussian or normal probability distribution with mean $\langle x \rangle$ and standard deviation Δx. Alternatively the eigenvalue problem may be solved in momentum space giving the eigenfunction

$$\varphi(p, 0) = \frac{1}{(2\pi(\Delta p)^2)^{1/4}} \exp\left\{ -\mathrm{i}\frac{p\langle x \rangle}{\hbar} - \frac{1}{4}\left(\frac{p - \langle p \rangle}{2\Delta p} \right)^2 \right\} \tag{6.56}$$

Thus $|\varphi(p, 0)|^2$ is also a Gaussian probability density with mean $\langle p \rangle$ and standard deviation Δp. One may of course choose $\langle x \rangle = 0$ without loss of generality, and a minimum uncertainty state is therefore characterized by two free parameters, i.e. $\langle p \rangle$ and either Δx or Δp.

It is instructive to observe that the Gaussian wavefunction (6.54) is exactly of the form postulated for the WKB solution in (6.34) although that function represents an approximation to an energy eigenfunction whereas the wave packet (6.54) is a superposition of momentum eigenstates, and hence of energy eigenstates in the case of a free particle. It is then easy to show that the condition that $\Psi(x, 0)$ be an acceptable WKB solution, i.e.

$$\frac{\partial^2}{\partial x^2} |\Psi(x, 0)| \ll (\langle k \rangle^2) |\Psi(x, 0)| \tag{6.57}$$

is satisfied for all points x in space such that

$$|x| \ll (\Delta x)\left|\frac{\langle p\rangle}{\Delta p}\right| = \hbar\,\frac{|\langle p\rangle|}{(\Delta p^2)^2} \tag{6.58}$$

The time-dependent wavefunction $\Psi(x, t)$ is now determined from (6.47)–(6.48) with the result

$$\Psi(x, t) = (2\pi(\Delta x)^2)^{-1/4}(1 + i\hbar t/2m(\Delta x)^2)^{-1/2}$$
$$\times \exp\left\{ i\left[\frac{p(0)x - E(0)t}{\hbar}\right] - \frac{1}{4}\left[\frac{(x - p(0)t/m)^2}{(\Delta x)^2 + i\hbar t/2m}\right]\right\}$$
$$-\pi < \arg(t) \leqslant \pi \tag{6.59}$$

where $p(0)$ is the average momentum and $E(0) = p^2(0)/2m$ the average energy in the initial state. Thus $\Psi(x, t)$ also represents a Gaussian wave packet which has propagated from the point $(0, p(0))$ in phase space, to the point

$$\langle x(t)\rangle = \frac{p_0 t}{m} \qquad \langle p(t)\rangle = p(0) \tag{6.60}$$

in agreement with the solution of the Ehrenfest relations (6.42) for $V(x) = 0$. The standard deviation in momentum $\Delta p(t) = \Delta p$ is also a constant of the motion but the wave packet has undergone a spreading in space as expressed by the new variance

$$(\Delta x(t))^2 = (\Delta x)^2 + \left(\frac{\hbar t}{2m\,\Delta x}\right)^2 = (\Delta x)^2\left\{1 + \left(\frac{\langle x(t)\rangle}{\Delta x}\right)^2\left(\frac{\Delta p}{p(0)}\right)^2\right\} \tag{6.61}$$

It follows that $\Psi(x, t)$ no longer describes a minimum uncertainty state and, as the wave packet spreads in space, the peak amplitude $\Psi(\langle x(t)\rangle, t)$ reduces in proportion to conserve the normalization.

Another way to exhibit the spreading of the wavefunction is to calculate the current (cf. section 2.3)

$$j(x, t) = |\Psi(x, t)|^2\left\{\frac{p_0}{m}\right\}\left\{1 + 2\left(\frac{\Delta p}{p(0)}\right)^2\left(\frac{\langle x(t)\rangle}{\Delta x(t)}\right)\left(\frac{x - \langle x(t)\rangle}{\Delta x(t)}\right)\right\} \tag{6.62}$$

This result shows that, in addition to the mean current $|\Psi(x, t)|^2 p(0)/m$, $j(x, t)$ also contains a component which is positive for $x > \langle x(t)\rangle$ and negative for $x < \langle x(t)\rangle$ and hence tends to spread the wavefunction. This spreading is not a purely quantum phenomenon since a classical ensemble of free particles characterized by a Liouville density function $f(x, p)$ centred at the point $x = 0$, $p = p(0)$ in phase space exhibits an identical spreading in configuration space. Furthermore $\Delta x(t)$ tends asymptotically towards the value $\langle x(t)\rangle(\Delta p/p(0))$ which would be reached at time t by a swarm of classical particles distributed normally about the origin with relative momentum dispersion $\Delta p/p(0)$. What these results show is that the quantum mechanical average in Ehrenfest's theorem is to be interpreted as an ensemble average in the transition to the classical limit (Holstein and Swift 1972).

6.2.3 Coherent states

In the case of a particle moving in a harmonic potential for which the Ehrenfest equations (6.42)–(6.43) have solutions which satisfy the equations of classical mechanics exactly, it is indeed possible to select a subset of minimum uncertainty states which has the required property that the wave packet does not spread in time. These states were discovered by Schrödinger who remarked that (Schrödinger 1926a, Steiner 1988)

> Also such a wave group will remain compact, in contrast, e.g. to a wave packet in classical optics, which is dissipated in the course of time. This distinction may originate in the fact that our wave group is built up of separate discrete harmonic components, and not out of a continuum of such.

The Schrödinger wave packet is constructed from a special linear combination of energy eigenfunctions

$$\Psi(x,t) \approx \sum_{n=0} \left(\frac{A}{2}\right)^2 \frac{1}{n} \Psi_n(x,t) \qquad (6.63)$$

where the simple harmonic oscillator eigenfunctions $\Psi_n(x, t)$ are defined in (6.31). and A is an arbitrary real number large in comparison with unity. The sum (6.63) represents a Poisson distribution over phonon number eigenstates with mean values

$$\langle n \rangle = \left(\frac{A^2}{2}\right) \qquad \langle E \rangle \approx \hbar\omega \langle n \rangle = A^2(\hbar\omega/2) \qquad (6.64)$$

It describes a minimum uncertainty Gaussian wave packet which propagates in phase space according to the classical motion

$$\langle x(t) \rangle = \sqrt{(2\langle E \rangle / m\omega^2)} \cos \omega t \qquad \langle p(t) \rangle = -\sqrt{(2m\langle E \rangle)} \sin \omega t \quad (6.65)$$

with standard deviations

$$\Delta x = \left(\frac{\hbar}{2m\omega}\right) \qquad \Delta p = \hbar/2\Delta x \qquad (6.66)$$

which do not vary in time.

The Schrödinger oscillator wave packets are identical with the 'coherent states' of the quantized radiation field (cf. section 2.1.3.1) used by Glauber (1963) in his analysis of the photon statistics of light beams, for which the phase φ rather than the photon number n is specified within the limitations imposed by the Heisenberg principle. The properties of these coherent states may be established most directly by deriving the minimum uncertainty states of the linear oscillator according to the prescription given in equations (6.50)–(6.53) above, and imposing the condition that the ground state ($n=0$), with $\langle x\rangle=\langle p\rangle=0$, should be a member of the set. This removes the freedom to select Δx arbitrarily and this number is now fixed according to (6.66). It follows that $\alpha=(m\omega)^{-1}$ (cf. 6.53), and the coherent states may be defined as the eigenstates of the (non-Hermitian) photon annihilation operator

$$a=\sqrt{\frac{m\omega}{2\hbar}}\left(x+\frac{\mathrm{i}p}{m\omega}\right) \tag{6.67}$$

The eigenstates constitute a complete, but not an orthogonal, set.

That the free particle wave packet (6.59) spreads in time whereas the Schrödinger wave packet (6.63) coheres is indeed related to the fact that the former is a superposition of a continuum of states, whereas the latter is constructed from a discrete set. This distinction was evidently appreciated by Schrödinger, but there is more to it than that. The important difference is that the oscillator energy levels are equally spaced whereas, even where a free particle is confined within a length L and periodic boundary conditions imposed, the energy levels are given by

$$E_n=\frac{\hbar^2}{2m}\frac{4\pi^2n^2}{L^2} \tag{6.68}$$

where n is integral, and those energy levels are not equally spaced. The same difficulty arises with the quantum Kepler problem since the states of the Schrödinger hydrogen atom are not equally spaced (Gerry 1984).

It is possible by imposing the property of equal spacing to construct generalized coherent states in a range of non-harmonic systems which propagate in time almost, but not quite, according to the laws of classical motion (Nieto and Simmons 1978, 1979). These states have been widely discussed in the theoretical literature, but in no one case have they been subjected to experimental study. Since intense beams of neutrons are now available and many potentials $V(r)$ can be approximated through some combination of interactions $-\boldsymbol{\mu}_\mathrm{n}\cdot\boldsymbol{B}(\boldsymbol{r})$ which a neutron experiences in a magnetic field, and the gravitational potential $m_\mathrm{n}gh$ which it experiences in a gravitational field, it seems that neutrons provide an ideal, if not indeed a unique tool to explore what is obviously a very interesting field.

Indeed one may construct coherent spin states which are the analogues of the coherent spin states of the harmonic oscillator discussed above (Radcliffe 1971). In the case of the neutron spin these states may be represented by Pauli spinors (cf. section 7.4.1)

$$|\theta,\varphi\rangle=\begin{pmatrix}\cos\theta/2\ \mathrm{e}^{-\mathrm{i}\varphi/2}\\ \sin\theta/2\ \mathrm{e}^{\mathrm{i}\varphi/2}\end{pmatrix}\qquad 0\leqslant\theta<\pi\qquad 0\leqslant\varphi<2\pi \tag{6.69}$$

That these states constitute a normalized complete set follows from the result

$$\frac{2}{\pi}\int |\theta, \varphi\rangle\langle\theta, \varphi| \sin\theta \, d\theta \, d\varphi = \sigma_0 \tag{6.70}$$

where σ_0 is the 2×2 unit matrix.

6.3 Neutron propagation in condensed media

6.3.1 Dispersion

6.3.1.1 Refractive index

The integral equation (2.36) describing the propagation of neutron waves in a uniform isotropic condensed medium containing spinless nuclei located at scattering centres $\boldsymbol{r}_j$ may be written as (Sears 1982)

$$\begin{aligned}\Psi(\boldsymbol{r}) &= \Psi_0(\boldsymbol{r}) + \sum_j \int G_0(\boldsymbol{r}-\boldsymbol{r}')V_j(\boldsymbol{r}'-\boldsymbol{r}_j)\Psi(\boldsymbol{r}')\, d^3r' \\ &= \Psi_0(\boldsymbol{r}) + \sum_j \int G_0(\boldsymbol{r}-\boldsymbol{r}')T(\boldsymbol{r}'-\boldsymbol{r}_j)\chi_j(\boldsymbol{r}')\, d^3r' \end{aligned} \tag{6.71}$$

where the sum is taken over the whole volume of the medium and $\Psi_0(\boldsymbol{r})$ is the incident wavefunction representing a free neutron coming from outside the medium. $V_j(\boldsymbol{r}-\boldsymbol{r}_j)$ is the potential at $\boldsymbol{r}$ produced by the nucleus at $\boldsymbol{r}_j$ and $T_j(\boldsymbol{r}-\boldsymbol{r}_j)$ is the transition operator associated with $V_j(\boldsymbol{r}-\boldsymbol{r}_j)$ through the integral equation (2.50). The function $\chi_j(\boldsymbol{r})$ appearing in (6.71) is the wave at $\boldsymbol{r}$ which is incident on the nucleus at $\boldsymbol{r}_j$. It represents the effective field which interacts with the nucleus at $\boldsymbol{r}_j$, and must therefore be equal to the total field at $\boldsymbol{r}$ less the field scattered from $\boldsymbol{r}_j$ to $\boldsymbol{r}$. Thus $\chi_j(\boldsymbol{r})$ satisfies the integral equation

$$\begin{aligned}\chi_j(\boldsymbol{r}) &\equiv \Psi(\boldsymbol{r}) - \int G_0(\boldsymbol{r}-\boldsymbol{r}')T_j(\boldsymbol{r}'-\boldsymbol{r}_j)\chi_j(\boldsymbol{r}')\, d^3r' \\ &= \Psi_0(\boldsymbol{r}) + \sum_{i\neq j} \int G_0(\boldsymbol{r}-\boldsymbol{r}')T_i(\boldsymbol{r}'-\boldsymbol{r}_i)\chi_i(\boldsymbol{r}')\, d^3r' \end{aligned} \tag{6.72}$$

The exact solutions $\Psi(\boldsymbol{r})$, $\chi_j(\boldsymbol{r})$ satisfying the complete array of equations (6.71)–(6.72) are functions not only of $\boldsymbol{r}$, but also of all the position vectors $\boldsymbol{r}_j$. However, not only are these microscopic solutions impossible to obtain; they are not even needed. What is required is the macroscopic solution $\Psi(\boldsymbol{r})$, which is an average of $\Psi(\boldsymbol{r})$ over the variables $\boldsymbol{r}_j$, which in real situations are fluctuating quantities, and also over other variables such as spins, velocities, scattering lengths, etc. which in general enter the problem.

If it is now assumed that the range of the neutron–nucleus interaction and the scattering length are each much less than the internuclear separation, then, for low energy neutrons, it is permissible to replace both the true scattering potential and its transition operator by Fermi's pseudo-potential (cf. section 2.5.5)

$$V_F(|\boldsymbol{r}-\boldsymbol{r}_j|)=\frac{2\pi\hbar^2}{\mu}\,a_j\,\delta^3(\boldsymbol{r}-\boldsymbol{r}_j) \tag{6.73}$$

where μ is the reduced mass of a neutron scattering from a free nucleus of mass $M(A, Z)$. In practice it is convenient to replace the reduced mass μ by the rest mass m_n in (6.73), and the free scattering length a by the bound scattering length b, where

$$b_j=a_j(1+m_n/M(A, Z)) \tag{6.74}$$

is the scattering length which would be observed if the target nucleus were infinitely massive (Koester and Steyerl 1977).

Carrying out appropriate averages over $\boldsymbol{r}_j$ we now find that

$$\begin{aligned}
&\sum_j\langle\delta^3(\boldsymbol{r}'-\boldsymbol{r}_j)\rangle=N\\
&\sum_j\langle b_j\,\delta^3(\boldsymbol{r}'-\boldsymbol{r}_j)\rangle=Nb\\
&\sum_j\langle T_j(\boldsymbol{r}'-\boldsymbol{r}_j)\,\delta^3(\boldsymbol{r}'-\boldsymbol{r}_j)\rangle=\frac{2\pi\hbar^2}{m_n}\,Nb\chi(\boldsymbol{r}')
\end{aligned} \tag{6.75}$$

where N, the mean number of scattering nuclei per unit volume, and b, the mean bound scattering length, are taken to be uniform over the medium. This assumption is of course not necessary, and is not true for example for ultra-cold neutrons confined in a bottle where, due to the gravitational potential, there is a density gradient which varies with height h like $\exp[-m_n gh(kT)]$. The function $\chi(\boldsymbol{r}')$ which is defined by (6.75), is the mean effective field for scattering at nuclei in the immediate vicinity of the point $\boldsymbol{r}'$. However carrying out the averaging process over the functions $\chi_j(\boldsymbol{r}', \boldsymbol{r}_j)$ is a very difficult process indeed, since this average depends strongly on the degree to which the positions of the individual nuclei are correlated. When there is no spatial correlation between scattering centres, as happens for example in a dilute gas, then $\chi(\boldsymbol{r})$ may be identified with $\Psi(\boldsymbol{r})$. However, in the general case the positions of the nuclei are correlated to some degree, so that $\chi(\boldsymbol{r})$ and $\Psi(\boldsymbol{r})$ are not identical, but are connected through a constitutive relation (Foldy 1945, Lax 1951, 1952, Sears 1982, 1989)

$$\chi(\boldsymbol{r})=c_m\Psi(\boldsymbol{r}) \tag{6.76}$$

where c_m is a factor determined by the thermodynamic state of the medium and its structure as expressed by the pair correlation function $g(\boldsymbol{r}-\boldsymbol{r}')$ (cf. section 8.3.4).

The situation desribed by (6.76) closely parallels that which arises in the electrodynamics of condensed media, where the effective electric field in the medium, which is the analogue of $\chi(\boldsymbol{r})$, is given as the sum of the true mean electric field $\boldsymbol{E}(\boldsymbol{r})$, which is the analogue of $\bar{\Psi}(\boldsymbol{r})$, and a term equal to $(4/3)\pi P(\boldsymbol{r})$ where $P(\boldsymbol{r})$ is the mean electric polarization. Hence the corresponding expression for c_m is $(\varepsilon+2)/3$, where ε is the dielectric constant, a result which is known as the Lorentz–Lorenz law. In the corresponding expression for neutron scattering in a simple cubic crystal it has been shown that c_m departs from unity by terms of the order of the ratio of the scattering length to the lattice parameter giving a correction of about $10^{-2}\%$ (Ekstein 1951, 1953). We may therefore set c_m equal to unity in all cases of practical interest, although to maintain the correct phase relations for neutrons propagating in perfect crystals it may be necessary to take into account a non-vanishing phase in c_m (cf. section 7.2.4.1).

We conclude therefore that the macroscopic wave $\bar{\Psi}(\boldsymbol{r})$ inside the medium satisfies a one-body Schrödinger equation

$$(\nabla^2+K^2)\bar{\Psi}(\boldsymbol{r})=0 \qquad K^2=k^2-4\pi Nb \tag{6.77}$$

This result implies that the macroscopic wave $\bar{\Psi}(\boldsymbol{r})$ satisfies the same equation in a discontinuous atomic medium as is satisfied by a microscopic wave $\Psi(\boldsymbol{r})$ in a continuous medium where there is a uniform potential

$$V_F=\frac{\hbar^2}{2m_n}4\pi Nb=\frac{2\pi\hbar^2 Nb}{m_n} \tag{6.78}$$

According to (6.21) the corresponding refractive index is

$$n=\frac{K}{k}=\sqrt{\left(1-\frac{4\pi Nb}{k^2}\right)}\simeq 1-\frac{2\pi Nb}{k^2} \tag{6.79}$$

Since it is known experimentally that the scattering length is normally positive, and this is true for reasons which are well understood (cf. section 2.5.1), it follows from (6.79) that most materials are neutron-optically less dense than vacuum. When the medium has a non-vanishing neutron capture cross-section σ_c this can be taken into account by assigning an imaginary component to V (cf. section 2.4.3). The net result is that an imaginary term $\mathrm{i}N\sigma_c/k$ is added to the expression for n given in (6.79). The magnitude of this term is of exceptional importance in assessing which materials are most suitable for ultra-cold neutron storage (cf. section 1.6.6).

The macroscopic wave $\bar{\Psi}(\boldsymbol{r})$ is formed from the superposition of the incident wave and these waves which are coherently scattered from the scattering centres. For this reason $\bar{\Psi}(\boldsymbol{r})$ is referred to as the 'coherent wave' in the medium, and it is simply written as $\Psi(\boldsymbol{r})$. No confusion arises as a result of using this notation just as no confusion arises in the study of electromagnetism when the same symbols are used for the microscopic fields in vacuum and the macroscopic fields in the medium. The results (6.78)–(6.79) cannot be applied as they stand to crystalline media (cf. section 6.5.3) since in this case the positions of the scattering centres are spatially correlated with the period of the lattice, and the existence of Bragg scattering requires that the refractive index should vary in space with the same period.

In arriving at the Schrödinger equation (6.77) which describes the propagation of coherent waves in a medium containing a very large number of scattering nuclei, we have treated the individual collisons in the impulse approximation, i.e. each neutron–nucleus collison is viewed as if the scattering nucleus was free. This approach is justified on the grounds that nuclear forces are vastly stronger than any chemical force which binds the nucleus in the medium. In representing both the nuclear potential and its transition operator in terms of Fermi's pseudo-potential (cf. section 2.5.5) we apply the Born approximation (cf. section 2.3.1) which means that multiple scattering of an incident neutron in the field of an individual nucleus is neglected. However, it is not permissible in general to ignore the multiple scattering of neutrons by many nuclei which is implied by applying the Born approximation to the solution of (6.77). Indeed, as the discussion of the extinction theorem in section 6.3.2 shows, to account for the phenomena of reflection and refraction it is necessary to solve (6.77) with full rigour and the Born approximation is not applicable.

However, for the treatment of diffraction phenomena it is indeed possible to use the Born approximation and view this process solely as an interference effect between the incident wave and waves generated at primary scattering events. This is the so-called kinematical theory of diffraction in which the contribution of waves produced in plural or multiple scattering processes is neglected. This theory is adequate to describe macroscopic diffraction by slits, edges, etc., or microscopic diffraction in thin crystals (cf. section 7.2.1–3), but in large crystals it fails since the theory violates conservation of the probability current (cf. 2.52) and predicts a diffracted flux which exceeds the incident flux. In these circumstances the kinematical theory must be replaced by a dynamical theory based on the exact solutions of (6.76) in the scattering medium. In the particular case of Bragg diffraction (cf. section 7.2.3) the dynamical description is of particular interest since it predicts new phenomena which lie outside the range of kinematical theory.

6.3.1.2 The influence of nuclear spin

If the neutron spin is denoted by $\boldsymbol{s}$, and the nuclear spin by $\boldsymbol{I}$, where both spins are measured in units of $\hbar$, then the joint system of neutron plus nucleus can have spin quantum numbers $I+\frac{1}{2}$ or $I-\frac{1}{2}$, corresponding to parallel and anti-parallel spins, respectively. In this case the scattering potential contains a spin–spin coupling term proportional to $\boldsymbol{s}\cdot\boldsymbol{I}$ and, ignoring tensor forces (cf. section 3.3.2) the Fermi pseudo-potential assumes the modified form

$$V_{\mathrm{F}}=\frac{2\pi\hbar^2}{m_{\mathrm{n}}}\{b+c\boldsymbol{s}\cdot\boldsymbol{I}\} \tag{6.80}$$

Since the operator $\boldsymbol{s}\cdot\boldsymbol{I}$ has eigenvalues $\frac{1}{2}I$ and $-\frac{1}{2}(I+1)$ in the spin states $(I+\frac{1}{2})$ and $(I-\frac{1}{2})$, respectively, the parameters b and c are given by

$$b=\alpha^+b^+ +\alpha^-b^- \qquad c=\frac{2(b^+-b^-)}{2I+1} \tag{6.81}$$

where $b^{\pm}$ are bound scattering lengths in the spin states $|I\pm\frac{1}{2}\rangle$ and

$$\alpha^+=\frac{I+1}{(2I+1)} \qquad \alpha^-=\frac{I}{(2I+1)} \tag{6.82}$$

are the corresponding relative statistical weights. The quantity b defined by (6.81)–(6.82) is called the coherent bound scattering length for the nucleus of spin I and reduces to the usual bound scattering length in the case of a spinless nucleus. Using it we may define a corresponding incoherent bound scattering length $b_{\mathrm{inc}}(I)$ through the relation

$$\begin{aligned} b_{\mathrm{inc}}^2(I)&=(\alpha^+|b^+|^2+\alpha^-|b^-|^2)-|(\alpha^+b^+ +\alpha^-b^-)|^2\\ &\equiv b_{\mathrm{s}}^2-b^2\\ &=I(I+1)(c/2)^2=\alpha^+\alpha^-(b^+-b^-)^2 \end{aligned} \tag{6.83}$$

where $\sigma_{\mathrm{s}}=4\pi b_{\mathrm{s}}^2$ is the average scattering cross-section.

The coherent and total scattering cross-sections for unpolarized neutrons in a chemically pure unpolarized substance composed of several isotopes may now be derived from (6.80)–(6.81) giving the results

$$\sigma_{\mathrm{coh}}=4\pi\left|\sum_i f_i(\alpha^+b^+ +\alpha^-b^-)_i\right|^2=4\pi\langle b\rangle^2\equiv 4\pi b_{\mathrm{coh}}^2$$

$$\sigma_{\mathrm{s}}=4\pi\sum_i f_i(\alpha^+|b^+|^2+\alpha^-|b^-|^2)_i=4\pi\langle b_{\mathrm{s}}^2\rangle \tag{6.84}$$

where f_i represents the relative abundance of the ith isotope in the medium. The corresponding incoherent scattering cross-section is now given by

$$\begin{aligned} \sigma_{\mathrm{inc}}&=\sigma_{\mathrm{s}}-\sigma_{\mathrm{coh}}=4\pi(\langle b_{\mathrm{s}}^2\rangle-\langle b\rangle^2)\\ &=4\pi(\langle b_{\mathrm{inc}}(I)^2\rangle+\langle b^2\rangle-\langle b\rangle^2) \end{aligned} \tag{6.85}$$

Thus, in addition to the spin contribution $4\pi\langle b_{\mathrm{inc}}(I)^2\rangle$, the incoherent scattering cross-section contains an isotopic contribution $4\pi(\langle b^2\rangle-\langle b\rangle^2)$ proportional to the variance of the distribution of bound scattering lengths, which is non-vanishing even when all of the contributing isotopes have zero spin.

For pure potential scattering b^+ and b^- are approximately equal and comparable in size with the nuclear radius so that, according to (6.83), the spin-incoherent scattering is weak. This conclusion does not hold in the neighbourhood of a resonance, which is always characterized by a definite spin value, so that one of b^+ or b^- is large and the incoherent scattering is much enhanced. In these circumstances the total scattering cross-section becomes (cf. section 2.3.2)

$$\sigma_s = 4\pi\alpha^{\pm} \left| \frac{\Gamma_s/2k}{(E-E_r)+\dfrac{\Gamma_t}{2}} + b^{\pm} \right|^2 + 4\pi\alpha^{\mp} |b^{\mp}|^2 \tag{6.86}$$

where the upper (lower) sign applies when the resonance has spin $(I \pm \frac{1}{2})$.

In the general case a scattering medium contains more than one isotope and indeed more than one element, and in the latter case the sum over f_i must be replaced by a double sum over f_{ij} where the index j distinguishes the different elements. One should note that $\sigma_{inc} \neq 0$ when more than one of the isotopic abundances f_i are non-vanishing even when $I=0$ and $b_{inc}=0$ for each of the contributing isotopes. Thus an isotopic incoherence exists in addition to a spin incoherence. Indeed there is only one element in nature, namely thorium, for which the isotopic and spin incoherence contributions both vanish, because the only naturally occurring isotope of thorium, ${}^{232}_{90}Th$, has $I=0$.

In every case the total cross-section σ_t can be derived from the imaginary part of the corresponding scattering length by means of the optical theorem as expressed in equation (2.100), where Imf0) is replaced by $-\text{Im}\, b \geqslant 0$. Thus, in the case of the three important (n, α) reactions listed in Table 1.3, Im b^+ is essentially zero for ${}^3_2He$ and ${}^{10}_5B$ and the reaction takes place only in states with channel spins 0^+ and $\frac{5}{2}^+$, respectively, whereas for ${}^6_3Li$ the reaction takes place in both states $I^{\pm}_{1/2}$, with channel spins $\frac{1}{2}^+, \frac{3}{2}^+$, respectively.

Compilations of coherent and incoherent scattering lengths and cross-sections are presented in Table 6.1 for the most abundant stable isotopes of the light elements, and in Table 6.2 for the very useful group of stable isotopes of the transition elements.

6.3.1.3 Magnetic birefringence

In the presence of a magnetic field $\boldsymbol{B}$, because of the magnetic interaction $-\boldsymbol{\mu}_n \cdot \boldsymbol{B}$ which introduces an energy splitting between the two spin-states of the neutron, it is necessary to take account of the spinor nature of the coherent wave $\Psi(\boldsymbol{r})$ and replace the scattering potential $V(\boldsymbol{r}-\boldsymbol{r}_j)$ by a 2×2 matrix. The magnetic scattering differs fundamentally from the nuclear scattering in two main respects: (i) since it arises from the interaction between the neutron magnetic moment and the spin and orbital magnetic moments of the scattering atoms it is a long-range and not a short-range

Table 6.1. Coherent and incoherent scattering lengths and cross-sections for stable light nuclei in the s, p and s–d shells, with natural abundances >1%. Data selected from AECL-8490 (Sears 1984). The table also includes data for the 12.26 y, β^--emitter ${}^3_1$H because of its importance as a neutron source, and for ${}^{36}_{18}$A (0.36%) which has an anomalously large coherent scattering length. Note that, although individual isotopes with $I=0$ have $b_{inc}\equiv 0$, $\sigma_{inc}\neq 0$ in general for a mixture of isotopes each with $I=0$. For example natural argon has $\sigma_{coh}=0.458\pm 0.003$b and $\sigma_{inc}=0.22\pm 0.02$b, even though all naturally occurring isotopes of argon have $I=0$

Nucleus	I^π	Abundance (%)	b_{coh} (fm)	b_{inc} (fm)	σ_{coh} (b)	σ_{inc} (b)
${}^1_1$H	$\frac{1}{2}^+$	99.985	-3.7423 ± 0.0012	25.217 ± 0.006	1.7599 ± 0.0011	79.91 ± 0.04
${}^2_1$H	1^+	0.015	6.674 ± 0.006	4.033 ± 0.032	5.597 ± 0.010	2.04 ± 0.03
${}^3_1$H	$\frac{1}{2}^+$	0.0	4.94 ± 0.08	0.00 ± 0.37	3.07 ± 0.10	0.00 ± 0.02
${}^3_2$He	$\frac{1}{2}^+$	1.3×10^{-4}	5.74 ± 0.07	-1.8 ± 0.6	4.42 ± 0.10	1.2 ± 0.3
${}^4_2$He	0^+	~ 100	3.26 ± 0.03	0	1.34 ± 0.02	0
${}^6_3$Li	1^+	7.4	2.0 ± 0.1	-1.79 ± 0.24	0.51 ± 0.05	0.41 ± 0.11
${}^7_3$Li	$\frac{3}{2}^-$	92.6	-2.22 ± 0.01	-2.49 ± 0.05	0.619 ± 0.006	0.78 ± 0.03
${}^9_4$Be	$\frac{3}{2}^-$	100	7.79 ± 0.01	0.20 ± 0.02	7.63 ± 0.02	0.005 ± 0.001
${}^{10}_5$B	3^+	18.7	-0.1 ± 0.4	-4.7 ± 0.3	0.14 ± 0.02	3.0 ± 0.4
${}^{11}_5$B	$\frac{3}{2}^-$	81.3	6.65 ± 0.04	-1.31 ± 0.17	5.56 ± 0.07	0.22 ± 0.06
${}^{12}_6$C	0^+	98.9	6.6535 ± 0.0014	0	5.563 ± 0.002	0
${}^{13}_6$C	$\frac{1}{2}^-$	1.1	6.19 ± 0.09	-0.52 ± 0.09	4.81 ± 0.14	0.034 ± 0.012
${}^{14}_7$N	1^+	99.6	9.37 ± 0.02	1.98 ± 0.17	11.03 ± 0.05	0.49 ± 0.08
${}^{16}_8$O	0^+	99.76	5.805 ± 0.005	0	4.235 ± 0.007	0
${}^{19}_9$F	$\frac{1}{2}^+$	100	5.654 ± 0.012	-0.082 ± 0.009	4.017 ± 0.017	0.0008 ± 0.0002

$^{20}_{10}Ne$	0^+	90.9	4.610 ± 0.012	0	2.671 ± 0.014	0
$^{22}_{10}Ne$	0^+	8.8	3.87 ± 0.01	0	1.88 ± 0.01	0
$^{23}_{11}Na$	$\frac{3}{2}^+$	100	3.63 ± 0.02	3.59 ± 0.03	1.66 ± 0.02	1.62 ± 0.03
$^{24}_{12}Mg$	0^+	78.6	5.68 ± 0.02	0	4.05 ± 0.03	0
$^{25}_{12}Mg$	$\frac{5}{2}^+$	10.1	3.62 ± 0.14	0.9 ± 0.3	1.65 ± 0.13	0.10 ± 0.07
$^{26}_{12}Mg$	0^+	11.3	4.92 ± 0.15	0	3.04 ± 0.19	0
$^{23}_{13}Al$	$\frac{5}{2}^+$	100	3.449 ± 0.005	0.26 ± 0.01	1.495 ± 0.004	0.0085 ± 0.0007
$^{28}_{14}Si$	0^+	92.2	4.106 ± 0.006	0	2.119 ± 0.006	0
$^{29}_{14}Si$	$\frac{1}{2}^+$	4.7	4.7 ± 0.1	-1.1 ± 0.2	2.78 ± 0.12	0.15 ± 0.06
$^{30}_{14}Si$	0^+	3.1	4.58 ± 0.08	0	2.64 ± 0.09	0
$^{31}_{15}P$	$\frac{1}{2}^+$	100	5.13 ± 0.01	0.22 ± 0.07	3.307 ± 0.013	0.006 ± 0.004
$^{32}_{16}S$	0^+	95.0	2.804 ± 0.002	0	0.9880 ± 0.0014	0
$^{34}_{16}S$	0^+	4.24	3.48 ± 0.03	0	1.52 ± 0.03	0
$^{35}_{17}Cl$	$\frac{3}{2}^+$	75.4	11.66 ± 0.02	6.0 ± 0.2	17.08 ± 0.06	4.5 ± 0.3
$^{37}_{17}Cl$	$\frac{3}{2}^+$	24.6	3.08 ± 0.06	0.02 ± 0.05	1.19 ± 0.05	0.0001 ± 0.0003
$^{36}_{18}A$	0^+	0.34	24.90 ± 0.07	0	77.9 ± 0.4	0
$^{40}_{18}A$	0^+	99.60	1.83 ± 0.05	0	0.42 ± 0.02	0
$^{39}_{19}K$	$\frac{3}{2}^+$	93.08	3.79 ± 0.02	1.4 ± 0.3	1.81 ± 0.02	0.25 ± 0.11
$^{41}_{19}K$	$\frac{3}{2}^+$	6.91	2.58 ± 0.06	< 0.2	0.84 ± 0.04	< 0.005
$^{40}_{20}Ca$	0^+	96.94	4.99 ± 0.03	0	3.13 ± 0.04	0
$^{44}_{20}Ca$	0^+	2.1	1.8 ± 0.1	0	0.41 ± 0.05	0

Table 6.2. Coherent and incoherent scattering lengths and cross-sections for stable nuclei of transition elements in the first long period of the Periodic Table. Data selected from AECL-8490 (Sears 1984). Excluding the light nuclei listed in Table 6.1 and a few examples of heavy nuclei, e.g. $^{113}_{48}Cd$, $^{149,152}_{62}Sm$, $^{162}_{66}Dy$ and $^{186}_{74}W$, this group includes all nuclei with $b_{coh}<0$. It also contains the important trio of group VIII elements, Fe, Co, Ni, whose isotopes can be mixed in various proportions to bring about effective cancellation of nuclear and magnetic scattering lengths for one sense of the spin

Nucleus	I^π	Abundance (%)	b_{coh} (fm)	b_{inc} (fm)	σ_{coh}(b)	σ_{inc}(b)
$^{45}_{21}Sc$	$\frac{7}{2}^-$	100	12.29 ± 0.11	-6.0 ± 0.3	19.0 ± 0.3	4.5 ± 0.5
$^{46}_{22}Ti$	0^+	8.0	4.73 ± 0.06	0	2.81 ± 0.07	0
$^{47}_{22}Ti$	$\frac{5}{2}^-$	7.3	3.49 ± 0.12	-3.5 ± 0.2	1.53 ± 0.11	1.54 ± 0.18
$^{48}_{22}Ti$	0^+	73.8	-5.84 ± 0.02	0	4.29 ± 0.03	0
$^{49}_{22}Ti$	$\frac{7}{2}^-$	5.5	1.00 ± 0.05	5.1 ± 0.2	0.13 ± 0.01	3.27 ± 0.26
$^{50}_{22}Ti$	0^+	5.4	5.93 ± 0.08	0	4.42 ± 0.12	0
$^{50}_{23}V$	6^+	0.25	7.6 ± 0.7	2 ± 2	7.3 ± 1.4	0.5 ± 1.0
$^{51}_{23}V$	$\frac{7}{2}^-$	99.75	-0.4024 ± 0.0021	6.419 ± 0.010	0.0203 ± 0.0002	5.178 ± 0.016
$^{50}_{24}Cr$	0^+	4.4	-4.50 ± 0.05	0	2.54 ± 0.06	0
$^{52}_{24}Cr$	0^+	83.7	4.920 ± 0.0010	0	3.042 ± 0.012	0
$^{53}_{24}Cr$	$\frac{3}{2}^-$	9.5	-4.20 ± 0.03	6.86 ± 0.01	2.22 ± 0.03	5.91 ± 0.19
$^{54}_{24}Cr$	0^+	2.4	4.55 ± 0.10	0	2.60 ± 0.11	0

$^{55}_{25}Mn$	$\frac{5}{2}^+$	100	-3.73 ± 0.02	1.79 ± 0.04	1.75 ± 0.02	0.40 ± 0.02
$^{54}_{26}Fe$	0^+	5.9	4.2 ± 0.1	0	2.2 ± 0.1	0
$^{56}_{26}Fe$	0^+	91.6	10.03 ± 0.07	0	12.64 ± 0.18	0
$^{57}_{26}Fe$	$\frac{1}{2}^-$	2.2	2.3 ± 0.1	2 ± 2	0.66 ± 0.06	0.5 ± 1.0
$^{58}_{26}Fe$	0^+	0.33	15 ± 7	0	28 ± 26	0
$^{59}_{27}Co$	$\frac{7}{2}^-$	100	2.50 ± 0.03	-6.2 ± 0.2	0.79 ± 0.02	4.8 ± 0.3
$^{58}_{28}Ni$	0^+	67.9	14.4 ± 0.1	0	26.1 ± 0.4	0
$^{60}_{28}Ni$	0^+	26.2	2.8 ± 0.1	0	0.99 ± 0.07	0
$^{61}_{28}Ni$	$\frac{3}{2}^-$	1.2	7.60 ± 0.06	4.0 ± 0.3	7.26 ± 0.11	2.0 ± 0.3
$^{62}_{28}Ni$	0^+	3.7	-8.7 ± 0.2	0	9.5 ± 0.4	0
$^{64}_{28}Ni$	0^+	1.0	-0.38 ± 0.07	0	0.018 ± 0.007	0
$^{63}_{29}Cu$	$\frac{3}{2}^-$	69.0	6.42 ± 0.15	0.22 ± 0.02	5.2 ± 0.2	0.0061 ± 0.0011
$^{65}_{29}Cu$	$\frac{3}{2}^-$	31.0	10.61 ± 0.19	1.79 ± 0.10	14.1 ± 0.5	0.40 ± 0.05
$^{64}_{30}Zn$	0^+	48.9	5.23 ± 0.10	0	3.44 ± 0.13	0
$^{66}_{30}Zn$	0^+	27.8	6.01 ± 0.12	0	4.54 ± 0.18	0
$^{67}_{30}Zn$	$\frac{5}{2}^-$	4.1	7.64 ± 0.15	2 ± 2	7.3 ± 0.3	0.5 ± 1.0
$^{68}_{30}Zn$	0^+	18.6	6.05 ± 0.12	0	4.60 ± 0.18	0
$^{70}_{30}Zn$	0^+	0.62	6 ± 1	0	4.5 ± 1.5	0

interaction and cannot in general be represented as a modification of the parameters of the Fermi pseudo-potential; (ii) the scattering is not isotropic but has an angular dependence characteristic of a dipole–dipole interaction, and again the description of the scattering in terms of a pseudo-potential is not possible.

When a neutron flips its spin on scattering from a magnetic atom the scattering is incoherent and even when the spin is not flipped the scattering is incoherent if the atomic magnetic moments are oriented at random as happens in a paramagnetic substance. The situation is quite different in a ferromagnet where the atomic moments are spontaneously polarized below the Curie temperature giving a net magnetic field $\boldsymbol{B}$ throughout a magnetic domain which is much larger than any individual atom. In this case the ferromagnet is neutron-optically birefringent with two values of the refractive index

$$n^{\pm}=\left\{1-\frac{4\pi}{k^2}\left(Nb\pm\left(\frac{2\pi\hbar^2}{m_n}\right)^{-1}\mu_n B\right\}^{1/2} \tag{6.87}$$

corresponding to the two possible orientations of the neutron spin relative to the local field. This result has been successfully employed in various ways to produce beams of polarized neutrons (cf. section 7.4.3).

6.3.1.4 *The effect of chemical binding*

When the scattering nucleus is bound by electromagnetic forces in a molecule the scattering of neutron on nucleus becomes a three-body rather than a two-body problem, and in general the motion of the centre of mass cannot be eliminated from the momentum equation using the concept of reduced mass. The problem is therefore posed of calculating the bound scattering length b, in terms of the free scattering length a, taking account of the influence of the chemical binding.

The calculation was first performed by Fermi (1936) who showed that the differential cross-section for elastic scattering in the Born approximation could be expressed in the form

$$\frac{d\sigma_s}{d\Omega}=\left(\frac{\mu'}{\mu}\right)^2\left(\frac{v_f}{v_i}\right)|a|^2\left|\sum_j\int e^{i\boldsymbol{q}\cdot\boldsymbol{r}_m}u_f^*u_i\,d^3r_m\right|^2 \tag{6.88}$$

where $u_m(\boldsymbol{r}_m, \xi_j)$ is the molecular wavefunction, ξ_j collectively describes the internal molecular coordinates, and

$$\boldsymbol{q}=\boldsymbol{k}_i-\boldsymbol{k}_f \tag{6.89}$$

represents the momentum transferred from the neutron to the molecule. The initial and final neutron velocities v_i, v_f are related to the corresponding momenta

$$v_i = \hbar k_i/\mu' \qquad v_f = \hbar k_f/\mu' \tag{6.90}$$

where

$$\mu' = \frac{m_n M_m}{m_n + M_m} \tag{6.91}$$

is the reduced mass of the neutron in its collision with the molecule of mass M_m.

It follows from (6.88) that, to determine the effect of chemical binding, it is necessary to know the molecular wavefunction $u(\boldsymbol{r}_m, \xi_j)$, and a large number of calculations of this type have been carried out using various simple models. Nevertheless there are two limiting situations where such information is not needed. The first limiting case is obviously the high energy limit (i.e. $E_n > 1$ eV), where the molecular binding energy of the target nucleus may be neglected in comparison with the incident kinetic energy. In this $\mu' = \mu$, $v_i = v_f$ and the overlap integral in (6.88) vanishes unless $\boldsymbol{q} \equiv 0$, i.e. momentum is conserved in the collision of neutron and nucleus. This corresponds to the normal situation of neutrons colliding with free nuclei. The second limiting case arises when the neutron energy is so low that molecular excitation is impossible. In this case also the scattering is elastic, i.e. $v_i = v_f$ and, setting $\boldsymbol{q} \cdot \boldsymbol{r}_n \simeq 0$, the overlap integral is unity since $u_i \equiv u_f$. The angular dependence therefore disappears from the integrand giving the result

$$\sigma_s = 4\pi \left(\frac{\mu'}{\mu}\right)^2 |a|^2 \tag{6.92}$$

Thus the relation (6.74) which connects bound and free scattering lengths is replaced by

$$b = \left(\frac{\mu'}{\mu}\right)\left(\frac{m_n}{\mu}\right) a \tag{6.93}$$

In general the correction introduced by (6.93) for neutron scattering on light nuclei is not a trivial one. For example it increases the zero-energy scattering cross-section for neutrons on protons bound in heavy molecules by a factor of about four as compared with neutrons scattering from atomic hydrogen (Breit 1947, Lippmann 1950).

6.3.2 The extinction theorem

The extinction theorem, which was first discussed in the context of electromagnetic wave propagation by Oseen (1915) and Ewald (1916), is generally applicable to wave phenomena in dispersive media, and is very easily proved for the de Broglie waves associated with neutrons. The theorem is usually interpreted as a demonstration of how an incident wave which penetrates a scattering medium is 'extinguished' by destructive interference with waves generated by surface sources, to be replaced by a transmitted ray which travels in a slightly different direction. In a sense therefore the extinction theorem may be said to 'explain' the phenomena of both refraction and reflection, since the same surface sources, emitting out of the medium rather than into it, account also for the reflected wave. These surface sources are, however, at least partly fictitious since, as we show below, they act even when there is no scattering medium, and are related to the secondary sources which play such an important role in the Huygens–Fresnel formulation of diffraction theory (cf. section 7.4).

To derive the extinction theorem it is convenient to divide the whole of space into two regions denoted respectively by M, the scattering medium, and $\bar{M}$, the vacuum. The scattering potential $V(\boldsymbol{r})$ may then be defined in the following way

$$\begin{aligned} V_{\mathrm{F}}(\boldsymbol{r}) &= V_{\mathrm{F}} \qquad \boldsymbol{r} \in M \\ &= 0 \qquad \boldsymbol{r} \in \bar{M} \end{aligned} \tag{6.94}$$

The coherent wave $\Psi(\boldsymbol{r})$ exists throughout space and satisfies a Schrödinger equation

$$(\nabla^2 + k^2)\Psi(\boldsymbol{r}) = \frac{2m_{\mathrm{n}}}{\hbar^2} V_{\mathrm{F}}(\boldsymbol{r})\Psi(\boldsymbol{r}) \tag{6.95}$$

whose solution behaves asymptotically like outgoing waves at infinity, and hence satisfies the integral equation (cf. section 2.3.1)

$$\Psi(\boldsymbol{r}) = \mathrm{e}^{\mathrm{i}\boldsymbol{k}\cdot\boldsymbol{r}} + \int \mathrm{d}^3 r' G_0(\boldsymbol{r}-\boldsymbol{r}')\Psi(\boldsymbol{r}')V_{\mathrm{F}}(\boldsymbol{r}') \tag{6.96}$$

where $\mathrm{e}^{\mathrm{i}\boldsymbol{k}\cdot\boldsymbol{r}}$ represents the incoming wave. The Green function $G_0(\boldsymbol{r})$, whose explicit form is given in (2.38) satisfies the inhomogeneous equation

$$(\nabla^2 + k^2)G_0(\boldsymbol{r}) = \frac{2m_{\mathrm{n}}}{\hbar^2}\delta^3(\boldsymbol{r}) \tag{6.97}$$

We now define a vector

$$\boldsymbol{N}(\boldsymbol{r}, \boldsymbol{r}') = \Psi(\boldsymbol{r}')\nabla' G_0(\boldsymbol{r}-\boldsymbol{r}') - G_0(\boldsymbol{r}-\boldsymbol{r}')\nabla'\Psi(\boldsymbol{r}') \tag{6.98}$$

and use equations (6.95) and (6.97), applied at points $\boldsymbol{r}' \in M$ to derive the result

$$\nabla' \cdot \boldsymbol{N}(\boldsymbol{r}, \boldsymbol{r}') = -\frac{2m_{\mathrm{n}}}{\hbar^2}\{G_0(\boldsymbol{r}-\boldsymbol{r}')V_{\mathrm{F}}(\boldsymbol{r}') - \delta^3(\boldsymbol{r}-\boldsymbol{r}')\}\Psi(\boldsymbol{r}') \tag{6.99}$$

Integrating (6.99) over all points $\boldsymbol{r}' \in M$, and noting that the δ-function makes a contribution only if $r \in M$, we find, using Gauss's theorem, that

$$\frac{\hbar^2}{2m_n}\int_\Sigma N(r, r')\cdot d\sigma' = e^{ik\cdot r} \qquad r \in M$$
$$= -\Psi_R(r) \qquad r \in \bar{M} \qquad (6.100)$$

where

$$\Psi_R(r) = \Psi(r) - e^{ik\cdot r} \qquad r \in \bar{M} \qquad (6.101)$$

is the reflected wave. The surface integral in (6.100) is carried out over points $(x', y', 0)$ as shown in figure 6.4.

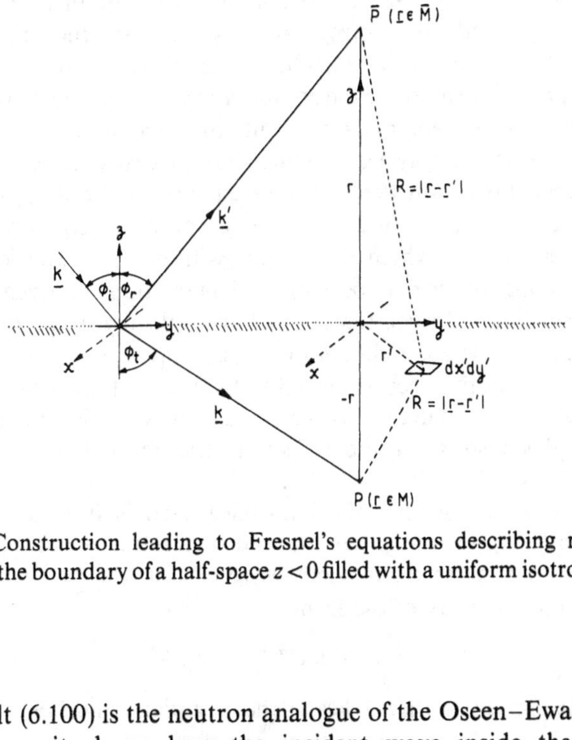

Figure 6.4. Construction leading to Fresnel's equations describing reflection and refraction at the boundary of a half-space $z < 0$ filled with a uniform isotropic scattering medium.

The result (6.100) is the neutron analogue of the Oseen–Ewald extinction theorem since it shows how the incident wave inside the medium is 'extinguished' by waves emitted from a surface source density $-(\hbar^2/2m_n)N_n$, where N_n is the outward drawn normal component of N, while at the same time a reflected wave is generated outside the medium. It should be emphasized, however, that if V_F is made to vanish identically throughout space, then the surface integral appearing in (6.100) becomes identical with the Helmholtz–Kirchoff diffraction integral defined in (7.11). The integral therefore take the value $e^{ik\cdot r}$ for $r \in M$, but vanishes for $r \in \bar{M}$, in accordance with the fact that $\Psi(r) = e^{ik\cdot r}$ everywhere. Thus for $r \in M$, the source density $-(\hbar^2/2m_n)N_n$ must contain a 'fictitious' component as discussed above.

6.3.3 Reflection and refraction

6.3.3.1 'Fresnel' equations

Taken in conjunction, equations (6.100) and (6.101) may be interpreted as coupled integral equations for $\Psi(\boldsymbol{r}')$ and $\partial\Psi(\boldsymbol{r}')/\partial n$ at points $\boldsymbol{r}'$ on the bounding surface Σ when the incident wave $e^{i\boldsymbol{k}\cdot\boldsymbol{r}}$ and the reflected wave $\Psi_R(\boldsymbol{r})$, are specified at points $\boldsymbol{r}\in\bar{M}$ outside the medium. Taken in sequence these equations may be used to derive the laws of reflection and refraction, by first postulating a solution of (6.95) for $\boldsymbol{r}\in M$ of the form $A_t\,e^{i\boldsymbol{K}\cdot\boldsymbol{r}}$, where $|K|=nk$, and then solving for the transmitted amplitude A_t and the direction cosines of $\boldsymbol{K}$ using (6.100). In the same way solutions are obtained for the reflected amplitude A_r, and the direction cosines for the reflected wave $\Psi_R(\boldsymbol{r})=A_r\,e^{i\boldsymbol{k}'\cdot\boldsymbol{r}}$ ($|\boldsymbol{k}'|=k$) using (6.100), where $\Psi_R(\boldsymbol{r})$ is a solution of (6.95) for $\boldsymbol{r}\in\bar{M}$. This procedure involves the computation of an integral over the surface sources which in the general case is quite intractable but, in the case of an infinite plane surface, may be evaluated rather easily using the method of steepest descents (Copson 1935). The results are valid only to the neglect of terms of order $(k|\boldsymbol{r}'-\boldsymbol{r}|)^{-1}$ and therefore do not take account of phenomena, e.g. evanescent waves, which may occur within a few wavelengths of the surface. The conclusions may be expressed as relations between the incident, refracted and reflected rays, represented symbolically by $(1, \boldsymbol{k})$, $(A_t, \boldsymbol{K})$ and $(A_r, \boldsymbol{k}')$, which are the analogues of the Fresnel equations for electromagnetic waves (with the electric vector normal to the plane of incidence (Goldberger and Seitz 1947)), although there is perhaps a somewhat closer correspondence with Rayleigh's results for the reflection and refraction of sound waves (Rayleigh 1929).

We therefore assume that the half-space $z<0$ is filled with a uniform isotropic scattering medium characterized by a refractive index n, as shown in figure 6.4 and, given an incident wave $e^{i\boldsymbol{k}\cdot\boldsymbol{r}}$ in the region $\boldsymbol{r}\in\bar{M}$ $(z>0)$, postulate a refracted ray of the form

$$\Psi(\boldsymbol{r})=A_t\,e^{i\boldsymbol{K}\cdot\boldsymbol{r}} \qquad \boldsymbol{r}\in M \tag{6.102}$$

The axes are chosen so that $\boldsymbol{K}$ lies in the x–z plane and has components

$$\boldsymbol{K}\equiv kn\cos\varphi_t \qquad 0 \qquad -kn\sin\varphi_t \qquad |\boldsymbol{K}|=kn \tag{6.103}$$

Anticipating the final result we also derive from $\boldsymbol{K}$ a new vector $\boldsymbol{\kappa}$, where

$$\boldsymbol{\kappa}\equiv k\cos\varphi \qquad 0 \qquad -k\sin\varphi \qquad |\boldsymbol{\kappa}|=k \tag{6.104}$$

and the angle φ is defined by

$$\sin\varphi=n\sin\varphi_t \tag{6.105}$$

We now apply (6.100) to derive a relation between the incident and refracted wave at the point $P(\boldsymbol{r})\equiv(0, 0, -\boldsymbol{r})$ inside the medium. Substituting

(6.101) into (6.100) and using (6.103)–(6.105) we find then that

$$A_t \frac{\sin(\varphi+\varphi_t)}{2\cos\varphi\sin\varphi_t} e^{i\boldsymbol{\kappa}\cdot\boldsymbol{r}} = e^{i\boldsymbol{k}\cdot\boldsymbol{r}} \qquad \boldsymbol{r}\in M \tag{6.106}$$

From this result we may deduce that $\boldsymbol{\kappa}=\boldsymbol{k}$ and $\varphi=\varphi_i$, where φ_i is the angle of incidence. Thus the refracted ray lies in the plane of incidence and the definition (6.105) becomes an expression of Snell's law. The transmitted amplitude is then given according to (6.106) and (6.105) by

$$A_t = \frac{2\cos\varphi_i \sin\varphi_t}{\sin(\varphi_i+\varphi_t)} = \frac{2\cos\varphi_i\sin\varphi_t}{\cos\varphi_i + \sqrt{(n^2-\sin^2\varphi_i)}} \tag{6.107}$$

To arrive at an expression for the reflected wave

$$\Psi_R(\boldsymbol{r}) = A_r\, e^{i\boldsymbol{k}'\cdot\boldsymbol{r}'} \qquad \boldsymbol{r}'\in\bar{M} \tag{6.108}$$

which is a solution of (6.95) for $\boldsymbol{r}'\in\bar{M}$ and $|\boldsymbol{k}'|=k$, we define a new vector $\boldsymbol{\kappa}'$, analogous to $\boldsymbol{\kappa}$, i.e.

$$\boldsymbol{\kappa}' \equiv k\cos\varphi_i \qquad 0 \qquad k\sin\varphi_i \qquad |\boldsymbol{\kappa}'|=k \tag{6.109}$$

Using (6.108) and evaluating the surface integral in (6.100) for the point $\bar{P}(\boldsymbol{r})=(0,0,r)$ outside the medium, we obtain the result

$$\frac{\sin(\varphi_i-\varphi_t)}{\sin(\varphi_i+\varphi_t)} e^{i\boldsymbol{\kappa}'\cdot\boldsymbol{r}'} = -A_r\, e^{i\boldsymbol{k}'\cdot\boldsymbol{r}'} \qquad \boldsymbol{r}\in\bar{M} \tag{6.110}$$

Thus $\boldsymbol{k}'\equiv\boldsymbol{\kappa}'$ and $\varphi_r=\varphi_i$ so that the reflected ray also lies in the plane of incidence and propagates at an equal angle. These are the familiar laws of reflection. According to (6.110) and (6.105) the reflected amplitude is

$$A_r = -\frac{\sin(\varphi_i-\varphi_t)}{\sin(\varphi_i+\varphi_t)} = \frac{\cos\varphi_i - \sqrt{(n^2-\sin^2\varphi_i)}}{\cos\varphi_i+\sqrt{(n^2-\sin^2\varphi_i)}} = A_t - 1 \tag{6.111}$$

We note that the result (6.111) implies that, if $n>1$, which is rather unusual for neutrons, then $A_r<0$ and there is a phase change on reflection at the boundary. Finally, to check on the consistency of (6.107) and (6.111) we must confirm that these results accord with the requirement that the probability current density $\boldsymbol{j}(\boldsymbol{r},t)$ be conserved. This demand is expressed as a continuity condition on the normal component of the current at the boundary, i.e.

$$k\cos\varphi_i - k'|A_r|^2\cos\varphi_r = K|A_t|^2\cos\varphi_t \tag{6.112}$$

Using the results derived above this may be rewritten as

$$|A_r|^2 + \sqrt{\left(\frac{n^2-\sin^2\varphi_i}{\cos^2\varphi_i}\right)}|A_t|^2 = R+T=1 \tag{6.113}$$

where R and T denote the reflection and transmission coefficients respectively.

The results (6.107) and (6.111), which are the neutron analogues of the Fresnel equations, can of course be derived directly by solving the boundary value problem for a potential step in two dimensions, and imposing the requirements that $\Psi(\boldsymbol{r})$, and the normal component of its gradient, be continuous across the boundary. Nevertheless the exercise provides the complete laws of reflection and refraction and demonstrates the power of the extinction theorem which may be applied to boundary surfaces of arbitrary shape without imposing any restrictions as to the number of dimensions or the separability of the wave equation.

6.3.3.2 Total external reflection

According to (6.111) the reflected amplitude A_r becomes unity at an angle of incidence

$$\varphi_c = \sin^{-1}(n) \tag{6.114}$$

and, when $\varphi > \varphi_i$, A_r is complex with unit modulus. Thus, introducing the glancing angle

$$\theta = \pi/2 - \varphi_i \tag{6.115}$$

total external reflection occurs for all $\theta < \theta_c$, where θ_c is a critical glancing angle which, for small θ_c, is given by

$$\theta_c = \lambda \sqrt{\left(\frac{Nb}{\pi}\right)} \tag{6.116}$$

For the phenomenon of total external reflection to occur it is necessary of course that the scattering length be positive, but this is the case for the great majority of substances.

Total external reflection was first observed by Fermi and Zinn (1946) using polychromatic neutrons, although they were unable to determine θ_c. This was first achieved by Fermi and Marshall (1947) using monochromatic neutrons selected by Bragg reflection in single crystals of calcium fluoride, and the procedure they employed forms the basis of most methods for determining scattering lengths. The technique is effective because it relies on mirror reflection and no corrections for temperature-dependent diffuse scattering are required, since the corresponding Debye–Waller factor is unity for forward scattering (cf. section 8.1).

In this context it should be emphasized that the analysis of section 6.3.1 applies only to isotropic media and in crystalline media mirror reflection occurs only at glancing angles θ satisfying Bragg's law (cf. section 7.2.3)

$$2d \sin\theta = m\lambda \tag{6.117}$$

where d is the appropriate crystal lattice spacing. The corresponding scattering length turns out to be infinite but this paradoxical result implies only that the Born approximation ceases to be valid since the incident wave is strongly attenuated as it penetrates the medium. Indeed in most crystals extinction is complete at depths of about 100 μm. This result is the basis of the method introduced by Hughes and Burgy (1951) for producing cold neutrons ($\lambda \geqslant 4$ Å) by filtering out fast neutrons with $\lambda < \lambda_B$ by Bragg scattering in polycrystalline samples of beryllium oxide and graphite. Here $\lambda_B = 2d$ is the Bragg cut-off wavelength above which Bragg's equation (6.115) cannot be satisfied for any value of θ.

The most important source of error in scattering length measurements is associated with the wavelength determination, and even a crystal monochromator has a finite resolving power due to the beam divergence and the mosaic block structure of the single crystal which is determined by its dislocation pattern. In Koester's gravity refractometer (cf. section 1.5.2.2) this source of error is eliminated at least in first order. The refractometer relies for its action on the fact that, in many substances, the Fermi potential V_F is comparable with the gravitational potential energy of the neutron ($\simeq 10^{-7}$ eV m^{-1}) (Maier–Leibnitz 1962). Thus neutrons falling from a height h onto a horizontal plane surface are totally reflected unless $h > h_c$, where h_c is a critical height defined by

$$m_n g h_c = V_F = 2\pi\hbar^2 Nb/m_n \tag{6.118}$$

The determination of h_c then permits a measurement of b according to (6.118).

Apart from its application to the determination of scattering lengths, total external reflection has also been successfully exploited for producing polarized neutrons (cf. section 7.4.1) for storing ultra-cold neutrons (cf. sections 1.6.6, 9.2.5.4) and in the manufacture of neutron guides of high transmission (Christ and Springer 1962, Maier-Leibnitz and Springer 1963). An approximate theory of the neutron guide based on (6.111) indicates that, if θ_c is fixed by the experimental conditions, the equation determines a critical wavelength λ_c such that, for $\lambda > \lambda_c$ neutrons are totally reflected at the guide walls and may be channelled to the experimental area, whereas for $\lambda < \lambda_c$, the neutrons penetrate the walls and are lost. The value of θ_c is determined by curving the guide and forcing each neutron to make at least one bounce during its passage. Thus, the shorter the radius of curvature R of the guide the larger are θ_c and λ_c. Neutron guides are conventionally constructed from nickel-coated boron glass, and have rectangular cross-sections of side typically about 10 cm. Thus a thermal neutron guide will have $R \simeq 2 \times 10^4$ cm, $\lambda_c = 1$ Å, and will be about 80 m long. For a very cold neutron guide $R \simeq 25$ m, $\lambda_c \simeq 30$ Å and the length will be about 10 m. An important feature in the use of a neutron guide is that the flux of γ-rays is attenuated in proportion to the square of the length.

6.3.3.3 Refraction

Neutron refraction was first observed by Harvey *et al* (1951) but, despite the huge array of neutron optical instruments which have been developed in solid state physics research, prisms and lenses have not been used as dispersing or focusing elements to anything like the same extent as in electromagnetic wave optics. This is because the refractive power $|n-1|$ is extremely small for thermal neutrons and even for very cold neutrons ($\lambda \simeq 20$ Å), is typically of order 10^{-4}. Nevertheless a number of instruments based on neutron refraction have been constructed from time to time and perhaps the most successful of these has been Shull's double crystal refractometer which is shown in figure 6.5 (Shull *et al* 1967, Schneider and Shull 1971).

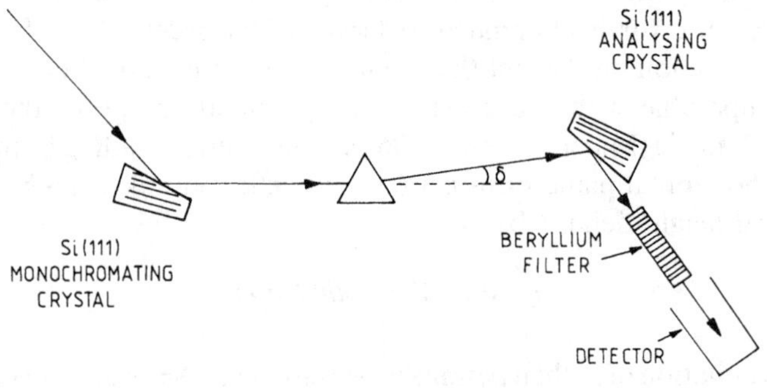

Figure 6.5. The double crystal refractometer for the determination of neutron scattering lengths. The refractive index of the prism material is derived from a measurement of the angle of deviation δ of a monochromatic neutron beam directed parallel to the prism base (Shull and Nathans 1967).

This refractometer is an instrument for determining scattering lengths by measuring the angle δ through which a monochromatic neutron beam is deflected on passage through a prism of apex half-angle α. The relevant connection is

$$\delta = 2(1-n)\tan\alpha \simeq \left(\frac{Nb\lambda^2}{\pi}\right)\tan\alpha \qquad (6.119)$$

and, since $n<1$, the deflection is towards the apex of the prism rather than away from it as in traditional prism devices. The prism is placed between two perfect silicon crystals aligned parallel, which function as monochromator and analyser respectively, and δ is derived from the position of the diffraction maximum in the rocking curve, i.e. the plot of analysed intensity versus angle, which is observed when the analyser is rotated with respect to the monochromator. By this means it has proved possible to measure scattering lengths to precisions of between 1% and 2×10^{-2}% in various prism materials, e.g. copper, germanium and quartz (Schneider and Shull 1971, Schneider 1973, 1976).

Neutron lenses have also found some uses in certain special experiments (cf. Gähler *et al* 1980, 1981, Kearney *et al* 1980) and again, because $n<1$, a converging (diverging) neutron lens is concave (convex). Like many other double slit interferometric devices based on division of wavefront a neutron interferometer designed essentially along the lines of Fresnel's bi-prism (Maier-Leibnitz and Springer 1962) has proved difficult to operate (Friedrich and Heintz 1978). Thus almost all successful developments in neutron interferometry have been brought about using the Bonse–Hart interferometer (cf. section 7.3.3) which, like the Michelson interferometer, is an instrument which relies on division of amplitude.

6.4 Phase relations and superposition of states

6.4.1 Coherence

The concept of coherence is a familiar one from optics. When we combine together waves whose phases are in some definite relation this addition is said to be coherent and the existence of interference, constructive or destructive, is the result of physical coherence in the constituent parts of the wave. The word 'coherent' is also applied to the sources of coherent waves. If the phases of waves to be combined are distributed at random, the waves are said to be incoherent and no interference can take place. When the amplitudes of waves from N sources emitting waves of equal amplitude A and equal phase δ are added together, the resultant wave has an intensity N^2A^2, whereas, if the phases are distributed at random, the total intensity is NA^2.

The same ideas are brought forward into the quantum-mechanical theory of scattering. For example, the differential cross-section for scattering of particles of definite energy and zero spin may be represented as the square of a sum of partial wave amplitudes, each amplitude corresponding to a definite angular momentum state in the scattered wave (cf. section 2.3). This set of partial waves is coherent. If, however, the particles are fermions having spin one-half, it is necessary to separate the beam into two incoherent component beams each having the same energy, but with spins pointing in opposite directions, and calculate the differential cross-section for each beam separately. The total cross-section is then obtained by adding the contributions of each of the component beams. Furthermore the incident wave with spin 'up' is coherent with the scattered wave with spin 'up', but incoherent with the scattered wave with spin 'down'. The spin 'flip' may be considered as removing all information regarding the phases and thus introduces a random phase change which destroys the coherence.

6.4.2 Mixed states in quantum mechanics

The idea of coherence in quantum mechanics is a little more complicated than in classical optics. In order to decide in any particular case whether or not coherence is present we shall use the following criterion: quantum mechanical waves associated with beams of particles are coherent provided they have the same energy and provided there exists no means by which this energy-degeneracy can be removed. In the case of spin one-half particles with finite magnetic moment considered above, the degeneracy is automatically removed by application of a magnetic field.

We now seek to establish a connection between the concept of coherence and the idea of a 'pure state' of a quantum mechanical system. The 'state' of a system may be defined as an undisturbed motion that is restricted by as many conditions as are theoretically possible without mutual interference or contradiction. This definition of a 'state' is not the most general one; rather it describes a 'pure state' in that it implies the existence of an experiment, uniquely corresponding to that state, for which the results are predictable with certainty. The uniqueness of the correspondence is essential, for, if there existed two distinct states possessing that property, it is clear that the full number of restrictions theoretically possible could not have been imposed. It is only to pure states that the full principle of superposition can be applied. This principle asserts that any pure state may be regarded as being formed from the coherent superposition of two or more pure states, where the description 'coherent' is understood in the same sense as above. There may however exist pure states carrying different eigenvalues of an absolutely conserved observable, which cannot be superposed because there exists no observable which has a non-vanishing matrix element between them. In this case a superselection rule is said to operate (Wick *et al* 1952). An example is the observable T^2 (where T is the time reversal operator discussed in section 3.3.1), which has eigenvalue $+1$ (-1) for boson (fermion) systems. Whether or not baryon number **B** is governed by a superselection rule has yet to be decided experimentally (cf. section 10.4.3).

States for which the full number of theoretically ascertainable data are not known, e.g. the state of a neutron of known energy and momentum for which the spin state is entirely unknown, are called 'mixed' states. By this we mean that, in order to calculate the probability of obtaining a certain result in an experiment on a system in a mixed state, we must calculate this probability for each of the pure states allowed to the system, and then average the result, assigning the appropriate weight to each state. In many problems in quantum mechanics it is quite sufficient to work with pure states only; in some instances, however, we are interested in the properties of systems in mixed states and the ordinary formalism, although adequate, is not always convenient for purposes of easy investigation. The usual method for studying systems in mixed states is by means of the density operator which we discuss immediately below.

6.4.3 The density operator

According to the principle of superposition a quantum mechanical system in the normalized pure state $|\Psi_n\rangle$ may be expressed as a linear combination of states selected from a complete set of orthornormal pure states $|\varphi_m\rangle$, i.e.

$$|\Psi_n\rangle=\sum_m |\varphi_m\rangle C_{nm}=\sum_m |\varphi_m\rangle\langle\varphi_m|\Psi_n\rangle \tag{6.120}$$

In general the states $|\varphi_m\rangle$ are simultaneous eigenstates of a complete set of commuting observables satisfying the closure condition

$$\sum_m |\varphi_m\rangle\langle\varphi_m|=1$$

In an experiment designed to measure an observable A of the system when it is the state $|\Psi_n\rangle$, the expectation value of A is given by the relation

$$\langle A_n\rangle\equiv\langle\Psi_n|A|\Psi_n\rangle=\sum_{lm} C^*_{nl}C_{nm}a_{lm} \tag{6.121}$$

where

$$a_{lm}\equiv\langle\varphi_l|A|\varphi_m\rangle \tag{6.122}$$

is the matrix element of A in the representation of basis vectors $|\varphi_m\rangle$. The result (6.121) may now be expressed in the form

$$\langle A_n\rangle=\sum_{lm} \rho^{(n)}_{ml}a_{lm}=\text{Tr}\,(\rho^{(n)}A) \tag{6.123}$$

where

$$\rho^{(n)}_{ml}=C_{nm}C^*_{nl} \tag{6.124}$$

are the matrix elements of a density operator (or density matrix) $\rho^{(n)}$ which describes the statistical properties of the system when it is in the pure state $|\Psi_n\rangle$.

From (6.124) we may now calculate the matrix elements of the operator $\rho^{2(n)}$ and we find

$$\rho^{2(n)}_{ml}\equiv\sum_j \rho^{(n)}_{mj}\rho^{(n)}_{jl}=\sum_j C_{nm}C^*_{nj}C_{nj}C^*_{nl}$$

$$=C_{nm}C^*_{nl}\sum_j |C_{nj}|^2=C_{nm}C^*_{nl}=\rho^{(n)}_{ml} \tag{6.125}$$

Thus $\rho^{2(n)}$ satisfies the idempotency condition

$$\rho^{2(n)}=\rho^{(n)} \tag{6.126}$$

showing that $\rho^{(n)}$ has the nature of a projection operator with eigenvalues 0 or 1.

When the system has been prepared in such a way that all the matrix elements $\rho_{ml}^{(n)}$ are not known precisely, then this system is said to be in a mixed state and it is necessary to set up a Gibbs ensemble with elements in pure states $|\Psi_n\rangle$ weighted according to the information which was available when the mixed state was prepared. In particular we may assume that the state preparation involves making a large number of observations on an observable A yielding a sampling average $\langle A\rangle$ such that

$$\langle A\rangle=\sum_n P_n\langle A_n\rangle \tag{6.127}$$

where the sum is taken over all the pure states $|\Psi\rangle$ available to the system and P_n represents the relative weighting of the pure state expectation value $\langle A_n\rangle$. The notation $\langle A\rangle$ introduced in (6.127) therefore incorporates two averaging processes, a quantum mechanical average as represented by $\langle A_n\rangle$, and a statistical or ensemble average as represented by the weighted sum over all averages $\langle A_n\rangle$ in the ensemble.

Since it is not now possible to write down an expression for the mixed state analogous to (6.120), it is necessary to find some other means of expressing in mathematical form the limited amount of information which is actually ascertainable. The most useful instrument for this purpose is the density operator ρ which is now re-defined by the relation

$$\rho=\sum_n |\Psi_n\rangle P_n\langle\Psi_n| \tag{6.128}$$

The matrix element of ρ in the basis $|\varphi\rangle$ is now given by

$$\begin{aligned}\rho_{lm}\equiv\langle\varphi_l|\rho|\varphi_m\rangle&=\sum_n\langle\varphi_l|\Psi_n\rangle P_n\langle\Psi_n|\varphi_m\rangle\\&=\sum_n P_n C_{nl}C_{nm}^*=\sum_n P_n\rho_{lm}^{(n)}\end{aligned} \tag{6.129}$$

We now find, by combining (6.129) and (6.125), that

$$\begin{aligned}\mathrm{Tr}\,(\rho A)&=\sum_{lm}\rho_{ml}a_{lm}=\sum_n P_n\sum\rho_{ml}^{(n)}a_{lm}\\&=\sum_n P_n\langle A_n\rangle=\langle A\rangle\end{aligned} \tag{6.130}$$

Thus the density operator ρ defined by (6.128) plays exactly the same role for the mixed state as the operator $\rho^{(n)}$ does for the pure state $|\Psi_n\rangle$ and the result (6.130) expresses its most important property. We establish some additional properties immediately below.

First we assume for convenience that each state $|\Psi_n\rangle$ is normalized so that

$$\langle\Psi_n|\Psi_n\rangle=1 \tag{6.131}$$

Since the probabilities P_n satisfy $P_n \leqslant 1$ and $\sum_n P_n = 1$ it follows that

$$\mathrm{Tr}\,(\rho) = \sum_n \sum_l P_n^* C_{nl} C_{ln} = \sum_n P_n = 1 \tag{6.132}$$

In the important case where a beam of neutrons of momentum $\hbar k$ impinges on a target, normalization of the incident beam according to (6.131) means that there is an incident flux of particles $\hbar k/m$ cm^{-2} s^{-1}. The beam scattered into the element of solid angle $\mathrm{d}\Omega$ through the angle θ has a flux $\mathrm{d}\sigma(\theta)/\mathrm{d}\Omega$ and for the scattered beam the result (6.132) must be replaced by

$$\mathrm{Tr}\,(\rho) = \frac{\mathrm{d}\sigma}{\mathrm{d}\Omega} \tag{6.133}$$

In general when (6.131) does not hold (as for the scattered beam above) the result (6.130) must be replaced by

$$\langle A \rangle = \mathrm{Tr}\,(\rho A)/\mathrm{Tr}(\rho) \tag{6.134}$$

Further important properties of the density operator ρ are that it is Hermitian and that is positive. The Hermiticity of ρ follows from the definition (6.128) which shows that ρ is a measureable quantity. By the description 'positive' we mean that ρ has a positive expectation value in any state $|\chi\rangle$. This result also follows from (6.128) and we find

$$\langle\chi|\rho|\chi\rangle = \sum_n \langle\chi|\Psi_n\rangle P_n \langle\Psi_n|\chi\rangle = \sum_n |\langle\chi|\Psi_n\rangle|^2 P_n \geqslant 0 \tag{6.135}$$

Thus $\langle\rho\rangle$ is the quantum mechanical analogue of the Liouville phase space density in classical statistical mechanics. Since for a normalized pure state $\mathrm{Tr}\,(\rho) = \mathrm{Tr}\,(\rho^2) = 1$, and the trace of a matrix is invariant under unitary transformations, it follows that the diagonal elements of ρ in any representation must each be less than, or equal to unity, and in general

$$\mathrm{Tr}\,(\rho^2) \leqslant 1 \tag{6.136}$$

where the equality sign applies if, and only if, the state described by ρ is a pure state.

Since the density operator provides a means of treating states for which certain information is missing, it is connected with the statistical concept of entropy. In classical statistical mechanics the entropy S of an ensemble of microscopic systems is defined by

$$S = -k \sum_n P_n \ln(P_n) \tag{6.137}$$

where k is Boltzmann's constant. The corresponding definition of S in quantum statistical mechanics is based on the definition of ρ given in terms of the P_n in (6.128). The modification consists in replacing the quantities P_n

in (6.137) by the matrix elements of ρ in the representation in which it is diagonal. In an arbitrary representation the procedure leads to a redefinition of S as the operator

$$S = -k\,\mathrm{Tr}\,(\rho \ln(\rho)) \tag{6.138}$$

where the operator $\ln(\rho)$ is defined by a relation analogous to (6.128), i.e.

$$\ln(\rho) = \sum_n |\Psi_n\rangle \, ln(P_n) \langle \Psi_n| \tag{6.139}$$

6.4.4 Time dependence of the density operator

6.4.4.1 The quantum Liouville equation

In general the pure states $|\Psi_n\rangle$ which make up the ensemble are time dependent, and develop in time according to the Schrödinger equation

$$i\hbar \frac{\partial}{\partial t}(|\Psi_n(\mathrm{t})\rangle) = H|\Psi_n(\mathrm{t})\rangle \tag{6.140}$$

where H is the total Hamiltonian. Multiplying (6.140) to the right by $\langle \Psi_n(t)|$ then yields the result

$$i\hbar \frac{\partial}{\partial t}(|\Psi_n(t)\rangle)\langle \Psi_n(t)| = H|\Psi_n\rangle\langle \Psi_n| \tag{6.141}$$

We now take the Hermitian adjoint of (6.140) and multiply to the left by $|\Psi_n(t)\rangle$. Using the Hermitian property of H this results in a relation similar to (6.141),

$$-i\hbar|\Psi_n(t)\rangle \frac{\partial}{\partial t}(\langle \Psi_n(t)|) = |\Psi_n\rangle\langle \Psi_n|H \tag{6.142}$$

Subtracting (6.142) from (6.141) then leads to the equation of motion of the pure state density operator $\rho^{(n)}(t)$

$$i\hbar \frac{\partial}{\partial t}\rho^{(n)}(t) = H\rho^{(n)}(t) - \rho^{(n)}(t)H = (H, \rho^{(n)}(t)) \tag{6.143}$$

where $(H, \rho^{(n)}(t))$ denotes the commutator of H and $\rho^{(n)}(t)$.

To obtain the time dependence of the mixed state density operator ρ, we multiply both sides of (6.143) by the weights $P_n(0)$ of the pure states $|\Psi_n(0)\rangle$ in the initial ensemble, and sum over all n. This procedure leads to the final result

$$i\hbar \frac{\partial \rho}{\partial t}(t) = (H, \rho(t)) \tag{6.144}$$

Equation (6.144) is the quantum mechanical analogue of the Liouville equation in classical statistical mechanics, and apart from a significant difference in sign, closely resembles the equation of motion of an operator $A(t)$ in the Heisenberg representation. For this reason it is well to emphasize that (6.144) applies in the Schrödinger representation; in the Heisenberg representation the state functions $|\Psi_n\rangle$ are independent of time and so also is the density operator ρ.

If ρ commutes with H, then ρ is constant in time and the corresponding mixed state is said to be 'stationary'. If the state is a pure state then it is an eigenstate of H and it is 'stationary' in the restricted sense originally applied by Bohr to the states of the hydrogen atom. In the general case it is a sufficient condition for the state to be stationary that ρ be a function of H (cf. section 6.4.4.3 below).

6.4.4.2 Wave packet ensembles

We can illustrate some of these ideas by first considering the case of a neutron in the pure state $|\Psi(t)\rangle$ represented by the wave packet (cf. 6.45–47)

$$\Psi(x,t)=\frac{1}{\sqrt{(2\pi\hbar)}}\int_{-\infty}^{\infty}\varphi(p)\,e^{i[kx-\omega(k)t]}\,dp \tag{6.145}$$

Since this is a positive energy state it has to be normalized in a volume of length L and $\Psi(x,t)$ contains a normalization factor proportional to $L^{-1/2}$. According to the definition (6.124) the matrix elements of the density operator ρ in the continuous representation of basis states $|x\rangle$ is given by

$$\rho(x,x';t)\equiv\langle x|\rho(t)|x'\rangle=\Psi(x,t)\Psi^*(x't) \tag{6.146}$$

Using the Fourier decomposition (6.145) we may now re-write this result as

$$\rho(x',x;t)=\frac{1}{2\pi\hbar}\int_{-\infty}^{\infty}\varphi(p)\,e^{ikx}\,dp\int_{-\infty}^{\infty}\varphi^*(p')\,e^{-ik'x'+i(\omega'-\omega)t}\,dp' \tag{6.147}$$

In a physically realizable system which is made up from many particles whose wavefunctions are in no fixed phase relationship with respect to each other, it is necessary to replace the matrix element $\rho(x',x;t)$ by its ensemble average. This may be achieved by inserting a random path length ξ in the phase of each wavefunction and then carrying out an average of ξ. This means that $\rho(x',x;t)$ acquires a factor $\exp[i(k-k')\xi]$ where

$$\langle\exp[i(k-k')\xi\rangle=\frac{1}{L}\int_{-L/2}^{L/2}e^{i(k-k')\xi}\,d\xi=\frac{2\pi}{L}\left\{\frac{\sin[(k-k')(L/2)]}{(k-k')}\right\} \tag{6.148}$$

Since, provided $L\to\infty$ ultimately, the factor in parenthesis in (6.148) can be replaced by $\delta(k'-k)$, the integral over p' in (6.147) can be written as

$$\int_{-\infty}^{\infty} \varphi^*(p')\, e^{-ik'x+i(\omega'-\omega)t}\left(\frac{2\pi}{L}\right)\left(\frac{dp'}{dk'}\right)\delta(k'-k)\, dk' = \frac{2\pi\hbar}{L}\,\varphi^*(p)\, e^{-ikx} \tag{6.149}$$

Inserting (6.149) in (6.147) and carrying out the appropriate limiting procedure leads to the final result

$$\langle \rho(x, x')\rangle = \operatorname*{Lim}_{L\to\infty} \int_{-\infty}^{\infty} \frac{|\varphi(p)|^2}{L}\, e^{ik(x-x')}\, dp \tag{6.150}$$

Going back to (6.146), and writing $x-x'=x$, we now see that $\langle \rho(x, x'; t)\rangle$ can be written in the form

$$\langle \rho(x, x'; t)\rangle = \langle \Psi(x'+x, t)\Psi^*(x't\rangle \tag{6.151}$$

Thus $\langle \rho(x, x'; t)\rangle$ can be interpreted as the spatial autocorrelation function of the wavefunction $\Psi(x, t)$, and the result (6.150) is an expression of the Wiener–Khinchine theorem (Wiener 1930, Khinchine 1934) applied in the space domain rather than in the time domain. It is consistent with the requirement (6.132) that $\mathrm{Tr}\,(\rho)=1$, for we now have

$$\begin{aligned}\mathrm{Tr}\,(\rho) &= \int_{-\infty}^{\infty} \langle \rho(x, x')\rangle\, dx = \operatorname*{Lim}_{L\to\infty} \frac{|\varphi(p)|^2\, dp}{L} \\ &= \int_{-\infty}^{\infty} |\varphi(p)|^2\, dp = 1\end{aligned} \tag{6.152}$$

The fact that the time no longer appears in $\langle \rho(x, x')\rangle$ shows of course that the ensemble describes a stationary system. The quantity

$$\operatorname*{Lim}_{L\to\infty} (|\rho(p)|^2/L)$$

can be interpreted as the density of neutrons per unit length per unit velocity interval. The ensemble average of the neutron current density $\boldsymbol{j}(\boldsymbol{r})$ defined in equation (2.51) can be arrived at by an argument identical with that which leads to (6.151) and we find then that

$$\langle j_x\rangle = \operatorname*{Lim}_{L\to\infty} \int_{-\infty}^{\infty} \frac{|\varphi(p)|^2 v\, dp}{L} \tag{6.153}$$

This permits us to interpret the quantity

$$\operatorname*{Lim}_{L\to\infty} (|\varphi(p)|^2 v/L)$$

as the mean flux of neutrons per unit length per unit velocity interval.

6.4.4.3 The equilibrium state

A stationary state is of course not identical with the equilibrium state which may be defined as in classical statistical mechanics as that state which maximizes the expectation value of the entropy operator defined in (6.138) subject to whatever conditions are imposed on the ensemble. In the case of a non-degenerate gas of neutrons moving in one dimension in the thermal column of a reactor at temperature T the appropriate ensemble is the canonical ensemble where the mean energy $\langle E \rangle = \frac{1}{2}kT$ is given. Thus $\langle S \rangle = \mathrm{Tr}\,(\rho S)$ must be maximized subject to the condition (6.132) that $\mathrm{Tr}\,(\rho) = 1$, and also that

$$\mathrm{Tr}\,(\rho H) = \langle E \rangle \tag{6.154}$$

The maximization of $\langle S \rangle$ is carried out in the usual manner using the method of Lagrange multipliers, i.e.

$$\delta\,\mathrm{Tr}\,(\rho \ln(\rho) + \lambda\rho + \mu(\rho H)) = \mathrm{Tr}\,(1 + \ln(\rho) + \lambda + \mu H)\,\delta\rho = 0 \tag{6.155}$$

Equation (6.155) gives the usual solution for ρ

$$\rho = \mathrm{e}^{-[1+\lambda+\mu H]} \tag{6.156}$$

Applying the condition $\mathrm{Tr}\,(\rho) = 1$ we can eliminate λ giving

$$\mathrm{e}^{(1+\lambda)} = \mathrm{Tr}\,(\mathrm{e}^{-\mu H}) = Z \tag{6.157}$$

where Z is the canonical partition function.

To obtain an explicit form for ρ we need to evaluate the matrix elements of $\mathrm{e}^{-\mu H}$ in the basis $|x\rangle$, i.e.

$$\begin{aligned}
\langle x|\mathrm{e}^{-\mu H}|x'\rangle &= \int_{-\infty}^{\infty} \mathrm{d}p \int_{-\infty}^{\infty} \mathrm{d}p' \langle x|p\rangle\langle p|\mathrm{e}^{-\mu H}|p'\rangle\langle p'|x'\rangle \\
&= \int_{-\infty}^{\infty} \mathrm{d}p'\, \mathrm{e}^{\mathrm{i}(kx - k'x') - \mu p'^2/2m} \\
&= \int_{-\infty}^{\infty} \mathrm{d}p\, \mathrm{e}^{\mathrm{i}k(x - x') - \mu p^2/2m}
\end{aligned} \tag{6.158}$$

The partition function Z is therefore given by

$$\begin{aligned}
Z = \mathrm{Tr}\,(\mathrm{e}^{-\mu H}) &= \lim_{L\to\infty} \int_{-L/2}^{L/2} \mathrm{d}x \langle x|\mathrm{e}^{-\mu H}|x\rangle \\
&= \lim_{L\to\infty} \int_{-\infty}^{\infty} \mathrm{d}p\, \mathrm{e}^{-\mu p^2/2m} = \lim_{L\to\infty} L \frac{2\pi m}{\mu}
\end{aligned} \tag{6.159}$$

We can now eliminate the Lagrange multiplier μ in the usual way using the result

$$\langle E\rangle=\frac{\mathrm{Tr}\,(H\,\mathrm{e}^{-\mu H})}{\mathrm{Tr}\,(\mathrm{e}^{-\mu H})}=-\frac{\partial}{\partial\mu}\ln(Z)=\frac{1}{2\mu}=\frac{kT}{2} \tag{6.160}$$

The final expression for $\rho(x, x')$ is then given by

$$\langle\rho(x, x')\rangle=\lim_{L\to\infty}\int_{-\infty}^{\infty}\frac{\mathrm{e}^{-1/2(p/\Delta p)^2+ik(x-x')}}{L\sqrt{(2\pi(\Delta p)^2)}}\,\mathrm{d}p \tag{6.161}$$

where

$$(\Delta p)^2=2m\langle E\rangle=mkT \tag{6.162}$$

Comparing the results (6.161)–(6.162) with the general result for a stationary ensemble given by (6.150) we see that the equilibrium wave packet has a momentum-space wavefunction with modulus

$$|\varphi(p)|=\frac{1}{(2\pi(\Delta p)^2)^{1/4}}\exp\left\{-\frac{1}{4}\left(\frac{p}{\Delta p}\right)^2\right\} \tag{6.163}$$

which is a minimum uncertainty Gaussian wave packet (cf. 6.56) of undetermined mean position $\langle x\rangle$ and mean momentum $\langle p\rangle$ equal to zero. The width of such a wave packet is

$$\Delta x=\frac{\hbar}{2\,\Delta p}=\left(\frac{\hbar^2}{4mkT}\right)=\frac{\lambda\!\!\!^{-}_{\mathrm{c}}}{2}\frac{mc^2}{kT} \tag{6.164}$$

where $\lambda\!\!\!^{-}_{\mathrm{c}}=\hbar/mc$ is the reduced Compton wavelength of the neutron. For a thermal neutron with $\langle E\rangle=0.025$ eV, the root mean square width of the wave packet is $\Delta x\simeq 3$ Å.

6.4.5 Berry's phase

6.4.5.1 Dynamical and topological phases

In the thermal equilibrium ensemble described above the density operator is given by (6.156) and (6.157) and, when H is just the magnetic potential given by (1.49), with $\boldsymbol{B}$ parallel to the z-axis, only σ_z has a non-vanishing expectation value and the other components $\langle\sigma_x\rangle$ and $\langle\sigma_y\rangle$ vanish. In the same way all operators A which are orthogonal to all powers of H have vanishing expectation values and all phase information disappears. Yet, quite generally, phases are important since they can carry information, e.g. in the neutron spin-echo spectrometer described in section 8.5.3.2. The difficulty has been that to date no completely acceptable definitions have been available for phase and angle variables which would make these acceptable as true observables (Carruthers and Nieto 1968). It has therefore been generally assumed that the phase could be fixed by convention and only phase differences are measurable. What meaning could be attached to the notion of the phase of a single particle has barely been considered.

All this has changed with the discovery of local gauge invariance symmetry which shows that the very existence of electromagnetic and weak interactions between elementary particles may be demanded by the requirement that all phases must be locally variable in an arbitrary way, the locality requirement being necessary to accord with relativity which demands that there be no causal relation between events with a space-like separation. For these reasons the discovery of the Berry phase, with its non-trivial experimental implications, has aroused a great deal of interest.

Consider a quantum system described by a Hamiltonian $H(\boldsymbol{R})$, where $\boldsymbol{R}$ represents a set of parameters which may in general be viewed as the components of a multi-dimensional vector, but which is most likely to be a vector in three-dimensional space. If we suppose that the system is prepared in an eigenstate $|\varphi_n(\boldsymbol{R})\rangle$ of $H(\boldsymbol{R})$ at zero time, then the state at time t is given by

$$|\psi(t)\rangle = |\varphi_n(\boldsymbol{R})\rangle \exp[-\mathrm{i}E_n t/\hbar] \tag{6.165}$$

Hence the initial state acquires a dynamical phase $\exp[-\mathrm{i}E_n t/\hbar]$ as it propagates in time.

Let us now suppose that the parameter set is permitted to vary in time, returning to its initial configuration $\boldsymbol{R}(0)$ after a period of time T. The state $|\psi(t)\rangle$ now satisfies the Schrödinger equation

$$\mathrm{i}\hbar \frac{\partial}{\partial t}|\psi(\mathrm{t})\rangle = H(\boldsymbol{R}(t))|\psi(t)\rangle \qquad \boldsymbol{R}(T)=\boldsymbol{R}(0) \tag{6.166}$$

and the adiabatic theorem of quantum mechanics predicts that the system remains in the instantaneous eigenstate $|\varphi_n(\mathrm{t})\rangle$, provided $\boldsymbol{R}(t)$ varies sufficiently slowly, and returns to its initial state $|\varphi_n(0)\rangle$ at $t=T$, multiplied by some phase-factor. Berry's theorem then states that, in addition to the dynamical phase $\varphi_n(T)$ defined by

$$\varphi_n(T) = -\frac{1}{\hbar}\int_0^T \mathrm{d}t\, E_n(\boldsymbol{R}(t))\,\mathrm{d}t \tag{6.167}$$

the state function picks up a geometric or topological phase $\gamma_n(C)$ whose value is determined only by the geometry of the circuit C over which the system is transported in parameter space, and is independent of the time T (Berry 1984, 1987).

For Berry's theorem to apply it is essential that the adiabatic condition be maintained throughout the whole cycle of change, which means, in simple terms, that at no point on its path must the system be subjected to Fourier components of the time-varying force fields in which it moves, of sufficiently high frequency to induce transitions between its instantaneous energy eigenstates. Formally this means that the angular frequency $\alpha_{nm}(t)$, which describes the rate at which the eigenstate $|\varphi_n\rangle$ is rotated into $|\varphi_m\rangle$, must satisfy the condition (Messiah 1964)

$$|\alpha_{nm}(t)|_{\max} \ll \omega_{nm}(t)_{\min} \tag{6.168}$$

where

$$\hbar\omega_{mn}(t) = E_n(t) - E_m(t) \tag{6.169}$$

To arrive at a formula for $\gamma_n(t)$ we postulate a solution of the form (cf. section 6.2.1)

$$|\psi(t)\rangle = \exp[\mathrm{i}(\varphi_n(t) + \gamma_n(t))] \tag{6.170}$$

and substitution into the Schrödinger equation (6.166) leads to the result

$$\frac{\mathrm{d}\gamma_n(t)}{\mathrm{d}t} = \mathrm{i}\langle\varphi_n(\boldsymbol{R}(t)|\nabla_R\varphi_n(\boldsymbol{R}(t))\rangle\cdot\frac{\mathrm{d}\boldsymbol{R}(t)}{\mathrm{d}t} \tag{6.171}$$

where ∇_R is the gradient operator in $\boldsymbol{R}$-space. This expression may now be integrated over the complete circuit C and, assuming the transport takes place infinitely slowly as required by the adiabatic theorem, integration by parts yields the result

$$\gamma_n(C) = \mathrm{i}\oint_C \langle\varphi_n(\boldsymbol{R}(0))|\nabla_R\varphi_n(\boldsymbol{R}(0))\rangle \boldsymbol{dR} \tag{6.172}$$

where the reality of $\gamma_n(C)$ follows from the normalization condition on $|\varphi_n\rangle$.

When $\boldsymbol{R}(t)$ is a vector in three-dimensional space the expression (6.172) may be transformed by Stokes' theorem into a surface integral over any open surface S bounded by C, i.e.

$$\gamma_n(C) = -\int_S \boldsymbol{V}_n(\boldsymbol{R})\cdot\boldsymbol{d\sigma} \tag{6.173}$$

where $\boldsymbol{V}_n(\boldsymbol{R})$ is the vector product

$$\boldsymbol{V}_n(\boldsymbol{R}) = \mathrm{Im}\sum_{m\neq n}\left\{\frac{\langle\varphi_n(\boldsymbol{R})|\nabla_R H(\boldsymbol{R})|\varphi_m(\boldsymbol{R})\rangle\times\langle\varphi_m(\boldsymbol{R})|\nabla_R H(\boldsymbol{R})|\varphi_n(\boldsymbol{R})\rangle}{(E_m(\boldsymbol{R}) - E_n(\boldsymbol{R}))^2}\right\} \tag{6.174}$$

The result (6.174) may be expressed in particularly simple form if the circuit C lies close to a point of degeneracy $\boldsymbol{R}^*$ of $H(\boldsymbol{R})$ such that $E_m(R^*) = E_n(R^*)$ for some range of m. For in this case the sum in (6.174) may be restricted to the set of states $|\varphi_m\rangle$ degenerate with $|\varphi_n\rangle$. A very important example of such a case arises for a neutron spin precessing in a magnetic field $\boldsymbol{B}$, where the two spin states become degenerate at $\boldsymbol{B}=0$, and the Berry phase is given by

$$\gamma_{\pm 1/2}(C) = \mp \tfrac{1}{2}\Omega(C) \qquad (6.175)$$

where $\Omega(C)$ is the solid angle subtended by the closed circuit C at the origin in $\boldsymbol{B}$-space. Thus, if B_z is the field component along the axis of quantization, and the transverse field component $B_1 = (B_x^2 + B_y^2)^{1/2}$ performs one complete cycle of revolution about the z-axis in the period T, the result (6.175) reduces to

$$\gamma_{\pm 1/2}(C) = 2\pi(1 - \cos\theta) \qquad \cos\theta = B_z/(B_1^2 + B_z^2)^{1/2} \qquad (6.176)$$

Since the appearance of Berry's original work similar ideas have been applied successfully in a number of fields where phases are important. These fields include the quantum Hall effect (Semenoff and Sodano 1986, Kuratsuji 1987) and the appearance of gauge anomalies in quantum field theory (Niemi and Semenoff 1985, 1986, Sonada 1986). A gauge-invariant generalization of Berry's phase defined for any cyclic evolution of a quantum system (Aharonov and Ananden 1987) has also been verified experimentally (Suter *et al* 1988).

6.4.5.2 Observation of the Berry phase using polarized light

The first experimental observation of a Berry phase was successfully carried out, not with polarized neutrons, but with linearly polarized laser light, for which the result (6.175) also applies with the spin components $\pm\frac{1}{2}$ replaced by ± 1 corresponding to the two helicity (or circular polarization) states $|\pm 1\rangle$ of the photon (Chiao and Wu 1986, Tomita and Chiao 1986). In this case the states of linear polarization along the $\hat{x}$ and $\hat{y}$ axes are defined as the superpositions of helicity states

$$|\hat{x}\rangle = \frac{1}{\sqrt{2}}(|1\rangle + |-1\rangle)$$

$$|\hat{y}\rangle = \frac{1}{\sqrt{2}}(|1\rangle - |-1\rangle) \qquad (6.177)$$

Thus, after the parameters governing the light propagation have been taken through a complete cycle C, these states acquire phases such that

$$|\hat{x}(C)\rangle = \frac{1}{\sqrt{2}}\{|1\rangle\, e^{i(\delta_+ + \gamma_1(C))} + |-1\rangle\, e^{i(\delta_- - \gamma_1(C))}\}$$

$$|\hat{y}(C)\rangle = \frac{1}{\sqrt{2}}\{|1\rangle\, e^{i(\delta_+ + \gamma_1(C))} - |-1\rangle\, e^{i(\delta_- - \gamma_1(C))}\} \qquad (6.178)$$

where

$$\delta_\pm = \pm\frac{2\pi}{\lambda}(n_+ - n_-)l \tag{6.179}$$

are the dynamical phases corresponding to a path length l. The quantities n_+ and n_- are the refractive indices for right and left circularly polarized light respectively, and are equal if the medium is not optically active. Thus in general the plane of polarization of the light is rotated through an angle Φ_l given by

$$\Phi_l = (\delta_+ - \delta_-) + 2\gamma_1(C) \tag{6.180}$$

In the experimental test of these relations, the light was transmitted along N turns of an optical fibre wound on a cylindrical former in the shape of a helical curve of pitch angle θ_p, for which the subtended solid angle is

$$\Omega(c) = 2\pi N(1 - \cos\theta_p) \tag{6.181}$$

In this example it is the direction of the axis of symmetry of the optical fibre which carries out N rotations in space and, since the photon momentum is adiabatically locked to this axis, the two helicity states acquire equal and opposite Berry phases. Since the initial state is some linear combination of $|\hat{x}\rangle$ and $|\hat{y}\rangle$, the presence of the Berry phase is manifested as a rotation of the plane of polarization. It may be shown that this result follows directly from application of the Gauss–Bonnet theorem in differential geometry to the tangent space of the twisted optical fibre (Ryder 1991).

6.4.5.3 Observation of the Berry phase using polarized neutrons

The formal theory of neutron polarization is developed in section 7.4.1, but since the neutron spin, like the photon polarization, is a two-state system, a common formalism may be applied, with differing physical interpretations, to both photons and neutrons (McMaster 1961, Byrne 1971). It is therefore not surprising that the first experimental search for the Berry phase using polarized neutrons was carried out using an arrangement which was almost the exact neutron analogue of the optical experiment described above (Bitter and Dubbers 1987, Dubbers 1988). In this experiment a beam of very cold neutrons in a state of longitudinal polarization (spin parallel or anti-parallel to momentum) was projected non-adiabatically into a region of space containing a weak longitudinal field B_z, and a strong transverse helical magnetic field B_1. In one cycle of this helical field as viewed from the rest-frame of the neutron, each of the helicity states of the neutron acquires equal and opposite dynamical and topological phases according to (6.167) and (6.175), since $E_{1/2} = -E_{-1/2} = \frac{1}{2}\hbar\omega_0$ where ω_0 is the angular frequency of Larmor spin precession (cf. section 7.5.2.1). Since $B = (B_1^2 + B_z^2)^{1/2}$ remains constant in magnitude, while B_1 carries out one rotation about the z-axis, the spin rotates in a full cycle through an angle Φ_n given by

$$\Phi_n = \omega_0 T + 2\gamma_{1/2}(C) \tag{6.182}$$

At the end of the cycle the neutrons are ejected non-adiabatically from the helical field and Φ_n is determined by measuring the helicity as a function of B_1 (cf. section 7.4.3). Since the adiabatic condition requires that $\omega_0 T \gg 2\gamma_{1/2}(C)$, a plot of Φ_n against B_1 approaches $\pm\omega_0(B_1)T$ as $B_1/B_z \to \pm\infty$, the asymptotes to this plot intercepting the ordinates at the points $\Phi_n = \pm 2\gamma_{1/2}(C)$, thereby allowing a determination of the Berry phase to be made. This effect is illustrated in figure 6.6 which shows both the longitudinal polarization f_z (cf. section 7.4.1.1) and the accumulated phase Φ_n as functions of applied helical magnetic field B_1. The existence of a Berry phase is proved, not only by the relative displacement of the two asymptotes, but also by the anomalous width of the polarization curve near $B_1 \simeq 0$, which, in the absence of a Berry phase, should obey a pure cosine law.

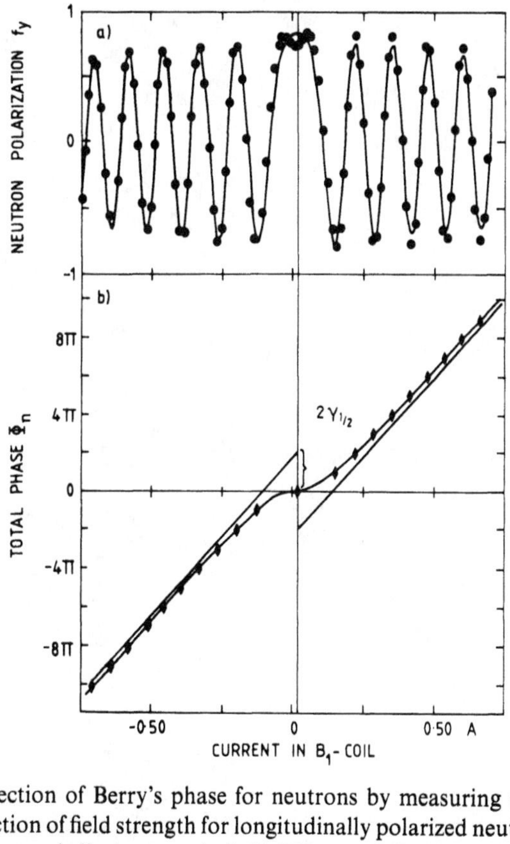

Figure 6.6. Detection of Berry's phase for neutrons by measuring the spin-rotation pattern as a function of field strength for longitudinally polarized neutrons transmitted through a transverse helical magnetic field (Bitter and Dubbers 1987).

In the experiment described above the neutron spins were subjected to one cycle of a helical magnetic field varying in space; subsequently the Berry phase phenomenon was studied in much greater detail using ultra-cold neutrons stored in a beryllium bottle (Richardson *et al* 1988). This experiment was carried out using a modified version of the neutron electric dipole apparatus shown in figure 9.8, in which neutrons were stored for a fixed time $T = 7.387 \pm 0.005$ s, and subjected to many cycles N of a range of magnetic fields generated by three mutually orthogonal and independent field coils. Polarized neutrons have also been used to study the generation of topological phases for incomplete cycles and under non-adiabatic conditions (Weinfurter and Badurek 1990).

7 Wave phenomena with neutrons

7.1 The Huygens–Fresnel principle for neutrons

7.1.1 Huygens' principle

The famous principle enunciated by the Dutch astronomer, Christiaan Huygens, in his *Traité de la Lumière* (1690), expressed an hypothesis concerning the nature of light which was opposed in its entirety to those corpuscular theories favoured by Huygens' more famous contemporary, Isaac Newton. Huygens held that the phenomena of reflection and refraction, and in particular the spectacular discovery of double refraction in Iceland Spar reported by Erasmus Bartholinus in 1669, could be most easily understood on the assumption that light was a wave disturbance. His principle asserts that a point source of light in an optically homogeneous medium generates a solitary spherical wave according to a step-by-step process, whereby each point on the wave surface at time t is the source of a secondary spherical wavelet, and the envelope of all such wavelets at time $t+\mathrm{d}t$ constitutes the new position of the wave surface. The waves envisaged by Huygens were not the periodic disturbances contemplated by Thomas Young, Augustin Fresnel, and other investigators of the early nineteenth century. Rather they may be regarded as isolated flashes of light whose envelope propagates in space according to the eikonal equation (6.4).

By carrying out an appropriate geometrical construction of the wavefronts at a plane boundary separating media with different optical properties Huygens was able to reproduce correctly the paths of both reflected and refracted rays and, assuming that light waves travel more slowly in the optically denser medium, to account for both Fermat's principle and Snell's law. Huygens' construction describing optical reflection and refraction at a plane boundary is shown in figure 7.1. The theory also accounted satisfactorily for a whole range of anomalous effects, familiar to astronomers, associated with refraction by the earth's atmosphere, whose optical properties may vary in space and time. The phenomenon of double refraction could be included in this scheme by postulating that there were two components to the secondary wavelet generated in anisotropic crystalline media, one a sphere and one an ellipsoid of revolution, whose envelopes could be identified with the wavefronts of the ordinary and extraordinary rays, respectively.

Huygens theory was faced with two major difficulties. In the first place the envelope of secondary wavelets emitted from points on the surface of a sphere is composed in general of two sheets, one diverging from the centre of the sphere which Huygens identified with the physical wavefront, and a second sheet converging toward the centre which he was compelled to assume did not exist. The second difficulty was that, in order to account for the observed rectilinear propagation of light, each secondary wavelet had to be

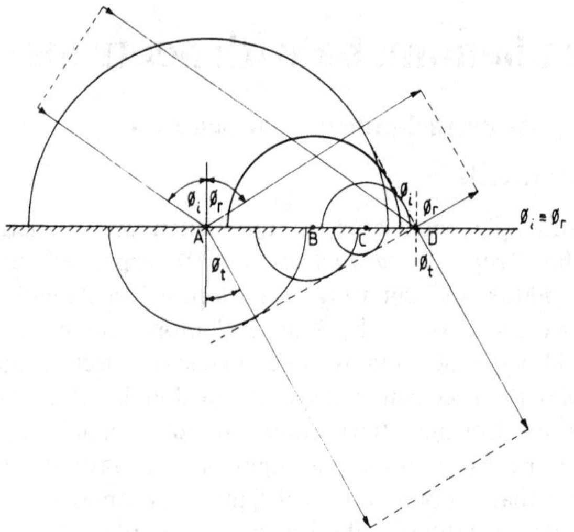

Figure 7.1. Reflection and refraction at a plane surface described by a construction based on Huygens' principle. The angles of incidence, reflection and refraction are denoted by φ_i, φ_r and φ_t respectively. The incident wavefront reaches the boundary surface first at point A, and subsequently at points B and C. The refracted wavefront is then the envelope of spherical waves emitted from these points in the forward direction while the reflected wave is the envelope of spherical waves emitted in the backwards direction. The diagram describes the situation where light travels fastest in the half-space of the incident wave.

treated as 'active' only at that point where it touched the envelope (Baker and Copson 1950). The first difficulty is connected with the fact that d'Alembert's scalar wave equation (6.6) is of second order in time so that both $\tilde{\Psi}(\boldsymbol{r}, 0)$ and $(\partial\tilde{\Psi}/\partial t)(\boldsymbol{r}, 0)$ may be specified arbitrarily. Thus for any given spherically symmetric optical disturbance at $t=0$, the solution at arbitrary t in general consists of an incoming and an outgoing wave, i.e.

$$r\tilde{\Psi}(\boldsymbol{r}, t)=\tilde{\varphi}_{\text{out}}(r-ct)+\tilde{\varphi}_{\text{in}}(r+ct) \tag{7.1}$$

If the initial wavefield

$$r\tilde{\Psi}(\boldsymbol{r}, 0)=\tilde{\varphi}_{\text{out}}(r) \qquad \tilde{\varphi}_{\text{in}}(r)=0 \tag{7.2}$$

contains outgoing waves only, in order to suppress the incoming wave for all time as required by Huygens' principle, it is necessary to impose the additional boundary condition

$$\frac{\partial\tilde{\Psi}}{\partial t}(\boldsymbol{r}, 0)+\frac{c}{r}\frac{\partial}{\partial r}(r\tilde{\Psi}(\boldsymbol{r}, 0))=0 \tag{7.3}$$

As an example, in the case of the initial conditions

$$\tilde{\Psi}(\mathbf{r}, 0) = \delta(r)/4\pi r \qquad \frac{\partial \tilde{\Psi}}{\partial t}(\mathbf{r}, 0) = -c\delta'(r)/4\pi r \tag{7.4}$$

Huygens' construction gives the correct solution, namely the retarded Green function

$$\tilde{\Psi}(\mathbf{r}, t) = \delta(r - ct)/4\pi r \tag{7.5}$$

provided it is assumed always that the secondary wave acts only at that point where it touches the envelope. This result was first demonstrated by Poisson (1819).

7.1.2 Fresnel's theory of diffraction

The principal defect in Huygens' formulation of his principle, namely its restriction to isolated spherical waves, was remedied in 1818 in a prize-winning memoir by Fresnel who supposed that a light source emitted, not discontinuous pulses of light like that which is described by (7.5), but continuous trains of waves of definite periodicity (Fresnel 1826). Fresnel considered the light-supporting ether as having the nature of an elastic solid so that each element on the wave surface of the primary wave can act as a secondary source because of the energy and momentum which has been communicated to it from the original source. The resultant amplitude at any point is then determined by interference between all the secondary waves which arrive there. Thus light is observed, not only at the envelope of the secondary waves, but at all points where the net effect of interference is constructive. In thus successfully uniting the concept of wave interference (Young 1802) with Huygens geometrical construction, Fresnel succeeded, not only in preserving the law of rectilinear propagation of an unobstructed light-beam, but also in accounting for departures from that law near the boundaries of opaque objects. These effects, which lead to the appearance of light in the geometrical shadow, constitute the phenomenon of diffraction, first discussed by Leonardo da Vinci in the fifteenth century as an element in the theory of colour and perspective, and subsequently rediscovered by Grimaldi, in a memoir published at Bologna in 1665.

In order to maintain the spirit of the original version of Huygens' principle it was necessary that Fresnel's system of interfering secondary wavelets should not only describe correctly the amplitude ahead of the leading edge of an advancing wavefront, but also give a null effect behind its trailing edge. He therefore postulated the existence at each secondary source of an inclination factor assumed to be a maximum in the forward direction and zero in the backwards direction. By employing a construction based on dividing the primary wave into zones such that the amplitudes contributed by successive zones at any point P ahead of the wavefront alternate in sign, he succeeded in reproducing the forward wave exactly with no returning wave. Success in this area required the introduction of additional postulates concerning the nature of the secondary sources, which were required to lead the primary wave by a quarter of a period, with an amplitude $1/\lambda$ times the primary wave amplitude. These postulates were ultimately shown to be consistent with the Helmholtz–Kirchhoff integral representation of the optical wavefield in the short-wavelength limit.

The treatment of diffraction problems in optics by the Huygens–Fresnel principle may be strictly justified only on the assumption that the optical wavefield satisfies d'Alembert's scalar wave equation, whereas the phenomenon of optical polarization, discovered by Huygens himself, shows that the optical field is indeed a vector field. The formulation of a vector version of the principle which satisfies the full panoply of Maxwell's equations, and is therefore valid for electromagnetic waves (Kottler 1923, Stratton and Chiu 1939) is a very difficult matter which is of no concern in the present context. The de Broglie waves associated with the neutron are, however, spinor waves, i.e. the wave has two components corresponding to the two directions of the neutron spin. At least in the absence of strong magnetic fields they may be treated as scalar waves described by a wave equation which is first order in time, to which the Huygens–Fresnel principle may be applied as readily as to the optical field in the scalar approximation. Indeed the principle is wholly valid when applied to the time-dependent Schrödinger equation since it may be applied without the additional restrictions imposed by (7.3). This conclusion follows from the fact that, if an initial Schrödinger wave function $\Psi(\boldsymbol{r}, 0)$ describes outgoing waves, so that its Fourier transform $\varphi(\boldsymbol{p}, 0)$ vanishes in that region of momentum space for which $\boldsymbol{p}\cdot\boldsymbol{r}<0$, the same is true for $\Psi(\boldsymbol{r}, t)$ at all times since $\varphi(\boldsymbol{p}, t)=\varphi(\boldsymbol{p}, 0)\exp(-\mathrm{i}\omega(k)t)$.

In the next section we consider briefly the quantitative theory of diffraction based on the Helmholtz–Kirchhoff integral representation of the scalar field, which is just the mathematical expression of the Huygens–Fresnel principle. However, because the vacuum is dispersive for de Broglie waves (cf. 6.48) the concept of retarded time which is central to Kirchhoff's general theory has no validity when applied to polychromatic neutrons. We shall therefore restrict attention to the Helmholtz version of the theory which is valid for monochromatic neutrons only.

7.1.3 The Helmholtz–Kirchhoff integral theorem

We shall confine the discussion of this theorem to the case of monochromatic neutrons of energy $E_{\mathrm{n}}=\hbar\omega(k)$, described by a wavefunction

$$\Psi(\boldsymbol{r}, t)=\Psi(\boldsymbol{r})\,\mathrm{e}^{-\mathrm{i}\omega(k)t} \tag{7.6}$$

where $\Psi(\boldsymbol{r})$ satisfies the scalar Helmholtz equation (6.19) for field-free propagation (i.e. $V(\boldsymbol{r})=0$), and k and $\omega(k)$ obey the free particle dispersion relation (6.48). The objective of the exercise is to find a representation of Ψ at a given point P, in terms of an integral over points Q on a surface Σ enclosing the real sources, which can be interpreted as a statement of the Huygens–Fresnel principle. The point of departure for the calculation is Green's Theorem

$$\int_v (\varphi\nabla^2\Psi-\Psi\nabla^2\varphi)\,\mathrm{d}v=\int_S\left(\varphi\frac{\partial\Psi}{\partial n}-\Psi\frac{\partial\varphi}{\partial n}\right)\mathrm{d}S \tag{7.7}$$

where φ and Ψ are solutions of (6.19) regular in a volume v enclosed by a surface S. All neutron sources are confined within a finite region of space and all singular source-points P′ are excluded from v by enclosing them in a surface Σ, which is further subdivided into a surface Σ_1 which is opaque to neutrons, and a surface Σ_2 which is transparent.

We choose P as the origin of coordinates and select $\varphi(\boldsymbol{r})$ as that solution which describes divergent spherical waves,

$$\varphi(\boldsymbol{r})=\frac{\mathrm{e}^{ikr}}{r}\qquad \boldsymbol{r}=r\hat{\boldsymbol{r}} \tag{7.8}$$

where $\hat{\boldsymbol{r}}$ is a unit vector directed outward from P. Because of the reciprocal relationship which exists between source point and field point, the function $\exp(\mathrm{i}kr)/r$ also represents the spherical wave amplitude at P which is emitted by a point source at $\boldsymbol{r}$. Since $\varphi(\boldsymbol{r})$ is singular at P, this point must also be excluded from v by enclosing it in a small sphere Σ_a of radius a. Finally, to exclude the point at infinity, both Σ and Σ_a are enclosed in a large sphere Σ_R of radius R. This system of surfaces is illustrated in figure 7.2.

The surface S is therefore composed of three sub-surfaces Σ, Σ_a and Σ_R and the outward drawn unit normal $\hat{\boldsymbol{n}}$ on S is an outward normal for Σ_R, and an inward normal for Σ and Σ_a. Since φ and Ψ are now regular solutions in the volume v interior to Σ_R and exterior to Σ and Σ_a, the volume integral in (7.7) vanishes and we find

$$\int_\Sigma\left\{\frac{\mathrm{e}^{ikr}}{r}\frac{\partial\Psi}{\partial n}-\Psi\frac{\partial}{\partial n}\left(\frac{\mathrm{e}^{ikr}}{r}\right)\right\}\mathrm{d}S-\lim_{a\to 0}\int\left\{\frac{\mathrm{e}^{ika}}{a}\frac{\partial\Psi}{\partial a}-\Psi\left(\frac{-\mathrm{e}^{ika}}{a^2}+\frac{ik}{a}\mathrm{e}^{ika}\right)\right\}r^2\,\mathrm{d}\Omega$$
$$+\lim_{R\to\infty}\int\left\{\frac{\mathrm{e}^{ikR}}{R}\frac{\partial\Psi}{\partial R}-\Psi\left(\frac{-\mathrm{e}^{ikR}}{R^2}+\frac{ik}{R}\mathrm{e}^{ikR}\right)\right\}R^2\,\mathrm{d}\Omega=0 \tag{7.9}$$

If we now impose the condition that $\Psi(\boldsymbol{r})$ should describe outgoing waves as $\boldsymbol{r}$ tends to infinity in any direction (cf. section 2.3), i.e. $R\Psi(R)$ is bounded and

$$\lim_{R\to\infty} R\left\{\frac{\partial\Psi}{\partial R}-\mathrm{i}k\Psi\right\}\to 0 \tag{7.10}$$

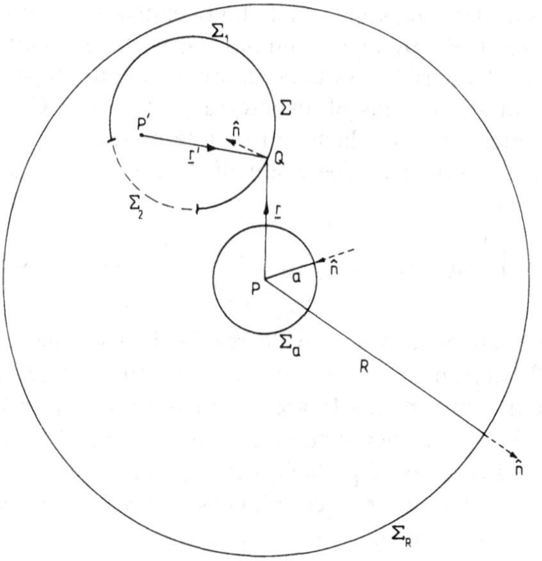

Figure 7.2. Helmholtz–Kirchhoff representation of the wave amplitude at P emitted by a point source at P′, as a superposition of wavelets arriving at P from secondary sources Q, distributed over a surface Σ enclosing P′. The surface Σ_1 is opaque, and the surface Σ_2 is transparent to radiations emitted from P′.

then the surface integral at infinity vanishes. The condition (7.10), which is known as Sommerfeld's '*Austrahlungsbedingung*' or radiation condition, permits the exclusion of standing wave solutions for $\Psi(\boldsymbol{r})$ which may be possible for certain characteristic values of k.

When the limiting procedure $a \to 0$ is carried out over the small sphere Σ_a surrounding P, there is only one surviving term, namely $\Psi(0)$, and we arrive at the result

$$\Psi(0) = \frac{1}{4\pi} \int_\Sigma \left\{ \frac{e^{ikR}}{r} \frac{\partial \Psi}{\partial n} - \Psi \frac{\partial}{\partial n} \left(\frac{e^{ikr}}{r} \right) \right\} dS \tag{7.11}$$

Thus the surface integral in (7.11) provides a representation of the wave amplitude $\Psi(0)$ at P as a superposition of wavelets e^{ikr}/r arriving from secondary sources Q distributed over a surface Σ which encloses the primary sources. There are evidently two types of secondary source, each proportional to the amplitude of the primary wave on Σ: a simple source with surface density $(1/4\pi)(\partial\Psi/\partial n)$, and a doublet source with surface density $-\Psi/4\pi$. Furthermore, if the point P is taken inside Σ, then since $\varphi(\boldsymbol{r})$ now has no singularities outside Σ, it is unnecessary to exclude any part of the volume exterior to Σ from the volume integral and the surface Σ_a is redundant. Thus the surface integral in (7.11) vanishes, i.e. there is no returning wave at any point inside Σ. It follows that (7.11) provides an exact analytical formulation of the Huygens–Fresnel principle subject to the condition (7.10), and the requirement that there be no sources at infinity.

In a sense (7.11) may be regarded as an existence theorem which validates the Huygens–Fresnel principle, but, if the result is to be of any practical value, it must be possible to evaluate the integral, and this requires a knowledge of Ψ and $\partial\Psi/\partial n$ on Σ. If these quantities are replaced by arbitrary functions χ, η defined on Σ the resultant surface integral is still a solution of the wave equation at P because the functions $\exp(ikr)/r$ and $(\partial/\partial n)(\exp(ikr)/r)$, considered as functions of the coordinates of P, are solutions. The difficulty is to find boundary functions χ, η which coincide with the limiting values of Ψ and $\partial\Psi/\partial n$ as the point P moves towards Σ. To illustrate this difficulty we consider the Kirchhoff boundary conditions, which, at first glance, seem highly plausible. These are

$$\Psi = \frac{\partial\Psi}{\partial n} = 0 \qquad \text{on } \Sigma_1 \text{ (opaque)}$$

$$\Psi = \frac{A\,e^{ikr'}}{r'} \qquad \frac{\partial\Psi}{\partial n} = \frac{\partial}{\partial n}\left(\frac{A\,e^{ikr'}}{r'}\right) \qquad \text{on } \Sigma_2 \text{ (transparent)} \tag{7.12}$$

where $\boldsymbol{r}' = r'\hat{\boldsymbol{r}}'$ is a vector directed from the source point P′ to the point Q on the surface. We then find that, combining (7.12) and (7.11),

$$\Psi(0) = \frac{A}{4\pi}\int_{\Sigma_2} \frac{e^{ik(r+r')}}{rr'}\left\{\left(ik - \frac{1}{r'}\right)\hat{\boldsymbol{n}}\cdot\hat{\boldsymbol{r}}' - \left(ik - \frac{1}{r}\right)\hat{\boldsymbol{n}}\cdot\hat{\boldsymbol{r}}\right\} dS \tag{7.13}$$

If we now assume that both source and field points are displaced from the surface Σ_2 by distances large in comparison with the reduced de Broglie wavelength $\lambda\!\!\!^{-} = k^{-1}$, i.e. $r \gg \lambda$, $r' \gg \lambda$, then (7.13) reduces to

$$\Psi(0) \simeq \frac{ikA}{4\pi}\int_{\Sigma_2} \frac{e^{ik(r+r')}}{rr'}\,\hat{\boldsymbol{n}}\cdot(\hat{\boldsymbol{r}}' - \hat{\boldsymbol{r}})\,dS \tag{7.14}$$

One immediate conclusion which may be drawn from (7.14) is that the relative amplitude of the secondary source in the first Fresnel zone for which $\hat{\boldsymbol{r}} = \hat{\boldsymbol{n}} = -\hat{\boldsymbol{r}}'$, is given by

$$\frac{-ik}{4\pi}\,2\hat{\boldsymbol{n}}\cdot\hat{\boldsymbol{n}} = e^{-i\pi/2}/\lambda \tag{7.15}$$

This result agrees with Fresnel's postulates, both as to the presence of the factor $1/\lambda$ and to the phase shift of a quarter of a period in advance, which is introduced when the time factor $e^{-i\omega t}$ is replaced by $e^{-i(\omega t + \pi/2)}$.

The difficulty is that the value of $\Psi(0)$ given by (7.13) does not match to the boundary conditions (7.12) as the point P moves towards the surface. Furthermore the boundary conditions (7.12) are mutually inconsistent in that, as first pointed out by Poincaré (1892), a solution of the wave equation which vanishes along with its normal derivative on a finite surface Σ_1, vanishes identically everywhere in direct contradiction to the boundary condition imposed at the surface Σ_2.

An alternative procedure is to replace the principal or unrestricted solution $\varphi(\boldsymbol{r})$ defined by (7.8) by the Green function $g(\boldsymbol{r})$ which describes the emission of spherical waves from P in the presence of an absorbing boundary at Σ. The function $g(\boldsymbol{r})$ therefore vanishes on Σ, tends towards $\varphi(\boldsymbol{r})$ as $r \to 0$, and satisfies the radiation condition at infinity. In these circumstances the first term in the surface integral in (7.7) is missing, and the final solution (7.11) is replaced by

$$\Psi(0) = -\frac{1}{4\pi}\int_{\Sigma} \Psi \frac{\partial g}{\partial n}\,\mathrm{d}S \tag{7.16}$$

In this case, since it is not necessary to specify both Ψ and $\partial\Psi/\partial n$ on Σ, a consistent solution may be arrived at by imposing the modified Kirchhoff boundary condition

$$\Psi = 0 \qquad \text{on } \Sigma_1 \qquad \Psi = A\frac{\mathrm{e}^{ikr'}}{r'} \qquad \text{on } \Sigma_2 \tag{7.17}$$

From various viewpoints, however, this is not a satisfactory conclusion since $g(\boldsymbol{r})$ has singularities both inside and outside Σ no matter where the point P is located. Thus when P is taken inside Σ the integral in (7.16) does not vanish; thus there is a returning wave and the result does not provide an analytic formulation of the Huygens–Fresnel principle. The other difficulty is that the Green function $g(\boldsymbol{r})$ has in general an infinite number of singularities and cannot be expressed in closed form. The exceptional case occurs where Σ is a plane surface, in which case $g(\boldsymbol{r})$ is given by

$$g(\boldsymbol{r}) = \frac{\mathrm{e}^{ikr}}{r} - \frac{\mathrm{e}^{ikr_{\mathrm{p}}}}{r_{\mathrm{p}}} \tag{7.18}$$

where $\boldsymbol{r}_{\mathrm{p}} = r_{\mathrm{p}}\hat{\boldsymbol{r}}_{\mathrm{p}}$ is the position vector of the field point $\boldsymbol{r}$, referred to an origin which is the mirror image of P in the surface Σ.

In the case of a plane surface Σ, the application of the boundary condition (7.17) to the representation (7.16) leads to the result

$$\Psi(0) = -\frac{1}{4\pi}\int_{\Sigma_2} A\frac{\mathrm{e}^{ikr'}}{r'}\,2\frac{\partial}{\partial r}\left(\frac{\mathrm{e}^{ikr}}{r}\right)(\hat{\boldsymbol{n}}\cdot\hat{\boldsymbol{r}})\,\mathrm{d}S$$

$$= -\frac{ikA}{2\pi}\int_{\Sigma_2} A\frac{\mathrm{e}^{ik(r+r')}}{rr'}\left(1+\frac{i}{kr}\right)(\hat{\boldsymbol{n}}\cdot\hat{\boldsymbol{r}})\mathrm{d}S \tag{7.19}$$

which, in the wave zone defined by $r \gg \lambda$, $r' \gg \lambda$, becomes approximately

$$\Psi(0) \simeq \frac{-ikA}{2\pi} \int_{\Sigma_2} \frac{e^{ik(r+r')}}{rr'} (\hat{n} \cdot \hat{r}) \, dS \tag{7.20}$$

The results (7.14) and (7.20) agree in detail in the first Fresnel zone where $\hat{r} + \hat{r}' = 0$ and it is only at higher zone numbers that divergences begin to appear.

It should be emphasized that the surface integrals (7.13) and (7.19) are both exact solutions of the scalar Helmholtz equation at P, which correspond to slightly different boundary conditions, and neither is an exact realization of the Huygens–Fresnel principle. Either may be used in the solution of diffraction problems but, where applicable, the Green function method is to be preferred since the solution finally arrived at does in fact satisfy the stated boundary condition (7.17).

7.1.4 Diffraction by an opaque circular disc in the geometric limit

We illustrate the results of the preceding discussion by applying the modified Helmholtz–Kirchhoff integral (6.139) to the problem of diffraction by an opaque circular disc of radius a in the limit of very short wavelength (Sommerfeld 1965). To this end we set up a system of Cartesian axes, with origin at the centre of the disc and z-axis aligned along the normal to the disc as shown in figure 7.3. If the straight line joining the source point P′ to the field point P crosses the disc plane $z=0$ at a point P_0 located a distance R_0 from the origin, then P lies inside or outside the shadow of the disc according as $R_0 < a$ or $R_0 > a$. Similarly if $Q(x, y)$ is any point on the plane $z=0$, the surface Σ_2 containing the secondary sources is defined by

$$\Sigma_2 \colon \sqrt{(x^2+y^2)} > a \tag{7.21}$$

To compute the integral in (7.20) we identify those 'surfaces of constant phase', for which the path length between P′ and P is constant, which form a system of confocal ellipsoids of revolution with foci at P′ and P, and may be described by the equation

$$r' + r = R' + R + l \tag{7.22}$$

where $R = PP_0$, $R' = P'P_0$ and l is a parameter having the dimensions of length which, in the limit $l \to 0$, selects the degenerate ellipsoid which envelops the straight line P′P. These surfaces cut the plane $z=0$ in a corresponding system of ellipses such that secondary wavelets emitted from sources in the annulus between neighbouring ellipses in the region exterior to the disc, interfere constructively at P. We may therefore write (7.20) in the form

$$\Psi(P) = \frac{-ik}{2\pi} A \frac{e^{ik(R'+R)}}{(R'+R)} \int_0^\infty 2\pi\eta(l) f(l) \, e^{ikl} \, dl \tag{7.23}$$

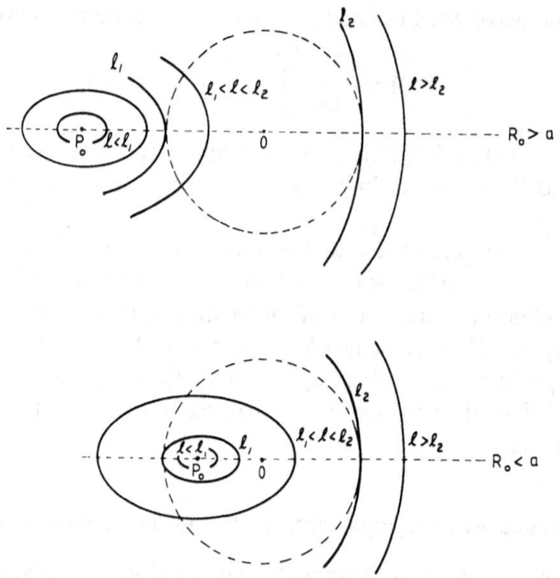

Figure 7.3. Diffraction by an opaque circular disc of radius a in the geometric limit. The line joining the source point P′ to the field point P crosses the plane of the disc at a point P_0, a distance R_0 from the centre of the disc, whose geometric shadow is indicated by the hatched circle. Ellipsoids of constant phase intersect the disc plane along ellipses whose positions depend on whether P lies outside ($R_0 > a$), or inside ($R_0 < a$), the geometric shadow of the disc.

where

$$2\pi\eta(l) = (R' + R)\frac{dS(l)}{dl}\left(\frac{\hat{n}\cdot\hat{r}}{r'r}\right) \qquad \eta(0) = 1 \tag{7.24}$$

and $f(l) = 1 - g(l)$ represents the fraction of the element of area $dS(l)$ which lies in the region $\sqrt{(x^2 + y^2)} > a$ outside the disc. With these definitions the integral may now be re-expressed in convenient form

$$\Psi(P) = -\Psi_0(P)\int_0^\infty \eta(l)f(l)\,d(e^{ikl}) \tag{7.25}$$

where

$$\Psi_0(P) = A\frac{e^{ik(R'+R)}}{(R' + R)} \tag{7.26}$$

is the wave amplitude which would be observed at P in the absence of the diffracting disc.

There are now two cases to consider. In case (1) $R_0 > a$ and P lies outside the geometrical shadow, and in this case it is more convenient to write (7.25) in terms of the function $g(l)$ which represents the fraction of the area $\mathrm{d}S(l)$ which is obscured by the disc. In this case we have the result

Case (1)

$$R_0 > a \quad \begin{array}{ll} l \leqslant l_1 & g(l) = 0 \\ l_1 < l < l_2 & 0 < g(l) < 1 \\ l \geqslant l_2 & g(l) = 0 \end{array} \tag{7.27}$$

where the significance of the limiting values l_1, l_2 may be understood from the shadow diagram shown in figure 7.3. In case (2) $R_0 < a$, P lies inside the geometrical shadow, and the conditions on $f(l)$ are

Case (2)

$$R_0 < a \quad \begin{array}{ll} l \leqslant l_1 & f(l) = 0 \\ l_1 < l < l_2 & 0 < f(l) < 1 \\ l \leqslant l_2 & f(l) = 1 \end{array} \tag{7.28}$$

In case (1) we may now write

$$\begin{aligned} \Psi(\mathrm{P}) &= -\Psi_0(\mathrm{P}) \left\{ \int_0^\infty \eta(l)\, \mathrm{d}(\mathrm{e}^{ikl}) - \int_{l_1}^{l_2} \eta(l) g(l)\, \mathrm{d}(\mathrm{e}^{ikl}) \right\} \\ &= \Psi_0(\mathrm{P}) \left\{ 1 + \int_0^\infty \eta'(l)\, \mathrm{e}^{ikl} \mathrm{d}l - \int_{l_1}^{l_2} \frac{\mathrm{d}}{\mathrm{d}l} (\eta(l) g(l))\, \mathrm{e}^{ikl} \mathrm{d}l \right\} \end{aligned} \tag{7.29}$$

following an integration by parts. Since both the Fourier transforms in (7.29) vanish in the short-wave limit, $k \to \infty$, this result shows that $\Psi(\mathrm{P}) \to \Psi_0(\mathrm{P})$ and the eikonal approximation is reached where the diffracting disc has no effect. For case (2) a similar analysis shows that

$$\begin{aligned} \Psi(\mathrm{P}) &= -\Psi_0(\mathrm{P}) \int_{l_1}^\infty \eta(l) f(l)\, \mathrm{d}(\mathrm{e}^{ikl}) \\ &= \Psi_0(\mathrm{P}) \int_{l_1}^\infty \frac{\mathrm{d}}{\mathrm{d}l} (\eta(l) f(l))\, \mathrm{e}^{ikl} \mathrm{d}l \end{aligned} \tag{7.30}$$

In this case $\Psi(\mathrm{P}) \to 0$ as $k \to \infty$, once again in agreement with the predictions of the eikonal approximation, that no radiation appears in the geometrical shadow.

There is, however, an important exception to (7.30), and this occurs when both P′ and P lie on the axis of the disc. In this case the parameters l_1 and l_2 coincide at the value

$$l_1 = l_2 = \sqrt{(R'^2 + a^2)} + \sqrt{(R^2 + a^2)} - (R' + R) \tag{7.31}$$

and (7.28) must be replaced by

$$\text{Case (3)} \qquad f(l)=0 \quad l<l_1 \quad \lim_{l\to l_1-} f(l)=0$$

$$f(l)=1 \quad l>l_1 \quad \lim_{l\to l_1+} f(l)=1 \tag{7.32}$$

Thus, in place of (7.30), we have the result

$$\Psi(\mathrm{P}) = -\Psi_0(\mathrm{P})\left\{\left|\eta(l)\,\mathrm{e}^{ikl}\right|_{l_1}^{\infty} - \int_{l_1}^{\infty} \eta'(l)\,\mathrm{e}^{ikl}\,\mathrm{d}l\right\}$$

$$\to \Psi_0(\mathrm{P})\eta(l_1)\,\mathrm{e}^{ikl_1} \qquad k\to\infty \tag{7.33}$$

where

$$\eta(l_1) = \frac{R(R'+R)}{(R^2+a^2)+\sqrt{(R^2+a^2)(R'^2+a^2)}} \tag{7.34}$$

In the limit $R\to 0$, i.e. immediately behind the diffracting disc, $\eta(l_1)\to 0$ in accord with the result expected from geometric optics. However $\eta(l_1)\to 1$ when $R\to\infty$, a result which shows that, independently of wavelength, the absolute amplitude on the disc axis is the same as it would have been in the absence of the disc. This conclusion totally contradicts the predictions of the eikonal approximation in the geometric limit.

These results were originally derived by Poisson in 1818, and were advanced by him as evidence against the Huygens–Fresnel theory. The subsequent observation by Fresnel and Arago of a bright spot on the axis in the shadow of an illuminated disc therefore served to forward the claims of the theory in a very dramatic manner. Formally the effect is associated with a discontinuity in $f(l)$, but physically it comes about because the edge of the disc lies on a surface of constant phase. It also falls on a surface of constant action S for neutrons emitted from the source at P′ and it is the existence of a corresponding discontinuity in the wave amplitude at the edge of the disc (cf. 6.34) which leads to the breakdown of the eikonal approximation (cf. sections 6.1.1 and 6.2.1). In the realm of fast neutron physics the phenomenon is of course very familiar since it is precisely this effect which gives rise to the pronounced shadow scattering effect discussed in section 2.5.4.

7.2 Neutron diffraction

7.2.1 Fraunhofer and Fresnel diffraction

7.2.1.1 Fraunhofer diffraction

Since the angular diameter subtended by a celestial body at an observation point on the earth is very small ($\approx 10^{-2}$ for sun or moon, $\approx 10^{-4}$ for a large planet and $\approx 10^{-7}$ for a giant star) the laws of geometric optics would predict that the image formed in the back focal plane of an astronomical telescope is a point. However, because the telescope has a finite aperture, what is actually observed is the diffraction pattern of this aperture. These effects were first investigated using the Huygens–Fresnel principle by Airy (1835), fifty years before Kirchhoff (1883) established the theory of diffraction on a secure basis, and he obtained the familiar formula for the light intensity diffracted at angle θ

$$I(\theta)=I(0)\left\{\frac{J_1(ka\sin\theta)}{ka\sin\theta}\right\} \tag{7.35}$$

where a is the radius of the diffracting aperture and $J_1(x)$ is the first-order Bessel function.

The Airy pattern described by (7.35) consists of a central disc of angular diameter $2.44\lambda/a$, surrounded by a system of concentric circular fringes. It provides an example of diffraction in the Fraunhofer limit, a description which is applied to any diffraction phenomenon when the source point and field point are so far removed from the diffracting obstacle that effects associated with the curvature of the wavefront may be neglected. In the case of the telescope the wavefront in the aperture is a plane wave and the diffraction pattern is at infinity. It is the lensing action of the objective which produces an image of the diffraction pattern in the back focal plane, but this does not alter the fact that the pattern at the focus is of the Fraunhofer type. Between the focus and the lens of course this is no longer true.

We may easily derive the condition for Fraunhofer diffraction by returning to figure 7.2 and equation (7.20). Assume that both P′ and P are so far removed from a plane aperture Σ_1 that only the complex exponential in the integrand varies significantly across Σ_1, then this equation may be rewritten in the form

$$\Psi(\mathrm{P})=\frac{-ikA}{2\pi}\frac{\mathrm{e}^{\mathrm{i}k(R'+R)}}{(R'+R)}\hat{n}\cdot\hat{r}\int_{\Sigma_1}\mathrm{e}^{\mathrm{i}k(r'+r-R'-R)}\,\mathrm{d}S_Q(x,y) \tag{7.36}$$

where $Q(x, y)$ is a point on the aperture. We may now expand r' in the form

$$\begin{aligned} r' &= \{(x'-x)^2+(y'-y)^2+A'^2\}^{1/2} \\ &\simeq R'+\frac{1}{2R'}\left\{-2X'x-2Y'y+(x^2+y^2)-\frac{(X'x+Y'y)^2}{R'^2}+\ldots\right\} \end{aligned} \tag{7.37}$$

with an exactly equivalent expansion for r. The exponent in the integrand may therefore be expressed as

$$(r'+r-R'-R)\simeq-\left\{\frac{X'x+Y'y}{R'}+\frac{Xx+Yy}{R}\right\}+\frac{1}{2}\left(\frac{1}{R'}+\frac{1}{R}\right)(x^2+y^2) \tag{7.38}$$

The Fraunhofer condition requires that the quadratic terms in (7.38) make a contribution to the total path difference across the aperture which is much less than λ, i.e.

$$\tfrac{1}{2}(\alpha'+\alpha)\ll\lambda/D \tag{7.39}$$

where $\langle x^2+y^2\rangle^2\simeq D^2$, D is a characteristic dimension of the aperture, and

$$\alpha'=\frac{D}{R'} \qquad \alpha=\frac{D}{R} \tag{7.40}$$

are the angular diameters of the aperture subtended at source point and field point respectively.

In the Fraunhofer limit the diffraction integral (7.36) therefore reduces to the result

$$\Psi(\mathrm{P})=\frac{-ik}{2\pi f}\Psi_0(\mathrm{P})\hat{n}\cdot\hat{r}\int_{\Sigma_1}\mathrm{e}^{ik[(l-l')x+(m-m')y]}\,\mathrm{d}S_Q(x,y) \tag{7.41}$$

where

$$\frac{X'}{R'}=-l' \qquad \frac{X}{R}=l \qquad \frac{Y'}{R'}=-m' \qquad \frac{Y}{R}=m \tag{7.42}$$

and f is a pseudo focal length defined by

$$\frac{1}{f}=\frac{1}{R'}+\frac{1}{R} \tag{7.43}$$

Thus $\Psi(\mathrm{P})$ is determined by the two-dimensional Fourier transform of a pupil function which is defined to be unity at points where the plane of diffraction is transparent, and zero where it is opaque. The argument of the transform is a function of the difference between the direction cosines $(l'm'n')$, (lmn) of the incident and diffracted rays respectively. It does not depend explicitly on R' and R, which may henceforth be regarded as 'infinite', and whose residual dependence is subsumed into a normalization factor which is the coefficient of the integral in (7.41).

To reproduce Airy's result (7.35) using (7.41) we assume that the incident light falls normally on a diffracting aperture of radius a so that $l'=m'=0$. The $x-y$ axes are then oriented so that the incident ray is in the $x-z$ plane; thus $m=0$ and $l=\sin\theta$ where θ is the angle of diffraction. The integral in (7.41) may now be evaluated explicitly using two dimensional polar coordinates in the $x-y$ plane giving the result

$$\int_0^{2\pi}\mathrm{d}\varphi\int_0^a \mathrm{e}^{-ik\rho\sin\theta\cos\varphi}\rho\,\mathrm{d}\rho=\pi a^2\left(\frac{2J_1(ka\sin\theta)}{ka\sin\theta}\right) \tag{7.44}$$

which, appropriately normalized, is just (7.35).

For easy reference we also write down expressions for (1) the single-slit and (2) and the N-slit diffraction patterns. These are

Single-slit $$I(\theta)=I(\theta)\left(\frac{\sin\beta}{\beta}\right)^2 \tag{7.45a}$$

N-slit $$I(\theta)=I(\theta)\left(\frac{\sin\beta}{\beta}\right)\left(\frac{\sin N\gamma}{\sin\gamma}\right)^2 \tag{7.45b}$$

where

$$l'=\sin\theta' \qquad l=\sin\theta \qquad \beta=\frac{ka}{2}(l-l') \qquad \gamma=\frac{kd}{2}(l-l') \tag{7.46}$$

a is the slit width, and d is the distance separating the centres of neighbouring slits. $I(\theta')$ represents the intensity falling on one slit at angle of incidence θ' and the intensity of the forward diffracted beam $(\theta+\theta'=0)$ is $N^2I(\theta')$ because at this angle the amplitudes arriving at P from all N slits add coherently. It is of course not necessary that each diffracting element be a slit; the essential point is that they be identical of dimension $a \ll d$, where d is the spatial period, or grating space, of the system of diffracting elements. In general the intensity factorizes into a diffraction factor, e.g. $(\sin\beta/\beta)^2$ for each element, and an interference factor $(\sin N\gamma/\sin\gamma)^2$ which describes the interference pattern which is observed in those directions in which the individual diffraction patterns overlap.

The interference factor in (7.45) is indeterminate when $\gamma=p\pi$, where p is a positive or negative integer, and application of L'Hospital's limit rule shows that these values represent principal maxima in the interference pattern. As an example, for the two-slit interference pattern, the interference factor reduces to $\cos^2\gamma$, which has a central maximum at $\gamma=0$, flanked by symmetric maxima at $\gamma=\pm p/\pi$, all of which will be observed so long as they fall within the diffraction envelope. This is illustrated for the case of 20 Å neutrons in figure 7.5(b). An almost identical picture is obtained for the Ramsey interference pattern in spin-space, which is observed using Ramsey's separated oscillatory field magnetic resonance technique for neutrons (cf. sections 7.5.3 and 9.3.2).

7.2.1.2 Fresnel diffraction

The case where the condition (7.39) fails is referred to as Fresnel diffraction, and in these circumstances it is necessary to take both the inclination factor $\hat{n}\cdot\hat{r}$ and the quadratic terms in the exponent under the integral sign. Up to this point the origin of coordinates has been chosen arbitrarily on the plane of the diffracting aperture and usually it is selected to exploit any symmetries of the aperture. However, when linear and quadratic terms in the exponent are comparable in magnitude it is convenient to choose the origin at the point P_0 where the line P′P crosses the diffracting plane $z=0$, and to reorient the $x-y$ axes so that the line P′P lies in the $x-z$ plane. This is shown in figure 7.4. On this redefinition, the length of the line P′P is just $(R'+R)$.

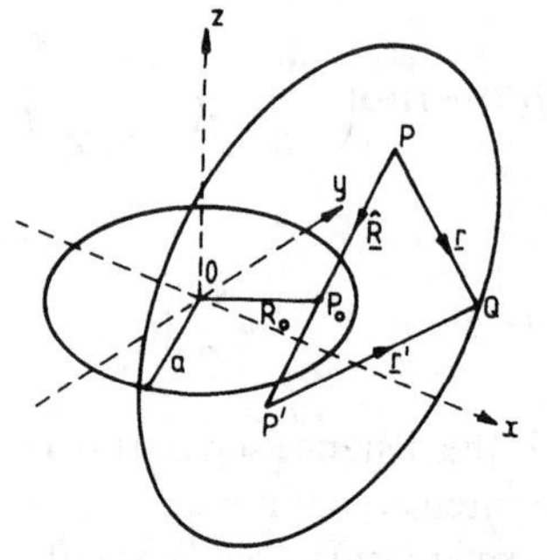

Figure 7.4. To illustrate Fresnel diffraction by a diffracting aperture the origin of coordinates is transferred to the point P_0, where the line joining the source point P′ to the field point P intersects the plane of the aperture. The axes are then rotated so that P′P lies in the $x-z$ plane. P_0 falls inside or outside the aperture according as P lies outside or inside the geometric shadow of the screen containing the aperture.

Thus in place of (7.42) we now write

$$Y'=Y=0 \qquad \frac{X}{R}=\frac{-X'}{R'}=\sin\delta \qquad \hat{n}\cdot\hat{r}=\cos\delta \tag{7.47}$$

and the diffraction integral (7.36) becomes

$$\begin{aligned}\Psi(\mathrm{P})&=\frac{-ik\Psi_0(\mathrm{P})}{2\pi f}\int_{\Sigma_1}\mathrm{e}^{\mathrm{i}(k/2f)(x^2\cos^2\delta+y^2)}\,\mathrm{d}(x\cos\delta)\,\mathrm{d}y\\&=\frac{-\mathrm{i}\Psi_0(\mathrm{P})}{2}\int_{\Sigma_1}\mathrm{e}^{\mathrm{i}(\pi/2)(u^2+v^2)}\,\mathrm{d}u\,\mathrm{d}v\end{aligned} \tag{7.48}$$

where

$$\frac{\pi u^2}{2}=\frac{k}{2f}(x^2\cos^2\delta) \qquad \frac{\pi v^2}{2}=\frac{k}{2f}y^2 \tag{7.49}$$

It should be emphasized that the origin of the dimensionless coordinates (u, v) located at P_0 is determined by the positions of P′ and P and there is nothing to guarantee that P_0 lies inside the diffracting aperture Σ_1 rather than on the opaque part of the diffracting plane Σ_2. In the latter instance P lies inside the geometric shadow and there is some danger that the error involved in neglecting the higher terms in the expansion (7.38) may no longer be negligible. Occasions may also arise when there is a point of inflection on the wave surface in the aperture so that the quadratic terms in (7.38) vanish identically. In this case the 'Fresnel integral' over quadratic terms in the complex exponent is replaced by an 'Airy integral' over cubic terms. An example of this phenomenon occurs in the theory of the rainbow.

It is evident from (7.48) that, in any particular instance, $\Psi(\mathrm{P})$ is determined by a product of functions of the form

$$F(z)=\int_0^z e^{i\pi z^2/2}\,dz \tag{7.50}$$

where

$$z=D/\sqrt{(\lambda f/2)} \tag{7.51}$$

and D is a characteristic dimension which depends on the positions of P′ and P, and on the size and shape of the diffracting aperture. $F(z)$ is given in terms of the confluent hypergeometric function, $F_1(a; b; z)$ by the relation (Abramowitz and Segun 1965)

$$F(x)=z_1F_1(\tfrac{1}{2};\tfrac{3}{2};\,i\pi z^2/2) \tag{7.52}$$

and has an asymptotic expansion

$$F(z)\to \pm\left\{\frac{1+i}{2}-i\,e^{i\pi z^2/2}/\pi|z|\right\} \qquad z\to\pm\infty \tag{7.53}$$

The real and imaginary parts of $F(z)$ are the familiar Fresnel integrals $C(z)$, $S(z)$ for which detailed tabulations are available.

The integral in (7.48) can be evaluated fairly easily in some simple, but experimentally interesting cases, e.g. the single slit and the straight edge. In both these examples the integral over the infinite y-dimensions gives the factor

$$F(\infty)-F(-\infty)=1+i \tag{7.54}$$

The simplest case arises when the source point P′ is at infinity and the neutrons are incident normally on the diffracting plane, so that R represents the distance of the plane of observation behind the diffracting plane. For a single slit of width a the result is

$$\Psi(\mathrm{P})=\left(\frac{1-i}{2}\right)\Psi_0(\mathrm{P})\{F(z_+)-F(z_-)\} \tag{7.55}$$

where

$$z_\pm=(\pm a/2-x(\mathrm{P}))/\sqrt{(\lambda R/2)} \tag{7.56}$$

and $x(\mathrm{P})$ is the distance of the point of observation measured from an origin in the observing plane in the centre of the slit. The corresponding result for a straight edge is

$$\Psi(\mathrm{P})=\left(\frac{1-\mathrm{i}}{2}\right)\Psi_0(\mathrm{P})\{F(z)-F(-\infty)\} \tag{7.57}$$

where

$$z=x(\mathrm{P})/\surd(\lambda R/2) \tag{7.58}$$

In this case the origin in the observing plane is set immediately behind the diffracting edge and $x>0$ ($x<0$) describes the illuminated (shadow) region. It follows immediately from (7.54), (7.57) and (7.58) that the intensity at the diffracting edge ($z=0$) is one quarter of its value in the illuminated region far from the edge ($z\to+\infty$).

7.2.2 Macroscopic diffraction

By the term 'macroscopic neutron diffraction' we mean to describe phenomena associated with the diffraction of slow neutrons ($\lambda=1$ Å → 1000 Å) by macroscopic objects whose dimensions are greater than the neutron wavelength by factors of order 10^2-10^4. A selection of experiments expressly designed to demonstrate macroscopic diffraction is listed in Table 7.1.

Table 7.1. Experiments demonstrating macroscopic diffraction with slow neutrons

Geometry	Diffraction type	Wavelength (Å)	Reference
Single slit	Fraunhofer	4.43	Shull (1969)
Single and double slit	Fraunhofer and Fresnel	20	Zeilinger *et al* (1981, 1988)
Ruled grating	Fraunhofer	2–4	Kurz and Rauch (1969)
Ruled grating	Fraunhofer	37	Graf *et al* (1979)
Ruled grating	Fraunhofer	$\simeq 600$	Scheckenhofer and Steyerl (1977)
Straight edge	Fresnel	4.33	Klein *et al* (1977)
Straight edge	Fresnel	20	Gähler *et al* (1981)
Zone plate	Fresnel	20	Kearney *et al* (1980)
Zone plate	Fresnel	20	Gähler *et al* (1980)
Zone plate	Fresnel	20	Klein *et al* (1981)
Zone plate	Fresnel	$\simeq 500$	Schütz *et al* (1980)

The first experiment of this type was carried out by Shull (1969) with the objective of confirming the single slit Fraunhofer diffraction pattern predicted by (7.45a). He used Bragg diffraction (cf. section 7.2.3) in a single crystal of pure silicon to define a parallel monochromatic beam of neutrons with

$\lambda = 4.43$ Å, which was incident normally on a slit of variable width $a > 4$ μm. The transmitted beam was analysed by a second silicon crystal of variable orientation to select those neutrons diffracted in a specified direction, and the results confirmed the theoretical prediction. More recently both the Fraunhofer and Fresnel patterns for both single and double slits have been studied to $<1\%$ precision by Zeilinger *et al* (1988) giving very good agreement with theory. The results obtained for the Fraunhofer patterns are shown in figure 7.5. The predicted form of the N-slit pattern (7.45b) has also been confirmed with both cold ($\lambda \approx 20$ Å) and ultra-cold ($\lambda \approx 600$ Å) neutrons.

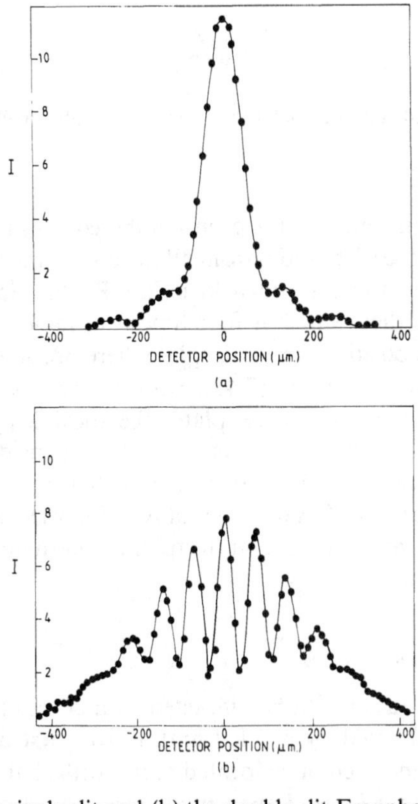

Figure 7.5. (a) The single slit and (b) the double slit Fraunhofer diffraction patterns observed with 20 Å neutrons (Zeilinger *et al* 1981, 1988).

In the case of Fresnel diffraction, apart from the single and double slit, the straight edge pattern has also been measured a number of times, and the results obtained by Gähler *et al* (1981) using 20 Å neutrons are shown in figure 7.6. However, an even more convincing demonstration of Fresnel diffraction has been achieved by means of the zone plate, of which a number have been constructed and operated successfully. A zone plate is a

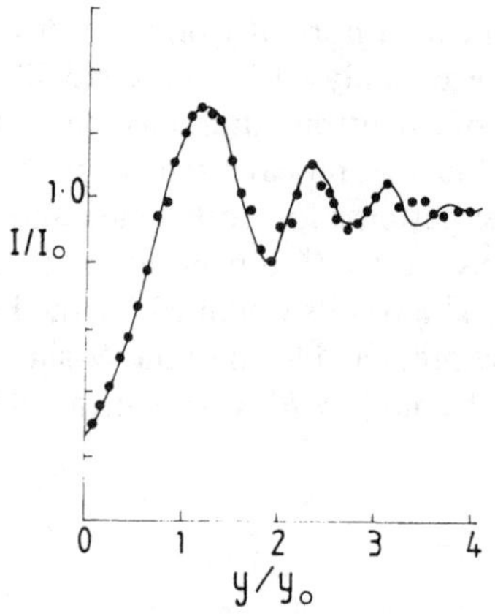

Figure 7.6. The straight edge Fresnel diffraction pattern observed with 20 Å neutrons (Gähler *et al* 1981).

circular diaphragm in which all the even-numbered Fresnel zones have been made opaque. Thus the amplitudes from all the odd numbered zones, defined with respect to given sources and field points P′, P, add coherently at P, yielding an intensity there which is four times the value it would have were the zone plate not in position. The zone plate therefore acts like a converging lens with focal length f given by (7.43). Since the radius $r_n = \sqrt{(n\lambda f)}$ of the nth zone is fixed for a given zone plate, the focal length f is inversely proportional to λ and the device is subject to considerable chromatic aberration. Nevertheless, in the system constructed by Gähler *et al* (1981), a 50 μm wide line source of 20 Å neutrons at $R' = 5$ m was successfully focused onto a 73 μm wide image at $R = 5$ m, using a 2 mm diameter zone plate of focal length $f = 2.5$ m.

7.2.3 Bragg diffraction

The proposal that X-rays might be diffracted by a crystal lattice in the same way that light is diffracted by a ruled grating was first advanced by Max von Laue, whose name is commemorated along with that of Paul Langevin in the title of the Institut Laue–Langevin at Grenoble (cf. section 1.6.4) (von Laue 1912, 1931, Darwin 1914, Ewald 1916, 1917, Batterman and Cole 1964). The effect is described by Bragg's law (1912).

$$2d \sin\theta = n\lambda \qquad (7.59)$$

where 2θ is the angle through which a beam of X-rays of wavelength λ is deflected by a crystal lattice of interplanar spacing d, which bisects the angle between incident and emergent beams as shown in figure 7.7, and n is the order of diffraction. The phenomenon was first observed by Friedrich *et al*

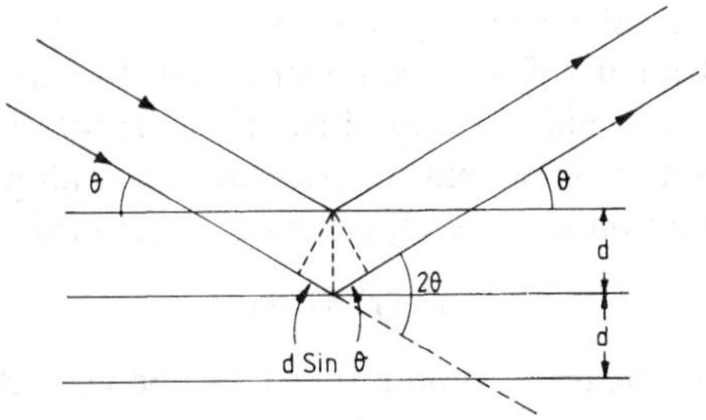

Figure 7.7. Symmetric Bragg diffraction at a crystal surface parallel to the diffracting planes.

(1913) using a continuous spectrum of X-rays, and subsequently Bragg and Bragg (1913) studied the diffraction of monochromatic X-rays from various crystal faces, and were able to determine the structures of some simple crystals and measure λ. The important part played by the technique of X-ray diffraction in Moseley's work on atomic structure has already been noted (cf. section 1.1).

Bragg (1912) considered X-ray diffraction as a problem involving the reflection of X-rays from a set of parallel crystal planes separated by a distance d, and he assumed that the usual laws of reflection applied. Bragg's law (7.59) then expresses the condition that multiple reflections at grazing angles θ from the system of planes should interfere constructively. In general the effect is observable only if λ and d are comparable in magnitude which explains why it is essential to use X-rays ($\lambda \simeq d \simeq 1$ Å) rather than visible light ($\lambda \approx 6000$ Å). However, to see how the result (7.59) is arrived at through an analysis in terms of three-dimensional Fraunhofer diffraction, it is necessary to make use of some simple notions from the theory of crystals.

A crystal is an object in which the atoms are distributed in three dimensions, not at random as in a gas, but in a regular array which is a periodic repetition of a fundamental pattern unit. It therefore exhibits long-range order whereas in a liquid there is only short-range order in that the positions of atoms are correlated only over short distances comparable with the separation of nearest neighbours. There also exist certain organic substances which show long-range order in one or two dimensions, and these are known as 'liquid crystals'.

Crystals are classified according to their symmetry properties under displacements and rotations, and the natural language to use in expressing such a classification is the language of group theory (Elliott and Dawber 1979). Two points or directions in a crystal are said to be equivalent if the geometric and physical properties of the crystal are identical at those two points or in those two directions. The set of all transformations which map every point onto an equivalent point, and every direction onto an equivalent direction, constitute the elements of the crystal group. The groups of relevance in this context are the space groups whose elements are obtained by combining the elements of a three-dimensional discrete translation group with the elements of a point group of rotations or rotation-inversions.

A translation is a displacement in space such that all points $\boldsymbol{r}$ are equally displaced. A translation may be represented in the form

$$\boldsymbol{r} \rightarrow \boldsymbol{r}' = \boldsymbol{r} + \boldsymbol{v} \tag{7.60}$$

and the elements of the translation group T can be viewed as a set of vectors $\boldsymbol{v}$, and the law of composition of the group is just vector addition. The group T is a three-dimensional discrete group if every group element $\boldsymbol{v}$ is restricted to have the form

$$\boldsymbol{v} = m_1\boldsymbol{d}_1 + m_2\boldsymbol{d}_2 + m_3\boldsymbol{d}_3 \tag{7.61}$$

where the integers m_1, m_2, m_3 can be positive, negative or zero, and the vectors $\boldsymbol{d}_j$ $(j=1-3)$ form a triad of non-coplanar basis vectors. The set of points $\boldsymbol{r}'$ obtained by performing all possible operations of the group T on an arbitrary point $\boldsymbol{r}$ in the crystal is called a Bravais lattice and 14 distinguishable lattices may be identified. Every lattice is characterized by a unit cell or repetition unit which is not uniquely defined, but an acceptable unit cell is the volume of the Bravais lattice constructed on the basis vectors.

We have already encountered the group R(3) of rotations in three-dimensional space, and when we adjoin to this group the operation of inversion, i.e. the 'parity' operation (cf. section 3.3.1), we obtain the full orthogonal group O(3) which plays an important role in the classification of states in the interacting boson model of the nucleus (cf. section 4.6.2). The point groups are the finite subgroups of O(3). Every translation group T has a symmetry group S of point transformations, such that every element of S transforms a given element of T into another element of T, and all groups T which have the same symmetry group S are said to belong to the same symmetry system. There are in total seven symmetry systems (or syngonies) which are designated as triclinic, monoclinic, orthorhombic, tetragonal, rhombohedral, hexagonal and cubic. The symmetry groups S together with their subgroups constitute the set of crystal direction groups and these number 32 in total. All crystals which have the same direction group are said to belong to the same crystal class.

In the present context we shall assume for simplicity that the crystal belongs to the orthorhombic system for which the basis vectors $\boldsymbol{d}_j$ are mutually orthogonal $(d_1 \neq d_2 \neq d_3)$ forming a natural Cartesian coordinate system, and which manifest two-fold symmetry in the sense that rotation about any basic axis through an angle π transforms the crystal into itself. Within the orthorhombic system we may distinguish three crystal classes corresponding to the symmetry group D_{2h} and its subgroups D_2 and C_{2v}, and four Bravais lattices. These are described as simple, base-centred, body-centred and face-centred and are shown in figure 7.8. Since no two equivalent directions

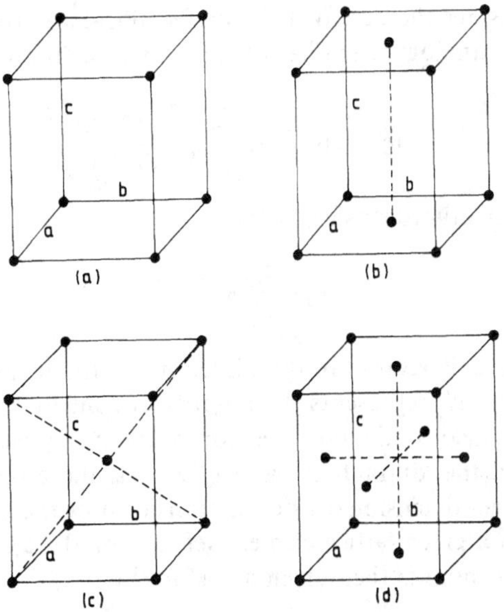

Figure 7.8. Orthorhombic crystals (a) simple, (b) base-centred, (c) body-centred and (d) face-centred.

can be identified in the orthorhombic crystals, these crystals are explicitly biaxial and display the phenomenon of conical refraction discussed in section 6.1.2.

Any lattice plane in a crystal which passes through the point which is chosen as the origin of coordinates may be represented by the equation

$$\frac{h_1x}{d_1}+\frac{h_2y}{d_2}+\frac{h_3z}{d_3}=0 \tag{7.62}$$

where the numbers h_j ($j=1-3$) are integers which contain no common factor, and which are known as the Miller indices. Since any plane parallel to the selected plane has an equation

$$\frac{h_1x}{d_1}+\frac{h_2y}{d_2}+\frac{h_3z}{d_3}=m \tag{7.63}$$

where the periodicity of the crystal requires that m be integral (positive or negative), the Miller indices are sufficient to characterize a whole system of parallel planes. The perpendicular distance $d(h_j)$ between neighbouring parallel planes, which is the distance d which appears in Bragg's law (7.59), is given by

$$\frac{1}{d(h_j)}=\left[\sum_{j=1}^{3}\left(\frac{h_j}{d_j}\right)^2\right]^{1/2} \tag{7.64}$$

We now consider the case of Fraunhofer diffraction by an orthorhombic crystal where, in analogy with (7.45), the intensity of diffracted radiation is

$$I(\boldsymbol{q})=I(0)F(\boldsymbol{q})\prod_{j=1}^{3}\left(\frac{\sin(N_j\gamma_j)}{\sin(\gamma_j)}\right)^2 \tag{7.65}$$

and $\boldsymbol{q}$, the momentum transfer, is given by

$$q_j=\frac{2\gamma_j}{d_j}k(\xi_j-\xi_j') \tag{7.66}$$

where $\boldsymbol{\xi}$, $\boldsymbol{\xi}'$ are unit vectors in the directions of the incident and diffracted rays respectively. N_j represents the number of contributing lattice elements which is determined by the cross-section of the X-ray beam and the size of the crystal, and the 'diffraction pattern' $F(\boldsymbol{q})$ is the atomic form factor or normalized three-dimensional Fourier transform of the charge distribution associated with a given lattice element (cf. section 3.2.5). The condition for a principal maximum is then given as before by

$$\xi_j-\xi_j'=n_j\frac{\lambda}{d_j} \qquad j=1-3 \tag{7.67}$$

and these equations are known as the Laue equations.

In the cases of one or two-dimensional Fraunhofer diffraction patterns the corresponding Laue equations are independent, but this is not true for the three-dimensional case described by (7.65) since $\boldsymbol{\xi}'$ and $\boldsymbol{\xi}$, being unit vectors, must satisfy the normalization condition

$$\boldsymbol{\xi}'^2=\boldsymbol{\xi}^2=1 \tag{7.68}$$

Thus (7.65) can be satisfied simultaneously only for a unique value of λ given by

$$\lambda=-2\left(\sum_{j=1}^{3}\xi_j'n_j/d_j\right)\bigg/\left(\sum_{j=1}^{3}(n_j/d_j)^2\right) \tag{7.69}$$

In particular, squaring and adding both sides of (7.67) we find that

$$\sum_{j=1}^{3}(\xi_j-\xi_j')^2=2(1-\boldsymbol{\xi}'\cdot\boldsymbol{\xi})=4\sin^2\theta=\lambda^2\sum_{j=1}^{3}(n_j/d_j)^2 \tag{7.70}$$

where $\cos 2\theta=\boldsymbol{\xi}'\cdot\boldsymbol{\xi}$ and 2θ is the angle between the incident and emergent beams as defined above. We now remove any common factor n from diffraction order numbers $n_j=nh_j$, where h_j are the Miller indices of the lattice plane which is normal to the plane of $\boldsymbol{\xi}'$ and $\boldsymbol{\xi}$, and (7.67) then reduces to Bragg's law (7.59).

Let us now write the Fermi potential (6.78) inside the crystal in the form of a compound Fourier series

$$V_{\mathrm{F}}(\boldsymbol{r})=\sum_{\boldsymbol{\mu}} V_{\mathrm{F}}(\boldsymbol{q}(\boldsymbol{\mu}))\,\mathrm{e}^{-\mathrm{i}\boldsymbol{q}(\boldsymbol{\mu})\cdot\boldsymbol{r}} \tag{7.71}$$

where the vector $\boldsymbol{q}(\boldsymbol{\mu})$ has the property that $V_{\mathrm{F}}(\boldsymbol{r})$ repeats itself whenever $\boldsymbol{r}$ is replaced by $\boldsymbol{r}+\boldsymbol{v}$ and $\boldsymbol{v}$ is a lattice vector which obeys a relation of the form (7.61). It follows then that $\exp(-\mathrm{i}\boldsymbol{q}(\boldsymbol{\mu})\cdot\boldsymbol{v})=1$ and

$$\boldsymbol{q}(\boldsymbol{\mu})\cdot\boldsymbol{v}=2\pi n \tag{7.72}$$

where n is an integer. This condition is clearly satisfied whenever $\boldsymbol{q}$ is a vector of the reciprocal lattice

$$\boldsymbol{q}(\boldsymbol{\mu})=\mu_1\boldsymbol{\delta}_1+\mu_2\boldsymbol{\delta}_2+\mu_3\boldsymbol{\delta}_3 \tag{7.73}$$

whose basis vectors obey the relation

$$\boldsymbol{\delta}_i\cdot\boldsymbol{d}_j=2\pi\,\delta_{ij} \qquad i,\ j=1-3 \tag{7.74}$$

and where the coefficients μ_j are integers.

We can now express the Laue equations (7.67) in the form

$$k(\boldsymbol{\xi}-\boldsymbol{\xi}')=\boldsymbol{k}-\boldsymbol{k}'=\sum_{j=1}^{3} n_j\boldsymbol{\delta}_j=\boldsymbol{q}(\boldsymbol{n}) \tag{7.75}$$

It follows that the Fraunhofer diffraction pattern of a perfect infinite crystal selects out a set of directions in $\boldsymbol{k}$-space, each direction corresponding to a vector $\boldsymbol{q}(\boldsymbol{n})$ of the reciprocal lattice.

The suggestion that Laue diffraction patterns could be observed using neutrons rather than X-rays was made by Elsasser (1936) and was immediately verified (Halban and Preiswerk 1936, Mitchell and Powers 1936). Neutron diffraction may be observed using either monochromatic or polychromatic neutrons with single crystals or by using the powder or Debye–Scherrer method with monochromatic neutrons in polycrystalline samples. More generally the description 'neutron diffraction' is understood to include all phenomena associated with coherent scattering of slow neutrons in condensed media, when the de Broglie wavelength is comparable with the spatial separations between the scattering centres.

Neutron diffraction differs from X-ray diffraction in three main respects. (1) An X-ray with $\lambda_x \simeq 1$ Å has an energy $E_x \simeq 10$ keV whereas a neutron with $\lambda_n \simeq 1$ Å has an energy $E_n \simeq 0.1$ eV. This energy is comparable with the thermal energy of vibration of the scattering centres and inelastic neutron scattering through the exchange of vibrational energy with the lattice may therefore be observed. (2) The form factor for X-ray diffraction is determined by the electronic charge distribution and the scattered intensity is approximately proportional to the square of the atomic number of the scattering centre. Neutrons on the other hand are primarily scattered by nuclei although there is an important magnetic component due to the interaction of the neutron and electron magnetic moments (cf. section 8.4.1). (3) The number of sources of incoherence in the scattered beam is much greater in the case of neutrons than it is for X-rays for which the principal source is the thermal motion of the scattering centre.

7.2.4 Dynamical diffraction theory

7.2.4.1 Structure factor and lattice sum

In section 7.2.3 we briefly summarized the kinematical description of X-ray Fraunhofer diffraction in a finite orthorhombic crystal, and deduced the familiar result (7.65) that the angular distribution in intensity is determined by the product of two factors; a 'diffraction factor' which characterizes the charge structure of the individual identical diffracting elements, and an 'interference factor' which is a function of the three-dimensional periodicity of the lattice. In the following we pursue a more general approach to the diffraction problem with the specific case of dynamical diffraction of neutrons in crystals kept in view (Goldberger and Seitz 1947, Kagan and Afazesev 1965, Sears 1978).

We begin by expressing the Fermi potential at a point $\boldsymbol{r}$ in the crystal at an arbitrary time t, in the form of a Fourier integral

$$V_{\mathrm{F}}(\boldsymbol{r}, t)=\int V_{\mathrm{F}}(\boldsymbol{q}, t)\,\mathrm{e}^{-\mathrm{i}\boldsymbol{q}\cdot\boldsymbol{r}}\,\mathrm{d}^3q \qquad (7.76)$$

where

$$V_{\mathrm{F}}(\boldsymbol{q}, t)=\left(\frac{1}{2\pi}\right)^3\int_\Omega V_{\mathrm{F}}(\boldsymbol{r}, t)\,\mathrm{e}^{\mathrm{i}\boldsymbol{q}\cdot\boldsymbol{r}}\,\mathrm{d}^3r$$

$$=\left(\frac{1}{2\pi}\right)^3\sum_j\int_\Omega\left(\frac{2\pi\hbar^2}{m_n}\right)c_{\mathrm{m}}b_j\cdot\delta^3(\boldsymbol{r}-\boldsymbol{r}_j(t))\,\mathrm{e}^{\mathrm{i}\boldsymbol{q}\cdot\boldsymbol{r}}\,\mathrm{d}^3r \qquad (7.77)$$

and the integration is over the volume Ω of the crystal. Anticipating the application of (7.77) in the dynamical theory of diffraction in crystals, where the effective field correction is important, each bound scattering length b_j in (7.77) is renormalized by the appropriate factor c_{m} introduced in section 6.3.1.

If the origin of a given unit cell is given by the lattice vector $\boldsymbol{r}_{\mathrm{m}}$ satisfying (7.61), then the position vector of the jth scattering nucleus in the cell may be written

$$\boldsymbol{r}_j(t)=\boldsymbol{r}_{\mathrm{m}}+\boldsymbol{r}_j+\boldsymbol{r}'_j(t) \qquad \langle\boldsymbol{r}'_j(t)\rangle=0 \qquad (7.78)$$

where the fluctuating vector $\boldsymbol{r}'_j(t)$ describes the thermal and zero-point oscillations of the nucleus about its mean position $\boldsymbol{r}_j$ in the lattice. Carrying out an appropriate time average this yields the result

$$\langle V_F(\boldsymbol{q}, t)\rangle_t \equiv v_F(\boldsymbol{q}) = \left(\frac{1}{2\pi}\right)^3 I(\boldsymbol{q}) \sum_{j=1}^{N_c} \left(\frac{2\pi\hbar^2}{m_n}\right) c_m b_j \, e^{i\boldsymbol{q}\cdot\boldsymbol{r}_j} \varphi_j(\boldsymbol{q}) \tag{7.79}$$

where

$$I(\boldsymbol{q}) = \sum_m e^{i\boldsymbol{q}\cdot\boldsymbol{r}_m} \tag{7.80}$$

is the lattice sum, which is computed over all the unit cells in the crystal, $N_c = N\Omega_c$ is the number of nuclei per unit cell, and

$$\varphi_j(\boldsymbol{q}) = \langle e^{i\boldsymbol{q}\cdot\boldsymbol{r}'_j(t)}\rangle_t \tag{7.81}$$

is the characteristic function for the distribution of the random variable $\boldsymbol{r}'_j(t)$. Assuming that the corresponding probability density is a Gaussian, then $\varphi_j(\boldsymbol{q})$ is also Gaussian, i.e.

$$\varphi_j(\boldsymbol{q}) = e^{-(1/2)q^2\langle r'_j(t)^2\rangle} = e^{-W_j} \tag{7.82}$$

The function e^{-W_j} is known as the Debye–Waller factor and the exponent W_j is proportional to the absolute temperature T for temperatures above the crystal Debye temperature. At absolute zero W_j tends to a finite limit representing the effect of the zero-point motion. It is convenient, although not necessary, to incorporate the Debye–Waller factor into the unit cell structure factor which is now defined by

$$F(\boldsymbol{q}) = \sum_{j=1}^{N_c} (c_m b_j) \, e^{i\boldsymbol{q}\cdot\boldsymbol{r}_j - W_j} = N_c c_m b_j \bar{F}(\boldsymbol{q}) \tag{7.83}$$

where $\bar{F}(\boldsymbol{q})$ is the normalized structure factor, or unit cell form factor which appears in (7.65), and is defined such that $\bar{F}(0) = 1$.

In the case of a perfect crystal containing $N_1 \times N_2 \times N_3$ unit cells arranged on a lattice described by basis vectors $\boldsymbol{d}_j$ $(j = 1-3)$, the lattice sum $I(\boldsymbol{q})$ defined in (7.80) can be calculated explicitly and we find

$$I(\boldsymbol{q}) = \prod_{j=1}^{3} \sum_{m_j=0}^{N_j} e^{im_j(\boldsymbol{q}\cdot\boldsymbol{d}_j)} = \prod_{j=1}^{3} \exp\left[\frac{i}{2}(N_j - 1)\boldsymbol{q}\cdot\boldsymbol{d}_j\right] \frac{\sin(\frac{1}{2}N_j\boldsymbol{q}\cdot\boldsymbol{d}_j)}{\sin(\frac{1}{2}\boldsymbol{q}\cdot\boldsymbol{d}_j)} \tag{7.84}$$

For a large crystal $(N_j \to \infty)$ the interference factor $\sin(\frac{1}{2}N_j\boldsymbol{q}_j\cdot\boldsymbol{d}_j)/\sin(\frac{1}{2}\boldsymbol{q}\cdot\boldsymbol{d}_j)$ vanishes unless $\boldsymbol{q}$ is identical with a vector $\boldsymbol{q}(\boldsymbol{\mu})$ of the reciprocal lattice, in which case it takes the value N_j. Thus, apart from a phase factor which may be omitted, $I(\boldsymbol{q}(\boldsymbol{\mu}))$ becomes equal to the number of unit cells and we may write for an infinite perfect crystal

$$I_\infty(\boldsymbol{q}) = \sum_\mu \int e^{i(\boldsymbol{q}-\boldsymbol{q}(\boldsymbol{\mu}))\cdot\boldsymbol{r}} \, d^3r/\Omega_c = \frac{(2\pi)^3}{\Omega_c} \sum_\mu \delta^3(\boldsymbol{q} - \boldsymbol{q}(\boldsymbol{\mu})) \tag{7.85}$$

For a finite non-perfect crystal (7.85) may be modified by inserting in the integral the unit step function $S(\boldsymbol{r})$ which is defined such that (in the notation of section 6.3.2)

$$S(\boldsymbol{r})=1 \qquad \boldsymbol{r}\in M \qquad S(\boldsymbol{r})=0 \qquad \boldsymbol{r}\in \bar{M} \tag{7.86}$$

In this case the lattice sum becomes

$$I(\boldsymbol{q})=\sum_{\mu}\int \mathrm{e}^{\mathrm{i}(\boldsymbol{q}-\boldsymbol{q}(\boldsymbol{\mu}))\cdot\boldsymbol{r}}\,S(\boldsymbol{r})\,\mathrm{d}^3r/\Omega_\mathrm{c}=\left(\frac{2\pi}{\Omega_\mathrm{c}}\right)^3\sum_{\mu}S(\boldsymbol{q}-\boldsymbol{q}(\boldsymbol{\mu})) \tag{7.87}$$

where $S(\boldsymbol{q})$ is the three-dimensional Fourier transform of $S(\boldsymbol{r})$.

We may now combine (7.87), (7.83) and (7.81) to give the final results

$$v_\mathrm{F}(\boldsymbol{q})=\left(\frac{2\pi\hbar^2}{m_\mathrm{n}}\right)(Nc_\mathrm{m}b)\sum_{\mu}\bar{F}(\boldsymbol{q})S(\boldsymbol{q}-\boldsymbol{q}(\boldsymbol{\mu})) \tag{7.88}$$

and

$$V_\mathrm{F}(\boldsymbol{r})\equiv\langle V_\mathrm{F}(\boldsymbol{r},t)\rangle=\left(\frac{2\pi\hbar^2}{m_\mathrm{n}}\right)(Nc_\mathrm{m}b)\sum_{\mu}\int \mathrm{e}^{-\mathrm{i}\boldsymbol{q}\cdot\boldsymbol{r}}\bar{F}(\boldsymbol{q})S(\boldsymbol{q}-\boldsymbol{q}(\boldsymbol{\mu}))\,\mathrm{d}^3q \tag{7.89}$$

In the case of a small crystal where beam attenuation and multiple scattering effects are negligible the kinematical theory is applicable and there is no effective field correction (i.e. $c_\mathrm{m}=1$ in (7.89)). In this situation the differential coherent scattering cross-section for the whole crystal is given by the Born approximation formula

$$\begin{aligned}\frac{\mathrm{d}\sigma}{\mathrm{d}\Omega}(\boldsymbol{q})&=\left|\frac{m_\mathrm{n}v_\mathrm{F}(\boldsymbol{q})}{2\pi\hbar^2}\right|^2=(Nb)^2|\bar{F}(\boldsymbol{q})|^2\left|\sum_{\mu}(2\pi)^2S(\boldsymbol{q}-\boldsymbol{q}(\boldsymbol{\mu}))\right|^2\\&=(Nb)^2|\bar{F}(\boldsymbol{q})|^2\sum_{\mu}|(2\pi)^2S(\boldsymbol{q}-\boldsymbol{q}(\boldsymbol{\mu}))|^2\end{aligned} \tag{7.90}$$

where the contribution of each site in the reciprocal lattice may be treated independently since there is no overlap in the corresponding resonance peaks in $S(\boldsymbol{q}-\boldsymbol{q}(\boldsymbol{\mu}))$. Also since $S(\boldsymbol{q})$ is proportional to the crystal volume, it follows that at each Bragg direction the waves scattered from every nucleus in the crystal combine coherently.

In the case of a large perfect crystal the lattice sum is given by (7.85), in which case the Fermi potential (7.89) reduces to the result (7.71) quoted earlier, where the coefficients $V_\mathrm{F}(\boldsymbol{q}(\mu))$ are defined by

$$V_\mathrm{F}(\boldsymbol{q}(\boldsymbol{\mu}))=\left(\frac{2\pi\hbar^2}{\mu_\mathrm{n}}\right)(Nc_\mathrm{m}b)\bar{F}(\boldsymbol{q}(\boldsymbol{\mu})) \tag{7.91}$$

7.2.4.2 Neutron propagation in perfect crystals

In discussing the dynamical theory it is convenient to restrict attention to the Laue case where the crystal (M) is confined to the region $0<x<t$, and

there are two vacuum spaces ($\bar{M}$), an entrance space $x<0$ and an exit space $x>t$. The configuration shown in figure 7.9 corresponds to symmetric Laue diffraction where the entrance space is normal to the reflecting planes and the diffracted wave emerges at the opposite face. In contrast figure 7.7 illustrates the case of symmetric Bragg diffraction where the entrance face is parallel to the reflecting planes and the diffracted wave emerges at the same face. Neglecting mirror reflection at the plane $x=0$, and internal reflection at $x=t$, the corresponding boundary conditions to be satisfied are

$$\begin{aligned}
&\text{(a) Laue} \quad \Psi(\boldsymbol{r})=\mathrm{e}^{\mathrm{i}\boldsymbol{k}\cdot\boldsymbol{r}} && x<0 \\
&\qquad\qquad =A_{\mathrm{B}}\,\mathrm{e}^{\mathrm{i}\boldsymbol{k}'\cdot\boldsymbol{r}}+A_{\mathrm{T}}\,\mathrm{e}^{\mathrm{i}\boldsymbol{k}''\cdot\boldsymbol{r}} \quad k'=k''=k && x>t \qquad (7.92)\\
&\text{(b) Bragg} \quad \Psi(\boldsymbol{r})=\mathrm{e}^{\mathrm{i}\boldsymbol{k}\cdot\boldsymbol{r}}+A_{\mathrm{B}}\,\mathrm{e}^{\mathrm{i}\boldsymbol{k}'\cdot\boldsymbol{r}} \quad k'=k && x<0 \\
&\qquad\qquad =A_{\mathrm{T}}\,\mathrm{e}^{\mathrm{i}\boldsymbol{k}''\cdot\boldsymbol{r}} \quad k''=k && x>t
\end{aligned}$$

For a homogenous non-crystalline medium the interior coherent wave is given by

$$\Psi(\boldsymbol{r})=A\,\mathrm{e}^{\mathrm{i}\boldsymbol{k}\cdot\boldsymbol{r}} \qquad (7.93)$$

where the relation (6.79) between $\boldsymbol{K}$ and $\boldsymbol{k}$ defines a dispersion surface

$$K=k\sqrt{(1-\xi(k))} \qquad (7.94)$$

with

$$\xi(k)=4\pi c_{\mathrm{m}} bN/k^2=1-n^2 \qquad (7.95)$$

and n is a single-valued refractive index. To conform with the optical theorem (2.90) the bound scattering length must have a small imaginary part, but we shall assume there is no capture cross-section so that the probability current is conserved. In this case the reflection form factor $\bar{F}_{\mu}\equiv\bar{F}(\boldsymbol{q}(\boldsymbol{\mu}))$, and the anti-reflection form factor $\bar{F}_{-\mu}\equiv\bar{F}(\boldsymbol{q}(-\boldsymbol{\mu}))$ satisfy the crossing relation known as Friedel's law

$$\bar{F}_{-\mu}=\bar{F}_{\mu} \qquad (7.96)$$

Since in a practical situation the neutron wavelength is matched to the interatomic spacing so that $Nk^{-3}=\mathrm{O}(1)$, it follows that $\xi\simeq b/\lambda\!\!\!^{-}\simeq 10^{-6}$ may be treated as an infinitesimal.

In a large perfect crystal the Fermi potential is modulated with the period of the lattice as expressed by (7.71), where, according to (7.91) and (7.96), the expansion coefficients may be written

$$V(\boldsymbol{q}(\boldsymbol{\mu}))=\xi(k)E_n\bar{F}_{\mu} \qquad (7.97)$$

and $E_n=\hbar^2k^2/2m_{\mathrm{n}}$ is the neutron energy. The amplitude of the coherent wave acquires a corresponding modulation

$$A(\boldsymbol{r})=\sum_{\boldsymbol{\mu}} A_{\boldsymbol{\mu}}\,\mathrm{e}^{-\mathrm{i}\boldsymbol{q}(\boldsymbol{\mu})\cdot\boldsymbol{r}} \qquad (7.98)$$

so that $\Psi(\boldsymbol{r})$ may be represented as a linear superposition of Bloch waves

$$\Psi(\boldsymbol{r})=\sum_{\boldsymbol{\nu}} A_{\boldsymbol{\nu}}\, \mathrm{e}^{\mathrm{i}\boldsymbol{K}(\boldsymbol{\nu})\cdot\boldsymbol{r}} \tag{7.99}$$

where

$$\boldsymbol{K}(0)-\boldsymbol{K}(\boldsymbol{\nu})=\boldsymbol{q}(\boldsymbol{\nu}) \tag{7.100}$$

The result (7.100) implies that the refracted wavevector $\boldsymbol{K}(0)$ in the medium can be Bragg diffracted into the direction $\boldsymbol{K}(\boldsymbol{\nu})$ only when conditions are such that the recoil momentum $\boldsymbol{q}(\boldsymbol{\nu})$ is identical with a vector of the reciprocal lattice which is normal to the corresponding refracting plane and can therefore be absorbed by the crystal as a whole. Forward scattering is of course always allowed, independently of the boundary conditions, since $\boldsymbol{q}(0)\equiv 0$ is a vector of the reciprocal lattice with $\nu_1=\nu_2=\nu_3=0$. Thus the result (7.100) may be interpreted as a reformulation of the Bragg–Laue equation (7.75) taking account of refraction.

To derive an expression for the dispersion surface the coherent wave (7.99) has to be substituted into the Schrödinger equation with potential given by (7.71) and (7.97), a procedure which, noting that $\boldsymbol{q}(\boldsymbol{\mu}-\boldsymbol{\nu})\equiv\boldsymbol{q}(\boldsymbol{\mu})-\boldsymbol{q}(\boldsymbol{\nu})$, leads to the following infinite set of coupled linear equations for the amplitudes $A_{\boldsymbol{\sigma}}$

$$[K^2-K^2(\boldsymbol{\sigma})]A_{\boldsymbol{\sigma}}=k^2\xi \sum_{\boldsymbol{\tau}\neq\boldsymbol{\sigma}} \bar{F}_{\boldsymbol{\sigma}-\boldsymbol{\tau}}A_{\boldsymbol{\tau}} \tag{7.101}$$

Since ξ is infinitesimal, the coefficient $[K^2-K^2(\boldsymbol{\sigma})]$ must also be $\mathrm{O}(\xi)$ for $A_{\boldsymbol{\sigma}}$ to be finite. This condition is always satisfied for the forward diffracted wave ($\boldsymbol{\sigma}=0$) but, for it to be fulfilled for at least one other amplitude $A_{\boldsymbol{\sigma}}$ ($\boldsymbol{\sigma}\neq 0$), the incident momentum $\boldsymbol{k}$ must meet the entrance surface at an angle very close to the corresponding Bragg angle. In this case there are two non-vanishing amplitudes in (7.101), which reduces to

$$\begin{pmatrix} K^2-K^2(0) & -k^2\xi\bar{F}_{-\boldsymbol{\sigma}} \\ -k^2\xi\bar{F}_{\boldsymbol{\sigma}} & K^2-K^2(\boldsymbol{\sigma}) \end{pmatrix}\begin{pmatrix} A_0 \\ A_{\boldsymbol{\sigma}} \end{pmatrix}=0 \tag{7.102}$$

The dispersion surface is then determined by the condition that (7.102) should have a non-trivial solution, which implies that the determinant should vanish, i.e.

$$(K^2-K^2(0))(K^2-K^2(\boldsymbol{\sigma}))=|k^2\xi\bar{F}_{\boldsymbol{\sigma}}|^2 \tag{7.103}$$

We can now solve (7.103) directly for forward scattering and we find that, in place of (7.95), the dispersion surface separates into two distinct branches

$$K_{\pm}(0)=K\sqrt{(1\mp(k/K)^2\xi|\bar{F}(\boldsymbol{\sigma})|)} \tag{7.104}$$

with a splitting equal to $|k\xi F(\boldsymbol{\sigma})|$ in first order. The corresponding solutions for the Bragg-diffracted vectors $K_{\pm}(\boldsymbol{\sigma})$ are then most easily deduced from

(7.100). This situation clearly arises as a direct consequence of the coupling term $|k^2\xi F(\boldsymbol{\sigma})|^2$ in the secular equation (7.103), whose presence signals the fact that, when $\boldsymbol{K}(0)$ satisfies the condition for Bragg diffraction at the reciprocal lattice site $\boldsymbol{q}(\boldsymbol{\sigma})$, $\boldsymbol{K}(\boldsymbol{\sigma})$ equally satisfies the same condition at $\boldsymbol{q}(-\boldsymbol{\sigma})$, and the scattered wave is rescattered into the forward direction. Under exact conditions the amplitudes of forward-scattered and Bragg-scattered waves will then be equal.

The situation described above has a close analogue in the band theory of solids where an energy gap $\Delta E = 2V(\boldsymbol{q}(\boldsymbol{\sigma}))$ develops at the Brillouin zone boundary defined by the Bragg condition (7.75), between the energies of those quasi-free electron states which exist at opposite sides of the zone boundary. For the case of neutron, as opposed to electron, propagation in perfect crystals, the total energy is of course a constant of the motion which is fixed by the incident wave energy, and the division of the dispersion surface described by (7.104) corresponds to a splitting in the kinetic energy given by

$$\Delta(\hbar^2K^2(0)/2m_{\rm n}) = 2V(\boldsymbol{q}(\boldsymbol{\sigma})) \tag{7.105}$$

When $\boldsymbol{K}(0)$ does not lie close to a zone boundary only forward scattering is permitted, and there can be no attenuation in a perfect crystal for which there is no nuclear capture cross-section. This is ensured by the effective field correction factor $c_{\rm m}$ which acquires such a phase as to compensate for any phase in b. Thus $(c_{\rm m}b)$ is real and the total cross-section vanishes. Precisely the opposite effect occurs when $\boldsymbol{K}(0)$ lies close to a Brillouin zone corner and the substantial mixing of states which then occurs enhances the non-forward scattering leading to an increased attenuation in the transmitted wave.

7.2.4.3 The Borrmann effect

Since Bragg reflection conserves the component of momentum parallel to the reflecting plane, while the normal component is reversed, the stable solutions at the exact Bragg condition correspond to progressive waves in parallel directions and standing waves in normal directions. This means that the energy is transported parallel to the reflecting plane. Thus, in the symmetric Laue case shown in figure 7.9, the perfect crystal acts like a waveguide, with wave patterns corresponding to the two discrete branches of the dispersion surface given by

$$\Psi^+ = \sin(q(\boldsymbol{\sigma})y/2)\, e^{i(K\cos\theta_{\rm B} + i\pi/\Delta)x}$$

$$\Psi^- = \cos(q(\boldsymbol{\sigma})y/2)\, e^{i(K\cos\theta_{\rm B} - i\pi/\Delta)x} \tag{7.106}$$

where Δ is the extinction length (also called the Pendellösung period) defined by

$$\Delta = 2\pi\cos\theta_{\rm B}/k\xi\bar{F}(\boldsymbol{\sigma}) = \pi\cos\theta_{\rm B}/\lambda_{\rm B}Nc_{\rm m}b\bar{F}(\boldsymbol{\sigma}) \tag{7.107}$$

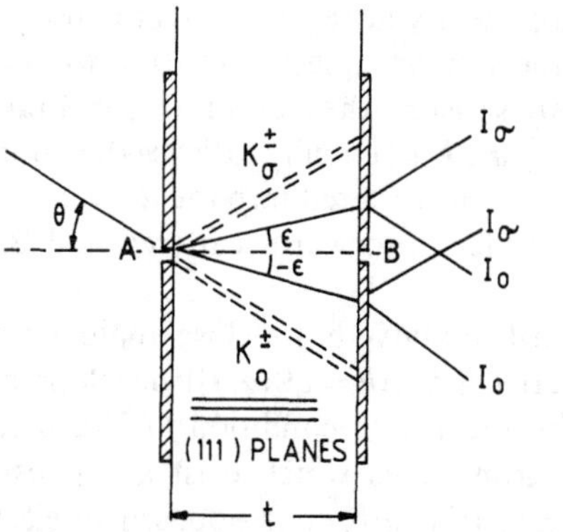

Figure 7.9. Demonstration of Pendellösung interference in perfect crystals of silicon by symmetric Laue diffraction at a surface normal to the (111) diffracting planes. Neutrons are incident at the entrance slit A and are detected at the scanning slit B. The amount by which the angle of incidence θ departs from the exact Bragg angle θ_B is denoted by ε. The two forward wave-vectors $K_0^\pm$, and the two diffracted wave-vectors $K_\sigma^\pm$ are predicted by dynamical diffraction theory.

and the phases correspond to the convention that the x-axis lies in a reflecting plane.

At the exit face ($x=t$) the interior waves Ψ^+ and Ψ^- recombine to produce transmitted (Ψ_0) and diffracted (Ψ_σ) waves

$$\Psi_0(\boldsymbol{r}) \approx \exp\left[\, i(k \cos \theta_B x - \frac{1}{2} q(\boldsymbol{\sigma}) y)\right] \qquad (7.108)$$

$$\Psi_\sigma(\boldsymbol{r}) \approx \exp\left[\, i(k \cos \theta_B x + \frac{1}{2} q(\boldsymbol{\sigma}) y)\right]$$

with relative intensities

$$I_0 \approx \cos^2\left(\frac{\pi}{4} + \frac{\pi t}{\Delta}\right) \qquad I_\sigma \approx \sin^2\left(\frac{\pi}{4} + \frac{\pi t}{\Delta}\right) \qquad (7.109)$$

The wave Ψ^+ vanishes at a reflecting plane ($q(\boldsymbol{\sigma})y = 2n\pi$) and its nodes are located at points mid-way between reflecting planes. As a result it propagates through the crystal essentially unhindered. In contrast the nodes of Ψ^- are located on the reflecting planes and the wave interacts strongly with nuclei in the crystal. The mean potential energy difference between the two states is thus equal to the 'energy gap' defined in (7.105). If now some of the nuclei have significant capture cross-sections the wave Ψ^- will be severely attenuated and a rapid variation in transmission is predicted in the vicinity of the Bragg condition. This curious behaviour is described as the 'Borrmann effect' in recognition of its discoverer, who first observed the phenomenon in X-ray diffraction (Borrmann 1941, 1950). Anomalous transmission effects with neutrons were first observed by Knowles (1956) who recorded a $2.5 \pm 0.6\%$ change in the yield of (n, γ) reactions in calcium, when a perfect crystal of calcium was rocked through the Bragg angle.

7.2.4.4 Pendellösung interference

When the angle of incidence θ departs by a small amount $\delta\theta = \theta_B - \theta$ from the Bragg angle, the energy no longer propagates parallel to the reflecting planes but is channelled into two streams flowing at angles $\pm\varepsilon$ with respect to the normal, where

$$\gamma = \frac{\tan \varepsilon}{\tan \theta_B} = \frac{\Delta\, \delta\theta/2d}{\sqrt{1 + (\Delta\, \delta\theta/2d)^2}} \tag{7.110}$$

In the case where the incident neutrons reach the crystal face through a fine slit of width $\simeq 0.1$ mm, so that the angular spectrum has an appreciable value within a width $\Delta\theta_{inc} > \Delta\theta$, where

$$\Delta\theta = 2\xi |\tilde{F}_\sigma| / \sin 2\theta_B \tag{7.111}$$

is the angular width of the single-crystal reflection, then it is a better approximation to represent the incident wavefield as a spherical rather than as a plane wave (Kato 1961). Under these circumstances each branch of the dispersion surface is uniformly excited and diffracted neutrons emerge from a finite area on the exit face, whose width is determined by t and ε $(-\theta_B < \varepsilon < \theta_B)$, with an intensity

$$I_\sigma(\gamma) \approx (1 - \gamma^2)^{-1/2} \sin^2\left(\frac{\pi}{4} + \frac{\pi t}{\Delta}(1 - \gamma^2)^{-1/2}\right) \tag{7.112}$$

Since for silicon $d \simeq 10^{-10}$ m and $\Delta \simeq 10^{-4}$ m it follows from (7.110) that $\varepsilon/\delta\theta \simeq 10^6$. Thus an infinitesimal variation $\delta\theta$ ($\simeq 10^{-5}$) is amplified to produce a finite variation in $I_\sigma(\gamma)$ with a corresponding variation in the transmitted intensity $I_0(\gamma)$ in anti-phase. This switching of the intensity between Bragg diffracted and forward scattered beams was described as 'Pendellösung' by Ewald (1917) on the analogy with the periodic transfer of energy between coupled pendula. It is caused by the beating of the Ψ^+ and Ψ^- waves (7.106) which occurs at various depths x in the crystal with a spatial period Δ.

Pendellösung interference with monochromatic neutrons was first investigated by Shull (1968) for Bragg diffraction in the (1 1 1) plane of perfect silicon crystals, using the arrangement shown in figure 7.9. Using a fixed narrow entrance slit A in combination with a movable scanning slit B he succeeded in verifying the intensity pattern (7.112) for values of γ in the range $|\gamma| < 0.6$. This is shown in figure 7.10(a) for an incident wavelength

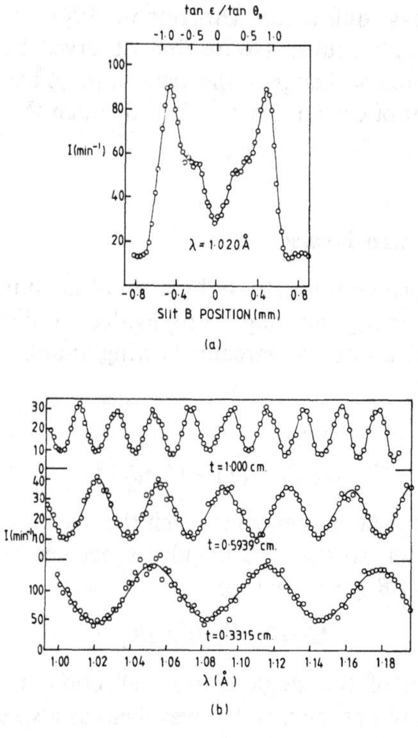

Figure 7.10. (a) Pendellösung interference observed with 1.020 Å neutrons incident on perfect crystals of silicon. The 0.13 mm entrance slit A is fixed while the scanning slit B is varied in position. (b) Pendellösung interference patterns observed with slits A and B both fixed, and λ and θ simultaneously varied to maintain the Bragg condition. Results are shown for three thicknesses of perfect silicon crystal (Shull 1968).

$\lambda = 1.020$ Å. An even more dramatic demonstration was brought about by fixing B symmetrically along the reflecting plane relative to A and varying λ, while synchronously stepping the crystal orientation to maintain the exact Bragg condition. By this means the Pendellösung interference pattern could be traced to fringe order numbers t/Δ as high as 55. The comparison between theory and experiment for three values of the crystal thickness is shown in figure 7.10(b).

It may be remarked that Borrmann anomalous transmission and Pendellösung interference are almost mutually exclusive phenomena. For X-ray diffraction where there is strong absorption the Borrmann effect dominates whereas with perfect crystals such as silicon, which are essentially transparent to neutrons, there is no Borrmann effect but a well developed Pendellösung fringe structure is observed (cf. section 7.4.2).

7.3 Neutron interference

7.3.1 Interferometric measures of coherence

7.3.1.1 Temporal coherence

In 1801 Thomas Young performed the famous double beam interference experiment which bears his name, and which was to become an important landmark in the long history of the wave theory of light and material particles. In its original version Young's experiment was carried out using sunlight falling on a tiny pinhole which acted as the primary light source S. The emergent light illuminated two pinholes $P'_1P'_2$ which became the secondary sources of the interfering beams. In modern versions of the experiment the pinholes are replaced by narrow slits as shown in figure 7.11. Here S represents either (a) a line source, or (b) an extended source, of quasi-monochromatic light which illuminates two narrow slits P'_1 and P'_2. Interference fringes parallel to the slits are then observed at points P on a distant observing screen.

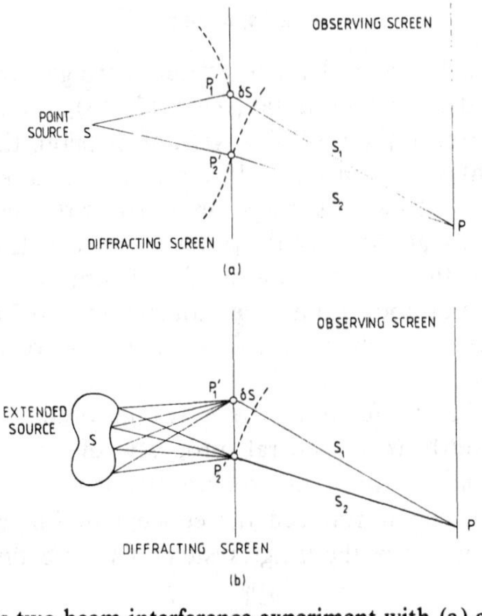

Figure 7.11. Young's two beam interference experiment with (a) a point source, and (b) an extended source. The quantity $\delta S = S_1 - S_2$ denotes the path difference for waves arriving at the field point P from illuminated pinholes at P'_1 and P'_2.

The classic Young's experiment uses a line source as shown in figure 7.11(a) and relies for its action on the division of the incident wavefront at the diffracting apertures P'_1 and P'_2 but, because the wavelength is small, the slits have to be made very narrow, and a substantial experimental effort must be mounted to observe the fringes, without even attempting to discern the details. A much more efficient method for producing interference makes use of the division of amplitude at the interface between two media, where the incident wave is spatially separated into reflected and transmitted components. A celebrated example of interference by division of amplitude is provided by the optical phenomenon which goes by the name of 'Newton's rings', although it was discovered independently by Boyle and by Hooke. Newton's rings are equal thickness circular fringes formed in the film of air between a convex lens and an optical flat which, like the interference fringes created in films of oil, are easily visible to the naked eye. Just as in Young's arrangement all these two-beam interference devices allow a determination of wavelength from the observed interference pattern, and their ultimate development is represented by Michelson's interferometer (1881) and successor instruments derived from it.

In a two-beam interference experiment interference is brought about by introducing a path difference $\delta s=(s_1-s_2)$ between the two beams and interference maxima (or minima in the case of Newton's rings) are observed for a given spectral component for path differences δs_n which satisfy

$$k\,\delta s_n=2n\pi \tag{7.113}$$

For strictly monochromatic light interference fringes are observed for all values of the order number n satisfying (7.113), no matter how large, assuming, of course, in the case of wavefront division, that they fall within the appropriate diffraction envelope. It was, however, noted by Fizeau (1862) that the system of Newton's rings observed with yellow sodium light vanished completely at values of the order number n close to $n_0 \simeq 490$, and then reappeared with full contrast at $n \simeq 2n_0$. From this observation Fizeau deduced correctly that sodium light was not characterized by a single spectral line but rather by a closely spaced doublet with a wavelength separation $\delta\lambda \simeq 0.1$ Å.

The fading of the interference pattern at high order numbers due to the finite spectral width is a general property of the radiation from a quasi-monochromatic primary line source. It was studied in great detail by Michelson (1891) who introduced the concept of fringe visibility V as a measure of the contrast in the fringe system. This is defined by

$$V(n)=\frac{I_{\max}-I_{\min}}{I_{\max}+I_{\min}} \tag{7.114}$$

where $I_{\max}$ and $I_{\min}$ denote the maximum and minimum intensities respectively in the nth order of interference. The fact that $V(n)$ vanishes ultimately shows that there is an upper limit to the time delay which can be inserted between two beams from the one primary source, while still giving rise to identifiable interference fringes, and $V(n)$ provides a measure of what is known as the temporal coherence of the radiation from this one source.

The maximum permitted time delay is called the coherence time since it determines the length of time it takes for a wavetrain associated with either of the two beams to pass the point of observation.

To understand the relationship which exists between the coherence time τ_c and the spectral bandwidth $\Delta_c\nu$ we may suppose first of all that the signal from the light source may be represented at the point of observation by the single pulse

$$\tilde{\Psi}(\boldsymbol{r}, t) = e^{i(\boldsymbol{k}\cdot\boldsymbol{r} - \omega t)} \qquad -\frac{\tau_c}{2} \leqslant t \leqslant \frac{\tau_c}{2} \tag{7.115}$$

where $\bar{\omega} = c\bar{k} = 2\pi\bar{\nu}$ and $\bar{\nu}$ is the mean frequency. Then the Fourier transform of $\Psi(\boldsymbol{r}, t)$ from the time domain to the frequency domain is given by

$$\tilde{\varphi}(\boldsymbol{r}, \nu) \equiv \int_{-\infty}^{\infty} \tilde{\Psi}(\boldsymbol{r}, t)\, e^{i\omega t}\, dt = \tau_c \left\{ \frac{\sin[(\omega - \bar{\omega})(\tau_c/2)]}{(\omega - \bar{\omega})(\tau_c/2)} \right\} \tag{7.116}$$

Thus, since $\tilde{\varphi}(\boldsymbol{r}, \nu)$ is negligible outside a spectral bandwidth $\Delta_c\nu$, defined such that

$$2\pi\, \Delta_c \nu \frac{\tau_2}{2} = \pi \tag{7.117}$$

the coherence time τ_c satisfies the relation

$$\tau_c = (\Delta_c \nu)^{-1} \tag{7.118}$$

The definition of coherence time given immediately above is clearly consistent with the energy-time uncertainty relation

$$\Delta E\, \Delta t \geqslant \hbar/2 \tag{7.119}$$

where $\Delta E = 2\pi\hbar\, \Delta\nu$ is the uncertainty, i.e. the standard deviation, in the energy of radiation emitted by a decaying source, and Δt is the uncertainty in the time during which the source continues to radiate. It should be emphasized, however, that the uncertainty relation (7.117) sets a lower limit, and not an upper limit, on the coherence time, and, even for a very broad band primary spectrum, interference fringes, 'white light' fringes, are in principle always observable. If $\Delta\nu \ll \bar{\nu}$ the source is said to be quasi-monochromatic and in this case, the coherence time τ_c can always be translated into a longitudinal coherence length l_c defined by

$$l_c = c\tau_c = \frac{c}{\Delta_c \nu} = \frac{\bar{\lambda}^2}{\Delta_c \lambda} \tag{7.120}$$

If $\Delta\nu \geqslant \bar{\nu}$, the shape of the wave packet alters continuously in time and no unambiguous meaning can be attached to the notion of coherence length.

7.3.1.2 Spatial coherence

Another, quite different, way to reduce the fringe visibility is to leave the spectral content of the radiation unaltered, but to replace the line source by an extended source as shown in figure 7.11(b). In this case the secondary sources P_1' and P_2' sample quite different spatial regions of the primary source, and the fringe visibility decreases quite rapidly as P_1' and P_2' are moved further apart, even though their average separations from S remain equal. In this case the fringe visibility is primarily a measure of the degree to which the phase of the radiation field is different at different points in space, i.e. it measures the spatial coherence of the radiation arriving at P from P_1' and P_2', although it is never possible totally to disentangle the effects of temporal and spatial incoherence. Indeed, for purely monochromatic radiation, which by definition has complete temporal coherence, there would be no effect on the fringe visibility due to spatial incoherence either.

These effects were first investigated by Verdet (1869) who showed that Young's interference pattern observed with sunlight disappeared if the separation between the two secondary pinholes was increased to about 50 μm. This happens because the sun is not a point source, but an extended source which subtends an angular diameter $\Delta\theta \simeq 10^{-2}$ at points S on the earth's surface, and each individual excited atom in the sun's disc radiates independently. It is precisely this effect which is exploited in Michelson's stellar interferometer (1920) to determine the angular diameters subtended by distant sources. According to the theory of the device the first zero in the fringe visibility, due to the overlapping of the totality of incoherent interference patterns, occurs when the separation d between the two slits is given by

$$d = \bar{\lambda}/\Delta\theta \tag{7.121}$$

a result which provides a completely adequate explanation of Verdet's observations for a mean wavelength of solar radiation $\bar{\lambda} = 5000$ Å.

In the application of these ideas to the de Broglie waves associated with neutrons the concept of spatial coherence can be adopted almost without change, but there are difficulties with the notion of temporal coherence because each spectral component travels with a different velocity and even the vacuum is dispersive. Thus in general the path lengths $P_1'P$ and $P_2'P$ are curvilinear and energy-dependent. If $\hbar k$ is the momentum of a neutron detected at P then we must define reduced path lengths

$$ks_1 = \int_{P_1'}^{P} k_1(s_1)\,ds_1 \qquad ks_2 = \int_{P_2'}^{P} k_2(s_2)\,ds_2 \tag{7.122}$$

and the phase difference $k\,\delta s$ appearing in (7.113) is in general frequency dependent. We may therefore expand it as a Taylor series about the mean frequency of the wavepacket, i.e.

$$k\,\delta s=(k\,\delta s)_{\nu=\bar{\nu}}+\frac{\mathrm{d}}{\mathrm{d}\nu}(k\,\delta s)_{\nu=\bar{\nu}}(\nu-\bar{\nu})+\ldots$$

$$=\overline{k\,\delta s}+\frac{2\pi}{\bar{v}_{\mathrm{g}}}\,\delta\bar{s}\{1+C(\bar{\nu})\}(\nu-\bar{\nu})+\ldots \tag{7.123}$$

where

$$\bar{v}_{\mathrm{g}}=\left(\frac{\mathrm{d}\omega}{\mathrm{d}k}\right)_{\nu=\bar{\nu}} \tag{7.124}$$

is the group velocity of the wavepacket and $C(\nu)$ is a chromatic coefficient (Gabor 1956) defined by

$$C(\nu)=\mathrm{d}\ln(\delta)/\mathrm{d}\ln(k) \tag{7.125}$$

The value of $C(\nu)$ depends on the means by which the splitting of the primary beam is brought about in the first place. If this is achieved by diffraction then $\delta s\propto\lambda$, $C(\nu)=-1$ and the resultant two-beam interferometer is achromatic.

7.3.1.3 A gravitational rainbow

An appealing example of an interference effect described as a 'falling neutron rainbow' is in principle observable with slow neutrons, although it has not yet proved possible to realize those conditions of spatial and temporal coherence which are necessary for its detection (Berry 1982). The notion is that a point source of monoenergetic neutrons emits isotropically in a gravitational field so that each neutron traverses a parabolic orbit. The envelope of all possible neutron orbits is then a paraboloidal caustic surface (cf. section 6.1.1) whose radius at the level of the source has the value $R=v^2/g$. This system is analogous to the optical rainbow, both with respect to the focusing action onto caustic surfaces, and to the effects of dispersion, which defines different caustic surfaces for different wavelengths.

Since every point in the interior of the caustic is crossed by two neutron orbits, an interference pattern develops such that, in the neighbourhood of the caustic, the fringe separation ΔR is independent of velocity. This result is unique to de Broglie waves and follows from the form of the gravity potential and the dispersion relation (6.15). The very small value $\Delta R\simeq 26\ \mu\mathrm{m}$ requires that the source be smaller still if spatial coherence is to be maintained. Furthermore, assuming $R\leqslant 10$ m, it is necessary that $v<10\ \mathrm{m\,s^{-1}}$; this means that the experiment can only be performed with ultra-cold neutrons. Finally, to fulfill the requirements of temporal coherence and produce visible fringes, it is necessary that the velocity dispersion satisfies the condition $\Delta v/v<10^{-6}$. Taken together these very stringent requirements present formidable difficulties for the performance of a successful experiment.

7.3.2 The mutual coherence function

An important distinction between the wavefield associated with light, and the de Broglie wavefield associated with the neutron, stems from the fact that the former is real whereas the latter is inherently complex (cf. section 6.2.1). Since the time-Fourier transform $\tilde{\varphi}_r(\boldsymbol{r}, \nu)$ of a real optical field $\tilde{\Psi}(\boldsymbol{r}, t)$ satisfies the crossing condition

$$\tilde{\varphi}_r(\boldsymbol{r}, \nu) = \tilde{\varphi}_r^*(\boldsymbol{r}, \nu) \tag{7.126}$$

there is no information carried by the negative frequency components ($\nu < 0$) which is not already present in the positive frequency components ($\nu > 0$). Thus no information is lost or gained when $\tilde{\Psi}_r(\boldsymbol{r}, t)$ is replaced by the complex field

$$\tilde{\Psi}(\boldsymbol{r}, t) = \int_0^\infty \tilde{\varphi}_r(\boldsymbol{r}, \nu)\, \mathrm{e}^{-i\omega t}\, \mathrm{d}\nu \tag{7.127}$$

which it usually proves very convenient to do. The function $\tilde{\Psi}(\boldsymbol{r}, t)$ which contains positive frequency components only was first introduced by Gabor (1946). It is referred to as an 'analytical signal', whose real and imaginary parts constitute a Hilbert transform pair which satisfy an appropriate set of dispersion relations (cf. 2.114).

In the case of a positive energy neutron wavefunction $\Psi(\boldsymbol{r}, t)$, e.g. a wavepacket such as that described by equations (6.47)–(6.48), the time-Fourier transform $\varphi(\boldsymbol{r}, \nu)$ contains positive frequency components only which means that Wolf's (1955) formulation of optical coherence theory based on the analytic signal can be taken over into neutron optics almost without change. In this context it should be noted that the optical pulse described by (7.115), although very useful for purposes of illustration, is not an analytic signal. Its usefulness arises from the fact that it is square integrable in time, and therefore possesses a Fourier transform. It is what is known in communications engineering as an energy signal. However, in most cases radiation from a source arrives in the form of a continuously fluctuating signal rather than as a sequence of discrete pulses. Thus the received signal is an incoherent superposition of waveforms which must be treated by statistical methods. To treat a continuous signal by the methods of Fourier transformation it is first necessary to choose a sample signal which is defined for a finite period of time T

$$\begin{aligned} \Psi_T(\boldsymbol{r}, t) &= \Psi(\boldsymbol{r}, t) \qquad -\frac{T}{2} \leqslant t \leqslant \frac{T}{2} \\ &= 0 \qquad |t| > \frac{T}{2} \end{aligned} \tag{7.128}$$

Since the truncated signal $\Psi_T(\boldsymbol{r}, t)$ is square integrable it has a time-Fourier transform

$$\varphi_T(\boldsymbol{r}, \nu) = \int_{-\infty}^{\infty} \Psi_T(\boldsymbol{r}, t)\, \mathrm{e}^{\mathrm{i}\omega t}\, \mathrm{d}t \tag{7.129}$$

with a power spectrum or spectral density function $G(\nu)$, where

$$G(\nu) = \underset{T\to\infty}{\mathrm{Lim}} \frac{1}{T} |\varphi_T(\boldsymbol{r}, \nu)|^2 \tag{7.130}$$

In general the limit (7.130) exists only when the emission process is stationary and ergodic. The condition of stationarity we have already encountered in section 6.4.4.1 and it means that the process is invariant with respect to displacement of the zero of time. An ergodic process is one in which ensemble averages may be replaced by time averages over a single member of the ensemble. Both conditions apply in most cases of experimental interest.

We now take (7.118) as the definition of coherence time for a continuously varying wavefunction $\Psi(\boldsymbol{r}, t)$, where the spectral or coherence width $\Delta_c\nu$ is given by

$$(\Delta_c\nu)^{-1} = \int_0^{\infty} g^2(\nu)\, \mathrm{d}\nu \tag{7.131}$$

and

$$g(\nu) = G(\nu) \Big/ \int_0^{\infty} G(\nu)\, \mathrm{d}\nu \tag{7.132}$$

is the normalized power spectrum.

We now return to a consideration of the two-beam interference experiment and assume that the times of travel of a neutron wavepacket from source points $\mathrm{P}_1'(\boldsymbol{r}_1)$ and $\mathrm{P}_2'(\boldsymbol{r}_2)$ to the observation point $\mathrm{P}(\boldsymbol{r})$ are given by t_1 and t_2 respectively. The wavefunction $\Psi(\boldsymbol{r}, t)$ at P may then be expressed as the superposition state

$$\Psi(\boldsymbol{r}, t) = \Psi(\boldsymbol{r}_1, t - t_1) + \Psi(\boldsymbol{r}_2, t - t_2) \tag{7.133}$$

with a resultant intensity

$$I(\boldsymbol{r}) = I_1 + I_2 + 2\,\mathrm{Re}\,\{\Gamma_{12}(\tau)\} \tag{7.134}$$

where $I_1(I_2)$ is the intensity at $\boldsymbol{r}_1(\boldsymbol{r}_2)$ and $\Gamma_{12}(\tau)$ is the mutual coherence function defined by

$$\begin{aligned}\Gamma_{12}(\tau) &\equiv \Gamma(\boldsymbol{r}_1, \boldsymbol{r}_2; \tau) \equiv \langle \Psi^*(\boldsymbol{r}_1, t - t_1)\Psi(\boldsymbol{r}_2, t - t_2)\rangle \\ &= \underset{T\to\infty}{\mathrm{Lim}} \frac{1}{T} \int_{-T/2}^{T/2} \Psi_T^*(\boldsymbol{r}_1, t)\Psi_T(\boldsymbol{r}_2, t + \tau)\, \mathrm{d}t \end{aligned} \tag{7.135}$$

where $\tau = (t_1 - t_2)$ is the time delay between the arrival times of the interfering wavepackets at P($\boldsymbol{r}$). In the case of light propagating in vacuum or in a medium which is not highly dispersive, the time difference τ is determined by the optical path difference inserted between the two beams and is therefore well defined. In the case of neutron wavepackets the time delay is determined by the path difference only when the mean momentum is well defined and in practice this means that the source must be quasi-monochromatic.

We can now rewrite (7.134) in terms of the complex degree of coherence $\gamma_{12}(\tau)$ defined by

$$\gamma_{12}(\tau) \equiv \gamma(\boldsymbol{r}_1, \boldsymbol{r}_2; \tau) = \Gamma_{12}(\tau)/\sqrt{(I_1 I_2)} \tag{7.136}$$

and the intensity at P($\boldsymbol{r}$) now becomes

$$I(\boldsymbol{r}) = I_1 + I_2 + 2\sqrt{(I_1 I_2)}\,\mathrm{Re}\,(\gamma_{12}(\tau)) \tag{7.137}$$

To interpret the result (7.137) in terms of the concept of fringe visibility introduced in (7.114), we must return to a description in the frequency domain, and rewrite the interference term $I_{12}(\tau)$ in (7.134) in the form

$$\begin{aligned} I_{12}(\tau) &= 2\,\mathrm{Re}\left\{ \lim_{T\to\infty} \frac{1}{T} \int_0^\infty \int_0^\infty \varphi_T^*(\boldsymbol{r}_1, \nu_1)\varphi_T(\boldsymbol{r}_2, \nu_2)\,\mathrm{e}^{-\mathrm{i}(\omega_1 t_1 - \omega_2 t_2)} \right. \\ &\qquad \left. \times \frac{\sin\left[(\omega_1 - \omega_2)\dfrac{T}{2}\right]}{(\omega_1 - \omega_2)}\,\mathrm{d}\nu_1\,\mathrm{d}\nu_2 \right\} \\ &= 2\,\mathrm{Re} \int_0^\infty G_{12}(\nu)\,\mathrm{e}^{-\mathrm{i}\omega\tau}\,\mathrm{d}\nu \end{aligned} \tag{7.138}$$

where

$$G_{12}(\nu) \equiv G(\boldsymbol{r}_1, \boldsymbol{r}_2; \nu) = \lim_{T\to\infty} \frac{\varphi_T^*(\boldsymbol{r}_1, \nu)\varphi_T(\boldsymbol{r}_2, \nu)}{T} \tag{7.139}$$

is the mutual spectral density function. When $\boldsymbol{r}_1 \equiv \boldsymbol{r}_2$, $G_{11}(\nu) \equiv G(\nu)$ where $G(\nu)$ is the spectral density function introduced in (7.130). Finally, comparing (7.138) and (7.134) we have the result

$$\Gamma_{12}(\tau) = \int_0^\infty G_{12}(\nu)\,\mathrm{e}^{-\mathrm{i}\omega\tau}\,\mathrm{d}\nu \tag{7.140}$$

which is a generalization of the Wiener–Khinchine theorem already encountered in section 6.4.4.2.

In the case of a quasi-monochromatic source the function $G_{12}(\nu)$ has a sharp peak in the spectral region $\nu \simeq \bar{\nu}$ and is vanishingly small outside the range $\bar{\nu} - \Delta_c\nu/2 < \nu < \bar{\nu} + \Delta_c\nu/2$ where $\Delta_c\nu \ll \bar{\nu}$. In this case the integral in (7.140) can always be evaluated approximately using the method of steepest descents

(Copson 1935) and we find that

$$\arg \Gamma_{12}(\tau) = -\bar{\omega}\tau + \theta_{12}(\tau) \tag{7.141}$$

where $\theta_{12}(\tau) = \theta(\boldsymbol{r}_1, \boldsymbol{r}_2; \tau)$ is a slowly varying function of τ. Under these circumstances the expression (7.137) for the intensity becomes

$$I(\boldsymbol{r}) = I_1 + I_2 + 2\sqrt{(I_1 I_2)} \cos(\theta_{12}(\tau) - \bar{\omega}\tau) \tag{7.142}$$

Since the fringe maxima and minima appear at order numbers n such that

$$\begin{aligned} I(\boldsymbol{r}) &= I_{\text{max}} \qquad \alpha_{12}(\tau) = \bar{\omega}\tau + 2n\pi \\ I(\boldsymbol{r}) &= I_{\text{min}} \qquad \alpha_{12}(\tau) = \bar{\omega}\tau + (2n+1)\pi \end{aligned} \tag{7.143}$$

the fringe visibility at $\boldsymbol{r}$ becomes

$$V(n) = \frac{2\sqrt{(I_1 I_2)}}{I_1 + I_2} |\gamma_{12}(\tau(n))| \tag{7.144}$$

Since the phase of $\gamma_{12}(\tau)$ is determined according to (7.142) by the positions of the maxima and minima in the interference pattern, and $|\gamma_{12}(\omega)|$ is given by the fringe visibility according to (7.144), the functions $\gamma_{12}(\tau)$ and $\Gamma_{12}(\tau)$ are both measurable. In particular, if $\boldsymbol{r}_1$ coincides with $\boldsymbol{r}_2$, and beam splitting is brought about by amplitude division at $\boldsymbol{r}_1$, then $\Gamma_{11}(\tau)$ represents the correlation at $\boldsymbol{r}_1$ for two different instants in time separated by an interval τ. Thus the fringe visibility in Michelson's interferometer (and neutron interferometers developed according to the same principle as discussed in section 7.3.3 below) yields information about the spectral density $G_{11}(\nu) = G(\nu)$. If, on the other hand, beam splitting is brought about by division of wavefront as in Michelson's stellar interferometer, then the fringe visibility near the centre of the pattern ($\tau \simeq 0$) measures $\Gamma_{12}(0)$, representing the correlation at $\boldsymbol{r}_1$ and $\boldsymbol{r}_2$ at the same instant in time, and the spatial Fourier transform of $\Gamma_{12}(0)$ yields a measure of the spatial variation in luminosity across the source.

7.3.3 Neutron interferometry

7.3.3.1 The perfect crystal neutron interferometer

The earliest operating neutron interferometer (Maier-Leibnitz and Springer 1962) was based on the principle of Fresnel's biprism but, essentially because the neutron refractive index is so close to unity, the beam separation achieved ($\simeq 60\ \mu$m) was insufficient to permit phase measurements when a sample was introduced into one of the arms (Landkammer 1966). This difficulty is circumvented in the triple Laue (LLL) perfect crystal interferometer first constructed by Rauch *et al* (1974) following the successful development of a similiar instrument for X-rays (Bonse and Hart 1965, 1966). In this scheme (figure 7.12), which resembles in many ways the Mach–Zehnder modification

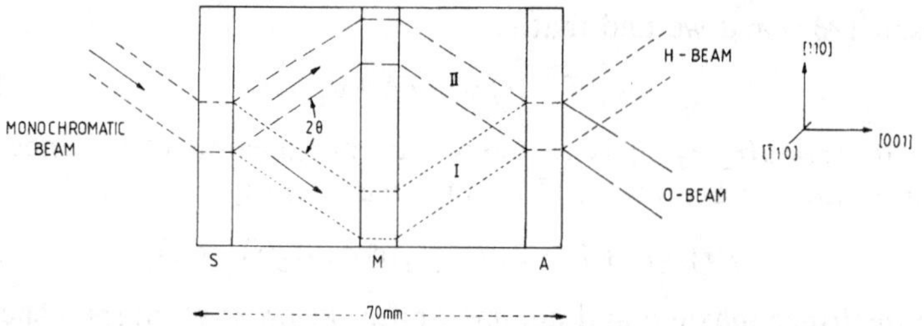

Figure 7.12. The symmetric triple Laue perfect crystal neutron interferometer (Rauch *et al* 1974). The instrument is constructed from a single silicon monolith, and interference patterns may be observed in both the undeviated (O) and deviated (H) beams. Perfect crystal interferometers operate over a wavelength range $0.1\ \text{Å} < \lambda < 6\ \text{Å}$. Under the best conditions fringe visibilities as high as 85% may be observed in the O-beam.

of Michelson's interferometer, an incident beam is divided in amplitude by Bragg reflection at a perfect silicon crystal S, which acts as a beam splitter producing two coherent beams with a finite relative deviation $2\theta_B$. Following reflection at a mirror crystal M, these are recombined at an analyser crystal A producing in the final state an undeviated beam (O), and a deviated beam (H).

The essential idea of the interferometer is that the exit wavefield is composed of two linear combinations of two wavefields traversing widely separated paths I and II. Thus when a sample of thickness D is inserted into one of these paths, a phase change $(n-1)kD$ is thereby introduced which, under idealized conditions, generates a corresponding intensity modulation in the O and H beams. The modulated intensities are then given by

$$I_O = \frac{1}{2}\{|\Psi_I|^2 + |\Psi_{II}|^2 + 2\,\mathrm{Re}\,(\Psi_I \Psi_{II}^* \, e^{i(n-1)kD}\}$$

$$I_H = \frac{1}{2}\{|\Psi_I|^2 + |\Psi_{II}|^2 - 2\,\mathrm{Re}\,(\Psi_I \Psi_{II}^* \, e^{i(n-1)kD}\} \qquad (7.145)$$

The version of the interferometer shown in figure 7.12 uses symmetric Laue diffraction at each of the three crystals but asymmetric and skew-symmetric versions are also possible (Bonse 1979).

Because of the correlation between wavelength and angle which is inherent in Bragg diffraction, the practical application of the instrument is restricted to a useful wavelength range between 6 Å and 0.1 Å. The long wavelength limit is determined by the maximum value $d = 3.135$ Å in the (1 1 1) planes of silicon and the short-wave limit is governed by the Debye–Waller function which falls below $\simeq 0.1$ at fringe order numbers approaching 10^3. In practice the instrument has been employed successfully at fringe order numbers of about 300 (Rauch 1979, Kaiser *et al* 1983).

Ideally the (LLL) interferometer should be constructed from three very thin perfect crystals, separated by precisely matched spacings, but in a practical system the deviating elements must have finite thickness. As a result the multiple-path system of possible routes through the interferometer needs to be carefully analysed, paying special attention to phase changes which depend on the total area enclosed by the loop. For example, the first interferometer of this type was cut from a perfect monolithic silicon crystal, 80 mm in diameter and 70 mm long, and Bragg reflection was brought about in the (2 2 0) planes of the three plates, each 4.3954 ± 0.0008 mm thick, separated by spacings each of length 27.2936 ± 0.0009 mm. Although polylithic X-ray interferometers have been built successfully, for neutrons it appears necessary to make use of monoliths only. This is because of the much more demanding stability criteria which must be met. These may ultimately be traced to the very small neutron velocity, as compared with the X-ray case, which make the instrument very sensitive to accoustic disturbances. In addition, crystal imperfections are that much more important because large crystals are needed, and because the reflection range is only half that available to X-rays.

Wave propagation in the three crystal system is a highly complex phenomenon requiring a complete re-working of the dynamical diffraction theory for neutrons (Petrascheck 1976, Bauspiess *et al* 1976). At each reflection both deviated and undeviated beams are spatially broadened to a complete Borrmann fan-width (Kikuta 1979) and, because silicon has no absorption cross-section, both Ψ^+ and Ψ^- waves are transmitted without attenuation and each of the separated beams (I) and (II) displays prominent Pendellösung structure. However, it is easy to show by ray tracing that, with respect to the number of deviations each beam experiences, there is a symmetry in the O-beam which is missing from the H-beam. As a result the conditions for high contrast cannot be maintained in the H-beam whose fringe visibility (7.114) barely reach 50%, whereas, in the O-beam it may easily exceed 85%.

7.3.3.2 Applications of neutron interferometry

To date the major interest in the perfect crystal interferometer has centred on its application to the systematic study of the foundations of quantum mechanics (Namiki 1988, Silverman 1988, Uffink and Hilgevoord 1988), and to the observation and interpretation of phase shifts produced in acceleration fields, of which gravity is, perhaps, the most important (Greenberger and Overhauser 1979, Greenberger 1983, Klein and Werner 1983, Werner and Klein 1984, Horne 1986). A selection of the relevant neutron interferometric experiments is assembled in Table 7.2. However, the technique also has valuable applications in nuclear and solid state physics. These applications include precise determinations of scattering lengths and resonance parameters (Bauspiess *et al* 1978), e.g. in ^{3}H (Kaiser *et al* 1979) and in ^{235}U (Kaiser *et*

Table 7.2. A selection of experiments in fundamental physics carried out using the perfect crystal neutron interferometer. The observability of 2π rotations for spinor fields has been the subject of much discussion over a long period (Bernstein 1967, Aharonov and Susskind 1967, Hegerfeldt and Kraus 1968, Moore 1970, Byrne 1978a, Zeilinger 1981, 1986)

Topic	Reference
2π spinor rotation	Rauch *et al* (1975)
	Werner *et al* (1975)
	Klein and Opat (1975, 1976)
Wigner spin superposition	Summhammer *et al* (1982, 1983)
Longitudinal coherence length	Kaiser *et al* (1983)
	Hamilton *et al* (1983)
Fizeau effect for neutrons	Klein *et al* (1981)
	Bonse and Rumpf (1986)
Non-linear quantum mechanics	Shull *et al* (1980)
Quaternions in quantum mechanics	Kaiser *et al* (1984)
Aharanov–Bohm effect	Greenberger *et al* (1981)
Aharanov–Casher effect	Cimmino *et al* (1989), Hagen (1990)
Gravitationally induced quantum interference	Colella *et al* (1975)
	Staudenmann *et al* (1980)
Acceleration induced quantum interference	Bonse and Wroblewski (1983)
Sagnac (Coriolis) effect	Atwood *et al* (1984)

al 1986), and the study of inhomogeneous materials, e.g. magnetic domain structures by neutron phase topography, a technique similar in principle to phase-contrast optical microscopy (cf. section 8.2.1.2).

At the present time a number of new experiments are under consideration which can, in principle, be performed with perfect crystal interferometers of conventional design, i.e. monoliths constructed from single perfect crystals of major dimension $\leqslant 20$ cm. Of these new studies perhaps the most intriguing is the Wheeler delayed choice experiment (Wheeler 1978, Miller and Wheeler 1983, Bernstein 1986), which is a version of Young's two slit interference experiment where the choice between single slit (no interference) or double slit (interference) is made after the particle has passed the screen containing the two slits. Another new experiment which is concerned with acceleration induced phase shifts is the search for the Mashoon effect (Mashoon 1987, 1988). This is a phase shift predicted to occur in a rotating reference frame associated with a coupling of the neutron spin to the axis of rotation. It is closely allied to the Sagnac effect (Sagnac 1913, Michelson *et al* 1925, Post 1967, Atwood *et al* 1984). Current proposals also envisage the construction of long baseline interferometers constructed from separated perfect crystals with stabilization by laser interferometry (Deslattes 1986). Such instruments would permit the performance with neutrons of a number of classical experiments, e.g. the Michelson–Morley experiment and the Cavendish experiment, although it is not entirely clear what scientific objective is being pursued by such studies.

7.4 Neutron polarization

7.4.1 The representation of neutron polarization

7.4.1.1 The Stokes parameters

The spin of the neutron may be defined as its angular momentum measured in its own rest-frame. For non-relativistic neutrons the spin operators $\boldsymbol{s}=\frac{1}{2}\hbar\boldsymbol{\sigma}$ act in a two-dimensional complex spinor space, and the expectation value $\langle\boldsymbol{\sigma}\rangle$ in a pure state is an axial vector called the 'polarization'. For relativistic neutrons described in the Pauli–Dirac four-component representation the 2×2 Pauli spin matrices are replaced by 4×4 matrices $\boldsymbol{\sigma}'$, which act in a four-dimensional complex spinor space, and which are the space–space components of the anti-symmetric tensor, or six-vector, $\sigma^{\mu\nu}=(1/2\mathrm{i})(\gamma^{\mu},\gamma^{\nu})$. In the present context we shall be concerned, not so much with the motion of the spin, although that is important in the Stern–Gerlach apparatus (cf. section 1.5.3.2), and in neutron magnetic resonance (cf. section 7.5), but with the average spin state of a beam of slow neutrons. Thus the term 'polarization' will be used in a slightly different, although related, sense to be explained below. The application of polarized slow neutrons in the study of fundamental interactions has been discussed in detail by Krupchitsky (1987).

For a relativistic free neutron it is possible to construct simultaneous eigenstates of the momentum and the helicity, i.e. the spin-component along the momentum. By convention this direction is taken as the z-axis, and we shall denote such eigenstates by $|\pm\frac{1}{2}\rangle$. The general mixed polarization state of a beam of neutrons may be represented as an incoherent superposition of any two orthogonal pure states characterized by arbitrary complex amplitudes u_1 and u_2

$$
\begin{aligned}
|a\rangle &= u_1|\tfrac{1}{2}\rangle + u_2|-\tfrac{1}{2}\rangle \\
|b\rangle &= u_2^*|\tfrac{1}{2}\rangle - u_1^*|-\tfrac{1}{2}\rangle
\end{aligned}
\tag{7.146}
$$

with probabilities P_a and P_b, respectively. The density operator describing the beam may then be written in the form of a 2×2 Hermitian matrix

$$
\rho=\sum_{i=a}^{b}|i\rangle P_i\langle i|=\tfrac{1}{2}(P_0\sigma_0+\boldsymbol{P}\cdot\boldsymbol{\sigma}) \tag{7.147}
$$

where σ_0 is the unit matrix and σ_j $(j=x, y, z)$ are the Pauli spin matrices,

$$\sigma_0=\begin{pmatrix}1 & 0\\ 0 & 1\end{pmatrix} \quad \sigma_x=\begin{pmatrix}0 & 1\\ 1 & 0\end{pmatrix} \quad \sigma_y=\begin{pmatrix}0 & -\mathrm{i}\\ \mathrm{i} & 0\end{pmatrix} \quad \sigma_z=\begin{pmatrix}1 & 0\\ 0 & -1\end{pmatrix} \tag{7.148}$$

The density matrix ρ is completely determined by the intensity P_0 and the three real numbers P_j $(j=x, y, z)$ which may be termed the Stokes parameters for the beam, because they are the direct analogues of the three real numbers introduced by Stokes (1852) to characterize a partially polarized beam of light. They are defined by the relations

$$\begin{aligned} P_x&=(P_a-P_b)(u_1u_2^*+u_1^*u_2)=P_0\langle\overline{s_x}\rangle(E_n/m_nc^2)/\tfrac{1}{2}\hbar \\ P_y&=(P_a-P_b)\mathrm{i}(u_1u_2^*-u_1^*u_2)=P_0\langle\overline{s_y}\rangle(E_n/m_nc^2)/\tfrac{1}{2}\hbar \\ P_z&=(P_a-P_b)(|u_1|^2-|u_2|^2)=P_0\langle\overline{s_z}\rangle/\tfrac{1}{2}\hbar \end{aligned} \tag{7.149}$$

where $\langle\bar{\boldsymbol{s}}\rangle$ represents an ensemble average of the spin expectation value $\langle\boldsymbol{s}\rangle$ carried out over all the neutrons in the beam.

It is evident from (7.149) that the 'polarization' $\boldsymbol{P}/P_0$ reduces to an axial vector in the non-relativistic limit $(E_n/m_nc^2\rightarrow 1)$ although this is not true for relativistic neutrons where, although P_z/P_0 always measures the mean longitudinal spin component, the mean transverse spin-components $\langle\overline{s_x}\rangle$ and $\langle\overline{s_y}\rangle$ vanish in the extreme relativistic limit, even when the corresponding polarization components P_x/P_0 and P_y/P_0, remain finite.

Since the four real number $(P_0, \boldsymbol{P})$ satisfy the relation

$$P_0^2-\boldsymbol{P}^2=P_0^2(1-f_n^2)\leqslant P_0^2 \tag{7.150}$$

where

$$f_n=\boldsymbol{P}/P_0 \tag{7.151}$$

is the neutron beam polarization vector (cf. section 7.4.5.1), the quantity $(P_0^2-\boldsymbol{P}^2)$ enters in a natural way into any treatment of the interactions of polarized neutron beams. It is therefore a useful notion to visualize $\boldsymbol{P}$ as a vector, the Stokes vector, in a real abstract three-dimensional polarization space, analogous to position space, and $(P_0, \boldsymbol{P})$ as a four vector in a real abstract four-dimensional polarization space of indefinite metric, analogous to four-dimensional Minkowski space-time.

A simple example of the use of the formalism introduced above is provided by the case of a slow neutron propagating in a region of space where there is a magnetic field $\boldsymbol{B}$, oriented along a unit vector $\hat{\boldsymbol{n}}$ specified by polar angles (θ, φ). If $|a\rangle$ and $|b\rangle$ are selected as those spin-states in which the neutron spin is oriented parallel or anti-parallel to $\boldsymbol{B}$, then the complex amplitudes

u_1, u_2 appearing in (7.146) are given by

$$u_1 = \cos\theta/2\, e^{-i\varphi/2} \qquad u_2 = \sin\theta/2\, e^{i\varphi/2} \tag{7.152}$$

This result indicates why these states may be described as coherent (cf. section 6.2.3) since they provide the closest quantum equivalent to the case of a classical spin pointing in the direction (θ, φ). If we now assume that the mixed state of the beam is an equilibrium state corresponding to a temperature T, then the probabilities P_a and P_b are proportional to the corresponding Boltzmann factors $\exp(\pm\mu_n B/kT)$ and the Stokes vector $\boldsymbol{P}$ becomes

$$\boldsymbol{P} = P_0 \tanh(\mu_n B/kT)\hat{\boldsymbol{n}} \tag{7.153}$$

We may then deduce from (7.153) that, even at liquid helium temperatures $T \approx 4$ K, in magnetic fields $\boldsymbol{B}$ as high as 5×10^3 G, the spontaneous neutron polarization has a value $P = \tanh(\mu_n B/kT)$ no larger than $\simeq 0.1\%$.

7.4.1.2 Transfer matrices

We now represent the pure polarization state $|a\rangle$ defined in equation (7.146) in 'spinor' form

$$\mathbf{u} = u_1\begin{pmatrix}1\\0\end{pmatrix} + u_2\begin{pmatrix}0\\1\end{pmatrix} \tag{7.154}$$

where we have chosen the basis states

$$\begin{pmatrix}1\\0\end{pmatrix} \text{ and } \begin{pmatrix}0\\1\end{pmatrix}$$

to be states of definite spin component with respect to the momentum. One must emphasize again that u is a spinor not in space-time but in the abstract polarization space introduced in (7.146) and $\mathbf{u}$ is in general not a state of definite spin-component in any direction. This is true only in the non-relativistic limit which for present purposes is the most important situation. Furthermore in this limit it is not even necessary to restrict the axis of quantization to be parallel to the momentum although this is very convenient in many cases because then u_1 and u_2 may be identified with the large components of the Dirac spinor (cf. section 9.2.1). Provided we restrict the discussion to processes involving coherent scattering or pure absorption only, the transformation of the pure polarization state of a neutron from $\mathbf{u}$ to $\mathbf{u}'$ may be described by the equation

$$\mathbf{u}' = M\mathbf{u} \tag{7.155}$$

where M is a 2×2 transfer matrix.

Since matrices generated from a given matrix by similarity transformations are equivalent, one is led to consider the invariants of such transformations, namely the trace T, the determinant Δ, and the matrix rank. In general 2×2 matrices can have rank 0, 1 or 2. Rank zero matrices are null matrices while rank 1 matrices have vanishing determinant and have the nature of projection operators. Such matrices are called polarizers since the output beam is fully polarized when either or both of the eigenvalues of M vanish, the sense of the polarization being determined by the non-vanishing eigenvalue.

An example of a neutron polarizer is given by the Stern–Gerlach apparatus described in section 1.5.3. If in the general case the magnetic axis of the system is aligned along the unit vector $\hat{\boldsymbol{n}}(\theta, \varphi)$ introduced above, then the action of the instrument acting on any pure polarization state $\boldsymbol{u}$ may be represented by the projection operator

$$M_{\text{p}} = \tfrac{1}{2}(1 + \boldsymbol{\sigma}\cdot\boldsymbol{n}) = \frac{1}{2}\begin{pmatrix} 1+\cos\theta & \sin\theta\, \mathrm{e}^{-\mathrm{i}\varphi} \\ \sin\theta\, \mathrm{e}^{\mathrm{i}\varphi} & 1-\cos\theta \end{pmatrix} \tag{7.156}$$

Since M_{p} satisfies the idempotency condition

$$M_{\text{p}}^2 = M_{\text{p}} \tag{7.157}$$

it has eigenvalues 1 and 0 corresponding to the pure states $|a\rangle$ and $|b\rangle$ defined by (7.146) and (7.152). It therefore projects out from an arbitrary state $\boldsymbol{u}$ that component in which the spin is oriented parallel to $\hat{\boldsymbol{n}}$, and suppresses the anti-parallel component.

In practice the construction of devices which achieve 100% polarization is very difficult to achieve (cf. section 7.4.3) and to discuss most polarization-sensitive processes it is necessary to extend the analysis of the transfer operators to matrices of rank 2, i.e. those for which Δ does not vanish. Since the unit matrix σ_0 falls in this category, and since each such matrix possesses an inverse, they constitute a group with respect to multiplication. Further, since $\Delta \neq 0$ we can divide each element of M by $\Delta^{1/2}$, thus reducing the number of real parameters required to specify the matrix from eight to six (Δ is complex in general). The resultant unimodular matrices can be expressed as elements of the Pauli ring

$$M_2 = \sum_{i=0}^{2} k_i\sigma_i = \begin{pmatrix} k_0 + k_z & k_x - ik_y \\ k_x + ik_y & k_0 - k_z \end{pmatrix} \tag{7.158}$$

where the unimodular condition is expressed as a relation between the complex coefficients

$$k_0^2 - (k_y^2 + k_y^2 + k_z^2) = 1 \tag{7.159}$$

The advantage of this formulation of the problem is that the unimodular group C_2 is homomorphic to the group L_{p} of proper homogeneous Lorentz transformations in space-time. The properties of this group are of course well known, and afford us a geometrical picture of the operations in the abstract polarization space of the various transfer matrices M_2 which we encounter. For example every Lorentz transformation is characterized by six real parameters, i.e. one parameter each to specify the degree of 'boost' and the degree of rotation and two parameters each to specify the directions of these operations. For the polarization transfer matrices these two 'directions' coincide in most cases of practical interest and only four real parameters are needed, i.e. two to specify a direction together with the real and imaginary parts of the scattering amplitude. Such matrices are said to be normal and are unitarily similar to diagonal matrices. If the scattering amplitude is real, three real parameters suffice and the transfer matrix is itself unitary.

To find how the four components of the Stokes four-vector $(P_0, \boldsymbol{P})$ transform under the matrix operation (7.158) we form the direct product $M_2 \times M_2^*$ and then subject it to a similarity transformation generated by the unitary transformation

$$U=\frac{1}{2}\begin{pmatrix} 1 & 0 & 0 & 1 \\ 0 & 1 & 1 & 0 \\ 0 & \mathrm{i} & -\mathrm{i} & 0 \\ 1 & 0 & 0 & -1 \end{pmatrix} \tag{7.160}$$

We then find that the matrix

$$M_4=U(M_2 \times M_2^*)U^{-1} \tag{7.161}$$

is the required transfer matrix. These matrices, which were first introduced into optics by Mueller (1943), constitute a 4×4 representation of L_p (a 'tensor' representation as opposed to the true 'spinor' representation M_2); they have an algebraic form which is identical with those matrices describing pure rotations and pure Lorentz transformations of four-vectors in space-time (Byrne 1971, 1979).

As an illustration of the above ideas we can write down the elements of the transfer matrix describing a rotation of the spinor $\boldsymbol{u}$ through an angle Ψ about a unit vector $\hat{\boldsymbol{n}}$ whose direction cosines are n_x, n_y, n_z $(n_x^2+n_y^2+n_z^2=1)$. This matrix is given by

$$\begin{aligned} M_2(\Psi) &= \cos\frac{\Psi}{2}(\sigma_0)-\mathrm{i}\sin\frac{\Psi}{2}(\boldsymbol{\sigma}\cdot\hat{\boldsymbol{n}}) \\ &= \begin{pmatrix} \cos\frac{\Psi}{2}-in_z\sin\frac{\Psi}{2} & -\mathrm{i}(n_x-in_y)\sin\frac{\Psi}{2} \\ -\mathrm{i}(n_x+in_y)\sin\frac{\Psi}{2} & \cos\frac{\Psi}{2}+in_z\sin\frac{\Psi}{2} \end{pmatrix} \end{aligned} \tag{7.162}$$

In an elementary spin-sensitive scattering process the angle Ψ is related to the real part of the scattering amplitude. If there is an absorptive part Ψ

is complex and $M_2(\Psi)$ is no longer unitary but still normal. The additional parameter describes a pure Lorentz transformation or 'boost' in the polarization space. An example of such a scattering process is described in section 7.4.4.

7.4.2 Spin rotators and spin flippers

As discussed in section 7.4.3, neutron polarizers of whatever description operate by transmitting only one sense of the spin component measured with respect to an applied magnetic field $\boldsymbol{B}_0$. However in many experiments with polarized neutrons it is required to observe Larmor precession of the spin about the applied field, and in many others the information is extracted from a counting rate asymmetry observed when the neutron polarization is reversed in direction. It is therefore necessary to construct instruments which rotate the neutron spin through angles of $\pi/2$ ($\pi/2$ coil), or π (π coil or spin-flipper). To describe spin-rotators of this nature by the transfer matrix $M_2(\Psi)$ given in (7.162), we suppose that the neutron travels with velocity v along the y-axis, normal to a magnetic field $\boldsymbol{B}_0$ pointing along the z-axis. In this case a $\pi/2$ coil requires $\Psi=\pi/2, n_z=0$, while a π coil requires $\Psi=\pi, n_z=0$.

The most widely used device for rotating the neutron spin is the radiofrequency (RF) resonance coil whose mode of operation is analysed in detail in section 7.5.2.1 (Weinfurter *et al* 1988, 1989). The coil is aligned with its axis along the beam direction and, at resonance, rotates the component P_z of the polarization through an angle directly proportional to the length of time the neutron spends inside it. This type of device was applied as a spin-flipper in the first ever measurement of the neutron magnetic moment (Alvarez and Bloch 1940). The layout is shown in figure 7.16 (cf. section 7.5.1). Subsequently it was used to study the polarizing and analysing properties of polycrystalline iron (Stanford *et al* 1954) and, in various combinations, for experiments on neutron scattering by spin-waves (Drabkin 1963, Drabkin *et al* 1965). Two $\pi/2$ coils separated by a quasi-uniform magnetic field form an integral part of the Ramsey magnetic resonance technique (cf. figure 7.17, section 7.5.3).

The RF coil is of course a dynamic system and a much simpler device is the static-field Mezei coil shown in figure 7.13(a) (Mezei 1972). In the Mezei spin rotator the neutron crosses the windings of a coil of rectangular cross-section aligned in the x-direction. Inside the coil, the field $\boldsymbol{B}_0+\boldsymbol{B}_1$ lies in the xy plane, at some selected angle θ with respect to the z-axis. The neutron spin cannot follow the non-adiabatic change in the field, but retains its initial orientation, and subsequently precesses around the new field $\boldsymbol{B}_0+\boldsymbol{B}_1$. After a time $t=d/v$ it exits from the coil, without changing its orientation, returning once again to its precessional motion about $\boldsymbol{B}_0$. Since the spin precesses about the unit vector $\hat{\boldsymbol{n}}=(\sin\theta, 0, \cos\theta)$ while inside the coil, then, if we choose $\boldsymbol{B}_1$, d and v so that the passage time is equal to half

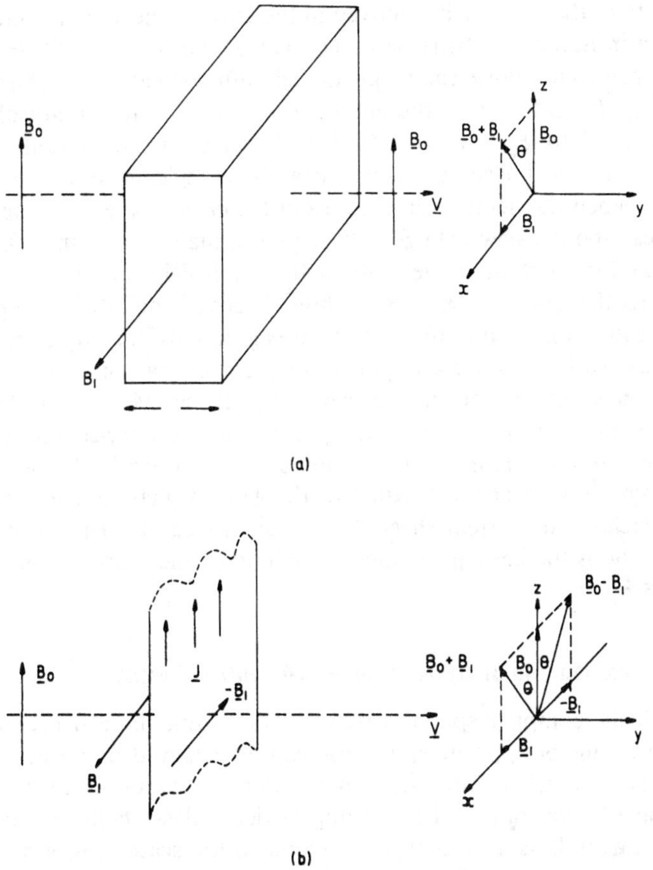

Figure 7.13. (a) The Mezei coil neutron spin rotator. The magnetic field $\boldsymbol{B}_0+\boldsymbol{B}_1$ inside the coil lies in the x–z plane and is set at an angle θ with respect to the field $\boldsymbol{B}_0$ outside the coil. (b) The current sheet neutron spin rotator. The magnetic fields $\boldsymbol{B}_0+\boldsymbol{B}_1$ and $\boldsymbol{B}_0-\boldsymbol{B}_1$ on either side of the current sheet are equal in magnitude and lie in the x–z plane at an angle 2θ with respect to each other.

of one Larmor period, the polarization component P_y will be reversed on emergence, while P_z and P_x will have rotated through angles 2θ and $2\theta-\pi$, respectively. This conclusion may be checked by setting $\Psi=\pi$, and $\hat{\boldsymbol{n}}=(\sin\theta, 0, \cos\theta)$ in (7.162), and calculating the components of $\boldsymbol{P}$ from (7.149). Choosing $\theta=\pi/4$ the system acts as a $\pi/2$ coil, while, with $\theta=\pi/2$, it acts as a spin-flipper.

Both the RF coil and the Mezei coil are very suitable for use with quasi-monochromatic neutrons, since the net rotation produced in both cases depends on the total phase angle accumulated by the neutron spin as it follows the magnetic field adiabatically along a fixed length of path. This sensitivity to the velocity is removed in the current sheet spin-rotator which is shown in figure 7.13(b) (Majorana 1932, Dabbs *et al* 1955). In this case the neutron spin follows the magnetic field adiabatically as it changes from $\boldsymbol{B}_0$ to $\boldsymbol{B}_0+\boldsymbol{B}_1$ just outside the current sheet so that an initial polarization P_z is reduced to $P_z \cos\theta$. In addition the precessional motion about the slowly varying field accumulates an extra phase angle $\delta(\omega_0 t)$ due to the extra field $\boldsymbol{B}_1$. When the neutron emerges from the current sheets the spin begins to process about a new field $\boldsymbol{B}_0-\boldsymbol{B}_1$, equal in magnitude (since $\boldsymbol{B}_0\cdot\boldsymbol{B}_1=0$), but rotated through an angle π about the z-axis. The spin component with respect to the direction $\boldsymbol{B}_0-\boldsymbol{B}_1$ is now $P_z\cos 2\theta$, and this component is adiabatically conserved as the field changes gradually back to its initial value $\boldsymbol{B}_0$. Hence, so far as the z-component of the polarization is concerned, the current sheet acts exactly in the same way as the Mezei coil. There is a difference, however, in respect of the polarization components normal to $\boldsymbol{B}_0$ for these lose an amount of phase angle $-\delta(\omega_0 t)$, while the field reduces away from the current sheet, equal to the gain in phase angle accumulated in approaching the current sheet. This result is true independent of neutron velocity, and is the basic phenomenon underlying the spin-echo principle (cf. section 8.5.3.2).

7.4.3 Production and analysis of polarized neutron beams

Although the complete spatial separation of an unpolarized neutron beam into two beams of opposite spin-sense can be achieved in a Stern–Gerlach apparatus, insufficient intensity can be attained by these means and the operation of most practical polarizing devices relies on the existence of a magnetic, as well as a nuclear, component in the scattering of neutrons in ferromagnetic material (Halpern and Johnson 1939). One then looks for conditions in which the nuclear and magnetic scattering amplitudes cancel for one sense of the spin and in some cases, e.g. Bragg reflection in single crystals of Fe_3O_4 (Shull 1951), polarizations approaching 100% can be achieved with monochromatic neutrons. Suppose B is the magnetic field in a magnetized ferromagnetic material such as iron, then the Fermi pseudo-potential V_F will be changed to $V_F-\boldsymbol{\mu}_n\cdot\boldsymbol{B}$; the effect is to replace the nuclear scattering length a_n by $(a_n \pm a_m)$, where the alternate signs correspond to the two direction of the spin relative to the local magnetic field and the magnetic scattering length a_m is given by

$$a_m=(m_n\mu_n/2\pi\hbar^3)\int B\,d^3r \qquad (7.163)$$

where the integral is taken over the volume of lattice occupied by the ferromagnetic atom.

It follows from (7.163) that the total cross-section $4\pi|a_n \pm a_m|^2$ differs for the two senses of the spin so that one spin state can be selectively filtered out by transmission through a slab of material magnetized to saturation (Bloch *et al* 1943). This technique suffers from the disadvantage, that, for path lengths exceeding about two mean free paths, the gain in polarizing or analysing power is offset by the loss of intensity. The conventional approach is to exploit the fact that in a ferromagnetic medium the refractive index is two-valued, i.e.

$$n^{\pm} \simeq 1 - \frac{\lambda^2}{2\pi} N(a_n \mp a_m) \tag{7.164}$$

Then, in a material for which there is near-cancellation of nuclear and magnetic scattering lengths for one spin state, the other spin state can be selected by total internal reflection at angles greater than the critical angle (Hughes 1953). Such a system is realised near grazing incidence at the surface of a totally reflecting Fe–Co magnetic mirror made from an alloy of iron and cobalt in the approximate ratio of one to two. For beams with a broad-band thermal spectrum such mirrors can produce 70–80% polarization at 1500 G and close to 90% at 2500 G (Mezei 1978).

The main problem with totally reflecting magnetic mirrors is that to reach high intensity with minimal divergence they need to be up to 5 m long. This situation can be remedied to a very large extent by using stacks of double-sided curved mirrors in a 'Soller' configuration (Abrahams *et al* 1966) and by this means 90% polarization can be achieved in a 25 cm polarizer, magnetised to 1500 G, with a total transmission efficiency of 15%. Further improvement of the output polarization to levels of 95% or better can be brought about by using reflecting layers evaporated onto thin glass (Drabkin 1976) or stretched plastic sheet (Hayter *et al* 1978). At longer wavelengths, polarizing 'supermirrors' based on the principle of the interference filter (Schoenborn *et al* 1974, Mezei 1976, 1978, Schäerpf 1989) have produced polarizations of the order of 97–98% for neutrons in the range 6–7 Å (Mezei and Dagleish 1977). In the ultra-cold range ($\lambda \approx 550$ Å) polarizing Fe–Co films deposited on aluminium or vanadium substrates have been successfully used at normal incidence where one spin state is totally reflected at the surface and the transmitted spin state is accepted (cf. section 6.3.1).

7.4.4 Schwinger scattering

The methods employed to produce polarized neutrons which are discussed in section 7.4.3 above rely for their effectiveness on the interaction $-\boldsymbol{\mu}_n \cdot \boldsymbol{B}$ between the neutron magnetic moment $\boldsymbol{\mu}_n$ and a locally generated magnetic field $\boldsymbol{B}$, and apply for slow neutrons only. Fast neutron polarization may be achieved either through the nuclear spin-orbit interaction, e.g. in the scattering of fast neutrons at the $p_{3/2}$ resonance in helium (cf. section 3.3.2.3), or through the spin-orbit interaction experienced by a fast neutron in the Coulomb field of a heavy nucleus. This latter process is called Schwinger scattering (Schwinger 1946, 1948, Lepore 1950), and it comes about because a neutron moving in an electric field sees a magnetic field in its own rest-frame. The corresponding induced magnetic interaction produces a degree of partial polarization in an initially unpolarized beam, and a left–right scattering asymmetry in an incident polarized beam which is proportional to its component of polarization normal to the plane of scattering. The process is entirely analogous to the Mott scattering of electrons by heavy nuclei (cf. section 3.2.5) which is the conventional analyser for electron polarization for energies up to 1 MeV.

We take as point of departure Schwinger's result that, when neutrons of momentum $\boldsymbol{k}_0$ are elastically scattered by spinless nuclei of atomic number Z into the momentum state $\boldsymbol{k}$, the scattering amplitude is given by

$$f(\theta)=f_0(\theta)+\mathrm{i}\gamma\cot(\theta/2)\boldsymbol{\sigma}\cdot\hat{\boldsymbol{n}} \tag{7.165}$$

where

$$\gamma=\tfrac{1}{2}\alpha Z\mu_n\lambda_n \tag{7.166}$$

α is the fine structure constant, and $\hat{\boldsymbol{n}}$ is a unit vector defined by

$$\boldsymbol{k}\times\boldsymbol{k}_0=\hat{\boldsymbol{n}}k^2\sin\theta \tag{7.167}$$

The function $f_0(\theta)$ is a spin-independent nuclear scattering amplitude. Expressed in the notation of section 7.4.2 the result (7.165) implies that the large components of the scattered and incident Dirac spinors are related by

$$\begin{pmatrix}u_1'\\u_2'\end{pmatrix}=\begin{pmatrix}f_0(\theta) & \gamma\cot(\theta/2)\\-\gamma\cot(\theta/2) & f_0(\theta)\end{pmatrix}\begin{pmatrix}u_1\\u_2\end{pmatrix} \tag{7.168}$$

where the y-axis is taken normal to the plane of scattering as shown in figure 7.14. The trace and determinant of the transfer matrix $M_2(\theta)$ defined in (7.168) are now given by

$$\mathrm{T}=2f_0(\theta)\neq 0 \qquad 0\leqslant\theta\leqslant\pi$$

$$\Delta=f_0^2(\theta)+\gamma^2\cot^2(\theta/2) \tag{7.169}$$

where

$$|\Delta|^2=(|f_0|^2+\gamma^2\cot^2(\theta/2))-4\gamma^2\cot^2(\theta/2)(\mathrm{Im}\,f_0(\theta))^2 \tag{7.170a}$$

$$=(|f_0|^2-\gamma^2\cot^2(\theta/2))^2+4\gamma^2\cot^2(\theta/2)(\mathrm{Re}\,f_0(\theta))^2 \tag{7.170b}$$

It follows from (7.170b) that $|\Delta|\neq 0$ so that the scattering process does not act as a perfect polarizer. However, according to (7.170a) we may expect a substantial degree of partial polarization when

$$\tan\frac{\theta}{2}\simeq\frac{2\gamma\,\mathrm{Im}\,f_0(\theta)}{|f_0(\theta)|^2}=\frac{2\gamma k\sigma_t}{4\gamma|f_0(0)|^2} \tag{7.171}$$

where σ_t is the total cross-section and we have made use of the optical theorem (cf. section 2.4.2).

To determine the Mueller matrix corresponding to (71.68) we only require expressions for the k-parameters defined in (7.158)–(7.159). These are

$$k_0 = f_0/\Delta^{1/2} \qquad k_y = \mathrm{i}\gamma \cot(\theta/2)/\Delta^{1/2} \qquad k_x = k_z = 0 \tag{7.172}$$

The transformation of the Stokes four-vector $(P_0, \boldsymbol{P})$, which corresponds to the spinor transformation (7.168), is then given according to (7.161) by the matrix equation

$$\begin{pmatrix} P_0' \\ P_x' \\ P_y' \\ P_z' \end{pmatrix} = |\Delta| \begin{pmatrix} \cosh\varepsilon & 0 & \sinh\varepsilon & 0 \\ 0 & \cos\eta & -\sin\eta & 0 \\ \sinh\varepsilon & 0 & \cosh\varepsilon & 0 \\ 0 & \sin\eta & \cos\eta & 0 \end{pmatrix} \begin{pmatrix} P_0 \\ P_x \\ P_y \\ P_z \end{pmatrix} \tag{7.173}$$

where the matrix elements of $M_4(\theta)$ are given by

$$\begin{aligned} |\Delta|\cosh\varepsilon &\equiv |\Delta|(|k_0|^2 + |k_y|^2) = (|f_0(\theta)|^2 + \gamma^2 \cot^2(\theta/2)) \\ |\Delta|\sinh\varepsilon &\equiv |\Delta|(2\,\mathrm{Im}\,(\mathrm{i}k_0 k_y^*)) = (2\gamma \cot(\theta/2)\,\mathrm{Im}\,(f_0(\theta)) \\ |\Delta|\cos\eta &\equiv |\Delta|(|k_0|^2 - |k_y|^2) = (|f_0(\theta)|^2 - \gamma^2 \cot^2(\theta/2)) \\ |\Delta|\sin\eta &\equiv |\Delta|(2\,\mathrm{Re}\,(\mathrm{i}k_0 k_y^*)) = (2\gamma \cot(\theta/2)\,\mathrm{Re}\,(f_0(\theta)) \end{aligned} \tag{7.174}$$

The reduction of the matrix $M_4(\theta)$ which appears on (7.173) into the direct sum of two 2×2 matrices which we may denote by $N_2(\varepsilon)$ and $N_2(\eta)$ is an expression of the fact that Schwinger scattering is sensitive only to the polarization component normal to the plane of scattering. Thus, in the system

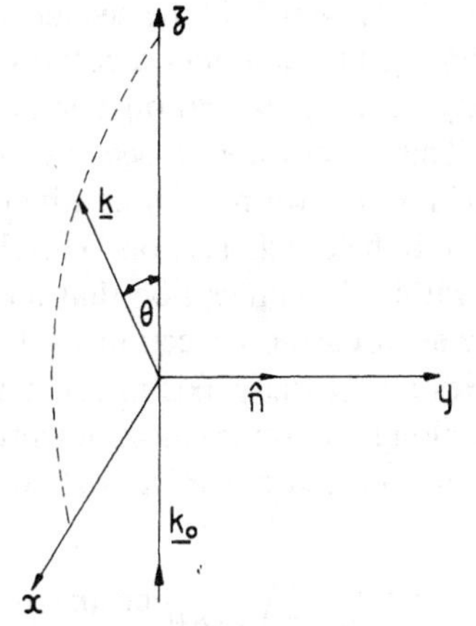

Figure 7.14. Geometrical construction appropriate to the description of Schwinger scattering of neutrons by a spin-orbit potential. The incident and scattered wavevectors are denoted by $\boldsymbol{k}_0$ and $\boldsymbol{k}$, respectively.

of coordinates used in figure 7.14, where the scattering occurs in the $x-z$ plane, k_y is the only non-vanishing parameter k_i $(i=xyz)$. The matrix $M_2(\theta)$ is therefore unitarily similar to a diagonal matrix and the direct product in (7.161) therefore reduces to a direct sum as noted.

The matrix $N(\varepsilon)$ has the characteristic form of a pure Lorentz transformation describing a 'boost' along the y-axis and, in the present context characterizes both the polarizing and analysing properties of the process. The polarizing efficiency, i.e. the polarization f' produced in an initially unpolarized beam, is then given by

$$f'=\frac{P'_y}{P_0}=\tanh\varepsilon=\frac{2\gamma\cot\theta/2\ \mathrm{Im}\ (f_0(\theta))}{|f_0(\theta)|^2+\gamma^2\cot^2\theta/2} \tag{7.175}$$

It follows immediately that, ignoring the term $\gamma^2\cot^2\theta/2$ in the denominator of (7.175), f' approaches unity at angles θ which satisfy (7.171). For the scattering of 1 MeV neutrons in lead ($Z=82$, $\gamma\simeq 0.12$ fm) the maximum polarization occurs at scattering angles of a few degrees and for this reason Schwinger scattering is sometimes referred to as 'small angle scattering'.

Adopting the system of axes shown in figure 7.14 the scattered neutrons are detected at the azimuthal angle $\varphi=0$ and, if instead they are detected at $\varphi=\pi$ for the same angle of scatter θ, the effect is the same as reversing the polarization components $P_x\to -P_x$, $P_y\to -P_y$. Thus any initial polarization P_y in the incident beam may be detected through a determination of the left–right counting rate asymmetry

$$A=\frac{P'_0(0)-P'_0(\pi)}{P'_0(0)+P'_0(\pi)}=P_x\tanh\varepsilon/P_0(0) \tag{7.176}$$

The unitary matrix $N_2(\eta)$, which is the second matrix in the reduction of $M_4(\theta)$, does not act on P_y, but describes a rotation about the normal to the scattering plane of any initial polarization components P_x, P_z which lie in the scattering plane. This comes about because the neutron spin precesses about the magnetic field in its own rest frame which is induced by the motion in the nuclear Coulomb field (cf. section 3.3.2.3). However, in order to maintain consistency within the convention that the basis states be eigenstates of the neutron helicity as discussed in section 7.4.1, it is necessary in addition to rotate the space axes about the y-axis in order to bring the z-axis into its new position aligned along the outgoing momentum $\boldsymbol{k}$. In this new system of axes P'_0 and P'_y are unchanged but P'_x and P'_y are transformed into P''_x and P''_z where

$$\begin{pmatrix}P''_x\\ P''_z\end{pmatrix}=\begin{pmatrix}\cos\theta & \sin\theta\\ -\sin\theta & \cos\theta\end{pmatrix}\begin{pmatrix}P'_x\\ P'_z\end{pmatrix}=|\Delta|\begin{pmatrix}\cos(\eta-\theta) & -\sin(\eta-\theta)\\ \sin(\eta-\theta) & \cos(\eta-\theta)\end{pmatrix}\begin{pmatrix}P_x\\ P_z\end{pmatrix} \tag{7.177}$$

The form of the spin-orbit scattering amplitude (7.165) has been verified experimentally by Shull (1963), where Z is replaced by $Z(1-f_e(q))$ in (7.166) and $f_e(q)$ is the atomic form factor corresponding to a momentum transfer $q=2k\sin\theta/2$. However, Schwinger scattering is not a very useful source of polarized fast neutrons because the scattering angle has to be fixed very precisely necessitating a high degree of collimation. The process is of interest principally as a source of spurious effects in experiments where precise determination of the degree and orientation of the neutron polarization is important (cf. sections 9.2.5.3 and 9.2.5.4).

Most practical sources of fast polarized neutrons rely on the spin-orbit force as manifested in nuclear reactions (cf. section 3.3.2), and a list of useful sources is provided in Table 7.3. The analysis of fast neutron polarization is usually carried out using resonance scattering in ^4He or ^{12}C as noted above.

Table 7.3. Sources of polarized fast neutrons (Walter 1971). E_i represents the energy of the bombarding particle and E_n the energy of the ejected neutron in the laboratory system. θ_L is the angle of emission in the laboratory at which the polarization is maximum

Nuclear reaction	E_i (MeV)	E_n (MeV)	θ_L (°)	Polarization (%)
$^7Li(p, n)^7Be$	2.9	1.0	50	31 ± 0.4
$^3H(p, n)^3He$	3.1	2.0	33	23 ± 1
$^3H(p, n)^3He$	7.8	5.9	40	19.8 ± 0.7
$^{12}C(d, n)^{13}N$	2.8	2.4	25	46.9 ± 0.6
$^2H(d, n)^3He$	8.4	10.0	32	23 ± 1
$^3H(d, n)^4He$	6.0	16.4	90	46 ± 2
$^3H(d, n)^4He$	7.0	17.7	82	41 ± 2
$^2H(t, n)^4He$	11.5	11.0	133	61 ± 1
$^9Be(\alpha, n)^{12}C$	2.6	3.4	45	53.5 ± 0.3
$^9Be(\alpha, n)^{12}C$	2.6	7.8	45	53 ± 0.3

7.4.5 Interaction of polarized neutrons with polarized nuclei

7.4.5.1 Nuclear orientation

An assembly of nuclei whose spins are not randomly distributed over all directions is said to be oriented, and the statistical properties of the system may be described by a Hermitian density matrix (cf. sections 6.4.3–6.4.4)

$$\rho_N=\sum_i |\Psi_i\rangle P_i\langle\Psi_i| \tag{7.178}$$

where P_i represents the probability that a given nucleus is found in the pure state $|\Psi_i\rangle$. For a stationary system ρ_N commutes with the Hamiltonian and the states $|\Psi_i\rangle$ are energy eigenstates. Also it is permissible to assume that the temperature is so low that only the magnetic substates of the nuclear ground state of spin I are populated. In general the electromagnetic field producing the nuclear orientation need not have an axis of symmetry and ρ_N is in general not diagonal in the basis of angular momentum eigenstates $|I, M\rangle$.

We may however expand ρ_N as a series of tensor operators $T_{LM}(\boldsymbol{I})$ each of which transforms according to an irreducible representation of the rotation group, i.e.

$$\rho_N = \sum_{L=0}^{2I} \sum_{M=-L}^{L} \alpha_{LM} T_{LM}(\boldsymbol{I}) \tag{7.179}$$

where the coefficients α_{LM} have simple interpretations which we derive below. The tensor operators $T_{LM}(\boldsymbol{I})$ are also known as polarized harmonics, since they may be constructed by applying the scalar operator $(\boldsymbol{I}\cdot\nabla)^L$ to the solid harmonics $r^L Y_{LM}(\theta, \varphi)$, a process which is equivalent to replacing $\boldsymbol{r}$ by $\boldsymbol{I}$ making due allowance for the non-commuting property of the components of $\boldsymbol{I}$ (Falkoff and Uhlenbeck 1950). They are Hermitian in the sense that the adjoint operators are defined by the rule (Brink and Satchler 1968)

$$T_{LM}(\boldsymbol{I}) = (-1)^{-M} T^{\dagger}_{L-M}(\boldsymbol{I}) \tag{7.180}$$

This does not mean that the spherical components of $T_{LM}(\boldsymbol{I})$ are individually Hermitian operators, but that there exist linear combinations of operators of the same rank L, e.g. the Cartesian components, which have this property. The polarized harmonics are normalized according to the rule

$$\mathrm{Tr}(T^{\dagger}_{LM} T_{L'M'}) = \delta_{LL'}\delta_{MM'} \tag{7.181}$$

Evaluating the expectation value $\langle T_{kq}(\boldsymbol{I})\rangle$ using (7.180) and (7.181) we find that

$$\rho_N = \alpha_{00} \sum_{L=0}^{2I} \sum_{M=-L}^{L} \langle T_{LM}(I)\rangle^* T_{LM}(\boldsymbol{I}) \tag{7.182}$$

where

$$\alpha_{00} = Tr(\rho_N) = \sum_i P_i = 1 \tag{7.183}$$

A simple formula for the statistical tensors $\langle T_{kq}\rangle$ expressed in terms of the matrix elements of ρ_N may be derived by application of the Wigner–Eckhart theorem and making use of the symmetry properties of the

Clebsch–Gordon coefficients (Fano 1951). Thus we find that

$$\langle T_{kq}\rangle = \sum_{mm'} \langle m|\rho_N|m'\rangle\langle m'|T_{kq}|m\rangle$$

$$= \sum_{mm'} \langle m|\rho_N|m'\rangle\langle Im, kq|Im'\rangle\langle I\|T_k\|I\rangle$$

$$= \sum_{mm'} \langle m|\rho_N|m'\rangle(-1)^{I-m}[(2I+I)/(2k+1)]^{1/2} \times\langle Im', I-m|kq\rangle\langle I\|T_k\|I\rangle \tag{7.184}$$

where $\langle I\|T_k\|I\rangle$ represents the reduced matrix element of $T_{kq}(\mathbf{I})$. This factor may be eliminated by replacing ρ_N by $T^{\dagger}_{kq}$ in (7.184) and using the normalization condition (7.181) in combination with the orthogonality property of the Clebsch–Gordon coefficients. This leads to the results

$$[(2I+1)/(2k+1)]|\langle I\|T_k\|I\rangle|^2 = 1 \tag{7.185}$$

and

$$\langle T_{kq}\rangle = \sum_{mm'} (-1)^{I-m}\langle m|\rho|m'\rangle\langle Im', I-m|kq\rangle \tag{7.186}$$

The simplest situation occurs when an axis of symmetry exists which may be chosen as axis of quantization, for in this case ρ_N is diagonal in the angular momentum basis and only $\langle T_{k0}\rangle$ has the non-zero value

$$\langle T_{k0}\rangle = \sum_m P_m(-1)^{I-m}\langle Im, I-m|k0\rangle \tag{7.187}$$

In these circumstances it is convenient to introduce the orientation parameters $f_k(I)$ (Tolhoek and Cox 1953), where

$$f_k(I) = \binom{2k}{k}^{-1} I^{-k}\left[\frac{(2I+k+1)!}{(2k+1)(2I-k)!}\right]^{1/2}\langle T_{k0}\rangle \tag{7.188}$$

Since the coefficient $\langle Im, I-m|k0\rangle$ gains a factor $(-1)^k$ when the quantum numbers m and $-m$ are interchanged, it follows that $f_k(I)$ vanishes identically for odd k whenever $P_{-m} = P_m$ in (7.187). In this case, although the z-axis is an axis of symmetry, the nuclear spins exhibit no preference with respect to the positive or negative z-axis, and the system is said to be aligned along the z-axis. Thus the first significant non-vanishing orientation parameter is $f_2(I)$, where

$$f_2(I) = g_N = I^{-2}\left(\sum_{m=-I}^{I} m^2P_m - \frac{1}{3}I(I+1)\right) \tag{7.189}$$

measures the degree of alignment.

Since $T_{q2}(\mathbf{I})$ is a quadrupole operator it cannot be detected by neutrons which, having spin $\frac{1}{2}$, have no quadrupole moment. To detect nuclear orientation by neutron scattering it is necessary that $P_m \neq P_{-m}$, in which case the nuclei are said to be polarized. The relevant orientation parameter is then

$$f_1(I) = f_N = I^{-1} \sum_{m=-I}^{I} mP_m \tag{7.190}$$

where f_N is the nuclear polarization and

$$\langle I_z \rangle = f_N I \tag{7.191}$$

In most cases polarization of the nuclear assembly is achieved by placing the nuclei in a strong magnetic field $\boldsymbol{B}$, in a low temperature environment, in which case f_N is given by

$$f_N = \frac{2I+1}{2I} \coth\left[\frac{2I+1}{2}\frac{\gamma_N B}{kT}\right] - \frac{1}{2I}\coth\left[\frac{\gamma_N B}{2kT}\right] \tag{7.192}$$

where γ_N is the nuclear gyromagnetic ratio.

7.4.5.2 Scattering of polarized neutrons by polarized nuclei

We may suppose that the nuclei have a non-vanishing mean spin component $\langle I_z \rangle$ given by (7.191) and that the neutron–nucleon interaction is represented by a Fermi potential (cf. section 6.3.1.2)

$$V_F = \frac{2\pi\hbar^2}{m_n} v_{nN} \tag{7.193}$$

where

$$v_{nN} = b\sigma_0 I_0 + \tfrac{1}{2} c\boldsymbol{\sigma}\cdot\boldsymbol{I} \tag{7.194}$$

is the sum of spin-independent and spin-dependent terms characterized by bound scattering amplitudes b and c defined in (6.81), each composed from the product of a 2×2 matrix operator acting in the neutron spin-space, and a $(2I+1)\times(2I+1)$ matrix operator acting in the nuclear spin space. σ_0 and I_0 represent the unit operators acting in the respective spin-spaces. The polarization state of the (non-relativistic) neutron beam defined with respect to an arbitrary direction is then described by the density matrix (cf. 7.147)

$$\rho_n = \tfrac{1}{2}(P_0\sigma_0 + \boldsymbol{P}\cdot\boldsymbol{\sigma}) \qquad \mathrm{Tr}\,(\rho_n) = P_0 \tag{7.195}$$

To find an expression for the scattering cross-section the operator $v^\dagger v$ has first to be averaged over the neutron spin-states giving the result

$$\begin{aligned}\langle v^\dagger v\rangle_{(n)} &= \mathrm{Tr}\,[\rho_n(v^\dagger v)]/\mathrm{Tr}\,(\rho_n)\\ &= \mathrm{Tr}\,[\tfrac{1}{2}(P_0\sigma_0 + \boldsymbol{P}\cdot\boldsymbol{\sigma})(b^*\sigma_0 + \tfrac{1}{2}c^*\boldsymbol{s}\cdot\boldsymbol{I})(b\sigma_0 + \tfrac{1}{2}c\boldsymbol{\sigma}\cdot\boldsymbol{I})]/P_0\\ &= \left[|b|^2 + \frac{|c|^2}{2}\boldsymbol{I}\cdot\boldsymbol{I} + \tfrac{1}{2}(b^*c + bc^* - |c|^2)\boldsymbol{f}_n\cdot\boldsymbol{I}\right]\end{aligned} \tag{7.196}$$

where

$$f_n = P/P_0 \tag{7.197}$$

is the neutron beam polarization vector defined in (7.151). Completing the average over the nuclear spin states is then straightforward since $\boldsymbol{I}\cdot\boldsymbol{I}$ is a scalar operator with the eigenvalue $I(I+1)$. We find therefore that

$$\langle v^\dagger v\rangle_{(\mathrm{n,N})} = \left[\,|b|^2 + \frac{|c|^2}{2} I(I+1) + (\mathrm{Re}\,(b^*c) - \tfrac{1}{2}|c|^2) f_n \cdot \langle I\rangle\,\right] \tag{7.198}$$

and, in place of (6.85) the incoherent scattering length becomes

$$b_{\mathrm{inc}} = \left\{\frac{||c|^2}{2} I(I+1) + [\mathrm{Re}\,(b^*c) - \tfrac{1}{2}|c|^2] f_n \cdot \langle I\rangle\right\}^{1/2} \tag{7.199}$$

7.4.5.3 Capture of polarized neutrons on polarized nuclei

We suppose first that the incident neutrons are prepared in the pure state (cf. 7.147)

$$|a\rangle = u_+|\tfrac{1}{2}, \tfrac{1}{2}\rangle + u_-|\tfrac{1}{2}, -\tfrac{1}{2}\rangle \tag{7.200}$$

referred to a quantization axis directed along the symmetry axis of the polarized nuclei. The compound nucleus will then be created in a state with spin $I_f = I \pm \frac{1}{2}$, whichever is the spin of the nearest resonance. Apart from a normalization factor the transition amplitudes for the states $|I_f, M \pm \frac{1}{2}\rangle$ are given by

$$A_\pm(I_f, M) = u_\pm \langle I_f, M \pm \tfrac{1}{2} | \tfrac{1}{2}, \pm\tfrac{1}{2}, IM\rangle \tag{7.201}$$

where the relevant Clebsch–Gordon coefficients are listed in table 7.4. The

Table 7.4. Explicit expressions for the Clebsch–Gordon coefficients which arise in the analysis of the polarized neutron–polarized nuclei interaction, expressed as functions of the spin projection quantum number of the neutron (m) and the target nucleus (M)

I_f \ m	$\frac{1}{2}$	$-\frac{1}{2}$
$I+\frac{1}{2}$	$\left(\frac{I+M+1}{2I+1}\right)^{1/2}$	$\left(\frac{I-M+1}{2I+1}\right)^{1/2}$
$I-\frac{1}{2}$	$-\left(\frac{I-M}{2I+1}\right)^{1/2}$	$\left(\frac{I+M}{2I+1}\right)^{1/2}$

total transition rate is then found to be

$$
\begin{aligned}
W(I_f, M) &= |A_+(M)|^2 + |A_-(M)|^2 \\
&= \tfrac{1}{2}(P_0 + P_z)|\langle I_f, M+\tfrac{1}{2}|\tfrac{1}{2}\tfrac{1}{2}, IM\rangle|^2 \\
&\quad + \tfrac{1}{2}(P_0 - P_z)|\langle I_f, M-\tfrac{1}{2}|\tfrac{1}{2}-\tfrac{1}{2}, IM\rangle|^2
\end{aligned} \tag{7.202}
$$

where we have removed the restriction that the neutrons be prepared in a pure state and $(P_0, \boldsymbol{P})$ are the Stokes parameters defined in (7.149).

We now take the first special case and set $I_f = I + \frac{1}{2}$. Averaging over the ensemble of polarized nuclei the mean transition rate is

$$
\begin{aligned}
W &= \sum_M P_M W(I+\tfrac{1}{2}, M) = \tfrac{1}{2}\sum_M P_M \left\{ \frac{P_0(I+1)}{2I+1} + \frac{P_z M}{2I+1} \right\} \\
&= \tfrac{1}{2} P_0 \left\{ \frac{I+1}{2I+1} + I f_{nz} f_N \right\}
\end{aligned} \tag{7.203}
$$

From this result the total polarized capture cross-section into the state with $I_f = I + \frac{1}{2}$ is immediately seen to be

$$
\sigma_c^+ = \sigma_{cu}^+ \left\{ 1 + \left(\frac{I}{I+1} \right) \boldsymbol{f}_n \cdot \boldsymbol{f}_N \right\} \tag{7.204}
$$

where σ_{cu}^+ is the capture cross-section for unpolarized neutrons.

Similarly, taking the appropriate Clebsch–Gordon coefficient from Table 7.3, the polarized cross-section for capture into the state with $I_f = I - \frac{1}{2}$ is found to be

$$
\sigma_c^- = \sigma_{cu}^- \{ 1 - \boldsymbol{f}_n \cdot \boldsymbol{f}_N \} \tag{7.205}
$$

The practical significance of the results (7.204)–(7.205) follows from the fact that the change in cross-section due to the polarization effect operates in opposite directions for $I_f = I \pm \frac{1}{2}$. This result permits the spin of the capturing resonance to be determined when I is known, and when the relative orientation of $\boldsymbol{f}_n$ and $\boldsymbol{f}_N$ is known, and from such observations the spin of the 1.458 eV resonance in ^{115}In was shown to take the value $I_f = 5$ (Dabbs *et al* 1955).

Finally, it is straightforward to show that the polarization of the compound nucleus is given by the formula

$$
(I+\tfrac{1}{2})\boldsymbol{f}_N^c = \frac{(I+\tfrac{3}{2})\left[\dfrac{\boldsymbol{f}_n}{3} + \dfrac{I}{I+1}\boldsymbol{f}_N \right] + \dfrac{I^2}{I+1} f_n g_N}{1 + \left(\dfrac{I}{I+1} \right) \boldsymbol{f}_n \cdot \boldsymbol{f}_N} \qquad I_f = I + \tfrac{1}{2} \tag{7.206}
$$

and

$$(I-\tfrac{1}{2})f_{\rm N}^{\rm c}=\frac{(I-\tfrac{1}{2})\left(f_{\rm N}-\dfrac{f_{\rm n}}{3}\right)-If_{\rm n}g_{\rm N}}{1-f_{\rm n}\cdot f_{\rm N}} \qquad I_{\rm f}=I-\tfrac{1}{2} \tag{7.207}$$

where g_N is the degree of initial nuclear alignment defined in (7.189). This parameter is identically zero of course when $I=\frac{1}{2}$, e.g. in the capture of polarized neutrons on polarized protons.

7.5 Neutron spin resonance

7.5.1 Nuclear magnetic resonance with neutrons

The spectroscopic technique described as 'nuclear magnetic resonance' or 'NMR' (Gorter 1936, Rabi *et al* 1938) was first applied to the problem of measuring nuclear magnetic moments (Rabi *et al* 1939, Kellogg *et al* 1939). It was subsequently extended to the study of the radiofrequency spectra of atoms and molecules (Kellogg *et al* 1940, Purcell *et al* 1946, 1948). The technique was applied in the very first measurement of the neutron magnetic moment using a version of the Rabi single coil resonance method as illustrated in figure 7.16 (Alvarez and Bloch 1940). The neutron magnetic moment was re-measured a number of times in the immediate post-war period using the Rabi technique (Arnold and Roberts 1947, Bloch *et al* 1948), resulting in improvements in accuracy by more than an order of magnitude. However, more recent measurements have exploited the Ramsey separated oscillatory coil technique (Ramsey 1949, 1950, Ramsey and Pound 1951, Ramsey and Silsbee 1951) yielding results of quite astonishing accuracy (Corngold *et al* 1956, Greene *et al* 1977, 1979). The same method has also been applied using ultra-cold neutrons to search for a neutron electric dipole moment. The discussion of these experiments is deferred to section 9.2.5.

The state of a neutron beam with an arbitrary degree of partial polarization may be characterized by its four Stokes parameters $(P_0, \boldsymbol{P})$, when P_0 is the total intensity and $\boldsymbol{P}/P_0$ is an axial vector in position space, which may be identified with the expectation value of the spin (cf. section 7.4.1). In a uniform magnetic field $\boldsymbol{B}_0$, the component of $\boldsymbol{P}$ normal to $\boldsymbol{B}_0$, precesses about $\boldsymbol{B}_0$ with Larmor angular frequency given by

$$\hbar\omega_0=-\hbar\gamma_{\rm n}B_0 \tag{7.208}$$

where γ_n (<0) is the gyromagnetic ratio. When a weak magnetic field $\boldsymbol{B}_1$ is superimposed in a plane normal to $\boldsymbol{B}_0$, rotating with angular frequency ω in the same sense as the free precession, the Stokes vector carries out a forced precession. Thus, when $\omega=\omega_0$, the neutron spin senses a constant weak field when viewed from a frame of reference rotating with angular velocity ω, and a resonance occurs which is similar to any classical resonance phenomenon.

The quantum mechanical description of the resonance phenomenon is much more akin to that of an optical interference experiment, for in this formulation each component of the Pauli spinor describing a neutron in a magnetic field picks up equal and opposite phases. As a consequence the superposition state varies in time giving rise to the phenomenon of Larmor precession. When a time-dependent oscillatory field is applied, the neutron spin also carries out a forced precession which, under resonance conditions, induces transitions between the stationary states of the spin. Thus, from this viewpoint the resonance is manifested as constructive (flop-in) or destructive (flop-out) interference. Pursuing this comparison a bit further, the original Rabi single coil resonance technique resembles the optical single slit interference experiment, whereas the more sophisticated Ramsey separated oscillatory coil method is the analogue in spin-space of the double slit optical interference experiment (cf. section 7.2.1.1).

7.5.2 The Rabi single coil resonance method

7.5.2.1 Determination of the resonance condition

The spin Hamiltonian of a neutron in a magnetic field $\boldsymbol{B}$ is given by

$$H_s = -\boldsymbol{\mu}_n \cdot \boldsymbol{B} = -\frac{\hbar\gamma_n}{2}\boldsymbol{\sigma}\cdot\boldsymbol{B} \tag{7.209}$$

and, since $\gamma_n < 0$, the spin state defined as 'spin-up' with respect to the magnetic field is the stationary state with higher energy. Thus if $E_1(E_2)$ denotes the energy of the spin-up (spin-down) state we have the results

$$E_1 = \tfrac{1}{2}\hbar\omega_0 \qquad E_2 = -\tfrac{1}{2}\hbar\omega_0 \tag{7.210}$$

where $\hbar\omega_0$ is the quantum energy absorbed by the neutron in flipping its spin from the down state to the up state.

In a magnetic resonance experiment a strong magnetic field B_0 is established in the z-direction, which is taken as the axis of quantization, and a weak rotating magnetic field is applied in the xy plane. The complete magnetic field is therefore

$$\boldsymbol{B} \equiv B_1 \cos\omega t,\ B_1 \sin\omega t,\ B_0 \qquad B_1 \ll B_0 \tag{7.211}$$

and the time-dependent Schrödinger equation for the Pauli spinor $\boldsymbol{u}$ is

$$\dot{\boldsymbol{u}}(t) + M(t)\boldsymbol{u}(t) = 0 \tag{7.212}$$

Here $M(t)$ is the transfer matrix

$$M(t)=\mathrm{i}\begin{pmatrix}\omega_0/2 & b\,\mathrm{e}^{-\mathrm{i}\omega t}\\ b\,\mathrm{e}^{\mathrm{i}\omega t} & -\omega_0/2\end{pmatrix} \tag{7.213}$$

where

$$2b=-\gamma_\mathrm{n}B_1\ll\omega_0 \qquad (b>0) \tag{7.214}$$

The time dependence of the off-diagonal elements in (7.213) may be removed by transforming to a frame of reference rotating with angular frequency ω about the z-axis (Rabi *et al* 1954). This transformation is

$$\boldsymbol{u}(t)=\begin{pmatrix}\mathrm{e}^{-\mathrm{i}\omega t/2} & 0\\ 0 & \mathrm{e}^{\mathrm{i}\omega t/2}\end{pmatrix}\boldsymbol{v}(t)=G(t)\boldsymbol{v}(t) \tag{7.215}$$

where the transformed spinor $\boldsymbol{v}(t)$ satisfies the equation

$$\dot{\boldsymbol{v}}(t)+T\boldsymbol{v}(t)=0 \tag{7.216}$$

with

$$T=\mathrm{i}\begin{pmatrix}\frac{1}{2}(\omega_0-\omega) & b\\ b & -\frac{1}{2}(\omega_0-\omega)\end{pmatrix} \tag{7.217}$$

The T matrix is unitary to within a constant (imaginary) factor and may be conveniently expressed in the form

$$T(\theta)=-\frac{\mathrm{i}a}{2}\begin{pmatrix}-\cos\theta & \sin\theta\\ \sin\theta & \cos\theta\end{pmatrix} \tag{7.218}$$

where we have introduced the notation (Ramsey 1956)

$$\cos\theta=(\omega_0-\omega)/a \qquad \sin\theta=-2b/a$$
$$a=\sqrt{((\omega_0-\omega)^2+(2b)^2)} \tag{7.219}$$

$T(\theta)$ may now be converted into a constant multiple of the Pauli matrix σ_z using the rotation matrix

$$R(\theta)=\begin{pmatrix}\cos(\theta/2) & \sin(\theta/2)\\ -\sin(\theta/2) & \cos(\theta/2)\end{pmatrix} \tag{7.220}$$

so that

$$\boldsymbol{v}(t)=R(\theta)\boldsymbol{w}(t) \tag{7.221}$$

where $\boldsymbol{w}(t)$ satisfies the differential equation

$$\dot{\boldsymbol{w}}(t)+\frac{\mathrm{i}a\sigma_z}{2}\boldsymbol{w}(t)+R^{-1}(\theta)\dot{R}(\theta)\boldsymbol{w}(t)=0 \tag{7.222}$$

Equation (7.222) separates into independent equations for the two components of $w(t)$ only in the special circumstance that $\dot{R}(\theta)\equiv 0$, and this is true only in two cases: (i) $B_1\equiv 0$ and $\theta\equiv 0$, i.e. no rotating field; and (ii) B_0 and B_1 are both truly uniform homogeneous fields. The point here is that, as time develops, the neutron spin samples different regions of the magnetic field so that B_0 and B_1 become implicitly weakly time dependent. To solve (7.222) in the case $B_1\neq 0$ we therefore have to assume that these fields are both perfectly homogeneous, which effectively limits the applicability of the resultant solution to short times or, for a beam of neutrons of speed v moving a distance $l=vt$ in time t, for distances l such that

$$\frac{l}{B_0}\frac{\partial B_0}{\partial l}\ll 1 \tag{7.223}$$

If this condition is fulfilled then (7.222) integrates immediately to

$$w(t)=\begin{pmatrix} e^{iat/2} & 0 \\ 0 & e^{-iat/2} \end{pmatrix} w(0)=S(t)w(0) \tag{7.224}$$

The complete solution for $\boldsymbol{u}(t)$ obtained using (7.215), (7.221) and (7.224) now becomes

$$\begin{aligned} \boldsymbol{u}(t) &= G(t)R(\theta)S(t)w(0) \\ &= G(t)R(\theta)S(t)R^{-1}(\theta)G^{-1}(0)\boldsymbol{u}(0) \\ &= G(t)S(\theta, t)\boldsymbol{u}(0) \end{aligned} \tag{7.225}$$

where

$$S(\theta, t)=\begin{pmatrix} \cos\left(\frac{at}{2}\right)-i\cos\theta\sin\left(\frac{at}{2}\right) & i\sin\theta\sin\left(\frac{at}{2}\right) \\ i\sin\theta\sin\left(\frac{at}{2}\right) & \cos\left(\frac{at}{2}\right)+i\cos\theta\sin\frac{at}{2} \end{pmatrix} \tag{7.226}$$

It follows from a comparison of (7.226) with (7.162) that $S(\theta, t)$ is a unitary matrix describing the rotation of the initial spin vector (or the initial Stokes vector $\boldsymbol{P}(0)$ in the case of partial polarization) about an axis in space whose direction cosines are $(-\sin\theta, 0, \cos\theta)$. When $\theta=-\pi/2$ this axis is oriented normal to the axis of quantization and the neutron spin senses a constant magnetic field in the rotating frame of reference. This describes the resonance condition. Hence a neutron spin originally polarized parallel or anti-parallel to the strong magnetic field rotates between these two states while it precesses about this constant field.

We may immediately derive the Rabi resonance condition (the flop-in formula) from (7.226). For if we suppose that the neutron is in the spin-down position at zero time, i.e. $u_1(0)=0$, $u_2(0)=1$, then the probability for a transition to the spin-up state at time t is given by

$$W_{12}(\theta, t)=|u_1(t)|^2=\sin^2\theta \sin^2 at/2$$

$$=\frac{(2b)^2 \sin^2\left[\frac{t}{2}\sqrt{((\omega_0-\omega)^2+(2b)^2)}\right]}{(\omega-\omega_0)^2+(2b)^2} \tag{7.227}$$

The similarity of the result (7.227) and the single slit diffraction pattern (7.45) is obvious. $W_{12}(\theta, t)$ is a product of two factors: an envelope function having the form of a Breit–Wigner resonance (cf. section 2.6.3.2) which reaches a maximum of unity when $\omega=\omega_0$, with a full width at half maximum $\Delta\omega_e=4b$, and an oscillating signal which, for $t\geqslant\pi/2b$, can also reach unity at resonance, in which case the neutron spin flips with unit probability. For $t<\pi/2b$, there is still a resonance at $\omega=\omega_0$ but the oscillating factor does not reach unity at resonance, and there is only partial flipping of the spin. This is because, under these conditions, insufficient time is available for the neutron spin to carry out a full precession cycle about the constant field in the rotating frame.

The function $W_{12}(\theta, t)$ is plotted in figure 7.15 for various values of the time. It may be noted that the width $\Delta\omega'$ of the central lobe of the pattern decreases in inverse proportion to the time t that the neutron spends in the oscillating field. This is a manifestation of the energy–time uncertainty relation

$$(\hbar\,\Delta\omega')t\geqslant\hbar \tag{7.228}$$

indicating that the main limitation on the accuracy with which the resonance angular frequency ω_0 may be determined using a single resonance coil as described above, is the restriction on the time t that the neutron spends in the resonating field. This time can clearly be increased using very cold neutrons, but the method is primarily limited by the technical difficulty of constructing a magnetic field which is strictly homogeneous over a large region of space.

The apparatus used by Alvarez and Bloch (1940) in their measurement of the neutron magnetic moment by the resonance technique is shown in figure 7.16. The experiment made use of neutrons from the reaction $^7_4\mathrm{Be(d, n)}^8_5\mathrm{B}$ polarized and analysed by transmission through magnetized iron. The result obtained, $|\mu_n|=1.93\pm0.02$ nm was in every way compatible with the measured magnetic moments of proton and deuteron, but quite incompatible with the picture of the neutron as a Dirac point particle. The sign of μ_n could not be determined because the experiment made use of a linear oscillating field rather than a rotating field so the sense of the spin precession was not measured. The significance of this variation is discussed immediately below.

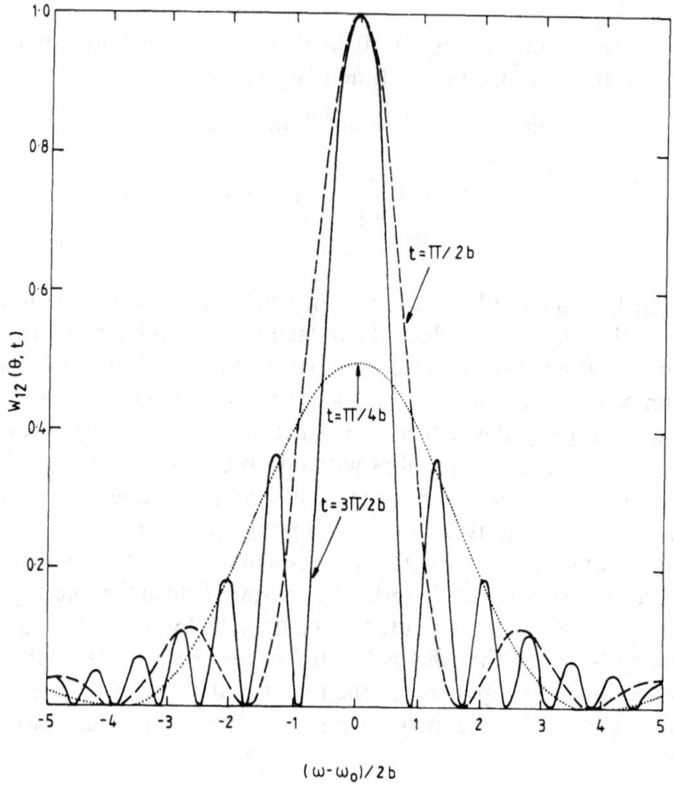

Figure 7.15. The spin-flipping transition probability $W_{12}(\theta, t)$ in a Rabi resonance coil plotted as a function of $\cot\theta = (\omega - \omega_0)/2b$, for three values of the time ($t = \pi/4b$), $\pi/2b$ and $3\pi/2b$.

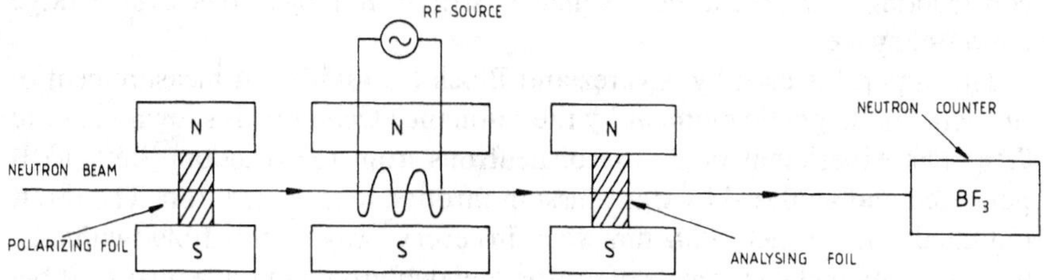

Figure 7.16. Apparatus used by Alvarez and Bloch (1940) to measure the magnetic moment of the neutron. Neutron polarization and analysis were performed by transmission through polarized iron. The resonance condition corresponding to maximum spin-flip probability was detected by observing a minimum in the transmitted intensity.

7.5.2.2 The Bloch–Siegert shift

In practice it turns out to be wholly inconvenient to work with a transverse rotating magnetic field $B_1 \, e^{i\omega t}$, which is replaced by a linear sinusoidal magnetic field $2B_1 \cos \omega t$ oriented along the x-direction. Thus each of the off-diagonal elements in the transfer matrix $M(t)$, defined in (7.213) is replaced by $2b \cos \omega t$. The corresponding matrix $T(\theta)$ given in (7.218) then goes over to

$$\bar{T}(\theta) = -\frac{ia}{2}\begin{pmatrix} -\cos\theta & \sin\theta[1+e^{2i\omega t}] \\ \sin\theta[1+e^{-2i\omega t}] & \cos\theta \end{pmatrix} \tag{7.229}$$

This result expresses the fact that, viewed from a frame of reference rotating about the z-axis with angular frequency ω, the neutron spin senses a counter-rotating magnetic field of angular frequency 2ω. The corresponding Pauli spinor $\boldsymbol{w}(t)$ defined in (7.221) then satisfies a modified equation

$$\dot{\boldsymbol{w}}(t) + \frac{ia}{2}[\sigma_z + \sin\theta\, \hat{\boldsymbol{n}}(t)\cdot\boldsymbol{\sigma}]\boldsymbol{w}(t) = 0 \tag{7.230}$$

where $\hat{\boldsymbol{n}}(t)$ is the unit vector

$$\hat{\boldsymbol{n}}(t) \equiv -\cos\theta\cos 2\omega t,\ \sin 2\omega t,\ \sin\theta\cos 2\omega t \tag{7.231}$$

Equation (7.230) is just the Schrödinger equation for the Hamiltonian

$$H = H_0 + H'(t) = \tfrac{1}{2}\hbar a\sigma_z - \hbar b\hat{\boldsymbol{n}}(t)\cdot\boldsymbol{\sigma} \tag{7.232}$$

in which the perturbing potential $H'(t)$ contains only terms which are periodic in time. It might therefore be assumed, on this account, that the action of the perturbation would average to zero over many periods of oscillation. This, however, is not the case; if either of the components of $\boldsymbol{w}(t)$ is eliminated from (7.230), the other component satisfies a second-order linear differential equation with periodic coefficients. This equation is of a general class of which Mathieu's equation provides the simplest example. It is well known that, in general, the solutions of Mathieu's equation and similar equations, are of the form $e^{\alpha t} f(t)$, where $f(t)$ is a periodic function, and in theories of planetary motion non-vanishing values of α are associated with secular motions as exemplified in Hill's classic investigation of the motion of the lunar perigee.

In the present context, secular terms in the exact solution of (7.230) renormalize the energies $\pm\frac{1}{2}\hbar a$ of the neutron spin states as observed in the rotating frame, thereby inducing a displacement in the position of the resonance which is described as the Bloch–Siegert shift (Bloch and Siegert 1940, Stevenson 1940, Ramsey 1955, Shirley 1963). Since it is not possible to solve (7.230) exactly, the value of the shift has to be deduced from a perturbation calculation.

The procedure adopted is to assume that the perturbation renormalizes the difference angular frequency $(\omega_0 - \omega)$ by an amount characterized by the small quantity ε, i.e.

$$(\omega_0-\omega)\to(\omega_0-\omega)(1+\varepsilon) \qquad \varepsilon=\varepsilon(\omega)\ll 1$$
$$a\to a'=((\omega_0-\omega)^2(1+\varepsilon)^2+4b^2)^{1/2}$$
$$\simeq a+\varepsilon(\omega_0-\omega)^2/a+O(\varepsilon^2) \qquad (7.233)$$

The quantity ε is now determined by the requirement that $(\omega_0-\omega)\varepsilon$, which, evaluated at the true resonant regular frequency ω_r, is just the Bloch–Siegert shift, should cancel any secular terms introduced into the spin motion by the perturbation. With the substitution (7.233), equation (7.232) may be rewritten in the form

$$H=\hat{H}_0+\hat{H}'(t)=\tfrac{1}{2}\hbar a'\sigma_z-\tfrac{1}{2}\hbar[(a'-a)\sigma_z+2b\hat{\boldsymbol{n}}(t)\cdot\boldsymbol{\sigma}] \qquad (7.234)$$

To obtain an approximate solution to (7.234) we transform to the interaction representation (cf. section 8.2.2) in which the time dependence is due entirely to the perturbation, i.e.

$$\boldsymbol{w}(t)\to\bar{\boldsymbol{w}}(t)=\exp[i\hat{H}_0t/\hbar]\boldsymbol{w}(t)=\exp\left(\frac{ia't}{2}\sigma_z\right)\boldsymbol{w}(t) \qquad (7.235)$$

Thus $\bar{\boldsymbol{w}}(t)$ satisfies the equation

$$\dot{\bar{\boldsymbol{w}}}(t)+\frac{ia}{2}\left[-\left(\frac{a'-a}{a}\right)\sigma_z+\sin\theta\,(\hat{n}_z(t)\sigma_z+\hat{n}_-(t)\,e^{ia't}\sigma_++\hat{n}_+(t)e^{-ia't}\sigma_-)\right]\bar{\boldsymbol{w}}(t)=0$$

$$(7.236)$$

where $\hat{n}_\pm(t)=\hat{n}_x(t)\pm i\hat{n}_y(t)$ are the spherical components of $\hat{\boldsymbol{n}}(t)$.

The solution of (7.236) may now be written as a perturbation series

$$\boldsymbol{w}(t)=\left[\sigma_0\frac{-i}{\hbar}\int_0^t\bar{H}'(t_1)\,dt_1+\left(\frac{i}{\hbar}\right)^2\int_0^t dt_1\int_0^{t_1}\bar{H}'(t_1)\bar{H}'(t_2)\,dt_2+\ldots\right]\boldsymbol{w}(0)$$

$$(7.237)$$

where σ_0 is the 2×2 unit matrix. It is evident from (7.236) that the second term in (7.237) can give rise to oscillatory terms only, and to obtain secular terms it is necessary to proceed to the next order. Secular terms are then introduced by integrating products of the form $\hat{n}_-(t_1)\hat{n}_+(t_2)\exp[ia'(t_1-t_2)]$ over t_2, with the end result

$$\left(\frac{a'-a}{a}\right)-\sin^2\theta\left(\frac{4\omega a\cos\theta}{4(4\omega^2-a'^2)}\right)=0 \qquad (7.238)$$

Close to resonance ($\omega\simeq\omega_0$, $2\omega\gg a$, $2\omega\gg a'$), the term proportional to a'^2 may be neglected in (7.238) with the result that, to first order in $\varepsilon(\omega)$,

$$(\omega_0-\omega)\left[\frac{(\omega_0-\omega)\varepsilon(\omega)}{a^2}-\frac{\sin^2\theta}{4\omega}\right]=0 \qquad \omega\neq\omega_0 \qquad (7.239)$$

Evaluating (7.239) at resonance we find that

$$\omega_r = \omega_0 + (\omega_0 - \omega_r)\varepsilon(\omega_r) = \omega_0 + \frac{b^2}{\omega_0} \tag{7.240}$$

which is the familiar expression for the Bloch–Siegert shift. Since $b/\omega_0 \simeq 10^{-3}$ in a typical experiment, the Bloch–Siegert shift introduces a correction at the level of $10^{-4}\%$.

7.5.3 The Ramsey separated oscillatory fields method

In the Rabi resonance method the single coil plays a dual role, for it provides both the strong uniform magnetic field in which the stationary states of the neutron spin are established, and the weak perturbing field which induces transitions between these states. During the first half of its passage time through the coil under resonance conditions, the neutron spin slowly rotates from its initial eigenstate into a superposition of energy eigenstates, completing the transition to the orthogonal state during the second half. At the halfway stage the spin lies in a plane orthogonal to the uniform field and precesses rapidly about that field. As noted above, the width of the resonance reduces in proportion to the total passage time through the system in accordance with the Heisenberg uncertainty principle. In the Ramsey separated oscillatory coil method, which is shown schematically in figure 7.17 the total time 2τ spent in the oscillating field is split in two, and between these two short times, the neutron spin freely precesses about the uniform field for a relatively long time T. Thus the width of the resonance is reduced approximately in the ratio $2\tau/(T+2\tau)$. The principle is similar to that applied

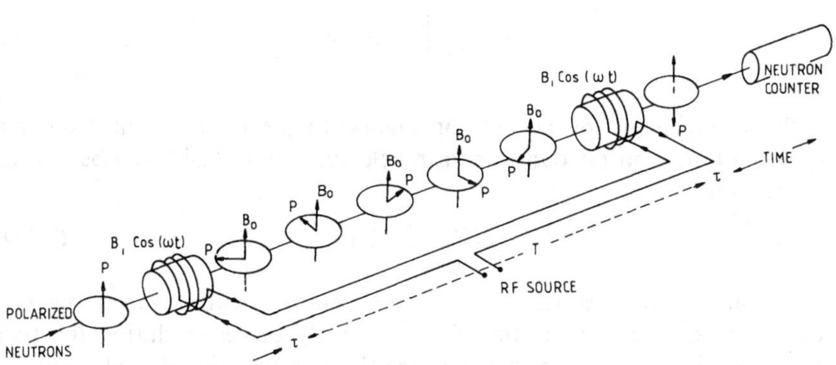

Figure 7.17. The Ramsey separated oscillatory fields resonance apparatus (Pendlebury and Smith 1980). The first $\pi/2$ coil turns the neutron spin from an orientation parallel to the uniform field $\boldsymbol{B}_0$ into the transverse plane. After a time T of free precession the second $\pi/2$ coil completes turning the neutron spin into an orientation anti-parallel to $\boldsymbol{B}_0$.

in Michelson's stellar interferometer where the angular resolution of stellar diameters is dramatically improved by increasing the separation between two light-collecting mirrors in a version of Young's double slit interference experiment (Michelson 1920, Michelson and Pease 1921).

The reason why the Ramsey configuration works is that it is not necessary for the strong field to be accurately homogeneous in the region of free precession, where the resonance frequency is rigorously determined by the average field sampled by the neutron spin during the passage time T. Of course such inhomogeneities as do exist cannot be too great or otherwise the transverse field components required to maintain the condition $\nabla \cdot \boldsymbol{B}=0$ will also display significant inhomogeneities, and every neutron will not sample the same average field. Thus the problem of field inhomogeneity can never be totally eliminated.

To prove the contention that only the average field is significant we start with equation (7.212) when the transfer matrix M is replaced by

$$\tilde{M}(t)=\mathrm{i}\begin{pmatrix} \omega_0(t)/2 & 0 \\ 0 & -\omega_0(t)/2 \end{pmatrix} \tag{7.241}$$

where $b \equiv 0$ but, because the neutron spin is moving in a spatially varying magnetic field, $\omega_0(t)$ is a slowly varying function of the time. With the substitution (7.241), the solution for the Pauli spin $\boldsymbol{u}(t)$ is replaced by

$$\begin{pmatrix} \tilde{u}_1(t) \\ \tilde{u}_2(t) \end{pmatrix}=\begin{pmatrix} \mathrm{e}^{-\mathrm{i}\bar{\omega}_0 t} & 0 \\ 0 & \mathrm{e}^{\mathrm{i}\bar{\omega}_0 t} \end{pmatrix}\begin{pmatrix} u_1(0) \\ u_2(0) \end{pmatrix} \tag{7.242}$$

where

$$\bar{\omega}_0=\frac{1}{t}\int_0^t \omega_0(t')\,\mathrm{d}t' \tag{7.243}$$

is the average value of the Larmor angular frequency. If the neutron spin spends a time T in the quasi-uniform field the result (7.242) can be written in the form

$$\boldsymbol{u}(T)=\tilde{S}(T)\boldsymbol{u}(0) \tag{7.244}$$

We now have a situation where the neutron spin traverses the first oscillating coil during the time $0 \leqslant t \leqslant \tau$, freely precesses during the time $\tau < t \leqslant (\tau+T)$, and completes its passage through the second oscillatory coil during the remaining time $\tau+T<t \leqslant 2\tau+T$. This sequence of events is illustrated in figure 7.17. The final emergent spinor then has the form

$$\begin{aligned} \boldsymbol{u}(T+2\tau) &= G(T+2\tau)S(\theta, T+2\tau)\tilde{S}(T)G(\tau)S(\theta,\tau)G^{-1}(0)\boldsymbol{u}(0) \\ &= G(T+2\tau)S_{\mathrm{R}}(\theta, T, \tau)\boldsymbol{u}(0) \end{aligned} \tag{7.245}$$

where $S_R(\theta, T, \tau)$, the transfer matrix for the combined system, is given by

$$S_R(\theta, T, \tau) = \begin{pmatrix} (S_{11})^2 e^{i\lambda T/2} - |S_{12}|^2 e^{-i\lambda T/2} & 2S_{12}Re(S_{11} e^{i\lambda T/2}) \\ -2S_{12}^* Re(S_{11} e^{i\lambda T/2}) & (S_{11}^{*2} e^{-i\lambda T/2} - |S_{12}| e^{i\lambda T/2} \end{pmatrix}$$

(7.246)

The quantities S_{ij} ($j=1, 2$) appearing in $S_R(\theta, T, \tau)$ are the matrix elements of the transfer matrix $S(\theta, \tau)$ defined in (7.226) and

$$\lambda = \bar{\omega}_0 - \omega \tag{7.247}$$

The transition probability, from the spin eigenstate $\binom{0}{1}$ to the orthogonal eigenstate $\binom{1}{0}$, is thus given by (Ramsey 1956)

$$W_{12}(T+2\tau) = 4|S_{12}|^2[Re(S_{11} e^{i\lambda T/2})]^2$$

$$= 4 \sin^2\theta \sin^2\frac{a\tau}{2}\left[\cos\frac{\lambda T}{2}\cos\frac{a\tau}{2} - \cos\theta \sin\frac{\lambda T}{2}\sin\frac{a\tau}{2}\right]^2$$

(7.248)

The correspondence between the expression for $W_{12}(T+2\tau)$ and the familiar intensity pattern for the optical double slit interference experiment may be demonstrated most convincingly by assuming that near-resonance conditions are in force and $\cos\theta \simeq 0$. In this case (7.248) reduces to

$$W_{12}(T+2\tau) \simeq \sin^2\theta \sin^2 a\tau \cos^2 \lambda T/2 \tag{7.249}$$

which is just the product of a 'diffraction factor' $\sin^2\theta \sin^2 a\tau$ corresponding to a total passage time 2τ in the oscillating field region (cf. 7.227) and an 'interference factor' $\cos^2 \lambda T/2$ introduced by virtue of a passage time T through the region of free precession. Since $T \gg \tau$, the full width at half maximum of the interference factor $\Delta\omega \approx 1/T$ determines the frequency resolution of the joint system. Since the resonance angular frequency in the oscillatory region is ω_0, whereas in the free precession region it is $\bar{\omega}_0$, and these two quantities do not coincide in general, the overall resonant angular frequency derived from (7.248) is slightly shifted from $\bar{\omega}_0$; a good approximation is given by (Code and Ramsey 1971)

$$\omega = \bar{\omega}_0 + \left(\frac{(l/L)\tan\frac{1}{2}a\tau}{\frac{1}{2}a\tau + (2l/L)\tan\frac{1}{2}a\tau}\right)(\omega_0 - \bar{\omega}) \tag{7.250}$$

where l and L are the lengths of each of the resonating field regions, and the quasi-homogeneous field region, respectively. Since $\tau = l/v$ and $T = L/v$ where v is the appropriate component of the neutron velocity, and since v has some distribution in general, the results (7.237) and (7.248) have to be averaged over neutron velocity in order to be applicable in real experiments (Kruse and Ramsey 1951, Ramsey 1956).

7.5.4 Dressed neutrons

In section 7.5.2.1 we treated the problem of a neutron spin in a strong magnetic field $\boldsymbol{B}_0$, interacting with a weak radiofrequency field $\boldsymbol{B}_1\, e^{i\omega t}$ rotating in the same sense as the precessing spin. The Hamiltonian describing this system may be written in the form

$$H_s = \frac{\hbar\omega_0}{2}\sigma_z + \hbar b[\sigma_-\, e^{i\omega t} + \sigma_+\, e^{-i\omega t}] \tag{7.251}$$

where $\sigma_\pm = \frac{1}{2}(\sigma_x \pm i\sigma_y)$ are the raising and lowering operators in spin space. However, if the restriction that the radiofrequency field be weak is relaxed, then H_s must be supplemented by a term describing the energy stored in this field. In this case, since radiofrequency photons are destroyed and created because of the interaction with the neutron spin, it is more convenient to treat the problem using the formalism of the quantized electromagnetic field (cf. section 2.1.3.1). We therefore make the replacements

$$b\, e^{i\omega t} \to \lambda q^\dagger \qquad b\, e^{-i\omega t} \to \lambda q$$

$$b^2 \to \lambda^2 \langle q^+ q \rangle = \lambda^2 \bar{n} \tag{7.252}$$

where $\bar{n}$ is the mean number of photons of angular frequency ω. The resultant Hamiltonian then becomes

$$H = \hbar\left[\omega q^\dagger q + \frac{\omega_0}{2}\sigma_z\right] + \hbar\lambda[\sigma_- q^\dagger + \sigma_+ q]$$

$$= H_0 + H_{\text{int}} \tag{7.253}$$

A neutron whose spin motion is described by H is said to be 'dressed', since the specification of the states of the joint system of neutron spin plus field requires a statement about the distribution of photon number as well as about the orientation of the spin. The description 'dressed neutron' is taken over from atomic physics where the states of atoms dressed with radiofrequency photons have been shown to differ in many important respects from the states of atoms in weak fields (Cohen-Tannoudji and Haroche 1969).

An exact solution of the Schrödinger–Pauli equation for the Hamiltonian H has been obtained for the case of the two-state maser (Jaynes and Cummings 1963), which may be readily adapted to describe the present problem (Louiselle 1964). We note first of all that the solution of the eigenvalue problem for the unperturbed Hamiltonian may be written down immediately

$$H_0|n, \pm\tfrac{1}{2}\rangle = \left(\hbar\omega n \pm \hbar\frac{\omega_0}{2}\right)|n, \pm\tfrac{1}{2}\rangle \tag{7.254}$$

where $|n, \pm\frac{1}{2}\rangle$ represents the state of the joint system with n photons and neutron spin-orientation $\pm\frac{1}{2}$. However, since H_{int} does not commute with

H_0, the addition of H_{int} to H_0 brings about a mixing of these eigenstates. It turns out that this mixing is of the simplest possible kind.

The Hamiltonian H is first split into two commuting parts

$$H = [H_0 - \tfrac{1}{2}\hbar(\omega_0 - \omega)\sigma_z] + [H_{\text{int}} + \tfrac{1}{2}\hbar(\omega_0 - \omega)\sigma_z] = H_1 + H_2 \quad (7.255)$$

where the eigenstates $|n, \pm\tfrac{1}{2}\rangle$ of H_0 are also eigenstates of H_1 but with different eigenvalues, i.e.

$$H_1|n, \pm\tfrac{1}{2}\rangle = \hbar\omega(n \pm \tfrac{1}{2})|n, \pm\tfrac{1}{2}\rangle \quad (7.256)$$

With the exception of the lowest state $|0, -\tfrac{1}{2}\rangle$, the eigenstates $|n, \tfrac{1}{2}\rangle$ and $|n+1, -\tfrac{1}{2}\rangle$ are degenerate. Thus, since H_2 commutes with H_1, it is possible to form orthogonal linear combinations of these degenerate eigenstates

$$|\varphi_+(n)\rangle = \cos(\tfrac{1}{2}\theta_n)|n+1, -\tfrac{1}{2}\rangle - \sin(\tfrac{1}{2}\theta_n)|n, \tfrac{1}{2}\rangle$$

$$|\varphi_-(n)\rangle = \sin(\tfrac{1}{2}\theta_n)|n+1, -\tfrac{1}{2}\rangle + \cos(\tfrac{1}{2}\theta_n)|n, \tfrac{1}{2}\rangle \quad (7.257)$$

which provide a basis in which H_2 is diagonal. An appropriate choice of the mixing angle θ_n is given by (cf. 7.219)

$$\cos(\theta_n) = (\omega_0 - \omega)/a_n$$

$$\sin(\theta_n) = -2\lambda(n+1)^{1/2}/a_n \quad (7.258)$$

where

$$a_n = [(\omega_0 - \omega)^2 + (2\lambda)^2(n+1)]^{1/2} \quad (7.259)$$

The states $|\varphi_\pm(n)\rangle$ are then eigenstates of H with eigenvalues

$$E^\pm(n) = \hbar\omega(n + \tfrac{1}{2}) \pm \tfrac{1}{2}\hbar a_n \quad (7.260)$$

To write down the complete solution of the time-dependent Schrödinger–Pauli equation we adopt the usual spinor notation (cf. section 7.5.2) with basis states

$$|n, \tfrac{1}{2}\rangle = \begin{pmatrix} 1 \\ 0 \end{pmatrix} \qquad |n+1, -\tfrac{1}{2}\rangle = \begin{pmatrix} 0 \\ 1 \end{pmatrix} \quad (7.261)$$

The solution is then given by

$$\boldsymbol{u}_n(t) = \exp[-\mathrm{i}Ht/\hbar]\boldsymbol{u}_n(0) = P_n(t)G(t)S(\theta_n, t)\boldsymbol{u}_n(0) \quad (7.262)$$

where

$$P_n(t) = \begin{pmatrix} \mathrm{e}^{-\mathrm{i}\omega nt} & 0 \\ 0 & \mathrm{e}^{-\mathrm{i}\omega(n+1)t} \end{pmatrix} \quad (7.263)$$

is a matrix which takes account of the phases associated with the states of the radiation field, $G(t)$ is the matrix defined in (7.215) which describes the transformation to the rotating frame of reference, and the matrix $S(\theta, t)$ is defined in (7.226). Thus the main addition provided by the quantized theory is that the angle θ defined in (7.219) is replaced by the angle θ_n defined in (7.258). These angles differ in only one respect, namely that b is replaced by $b\sqrt{((n+1)/\bar{n})}$, and the quantized theory takes account of the contribution from spontaneous transitions, as implied by the appearance of the factor $(n+1)$ rather than n.

Since the eigenstates of H are eigenstates of H_1 the quantum number n is a good quantum number and $(n+\frac{1}{2})\hbar$ may be interpreted as the value of the projection of the angular momentum along the z-axis. This follows from the fact that each photon in the right circularly polarized radio-frequency field carries unit component of angular momentum. In this context it is worth noting that the matrix $P_n(t)G(t)$ is simply the unit matrix multiplied by the phase factor $\exp[-i(n+\frac{1}{2})\omega t]$. Since the states $|\varphi_\pm(n)\rangle$ are not eigenstates of H_0 the quantum number n cannot be interpreted in general as the number of photons in this state. This happens only when b is reduced to zero, in which case n is the number of photons for the spin-up case, whereas $(n+1)$ is the photon number when the spin is down.

Since the function a_n^2 is a quadratic function of $(\omega_0-\omega)$, it has a minimum value when $\omega_0=\omega$, and there is one value of ω_0 at which the energy levels $E_+(n)$ and $E_-(n+)$ become equal. This is the phenomonon described as 'level-crossing'. In the 'resonant' radiofrequency field described by (7.253) there is only one crossing point, but, more generally, states of opposite spin orientation can cross provided the difference in their quantum numbers n is an odd number, and coherent mixing occurs at the crossing point. States of opposite spin orientation whose quantum numbers n differ by an even number never cross, and instead exhibit the phenomenon of anti-crossing. For example, in the resonant case, the energy levels $E_+(n)$ and $E_-(n)$ anti-cross at their extremum point $\omega_0=\omega$.

The solution given above breaks down of course when the rotating field $\boldsymbol{B}_1\,e^{i\omega t}$ is supplemented by a counter-rotating field $\boldsymbol{B}_1^{-i\omega t}$, so that the neutron spin is acted on by a radiofrequency field in a state of linear polarization as described in section 7.5.2.2. In this case the Hamiltonian is supplemented by a term

$$H_3=\hbar\lambda(\sigma_-q+\sigma_+q^\dagger) \tag{7.264}$$

which connects states differing by two units of angular momentum measured with respect to the z-axis. The total Hamiltonian is then given by

$$H=\hbar\left[\omega q^\dagger q+\frac{\omega_0\sigma_z}{2}+\lambda\sigma_x(q^\dagger+q)\right] \tag{7.265}$$

The addition of the term H_3 means that the new eigenstates $|\varphi_\pm^m(n)\rangle$ are now constructed from admixtures of states $|n+1+m, -\frac{1}{2}\rangle$ and $|n+m, \frac{1}{2}\rangle$ where $m=0\pm 1$, ± 2, etc, so that there is an infinite number of crossings and anti-crossings of the resultant energy level scheme.

The eigenfrequencies of the dressed neutron system described by (7.265) have been determined by measuring the spin-rotation pattern for a beam of longitudinally polarized monochromatic cold neutrons, transmitted along the axis of a coil generating an axial oscillating field $\boldsymbol{B}_1 \cos \omega t$ transverse to a constant magnetic field $\boldsymbol{B}_0$ (Muskat *et al* 1987). For the rotating oscillating field given in (7.253), the corresponding axial component of the polarization is given by

$$f_x(t) = \cos^2 \theta_n + \cos(a_n t) \sin^2 \theta_n \tag{7.266}$$

for $f_x(0) = 1$. Thus the extrema in the spin rotation pattern for a fixed transit

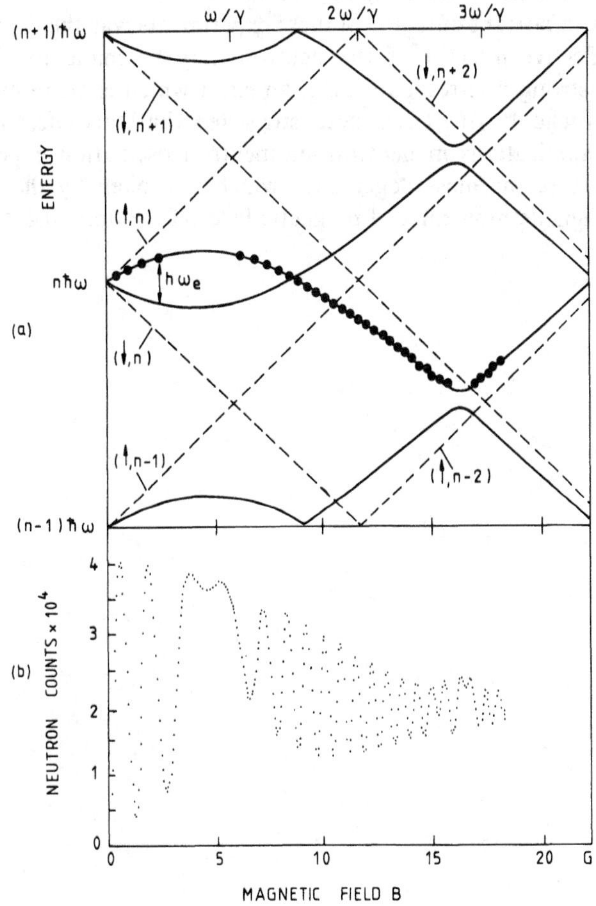

Figure 7.18. The spin-rotation pattern of polarized neutrons 'dressed' by photons of a high intensity linearly polarized radiofrequency field. The full lines denote energy levels calculated for $B_x = 1G$ while the points are experimental values (Muskat *et al* 1987).

time T are given at values of ω_0 such that

$$a_n T = N\pi \qquad N = 0 \pm 1 \ldots \tag{7.267}$$

and this relationship permits a determination of the transition energies $\hbar a_n$ given in (7.260).

In the case of the linearly oscillating field the angular eigenfrequencies $\omega_n(B_0)$ analogous to $a_n(B_0)$ were similarly determined by setting $B_1 = 1G$, $\omega = 113 \times 10^3\ \mathrm{s}^{-1}$. The spin rotation pattern observed was as shown in figure 7.18(b), and this was interpreted in terms of the crossings and anti-crossings in the energy level diagram shown in figure 7.18(a). These were determined by truncating the set of contributing states at some appropriate point and carrying out the necessary diagonalization.

A further important outcome of these experiments was the demonstration that the effective g-factor of the neutron may be made to vanish in a sufficiently strong non-resonant field, an effect which is also observed with atoms (Haroche 1970). It has been suggested that this effect could have important applications in neutron–antineutron oscillation experiments in order to restore the mass degeneracy which is broken by the interaction between magnetic moment and magnetic field (cf. section 10.4.2).

8 The neutron and condensed states of matter

8.1 Applied neutron optics

8.1.1 Neutron radiography

8.1.1.1 Radiography with X-rays and neutrons

X-radiography was invented by Röntgen in 1895, shortly after he had discovered X-rays, when he constructed the first radiogram showing a zinc plate weld. The feature which distinguishes radiography from photography is that, in the former technique, the latent image is created by radiation transmitted through, rather than reflected from, the object under study. Although the first neutron radiograms were recorded by Kalmann and Kuhn as long ago as 1935, using neutrons generated by a small accelerator (Kalmann 1948, Peter 1946), the modern technology of neutron radiography, which relies primarily, although not entirely, on intense beams of neutrons from nuclear reactors, originated at AERE Harwell (Thewles and Derbyshire 1956, Thewles 1956). These techniques were subsequently developed by Berger and his associates at the Argonne National Laboratory into a major diagnostic tool for non-destructive testing of highly radioactive components for the nuclear energy industry (Berger 1971).

Thermal neutrons and X-rays share the common properties that neither is charged, and both have wavelengths comparable with the mean separations of atoms in solids. However, in other important aspects their properties contrast rather than compare, e.g. whereas light elements are transparent to X-rays and heavy elements are relatively opaque, several light elements such as hydrogen, lithium and boron are highly neutron absorbant, whereas many heavy elements, e.g. lead and bismuth, transmit neutrons freely. Another difference is that neighbouring elements in the periodic table, e.g. boron and carbon, or cadmium and tin, virtually indistinguishable from the point of view of X-ray transmission, are easily separated by neutron transmission, and the same result is true even for different isotopes of the same element, e.g. ^{1}H and ^{2}H, or ^{3}He and ^{4}He. The significance of these and other differences is discussed further in sections 8.2.1 and 8.4.1.

In neutron radiography, as in X-radiography, the conditions necessary for the validity of geometrical optics apply to a good approximation, and neutrons in free space are assumed to travel in straight lines (cf. sections 6.1 and 6.2). In constructing a neutron radiogram, the transmission of a narrow beam of thermal (or epithermal) neutrons is measured as a function of the position and orientation of the object under examination, and variations are observed due to large-scale inhomogeneities, e.g. material or geometric imperfections, which influence both the absorption and the large-angle scattering of the incident beam. The smallest inhomogeneity resolvable by neutron radiographic methods is about 1 μm.

As a general technique for non-destructive testing with neutrons, radiography complements that of small-angle neutron scattering or SANS (cf. section 8.5.1.4), which relies on the wave properties of the neutron, and can readily resolve inhomogeneities on the level of $(1-10^3)$ Å. However, in recent years, this scale has been extended upwards to about 1 μm, so that inhomogeneities of all sizes can now be detected by one technique or the other.

Neutron radiography is applied on a large scale in the nuclear energy industry where it is used to locate voids, cracks, swelling, and radiation damage generally, in fuel elements and control rods. In these applications it has the great advantage over X-radiography that it can readily distinguish between ^{235}U and ^{238}U. Most non-nuclear uses exploit the fact that hydrogen is a strong neutron absorber, and many materials of commercial importance such as rubber and plastics contain hydrogen. Neutron radiography of explosives by the transfer technique (cf. section 8.1.1.3) is also an important application. The existence of a magnetic interaction between neutrons and matter (cf. section 8.4.1) has also been put to advantage in neutron radiography, where studies of neutron transmission through single ferromagnetic crystals for different crystallographic directions, have been used to obtain information about average domain wall inclinations and separations (Shil'shtein *et al* 1973). More recently, through a development described as neutron radiography with refraction contrast, it has proved possible to image individual domain walls (Kvardakov *et al* 1987, Podurets *et al* 1989).

8.1.1.2 Neutron beam characteristics

In addition to the nuclear reactor, all the neutron sources discussed in sections 1.6–1.8 have been used, in conjunction with suitable moderators, as slow neutron sources for neutron radiography. Demineralized water or polyethylene are most widely employed as moderators. Because of exposure time limitations high intensity is the most important requirement, and, in practice, capture fluxes $\varphi_c \geqslant 10^5$ cm^{-2} s^{-1} represent the minimum necessary. In the context of given source and given moderator the figure of merit is the thermalization factor f_{th}, which is defined as the ratio of the source yield (cf. Tables 1.1 and 1.2) to the peak thermal flux. Because they operate with low energy deuterons (cf. section 1.6.4) accelerator sources based on the reaction ^{3}H(d, n)^{4}He have been favoured, although much higher peak fluxes are achievable using the (d, n) reaction based on ^{242}Cm ($f_{th} \approx 100$), or the γ–n reaction based on ^{124}Sb ($f_{th} \approx 50$). The spontaneously fissioning isotope ^{252}Cf has proved a popular choice, with or without a sub-critical multiplying assembly, which can multiply the slow neutron yield by factors up to 100 (Miller *et al* 1971).

Apart from its intensity, the other important properties of the neutron beam are its spectral composition, in particular its 'hardness' as determined by its cadmium ratio, since this quantity controls the sample penetration, and the beam quality, which is defined as the neutron-to-gamma ratio. The latter index is of particular significance where γ-rays as well as neutrons contribute to the final image, since a high efficiency for γ-ray detection spoils the spatial resolution in the resultant radiogram. Having selected a neutron source, the next requirement is to produce a parallel beam, and this is most easily achieved by passing the neutron down a cylindrical tube of length L and diameter D. Then D/L defines the angular divergence, and the emergent flux is (Matfield 1971)

$$\varphi(L) \simeq \varphi(0)[L/4D]^2 \tag{8.1}$$

An improved angular divergence is obtained using an array of small tubes or parallel plates in a Soller configuration (cf. section 8.5.1.2), but this system has the disadvantage that the resultant radiogram contains an additional pattern of circles or lines which tends to confuse the interpretation. A commonly used system is the divergent collimator which uses a small entrance hole and a collimating cylinder diverging uniformly along its length. This has the advantage that the angular divergence is determined only by the length and the source size.

A practical layout suitable for neutron radiography is illustrated in figure 8.1. This illustrates the significance in relation to image quality of the geometric unsharpness defined by

$$U = DL_{\text{I}}/L_{\text{O}} \tag{8.2}$$

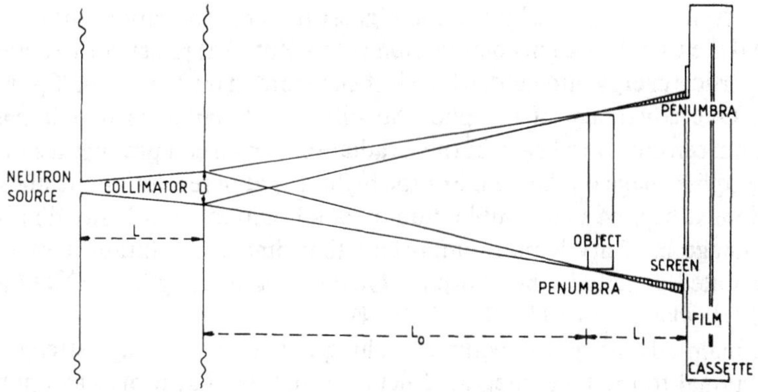

Figure 8.1. Neutron radiography by the direct technique using ^{6}Li, ^{10}B or ^{155}Gd detecting screens. The quality of the image recorded on X-ray film is determined by the geometric unsharpness U, defined equal to $DL_{\text{I}}/L_{\text{O}}$.

where L_O is the source-to-object distance, and L_I the object-to-image distance. Clearly U is mininized by reducing the ratio D/L_O, with a corresponding reduction in the neutron flux intercepted by the sample. Thus the usual compromise has to be made between resolution and intensity, which is common to all imaging systems without exception. Because neutron radiography relies on the action of a detecting screen which converts the neutrons into secondary radiations which are subsequently recorded, the conversion process introduces additional unsharpness, which increases with the range of the secondary radiation and the thickness of its detector. If only for this reason, neutron screens based solely on the (n, γ) reaction are unsuitable for neutron radiography.

8.1.1.3 Neutron image detectors

There are two techniques in general use for recording the neutron image, and these are termed the direct technique and the transfer technique. In the direct technique the transmitted neutron is absorbed in a screen which promptly emits a secondary radiation which is immediately recorded. Prompt emissions of interest in this respect are α-particles and internal conversion elections. In the transfer technique the absorbing screen becomes β-radioactive, and the latent image stored in the radioactive screen is removed from the neutron beam to be counted elsewhere, and at a later time. Table 8.1 lists all the target nuclei which have been used at one time or another as neutron detectors for neutron radiography, although, in practice, only six of them, ^{6}Li, ^{10}B, ^{155}Gd, ^{157}Gd, ^{115}In and ^{164}Dy are used in normal applications.

For direct radiography the obvious choices are ^{6}Li and ^{10}B, with ^{155}Gd especially valuable, firstly because of its high cross-section for thermal neutrons, but mainly because it emits 70 keV conversion electrons which have very short range leading to the highest possible spatial resolution. ^{6}Li and ^{10}B are usually used in combination with a ZnS(Ag) phosphor to convert the α-particle energy into light, which is then recorded on X-ray film (Spowart 1971). One advantage of the phosphor-film combination is that it has a minimum sensitivity to beam-borne γ-radiation, which can produce a second radiographic image on the film. For the highest spatial resolution the nucleus ^{155}Gd must be used in a combination of gadolinum foil and X-ray film but, when image intensity is more important than image resolution it may be incorporated directly in the phosphor Gd_2O_2S, containing both ^{157}Gd and ^{155}Gd (Hawkesworth 1973, Swinth 1974).

The main advantage of the direct technique is that the image strength is proportional to the time-integrated neutron flux, so that it may be applied even when the available neutron flux is quite low. The principal disadvantage is that γ-rays as well as neutrons leave an image, with a corresponding reduction in spatial resolution.

Table 8.1. Characteristics of target nuclei suitable for neutron conversion in neutron radiography screens. The symbol 'm' on the daughter nucleus signifies an isomeric state.

Target nucleus	Reaction products	σ (barn)	$E_{max}(\alpha, \beta, \gamma)$ (MeV)	$T_{1/2}$ Direct	Transfer
^{6}Li	^{3}H$+\alpha$	941	2.73(α)	Prompt	
^{10}B	^{7}Li$+\alpha$	3838	1.78(α)	Prompt	
^{103}Rh	^{104}Rh^{m}(γ)	11	0.05(γ)		4.5 min
	^{104}Rh (β)	139	2.44(β)		42 s
^{107}Ag	^{108}Ag (β)	35	1.70(β)		2.3 min
^{109}Ag	^{110}Ag (β)	91	2.87(β)		24 s
^{113}Cd	^{114}Cd$+\gamma$	2×10^{4}		Prompt	
^{115}In	^{116}In^{m} (γ, β)	157	1.0(β)		54 min
	^{116}In (β)	42	3.0(β)		14 s
^{149}Sm	^{150}Sm$+\gamma$	4.1×10^{4}		Prompt	
^{152}Sm	^{153}Sm (β)	210	0.8(β)		47 h
^{155}Gd	^{156}Gd$+\gamma$	6.1×10^{4}		Prompt	
	^{156}Gd$+e^{-}$		0.07(e^{-})	Prompt	
^{157}Gd	^{158}Gd$+\gamma$	2.54×10^{5}		Prompt	
^{164}Dy	^{165}Dy^{m} (γ)	2200	0.11(γ)		1.25 min
	^{165}Dy (β)	800	1.25(β)		138 min
^{197}Au	^{198}Au (β)	98.65	0.97(β)		2.7 days

The advantages of the transfer technique are that it is ideal for use in hostile environments, e.g. for studies of highly radioactive fuel rods, and that it does not detect any local γ-radiation. The disadvantage is that the image strength cannot be increased by prolonging the exposure. This is because, after a couple of half-lives, the screen activity reaches saturation, at which point the numbers of radioactive atoms created and decaying in unit time are equal. For transfer radiography the requirements are for a neutron-absorbant material producing a β-emitting isotope, with a half-life that is long enough that a high saturation activity can be reached, but not so long that a large interval of time must elapse before re-use. The favourite candidates are ^{164}Dy and ^{115}In, the former because of its relatively long half-life, and the latter because of its strong resonance at 1.43 eV which makes it a powerful detector of epithermal neutrons. All of the various neutron screens listed above are conventionally used with X-ray film to store the image, or with etched plastic film which is less sensitive to γ-radiation. Alternatives to photosensitive films are provided by image intensifiers of various kinds, and indeed the Anger camera (cf. section 8.3.6) has also been called into service as an image recorder (Holland and Pain 1971).

8.1.2 Neutron reflectometry

The reflection of neutrons at a plane surface was discussed in section 6.3.3 where it was shown that the reflected amplitude was given by the Fresnel formula (6.111). Applying (6.114)–(6.116), this result can be written in the form

$$A_r = -\left(\frac{1-[1-(\sin\theta_c/\sin\theta)^2]^{1/2}}{1+[1-(\sin\theta_c/\sin\theta)^2]^{1/2}}\right) \tag{8.3}$$

where θ_c, the critical glancing angle, has values typically in the range of 3–4 milliradian. The condition $\theta \leqslant \theta_c$ defines the regime of total external reflection, a phenomenon which is exploited to great advantage in the gravity refractometer (cf. sections 1.5.2.2 and 6.3.3.2). Since in this case we have

$$(\sin\theta_c/\sin\theta) \simeq (\theta_c/\theta) \simeq (h_c/h)^{1/2} \tag{8.4}$$

where h is the height of fall under gravity, and, according to (6.118) the critical height h_c is independent of λ, the operation of the instrument is independent of λ to first order, a feature which accounts for its great success.

When $\theta \gg \theta_c$ we may expand A_r to lowest order in $(\sin\theta_c/\sin\theta)$ giving the result

$$A_r \simeq -(\sin\theta_c/2\sin\theta)^2 \simeq -2m_n V_F/(\hbar q)^2 \tag{8.5}$$

where V_F is the Fermi pseudo-potential defined in (6.78) and

$$q = 2k\sin\theta \tag{8.6}$$

is the component of momentum transfer normal to the surface. For true mirror or specular reflection there is no other component. However, specular reflection is always accompanied by some diffuse scattering which transfers momentum in the plane of the reflecting surface. For sufficiently large values of θ the diffuse scattering eventually dominates the specular component.

Since the result (8.5) is linear in V_F it must represent the Born approximation to A_r (cf. section 2.3.1), and we may apply (2.55) to rewrite it in the form

$$A_r = -2m_n V_F/(\hbar q)^2 \int_{-\infty}^{\infty} e^{iqz}\,\delta(z)\,dz \tag{8.7}$$

where the function $\delta(z)$ is inserted to take account of the fact that the virtual sources responsible for extinguishing the normal component of the incident wave lie on the boundary surface. Integrating (8.7) by parts we may write it in the form

$$A_r = -2im_n V_F/\hbar^2 q \int_{-\infty}^{\infty} e^{iqz}\,\theta(-z)\,dz \tag{8.8}$$

where $\theta(-z)$ is unity for $z<0$ and vanishes for $z>0$.

Let us now suppose that the scattering centres are not uniformly distributed over the scattering medium but are described by a density distribution $\rho(z)$ which may, for convenience, be normalized such that $\rho(-\infty)$ is unity, i.e. the V_F appearing in (8.8) is that appropriate to the bulk material. Then we may replace $\theta(-z)$ by $\rho(z)$ in (8.8) and eventually recover (8.7) with $\delta(z)$ replaced by the derivative $\rho'(z)$. The net result is that the reflectivity in the kinematic approximation may be expressed in the form

$$R(q) = R_F(q)|\hat{\rho}_1(q)|^2 \tag{8.9}$$

where $R_F(q)$ is the Fresnel reflectivity given by (8.3) and $\hat{\rho}_1(q)$ is the one-dimensional Fourier transform of $\rho'(z)$. The factor $|\hat{\rho}_1(q)|^2$ then describes the influence of any inhomogeneties, normal to the surface, on the specular reflection (Als-Nielsen 1984, 1985). The detailed theory of specular reflection of neutrons from inhomogeneous surface layers has been given by Steyerl *et al* (1991) and by Leckner (1991), and the application of polarized neutron reflectometry have been discussed by Majkrzak (1991).

Although the study of surfaces and interfaces by X-ray reflectometry is a well-established technique (Parratt 1954, Marra *et al* 1979) the application of neutron reflectometry to problems in the same field is a relatively recent development (Penfold and Thomas 1980, Felcher 1981, Felcher *et al* 1984, Als-Nielsen 1986, Stamm *et al* 1989). In spite of the substantially lower relative brightness of available neutron sources, neutrons offer certain advantages as compared with X-rays, particularly in relation to their sensitivity to magnetism, and the potential for altering the contrast by substituting deuterium for hydrogen (cf. sections 8.2.1 and 8.4.1). Perhaps for these reasons the growth of activity in neutron reflectometry in recent years has been spectacular (Felcher and Russell 1991).

An important requirement for a neutron reflectivity experiment is that the neutron beam be well collimated in order to distinguish clearly between specular and diffuse reflection. With this condition assured two main techniques are employed. In the first of these the neutron wavelength is kept constant and q is varied by scanning over a wide range of θ. Thus the response pattern is developed on a point-by-point basis (Penfold 1991). This is a technique which is suitable for reactor based instruments such as the neutron reflectometer TOREMA at Jülich (Stamm *et al* 1989, 1991) or the low resolution diffractometer/small angle scattering spectrometer D17 at Grenoble. The former instrument operates at a fixed wavelength of 4.3 Å over a range of reflectivities between 1 and 10^{-4}, while the latter can operate at selected wavelengths in the range 8 Å$\leqslant \lambda \leqslant$20 Å with a wavelength resolution between 5 and 10%.

The second technique makes use of a white incident neutron beam and the reflected beam is momentum analysed by time-of-flight techniques. This makes it more suitable for use with pulsed sources and it has the additional advantage that it samples the whole range of the q-spectrum at the same time. An instrument which operates in this mode is the reflectometer CRISP at the spallation neutron source ISIS (cf. section 1.6.7) (Penfold *et al* 1987, Penfold 1991). This covers the momentum transfer range 3×10^{-3} Å$^{-1} < q <$ 0.65 Å^{-1} and is primarily applied to the study of surface chemistry, surface magnetism, polymers, solid films and biological materials. A feature of the time of flight reflectometer POSY II at the Argonne National Laboratory's IPNS facility is that it measures reflectivities as low as 10^{-6} over a momentum transfer range $0 < q < 0.1$ Å^{-1} (Karim *et al* 1991).

8.1.3 Neutron topography

8.1.3.1 Neutron diffraction topography

The construction of highly magnified images in optical and electron microscopes is possible only because beams of optical photons or electrons can be readily made to deviate, using dispersive media or electromagnetic fields. This is not so easy to achieve using beams of X-rays or thermal neutrons, since neither is electrically charged and the corresponding refractive indices in dispersive media are each very close to unity. Nevertheless several clever techniques have been devised to avoid these difficulties; e.g. it is possible by radiographic means to construct three-dimensional images, a procedure known as tomography, by combining a series of shadowgraphs taken at various relative inclinations and rotations. By using the magnetic force acting on the spin one may split a beam of cold neutrons into two beams of opposite polarization in a Stern–Gerlach apparatus (cf. section 1.5.3.2), and the same force has been successfully applied to confine ultra-cold neutrons of one spin sense in a magnetic bottle (cf. section 9.3.7.3). Similarly cold neutrons ($\lambda \simeq 20$ Å) may be focused using a zone plate (cf. section 7.2.2) and a working microscope has been constructed for use with ultra-cold neutrons ($\lambda \simeq 700$ Å) (Kerrmann *et al* 1985). However, these techniques rely critically on having an exceptionally long neutron wavelength, and cannot be made to work with thermal neutrons ($\lambda \leqslant 10$ Å), which alone are suitable for the detailed study of matter in the condensed state.

There is, however, one phenomenon which does introduce a finite deviation between incident and emergent beams, and which is common to both X-ray and neutron propagation. This is the phenomenon of Bragg diffraction in near-perfect crystals (cf. sections 7.2.3–7.2.4). In its simplest and most direct application, Bragg diffraction provides the basis for a method of image formation known as diffraction topography. However, used as a technique for beam splitting by division of amplitude in X-ray or neutron interferometry (cf. section 7.3.3), it also indirectly underpins an essentially distinct technique for image formation known as phase topography.

Since the angular width of an X-ray beam, Bragg diffracted from a thick crystal, is proportional to the structure function $F(\boldsymbol{q})$ (cf. section 7.2.4.1), and therefore has values of order 10^{-5} radian, even very small misorientations in the lattice planes can produce major changes in reflectivity. The detection of these changes is the objective of diffraction topography, which provides two-dimensional images of bulk defects in near perfect crystals, produced by strain fields, impurities or other imperfections. The basic idea of a topograph is illustrated in figure 8.2, where the sample S, and the photographic plate P, are kept as close together as possible, limited only by the requirement that the plate should not intercept the directly transmitted beam. The essential difference between a diffraction topograph and a diffraction pattern is that, in the latter case, the detector is placed a long way from the sample. Its response therefore represents an average over the whole volume of the sample, providing an image of the reciprocal space only.

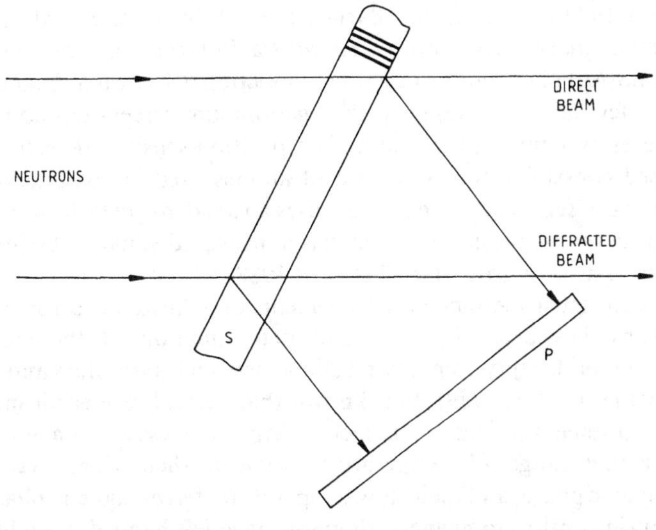

Figure 8.2. The principle of neutron diffraction topography applied to the location of imperfections in near-perfect crystals. The sample S and the photographic plate P are kept as close as possible, consistent with the requirement that the plate must not intercept the direct beam (Baruchel *et al* 1981).

Because it is possible with X-rays to adjust the diffracting crystal so that the wavelength at exact Bragg incidence coincides with an X-ray wavelength characteristic of the target in the X-ray tube, low beam divergences and high intensites are available, leading to high spatial resolution in the image. Thus several modes of operation are possible with different scans of the sample providing 'projection topographs', suitable for use when the cross-section of the X-ray beam is greater than the sample thickness (Lang 1958, 1959), or 'section topographs' which require narrow beams and thick crystals (Kato and Lang 1959, Kato 1969). However, because of intensity limitations on neutron beams, larger beam divergences and wavelength spreads must be tolerated. Thus only projection topographs may be constructed with poor spatial resolutions in the range $\geqslant 50\ \mu m$.

Neutron diffraction topography was first developed in Japan, where it was applied to the study of Pendellösung fringes (cf. section 7.2.4.4) in wedge-shaped single crystals of silicon (Kikuta *et al* 1971), and to the detection of strain fields in single crystals of germanium (Doi *et al* 1971). Spin-waves in antiferromagnetic chromium at low temperatures provided the first example of a neutron topographic study of a magnetic phenomenon (Ando and Hosoya 1972, 1978). Subsequently a dedicated neutron topography facility was established on the double axis neutron diffractometer S20 at the ILL Grenoble. The diffractometer operates with thermal neutrons and may be used with both unpolarized and polarized beams (Baruchel and Schlenker 1980, Schlenker and Baruchel 1986). In the standard operational mode preliminary images are taken with a Polaroid camera coupled to a ^{6}LiF–ZnS(Ag) phosphor as described in section 8.1.1, while final exposures lasting a few hours are taken with a gadolinium screen-film combination. Quite recently a 'multi-photo-multiplier' position-sensitive detector has been developed consisting of a 4×4 grid of anodes used in conjunction with a standard ^{6}LiF scintillator. The system uses an axial magnetic field to maintain a moderate spatial resolution which has a measured value $\simeq 200\ \mu m$ over an $8 \times 8\ mm^2$ sensitive area (Baruchel *et al* 1988).

One of the most spectacular achievements of polarized neutron diffraction topography has been the successful determination of the shapes and orientations of chirality domains in helimagnets such as terbium and holmium (Baruchel *et al* 1981, 1989). It is known that several rare earth metals and alloys, and some other materials such as MnP, possess a helimagnetic phase in a narrow range of temperature between their (high temperature) paramagnetic phase, and their (low temperature) ferromagnetic phase, where neighbouring anti-ferromagnetic domains have left handed or right handed spiral structure, a condition described as chirality (cf. section 8.4.1.3). Although the ratios of the volumes occupied by each chirality can be determined by polarized neutron diffraction (Felcher *et al* 1971, Siratori and Kita 1980) yielding values usually of the order of unity but occasionally as high as three, polarized neutron diffraction topography provides a unique means for determining their sizes, shapes and orientations.

It turns out that the domain shape is highly dependent on sample history and, although domains produced in the transition from the paramagnetic phase have no simple crystallographic orientation, domains produced from the ferromagnetic phase lie in planes parallel to the helix axis (Patterson *et al* 1985). An even more remarkable discovery is that of a memory effect, whereby domains are located in the same place, with the same chirality, after successive thermal cycles. The physical size of chirality domains is strongly dependent on sample purity, with dimensions of order 1 mm when purity is high. However, with decreasing purity the domain size falls below the minimum detectable by topographic techniques (Palmer *et al* 1986).

8.1.3.2 Neutron phase topography

Whereas in neutron diffraction topography the introduction of defects into an otherwise perfect crystal is accompanied by local changes in the Bragg reflectivity, in neutron phase topography the objective is to detect local inhomogeneities by the phase shifts they impress on a transmitted neutron beam. These phase shifts are then detected by placing the sample in one arm of a neutron interferometer (cf. section 7.3.3). Although neutron interferometry has been applied to measure concentrations of hydrogen and deuterium in crystals of vanadium and holmium (Rauch *et al* 1978), phase imaging of bulk specimens has not been widely applied in practice, essentially because of the presence of internal phase patterns which makes interpretation of the experimental data difficult. The main problem is the reduction in resolution caused by the width of the Borrmann fan in each plate, an effect which can be eliminated only by the use of thin diffracting plates.

The layout of a skew symmetric LLL neutron interferometer with well separated beams, which has been used to construct neutron phase topograms is shown in figure 8.3 (Bonse 1979). The 0.6 mm thick crystal plates are cut from a monolithic single silicon crystal in which O and H beams perform three reflections in symmetrical Laue geometry, and images are observed in both emergent beams using standard screen-film combinations. Of the two distinct forms of contrast in a phase topogram the most important is 'area contrast'; this is related to variations in the thickness of the sample crystal and shows up as equal thickness fringes in the case of a wedge-shaped film. The second kind of contrast has been described as 'contour contrast', and this remains even when one of the interfering beams is blocked off by a cadmium sheet. Contour contrast is an internal interference effect which occurs when a sharp step on the surface of the sample introduces a phase shift between different portions of the Borrmann fan, between the second pair of plates M_2 and A in figure 8.3. Similar effects have been observed in a two-crystal interferometer which uses wave spreading in the Borrmann fan to create two coherent beams (Kikuta *et al* 1975).

When ferromagnets are used as phase objects, the effect on the transmitted neutron beam is to rotate the spin as well as to alter the phase, a phenomenon which permits the phase-contrast imaging of magnetic domain walls using polarized neutrons. The technique has been applied to the study of ferromagnetic domains in Fe–3% Si crystals, where domain contrast was observed in the interference pattern in the presence of a magnetic field, and wall contrast was detected in the absence of a field (Schlenker *et al* 1980).

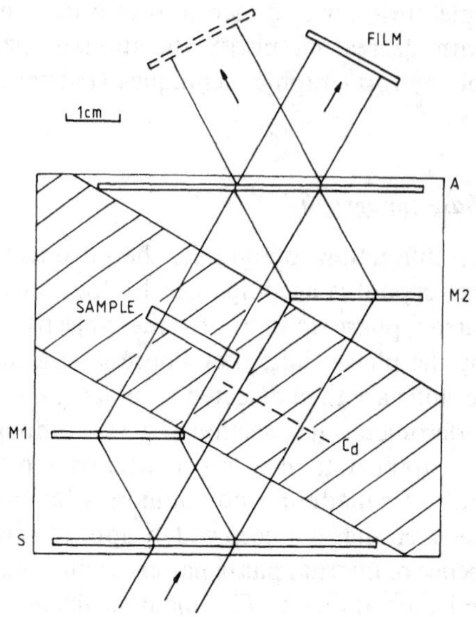

Figure 8.3. The principle of neutron phase topography whereby an image of an inhomogeneous sample is formed through the locally varying phase shift it produces in a transmitted beam. A skew-symmetric triple Laue neutron interferometer with four mirror wafers S, M_1, M_2, A, is used to allow acceptance of large samples without intercepting both interfering beams (Bonse 1979).

8.2 Neutron scattering in condensed systems

8.2.1 The neutron in condensed matter research

The most familiar application of the intense neutron fluxes present in high power nuclear reactors is the direct generation from fissionable fuel of nuclear energy for commercial purposes (cf. chapter 5). Radioisotopes constitute an important side-product of the fission process and these have found practical applications in all manner of activities including mechanics, geology, archaeology and industry of all types. The high neutrino fluxes generated by power reactors have also been applied towards scientific ends, including the first direct detection of the electron anti-neutrino (Cowan *et al* 1956), and the continuing search for neutrino oscillations (Mössbauer 1984). However the primary motivation for constructing fission reactors of intermediate power in the range 10–100 MW has been to extract intense beams of neutrons, with source fluxes at the level of 10^{15} cm^{-2} s^{-1}, to be applied to the determination of structure and dynamics in condensed matter systems. This is the realm of neutron scattering, a vast and continuously growing field of research with its own language and discipline, to which it is impossible to do justice in the limited space available. Fortunately the field is covered in depth by a number of excellent text books (Lovesey 1984, Bée 1988, Newport *et al* 1988) and innumerable reports and conference proceedings (Lander and Robinson 1986, Birgenau *et al* 1986, Gläser *et al* 1989, McEwan *et al* 1992).

Condensed matter can assume a wide range of forms, crystalline, liquid, amorphous, etc., depending on the degree of thermodynamic ordering and the conditions of temperature and pressure to which it is subjected. Within these broad categories there exists a bewildering variety of substates or phases, e.g. magnetic phases, superconductors, superfluids, liquid crystals, polymers, macromolecules — the list is endless — all of whose properties can be elucidated to a greater or lesser extent by the non-destructive techniques of neutron scattering. Of the wide range of condensed state phenomena which have been investigated by neutron scattering, and the even vaster catalogue of specific materials, a very restricted sample is listed in Table 8.2. Each and every one of these topics has its own particular difficulties and features of interest which cannot be pursued here, and the details must be sought in the specialist literature.

In the present and subsequent sections we shall only attempt to summarize the principles of neutron scattering from many-particle systems, and to illustrate these principles by looking in a little more detail at some idealized systems, e.g. phonon propagation in an infinite perfect crystal with Bravais lattice (cf. section 8.3.5). This particular system is made relatively simple because the periodicity of the crystal allows a reduction of the many-body aspects of the problem to one of counting normal modes, while translational invariance permits us to restrict attention to the properties of the unit cell. Another problem we shall discuss in some detail although at an elementary level, is magnetic scattering from ordered systems (cf. section 8.4). This topic has been chosen, not only because of the scientific and technological importance of magnetic materials, but because it allows the possibility of demonstrating the value and potential of polarized neutron scattering in condensed matter research.

Considered as a probe of the behaviour of matter in its various macroscopic states, the principal advantages the neutron enjoys in comparison with other projectiles, e.g. protons, deuterons, α-particles, is that it is uncharged, a feature whose significance was emphasized by Rutherford many years before the neutron was discovered (cf. section 1.2). Neutrons hold this property in common with X-rays although, unlike X-rays, they do not cause ionization, and it is primarily through their nuclear interactions that they sample the static and dynamic states of matter. In addition neutrons have magnetic moments which means they can be used to determine the magnitude and direction of the magnetization in the sample. A second important property, shared with X-rays, is that the neutron–matter interaction can be treated as a perturbation and the corresponding scattering process described in Born

Table 8.2. A selection of topics in physics, chemistry and biology which have been studied by means of elastic and inelastic neutron scattering. The materials or phenomena included are by no means exhaustive, and are chosen to illustrate the range and power of the neutron scattering technique (cf. Proceedings of International Conferences on Neutron Scattering: Lander and Robinson 1986, Birgenau *et al* 1986, Gläser *et al* 1989, McEwan *et al* 1992)

Crystallography	Magnetism
Crystal structure (ionic, covalent, molecular)	Paramagnetism
Long range order	Ferromagnetism
Location of hydrogen atoms in crystals	Anti-ferromagnetism
Liquid crystals (translational disorder)	Helimagnetic order
Plastic crystals (orientational disorder)	Ferrimagnetism
Alloys	Superexchange
Ferroelectrics	Magnetic structures
Incommensurate structures	Spin densities and form factors
Low-dimensional solids	Magnetic excitations
High T_c superconductors	Spin waves
Heavy fermions	Magnon dispersion curves
Lattice dynamics	Magnetic alloys (FeCo, NiMo)
Phonon dispersion curves	Magnetic superconductors
Phonon interactions	Spin glasses
Structural phase transitions	Phase transitions
Critical phenomena	Magnetic critical scattering
Disordered materials	**Chemistry**
Intermediate range order	Molecules in aqueous solution
Defects in solids	Hydrogen bonding
Glasses	Metal hydrogen bonds
Liquid–glass phase transition	Molecular dynamics (vibration, rotation, diffusion)
Fractals	Rotational tunnelling
Amorphous semiconductors	Dynamics of adsorbed systems
Critical scattering of fluids	Polymers and gels
Diffusion	Micelles
Chaos	
Liquids	**Biology**
Structure and dynamics of liquids	Biological crystals
Short range order	Macromolecules
Liquid H_2 and D_2	Protein crystallography
Molecular liquids	DNA and nucleo proteins
Binary liquid alloys	Haemoglobin
Multicomponent liquids	Cell walls and membranes
Ionic solutions	Bacteria
Quantum liquids (^4He, ^3He)	Viruses

approximation. For reasons discussed in section 2.3.1, this result remains true despite the fact that individual encounters between neutrons and nuclei are not amenable to a perturbation treatment.

Excluding ultra-cold neutrons which provide the subject matter for a separate discussion (cf. section 1.6.6), slow neutrons used in condensed matter research cover a wide spectrum of wavelengths in the range $0.3\ \text{Å} \leqslant \lambda \leqslant 20\ \text{Å}$, with a typical thermal value $\lambda \simeq 1.8\ \text{Å}$ which is comparable with the mean interatomic separation in most substances. The corresponding neutron wavevectors $\boldsymbol{k}$ therefore match the dimensions of Brillouin zones in crystals, which makes thermal neutrons ideal for determining three-dimensional structures by diffraction. Broadly speaking neutron scattering can distinguish details of structure over a very wide spatial range, stretching from as little as $10^{-2}\ \text{Å}$ to perhaps $10^3\ \text{Å}$. The corresponding energy of 24 meV for a thermal neutron is also comparable with the energies of quantized excitations in condensed matter, e.g. phonons, magnons, plasmons, etc., which means that neutron scattering also has a spectroscopic function (Weinstock 1944). Thus dynamics rather than structure provides the motivation for most inelastic neutron scattering experiments. In this respect neutron scattering contrasts favourably with X-ray scattering where the $\simeq 67$ keV energy carried by an X-ray quantum with $\lambda \approx 1.8\ \text{Å}$ is six orders of magnitude higher than a characteristic excitation in a condensed medium. This situation could however change in the future when inelastic scattering experiments may prove feasible at new intense sources of synchrotron radiation with energy transfers of 50 meV or greater.

There are three further advantages that neutrons possess in comparison with X-rays; the first is that, whereas X-ray cross-sections increase in proportion to Z^2, there is no systematic variation with Z for neutron-nuclear cross-sections, which may indeed display marked differences between isotopes of the same chemical element, e.g. ^{1}H and ^{2}H. This means that the positions of light atoms, the hydrogen atom in particular, may be established just as precisely as the positions of heavy atoms. Second, the ratio of absorption to coherent scattering is in general greater than unity for X-rays, whereas the reverse is usually the case for neutrons. It follows that extinction of the beam is less significant for neutrons than it is for X-rays. Thirdly there is no Compton background as in X-ray scattering although it is necessary to take account of the Doppler shift produced by nuclear recoil.

Since nuclear scattering lengths are typically four orders of magnitude less than the de Broglie wavelength of a thermal neutron, it follows that there is no q-dependent form factor, analogous to the atomic form factor, for scattering at an individual nucleus. Thus the cross-section for scattering from a many-particle system may be separated into two factors, namely the square of the relevant nuclear scattering length and a function $S(\boldsymbol{q}, \omega) > 0$ which depends only on the momentum $\hbar\boldsymbol{q}$, and the energy $\hbar\omega$, transferred between the neutron and the scattering medium. The so-called 'scattering law' $S(\boldsymbol{q}, \omega)$ is determined solely by the dynamics of the scattering medium, as expressed by the fluctuation-dissipation theorem of non-equilibrium thermodynamics (cf. section 8.2.3). This theorem relates the first-order response of a perturbed system to the random fluctuations of the generalized coordinates of the system about their equilibrium values. In the case of nuclear scattering it is the particle density which is perturbed by the bombarding neutron, and $S(\boldsymbol{q},\omega)$ is interpreted as the spectrum of density fluctuations in the density mode characterised by the wave vector $\boldsymbol{q}$. In the Van Hove formulation (cf. section 8.3) $S(\boldsymbol{q}, \omega)$ translates into the space time function $G(\boldsymbol{r}, t)$, which measures the correlation between two particles displaced in space-time by an amount $(\boldsymbol{r}, t)$. When these two particles are the same particle the scattering is described as incoherent, when they are different particles the scattering is coherent.

The interpretation of magnetic scattering is somewhat more complicated in that, in addition to $\boldsymbol{q}$, the scattering also depends on the directions of the local magnetization and the neutron spin polarization. Furthermore, since the magnetization is distributed over a volume comparable with atomic rather than nuclear dimensions, the magnetic scattering length is $\boldsymbol{q}$-dependent in general. Nevertheless, the relation between the observed scattering cross-sections and fluctuations in the spin systems within the medium, remains as before.

8.2.2 Neutron scattering in the interaction representation

In section 2.2 we discussed the collision of two particles interacting through a central potential $V(r)$, suitably restricted so that asymptotically incoming and outgoing states behave like eigenstates of the unperturbed Hamiltonian, of the same energy as the total energy of the interacting system. Evidently this description cannot be quite accurate since, in order to bring about the required spatial separation of the two particles in the remote past and in the distant future, incoming and outgoing states must be represented by wavepackets (cf. section 6.4.4.2). Nevertheless, even for complex collisions, e.g. neutron collisions with atoms or solids, this procedure may be legitimized by a device (Lippmann and Schwinger 1950) whereby the interaction is switched on and off again adiabatically, i.e. so slowly that no transitions are generated by the switching process.

In time-dependent scattering theory the switching process is realized by writing the total Hamiltonian in the non-conservative form

$$H = H_0 + H_{\text{int}}\, e^{-\varepsilon t/\hbar} \qquad \varepsilon > 0 \tag{8.10}$$

where ε is ultimately allowed to tend to zero so that energy conservation is recovered. The parameter ε has the dimensions of energy and evidently provides a measure of the uncertainty in the energy of the interaction which is associated with the finite duration $2\hbar/\varepsilon$ of the collision. In the time-independent formulation of scattering theory the finite energy width ε

is represented by the appearance of small imaginary terms $\pm i\varepsilon$ in the energy, where the selection of sign serves to distinguish between incoming (−) and outgoing (+) solutions (cf. section 2.3.1).

Up to this point we have chosen to describe all scattering processes in the Schrödinger representation of quantum states, where the state vectors $|\Psi_s(t)\rangle$ describing pure states, or density matrices $\rho_s(t)$ describing mixed states (cf. section 6.4.3), have a wave-like time dependence, whereas the Hermitian operators A_s which characterize the observables of the system are time independent. The Heisenberg representation, on the other hand, more closely resembles the classical description in that the state vectors $|\Psi_H\rangle$ are constant in time, whereas the operators $A_H(t)$ vary in time according to the rule

$$i\hbar \frac{\partial}{\partial t} A_H(t) = -[H, A_H(t)] \tag{8.11}$$

which is the quantum equivalent of Hamilton's equations for their classical analogues. There also exists an intermediate representation, the so-called interaction representation, which is peculiarily suited to the description of complex collisions, since it specifically places the emphasis on those changes in the total system which are brought about by the interaction between its component parts.

In the interaction representation the operators $\bar{A}(t)$ vary in time as if there were no interaction present, i.e.

$$i\hbar \frac{\partial}{\partial t} \bar{A}(t) = -[H_0, \bar{A}(t)] \tag{8.12}$$

whereas the state vectors $|\bar{\Psi}(t)\rangle$, and their basis vectors, have a time dependence determined by the interaction Hamiltonian alone

$$i\hbar \frac{\partial}{\partial t} |\bar{\Psi}(t)\rangle = \bar{H}_{\text{int}}(t)|\bar{\Psi}(t)\rangle \tag{8.13}$$

For unperturbed free motion the interaction representation reduces to the Heisenberg representation.

As may be verified by direct substitution the interaction representation may be derived from the Schrödinger representation by the following transformations

$$|\bar{\Psi}(t)\rangle = e^{iH_0t/\hbar}|\Psi_s(t)\rangle \tag{8.14}$$

$$\bar{A}(t) = e^{iH_0t/\hbar} A\, e^{-iH_0t/\hbar} \tag{8.15}$$

Thus $\bar{H}_0 \equiv H_0$ is time independent and $\bar{H}_{\text{int}}(0)$ may be identified with the corresponding Schrödinger operator. It may be shown immediately using (8.14–8.15) that $\bar{\rho}(t)$, the density matrix in the interaction representation, varies in time according to

$$i\hbar \frac{\partial}{\partial t} \bar{\rho}(t) = [\bar{H}_{\mathrm{int}}(t), \bar{\rho}(t)] \tag{8.16}$$

This result replaces (6.124) which describes the time behaviour of $\rho_s(t)$ and is equivalent to the integral equation

$$\bar{\rho}(t) = \bar{\rho}(-\infty) + \frac{1}{i\hbar} \int_{-\infty}^{t} [\bar{H}_{\mathrm{int}}(t'), \bar{\rho}(t')] \, \mathrm{d}t' \tag{8.17}$$

8.2.3 The fluctuation-dissipation theorem

Determination of structure in condensed media using the techniques of X-ray or neutron scattering requires that we observe the behaviour of an elementary particle in a well-defined, if not necessarily pure, state, interacting with a complex many-particle system possessing an enormous number of degrees of freedom, which is prepared in a state of thermodynamic equilibrium. The experimental problem consists first of all in deducing how the complex system responds to the stimulus provided by the scattered particle, and second, in analysing the relationship between the response and the structure of the system in its equilibrium state. This is a very familiar scenario which is frequently encountered in many areas of physics, and it is the central problem which is addressed in non-equilibrium thermodynamics.

An intuitively appealing solution to this problem, whose relevance to the interpretation of X-ray and neutron scattering data appears to have been first appreciated by Van Hove (1958) is expressed by the fluctuation-dissipation theorem (Callen and Welton 1951, Lax 1960, Kubo 1966, Landsberg and Cole 1967) which establishes a relation between the first-order response of a macroscopic quantum system coupled to an elementary signal generator and the fluctuations of the system in equilibrium.

The origins of this theorem may be traced to Einstein's classic paper (Einstein 1906), where he predicts that Brownian motion of the charge carriers should generate a fluctuating voltage across the ends of an electrical resistance R, maintained in thermal contact with a heat bath at temperatures T. Following Johnson's (1928) experimental verification of this prediction, Nyquist (1928) derived an expression for the power spectrum of the voltage fluctuations

$$w(\nu) = 4kTR(\nu) \tag{8.18}$$

This relation connects the linear response of the system when subjected to an applied voltage of frequency ν, i.e. the induced current as characterized by the dissipative element $R(\nu)$ according to Ohm's law, to the mean power per unit frequency interval, $w(\nu)$, generated by the voltage fluctuations in the equilibrium state.

In fact Nyquist's law is of much greater generality than this simple application would suggest for, if the voltage generator is replaced by a classical particle undergoing sychrotron oscillations at frequency ν, whose energy is radiatively dissipated through its coupling to the quantum electrodynamic vacuum, then (8.18) may be rewritten in the form

$$w(\nu)=\langle E_i^2(\nu)\rangle=4kT\cdot 2\omega^2/3c^3 \tag{8.19}$$

where $2\omega^2/3c^3$ is the radiation resistance and $E_i(\omega)$ is the ith component of the fluctuating electric field with $\langle E_i(\omega)\rangle=0$. Equation (8.19) may now be converted into an expression for the energy density $u(\omega)$, by multiplying both sides by $6/8\pi$ to take account of the six components of the electromagnetic field. The resultant relation reduces to a re-statement of Planck's law when the equipartition value kT for the mean energy of an oscillator is replaced by the quantum value

$$E(\omega, T)=\hbar\omega/(e^{\hbar\omega/kT}-1) \tag{8.20}$$

To state the fluctuation-dissipation theorem in a form which may be applied to the interpretation of neutron scattering by condensed media in the Born approximation, we may suppose that the neutron–medium interaction Hamiltonian is expressible in the form

$$\bar{H}_{\text{int}}(t)=-\sum_j f_j(t)\bar{q}_j^\dagger(t) \tag{8.21}$$

where $\bar{q}_j^\dagger(t)$ are interaction representation operators characterizing macroscopic properties, e.g. density and magnetization, of the medium and the $f_j(t)$ are classical functions representing the forces exerted on the medium by the impressed signal. For example in the case of an incident electromagnetic wave these forces would be the components of the electric field $\boldsymbol{E}$ acting on the induced electric dipole polarization $\boldsymbol{p}_e$, and the $\bar{q}_j^\dagger(t)$ would be interpreted as creation operators describing the creation of the corresponding electric dipole excitations (cf. section 2.1.3). In general the operators $\bar{q}_j^\dagger(t)$ are non-Hermitian but $(\bar{q}_j^\dagger(t)+\bar{q}_j(t))$ and $i(\bar{q}_j^\dagger(t)-\bar{q}_j(t))$ are Hermitian operators, as indeed are the anti-commutator and (i times) the commutator of any two operators $\bar{q}_j^\dagger(t)$ and $\bar{q}_i(t')$.

In the case of an incoming neutron the $f_j(t)$ would be scattering lengths or components of the Stokes vector describing the polarization state of the neutron in the case of spin-dependent scattering. In any case the quantum properties of the 'signal generator', whether X-ray or neutron, are essentially irrelevant to the description of the interaction process.

For simplicity it is convenient to define the operators $\bar{q}_i(t)$ such that each has mean value zero in the equilibrium state, i.e.

$$\langle\bar{q}_i(-\infty)\rangle\equiv\langle\bar{q}_i\rangle^{(0)}=\text{Tr}\,(\bar{\rho}(-\infty)\bar{q}_r(-\infty))=0 \tag{8.22}$$

where, according to (6.156) and (6.157)

$$\bar{\rho}(-\infty)\equiv\bar{\rho}^{(0)}=\mathrm{e}^{-\mu H_s}/Z \tag{8.23}$$

and Z is the canonical partition function for the unperturbed medium.

To derive the first-order response of the medium to the applied signal we solve (8.17) to first order by inserting $\bar{\rho}^{(0)}$ for $\bar{\rho}(t')$ in the integral, i.e.

$$\bar{\rho}(t)^{(1)}=\bar{\rho}^{(0)}+\frac{1}{i\hbar}\int_{-\infty}^{t}[\bar{H}_{\text{int}}(t'),\bar{\rho}^{(0)}]\,\mathrm{d}t' \tag{8.24}$$

We then find, multiplying (8.24) to the right by $\bar{q}_j(t)$ and taking the trace,

$$\langle\bar{q}_j(t)\rangle^{(1)}\equiv\mathrm{Tr}\,(\bar{\rho}^{(1)}(t)\bar{q}_j(t))=\frac{1}{i\hbar}\sum_i\int_{-\infty}^{t}f_i(t')\langle\bar{q}_i^{\dagger}(t'),\bar{q}_j(t)]\rangle^{(0)}\mathrm{d}t' \tag{8.25}$$

a result which is often referred to as Kubo's equation (Kubo 1957).

Since the response of a linear system subjected to a superposition of perturbations is just the sum of the separate responses, the solution of (8.25) for arbitrary $f_i(t)$ is determined by the retarded Green function

$$G_{ij}^{\mathrm{r}}(t)=\langle[\bar{q}_i^{\dagger}(0),\bar{q}_j(t)]/i\hbar\rangle\theta(t) \tag{8.26}$$

where

$$\theta(t)=1\qquad t\geqslant 0\qquad \theta(t)=0\qquad t<0 \tag{8.27}$$

$G_{ij}^{\mathrm{r}}(t)$ is the response in $\bar{q}_j$ at time t to a unit impulse applied to $\bar{q}_i$ at zero time, brought about by the coupling between all the operators of the condensed system, and its restriction to $t\geqslant 0$ imposed by (8.27) is a consequence of causality which requires that there be no present response to future inputs.

Because $\bar{\rho}(0)$ describes a stationary ensemble the real matrix commutator

$$\varphi_{ij}(t-\tau)=\langle[\bar{q}_i^{\dagger}(\tau),\bar{q}_j(t)]/i\hbar\rangle \tag{8.28}$$

is a function of $(t-\tau)$ only. It is defined for all values of t and its transformation under reversal of the time is given by

$$\varphi_{ij}(-t)=-\varphi_{ji}(t) \tag{8.29}$$

The general solution (8.25) now reduces to

$$\langle\bar{q}_j(t)\rangle^{(1)}=\sum_i\int_0^{\infty}f_i(t-t')\varphi_{ij}(t')\,\mathrm{d}t' \tag{8.30}$$

In particular if we suppose that the applied input $f_i(t)$ is the continuous harmonic signal $\mathrm{e}^{i\omega t}$, then the response $\langle\bar{q}_j(t)\rangle^{(1)}$ is just $\mathrm{e}^{i\omega t}\chi_{ij}(\omega)$ where

$$\chi_{ij}(\omega)=\frac{1}{2\pi}\int_0^{\infty}\mathrm{e}^{-i\omega t'}\varphi_{ij}(t')\,\mathrm{d}t' \tag{8.31}$$

is the complex susceptibility of the system. More generally taking the Fourier transform of both sides of (8.30) we have the result

$$\bar{q}_j(\omega)^{(1)} = \sum_i \chi_{ij}(\omega) f_i(\omega) \tag{8.32}$$

where $\bar{q}_j(\omega)^{(1)}$ and $f_i(\omega)$ are the Fourier transforms of $\langle \bar{q}_j(t) \rangle^{(1)}$ and $f_i(t)$, respectively.

The characterization of the response in terms of a susceptibility is clearly appropriate when the driven operator and its response are of the same general nature, e.g. when a system of magnetic moments is perturbed by an applied magnetic field of intensity $\boldsymbol{H}$, and an induced magnetization is observed. However in many cases, e.g. when 'charges' $\langle \bar{q}_i^{(0)} \rangle$ are driven by 'electric fields' $f_i(t)$, it may be convenient to describe the response in terms of induced 'currents' $\langle (\mathrm{d}/\mathrm{d}t)\bar{q}_j(t) \rangle^{(1)}$, in which case (8.32) is replaced by

$$\bar{i}_j(\omega)^{(1)} = \frac{1}{2\pi} \int_{-\infty}^{\infty} \mathrm{e}^{-i\omega t} \frac{\mathrm{d}}{\mathrm{d}t} \langle \bar{q}_j(t) \rangle^{(1)} \, \mathrm{d}t = \sum_i Y_{ij}(\omega) f_i(\omega) \tag{8.33}$$

where, by an obvious analogy,

$$Y_{ij}(\omega) = \mathrm{i}\omega \chi_{ij}(\omega) \tag{8.34}$$

is termed the 'admittance matrix' of the system. Since both $\chi_{ij}(\omega)$ and $Y_{ij}(\omega)$ are Fourier transforms which vanish for $t<0$ they satisfy dispersion relations of the type described in section 2.4.4.

To relate the first-order response to the equilibrium fluctuations in the manner of Nyquist's theorem (8.18), we need to establish a relation between $Y_{ij}(\omega)$ and the Fourier transform of $\varphi_{ij}(t)$ defined by

$$\Phi_{ij}(\omega) = \frac{1}{2\pi} \int_{-\infty}^{\infty} \mathrm{e}^{-i\omega t} \varphi_{ij}(t) \, \mathrm{d}t \tag{8.35}$$

Using the time reversal transformation (8.29) the connection is easily shown to be

$$\mathrm{i}\omega \Phi_{ij}^{(\mathrm{s})}(\omega) = 2 \operatorname{Re} Y_{ij}^{(\mathrm{s})}(\omega) \tag{8.36a}$$

$$\mathrm{i}\omega \Phi_{ij}^{(\mathrm{a})}(\omega) = 2\mathrm{i} \operatorname{Im} Y_{ij}^{(\mathrm{a})}(\omega) \tag{8.36b}$$

where the superscripts (s) and (a) refer to the symmetric and anti-symmetric parts respectively with respect to interchange of the indices (i, j).

If all the operators $\bar{q}_j(t)$ are assumed to be even under reversal of the time, i.e. all magnetic and Coriolis forces are disregarded, then it follows using (8.29) that

$$\varphi_{ij}(-t) = \varphi_{ij}(t) = -\varphi_{ji}(t) \tag{8.37}$$

and the anti-symmetric part of $\varphi_{ij}(t)$ vanishes. Thus $Y_{ij}^{(\mathrm{a})}(\omega) = 0$, a result which is familiar from elementary circuit analysis in the absence of mutual inductive coupling.

Although the operators $\bar{q}_i(0)$ and $\bar{q}_j(t)$ are Hermitian, their product $\bar{q}_i(0)\bar{q}_j(t)$ is not Hermitian since the two operators do not commute in general. However the expectation value $\langle \bar{q}_i(0)^\dagger \bar{q}_j(t)\rangle^{(0)}$ can always be represented as a complex number whose real and imaginary parts are themselves the expectation values of Hermitian operators, i.e.

$$\langle \bar{q}_i(0)^\dagger \bar{q}_j(t)\rangle^{(0)} = \psi_{ij}(t) + \frac{i\hbar}{2}\varphi_{ij}(t) \tag{8.38}$$

where $\varphi_{ij}(t)$ is defined in (8.28) and

$$\psi_{ij}(t) = \frac{1}{2}\langle \bar{q}_i^\dagger(0)\bar{q}_j(t) + \bar{q}_j(t)\bar{q}_i^\dagger(0)\rangle^{(0)} = \psi_{ji}(-t) \tag{8.39}$$

defines a symmetrized cross-correlation function whose classical analogue is simply the cross-correlation coefficient $\langle q_i(0)q_j(t)\rangle^{(0)}$ of the classical functions $q_i(0)$ and $q_j(t)$. The classical analogue of $\varphi_{ij}(t)$ on the other hand is the mean value of the Poisson bracket of $q_i(0)$ and $q_j(t)$. This has the value $\langle q_i(0)\dot{q}_j(t)\rangle/kT$ but its physical interpretation is not immediately evident.

$\Psi_{ij}(\omega)$, the Fourier transform of $\psi_{ij}(t)$ may be linked directly to $\Phi_{ij}(\omega)$ through the identity

$$\langle \bar{q}_i^\dagger(0)\bar{q}_j(t)\rangle^{(0)} \equiv \langle \bar{q}_j(t - i\mu\hbar)\bar{q}_i^\dagger(0)\rangle^{(0)} \tag{8.40}$$

a result which follows from the invariance of the trace of a product of operators with respect to any cyclic permutation of those operators, when applied to products of interaction representation operators in the equilibrium ensemble. Taking the Fourier transform of both sides of (8.40) and forming the commutator and anti-commutator of the transformed equation leads immediately to the result

$$\Phi_{ij}(\omega) = -i\omega\Psi_{ij}(\omega)/\tilde{E}(\omega, T) \tag{8.41}$$

where

$$\tilde{E}(\omega, T) = \frac{\hbar\omega}{2}\coth\left(\frac{\mu\hbar\omega}{2kT}\right) = \frac{\hbar\omega}{2} + E(\omega, T) \tag{8.42}$$

is the mean thermal energy of a quantum oscillator including the zero-point energy. Expressed in terms of $Y_{ij}(\omega)$ through the relations (8.36) the symmetric and anti-symmetric parts of $\Psi_{ij}(\omega)$ now become

$$\Psi_{ij}^{(s)}(\omega) = 2\tilde{E}(\omega, T)\,\mathrm{Re}\,Y_{ij}^{(s)}(\omega)/\omega^2 \tag{8.43a}$$

$$\Psi_{ij}^{(a)}(\omega) = 2i\tilde{E}(\omega, T)\,\mathrm{Im}\,Y_{ij}^{(a)}(\omega)/\omega^2 \tag{8.43b}$$

Since according to the Wiener–Khinchine theorem the function $\Psi_{ij}(\omega)$ represents the spectral density or power spectrum of the correlated fluctuations in the equilibrium state, the results (8.43) constitute the generalization of Nyquist's fluctuation-dissipation theorem (8.18).

Confining attention to the symmetric equation (8.43a) we note that $\tilde{E}(\omega, T)$ is an even function of ω which reduces to kT in the classical limit, and Re $Y_{ij}(\omega)/\omega^2$ is also even because $\varphi_{ij}(t)$ is real. It follows therefore that $\Psi_{ij}^{(s)}(\omega)$ is determined by its positive frequency components only and

$$\int_{-\infty}^{\infty} \Psi_{ij}^{(s)}(\omega)\,\mathrm{d}\omega = 2\int_{0}^{\infty} \Psi_{ij}^{(s)}(\omega)\,\mathrm{d}\omega = 4\int_{0}^{\infty} \tilde{E}(\omega, T)\,\mathrm{Re}\,Y_{ij}^{(s)}(\omega)\,\mathrm{d}\omega/\omega^2 \tag{8.44}$$

In the case of a simple resistor R, the single operator $\bar{q}(t)$ represents the fluctuating charge and the real part of the admittance, Re $Y(\omega)$, is just the reciprocal of R. The spectral density $\Psi(\omega)$ of the charge fluctuations corresponds to the spectral density $\omega^2 R^2 \Psi(\omega)$ of the voltage fluctuations across the resistor. Thus, replacing $\tilde{E}(\omega, T)$ by its classical approximation kT, and confining attention to positive frequencies only, the simple Nyquist relation (8.18) is immediately recovered.

8.2.4 Neutron scattering cross-sections

We consider the case of a free neutron in an initial state $|\boldsymbol{k}_0, s_0\rangle$ of momentum and spin, which undergoes single scattering from a condensed medium composed of N atoms, into a final state $|\boldsymbol{k}, s'\rangle$. We suppose that the medium has available to it a complete set of states $|m\rangle$ whose energy levels ε_m are the eigenvalues of the unperturbed system Hamiltonian H_s. At this stage we consider nuclear scattering only and assume that the neutron–medium interaction can be represented as a Fermi pseudo-potential

$$V_F(\boldsymbol{r}, \boldsymbol{s}; \boldsymbol{r}_j, \boldsymbol{I}_j) = \frac{2\pi\hbar^2}{m_n}\sum_{j=1}^{N} \{(b_j + c_j \boldsymbol{s}\cdot\boldsymbol{I}_j)\,\delta^3(\boldsymbol{r}-\boldsymbol{r}_j)\} \tag{8.45}$$

where $\boldsymbol{r}_j$, $\boldsymbol{I}_j$ are the position and spin vectors respectively of the jth nucleus and b_j and c_j are bound scattering lengths defined in (6.80)–(6.82).

The first-order transition probability per unit time of the interacting system of neutron plus medium is now given by

$$w_{fi} = \frac{2\pi}{\hbar}|\langle f|V_F|i\rangle|^2\,\delta(E_f - E_i) \tag{8.46}$$

where

$$E_i = \varepsilon_i + \frac{\hbar k_0^2}{2m_n} \qquad E_f = \varepsilon_f + \frac{\hbar^2 k^2}{2m_n} \tag{8.47}$$

are the initial and final energies, respectively, of the joint system corresponding to the transfer of momentum $\boldsymbol{q} = \boldsymbol{k}_0 - \boldsymbol{k}$, and energy $E = E_f - E_i$ from the neutron to the medium.

To compute the matrix element $\langle f|V_F|i\rangle$ we assume that the interaction takes place in a quantization volume L^3, so that the general neutron wavefunction has the form

$$\Psi(\boldsymbol{r}, s) = L^{-3/2}\, e^{i\boldsymbol{k}\cdot\boldsymbol{r}}\, \boldsymbol{u}(s) \tag{8.48}$$

It proves convenient to perform the integration over $\boldsymbol{r}$ first and write

$$\langle f|V_F|i\rangle = \langle f, s'|\int V_F(\boldsymbol{r})\, e^{i\boldsymbol{q}\cdot\boldsymbol{r}}\, d^3r|i, s_0\rangle \tag{8.49}$$

where, from this point onwards, the symbols '*f*' and '*i*' refer only to the medium. With the substitution (8.49) the transition rate becomes

$$w_{fi} = \frac{2\pi}{\hbar}\left(\frac{2\pi\hbar^2}{m_n L^3}\right)^2 \left|\langle f, s'|\sum_{j=1}^{N}(b_j + c_j \boldsymbol{s}\cdot\boldsymbol{I}_j)\, e^{i\boldsymbol{q}\cdot\boldsymbol{r}_j}|i, s_0\rangle\right|^2 \delta(E + \varepsilon_i - \varepsilon_f) \tag{8.50}$$

where

$$E = \hbar\omega = \frac{\hbar^2}{2m_n}(k_0^2 - k^2) \tag{8.51}$$

To express the result (8.50) in the form of a differential scattering cross-section, w_{fi} must be divided by the incident neutron flux (cf. section 2.4.1) and multiplied by the number $dn(k)$ of neutron states in a volume L^3 and in a range of d^3k about $\boldsymbol{k}$ where

$$dn(k) \equiv \left(\frac{L}{2\pi}\right)^3 d^3k = \left(\frac{L}{2\pi}\right)^3 \left(\frac{m_n}{\hbar}\right) k\, d\Omega\, d\omega \tag{8.52}$$

Further, since initially the medium is in a state of thermal equilibrium, we must average over its initial microscopic states $|i\rangle$ with probabilities

$$P_i = e^{-\varepsilon_i/kT}/Z \tag{8.53}$$

and sum over all unobserved final states $|f\rangle$.

Since the inclusion of the factor $\delta(E + \varepsilon_i - \varepsilon_f)$ in (8.50) automatically eliminates the contribution of those final states which do not satisfy energy conservation, the sum over final states may be carried out over all the internal states of the medium. These constitute a complete set and hence satisfy the closure property

$$\sum_f |f\rangle\langle f| = 1 \tag{8.54}$$

The resultant cross-section per atom expressed in the variables $(\boldsymbol{q}, \omega)$ is then found to be

$$\frac{d^2\sigma}{d\Omega\, d\omega}(\boldsymbol{q}, \omega) = N^{-1}\left(\frac{k}{k_0}\right)\sum_i P_i\left\{\sum_f |\langle f, \boldsymbol{s}'|\sum_{j=1}^{N}(b_j + c_j \boldsymbol{s}\cdot\boldsymbol{I}_j)\, e^{i\boldsymbol{q}\cdot\boldsymbol{r}_j}|i, \boldsymbol{s}_0\rangle|^2 \right.$$
$$\left. \times\, \delta(\omega + (\varepsilon_i - \varepsilon_f)/\hbar)\right\} \tag{8.55}$$

Each nucleus is characterized by position and spin operators ($\boldsymbol{r}_j$, $\boldsymbol{I}$) and by scattering parameters (b_j, c_j). All these variables are in general correlated but, in certain conditions, when carrying out sums and averages they may be treated as if they were independent.

We first assume that the medium is composed from atoms of a unique chemical species containing a single isotope without nuclear orientation (Turchin 1965). For unpolarized neutrons this last condition ensures that there is no correlation between initial and final spin states of the scattering medium. Thus, in a natural extension of the treatment of unpolarized spin-dependent scattering presented in section 8.3.1.2, we may sum over $\boldsymbol{s}'$ and average over $\boldsymbol{s}_0$. The resultant double differential scattering cross-section per atom then splits into coherent and incoherent cross-sections given by

$$\frac{d^2\sigma}{d\Omega\, d\omega}(\boldsymbol{q}, \omega)_{\text{coh}} = N^{-1}\left(\frac{k}{k_0}\right)\sum_i P_i\left\{\sum_f \left|\langle f|\sum_{j=1}^{N} b_j\, e^{i\boldsymbol{q}\cdot\boldsymbol{r}_j}|i\rangle\right|^2 \delta(\omega + (\varepsilon_i - \varepsilon_f)/\hbar\right\} \tag{8.56}$$

and

$$\frac{d^2\sigma}{d\Omega\, d\omega}(q, \omega)_{\text{inc}} = N^{-1}\left(\frac{k}{k_0}\right)\sum_i P_i\left\{\sum_f \sum_{j=1}^{N} |\langle f|b_{\text{inc}}(I_j)\, e^{i\boldsymbol{q}\cdot\boldsymbol{r}_j}|i\rangle|^2 \right.$$
$$\left. \times\, \delta(\omega + (\varepsilon_i - \varepsilon_f)/\hbar)\right\} \tag{8.57}$$

where $b_{\text{inc}}(I_j)$ is the bound spin-dependent incoherent scattering length defined in 6.83.

To break the correlation between $\boldsymbol{r}_j$ and b_j where more than one isotope of a given chemical species is present we need only assume that these isotopes are distributed at random over the sample. Under these circumstances we may average over the products of scattering lengths in the coherent (double) sum (8.56) with the result

$$\begin{aligned} \langle b_j^* b_k\rangle &= \langle b_j^*\rangle\langle b_k\rangle = \langle |b|\rangle^2 = b_{\text{coh}}^2 \qquad j \neq k \\ &= \langle |b|^2\rangle \qquad j = k \end{aligned} \tag{8.58}$$

Thus in general we may write

$$\langle b_j^* b_k\rangle = b_{\text{coh}}^2 + (\langle |b|^2\rangle - \langle |b|\rangle^2)\,\delta_{jk} \tag{8.59}$$

where the term proportional to δ_{jk} is to be added to the average of the incoherent sum (8.57). In this case the total incoherent scattering length is given by

$$b_{\text{inc}}^2 = \langle b_{\text{inc}}^2(I) \rangle + (\langle |b|^2 \rangle - \langle |b| \rangle^2) \tag{8.60}$$

in agreement with the definition (6.85). The total cross-section may therefore be written in the form

$$\frac{d^2\sigma}{d\Omega\, d\omega}(\boldsymbol{q}, \omega) = \frac{1}{4\pi}\left[\frac{k}{k_0}\right][\sigma_{\text{coh}} S(\boldsymbol{q}, \omega) + \sigma_{\text{inc}} S_s(\boldsymbol{q}, \omega)] \tag{8.61}$$

where the coherent scattering law $S(\boldsymbol{q}, \omega)$ is defined by

$$S(\boldsymbol{q}, \omega) = N^{-1} \sum_{j,k=1}^{N} \sum_i P_i \left\{ \sum_f \langle i | e^{-i\boldsymbol{q}\cdot\boldsymbol{r}_j} | f \rangle \langle f | e^{i\boldsymbol{q}\cdot\boldsymbol{r}_k} | i \rangle \delta(\omega + (\varepsilon_i - \varepsilon_f)/\hbar) \right\} \tag{8.62}$$

The coherent scattering law $S(\boldsymbol{q}, \omega)$ depends on the momentum and energy $(\boldsymbol{q}, \omega)$ transferred by the scattered neutron to the medium, but is otherwise independent of the details of the scattering process. It is therefore a function only of the structure and thermodynamic state of the medium, and it is through the detailed study of $S(\boldsymbol{q}, \omega)$ that the crystalline, chemical or biological properties of the scattering medium may be determined. The incoherent scattering law $S_s(\boldsymbol{q}, \omega)$, which characterizes the incoherent components of the total scattering cross-section (8.61), may be derived from $S(\boldsymbol{q}, \omega)$ by retaining the diagonal terms only in (8.62).

Under the same assumption as to the random distribution of the corresponding isotopes in the sample, equation (8.62) may also be applied to describe neutron scattering in a polyatomic medium, provided only the substitution is made

$$\sigma_{\text{coh}} S(\boldsymbol{q}, \omega) \rightarrow 4\pi \sum_{l=1}^{N_l} \sum_{m=1}^{N_m} (f_l f_m)^{1/2} b_{\text{coh}}^l b_{\text{coh}}^m S^{lm}(\boldsymbol{q}, \omega) \tag{8.63}$$

where $f_n = N_n/N$ is the proportion of the Nth atomic species in the sample and

$$S^{lm}(\boldsymbol{q}, \omega) = (N_l N_m)^{-1/2} \sum_{j_l=1}^{N_l} \sum_{k_m=1}^{N_m} \sum_i \left\{ \sum_f \langle i | e^{-\boldsymbol{q}\cdot\boldsymbol{r}_{j_l}} | f \rangle \langle f | \times e^{i\boldsymbol{q}\cdot\boldsymbol{r}_{k_m}} | i \rangle \, \delta(\omega + (\varepsilon_i - \varepsilon_f)/\hbar) \right\} \tag{8.64}$$

The corresponding incoherent scattering law $S_s^{lm}(\boldsymbol{q}, \omega)$ is derived from $S^{lm}(\boldsymbol{q}, \omega)$ by retaining only the diagonal terms as before.

Because of the relatively large mass differences between the three isotopes of hydrogen, equations (8.63)–(8.64) are also applicable to deuterated or tritiated hydrogenous materials for which incoherent scattering from normal hydrogen is the dominant scattering process (cf. table 6.1).

8.2.5 Scattering by a monatomic perfect gas

Excluding the important case of scattering by hydrogen atoms noted immediately above, $\sigma_{\text{coh}} \gg \sigma_{\text{inc}}$ in general, and in a perfect crystal the scattering is essentially coherent in nature. Indeed in these cases σ_{inc} vanishes identically provided the incident momentum does not lie close to a Brillouin zone boundary. Precisely the opposite situation occurs for the case of neutron scattering in a monatomic perfect gas where, because the positions of the scattering centres are totally uncorrelated, the scattering is essentially incoherent in nature, excluding of course forward scattering with $\boldsymbol{q} \equiv 0$ which is always coherent. Thus every nucleus makes an identical contribution to $S(\boldsymbol{q}, \omega)$ and the corresponding spatial matrix element for the nucleus may be directly calculated with the result

$$\langle i|e^{-i\boldsymbol{q}\cdot\boldsymbol{r}}|f\rangle = \int e^{i(\boldsymbol{K}_f - \boldsymbol{K}_i - \boldsymbol{q})\cdot\boldsymbol{r}}\, d^3r = \delta^3(\boldsymbol{K}_f - \boldsymbol{K}_i - \boldsymbol{q}) \tag{8.65}$$

where $\boldsymbol{K}_i$ ($\boldsymbol{K}_f$) is the initial (final) momentum of the struck nucleus.

Equation (8.65) expresses the conservation of momentum at each collision with a free nucleus and shows that, for given $\boldsymbol{K}_i$ and $\boldsymbol{q}$, $\boldsymbol{K}_f$ is unique. The corresponding energy transfer is thus given by

$$\varepsilon_f - \varepsilon_i = \frac{\hbar^2}{2M}(K_f^2 - K_i^2) = \frac{\hbar^2 q^2}{2M} + \frac{\hbar^2}{M}\boldsymbol{K}_i\cdot\boldsymbol{q} \tag{8.66}$$

and the (incoherent) scattering law assumes the form

$$S_s(\boldsymbol{q}, \omega) = \sum_i P_i \delta\left(\omega - \frac{\hbar^2 q^2}{2M} - \frac{\hbar \boldsymbol{K}_i}{M}\cdot\boldsymbol{q}\right) \tag{8.67}$$

To obtain the total scattering law $S(\boldsymbol{q},\omega)$ we have to add to $S_s(\boldsymbol{q}, \omega)$ the elastic contribution from coherent forward scattering

$$S_{\text{el}}(\boldsymbol{q}, \omega) = (2\pi)^3 \delta^3(\boldsymbol{q})\delta(\omega) n \tag{8.68}$$

where n is the particle number density (cf. 8.111).

Since a free nucleus in a gas has a continuous spectrum of energies the discrete probability P_i appearing in (8.67) must be replaced by the Maxwell–Boltzmann distribution of energies appropriate to an ideal classical gas

$$P_i \to p(\varepsilon_i)\, d\varepsilon_i = \frac{2\pi}{(MkT)^{3/2}} e^{-\varepsilon_i/kT} \varepsilon_i^{1/2}\, d\varepsilon_i \tag{8.69}$$

The resultant expression for $S(\boldsymbol{q}, \omega)$ then reduces to the Gaussian form

$$S_s(\boldsymbol{q}, \omega| = [2\pi(\Delta\omega)^2]^{-1/2} \exp\left\{-\frac{1}{2}\left(\frac{\omega - \omega_r}{\Delta\omega}\right)^2\right\} \tag{8.70}$$

which, for given q^2, represents a normal distribution in the energy transfer $\hbar\omega$, with a mean at $E_r = \hbar\omega_r$, and a standard deviation $\hbar\,\Delta\omega = \sqrt{(2E_r kT)}$, where $E_r = \hbar^2 q^2/2M$ is the recoil energy absorbed by a struck particle initially at rest.

The result (8.69) shows that, in this particularly simple example, $S(\boldsymbol{q}, \omega)$ is a real function which may be interpreted as the spectrum of the absorbed excitation. This is a line spectrum centred at $\omega = \omega_r$, on which is superimposed Gaussian noise due to the random motion of the atoms in the gas, whose presence is detected through the fluctuating Doppler shift $\hbar \boldsymbol{K}_i \cdot \boldsymbol{q}/M$ which contributes to the energy transfer in (8.66). Indeed (8.69) may be inferred directly from the well-known expression for the Doppler broadening of a spectral line

$$\frac{\Delta\omega_D}{\omega} = \sqrt{\left(\frac{kT}{Mc^2}\right)} \tag{8.71}$$

on replacing c, the velocity of light, by the phase velocity of c_p a de Broglie wave propagating with momentum q

$$c_p = \frac{\omega_r}{q} = \frac{\hbar q}{2M} \tag{8.72}$$

8.3 The Van Hove theory of space–time correlations

8.3.1 The intermediate scattering function

The scattering law $S(\boldsymbol{q}, \omega)$ is defined in energy-momentum space by (8.62) and, to determine it, we require to know the complete set of eigenfunctions of the scattering system, which is a demand impossible to meet in general. For purposes of calculating $S(\boldsymbol{q}, \omega)$ in certain models, an intuitively simpler approach may be based on the space–time correlation function $G(\boldsymbol{r}, t)$ introduced by Van Hove (1954), and defined as the four-dimensional Fourier transform of $S(\boldsymbol{q}, \omega)$

$$G(\boldsymbol{r}, t) = \left(\frac{1}{2\pi}\right)^3 \int e^{-i(\boldsymbol{q}\cdot\boldsymbol{r} - \omega t)} S(\boldsymbol{q}, \omega)\, d\omega\, d^3q \tag{8.73}$$

To carry out the transformation we first define an intermediate scattering function

$$\begin{aligned} I(\boldsymbol{q}, t) &= \int e^{i\omega t} S(\boldsymbol{q}, \omega)\, d\omega \\ &= N^{-1} \sum_i P_i \left\{ \sum_f \left(\sum_{j,k=1}^{N} \langle i | e^{-\boldsymbol{q}\cdot\boldsymbol{r}_j} | f \rangle \langle e^{i\varepsilon_f t/\hbar} e^{i\boldsymbol{q}\cdot\boldsymbol{r}_k} e^{-i\varepsilon_f t/\hbar} | i \rangle \right) \right\} \\ &= N^{-1} \left\langle \sum_{j,k=1}^{N} e^{-\boldsymbol{q}\cdot\bar{\boldsymbol{r}}_j(0)} e^{i\boldsymbol{q}\cdot\bar{\boldsymbol{r}}_k(t)} \right\rangle^{(0)} \end{aligned} \tag{8.74}$$

where

$$\bar{r}_k(t) = e^{iH_s t/\hbar} r_k e^{-iH_s t/\hbar} \tag{8.75}$$

is the time-dependent interaction representation operator introduced in (8.15), and the set of final states $|f\rangle$ has been eliminated through application of the closure property (8.54).

Because of the presence of the momentum operators in the system Hamiltonian

$$H_s = \sum \frac{p_j^2}{2m} + V(r_1, r_2, \ldots, r_N) \tag{8.76}$$

the operators $\bar{r}_k(t)$ and $\bar{r}_j(0)$ do not commute in (8.74), except at zero time when they each reduce to the corresponding Schrödinger operator. Accepting only the diagonal terms in (8.74) we may also define an incoherent intermediate scattering function

$$I_s(q, t) = N^{-1} \left\langle \sum_{j=1}^{N} e^{-iq \cdot \bar{r}_j(0)} e^{iq \cdot \bar{r}_j(t)} \right\rangle^{(0)} \tag{8.77}$$

To establish the relationship between $I(q, t)$ and the density correlations of the system we introduce the δ-function operator $\delta^3(\bar{r}_j(t))$ through the definition

$$\delta^3(\bar{r}_j(t)) = \left(\frac{1}{2\pi}\right)^3 \int e^{-iq \cdot \bar{r}_j(t)} d^3 q \tag{8.78}$$

and use it to write the particle density operator in the Fermi potential (8.45)

$$n(r) = \sum_{j=1}^{N} \delta^3(r - r_j) \tag{8.79}$$

in the interaction representation

$$\bar{n}(r, t) = \sum_{j=1}^{N} \delta^3(r - \bar{r}_j(t)) = \left(\frac{1}{2\pi}\right)^3 \int e^{-iq \cdot r} \bar{n}(q, t) d^3 q \tag{8.80}$$

where

$$\bar{n}(q, t) = \bar{n}^\dagger(-q, t) = \sum_{j=1}^{N} e^{iq \cdot \bar{r}_j(t)} = \int e^{iq \cdot r} \bar{n}(r, t) d^3 r \tag{8.81}$$

is the particle density operator in momentum space. In this notation $I(q, t)$ becomes

$$I(q, t) = N^{-1} \langle \bar{n}^\dagger(q, 0) \bar{n}(q, t) \rangle^{(0)} = I^*(-q, t) \tag{8.82}$$

and may therefore be viewed as the time auto-correlation function of the spatial density mode labelled by q.

8.3.2 The detailed balanced condition

Following (8.18) we may now split $I(\boldsymbol{q}, t)$ into real and imaginary parts

$$I(\boldsymbol{q}, t)=\psi(\boldsymbol{q}, t)+\frac{i\hbar}{2}\,\varphi(\boldsymbol{q}, t) \tag{8.83}$$

where the real functions $\psi(\boldsymbol{q}, t)$ and $\varphi(\boldsymbol{q}, t)$ are the analogues of the diagonal matrix functions $\psi_{ii}(t)$ and $\varphi_{ii}(t)$ defined in (8.28) and (8.39). By Fourier inversion of (8.74) $S(\boldsymbol{q}, \omega)$ may now be expressed in the form

$$S(\boldsymbol{q}, \omega)=\psi(\boldsymbol{q}, \omega)+\frac{i\hbar}{2}\,\varphi(\boldsymbol{q}, \omega) \tag{8.84}$$

where

$$\psi(\boldsymbol{q}, \omega)=\frac{1}{2\pi}\int e^{-i\omega t}\,\psi(\boldsymbol{q}, t)\,dt=\psi(-\boldsymbol{q}, -\omega) \tag{8.85}$$

is the real power spectrum of the density fluctuation in the mode $\boldsymbol{q}$ at equilibrium, and

$$\varphi(\boldsymbol{q}, \omega)=\frac{1}{2\pi}\int e^{-i\omega t}\,\varphi(\boldsymbol{q}, t)\,dt=-\varphi(-\boldsymbol{q}, -\omega) \tag{8.86}$$

describes the first-order disturbance in this mode under a perturbation of the system according to (8.25). Since $\varphi(\boldsymbol{q}, \omega)$ is pure imaginary $S(\boldsymbol{q}, \omega)$ is a real function as it must be according to its definition (8.62).

The behaviour of $\psi(\boldsymbol{q}, \omega)$ and $\varphi(\boldsymbol{q}, \omega)$ under inversion of $(\boldsymbol{q}, \omega)$ which corresponds to the case that the neutron absorbs energy and momentum in its interaction with the sample, rather than conversely, is determined by (8.82) and the transformations of $\psi(\boldsymbol{q}, t)$ and $\varphi(\boldsymbol{q}, t)$ under time reversal given in (8.39) and (8.37).

Applying (8.38) and (8.41) we may now write

$$\psi(\boldsymbol{q}, \omega)=\frac{1}{2}\,(1+e^{-\mu\hbar\omega})S(\boldsymbol{q}, \omega) \tag{8.87a}$$

$$\frac{i\hbar}{2}\,\varphi(\boldsymbol{q}, \omega)=\frac{1}{2}\,(1-e^{-\mu\hbar\omega})S(\boldsymbol{q}, \omega) \tag{8.87b}$$

and, by subtraction, obtain the relation

$$S(-\boldsymbol{q}, -\omega)=e^{-\mu\hbar\omega}\,S(\boldsymbol{q}, \omega) \tag{8.88}$$

This result is a statement of the principle of detailed balance, which requires that the probability that a neutron absorbs momentum and energy $(\boldsymbol{q}, \omega)$ in a collision, should be less on average than the probability that it should give up this momentum and energy, by the Boltzmann factor $e^{-\mu\hbar\omega}$, which gives the relative populations of states, differing in energy by $\hbar\omega$, for a system in thermal equilibrium. We note that the detailed balance condition is satisfied by the scattering law (8.69) describing neutron scattering in a monatomic perfect gas. The result (8.88) holds independently for $S(\boldsymbol{q}, \omega)$ and $S_s(\boldsymbol{q}, \omega)$.

Finally, we may combine (8.87b) with (8.41) and (8.43a) to express $S(\boldsymbol{q}, \omega)$ in terms of the functuation dissipation theorem

$$S(\boldsymbol{q}, \omega) = -\frac{2\hbar \operatorname{Im} \psi(\boldsymbol{q}, \omega)}{1 - e^{-\mu\hbar\omega}} \tag{8.89}$$

where $\operatorname{Im} \psi(\boldsymbol{q}, \omega)$ is the dissipative part of the dynamic susceptibility of the system defined by (8.31). Thus in principle $S(\boldsymbol{q}, \omega)$ may be determined by the response of the system to any perturbation which couples to the density.

8.3.3 The frequency sum rule

Since by definition $S(\boldsymbol{q}, \omega)$ is real and positive, it may be viewed as an unnormalized probability density function for ω. Thus, by (8.74), $I(\boldsymbol{q}, t)$ is the corresponding characteristic function, and, provided it behaves suitably at zero time, the moments of the distribution in ω are given by

$$S^{(n)}(\boldsymbol{q}) = \int \omega^n S(\boldsymbol{q}, \omega)\, d\omega = i^n \frac{d^n}{dt^n} I(\boldsymbol{q}, t)\bigg|_{t=0} \tag{8.90}$$

We now expand the operator $\bar{r}_k(t)$ appearing in (8.74) to first order in t

$$\bar{r}_k(t) = \bar{r}_k(0) + \dot{\bar{r}}_k(0)t + \ldots \tag{8.91}$$

where

$$\bar{r}_k(0) = r_k \qquad \dot{\bar{r}}_k(0) = \frac{i}{\hbar}[H_s, \bar{r}_k(0)] = p_k/M = \hbar K_k/M \tag{8.92}$$

Since the operators $\boldsymbol{r}_k$ and $\boldsymbol{p}_k$ do not commute, but each commutes with their commutator, we make use of the identity

$$e^{i\boldsymbol{q}\cdot(\boldsymbol{r}_k + \boldsymbol{p}_k t/M)} = e^{i\boldsymbol{q}\cdot\boldsymbol{r}_k}\, e^{i\boldsymbol{q}\cdot\boldsymbol{p}_k t/M}\, e^{(t/2M)[\boldsymbol{q}\cdot\boldsymbol{r}_k, \boldsymbol{q}\cdot\boldsymbol{p}_k]} \tag{8.93}$$

where

$$[\boldsymbol{q}\cdot\boldsymbol{r}_k, \boldsymbol{q}\cdot\boldsymbol{p}_k] = i\hbar q^2 \tag{8.94}$$

Thus the short-time behaviour of $I(\boldsymbol{q}, t)$ reduces to the form

$$I(\boldsymbol{q}, t) = N^{-1}\left\langle \sum_{j,k=1}^{N} e^{-i\boldsymbol{q}(\boldsymbol{r}_j - \boldsymbol{r}_k)}\left(1 + \frac{i\hbar}{M}\boldsymbol{q}\cdot\left(\boldsymbol{K}_k - \frac{\boldsymbol{q}}{2}\right)t + O(t^2)|\right)\right\rangle^{(0)} \tag{8.95}$$

The zeroth and first moments are immediately obtained from (8.95) with the results

$$S^{(0)}(\boldsymbol{q})=N^{-1}\left\langle\sum_{j,k=1}^{N} \mathrm{e}^{-\mathrm{i}\boldsymbol{q}\cdot(\boldsymbol{r}_j-\boldsymbol{r}_k)}\right\rangle^{(0)} \tag{8.96a}$$

$$S_{\mathrm{s}}^{(0)}(\boldsymbol{q})=1 \tag{8.96b}$$

and

$$S^{(1)}(\boldsymbol{q})=N^{-1}\frac{\hbar\boldsymbol{q}}{M}\cdot\left\langle\sum_{j,k=1}^{N} \mathrm{e}^{-\mathrm{i}\boldsymbol{q}\cdot(\boldsymbol{r}_j-\boldsymbol{r}_k)}\left(\frac{\boldsymbol{q}}{2}-\boldsymbol{K}_k\right)\right\rangle^{(0)} \tag{8.97a}$$

$$S_{\mathrm{s}}^{(1)}(\boldsymbol{q})=\frac{\hbar q^2}{2M} \tag{8.97b}$$

We note that the result (8.97b) agrees with the equivalent result (8.70) for the mean energy transferred per particle of a perfect gas in thermal equilibrium.

Since the chemical forces do not contribute any operators in $S^{(1)}(\boldsymbol{q})$, although they generate terms proportional to $\nabla_k V$ in $S^{(2(}(\boldsymbol{q})$, they can influence $S^{(1)}(\boldsymbol{q})$ only through the wavefunction dependence on V. However when the relevant wavefunctions are real, and this includes the important case that V is a quadratic function of $|\boldsymbol{r}_j-\boldsymbol{r}_k|$, it may be shown directly that (Placzek 1952)

$$\langle i|\mathrm{e}^{\mathrm{i}\boldsymbol{q}\cdot\boldsymbol{r}_k}\boldsymbol{K}_k|i\rangle=\frac{1}{2}\boldsymbol{q}\langle i|\mathrm{e}^{\mathrm{i}\boldsymbol{q}\cdot\boldsymbol{r}_k}|i\rangle \tag{8.98}$$

in which case the interference terms vanish in (8.97a) and this relation may be rewritten as

$$S^{(1)}(\boldsymbol{q})=\frac{\hbar q^2}{2M} \tag{8.99}$$

The sum-rule (8.99) expresses the fact that, even for a system of chemically bound particles under the stated conditions, the mean recoil energy accepted is the same as if the particles were free. One should note that the relevant term $\hbar q^2/2M$ appearing in (8.97) is of purely quantum mechanical origin, and the sum-rule is necessary to ensure that the recoil energy absorbed by the medium is properly taken into account.

8.3.4 The space–time correlation functions

The space–time correlation function $G(\boldsymbol{r}, t)$ defined in (8.73) may now be derived from the momentum space Fourier transform of $I(\boldsymbol{q}, t)$ with the result

$$G(\boldsymbol{r},t)=N^{-1}\sum_{j,k=1}^{N}\left(\frac{1}{2\pi}\right)^3\int \mathrm{e}^{-\mathrm{i}\boldsymbol{q}\cdot\boldsymbol{r}}\langle \mathrm{e}^{-\mathrm{i}\boldsymbol{q}\cdot\tilde{\boldsymbol{r}}_j(0)}\,\mathrm{e}^{\mathrm{i}\boldsymbol{q}\cdot\tilde{\boldsymbol{r}}_k(t)}\rangle^{(0)}\,\mathrm{d}^3q \tag{8.100}$$

Applying the convolution theorem of Fourier transforms to the operator product in the integrand of (8.100), and using (8.78), this result can be recast in the form

$$G(\boldsymbol{r}, t)=N^{-1} \sum_{j,k=1}^{N} \int \langle \delta^3(\boldsymbol{r}+\bar{\boldsymbol{r}}_j(0)-\boldsymbol{r}')\,\delta^3(\boldsymbol{r}'-\bar{\boldsymbol{r}}_k(t))\rangle^{(0)}\,\mathrm{d}^3r' \qquad (8.101)$$

By considering only the diagonal terms of the sum in (8.101) an incoherent or self-correlation function $G_s(\boldsymbol{r}, t)$ can be extracted from $G(\boldsymbol{r}, t)$ and by direct calculation from (8.100), the two functions may be shown to satisfy the integral relations

$$\int G(\boldsymbol{r}, t)\,\mathrm{d}^3r=N \qquad \int G_s(\boldsymbol{r}, t)\,\mathrm{d}^3r=1 \qquad (8.102)$$

Since the operators $\bar{\boldsymbol{r}}_j(0)$ and $\bar{\boldsymbol{r}}_k(t)$ do not commute for $t\neq 0$, the integral in (8.101) cannot be computed in general, except at $t=0$ when

$$G(\boldsymbol{r}, 0)=G_s(\boldsymbol{r}, 0)+N^{-1} \sum_{j\neq k=1}^{N} \langle \delta^3(\boldsymbol{r}+\boldsymbol{r}_j-\boldsymbol{r}')\,\delta^3(\boldsymbol{r}'-\boldsymbol{r}_k)\rangle^{(0)}$$

$$=G_s(\boldsymbol{r}, 0)+G_d(\boldsymbol{r}, 0)=\delta^3(\boldsymbol{r})+g(\boldsymbol{r}) \qquad (8.103)$$

where

$$g(\boldsymbol{r})=N^{-1} \sum_{j\neq k=1}^{N} \langle \delta^3(\boldsymbol{r}+\boldsymbol{r}_j-\boldsymbol{r}_k)\rangle^{(0)} \qquad (8.104)$$

is the static pair distribution function, representing the probability density for finding a particle at $\boldsymbol{r}$ given that there is a particle at the origin. Quite generally the separation of $G(\boldsymbol{r}, t)$ into a term $G_s(\boldsymbol{r}, t)$ describing the space–time correlations for a single particle, and a term $G_d(\boldsymbol{r}, t)$ describing the space–time correlations for distinct pairs of particles, can be maintained for $t>0$ provided the reduced wavelength $\lambda\!\!\!^{-}$ is much less than the mean separation between particles. Under these conditions $G_d(\boldsymbol{r}, t)$ is purely real and $G_s(\boldsymbol{r}, t)$ is real for $r\gg\lambda\!\!\!^{-}$. The spatial behaviour of these functions is shown in figure 8.4 for times $t\ll t_0$, $t\simeq t_0$ and $t\gg t_0$, where t_0 is the mean time it takes for a particle of the system to traverse the mean interparticle separation.

The static approximation may be applied to the interpretation of neutron scattering data whenever $\hbar\omega\ll(\hbar k_0)^2/2m_n$, for in this case we may set (k/k_0) equal to unity in the double differential scattering cross-section given in (8.61). Integrating over ω and using (8.90) we then obtain the result

$$\frac{\mathrm{d}\sigma}{\mathrm{d}\Omega}(\boldsymbol{q})=\frac{1}{4\pi}[\sigma_{\mathrm{coh}}I(\boldsymbol{q}, 0)+\sigma_{\mathrm{inc}}I_s(\boldsymbol{q}, 0)]$$

$$=\frac{1}{4\pi}[\sigma_s+\sigma_{\mathrm{coh}}\gamma(\boldsymbol{q})] \qquad (8.105)$$

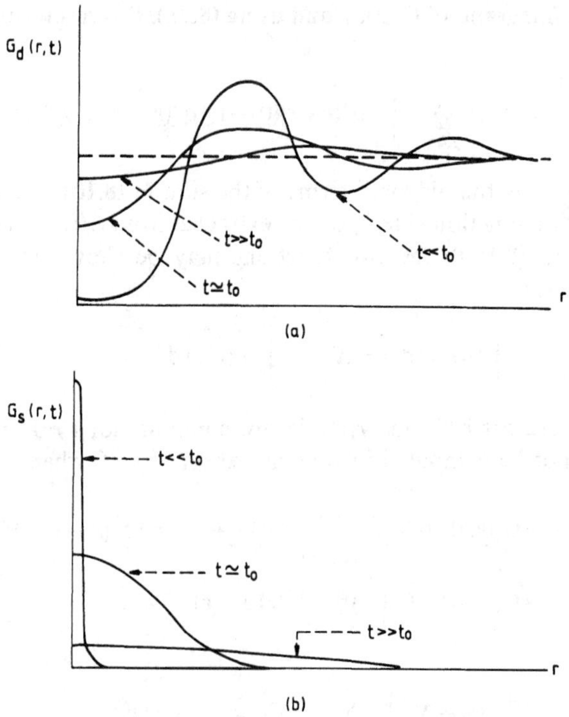

Figure 8.4. (a) The space–time correlation function $G_d(r, t)$ for pairs of distinct particles and (b) the space time correlation function $G_s(r, t)$ for one and the same particle, presented as functions of r for $t \ll t_0$, $t \simeq t_0$ and $t \gg t_0$ (Van Hove 1954). Here t_0 represents the average time required for a particle of the system to transverse the mean inter-particle separation. The function $G_s(r, t)$ is complex for $r \leqslant \lambda\!\!\!^{-}$, the reduced de Broglie wavelength.

where

$$\gamma(\boldsymbol{q}) = \int e^{i\boldsymbol{q}\cdot\boldsymbol{r}} g(\boldsymbol{r})\, d^3r \tag{8.106}$$

and we have used the identity

$$I(\boldsymbol{q}, 0) = \int e^{i\boldsymbol{q}\cdot\boldsymbol{r}} G(\boldsymbol{r}, 0)\, d^3r \tag{8.107}$$

Hence the (elastic plus inelastic) coherent scattering with momentum transfer $\boldsymbol{q}$ provides a direct measure of the pair distribution function $g(\boldsymbol{r})$ exactly as for X-ray scattering.

The other limit where $G(\boldsymbol{r}, t)$ can be evaluated exactly is at large times $|t| \to \infty$ where, because of the long time interval between observations, $\bar{r}_j(0)$

and $\bar{r}_k(\infty)$ are uncorrelated, and we may define an asymptotic correlation function

$$G(\boldsymbol{r}, \infty) = N^{-1} \int \langle n(\boldsymbol{r}' - \boldsymbol{r}) \rangle^{(0)} \langle n(\boldsymbol{r}') \rangle^{(0)} \mathrm{d}^3 r' \qquad (8.108)$$

Using this result we may now derive an expression for the scattering law for elastic scattering

$$\begin{aligned} S_{\mathrm{el}}(\boldsymbol{q}, \omega) &= N^{-1} \delta(\omega) \int \mathrm{e}^{\mathrm{i}\boldsymbol{q}\cdot\boldsymbol{r}} \langle n(\boldsymbol{r}' - \boldsymbol{r}) \rangle^{(0)} \langle n(\boldsymbol{r}') \rangle^{(0)} \mathrm{d}^3 r \, \mathrm{d}^3 r' \\ &= N^{-1} \delta(\omega) \int \mathrm{e}^{-\mathrm{i}\boldsymbol{q}\cdot\boldsymbol{r}''} \langle n(\boldsymbol{r}'') \rangle^{(0)} \mathrm{d}^3 r'' \int \mathrm{e}^{\mathrm{i}\boldsymbol{q}\cdot\boldsymbol{r}'} \langle n(\boldsymbol{r}') \rangle^{(0)} \mathrm{d}^3 r \\ &= N^{-1} \delta(\omega) |\langle n(\boldsymbol{q}) \rangle^{(0)}|^2 \end{aligned} \qquad (8.109)$$

In the case of a uniform infinite medium (of volume $\Omega \to \infty$), where the mean density is constant in space we have the result

$$\langle n(\boldsymbol{q}) \rangle^{(0)} = (2\pi)^3 (N/\Omega) \delta^3(\boldsymbol{q}) = (2\pi)^3 \Omega_{\mathrm{c}}^{-1} \delta^3(\boldsymbol{q}) \qquad (8.110)$$

where $\Omega_{\mathrm{c}} = (\Omega/N)$ is the volume of a unit cell and

$$\begin{aligned} S_{\mathrm{el}} &= N^{-1} \delta(\omega) (N/\Omega)^2 |(2\pi)^3 \delta^3(\boldsymbol{q})|^2 = N \delta(\omega) |(2\pi)^3 \delta^3(\boldsymbol{q})/\Omega|^2 \\ &= N \delta(\omega) |(2\pi)^3 \delta^3(\boldsymbol{q})/\Omega| \\ &= (2\pi)^3 \Omega_{\mathrm{c}}^{-1} \delta(\omega) \delta^3(\boldsymbol{q}) \end{aligned} \qquad (8.111)$$

An expression for $S_{\mathrm{el}}(\boldsymbol{q}, \omega)$ can also be obtained for an infinite perfect crystal where $\langle n(\boldsymbol{r}) \rangle^{(0)}$ can be represented as a Fourier series

$$\langle n(\boldsymbol{r}) \rangle^{(0)} = (N/\Omega) \sum_{\boldsymbol{\mu}} \langle \bar{F}(\boldsymbol{q}(\boldsymbol{\mu})) \rangle^{(0)} \mathrm{e}^{-\mathrm{i}\boldsymbol{q}(\boldsymbol{\mu})\cdot\boldsymbol{r}} \qquad (8.112)$$

Here $\boldsymbol{q}(\boldsymbol{\mu})$ is any vector of the reciprocal lattice and $\bar{F}(\boldsymbol{q}(\boldsymbol{\mu}))$ is the normalized atomic form factor defined in (7.37). The resultant expression for $S_{\mathrm{el}}(\boldsymbol{q}, \omega)$ is

$$S_{\mathrm{el}}(\boldsymbol{q}, \omega) = (2\pi)^3 \Omega_{\mathrm{c}}^{-1} \delta(\omega) \sum_{\boldsymbol{\mu}} |\langle \bar{F}(\boldsymbol{q}(\boldsymbol{\mu})) \rangle^{(0)}|^2 \delta^3(\boldsymbol{q} - \boldsymbol{q}(\boldsymbol{\mu})) \qquad (8.113)$$

This result just reproduces the predictions of kinematical diffraction theory (cf. equation (7.75) in the limit $N_j \to \infty$) applied to a perfect crystal in thermal equilibrium.

In general, classical conditions apply at large distances and long times, i.e. when $\hbar^2 q^2/2M \ll \frac{1}{2}kT$ and $\hbar\omega \ll \frac{1}{2}kT$. In this case the operators $\boldsymbol{r}_j$ and $\boldsymbol{r}_k$ can be treated as ordinary numbers. If in addition we assume that all particles are dynamically equivalent then we obtain for the classical limits

$$G^{c}(\boldsymbol{r}, t)=N^{-1} \sum_{j,k=1}^{N} \int \langle \delta^3(\boldsymbol{r}+\boldsymbol{r}_j(0)-\boldsymbol{r}_k(t))\rangle^{(0)} \mathrm{d}^3 r$$

$$= \sum_{k=1}^{N} \langle \delta^3(\boldsymbol{r}+\boldsymbol{r}_j(0)-\boldsymbol{r}_k(t))\rangle^{(0)} \tag{8.114}$$

and

$$G_s^{c}(\boldsymbol{r}, t)=\langle \delta^3(\boldsymbol{r}+\boldsymbol{r}_j(0)-\boldsymbol{r}_j(t))\rangle^{(0)} \tag{8.115}$$

Thus $G^{c}(\boldsymbol{r}, t)$ represents the probable density of particles at $(\boldsymbol{r}, t)$ given that there is a particle at the origin at zero time, while $G_s^{c}(\boldsymbol{r}, t)$ represents the probability for finding a particle at $(\boldsymbol{r}, t)$ given that the same particle was at the origin at zero time.

The advantage of having clear physical interpretations of $G^{c}(\boldsymbol{r}, t)$ and $G_s^{c}(\boldsymbol{r}, t)$, is that these functions may be evaluated for simple models in cases where the quantum mechanical calculations would prove wholly intractable. By Fourier inversion we may then compute the corresponding classical scattering laws $S^{c}(\boldsymbol{q}, \omega)$ and $S_s^{c}(\boldsymbol{q}, \omega)$. The difficulty with this approach is that, according to (8.87), in the classical limit $\psi^{c}(\boldsymbol{q}, \omega)\rightarrow S^{c}(\boldsymbol{q}, \omega)$ and $\varphi^{c}(\boldsymbol{q}, \omega)\rightarrow 0$; thus $S^{c}(\boldsymbol{q}, \omega)$ no longer satisfies the detailed balanced condition (8.88). Also since $\varphi^{c}(\boldsymbol{q}, t)\rightarrow 0$, $I^{c}(\boldsymbol{q}, t)$ is purely real and even in t, and it is no longer possible to satisfy the sum rule (8.99).

A way out of these difficulties may be found by defining a symmetrized scattering law (Schofield 1960)

$$\tilde{S}(\boldsymbol{q}, \omega)=\mathrm{e}^{-(1/2)\mu\hbar\omega} S(\boldsymbol{q}, \omega)=\tilde{S}(-\boldsymbol{q}, -\omega) \tag{8.116}$$

which behaves similarly to $S^{c}(\boldsymbol{q}, \omega)$ under inversion of $(\boldsymbol{q}, \omega)$. Thus $\tilde{S}(\boldsymbol{q}, \omega)$ may be approximated by $S^{c}(\boldsymbol{q}, \omega)$ while still maintaining the detailed balance condition. Corresponding to $\tilde{S}(\boldsymbol{q}, \omega)$ there is a symmetrized intermediate scattering function

$$\tilde{I}(\boldsymbol{q}, t)=I\left(\boldsymbol{q}, t+\frac{1}{2}\mathrm{i}\mu\hbar\right)=I^*\left(-\boldsymbol{q}, t+\frac{1}{2}\mathrm{i}\mu\hbar\right) \tag{8.117}$$

The relation (8.117) may now be written in the operator form

$$I(\boldsymbol{q}, t)=\exp\left[-\mathrm{i}\mu\hbar\frac{\partial}{\partial t}\right]\tilde{I}(\boldsymbol{q}, t) \tag{8.118}$$

yielding a relation between the real and imaginary parts of $I(\boldsymbol{q}, t)$

$$\mathrm{Im}\,[I(\boldsymbol{q}, t)] = -\tan\left(\frac{1}{2}\mu\hbar\frac{\partial}{\partial t}\right)\mathrm{Re}\,[I(\boldsymbol{q}, t)] \tag{8.119}$$

which is an expression of the fluctuation-dissipation theorem for $I(\boldsymbol{q}, t)$.

To satisfy the symmetry conditions imposed on $I(\boldsymbol{q}, t)$ by (8.82), (8.91) and (8.102), $\tilde{I}(\boldsymbol{q}, t)$ must be a real and even function of t. This requirement is automatically fulfilled when $\tilde{I}(\boldsymbol{q}, t)$ is required to be a real and even function of the variable $y=\sqrt{(t^2+\mu^2/4)}$. In particular if we approximate $\tilde{I}(\boldsymbol{q}, t)$ by a classical function $I^c(\boldsymbol{q}, t)$ such that

$$\tilde{I}_s^c(\boldsymbol{q}, t)=1-\frac{q^2y^2}{2\mu M}+O(y^4)$$

$$\tilde{I}^c(\boldsymbol{q}, t)=1+\gamma(\boldsymbol{q})-\frac{2^2y^2}{2\mu M}+O(y)^4 \tag{8.120}$$

where $\gamma(\boldsymbol{q})$ is defined in (8.106), then both expressions for the zeroth moments given in (8.33), and the frequency sum rule (8.99), are correctly retained.

One may easily confirm that these relations are satisfied for neutron scattering in a perfect gas (cf. (8.69)–(8.70)). For in this case

$$I_s(\boldsymbol{q}, t)=e^{-i\omega_r t-(\Delta\omega)^2t^2/2} \qquad (\Delta\omega)^2=\frac{2\omega_r}{\mu\hbar}=\frac{q^2}{\mu M} \tag{8.121a}$$

$$\tilde{I}_s(\boldsymbol{q}, t)=e^{-q^2y^2/2\mu M}\simeq 1-\frac{q^2y^2}{2\mu M} \tag{8.121b}$$

The classical limit $I_s^c(\boldsymbol{q}, t)$, obtained by letting $\hbar\to 0$ in (8.116), satisfies the identity

$$I_s^c(\boldsymbol{q}, y)\equiv\tilde{I}_s(\boldsymbol{q}, t) \tag{8.122}$$

so that, in the simple example given above, the exact expression for $I_s(\boldsymbol{q}, t)$ may be regained when the classical limit $I_s^c(\boldsymbol{q}, t)$ is known.

8.3.5 Neutron scattering in harmonic crystals

8.3.5.1 Crystal dynamics

Consider a perfect crystal consisting of N_c unit cells each containing p atoms, then the position of the jth atom in the mth unit cell at time t is given by

$$\boldsymbol{r}_{j,m}(t)=\boldsymbol{R}_m+\boldsymbol{r}_{j,m}+\boldsymbol{r}'_{j,m}(t) \qquad \langle\boldsymbol{r}'_{j,m}(t)\rangle=0 \qquad j=1,\ldots,p \tag{8.123}$$

$\boldsymbol{R}_m$ is a lattice vector satisfying (7.41), where the index m represents the triad $(m_1\ m_2\ m_3)$, and $\boldsymbol{r}_{j,m}$ is the equilibrium position of the atom within this unit cell. The coordinate $\boldsymbol{r}'_{j,m}(t)$ describes the fluctuating position of the atom around its equilibrium position. Assuming the atoms of the crystal interact via harmonic forces only, the classical equations of motion can be solved as a system of $3N_cp$ normal modes, each mode corresponding to one degree of freedom. Because of the translational symmetry of the crystal the normal coordinates are plane waves and, by imposing periodic boundary conditions at the surface of the crystal, $r'_{j,m}(t)$ may be expressed as a superposition of solutions of the type

$$r'_{j,m}(\boldsymbol{q}, s, t)=\frac{\varepsilon_j(\boldsymbol{q}_p, s)}{\sqrt{M_j}}\exp\{\mathrm{i}(\boldsymbol{q}_p\cdot[\boldsymbol{R}_m+\boldsymbol{r}_{j,m}]-\omega(\boldsymbol{q}_p, s)t)\} \tag{8.124}$$

which represents a wave propagating with wavevector $\boldsymbol{q}_p$, angular frequency $\omega(\boldsymbol{q}_p, s)$ and complex polarization vector $\varepsilon_j(\boldsymbol{q}_p, s)$. There are N_c physically distinguishable wavevectors $\boldsymbol{q}_p$, uniformly distributed over the first Brillouin zone, and both $\omega(\boldsymbol{q}_p, s)$ and $\varepsilon_j(\boldsymbol{q}_p, s)$ are periodic in $\boldsymbol{q}_p$ with the periodicity of the reciprocal lattice.

For each wavevector $\boldsymbol{q}_p$ there are $3p$ independent modes or branches characterized by an index s, and these may always be chosen to be orthogonal so that we may impose on the polarization vectors the orthonormal condition

$$\sum_{j=1}^{3p}\varepsilon_j(\boldsymbol{q}_p, s)\varepsilon_j^*(\boldsymbol{q}_p, s')=\delta_{ss'} \tag{8.125}$$

The corresponding $3p$ real values of $\omega^2(\boldsymbol{q}_p, s)$, which are the roots of the determinantal equation, are functions of q_p^2 only expressing the fact that plane waves travelling in opposite senses are degenerate in energy. For the same reason the complex polarization vectors $\varepsilon_j(\boldsymbol{q}_p, s)$ satisfy the crossing relation

$$\varepsilon_j(-\boldsymbol{q}_p, s)=\varepsilon_j^*(\boldsymbol{q}_p, s) \tag{8.126}$$

Three of the normal modes are acoustic modes which satisfy the dispersion relation

$$\omega(\boldsymbol{q}_p, s)=c(\boldsymbol{q}_p, s)q_p \tag{8.127}$$

where $c(\boldsymbol{q}_p, s)$ is the velocity of a sound wave with polarization index s. For wavevectors $\boldsymbol{q}_p$ which are simply related to the orientation of the crystal planes, the acoustic modes may be interpreted as two transverse and one longitudinal oscillation. For $p>1$ there are in addition $3(p-1)$ optical modes for which $\omega(\boldsymbol{q}_p, s)$ remains finite in the long-wave limit. Only acoustic modes are permitted in a Bravais lattice for which $p=1$, to which we subsequently confine attention. Here we may set $\boldsymbol{r}_{j,m}\equiv 0$ and discard the index j.

The most general motion of the atom in the mth unit cell of a Bravais lattice may now be written in the form

$$r'_m(t)=\sum_{\boldsymbol{q},s}\frac{Q(\boldsymbol{q}, s, t)}{\sqrt{NM}}\,\varepsilon_m(\boldsymbol{q}_p, s)\exp(\mathrm{i}\boldsymbol{q}_p\cdot\boldsymbol{R}_m) \tag{8.128}$$

where $Q(\boldsymbol{q}_p, s, t)$ is a collective generalized coordinate which, to preserve the reality of $\boldsymbol{r}'_m(t)$, also satisfies a crossing relation

$$Q(-\boldsymbol{q}_p, s, t)=Q^*(\boldsymbol{q}_p, s, t) \tag{8.129}$$

We may now reconstruct the system Lagrangian L_s in terms of the $3N/2$ complex collective coordinates $Q(\boldsymbol{q}_p, s, t)$, in place of the $3N$ real particle coordinates $r'_m(t)$, and use this Lagrangian to define a canonical momentum

$$P(\boldsymbol{q}_p, s, t) = \partial L_s / \partial \dot{Q}(\boldsymbol{q}_p, s, t) \tag{8.130}$$

With the normalization condition (8.124) the system Hamiltonian now becomes

$$H_s = \frac{1}{2} \sum_{\boldsymbol{q},s} (|P(\boldsymbol{q}_p, s)|^2 + \omega^2(\boldsymbol{q}_p, s)|Q(\boldsymbol{q}_p, s)|^2) \tag{8.131}$$

The quantum conditions are now introduced exactly as for the corresponding Fourier decomposition of the radiation field (cf. section 2.1.3.1) by replacing $Q(\boldsymbol{q}_p, s, t)$, $Q^*(\boldsymbol{q}_p, s, t)$ and $P(\boldsymbol{q}, s, t)$, $P^*(\boldsymbol{q}_p, s, t)$ by Schrödinger operators $Q(\boldsymbol{q}_p, s)$, $Q^\dagger(\boldsymbol{q}_p, s)$ and $P(\boldsymbol{q}_p, s)$, $P^\dagger(\boldsymbol{q}_p, s)$ which satisfy the usual commutation relations. In the same way we transform to the occupation number representation by introducing annihilation and creation operators

$$a(\boldsymbol{q}_p, s) = (\omega(\boldsymbol{q}_p, s)/2\hbar)^{1/2}[Q(\boldsymbol{q}_p, s) + (\mathrm{i}/\omega)P^\dagger(\boldsymbol{q}_p, s)]$$

$$a^\dagger(\boldsymbol{q}_p, s) = (\omega(\boldsymbol{q}_p, s)/2\hbar)^{1/2}[Q^\dagger(\boldsymbol{q}_p, s) + (\mathrm{i}/\omega)P(\boldsymbol{q}_p, s)] \tag{8.132}$$

in which representation the Hamiltonian may be written

$$H_s = \sum_{\boldsymbol{q}_p,s} \hbar\omega(\boldsymbol{q}_p, s)\left[n_p(\boldsymbol{q}_p, s) + \frac{1}{2}\right] \tag{8.133}$$

where

$$n_p(\boldsymbol{q}_p, s) = a^\dagger(\boldsymbol{q}_p, s)a(\boldsymbol{q}_p, s) = a(\boldsymbol{q}_p, s)a^\dagger(\boldsymbol{q}_p, s) + 1 \tag{8.134}$$

is the operator for the number of phonons in the mode $(\boldsymbol{q}_p, s)$.

The time-dependent harmonic coordinate operator $\bar{r}'_m(t)$ may now be expressed in the form

$$\bar{r}'_m(t) = \sum_{\boldsymbol{q}_p,s} \left(\frac{\hbar}{2MN\omega(\boldsymbol{q}_p, s)}\right)^{1/2} \boldsymbol{\varepsilon}_m(\boldsymbol{q}_p, s)\, \mathrm{e}^{\mathrm{i}\boldsymbol{q}_p\cdot\boldsymbol{R}m}(\bar{a}(\boldsymbol{q}_p, s, t) + \bar{a}^\dagger(-\boldsymbol{q}_p, s, t)) \tag{8.135}$$

where

$$\bar{a}(\boldsymbol{q}_p, s, t) = a(\boldsymbol{q}_p, s)\,\mathrm{e}^{-\mathrm{i}\omega(\boldsymbol{q}_p,s)t} \qquad \bar{a}^\dagger(\boldsymbol{q}_p, s, t) = a^\dagger(\boldsymbol{q}_p, s)\,\mathrm{e}^{\mathrm{i}\omega(\boldsymbol{q}_p,s)t} \tag{8.136}$$

are annihilation and creation operators in the interaction representation. Using (8.134) and (8.135) one may now show in a straightforward way that all commutators of the type $[\hat{y}'_m(0), \hat{z}'_n(t)]$ are c-numbers.

The conservation of energy and momentum when a neutron is scattered from the state $|\boldsymbol{k}_0\rangle$ to the state $|\boldsymbol{k}\rangle$ may now be written as

$$\frac{\hbar^2}{2M_n}(k_0^2-k^2)=\pm\hbar\omega(\boldsymbol{q}_p,\omega)$$

$$\hbar(\boldsymbol{k}_0-\boldsymbol{k})=\hbar\boldsymbol{q}=\pm\hbar\boldsymbol{q}_p+\hbar\boldsymbol{q}(\boldsymbol{\mu}) \tag{8.137}$$

where $\boldsymbol{q}(\boldsymbol{\mu})$ is any vector of the reciprocal lattice. The creation or annihilation of a phonon of energy $\hbar\omega(\boldsymbol{q}_p, s)$ requires that there exists a mode whose angular frequency $\omega(\boldsymbol{q}_p, s)$ satisfies (8.137).

The case where $\boldsymbol{q}(\boldsymbol{\mu})\neq 0$ describes a situation where the crystal as a whole absorbs momentum in the interaction, while simultaneously a phonon is created or destroyed. In particular when $\omega(\boldsymbol{q}_p, s)\equiv 0$, (8.137) reduces to Bragg's law in the form (7.66).

8.3.5.2 Determination of the scattering law

In order to determine the scattering law for neutrons scattering in the crystal, we first write the intermediate scattering function $I(\boldsymbol{q}, t)$ in the form

$$I(\boldsymbol{q},t)=N^{-1}\sum_{j,k=1}^{N} I_{jk}(\boldsymbol{q},t)\,\mathrm{e}^{-\mathrm{i}\boldsymbol{q}\cdot(\boldsymbol{R}_j-\boldsymbol{R}_k)} \tag{8.138}$$

where

$$I_{jk}(\boldsymbol{q},t)=\langle \mathrm{e}^{-\boldsymbol{q}\cdot\bar{\boldsymbol{r}}_j'(0)}\,\mathrm{e}^{\mathrm{i}\boldsymbol{q}\cdot\bar{\boldsymbol{r}}_k'(t)}\rangle^{(0)} \tag{8.139}$$

Since the operators $\boldsymbol{q}\cdot\bar{\boldsymbol{r}}_j'(0)$ and $\boldsymbol{q}\cdot\bar{\boldsymbol{r}}_k'(t)$ each commutes with its commutator, we may make use of the identity (cf. 8.55)

$$\mathrm{e}^{-\boldsymbol{q}\cdot\bar{\boldsymbol{r}}_j'(0)}\,\mathrm{e}^{\mathrm{i}\boldsymbol{q}\cdot\bar{\boldsymbol{r}}_k'(t)}=\exp(-\mathrm{i}\boldsymbol{q}\cdot(\bar{\boldsymbol{r}}_j'(0)-\bar{\boldsymbol{r}}_k'(t))+\tfrac{1}{2}[\boldsymbol{q}\cdot\bar{\boldsymbol{r}}_j'(0),\boldsymbol{q}\cdot\bar{\boldsymbol{r}}_k'(t)]) \tag{8.140}$$

where, according to (8.134) and (8.135) the commutator $[\boldsymbol{q}\cdot\bar{\boldsymbol{r}}_j'(0), \boldsymbol{q}\cdot\bar{\boldsymbol{r}}_k'(t)]$ is a c-number. We may therefore write

$$I_{jk}(\boldsymbol{q},t)=\langle\exp\{-\mathrm{i}\boldsymbol{q}\cdot(\bar{\boldsymbol{r}}_j'(0)-\bar{\boldsymbol{r}}_k'(t))\}\rangle^{(0)}\exp\left\{\frac{1}{2}[\boldsymbol{q},\bar{\boldsymbol{r}}_j'(0),\boldsymbol{q}\cdot\bar{\boldsymbol{r}}_k'(t)]\right\} \tag{8.141}$$

Since each Cartesian component of $\bar{\boldsymbol{r}}_j'(0)$ or $\bar{\boldsymbol{r}}_k'(t)$ is an independent harmonic coordinate we may quote Bloch's theorem to the effect that any linear combination of such coordinates is normally distributed in the thermal equilibrium state (Bloch 1932). The joint characteristic function therefore has the form of a Gaussian, i.e.

$$\langle\exp\{-\mathrm{i}\boldsymbol{q}\cdot(\bar{\boldsymbol{r}}_j'(0)-\bar{\boldsymbol{r}}_k'(t)\}\rangle^{(0)}=\exp\left\{-\frac{1}{2}q^2\langle(\bar{\boldsymbol{r}}_j'(0)-\bar{\boldsymbol{r}}_k'(t))^2\rangle^{(0)}\right\} \tag{8.142}$$

Also the commutator in (8.141) may be conveniently written as

$$[\boldsymbol{q}\cdot\bar{\boldsymbol{r}}_j'(0), \boldsymbol{q}\cdot\bar{\boldsymbol{r}}_k'(t)] = \langle \boldsymbol{q}\cdot\bar{\boldsymbol{r}}_j'(0)\boldsymbol{q}\cdot\bar{\boldsymbol{r}}_k'(t) - \boldsymbol{q}\cdot\bar{\boldsymbol{r}}_k(t)\boldsymbol{q}\cdot\bar{\boldsymbol{r}}_j'(0)\rangle^{(0)} \tag{8.143}$$

Combining (8.141), (8.79) and (8.143) we can now write $I_{jk}(\boldsymbol{q}, t)$ as an exponential function of the average of time-dependent operators

$$I_{jk}(\boldsymbol{q}, t) = \exp\left\{-\frac{1}{2}\langle (\mathrm{q}\cdot\bar{\boldsymbol{r}}_j'(0))^2 + (\boldsymbol{q}\cdot\bar{\boldsymbol{r}}_k'(t))^2 + 2(\boldsymbol{q}\cdot\bar{\boldsymbol{r}}_j'(0)\boldsymbol{q}\cdot\bar{\boldsymbol{r}}_k'(t))\rangle^{(0)}\right\} \tag{8.144}$$

where the time-ordering of the operators is correctly maintained.

To express $I(\boldsymbol{q}, t)$ in a more compact form we now introduce the Cartesian tensor $T_{\alpha\beta}^{jk}(t)$ (cf. section 3.3.1) through the definition

$$T_{\alpha\beta}^{jk}(t) = \{M_{\alpha\beta}^{jj}(0) + M_{\alpha\beta}^{kk}(0) - 2M_{\alpha\beta}^{jk}(t)\} \tag{8.145}$$

where $M_{\alpha\beta}^{jk}(t)$ is the symmetric correlation tensor

$$M_{\alpha\beta}^{jk}(t) = \langle \bar{r}_{j\alpha}'(0)\bar{r}_{k\beta}'(t)\rangle^{(0)} \qquad \alpha, \beta = x, y, z \tag{8.146}$$

Applying (8.138), (8.144) and (8.145) we now arrive at the relation

$$I(\boldsymbol{q}, t) = N^{-1}\sum_{j,k=1}^{N} \exp\left\{-\frac{1}{2}T_{\alpha\beta}^{jk}(t)q_\alpha q_\beta - \mathrm{i}\boldsymbol{q}\cdot(\boldsymbol{R}_j - \boldsymbol{R}_k)\right\} \tag{8.147}$$

where we have used the usual summation convention applied to tensor indices. The corresponding expression for $G(\boldsymbol{r}, t)$ given by Fourier inversion of $I(\boldsymbol{q}, t)$ is

$$G(\boldsymbol{r}, t) = N^{-1}\sum_{j,k=1}^{N}\left\{\frac{\Delta^{jk}(t)}{8\pi^3}\right\}^{1/2}\exp\left\{-\frac{1}{2}(T_{\alpha\beta}^{jk}(t))^{-1}(\boldsymbol{r}+\boldsymbol{R}_j-\boldsymbol{R}_k)_\alpha(\boldsymbol{r}+\boldsymbol{R}_j-\boldsymbol{R}_k)_\beta\right\} \tag{8.148}$$

where $\Delta^{jk}(t)$ is the determinant of the inverse matrix $(T_{\alpha\beta}^{jk}(t))^{-1}$.

To express the formulae in more easily interpretable form we assume that the crystal is effectively infinite, in which case all unit cells labelled by the index j are equivalent to the unit cell with $R_j \equiv 0$ located at the origin of coordinates. In this limit the sum over j in (8.147) reduces to a sum over N equal terms, i.e.

$$I(\boldsymbol{q}, t) = \sum_{k=1}^{N} \exp\left\{-\frac{1}{2}T_{\alpha\beta}^{0k}(t)q_\alpha q_\beta + \mathrm{i}\boldsymbol{q}\cdot\boldsymbol{R}_k\right\} \tag{8.149}$$

We now make use of the symmetry property of a simple cubic crystal

$$M_{\alpha\beta}^{00}(t) = M(t)\delta_{\alpha\beta} = \frac{1}{3}\langle \bar{r}_0'(0)\bar{r}_0'(t)\rangle^{(0)}\delta_{\alpha\beta} \tag{8.150}$$

which allows us to write the incoherent scattering functions in simple form

$$I_s(\boldsymbol{q}, t) = \exp\left\{-\frac{1}{2}q^2\Gamma(t)\right\}$$

$$G_s(\boldsymbol{r}, t) = [2\pi\Gamma(t)]^{-3/2} \exp\left\{-\frac{1}{2}r^2/\Gamma(t)\right\} \tag{8.151}$$

where

$$\Gamma(t) = 2(M(0) - M(t)) \tag{8.152}$$

Thus both $I_s(\boldsymbol{q}, t)$ and $G_s(\boldsymbol{r}, t)$ are three-dimensional Gaussian distributions in momentum and position space, respectively.

The auto-correlation function $M(t)$ may be derived directly from (8.135) and (8.136) making use of the Hermitian property of the harmonic coordinates and applying the orthogonality relations (8.125). The result is

$$M(t) = (1/3N) \sum_{\boldsymbol{q}_p, s} \left(\frac{\hbar}{2M\omega(\boldsymbol{q}_p, s)}\right)$$

$$\times \{\langle n(\boldsymbol{q}_p, s) + 1\rangle^{(0)} e^{i\omega t} + \langle n(\boldsymbol{q}_p, s)\rangle^{(0)} e^{-i\omega t}\} \tag{8.153}$$

where $\langle n(\boldsymbol{q}_p, s)\rangle^{(0)}$ is given by the familiar Planck formula

$$\langle n(\boldsymbol{q}_p, s)\rangle^{(0)} = [\exp(\mu\hbar\omega(\boldsymbol{q}_p, s)) - 1]^{-1} \tag{8.154}$$

The conclusion that the coefficient of $e^{i\omega t}\langle n(\boldsymbol{q}_p, s) + 1\rangle^{(0)}$ is increased by unity as compared with the coefficient of $e^{-i\omega t}\langle n(\boldsymbol{q}_p, s)\rangle^{(0)}$ in (8.154) is an expression of the fact that, because of the vacuum fluctuations, there is a spontaneous as well as an induced contribution to phonon creation. Equivalently, since by (8.150) $M(0)$ gives the mean square deviation in any direction of a nucleus about its equilibrium position in the lattice, the same effect describes the zero point fluctuations of the nucleus which remain finite at absolute zero where $\langle n(\boldsymbol{q}_p, s)\rangle^{(0)}$ vanishes.

To compute the sum (8.153) for any particular crystal we need to know the frequency spectrum of the phonons. If, for simplicity, we assume that this spectrum is characterized by a continuous density of states

$$dN(\omega)/d\omega = 3Ng(\omega) \qquad \int_0^\infty g(\omega)\, d\omega = 1 \tag{8.155}$$

then $M(t)$ can be expressed as an integral

$$M(t) = \frac{\hbar}{2M}\int_0^\infty \frac{g(\omega)}{\omega}\{\langle n(\omega) + 1\rangle^{(0)} e^{i\omega t} + \langle n(\omega)\rangle^{(0)} e^{-i\omega t}\}\, d\omega \tag{8.156}$$

Further, since $g(\omega)$ is an even function of ω, $M(t)$ can be recast as a Fourier integral

$$M(t)=\frac{\hbar}{2M}\int_{-\infty}^{\infty}\frac{g(\omega)}{\omega}\{\langle n(\omega)+1\rangle^{(0)}\}\,\mathrm{e}^{\mathrm{i}\omega t}\,\mathrm{d}t \tag{8.157}$$

a result which we employ immediately below.

In the case of a simple cubic crystal a suitable approximation to $g(\omega)$ is given by the Debye spectrum (Debye 1912)

$$\begin{aligned} g(\omega)&=3\omega^2/\omega_m^3 \qquad \omega\leqslant\omega_m \qquad \hbar\omega_m=k\theta_\mathrm{D}\\ &=0 \qquad\qquad\quad \omega>\omega_m \end{aligned} \tag{8.158}$$

where θ_D is the Debye temperature.

To complete the calculation of the scattering law we return to (8.151) which now yields the results

$$\begin{aligned} S_\mathrm{s}(\boldsymbol{q},\omega)&=\exp(-q^2M(0))\left(\frac{1}{2\pi}\right)\int_{-\infty}^{\infty}\mathrm{e}^{-\mathrm{i}\omega t}\exp[q^2M(t)]\,\mathrm{d}t\\ &=\exp(-q^2M(0))\left(\frac{1}{2\pi}\right)\sum_{n=0}^{\infty}\int_{-\infty}^{\infty}\mathrm{e}^{-\mathrm{i}\omega t}([q^2M(t)]^n/n!)\,\mathrm{d}t\\ &=\exp(-q^2M(0)\left\{\delta(\omega)+q^2M(0)S_{1\mathrm{s}}(\omega)+\left(\frac{q^2M(0)}{2!}\right)^2S_{2\mathrm{s}}(\omega)+\dots\right\} \end{aligned} \tag{8.159}$$

The first term in this expansion describes elastic scattering modified by the temperature-dependent Debye–Waller factor $\exp[-q^2M(0)]$. Subsequent terms correspond to one phonon, two phonon and multiple phonon inelastic processes. Of these clearly the most important term is the one phonon (creation or annihilation) process described by the normalized coefficient $S_{1\mathrm{s}}(\omega)$. By Fourier inversion of (8.157) this is given by

$$S_{1\mathrm{s}}(\omega)=(g(\omega)/\omega)\langle n(\omega)+1\rangle^{(0)}/M(0) \tag{8.160}$$

From the convolution theorem of Fourier transforms it follows that $S_{2\mathrm{s}}(\omega)$ is just the convolution of $S_{1\mathrm{s}}(\omega)$ with itself, and each successive term $S_{n\mathrm{s}}(\omega)$ is the convolution of $S_{(n-1)\mathrm{s}}$ with $S_{1\mathrm{s}}(\omega)$. Hence the complete scattering law $S_\mathrm{s}(\boldsymbol{q},\omega)$ is determined once the frequency spectrum $g(\omega)$ is known. The generalization of these results to take account of coherent scattering is straightforward if cumbersome. Finally we note from (8.151), (8.152) and (8.156) that

$$\left.\frac{\mathrm{d}I_s}{\mathrm{d}t}(\boldsymbol{q},t)\right|_{t=0}=\frac{-q^2}{2}\Gamma'(0)=q^2M'(0)=\frac{\mathrm{i}\hbar q^2}{2M} \tag{8.161}$$

Hence we confirm that

$$S_\mathrm{s}^{(1)}(\boldsymbol{q})=\hbar^2q^2/2M \tag{8.162}$$

in accordance with the sum rule (8.99).

8.3.6 Small-angle neutron scattering

When neutrons of wavenumber k_0 are elastically scattered through an angle φ the momentum transfer $\boldsymbol{q}$ has magnitude

$$q = 2k_0 \sin(\varphi/2) = \frac{4\pi}{\lambda_0} \sin(\varphi/2) \tag{8.163}$$

In a perfect crystal elastic scattering is observed only when the condition of Bragg resonance is satisfied, i.e.

$$n\lambda_0 = 2d \sin\theta \qquad \theta = \varphi/2 \tag{8.164}$$

in which case $\boldsymbol{q}_n$ is a vector of the reciprocal lattice with

$$q_n = 2\pi n/d \tag{8.165}$$

Clearly the minimum possible momentum transfer corresponds to $n = 1$ with $q_1 = 2\pi/d$.

Real crystals are not, however, perfect in general, but contain defects and inhomogeneities of linear dimension greater than d, whose size and distribution may be studied by neutron scattering with momentum transfers less than $2\pi/d$, without the results being contaminated by Bragg scattering (excluding of course the case of forward scattering with $q = 0$). It is not of course necessary for the medium to be crystalline; in any homogeneous medium, for which the mean separation between scattering centres is d, we may determine a scattering regime for which

$$0 < q < \pi/d \tag{8.166}$$

This condition defines the domain of 'small angle neutron scattering', or 'SANS', which is suitable for the study of inhomogeneities in the medium. Typical values of q fall within the range 3×10^{-3}–0.6 Å^{-1}.

Examples of the inhomogeneities to which the SANS technique may be applied, which we shall loosely describe as 'diffraction objects', are lattice defects and dislocations, voids, aggregates of atoms of linear dimension in the range 5–5000 Å, precipitates in solution, polymers and biological macromolecules. A particularly important application of SANS with neutrons has been developed in the shale-oil industry in the United States where SANS is used to locate voids in the shale. The point is that, since hydrogen has a negative scattering length, hydrocarbons (and water) have a mean scattering length approaching zero. Thus SANS locates the voids irrespective of whether they are full or not. X-ray scattering on the other hand only senses the empty voids, and a combination of the two techniques permits an estimate to be made of the potential of the shale as a source of petroleum.

Since, for given q, the scattering angle φ increases with λ, SANS experiments are usually performed using very slow neutrons with wavelengths in the range 2–20 Å. The scattering angle (which may indeed be quite large in spite of the description 'small') is precisely determined using a position sensitive detector (PSD). The most widely used instrument of this type is the Anger camera, which determines the exact position at which a slow neutron is absorbed in a ^{6}Li-loaded glass scintillator, by recording the scintallations in separated photomultipliers, looking from different directions (Anger 1958).

The amplitude for elastic scattering of unpolarized neutrons, in a medium at thermal equilibrium containing N atoms in a volume Ω, may be extracted from the general cross-section formula (8.55) with the result

$$f(\boldsymbol{q})=\int_{\Omega} \beta(\boldsymbol{r})\,\mathrm{e}^{\mathrm{i}\boldsymbol{q}\cdot\boldsymbol{r}}\,\mathrm{d}^3r \tag{8.167}$$

where

$$\beta(\boldsymbol{r})=\left\langle \sum_{j=1}^{N} b_j\,\delta^3(\boldsymbol{r}-\boldsymbol{r}_j)\right\rangle^{(0)} \tag{8.168}$$

is the 'scattering length density' and the symbol (0) indicates that the scattering atoms are at their equilibrium average values $\boldsymbol{r}_j$ (cf. section 8.2.3). In general $\beta(\boldsymbol{r})$ can be represented as a Fourier integral in $\boldsymbol{q}$-space (cf. section 8.3.8) but, since according to (8.166) q is confined to a wavenumber range which excludes the smallest wavenumber which contributes significantly to the spatial fluctuations in $\beta(\boldsymbol{r})$, it is sufficient to replace $\beta(\boldsymbol{r})$ by its average value, i.e.

$$\beta(\boldsymbol{r})\rightarrow\langle\beta\rangle=\sum_{j=1}^{N} b_j/\Omega=n\langle b\rangle \tag{8.169}$$

where $\langle b\rangle$ is the mean scattering length, and n is the number of atoms per unit volume.

Let us now suppose that the medium is divided into two regions, consisting of a small-scale diffracting object (M), where $\beta(\boldsymbol{r})=\langle\beta_0\rangle$, embedded in a homogeneous background ($\bar{M}$), where $\beta(\boldsymbol{r})=\langle\beta_{\mathrm{m}}\rangle$. In this case $\beta(\boldsymbol{r})$ acquires a large-scale spatial dependence given by

$$\begin{aligned}\beta(\boldsymbol{r})&=\langle\beta_{\mathrm{m}}\rangle(1-S(\boldsymbol{r}))+\langle\beta_0\rangle S(\boldsymbol{r})\\&=\langle\beta_{\mathrm{m}}\rangle+(\langle\beta_0\rangle-\langle\beta_{\mathrm{m}}\rangle)S(\boldsymbol{r})\end{aligned} \tag{8.170}$$

where $S(\boldsymbol{r})$ is the shape-function defined in (7.86), i.e.

$$S(\boldsymbol{r})=1 \qquad \boldsymbol{r}\in M \qquad S(\boldsymbol{r})=0 \qquad \boldsymbol{r}\in\bar{M} \tag{8.171}$$

According to (8.167) the scattering amplitude is now given by

$$f(\boldsymbol{q})=(2\pi)^3\langle\beta_{\mathrm{m}}\rangle\delta^3(\boldsymbol{q})+(\langle\beta_0\rangle-\langle\beta_{\mathrm{m}}\rangle)\int_{\Omega}\mathrm{e}^{\mathrm{i}\boldsymbol{q}\cdot\boldsymbol{r}}S(\boldsymbol{r})\,\mathrm{d}^3r \tag{8.172}$$

The first term in (8.172) represents the contribution from forward scattering in the medium. It is written as a δ-function in $\boldsymbol{q}$-space, although, strictly speaking, it has a finite width of order Ω^{-1}. We shall however assume that this component is indistinguishable from the transmitted beam and we shall neglect it. The second term in (8.172) describes the contribution from the diffracting object. If we suppose that the medium contains N_0 objects of this type, the jth diffracting object having a volume ω_j, the small-angle scattering amplitude ($q>0$) may be written in the form

$$f(\boldsymbol{q})=(\langle\beta_j\rangle-\langle\beta_{\mathrm{m}}\rangle)\sum_{j=1}^{N_0} \mathrm{e}^{\mathrm{i}\boldsymbol{q}\cdot\boldsymbol{r}_j}F_j(\boldsymbol{q}) \tag{8.173}$$

where

$$\bar{\boldsymbol{r}}_j=\omega_j^{-1}\int_{\Omega}\boldsymbol{r}S(\boldsymbol{r})\,\mathrm{d}^3r \tag{8.174}$$

is the position vector of the centre of mass, and

$$F_j(\boldsymbol{q})=\int_{\omega_j}\mathrm{e}^{\mathrm{i}\boldsymbol{q}\cdot\boldsymbol{r}'}\,\mathrm{d}^3r' \qquad \boldsymbol{r}'\equiv\boldsymbol{r}-\bar{\boldsymbol{r}}_j \tag{8.175}$$

is the form factor of the jth diffracting object. The total differential cross-section is then given by

$$\frac{\mathrm{d}\sigma}{\mathrm{d}\Omega}=|f(\boldsymbol{q})|^2=N_0(\langle\beta_0\rangle-\langle\beta_{\mathrm{m}}\rangle)^2\left\{\frac{1}{N_0}\sum_{j,k=1}^{N_0}\mathrm{e}^{-\mathrm{i}\boldsymbol{q}\cdot(\bar{\boldsymbol{r}}_j-\bar{\boldsymbol{r}}_k)}F_j^*(\boldsymbol{q})F_k(\boldsymbol{q})\right\} \tag{8.176}$$

We now introduce into (8.176), in place of each individual form factor $F_j(\boldsymbol{q})$, an average normalized form factor $\bar{F}(\boldsymbol{q})$ through the prescription

$$F_j(\boldsymbol{q})\equiv\omega_j\bar{F}_j(\boldsymbol{q})\rightarrow\omega_0\bar{F}(\boldsymbol{q}) \tag{8.177}$$

where ω_0 is the average volume. In this approximation (8.176) may be rewritten as

$$\frac{\mathrm{d}\sigma}{\mathrm{d}\Omega}=N_0\omega_0^2(\langle\beta_0\rangle-\langle\beta_m\rangle)^2 I(\boldsymbol{q})|\bar{F}(\boldsymbol{q})|^2 \tag{8.178}$$

where

$$I(\boldsymbol{q})=\frac{1}{N_0}\sum_{j,k=1}^{N_0}\mathrm{e}^{-\mathrm{i}\boldsymbol{q}\cdot(\bar{\boldsymbol{r}}_j-\bar{\boldsymbol{r}}_k)} \tag{8.179}$$

is immediately recognizable as the intermediate scattering function (cf. 8.74) for elastic scattering from the N_0 diffracting objects embedded in the homogeneous medium.

Clearly the factor $I(\boldsymbol{q})$ describes the interference pattern for waves scattered from the N_0 diffracting centres whose diffraction patterns are described by the normalized form factor $F(\boldsymbol{q})$. However interference may be neglected if $qD \geqslant 6$, where D is the mean separation between neighbouring objects. In this limit $I(\boldsymbol{q})$ may be replaced by unity, an approximation that implies, according to (8.166), that the inhomogeneities that form the scattering centres occupy no more than 5–10% of the total volume. Note that this is precisely the opposite situation to Bragg scattering where the angular dependence is a reflection of the interference pattern for scattering from nuclei, each of which is characterized by a unit normalized form factor, indicative of the fact that nuclear dimensions are infinitesimal in comparison with the wavelengths of cold neutrons. A much closer analogue to small-angle neutron scattering is familiar from classical optics, where the scattering of visible light from small droplets of water is responsible for the lunar halo which may be observed in foggy weather.

All the available information about the size and structure of the inhomogeneities themselves is contained in $\bar{F}(\boldsymbol{q})$ which may be evaluated once the geometry is known. In general, for a given orientation of the object relative to $\boldsymbol{q}$, $\bar{F}(\boldsymbol{q})$ may be written as the single integral

$$\bar{F}(\boldsymbol{q}) = \frac{1}{\omega_0} \int e^{iqz} \sigma(z) \, dz \tag{8.180}$$

where $\sigma(z)$ is the cross-section of the object cut-off by a plane normal to $\boldsymbol{q}$, a distance z from the centre of mass, as shown in figure 8.5. For the rather special case of a sphere of radius a, for which the notion of 'orientation' is meaningless, $\bar{F}(\boldsymbol{q})$ can be evaluated exactly with the result

$$\bar{F}(\boldsymbol{q}) = 3[\sin(qa) - qa\cos(qa)]/(qa)^3 \tag{8.181}$$

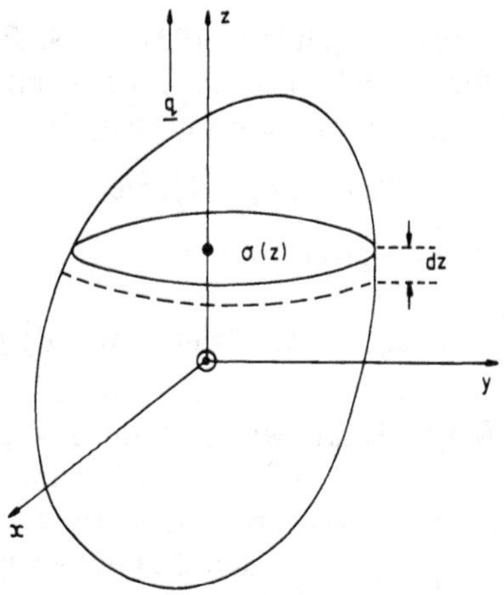

Figure 8.5. Evaluation of the normalized form factor $\bar{F}(\boldsymbol{q})$ describing small-angle neutron scattering (SANS) from an object whose cross-sectional area $\sigma(z)$, normal to $\boldsymbol{q}$, is known as a function of z.

In the general case, since $\sigma(z)/\omega_0$ is a probability density in z, $\bar{F}(\boldsymbol{q})$ is a characteristic function, and may be expressed in terms of a cumulant generating function $\kappa(q)$, i.e.

$$\bar{F}(\boldsymbol{q})=\exp[\kappa(q)] \tag{8.182}$$

where

$$\kappa(q)=\sum_{r=0}^{\infty} k_r(\mathrm{i}q)^r/r\,!$$

with cumulants

$$k_0=0 \qquad k_1=\langle z\rangle \qquad k_2=\langle z^2\rangle-\langle z\rangle^2,\ \text{etc.} \tag{8.183}$$

The moments of the distribution in z are defined in the usual way

$$\langle z^n\rangle=\frac{1}{\omega_0}\int z^n\sigma(z)\,\mathrm{d}z \tag{8.184}$$

The first cumulant k_1 vanishes in (8.181) because all moments $\langle z^n\rangle$ are evaluated with respect to the centre of mass. In any case it is only the even-order cumulants which are of interest because these add in the product $|F(q)|^2$ whereas the odd-order cumulants cancel. The most important cumulant is therefore $k_2=\langle z^2\rangle$, and this may be replaced by $R_{\mathrm{g}}^2/3$ after a further average over all orientations of the object, where R_g is the radius of gyration defined by

$$R_{\mathrm{g}}^2=\frac{1}{\omega_0}\int_{\omega_0} r^2\,\mathrm{d}^3r \tag{8.185}$$

The Guinier approximation (Guinier 1937, 1939, Porod 1948) amounts to the neglect of the fourth and all higher order cumulants in (8.180), and applies whenever $|qR_g|\leqslant 2$. In this case $\bar{F}(\boldsymbol{q})$ may be written

$$\bar{F}(q)=\exp[-R_{\mathrm{g}}^2q^2/6] \tag{8.186}$$

and the total cross-section is given by

$$\frac{\mathrm{d}\sigma}{\mathrm{d}\Omega}=N_0(\langle\beta_0\rangle-\langle\beta_{\mathrm{m}}\rangle)^2\omega_0^2\exp[-R_{\mathrm{g}}^2q^2/3] \tag{8.187}$$

The coefficient of $|\bar{F}(\boldsymbol{q})|^2$ is known as Guinier's constant; it may be determined simultaneously with R_{g}^2 by plotting the logarithm of $\mathrm{d}\sigma/\mathrm{d}\Omega$ against q^2. A comparison between the exact expression for $|\bar{F}(\boldsymbol{q})|^2$ for a sphere of radius a, with $R_{\mathrm{g}}=\sqrt{(3/5)}a$, and the Guinier approximation, shown in figure 8.6, indicates that the correspondence is very close for $|qR_{\mathrm{g}}|\leqslant 2$. The Guinier approximation does not of course reproduce the subsidiary weak maximum in $|\bar{F}(\boldsymbol{q})|^2$ at $(qa)\approx 5.5$. For $|qR_{\mathrm{g}}|\geqslant 3$ an expansion of $\bar{F}(q)$ in

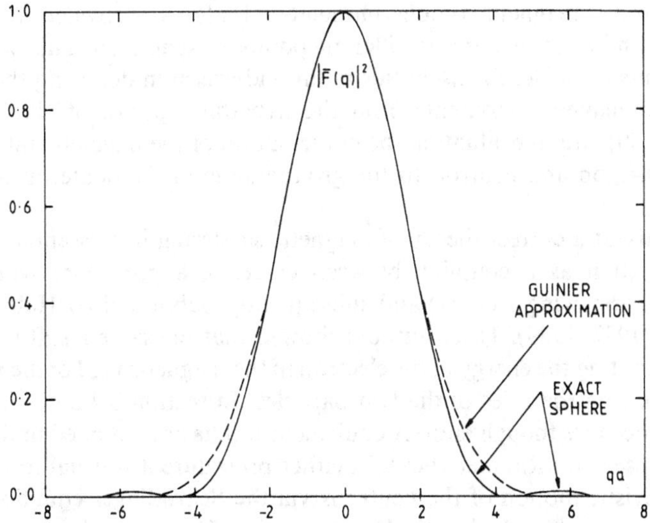

Figure 8.6. The Guinier approximation to $|\bar{F}(\boldsymbol{q})|^2$ for a sphere of a radius a. This approximation neglects the fourth and higher order cumulants in the Fourier integral defining $\bar{F}(\boldsymbol{q})$. It is valid provided $|qR_g|<2$, where R_g is the radius of gyration of the scattering object.

ascending powers of q is not appropriate and, as is evident from (8.178) for the case of a sphere, $\bar{F}(\boldsymbol{q})$ behaves asymptotically as q^{-2} (Porod 1982). However, this range of q is generally of less interest from an experimental angle.

8.4 Neutron scattering in magnetic media

8.4.1 The neutron as a probe of magnetic structure

8.4.1.1 The neutron–electron magnetic interaction

The suggestion that the neutron magnetic moment might be measured through observation of scattering by magnetic atoms was first advanced by Bloch (Bloch 1936, 1937). This was at a time when the existence of a finite magnetic moment in the neutron could only be inferred from the measured magnetic moments of proton and deuteron (Bethe and Bacher 1936). Bloch also proposed that magnetic scattering could be exploited to produce polarized neutrons and to investigate the structure of ferromagnetic materials.

Unfortunately the theory as first presented was incorrect since it assumed a classical dipole–dipole interaction between neutrons and electrons. This approach leads to a correct result for both orbital and spin contributions to the electronic magnetic dipole moments of atoms, provided only that neutrons and electrons are at different points in space, but fails whenever these points coincide. Precisely this point had arisen in deducing the values of nuclear magnetic moments from the hyperfine spectra of alkali atoms (Fermi 1930) and in evaluating the contribution of the magnetic interaction between proton and neutron to the ground state of the deuteron (Casimir 1936).

To arrive at a correct theory of magnetic scattering it is essential to treat the interaction as a coupling between currents, a procedure which was adopted by Schwinger (1937) and subsequently elaborated by Halpern and Johnson (1937, 1939). These authors showed that the same result is arrived at by computing the energy of the electron in the magnetic field of the neutron or by reversing the roles of the two particles. In section 8.4.2 we apply the former procedure though entirely equivalent results are obtained in the latter case (Lovesey and Rimmer 1969). In either procedure it is usual to treat the non-relativistic motion of the neutrons via the Schrödinger equation while the electrons are described by the Dirac equation. However, when the neutron motion is described by a Dirac equation modified by a phenomenological tensor coupling to the electromagnetic field tailored to reproduce the observed magnetic moment (Pauli 1941), an additional non-magnetic electron–neutron interaction is revealed (Fermi and Marshall 1947, Foldy 1951, 1952) whose origins and significance will be explored in section 9.2.2.2.

Concerning the effective magnetic potential acting on neutrons propagating in condensed media, what the resolution of the original controversy established was that this interaction energy is given by $-\boldsymbol{\mu}_n \cdot \boldsymbol{B}$ (Schwinger) rather than by $-\boldsymbol{\mu}_n \cdot \boldsymbol{H}$ (Bloch), where $\boldsymbol{B}$ and $\boldsymbol{H}$ are the usual macroscopic fields (Ekstein 1949, 1950). In this respect the observation that ferromagnetic materials are neutron optically bi-refringent (Hughes and Burgy 1949, 1951) was decisive. It has also been remarked (Hamermesh and Eisner 1950) that the same observations provide independent, and quite convincing, evidence, that the neutron is a spin $\frac{1}{2}$ particle (cf. section 1.5.3).

8.4.1.2 Paramagnetism and ferromagnetism

Atoms subjected to an applied magnetic field $\boldsymbol{H}$ acquire orbital angular momenta and induced magnetic moments proportional to H but oppositely directed. This diamagnetic effect, which is present in all substances, is a direct consequence of Faraday's law of electromagnetic induction. In paramagnetic substances, e.g. transition elements, rare earths, actinides, the atoms possess permanent magnetic dipole moments due to the presence of unpaired electron spins in incomplete electronic shells. In the absence of an applied field these magnetic moments are randomly oriented so there is no net macroscopic magnetization.

Paramagnetic atoms in gaseous form can interact via dipole–dipole forces as described in section 8.4.1.1 but, in condensed media at close separation these forces are dominated by exchange forces described approximately by the Heisenberg exchange Hamiltonian $H_s = -\sum_{jk} J_{jk} \boldsymbol{S}_j \cdot \boldsymbol{S}_k$ (cf. section 8.4). The exchange energy originates in the quantum mechanical law which requires the total electronic wavefunction to be anti-symmetric with respect to the exchange of position and spin coordinates between any pair of electrons, and the exchange integral J_{jk}, which may in principle be positive or negative, is a measure of the overlap of electronic wavefunction of atoms at neighbouring sites (cf. section 8.4.4). To a good approximation each atomic spin experiences an internal field $\boldsymbol{H}_i$, proportional to the local magnetization $\boldsymbol{M}$, exactly as postulated by Weiss (1907) to account for the phenomenon of ferromagnetism, namely the generation of a strong co-operative magnetization in a weak (or even zero) applied field, at temperatures below a critical (Curie) temperature T_c.

For ferromagnetism to occur it is essential that J_{jk} be positive so that, for $T < T_c$, the states of high spin multiplicity corresponding to a parallel orientation of atomic spins, should have the lowest energy. For $T > T_c$, the spontaneous magnetization disappears and the substance reverts to the normal paramagnetic state. For the most famous of all magnetic materials Fe_3O_4 magnetite or 'loadstone', $T_c \equiv 893$ K and this substance is spontaneously, if weakly, magnetized at room temperatures. Elastic (Bragg) scattering of neutrons by ferromagnetic salts, both above and below the Curie temperature T_c, was one of the earliest phenomena to be investigated by Halpern and his collaborators, and this technique may be applied both to locate the positions of magnetic atoms in crystals, and to measure the spatial distributions of spin densities inside these atoms. In recent years elastic scattering measurements have come to rely more and more on the use of polarized neutrons, and for this reason we postpone a discussion of Bragg scattering in ferromagnetic crystals to section 8.4.5, where it is treated in the general context of polarized neutron scattering.

The use of the bi-refringent properties of magnetized iron to produce polarized neutrons by selective reflection from magnetic mirrors characterized by two critical glancing angles (Achieser and Pomeranchuk 1948, Hughes *et al* 1949) (cf. section 7.4.3)

$$\theta_c = \lambda(N/\pi)[b_{coh} \pm p(0)]^{1/2} \tag{8.188}$$

was among the first applications of magnetic scattering to be explored experimentally, and this is discussed briefly in section 8.4.3. As a general rule, for a ferromagnet at a finite temperature $T < T_c$, in the absence of an applied field the spins within a given domain are established in a parallel configuration, but the individual domains are themselves oriented at random, and there is no net magnetization. A simple crystal (body centred cubic) of iron, magnetized along the preferred (1 1 0) direction will reach its saturation flux density $B_s = 21.6$ kG in an applied field of 3 kOe, whereas at least ten times this field is required to saturate (face-centred-cubic) crystals of nickel ($B_s = 6.05$ kG) or cobalt ($B_s = 17.7$ kG). This is unfortunate because in cobalt the nuclear scattering length is less than the magnetic scattering length. Thus only one index of refraction is less than unity and only one spin component is critically reflected at all wavelengths (Hamermesh 1949).

A major application of neutron scattering in ferromagnets is the study of inelastic processes, of which the simplest are the collective excitations of Heisenberg ferromagnets in the ground state. These excitations arise when atomic spins are re-oriented after a neutron interaction and, because of the exchange coupling with nearest neighbours, the subsequent disturbance is propagated throughout the crystal. This is a quantized excitation whose quantum, the magnetic analogue of the lattice phonon, is called a magnon or spin wave. It describes a precessional motion of the atomic spins with a phase which varies coherently in space and time according to a definite wavelength and frequency. The treatment of spin waves given in section 8.4.4 is confined to Heisenberg ferromagnets at low temperatures, but spin waves can be generated in other transition elements and in rare earths, and also in more exotic magnetic materials such as antiferromagnets and ferrites.

8.4.1.3 Antiferromagnetism

Neutron magnetic scattering from ferromagnets depends on the relative orientation of the atomic spin $\boldsymbol{S}$ and the momentum transfer $\boldsymbol{q}$ (cf. section 8.4.3) and, because of the difficulty involved in aligning the magnetic domains, this is always uncertain to some degree (Shull *et al* 1951b). A more suitable test of the Halpern–Johnson theory is provided by paramagnetic materials for which the theory predicts the scattering to be completely incoherent with an angular distribution determined by a calculable magnetic form factor $F(\boldsymbol{q})$. These predictions were exhaustively tested over the period (1948–52) at the Oak Ridge National Laboratory by Shull and his collaborators, who carried out a series of experiments with 1.058 Å neutrons scattered by anhydrous polycrystalline salts of the Group VIII transition element manganese. This element is monatomic having a single stable isotope $^{55}_{22}$Mn with $I^{\pi} = 5/2^{+}$, and a negative bound coherent scattering length $b_{\text{coh}} = -3.73$ fm (cf. Table 6.2). The divalent ion Mn^{++} has five electrons in the 3d-shell with parallel spins giving a total spin $S = 5/2$. The corresponding magnetic scattering length is $p(0) = 13$ fm.

The salts studied were MnF_2, $MnSO_4$ and MnO. Since the element fluorine is also monatomic and $^{19}_{9}$F has $I^{\pi} = \frac{1}{2}^{+}$ with $b_{\text{coh}} = 5.654$ fm, the nuclear and magnetic components in the incoherent scattering from MnF_2 are expected to be comparable in intensity, with no isotopic-incoherent nuclear scattering and most of the spin-incoherent nuclear scattering coming from the ^{55}Mn. There was no difficulty in separating out the nuclear coherent components since these were confined to Debye–Scherrer diffraction peaks directed along selected canonical surfaces defined with respect to the incident neutron momentum. The experimental results for MnF_2, and less convincingly for the magnetically dilute salt $MnSO_4$, confirmed the expected form factor angular dependence and allowed the radial distribution of the 3d-electrons to be evaluated, yielding results which compared favourably with theory (Shull and Smart 1949, Shull *et al* 1951).

The experimental results for MnO were quite different, for instead of the broad maximum in the forward direction predicted by theory for room temperature scattering ($T_c = 120$ K for MnO), significant peaking was observed near 10° reminiscent of the patterns observed for X-ray scattering in liquids indicative of position short-range order. In addition, strong nuclear coherent scattering at the (1 1 1) and (3 1 1) positions was observed, the all-odd nature of the indices being associated with the sign difference for the scattering lengths in ^{55}Mn and ^{16}O. When these experiments were repeated at 80 K, four additional peaks were identified in the spectrum, which is illustrated in figure 8.7. These were subsequently identified as purely antiferromagnetic Bragg reflections.

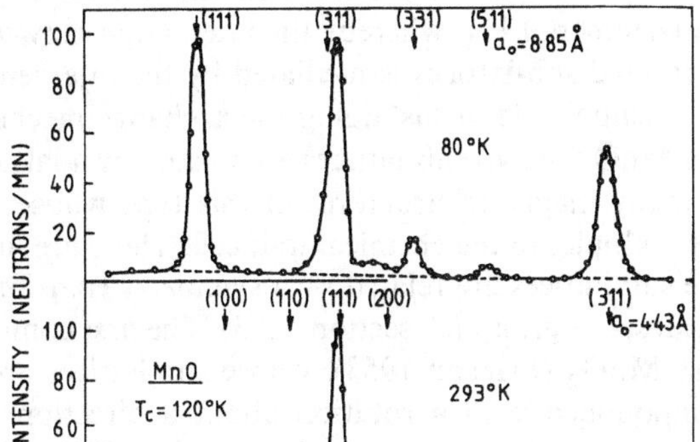

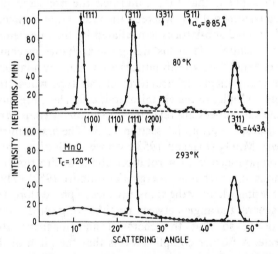

Figure 8.7. Neutron diffraction patterns observed in MnO at 293 K and 80 K. The nuclear scattering peaks observed at the (1 1 1) and (3 1 1) positions at room temperature are supplemented by the antiferromagnetic Bragg peaks at the (1 1 1), (3 1 1), (3 3 1) and (5 1 1) positions, when the temperature is reduced below the Néel point (Shull and Smart 1949).

Antiferromagnetism was first proposed by Néel (Néel 1932, 1936) who predicted that, in certain conditions involving a high concentration of magnetic atoms, the exchange integral J_{jk} becomes negative, in which case the lowest energy state corresponds to an anti-parallel, rather than a parallel orientation of atomic magnetic moments. An essential feature of the proposed scheme was that an antiferromagnetic crystal has a structure composed from two identical sub-lattices each of which would separately behave like a ferromagnet below the Curie temperature. However, the two sub-lattices interlock in such a way that atoms with one spin orientation on the first sub-lattice always have nearest neighbours with an antiparallel spin orientation on the other sub-lattice. Thus below the Curie point (or in this case renamed the Néel point) there is zero magnetization in the absence of an applied field.

A characteristic feature of the antiferromagnetic crystal is that the magnetic unit cell has a linear dimension twice that of the chemical unit cell, a consequence of the fact that the direction of magnetization alternates from one chemical unit cell to the next. This is of course precisely what is observed for neutron scattering in MnO, where the purely magnetic Debye–Scherrer peaks can be indexed only on the assumption of a twice-enlarged magnetic unit cell. On this basis, assuming a body-centred-cubic lattice for MnO as indicated by X-ray scattering, there should be a near coincidence of the (3 1 1) antiferromagnetic reflection with the (1 1 1) nuclear reflection exactly as is observed. A further important deduction from the neutron scattering experiments in MnO is that they confirmed the predicted phenomena of superexchange (Kramers 1934) whereby the interaction between two Mn^{++} ions on the separated sub-lattices is mediated by the intervening O^{--} ions which play an essential role in sustaining the exchange mechanism.

In the case of MnO the two sub-lattices are related by a lattice translation, and there are many magnetic structures of this type whose magnetic unit cells are simple multiples of the chemical unit cell. There are, however, other cases where the sub-lattices are related by a symmetry transformation of the relevant chemical space group (cf. section 7.2.3). The first confirmed example of this type was MnO_2 (Erikson 1953), whose sub-lattices are related by a translation compounded with a rotation about a direction in the crystal which minimizes the exchange energy (Yoshimori 1959). This leads to a helical spin ordering whereby the spins advance in phase along this minimum direction. In general the angle between neighbouring spins is not an integral submultiple of 2π so that the chemical and magnetic structures are incommensurate. A further complication is that the pitch of the spin helix may be temperature dependent.

8.4.1.4 Ferrimagnetism

Finally, we take brief note of an intermediate type of co-operative magnetism, namely ferrimagnetism, in which the interlocking sub-lattices are not equivalent, but may differ, either in the nature of the crystallographic site, or of the magnetic ion, or in both these characteristics (Néel 1948). Magnetite (Fe_3O_4) was the first ferrimagnetic substance to be studied by neutron scattering, with results which confirmed the predictions of Néel's proposed magnetic structure in quite remarkable detail (Shull 1951a).

Magnetite has the same crystal structure as spinel ($MgAl_2O_4$) where the oxygen ions form a cubic array with the metallic ions located in the interstices. These are of two types; in the first type each metallic ion is surrounded by four oxygen ions at the vertices of a tetrahedron, and in the second type six oxygen ions are located at the vertices of an octahedron. In magnetite the octahedral sites are occuped by equal numbers of divalent (F_e^{++}) and trivalent (F_e^{+++}) ions, while the tetrahedral sites are occupied by F_e^{+++} ions only. The iron ions at tetrahedral positions are coupled anti-ferromagnetically to those at octahedral positions, which, being in the majority, are magnetically oriented parallel to an applied field. The tetrahedral ions are oriented anti-parallel, and the net result is a weak ferromagnetism.

Since the positions of the iron and oxygen nuclei are known from X-ray diffraction data, the contribution of the nuclear scattering can be evaluated quite accurately. It follows that the neutron scattering pattern provides a detailed picture of the magnetic structure, bearing in mind that, because of the antiferromagnetic coupling, the effective magnetic scattering length of F_e^{+++} ions at tetrahedral sites is negative. As remarked briefly in section 8.4.5.2, there is almost complete cancellation between nuclear and magnetic scattering lengths in the (2 2 0) reflection from saturated single crystals of Fe_3O_4 so that the scattered neutrons are almost fully polarized (Shull 1950). This result contrasts sharply with the result for Bragg scattering at the (1 1 0) position in saturated iron, where the scattered polarization is much reduced and has the opposite sign. The explanation is that, whereas in pure iron the effective value of S is about 2.1, and the magnetic scattering length is positive, in Fe_3O_4 only the Fe^{+++} ions at tetrahedral sites contribute to the scattering at the (2 2 0) position. In this case $S=5/2$ and the effective magnetic scattering length is about -13 fm.

The investigation of magnetic structures described briefly above belongs to the classical era of neutron scattering, but in recent times new exotic materials and more general questions relating to the understanding of condensed matter systems have dominated the stage. For example magnetic phenomena in low-dimensional arrays have excited great interest and inelastic neutron scattering in the one-dimensional Heisenberg antiferromagnets $CsNiF_3$ and tetramethal ammonium manganese chloride (TMMC) has been intensively studied. At the fundamental level magnetic phase transitions at temperature $T \sim T_c$ have been explored by neutron scattering, but these are all complex matters which fall quite outside the range of an elementary discussion.

8.4.2 The magnetic scattering amplitude

A slow neutron at the point $\boldsymbol{r}'$ carrying a magnetic moment $\boldsymbol{\mu}_n = \kappa_n \mu_N \boldsymbol{\sigma}$, where $\mu_N = e\hbar/2m_p c$ is the nuclear magneton and $\kappa_n = -1.913$, creates a vector potential $A(r, r')$ at the point r given by

$$A(r, r') = \mu_n \times (r - r')/|r - r'|^3 \tag{8.189}$$

The interaction potential with an unpaired non-relativistic electron at r_j is given by the usual expression (cf. 2.22)

$$V_m(r_j) = -\mu_e \cdot \nabla_j \times (A(r_j)) + \frac{e}{2m_e c}(p_j \cdot A(r_j) + A(r_j) \cdot p_j) \tag{8.190}$$

where $\mu_e = 2\mu_B S_j$ is the magnetic moment of the electron and $\mu_B = e\hbar/2m_e c$ is the Bohr magneton.

Following the procedure adopted in section 8.2.4 we include the initial and final quantum numbers s_0, s' of the neutron spin-state along with the quantum numbers specifying the initial and final states of the scattering atom, and evaluate the matrix elements of $V_m(r_j)$ between initial and final planewave states of the scattered neutron. We then find that

$$\begin{aligned}\langle k|A(r, r')|k_0\rangle &= \int e^{iq\cdot r'} A(r, r')\, d^3r' \\ &= \mu_n \times \int e^{-iq\cdot(r''-r)}(r''/r''^3)\, d^3r'' \\ &= -4\pi i \mu_n \times (e^{iq\cdot r} q/q^2)\end{aligned} \tag{8.191}$$

When there are n unpaired electrons in the scattering atom the matrix element of V_m becomes

$$\begin{aligned}\langle k|V_m|k_0\rangle = 8\pi\kappa_n\mu_N\mu_B \sum_{j=1}^{n} \{&S_i \cdot \hat{q} \times (\sigma \times \hat{q})\, e^{iq\cdot r_j} \\ &-(i/2\hbar q)(p_j \cdot (\sigma \times \hat{q})\, e^{iq\cdot r_j} + (\sigma \times \hat{q})\, e^{iq\cdot r_j} \cdot p_j)\}\end{aligned} \tag{8.192}$$

where $\hat{q} = q/q$ is a unit vector in the direction of the momentum transfer $q = k_0 - k$. Evaluating the commutator

$$[p_i, e^{iq\cdot r_j}] = \hbar q_j \delta_{ij} \tag{8.193}$$

and using the identity $\hat{q}\cdot(\sigma \times \hat{q}) \equiv 0$, the matrix element (8.192) reduces to the form

$$\langle k|V_m|k_0\rangle = 8\pi\kappa_n\mu_N\mu_B \left\{ \hat{q} \times (\sigma \times q) \cdot \sum_{j=1}^{n} e^{iq\cdot r_j} S_j - (i/\hbar q)\sigma \times q \cdot \sum_{j=1}^{n} e^{iq\cdot r_j} p_j \right\} \tag{8.194}$$

which is now an operator acting in the space of the scattering atom (and the neutron spin).

We note first of all that the term in (8.194)

$$\boldsymbol{T}(\boldsymbol{q}) = \sum_{j=1}^{n} e^{i\boldsymbol{q}\cdot\boldsymbol{r}_j}\boldsymbol{S}_j \tag{8.195}$$

which acts on the spin degrees of freedom, satisfies the commutation relation

$$\boldsymbol{T}\times\boldsymbol{S} = i\boldsymbol{T} \tag{8.196}$$

with the total atomic spin

$$\boldsymbol{S} = \sum_{j=1}^{n} \boldsymbol{S}_j$$

Thus, assuming that initial and final atomic states are uncoupled eigenstates of orbital and spin angular momenta $|L\, m_L; Sm_S\rangle$ which have the same quantum numbers L, S but which may differ in their magnetic quantum numbers m_L, m_S, we may apply the decomposition theorem (Condon and Shortley 1935)

$$\langle Jm'_J|\boldsymbol{T}|Jm_J\rangle = \langle Jm'_J|\boldsymbol{J}|Jm_J\rangle\{\langle J\|\boldsymbol{T}\cdot\boldsymbol{J}\|J\rangle/J(J+1)\} \tag{8.197}$$

where $\boldsymbol{J}$ is any angular momentum (such as $\boldsymbol{S}$) which satisfies the commutation relation (8.196) with respect to $\boldsymbol{T}$, and $\langle J\|\boldsymbol{T}\cdot\boldsymbol{J}\|J\rangle$ is the reduced matrix element of $\boldsymbol{T}\cdot\boldsymbol{J}$ in the basis $|J, m_J\rangle$. The theorem (8.197) then states that, to within a factor given by the reduced matrix element of $\boldsymbol{T}\cdot\boldsymbol{J}$, the projection of $\boldsymbol{T}$ on $\boldsymbol{J}$, the matrix elements of $\boldsymbol{T}$ are the same as those of $\boldsymbol{J}$. Thus $\boldsymbol{T}$ is replaced by an effective vector $\boldsymbol{T}_{\text{eff}}$, since the components of $\boldsymbol{T}$ normal to $\boldsymbol{J}$ average to zero in a state where J is a good quantum number. We then find that

$$\langle Sm_{S'}|\boldsymbol{T}(\boldsymbol{q})|Sm_{S_0}\rangle = \langle Sm_{S'}|\boldsymbol{S}|Sm_{S_0}\rangle F(\boldsymbol{q}) \tag{8.198}$$

where

$$F(\boldsymbol{q}) = \langle S, \alpha\| \sum_{j=1}^{n} e^{i\boldsymbol{q}\cdot\boldsymbol{r}_j}\boldsymbol{S}_j\cdot\boldsymbol{S}/S(S+1)\|S, \alpha\rangle \tag{8.199}$$

is a normalized magnetic form factor describing the net effect of interference between waves scattered from all the unpaired spins within the atom, and the symbol α represents all other conserved quantum numbers of initial (and final) states. We may therefore write

$$F(\boldsymbol{q}) = \int \Psi^*\left\{\sum_{j=1}^{n} \langle S\|e^{i\boldsymbol{q}\cdot\boldsymbol{r}_j}\boldsymbol{S}_j\cdot\boldsymbol{S}/S(S+1)\|S\rangle\right\}\Psi\, d^3r_1\, d^3r_2 \ldots d^3r_n \tag{8.200}$$

where $\Psi(\boldsymbol{r}_1\, \boldsymbol{r}_2 \ldots \boldsymbol{r}_n)$ is the orbital wavefunction of the atom, which, by hypothesis, is the same for initial and final states.

To evaluate $F(\boldsymbol{q})$ in the special case that only the unpaired electron spins contribute to the scattering and we may neglect the orbital contribution described by the second term in (8.194), we expand the operator $e^{i\boldsymbol{q}\cdot\boldsymbol{r}_j}$ in a series of spherical standing waves of orbital angular momentum λ and parity $(-1)^\lambda$

$$e^{i\boldsymbol{q}\cdot\boldsymbol{r}_j}=4\pi\sum_{\lambda=0}^{\infty}\sum_{\mu=-\lambda}^{\lambda} i^\lambda Y_\lambda^{\mu}(\hat{\boldsymbol{r}}_j)Y_\lambda^{\mu}(\hat{\boldsymbol{q}})j_\lambda(qr_j) \tag{8.201}$$

where $\hat{\boldsymbol{r}}_j$ is the unit vector $\boldsymbol{r}_j/r_j$ and $j_\lambda(qr_j)$ is the spherical Bessel function of order λ. Since $\boldsymbol{S}_j$ is an axial vector, it follows that only the even valued operators $Y_\lambda^{\mu}(\hat{\boldsymbol{r}}_j)$ in (8.201) can have non-vanishing matrix elements between common states $|\Psi\rangle$ having definite parity. In particular, if we adopt the dipole approximation which limits consideration to spin 1 tensor operators connecting initial and final states, we need take into account only the term with $\lambda=0$ in (8.201). Thus, assuming that all electrons belong to the same orbital $Y_l^{m_l}(\hat{\boldsymbol{r}})R_l(r)$, and are distinguished only by their projections m_l, $F(\boldsymbol{q})$ may be evaluated immediately with the result (Lovesey and Rimmer 1969)

$$F(\boldsymbol{q})=\bar{j}_0^{(l)}(q) \tag{8.202}$$

where

$$\bar{j}_\lambda^{(l)}(q)=\int_0^\infty j_\lambda(qr)|P_l(r)|^2 r^2\,dr \tag{8.203}$$

Extending the discussion to take account of scattering by orbital magnetic moments we must consider the momentum dependence which is described by the second term in (8.194). The operator $\boldsymbol{p}_j$ is of course a polar vector, which itself only has matrix elements between states of opposite parity. It must therefore be combined with odd parity terms in (8.201) to generate a total operator of even parity. Within the dipole approximation only the term with $\lambda=1$ need be included, and we make the replacement

$$e^{i\boldsymbol{q}\cdot\boldsymbol{r}_j}\boldsymbol{p}_j\rightarrow 3i\hat{\boldsymbol{r}}_j\cdot\hat{\boldsymbol{q}}j_1(qr)\boldsymbol{p}_j=i\boldsymbol{r}_j\cdot\boldsymbol{q}(j_0(qr)+j_2(qr))\boldsymbol{p}_j \tag{8.204}$$

where we have made use of the recurrence relation

$$j_\lambda(x)=(x/(2\lambda+1))(j_{\lambda-1}(x)+j_{\lambda+1}(x)) \tag{8.205}$$

Combining (8.204) with (8.194) we see that the jth electron contributes a term proportional to $(\boldsymbol{r}_j\cdot\boldsymbol{q})\,[(\boldsymbol{\sigma}\times\hat{\boldsymbol{q}})\cdot\boldsymbol{p}_j]$, which is just the contraction of the second rank Cartesian tensors $r_{j\alpha}p_{j\beta}$ and $(\boldsymbol{\sigma}\times\hat{\boldsymbol{q}})_\alpha q_\beta$. When these tensors are expressed in terms of spherical tensors according to the prescription given in section 3.3.1 we see that there is of course no electric monopole interaction (the neutron is uncharged) since $(\sigma\times\hat{q})_\alpha q_\beta$ is a traceless tensor, and we are left with magnetic dipole and electric quadrupole interactions described by the contractions of the anti-symmetric and symmetric parts respectively of the two Cartesian tensors.

The orbital magnetic dipole interaction is therefore given by

$$2\left[\frac{1}{2}(\boldsymbol{\sigma}\times\hat{\boldsymbol{q}})\times\boldsymbol{q}\right]\cdot\left[\frac{1}{2}(\boldsymbol{r}_j\times\boldsymbol{p}_j)\right]=-\hbar q[\hat{\boldsymbol{q}}\times(\boldsymbol{\sigma}\times\hat{\boldsymbol{q}})]\cdot\boldsymbol{l}_j \tag{8.206}$$

where $\boldsymbol{l}_j$ is the orbital angular momentum of the jth electron (in units of $\hbar$). According to (8.194) this means that the vector operator $\bar{j}_o(q)^{(l)}\sum_j \boldsymbol{S}_j$ must be replaced by

$$\bar{j}_0(q)^{(l)}\sum_j \boldsymbol{S}_j \rightarrow \sum_j \{\bar{j}_0(q)^{(l)}\boldsymbol{S}_j-(\mathrm{i}/\hbar q)(\mathrm{i}\hbar q/2)(\bar{j}_0(q)^{(l)}+\bar{j}_2(q)^{(l)})\boldsymbol{l}_j\}$$

$$=\bar{j}_0(q)^{(l)}\boldsymbol{S}+\frac{1}{2}(\bar{j}_0(q)^{(l)}+\bar{j}_2(q)^{(l)})\boldsymbol{L} \tag{8.207}$$

A description in terms of the operators $\boldsymbol{S}$ and $\boldsymbol{L}$ separately is suitable only when the atom is adequately described in terms of uncoupled angular momentum eigenstates $|Lm_L; Sm_S\rangle$. In general because of the spin-orbit interaction (cf. section 3.3.2.3) a description in terms of the coupled states $|L\,S\,J\,m_J\rangle$ is more accurate. In this case, since both $\boldsymbol{L}$ and $\boldsymbol{S}$ satisfy commutation relations of the type (8.196) with respect to $\boldsymbol{J}$, these operators can be replaced by their corresponding effective operators and we find

$$\langle Jm'_J|2\bar{j}_0^{(l)}(q)\boldsymbol{S}+(\bar{j}_0^{(l)}(q)+\bar{j}_2^{(l)}(q)\boldsymbol{L}|Jm_J\rangle=gF(\boldsymbol{q})\langle Jm'_J|\boldsymbol{J}|Jm_J\rangle \tag{8.208}$$

where

$$g=1+[J(J+1)+S(S+1)-L(L+1)]/2J(J+1) \tag{8.209}$$

is the Landé g-factor and

$$F(\boldsymbol{q})=\bar{j}_0^{(l)}(q)+(2/g-1)\bar{j}_2^{(l)}(q) \tag{8.210}$$

is the normalized magnetic form-factor.

According to (8.208) and (8.209) the magnitude of the magnetic moment of a free paramagnetic ion is just $g\sqrt{(J(J+1))}\mu_B$, a result which has been tested for paramagnetic ions in the crystalline state by carrying out magnetic susceptibility or paramagnetic resonance experiments on dilute (e.g. hydrated) salts, where the magnetic coupling between neighbouring ions is minimized. It turns out that, for the 3d transition elements, the results approximate very closely to the value $2\sqrt{(S(S+1))}\mu_B$, indicating that the orbital angular momentum has been 'quenched'. This effect is due to the electrostatic interaction of the electrons in the 3d shell, in this case the outermost shell, with the crystalline electron field, which exhibits the full point group symmetry of the crystal lattice. When this is the dominant interaction, the single electron orbitals are labelled by the irreducible representations of the corresponding point group and L is no longer a good quantum number (Elliott and Dawber 1979). In the ground state the operator $\boldsymbol{L}$ has a zero expectation value whereas the spin operator $\boldsymbol{S}$ is unaffected by the crystalline field interaction.

The maximum number n of electrons in the 3d shell is 10 and the spin–orbit coupling changes sign when the shell is half-full. Thus, at $n=5$ (e.g. M_n^{++}, F_e^{++}), $L=0$ and there is no spin–orbit coupling in the free ion. For $n \leqslant 4$ the effect of the residual spin–orbit coupling is to reduce slightly the (quenched) magnetic moment whereas, for $n \geqslant 6$ the effect is to increase it. Thus for the divalent ions of Fe, Co and Ni the effective g-factor is slightly greater than 2. In the case of the rare earth paramagnetic ions the incomplete 4f shell is enclosed by two full shells; the spin–orbit interaction is therefore dominant and the orbital angular momentum is unquenched.

We have neglected the electric quadrupole interaction in the above analysis, first because it goes beyond the dipole approximation, but also because this interaction plays no role in elastic scattering. This is obvious from the fact that there is no electric field associated with the neutron magnetic moment when it is in a stationary state, but can easily be demonstrated directly. Carrying out the appropriate contraction of the symmetric tensors immediately identifies the term $[(\boldsymbol{\sigma}\times\hat{\boldsymbol{q}})\cdot\boldsymbol{r}_j][\boldsymbol{q}\cdot\boldsymbol{p}_j]$, in which we make the replacement

$$\boldsymbol{p}_j = m_e \frac{d\boldsymbol{r}_j}{dt} = m_e[H_a, \boldsymbol{r}_j]/i\hbar \tag{8.211}$$

where H_a is the atomic Hamiltonian. We are therefore led to compute the matrix element

$$\langle f|[(\boldsymbol{q}\cdot\boldsymbol{r}_j)\boldsymbol{r}_j, H_a]|i\rangle = \hbar\omega\langle f|(\boldsymbol{q}\cdot\boldsymbol{r}_j)\boldsymbol{r}_j|i\rangle \tag{8.212}$$

where $\hbar\omega=(\varepsilon_i-\varepsilon_f)$ is the energy taken up by the atom. Hence this term vanishes for elastic scattering when $\omega=0$. More generally it is true that there is no electric multipole contribution to elastic scattering and, after the magnetic dipole interaction, the next most important term derives from the magnetic octupole interaction which vanishes unless $J \geqslant 3/2$.

Combining the results (8.194), (8.199) and (8.208) we can now write the scattering matrix element as

$$\langle f|V_m(\boldsymbol{q})|i\rangle = 4\pi\kappa_n\mu_N\mu_B gF(\boldsymbol{q})\langle m_{Jf}s'|\hat{\boldsymbol{q}}\times(\boldsymbol{\sigma}\times\hat{\boldsymbol{q}})\cdot\boldsymbol{J}|m_{Ji}s_0\rangle \tag{8.213}$$

where the operator

$$\hat{\boldsymbol{q}}\times(\boldsymbol{\sigma}\times\hat{\boldsymbol{q}})\cdot\boldsymbol{J} = \boldsymbol{\sigma}\cdot\boldsymbol{J} - (\boldsymbol{\sigma}\cdot\hat{\boldsymbol{q}})(\boldsymbol{J}\cdot\hat{\boldsymbol{q}}) = \hat{\boldsymbol{q}}\times(\boldsymbol{J}\times\hat{\boldsymbol{q}})\cdot\boldsymbol{\sigma} \tag{8.214}$$

is symmetric in $\boldsymbol{\sigma}$ and $\boldsymbol{J}$. Indeed we may use this result to rewrite the matrix element in the form

$$\langle f|V_m(\boldsymbol{q})|i\rangle = -\boldsymbol{\mu}_n\cdot\tilde{\boldsymbol{B}}(\boldsymbol{q}) \tag{8.215}$$

where

$$\tilde{\boldsymbol{B}}(\boldsymbol{q}) = 4\pi g\mu_B F(\boldsymbol{q})[-\boldsymbol{J}+\hat{\boldsymbol{q}}(\hat{\boldsymbol{q}}\cdot\boldsymbol{J})] \tag{8.216}$$

is the spatial Fourier transform

$$\tilde{\boldsymbol{B}}(\boldsymbol{q}) = \int e^{i\boldsymbol{q}\cdot\boldsymbol{r}} \boldsymbol{B}(\boldsymbol{r})\, d^3r \tag{8.217}$$

of the magnetic field $\boldsymbol{B}(\boldsymbol{r}) = \boldsymbol{H}(\boldsymbol{r}) + 4\pi\boldsymbol{M}(\boldsymbol{r})$ acting on the neutron spin. On this interpretation the first term in (8.216) comes from the Fourier transform of the magnetization within the scattering atom, i.e.

$$\tilde{M}(\boldsymbol{q}) = \int e^{i\boldsymbol{q}\cdot\boldsymbol{r}} M(\boldsymbol{r})\, d^3r = \int e^{i\boldsymbol{q}\cdot\boldsymbol{r}} \sum_{j=1}^{n} (-g\mu_B \boldsymbol{J}\delta^3(\boldsymbol{r}-\boldsymbol{r}_j))\, d^3r$$

$$= -g\mu_B F(\boldsymbol{q})\boldsymbol{J} \tag{8.218}$$

The second term in (8.216) is the Fourier transform of the magnetic intensity vector $\boldsymbol{H}(\boldsymbol{r})$ whose source is the 'magnetic pole density' $-\nabla\cdot\boldsymbol{M}(\boldsymbol{r})$. The result $\boldsymbol{q}\cdot\tilde{\boldsymbol{B}}(\boldsymbol{q}) = 0$, which is satisfied identically from (8.216), is a consequence of the fact that $\boldsymbol{B}(\boldsymbol{r})$ is a transverse vector with $\nabla\cdot\boldsymbol{B}(\boldsymbol{r}) = 0$. It follows that there is no magnetic scattering with momentum transfer $\boldsymbol{q}$ normal to the magnetic scattering vector $\boldsymbol{h}(\boldsymbol{q})$ defined by

$$\boldsymbol{h}(\boldsymbol{q}) = -\boldsymbol{J} + \hat{\boldsymbol{q}}(\hat{\boldsymbol{q}}\cdot\boldsymbol{J}) \tag{8.219}$$

Finally, to obtain an expression for the magnetic scattering amplitude $a_m(\boldsymbol{q})$ in Born approximation, we multiply $\langle f|V_m(\boldsymbol{q})|i\rangle$ by $-m_n/2\pi\hbar^2$ (cf. 2.48) and, ignoring the difference between neutron and proton masses, arrive at the result

$$a_m(\boldsymbol{q}) = \kappa_n r_e (g/2) F(\boldsymbol{q}) \langle m'_J s' | \boldsymbol{\sigma}\cdot\boldsymbol{h} | m_{J0} s_0 \rangle \tag{8.220}$$

where $r_e = 2.82$ fm is the classical electron radius defined in section 3.2.4. Since the radius r_e has a value comparable with those of nuclear radii, it follows that magnetic and nuclear cross-sections are also comparable in magnitude.

8.4.3 Paramagnetic and ferromagnetic atoms

To calculate the differential cross-section in Born approximation for scattering of slow neutrons from a single magnetic atom with zero nuclear spin, we have to add the nuclear and magnetic amplitudes to give a total scattering amplitude

$$A(\boldsymbol{q}) = -\langle m'_J s' | b + p(\boldsymbol{q})\boldsymbol{\sigma}\cdot\boldsymbol{h} | m_{J0} s_0 \rangle \tag{8.221}$$

where b denotes the bound coherent nuclear scattering amplitude defined in section 6.3.1.2 and

$$p(\boldsymbol{q}) = -\kappa_n r_e (g/2) F(\boldsymbol{q}) \tag{8.222}$$

is a magnetic scattering length defined such that $p(0) > 0$ for forward scattering. To determine the elastic cross-section we introduce a density matrix ρ_n to describe the polarization state of the incident neutron beam and evaluate the average $\langle A^{\dagger}(\boldsymbol{q})A(\boldsymbol{q})\rangle_{(n)}$ exactly as was performed for spin-dependent neutron–nucleus scattering in section 7.4.5.2. The result, analogous to (7.196) derived in that case, is then

$$\langle A(\boldsymbol{q})^{\dagger}A(\boldsymbol{q})\rangle_{(n)}=|b|^2+2\,\mathrm{Re}\,(bp^*(\boldsymbol{q}))\boldsymbol{f}_n\cdot\boldsymbol{h}+|p(\boldsymbol{q})|^2(\boldsymbol{h}^2-(\boldsymbol{f}_n\cdot\hat{\boldsymbol{q}})(\hat{\boldsymbol{q}}\cdot\boldsymbol{J})) \tag{8.223}$$

where $4\pi|b|^2$ is the purely nuclear scattering cross-section (cf. 6.83), and $\boldsymbol{f}_n$ is the neutron polarization vector defined in (7.259).

At this point we need to distinguish between paramagnetic atoms for which $\langle \boldsymbol{J}\rangle\equiv 0$ and ferromagnetic atoms for which this is decidedly not the case. For paramagnetic atoms we then have the result

$$\langle J|\boldsymbol{h}\cdot\boldsymbol{h}|J\rangle\equiv\langle J|\boldsymbol{J}^2-(\hat{\boldsymbol{q}}\cdot\boldsymbol{J})(\hat{\boldsymbol{q}}\cdot\boldsymbol{J})|J\rangle$$
$$=J(J+1)-\sum_{\alpha}\frac{1}{3}q_{\alpha}^2\langle J|\boldsymbol{J}^2|J\rangle=\frac{2}{3}J(J+1) \tag{8.224}$$

The resultant elastic differential cross-section is then given by

$$\left(\frac{\mathrm{d}\sigma}{\mathrm{d}\Omega}(\boldsymbol{q})\right)_p=|b|^2+\frac{2}{3}|p(\boldsymbol{q})|^2J(J+1) \tag{8.225}$$

In this case therefore there is a magnetic contribution to the incoherent cross-section but there is no sensitivity to neutron polarization.

A paramagnetic substance below the Curie point becomes a ferromagnetic substance in that it acquires a spontaneous magnetization even in the absence of an applied magnetic field. There is of course no sense in which we may properly consider the properties of ferromagnetic atoms in isolation from their environment because, although the Hamiltonian is spherically symmetric, the ground state of a ferromagnet is not, because the spins are polarized along the local magnetization vector which, at least in an infinite ferromagnet, can point in an arbitrary direction. This is an example of 'spontaneous symmetry breaking', a concept which in recent years has proved so fruitful in arriving at an understanding of the weak interaction (cf. chapter 9). In the present context it means that the spin projection measured with respect to the spontaneous magnetization is always a good quantum number. In the ground state of a ferromagnet $\langle S_z\rangle=S$, $\langle S_x\rangle=\langle S_y\rangle=0$, where the axis of quantization is taken parallel to the magnetization. The differential cross-section for unpolarized neutrons is therefore given by

$$\frac{\mathrm{d}\sigma(\boldsymbol{q})}{\mathrm{d}\Omega}=|b|^2+(1-\hat{q}_z^2)|p(\boldsymbol{q})|^2S^2 \tag{8.226}$$

and there is a contribution to the incoherent scattering as before.

When the neutrons are polarized the interference term between nuclear and magnetic coherent scattering in (8.222) comes into play, yielding a contribution to the elastic cross-section which changes sign when the neutron polarization is reversed. There is also a purely magnetic polarization-dependent term in (8.223) which comes from the usual expansion of the operator product $(\boldsymbol{\sigma}\cdot\boldsymbol{h})(\boldsymbol{\sigma}\cdot\boldsymbol{h})$ (cf. 8.249); however, since it depends on the component of polarization in the plane of scattering and is therefore a spin flipping term (cf. section 8.4.5.1), it makes no contribution to the coherent cross-section. Specializing to the case of forward scattering of fully polarized neutrons the differential cross-section becomes

$$\frac{\mathrm{d}\sigma(0)}{\mathrm{d}\Omega}=|b\pm a_{\mathrm{m}}(0)|^2 \qquad \boldsymbol{f}_{\mathrm{n}}\cdot\boldsymbol{h}(0)=\pm h(0) \tag{8.227}$$

where $a_{\mathrm{m}}(0)$ is the coherent forward magnetic scattering amplitude defined by

$$a_{\mathrm{m}}(0)=-p(0)\boldsymbol{f}_{\mathrm{n}}\cdot\boldsymbol{h}(0)=(m_{\mathrm{n}}\mu_{\mathrm{n}}/2\pi\hbar^2)\boldsymbol{f}_{\mathrm{n}}\cdot\int\boldsymbol{B}(\boldsymbol{r})\,\mathrm{d}^3r \tag{8.228}$$

where we have used (8.216) and (8.217), and the integral in (8.228) is extended over the volume occupied by the ferromagnetic atom. The significance of this result for developing methods of producing polarized slow neutrons has already been discussed in section 7.4.3.

8.4.4 Spin waves

In the ground state of a ferromagnetic crystal at low temperature, each atomic spin $\boldsymbol{S}_j$ is almost fully polarized along the effective field $\boldsymbol{B}_{\mathrm{eff}}$, with $\langle S_{jz}\rangle=S$ and $\langle S_{jx}\rangle=\langle S_{jy}\rangle=0$. A magnetically scattered neutron can disturb this orientation of the spin, which then precesses about the effective field. However, because of the exchange energy linking atomic spins at neighbouring lattice sites, this disturbance can propagate with various wavevectors $\boldsymbol{q}_s$, determined by the crystal symmetry. The normal modes of the spin system are called 'spin waves' or 'magnons' and, like phonons, are quantized excitations of the crystal as a whole (Bloch 1930, Elliott and Lowde 1955, Van Kranendonk and Van Vleck 1958). The propagation of a spin wave in a ferromagnet is illustrated in figure 8.8.

The point of departure for a theory of spin waves is the Heisenberg Hamiltonian for an ordered ferromagnet

$$H_{\mathrm{s}}=-\sum_{j,k=1}^{N} J_{jk}\boldsymbol{S}_j\cdot\boldsymbol{S}_k \tag{8.229}$$

where N is the number of atoms in the crystal and

$$J_{jk}\equiv J(\boldsymbol{R}_j-\boldsymbol{R}_k) \tag{8.230}$$

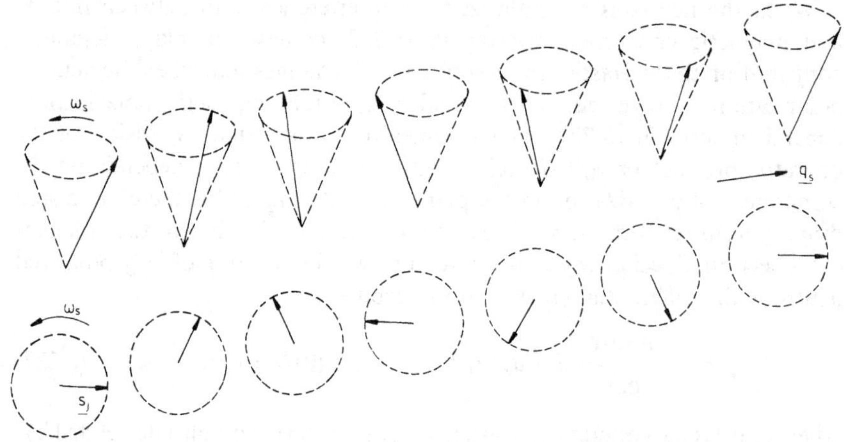

Figure 8.8. Propagation of a spin wave of energy $\hbar\omega_s$ and momentum $\hbar \boldsymbol{q}_s$ in a ferromagnet.

is the exchange parameter for atoms at the lattice sites $\boldsymbol{R}_j$ and $\boldsymbol{R}_k$. The symmetry of the crystal is reflected in the symmetry of $J(\boldsymbol{R})$ which can be expressed as a three-dimensional Fourier series

$$J(\boldsymbol{R}) = N^{-1} \sum_s \tilde{J}(\boldsymbol{q}_s)\, \mathrm{e}^{\mathrm{i}\boldsymbol{q}_s \cdot \boldsymbol{R}} \tag{8.231}$$

where the $\boldsymbol{q}_s$ form a set of N_c wavevectors distributed quasi-continuously over the first Brillouin zone. We shall restrict attention to Bravais lattices ($N_c = N$) where both $J(\boldsymbol{R})$ and $\tilde{J}(\boldsymbol{q}_s)$ are even functions of their arguments. Since $J(\boldsymbol{R})$ is a periodic function of the lattice, $\tilde{J}(\boldsymbol{q}_s) = \tilde{J}(\boldsymbol{q}_s + \boldsymbol{q}(\boldsymbol{v}))$ where $\boldsymbol{q}(\boldsymbol{v})$ is any vector of the reciprocal lattice.

The equations of motion of the jth spin are immediately found from (8.229) to be

$$\frac{\mathrm{i}\hbar\, \partial \boldsymbol{S}_j}{\partial t} = [H_s, \boldsymbol{S}_j] = \sum_k J_{jk} \boldsymbol{S}_j \times \boldsymbol{S}_k \tag{8.232}$$

These non-linear equations are symmetric in the Cartesian components of each spin $\boldsymbol{S}_j$, and are to be solved perturbatively as a small departure from the ground state of the crystal, where that symmetry is spontaneously broken. To carry out the required linearization it is useful to rewrite them in a form which explicitly recognizes the special status of the z-axis. We therefore introduce raising and lowering operators

$$S_j^{\pm} \equiv S_{jx} \pm \mathrm{i} S_{jz} \tag{8.233}$$

which satisty the commutation relations

$$[S_j^+, S_k^-] = 2\delta_{jk} S_{jz} \qquad [S_{jz}, S_k^{\pm}] = \pm \delta_{jk} S_j^{\pm} \tag{8.234}$$

Since the ground state is a simultaneous eigenstate of $\boldsymbol{S}_j^2$ and S_{jz} for all j, with eigenvalues $S(S+1)$ and S, respectively, it follows from the identity

$$\boldsymbol{S}_j^2 \equiv S_{jz}^2 + S_{jz} + S_j^- S_j^+ \tag{8.235}$$

that the ground state is also an eigenstate of $S_j^- S_j^+$ with eigenvalue zero. More generally, since the total spin component $\sum_j S_{jz}$ is a constant of the motion in all eigenstates of H_s, it follows that

$$\langle S_j^+ S_k^+ \rangle = \langle S_j^- S_k^- \rangle = 0 \tag{8.236}$$

for all pairs of spins in eigenstates of H_s. Finally we may deduce from (8.234) that S'_{jz}, the small change in S_{jz}, is given approximately by

$$S'_{jz} \equiv S_{jz} - S \simeq -S_j^- S_j^+ / 2S \tag{8.237}$$

The linearized form of the equations of motion (8.232) now becomes

$$i\hbar \frac{\partial S'_{jz}}{\partial t} \simeq \sum_k J_{jk}(S_k^- S_j^+ + S_k^+ S_j^-) \simeq 0 \tag{8.238}$$

and

$$i\hbar \frac{\partial S_j^\pm}{\partial t} \simeq 2S \sum_k J_{jk}(S_j^\pm - S_k^\pm) \tag{8.239}$$

We may deduce from (8.238) that, to first order, S'_{jz} remains constant at its value given by (8.237); thus the longitudinal perturbations of the ground state contribute to elastic scattering only and we may henceforth disregard these degree of freedom. Within this approximation the modified Hamiltonian becomes

$$H'_s = -NS^2\tilde{J}(0) - \sum_{j,k=1}^{N} J_{jk} S_j^+ S_k^- \tag{8.240}$$

It may be diagonalized in the usual way by introducing normal modes

$$\xi_s = \xi(\boldsymbol{q}_s) = (2NS)^{-1/2} \sum_j \mathrm{e}^{-\mathrm{i}\boldsymbol{q}_s \cdot \boldsymbol{R}_j} S_j^+$$

$$\xi_s^\dagger = \xi^\dagger(\boldsymbol{q}_s) = (2NS)^{-1/2} \sum_k \mathrm{e}^{\mathrm{i}\boldsymbol{q}_s \cdot \boldsymbol{R}_k} S_k^- \tag{8.241}$$

where

$$S_j^+ = (2S/N)^{-1/2} \sum_s \xi_s \mathrm{e}^{\mathrm{i}\boldsymbol{q}_s \cdot \boldsymbol{R}_j} \qquad S_k^- = (2S/N)^{-1/2} \sum_s \xi_s^\dagger \mathrm{e}^{-\mathrm{i}\boldsymbol{q}_s \cdot \boldsymbol{R}_k} \tag{8.242}$$

It follows from (8.241) and (8.234) that $\xi_s^\dagger$ and ξ_s satisfy the commutation relations

$$[\xi_s^\dagger, \xi_{s'}] = \delta_{ss'} \tag{8.243}$$

and may be interpreted in the usual way as creation and annihilation operators respectively for magnons in the sth mode. Rewriting the crystal Hamiltonian in the form

$$H'_s = -NS^2\bar{J}(0) - 2S\sum_s (\bar{J}(\boldsymbol{q}_s) - \bar{J}(0))(\xi_s^\dagger \xi_s + \tfrac{1}{2}) \qquad (8.244)$$

It follows that the spin-wave contribution to the Hamiltonian is given by

$$H_{\rm sw} = \sum_s \hbar\omega_s(\xi_s^\dagger \xi_s + \tfrac{1}{2}) \qquad (8.245)$$

where

$$\hbar\omega_s \equiv \hbar\omega(\boldsymbol{q}_s) = 2S(\bar{J}(0) - \bar{J}(\boldsymbol{q}_s)) \qquad (8.246)$$

is the corresponding dispersion relation relating angular frequency and wave number. This relation was first checked experimentally for the case of a face-centred cubic cobalt alloy using neutron spectrometric techniques (cf. section 8.5.2) (Sinclair and Brockhouse 1960).

To determine the scattering law for inelastic scattering of unpolarized neutrons with creation or annihilation of magnons, ignoring for simplicity phonon interactions and incoherent nuclear scattering which can of course take place simultaneously, we introduce a dimensionless magnetic intermediate scattering function in analogy with (8.74)

$$I_{\rm m}(\boldsymbol{q}, t) = N^{-1}\left\langle \sum_{j,k=1}^{N} e^{-i\boldsymbol{q}\cdot\bar{\boldsymbol{r}}_j(0)}\boldsymbol{\sigma}\cdot\bar{\boldsymbol{h}}_j(0)\boldsymbol{\sigma}\cdot\bar{\boldsymbol{h}}_k(t)\, e^{i\boldsymbol{q}\cdot\bar{\boldsymbol{r}}_k(t)}\right\rangle^{(0)} \qquad (8.247)$$

where $\bar{\boldsymbol{r}}_j(t)$, $\bar{\boldsymbol{h}}_j(t)$, etc. are operators in the interaction representation. If we disregard any coupling between the spin and position operators in (8.247), then, in the assumed absence of any inelastic processes involving the lattice, we may factorize the joint correlation function in (8.247) into the product of independent correlation functions and extract the position operators to determine a Debye–Waller factor exactly as in section 8.3.5.2. The result is

$$\langle e^{i\boldsymbol{q}\cdot\bar{\boldsymbol{r}}_j(0)}\, e^{i\boldsymbol{q}\cdot\bar{\boldsymbol{r}}_k(t)}\rangle^{(0)} = e^{-i\boldsymbol{q}\cdot(\boldsymbol{R}_j - \boldsymbol{R}_k)}\, e^{-2q^2M(0)} \qquad (8.248)$$

where $M(0)$ is the integral over the phonon spectrum defined is (8.157). Thus, using the identity for unpolarized neutrons ($\langle\boldsymbol{\sigma}\rangle \equiv 0$)

$$\boldsymbol{\sigma}\cdot\bar{\boldsymbol{h}}(0)\boldsymbol{\sigma}\cdot\bar{\boldsymbol{h}}_k(t) \equiv \bar{\boldsymbol{h}}_j(0)\cdot\bar{\boldsymbol{h}}_k(t) + i\boldsymbol{\sigma}\cdot\bar{\boldsymbol{h}}_j(0)\times\bar{\boldsymbol{h}}_k(t) = \bar{\boldsymbol{h}}_j(0)\cdot\bar{\boldsymbol{h}}_{jk}(t) \qquad (8.249)$$

$I_{\rm m}(\boldsymbol{q}, t)$ reduces to the form

$$I_{\rm m}(\boldsymbol{q}, t) = e^{-q^2M(0)}N^{-1}\left\langle \sum_{j,k=1}^{N} e^{-i\boldsymbol{q}\cdot(\boldsymbol{R}_j - \boldsymbol{R}_k)} \sum_{\alpha\beta} (\delta_{\alpha\beta} - \hat{q}_\alpha\hat{q}_\beta)\bar{S}_{j\alpha}(0)\bar{S}_{k\beta}(t)\right\rangle^{(0)}$$

$$(8.250)$$

The sum over the Cartesian indices $(\alpha\beta)$ in (8.250) is confined to the components (x, y) alone, since only the transverse magnon modes contribute to the inelastic scattering. This sum therefore reduces to

$$\sum_{\alpha\beta=x,y} (\delta_{\alpha\beta}-\hat{q}_\alpha\hat{q}_\beta)\bar{S}_{j\alpha}(0)\bar{S}_{k\beta}(t)=\frac{1}{4}(1+\hat{q}_z^2)(\bar{S}_j^-(0)\bar{S}_k^+(t)+\bar{S}_j^+(0)\bar{S}_k^-(t)) \tag{8.251}$$

where, following (8.236), we have omitted all operators constructed from two lowering or two raising operators. Transforming to normal modes using (8.241) and (8.242) the intermediate scattering function becomes

$$I_{\mathrm{m}}(\boldsymbol{q}, t)=(1+\hat{q}_z^2)\,\mathrm{e}^{-2q^2M(0)}N^{-1}$$

$$\times\left\langle \sum_{j,k=1}^{N} \mathrm{e}^{-\mathrm{i}\boldsymbol{q}\cdot(\boldsymbol{R}_j-\boldsymbol{R}_k)}\frac{1}{4}(\bar{S}_j^-(0)\bar{S}_k^+(t)+\bar{S}_j^+(0)\bar{S}_k^-(t))\right\rangle^{(0)}$$

$$=(1+\hat{q}_z^2)\,\mathrm{e}^{-2q^2M(0)}\frac{SN^{-2}}{2}\left\langle\sum_{j,k=1}^{N}\sum_{s}(\mathrm{e}^{-(\boldsymbol{q}+\boldsymbol{q}_s)\cdot(\boldsymbol{R}_j-\boldsymbol{R}_k)-\mathrm{i}\omega_s t}\xi_s^\dagger\xi_s\right.$$

$$\left.+\mathrm{e}^{-(\boldsymbol{q}-\boldsymbol{q}_s)\cdot(\boldsymbol{R}_j-\boldsymbol{R}_k)+\mathrm{i}\omega_s t}[\xi_s^\dagger\xi_s+1])\right\rangle^{(0)} \tag{8.252}$$

where the interaction representation operators $\xi_s^\dagger(t)$, $\xi_s(t)$ are given by (cf. 8.136)

$$\xi_s(t)=\xi_s\,\mathrm{e}^{-\mathrm{i}\omega_s t} \qquad \xi_s^\dagger(t)=\xi_s^\dagger\,\mathrm{e}^{\mathrm{i}\omega_s t} \tag{8.253}$$

We note that, because of the periodicity of the crystal, sums of the form

$$\sum_{j,k=1}^{N}\exp[-\mathrm{i}\boldsymbol{q}\cdot(\boldsymbol{R}_j-\boldsymbol{R}_k)]$$

which appear in (8.252), take the value N^2 when $\boldsymbol{q}=\boldsymbol{q}(\boldsymbol{\nu})$ where $\boldsymbol{q}(\boldsymbol{\nu})$ is a vector of the reciprocal lattice, but vanish otherwise. Therefore for a large crystal they may be represented in the form (cf. 8.110–8.112)

$$\sum_{j,k=1}^{N}\mathrm{e}^{-\mathrm{i}\boldsymbol{q}\cdot(\boldsymbol{R}_j-\boldsymbol{R}_k)}=\left|\sum_{j=1}^{N}\mathrm{e}^{-\mathrm{i}\boldsymbol{q}\cdot\boldsymbol{R}_j}\right|^2=(2\pi)^3(N/\Omega_{\mathrm{c}})\sum_{\boldsymbol{\nu}}\delta^3(\boldsymbol{q}-\boldsymbol{q}(\boldsymbol{\nu})) \tag{8.254}$$

where Ω_{c} is the volume of a unit cell. Using this result we then obtain a final expression for $I_{\mathrm{m}}(\boldsymbol{q}, t)$

$$I_{\mathrm{m}}(\boldsymbol{q}, t)=(1+\hat{q}_z^2)\,\mathrm{e}^{-2q^2M(0)}(S/2\Omega_{\mathrm{c}})\sum_{\boldsymbol{\nu}}(N^{-1})\sum_{s}\{\delta^3(\boldsymbol{q}+\boldsymbol{q}_s-\boldsymbol{q}(\boldsymbol{\nu}))$$

$$\times\,\mathrm{e}^{-\mathrm{i}\omega_s t}\langle n_s\rangle^{(0)}+\delta^3(\boldsymbol{q}-\boldsymbol{q}_s-\boldsymbol{q}(\boldsymbol{\nu}))\,\mathrm{e}^{\mathrm{i}\omega_s t}\langle n_s+1\rangle^{(0)} \tag{8.255}$$

where

$$\langle n_s \rangle^{(0)} = \langle \xi_s^\dagger \xi_s \rangle^{(0)} = [\exp(\mu\hbar\omega_s) - 1]^{-1} \tag{8.256}$$

is the mean number of magnons in the sth spin-wave mode at the temperature of the crystal. Thus there is a contribution to $I_{\mathrm{m}}(\boldsymbol{q}, t)$ from each spin wave, labelled by s, and from each Bragg peak, labelled by $\boldsymbol{\nu}$.

The magnetic scattering law now becomes

$$\begin{aligned} S_{\mathrm{m}}(\boldsymbol{q}, \omega) &= \left(\frac{1}{2\pi}\right) \int \mathrm{e}^{-\mathrm{i}\omega t} I_{\mathrm{m}}(\boldsymbol{q}, t)\, \mathrm{d}t \\ &= (1 + \hat{q}_z^2)\, \mathrm{e}^{-2q^2 M(0)} (S/2\Omega_{\mathrm{c}})(2\pi)^3 \sum_{\boldsymbol{\nu}} (N^{-1}) \sum_s \{\delta^3(\boldsymbol{q} - \boldsymbol{q}_s - \boldsymbol{q}(\boldsymbol{\nu})) \\ &\quad \times \delta(\omega + \omega_s) \langle n_s \rangle^{(0)} + \delta^3(\boldsymbol{q} - \boldsymbol{q}_s(\boldsymbol{\nu}))\, \delta(\omega - \omega_s) \langle n_s + 1 \rangle^{(0)}\} \end{aligned} \tag{8.257}$$

The corresponding expression for the inelastic differential scattering cross-section per target atom is then given by

$$\begin{aligned} \frac{\mathrm{d}^2 \sigma_{\mathrm{m}}^{\pm}}{\mathrm{d}\Omega\, \mathrm{d}\omega}(\boldsymbol{q}, \omega) &= \left(\frac{k}{k_0}\right) (1 + \hat{q}_z^2)\, \mathrm{e}^{-2q^2 M(0)} (2\pi)^3 \sum_{\boldsymbol{\nu}} (N^{-1} |p(\boldsymbol{q})|^2 \\ &\quad \times \sum_s \left[\left\langle n_s + \frac{1}{2} \pm \frac{1}{2} \right\rangle^{(0)} \delta^3(\boldsymbol{q} \mp \boldsymbol{q}_s - \boldsymbol{q}(\boldsymbol{\nu})\, \delta(\omega \mp \omega_s) \right] \Bigg) \end{aligned} \tag{8.258}$$

where the magnetic scattering length has been taken from (8.222) and the upper (lower) sign refers to magnon creation (annihilation) respectively.

It should be noted that it is always possible in principle to distinguish magnetic from nuclear scattering, by varying the orientation parameter $(1 + \hat{q}_z^2)$ over its allowed range between 1 and 2. This is because $\hat{q}_z$ is just the projection of the normalized momentum transfer along the direction of the effective internal magnetic field $\boldsymbol{B}_{\mathrm{eff}}$. This field can be controlled by applying an external magnetic field $\boldsymbol{H}_0$ in an arbitrary direction. Under these circumstances an additional term $-2\mu_{\mathrm{B}} H_0 \sum_j S_{jz}$ is added to H_s but the spin wave Hamiltonian H_{sw} is unaltered in form.

8.4.5 Polarized neutron scattering

8.4.5.1 Spin-flip and non-spin-flip scattering

Taking the z-axis as the quantization axis, we consider an incident neutron beam in either of the pure polarization states $|\pm\frac{1}{2}\rangle$ (cf. section 7.4.1). When the neutron scatters from an atom for which the bound coherent nuclear scattering length is b, and the nuclear spin-dependent scattering length is c (cf. section 6.3.1.2), there are two non-spin-flipping matrix elements which we denote by (Moon *et al* 1969),

$$\langle s'|V_{\rm m}|s_0\rangle_{\rm NSF}=\left\langle \pm\frac{1}{2}\middle|b+p(\boldsymbol{q})\boldsymbol{h}\cdot\boldsymbol{\sigma}+\frac{c}{2}\boldsymbol{I}\cdot\boldsymbol{\sigma}\middle|\pm\frac{1}{2}\right\rangle$$

$$=b\pm ph_z(\boldsymbol{q})\pm\frac{c}{2}I_z \tag{8.259}$$

and two corresponding spin-flip matrix elements

$$\langle s'|V_{\rm m}|s_0\rangle_{\rm SF}=\left\langle \mp\frac{1}{2}\middle|\boldsymbol{b}+p(\boldsymbol{q})\boldsymbol{h}\cdot\boldsymbol{\sigma}+\frac{c}{2}\boldsymbol{I}\cdot\boldsymbol{\sigma}\middle|\pm\frac{1}{2}\right\rangle$$

$$=ph^{\pm}(q)+\frac{c}{2}I^{\pm} \tag{8.260}$$

Here $h^{\pm}$, $I^{\pm}$ are raising and lowering operators defined in the usual way (cf. 8.253), and $\sigma^{\pm}$, which are the corresponding operators for a spin-$\frac{1}{2}$ particle, obey the relations (cf. section 3.4.2)

$$\sigma^{\pm}\equiv\frac{1}{2}(\sigma_x\pm \mathrm{i}\sigma_y)\qquad \sigma^{\pm}\left|\mp\frac{1}{2}\right\rangle=\left|\pm\frac{1}{2}\right\rangle\qquad \sigma^{\pm}\left|\pm\frac{1}{2}\right\rangle=0 \tag{8.261}$$

We first consider the implications of (8.259) and (8.260) as applied to the nuclear component in the scattering. In the first place it is clear that there is no spin-flip in the coherent nuclear scattering, and the same conclusion is obviously true for incoherent nuclear scattering generated by isotopic mixing (cf. section 8.2.4). The nuclear spin-dependent scattering is independent of $\boldsymbol{q}$ and therefore, in the absence of any nuclear polarization, two-thirds of the nuclear spins will on average be oriented normal to the neutron spin and one-third oriented parallel or anti-parallel to the neutron spin. It follows therefore that the spin-dependent nuclear incoherent scattering will be divided into spin-flipping (cf. 8.260) and non-spin-flipping (cf. 8.259) in the same ratio. The situation is entirely different in relation to magnetic scattering, since there is no spin-flip scattering when the scattering vector $\boldsymbol{h}(\boldsymbol{q})$ is parallel or anti parallel to the neutron polarization $\boldsymbol{P}$, and the scattering is totally spin-flip when $\boldsymbol{h}(\boldsymbol{q})$ is normal to $\boldsymbol{P}$.

Unless the target is itself polarized we cannot know the direction of $\boldsymbol{h}$, since each target spin $\boldsymbol{S}$ is oriented at random in space. We can however measure the direction of $\hat{\boldsymbol{q}}$ and, since $\hat{\boldsymbol{q}}\cdot\boldsymbol{h}\equiv 0$, select the plane in which $\boldsymbol{h}(\boldsymbol{q})$ lies. This circumstance allows us to identify two special cases which are illustrated in figure 8.9. In case (i) (figure 8.9(a)) we set $\boldsymbol{P}\cdot\hat{\boldsymbol{q}}=0$ and, choosing the x-axis to point along $\hat{\boldsymbol{q}}$, $\boldsymbol{h}$ is restricted to be in the y–z plane. Thus, in general, both h_z (non-spin-flip) and h_y (spin-flip) are non-zero, and, averaging over all directions of $\boldsymbol{S}$, the observed magnetic scattering will on average be equally divided between non-spin-flip and spin-flip modes. In case (ii) (figure

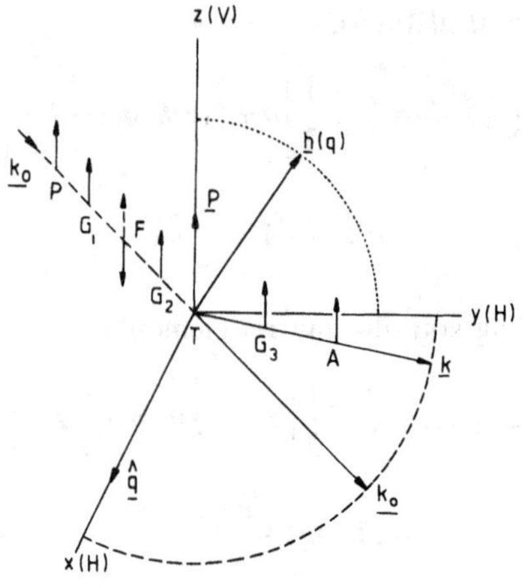

(a)

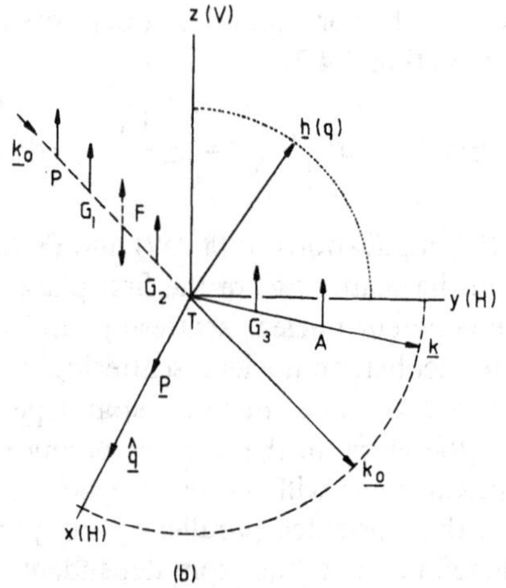

(b)

Figure 8.9. Magnetic scattering of polarized neutrons from unpolarized targets (T). (a) In case (i) the neutron polarization $\boldsymbol{P}$ is normal to the direction of the momentum transfer $\hat{\boldsymbol{q}}$; (b) in case (ii) $\boldsymbol{P}$ and $\hat{\boldsymbol{q}}$ are parallel or anti-parallel. In each case the polarizer P, analyser A, neutron spin guide-fields G and flipper coils F are aligned in the vertical (V), while the scattering is observed in the horizontal (H). In case (i) the magnetic field at the target is vertical while in case (ii) it is horizontal; this means that the pure neutron spin state changes adiabatically from an eigenstate of σ_z at P, to an eigenstate of σ_x at T, back to an eigenstate of σ_z at A.

8.9(b)) we set $\boldsymbol{P}\cdot\hat{\boldsymbol{q}} = \pm 1$ and, since $h_z \equiv 0$, the magnetic scattering occurs totally in the spin–flip mode. We should emphasize that case (ii) does not provide a suitable configuration for magnetic scattering from polarized targets since, to preserve the neutron spin in an (adiabatic) stationary state, the scattering must take place with $\boldsymbol{P}\cdot\boldsymbol{S}=\pm|\boldsymbol{P}|S$ and hence $\hat{\boldsymbol{q}}\cdot\boldsymbol{S}=\pm S$. It follows that $\boldsymbol{h}(\boldsymbol{q})\equiv 0$ and there is no magnetic scattering.

For the study of magnetic scattering in unpolarized targets (Ziebeck and Brown 1980, Steinsvoll *et al* 1984) the most effective procedure is to observe spin-flip scattering only, since this eliminates all but the spin-incoherent component of nuclear scattering. In case (i) 50% of the magnetic scattering is observed and in case (ii) the proportion is 100%. Since the amount of nuclear scattering is the same in either case, the difference is a measure of 50% of the magnetic scattering. In both cases the non-spin-flip component is eliminated by placing a spin flipper either before or after the target as convenient, so that the analyser transmits the same sense of the neutron spin as does the polarizer.

8.4.5.2 Bragg scattering from ferromagnetic crystals

In section 8.4.3 we touched briefly on the topic of spin-dependent forward scattering from a single ferromagnetic atom. In the more complex situation which arises with spin-dependent elastic scattering from a saturated ferromagnetic crystal we have to evaluate the correlation function

$$K_{\mathrm{B}}(\boldsymbol{q})=N^{-1}\sum_{j,k=1}^{N}\langle \mathrm{e}^{-\mathrm{i}\boldsymbol{q}\cdot\boldsymbol{r}_j}A_j^{\dagger}(\boldsymbol{q})A_k(\boldsymbol{q})\,\mathrm{e}^{\mathrm{i}\boldsymbol{q}\cdot\boldsymbol{r}_k}\rangle^{(0)} \tag{8.262}$$

where $A_j(\boldsymbol{q})$ is the total scattering amplitude defined in (8.221). In contrast to the dimensionless intermediate scattering function $I_{\mathrm{m}}(\boldsymbol{q},t)$, the function $K_{\mathrm{B}}(\boldsymbol{q})$ has the dimensions of a cross-section. Expanding the operator product in (8.262) we may identify nuclear, magnetic and interference terms in $K_{\mathrm{B}}(\boldsymbol{q})$, i.e.

$$K_{\mathrm{B}}(\boldsymbol{q})=N^{-1}\sum_{j,k=1}^{N}\langle \mathrm{e}^{-\mathrm{i}\boldsymbol{q}\cdot(\boldsymbol{r}_j-\boldsymbol{r}_k)}\{b_{\mathrm{coh}}^2+2b_{\mathrm{coh}}(p^*(\boldsymbol{q})\boldsymbol{\sigma}\cdot\boldsymbol{h}_j+p(\boldsymbol{q})\boldsymbol{\sigma}\cdot\boldsymbol{h}_k)$$
$$+|p(\boldsymbol{q})|^2(\boldsymbol{\sigma}\cdot\boldsymbol{h}_j)(\boldsymbol{\sigma}\cdot\boldsymbol{h}_k)\} \tag{8.263}$$

where b_{coh} is the coherent scattering length defined in (8.58). For simplicity we shall maintain the assumption that the crystal is a Bravais lattice in which case $F(\boldsymbol{q})$ is an even, and hence a real function of $\boldsymbol{q}$.

According to (8.249) the operator product $(\boldsymbol{\sigma}\cdot\boldsymbol{h}_j)(\boldsymbol{\sigma}\cdot\boldsymbol{h}_k)$ gives rise to a spin-dependent term $\mathrm{i}\boldsymbol{\sigma}\cdot\boldsymbol{h}_j\times\boldsymbol{h}_k$ which, in the most general case, is non-vanishing. For example in certain anti-ferromagnetic crystals the spin $\boldsymbol{S}_j$ does not point in a fixed direction but undergoes a helical variation in space which induces a precession of the neutron spin of the same screw sense about the vector $\boldsymbol{h}_j\times\boldsymbol{h}_k$. We shall however assume that $\boldsymbol{S}_j$ is constant everywhere, i.e.

$$\boldsymbol{S}_j=-S\hat{\boldsymbol{f}}_{\mathrm{m}} \qquad \text{for all } j \tag{8.264}$$

where $\hat{f}_m$ is a unit vector directed along the internal magnetization, and hence anti-parallel to the atomic spin. In this case all operators $\boldsymbol{h}_j$ are equivalent and each is given by

$$\boldsymbol{h}_j(\boldsymbol{q})=S(\hat{f}_m-\hat{\boldsymbol{q}}(\hat{\boldsymbol{q}}\cdot\hat{f}_m))=S\sqrt{(1-(\hat{\boldsymbol{q}}\cdot\hat{f}_m)^2)}\hat{\boldsymbol{h}}\cdot(\boldsymbol{q}) \tag{8.265}$$

where $\hat{\boldsymbol{h}}(\boldsymbol{q})$ is a unit vector along $\boldsymbol{h}_j(\boldsymbol{q})$, with

$$\mathrm{i}\boldsymbol{\sigma}\cdot(\boldsymbol{h}_j(\boldsymbol{q})\times\boldsymbol{h}_k(\boldsymbol{q}))=-(\boldsymbol{\sigma}\cdot\hat{\boldsymbol{q}})(\hat{\boldsymbol{q}}\cdot S\hat{f}_m)\delta_{jk} \tag{8.266}$$

This term vanishes when $\hat{\boldsymbol{q}}\cdot\hat{f}_m=0$, which is the conventional configuration for polarized neutron scattering from a polarized target.

We may now extract the space-dependent operators from $K_B(\boldsymbol{q})$ exactly as for the spin wave analysis in section 8.4 and the remainder of the calculation proceeds in similar fashion. The final result for the differential cross-section per atom in Bragg scattering is then given by

$$\left(\frac{\mathrm{d}\sigma(\boldsymbol{q})}{\mathrm{d}\Omega}\right)_B=\frac{(2\pi)^3}{\Omega_c}\sum_{\nu}\mathrm{e}^{-2q^2M(0)}\delta^3(\boldsymbol{q}-\boldsymbol{q}(\boldsymbol{\nu}))$$
$$\times\{b_{coh}^2+2b_{coh}Sp(\boldsymbol{q})\sqrt{(1-(\hat{\boldsymbol{q}}\cdot\hat{f}_m)^2)}\boldsymbol{\sigma}\cdot\hat{\boldsymbol{h}}(\boldsymbol{q})+(1-(\hat{\boldsymbol{q}}\cdot\hat{f}_m)^2)S^2p^2(\boldsymbol{q})\} \tag{8.267}$$

If the incident neutron beam is prepared in an eigenstate of the operator $\boldsymbol{\sigma}\cdot\hat{f}_m$, i.e.

$$\hat{f}_n\cdot\hat{f}_m=\pm 1 \tag{8.268}$$

then the function given in (8.267) can be rewritten as a two-valued function for each Bragg peak

$$\left(\frac{\mathrm{d}\sigma^{\pm}}{\mathrm{d}\Omega}(\boldsymbol{q}(\boldsymbol{\nu}))\right)_B=\frac{(2\pi)^3}{\Omega_c}\mathrm{e}^{-2q^2(\nu)M(0)}\{b_{coh}\pm Sp(\boldsymbol{q}(\boldsymbol{\nu}))\}^2 \tag{8.269}$$

The result (8.269) may be put to a number of uses the most straightforward of which is in the production of polarized neutrons. For, provided $|b_{coh}|\leqslant Sp(\boldsymbol{q})$ it may be possible to select a particular Bragg angle at which one or other of the cross-sections vanishes, in which case the diffracted neutrons are fully polarized in the opposite spin-sense. The effect was first observed in the (2 2 0) diffraction in crystalline magnetite, Fe_2O_3 (Shull 1951, Shull *et al* 1951). Since diffraction can only produce polarization at wavelengths less than the crystal cutoff it is of less general application than mirror reflection (cf. section 7.4.3), which is also effective at longer wavelengths. A more important application of (8.269) is in the experimental determination of spin densities (Nathans *et al* 1959).

For a Bravais lattice where $\boldsymbol{S}$ is the atomic spin at a given lattice site, and there are n unpaired electrons per atom, the normalized spin density $s(\boldsymbol{r})$

may be defined by

$$s(\boldsymbol{r})=\left\langle S\left|\sum_{i=1}^{n} \boldsymbol{S}_i\delta^3(\boldsymbol{r}-\boldsymbol{r}_i)\cdot\boldsymbol{S}/S(S+1)\right|S\right\rangle \tag{8.270}$$

Thus, according to (8.199), $s(\boldsymbol{r})$ may be written as the Fourier transform

$$s(\boldsymbol{r})=\left(\frac{1}{2\pi}\right)^3\int_{\Omega_c} e^{-i\boldsymbol{q}\cdot\boldsymbol{r}}F(\boldsymbol{q})\,d^3q \tag{8.271}$$

and a measurement of $F(\boldsymbol{q})$ for a range of vectors $\boldsymbol{q}(\boldsymbol{\nu})$ provides a mapping of $s(\boldsymbol{r})$ throughout the unit cell. Returning to (8.269) if we set

$$\hat{\boldsymbol{q}}\cdot\boldsymbol{f}_m=0 \qquad \boldsymbol{f}_n\cdot\hat{\boldsymbol{f}}_m=\pm 1 \tag{8.272}$$

then, by carrying out diffraction measurements with both senses of neutron polarization, we may determine the flipping ratio (Moon 1986)

$$R(\nu)=\frac{d\sigma^+(\boldsymbol{q}(\boldsymbol{\nu}))}{d\sigma^-(\boldsymbol{q}(\boldsymbol{\nu}))}=\left\{\frac{b_{\mathrm{coh}}+p(\boldsymbol{q}(\boldsymbol{\nu}))}{b_{\mathrm{coh}}-p(\boldsymbol{q}(\boldsymbol{\nu}))}\right\}^2 \tag{8.273}$$

whose measurement over a range of Bragg angles permits a determination of $F(\boldsymbol{q}(\boldsymbol{\nu}))$ and hence of $s(\boldsymbol{r})$. The result (8.273) is independent of the Debye–Waller factor, and of attenuation in the crystal, and has provided a highly sensitive method for determining $F(\boldsymbol{q})$ in a range of magnetic materials. The method is applied using the vertical configuration shown in figure 8.9(a), with a spin–flipper set up between the polarizer and the sample, and does not require polarization analysis of the diffracted beam.

8.5 Neutron scattering instrumentation

8.5.1 Conventional diffractometry

8.5.1.1 The double-axis diffractometer

The first instruments which exploited Bragg scattering in crystals for the production of monochromatic neutrons were specifically designed for investigations in nuclear physics. They were used to measure total cross-sections, to investigate the detailed shapes of neutron resonances, and to verify the $1/v$ law of neutron capture (cf. section 1.7.1.1). These instruments were of the single-axis type, using a monochromating crystal rotatable about an axis normal to the scattering plane and a single BF_3 neutron counter located immediately behind the sample (Zinn 1947, Stürm 1947). Subsequently Fermi and Marshall (1947) developed a double-axis version, with a counter which could be rotated about the sample permitting the determination of angular distributions. This work initiated the application of neutron diffractometry in the study of condensed states of matter.

The techniques of double-axis diffractometry are applicable in these situations where the scattering law $S(\boldsymbol{q}, \omega)$ assumes large values only in a range of ω for which $\hbar\omega$ is much less than the neutron energy. This condition, which always holds good for X-ray diffraction, is generally valid for Bragg scattering from crystals. In this case the scattered intensity is proportional to $\int S(\boldsymbol{q}, \omega)\mathrm{d}\omega$, and provides information about the lattice structure of the crystal. Since there is no energy selection of the scattered neutrons conventional diffractometry must be distinguished from inelastic diffractometry which requires the use of triple-axis or time of flight spectrometers (cf. section 8.5.2).

The essential elements of a double-axis diffractometer are (1) a monochromator which accepts neutrons from a polychromatic source and directs a monochromatic beam onto the sample; (2) a set of collimators to determine the directions of the momentum $\boldsymbol{k}_\mathrm{i}$ incident on the monochromator, the momentum $\boldsymbol{k}_0$ delivered by the monochromator to the sample, and the momentum $\boldsymbol{k}$ scattered by the sample; and (3) a detector to record the neutrons scattered through a specific angle φ. The layout of such a system is illustrated in figure 8.10. In a modern instrument the neutrons are recorded

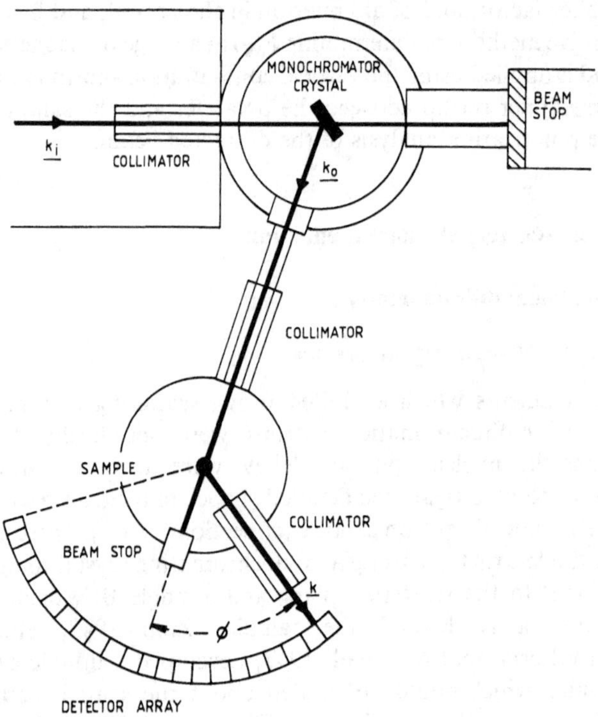

Figure 8.10. Principle of operation of a double-axis neutron diffractometer.

in ^3He counters or position-sensitive detectors (cf. section 8.3.6), or in one of an array of many counters set at different angles φ.

8.5.1.2 The single-crystal diffractometer

In the two-circle instrument, used to study liquids and amorphous materials, the scattering angle may be varied by swinging the counter arm through an angle φ with respect to $\boldsymbol{k}_0$, in the range $0<\varphi<2\pi$ and by rotating the sample through an angle ψ about an axis normal to the scattering plane, in the range $-\pi<\psi<\pi$. To study Bragg scattering in crystals, where $\varphi=2\theta$, a four-circle instrument is used, which incorporates an 'Eulerian cradle' to support the sample at any specified orientation in space. This addition permits an axis of symmetry of the sample to be fixed at an angle χ with respect to the scattering plane $(-\pi<\chi<\pi)$, and the sample to be rotated through an angle φ about this axis. Thus ψ, $\pi-\chi$ and η constitute a set of Euler angles defining the orientation of the sample.

In most diffractometers the monochromator is a large single crystal oriented with respect to $\boldsymbol{k}_i$ in such a way that only those neutrons incident at a glancing angle θ_m which satisfies Bragg's law (8.164) are Bragg diffracted through a 'take-off' angle $2\theta_m$. As with most dispersing instruments the design problem presented reduces to one of finding a suitable compromise to the conflicting demands of maximizing the intensity, and reducing the resolution width ΔE, of the transmitted neutrons. Unfortunately the kinematical theory of Bragg scattering is inapplicable to monochromators since the expression (7.90) for the differential cross-section applies only for small perfect crystals, where multiple scattering may be neglected. For large perfect crystals it is necessary to apply the dynamical theory (cf. section 7.2.4) where the decay of the neutron wave amplitude in the interior is characterized by the extinction length Δ given in (7.107). This parameter measures the degree of primary extinction whereby the outer layers of the crystal shield the inner regions from participating in the reflection (Bacon and Lowde 1948, Hamilton 1957, 1958). Typical values of Δ fall in the range 10^5-10^6 Å. In any case perfect crystals cannot be used efficiently as monochromators since the angular width they can accept is very much less than the beam divergences from available sources.

These difficulties may be overcome in practice by constructing monochromators from mosaic crystals, i.e. crystals formed from mosaic blocks typically about 10^4 Å in size, each of which behaves as a small perfect crystal. In order to reduce the effect of secondary extinction which arises because a given mosaic block may shield its neighbour to some degree, it is necessary that the coherence of the neutron wave be destroyed at the interfaces between neighbouring blocks. This is achieved by arranging the blocks with a slight misorientation, so that the Bragg angle θ is no longer unique, but is distributed with a Gaussian density (Darwin 1922)

$$p(\theta)=(2\pi\eta^2)^{-1/2}\exp[-\tfrac{1}{2}((\theta-\theta_m)/\eta)^2] \tag{8.274}$$

The standard deviation η has values falling typically in the range 0.3–3.0 milliradian, and is known as the mosaic spread. With this range of η, neutrons scatter incoherently from the mosaic blocks, and neutron propagation in the crystal may be treated as a problem in neutron transport (Sears 1977).

Monochromators are constructed from single crystals with a large mosaic spread using materials featuring reduced incoherent scattering and capture cross-sections. The most commonly used elements are Be, Cu, Ge, Si and pyrolytic graphite which, using a variety of reflecting planes, together cover the wavelength range 0.4–6.0 Å. In some instruments squashed germanium crystals have been employed successfully to produce focusing normal to the scattering plane.

If λ_0 is the wavelength selected in the first order, the transmitted beam will be contaminated with faster neutrons of wavelength λ_0/n, with $n \geqslant 2$, and momentum transfers q_n given in (8.165). However, since the higher momentum transfers are suppressed by the Debye–Waller factor (cf. section 8.3.5.2), it is usually only necessary to take account of higher order reflections with $n=2$ or 3. A convenient method for removing fast neutrons is to use a polycrystalline BeO filter which removes all neutrons with wavelengths less than the Bragg cut-off at 3.96 Å. Another useful technique makes use of 'resonance filters', whereby λ_0 is selected so that $\lambda_0/2$ coincides with a strong absorption resonance in a particular isotope. Convenient resonances exist for short wavelength neutrons with $\lambda=0.1$–0.3 Å in certain actinide elements of which the most important is the plutonium isotope ^{239}Pu which has a resonance at 290 meV corresponding to the second-order reflection at $\lambda=1.06$ Å. For longer wavelength neutrons with λ in the range 0.3–0.9 Å, a selection of absorbing resonances is available among isotopes of rare earth elements.

The intrinsic energy resolution for a monochromating crystal is given by

$$\Delta E/E=\cot\theta_m\,\Delta\theta_m \tag{8.275}$$

where

$$\Delta\theta_m=\left\{\frac{(\alpha_i^2+\alpha_0^2)\eta^2+\alpha_i^2\alpha_0^2}{(\alpha_i^2+\alpha_0^2)+4\eta^2}\right\}^{1/2} \tag{8.276}$$

Here α_i and α_0 are respectively the in-plane acceptance angles for the beam incident on, and the beam reflected by, the monochromator. These angles are usually limited by Soller collimators, each consisting of an array of neutron absorbing discs (e.g. gadolinium deposited on Mylar sheet) arranged parallel to the selected direction (Soller 1924). If each disc is of length l, separated from its neighbour by a distance s, the acceptance angles are restricted to values less than $\alpha=s/l$.

It follows from (8.275) that the best resolution is achieved when $\theta_m = \pi/2$, i.e. backscattering, although such a condition is not achievable in general. Thus most designs aim at maximizing the take-off angle. To achieve maximum intensity with minimum background at a given resolution, α_i and α_0 should each be set equal to 2η so as to accept that range of wavelength determined by the mosaic spread.

8.5.1.3 The powder diffractometer

In single-crystal diffraction experiments the individual Bragg reflections are well separated in angle, and may be studied independently of each other by rotating the detection system and altering the orientation of the sample (Peterson and Levy 1951, 1952, Bacon 1951, Busing and Levy 1957). The crystal structure thus revealed is refined by least-squares minimization of the weighted sums of squares of residuals, determined by subtracting the measured integrated intensities in the Bragg line, from theoretical intensities calculated on the basis of some hypothetical structure. However, problems of primary or secondary extinction make this technique difficult to apply in the case of large crystals where the Debye–Scherrer or powder technique must be used. Here the sample is composed from a powder of small crystals, whose orientations are distributed isotropically in space, and the three-dimensional crystallographic diffraction pattern characterized by a spacing $d(h_1h_2h_3)$, is projected on to a Debye–Scherrer cone given by

$$\lambda = 2d(h_j)\sin[\theta(h_j)] \qquad j = 1, 2, 3 \tag{8.277}$$

The powder samples are usually contained in cylindrical cells made from vanadium for which the coherent scattering cross-section is vanishingly small (cf. Table 6.2).

The difficulty with this technique is, of course, that the information contained in the three-dimensional pattern is compressed into one dimension with the inevitable overlap of reflections implied by that reduction. The number of equivalent sets of crystal planes characterized by the same $d(h_j)$ is known as the multiplicity; it depends on the indices h_j and in general is larger for crystals with higher symmetry. For triclinic crystals with no symmetry transformations the multiplicity is 2 whereas for orthorhombic crystals it is 8 in general, reducing to 4 when one index is zero, and to 2 when two indices are zero. In cubic crystals sets of crystal planes occur, characterized by quite distinct sets of indices, which have the same spacing d, and which are therefore quite indistinguishable in the powder spectrum.

The first powder diffractometer for neutrons was constructed by Wollan and Shull (1948) who used it to locate the positions of the hydrogen atoms in NaH. Indeed the location of hydrogen, and the heavier atoms C, N, O in complex organic molecules, has continued to provide a useful field of application for powder diffractometers. The transition metal oxides were among the first specimens to be investigated by the powder method, since magnetic diffraction occurs at low angles where resolution is less critical. Here it was noted that the azimuthal symmetry in the diffracted intensity around a Debye–Scherrer ring could be destroyed by application of a magnetic field, an observation which culminated eventually in the identification of the anti-ferromagnetic and ferrimagnetic phases (cf. sections 8.4.1.3 and 4). However the application of the integrated intensity procedure to the refinement of complex crystallographic structures by powder diffractometry was severely limited by the overlap problem, and attendant difficulties associated with extinction and sample orientation.

The principal weaknesses of the powder method were eventually overcome by the profile analysis method introduced by Rietveld, which treated the entire pattern as a single entity without attempting to resolve and identify individual lines (Rietveld 1966, 1967, 1969). The Rietveld procedure provides an algorithm based on least squares fitting of the complete spectrum without the introduction of intermediate steps involving the extraction of integrated intensities according to the traditional procedure. At the root of the Rietveld method was the recognition that the observed line shape for a given reflection, representing the convolution of many independent line shapes for the crystallites making up the sample, is truly a Gaussian in conformity with the central limit theorem of statistics. The introduction of Rietveld profile analysis enormously extends the potential of the powder method allowing the investigation of systems of low symmetry and the study of structural phase transitions.

These theoretical initiatives have motivated matching developments on the technical side, resulting in the construction of powder diffractometers with angular resolutions approaching the theoretical limit $\approx \lambda/D$, when D is the characteristic size of the mosaic block in the crystalline making up the sample (Hewat 1975, Hewat and Bailey 1976). The corresponding reduction in intensity is then compensated by introducing the multidetector array, simultaneously sampling a wide range of scattering angle 2θ, with the ultimate resolution achieved at $\theta = \theta_m$. This is the so-called parallel orientation which is represented schematically in figure 8.11.

8.5.1.4 The small-angle diffractometer

Small-angle neutron scattering was first applied at the Argonne National Laboratory to study magnetic refraction at domain boundaries in the interiors of ferromagnetic crystals (Hughes *et al* 1949). The small angular deviations observed were then interpreted in terms of variations in the refractive index caused by changes in the direction of magnetization (cf. section 8.4.1.2). Similar techniques were also employed to investigate multiple refraction and diffraction effects produced by powders suspended in liquid solution (Kruger *et al* 1950, Weiss 1951). The requirement to increase the

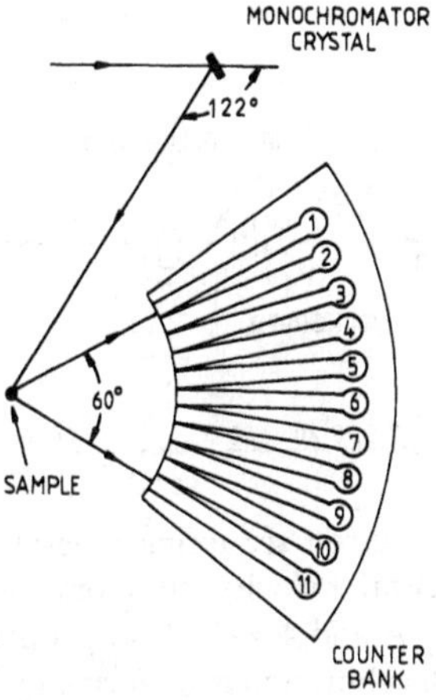

Figure 8.11. A high-resolution neutron powder diffractometer focused for parallel geometry ($2\theta = 122°$). The bank of ^{3}He counters can be swept from $2\theta = 0°$ to $2\theta = 160°$ in steps of 0.05° (Hewat and Bailey 1976).

relative intensity of 4 Å neutrons for small-angle scattering on liquid helium at AERE Harwell over the period 1952–53, provided the main impetus for constructing the first liquid hydrogen cooled moderator, which was inserted in the reactor BEPO in 1956. Subsequently all SANS diffractometers have used long-wave neutrons from cold sources with velocity selection replacing crystal monochromators to produce approximately monoenergetic neutrons (Dunning *et al* 1935).

The essential requirement of a SANS instrument is that the incident and scattered momenta be very precisely defined and the first diffractometer designed specifically for SANS experiments was constructed in the early 1970s at Jülich. This displayed all the characteristic features of modern instruments of this type, namely a neutron guide to select slow neutrons from a cold source, velocity selection to produce roughly monochromatic neutrons with a wavelength spread of order 9–20%, and a movable position-sensitive detector to count the neutrons scattered into a variable range of q (Schmatz *et al* 1974). Among the first applications of the SANS technique was the study of dislocation lines in copper and flux lines in Type II superconductors (Christen *et al* 1977).

Figure 8.12 is a schematic representation of the SANS diffractometer D11 at the Institut Laue Langevin which is a development of the original Jülich instrument. Exploiting the fact that the intensity of scattered neutrons increases in proportion to the square of a typical dimension, this is a very large machine, over 100 m in length at its maximum extension, and covering a range of q from 10^{-3} Å^{-1} to 0.5 Å^{-1}. Neutrons are accepted from a 25 l liquid deuterium cooled moderator maintained at 25 K, and subsequently

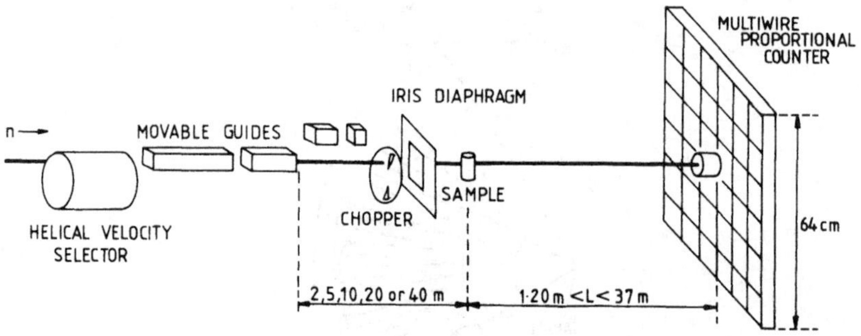

Figure 8.12. The small-angle neutron scattering time-of-flight spectrometer D11 at the ILL Grenoble. Two helical velocity selectors can provide neutrons with wavelengths in the range 2–20 Å and 4.5–20 Å with wavelength spreads of 9% and 50%, respectively. Wavelength calibration is performed with a chopper and neutrons are counted in a multiwire BF_3 proportional counter containing 3808 active elements.

monochromatized in one of a pair of helical velocity selectors. The first of these delivers 4.5–20 Å neutrons at a wavelength resolution of 9%; the second covers a large range of wavelength between 2 Å and 20 Å but at a resolution of $\approx 50\%$. With the maximum separation of 40 m between velocity selector and sample a high degree of primary collimation corresponding to a beam divergence of ≈ 0.2 milliradian is achievable.

The neutrons are detected in a 64×64 array of BF_3 counters with a ≈ 1 cm^2 mesh, which can be set at a distance L from the sample variable between 1.2 m and 37 m. Under these conditions the accessible range of q is approximately kd/L, where d is the distance from the counting cell to the point at which the direct beam intercepts the detector system. At the maximum value of L, q can be varied between 10^{-3} Å^{-1} and 10^{-2} Å^{-1}, and at the minimum separation between 2×10^{-2} Å^{-1} and 2×10^{-1} Å^{-1}. There are also facilities for increasing q to about 3 Å^{-1} by adding additional detectors at large angles.

8.5.2 Inelastic scattering

8.5.2.1 The triple-axis spectrometer

The general objective of coherent inelastic scattering experiments in single crystals is the determination of the energy transfer $\hbar\omega$ as a function of the momentum transfer $\hbar\boldsymbol{q}$ (cf. section 8.3.5.1). In general $\hbar\boldsymbol{q}$ is divided between the momentum $\hbar\boldsymbol{q}(\boldsymbol{\mu})$ absorbed by the lattice as a whole, and the momentum $\hbar\boldsymbol{q}_p$ transferred to an individual lattice phonon. The region of interest in $(\boldsymbol{q}, \omega)$ space is much reduced by the crystal symmetry and, for studies of lattice dynamics, interest tends to be concentrated on high symmetry points and lines in $\boldsymbol{q}$-space. Phonon modes, for example tend to be degenerate on high symmetry lines. For studies of structural and magnetic phase transitions, it is required to study the inelastic spectrum as a function of applied pressure, temperature or magnetic field, and in these cases, ω must be measured with high resolution along specified directions in $\boldsymbol{q}$-space. For these types of rather specific measurement the triple-axis spectrometer tends to be preferred to the time-of-flight spectrometer which is more suitable for incoherent scattering and broad scans over large regions of $(\boldsymbol{q}, \omega)$ space.

The triple-axis spectrometer has developed from the double axis diffractometer by the addition of a third axis which carries a second crystal, where the scattered neutrons are energy analysed by Bragg scattering prior to detection. The first instrument of this type was constructed by Brockhouse at the Chalk River Nuclear Laboratories, where it was used to observe neutron–phonon scattering in single crystals of aluminium (Placzek and Van Hove 1954, Brockhouse and Stewart 1955, 1958, Brockhouse and Iyengar 1958). These experiments confirmed the expectation that neutrons could gain as well as lose energy in interaction with lattice phonons (Egelstaff 1951, Cassels 1951, Brockhouse and Hurst 1952) and were in accord with dynamical effects observed earlier in boron absorption spectroscopy of single crystals of iron (Lowde 1952, 1954). Most triple-axis spectrometers operate with thermal or cold neutrons and many are now equipped for polarization analysis in incident and scattered beams (cf. section 8.4.5). For studies of spin waves where it is desirable to sample large values of ω at small values of q, so that the magnetic form factor should not become too large (cf. sections 8.4.3–8.4.4) spectrometers operating with hot neutrons, of wavelength in the range 0.3–1.2 Å, have proved very useful.

The layout of a conventional triple-axis spectrometer is illustrated in figure 8.13. The directions of all three axes must be subject to independent control, with three further controls fixing the orientations of the monochromator, the sample and the analyser about axes normal to the scattering plane. In most modern instruments a fission chamber monitor, which is relatively transparent to neutrons, is placed between the monochromator and the sample, so that allowance may be made for fluctuations in neutron intensity from the source. A second monitor is placed between the sample and the analyser to detect any spurious signal produced by Bragg scattering from the sample into the direction of the inelastically scattered beam, subsequently incoherently scattered from the analyser into the detector. Such a process produces an anomalously high signal in the second monitor, as indeed does any background in the incident beam, with energy equal to the twice scattered energy E, which is then twice Bragg reflected from the sample and from the analyser into the detector.

The resolution of the analyser is described by equations which are the analogues of (8.275)–(8.276) with $\theta_M \to \theta_A$, $\alpha_i \to \alpha'$ and $\alpha_0 \to \alpha$, where α' and α are the acceptance apertures for neutrons incident on, and Bragg scattered

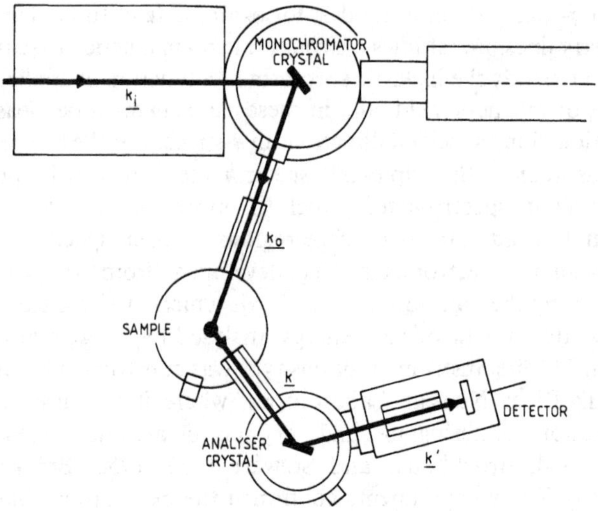

Figure 8.13. Principle of operation of the triple-axis neutron spectrometer.

by, the analyser. Both ω and $\boldsymbol{q}$ must be defined very closely, but the degree of precision required depends both on the size of the unit cell in the reciprocal space of the sample, and the rate of change of the phonon frequency with wave number. A 1.5 Å unit cell corresponds to a unit cell of order 4 Å^{-1} in reciprocal space for which a wavelength resolution of order 0.1 Å^{-1} would generally be necessary. Near high symmetry lines in the reciprocal lattice the rate of change of frequency with wave number is frequently reduced permitting a relaxation in the conditions of high resolution.

Since there are in general four variables at the disposal of the experimenter, namely the incident energy E_0, the scattered energy E, the angle of scatter φ and the sample orientation ψ, which may be defined as the angle between $\boldsymbol{k}_0$ and a selected crystal axis, whereas to view a particular point in $(\boldsymbol{q}, \omega)$ space only three numbers are needed since $\boldsymbol{q}$ always lie in the scattering plane, it follows that the problem is over-determined. This feature contributes greatly to the versatility of the instrument, permitting a number of quite distinct experimental scans in which one variable is maintained constant. For example in a conventional scan $\boldsymbol{k}_0$ and φ are kept fixed and ψ is varied by rotating the sample; this means that $\boldsymbol{q}$ sweeps out a restricted region in reciprocal space and may reveal the presence of groups of phonons in $(\boldsymbol{q}_p, \omega)$ space which satisfy the conditions of energy and momentum conservation given in (8.137). To search for phonons of given momentum $\boldsymbol{q}_p$, a specific reciprocal lattice vector $\boldsymbol{q}(\boldsymbol{\mu})$ must be selected with $\boldsymbol{q}$ fixed at a value given by

$$\boldsymbol{q} = \boldsymbol{q}_p + \boldsymbol{q}(\boldsymbol{\mu}) \tag{8.278}$$

To perform such a constant-$\boldsymbol{q}$ scan either E_0 may be kept constant and E varied, or conversely, with φ and ψ varied in such a way as to maintain $\boldsymbol{q}$ constant.

Keeping E_0 constant has the advantage that the sample position is fixed and the background is constant; the disadvantage is that the analyser resolution varies rapidly with variable θ_A. Thus the low frequency phonons are less accurately located in $(\boldsymbol{q}_p, \omega)$ space than are the high frequency phonons. The more usual procedure this is to keep E fixed and vary E_0. This technique has the added advantage that, if data are accumulated, not over fixed periods of time, but over fixed numbers of counts in the monitor between monochromator and sample, assumed to be a $1/v_0$ (fission) detector, it provides a direct measure of $S(\boldsymbol{q}, \omega)$, since it effectively removes the denominator k_0 from the factor (k/k_0) appearing before $S(\boldsymbol{q}, \omega)$ in (8.61) (Dorner 1972).

8.5.2.2 The time-of-flight spectrometer

A time-of-flight (TOF) neutron spectrometer contains two principal elements; a primary spectrometer or monochromator which determines the mean energy and energy spread of neutron pulses reaching the sample, as well as the time interval between the pulses, and a secondary spectrometer which records the energy spectrum of the scattered neutrons by measuring their flight times over a fixed path length (Buras and Leciejewicz 1963, Buras 1963). Time-of-flight spectrometers are suitable for detecting energy gains or losses in the range 10^{-5} eV to 10^{-1} eV, and momentum transfers in the range 10^{-2} Å^{-1} to 10 Å^{-1}.

Although several variations of the primary spectrometer are in current use, the choice depending on the specific experiments envisaged and the range of wavelength deployed, the principle of operation of the secondary spectrometer is the same in every case. There are of course differences in detail, e.g. the length of flight path, the numbers and types of neutron counter employed, and the range of scattering angle covered. However in a typical time-of-flight system there will be several hundred ^{3}He counters in variable configurations covering a range of angles between 10° and 130° with flight paths between 0.4 m (cold neutrons) and 4.0 m (thermal neutrons). Neutron pulse lengths and pulse separations in the ms range are typical, and counter outputs will be sorted into 10 μs time bins allowing a best energy resolution of about 10 μeV. Three versions of the primary spectrometer are in operation at the ILL (continuous) neutron source, the multichopper monochromator (IN5), the time-focusing crystal monochromator (IN6) and the rotating crystal monochromator (IN4). At the ISIS (pulsed) neutron source (cf. figure 1.6), the spectrometer HET, which operates with epithermal neutrons, uses a single chopper synchronized to the natural source frequency.

In the multichopper arrangement illustrated in figure 8.14, all four

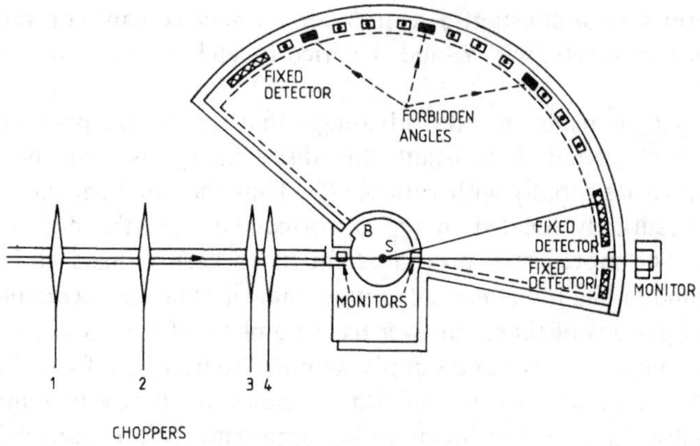

Figure 8.14. The multi-chopper time of flight spectrometer IN5 at the ILL Grenoble (Lechner *et al* 1973). The velocity of the transmitted neutrons is determined by the phase angle between choppers 1 and 4. Slow neutrons whose flight times exceed the selected flight time by an integral number of rotation periods are eliminated by chopper 2. The rate at which neutron pulses with the requisite velocity spread arrive at the target is controlled by chopper 3.

mechanical choppers spin about a common axis parallel to the incident beam, and each contains a radial transmission slot matched to the cross-section of the neutron guide transporting the neutrons through the chopper system. The first, second and fourth chopper spin at the same rate which is variable between 6×10^3 rpm and 2×10^4 rpm. The third chopper spins at a reduced rate equal to some simple fraction of the fundamental frequency. The first chopper produces a burst of white neutrons which disperse in space according to the range of relative velocities in the beam, and only those neutrons moving with such a velocity that they are transmitted by the fourth chopper, which spins at a fixed phase angle with respect to the first, will impinge on the target. This neutron pulse has a resolution function which is almost exactly triangular in shape.

If the selected group of neutrons arrives at the fourth chopper, a time τ after leaving the first, then it is clear that slower neutrons arriving at times $t_n=\tau+nT$, where T is the rotation period, will also be transmitted. It is the function of the second chopper to eliminate these higher order contaminants from the final pulse. Finally, the third chopper, running at the slower speed, regulates the rate at which neutron pulses arrive at the fourth chopper. It therefore eliminates the effect of frame overlap, whereby fast neutrons which have gained energy from the scattering process, overtake slow neutrons in the preceding pulse which have lost energy. The time-of-flight analysis is initiated by a pick-up signal sent from the fourth chopper, with calibration of the incident neutrons provided by observations on the elastic peak.

In the time focusing primary spectrometer shown in figure 8.15 the beam is extracted from the neutron guide by a system of three composite pyrolytic graphite crystals, which, acting in combination with a liquid nitrogen cooled beryllium filter to suppress the second-order reflection in graphite, focus the Bragg diffracted beam onto the sample. The focusing action increases the flux by a factor of about three over that transmitted by the multichopper but, because neutrons diffracted through the largest Bragg angle move more slowly than those diffracted through the smallest Bragg angle, this improvement in flux is acquired at the cost of a disimprovement in energy resolution. However, this loss in energy resolution is regained by pulsing the beam using a Fermi chopper, i.e. a chopper containing transmission slots running parallel to the axis of rotation which is perpendicular to the neutron beam (Fermi *et al* 1947) (cf. figure 1.3). The sense of rotation of this chopper is arranged so that the slowest neutrons are transmitted first and time focusing is brought about by setting the time delay between the times of transmission of the slowest and the fastest neutrons exactly equal to the difference in their flight times. The function of the second Fermi chopper, which precedes the time-focusing chopper, is to eliminate frame overlap as described above.

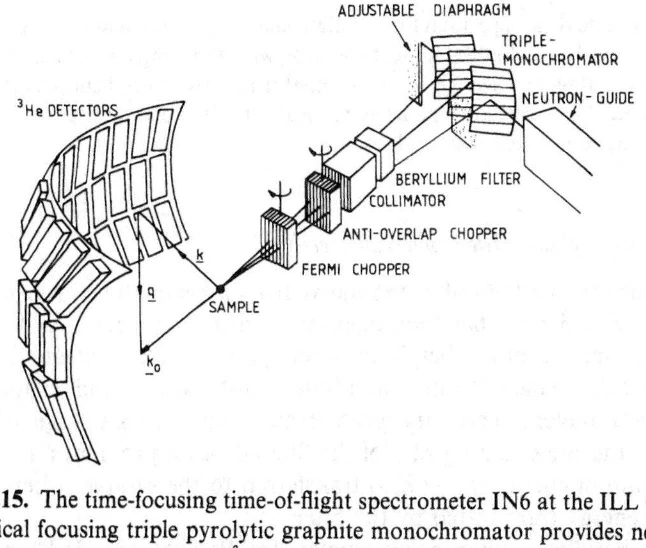

Figure 8.15. The time-focusing time-of-flight spectrometer IN6 at the ILL Grenoble. The vertical focusing triple pyrolytic graphite monochromator provides neutrons of wavelength 4.1, 4.6, 5.1 and 5.9 Å for studies of quasi-elastic and inelastic scattering. The time-focusing condition is achieved by a Fermi chopper with a suppressor chopper to eliminate frame overlap.

A third form of primary spectrometer, which is suitable for use with faster neutrons, makes use of a rotating crystal to extract neutrons from the guide, so that every time the Bragg condition is satisfied with respect to a direction specified by a collimator, a burst of monochromatic neutrons arrives at the sample. This monochromating crystal is used in conjunction with a second crystal to eliminate the background of thermal neutrons, and, when the second crystal also rotates, only one order of reflection can be Bragg diffracted. This system is in use at the spectrometer IN4 at the ILL, the complete layout of which is shown in figure 8.16.

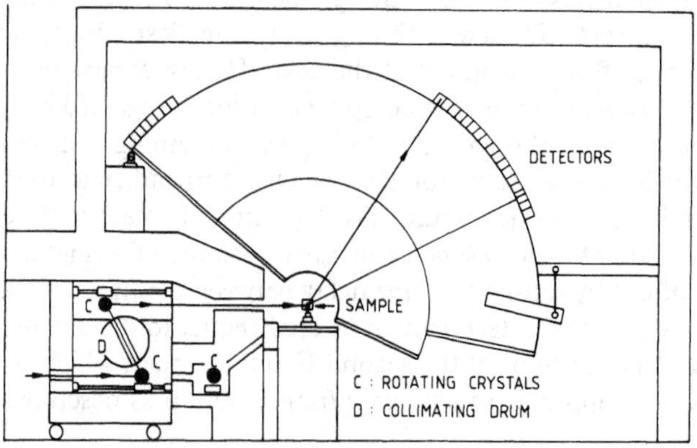

Figure 8.16. The rotating crystal time of flight spectrometer IN4 at the ILL Grenoble. The instrument has a double monochromator with two single crystals, at least one of which is rotating, and a single monochromator plus two Fermi choppers for a fixed take-off angle. The scattering plane is vertical and ^{3}He counters cover a range of scattering angles between $\approx 9°$ and 140°.

8.5.2.3 The beryllium filter spectrometer

The routine use of a BeO filter to remove from a beam all fast neutrons with $\lambda < 3.96$ Å ($E_n > 5$ meV) has been mentioned in section 8.5.1.2, and the same principle is applied in the beryllium filter spectrometer to create an energy window $0 < E_n < 5$ meV (Danner and Stiller 1961, Duggal and Thaper 1962). In this spectrometer the primary spectrometer scans across a range of incident energy E_i, and the counting rate of the filtered beam provides a measure of the spectrum of energy $(E_i - \langle E_f \rangle)$ transferred to the sample, where $\langle E_f \rangle$ is the mean energy transmitted by the filter.

In the beryllium filter spectrometer IN1BeF at the ILL, a cooled beryllium-graphite combination filter is used, with a fixed energy window $0 < E_n < 2$ meV, with variable incident energy provided by a set of three monochromators; Cu200 (16–227 meV), Cu220 (41–476 meV) and Cu331 (66–1000 meV). The instrument is used with hot neutrons with a capture flux $\varphi_i \simeq 2 \times 10^{10}$ cm^{-2} s^{-1} peaked near 1.2 Å, and is applied to the study of incoherent inelastic scattering in the measurement of the vibrational spectra of hydrides and molecular compounds in general.

8.5.3 Quasi-elastic scattering

8.5.3.1 Back-scattering spectrometers

The description 'quasi-elastic' is applied to neutron scattering in which very small amounts of energy $|\Delta E| \leqslant 2$ meV are exchanged between the incident neutron and the sample. Such small energy transfers occur in scattering from particles undergoing rotational or translational diffusive motions in atomic and molecular systems, e.g. in liquid or plastic crystals, and their detection requires a spectrometer with the highest possible energy resolution. Although direct geometry instruments, such as the time-of-flight spectrometers discussed above, can detect energy differences as low as 10 μeV in special circumstances, values at the level of 100 μeV are more typical. This is because, with a fixed incident energy, the ultimate resolution is determined by the precision with which the neutron velocity can be measured, over a broad spectrum of neutron energies.

The highest energy resolution can be achieved in a system which is specifically designed to detect scattered neutrons of a fixed energy, and such instruments are described as inverted geometry instruments, since it is the incident energy which is scanned across a pre-determined range. The beryllium filter spectrometer described in section 8.5.2.3 comes into this class. It may be observed that an inverted geometry spectrometer is analogous in many ways to the superheterodyne system of radio reception with the resonant level in the sample playing the role of carrier frequency, the incident neutron energy as local frequency, and the scattered energy as intermediate frequency.

Since the most precise way of measuring a neutron energy is by application of Bragg's law (cf. 7.59), the corresponding energy resolution may be estimated by evaluating the differential

$$\frac{\delta E}{E} \equiv 2\frac{\delta \lambda}{\lambda} = 2\left\{\frac{\delta d}{d} + \cot \theta_B \delta\theta_B\right\} \tag{8.279}$$

From this result we concluded that $\delta E/E$ is minimized when $\cot \theta_B = 0$, i.e. $\theta_B = \pi/2$. This corresponds to a fixed 'take-off' angle $2\theta_B = \pi$, i.e. back-scattering (Birr *et al* 1971). There are two back-scattering spectrometers of this type in use at the ILL, i.e IN10 and IN13, while a third, IRIS, is in operation at ISIS.

The spectrometer IN10 operates with capture fluxes $\varphi_c \simeq 10^3$–2×10^4 of cold neutrons centred around 6 Å, and high energy resolution is achieved by back-scattering at both monochromating and analysing crystals. The energy of the primary beam is scanned through a finite energy range varying between $\simeq 15\ \mu$eV and $\simeq 150\ \mu$eV, depending on the choice of crystal combination, through Doppler motion of the monochromator which is mounted on a velocity drive. A range of Si, Al_2O_3, CaF_2 and Ge crystals is available, allowing a selection of five modulated incident wavelengths covering the range between 3.27 Å and 6.53 Å.

The neutrons in the primary beam are directed on to the sample by Bragg reflection from a graphite crystal, and pulsing is brought about by inserting a chopper in front of the sample, phased so that neutrons scattered directly into the ^{3}He detectors are not recorded. Experiments can be carried out simultaneously for eight values of the momentum transfer between 0.1 $Å^{-1}$ and 5 $Å^{-1}$, allowing the measurement of energy transfers in the range 1 $\mu eV < \Delta E < 100\ \mu eV$, with energy resolutions between 0.3 μeV and 4 μeV. The layout of the system is shown in figure 8.17.

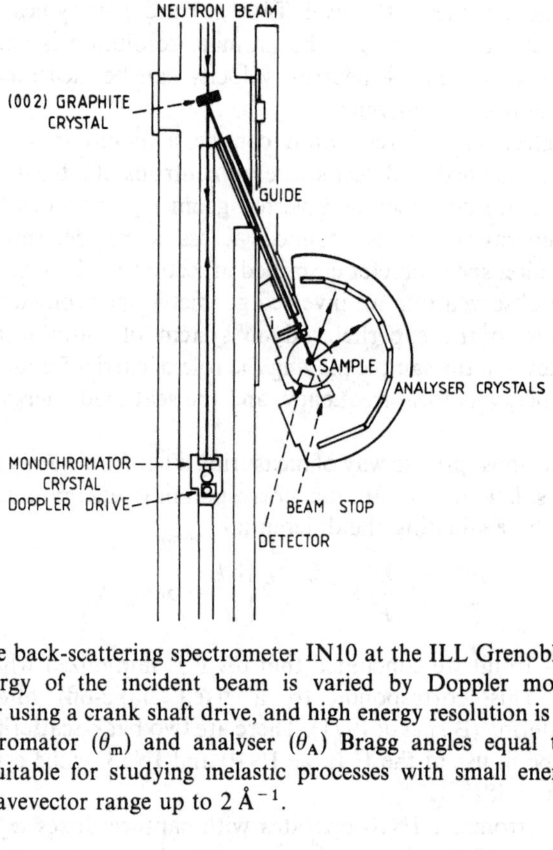

Figure 8.17. The back-scattering spectrometer IN10 at the ILL Grenoble (Birr *et al* 1971). The energy of the incident beam is varied by Doppler motion of the monochromator using a crank shaft drive, and high energy resolution is achieved by setting monochromator (θ_m) and analyser (θ_A) Bragg angles equal to 90°. The instrument is suitable for studying inelastic processes with small energy changes >0.1 eV in a wavevector range up to 2 $Å^{-1}$.

The spectrometer IN13 uses a CaF_2 monochromator to select neutrons from a thermal beam centred near 2.23 Å, with modulation brought about by temperature variation which changes the lattice parameter. The scattered neutrons are energy analysed by a set of six spherically curved composite crystal analysers in back-scattering mode, set to cover a range of momentum transfers $0.3\,\text{Å}^{-1}<q<5.5\,\text{Å}^{-1}$. Simultaneous time-of-flight analysis is used to eliminate neutrons scattered directly into the ^{3}He detectors. The spectrometer can detect energy exchanges over the range $-400\,\mu\text{eV}\leqslant\Delta E\leqslant 8\,\mu\text{eV}$, at an energy resolution of approximately 8 μeV. In the IRIS spectrometer at ISIS the incident energy is selected from the pulsed beam by a rotating disc chopper, and the scattered energy analysis is carried out using a pyrolytic graphite crystal close to the back-scattering orientation ($2\theta_B=175°$). This yields an energy resolution of approximately 15 μeV for energy exchanges in the range $-1\,\text{meV}\leqslant\Delta E\leqslant 4\,\text{meV}$.

8.5.3.2 The neutron spin-echo spectrometer

Direct geometry spectrometers determine the energy by measuring the velocity, and inverted geometry spectrometers perform the same function by measuring the wavelength. The neutron spin echo (NSE) spectrometer is in a class of its own, since it relies on a totally different phenomenon, namely that a neutron spin carries out a countable number of Larmor precessions about a uniform magnetic field, which, for a fixed path length, is inversely proportional to the velocity. In a sense therefore the neutron measures its own time of flight, using the precession of its own spin as a clock.

There is, however, a second phenomenon involved, namely that of 'spin echo', or spin refocusing, whereby the neutron spin winds itself up during the first half of its flight time, and unwinds during the second, so that all neutrons complete their journey with the same precession angle with which they began, irrespective of their velocity. If, however, a scattering takes place, at the passage mid-point, the resultant change in velocity is reflected in the final precession angle, which may be determined with high precision. Although the spin echo principle has been successfully applied for many years in NMR (Hahn 1950), its reincarnation as a major tool in quasi-elastic neutron spectroscopy is due to Mezei (Mezei 1972, 1980, 1983).

The principles underlying the NSE techniques may be understood by reference to figure 8.18(a). A neutron, initially polarized along the x-axis moves, with velocity v_0, a distance l_0 towards the origin in the positive z-direction, parallel to a uniform magnetic field B_0. Thus, by the time it reaches the origin, the spin has processed through an angle φ_0 given by

$$\varphi_0=\gamma_n l_0 B_0/v_0 \tag{8.280}$$

where γ_n is the gyromagnetic ratio (cf. section 7.5.2). If, after scattering at the origin it resumes its motion with velocity v_1, along the positive z-axis where there is a magnetic field component $-B_1$, by the time it has travelled a further distance l_1, the net angle of precession accumulated over the total path (l_0+l_1) is given by

$$\varphi(v_0, v_1)=\varphi_0-\varphi_1=\gamma_n[l_0B_0/v_0-l_1B_1/v_1] \tag{8.281}$$

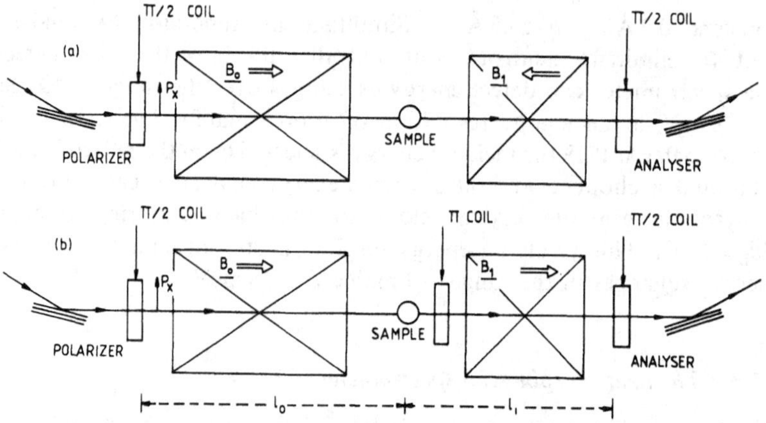

Figure 8.18. The neutron spin echo principle. The arrangement (a) with $\boldsymbol{B}_0$ and $\boldsymbol{B}_1$ anti-parallel is unsuitable in practice, since the neutrons are forced to traverse a region of zero field where the neutron spin states become degenerate. This difficulty is avoided in arrangement (b) where $\boldsymbol{B}_0$ and $\boldsymbol{B}_1$ are parallel, and the neutron spins are flipped after passing through the sample.

If the scattering is elastic so that $v_1 = v_0 = v$, the resultant x-polarization, for a given velocity distribution characterized by the probability density function $p(v)$, is given by

$$P_x = \langle \cos \varphi \rangle = \int_0^\infty \cos\left[\frac{\gamma_n}{v}[l_0 B_0 - l_1 B_1]\right] p(v)\, \mathrm{d}v \qquad (8.282)$$

Thus the NSE signal provided by P_x is proportional to the real part of the characteristic function describing the distribution of $(1/v)$, i.e. of the wavelength λ. In particular if conditions are chosen so that

$$l_0 B_0 = l_1 B_1 \qquad (8.283)$$

P_x returns to its initial value of unity for all values of v. If P_x is observed not to be unity, then this result is indicative of an inelastic component in the scattering and herein lies the potential of the NSE technique as a spectrometer tool.

A serious difficulty with the system as envisaged in figure 8.18(a) is that, if the field reversal is to be carried out in a continuous way, the neutrons reaching the origin will find themselves in zero magnetic field, where the spin states become degenerate, with all the attendant difficulties that situation generates (Majorana 1932) (cf. section 6.2.1). Thus the field reversal has to be carried out non-adiabatically, for example using the current sheet system described in section 7.4.2. In that case, however, the fringing fields associated with the current sheet creates problems in other parts of the apparatus. A far simpler arrangement, and the one which is universally adopted, is to maintain $\boldsymbol{B}_1$ parallel to $\boldsymbol{B}_0$, and insert a spin-flipper, other than the current sheet, at the origin, which replaces the component P_x by $-P_x$. It then follows, according to figure 8.18(b), that those spins most advanced in phase, which are associated with the slowest neutrons, exchange places with the most retarded spins associated with the fastest neutrons. Thus when the second stage of the journey of length l_1 has been completed, all spins come back into phase as before, and P_x returns to (minus) its starting value.

We now consider the more interesting case where the scattering is inelastic, so that the phase shift $\varphi(v_0, v_1)$ is accompanied by an energy transfer

$$\hbar\omega(v_0, v_1) = \tfrac{1}{2}m_{\mathrm{n}}(v_2^2 - v_1^2) \tag{8.284}$$

The basic objective now is to establish conditions under which $\hbar\omega(v_0, v_1)$ may be deduced from a measurement of $\varphi(v_0, v_1)$. We therefore introduce average angles of precession $\bar{\varphi}$, and average energy exchanges $\hbar\bar{\omega}$ according to the definitions

$$\bar{\varphi} = \varphi(\bar{v}_0, \bar{v}_1) \qquad \bar{\omega} = \omega(\bar{v}_0, \bar{v}_1) \tag{8.285}$$

and postulate a linear relation of the form

$$\varphi - \bar{\varphi} = t(\omega - \bar{\omega}) \tag{8.286}$$

where t is a quantity having the dimensions of time, which is independent of v_0 and v_1, but is, of course, a function of $\bar{v}_0$ and $\bar{v}_1$. In these circumstances a measure of φ provides a measure of ω, and the basis has been established for devising a spectrometer.

If we may assume that the velocity distributions $p_0(v_0)$ and $p_1(v_1)$ are so strongly peaked about their means that

$$\delta v_0 \equiv v_0 - \bar{v}_0 \ll \bar{v}_0 \qquad \delta v_1 \equiv v_1 - \bar{v}_1 \ll \bar{v}_1 \tag{8.287}$$

then (8.286) can be satisfied to first order in δv_0 and δv_1, provided t is defined such that

$$\frac{m_{\mathrm{n}}t}{\hbar\gamma_{\mathrm{n}}} = \frac{l_0 B_0}{v_0^3} = \frac{l_1 B_1}{v_1^3} \tag{8.288}$$

Under these circumstances the NSE signal is determined not by the characteristic functions corresponding to $p_0(v_0)$ and $p_1(v_1)$, but by the characteristic function which corresponds to the probability density function for energy transfer. Since this is proportional to the coherent scattering law defined in (8.62), we have the result

$$\begin{aligned} P_x &= \langle\cos(\varphi - \bar{\varphi})\rangle = \int_{-\infty}^{\infty} S(\boldsymbol{q}, \omega)\cos[(\omega - \bar{\omega})t]\,\mathrm{d}\omega \Big/ \int_{-\infty}^{\infty} S(\boldsymbol{q}, \omega)\,\mathrm{d}\omega \\ &= \mathrm{Re}\,\{\mathrm{e}^{-\mathrm{i}\bar{\omega}t}\,I(\boldsymbol{q}, t)/I(\boldsymbol{q}, 0)\} \end{aligned} \tag{8.289}$$

where $I(\boldsymbol{q}, t)$ is the intermediate scattering function defined in section 8.3.1.

In the general case the NSE signal differs from P_x by a factor P_S $(-1 \leq P_S \leq 1)$ which takes into account any explicit spin-dependence of the scattering mechanism.

A sketch of the NSE spectrometer IN11 at the ILL is shown in figure 8.19. Neutrons in the wavelength range $4\ \text{Å} < \lambda < 8\ \text{Å}$ are pulsed by a single chopper with a helical transmission slot which acts as velocity selector with energy resolution in the range 12–25%. These are longitudinally polarized by an Fe–Ag supermirror in a Soller type mirror configuration, and switched into a transverse polarization state P_x in the plane of the diagram, by a Mezei-type $\pi/2$ coil, located immediately in front of the first precession field B_0. After scattering, spin refocusing is carried out by reversing P_x in a Mezei-type π coil placed immediately after the sample. To detect spin-flip transitions in paramagnetic or magnetically isotropic samples (cf. section 8.4.5), the π coil may be omitted since the sample produces its own spin-echo action. Finally when the full cycle of precession has been completed, the component P_x is switched back into longitudinal mode by a second $\pi/2$ coil. The neutrons are then spin analysed by a second supermirror before being counted. When a graphite analyser crystal is placed before the detector the energy resolution of the detected neutrons is improved to between 1.5% and 6%. The instrument can detect energy transfers across the range 0.4–400 μeV with momentum transfers 0.03–2.7 Å^{-1} at $\lambda = 4$ Å, and 0.015–1.35 Å^{-1} at $\lambda = 8$ Å.

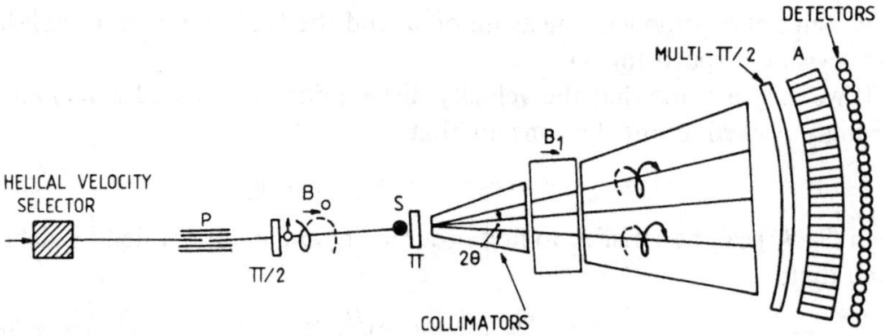

Figure 8.19. The neutron spin echo spectrometer IN11 at the ILL Grenoble. The instrument measures the Fourier transform $I(\boldsymbol{q}, t)$ of the sample scattering function over a range of time between 1.5×10^{-12} s at 4 Å and 1.5×10^{-8} s at 8 Å.

An early achievement of the spin-echo technique was the successful determination of the magnetic intermediate scattering function $I_m(\boldsymbol{q}, t)$ for a Cu–Mn spin glass alloy over the time range $10^{-12} < t < 10^{-9}$ s (Mezei and Murani 1979). Quite recently a new form of NSE spectrometer has been tested in which there are no guide fields and hence no significant precession of the neutron spin along its flight path (Golub and Gähler 1987, Gähler and Golub 1987, 1988, Dubbers *et al* 1989). In this case the periodic phenomenon used for purposes of time-keeping is the RF field used in the spin flipping coils, exactly as in the Ramsey separated oscillating field magnetic resonance technique (cf. section 7.5.3). The new instrument has the potential to employ much longer flight paths with a corresponding improvement in resolution.

9 The neutron as a composite system

9.1 Strong interactions and SU(3) flavour symmetry

9.1.1 Nucleon sizes

The science of atomic physics may be said to have begun with Mendeleev's discovery of the period table of the chemical elements (cf. section 1.1.1). Subsequent work by Thomson, Rutherford, Bohr and their associates showed that atoms were composite structures, with dimensions of the order of 10^{-10} m (1 Å), which could be understood as bound states of negative electrons and positive nuclei, whose electromagnetic interactions were mediated by a massless neutral boson, the photon. The simplest nucleus, namely that of hydrogen, consisted of a single massive positive particle, the proton, and further investigations by Chadwick, Pauli and others, revealed the presence of neutral partners to both proton and electron, namely the neutron and the neutrino. In terms of these four 'elementary' fermions (p, n, e^- and ν_e), together with their anti-particles, the structure and disintegration schemes of all nuclei could be explained.

Nuclei have dimensions of order 10^{-15} m (1 fm), and, broadly speaking, the sizes of nuclei can be explained in terms of the measured masses of the triplet of pions ($\pi^-\pi^0\pi^+$), charged bosons whose primary role it is to mediate the strong forces holding nuclei together. The heavy electron known as the muon (μ^-) and its anti-particle (μ^+) coexisted for a long time somewhat uneasily with this rather satisfactory scheme, to which they appeared to make no significant contribution. We now know, of course, that there are indeed three families of leptons, particles which, like the electron, have no strong interactions, namely (e^-, ν_e), (μ^-, ν_μ) and (τ^-, ν_τ), although the tau neutrino ν_τ has yet to be positively identified.

All known particles interact weakly and, excluding the neutrinos which are uncharged and (so far as is known) have no magnetic moments, all interact electromagnetically. However, only nucleons and pions among the particle species enumerated above have strong interactions, and this property is distinguished by grouping them together in the class of hadrons. The strong interactions are invariant under the group SU(2) of isospin transformations so that each hadron is assigned a pair of isospin quantum numbers (T, T_3). As discussed in some detail in sections 3.3 and 3.4, the nucleons (p, n) transform under operations of the group like the components of an isospinor ($T=\frac{1}{2}$, $T_3=\pm\frac{1}{2}$). All the hadrons couple to the electromagnetic field with charges Q given (in units of the proton charge $+e$) by (cf. section 3.5.15)

$$Q = T_3 + \tfrac{1}{2}B \qquad (9.1)$$

where the baryon number B takes the values $B=1(-1)$ for nucleons (anti-nucleons) and $B=0$ for pions.

Within a century of the development of the study of the structure of matter, first atoms, and then atomic nuclei, were shown not to be elementary particles, but composites of simpler entities. On the basis of evidence available to date the opposite conclusion must be reached about the nature of the charged leptons (at least the $\mu^{\pm}$ and $e^{\pm}$). Based on highly precise measurements of their magnetic moments, which agree up to eight significant figures with the values calculated from the Dirac equation, with radiative corrections evaluated to fourth order in α using well-established rules of quantum electrodynamic perturbation theory, the conclusion must be drawn that these are indeed point particles without an internal structure. The question we must now consider is: are protons and neutrons, and perhaps pions and other mesons, point particles, or do they have finite sizes and internal structures made up from still smaller objects?

Protons and neutrons have magnetic moments which do not agree even approximately with those predicted on the basis of the Dirac equation but display anomalous (Pauli) moments $\kappa_p=1.78$ nm, and $\kappa_n=-1.91$ nm where

$$\mu_p=(1+\kappa_p)\mu_N \qquad \mu_n=\kappa_n\mu_N \tag{9.2}$$

Note that in addition to their Pauli moments the (charged) proton has unit Dirac moment, and the (uncharged) neutron has zero Dirac moment. Since a tensor coupling of the nucleon spin to the electromagnetic field (cf. section 3.3.1) characterized by a Pauli moment of arbitrary magnitude and spin can be incorporated into the Dirac equation in a relativistically invariant way (cf. section 9.2.1), the existence of Pauli moments for the nucleons cannot of itself automatically exclude the possibility that neutrons and protons are truly point particles, although it can hardly be without significance that the measured Pauli moments are equal and opposite to within about 5%. These findings suggest the presence of internal electric currents which are also equal in magnitude but opposite in sign.

Fortunately it is not necessary to speculate on these matters, for compelling evidence has existed for more than 35 years that nucleons indeed have finite radii of order 1 fm, comparable with those of other light nuclei. The crucial observations were made by Hofstadter and his associates at Stanford University who measured the charge form factors by scattering electrons in the energy range 100–500 MeV on hydrogen and deuterium, and deduced values for both 'Dirac' and 'Pauli' root mean square radii in the proton on the order of 0.74 ± 0.24 fm (Hofstadter 1956, Bernstein and Goldberger 1958, Hofstadter *et al* 1958). The precise meanings to be attached to these radii will be deferred until section 9.2.3, where we discuss the electromagnetic properties of the neutron.

9.1.2 Hadron spectroscopy

9.1.2.1 Excited states of the nucleon

Since individual protons and neutrons have internal structures, then the obvious question arises as to what are their constituents, and whether also, like atoms and nuclei, they possess systems of excited states. The structure of the excited states of nuclei was discussed in some detail in section 4.7, where broadly three types of nuclear excited states were identified. These were (i) bound states with energies of the order of a few keV to a few MeV, and mean widths $\langle\Gamma\rangle$ of the order of a few meV, much less than the mean separation $\langle D\rangle$ between states, and (ii) resonance states with energies of order 10 MeV and mean widths $\langle\Gamma\rangle \approx 1$ eV, with $\langle\Gamma\rangle \ll \langle D\rangle$. At excitation energies $\geqslant 20$ MeV we enter the continuum of states (iii), where individual discrete states are not resolvable, but where the structure observed at lower energies is still discernible in the characteristics and locations of giant resonances which are prominent in this energy region. Experimentally it turns out that excited states of the nucleon can be produced in a number of ways, most simply by bombarding nucleons with pions, but all the states so generated are resonance states with lifetimes of order 10^{-24} s and mean widths $\langle\Gamma\rangle$ of order 100 MeV, comparable with the mean separation $\langle D\rangle$ between states. In particular no continuum of states is observed.

Since isospin is conserved in the interactions of pions with nucleons (cf. section 3.4–3.5), these particles can form compound states characterized by total isospin quantum numbers $T=\frac{3}{2}$, or $T=\frac{1}{2}$ corresponding to the isospin addition rule

$$\boldsymbol{T}=\boldsymbol{T}_{\mathrm{N}}+\boldsymbol{T}_{\pi} \qquad T_{\mathrm{N}}=\tfrac{1}{2} \qquad T_{\pi}=1 \tag{9.3}$$

The coupled isospin states $|(T_{\mathrm{N}}T_{\pi}); TT_3\rangle$ are thus expressed in terms of the uncoupled isospin states $|(T_{\mathrm{N}})T_{3\mathrm{N}}, (T_{\pi})T_{3\pi}\rangle$, where the quantum numbers $(T_{\mathrm{N}}, T_{\pi})$ are suppressed, by inserting the appropriate values of the Clebsch–Gordon coupling coefficients. We then obtain the results

$$T=3/2: \quad \begin{aligned} |3/2, 3/2\rangle &= |\mathrm{p}, \pi^+\rangle \\ |3/2, 1/2\rangle &= \surd(2/3)\,|\mathrm{p}, \pi^0\rangle + \surd(1/3)|\mathrm{n}, \pi^+\rangle \\ |3/2, -1/2\rangle &= \surd(2/3)\,|\mathrm{n}, \pi^0\rangle + \surd(1/3)|\mathrm{p}, \pi^-\rangle \\ |3/2, -3/2\rangle &= |\mathrm{n}, \pi^-\rangle \end{aligned} \tag{9.4}$$

and

$$T=1/2: \quad \begin{aligned} |1/2, 1/2\rangle &= -\surd(1/3)\,|\mathrm{p}, \pi^0\rangle + \surd(2/3)\,|\mathrm{n}, \pi^+\rangle \\ |1/2, -1/2\rangle &= \surd(1/3)\,|\mathrm{n}, \pi^0\rangle - \surd(2/3)\,|\mathrm{p}, \pi^-\rangle \end{aligned} \tag{9.5}$$

The group of states labelled with the quantum number $T=3/2$ are called Δ-particles, and they come in four charge states, Δ^{++} $(T_3=\frac{3}{2}, Q=2)$, Δ^+

($T_3=\frac{1}{2}$, $Q=1$), Δ^0 ($T_3=-\frac{1}{2}$, $Q=1$) and Δ^{-1} ($T_3=-\frac{3}{2}$, $Q=-1$) where the charges Q are determined by putting $B=1$ in (9.1). Since the pions and nucleons can collide with orbital angular momenta $l=0, 1, 2, \ldots$, there are s, p, d,... states of the Δ with even-parity states corresponding to odd values of l, since the pion is a pseudo-scalar particle with negative intrinsic parity. The group of states labelled with the quantum number $T=\frac{1}{2}$ are called N-particles, in recognition of the fact that the nucleon is the ground state of this system. There are clearly two charge states N^+, N^0.

Since it has not yet proved possible to construct targets of free neutrons in the laboratory, although neutrons bound in the deuteron are available, and there are no beams of free π^0s since these decay spontaneously into two photons with a lifetime of about 10^{-16} s (cf. section 3.5.1), $p+\pi^{\pm}$ collisions are the only ones accessible to experimental study. Thus the only excited states of the nucleon easily generated are the $|3/2, 3/2\rangle$ state from $p+\pi^+$ collisions, and the states $|3/2, -\frac{1}{2}\rangle$ and $|\frac{1}{2}-\frac{1}{2}\rangle$ from $p+\pi^-$ collisions. The $p+\pi^{\pm}$ cross-sections are shown as functions of the total energy in the centre of mass in figure 9.1 where we may identify Δ^{++} resonances at 1232 MeV, 1920 MeV and 2420 MeV in the $p+\pi^+$ cross-section, with a Δ^0 resonance at 1232 MeV, and N^0 resonances at 1520 MeV and 1680 MeV in the $p+\pi^-$ cross-section. The system of known resonances in the pion–nucleon system below 2.5 GeV is shown in figure 9.2.

The very strong resonance at 1232 MeV is known as the (3, 3) resonance ($T=I=3/2$), and was discovered by Fermi and his associates (Anderson *et al* 1952). It is also excited strongly in the photo–pion production process $p(\gamma, \pi^0)p$ which, being an electromagnetic interaction, conserves T_3 but violates T. The isospin–spin assignment for the (3, 3) resonance was first proposed by Brueckner (Brueckner 1952). Quite generally the spin of the resonance is determined by measuring the cross-section at maximum and applying (4.109) and (4.110) to determine the statistical factor g_I. For the pion–nucleon system this takes the value $(I+\frac{1}{2})$.

9.1.2.2 Baryons and mesons

The Λ^0 was the first hyperon, or heavy nucleon, to be discovered. This was detected in the cosmic radiation (Rochester and Butler 1947) and was subsequently produced in the laboratory via the reaction

$$\pi^- + p \rightarrow \Lambda^0 + K^0 \tag{9.6}$$

The production cross-section has a value in the millibarn range indicating that the reaction (9.6) is a strong one. The Λ^0 is a spin $\frac{1}{2}$ particle of mass 1115.6 MeV which decays into a nucleon

$$\Lambda^0 \rightarrow p + \pi^- \tag{9.7}$$

with a lifetime $\tau \approx 2.5 \times 10^{-10}$ s, a result which indicates that (9.7) is a weak

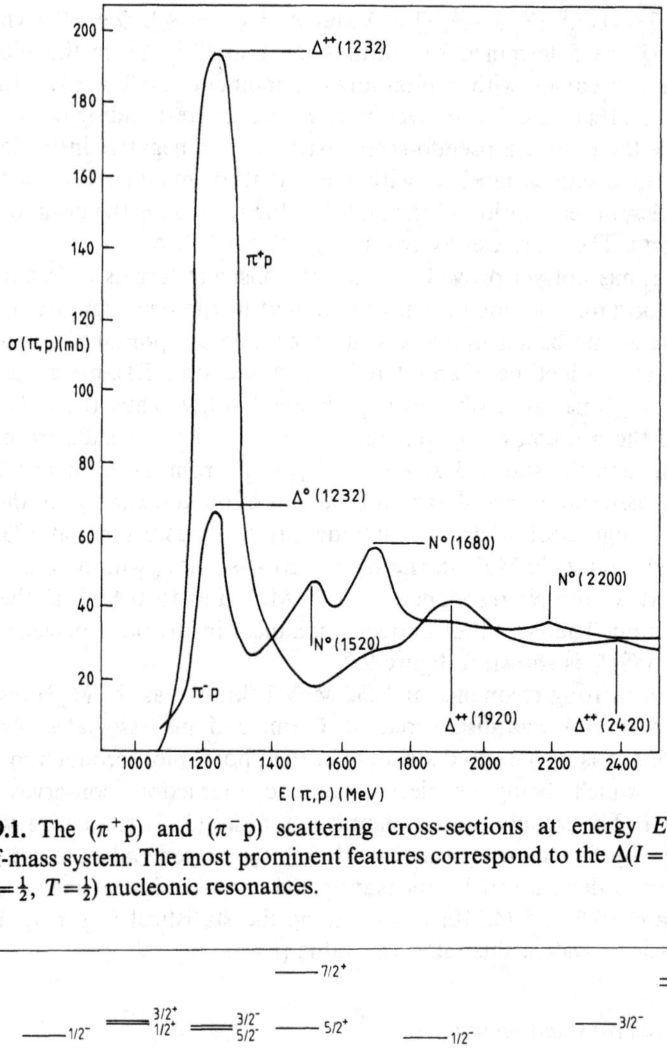

Figure 9.1. The (π^+p) and (π^-p) scattering cross-sections at energy E in the centre-of-mass system. The most prominent features correspond to the $\Delta(I=\frac{3}{2}, T=\frac{3}{2})$ and $N(I=\frac{1}{2}, T=\frac{1}{2})$ nucleonic resonances.

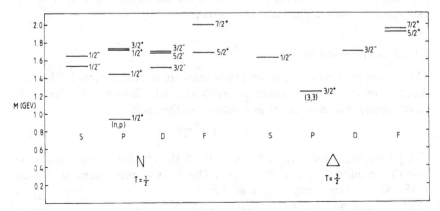

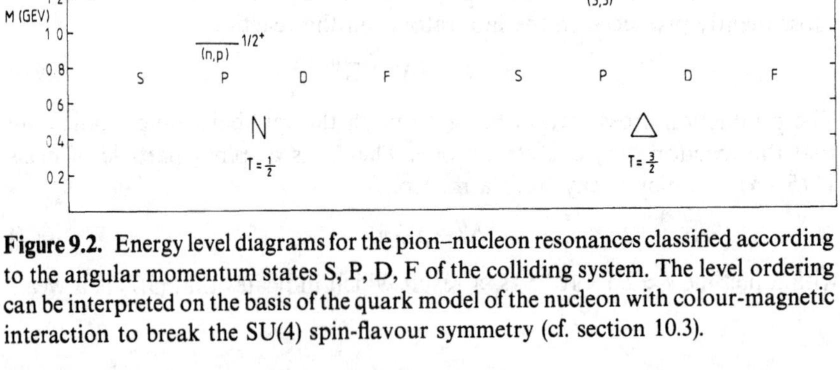

Figure 9.2. Energy level diagrams for the pion–nucleon resonances classified according to the angular momentum states S, P, D, F of the colliding system. The level ordering can be interpreted on the basis of the quark model of the nucleon with colour-magnetic interaction to break the SU(4) spin-flavour symmetry (cf. section 10.3).

interaction. The K^0 which accompanies it in (9.6) is a pseudo-scalar spin 0 meson, of mass 497.7 MeV, which decays weakly with a lifetime also in the range 10^{-10} s. The K^0-decay is famous for the fact that it provides the only known example of a *CP*-violating (and by implication *T*-violating) decay mode, which occurs with a branching ratio of about 0.2% (Christenson *et al* 1964). Since a lifetime of order 10^{-10} s corresponds to a width of about 10^{-5} eV it is conventional to describe Λ^0 and K^0 as 'stable' particles, in the same sense that the neutron is 'stable'. However, allowing for the much larger phase space available in Λ^0 and K^0 decay the value of the corresponding weak coupling constant is only about 10% of that describing neutron β-decay.

The terms 'hyperon' used to describe the Λ^0 has now largely been replaced by the term 'baryon', since the Λ^0 along with its spin $\frac{1}{2}$ partners, the sigma particles, Σ^-, Σ^0, Σ^+ and the cascade particles Ξ^-, Ξ^0, has baryon number $B=1$. The baryon number seems to be absolutely conserved, and it appears to be the fate of all baryons ultimately to decay into nucleons.

Because they were produced strongly yet decayed weakly the new baryons and mesons presented a theoretical puzzle, the clue to whose eventual solution was provided by the observation that they were always created in pairs, a phenomenon described as 'associated production' (Pais 1952). To account for the observed patterns of production and decay a new quantum number *S*, the 'strangeness' was introduced, which was assigned the value zero for pions, nucleons and nucleonic resonances, and which, like T and T_3, was assumed to be conserved in strong interactions, but violated in weak interactions (Gell-Mann 1953, Nakano and Nishijima 1953). Strange quantum numbers were assigned to baryons and mesons by generalizing the charge formula (9.1) to

$$Q=T_3+\tfrac{1}{2}(B+S)\equiv T_3+\tfrac{1}{2}Y \tag{9.8}$$

where the hypercharge quantum number Y is defined by this relation. The isospin quantum number T_3 could be measured experimentally by associating the new particles in isospin multiplets of approximately equal mass. In this way the Λ^0 was identified as an isospin singlet ($T=0$), the $\Sigma^-\Sigma^0\Sigma^+$ as a triplet ($T=1$), and the $\Xi^-\Xi^0$ as a doublet ($T=\frac{1}{2}$) analogous to the nucleon pair.

The resulting scheme of spin $\frac{1}{2}$ baryons is shown in figure 9.3(a). It consists of an octet of eight spin $\frac{1}{2}$ particles known as the 'eightfold-way' organized into four isospin multiplets $|T=\frac{1}{2}, Y=1\rangle$, $|T=0, Y=0\rangle$, $|T=1, Y=0\rangle$ and $|T=\frac{1}{2}, Y=-1\rangle$. An interesting feature of the scheme is that the charge quantum numbers Q given by (9.8) can assume exactly the same values as the hypercharge Y. In parallel with the baryon octet there is of course a meson octet with the same quantum numbers made up from the isospin multiplets (K^0K^+), η^0, ($\pi^-\pi^0\pi^+$) and ($K^-\bar{K}^0$).

The proposal that the baryons and mesons of the eightfold way could be interpreted as states of an eight-dimensional representation of a new SU(3) group, now known as flavour SU(3), allowed a natural explanation of the observed symmetries obeyed by these particles. Since SU(3) has eight

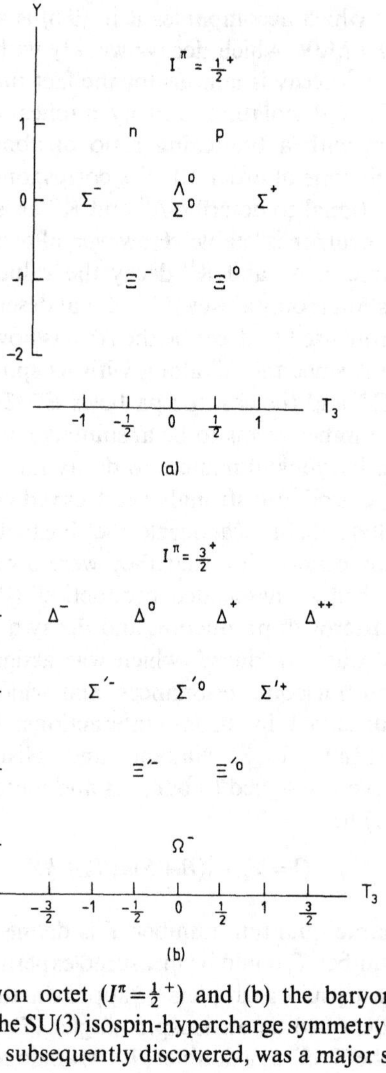

Figure 9.3. (a) The baryon octet ($I^\pi = \frac{1}{2}^+$) and (b) the baryon decuplet ($I^\pi = \frac{3}{2}^+$) interpreted according to the SU(3) isospin-hypercharge symmetry scheme for hadrons. The prediction of the Ω^-, subsequently discovered, was a major success of the theory (Barnes *et al* 1964).

generators the octet representation is the regular or adjoint representation $D^{(11)}$ (cf. section 4.6.2). Within the baryon octet the masses of particles in each isospin multiplet are equal to within about 1%, whereas the mean masses for multiplets differing by one unit in Y vary by about 15%. Hence the flavour $SU(3)_F$ symmetry is a much more broken symmetry than isospin $SU(2)_T$.

We have already encountered the group SU(3) in sections 4.6.2 and 4.6.3, during our discussion of the rotation–vibration states of deformed nuclei, and many considerations touched on there are of equal relevance in the treatment of flavour SU(3). If, for example, we take any two-dimensional subspace of the three-dimensional complex space on which the SU(3) operations are defined, then we may select three SU(2) subgroups, one of which may be identified with the isospin subgroup $SU(2)_T$, whose generators are T_1, T_2, T_3. Since the eighth generator commutes with all three isospin generators, we may identify this generator with hypercharge, and construct the subgroup $SU(2)_T \times U(1)_Y$. We performed a similar construction in section 4.6.2 in identifying the subgroup $SU(2)_\Lambda \times U(1)$. However, the true analogue of $SU(2)_T$ is not $SU(2)_\Lambda$ but $SU(2)_L$, for which the analogue of the formula (4.102) for the energy levels of a rotating nucleus, is the formula for the mass splitting in the baryon octet (Okubo 1962)

$$M(Y, T) = a + bY + c[T(T+1) - \tfrac{1}{4}Y^2 - 1] \qquad (9.9)$$

Here the term proportional to $T(T+1)$ matches the term proportional to $L(L+1)$ in (4.102) which is characteristic of nuclear rotational spectra.

The two other SU(2) sub-groups within $SU(3)_F$ for which quantum numbers may be assigned to the baryons in the octet, namely U-spin and V-spin, are of lesser significance than isospin $SU(2)_T$, since the masses show a much wider range of variation than within an isospin multiplet. Nevertheless the U-spin symmetry is of some interest since the generator U_3 satisfies the relation

$$U_3 = Y - \tfrac{1}{2}Q \qquad (9.10)$$

which may be obtained from (9.8) by the replacements $T_3 \rightarrow U_3$, $Y \leftrightarrows Q$. Thus members of a U-spin multiplet, e.g. $(p\Sigma^+)$ have the same charge, and (very approximately) the same magnetic moments.

The Δ particles ($T = \frac{3}{2}$, $I = \frac{1}{2}$) clearly cannot be classified according to an SU(3) octet; they can however be organized as members of an SU(3) decuplet as shown in figure 9.3(b). At the time this classification was first proposed the tenth particle, the Ω^- ($T = 0$, $Y = -2$), had yet to be found; its identification as a reaction product in $K^- + p$ collisions set the final seal of approval on the $SU(3)_F$ symmetry scheme (Barnes *et al* 1964).

9.1.3 Quarks

9.1.3.1 The quark model of Zweig and Gell-Mann

We now turn to the question of whether there exist in nature $SU(3)_F$ multiplets of dimensionality lower than eight, from which the observed octets and decuplets can be constructed by expansion of direct products, in exactly the same way as singlet and triplet representations of SU(2) are built up from direct products of spin $\frac{1}{2}$ representations. Indeed this question predates the advent of $SU(3)_F$, in the form of an earlier, failed, attempt to base a theory of elementary particles on the fundamental triplet p, n, Λ^0, combined with the triplet $\bar{p}$, $\bar{n}$ and $\bar{\Lambda}^0$ (Sakata 1956). In the original version of the quark model (Zweig 1964, Gell-Mann 1964) the basic building blocks are spin $\frac{1}{2}$ particles called quarks, which come in three 'flavours' listed as up (u), down (d) and strange (s). The u and d quarks constitute an isospin doublet with $T=\frac{1}{2}$, $T_3=\frac{1}{2}(u)$ and $T_3=-\frac{1}{2}(d)$; the s quark is an isospin singlet with $T=T_3=0$. The quarks (q) are assigned to a $D^{(10)}$ representation of $SU(3)_F$, while the antiquarks ($\bar{q}$) are assigned to a $D^{(01)}$ representation.

It is well-known that quarks are predicted to have non-integral values of the charge Q, and no free quark has been detected in nature (Smith 1989). Quarks are, therefore, said to be 'confined'. Furthermore, the original multiplet of three light quarks has been expanded to include three heavy quarks, the charmed (c), bottom (b) and top (t) quarks. According to this model there are six flavours of quark, and these are organized rather differently, appearing as three generations of doublets, i.e. (u, d), (c, s) and (t, b) each with charges ($\frac{2}{3}$, $-\frac{1}{3}$). The charmed quark therefore provides a partner for the strange quark, and for this reason occupies a place of some significance for low energy weak phenomena. Indeed it was to solve a problem in weak interaction physics, namely the unexplained absence of strangeness changing neutral currents, that the charmed quark was predicted in the first place (cf. section 9.3.6.2).

Subsequently first the J/ψ, and then the γ-meson were discovered and their properties interpreted as quasi-bound states of the heavy quarkoniums $c\bar{c}$ and $b\bar{b}$ respectively (Aubert *et al* 1974, Augustin *et al* 1974, Herb *et al* 1977). The top quark has yet to be identified as a constituent of any known particle.

The charge of any composite of quarks and antiquarks is given by the formula

$$Q=T_3+\tfrac{1}{2}(B+S+c+b+t) \tag{9.11}$$

where c, b, and t are quantum numbers which have non-zero values only for charmed, bottom and top quarks respectively. This formula is a further generalization of (9.8) and suggests the presence of a deeper $SU(6)_F$ symmetry. However, this symmetry is so badly broken by the mass differences exhibited by the heavy quarks, that only the subgroup $SU(3)_F$ of isospin and strangeness retains a practical utility. The quantum numbers of all six quarks are listed in table 9.1.

We may now assign the observed baryon octet and decuplet to $D^{(11)}$ and $D^{(30)}$ representations respectively, arising from the decomposition of the direct product of three $D^{(10)}$ (quark) representations (Elliott and Dawber 1979)

$$D^{(10)}\times D^{(10)}\times D^{(10)}=D^{(30)}+2D^{(11)}+D^{(00)} \tag{9.12}$$

Similarly the meson octet, and the meson singlet $\eta'(955)$ ($T=0$, $I^\pi=0^-$), can be assigned respectively to $D^{(11)}$ and $D^{(00)}$ representations of $SU(3)_F$, arising from the direct product of quark–antiquark representations

$$D^{(10)}\times D^{(01)}=D^{(11)}+D^{(00)} \tag{9.13}$$

Table 9.1 Quark quantum numbers. The anti-quarks $\bar{u}$ and $\bar{d}$ have $T=\frac{1}{2}$ as have u and d, but otherwise the quantum numbers of all anti-quarks have equal and opposite values to those of the quarks

Quark	B	T	T_3	Y	S	c	b	t	Q
u	$\frac{1}{3}$	$\frac{1}{2}$	$\frac{1}{2}$	$\frac{1}{3}$	0	0	0	0	$\frac{2}{3}$
d	$\frac{1}{3}$	$\frac{1}{2}$	$-\frac{1}{2}$	$\frac{1}{3}$	0	0	0	0	$-\frac{1}{3}$
c	$\frac{1}{3}$	0	0	$\frac{4}{3}$	0	1	0	0	$\frac{2}{3}$
s	$\frac{1}{3}$	0	0	$-\frac{2}{3}$	-1	0	0	0	$-\frac{1}{3}$
t	$\frac{1}{3}$	0	0	$\frac{4}{3}$	0	0	0	1	$\frac{2}{3}$
b	$\frac{1}{3}$	0	0	$-\frac{2}{3}$	0	0	-1	0	$-\frac{1}{3}$

In this way we may easily establish the quark content of the non-strange hadrons

$$p=uud \qquad Q=(2/3)+(2/3)+(-1/3)=1$$

$$n=udd \qquad Q=(2/3)+(-1/3)+(-1/3)=0 \tag{9.14}$$

and

$$\pi^- =\bar{u}d \qquad Q=(-2/3)+(-1/3)=-1$$

$$\pi^0 =\surd(\tfrac{1}{2})(u\bar{u}-d\bar{d}) \qquad Q=(\tfrac{1}{2})[(2/3-2/3)+(-1/3+1/3)]=0$$

$$\pi^+ =u\bar{d} \qquad Q=(1/3)+(2/3)=1 \tag{9.15}$$

A difficulty arises when we note that, whereas the baryon octet has $I=\frac{1}{2}$, the decuplet has $I=\frac{3}{2}$ (an association which is explored further in section 10.3.1). The problem is that, since the decuplet representation of $SU(3)_F$ and the $I=\frac{3}{2}$ representation of $SU(2)_I$ (spin) are each symmetric with respect to quark interchange, so also is the product representation $SU(2)_I \times SU(3)_F$. As a result the states $\Delta^{++}=uuu$, $\Delta^-=ddd$ and $\Omega^-=sss$, which have zero internal orbital angular momentum, would appear to be totally symmetric with respect to exchange of identical quarks. Since quarks are spin $\frac{1}{2}$ particles this leads to severe difficulties associated with quark statistics and the Pauli principle.

9.1.3.2 Coloured quarks

The solution proposed to solve the problem of quark statistics hinges on the introduction of a new tri-valued quantum number called colour, analogous to the two-valued quantum number associated with electric charge (Greenberg 1964, Greenberg and Resnikoff 1968). The colour theory rests on two basic assumptions, firstly that quarks of every flavour can exist in three quantum states distinguished by their 'colour'; the convention is then to describe quarks as red, blue or green. The number of colours required to accord with experiment may be determined e.g. by measuring the decay rate of the π^0 into two photons. The second assumption is that all physically observable states of quarks have zero colour. While this assumption takes account of the fact that free quarks have never been observed, it goes further in hypothesizing that the strong forces between quarks are related to their colour rather than to their flavour. This brings the theory in line with the observed fact that all quantum numbers T_3, S, c, etc., associated with quark flavour, are conserved in strong interactions. The central idea is that any two colours are equivalent to the complement of the third; thus red plus blue is equivalent to anti-green, and three quarks, or a quark–antiquark pair, will bind together in a state of zero colour.

The theory of coloured quarks is described as quantum chromodynamics or QCD, in analogy with quantum electrodynamics or QED. The proposal is that the three-dimensional colour charge is related to an $SU(3)_C$ colour group, and the physical states of hadronic matter carrying zero colour correspond to the completely anti-symmetric representation $D^{(00)}$ of this group (cf. 9.12). This assumption accounts for the fact that the only observed hadron multiplets correspond to totally symmetric representations of the product group $SU(2)_I \times SU(6)_F$, and explains the origin of the term 'colour singlet' applied to physical hadrons.

Unlike $SU(6)_F$, which describes unitary transformations of fields over a complex flavour space of six dimensions, $SU(3)_C$ is a gauge group. This means that it describes generalized phase transformations on fields $\psi(x)$ of the form

$$\psi(x) \rightarrow \psi'(x) = \exp[\mathrm{i}\boldsymbol{T} \cdot \boldsymbol{\lambda}(x)]\psi(x) \tag{9.16}$$

where $\lambda(x)$ are functions defined arbitrarily at each space–time point x. The properties of the group are defined by the commutation relations satisfied by the group generators $\boldsymbol{T}$

$$[T_i, T_k] = \mathrm{i} f^l_{jk} T_l \tag{9.17}$$

where f^l_{jk} are the structure functions of the group, and the convention that a repeated index is summed over all values is applied in (9.17). The essential feature of a gauge transformation, is that, in order that it should describe an exact symmetry, the derivatives of the field must transform in the same way as the fields themselves. In effect this means that the derivative $\partial_\mu \psi(x)$ must be replaced by the covariant derivative

$$D_\mu(x) = \partial_\mu \psi(x) - \mathrm{i}g\boldsymbol{T} \cdot \boldsymbol{G}_\mu(x) \tag{9.18}$$

where g is an arbitrary real number analogous to electric charge e, which defines the coupling of the field $\psi(x)$ to the massless vector gauge field $\boldsymbol{G}_\mu(x)$. The number of these is equal to the number of generators of the group, and they transform under the gauge transformation according to the rule

$$G^j_\mu(x) \to G^{j'}_\mu(x) = G^j_\mu(x) - f^j_{kl} G^k_\mu(x) \lambda^l(x) + g^{-1} \partial_\mu \lambda^j(x) \tag{9.19}$$

The simplest gauge group we know is the U(1) gauge group associated with electric charge. It has a single generator and a single massless gauge field $A_\mu(x)$ whose quanta are photons. The gauge symmetry is exact, and associated with this gauge symmetry we have the conserved electromagnetic current. In the case of $SU(3)_C$ colour symmetry, also hypothesized to be exact, there are eight massless vector fields $G^j_\mu(x)$ ($j=1, 2, \ldots, 8$) whose quanta are known as gluons, and eight conserved currents. In analogy with the antisymmetric electromagnetic field tensor $F_{\mu\nu}(x)$, whose six independent components give the electric and magentic field strengths, we may define colour electric and colour magnetic field tensors, with eight indices ($j=1, 2, \ldots, 8$)

$$F^j_{\mu\nu}(x) = \partial_\mu G^j_\nu(x) - \partial_\nu G^j_\mu(x) + g f^j_{kl} G^k_\mu(x) G^l_\nu(x) \tag{9.20}$$

In the case of the electromagnetic field there is no analogue to the g-dependent term in (9.20), an expression of the fact that photons carry no electric charge. However, because of this term the gluons are themselves coloured; this is a consequence of the fact that the generators T_i do not commute and the group SU(3) is said to be non-Abelian. In this property gluons behave more like the spin 2 gravitons which transmit the gravitational force, since gravitons carry energy which is the source of the gravitational field. It is the non-Abelian nature of the colour charge which is responsible for quark binding in hadrons, and for that property of quarks described as 'asymptotic freedom', whereby quarks undergoing close encounters appear to interact almost as free particles. Thus the effective colour charge carried by a quark appears to be reduced at close distances or at high momentum transfers. Thus we may visualize the colour charge carried by a quark as being spread out over the space which surrounds it. It is also generally accepted that the non-Abelian property is responsible for quark confinement although this hypothesis has not been proved rigorously in QCD.

The question of the direct observation of quarks within nucleons is discussed in section 9.2.4 below, and some more general properties of QCD are taken up briefly in section 10.2.2.

9.2 Electromagnetic interactions

9.2.1 Dirac equations for a spin $\frac{1}{2}$ fermion

9.2.1.1 The free particle wave equation

Nucleons are spin $\frac{1}{2}$ fermions and hence, were they point particles, would be described by four-component spinor wave-functions ψ satisfying the Dirac equation for a particle of mass m

$$H_0\psi=[c(\boldsymbol{\alpha}\cdot\boldsymbol{p})+\beta mc^2]\psi=\mathrm{i}\hbar\frac{\partial\psi}{\partial t} \tag{9.21}$$

where $p=-\mathrm{i}\hbar\nabla$, and $\boldsymbol{\alpha}$, β are, respectively, polar vector and scalar 4×4 Dirac matrices. To find a physical interpretation for the operator $\boldsymbol{\alpha}$ we may evaluate the velocity operator in the Heisenberg representation

$$\boldsymbol{v}(t)=\frac{\mathrm{d}\boldsymbol{r}(t)}{\mathrm{d}t}=\frac{\mathrm{i}}{\hbar}[H_0,\boldsymbol{r}(t)]=c\boldsymbol{\alpha}=\boldsymbol{v}(0) \tag{9.22}$$

a result which shows that the velocity operator $\boldsymbol{v}=\boldsymbol{v}(0)$ in the Schrödinger representation is also equal to $c\boldsymbol{\alpha}$. Since $\boldsymbol{\alpha}$ has eigenvalues ± 1, $\boldsymbol{v}$ must have eigenvalues $\pm c$. This result is usually taken to mean that the particle undergoes a very rapid random motion described as 'zitterbewegung' or 'trembling' (Schrödinger 1930). On the other hand the expectation value of the operator $c\boldsymbol{\alpha}$ for a free particle in the non-relativistic limit is equal to $\boldsymbol{p}/m$ which agrees with the normal interpretation of the momentum.

The Dirac equation (9.21) may be written in relativistically covariant form, assuming the usual convention that repeated Greek indices are summed from 1 to 4

$$[\gamma_\mu\partial_\mu+mc/\hbar]\psi=0 \tag{9.23}$$

Here x_μ, ∂_μ and γ_μ are four-vector operators defined by

$$x_\mu=[\boldsymbol{r},\mathrm{i}ct) \qquad p_\mu\equiv\mathrm{i}\hbar\frac{\partial}{\partial x_\mu}=\mathrm{i}\hbar\partial_\mu \qquad \gamma_\mu=(-\mathrm{i}\beta\boldsymbol{\alpha},\beta) \tag{9.24}$$

The equation for the adjoint wave function $\bar{\psi}$ corresponding to ψ is

$$[\gamma_\mu\partial_\mu-mc/\hbar]\bar{\psi}=0 \tag{9.25}$$

and both (9.23) and (9.25) may be derived by applying Lagrange's equations for fields to the free field Lagrangian density

$$L_0=-\hbar c[\bar{\psi}\gamma_\mu\partial_\mu\psi+(mc/\hbar)\bar{\psi}\psi] \tag{9.26}$$

Finally, by multiplying (9.23) on the left by $\bar{\psi}$, and (9.25) on the right by ψ, and adding the resulting equations, we may construct the conserved current density

$$j_\mu=\bar{\psi}\gamma_\mu\psi \tag{9.27a}$$

$$\partial_\mu j_\mu=0 \tag{9.27b}$$

9.2.1.2 Particle in an electromagnetic field

In classical electrodynamics the equations of motion satisfied by a particle carrying electric charge $+e$, moving in an electromagnetic field $A_\mu(x)$

are obtained by making the 'minimal substitution' in the Hamiltonian (cf. section 2.1.3)

$$p_\mu \to p_\mu - \frac{eA_\mu(x)}{c} \qquad A_\mu \equiv \mathbf{A}, \, \mathrm{i}\varphi/c \tag{9.28}$$

Carrying out the same prescription on the operator p_μ defined in (9.24) the Dirac equation (9.23) becomes

$$\left[\gamma_\mu\left(\partial_\mu - \frac{-\mathrm{i}e}{\hbar c}A_\mu\right) + mc/\hbar\right]\psi = 0 \tag{9.29}$$

If we make the corresponding change to L_0, and add to it the Lagrangian density for the free electromagnetic field

$$L_{\mathrm{em}} = -\tfrac{1}{4}F_{\mu\nu}F_{\mu\nu} \tag{9.30}$$

where $F_{\mu\nu}$ is the anti-symmetric electromagnetic field tensor

$$F_{\mu\nu} = \partial_\mu A_\nu - \partial_\nu A_\mu \tag{9.31}$$

we obtain the Lagrangian density L for the combined system of field-plus-particle

$$L = L_0 + \mathrm{i}ej_\mu A_\mu + L_{\mathrm{em}} \tag{9.32}$$

Thus the coupling of the particle to the electromagnetic field is represented in (9.32) by the term

$$L_1 = \mathrm{i}ej_\mu A_\mu \tag{9.33}$$

which is described as a vector interaction, since it is given by the contraction of two four-vectors. Moreover it is a parity conserving interaction since j_μ and A_μ are both polar four-vectors, i.e. their space components each form a Cartesian polar vector, and their time components are true scalars.

We can, however, also form an axial four vector $\gamma_5\gamma_\mu$ by making use of the pseudo-scalar operator

$$\gamma_5 = \gamma_1\gamma_2\gamma_3\gamma_4 \tag{9.34}$$

whose pseudo-scalar property follows from the third part of definition (9.24). To complete the selection of useful operators we may construct the antisymmetric tensor operator

$$\sigma_{\mu\nu} = (1/2\mathrm{i})[\gamma_\mu, \gamma_\nu] \tag{9.35}$$

The three space-space components $\sigma_{jk} = -\mathrm{i}\alpha_j\alpha_k$ are the component of the axial vector operator $-\mathrm{i}\boldsymbol{\alpha} \times \boldsymbol{\alpha}$, which is just the spin operator in units of $\frac{1}{2}\hbar$, whereas the three space-time component $\sigma_{j4} = \alpha_j$ are just the component of the polar vector operator $\boldsymbol{\alpha}$.

9.2.1.3 Gauge invariance and charge conservation

We may now show by direct substitution that the Lagrangian density L defined in (9.32) is invariant with respect to the local gauge transformations

$$\psi(x)\rightarrow\psi'(x)=\exp[i\lambda(x)]\psi(x) \tag{9.36}$$

and

$$A_\mu(x)\rightarrow A'_\mu(x)=A_\mu(x)+\left(\frac{1}{e}\right)\partial_\mu\lambda(x) \tag{9.37}$$

In contrast with the colour gauge transformations given in (9.16) and (9.19) which have eight generators $\boldsymbol{T}$, the electromagnetic gauge transformation only has one generator and the corresponding U(1) gauge group is Abelian. This is reflected in the definition of the electromagnetic field tensor in (9.31) which, unlike the corresponding colour field tensor defined in (9.20), is independent of the charge.

It is well known from classical electrodynamics that the electromagnetic field tensor $F_{\mu\nu}$ is gauge invariant and hence the term L_{em} in (9.33) is independently gauge-invariant. However only the combination L_0+L_1 is gauge-invariant rather than each term separately, and therefore the coupling of the particle to the massless gauge field $A_\mu(x)$ is necessary to maintain overall gauge invariance. The corresponding symmetry is expressed in the conservation of the electromagnetic current $j_\mu=\boldsymbol{j}, ic\rho$, defined in (9.27a). We can show this in the following way. For the U(1) group of gauge transformations (9.36) has a single generator, namely the unit operator which, assuming exact invariance under the group, is to be identified with a conserved 'charge' operator

$$Q(t)\equiv\int\rho(\boldsymbol{r},t)\,d^3r=1 \tag{9.38}$$

That this condition is fulfilled by the probability charge density $\rho(\boldsymbol{r}, t)$ defined in (9.27a) follows from the conservation law (9.27b). For this gives us the result

$$\frac{\partial}{\partial t}\int\rho(\boldsymbol{r},t)\,d^3r=-\int\nabla\cdot\boldsymbol{j}(\boldsymbol{r},t)\,d^3r=-\int\boldsymbol{j}(\boldsymbol{r},t)\cdot d\boldsymbol{\sigma}=0 \tag{9.39}$$

where the surface integral is taken over the surface at infinity. The result (9.38) follows immediately.

It is clear that, although (9.38) holds in the case of the proton, where Q is the charge measured in units of $+e$, it cannot apply to the neutron where $Q=0$, even though (ignoring weak interactions) the space integral of the probability density $\rho(\boldsymbol{r}, t)$ integrates to unity. The reason is that, supposing ψ to represent the wavefunction of a nucleon, the charge Q is given, not by (9.38), but by (9.1). Hence the electromagnetic current for the nucleon must be defined, not by (9.27a) but rather by

$$J_\mu = \bar{\psi}\gamma_\mu \tfrac{1}{2}(1+\tau_3)\psi = \tfrac{1}{2}\bar{\psi}\gamma_\mu\psi + \tfrac{1}{2}\bar{\psi}\gamma_\mu\tau_3\psi \tag{9.40}$$

where τ_3 is the isospin operator defined in (3.74). Thus in place of (9.27) we have a conserved isoscalar current

$$J^S_\mu = \tfrac{1}{2}\bar{\psi}\gamma_\mu\psi \tag{9.41a}$$

$$\partial_\mu J^S_\mu = 0 \tag{9.41b}$$

which is an expression of the conservation of baryon number *B*. This is a consequence, not of local gauge invariance, but of a U(1) global gauge invariance, corresponding to a transformation similar to (9.36), where the function $\lambda(x)$ is replaced by a constant λ.

Because of virtual strong processes of the type $p \rightleftarrows n+\pi^+$, $n \rightleftarrows p+\pi^-$ the isovector current

$$J^V_\mu = \tfrac{1}{2}\bar{\psi}\gamma_\mu\tau_3\psi \tag{9.42}$$

is not conserved, and to maintain exact conservation of the electromagnetic current, it is necessary to supplement J^V_μ by additional terms describing the contribution of pions, strange particles, etc., in virtual states. The conclusion must be, therefore, that nucleons cannot be described by Dirac equations, except in the non-relativistic limit, where strong interactions may be taken into account as the sources of certain induced electromagnetic interactions.

9.2.1.4 *The non-relativistic limit*

Suppose that ψ_0 is an energy eigensolution of (9.29) with eigenvalue $E_0>0$, then, reverting to a non-covariant notation, this equation may be rewritten in the form

$$\frac{1}{2mc^2}\{(E_0-e\varphi)^2-(mc^2)^2\}\psi_0 = \left\{\frac{1}{2m}\left(\boldsymbol{p}-\frac{e\boldsymbol{A}}{c}\right)^2 - \frac{e}{2mc}\left(\frac{\hbar}{2}\right)\sigma_{\mu\nu}F_{\mu\nu}\right\}\psi_0 \tag{9.43}$$

If we now apply this result in the non-relativistic limit when $E_0 \simeq mc^2$, the left-hand side reduces to $-e\varphi\psi_0$ approximately, and we are left with an equation which is remarkably similar to the Schrödinger–Pauli equation, except of course that ψ_0 is a four-component wavefunction, and there is an additional term in the Hamiltonian

$$-\frac{e}{2mc}\frac{1}{2}\hbar\sigma_{\mu\nu}F_{\mu\nu} = -\frac{e\hbar}{2mc}\{\boldsymbol{\alpha}\cdot\boldsymbol{B} - \mathrm{i}\boldsymbol{\alpha}\cdot E\} \tag{9.44}$$

where $\sigma_j \equiv \sigma_{kl}$. The first term in this equation is just the energy of a Dirac magnetic moment in a magnetic field $\boldsymbol{B}$ i.e.

$$H_{\text{mag}} = -\frac{e\hbar}{2mc}\boldsymbol{\sigma}\cdot\boldsymbol{B} = -\boldsymbol{\mu}_D\cdot\boldsymbol{B} \tag{9.45}$$

The second term in (9.44) vanishes in the extreme non-relativistic limit as we may understand by recalling that $\boldsymbol{\alpha} \simeq \boldsymbol{v}/c$. However, by constructing explicit representations for $\boldsymbol{\alpha}$, β which allow us to separate ψ_0 into its large components u_1, u_2 and its small components u_3, u_4, we can show that this term leads to a spin–orbit interaction

$$H_{\rm so} = -\frac{e\hbar}{4m^2c^2}\boldsymbol{\alpha}\cdot(\boldsymbol{E}\times\boldsymbol{p}) \simeq -\frac{1}{2}\mu_{\rm D}\cdot\frac{\boldsymbol{E}\times\boldsymbol{v}}{c} \tag{9.46}$$

In fact, apart from the factor $\frac{1}{2}$, $H_{\rm so}$ is just the energy of a magnetic moment $\mu_{\rm D}$, in the magnetic field $\boldsymbol{B} = (\boldsymbol{E}\times\boldsymbol{v})/c$ which it senses when moving with velocity $\boldsymbol{v}$ in an electric field $\boldsymbol{E}$. The factor $\frac{1}{2}$ represents the counter-effect caused by the Thomas precession discussed in section 3.3.1. When a proper expansion of the Hamiltonian in (9.43) is carried out correct to terms of order $(mc/\hbar)^2$, apart from terms expressing the relativistic mass increase, a further term is revealed, namely the Darwin term (Darwin 1928)

$$H_{\rm D} = -\frac{e\hbar^2}{8m^2c^2}\nabla\cdot\boldsymbol{E} \tag{9.47}$$

This term takes account of the fact that, because of the 'zitterbewegung' motion, the charge on the particle is not concentrated at a point, but is spread out over a region of space surrounding the particle, of dimension approximately equal to the reduced Compton wavelength $\hbar/mc$.

9.2.2 Induced electromagnetic effects

9.2.2.1 The Pauli magnetic moments

According to (9.45) a point proton has a magnetic moment equal to one nuclear magneton, and a point neutron has zero magnetic moment. Experimentally protons and neutrons have anomalous magnetic moments equal to 1.79 n m and −1.91 n m, respectively. These moments are now interpreted as induced magnetic moments which owe their origins to the motion of confined quarks. It has, however, been known for a long time that it is still possible to describe the motions of nucleons in not too strong electromagnetic fields by a modified Dirac equation (Pauli 1941)

$$\left[\gamma_\mu\left(\partial_\mu + \frac{ie}{\hbar c}A_\mu\right) + (\mu_{\rm P}/\hbar c)\tfrac{1}{2}\sigma_{\mu\nu}F_{\mu\nu} + mc/\hbar\right]\psi = 0 \tag{9.48}$$

where $\mu_{\rm P}$ is an arbitrary real number, without disturbing the relativistic covariance of the equation. In this way it is possible to assign to any spin $\frac{1}{2}$ fermion a Pauli moment $\mu_{\rm P}$, in addition to any Dirac moment it may carry.

The modification introduced in (9.48) is equivalent to adding to the

Lagrangian density (9.32) a Pauli term

$$L_P = -\frac{\mu_P}{2}(\bar{\psi}\sigma_{\mu\nu}\psi)F_{\mu\nu} \tag{9.49}$$

which is described as a 'tensor interaction' since it is constructed from the contraction of two anti-symmetric tensors. Since L_P is itself gauge invariant, the addition of L_P to L does not disturb gauge invariance or charge conservation. This result is familiar from classical electrodynamics, where the addition of a magnetisation current $c(\nabla \times \boldsymbol{M})$, to the conduction current $\rho\boldsymbol{v}$, does not disturb charge conservation since $\nabla \cdot (\nabla \times \boldsymbol{M}) \equiv 0$. To maintain agreement with experiment it is only necessary to assign anomalous Pauli moments to proton and neutron

$$\mu_{Pp} = 1.79\mu_N \qquad \mu_{Pn} = -1.91\mu_N \tag{9.50}$$

The expression (9.43) for the spin–orbit interaction is then retained provided $\boldsymbol{\mu}_D$ is replaced by the empirical magnetic moment $\boldsymbol{\mu}_D + \boldsymbol{\mu}_P$.

9.2.2.2 The Foldy potential

In exactly the same way that the Dirac magnetic moment $\boldsymbol{\mu}_D$ is associated with the Darwin term (9.47) one might expect an equivalent term to appear in conjunction with the Pauli moment $\boldsymbol{\mu}_P$, and indeed this turns out to be the case. The relevant term was first identified by Foldy (Foldy 1951, 1952, 1958)

$$H_F = -\tfrac{1}{2}\mu_P(\hbar/mc)\nabla \cdot \boldsymbol{E} \tag{9.51}$$

The Foldy potential owes its origin to the mean dipole interaction $\langle \boldsymbol{\mu}_P \times \boldsymbol{\alpha} \cdot \boldsymbol{E} \rangle$ induced by the zitterbewegung motion averaged over a volume of space having dimensions of order $(\hbar/mc)$. This term is of particular importance in the scattering of slow neutrons by electrons bound in spinless atoms for which the magnetic interaction vanishes (Fermi and Marshall 1947). In this case the Foldy term reduces to the contact potential

$$H_{ne} = -\tfrac{1}{2}\kappa_n\mu_N(\hbar/2m_n c)(4\pi e)\delta^3(\boldsymbol{r}) \tag{9.52}$$

which gives rise to a bound scattering length

$$b_F = \kappa_n\left(\frac{m_e}{2m_n}\right)r_e = -1.468 \times 10^{-3}\ \text{fm} \tag{9.53}$$

where $r_e = e^2/m_e c^2$ is the classical electron radius.

An alternative description of H_{ne} expresses it in terms of an equivalent spherical potential well of depth V_F and radius r_e (Fermi and Marshall 1947). Thus V_F is defined by

$$\frac{4}{3}\pi r_e^3 V_F = -\int H_{ne}\, d^3r = -\frac{1}{2}\kappa_n\mu_N\left(\frac{\hbar}{2m_n c}\right)(4\pi e) \tag{9.54}$$

leading to the result

$$V_F = (3/4)(-\kappa_n/\mu_N)(m_e c^2)(m_e/\alpha m_n)^2 = 4072.73 \text{ eV} \tag{9.55}$$

where $\alpha = e^2/\hbar c$ is the fine structure constant.

9.2.2.3 Electric dipole moment

The last possibility we consider, which is of considerable contemporary interest, is the construction of a contribution to the Lagrangian density of the form

$$L_{\text{edm}} = -\tfrac{1}{2} p_N (\bar{\psi} \sigma_{\mu\nu} \psi) \tilde{F}_{\mu\nu} \tag{9.56}$$

where

$$\tilde{F}_{\mu\nu} = -\frac{i}{2} \delta_{\mu\nu\rho\sigma} F_{\rho\sigma} \tag{9.57}$$

is the pseudo-tensor dual to $F_{\mu\nu}$ and $\delta_{\mu\nu\rho\sigma}$ is the completely anti-symmetric Levi–Civita symbol (Møller 1952). In the present context $\tilde{F}_{\mu\nu}$ is constructed most simply by making the replacements

$$\boldsymbol{E} \to \boldsymbol{B} \qquad \boldsymbol{B} \to -\boldsymbol{E} \tag{9.58}$$

The comparison of (9.56) with (9.49) shows that the interaction (9.56) takes into account any nucleon electric dipole moment $p_N = e d_N$, where d_N is referred to as the dipole length.

The interaction L_{edm} is invariant under all proper Lorentz transformations but changes sign for improper transformations which invert the coordinate system or reverse the direction of the time. The postulated interaction is therefore simultaneously P-violating and T-violating (cf. section 3.3.1). Since the neutron is uncharged it provides the most suitable candidate in which to search for a finite electric dipole moment which, were it detected, would be the first T-violating effect observed in nature outside the K^0–$\bar{K}^0$ system. This equation is discussed further in section 9.2.4.4.

9.2.3 Electromagnetic form factors

9.2.3.1 Electron–neutron scattering

The determination of nuclear size and charge distributions by elastic scattering of relativistic electrons was discussed in section 3.2.9 where the differential cross-section was expressed in the form (cf. 3.20)

$$\left(\frac{d\sigma}{d\Omega}\right) = \left(\frac{d\sigma}{d\Omega}\right)_M |F_c(q)|^2 \tag{9.59}$$

The function $F_c(q)$ is the charge form factor defined in (3.21), and the momentum transfer $\hbar\boldsymbol{q}$ may be interpreted as the momentum carried by the space-like virtual photon which is exchanged between electron and nucleus. The cross-section $(d\sigma/d\Omega)_M$ describes Mott scattering from a spinless massive nucleus carrying electric charge Ze. For a point charge $F_c(q)$ is equal to unity, but, for an extended object, $F_c(q)$ decreases rapidly with q, expressing the fact that a large object tends to break up into its constituent parts, and is therefore less efficient in absorbing momentum.

The description of electron scattering based on (9.59) is adequate for energies up to ≈ 100 MeV, but, to investigate the electromagnetic structures of neutrons and protons, energies of 0.5 GeV and upwards are required, and a proper relativistic description of the scattering process is necessary. We therefore consider electron scattering in the one-photon exchange approximation represented by the Feynman diagram shown in figure 9.4, where all energies and momenta indicated are measured in the laboratory system. The momentum and energy transferred in the collision are then given by

$$\hbar\boldsymbol{q} = \boldsymbol{p}_e - \boldsymbol{p}'_e \qquad \hbar\omega = E_e - E'_e \tag{9.60}$$

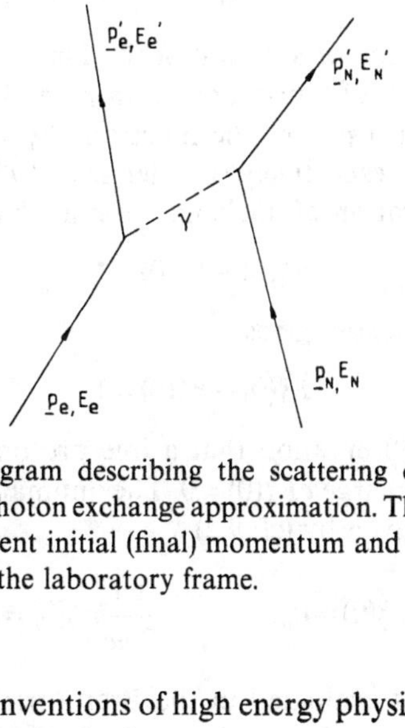

Figure 9.4. Feynman diagram describing the scattering of high energy electrons from nucleons in the one photon exchange approximation. The quantities $\boldsymbol{p}_e$, E_e ($\boldsymbol{p}'_e$, E'_e) and $\boldsymbol{p}_N$, E_N ($\boldsymbol{p}'_N$, E'_N) represent initial (final) momentum and energy, for electrons and nucleons, respectively, in the laboratory frame.

To conform with the conventions of high energy physics we write the squared four-momentum transfer in the form

$$Q^2 = -[\omega^2 - \boldsymbol{q}^2] \tag{9.61}$$

We now assume that the nucleon is coupled to the electromagnetic field via a polar vector interaction of the form (9.33), i.e.

$$L_1 = ie\hbar J_\mu(x) A_\mu(x) \tag{9.62}$$

where $A_\mu(x)$ is an electromagnetic field operator discussed in section 2.1.1 and $J_\mu(x)$ is a conserved nucleon four-current operator, i.e.

$$\partial_\mu J_\mu(x)=0 \tag{9.63}$$

For example a nucleon constructed entirely of u and d valence quarks would have a four-current of the form

$$J_\mu(x)=\sum_{j=1}^{2}\frac{1}{2}\left[\bar{\psi}_j(x)\left(\frac{1}{3}+\tau_{3j}\right)\gamma_\mu\psi_j(x)\right] \tag{9.64}$$

where $\psi_j(x)$ ($\bar{\psi}_j(x)$) is the annihilation (creation) operator for a quark of the jth flavour.

In any case the only charge-conserving parity-conserving relativistically covariant transition current matrix element which may be constructed from the available operators γ_μ, q_μ, etc., must have the form

$$\begin{aligned}\langle N'|eJ_\mu(0)|N\rangle=\bar{\boldsymbol{u}}'(p'_{\rm N}S'_z)\chi'\Bigg\{&\left[eF_1^{\rm p}(Q^2)\left(\frac{1+\tau_3}{2}\right)\gamma_\mu+eF_1^{\rm n}(Q^2)\left(\frac{1-\tau_3}{2}\right)\gamma_\mu\right]\\ &-\left[\frac{e\hbar}{2mc}F_2^{\rm p}(Q^2)\left(\frac{1+\tau_3}{2}\right)\mathrm{i}\sigma_{\mu\nu}q_\nu+\frac{e\hbar}{2mc}F_2^{\rm n}(Q^2)\left(\frac{1-\tau_3}{2}\right)\mathrm{i}\sigma_{\mu\nu}q_\nu\right]\Bigg\}\boldsymbol{u}(p_{\rm N}S_z)\chi\end{aligned} \tag{9.65}$$

where $\boldsymbol{u}(p_{\rm N}S_z)$ is a four-spinor solution of the Dirac equation describing the motion of a free nucleon of four-momentum $p_{\rm N}$ and spin projection S_z and χ is a two-component isospinor. The functions $F_1^{\rm p,n}(Q^2)$, $F_2^{\rm p,n}(Q^2)$ are Dirac and Pauli form factors respectively (Hohler *et al* 1976).

To maintain conservation of nucleons we must have

$$F_1^{\rm p}(0)+F_1^{\rm n}(0)=1 \tag{9.66}$$

while conservation of isospin gives

$$F_1^{\rm p}(0)-F_1^{\rm n}(0)=1 \tag{9.67}$$

corresponding to the observation that a free proton has charge $eF_1^{\rm p}(0)=\mathrm{e}$, and a free neutron has charge $eF_1^{\rm n}(0)=0$. The anomalous moments of proton and neutron are now given by (cf. 9.2)

$$\frac{e\hbar}{2mc}F_2^{\rm p}(0)=\kappa_{\rm p}\mu_{\rm N}\qquad\qquad \frac{e\hbar}{2mc}F_2^{\rm n}(0)=\kappa_{\rm n}\mu_{\rm N} \tag{9.68}$$

When $\langle N'|J_\mu(0)|N\rangle$ is evaluated in the Breit (brick-wall) frame, which is defined such that $\boldsymbol{p}_{\rm N}+\boldsymbol{p}'_{\rm N}=0$ and therefore such that $\hbar\omega=0$, the nucleon transition matrix elements become ($J_4=\mathrm{i}J_0=\mathrm{i}c\rho$)

$$\langle N'|eJ_0(0)|N\rangle=\frac{\delta_{S_zS_{z'}}}{\sqrt{1+(\hbar Q/2mc)^2}}eG_{\rm E}^{\rm p,n}(Q^2) \tag{9.69a}$$

$$\langle N'|e\boldsymbol{J}(0)|N\rangle=\frac{\langle S'_z|(\boldsymbol{\sigma}\times\boldsymbol{q})|S_z\rangle}{2mc\sqrt{1+(\hbar Q/2mc)^2}}eG_{\rm M}^{\rm p,n}(Q^2) \tag{9.69b}$$

where $G_E(Q^2)$ and $G_M(Q^2)$ are the electric and magnetic Sachs form factors defined by

$$G_E^{p,n}(Q^2)=F_1^{p,n}(Q^2)-\left(\frac{\hbar Q}{2mc}\right)F_2^{p,n}(Q^2) \tag{9.70a}$$

$$G_M^{p,n}(Q^2)=F_1^{p,n}(Q^2)+F_2^{p,n}(Q^2) \tag{9.70b}$$

We then find from (9.66) and (9.67)

$$\begin{aligned} &G_E^p(0)=1 \qquad G_E^n(0)=0 \\ &G_M^p(0)=1+\kappa_p=\mu_p \qquad G_M^n(0)=\kappa_n=\mu_n \end{aligned} \tag{9.71}$$

The differential cross-section is now determined in first-order Born approximation for the scattering of point electrons from the distributions of electric charge and current

$$\rho(\boldsymbol{r})=\left(\frac{1}{2\pi}\right)^3\int G_E(Q^2)e^{-i\boldsymbol{q}\cdot\boldsymbol{r}}\,d^3q \tag{9.72a}$$

$$\boldsymbol{j}(\boldsymbol{r})=\left(\frac{1}{2\pi}\right)^3\int i(\boldsymbol{\sigma}\times\boldsymbol{q})G_M(Q^2)e^{-i\boldsymbol{q}\cdot\boldsymbol{r}}\,d^3q \tag{9.72b}$$

resulting in the scattering formula (Rosenbluth 1950)

$$\left(\frac{d\sigma}{d\Omega}\right)=\left(\frac{d\sigma}{d\Omega}\right)_M\left(\frac{E_e'}{E_e}\right)\left\{\frac{G_E^2+(\hbar Q/2mc)^2G_M^2}{1+(\hbar Q/2mc)^2}+2\left(\frac{\hbar Q}{2mc}\right)^2G_M^2\tan^2\theta/2\right\} \tag{9.73}$$

Thus $G_E(Q^2)$ and $G_M(Q^2)$ can be determined by measuring $(d\sigma/d\Omega)$ as a function of $\tan^2\theta/2$ for fixed $(\hbar Q/2mc)^2$.

Electron scattering experiments on targets of hydrogen and deuterium indicate that three of the four Sachs form factors satisfy, at least to a good approximation, the dipole fit

$$G_E^p(Q^2)=\frac{G_M^p(Q^2)}{\kappa_p}=\frac{G_M^n(Q^2)}{\kappa_n}=\left\{1+\left(\frac{hQ}{m_v c}\right)^2\right\}^{-2} \tag{9.74}$$

a result which corresponds to scattering via the exchange of a vector meson of mass $m_v c^2=843$ MeV, having the same quantum numbers as the photon. In this respect the ρ-meson ($m_\rho c^2=770$ MeV) makes a good candidate. The results (9.74) might be interpreted to mean that those entities which carry the charge in the nucleon, also carry the magnetization. However recent measurements on the quasi-elastic reaction $^2H(e,e'n)^1H$ indicate an enhancement at low Q^2 of G_M^n relative to the dipole parametrization (9.74) (Markowitz *et al* 1993). By Fourier inversion of (9.72) and application of (3.21) we find for the mean square radii characterizing the distributions of electric charge and magnetization

$$\frac{1}{6}\langle r_c^2\rangle=-\frac{dG_E(0)}{dQ^2}\simeq-F_1'(0)+\left(\frac{\hbar}{2mc}\right)^2F_2(0) \tag{9.75a}$$

$$\frac{1}{6}\langle r_{\mathrm{m}}^2\rangle = -\frac{\mathrm{d}G_{\mathrm{M}}(0)}{\mathrm{d}Q^2} \simeq -F'_1(0) - F'_2(0) \tag{9.75b}$$

The size of the proton may be determined to quite high accuracy by studying e–H scattering at low momentum transfer, giving the result (Simon *et al* 1980)

$$\langle r_{\mathrm{c}}^2\rangle_{\mathrm{p}} = 0.862 \pm 0.012\ \mathrm{fm}^2 \tag{9.76}$$

However, largely, if not entirely, because of difficulties associated with the structure of the deuteron (Kukulin *et al* 1972), it is difficult to deduce a precise value for the electric form factor $G_{\mathrm{E}}^{\mathrm{n}}(Q^2)$ from measurements of e–D scattering (Bumiller *et al* 1970, Galster *et al* 1971, Berard *et al* 1973). Thus the error to be assigned to one published value for $\langle r_{\mathrm{c}}^2\rangle_{\mathrm{n}}$ based on the e–D scattering data (Trubnikov 1981)

$$\langle r_{\mathrm{c}}^2\rangle_{\mathrm{n}} = -0.113\ \mathrm{fm}^2 \tag{9.77}$$

is rather difficult to assess. However, this is a matter of no great importance since $\langle r_{\mathrm{c}}^2\rangle_{\mathrm{n}}$ may be determined to much higher precision from the very accurate measurements which have been made, over many years, on the coherent scattering of slow neutrons by spinless atoms. These are discussed in section 9.2.3.2, where it is shown that most of the measured charge radius can be accounted for by the contribution of the Pauli term in (9.75a). This is given by

$$\left(\frac{\hbar}{2m_{\mathrm{n}}c}\right)^2 F_2(0) = \alpha^{-1}\left(\frac{\hbar}{2m_{\mathrm{n}}c}\right) b_{\mathrm{F}} = -0.02115\ \mathrm{fm}^2 \tag{9.78}$$

where b_{F} is the Foldy scattering length defined in (9.53).

Expressions have been derived for both $G_{\mathrm{E}}^{\mathrm{p}}(Q^2)$ and $G_{\mathrm{E}}^{\mathrm{n}}(Q^2)$ using the harmonic oscillator quark model of the nucleon with colour hyperfine interaction (cf. section 10.3.2) (Carlitz *et al* 1977, Isgur 1977, Isgur and Karl, 1977, 1978)

$$G_{\mathrm{E}}^{\mathrm{p}}(Q^2) = e^{-Q^2/6\eta^2} \tag{9.79a}$$

$$G_{\mathrm{E}}^{\mathrm{n}}(Q^2) = -\frac{1}{6}\left\langle \sum_i e_i r_i^2 \right\rangle_{\mathrm{n}} Q^2 e^{-Q^2/6\eta^2} \tag{9.79b}$$

where

$$\eta^{-2} = \left\langle \sum_i e_i r_i^2 \right\rangle_{\mathrm{p}} \tag{9.80}$$

and e_i is the charge on the ith quark. According to these theories $\langle r_{\mathrm{c}}^2\rangle_{\mathrm{n}}/\langle r_{\mathrm{c}}^2\rangle_{\mathrm{p}} \simeq -0.16$, the minus sign resulting from the fact that, while the two *u–d* quark pairs in the neutron attract each other, the *d–d* pair repel. The value of $\langle r_{\mathrm{c}}^2\rangle_{\mathrm{n}}$ calculated from (9.76), and to be compared with (9.77), is then

$$\langle r_{\mathrm{c}}^2\rangle_{\mathrm{n}} = -0.138\ \mathrm{fm}^2 \tag{9.81}$$

9.2.3.2 Neutron–electron scattering

The first experimental searches for a possible non-magnetic term in the neutron–electron interaction were carried out almost simultaneously by Havens *et al* (1947), and by Fermi and Marshall (1947). These investigations were motivated by suggestions that the neutron might be surrounded by a cloud of virtual mesons, and predated both the discovery of the π-meson and the identification of the Foldy term (cf. section 9.2.2.2). It might have been supposed, for example, that the neutron could be viewed as a point proton enveloped in a charge density $-e\rho(r)$, in which case the equivalent well depth V_F for a radius r_e would be given by (cf. 9.54)

$$\frac{4}{3}\pi r_e^3 V_F \simeq \left(\frac{2\pi e}{3}\right)\int r^2\rho(r)\,\mathrm{d}r \tag{9.82}$$

Hence a measurement of V_F would provide an estimate of $\langle r_c^2 \rangle_n$.

As first pointed out by Condon (1936) conditions for observing a neutron–electron interaction are much more favourable using electrons bound in atoms rather than free electrons, since this increases the reduced mass by a factor of order (m_n/m_e), and the minimum detectable value of V_F by about the same amount. Also, to suppress the magnetic interaction, it is necessary to use spinless atoms, and preferably spinless nuclei if the spin-incoherent scattering is to be kept to a minimum. In these circumstances the total scattering amplitude may be written

$$f(q) = -\{b_N + ZF(q)b_{ne}\} \qquad q = 2k\sin(\theta/2) \tag{9.83}$$

where b_N and b_{ne} are nuclear and electronic bound scattering lengths, respectively, and $F(q)$ is an atomic form factor. The essential feature of (9.83) which has been exploited in all measurements of b_{ne} carried out to date, is that $F(q)$ is approximately unity for low energy neutrons with $E_n \leqslant 0.1$ eV, and is essentially zero for $E_n \geqslant 10$ eV. Thus any measured difference between high energy and low energy scattering cross-sections, by whatever method detected, may be used to isolate the interference term $2b_N b_{ne} ZF(q)$ which contains the required information.

The experiments may be separated into two distinct streams; those based on neutron scattering in heavy liquid metals, and those which use free rare gas atoms as targets (Hughes 1953, Halpern 1964). The earliest experiments in the first stream studied the variation with neutron wavelength of slow neutron transmission through targets of molten lead and bismuth (Havens *et al* 1947, 1949, 1951). After applying various corrections associated with target motion and liquid diffraction, they achieved the first significant result $V_F = -(5300 \pm 1000)$ eV. In a much improved version of the same technique, developed at Brookhaven National Laboratory, the precision was subsequently improved giving a result $V_F = -(4340 \pm 140)$ eV (Melkonian *et al* 1956, 1959).

A particularly ingenious variation on the same basic idea relies on the expression for the refractive index (cf. 6.79)

$$n^2 = 1 - \frac{N\lambda^2}{\pi}[b_N + Zb_{ne}] \tag{9.84}$$

where b_{ne} is deduced from simultaneous determinations of n^2 and b_N (Hughes *et al* 1953). One method for measuring the relative refractive index of two media is to determine the critical angle θ_c for total reflection which, for $\theta_c \ll 1$, is essentially equal to the increment in the refractive index. In these experiments neutron scattering at a bismuth–liquid oxygen interface was studied and, since the nuclear scattering amplitudes in this case were almost equal, virtually the whole of the difference could be attributed to the electron amplitude. The corrections for the nuclear contribution were measured in the usual way by studying the variation of the total cross-section in a region of energy where $F(q) \simeq 0$. The final results deduced a value $V_F = -(3860 \pm 370)$ eV.

More recently these methods have been refined into a very precise technique by Koester and his associates at Munich, reaching such a sensitivity that it has proved possible to isolate a contribution to the scattering length associated with the electric polarizability of the neutron in the Coulomb field of the nucleus (Koester *et al* 1976, 1976a, Waschkowski and Koester 1976, Koester *et al* 1986). In these experiments the total scattering lengths in liquid bismuth and liquid lead were determined to a high degree of accuracy using the neutron gravity refractometer described in section 1.5.2.2. These data were then combined with total cross-section measurements at well defined energies selected by resonances in ^{103}Rh (1.26 eV), ^{109}Ag (5.19 eV), ^{186}W (18.8 eV) and ^{59}Co (132 eV). After carrying out a series of corrections associated with finite range effects, resonance–potential scattering interference, and spin-incoherent scattering in ^{209}Bi (for which $I = 9/2$) a final result was obtained for the neutron–electron scattering length

$$b_{ne} = -(1.32 \pm 0.04) \times 10^{-3} \text{ fm} \tag{9.85}$$

The second experimental stream, involving the scattering of slow neutrons in rare gases, was initiated by Fermi and Marshall (1947). Rare gases were selected because they are monatomic and therefore cannot display the interference effects for scattering from atoms in the same molecule which were so crucial to the first determination of the neutron spin (cf. section 1.5.3.1). In these early experiments neutrons were scattered from xenon atoms at angles of 45° and 135° in order to isolate any interference term, which, depending as it does on $q = 2k \sin(\theta/2)$, is not isotropic in the centre-of-mass system, in contrast to the nuclear scattering. To detect such an asymmetric term it is of course necessary to separate the much larger Doppler spread associated with the thermal motion of the centre-of-mass. Although the early measurements were inconclusive, giving a value $V_F = 300 \pm 5000$ eV (Fermi and Marshall 1947), and $V_F = 4100 \pm 1000$ eV (Hammermesh *et al* 1952) the technique was subsequently developed to a high state of perfection by Krohn and Ringo (1960, 1973), who studied scattering in argon, neon, krypton and xenon. In this work the angular asymmetry associated with the Doppler motion was determined from accurate temperature measurements combined with calculations for each contributing isotopic species. A feature of the technique was the very high level of purity attained, thereby eliminating any contamination from hydrocarbons. The final result obtained was

$$b_{\mathrm{ne}} = -(1.30 \pm 0.03) \times 10^{-3}\ \mathrm{fm} \tag{9.86}$$

Since the two most precise results (9.85) and (9.86) agree with each other within the errors, we may compute a weighted mean value

$$\langle b_{\mathrm{ne}} \rangle = -(1.307 \pm 0.024) \times 10^{-3}\ \mathrm{fm} \tag{9.87}$$

The corresponding value for $\langle r_c^2 \rangle_n$ is then given by (cf. 9.78)

$$\langle r_c^2 \rangle_n = (6/\alpha)(\hbar/2m_n c) b_{\mathrm{ne}} = -(0.113 \pm 0.002)\ \mathrm{fm}^2 \tag{9.88}$$

a result which, coincidentally, agrees exactly with the value derived from neutron–deuteron scattering experiments (cf. 9.27) but with an error of order 2%. There is also a small but significant departure from the value of the Foldy scattering length b_F given in (9.53). The difference is

$$\Delta b = b_{\mathrm{ne}} - b_F = (0.161 \pm 0.024) \times 10^{-3}\ \mathrm{fm} \tag{9.89}$$

This may be associated with a non-vanishing derivative $-F_1'^n(0)$ of the Dirac form factor according to (9.75). We may also choose to define a corresponding Dirac charge radius

$$\langle r_c^2 \rangle_n^D \equiv -6F_1'^n(0) = (9.5 \pm 2.2) \times 10^{-3}\ \mathrm{fm}^2 \tag{9.90}$$

although it is not clear that a very precise meaning can be attached to such a concept. Indeed, if the value of b_{ne} given in (9.86) is replaced by the somewhat larger values measured by the Dubna group (Aleksandrov *et al* 1985, 1986), the corresponding values of $\langle r_c^2 \rangle_n^D$ are negative. It is not obvious, however, that this difference is significant. The chiral bag model of the nucleon (cf. section 10.3.4) predicts a value $\langle r_c^2 \rangle_n^{CB} = -0.130\ \mathrm{fm}^2$ (Thomas *et al* 1981) but, because the model contains no zitterbewegung effect, there is no Foldy term. Hence $\langle r_c^2 \rangle_n^{CB}$ should be compared with $\langle r_c^2 \rangle_n$ rather than with $\langle r_c^2 \rangle_n^D$ and there is excellent agreement.

9.2.4 Electromagnetic structure of the nucleon

9.2.4.1 Deep inelastic scattering

Thirty years ago the highest electron energy available for nuclear physics investigations was about 1 GeV, which was more than adequate to excite all nuclear processes of interest, and sufficient to determine the Sachs form factors of nucleons, and to generate the Δ and N nucleonic resonances. Over the succeeding decades the available energy has been raised by two orders of magnitude, and currently there are two principal sources of charged leptons. These are (i) the Stanford linear accelerator (SLAC) operating at energies between 5 and 20 GeV and (ii) the CERN muon beam which operates between 100 and 250 GeV. Both systems generate hadronic excitations via the exchange of space-like ($Q^2 > 0$) virtual photons.

It is possible at the present time to produce electron fluxes of order $10^{13}\ s^{-1}$ at 0.1% momentum resolution, and muon fluxes of order $10^6\ s^{-1}$. The analysis of muon scattering data is somewhat less susceptible to uncertainties associated with radiative corrections as compared with electron scattering data. The study of processes involving the exchange of time-like virtual photons ($Q^2 < 0$) is made possible by the availability of e^+e^- colliding beam machines leading to the creation of lepton pairs $\mu^+\mu^-$, $\tau^+\tau^-$, or hadrons, in the final state. This reaction is shown in figure 9.5. The work is further supplemented by data provided by scattering experiments using beams of high energy neutrinos ν_μ and $\bar{\nu}_\mu$ at CERN, involving the exchange of vector mesons $W^\pm$, Z^0 through the weak interaction.

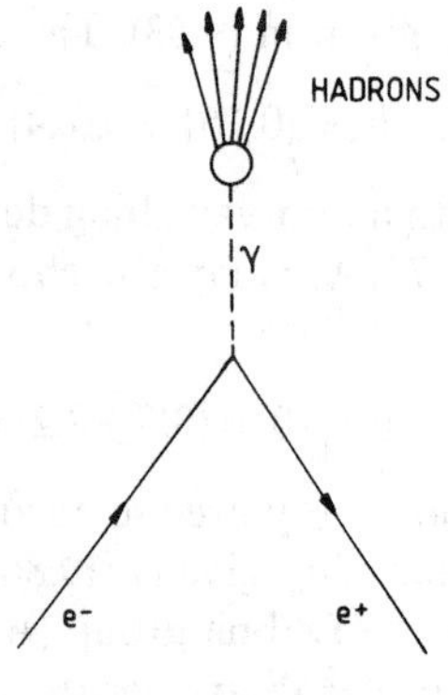

Figure 9.5. Creation of hadrons by time-like virtual photons produced by e^+–e^- annihilation in colliding beams.

At energies of order 0.1–0.2 GeV, and small scattering angles corresponding to low momentum transfers $Q^2 \approx 0.05\ (\text{GeV}/c)^2$, electron–nucleus scattering features an elastic peak corresponding to coherent scattering from the nucleus as a whole, together with the giant resonance at low energies due to the excitation of internal degrees of freedom (cf. 4.109). At backward angles corresponding to high momentum transfers of order 0.2 $(\text{GeV}/c)^2$, the elastic peak is strongly reduced and is replaced by a quasi-elastic peak at lower energies corresponding to elastic scattering from individual nucleons. This is because high momentum transfer corresponds to a shorter wavelength for the virtual photon, and increased spatial resolution, where the peak is Doppler broadened due to the Fermi motion of nucleons in the nucleus. When the energy is raised by one to two orders of magnitude the same general behaviour of the cross-section is reproduced with the nucleus now giving way to the nucleon, i.e. elastic nucleon scattering and nucleonic resonance generation for $Q^2 \approx 1$ (GeV/c)^2, and quasi-elastic scattering from individual 'parts' of the nucleon at momentum transfers $Q^2 \geqslant 5$ (GeV/c)^2. These particular nucleonic constituents are now described as 'partons' (Bjorken 1969, Feynman 1969). According to QCD theory the partons are either quarks or gluons.

The internal structure of the nucleon is most directly investigated in the deep inelastic experimental regime defined by

$$\hbar\omega \gg mc^2 \qquad (\hbar Q) \gg (mc) \tag{9.91}$$

where the double differential scattering cross-section, in analogy with (9.73), is given by (Close 1979)

$$\frac{\mathrm{d}^2\sigma}{\mathrm{d}E'_e\,\mathrm{d}\Omega} = \left(\frac{\mathrm{d}\sigma}{\mathrm{d}\Omega}\right)_{\mathrm{M}} \left\{\cos^2(\theta/2)\frac{F_2(Q^2,\omega)}{\hbar\omega} + 2\sin^2(\theta/2)\frac{F_1(Q^2,\omega)}{\hbar\omega}\right\} \tag{9.92}$$

and

$$\left(\frac{\mathrm{d}\sigma}{\mathrm{d}\Omega}\right)_{\mathrm{M}} = \left(\frac{4r_e^2 m_e^2 E_e'^2}{(\hbar Q)^4}\right) \tag{9.93}$$

is the usual Mott cross-section describing scattering from a spinless point charge (cf. 3.19). The structure functions $F_2(Q^2, \omega)$ and $F_1(Q^2, \omega)$ are form factors describing the scattering from the partons, where $F_2(Q^2, \omega)$ is related to the distribution of internal momentum and $F_1(Q^2, \omega)$ is related to the distribution of spin.

A formula very similar to (9.92) describes deep inelastic scattering of high energy neutrinos, where the principal modification is represented by the addition of a third structure function associated with the weak axial current and parity violation (cf. section 9.5). This term has opposite signs for scattering of ν_μ and $\bar{\nu}_\mu$. In addition the Mott cross-section (9.93) is replaced by the function $G^2 mE'/\pi$, where G is the weak Fermi coupling constant (cf. section 9.3). The electromagnetic and weak scattering cross-sections become comparable when $Q^2 \approx 10^7$ (GeV/c)^2.

9.2.4.2 Parton model of the nucleon

The first direct evidence for the presence of free or quasi-free partons in the nucleon was obtained through the study of inclusive reactions of the type

$$e + N \rightarrow e' + X \tag{9.94}$$

where X consist of a debris of hadronic matter emitted following the collision. The critical finding was the observation that, although the cross-section at the coherent elastic peak dropped by two orders of magnitude over the range

$(1<(\hbar Q)^2<5)$ $(\text{GeV}/c)^2$, the deep inelastic cross-section remained approximately constant (Breidenbach *et al* 1969). Furthermore it emerged that, whereas in principle, F_1 and F_2 may depend on two variables, e.g.

$$x=\hbar Q^2/2m\omega \qquad y=\hbar\omega/E_e \tag{9.95}$$

experimentally they were found to depend only on x. This is the phenomenon known as scaling, a certain indication of scattering from quasi-free point-like objects (Bjorken 1966, 1969, Bjorken and Paschos 1969), since it implies that the characteristics of the scattering process should be identical under experimental conditions defined by (Q^2, ω) or (Q'^2, ω'), provided only that the kinematic variables scale according to the rule

$$(\omega/\omega')=(Q/Q')^2 \tag{9.96}$$

It was the phenomenon of scaling, now firmly established (Panofsky 1968, Friedman and Kendal 1972, Bodek 1979), which led to the introduction of the parton model of nucleons. The model assumes that a lepton interacting with a hadron under conditions of high momentum transfer sees the hadron in the Breit frame as an assembly of free point particles, the partons, so that the observed inelastic spectrum is interpreted as the incoherent sum of elastic scattering events from individual partons. On this view the variable x is nothing more than the fraction of the hadron momentum in the Breit frame carried by the corresponding parton.

Comparing the hadron to a nucleus, the analogues of the partons are the nucleons, which are ejected from the nucleus either singly or in combinations e.g. deuterons, tritons, or α-particles. In remarkable contrast, partons are not ejected from the hadron as free particles, but only in combinations which exit from the nucleon in the form of hadronic material, by a process which is not yet understood. If, as many believe, hadrons are constructed from quarks and gluons confined through QCD forces, then it is natural to identify partons with quarks and gluons, and to interpret the non-emergence of free partons as just another example of confinement of coloured objects.

9.2.4.3 Quarks and gluons in the nucleon

If it is assumed that the partons picked out by charged (and neutral) leptons are indeed spin $\frac{1}{2}$ quarks then the parton model makes very specific predictions for the structure functions F_1 and F_2, i.e.

$$F_1(x)=\frac{1}{2}\sum_i e_i^2 f_i(x) \tag{9.97a}$$

$$F_2(x)=\sum_i e_i^2 x f_i(x) \tag{9.97b}$$

where $f_i(x)$ is the probability density that the ith quark with charge e_i should carry a fraction x of the total nucleon momentum. These results may now be combined into the relationship (Callan and Gross 1969)

$$F_2(x) = 2xF_1(x) \tag{9.98}$$

Since, according to (9.92) $F_1(x)$ makes no contribution to the scattering near forward angles which is the easiest case given to study, the result (9.98) is not easy to confirm experimentally. Nevertheless, it is sufficiently well established to prove that the scattering centres are not spinless objects and indeed the results are quite consistent with their being spin $\frac{1}{2}$ quarks. They cannot, of course, be gluons, since leptons do not participate in the strong interaction and carry no colour. However, summing the results over all values of x shows that the scattering objects, which we take to be quarks, cannot account for more than about 50% of the momentum in a nucleon, and the presumption must be that the missing momentum is carried by the gluons. Thus experiment agrees with QCD predictions that the partons are indeed quarks and gluons.

To determine whether the quarks have the same charges and colours postulated to account for the static properties of baryons we have to look to the results of e^+e^- collisions where we measure the ratio

$$R = \sigma(e^+e^- \to \text{hadrons})/\sigma(e^+e^- \to \mu^+\mu^-) = \sum_i e_i^2 \tag{9.99}$$

evaluated on the assumption that the scattering centres are indeed spin $\frac{1}{2}$ quarks. Assuming the collision takes place below the threshold for charmed quark production ($m_c c^2 \simeq 1$ GeV), and allowing for a correction due to $\tau^+\tau^-$ production, the ratio takes the value

$$R = 3\{4/9 + 1/9 + 1/9\} = 2 \tag{9.100}$$

assuming that up, down, and strange quarks are produced in equal measure. The factor 3 entering into (9.100) takes account of the fact that each quark comes in three colours.

Experimentally $R \approx 2.5$ below the charm threshold and $R \approx 4.6$ above it, in agreement with (9.100), and with the assignment of a charge equal to 2/3 the proton charge for the charmed quark given in table 9.1. The experimental results therefore do confirm the predictions of the quark–gluon nucleon model.

However, the quarks involved in deep inelastic scattering cannot be merely the valence quarks which define the SU(3) flavour properties of the baryon octet, and for the following reason. Integration of (9.97b) over x gives the result

$$\operatorname*{Lim}_{x\to 0} \int_x^1 F_2(x)\frac{dx}{x} = \sum_i e_i^2 \left\{ \operatorname*{Lim}_{x\to 0} \int_x^1 f_i(x)\, dx \right\} = \sum_i e_i^2 \tag{9.101}$$

where, for $x \leqslant 0.2$, $F_2(x)$ approaches a constant value in the range $0.1 < F_2(x) < 1$, i.e. as $\hbar\omega \to \infty$ at fixed Q^2. Hence $\sum_i e_i^2$ becomes infinite in this limit and the number of quarks contributing to the scattering is unbounded. We interpret this result on the basis of QCD theory by assuming that each valence quark in the target nucleon is surrounded by its gluon cloud. Each gluon can be resolved into a virtual $q\bar{q}$ pair and these pairs form a sea which is a singlet under SU(3). Thus large values of x (small $\hbar\omega$ at fixed Q^2) correspond to scattering from valence quarks, while small values of x (large $\hbar\omega$ at fixed Q^2) correspond to scattering from 'sea-quarks'. Since the valence quark contribution to the sum $\Sigma_i e_i^2$ takes the value 1(2/3) for the proton (neutron), and the sea, being isotopically neutral, makes the same contribution to the integral in (9.101) for protons and neutrons alike, the difference between these integrals is predicted to have the value 1/3. This result is known as Gottfried's sum rule (Gottfried 1967). The measured difference has the value 0.240 ± 0.016, indicating, perhaps, that the sea is flavour asymmetric, with an excess of $d\bar{d}$ sea-quarks in the proton (Amaudruz *et al* 1991, Foudas *et al* 1990). A possible mechanism stems from the suppression of $u\bar{u}$ pairs in the sea due to the additional valence u-quark in the proton (Kumano and Londergan 1991).

An important property of QCD is that it is only asymptotically free, which means that the parton model is valid only for values of $Q^2 \gg \Lambda^2$, where Λ is a QCD scale parameter to be determined by experiment. For $Q^2 \gg \Lambda^2$ the strong coupling constant $\alpha_s(Q^2)$, analogous to the fine structure constant $\alpha = e^2/\hbar c$ which plays a similar role in QED, is given by

$$\alpha_s(Q^2) = \frac{12\pi}{(33 - 2N_f)\ln(Q^2/\Lambda^2)} \qquad Q^2 \gg \Lambda^2 \tag{9.102}$$

where $N_f(=6)$ is the number of quark flavours. Experimentally $\Lambda \simeq 0.2$ (GeV/c) so that Λ^{-1} is a length scale close to the size of the nucleon. Thus the QCD forces described by the running coupling constant $\alpha_s(Q^2)$ look like strong forces when the question of quark confinement inside an object the size of a nucleon comes up for consideration, but are comparable with electromagnetic forces at values of $(\hbar Q)^2 \gg 0.04$ (GeV/c)^2 applicable to deep inelastic scattering. This explains why perturbative QCD and the notion of scaling work so well.

However, because $\alpha_s(Q^2)$ does depend on Q^2, even though the dependence is only logarithmic, scaling must break down at some point, and indeed experiments carried out with muons at fixed x over the range $(5 < (\hbar Q)^2 < 200)$ (GeV/c)^2 confirm that scaling is indeed violated. The experimental results agree with the QCD prediction that, for increasing Q^2, and all quark flavours, $f_i(x)$ is enhanced at low x representing an increased contribution from the sea-quarks. Thus, whereas a static nucleon looks to be made from valence quarks where neutrons are easily distinguished from protons, at the highest value of Q^2 the sea-quarks dominate and neutrons and protons are indistinguishable. The main difficulty is to bridge the gap between nucleon models based on confined valence quarks valid in the static limit $Q \approx \Lambda$ where perturbative QCD does not apply, and the quasi-free (although still confined) valence and sea-quarks of the parton model which may be described by perturbative QCD at high values of Q^2.

9.2.4.4 *Anomalous effects*

There are two anomalous effects which have been observed in deep inelastic scattering which have aroused wide interest and not a little controversy. The first of these is the so-called EMC effect, which is the term used to describe the discovery by the European Muon Collaboration that the structure function $F_2(x)$, as measured in ^2H differs significantly from the same structure function measured in ^{56}Fe (Aubert *et al* 1983, 1985, 1986). The effect was subsequently confirmed in electron scattering at much lower energies (Bodek *et al* 1983, Arnold *et al* 1984); it has also been seen in other nuclei, e.g. Cu, C, Xe and Ca (Arneodo *et al* 1988, 1990). The essential observation is that $F_2(x)$ is enhanced for $x<0.1$, and depleted for $x>0.1$, for nucleons bound in nuclei as compared with free nucleons. The effect has also been detected in the comparison of results for hydrogen and deuterium, and has been used to determine the structure function $F_2(x)$ for the neutron (Aubert *et al* 1987, Benvenuti *et al* 1990, Allasia *et al* 1990).

There is no universally accepted explanation for the EMC effect, which has been linked to the presence of π-mesons and Δ-isobars in nuclei, and to suggestions that quark confinement in less complete in nuclei as compared with nucleons (Szwed 1983, Close *et al* 1983). More generally the effect has been related to a description of the gluon cloud surrounding a quark, in terms of a colour dielectric constant which is non-zero outside a bound nucleon, allowing gluons and quarks to leak into the nucleus (Nachtmann and Perner 1984).

The second area where surprising results have been observed concerns the scattering of polarized charged leptons on polarized proton targets (Bourrely *et al* 1980). If $f_j^{\uparrow}(x)$ ($f_j^{\downarrow}(x)$) is the number density with respect to x for quarks of flavour type j with helicity parallel (anti-parallel) to the helicity of the target nucleon, and $\bar{f}_j^{\uparrow}(x)$ ($\bar{f}_j^{\downarrow}(x)$) is similarly defined for anti-quarks, then polarization effects are contained within the structure function

$$g_1(x) = A(x)F_1(x) \tag{9.103}$$

where $A(x)$ is the polarization asymmetry function defined by

$$A(x) = \sum_j e_j^2\{(f_j^{\uparrow}(x) - f_j^{\downarrow}(x)) + (\bar{f}_j^{\uparrow}(x) - \bar{f}_j^{\downarrow}(x))\} \Bigg/ \sum_j e_j^2\{f_j^{\uparrow}(x) + f_j^{\downarrow}(x) + \bar{f}_j^{\uparrow}(x) + \bar{f}_j^{\downarrow}(x)\} \tag{9.104}$$

The prediction of perturbative QCD is that, for $x \geqslant 0.2$, when scattering from valence quarks dominates, $0 \ll A(x) < 1$ for protons and $A(x) \simeq 0$ for neutrons, while

$$\lim_{x\to 0} A(x) = 0$$

for protons and neutrons (Kuti and Weisskopf 1971, Close 1974). The predictions are well satisfied for protons for $x \geqslant 0.2$, but the measured value of $A(x)$ falls off more rapidly than predicted for $x < 0.2$ (Baum *et al* 1983, Hughes and Kuti, 1983, Ashman *et al* 1988). Detailed comparison between experiment and theory then implies a vanishing result for the integrated quark polarization

$$\sum_j \int \mathrm{d}x\{f_j^{\uparrow}(x) - f_j^{\downarrow}(x) + \bar{f}_j^{\uparrow}(x) - \bar{f}_j^{\downarrow}(x)\}\ \mathrm{d}x \simeq 0 \qquad (9.105)$$

The somewhat surprising conclusion (9.105) has been taken by many to imply that the spin of the proton does not reside on the quarks but on something else, e.g. orbital angular momentum or polarized gluons (Sehgal 1974, Altarelli and Ross 1988, Carlitz *et al* 1988, Brodsky *et al* 1988, Leader and Anselmino 1988). It has, however, been emphasized that the theoretical input leading to this conclusion leans very heavily on the value assigned to two experimental numbers (Close 1990). These are (i) the g_A/g_V value for the β-decay of free neutrons (cf. sections 9.3.7 and 9.3.8) and (ii) the ratio F/D of the anti-symmetric and symmetric weak form factors in the lowest baryon SU(3) octet (Bourquin *et al* 1983, Roos 1990) (cf. section 9.3.6.1.) Nevertheless recent experimental evidence does lend support to the conclusion that, by some criteria, the valence quarks contribute very little to the proton spin, which resides primarily on the strange quark content of the sea (Ashman *et al* 1989). The neutron spin, on the other hand, is carried mainly by the valence quarks. However, the precise meaning to be attached to such statements remains unclear (Bass and Thomas 1993).

9.2.5 Electromagnetic moments of the neutron

9.2.5.1 Electric charge

An electromagnetic multipole operator T_l^m is an irreducible spherical tensor operator whose matrix elements $\langle IM|T_l^m|IM\rangle$ in a given angular momentum state $|IM\rangle$ vanish unless $0 \leqslant l \leqslant 2I$. Therefore the neutron, which has spin $I = \frac{1}{2}$, can only have monopole ($l = 0$) or dipole ($l = 1$) moments if angular momentum is to be conserved. Since electric (magnetic) multipole operators carry parity $(-1)^L((-1)^{L+1})$ it follows that, if the neutron considered as a composite system is a parity eigenstate, it can have an electric monopole moment (i.e. electric charge), or a magnetic dipole moment, but magnetic monopole and electric dipole moments violate parity conservation.

That the neutron is a neutral particle having zero electric charge has been checked many times in a series of experiments going right back to its first discovery (Dee 1932). Figure 9.6 illustrates the most recent measurement of this type, where the deflection of a focused beam of 200 m s^{-1} neutrons was sought in an electric field ± 6 kV mm^{-1} applied over a distance of 9 m (Gähler *et al* 1982, Baumann *et al* 1988). The most recent result

$$q_n = (-0.4 \pm 1.1) \times 10^{-21}\, e \qquad (9.106)$$

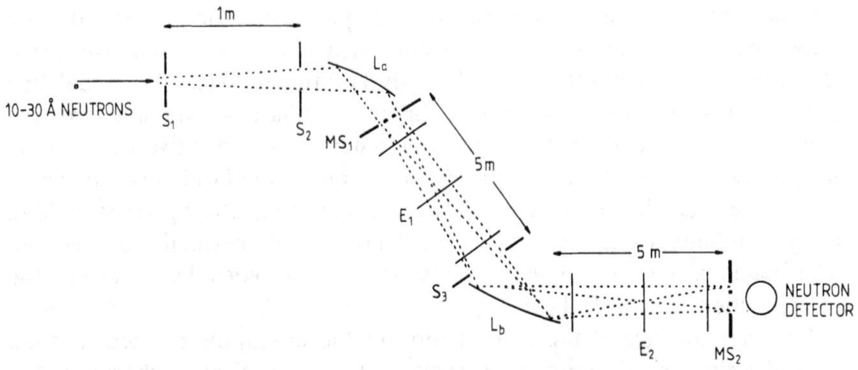

Figure 9.6. Search for a neutron electric charge by measuring the lateral deflection of a 200 m s^{-1} neutron beam over a length of 9 m in a transverse electric field of 6 kV mm^{-1}. L_a and L_b are 2.5 m focal length concave mirrors while MS_1 and MS_2 are multislits each containing 31 slits 30 μm wide (Baumann *et al* 1988).

was zero within experimental error, and showed an improvement in sensitivity by more than two orders of magnitude over earlier methods (Shapiro and Estulin 1957, Shull *et al* 1967). Since, in order to conserve electric charge in neutron β-decay, a non-vanishing neutron charge would imply that electron and proton charges were not equal and opposite, all atoms would accordingly carry some residual charge. It has been shown that the resultant electrostatic repulsion between atoms would be sufficient to account for the observed expansion of the universe provided $q_n \simeq 2 \times 10^{-18}|q_e|$ (Lyttleton and Bondi 1959, 1960, Hoyle 1960). According to the value of q_n given in (9.106) this conclusion is now ruled out.

A further point of interest is that a study of the β-decay

$$^{87}_{37}\mathrm{Rb} \rightarrow {}^{87}_{38}\mathrm{Sr} + e^- + \tilde{\nu}_e \qquad (\tau \approx 5 \times 10^{10} \text{ years}) \tag{9.107}$$

has established that the charge non-conserving decay $n \rightarrow p +$ neutrals has a lifetime $> 1.8 \times 10^{18}$ years (Norman and Seamster 1979).

9.2.5.2 Magnetic monopole moment

Although there are no (parity-violating) magnetic monopole charge and current density terms in Maxwell's equations of the classical electromagnetic field, no difficulty arises in principle by introducing them. Problems arise, however, in the Hamiltonian formulation of quantum mechanics based on (2.22) since it is no longer possible to apply (2.8) to express the electromagnetic field vectors in terms of potential functions ($\boldsymbol{A}$, φ) which are defined at all points in space. The difficulty is that the existence of a single-valued vector potential $\boldsymbol{A}$ implies the vanishing of the magnetic flux through any closed surface which clearly is not possible for any surface enclosing a magnetic monopole.

In the monopole theory of Dirac (1931) the magnetic potential may be defined for the field of a monopole except at points on a nodal line radiating outward from the monopole in a direction which may be chosen arbitrarily. By imposing the condition that the only additional effect of a nodal line crossing a small closed curve is to alter the phase accumulated by the wavefunction around this loop by $2\pi n$, where n is a positive or negative integer, Dirac showed that the monopole moment could only take the value $ne/2\alpha$. The nodal line could be viewed as a string of magnetic dipoles stretching away to infinity, or equivalently as a solenoid of infinitesimal cross-section. The magnetic field singularity on the string is avoided by requiring the wavefunction to vanish there.

One consequence of the quantization of the monopole moment is that, even if only a single monopole exists, it forces all electric charge in the universe to be quantized in units of e. Although such quantization is an experimental fact, it is, nevertheless, not a requirement of the standard model of elementary particles (cf. section 10.2.1). Neither the mass m_0, nor the spin S_0 are determined in the Dirac theory. However, when $S_0 = \frac{1}{2}$, the monopole must also acquire an electric dipole moment $(ne/2\alpha)(\hbar/2m_0c)$ in accordance with (9.25).

In a later paper Dirac (1948) generalized the discussion to include the interaction of a quantized magnetic monopole field in interaction with an electromagnetic field and, in so doing, revealed a number of severe difficulties which are still subject to investigation. In particular it emerged that magnetic monopoles could not be accommodated within the current theory of electromagnetic interactions with a single gauge field (Hagen 1965, Zwanziger 1965).

Although many failed attempts to detect magnetic monopoles have been reported (Bradner and Isbell 1959, Fidecaro *et al* 1961, Petukhov and Yakimenko 1963, Goto *et al* 1963), the interest in magnetic monopoles has revived in recent years because they arise naturally in spontaneously broken gauge theories, and are an intrinsic feature of grand unified theories where they are predicted to have masses near 10^{16} GeV, i.e. about the mass of a lead pellet 60 μm in diameter (t'Hooft 1974). There has, indeed, been one report of a monopole crossing the 20 cm^2 area of a superconducting loop (Cabrera 1982). However, since one event was detected in 151 days, this would imply a flux $\simeq 2 \times 10^{-10}\ cm^{-2}\ s^{-1}\ sr^{-1}$, i.e. approximately 4×10^6 times greater than the Parker limit set by the $\simeq 1\ \mu$G galactic magnetic field (Parker 1970). However, more detailed analysis indicated that the original identification was mistaken, and the final result sets a limit on the magnetic monopole flux less than $3.7 \times 10^{-11}\ cm^{-2}\ s^{-1}\ sr^{-1}$ (Cabrera *et al* 1983).

A single experiment has been carried out to date to search for a neutron magnetic monopole moment, in which a deflection was sought in a neutron beam propagating in a perfect silicon crystal in symmetric Laue geometry, transverse to an applied magnetic field, with a second silicon crystal acting as a sensitive deflection detector (Finkelstein *et al* 1986). The experiment exploited the fact that, according to dynamical diffraction theory, a neutron in a crystal has an effective mass $m_n^* \simeq \pm m_n \times 10^{-5}$, the sign depending on which of the two wavefields is excited in the crystal (cf. section 7.2.4.2). A small effective mass then results in an amplification in the deflection. The null result obtained

$$q_{mn} = [0.85 \pm 2.2] \times 10^{-20} \, (e/\alpha) \tag{9.108}$$

shows an improvement in sensitivity by six orders of magnitude over the previous best estimate (Ramsey 1982).

9.2.5.3 Magnetic dipole moment

The magnetic dipole moment is the only multipole moment of the neutron which is known to be definitely non-zero. The most recent measurement, carried out using the Ramsey technique of separated magnetic resonance coils (cf. section 7.5.3) has given the result (Greene *et al* 1979)

$$\mu_n = -(1.913\,043\,08 \pm 0.000\,000\,58)\mu_N \tag{9.109}$$

which is ≈ 100 times more accurate than the previous best measurement (Corngold *et al* 1956), and places this important quantity among the most accurately determined numbers in nuclear physics.

The main problem to be faced in making an accurate measurement using the technique of separated resonance coils, is how to carry out a precise averaging of the magnetic field in the quasi-uniform field between the coils. In the method under discussion here, shown in figure 9.7, the field is averaged

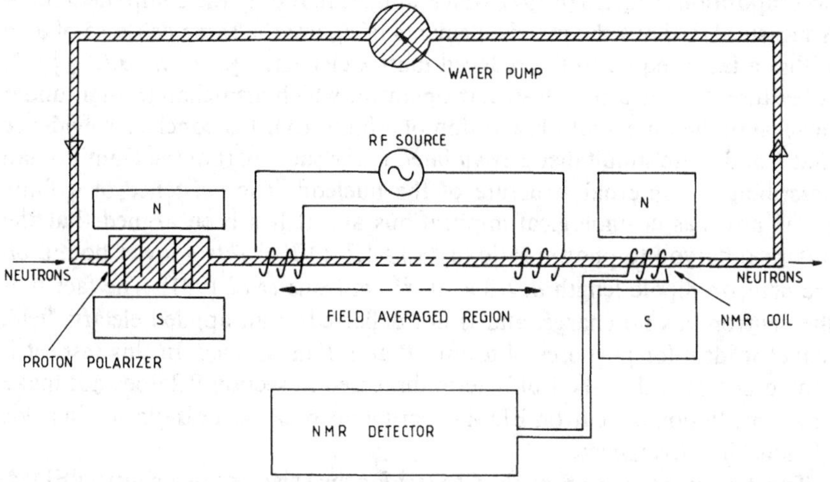

Figure 9.7. Measuring the neutron magnetic moment using the Ramsey separated oscillatory fields magnetic resonance technique (Greene *et al* 1979). The average magnetic field in the free precession region is determined by observing proton resonance in flowing water which samples approximately the same region.

by observing proton resonance in flowing water, which is polarized to near saturation in a 2 kG magnetic field, and which samples almost the same magnetic field as that traversed by the neutrons. Hence the experiment measures, in effect, the ratio of neutron to proton magnetic moment which is given by

$$\mu_n/\mu_p = -0.684\,979\,35 \pm 0.000\,000\,17 \tag{9.110}$$

An early success of the SU(6) spin-flavour symmetry scheme (cf. section 10.3.1) was its prediction that $\mu_n/\mu_p = -2/3$, and the result obtained for μ_n in the improved simple harmonic oscillator model of the nucleon is correct to four significant figures (cf. table 10.1). The MIT bag model (cf. section 10.3.3) preserves the SU(6) μ_n/μ_p ratio and also correctly predicts the magnetic moment of the Λ^0 (de Rujula *et al* 1975). The model does not do so well for the other strange baryons and greater success is obtained with the chiral bag model or 'little bag' (cf. section 10.3.4) where pions play the role of Goldstone bosons. In this model, where the pions exist outside the bag rather as in the early Yukawa model (Brown and Rho 1979), μ_n is predicted to have the value $-1.89\,\mu_N$ which is accurate to 1.6% (Brown and Myhrer 1983).

9.2.5.4 Electric dipole moment

The relation of a finite nucleon electric dipole moment (EDM) to *P* and *T* violation has been noted in section 9.2.1, and in this context it is important to remember that $\boldsymbol{d}_N$ is a vector operator, and is therefore subject to the decomposition theorem (8.197), which implies that only the component $\boldsymbol{d}_N \cdot \boldsymbol{\sigma}$ is measurable. Thus the matrix elements of $\boldsymbol{d}_N$ are the same as those of $\boldsymbol{\sigma}$ to within a factor equal to the reduced matrix element $\langle \frac{1}{2}, \pm\frac{1}{2} | \boldsymbol{d}_N \cdot \boldsymbol{\sigma} | \frac{1}{2}, \pm\frac{1}{2} \rangle$. It is because $\boldsymbol{d}_N \cdot \boldsymbol{\sigma}$ is a pseudo-scalar operator, which also changes sign under reversal of the time, that observation of a finite EDM is conclusive evidence that *P* and *T* are simultaneously violated in the same part of the Hamiltonian describing the internal structure of the nucleon. The existence of a finite EDM also has cosmological implications and it has been argued that the observed baryon to photon ratio $(n_B/n_\gamma) \geqslant 1.3 \times 10^{-10}$ sets a lower bound on the neutron dipole length $d_n \geqslant 3 \times 10^{-28}$ cm (Ellis *et al* 1981). The fact that the neutron has no charge, and is not deflected in an applied electric field, makes it ideal for purposes of testing *P* and *T* invariance. In this respect it should be noted that the Foldy term discussed in section 9.2 does not make any contribution in neutron EDM experiments because it disappears in fields of vanishing divergence.

Two techniques have been used to test for the existence of a neutron EDM. The first method seeks to establish a shift in the position of a magnetic resonance line caused by the presence of a strong electric field (Purcell and Ramsey 1950, Smith *et al* 1957, Miller *et al* 1967, Dress *et al* 1968, 1977) while the second method looks for a contribution of the electric interaction to the scattering of neutrons in the Coulomb fields of atoms (Shull and Nathans 1967). The availability in recent years of ultra-cold neutrons has led to an improvement in sensitivity by two orders of magnitude for NMR methods, and two experiments of this general class have been in progress for more than a decade. The first of these is at the Leningrad Nuclear Physics Institute (Altarev *et al* 1986, 1992) and the second at the ILL Grenoble (Pendlebury *et al* 1984, Smith *et al* 1990). Although differing in detail the principle of operation is the same in each case, in that both make use of ultra-cold neutrons (cf. section 1.6.6) in combination with the Ramsey separated oscillatory coils technique (cf. section 7.5.3).

The layout of the ILL experiment is shown in figure 9.8. Ultra-cold neutrons are polarized by passage through an Fe–Co film polarized in a 1 kG field, where the isotopic mixture is such that one polarization state is transmitted and one totally reflected (cf. section 7.4.3). The neutron spins remain locked adiabatically to the local magnetic field as it reduces from ≈ 0.2 G outside, to ≈ 0.01 G inside a five-layer mu-metal magnetic shield. The neutrons enter the storage volume through a neutron valve, and are stored for periods of time of order 100 s. Inside the storage volume there are uniform magnetic and electric fields $\boldsymbol{B}$, $\boldsymbol{E}$ whose relative orientation may be changed from parallel to anti-parallel by reversing the direction of $\boldsymbol{E}$. Pulses of RF field,

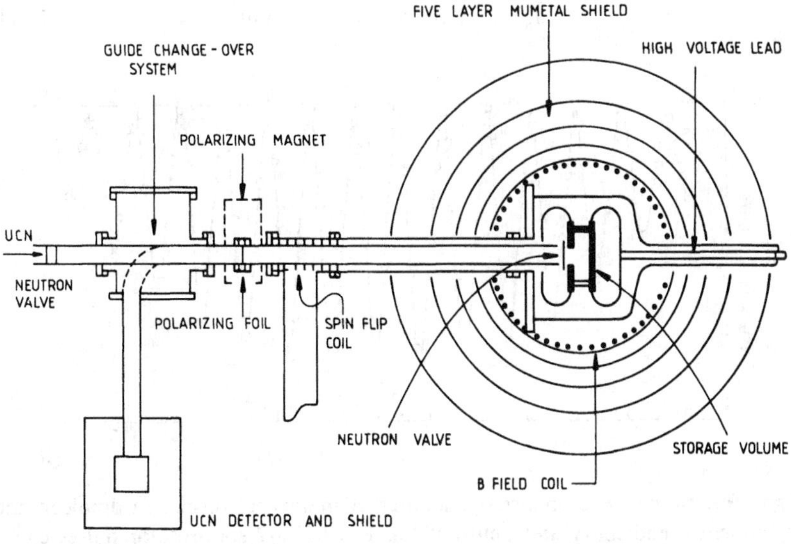

Figure 9.8. Search for a neutron electric dipole moment using polarized ultra-cold neutrons stored in a beryllium oxide bottle for periods of order 10^2 s. The experiment consists in looking for a shift in the resonance signal when the sense of the electric field is reversed with respect to the parallel magnetic field (Smith *et al* 1990).

applied ≈ 4 s after entry, and ≈ 4 s before exit from the storage volume, act as the required $\pi/2$ spin rotators, so this is a variant of the Ramsey technique in which the spin rotations are separated in time rather than in space. After exiting from the storage volume the neutron polarization is analysed by the same magnetized foil which acted as polarizer. However, instead of returning to the source the neutrons are deflected under gravity into a ^{3}He counter.

The resonance curve observed for a 10 s storage time in the 10^{-2} G field is shown in figure 9.9, and the search for a neutron EDM is carried out by seeing whether the central resonance peak shifts when the relative orientation of $\boldsymbol{E}$ and $\boldsymbol{B}$ is reversed from parallel to anti-parallel. This corresponds to the fact that the spin Hamiltonian (7.209) is changed to

$$H_s = -\boldsymbol{\mu}_n \cdot \boldsymbol{B} \mp e\boldsymbol{d}_n \cdot \boldsymbol{E} \tag{9.111}$$

depending on whether $\boldsymbol{E} \cdot \boldsymbol{B} = \pm |EB|$. One of the main difficulties with EDM experiments is the so-called '$\boldsymbol{E} \times \boldsymbol{V}$ effect' whereby a neutron moving with velocity $\boldsymbol{V}$ in an electric field $\boldsymbol{E}$, sees a magnetic field $(\boldsymbol{E} \times \boldsymbol{V})/2c$ in its own rest-frame (cf. section 9.2.1.4). Since the term $\boldsymbol{\mu}_n \cdot [\boldsymbol{E} \times \boldsymbol{V}]/2c$ changes sign when $\boldsymbol{E}$ changes sign, the term can mimic a genuine EDM. In a storage experiment the term averages to zero since positive and negative values of $\boldsymbol{V}$ are equally likely. In practice a more serious source of spurious effects is the stray magnetic field associated with leakage currents, which changes sign when the electric field is reversed.

To date the best limit obtained in the ILL experiment is (Smith *et al* 1990)

$$d_n = (-0.3 \pm 0.5) \times 10^{-25} \text{ cm} \tag{9.112}$$

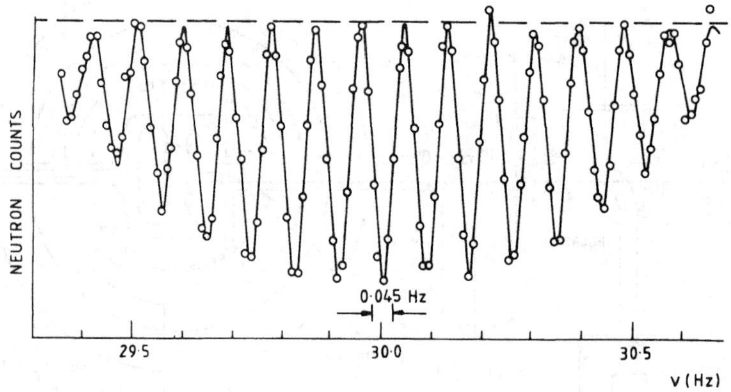

Figure 9.9. Magnetic resonance signal observed in the neutron electric dipole moment experiment (Pendlebury and Smith 1980). The RF $\pi/2$ spin-rotator pulses each of length τ are separated in time by a period $T \gg \tau$ of free precession in the system of parallel (anti-parallel) magnetic and electric fields. The peak width varies as T^{-1} while the width of the envelope is proportional to τ^{-1}. In these features the magnetic resonance signal closely resembles the double-slit optical interference pattern.

while the corresponding result for the LNPI experiment is (Altarev *et al* 1992)

$$d_n = (0.26 \pm 0.42 \pm 0.16) \times 10^{-25} \text{ cm} \tag{9.113}$$

where the first quoted error is statistical and the second systematic.

9.2.6 Electric and magnetic polarizabilities

The scattering of photons by a system of electric charges and magnetic moments is characterized by a scattering tensor, which in turn is related to the polarizability tensor of the system. In the case of the nucleon this is represented by dynamic electric polarizabilities $\bar{\alpha}_p$, $\bar{\alpha}_n$, and dynamic magnetic polarizabilities $\bar{\beta}_p$, $\bar{\beta}_n$. These quantities may be measured directly from Compton scattering experiments for low energy ($\approx$ 50–100 MeV) photons on nucleons (Baranov *et al* 1974), or indirectly through the application of dispersion relations (cf. section 2.4.4) to the total measured photoabsorption cross-section (Schröder 1980). The static polarizabilities α, β, on the other hand, are defined by the total potential energy $V(\boldsymbol{r})$ of the system in an applied electromagnetic field ($\boldsymbol{E}$, $\boldsymbol{B}$), i.e.

$$V(\boldsymbol{r}) = -\tfrac{1}{2}\alpha \boldsymbol{E}^2 - \tfrac{1}{2}\beta \boldsymbol{B}^2 \tag{9.114}$$

$V(\boldsymbol{r})$ does not, of course, include the contributions from intrinsic magnetic or electric dipole moments.

In the case of nucleons $\bar{\alpha}$ and α are related by (Shekter 1968)

$$\bar{\alpha}_N = \alpha_N + \left(\frac{e^2}{\hbar c}\right)\left\{\frac{\hbar}{3mc}\langle r_c^2\rangle + \frac{1}{4}\left(\frac{\hbar}{2mc}\right)^3 \kappa_N\right\} \tag{9.115}$$

where κ_N is the anomalous nucleon magnetic moment expressed in nuclear magnetons. The values of $\bar{\alpha}_N$ and $\bar{\beta}_N$ are then connected according to the sum rule (Schröder 1976, 1980)

$$\bar{\alpha}_N + \bar{\beta}_N = 2c \int_{\omega_0}^{\infty} \frac{d\omega}{\omega^2} \sigma_T^{\omega N}(\omega) \tag{9.116}$$

where $\sigma_T^{\omega N}(\omega)$ is the total photoabsorption cross-section and $\hbar\omega_0$ is the threshold energy.

Some notion of how α_p and α_n characterize the intrinsic nucleon structure may be gained from elementary optical dispersion theory. The static atomic polarizability due to a single dispersing electron may be expressed as a product of three factors: the fine structure constant, an atomic volume represented by the cube of the electron reduced Compton wavelength, and a factor representing the square of the ratio of electron rest mass to a characteristic atomic energy. For atoms this factor provides significant amplification; the ratio is typically about 5×10^4. But in considering the nucleon in such terms one replaces the electron by the pion; the characteristic dimension is the pion reduced Compton wavelength and, most importantly, the ratio of the pion rest mass to the excitation energy of the nearest nucleonic resonance is a mere 0.25. Thus it is that the small value of α_n mirrors the strong binding of the nucleon compared to the weak binding of the atom.

In the case of the proton the dynamic polarizabilities $\bar{\alpha}_p$, $\bar{\beta}_p$ have been determined from Compton scattering experiments on protons (Baranov *et al* 1974, L'vov 1981). After allowing for Thomson and Rayleigh scattering corrections the results obtained

$$\bar{\alpha}_p=(10.7\pm1.1)\times10^{-4}\ \mathrm{fm}^3 \qquad \bar{\beta}_p=(-0.7\pm1.6)\times10^{-4}\ \mathrm{fm}^3 \qquad (9.117)$$

are in reasonable accord (at least as far as $\bar{\alpha}_p$ is concerned) with the value derived by application of dispersion relations to the forward Compton amplitude (Schröder 1980, Bernard *et al* 1992)

$$\begin{aligned} \bar{\alpha}_p&=11.5\times10^{-4}\ \mathrm{fm}^3 \qquad & \bar{\beta}_p&=3.2\times10^{-4}\ \mathrm{fm}^3 \\ \bar{\alpha}_n&=8.5\times10^{-4}\ \mathrm{fm}^3 \qquad & \bar{\beta}_n&=4.8\times10^{-4}\ \mathrm{fm}^3 \end{aligned} \qquad (9.118)$$

The corresponding values for the static electric polarizabilities are

$$\alpha_p=7.6\times10^{-4}\ \mathrm{fm}^3 \qquad \alpha_n=8.5\times10^{-4}\ \mathrm{fm}^3 \qquad (9.119)$$

The value $\alpha_p=3.1\times10^{-4}\ \mathrm{fm}^3$, determined from a potential model of valence quarks (Schöberl and Leeb 1986) compares unfavourably with the experimental results given in (9.117)–(9.118), which indeed are much closer to the results $\alpha_p=\alpha_n=(8.6\pm2.6)\times10^{-4}\ \mathrm{fm}^3$ calculated on the basis of the chiral bag model. The implication would appear to be that the most significant contribution to the electric polarizability comes from the pion cloud (Weiner and Weise 1985, Schröder 1972). On the other hand most of the magnetic polarizability comes from the quark spin–flip transition to the Δ-isobar giving $\beta_p\leqslant2\times10^{-3}\ \mathrm{fm}^3$ with a value of β_n larger by about 10%.

Up to very recently a single measured value $\alpha_n=(13\pm8)\times10^{-3}\ \mathrm{fm}^3$ existed for the static electric polarizability of the neutron (Aleksandrov 1983) which was not consistent with the estimate $\alpha_n=(3\pm4)\times10^{-3}\ \mathrm{fm}^3$ obtained from neutron-electron scattering (cf. section 9.2.3.2) or with the estimate $\alpha_n\leqslant6\times10^{-3}\ \mathrm{fm}^3$ obtained from the backward–forward asymmetry observed in the scattering of fast neutrons on heavy nulei (Samosvat 1984). The theoretical estimates suggest a value $\alpha_n\approx10\times10^{-4}\ \mathrm{fm}^3$ comparable with the value for the proton. However a new value of α_n has now been derived from a measurement at the 136 MeV electron linac neutron source HELIOS at AERE Harwell (Schmiedmayer *et al* 1988). The experimental arrangement is shown in figure 9.10. Short pulses of neutrons of energy ranging between 50 eV and 50 keV were projected at a natural lead target located 50 m from the source; the unscattered neutrons were recorded in a ^{10}B detector located a further 100 m downstream. A range of targets was studied with transmission varying from 2% to 50%, and the scattering cross-section determined by time-of-flight analysis at about 50 neutron energies, each to a precision of 0.2%.

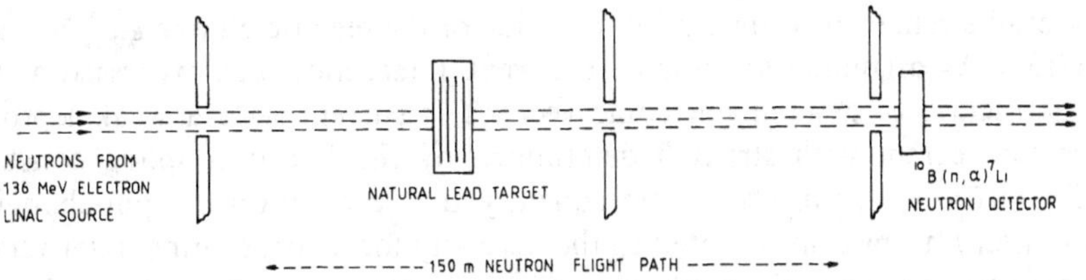

Figure 9.10. Experiment for measuring the electric polarizability of the neutron by detecting an interference term proportional to wave number between nuclear and electromagnetic scattering amplitudes from a high Z target (Schmiedmayer *et al* 1988). A null effect was observed in low Z targets.

The specific objective of the experiment was to demonstrate and measure an interference term between the nuclear scattering amplitude, proportional to the nuclear radius $R \propto Z^{1/3}$, and the electromagnetic amplitude $\propto Z^2$, which would vary linearly with the wave number $k \propto E_n^{1/2}$ of the incoming neutron. Such characteristic behaviour provides a unique signature for a finite α_n since it points to a long-range electromagnetic component of the scattering potential which cannot be associated with the short-range nuclear force (cf. section 2.3.2). The determination of α_n from the k-dependence of the cross-section is a difficult and complex procedure, since allowance must be made for potential and resonance contributions to the nuclear amplitude, and for neutron–atomic electron interactions and relativistic effects in the electromagnetic scattering amplitude. To confirm that the observed effect was genuine the experiment was repeated with carbon nuclei ($Z = 6$) for which the measured coefficient of k was zero within the errors.

The complete analysis yielded the result

$$\alpha_n = (12 \pm 10) \times 10^{-4} \text{ fm}^3 \tag{9.120}$$

a result which, in spite of its large error, compares well with the chiral bag model prediction quoted above. The experiment has recently been repeated to a much higher precision, arriving at the value (Schmiedmayer *et al* 1991)

$$\alpha_n = (12.0 \pm 1.5 \pm 2.0) \times 10^{-4} \text{ fm}^3 \tag{9.121}$$

9.3 Weak interactions

9.3.1 The current–current weak interaction

The theory of the weak interaction responsible for nuclear β-decay was developed by Fermi (Fermi 1934) on the basis of Pauli's neutrino hypothesis (Pauli 1930, 1934) and by analogy with quantum electrodynamics (QED), where the interaction is described by the coupling of a vector current of charged particles to the vector potential of the electromagnetic field, with overall strength determined by the value of the electric charge $e(e^2/\hbar c = \alpha \approx 1/137)$. As originally formulated the weak interaction was expressed as the contraction of a weak nucleonic vector (V) current with a weak leptonic vector current with strength determined by the Fermi coupling constant $G = 1.0270 \times 10^{-5}\, m_p c^2)^{-2}$. Subsequently the recognition of spin-changing allowed β-transitions pointed to the necessity for incorporating axial vector (A) contributions to the weak interaction (Gamow and Teller 1936), although the presence of other Lorentz invariant interactions (Bethe and Bacher 1936, Pauli 1936, 1941) was subsequently ruled out by experiment.

Following the discovery of parity violation in weak processes (Lee and Yang 1956, Wu *et al* 1957, Garwin *et al* 1957, Frauenfelder *et al* 1957) the whole subject of the weak interaction was thrown into a state of deep confusion. The difficulty was that, at least in the case of weak interactions involving leptons, the experimental data could equally well be described in terms of an S, T combination of interactions and a right-handed neutrino, as with a V, A combination and a left-handed neutrino. In addition there was a considerable amount of contradictory experimental evidence, e.g. from electron–neutrino angular correlation studies in the superallowed β-transitions of ^{6}He and ^{35}A. Fortunately the direct demonstration that the neutrino was a left-handed particle effectively settled the matter (Goldhaber *et al* 1958). At about the same time the theory was subjected to a major revision culminating in the successful (V–A) current–current interaction (Feynman and Gell-Mann 1958, Sudarshan and Marshak 1958, Sakurai 1958). In this formulation the weak interaction is represented by the Lagrangian density

$$L_w(x) = \frac{G_F}{\sqrt{2}} J_\mu(x) J_\mu^\dagger(x) \tag{9.122}$$

where the self-interacting weak current $J_\mu(x)$ contains both leptonic and hadronic contributions and has a space-time structure which is a left-handed admixture of polar vector and axial vector components. The weak current $J_\mu(x)$ is described as 'charged' in the sense that it simultaneously emits and absorbs particles which differ in electric charge by one unit. In the same sense the electromagnetic vector current $J_\mu^{EM}(x)$ is a 'neutral' current. The result (9.122) extends to all fermions the notion of a universal weak interaction first applied to the 'Puppi triangle' of $(\bar{\nu}_e e^-)$, $(\bar{\nu}_\mu \mu^-)$ and $(\bar{p}n)$ (Puppi 1948, Klein 1948, Tiomno and Wheeler 1949).

Weak decays of leptons and hadrons, of which the prototypes are muon decay (purely leptonic) and neutron decay (semi-leptonic)

$$\mu^- \rightarrow \nu_\mu + e^- + \bar{\nu}_e \tag{9.123a}$$

$$n \rightarrow p + e^- + \bar{\nu}_e \tag{9.123b}$$

arise as first-order terms in the perturbation series based on $L_w(x)$, which can account for muon decay, pion decay and most baryon decays even those not involving lepton creation. The interaction does not conserve C or P and involves only left-handed fermions and right-handed anti-fermions. It does conserve CP, at least to a high degree of approximation, and the small ($\simeq 0.2\%$) CP-violating effects observed in neutral kaon decay (Christenson *et al* 1964) are of uncertain origin (cf. section 9.3.6.3). The weak interaction does not conserve isospin or hypercharge quantum numbers which are associated with global symmetries of the strong interactions.

The additional element required to unify the description of semi-leptonic, leptonic and non-leptonic hadron decays was the quark degree of freedom (Gell-Mann 1964, Zweig 1964). This was originally introduced in the context of the SU(3) classification of hadrons, with baryons constructed from three quark 'flavours' up (u), down (d) and strange (s), and mesons from quark-anti quark pairs (cf. section 9.1.3). So far as weak decays within the lowest baryon and meson SU(3) octets are concerned (which include the nucleon and the pion), neither the heavy charmed (c), bottom (b), top (t) quarks nor the heavy τ-lepton play any role. It is therefore sufficient to work with the weak current

$$J_\mu = \bar{e}\gamma_\mu(1-\gamma_5)\nu_e + \bar{\mu}\gamma_\mu(1-\gamma_5)\nu_\mu + (\bar{d}\cos\theta_c + \bar{s}\sin\theta_c)\gamma_\mu(1-\gamma_5)u \tag{9.124}$$

In this expression (e^-, ν_e) and (μ^-, ν_μ) are the light leptons and u, d and s represent mass eigenstates of the light quarks. The operator $(1-\gamma_5)$ projects out the left-handed component of the current as required in the (V–A) interaction; right-handed currents projected out by the operator $(1+\gamma_5)$ would imply an admixture of $(V+A)$ interaction (cf. section 10.4.1). The Cabibbo angle θ_c (Cabibbo 1963) describes a mixing of the d and s quarks in the weak interaction and provides a measure of the relative strengths of hypercharge conserving ($\cos^2\theta_c$) and hypercharge non-conserving ($\sin^2\theta_c$) weak processes (cf. section 9.3.6).

9.3.2 Transition amplitudes for neutron β-decay

The weak current can be divided into a hadronic part $J_\mu^h(x)$ and a leptonic part $J_\mu^l(x)$ and we know from the study of nuclear β-decay, and from muon decay, that the leptonic current has the form

$$J_\mu^l(x) = \bar{\psi}_l(x)\gamma_\mu(1-\gamma_5)\psi_{\nu_l}(x) \tag{9.125}$$

which is an equal mixture of polar vector (Fermi) $\bar{\psi}_e\gamma_\mu\psi_\nu$ and axial vector (Gamow–Teller) $\bar{\psi}_e\gamma_\mu\gamma_5\psi_\nu$ currents. The leptonic matrix element for neutron β-decay is

$$\langle e^-\bar{\nu}_e|J_\mu^l(0)|0\rangle = \langle\bar{u}_e|\gamma_\mu(1-\gamma_5)|u_{\nu_e}\rangle \tag{9.126}$$

Since we have no field theoretic description of strongly interacting nucleons we cannot represent the hadronic current in the simple form (9.125) although it is useful to consider the symmetry properties of the 'bare' nucleonic vector

$\bar{\psi}_p\gamma_\mu\psi_n$ and axial vector $\bar{\psi}_p\gamma_\mu\gamma_5\psi_n$ currents. To stress the connection with isospin it is also convenient to represent the bare nucleonic currents in the form $\bar{\psi}\gamma_\mu\tau^+\psi$ and $\bar{\psi}\gamma_\mu\gamma_5\tau^+\psi$ where τ_i are the usual isospin operators and ψ represents the nucleon field.

What we can do is write down the most general form of the matrix elements of the true vector $V_\mu(x)$ and axial vector $A_\mu(x)$ hadronic currents, consistent with Lorentz invariance, and we find for neutron β-decay in particular (Holstein 1974, Grenacs 1985)

$$\langle \mathrm{p}|V_\mu|\mathrm{n}\rangle = \left\langle \bar{u}_\mathrm{p}|g_\mathrm{V}\gamma_\mu - \mathrm{i}\frac{\hbar(g_\mathrm{M}-g_\mathrm{V})}{2mc}\sigma_{\mu\nu}q_\nu + \frac{\hbar g_\mathrm{S}}{2mc}q_\mu|u_\mathrm{n}\right\rangle \tag{9.127a}$$

$$\langle \mathrm{p}|A_\mu|\mathrm{n}\rangle = \left\langle \bar{u}_\mathrm{p}|g_\mathrm{A}\gamma_\mu\gamma_5 - \mathrm{i}\frac{\hbar g_\mathrm{T}}{2mc}\sigma_{\mu\nu}q_\nu\gamma_5 + \frac{\hbar g_\mathrm{P}}{2mc}q_\mu\gamma_5|u_\mathrm{n}\right\rangle \tag{9.127b}$$

where $m = \frac{1}{2}(m_\mathrm{n} + m_\mathrm{p})$ and the invariant form factors $g_i(q)$ describe vector (g_V), axial vector (g_A), induced weak magnetism (g_M), tensor (g_T), scalar (g_S) and pseudo-scalar (g_P) respectively.

Under the discrete symmetry transformations P, C and T we find that the matrix elements (9.126) and (9.127) transform according to the rules

$$P\colon g_i \to g_i,\ \gamma_5 \to -\gamma_5 \qquad C\colon g_i \to g_i^*,\ \gamma_5 \to -\gamma_5 \qquad T\colon g_i \to g_i^*,\ \gamma_5 \to \gamma_5 \tag{9.128}$$

We conclude that, because of the mixing of vector and axial terms in the matrix elements, P and C symmetries are each violated in the weak interaction. The (unitary) symmetry CP and the (anti-unitary) symmetry T are violated if any of the $g_i(q)$ are complex. Thus a non-zero value for the triple correlation coefficient D defined in (9.130) below would be evidence for time-reversal non-invariance provided final state interactions could be ignored (Callan and Treiman 1967).

Since for neutron decay $|q/mc| \ll 1$, the most significant contribution comes from the vector and axial vector form factors associated with the 'bare' nucleonic currents. Concerning these quantities the most significant information is derived from measuring the differential probability that a polarized neutron should decay with emission of leptons into specified momentum states (Jackson *et al* 1957, 1957a, Alder *et al* 1957, Linke 1969)

$$\mathrm{d}^3W(p_\mathrm{e}, p_{\bar\nu}) = \mathrm{d}W(p_\mathrm{e})\,\mathrm{d}\Omega_\mathrm{e}\,\mathrm{d}\Omega_{\bar\nu}\{1 + a\boldsymbol{p}_\mathrm{e}\cdot\boldsymbol{p}_{\bar\nu}/E_\mathrm{e}E_{\bar\nu} + \langle \boldsymbol{I}_\mathrm{n}/I_\mathrm{n}\rangle(A\boldsymbol{p}_\mathrm{e}/E_\mathrm{e} + B\boldsymbol{p}_{\bar\nu}/E_{\bar\nu} + D\boldsymbol{p}_\mathrm{e}\times\boldsymbol{p}_{\bar\nu}/E_\mathrm{e}E_{\bar\nu})\} \tag{9.129}$$

where, to lowest order in momentum transfer q,

$$\begin{aligned} a &= (1-|\lambda|^2)/(1+3|\lambda|)^2 \\ A &= -2(|\lambda|^2 + |\lambda|\cos\varphi)/(1+3|\lambda|^2) \\ B &= 2(|\lambda|^2 - |\lambda|\cos\varphi)/(1+3|\lambda|^2) \\ D &= 2(|\lambda|\sin\varphi/(1+3|\lambda|^2) \end{aligned} \tag{9.130}$$

" and, *according to (V-A) theory, the coupling constant ratio*

$$\lambda = |\lambda| e^{i\varphi} = G_A/G_V = -g_A(0)/g_V(0) \qquad (9.131)$$

Integration of (9.129) over all momenta gives information on the absolute value of $|g_V|^2+3|g_A|^2$ in the form of the *ft*-value, where $t \ln 2$ is the neutron lifetime and f is the integrated Fermi phase space factor as determined from the kinetic energy release of 782.332±0.017 keV (Greenwood and Chrien 1980) and

$$ft=(2\pi^3 \ln 2\, \hbar^7/m_e^5c^4)/G_F^2 g_V^2 \cos^2\theta_c(1+3|\lambda|^2) \qquad (9.132)$$

Allowing for outer radiative corrections $f=1.714\,65\pm0.000\,15$ (Wilkinson 1981, 1982). Both the *ft* value (9.132) and the correlation coefficients (9.130) must be subjected to model independent (outer) and model dependent (inner) radiative corrections of the order of a few per cent (Sirlin 1967, 1974, Shann 1971, Angerson 1974, Garcia and Maya 1978, Garcia 1982, 1983, Glück and Tóth 1990).

Apart from a small correction to the Gamow–Teller matrix element, and making allowance for the fact that ^{3}H has a weak decay branch into an antineutrino and a neutral helium atom, the result (9.132) also applies to the β-decay of ^{3}H (Byrne 1984, Wilkinson 1991). The equivalent decay of a neutron into an antineutrino and a neutral hydrogen atom has an estimated branching ratio of 4.2×10^{-4}% (Kabir 1967, Nemenoff 1980, Nemenoff and Ovchinnikov 1980). This decay branch has not yet been observed (Green and Thompson 1990).

9.3.3 The conserved vector current (CVC) theory

If the strong interactions could be switched off the 'bare' nucleonic electromagnetic current would have the form (cf. 9.40)

$$J_\mu^{b,EM}(x)=\bar\psi\gamma_\mu\tfrac{1}{2}(1+\tau_3)\psi \qquad (9.133)$$

which subdivides into an isoscalar term $\frac{1}{2}\bar\psi\gamma_\mu\psi$ and an isovector term $\frac{1}{2}\bar\psi\gamma_\mu\tau_3\psi$. Supplemented by additional terms describing strange particles in virtual states the isoscalar current would be conserved from baryon conservation. In the same way, to get a conserved isovector current the 'bare' current would need to be supplemented by mesonic terms, etc. On the other hand the matrix element of the true hadronic electromagnetic current between nucleon states is known (cf. 9.65 and 9.68), i.e.

$$\langle N|J_\mu^{EM}|N\rangle=\langle\bar u|(\gamma_\mu-\kappa_p\mu_N i\sigma_{\mu\nu}q_\nu)(1+\tau_3/2)-\kappa_n\mu_N i\sigma_{\mu\nu}q_\nu(1-\tau_3/2)|u\rangle \qquad (9.134)$$

where κ_p and κ_n are the anomalous magnetic moments of proton and neutron. The isovector contribution is

$$\langle N|J_{\mu,3}^{EM}|N\rangle=\langle\bar u|\gamma_\mu-(\kappa_p-\kappa_n)\mu_N i\sigma_{\mu\nu}q_\nu|u\rangle \qquad (9.135)$$

That this is indeed a conserved current, i.e. $\partial_\mu J_{\mu,3}^{EM}=0$, is shown by the fact that the coefficient of γ_μ in (9.135) is unity expressing the fact that the electric charge on the physical proton (neutron) is always unity (zero).

Experimentally $g_V \simeq 1$ in (9.127) which is evidence for the conservation of the vector current $V_\mu(x)$. However since the 'bare' electromagnetic and weak vector currents are evidently members of an isotriplet it has been proposed that this property remains true for $J_\mu^{EM}(x)$ and $V_\mu(x)$ with full allowance for the strong interactions. This is the strong conserved vector current (CVC) hypothesis (Gershstein and Zel'dovich 1957, Feynman and Gell-Mann 1958). It implies that, to the extent that isospin is a good symmetry of the strong interactions, $V_\mu(x)$ is conserved because $J_\mu^{EM}(x)$ is conserved. Further, we may invoke the Wigner–Eckart theorem (Condon and Shortley 1935) to deduce that the matrix elements of the two currents are the same to within a factor given by the relevant Clebsch–Gordon coefficient. We find then that

$$g_S = 0 \qquad g_M(0) - g_V(0) = g_V(0)(\kappa_p - \kappa_n) \tag{9.136}$$

A convincing test of the CVC hypothesis in action outside the regime of baryon decays is the observed rate of the pure Fermi $(0^- \to 0^-)$ π^+ β-decay, $\pi^+ \to \pi^\circ + e^+ + \nu_e$, for which CVC theory makes a precise prediction of $0.4027 \pm 0.0018\ \mathrm{s}^{-1}$. This compares very favourably with the measured value of $0.394 \pm 0.015\ \mathrm{s}^{-1}$ (McFarlane *et al* 1985).

9.3.4 First and second-class currents

The CVC hypothesis predicts a simple relationship between the weak vector form factors and their electromagnetic analogues. In particular it says that $g_S = 0$ which raises the question why there is no induced scalar in the electromagnetic matrix element. The reason is that C-invariance of the strong interactions rules out the corresponding current $\partial_\mu(\bar{\psi}\psi)$ which has opposite transformation properties to the bare current $\bar{\psi}\gamma_\mu\psi$. A similar analysis may be applied to the induced weak form factors with this minor change that, since the weak current is charged, an additional n$\rightleftarrows$p interchange is required. The relevant symmetry is termed G parity, where

$$G = C \exp[-i\pi T_2] \tag{9.137}$$

and T_2 generates a rotation about the 2-axis in isospin space. It turns out that the 'bare' Fermi and Gamow–Teller nucleonic currents transform under the G parity transformation according to the rules

$$V_\mu^b \to G V_\mu^b G^{-1} = V_\mu^b \qquad A_\mu^b \to G A_\mu^b G^{-1} = -A_\mu^b \tag{9.138}$$

and these currents are described as 'first class' (Weinberg 1958). Currents which transform according to the opposite rule that V_μ changes sign while A_μ remains invariant are described as 'second class'. Second-class currents are ruled out on the grounds that G-parity is conserved in the strong interactions and the experimental evidence supports this view (Wilkinson

1977, Szybisz 1978, Klippinger *et al* 1978). Excluding second-class currents we then find that, for neutron decay, and for decays within an isotopic multiplet

$$g_S = g_T = 0 \tag{9.139}$$

There are two important addenda to the above argument: (i) it fails if the 'bare' current contains second class contributions or (ii) if the lepton and antilepton currents couple to hadronic currents which, like the 'bare' hadronic currents, are members of the same isospin multiplet, then $g_i = g_i^*$ for first class and $g_i = -g_i^*$ for second-class form factors (Cabibbo 1964). It follows that observation of T-violation in neutron β-decay would imply, either the presence of second class currents, or a breakdown of charge symmetry, or both. One further important consequence of the exclusion of second-class currents is that any residual corrections to the vector matrix elements in neutron decay are of order $[(m_n - m_p)/m_p]^2$ at worst (Behrends and Sirlin 1960).

CVC theory and strong-interaction G parity conservation predict the values (9.136) and (9.139) for weak magnetism (induced pseudo-tensor) and induced tensor form factors respectively. To the question: 'is it possible to test these predictions?', the answer is that a complete separation of $(g_M - g_V)$ and g_T may be brought about by measuring the energy dependence of the correlation coefficients $a(E_e)$ and $A(E_e)$ whose zero-momentum transfer values are given in (9.130) (Holstein 1974). However, the relevant terms are only a few per cent of the leading terms in these correlations and no successful experiment has been reported.

9.3.5 The Goldberger–Treiman relation

The Goldberger–Treiman relation provides a connection between the axial vector coefficient $g_A(0)$ in neutron decay, the pion decay constant $f_\pi(=93.24 \pm 0.09$ MeV) and the pion–nucleon coupling constant $g_{\pi NN}(=13.396 \pm 0.084)$

$$g_A(0) = f_\pi g_{\pi NN}/mc^2 \tag{9.140}$$

The relation was originally obtained by dispersion techniques (Goldberger and Treiman 1958) and subsequently re-derived in various forms of partially conserved axial current (PCAC) theory (Gell-Mann and Levy 1960). It is currently viewed as a consequence of an exact chiral invariance of the strong interactions (Pagels 1975, Henley *et al* 1992). The point of departure in the derivation of the Goldberger–Treiman relation, from this viewpoint, is the recognition that the Lagrangian describing the interaction of massless u and d quarks with the QCD field is not only isospin invariant with a conserved vector current $\frac{1}{2}\psi\gamma_\mu\tau\psi$ (cf. section 9.3.2 and 9.3.3) but isoaxially invariant with a conserved axial current $\frac{1}{2}\psi\gamma_\mu\gamma_5\tau\psi$. The corresponding chiral symmetry group factorizes into the product group $SU(2)_L \times SU(2)_R$ of right-handed and left-handed SU(2) groups. If, in addition the s quark is massless the chiral group is extended to $SU(3)_L \times SU(3)_R$ (cf. section 10.2.2). The argument now assumes that the conserved V and A currents are those which participate in the weak interactions.

Although $m_u \neq m_d \neq 0$ the vacuum is isosopin invariant, i.e. $V_\mu|0\rangle \equiv 0$, the vector current V_μ is conserved to a very good approximation (cf. section 9.2.3) and isospin symmetry is directly manifested in the observed isospin multiplets. None of these statements is true of the isoaxial symmetry; the non-vanishing of the pion decay amplitude $\langle 0|A_\mu|\pi\rangle = \mathrm{i}f_\pi q_\mu/(\hbar c)^2$ shows that the vacuum is not axially invariant, the axial current A_μ is not conserved since $\langle 0|\partial_\mu A_\mu|\pi\rangle = f_\pi(m_\pi c^2)/(\hbar c)^4 \neq 0$. Finally there is no trace of axial isospin symmetry in hadronic spectra, which would require each member of an isospin multiplet to be accompanied by a degenerate partner of opposite parity.

This situation is described by the statement that, although isospin symmetry is realized in the conventional (Wigner–Weyl) mode, the axial symmetry is realized in the Nambu–Goldstone mode where the vacuum state does not satisfy the symmetry of the Lagrangian. This circumstance, which we have already encountered in connection with the ferromagnet, whose ground state does not satisfy the rotational symmetry of its Lagrangian (cf. section 8.4.3) is described as spontaneous symmetry breaking. Although $A_\mu|0\rangle \neq 0$, we may still have $\partial_\mu A_\mu = 0$ provided $m_\pi = 0$; this is an example of Goldstone's theorem that to each generator of a spontaneously broken symmetry there corresponds a massless boson. Corresponding to the three generators of axial isospin the massless Goldstone bosons have the quantum numbers of the pions, and, according to present ideas, are to be identified with the pions.

The pions are not in fact massless and the axial current cannot be exactly conserved since $\langle 0|\partial_\mu A_\mu|\pi\rangle \neq 0$. However applying the conservation law $\partial_\mu A_\mu = 0$ in the limit $m_\pi \to 0$ to the axial matrix element (9.127b), and using the Dirac equation (9.23) gives the result

$$2g_A(0) = g_P(0)(\hbar q/mc)^2 \tag{9.141}$$

Since $g_A(0)$ is non-zero, $g_P(0)$ must have a pole at $q^2 = 0$ arising from the Goldstone pion with residue $2f_\pi g_{\pi NN} mc^2/(\hbar c)^2$. This is just the Goldberger–Treiman relation given above. Its principal importance is that, using the value of $g_A(0)$ determined from neutron decay, it provides a measure of departures from exact chiral symmetry through the parameter (Dominguez 1982)

$$\Delta = 1 - mc^2 g_A(0)/f_\pi g_{\pi NN} \tag{9.142}$$

which has a value of order (m_q/m) where m_q is the (current) quark mass (Gasser and Leutwyler 1982). One may use the Goldberger–Treiman relation to show that, since $\hbar|q| \simeq m_e c$ in neutron β-decay the effective induced pseudoscalar coupling coefficient $\simeq(2mm_e/m_\pi^2)g_A=0.05g_A$. Since pseudoscalar coupling does not contribute to allowed β-decay, and enters only as a second forbidden correction, its role in neutron β-decay is negligible.

9.3.6 Cabibbo theory and weak neutral currents

9.3.6.1 SU(3) symmetry of weak hadronic currents

The notion of 'universality' has been an important one in the development of weak interaction theory, e.g. the observed near-equality of the polar–vector coupling constants for muon and neutron decay led directly to the concept of a conserved vector current. However, this universality is approximate only since the strangeness non-conserving weak decays of hyperons are governed by transition amplitudes not more than 25% in absolute magnitude of those observed in nucleon or pion decay. Furthermore G_μ and $G_{\beta V}$, although close, differ by about $1.60\pm0.04\%$, a difference which, though small, is significant.

A solution to these problems came through the realization that the weak hadronic current satisfied a symmetry and, further, that the symmetry was a broken one. The strangeness-conserving ($\Delta S=0$) weak hadronic current is charged ($\Delta Q=\pm1$) and has the transformation properties of an isovector ($T=1$, $T_3=\pm1$); it therefore has all the quantum numbers of a charged pion. In the same way the strangeness non-conserving weak hadronic current ($\Delta S=\pm1$; $\Delta Q=\pm1$) has the quantum number of a charged kaon. It was therefore suggested that the weak hadronic currents transformed according to the octet representation of SU(3), each distinct current being identified by the quantum numbers of the charged members of the pseudoscalar meson octet (Cabibbo 1963). The total charged hadronic current is now expressed in a form which automatically brings about the observed selection rules for charge, strangeness and isospin:

$$J_\mu^{V,A}=\cos\theta_c J_\mu^{V,A}(\Delta S=0)+\sin\theta_c J_\mu^{V,A}(\Delta S=\pm1) \tag{9.143}$$

where the free parameter θ_c, known as the Cabibbo angle, fixes the degree of SU(3) symmetry breaking.

The assumption that the weak hadronic charged currents satisfy SU(3) symmetry implies that we may express these currents in the form

$$J_\mu^h(\Delta S=0)=(V_{\mu1}+iV_{\mu2})+(A_{\mu1}+iA_{\mu2}) \tag{9.144a}$$

$$J_\mu^h(\Delta S=1)=(V_{\mu4}+iV_{\mu5})+(A_{\mu4}+iA_{\mu5}) \tag{9.144b}$$

where $V_{\mu j}$ and $A_{\mu j}$ ($j=1,\ldots,8$) are respectively polar vector and axial vector currents which are members of an octet representation of SU(3). Adopting the view that the u, d and s quarks are the elementary hadrons, the total weak hadronic current may be taken to have the form

$$J_\mu^h=\bar{u}\gamma_\mu(1+\gamma_5)[d\cos\theta_c+s\sin\theta_c] \tag{9.145}$$

An important feature of the Cabibbo theory is that it generalizes the strong CVC hypothesis (cf. section 9.3.3) by postulating that the electromagnetic current is in the same octet as J_μ^V. We may therefore write, in addition to (9.144)

$$J_\mu^{EM} = V_{\mu 3} + \frac{V_{\mu 8}}{\sqrt{3}} \tag{9.146}$$

corresponding to the fact that J_μ^{EM} has an isovector part with the quantum numbers of the π^0, and an isoscalar part with the quantum numbers of the η^0 (cf. section 9.2.1.3). A feature of the theory is that, assuming any breaking of the SU(3) symmetry is due to a term which transforms like the eighth component of an octet, all vector coupling constants are unrenormalized to first order in this symmetry-breaking interaction (Ademollo and Gatto 1964). This theorem generalizes a similar result noted in section 9.3.4.

There is a slight complication in SU(3) as compared with SU(2) in that, in place of the single reduced matrix element in SU(2) for each piece of the current, there are two reduced matrix elements in SU(3). Thus the equivalent of the Wigner–Eckart theorem for SU(3) may be written

$$\langle i|\mathrm{O}_j|k\rangle = \mathrm{i}f_{ijk}f + d_{ijk}d \qquad i, j, k = 1, \ldots, 8 \tag{9.147}$$

where $|i\rangle$ and $|k\rangle$ are octet states, O_j is an octet operator, and f_{ijk} and d_{ijk} are SU(3) Clebsch–Gordon coefficients, respectively, anti-symmetric and symmetric in their indices. The real reduced matrix elements f and d correspond, respectively, to the symmetric and anti-symmetric coupled octets which are generated by the direct product of two octets (Bender *et al* 1968).

We may apply the formalism given above to determine the relationship between the vector (f_V, d_V) and weak magnetism (f_M, d_M) form factors in neutron β-decay, and the electromagnetic form factors of the proton and the neutron (cf. section 9.2.3). Making use of the SU(3) representations

$$|\mathrm{p}\rangle = \frac{1}{\sqrt{2}}(|B_4\rangle + \mathrm{i}|B_5\rangle) \qquad |\mathrm{n}\rangle = \frac{1}{\sqrt{2}}(|B_6\rangle + \mathrm{i}|B_7\rangle) \tag{9.148}$$

where $|B_j\rangle$ $(j = 1, \ldots, 8)$ are baryon octet states, and, substituting the appropriate values of f_{ijk} and d_{ijk} in (9.147), we find that

$$\langle \mathrm{p}|J_\mu^V(\Delta S = 0)|\mathrm{n}\rangle = f + d \tag{9.149a}$$

$$\langle \mathrm{p}|J_\mu^{EM}|\mathrm{p}\rangle = \tfrac{1}{2}\{(f + d) + (f - d/3)\} \tag{9.149b}$$

$$\langle \mathrm{n}|J_\mu^{EM}|\mathrm{n}\rangle = \tfrac{1}{2}\{-(f + d) + (f - d/3)\} \tag{9.149c}$$

where we have separated the isovector and isoscalar form factors in (9.149b) and (9.149c). We then find

$$g_V(0) = f_V(0) + d_V(0) = F_1^p(0) - F_1^n(0) = 1 \tag{9.150a}$$

$$[g_M(0) - g_V(0)] = f_M(0) + d_M(0) = F_2^p(0) - F_2^n(0) = \kappa_p - \kappa_n = 3.7 \tag{9.150b}$$

Of course the result (9.150) may be obtained from CVC theory and SU(2) symmetry and provides no test of SU(3), and it is necessary to study all the weak decays within the baryon and meson octets to put the Cabibbo formalism to a proper test.

There is no connection between the weak axial current $J_\mu^A(\Delta S=0)$ and the electromagnetic current, and the weak axial form factor

$$g_A(0)=F+D \tag{9.151}$$

cannot be related to the electromagnetic form factors. Experimentally $g_A(0)$ is best determined from a measurement of the electron momentum – neutron spin correlation coefficient A in the β-decay of polarized neutrons (cf. section 9.3.8.2). The most accurate value obtained to date is (Bopp *et al* 1986, Klemt *et al* 1988) (cf. section 9.3.9 for a competing result)

$$g_A(0)=1.262\pm0.005 \tag{9.152}$$

To determine F and D separately we need to survey all the weak baryon decays which yield the result (Bourquin *et al* 1983)

$$D/(F+D)=0.61 \tag{9.153}$$

9.3.6.2 Weak neutral currents

In its original formulation the Cabibbo theory predated the discovery of the charmed quark whose existence had first been postulated to provide a partner for the strange quark (Bjorken and Glashow 1964). This proposal was motivated by notions of quark–lepton symmetry arising from the gradual recognition that, like the lepton currents, the quark currents also have a space-time (V–A) structure. Thus, to match the left-handed lepton doublets

$$\begin{pmatrix} \nu_e \\ e^- \end{pmatrix}_L \quad \text{and} \quad \begin{pmatrix} \nu_\mu \\ \mu^- \end{pmatrix}_L$$

symmetry considerations suggested the introduction of a second quark doublet

$$\begin{pmatrix} c \\ s' \end{pmatrix}_L$$

in addition to the well-known quark doublet

$$\begin{pmatrix} u \\ d' \end{pmatrix}_L$$

Here d' and s' are the mutually orthogonal Cabibbo admixtures of quark states

$$d'=d\cos\theta_c+s\sin\theta_c \tag{9.154}$$

and

$$s'=-d\sin\theta_c+s\cos\theta_c \tag{9.155}$$

However there was also the major practical difficulty that, prior to the introduction of the charmed quark, no suitable symmetry assignment of quark fields could be found which did not give rise to terms of the form $\bar{s}\gamma_\mu(1+\gamma_5)d$, describing strangeness violating neutral members of the current SU(3) octet which are not observed experimentally (Kline 1986). The problem was not so much the apparent absence of strangeness conserving neutral currents of the form $\bar{d}\gamma_\mu(1+\gamma_5)d$, which are difficult to detect in the presence of overwhelming competition from electromagnetic currents; the difficulty was the pronounced absence of strangeness non-conserving neutral current processes like $K^+ \to \pi^+ + e^+ + e^-$ or $K^0 \to \mu^+ + \mu^-$ whose measured branching ratios are $\leqslant 10^{-8}$. The true significance of the fourth quark did not therefore become apparent until it was realized that this was exactly what was required to rid the theory of the unwanted neutral current processes. The suppression is described as the GIM mechanism (Glashow *et al* 1970) and works in the following way.

The assumption of a three-quark charged hadronic current of the form (9.143), automatically introduces into the effective Lagrangian density $L_w(x)$ defined in (9.122) terms of the form

$$L'_w(x) = \frac{G_F}{\sqrt{2}}\{\bar{d}'\gamma_\mu(1-\gamma_5)d[\bar{u}\gamma_\mu(1-\gamma_5)u + \bar{d}'\gamma_\mu(1-\gamma_5)d']\} \tag{9.156}$$

which generate terms proportional to $\cos\theta_c \sin\theta_c\,(\bar{d}s + d\bar{s})$ describing neutral current ($\Delta Q = 0$), strangeness violating ($|\Delta S| = 1$, 2) processes in the lowest baryon and meson SU(3) octets. If, however, the superposition state s' defined in (9.155) is assumed to enter the theory on the same level as d', the term $\bar{s}'\gamma_\mu(1-\gamma_5)s'$ must be added to $\bar{d}'\gamma_\mu(1-\gamma_5)d'$ in (9.156) and the resultant sum contains no cross-products proportional to $\cos\theta_c \sin\theta_c$ and the unwanted neutral current processes disappear. It is clear, of course, that the addition of the charmed quark takes us beyond the SU(3) symmetry scheme of the original Cabibbo hypothesis, into an even more highly broken SU(4) symmetry. However, the charmed quark is so heavy that, apart from the suppression of the strangeness-violating neutral current processes, the effect of charm is minimal at low energy.

The complete neutral current constructed from the left-handed components of u, d', c and s' (and hence parity violating) is then given by

$$J_\mu^{3w} = \bar{u}\gamma_\mu\frac{(1-\gamma_5)}{2}u + \bar{c}\gamma_\mu\frac{(1-\gamma_5)}{2}c - \bar{d}\gamma_\mu\frac{(1-\gamma_5)}{2}d - \bar{s}\frac{(1-\gamma_5)}{2}s \tag{9.157}$$

where the superscript '3w' is associated with the 3-direction in weak isospin space (cf. section 9.4.1). In addition there is, of course, the parity-conserving (neutral) electromagnetic current constructed from both the left-handed and right-handed components of u, d', c and s'

$$J_\mu^{EM} = \tfrac{2}{3}\bar{u}\gamma_\mu u + \tfrac{2}{3}\bar{c}\gamma_\mu c - \tfrac{1}{3}\bar{d}\gamma_\mu d - \tfrac{1}{3}\bar{s}\gamma_\mu s \tag{9.158}$$

Parity violating weak neutral currents of the form (9.148) were first detected in the leptonic sector via inelastic electron–neutrino scattering (Hasert *et al* 1973) and in the hadronic sector through single pion production by neutrinos (Barish *et al* 1974), and subsequently as competitors to the electromagnetic interaction in the inelastic scattering of polarized electrons by deuterons (Prescott *et al* 1978).

9.3.6.3 The Cabibbo–Kobayashi–Maskawa matrix

In the Cabibbo theory with two quark generations there is a single mixing parameter, namely the Cabibbo angle θ_c. When the theory is extended to three quark generations it is necessary to take account of mixing among all left-handed quarks which carry the same charge, and this mixing is described by what is now known as the CKM matrix (Kobayashi and Maskawa 1973). This is a unitary matrix which rotates the mass eigenstates d, s, b of quarks with electric charge equal to $-1/3$, into the weak eigenstates d', s', b'. In general this transformation is written in the form

$$\begin{pmatrix} d' \\ s' \\ b' \end{pmatrix} = \begin{pmatrix} V_{ud} & V_{us} & V_{ub} \\ V_{cd} & V_{cs} & V_{cb} \\ V_{td} & V_{ts} & V_{td} \end{pmatrix} \begin{pmatrix} d \\ s \\ b \end{pmatrix} \tag{9.159}$$

The nine elements of the CKM matrix can be expressed in terms of four real parameters, namely three real angles and a phase.

A number of parameterizations of the CKM matrix have been proposed, depending on the choice of quark phase, but the standard choice is the following (*Review of Particle Properties* 1992)

$$V_{\mathrm{CKM}} = \begin{pmatrix} c_{12}c_{13} & s_{12}c_{13} & s_{13}e^{-i\delta_{13}} \\ -s_{12}c_{23} - c_{12}s_{23}s_{13}e^{i\delta_{13}} & c_{12}c_{23} - s_{12}s_{23}s_{13}e^{i\delta_{13}} & s_{23}c_{13} \\ s_{12}s_{23} - c_{12}c_{23}s_{13}e^{i\delta_{13}} & -c_{12}s_{23} - s_{12}c_{23}s_{13}e^{i\delta_{13}} & c_{23}c_{12} \end{pmatrix} \tag{9.160}$$

where

$$c_{ij} = \cos\theta_{ij} \qquad s_{ij} = \sin\theta_{ij} \tag{9.161}$$

When $\theta_{23} = \theta_{13} = 0$, the third quark generation decouples from the first two quark generations and the remaining angle θ_{12} becomes identical with the Cabibbo angle θ_c.

The most intriguing feature of the CKM matrix is that the existence of the phase δ_{13} ($\neq n\pi$, $n = 0 \pm 1 \ldots$) allows for the possibility of accounting for *CP* violation, or *T* violation if the *CPT* theorem is assumed to hold good (Luders 1957, Feinberg 1957). Such an effect requires that at least three quark generations contribute to the mixing matrix. It is now well established that *CP* violation does occur to a small degree in the decay of the neutral kaon (Christenson *et al* 1964) (cf. section 10.4.2).

In the present context only the light u, d and s quarks are of interest and hence we direct attention to the matrix elements $V_{ud}(\simeq \cos\theta_c)$ and $V_{us}(\simeq \sin\theta_c)$. The matrix element V_{ud} has been determined experimentally by two independent methods. The first method is based on a measurement of the vector coupling constant $G_{\beta V}$ in nuclear β-decay, as derived from the ft-values of superallowed pure Fermi $0^+ \rightarrow 0^+$ decays, allowing for Coulomb effects which bring about a mismatch of initial and final nuclear wavefunctions, and for radiative corrections (Hardy *et al* 1990, Towner 1992, Barker *et al* 1992). The essential relationship is

$$G_{\beta V} = V_{ud} G_\mu (1 + \Delta\beta - \Delta\mu)^{1/2} \tag{9.162}$$

where G_μ is the weak Fermi coupling constant as derived from the lifetime of the muon (cf. section 10.2.1), and $\Delta\beta$ and $\Delta\mu$ are beta and muon radiative corrections (Marciano and Sirlin 1981, Bourquin *et al* 1983). The second method for determining V_{ud} also relies on a value of $G_{\beta V}$ as derived from a measurement of the neutron lifetime (τ_n), and the angular polarization coefficient A in the β-decay of polarized neutrons (cf. sections 9.3.7 and 9.3.8).

The matrix element V_{us} has been determined experimentally from studies of the semi-leptonic decays of the K-mesons K^+_{e3} and K^0_{e3} and from analysis of semi-leptonic hyperon β-decays, allowing for the breaking of SU(3) symmetry in all cases. Experimental results for both V_{ud} and V_{us} are assembled in Table 9.2.

Since the unitarity of the CKM matrix requires that

$$|V_{ud}|^2 + |V_{us}|^2 + |V_{ub}|^2 = 1 \tag{9.163}$$

we find that, applying the global values for V_{ud} and V_{us} from Table 9.2,

Table 9.2. Experimental values of V_{ud} and V_{us} for the CKM matrix

Matrix element	Method	Reference
V_{ud}		
0.9744 ± 0.0010	$(ft)0^+ \rightarrow 0^+$	Sirlin (1987)
0.9737 ± 0.0010	$(ft)0^+ \rightarrow 0^+$	Ormand and Brown (1989)
0.9750 ± 0.0004	$(ft)0^+ \rightarrow 0^+$	Wilkinson (1991)
0.9778 ± 0.0029	τ_n/A	Dubbers *et al* (1990)
0.9744 ± 0.0010	Global	*Review of Particle Properties* (1992)
V_{us}		
0.2196 ± 0.0023	K-decay	Leutwyler and Roos (1984)
0.220 ± 0.003	hyperon decay	Donoghue *et al* (1987)
0.218 ± 0.003	hyperon decay	Yamaguchi *et al* (1989)
0.2196 ± 0.0023	Global	*Review of Particle Properties* (1992)

$|V_{ub}| \simeq (4.3 \pm 2.0) \times 10^{-3}$. The most recent experimental estimate for this matrix element lies in the range $(4.3 \pm 1.5) \times 10^{-3}$ (*Review of Particle Properties* 1992).

9.3.7 Experimental determination of the neutron lifetime

9.3.7.1 Measurement techniques

That the free neutron would be β active was predicted by Chadwick and Goldhaber (1935) when they made the first precise measurement of the neutron mass. The process was first detected by Snell and his associates at Oak Ridge National Laboratory (Snell and Miller 1948, Snell *et al* 1950). Neutron decays were identified by observing triple coincidences between electron detectors and pulses in secondary electron multipliers caused by focused protons accelerated through a potential of 8 kV. They estimated a half-life in the range 10–30 min. Estimates by Robson (1950, 1950a) of 9–25 min, and by Spivak *et al* (1955b) giving 8–15 min were in agreement. The first definitive determination of the neutron lifetime was also carried out by Robson (Robson 1951).

Traditionally two quite different techniques have been employed to measure τ_n, which may be summarized as 'beam methods' and 'bottle methods'. In beam methods the number of decays in a (replenished) neutron ensemble is recorded, a technique which, on the whole, has tended to over-estimate τ_n. These methods are based on the differential equation

$$\frac{dN(t)}{dt} = \frac{-N(t)}{\tau_n} \tag{9.164}$$

where $dN(t)/dt$ is the measured decay rate, and $N(t)$ is the number of neutrons present at time t in the source volume V. In practice this relation has to be supplemented by two further relations

$$\left\langle \left| \frac{dN(t)}{dt} \right| \right\rangle = n_d/\Omega\eta \qquad \langle N(t) \rangle = \rho V \tag{9.165}$$

where n_d is the number of decay particles recorded per unit time in a detector for which the relative solid angle is Ω and the efficiency is η, and ρ is the mean neutron density in the beam. In general of course Ω and ρ are functions of position within the beam and it is necessary to carry out an integration.

'Bottle' experiments are based on the integral relation

$$N(t) = N(0) \exp[-t/\tau_n] \tag{9.166}$$

where τ_n is determined by counting the number of neutrons surviving to time t, as a function of t, for a fixed number $N(0)$ present at zero time. Such methods have the advantage that absolute neutron counting is unnecessary, but suffer the corresponding disadvantage of having to account for all neutrons lost through interactions other than β-decay. Since every unaccounted-for neutron is recorded as a β-decay, these methods have tended to underestimate τ_n. Only recently have the two quite different techniques produced results which are consistent within the stated errors. Two versions of the 'neutron bottle' have been employed to date, both using ultra-cold neutrons with energies in the range $E_n \leqslant 2 \times 10^{-7}$ eV. These will be considered in turn below. Finally one should take note of a proposed technique, yet to be attempted, which falls into neither of the two classes discussed above. This exploits the fact that neutrons created in the atmosphere with energies less than 0.65 eV are gravitationally bound, allowing τ_n to be determined from a measure of the ratio of the upward and downward neutron currents at the top of the atmosphere (Feldman *et al* 1990).

In the four decades which have elapsed since neutron β-decay was first detected a large number of measurements of the neutron lifetime have been reported, with accuracies quoted at the level of a few per cent, many of them with highly discrepant results. These results are listed in table 9.3. Their most

Table 9.3. Published measured values of the neutron lifetime. The symbol * represents a revised result

Source volume	*n*-counter	τ_n (s)	Reference
Beam	^{55}Mn	1108 ± 216	Robson (1951)
Beam	^{197}Au	1039 ± 130	Spivak *et al* (1956)
Beam + cloud chamber	^{197}Au	1100 ± 165	D'Angelo (1959)
Beam	^{197}Au	1013 ± 26	Sosnovski *et al* (1959)
Beam (4π)	^{3}He	935 ± 14	Christensen *et al* (1967)
		919 ± 14*	Christensen *et al* (1972)
Magnetic UCN bottle	^{3}He	909 ± 69	Kugler *et al* (1978), Paul and Trinks (1978)
Beam	^{197}Au	877 ± 8	Bondarenko *et al* (1978)
		891 ± 9*	Spivak (1988)
Beam (4π)	^{10}B	937 ± 18	Byrne *et al* (1978, 1980)
Al UCN bottle	^{3}He	875 ± 95	Kostvintsev *et al* (1976, 1980, 1982)
Al UCN bottle	^{3}He	903 ± 13	Kostvintsev *et al* (1986)
D_2O UCN bottle	^{3}He	893 ± 20	Kostvintsev *et al* (1987)
		900 ± 11*	Morozev (1989)
Pulsed beam (4π)	^{197}Au, ^{59}Co	876 ± 21	Last *et al* (1988)
Magnetic UCN bottle	^{3}He	877 ± 10	Paul *et al* (1989)
Fluid walled UCN bottle	^{3}He	887.6 ± 3	Mampe *et al* (1989)
Pulsed beam (4π)	^{3}He	878 ± 31	Kossakowski *et al* (1989)
Beam (4π)	^{10}B	893.5 ± 5.3	Byrne *et al* (1989, 1990)
Gravitational UCN bottle	^{3}He	888.4 ± 2.9	Alfimenkov *et al* (1990)
Gravitational UCN bottle	^{3}He	888.4 ± 3.3*	Nesvizhevskii *et al* (1992)
Fluid walled UCN bottle	^{3}He	882.6 ± 2.7	Mampe *et al* (1993)

obvious feature is the monotonic decline in the accepted value of the neutron lifetime over the period this work was carried out.

9.3.7.2 Neutron lifetime experiments in a neutron beam

The conventional 'beam' method for measuring τ_n employs thermal or cold neutrons with a continuous velocity spectrum, a detector for electrons or protons, and a thin neutron detector foil which absorbs only a tiny fraction of the neutrons which impinge on it. Assuming a 4π collection solid angle, as in all recent variants of the technique, and unit detection efficiency for recording the number of decays N_d occurring per second in a length L of beam, the value of τ_n is given by

$$\tau_n = N_n L / N_d \sigma_0 v_0 (\rho x) \tag{9.167}$$

Here N_n is the number of nuclear reactions recorded per unit time in the neutron counter, σ_0 is the reaction cross-section at some standard velocity v_0 (usually 2200 m s^{-1}), and (ρx) is the surface density of neutron detector isotope. This equation does not depend on the neutron velocity provided $\sigma(v)$ scales accurately as $(1/v)$. A list of suitable detector isotopes is assembled in tables 1.4 and 1.5.

A variant of the conventional beam technique makes use of a chopped neutron beam in combination with a system for recording decays from a finite pulse of N neutrons contained entirely within the sample volume. The number $\langle N \rangle$ is then determined by absorbing all of these neutrons in a 'black' neutron detector, thereby avoiding the necessity of determining L (or V) and ρ separately. This technique has been employed into two recent experiments (Last *et al* 1988, Kössakowski *et al* 1989).

In carrying out this kind of measurement we may identify four distinct problems which must be solved. (i) Should electrons or protons be counted and which detectors should be used? (ii) How are the genuine decay events to be distinguished from the background? (iii) How is the neutron density to be determined? (iv) How is the source volume to be defined? In the very first published measurement of the neutron lifetime carried out by Robson at Chalk River both electrons and protons were counted in coincidence and the observed electron spectrum was shown to be consistent with that expected for neutron β-decay (Robson 1951). In this method problems (i) and (ii) were solved simultaneously because coincidence counting suppresses the background. The neutron density (iii) was determined by measuring the activity generated by neutron capture in assayed manganese foils. However, determination of the neutron source volume (iv) remained a difficulty because the coincidence counting efficiency was a sensitive function of the position of the decaying neutron. This source of uncertainty is reflected in the 20% error quoted on the final result $\tau_n = 1108 \pm 216$ s.

The problems associated with coincidence counting were avoided in the series of experiments carried out much later on by D'Angelo who observed the individual decays directly in a diffusion cloud chamber (D'Angelo 1959). However, the final result $\tau_n = 1100 \pm 165$ s showed no marked improvement in accuracy mainly because of the increased background produced by nuclear reactions in the counter gas.

In subsequent experiments designed to measure the lifetime of neutrons decaying in a beam, attempts to isolate specific decay events were abandoned in favour of absolute counting of electrons or protons, combined with precise determination of the background. The crucial importance of reducing the background may be judged from the fact that a thermal flux of $\approx 2 \times 10^8$ cm^{-2} s^{-1} produces only about one neutron decay per cm^3 of beam per second, reflecting the fact that β-decay is an extremely rare event indeed in the career of a neutron.

The first programme established with the specific aim of measuring the neutron lifetime to an accuracy of a few per cent was initiated by Spivak and his collaborators at the I.Y. Kurchatov Atomic Energy Institute in Moscow (Sosnovski *et al* 1959). In this experiment a collimated beam of neutrons emerged from a reactor parallel to the axis of a hollow cylindrical electrode maintained at 20 kV, to which was attached at right angles a second cylindrical tube terminated by a grid. Decay protons generated in the beam travelled in a field-free region towards this grid, which determined the acceptance solid angle. They were then accelerated in a converging electric field towards a second spherical grid maintained at ground, behind which sat a ball proportional counter with a 0.07 μm collodion film window supported on a third grid. The sensitive volume of the counter was shielded by a fourth grid from charges collected on the window. Absolute and relative neutron density measurements were carried out by activation of gold and sodium foils. The results were found to be consistent within 0.5% with an overall precision of 1.8%. The final result was $\tau_n = 1013 \pm 26$ s, where the error quoted included contributions from counting statistics and uncertainties in the efficiencies for counting neutrons and protons.

The experiment was subsequently repeated using identical apparatus but obtaining a much reduced result i.e. $\tau_n = 877 \pm 8$ s (Bondarenko *et al* 1978). The discrepancy between the results of the two experiments was accounted for as arising from excessive background gas pressure in the earlier work. Whatever the origins of the problem the outcome serves to emphasize the difficulty of eliminating unsuspected systematic effects in absolute measurements of this type. The conclusions of this experiment have recently been subject to further revision leading to a final result (Spivak 1988, Gaponev *et al* 1990).

$$\tau_n = 885 \pm 9 \text{ s} \tag{9.168}$$

The major weakness in the technique described above appears to be the very small collection solid angle, a problem which was completely solved in the experiment carried out at the Risö reactor in Denmark (Christensen *et al* 1967, 1972). In this system the neutron beam passed transverse to a magnetic field which forced decay electrons generated in the beam to spiral about the magnetic field-lines on their passage to a large plastic scintillator sitting above the beam. The resultant light pulses were then fed via a light pipe to a photomultiplier located outside the magnetic field region. If an electron was scattered from the scintillator it reversed its helical trajectory back through the beam to be recorded in a second scintillator placed underneath. The two scintillation counters therefore acted as a 4π detector and the active volume was determined essentially by the length of beam cut off by the magnetic-field lines joining the detector apertures. Neutron density determination was made by counting neutrons in an ^{3}He proportional counter (Als-Nielsen *et al* 1967).

Subsequent experiments based on the beam technique have all used a 4π counter for detecting decay events and two of these have also made use of a pulsed beam as described above (Last *et al* 1988, Kossakowski *et al* 1989). To date the most precise experiment of this type made use of the Penning trap technique (Penning 1936, Byrne and Farago 1965, Byrne 1978) which is analogous to the Christensen method except that protons rather than electrons are counted (Byrne *et al* 1980, 1989, 1990). The operation of the system in its original version may be described as follows. The beam of very cold neutrons emerged from the reactor normal to a magnetic field of the order of 12 kG produced by a superconducting magnet. In such a field both electrons and protons move in tight spiral orbits of maximum radius $\leqslant 3.5$ mm. The beam passed transverse to a hollow cylindrical electrode aligned parallel to the magnetic axis and maintained at ground potential. On either side of this central electrode two coaxial electrodes were maintained at potentials of 1 kV, the whole combination of electric and magnetic fields acting as a potential well for the low-energy ($\leqslant 0.75$ keV) protons from neutron decay. Thus the protons could be trapped for significant periods of time (10^{-3}–10 s) before being released by suitable pulses and detected in a silicon surface barrier maintained at -30 kV.

The essential point of the device described above is that, if protons are stored for a time τ_s and the output spectrum is sampled for a period $\tau_c \ll \tau_s$ following the application of a release pulse, the background is reduced by a factor τ_c/τ_s; in practice values of the order of 10^{-4} were achieved for this ratio and the proton spectrum was observed essentially free of background. Neutron density determination was carried out by counting α particles emitted in the reaction ^{10}B(n, α)^{7}Li.

The first series of experiments carried out using this method gave the result $\tau_n = 937 \pm 18$ s, a value which must now be regarded as a substantial over-estimate. The same conclusion applies to the value $\tau_n = 914 \pm 17$ s obtained by the Christensen method. There were three major weaknesses to the Penning trap method as originally conceived. (i) The use of an inhomogeneous magnetic field to enhance the collecting power of the system, varying from 12 kG in the beam to 4 kG at the proton detector meant that the collection solid angle became position dependent. Thus an accurate assessment of the trapping volume was difficult. In addition the magnetic mirror effect played a dominant role. (ii) The neutron beam was projected normal to the magnetic field which meant that the trapping volume could not be systematically varied to eliminate edge effects. (iii) The quoted ^{10}B content and uniformity of the neutron counting standards proved unreliable (Lamberty *et al* 1990).

A much improved version of the Penning trap method is shown in figure 9.11 (Byrne *et al* 1989, 1990). Here the neutron beam enters the Penning trap parallel to the 50 kG magnetic field which bends through a 9° angle in such a way as to direct the decay protons onto a surface barrier detector located outside the neutron beam. Since the magnetic field at the detector is about 5% less than the magnetic field in the trap there is no magnetic mirror effect. In the system as used there were 20 electrodes whose potentials were controllable externally and independently. Thus by systematically varying the length of the trap the number of decays per unit length of beam could be measured. The neutron density was determined by α-particle counting as before, using boron targets deposited on single crystal silicon wafers, whose ^{10}B content was determined by isotope dilution mass spectroscopy (Pauwels *et al* 1991, Scott *et al* 1992). The current result based on this method is

$$\tau_n = 893.5 \pm 5.3 \text{ s} \qquad (9.169)$$

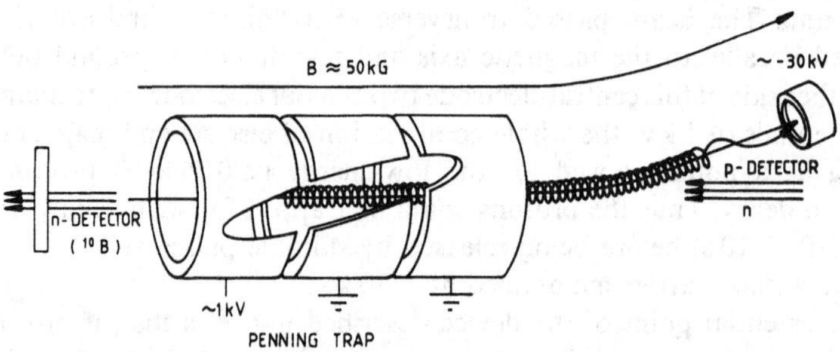

Figure 9.11. Determination of the neutron lifetime using the beam method, by counting protons from neutron decay stored in a Penning trap (Byrne *et al* 1990). The neutron density in the beam is measured by counting α-particles from the reaction ^{10}B(n, α) ^{7}Li generated in an assayed ^{10}B target.

9.3.7.3 Neutron lifetime experiment in a magnetic bottle

Since the neutron is uncharged the only available mechanism of electromagnetic confinement is through the force exerted on its magnetic moment

$$\boldsymbol{F} = -\nabla(\boldsymbol{\mu}_n \cdot \boldsymbol{B}(\boldsymbol{r})) \qquad (9.170)$$

in an inhomogeneous magnetic field $B(r)$. An m-pole magnetic field is of course not cylindrically symmetric although $|\boldsymbol{B}(\boldsymbol{r})| \approx r^{(m/2)-1}$. Thus, to produce a harmonic force on the spin moment, at least a sextupole field ($m=6$) is required. Since the sense of the force depends on the sign of the spin quantum number, only one neutron spin state may be confined. To achieve a stable orbit the inwardly directed radial magnetic force must be balanced by the centrifugal force, a requirement which limits the range of velocities which may be stored.

Magnetic storage was first achieved by Paul and his associates in 1978 using the toroidal storage ring NESTOR shown in figure 9.12 (Paul and Trinks 1978, 1978a, Kugler *et al* 1978, 1985). In the latest version of these early experiments, the storage ring was installed on the ultra-cold neutron vertical guide at the ILL, Grenoble (cf. figure 1.4), and neutrons in the velocity range 6–20 m s^{-1} were injected into the ring using a system of totally reflecting nickel mirrors. The maximum field in the ring had the value $B_{max} = 35$ kG, and the radial extent of the storage volume was defined by a pair of boron

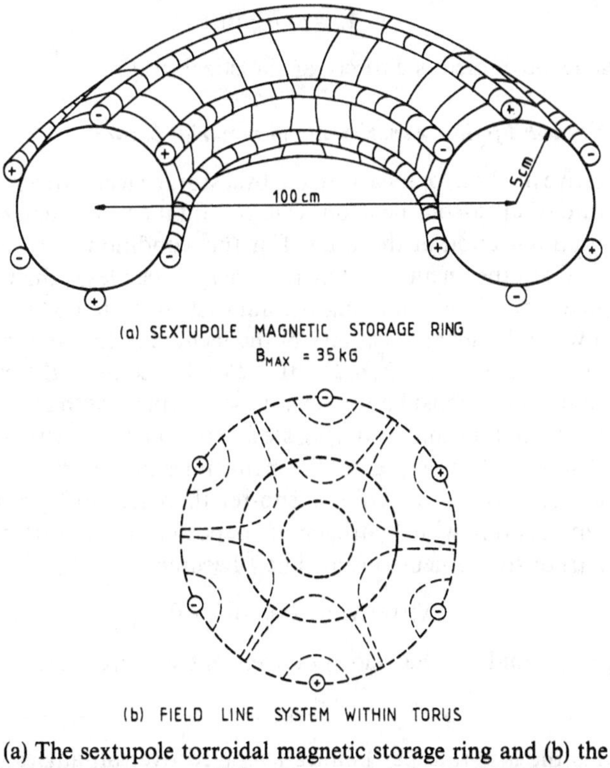

Figure 9.12. (a) The sextupole torroidal magnetic storage ring and (b) the system of magnetic field lines within the torus for the magnetic bottle NESTOR used to store ultra cold neutrons of one spin sense. The neutron lifetime was determined by measuring the decay rate of the stored neutrons (Paul *et al* 1989).

carbide 'beam scrapers'. Prior to injection the neutron flux was monitored by a ^{3}He counter, which was also used to count the number of neutrons surviving after a storage time t.

As with charged particles in a cyclic accelerator, stored neutrons undergo radial and axial betatron oscillations about a stable orbit, with the difference that, in the neutron case, resonance coupling between betatron modes takes a very severe form. Thus, to control the oscillation amplitude a decupole component has to be incorporated in the field configuration. An additional problem, peculiar to stored neutrons, is that, if the neutron spin senses a time varying magnetic field such that $\dot{B}/B \geqslant \omega_0$, the angular frequency of Larmor spin precession, the spin may flip and the neutron is ejected from the ring.

In the most recent series of experiments neutrons were stored for periods of time up to one hour and, since detailed analysis of the beam dynamics indicated that neutrons were lost by resonance coupling only during the first 450 s, the data accepted were limited to that obtained for storage times in the range 450 s$<t<$3600 s. The final result obtained was (Paul *et al* 1989)

$$\tau_n = 877 \pm 10 \text{ s} \tag{9.171}$$

where the error quoted is entirely of statistical origin.

9.3.7.4 Neutron lifetime experiments in material bottles

An alternative method for measuring τ_n by a storage technique is to determine the loss rate of ultra-cold neutrons (UCN) confined in a material bottle by total internal reflection at the walls. For this condition to be achieved it is necessary that the neutron kinetic energy be less than the Fermi pseudo-potential of the confining medium (cf. section 1.6.6), and it is this condition which limits applicability of the technique to the ultra-cold range of neutron energies with $E_n \leqslant 2 \times 10^{-7}$ eV. To confine the neutrons the material of the bottle must have a positive scattering length and quartz, glass, beryllium, aluminium and stainless steel have all been used with varying degrees of success. In practice confinement times in material bottles tend to have values an order of magnitude shorter than τ_n, most probably due to inelastic interactions with hydrogenous deposits on the surface. Thus, the number $N(t)$ of stored neutrons at time t becomes

$$N(t) = N(0) \exp[-(\theta_\beta + \theta_w)t] \tag{9.172}$$

where $\theta_\beta = \tau_n^{-1}$ and θ_w, the velocity-dependent wall loss-rate, is given by

$$\theta_w = \bar{\mu}(v)v/\lambda \tag{9.173}$$

Here $\bar{\mu}(v)$ is the loss rate per bounce averaged over all angles of incidence, and λ is the mean free path.

Three quite separate investigations have been carried out to date based on these equations, the essential idea underlying all three being that, although neutrons can disappear from the bottle either by β-decay or by nuclear interactions on the walls, the latter loss mechanism can be controlled to some degree by altering either the bottle geometry or the spectrum of confined neutrons or both. A major complication is that the spectrum changes with the passage of time as the faster neutrons are progressively removed from the storage volume.

The first attempt to measure τ_n by this technique was initiated by Morozev and his associates at JINR, Dubna, the basic idea being to maintain the volume of the bottle essentially constant while progressively increasing the surface area and hence the collision frequency γ. They obtained the preliminary result $\tau_n = 875 \pm 95$ s (Kostvintsev *et al* 1980, 1982). In the further development of this technique, experiments were carried out for different temperature ranges and with bottles made from aluminium, or coated with heavy water ice which has a much reduced total cross-section. By varying the collision frequency γ as described above, and extrapolating to zero γ, the results obtained were $\tau_n = 903 \pm 13$ s (Al) and $\tau_n = 893 \pm 20$ s (D_2O), leading to a mean value

$$\tau_n = 900 \pm 11 \text{ s} \tag{9.174}$$

In the second variant of the bottle technique which has been investigated, a dramatic improvement in accuracy has been achieved through the introduction of three important changes in technique: (i) altering the collision frequency by varying the volume; (ii) maintaining constant surface conditions by continuously renewing the surface (Bates 1983); (iii) scaling the storage time in proportion to the mean free path (Ageron *et al* 1986). The apparatus employed is shown in figure 9.13. If we suppose that $\theta_m(\lambda)$ is the measured

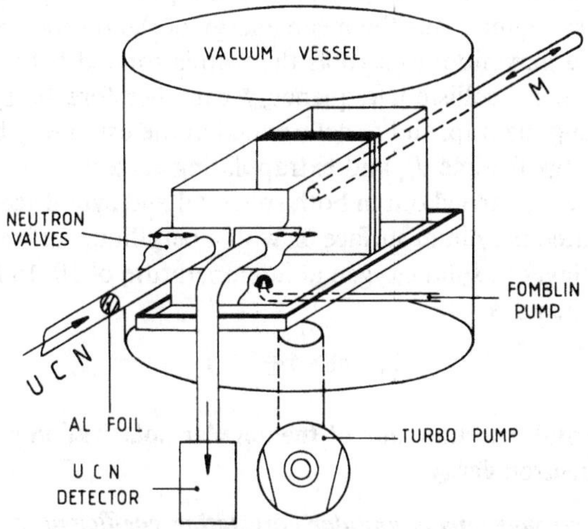

Figure 9.13. Experiment to measure the neutron lifetime by observing the decay rate of ultra-cold neutrons stored in a material bottle with a renewable fomblin oil surface (Mampe *et al* 1989).

loss rate for a given mean free path λ, then, by carrying out experiments for volumes corresponding to two different values of λ at a given velocity, we may extract a value for θ_β

$$\theta_\beta = [\theta_m(1)\lambda_1 - \theta_m(2)\lambda_2]/[\lambda_2 - \lambda_1] \tag{9.175}$$

The factor $\bar{\mu}(v)v$ can be eliminated from the equation only if the bottles for λ_1 and λ_2 have identical loss characteristics. This is ensured by using a glass bottle whose surface is continuously coated with the perfluoro polyether fluid 'fomblin'. At the same time, to apply the expression for θ_β given above to a wider range of velocities the storage times $[t_1(\lambda) - t_2(\lambda)]$ for the two values of λ must be scaled as

$$t_1(1)/t_1(2) = t_2(1)/t_2(2) = [t_1(1) - t_1(2)]/[t_2(1) - t_2(2)] = \lambda_1/\lambda_2 \tag{9.176}$$

The final value of τ_n is arrived at by plotting the measured inverse lifetime θ_m against inverse mean free path λ^{-1} for a range of scaled storage times, and extrapolating to zero λ^{-1}. The value obtained (Mampe *et al* 1989)

$$\tau_n = 887.6 \pm 3 \text{ s} \tag{9.177}$$

is one of the most accurate experimental values of τ_n achieved to date.

The most recent version of the bottle technique is the gravitational ultracold neutron trap constructed by Serebrov and his associates at Leningrad (Alfimenkov *et al* 1990). The apparatus consists essentially of a cryogenic trap for neutrons which simultaneously behaves as a gravitational spectrometer. The spectrometer action comes about because neutrons with a kinetic energy greater than a critical gravitational potential energy can escape through a hole in the trap. The trap is rotatable about a horizontal axis from a position with the hole at the highest point to a position with the hole at the lowest point. Thus the mean energy of the trapped neutrons may be varied from a maximum to zero as the trap is rotated between these two positions. The mean collision frequency γ can therefore be systematically varied by rotating the trap, and τ_n determined in the usual way by measuring the inverse storage lifetime θ_m and extrapolating to zero γ.

Experiments were carried out in both spherical and cylindrical bottles with either an untreated beryllium surface or with a beryllium surface coated with a 3000–5000 Å layer of solid oxygen at a temperature of 10–15 K. The result of these experiments is

$$\tau_n = 888.4 \pm 2.9 \text{ s} \tag{9.178}$$

9.3.8 Experimental determination of the angular polarization correlation coefficients in neutron decay

9.3.8.1 The electron–neutrino angular correlation coefficient 'a'

In an allowed β transition the shape of the electron spectrum is independent of the precise structure of the weak interaction responsible. There is some dependence on the form of the interaction in the shapes of forbidden spectra, but in practice it has proved difficult to give a detailed interpretation of the observed effects. Thus, until the discovery of the non-conservation of parity, the electron–neutrino angular correlation coefficient was the only measurable parameter which was sensitive in any significant degree to the details of the interaction. This result, which was first pointed out by Bloch and Møller (1935), explains the important role this coefficient has played in establishing the $(V-A)$ character of the weak interaction. The $(V-A)$ formulae for all four coefficients a, A, B and D are listed in section 9.3.2.

It should be emphasized that, since the angular correlation term $\boldsymbol{p}_e \cdot \boldsymbol{p}_{\bar{\nu}}$ is a scalar, this correlation does not originate in an interference effect between the parity conserving and parity non-conserving terms in the interaction. Neither does it come from interference between Fermi and Gamow–Teller terms because $\boldsymbol{p}_e \cdot \boldsymbol{p}_{\bar{\nu}}$ commutes with the operator for the total angular momentum of the leptons and does not therefore mix the singlet and triplet amplitudes.

The presence of a correlation between electron and neutrino momenta may be understood from the following arguments. If the neutrino emitted by an unpolarized neutron is not observed there is no preferred direction in space, and the system of electron plus proton is created into a state which may be regarded as an incoherent superposition of magnetic substates when referred to an arbitrary direction. Although each magnetic substate in general gives an anisotropic electron emission, this anisotropy vanishes upon averaging over the incoherent mixture. If, however, the neutrino is detected, the measurement forces it into a momentum eigenstate which means that the electron plus proton system is now in a pure state which may be represented as a coherent superposition of magnetic substates. In this case the anisotropy does not vanish in general because of the presence of the interference terms.

In the case of a pure Fermi $(0^+ \rightarrow 0^+)$ transition, a is given by the expression

$$a = \frac{|G_V|^2 + |G'_V|^2 - |G_S|^2 - |G'_S|^2}{|G_V|^2 + |G'_V|^2 + |G_S|^2 + |G'_S|^2} \tag{9.179}$$

where the constants G_i, G'_i, $i = V, S, T$ and A are parity conserving and parity violating coupling constants respectively for vector, scalar, tensor and axial vector couplings. In the $(V-A)$ weak interaction only V and A couplings contribute. Thus the experimental result $a = 0.97 \pm 0.14$ in the decay of ^{14}O (Allen *et al* 1959) establishes the essentially vector character of the Fermi operator. For a pure Gamow–Teller decay $(\Delta I = \pm 1)$ a is given by

$$a = -\left(\frac{1}{3}\right)\frac{|G_A|^2 + |G'_A|^2 - |G_T|^2 - |G'_T|^2}{|G_A|^2 + |G'_A|^2 + |G_S|^2 + |G'_S|^2} \tag{9.180}$$

and the experimental value $a = -0.3343 \pm 0.0030$ in the decay of ^6He (Johnston *et al* 1963) shows that the Gamow–Teller operator is predominantly of axial-vector type. To determine the relative magnitude of the vector and axial-vector coupling constants, a must be measured in a mixed transition like neutron decay although such a measurement can give no information about the relative sign of these coupling constants.

It is not, of course, practicable to observe the angular correlation between electron and neutrino directly since neutrino detectors of high efficiency do not exist. What is required is some experimentally accessible parameter which is a sensitive function of a and here three candidates merit serious consideration. These are (a) the distribution of decay events as a function of the angle between electron and recoiling proton similar to the distribution studied in the decay of ^{6}He (Rustad and Ruby 1955, Ridley 1961, Johnston *et al* 1963); (b) the momentum spectrum of electrons emitted into a given range of angles referred to the proton momentum (Kotani *et al* 1952, Reynolds *et al* 1953, Rose 1953) and (c) the spectrum of recoiling protons either in coincidence with selected electrons or not (Kofoed-Hansen 1948, 1954, Kotani *et al* 1952, Bilen'kii *et al* 1960, Riehs 1968, Nachtmann 1968).

No experiment of type (a) has been performed for neutron decay although there have been two experiments of type (b) (Robson 1958, Trebukhovskii *et al* 1959, Vladimirskii *et al* 1961). The main feature of type (b) experiments is the existence of a strong back-to-back correlation between electron and proton momenta, resulting from momentum conservation but, because an integration over two detector solid angles is necessary, it is difficult to achieve high precision when the correlation coefficient is small as indeed it turns out to be. In fact the only results of high accuracy published to date have been based on method (c) (Grigor'ev *et al* 1967, Dobrozemski *et al* 1975, Stratowa *et al* 1978). These are listed in table 9.4 along with the polarization–dependent correlation coefficients A, B and D.

9.3.8.2 The angular-polarization coefficients A and B

In a pure Fermi ($0^+ \rightarrow 0^+$) transition there can be no nuclear polarization and the leptons are emitted into a singlet ($I=0$, $M=0$) state. In a pure Gamow–Teller transition nuclear polarization is possible and the leptons are emitted into a triplet state ($I=1$, $M=0$, ± 1). However, the $M=0$ substate defined with respect to the polarization does not contribute to the angular correlation which derives from the $M=\pm 1$ substates only. This is the origin of the parity-violating angular symmetry first observed in the β decay of polarized ^{60}Co (Wu *et al* 1957).

In a mixed transition such as neutron decay there is an additional correlation arising from interference between singlet and triplet substates with $M=0$, and either the electron or the neutrino asymmetry will be enhanced in comparison with the pure Gamow–Teller transition, the other being reduced in proportion. Which is the larger depends on the relative phase of the interfering Fermi and Gamow–Teller amplitudes. Since it turns

Table 9.4. Experimental values for the angular-polarization correlation coefficients in free neutron decay

a	A	B	D	Reference
0.07 ± 0.12				Robson (1958)
	-0.114 ± 0.019	0.88 ± 0.15	-0.04 ± 0.05	Burgy *et al* (1960)
-0.06 ± 0.13				Vladimirskii *et al* (1961)
	-0.09 ± 0.05	0.96 ± 0.40	-0.14 ± 0.20	Clark and Robson (1961)
-0.091 ± 0.039				Grigor'ev *et al* (1967)
			-0.01 ± 0.01	Erozolimskii *et al* (1968, 1969, 1970)
		0.993 ± 0.034		Erozolimskii *et al* (1970a, 1971)
	-0.115 ± 0.009	1.01 ± 0.05		Christensen *et al* (1970)
	-0.118 ± 0.010			Erozolimskii *et al* (1971b)
		0.995 ± 0.038		Erozolimskii *et al* (1974)
			$-(1.1 \pm 1.7) \times 10^{-3}$	Steinberg *et al* (1974)
-0.099 ± 0.011				Dobrozemski *et al* (1975)
	-0.113 ± 0.006			Krohn and Ringo (1975)
			$+(2.2 \pm 3.0) \times 10^{-3}$	Erozolimskii *et al* (1978)
-0.1017 ± 0.0051				Stratowa *et al* (1978)
	-0.115 ± 0.006			Erozolimskii *et al* (1976)
	-0.114 ± 0.005			Erozolimskii *et al* (1979)
	-0.1146 ± 0.0019			Bopp *et al* (1986)
	-0.1131 ± 0.0014			Erozolimskii *et al* (1990, 1991)

out that the phase angle $\varphi = 180°$ (cf. 9.131) the neutrino correlation coefficient B is enhanced and the electron coefficient A reduced; it is the latter coefficient which is important therefore, since its value is proportional to the difference between the polar-vector and axial-vector coupling constants, and provides a direct measure of the renormalization of the axial-vector coupling constant by the strong interactions. In this sense therefore the very existence of a non-vanishing A in neutron decay has the same significance for the weak interaction as the Lamb shift or the electron g-factor anomaly have for the electromagnetic interaction. The value of B, on the other hand, is not sensitive to small changes in the ratio of the coupling constants and its precise value is of lesser significance.

Experimental investigations of the angular-polarization correlation coefficients in polarized neutron decay have been in progress for more than three decades. These were initiated at the Argonne National Laboratory and Chalk River Nuclear Laboratories and further developed at the Kurchatov Institute of Atomic Energy and, more recently at the ILL Grenoble. The methods adopted by the various groups have been reviewed in a number of places (Erozolimskii 1975, Byrne 1978, 1982). These have followed broadly similar lines, with some differences in detail, and, with one significant exception, all have used coincidence counting of protons and electrons to reduce the background. The exceptional case is the Grenoble experiment for the measurement of A, which dispensed with coincidence counting, thereby eliminating the systematic error associated with the neutron spin-proton momentum correlation (Bopp *et al* 1986). This experiment is described briefly below.

In this general class of experiments a particular orientation of the coplanar momentum vectors $\boldsymbol{p}_e$, $\boldsymbol{p}_{\bar{\nu}}$ and $\boldsymbol{p}_p$ is made, relative to the neutron spin $\boldsymbol{I}_n$ where $\boldsymbol{p}_e$ and $\boldsymbol{p}_p$ are set at an oblique angle so that $\boldsymbol{p}_e$ and $\boldsymbol{p}_{\bar{\nu}}$ are orthogonal. When the neutrons are polarized parallel to $\boldsymbol{p}_e$, $\boldsymbol{I}_n \cdot \boldsymbol{p}_e$ is maximized and $\boldsymbol{I}_n \cdot \boldsymbol{p}_{\bar{\nu}}$ is zero; thus the coefficient A can be determined by observing the relative change in the coincidence counting rate of electrons and protons when the neutron spin is reversed. In the same way polarizing the neutrons in the plane of $\boldsymbol{p}_e$ and $\boldsymbol{p}_p$, but normal to $\boldsymbol{p}_{\bar{\nu}}$, allows a determination of B. Finally to determine the triple correlation coefficient D, the neutrons must be polarized normal to the plane of $\boldsymbol{p}_e$ and $\boldsymbol{p}_{\bar{\nu}}$. When the directions of electron and proton momenta are selected in this way, the neutrino direction is determined within a certain finite range irrespective of the energy acceptance of the two detectors and these were essentially the experimental conditions established in the earlier generation of experiments (Burgy *et al* 1960, Clark and Robson 1961). However, such a procedure is very uneconomical when high precision is sought and counting statistics become important, since far too many useful events are discarded. In the more recent experiments the aim has been rather to extend the range of acceptable events by accurate labelling of the energy and momentum parameters.

The superconducting electron spectrometer PERKEO, first used to determine the correlation coefficient A (Bopp *et al* 1986, 1988, Klemt *et al* 1988), and subsequently modified to perform a neutron lifetime experiment (Last *et al* 1988) is shown in figure 9.14. The spectrometer is made from a 1.7 m long, 20 cm diameter superconducting solenoid which generates a 15 kG uniform magnetic field in its interior. In this field electrons of energy $\leqslant 782$ keV, created in neutron β-decay, are constrained to move in helical orbits less than 1 cm in diameter. The neutron beam passes parallel to the solenoid axis and trim coils arranged at each end of the solenoid force the magnetic field lines to bend vertically upwards so that the emergent electrons are detected in plastic scintillators placed outside and above the neutron beam. In this geometry an electron back-scattered from a scintillator at one end of the solenoid is recorded in the scintillator placed at the other end.

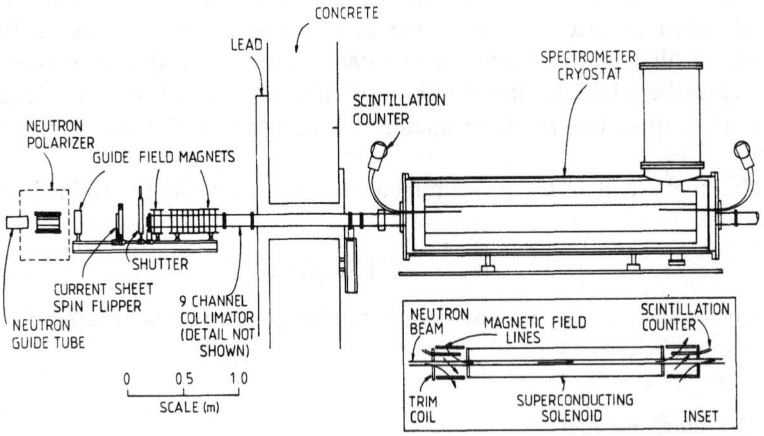

Figure 9.14. The superconducting electron spectrometer PERKEO used to measure both the electron momentum—neutron spin correlation coefficient A in the decay of longitudinally polarized neutrons and the neutron lifetime (Bopp *et al* 1986, Last *et al* 1988).

Each scintillator is coupled by light pipes to two photomultiplier tubes operated in coincidence, so that only those events which produce measurable pulses in both photo-tubes are recorded, thereby reducing the background. The electron energy is determined by the sum of the outputs of both photo-tubes with energy calibration provided by internal conversion sources of ^{109}Cd, ^{113}Sn and ^{207}Bi. When counters at both ends of the spectrometer are struck within a short time interval, the identity of the counter struck first is determined by timing, a feature which permits correcting for the $\simeq 1\%$ of events where an electron is first detected in one counter and is then back-scattered into the other.

The neutron beam is collimated by a 1.7 m long 9-channel Soller collimator made from ^{6}Li plates, and is polarized to a value of 97.4±0.5% by a supermirror polarizer. After leaving the polarizer the neutron spin is rotated by a series of guide fields so that it is oriented parallel to the longitudinal field in the spectrometer. The neutron polarization is periodically reversed by a non-adiabatic current sheet spin-flipper (cf. section 7.4.2), and the correlation coefficient $A(E_e)$ is thus determined at each electron energy E_e through the measured counting rate asymmetry

$$\frac{N_i^{\uparrow}(E_e)-N_i^{\downarrow}(E_e)}{N_i^{\uparrow}(E_e)+N_i^{\downarrow}(E_e)}=\frac{1}{2}\left(\frac{v_e}{c}\right)A(E_e)(1+\eta)S \qquad i=1,2 \qquad (9.181)$$

where $N_i^{\uparrow}(E_e)$ ($N_i^{\downarrow}(E_e)$) is the electron count-rate in the ith counter for the neutron spin in the up (down) orientation. The term $\eta \simeq 0.985 \pm 0.001$ is a correction introduced to allow for incomplete spin-reversal and $S=0.883\pm0.005$ is a correction for the magnetic mirror effect. This comes about when charged particles move slowly from a region of low field to a region of high field and are reflected backwards due to the requirement to conserve the adiabatic invariants. The effect has also been significant in several neutron lifetime experiments (Christensen *et al* 1972, Byrne *et al* 1980).

The final result of the measurement of the A coefficient described above was

$$A(0)=-0.1146\pm0.0019 \qquad (9.182)$$

which implies a value for the coupling constant ratio $\lambda=-1.262\pm0.005$ (cf. section 9.3.7.4).

9.3.8.3 Weak magnetism

According to (9.181) the counting rate asymmetry depends on energy primarily through the factor (v_e/c) which was indeed observed in this particular experiment for the first time. However there is also a weak dependence on E_e through the correlation coefficient $A(E_e)$ which is given to first order in E_e/m_nc^2 by (Holstein 1974)

$$A(E_e)=A(0)\left\{1-\frac{\frac{2}{3}E_e/m_nc^2}{A(0)(1+3|\lambda|^2)}[11|\lambda|^2-7|\lambda|+(5|\lambda|-1)(g_M/g_V)\right.$$
$$\left.+3A(0)(1+5|\lambda|^2+2|\lambda|(g_M/g_V))]\right\} \qquad (9.183)$$

Here g_M is the weak magnetism form factor defined in (9.127), whose predicted value according to CVC theory is given by (9.136), and second-class currents have been assumed to vanish. Thus in principle the CVC prediction could be tested in neutron β-decay by measuring the energy dependence of $A(E_e)$.

Unfortunately the energy release is so small that $(E_e/A(0)(\mathrm{d}A(0)/\mathrm{d}E) \leqslant 2\%$ and the experiment would be difficult to perform even with intense beams of highly polarized neutrons.

However, the weak magnetism term has been detected by measuring the shape factors $S(E_e)$ for electron and positron emission by unpolarized parent nuclei as originally proposed for the system of $A=12$ nuclei (Gell-Mann 1958). Subject to electromagnetic and other corrections (Calaprice and Holstein 1978) the approximate relationship is

$$S(E_e) = \gamma^2 \pm (4E_e/3Mc^2)\beta\gamma \tag{9.184}$$

for a β-transition between initial and final states which are not isotopic analogues. Here the $(+)$ and $(-)$ signs refer to electron and positron emission respectively, γ is a term which may be measured from the (pure Gamow–Teller) decay rate, and β is the weak magnetism term predicted by CVC theory as a function of the lifetime of the analogue $M1$ transition (cf. figure 3.5). For the $A=12$ system the predicted value of $(\mathrm{d}S/\mathrm{d}E)^{-}$-$(\mathrm{d}S/\mathrm{d}E)^{+}$ is equal to 0.93 (Holstein 1974), in good agreement with the measured values of (1.00 ± 0.13) (Lee *et al* 1963, Wu 1964, Wu *et al* 1977) and (0.98 ± 0.14) (Kaina *et al* 1977).

9.3.8.4 The triple angular-polarization correlation coefficient D

As has been stated in section 9.3.2 the existence of a non-vanishing coefficient D for the triple scalar product $\langle \boldsymbol{I}_n \rangle \cdot \boldsymbol{p}_e \times \boldsymbol{p}_{\bar{\nu}}$ is evidence for a violation of the principle of time reversal invariance in β-decay. In order to establish a violation of this principle in a decay process it is necessary to show that some feature of the decay depends on a term which changes sign under time reversal, but not under space inversion. In the decay of a polarized neutron the term $\langle \boldsymbol{I}_n \rangle \cdot (\boldsymbol{p}_e \times \boldsymbol{p}_{\bar{\nu}})$ possesses the required property and, within certain limits discussed briefly below, the observation of a dependence of the decay rate on this term may be regarded as conclusive for a violation of the postulated invariance. The earliest determinations of the triple correlation coefficient D (Burgy *et al* 1960, Clark and Robson 1961) gave null results showing that the β-decay interaction was time-reversal-invariant to within 5–10%. The most recent measurements continue to show null results but the precision has been refined to a level approaching 0.1% (Steinberg *et al* 1974, Erozolimskii *et al* 1978).

Since the aim of any technique designed to determine D must be to maximize $|\boldsymbol{I}_n \cdot \boldsymbol{p}_e \times \boldsymbol{p}_{\bar{\nu}}|$ an obvious procedure is to use longitudinally polarized neutrons and to detect electrons and protons in counters whose axes are normal to the beam in a configuration in which $\boldsymbol{I}_n, \boldsymbol{p}_e$ and $\boldsymbol{p}_{\bar{\nu}}$ constitute a triad of mutually orthogonal vectors. Figure 9.15 shows how this was achieved in the experiment carried out by Erozolimskii and associates (Erozolimskii *et al* 1968, 1978). In this arrangement neutrons with 85% polarization were obtained by reflection at a cobalt mirror and the polarization was maintained

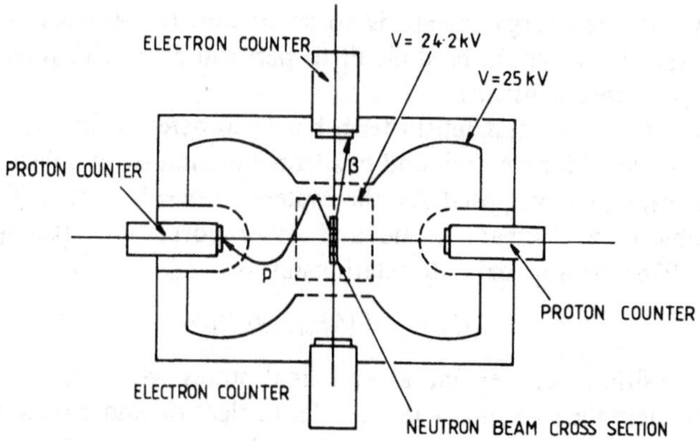

Figure 9.15. Apparatus used to search for a finite T-violating triple correlation coefficient D in the decay of polarized neutrons (Erzolimskii *et al* 1978).

parallel to the beam axis using a 2 G guide field. Non-adiabatic spin reversal was brought about using a current sheet spin flipper. The neutron decays were sampled from a field-free region within a gridded cylinder at 24.2 kV whose axis was normal to the common axis of two electron detectors placed above and below the beam in a symmetrical configuration. On leaving this cylinder through grids at either end, the protons entered spherical fields where they were focused onto two symmetrically located 20 μm thick CsI detectors at ground potential. The field-free region was surrounded by an additional coaxial gridded cylinder set at 25 kV, the gap between the two cylinders acting as a mirror to reflect protons back into the field-free region without altering their velocity components along the proton detector axis.

The value of D was derived from the counting rate asymmetry R through the formula

$$R=(N^+-N^-)/(N^++N^-)=Kf_nD \tag{9.185}$$

where K is an instrumental coefficient obtained by averaging over energy and solid angle, which had the value $K=0.47\pm0.03$. The counting rates N^+ and N^- were defined by the relations

$$\begin{aligned} N^+ &= N^{\uparrow}_{\beta_1 p_1}+N^{\uparrow}_{\beta_2 p_2}+N^{\downarrow}_{\beta_1 p_2}+N^{\downarrow}_{\beta_2 p_1} \\ N^- &= N^{\downarrow}_{\beta_1 p_1}+N^{\downarrow}_{\beta_2 p_2}+N^{\uparrow}_{\beta_1 p_2}+N^{\uparrow}_{\beta_2 p_1} \end{aligned} \tag{9.186}$$

where $N^{\uparrow}_{\beta_i p_j}$ $(N^{\downarrow}_{\beta_i p_j})$ denote coincidence counting rates in the ith electron counter and the jth proton counter for the two possible directions of the neutron spin signified by the arrows (↑↓). The final value obtained for D was

$$D=-(2.2\pm3.0)\times10^{-3} \tag{9.187}$$

A geometrical arrangement very similar to that shown in figure 9.15 was employed in the measurement of the triple correlation coefficient at the ILL Grenoble (Steinberg *et al* 1974), the principal difference being that two symmetrical counting systems were employed incorporating eight detectors in all. Cold neutrons ($\lambda \approx 2.8$ Å) were used and both polarizer and analyser were made from totally reflecting curved guides. The initial polarization was transverse to the beam direction but was turned into the longitudinal mode using a two-coil spin flipper and maintained there by a 3 G guide field. The electrons were detected in plastic scintillators and the protons in 4000 Å thick layers of NaI.

The value of D obtained in this work was derived from the formula

$$R=\frac{N^{\downarrow}_{\beta_1 \mathrm{p}_1} N^{\uparrow}_{\beta_1 \mathrm{p}_2} N^{\uparrow}_{\beta_2 \mathrm{p}_1} N^{\downarrow}_{\beta_2 \mathrm{p}_2}}{N^{\uparrow}_{\beta_1 \mathrm{p}_1} N^{\downarrow}_{\beta_1 \mathrm{p}_2} N^{\downarrow}{}_{\beta_2 \mathrm{p}_1} N^{\uparrow}_{\beta_2 \mathrm{p}_2}}=\frac{(1+Kf_{\mathrm{n}}D)^2}{(1-Kf_{\mathrm{n}}D)^2} \approx 1+4Kf_{\mathrm{n}}D \tag{9.188}$$

where the symbols are as defined above and the instrumental coefficient K had the value $K=0.45 \pm 0.05$. The final result was

$$D=-(1.1 \pm 1.7) \times 10^{-3} \tag{9.189}$$

The complete set of measurements of D published to date is given in table 9.4 together with values for the other correlation coefficients in neutron decay.

It may be observed that electromagnetic interactions of the particles in the final state may generate terms in the correlation proportional to the triple scalar product which in some circumstances can simulate a failure of time reversal invariance. These effects are due to the fact that time reversal is an anti-unitary symmetry transformation unlike space inversion which is unitary. Thus only terms odd under T which are first order in the weak interaction provide evidence of the breakdown of time reversal symmetry. Terms which are second order in the electromagnetic interaction, but which are odd under T, have no such implication. In the case of neutron decay the contribution to D of these effects is estimated at about 2×10^{-5} which still lies well beyond the accuracy which has been achieved to date (Callan and Treiman 1967).

The T-odd P-even triple correlation coefficient D receives T-violating contributions which are only of second order in the phase δ_{13} of the CKM matrix (cf. section 9.3.1.3) but which are of first order in the mixing angle ϕ of left–right symmetric models (cf. section 10.3.1). There are, however, also T-odd P-odd correlations e.g. (i) $\boldsymbol{\sigma}_{\mathrm{e}} \cdot (\boldsymbol{I}_{\mathrm{n}} \times \boldsymbol{p}_{\mathrm{e}})/E_{\mathrm{e}}$ and (ii) $\boldsymbol{\sigma}_{\mathrm{e}} \cdot (\boldsymbol{p}_{\mathrm{e}} \times \boldsymbol{p}_{\bar{\nu}})/E_{\mathrm{e}}E_{\bar{\nu}}$, where $\boldsymbol{\sigma}_{\mathrm{e}}$ is the electron spin matrix, which like the neutron electric dipole moment, can receive first-order contributions involving a scalar-type coupling in models involving more than one Higgs doublet. Since the measurement of electron polarization $\langle \boldsymbol{\sigma}_{\mathrm{e}} \rangle$ is notoriously inefficient ($\leqslant 10^{-3}$), the correlation (ii), which involves the simultaneous determination of the proton recoil momentum, is probably unobservable with present fluxes of neutrons, even though polarized neutrons are not required. The correlation (i) has yet to be studied in neutron decay, although it has been measured in ^{19}Ne decay where the corresponding coefficient takes the value $R=0.079\pm0.053$ (Schneider *et al* 1983), and in ^{8}Li where $R=0.004\pm0.014$ (Allet *et al* 1992).

9.3.9 Status of neutron decay data

It is apparent that the discrepancies between the various measured and 'theoretical' values of τ_n, which have existed for a very long time, have now been resolved to a very large extent. For example the world average experimental value of the neutron lifetime, $\tau_n=889.6\pm2.1$ s agrees very well with the value $\tau_n=890.7\pm1.8$ s based on the global value of V_{ud} listed in table 9.2, which is based on neutron decay data and on data from superallowed Fermi decays (Hardy *et al* 1990, Wilkinson 1990). The value $\tau_n=897\pm3$ s derived from the measured lifetime of ^{3}H is somewhat larger but in reasonably close accord (Budick *et al* 1991). However, some discrepancies remain; for

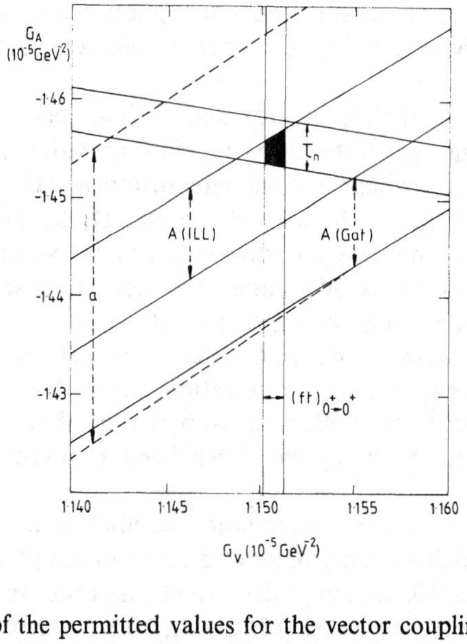

Figure 9.16. Plots of the permitted values for the vector coupling constant G_V and the axial vector coupling constant $G_A=\lambda G_V$, derived from measured values of the neutron lifetime, the *ft*-values of superallowed Fermi $0^+\rightarrow0^+$ β-transitions, the electron–neutron spin asymmetry coefficient A and the electron–neutrino angular correlation coefficient a. The allowed bands correspond to one standard deviation on either side of the mean value and all values have been subjected to radiative corrections up to order α (Wilkinson 1991, Towner 1992). It is clear that, whereas the value A (ILL) measured at the Institut Laue Langevin (Bopp *et al* 1988) provides a consistent solution within the standard model, the Gatchina value A (Gat) (Erozolimskii 1990, 1991) is discrepant in this respect.

example if the currently accepted value $\lambda = 1.262 \pm 0.005$ (Bopp *et al* 1986, Klemt *et al* 1988) (cf. section 9.3.8.2) is replaced by the significantly different value $\lambda = 1.2544 \pm 0.0036$ based on a recent measurement of the angular-polarization correlation coefficient A (Erozolimskii *et al* 1990, 1991), the 'predicted' value of τ_n increases to 906.4 ± 4.0 s, leading to a discrepancy with the measured value of τ_n well outside the quoted experimental error (Gaponov *et al* 1990, Erozolimskii and Mostovoi 1991). Thus this measured value of A does not lead to values of the weak coupling constants which are mutually consistent within the standard model (Wilkinson 1991). These results are illustrated in figure 9.16. It may, however, be unwise at this stage to attach too much importance to a two standard deviation difference between the two measured values of A. Indeed it may be observed that the two values of the neutron lifetime derived from measurements in a fluid-walled UCN bottle (Mampe *et al* 1989, 1993) also differ by close to two standard deviations. Thus the experimental situation in relation to neutron decay remains somewhat confused and further experiments are needed.

10 Fundamental processes and the role of the neutron

10.1 Parity violation in the nucleus

10.1.1 The weak interaction between nucleons

In the Cabibbo formulation of weak interaction theory the hadronic component of the charged weak current is given according to (9.143), where the strangeness-conserving ($\Delta S = 0$) and strangeness non-conserving ($\Delta S = \pm 1$) currents are constructed from vector and axial-vector octets. Thus there is a weak interaction between hadrons which is given for nucleons by the strangeness-conserving part of the contraction of J_μ^{h} with itself. This results in a charged-current weak internucleon Hamiltonian density

$$H_{\mathrm{W}}^{\mathrm{c}} = \frac{G_{\mathrm{F}}}{\sqrt{2}}[\cos^2\theta_{\mathrm{c}} J_\mu^h(\Delta S = 0) J_\mu^{\mathrm{h}\dagger}(\Delta S = 0) + \sin^2\theta_{\mathrm{c}} J_\mu^h(\Delta S = \pm 1) J_\mu^{\mathrm{h}\dagger}(\Delta S = \pm 1)] \tag{10.1}$$

In the Cabibbo theory of course the neutral currents are missing so that $H_{\mathrm{W}}^{\mathrm{c}}$ must be supplemented by the neutral-current strangeness-conserving contribution

$$H_{\mathrm{W}}^{\mathrm{n}} = \frac{G_{\mathrm{F}}}{\sqrt{2}} 2\rho J_\mu^{\mathrm{h}}(\Delta S = 0) J_\mu^{\mathrm{h}\dagger}(\Delta S = 0) \tag{10.2}$$

where the parameter ρ has the value unity in the Weinberg–Salam model (cf. section 10.2).

To lowest order in G_{F} the weak internucleon potential V_{w} is given by diagrams involving one weak and one strong vertex as illustrated in figure 10.1. Since, according to the Cabibbo hypothesis $J_\mu^{\mathrm{h}}(\Delta S = 0)$ is an isovector and $J_\mu^{\mathrm{h}}(\Delta S = \pm 1)$ an isospinor, the $\cos^2\theta_{\mathrm{c}}$ and $\sin^2\theta_{\mathrm{c}}$ terms in (10.1) have isospin selection rules $\Delta T = 0,1,2$ and $\Delta T = 0,1$, respectively. However, analysis of the behaviour of the $\cos^2\theta_{\mathrm{c}}$ term under the operation of charge symmetry shows that it is invariant, under the usual assumption that the charge-raising and charge-lowering operators belong to the same isovector. However, all $T = 1$ parity-violating couplings corresponding to π,ρ,ω exchange, etc., change sign under this operation and we conclude therefore that $\Delta T = 1$ processes are excluded from the $\cos^2\theta_{\mathrm{c}}$ term in (10.1). Furthermore CP conservation requires that single pion exchange contributes only to the $T = 1$ coupling (and that neutral π and ρ exchange be excluded altogether). Thus the $\cos^2\theta_{\mathrm{c}}$ term is dominated by vector meson exchange leading to a Yukawa-type potential with a range characteristic of the ρ-meson Compton wavelength. There may also be a significant contribution from 2π exchange diagrams.

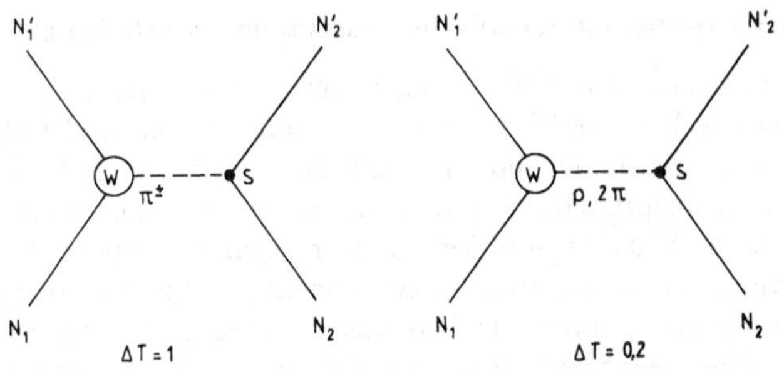

Figure 10.1. Feynman diagrams representing the weak interaction between nucleons in lowest order. The symbols W, S denote weak and strong vertices respectively. The $\Delta T=1$ potential is responsible for any circular polarization P_c of the γ-ray emitted in the p(n,γ)d reaction while the $\Delta T=0,2$ potential generates the photon momentum – neutron spin angular correlation coefficient A_γ.

Even though $\theta_c \simeq 0.22$ the $\sin^2 \theta_c$ term in (10.1) cannot be neglected since its isospin selection rule $\Delta T=1$ implies that the term is dominated by single charged-pion exchange; the corresponding Yukawa potential therefore has a comparatively long range, comparable with the Compton wavelength of the π meson. A further significance to this term is that neutral currents can lead to enhancement factors of the order of 40 whereas the likely enhancement of the $\cos^2\theta_c$ term by neutral currents is of the order of 2. The implication is that neutral-current effects are expected to be more prominent in processes for which $\Delta T=1$ than in those for which $\Delta T=0,2$.

The existence of the weak internucleon potential may be established experimentally by seeking out the characteristic fingerprint of weak processes, namely that they do not conserve parity. One of the most successful techniques employed to date to demonstrate the existence of parity-violating nuclear forces has been the study of polarized neutron reactions with complex nuclei (Krupchitsky 1989) and these are discussed in section 10.1.2 immediately below. Ideally, of course, one would like to study the two-nucleon system where there are essentially four parameters available for experimental investigation (Danilov 1965, 1971, 1972, McKellar 1975). These are (i) the analysing power A_p for proton longitudinal polarization in p–p scattering, (ii) the analysing power A_n for neutron longitudinal polarization in n–p scattering, (iii) the circular polarization P_c in radiative capture of unpolarized thermal neutrons by protons and (iv) the asymmetry coefficient A_γ of γ emission with respect to the neutron spin in the radiative capture of polarized thermal neutrons by protons.

A fifth process, coherent parity violation in the elastic forward scattering of transversely polarized neutrons by protons, is related to (ii) and provides the same information; this phenomenon is described in some detail in section 10.1.4.

10.1.2 Parity violation in radiative neutron capture, scattering and fission

Nuclear states are known to be parity eigenstates, at least to a very good approximation. However, if the self-interacting current model of the weak interaction is a good one, and all recent theories reduce to this form in the limit of zero momentum transfer, then the nuclear force must contain a parity violating term. Thus, in addition to their regular component ψ_r nuclear wavefunctions contain an irregular component ψ_i, of parity opposite to that of ψ_r, and of relative amplitude F. It was observed in the very early days of parity-violating phenomena that, to isolate the irregular component in the wavefunction of a complex nucleus, essentially three classes of phenomena can be investigated (Wilkinson 1958). These are (i) violation of absolute parity selection rules giving an effect of order F^2, (ii) observation of photon circular polarization or particle longitudinal polarization in radiation from an ensemble of unpolarized nuclei which is an effect proportional to F and (iii) observation of odd powers of $\cos\theta$ in the angular distribution of nuclear radiation relative to the momentum of an incident unpolarized particle, again giving an effect proportional to F.

Parity non-conserving processes in classes (i) and (iii) are exceedingly difficult to detect because their relative intensity is of the order of 10^{-12}. Nevertheless one such effect has been observed, namely the parity-forbidden α decay in the 2^- state in ^{16}O (Hättig *et al* 1970) which has been detected at a level consistent with theoretical prediction. The parity violating effect most commonly encountered in nuclei is the class (ii) phenomenon of circular polarization of γ radiation from unpolarized heavy nuclei (e.g. ^{41}K, ^{181}Ta, ^{175}Lu, ^{180}Hf) which is an interference effect between $M1$ and $E1$ components of the radiation field which have opposite parities. The parity violating effects which are most prominent in the interactions of neutrons with complex nuclei are class (ii) processes according to the Wilkinson classification, and are therefore proportional to F. These are (a) the angular asymmetry A_γ in the distribution of γ rays referred to the neutron spin in the radiative capture of polarized thermal neutrons and (b) the angular asymmetry in the distribution of the light fragment A_l, or of the fission neutrons A_{fn}, with respect to the spin of the incident neutron, in the slow-neutron-induced fission of heavy nuclei with polarized neutrons. Non-vanishing values of A_γ were first observed in complex nuclei and two such reactions, ^{113}Cd(n,γ)^{114}Cd and ^{117}Sn(n,γ)^{118}Sn have to date been studied in some detail. In both cases the target nucleus has $I=\frac{1}{2}$ so that capture can take place in either the singlet or the triplet state. Only the triplet state, of course, can be polarized. The finite value of A_γ is then a result of $E1$–$M1$ interference between the regular singlet transition and the irregular triplet transition and conversely.

In an extensive series of experiments with longitudinally polarized epithermal neutrons carried out at LAMPF, very large ($\simeq 10\%$) transmission asymmetries have been observed at the centres of $p_{1/2}$-wave resonances in

^{232}Th and ^{238}U (Alfimenkov *et al* 1983, Masuda *et al* 1989, Bowman *et al* 1990). The problem has been not so much to explain the magnitude of these effects, which may be put down to a magnification of the parity-violating matrix element $\langle S_{1/2}|V_w|P_{1/2}\rangle$ associated with the complex nature of the compound states (Sushkov and Flambaum 1982), but that the overwhelming majority of the measured asymmetries have the same sign (Frankle *et al* 1991, 1992, Zhu *et al* 1992). These observations contrast sharply with the description 'quantum chaos' applied to the distribution of resonances in these nuclei (cf. section 4.7.3.2). Recent attempts to account for the phenomenon (Bowman *et al* 1992) are in conflict with earlier analyses of parity-violating neutron scattering in heavy nuclei (Noguera and Desplanques 1986).

The proposal to look for parity-violating parameters in nuclear fission is quite an old one (Andreev and Vladimirskii 1961, Budnik and Rabatov 1973). These are also interference effects but the very complexity of the fission process itself makes the theoretical description of those processes that much more difficult (Danilyan 1980). Parity violating effects in nuclei are cumulative and are therefore more readily detectable in heavy nuclei; furthermore they can be enhanced by choosing for investigation those nuclei where there is a selection rule operating which hinders the regular transition in comparison with the irregular one. Conversely, because enhancement is a nuclear structure effect, it is difficult to disentangle the contribution of parity mixing in the static nuclear state and parity mixing in the neutron–nucleus interaction. This makes the interpretation of the experimental data in terms of the properties of the weak current very difficult (Vanhoy *et al* 1988).

10.1.3 Parity violation in the neutron–proton system

Because there exists the possibility of interpreting the experimental data in terms of specific models of the nucleonic weak interaction the n–p system is clearly of outstanding importance. This importance is reflected in the fact that, although for a long time only two searches for parity violation in the p(n, γ)d capture reaction had been reported, one on the circular polarization P_c (which is now known to be in error) (Lobashev *et al* 1972) and one on the γ-ray asymmetry A_γ (Cavaignac *et al* 1977), these experiments aroused enormous theoretical interest. An essential feature of the theoretical interpretation of this work is that, whereas A_γ is sensitive to the $\Delta T=1$ part of the weak interaction, P_c provides information on the $\Delta T=0,2$ contribution. The experimental arrangements for these experiments are illustrated in figure 10.2.

Apart from a $\simeq 4\%$ admixture of $^3D_1$ state as evidenced by the observation of a finite electric–quadrupole moment, the ground state of the deuteron has a $^3S_1$ configuration; there are therefore two S-wave capture states in the continuum of the n–p system, namely $^3S_1$ and $^1S_0$. Thus, since all the states involved have even parity and $I=0$ or 1, the regular γ-ray transition to the

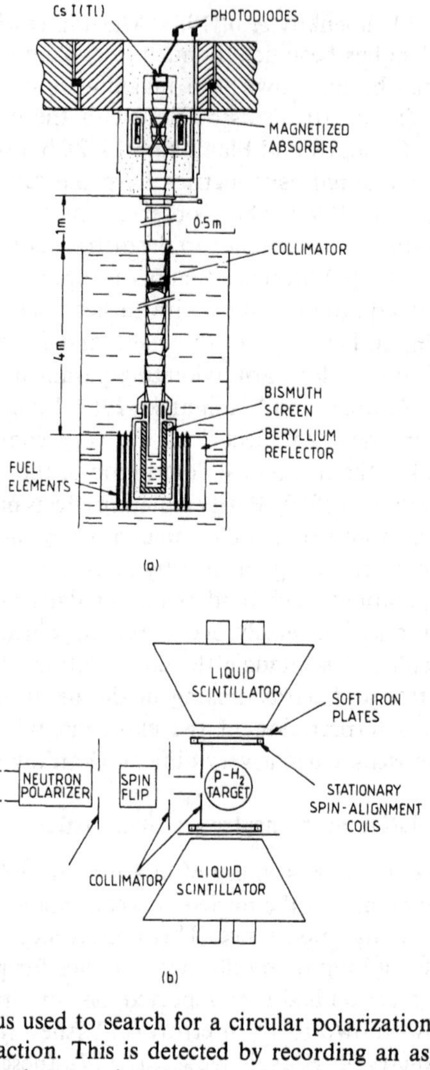

Figure 10.2. (a) Apparatus used to search for a circular polarization P_c of the γ-ray emitted in the p(n,γ)d reaction. This is detected by recording an asymmetry in the counting rates of transmitted γ-rays when the magnetization in the absorber is reversed with respect to the γ-ray momentum (Lobashev *et al* 1972). (b) Apparatus used to detect a parity violating angular correlation coefficient A_γ in the same reaction (Cavaignac *et al* 1977).

ground state is of magnetic-dipole type. The transition rate is therefore determined by the matrix element of the spin magnetic moment operator between the pairs of S states involved. However, this operator acts on the spin coordinates only, and the spatial parts of the wavefunctions for the $^3S_1$ states, being eigenstates of the triplet potential for the deuteron, are orthogonal. It follows that, excluding velocity-dependent terms in this potential (Breit and Rustgi 1971), the $M1$ matrix element connecting the $^3S_1$ states vanishes and capture is confined to the $^1S_0$ state. A theoretical calculation of the thermal capture cross-section for the $^1S_0$ state gives the value $\sigma_c = (302.5 \pm 4.0) \times 10^{-3}$ b (Noyes 1965, 1967) whereas the experimental value is $\sigma_c = (334.2 \pm 0.5) \times 10^{-3}$ b (Cox *et al* 1965), a discrepancy which is usually explained in terms of meson exchange currents involving the D-state component of the ground state, rather than to $^3S_1$ state capture (Riska and Brown 1972).

Applying the rule that, for states of the two-nucleon system, the sum of the quantum numbers $l+S+T$ must be an odd integer, we find that the $^3S_1$ round state of the deuteron has $T=0$ whereas the $^1S_0$ capture state has $T=1$. The existence of a parity violating component in the nucleon–nucleon interaction implies that either or both of the initial and final states contain admixtures of P states. This means that the irregular components $^1P_1$ ($T=0$) and $^3P_0$ ($T=1$) admixed into the $^3S_1$ ($T=0$) and $^1S_0$ ($T=1$) states, respectively, come from the $\Delta T = 0,2$ (ρ exchange) weak scalar potential, whereas the $^3P_1$ ($T=1$) irregular component in the $^3S_1$ states comes from the $\Delta T=1$ (π exchange) weak scalar potential. Thus the weak n–p interaction is governed by three weak mixing amplitudes $\langle ^3S_1 | V_w | ^1P_1 \rangle$ (isoscalar), $\langle ^1S_0 | V_w | ^3P_0 \rangle$ (isoscalar, isotensor) and $\langle ^3S_1 | V_w | ^3P_1 \rangle$ (isovector).

The spin and isospin assignments for the bound and low-energy continuum states of the n–p system are illustrated in figure 10.3. The figure also shows the regular and irregular electromagnetic transitions together with their multipolarity classifications. Here $m(I)$ and $e(I)$ are the electric and magnetic multipole reduced matrix elements labelled by the total spin of the initial

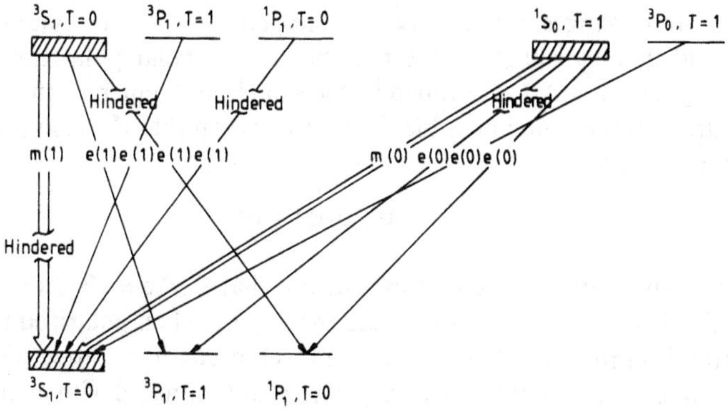

Figure 10.3. Diagram showing the spin and isospin assignments for the bound and low energy continuum levels of the n–p system and the electromagnetic dipole matrix elements $e(I_i)$, $m(I_i)$ linking them. A circular polarization P_c in the reaction p(n,γ)d is determined by $e(0)m(0)$ interference and the correlation coefficient A_γ by $e(1)m(0)$ interference.

channel. In the same notation P_c and A_γ are given by (Gari and Schlitter 1975)

$$P_c = 2[m(0)e(0) + m(1)e(1)]/m(0)^2 \simeq 2(e(0)/m(0))$$
$$A_\gamma = -\surd 2[m(0)e(1) + m(1)e(0)]/m(0)^2 \simeq -\surd 2(e(1)/m(0)) \qquad (10.3)$$

Of the irregular $E1$ transitions shown in figure 10.3 three are singlet-triplet intercombination lines and for these the matrix element is hindered from orthogonality of the spin states since the electric-dipole operator $\boldsymbol{r}$ does not act on the spin. Thus all the electromagnetic transitions with $\Delta T = 0$ are hindered, a result which is a general one for self-conjugate nuclei such as the deuteron (Morpurgo 1954). Intercombination lines can, however, arise through the relativistic operator $\boldsymbol{r} \times \boldsymbol{\sigma}$ and ultimately these contributions may prove important (Morioka and Ueda 1978, Kopeliovich 1981). The surviving electromagnetic transitions are now ${}^1S_0 \to {}^3S_1$, $m(0)$ (regular) and ${}^1S_0 \to {}^1P_1$ $e(0)$, ${}^3P_0 \to {}^3S_1$ $e(0)$ and ${}^3S_1 \leftrightarrow {}^3P_1$ $e(1)$ (irregular). It follows from (10.3) that the $\Delta T = 0,2$ weak potential does not contribute to A_γ nor the $\Delta T = 1$ weak potential to P_c.

Figure 10.2(a) shows a diagram of the apparatus used by Lobashev and his collaborators to measure the circular polarization P_c of the 2.2 MeV γ-ray emitted in the p(n,γ)d capture process with unpolarized neutrons (Lobashev *et al* 1972). The action of the γ-ray circular polarimeter was conventional in the sense of being based on the existence of a spin-dependent term in the Compton scattering of helical photons on polarized electrons (Schopper 1958, 1963). The essential technical innovation introduced in this experiment stemmed from the recognition that, to detect a circular polarization in the range $P_c \simeq 10^{-7}$–10^{-6}, γ-rays would have to be recorded at a rate which could not be resolved as differentiated pulses in the counters available. Thus, instead of recording individual pulses, an integral detector current was measured, and changes in the γ-ray intensity, correlated with magnetization reversal and thereby signalling the presence of a circular polarization, were observed by a resonant device tuned to the switching frequency, thus making accumulation of the signal possible. It is now known that the result obtained in this experiment

$$P_c = -(1.30 \pm 0.45) \times 10^{-6} \qquad (10.4)$$

is incorrect and more recent studies indicate that $P_c \leqslant 5 \times 10^{-7}$ (Knyaz'kov *et al* 1983). However it remains the case that $P_c \neq 0$ when polarized neutrons are captured on protons (Vesna *et al* 1981). One possible source of error in the original result could be that the target was contaminated with an impurity, e.g. Cl, for which $P_c \neq 0$ has been confirmed with unpolarized neutrons (Vesna *et al* 1982).

The γ-ray asymmetry A_γ in radiative capture of polarized neutrons has been determined using the apparatus illustrated in figure 10.2(b) (Cavaignac *et al* 1977). The neutrons were moderated in a liquid deuterium cold source to energies of the order of 1.5×10^{-2} eV which is the threshold at which normal molecular hydrogen, ortho-hydrogen ($I=1$), converts to para-hydrogen ($I=0$) in the presence of a catalyst. The point here is that atomic hydrogen has a capture cross-section for thermal neutrons $\sigma_c \simeq 0.3$ b which is small in comparison with the spin-dependent scattering cross-section $\sigma_s \simeq 95$ b. Thus an initial neutron spin polarization is rapidly dissipated by scattering in the target. However, in para-hydrogen $\sigma_s \simeq 4$ b and the scattering is much reduced; furthermore there is no depolarization since the para-hydrogen molecule has zero spin. The reduction in the scattering cross-section is an interference effect which becomes effective at energies such that the neutron wavelength is much greater than the separation of the protons in the para-hydrogen molecule (Schwinger and Teller 1937, Hamermesh and Schwinger 1947). The result of this experiment was

$$A_\gamma = (6.0 \pm 21) \times 10^{-8} \tag{10.5}$$

10.1.4 Coherent parity violation

By 'coherent parity violation' we understand parity violating effects associated with forward elastic scattering, the most prominent of these being the rotation of the neutron spin in nuclear scattering of transversely polarized neutrons, resulting from an effective weak interaction potential proportional to $\boldsymbol{\sigma}.\boldsymbol{p}$ between an incident neutron and an unpolarized target nucleus (Michel 1964, Stodolsky 1974). The fact that the scattering of the neutrons is coherent carries with it the implication that the atoms of the medium do not alter their states. Thus the medium contributes only to the spin-independent part, i.e. the vector part, and the neutrons to the axial-vector part of the weak current whose self-interaction determines the parity violating scattering potential.

To describe the weak rotation of the neutron spin we adopt the neutron optics approximation and assume that a neutron propagates in a medium of refractive index $n(\omega)$ with wavenumber $k = n(\omega)\omega/c$ where, for $|n^2(\omega) - 1| \ll 1$

$$n(\omega) \simeq 1 + 2\pi(c/\omega)^2 N f(\omega) \tag{10.6}$$

Here N is the number of scattering centres per unit volume and $f(\omega)$ is the coherent forward-scattering amplitude. If there is a weak potential present proportional to $\boldsymbol{\sigma}.\boldsymbol{p}$

$$f(\omega) = f_0(\omega) + p^{-1}\boldsymbol{\sigma}.\boldsymbol{p} f_1(\omega) \tag{10.7}$$

then there are two indices of refraction

$$n^{\pm}(\omega) = n_0(\omega) \pm n_1(\omega) \tag{10.8}$$

where the $(\pm)$ signs refer to positive and negative helicity states, respectively, for the incoming neutron.

One immediate consequence of the difference in the refractive indices of the two helicity states is that the neutron spin for a transversely polarized beam is rotated through an angle

$$\phi = -4\pi(c/\omega)\mathrm{Re}\,f(\omega)z \tag{10.9}$$

on transmission through a length z of the refracting medium. This rotation is analogous to the rotation of the plane of polarization of a beam of light transmitted through an optically active medium and the phenomenon is therefore termed 'weak optical activity'.

The imaginary part $\mathrm{Im}\,f(\omega)$ of the coherent forward-scattering amplitude is related to the total cross-section σ_t through the optical theorem (cf. section 2.4.2)

$$\sigma_t = 4\pi(c/\omega)\mathrm{Im}\,f(\omega) = \sigma_0(\omega) \pm \sigma_1(\omega) \tag{10.10}$$

Thus, in addition to the rotary power $\phi' = \mathrm{d}\phi/\mathrm{d}z$ there is an analysing power for neutron longitudinal polarization

$$A_n = \sigma_1(\omega)/\sigma_0(\omega) \tag{10.11}$$

an effect which may be termed 'weak circular dichroism'.

Of course, there are many other mechanisms by which a transversely polarized neutron can rotate its spin; if the medium contains a nuclear polarization for example. Neutron spin rotation in an anti-ferromagnet is a similar effect (cf. section 8.4.1.3). Another process which has excited some interest occurs when the medium itself is optically active; in this case there exists a resonant mechanism whereby the polarized neutron is first absorbed in an active nucleus and, while residing in the nucleus, the handedness of the orbital zero-point energy leads to a rotation of the neutron spin (Kabir *et al* 1974, Ritchie 1979). A non-resonant scattering based on spin-orbit coupling in the same system is less important (Kabir *et al* 1975). However, all these processes relate to the fact that the medium itself has a screw sense and have nothing to do with the weak interaction. Some or all of them may, however, be present in any experiment designed to look for weak spin rotation and they have to be taken into account.

As was first pointed out by Stodolsky (Stodolsky 1974, Harris and Stodolsky 1979) the weak neutron spin rotation can be induced not only by the neutron–nucleus interaction due to the charged current but also by the neutron–nucleus and neutron–electron weak interactions caused by neutral currents (Michel 1965, Bouchiat and Bouchiat 1974). Since the weak spin rotation is a parity-violating effect it arises as a result of interference between the vector and axial-vector parts of the weak current. However, in low-energy neutron scattering only the axial-vector part of the nucleonic current is spin-dependent and spin rotation therefore involves the vector part of the electronic weak neutral current; in the Weinberg–Salam model the relevant form factor $g_V^0 = -\frac{1}{2} + 2\sin^2\theta_w$ is very small since $\sin^2\theta_w \simeq 0.21$. Thus the contribution of the atomic electrons to the spin rotation is not large.

According to Stodolski (1974) the weak rotary power may be expressed

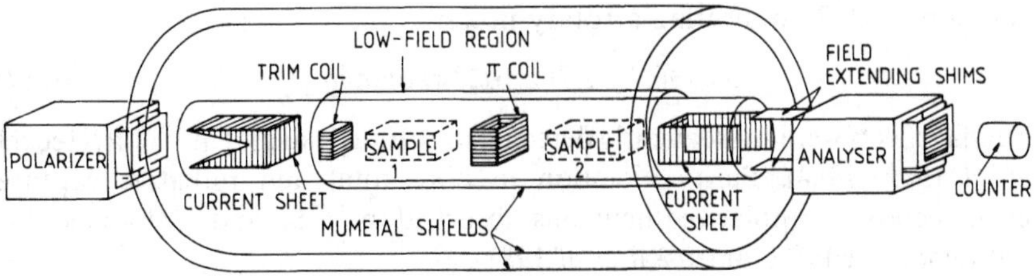

Figure 10.4. Apparatus used to measure coherent parity violation in the neutron–nucleus interaction by observing a small rotation of the neutron spin in forward scattering of transversely polarized neutrons from unpolarized nuclei. The imaginary part of the same parity violating amplitude is measured by detecting a differential absorption in the transmission of neutrons for opposite states of longitudinal polarization (Forte *et al* 1980).

in the form

$$\phi=\sqrt{2}NGF\{ZC_{\mathrm{ne}}+\eta[ZV_{\mathrm{np}}+(A-Z)C_{\mathrm{nn}}]\} \tag{10.12}$$

where GF is the Fermi coupling constant and C_{ne}, C_{np} and C_{nn} are weak charge numbers characterizing the neutron–electron, neutron–proton and neutron–neutron weak interactions, respectively, and η is a factor of the order of unity in the general case but which may be positive or negative. η represents a modification due to the strong nuclear forces which has an estimated value

$$\eta=1-2a/R-2b/R^3+ab/R^4 \tag{10.13}$$

where R is the nuclear radius, a is the neutron s-wave scattering length and b is a parameter characterizing the p-wave scattering.

The apparatus used to test for the presence of resonantly enhanced weak spin rotation in ^{124}Sn is illustrated in figure 10.4 (Forte *et al* 1980). The monochromatic 7 Å neutron beam with a flux of 10^7 neutrons $\mathrm{cm}^{-2}\,\mathrm{s}^{-1}$ was transversely polarized to about 91% by reflection from magnetically saturated Fe–Co film evaporated on to TPX (CH_2) plastic sheets. Upon leaving the polarizer the beam entered an open-ended input coil providing a 10 G guide field and exited from the region non-adiabatically through a current sheet into a low field ($<5\times10^{-3}$ G) region containing the target. On leaving the target the neutron spins were rotated through 180° in a π coil, and the neutron polarization was measured in an analyser identical to the polarizer. The neutrons were ultimately detected in a ^{6}Li-doped glass scintillator. The role of the π coil is understood once it is appreciated that any rotation of the neutron spin about the beam axis caused by the sample is reversed in sign when the positions of the sample and the π coil are interchanged. Thus rotations produced by residual magnetic fields are cancelled to first order.

The results of these measurements confirmed the existence of weak spin-rotation in ^{117}Sn indicating a rotary power

$$\phi=(36.7\pm2.7)\times10^{-4}\ \text{radian m}^{-1} \tag{10.14}$$

Similar effects were subsequently detected in natural lead and in ^{139}La (Heckel *et al* 1982, 1984). Parity violation in both total and radiative capture cross-sections for polarized neutrons absorbed in ^{117}Sn and ^{139}La has also been confirmed (Kolomenskii *et al* 1981).

10.2 Gauge theories of quarks and leptons

10.2.1 Electro-weak unification

The close similarity between the weak fermion current (9.124) and the corresponding electromagnetic current

$$J_\mu^{\mathrm{EM}}=-\bar{e}\gamma_\mu e-\bar{\mu}\gamma_\mu\mu+\tfrac{2}{3}\bar{u}\gamma_\mu u-\tfrac{1}{3}\bar{d}\gamma_\mu d-\tfrac{1}{3}\bar{s}\gamma_\mu s \tag{10.15}$$

suggests the hypothesis that the weak interactions may be mediated by charged massive vector bosons ($W_\mu^\pm$), in the same way that electromagnetic interactions are mediated by neutral massless photons (A_μ). From this viewpoint weak decays are second-order processes in the Lagrangian

$$\mathscr{L}_{\mathrm{w}}'=\frac{ig}{2\sqrt{2}}(J_\mu W_\mu^-+J_\mu^\dagger W_\mu^+) \tag{10.16}$$

which, in the limit of low momentum transfer ($|\hbar q|\ll M_{\mathrm{w}}c$) reduces to the four-fermion description (9.122), provided the coupling constant g is given in terms of the Fermi coupling constant G_{F}, through the relation

$$\frac{G_{\mathrm{F}}/(\hbar c)^3}{\sqrt{2}}=\frac{g^2/\hbar c}{8(M_{\mathrm{w}}c^2)^2} \tag{10.17}$$

So far as neutron β-decay is concerned the difference between the four-fermion interaction (9.122) and the Lagrangian (10.16) is represented in the Feynman diagrams shown in figure 10.5.

There are, however, great differences as well as great similarities between weak and electromagnetic interactions; electromagnetic forces are long range since the photon is massless, whereas the weak forces are short range and, on the current–current model given by (9.122), the currents interact at a point. Thus intermediate vector bosons, introduced to mediate the weak interactions, must be very massive. Indeed if we assume that $g\simeq e$, we find $M_{\mathrm{w}}c^2\simeq 40$ GeV. Fortunately it is no longer necessary to speculate about these matters since the creation of the heavy bosons $W_\mu^\pm$ (Arnison *et al* 1983, Banner *et al* 1983) and Z_μ^0 (Arnison *et al* 1983a, Bagnaia *et al* 1983) in high-energy p–p̄ collision has put the question beyond dispute. These particles, which mediate charged and neutral current processes respectively, all have masses in excess of 80 GeV. Thus, no matter what the details of an intermediate vector boson theory may

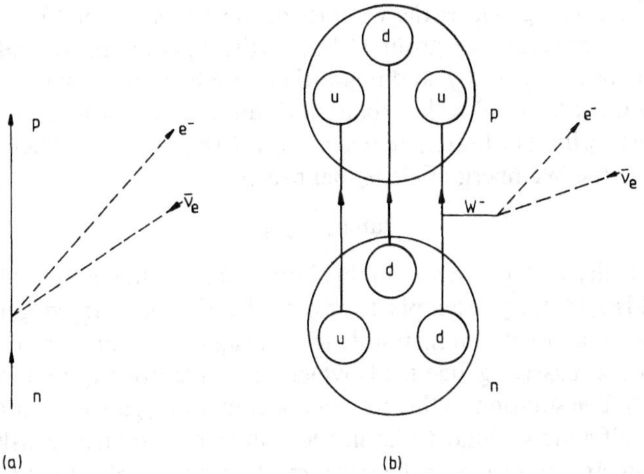

Figure 10.5. Feynman diagrams describing neutron β-decay according to (a) the weak current–current point interaction of four fermions and (b) the standard model according to which a d quark decays into a u quark with emission of a W^- intermediate vector boson, which decays immediately into an $e^-\bar{\nu}_e$ pair. Because the canonical Gamow–Teller operator is defined as $i\gamma_\mu\gamma_5$, rather than $\gamma_\mu\gamma_5$, contraction of the two axial currents introduces a factor of (-1). Thus the conventional V and A coupling constants differ in sign, which is why the interaction is described as V–A (cf. sections 9.3.1 and 10.4.1).

be, in low energy processes such as neutron decay, non-local effects associated with the exchange of massive vector bosons are unobservable and we may continue to work with the zero-range Lagrangian (9.122).

It has long been known that the 'minimal' electromagnetic interaction follows from demanding local gauge invariance under the symmetry group U(1) whose single generator is associated with the conserved electric charge (cf. section 9.2.1.3). For the symmetry to be exact the photon has to be massless. In the same way the charged current interaction (10.16), with massless gauge bosons $W_\mu^\pm$, may be obtained by assuming local gauge invariance under the non-Abelian group $SU(2)_w$, where the weak charges associated with the Cabibbo currents J_μ, $J_\mu^\dagger$ are identified with the group generators $T_w^\pm = T_{1w} \pm iT_{2w}$. However, SU(2) symmetry requires the presence of a neutral current whose weak charge is identified with the remaining generator T_{3w} and this cannot be the electromagnetic current which is parity conserving. In the Glashow–Weinberg–Salam theory ('the standard model') of weak interactions (Glashow 1961, Weinberg 1967, Salam 1968, Bilen'kii and Hosek 1982) this difficulty is overcome by taking the product group $SU(2)\times U(1)$ to be the gauge group, thereby bringing in a second neutral current whose charge is associated with the generator Y of the U(1) factor. The electric charge operator Q is now taken to be the generator (cf. 9.8)

$$Q = T_{3w} + \tfrac{1}{2}Y_w \tag{10.18}$$

Since the gauge group is the product of the 'weak isospin' group $SU(2)_w$ and the 'weak hypercharge' group $U(1)_w$, the theory contains two independent coupling constants g and g' and might therefore be more properly described as electro-weak 'mixing' rather than 'unification'. Since g is constrained by the low-energy limit (9.122) there remains one free parameter which is usually taken to be the Weinberg angle θ_w defined by

$$\tan \theta_w = g'/g \tag{10.19}$$

The difficulty with the theory in this form is that it applies only to massless gauge fields and early attempts to invent theories of charged gauge fields applicable to the weak interaction failed through their apparent inability to cope with the massive gauge fields which are evidently required (Yang and Mills 1954). The solution to the problem of combining gauge invariance with mass was ultimately found to lie in the notion of a spontaneously broken gauge symmetry coupled with the Higgs mechanism (Nambu 1960, Goldstone 1961, Higgs 1964, 1964a, 1966). In its simplest version such a theory visualizes a complex self-coupled scalar field on which is imposed invariance under a one-parameter group of gauge transformations; the gauge symmetry is spontaneously broken, as a consequence of which the gauge particles and one component of the scalar field acquire a mass while the other component decouples and disappears (Iliopoulos 1980). The net result is that one component of the scalar field provides the additional degree of freedom in the gauge field necessary for the gauge field to be massive (Bass and Schrödinger 1955, Baker and Glashow 1962).

In the most successful application of these ideas to construct a renormalized unified gauge theory of weak and electromagnetic interactions (Ross 1981) there are four gauge particles corresponding to the product group $SU(2) \times U(1)_w$ of gauge transformations. These enter the theory as a weak isotriplet $W_\mu^+ W_\mu^0 W_\mu^-$ and as a weak isosinglet B_μ^0; the neutral members W_μ^0 and B_μ^0 then mix to produce the photon A_μ and the neutral vector boson Z_μ^0 to mediate the weak neutral currents, i.e.

$$\begin{aligned} A_\mu &= -W_\mu^0 \sin \theta_w + B_\mu^0 \cos \theta_w \\ Z_\mu^0 &= W_\mu^0 \cos \theta_w + B_\mu^0 \sin \theta_w \end{aligned} \tag{10.20}$$

where the Weinberg angle θ_w is a free parameter to be determined from experiment. Although the gauge fields transform according to the adjoint representation of the group, i.e. the number of gauge bosons equals the number of generators, in this case four, the representation of fermions and Higgs scalars are not so constrained.

Thus parity violation can be introduced *ab initio* by assigning left- and right-handed fermion fields to doublet and singlet representations respectively of $SU(2)_w$. Fermion bare mass terms are not gauge invariant and for the fermions to aquire mass they must be coupled to Higgs scalars. This requirement necessitates the presence of at least one Higgs SU(2) doublet and in the standard model the minimal assumption is made that only one Higgs doublet is involved. This restriction leads to the unique predictions

$$e = g \sin\theta_w \qquad M_z = M_w^{\pm}/\cos\theta_w \tag{10.21}$$

In addition, in the minimal theory, charged and neutral current interactions have the same strengths so that, at low momentum transfer, the weak Lagrangian (9.122) holds equally for the neutral current J_μ^n as for the charged current. Apart from the electromagnetic and weak coupling constants e and G_F, the theory contains three additional measurable parameters M_w, M_z and θ_w, and all five constants are related through (10.17) and (10.21). A set of values derived from experiment which are consistent within the standard model is given in *Review of Particle Properties* (1992)

$$M_w c^2 = 80.22 \pm 0.26\ \text{GeV} \tag{10.22a}$$

$$M_z c^2 = 91.173 \pm 0.020\ \text{GeV} \tag{10.22b}$$

$$\sin^2\theta_w = 0.2337 \pm 0.0003 \tag{10.22c}$$

In its original version the standard model was applicable only to leptons, and attempts to incorporate the light u, d and s quarks encountered difficulties associated with the weak isospin assignments. For, although the left-handed linear combination of quark fields u, $d' = d\cos\theta_c + s\sin\theta_c$ which enter into the current–current weak interaction (9.122) could be assigned to a weak isospin doublet, the quark field $s' = -d\sin\theta_c + s\cos\theta_c$, orthogonal to d', had to be assigned to a singlet representation. As we have seen in section 9.3.6.2 the correct solution to this problem was the prediction of the c quark, to provide a partner for s' in a second weak isospin doublet. Both the resultant parity-violating neutral current J_μ^n, and the parity conserving electromagnetic current J_μ^{EM}

$$J_\mu^n = J_\mu^{3w}\cos^2\theta_w - \tfrac{1}{2}J_\mu^{Yw}\sin^2\theta_w \tag{10.23a}$$

$$J_\mu^{EM} = J_\mu^{3w} + \tfrac{1}{2}J_\mu^{Yw} \tag{10.23b}$$

are diagonal in all fermion mass eigenstates.

Weak processes observed at low energy involve the participation of four leptons and four quarks which are grouped into two generations or families (ν_e, e^-; u, d) and (ν_μ, μ^-; c, s). Apart from its important role in excluding strangeness-violating neutral processes the heavy charmed quark is not an effective participant, and the third fermion generation (ν_τ, τ^-; t, b), whose presence has been established by studies of the weak interaction at high energy, can also be neglected in the low energy domain. The leptons have no strong interactions and the process of muon decay (9.123b) is therefore of prime importance in that it allows us to study the electro-weak interaction in isolation from strong interaction effects. In particular the precise measurement of the muon lifetime τ_μ as given by the relation

$$\tau_\mu^{-1} = \frac{G_\mu^2 m_\mu^5 c^4}{192\pi^3\hbar^7}\left(1 - 8\frac{m_e^2}{m_\mu^2}\right)\left(1 + \frac{3}{5}\frac{m_\mu^2}{M_W^2}\right) \tag{10.24}$$

combined with the appropriate radiative correction on G_μ (Blin-Stoyle and Freeman 1970, Ross and Taylor 1973), permits us to make an accurate determination of the Fermi coupling constant (*Review of Particle Properties* 1992)

$$G_F/(\hbar c)^3 = 1.166\,39 \pm 0.000\,02 \times 10^{-5}\ \text{GeV}^{-2} \tag{10.25}$$

In contrast to the leptons, the quarks have strong interactions and this property is now directly linked to the observation that each quark flavour comes in three 'colours' (in this language leptons are colour singlets). Originally the colour degree of freedom was introduced to solve the statistics problem in the quark classification of hadron states (cf. section 9.1.3.2). However, at a more subtle level, the tri-valued colour quantum number is also necessary to preserve the renormalizability of electro-weak theory. This requires that the vector and axial vector currents, which couple to the gauge bosons, be conserved in the limit of zero spontaneous symmetry breaking, which in turn necessitates that the so-called 'triangle anomalies' which contribute to the divergence of axial currents should ultimately cancel (Fritzsch and Minkowski 1981). In the standard model this requirement takes the form that the total electric charge should vanish within each fermion generation, a criterion which is evidently fulfilled with tri-coloured quarks, e.g.

$$Q_{\nu_e} + Q_e + 3(Q_u + Q_d) = 0 \tag{10.26}$$

10.2.2 Strong interactions and quantum chromodynamics (QCD)

The modern theory of strong interactions (quantum chromodynamics or QCD) is conceptually quite different from the traditional approach which viewed the strong interaction as a short-range force between nucleons and other baryons, mediated by heavy mesons (π, ρ ,ω, etc.). It is instead similar in its essentials to the theory of quantum electrodynamics (QED) with the conserved colour quantum numbers, by convention red, blue and green, the analogues of the conserved electric charge. Like QED it is a renormalizable local gauge theory but based instead on the gauge group SU(3). However, in contrast with electro-weak theory, the gauge invariance is assumed to be unbroken as required to account for the observation of 'scaling' in deep inelastic lepton–nucleon scattering.

There are eight massless gauge bosons (gluons) in the theory which transform according to the adjoint representation of the gauge group. The coupling of quarks to gluons is independent of flavour and is non-chiral, i.e. parity conserving, and each quark of a definite flavour transforms according to the triplet representation (self-representation) of colour SU(3). The SU(3) operators act on the quark colour index and, because the gauge group is non-Abelian, the gluons must themselves carry colour, or rather they must carry colour and anti-colour simultaneously in order to conserve colour in interactions with quarks. One important consequence is that the effective quark–quark coupling depends on the distance between quarks, i.e. on the momentum carried by the gluon. Thus it is believed that, at short distances and high momentum transfer, the effective interaction between quarks is substantially reduced and may be treated as a perturbation. It is this feature of the theory, known as 'asymptotic freedom' which provides an explanation for the phenomenon of scaling.

At large separations, i.e. low momentum transfer, the colour force between quarks is believed to become very strong which would account for the observational fact of quark confinement within colour singlets. Since gluons also carry colour it is reasonable to suppose that they too are confined and cannot propagate in free space to transmit long-range forces like electromagnetism or gravity which are also mediated by massless gauge bosons. On this view hadrons can interact only via the exchange of colour singlets, e.g. $q\bar{q}$ pairs, giving rise to forces similar to the Yukawa nuclear force or to the Van der Waal's force between atoms. One of the main problems in applying the QCD theory to the nucleon is that there is no small parameter which could be used as the basis for an expansion. One possible parameter is the inverse of the number of colours N_c. Since there are (N_c^2-1) gluons in a model with $N_c \geqslant 3$, quarks would play a relatively unimportant role in a model with infinite N_c. Such a model is described as a 'skyrmion' (Skyrme 1958a, 1959, 1961, Balachandran *et al* 1983, Adkins *et al* 1983), and may perhaps be considered as an extreme form of the chiral bag model of the nucleon, in which the pion cloud is dominant (cf. section 10.3.4). It is assumed that there are no strongly interacting fundamental scalar fields and quark mass terms generated by interactions with Higgs scalars in the electro-weak sector are therefore colour independent and may be inserted as a bare mass matrix in the QCD Lagrangian exactly as bare lepton masses appear in the QED Lagrangian (cf. 9.26). Further, since the quark kinetic energy terms in the QCD Lagrangian are invariant with respect to those independent unitary transformations on the left-handed and right-handed quark fields in the electro-weak Lagrangian, which are required to transform the bare mass matrix into real diagonal form without γ_5-terms, one may assume that the QCD Lagrangian is expressed in the basis of quark mass eigenstates.

The QCD Lagrangian possesses an exact U(1) global gauge symmetry representing invariance under the transformations $q_r \rightarrow q_r e^{i\eta}$ for each quark flavour. This invariance is a statement of the law of baryon number conservation (cf. section 10.5.2.1) as expressed by conservation of the neutral vector current

$$J_\mu^{V,n} = \sum_{r=1}^{N} \bar{q}_r \gamma_\mu q_r \tag{10.27}$$

for N quark flavours. In the limit of vanishing quark mass there is also a U(1) axial symmetry $q_r \to q_r e^{i\eta\gamma_5}$ with a conserved axial current

$$J_\mu^{A,n} = \sum_{r=1}^{N} \bar{q}_r \gamma_\mu \gamma_5 q_r \qquad (10.28)$$

There is, however, no observational evidence to support this prediction, e.g. the vanishing of the axial divergence would forbid the decay $\eta \to 3\pi$ which experimentally accounts for about 50% of the decay-width of the η-meson. The difficulty is that, assuming the axial generator is spontaneously broken, and allowing for the fact that the axial symmetry is explicitly broken by the quark mass matrix, the corresponding massive isoscalar Goldstone boson is not observed experimentally (Weinberg 1975).

The violation of the U(1) axial symmetry is now believed to be associated with the existence of 'instanton' solutions of pure QCD whose presence raises very delicate questions about the nature of the vacuum. Instantons are also believed to play an essential role in quark confinement, quark condensation and *CP*-violation (cf. section 10.5.1) (Callan *et al* 1977, 1978). The instanton solutions are interpreted as tunnelling between an infinite number of degenerate classical vacua as a result of which one can define an infinity of 'true' vacua, in general inequivalent, each of which is a coherent superposition of classical vacua labelled by a continuous parameter θ ($|\theta| \leqslant \pi$). The choice of true vacuum is expressed by entering an additional term in the Lagrangian

$$L_\theta = \frac{\theta g^2}{64\pi^2} \mathrm{Tr}(F^\alpha_{\mu\nu} \tilde{F}^\alpha_{\mu\nu}) \qquad (10.29)$$

where $F^\alpha_{\mu\nu}$ is the colour field gauge tensor (analogous to the electromagnetic field tensor $F_{\mu\nu} = \partial_\nu A_\mu - \partial_\mu A_\nu$ in QED) and g is the QCD coupling constant. When the interaction with massless fermions is admitted into the Lagrangian the axial current J_μ^{An} acquires a non-vanishing divergence

$$\partial_\mu J_\mu^{A,n} = \frac{N_g^2}{32\pi^2} \mathrm{Tr}(F^\alpha_{\mu\nu} \tilde{F}^\alpha_{\mu\nu}) \qquad (10.30)$$

through the Alder–Bell–Jackiw triangle anomaly (Adler 1969, Bell and Jackiw 1969, t'Hooft 1976) and each classical vacuum becomes an eigenstate of axial baryon number with a different eigenvalue. Thus axial baryon number is no longer a symmetry of QCD constructed over the true vacuum.

Since the quark–gluon interaction is uniquely determined by gauge invariance and no explicit symmetry violating terms can be introduced without destroying renormalizability, all quantum numbers associated with individual quark flavours, e.g. isospin, hypercharge, charm, etc., are absolutely conserved in the strong interactions. Thus, if all N quark masses were equal the QCD Lagrangian would be invariant under the global gauge group SU(N). However the c, b and t quarks are so massive that this symmetry (with $N=6$) is broken down to an approximate flavour symmetry $SU(3)_F$ of isospin and hypercharge. The corresponding approximately conserved vector currents $\bar{q}_i\gamma_\mu q_j$ ($i, j=$ up, down, strange) are exactly those weak currents which participate in the charged current weak interactions according to the standard Cabibbo theory (cf. section 9.3.6). Thus QCD provides the ultimate justification for the conserved vector current hypothesis (CVC) which, in identifying the hyper-charge conserving weak (charged) vector currents, and the isovector electromagnetic (neutral) vector current as members of an isospin triplet, launched the first stage of electro-weak unification (Gershstein and Zel'dovich 1957, Feynman and Gell-Mann 1958).

In the limit where the masses of u, d and s quarks may be ignored, the QCD Lagrangian exhibits a chiral $SU(3)_L \times SU(3)_R$ symmetry although no remnant of this symmetry is observed in hadron spectra, e.g. no parity doubling is observed within SU(3) multiplets. The presumption is therefore that the eight isoaxial generators of the chiral group are spontaneously broken giving rise to eight pseudoscalar Goldstone bosons (a similar explanation was originally offered for the violation of the U(1) axial symmetry discussed above). Since QCD contains no fundamental scalar fields, those fields which acquire non-zero expectation values must be $q\bar{q}$ composites, with Goldstone bosons which are bound states of q and $\bar{q}$. An explanation for the dynamical symmetry breaking has been advanced which traces it to an effective interaction between quarks of opposite helicity generated by instantons which makes the vacuum unstable against the condensation of $q\bar{q}$ pairs (Callan *et al* 1977, 1978).

Since the chiral symmetry is explicitly broken by the quark mass matrix the Goldstone bosons acquire mass and may be identified with the members of the lowest SU(3) pseudo-scalar meson octet (Pagels 1975). Restricting attention to the u and d quarks which are the constituents of the nucleon, we have SU(2) isospin symmetry with conserved vector currents $\bar{q}\gamma_\mu \tau q$ where the operators τ are the generators of isospin. The Goldstone bosons associated with the spontaneous breaking of the chiral symmetry may be identified with the pion triplet and the divergence of the axial current $\bar{q}\gamma_\mu\gamma_5\tau q$ vanishes only in the limit of vanishing pion mass. However the pion is the lightest hadron and the axial currents may be regarded as almost conserved. This approximation is called the partially conserved axial current (PCAC) hypothesis (Gell-Mann and Levy 1960) and is important in the analysis of neutron β-decay in relation to the renormalization of the axial vector coupling constant through the strong interactions, since it permits the establishment of relations linking weak and strong interaction parameters by methods which are almost model independent (Goldberger and Treiman 1958, Weissberger 1965, Adler 1965) (cf. section 9.3.5).

10.3 Quark models of the nucleon

10.3.1 SU(6) (spin-flavour) symmetry

SU(2) (isospin) symmetry, originally introduced to represent neutron and proton as components of a nucleon doublet but later extended to all hadrons, is broken by the electromagnetic interaction, and the connection between electric charge Q and third component of isospin T_3 is given by the relation (cf. 9.8).

$$Q = T_3 + \tfrac{1}{2}Y \tag{10.31}$$

where the hypercharge Y is an additive quantum number which is the same for all members of an isospin multiplet. Formally (10.31) is the same as (10.18) where, in the latter case, I_{3w} and Y_w represent 'weak isospin' and 'weak hypercharge', respectively, and indeed the smallest group which can accommodate the commuting operators T_3 and Y is the product group $SU(2) \times U(1)$ which is just the gauge group of the electro-weak theory. However, such a group requires no relationship to exist between the quantum numbers of T_3 and Y contrary to observation in the context of the strong interactions. Such considerations led Gell-Mann and Ne'eman (1964) to enlarge the symmetry group to SU(3), showing that the isospin-hypercharge quantum numbers of the lowest mass spin $\frac{1}{2}$ baryons corresponded exactly to those of the $D^{(11)}$ octet representation of SU(3). Ultimately it was the identification of the spin $\frac{3}{2}$ particle Ω^- as the missing member of the $D^{(30)}$ decouplet which led to the general recognition of what is now called flavour SU(3) as an approximate symmetry of the strong interactions (Barnes *et al* 1964) (cf. section 9.1.2).

Subsequently it was shown (Gell-Mann 1964, Zweig 1964) that the observed baryons could be fitted into representations constructed from the product $D^{(10)} \times D^{(10)} \times D^{(10)}$ of fundamental SU(3) triplets leading to the suggestion that physical baryons were composed from three spin $\frac{1}{2}$ quarks which we now label as 'up', 'down' and 'strange'. The most prominent feature of quarks is that they come in states of fractional charge although this is not an absolute requirement with tricoloured quarks, which may carry integral charge determined by flavour and colour (Han and Nambu 1965). However, this possibility would require that electromagnetism break the colour symmetry which circumstance is not favoured for reasons given above. Furthermore the oberved decay characteristics of $c\bar{c}$ and $b\bar{b}$ bound states produced in high energy e^+e^- collisions is consistent with fractional quark charge (Schwitters and Strauch 1976, Chinowski 1977, Franzini and Lee-Franzini 1983).

Following the same line of reasoning as above, the observation that, whereas the baryon octet has spin $\frac{1}{2}$, the decouplet has spin $\frac{3}{2}$, suggests the construction of the product group SU(2) (spin) $\times$ SU(3) (flavour) and subsequent enlargement to SU(6) (spin-flavour) to see whether the observed spin-flavour correlation arises naturally from irreducible representations of SU(6) as indeed it does (Wigner 1937, Sakita 1964, Gursey and Ridicati 1964). Finally, in order to construct an anti-symmetric three-quark baryon wavefunction, it is necessary to insert a factor related to SU(3) (colour), and the $D^{(00)}$ colour singlet representation provides the only suitable candidate. It has not been proved in QCD that three quarks in a colour singlet form a bound state although such a result could be understood intuitively as the binding of a red quark say, to the anti-red combination of a blue and green quark. There is, however, some evidence from lattice gauge theory, which is a computational technique borrowed from solid state physics, that the colour singlet inter-quark potential does indeed reduce to a linear, and hence confining, form at large separation (Wilson 1974, Creutz 1980). However this result is valid only at zero temperature and, at temperatures above 10^{12}K, lattice QCD undergoes a first-order phase transition to a quark–gluon plasma, where Debye screening of the quark colour by coloured gluons suppresses the long-range confining potential. Such a phase change is believed to have occurred in the first 10^{-6} s of the early Universe.

To summarize then, quark models of baryons and mesons will satisfy approximate SU(6) (spin-flavour) symmetry, valid at low energy, provided the wavefunction separates into spin, space and colour factors, and the inter-quark force is spin-independent in leading order (Myhrer and Wroldsen 1988). The length parameters must then be chosen to reproduce the observed proton radius. Since the colour force is flavour independent the different quark flavours are distinguished by their masses and their electric charges. In the case of the nucleon, where only *u* and *d* quarks are involved, the SU(6) symmetry reduces to SU(4) (spin-isospin) symmetry and the predictions are fairly simple. In the case of the neutron the two *d* quarks can couple to $S_{\mathrm{dd}}=0$ or $S_{\mathrm{dd}}=1$; however the $S_{\mathrm{dd}}=0$ state is disallowed by the Pauli principle and coupling in the *u* quark results in $I=\frac{1}{2}$ and $I=\frac{3}{2}$ states which would be degenerate were the SU(4) symmetry exact. The colour-magnetic interaction breaks the SU(4) symmetry leading to an $I=\frac{1}{2}$ ground state as observed. The $I=\frac{3}{2}$ state is the Δ° isobar. A similar analysis applied to the proton shows that it has $I=\frac{1}{2}$ in the ground state with $I=\frac{3}{2}$ in the Δ^{+} isobar state. There are two further excited states Δ^{++} and Δ^{-} which correspond to *uuu* and *ddd* configurations of non-strange quarks in good agreement with the model (cf section 9.2.1.1). Indeed the very simplicity of the observed spectrum of nucleon excited states is itself evidence in favour of quark confinement since in these circumstances the continuum of excited states is eliminated.

10.3.2 The simple harmonic oscillator model

A non-relativistic (constituent) quark model of baryons without a continuum of states may be based on the three-dimensional harmonic oscillator potential with mass parameters $m_{\mathrm{u}}=m_{\mathrm{d}}=0.33$ GeV, $m_{\mathrm{s}}=0.55$ GeV (Isgur and Karl

1977, 1978). These 'constitutent masses' are not the same as the 'bare' or 'current' masses which appear in the QCD Lagrangian or current algebras, but are generated by the spontaneous breaking of the $SU(3)_L \times SU(3)_R$ chiral symmetry discussed briefly in section 10.2.2. They may be viewed as the masses the quarks would have were they not confined, and have estimated values $m_u = 5.6 \pm 1.1$ MeV, $m_d = 9.9 \pm 1.1$ MeV, $m_s = 199 \pm 33$ MeV, $m_c = 1.35 \pm 0.05$ GeV (*Review of Particle Properties* 1992). In this model the oscillator potential must be supplemented by a colour magnetic hyperfine interaction of the form

$$H_{chf} = \sum_{i>j} f(r_i r_j)\boldsymbol{\sigma}_i \cdot \boldsymbol{\sigma}_j \qquad f(r_i r_j) > 0 \tag{10.32}$$

to break the spin degeneracy of the oscillator states (cf. figure 4.1).

This model is quite successful in describing baryon spectra and magnetic moments (Beg *et al* 1964, Lipkin 1978, Schachinger *et al* 1978) and has the advantage, in common with non-relativistic models in general, that it is translationally invariant and allows separation of the centre-of-mass motion which causes difficulties in fully relativistic models. A serious problem with non-relativistic models is that the small components of the Dirac spinor for a bound quark vanish identically, one consequence of which is that the weak axial form factor is always over-estimated (Donoghue *et al* 1986). Relativistic corrections may, however, be incorporated by explicitly introducing a small Dirac wavefunction (Bozoian and Weber 1984).

The non-relativistic model is suitable as a basis for calculating magnetic moments in the baryon octet provided each quark is assumed to carry a Dirac magnetic moment

$$\mu_q = \frac{e\hbar}{2m_q c} \tag{10.33}$$

with $m_d \simeq m_u \simeq m_p/3$. To account for the measured magnetic moments of the strange baryons a slightly larger mass $m_s \simeq \frac{1}{2}m_p$ has to be given to the strange quark, thereby breaking the SU(6) spin-flavour symmetry (cf. section 10.3.1). The computed results are listed in table 10.1 both for the simplest version of the model (Isgur and Karl 1980), and for an improved version in which relativistic, pionic and quark confinement corrections are taken into account (Cohen and Weber 1985).

The spectra of the nucleonic resonances shown in figure 9.1 are grouped in terms of the angular momentum states s, p, d, f of the pion–nucleon collision in which these resonances are observed. However, the observed structure of these states can be interpreted in a fairly straightforward way as excited states of three quarks bound in a simple harmonic oscillator potential modified by the colour magnetic interaction (10.32). For if we assume, to be definite, that the nucleon is a neutron, then the addition of one unit of orbital angular momentum means that the singlet state of two *d* quarks, $S_{dd} = 0$, is

Table 10.1. Magnetic moments (in units of μ_N) in the baryon octet calculated on the basis of the simple harmonic oscillator model (μ(SHM)), and with the same model supplemented by a small Dirac quark wavefunction (μ(SHM*)). The Σ^0 decays electromagnetically to the Λ^0 with a lifetime of $7.4 \pm 0.7 \times 10^{-20}$ s so its magnetic moment cannot be measured. The experimental results are taken from the *Review of Particle Properties* (1992)

Baryon	μ(SHM)	μ(SHM*)	μ(Expt)
n	-1.80	-1.913	$-1.913\,042\,7 \pm 0.000\,000\,5$
p	2.70	2.618	$2.792\,847\,39 \pm 0.000\,000\,06$
Λ^0	-0.60	-0.614	-0.613 ± 0.004
Σ^-	-1.01	-1.088	-1.160 ± 0.025
Σ^0	–	–	–
Σ^+	2.59	2.475	2.42 ± 0.05
Ξ^-	-0.46	-0.552	-0.6507 ± 0.0025
Ξ^0	-1.36	-1.365	-1.250 ± 0.014

no longer forbidden by the Pauli principle. Therefore odd parity states with $I^\pi = \frac{1}{2}^-, \frac{3}{2}^-$ will be generated when the orbital angular momentum is added to the spin of the u quark. These may be identified with the $\frac{1}{2}^-$ and $\frac{3}{2}^-$, $T = \frac{1}{2}$ lowest odd parity states in figure 9.1. The triplet state of the two d quarks with $S_{dd} = 1$ will generate five odd parity states with $I^\pi = \frac{1}{2}^-, \frac{1}{2}^-, \frac{3}{2}^-, \frac{3}{2}^-, \frac{5}{2}^-$, of which the states with $I^\pi = \frac{1}{2}^-, \frac{3}{2}^-$ and $\frac{5}{2}^-$ have $T = \frac{1}{2}$, while the states with $I^\pi = \frac{1}{2}^-$ and $\frac{3}{2}^-$ have $T = \frac{3}{2}$. All these states fall somewhat higher on the energy level diagram for N and Δ resonances.

Note that it is the colour magnetic interaction (10.32) which splits the nucleon from the (3,3) Δ resonance, and which also splits the nucleonic excited states derived from the singlet state of two d quarks, from those derived from the triplet state. Exactly the same interaction lowers the pion mass below that of the ρ mass (cf. section 3.5.2).

10.3.3 The MIT 'bag' model

An advance on the simple harmonic oscillator model is provided by the MIT 'bag' model which does not rely on the concept of a potential (Chodos *et al* 1974, DeGrand *et al* 1975, Johnson 1975, 1978, Donoghue *et al* 1975, Donoghue and Johnson 1980, De Tar and Donoghue 1983). The basic assumption of the model is that the physical vacuum can exist in two phases, a normal phase in which coloured objects cannot propagate and a 'bag' phase in which quarks and gluons can exist and interact via QCD forces. Thus the model incorporates both confinement and asymptotic freedom, which are features characteristic of QCD. To create a bag of volume V against the pressure exerted by the external vacuum an energy BV is required, where the energy density B is termed the 'bag constant'. Quark confinement is enforced by the boundary condition on the quark wavefunctions which vanish at the bag radius R.

In QCD, as in QED, calculations based on bare masses and coupling constants encounter infinities associated with self-energy terms, and these are eliminated by working only with renormalized masses and coupling constants. In the simplest version of the bag model with massless quarks interacting through QCD forces, the only momentum-dependent input is the effective or running coupling constant $\alpha_s(Q^2)$ where the Q^2-dependence is fixed by the dominant perturbative corrections, and is expressed in terms of an arbitrary momentum Λ which fixes the mass scale (cf. 9.102). In the bag model this scale is set by the value of B and the energy of a baryon containing three massless quarks is then given by

$$E_b = \tfrac{4}{3}\pi R^3 B + 3(2.043)/R \tag{10.34}$$

B is then determined by minimizing E_b with respect to R and setting the resulting value of E_b equal to the baryon mass. In this way we find $R \simeq 1.5$ fm and $B \simeq 0.02$ GeV fm^{-3}.

These conclusions are altered only in detail when small 'current' masses are given to the quarks. In computing baryonic magnetic moments, for example, it is the quark energies which appear in the denominators exactly as the constituent quark masses do in non-relativistic potential models.

10.3.4 The chiral bag model

In the MIT bag model the weak vector current J_μ^V is conserved since the normal component of the current vanishes at the bag surface. This is not the case for the weak axial current J_μ^A and indeed no remnant of the original QCD chiral symmetry survives in the bag model, nor indeed in any of the non-relativistic potential models. This observation is the point of departure for the development of 'chiral bag' models which assume that, although $\partial_\mu J_\mu^A = 0$ everywhere, the chiral symmetry is implemented in different modes inside and outside the bag (Chodos and Thorn 1975, Callan *et al* 1978, Brown and Rho 1979, Brown *et al* 1980, Miller *et al* 1980, 1981, Théberge *et al* 1980, Thomas *et al* 1981). Specifically it is assumed that, inside the bag the symmetry is realized in the normal Wigner–Weyl mode, whereas outside the bag the symmetry is spontaneously broken and is implemented in the Nambu–Goldstone mode (Pagels 1975).

To consider the special case of the nucleon the axial current inside the bag is given by its normal expression

$$J_\mu^A = \sum_{k=1}^{3} \bar{q}_k \gamma_\mu \gamma_5 \tau q_k \tag{10.35}$$

whereas outside the bag the axial current is carried by a pion field

$$J_\mu^A = f_\pi D_\mu \phi_\pi \tag{10.36}$$

where D_μ is the non-linear differential operator $[1+\phi_\pi^2/f_\pi^2]^{-1}\partial_\mu$ and f_π is the pion decay constant (i.e. the matrix elements of the axial current connecting the vacuum to the one-pion state). Conservation of the axial current is enforced by matching the internal and external normal components of the axial current at the surface of the bag.

The chiral bag is somewhat smaller than the conventional bag but acts as the source of the surrounding pion cloud which is similar in many respects to that postulated in the Yukawa theory of pion–nucleon coupling. The model is composite in the sense that the pion is treated as a fundamental scalar field rather than as a $q\bar{q}$ bound state. However, this approximation does not produce any real conflict with the QCD theory since the pion radius is small in comparison with its Compton wavelength (the pion is the only physical hadron which has this property) and in the long wavelength limit the pion may be treated as elementary.

10.4 Neutron physics beyond the standard model

10.4.1 Right-handed weak currents and neutron decay

In section 10.2.1 parity violation was incorporated in the standard model of electro-weak interactions at the Lagrangian level by postulating that left and right-handed fermion fields be assigned to doublet and singlet representations of weak SU(2), respectively. This was necessary to accord with experimental observation that parity violation is maximal in all weak interactions in the low energy domain. However, no convincing reason has yet been advanced to support the notion that physical space should be left–right asymmetric at a fundamental level, and, for this reason, various models of the weak interaction have been proposed in which the weak Lagrangian is manifestly left–right symmetric, and parity violation comes about when reflection symmetry is spontaneously broken. In these models parity violation is a low-energy phenomenon and reflection invariance is restored at high energies and short distances.

A popular version of such a model is the left–right symmetric model based on the gauge group $SU(2)_L \times SU(2)_R \times U(1)$ (Mohapatra and Pati 1975, Mohapatra and Sidhu 1978). This model postulates a weak interaction Lagrangian which is the analogue of (10.16) extended to include right-handed currents, i.e.

$$L_W^{s'} = \frac{g}{2\sqrt{2}}\{[J_\mu^V + J_\mu^A]W_{\mu L}^- + [J_\mu^V - J_\mu^A]W_{\mu R}^- + [J_\mu^V + J_\mu^A]^\dagger W_{\mu L}^+ + [J_\mu^V - J_\mu^A]^\dagger W_{\mu R}^+\} \tag{10.37}$$

Although right and left-handed currents $J_\mu^V \mp J_\mu^A$ enter (10.37) on an equal footing, it is now assumed that the left–right symmetry is spontaneously broken generating an asymmetric vacuum even though the Higgs potential responsible for breaking the symmetry is reflection symmetric. Then, not only do the gauge bosons acquire mass in the usual way, but the mass eigenstates $W_{\mu1}$ and $W_{\mu2}$ appear as linear combinations of $W_{\mu L}$ and $W_{\mu R}$

$$W_{\mu1} = W_{\mu L} \cos \zeta - W_{\mu R} \sin \zeta$$
$$W_{\mu2} = W_{\mu L} \sin \zeta + W_{\mu R} \cos \zeta \tag{10.38}$$

where ζ is a real mixing angle. In general the masses m_1, m_2 will be unequal with squared ratio given by the free parameter

$$\delta = (m_1/m_2)^2 \tag{10.39}$$

The predictions of the standard model are therefore recovered in the limit $\zeta \to 0$, $\delta \to 0$, corresponding to an infinitely massive right-handed boson.

In analogy with the four-fermion interaction L_W given in (9.122) which is the low-energy approximation to L'_w given in (10.16), the corresponding left–right symmetric four-fermion Lagrangian is (Beg *et al* 1977)

$$L_w^s = \frac{GV}{\sqrt{2}} \{J_\mu^{V\dagger} J_\mu^V + \eta_{AA} J_\mu^{A\dagger} J_\mu^A + \eta_{AV}[J_\mu^{V\dagger} J_\mu^A + J_\mu^{A\dagger} J_\mu^V]\} \tag{10.40}$$

where

$$\frac{GV}{\sqrt{2}} = g^2 \left\{ \left[\frac{\cos\zeta - \sin\zeta}{m_1 c^2}\right]^2 + \left[\frac{\cos\zeta + \sin\zeta}{m_2 c^2}\right]^2 \right\} \tag{10.41}$$

and

$$\eta_{AA} = (\delta^2 + \varepsilon^2)/(1 + \varepsilon^2\delta^2)$$
$$\eta_{AV} = -\varepsilon(1 - \delta^2)/(1 + \varepsilon^2\delta^2)$$
$$\varepsilon = (1 + \tan\zeta)/(1 - \tan\zeta) \tag{10.42}$$

For purposes of making direct contact between theory and experiment a convenient notation is given by (Holstein and Treiman 1977)

$$x = \frac{1 + \eta_{VA}}{1 - \eta_{VA}} \simeq \delta - \zeta \qquad y = \frac{\eta_{AA} + \eta_{VA}}{\eta_{AA} - \eta_{VA}} \simeq \delta + \zeta \tag{10.43}$$

In the first comparison of the predictions of the model with experiment a number of tests were proposed of which the most sensitive to the parameters δ and ζ were (i) the longitudinal polarization of electrons emitted in pure Gamow–Teller nuclear β-decay (e.g. $^{60}Co \to {}^{60}Ni$) and (ii) the value of the Michel parameter ρ which determines the shape of the electron spectrum in the decay of the muon (Beg *et al* 1977). According to the model these quantities are given by the formulae

$$f_l = \langle \boldsymbol{\sigma}_e \cdot \boldsymbol{p}_e / p_e \rangle = (v_e/c)[(1-y^2)/(1+y^2)] \tag{10.44}$$

and

$$\rho = (\tfrac{3}{4})\{1 - \tfrac{1}{3}(x-y)^2[(1+y^2)(1+x^2) - (x+y)(1+xy)]^{-1}\} \tag{10.45}$$

From the experimental data it then proves possible to set limits $|\zeta| < 0.06$, $\delta < (2.78)^2$, a result which indicates that the mass of the heavy boson must satisfy $m_2 c^2 > 225$ GeV.

More recently attention has shifted to the question as to what limits may be set on ζ and δ from precise measurements in nuclear β-decay with particular reference to ^{19}Ne and the neutron (Holstein and Treiman 1977, Carnoy *et al* 1988). In the case of the neutron the interesting parameters are, as always, the neutron spin–electron momentum correlation coefficient A and the lifetime, coupled with the ft-values of pure Fermi $0^+ \to 0^+$ superallowed β-transitions. In the left–right symmetric model these are given by

$$A = -2\left\{\frac{|\lambda|^2(1-y^2) - |\lambda|(1-xy) + T_1}{(1+x^2) + 3|\lambda|^2(1+y^2) + T_2}\right\} \tag{10.46}$$

$$R = (ft)_{0^+ \to 0^+}/(ft)_n = \frac{(1+x^2) + 3|\lambda|^2(1+y^2) + T_3}{2(1+x^2) + T_4} \tag{10.47}$$

where T_1, T_2, T_3 and T_4 are small terms of recoil order which may be neglected.

There is no unique presciption for applying (10.46)–(10.47) for establishing an allowed region in the $\delta - \zeta$ plane, but a convenient method is to eliminate $|\lambda|^2$ between these equations and consider the experimental limits on A and R. A preliminary analysis along these lines emphasizes the importance of the neutron decay data in establishing significant constraints on the contribution of right-handed currents (Carnoy *et al* 1988, 1991, 1992, Quin 1993).

10.4.2 *CP*-violation

Because very intense sources of neutrons are available, and because it is uncharged, the neutron provides a very favourable system on which to test for violations of discrete space–time symmetries and in particular for time-reversal invariance. For example very substantial efforts have been expended on experimental searches for a non-vanishing triple correlation coefficient in neutron β-decay (cf. section 9.3.8.4) and for a finite neutron electric dipole moment (cf. section 9.2.5.4). In unbroken local gauge theories describing the interactions of massless fermions and gauge bosons *CP* is always a good symmetry provided an appropriate choice of phase is made for the fermion fields. Because of the *CPT* theorem (Lüders 1957, Feinberg 1957), this means that all such theories are invariant under the operation of time reversal (T). However, having fixed the phase convention, when the gauge symmetry is spontaneously broken through the interactions with Higgs scalars, *CP*-violating phases are introduced into the fermion mass matrix which may be eliminated in their turn when the fermion phases are redefined in such a way as to make the fermion mass-matrix real, diagonal and free from γ_5-terms. In general this redefinition is accompanied by the re-appearance of *CP*-violating phases in that part of the Lagrangian which describes the interaction between fundamental fields.

In the electro-weak sector spontaneous breaking of the gauge symmetry can generate *CP*-violation via the interactions of fermions either with gauge bosons or with Higgs scalars. With two fermion generations the *CP*-violating phases in the quark–gauge boson interaction can always be transformed away, but with three fermion generations there is a residual *CP*-violating phase represented by the angle δ_{13} in the CKM matrix (cf. section 9.3.6.3). However, the dipole length predicted on the basis of a non-vanishing δ_{13} is always very small, $d_n \simeq 10^{-33} - 2 \times 10^{-31}$ cm, and lies far outside the range of values detectable using current techniques. In the minimal electro-weak theory with only one Higgs doublet there is no *CP*-violation in the interaction with Higgs scalars and at least two Higgs doublets are necessary to generate such an effect. However, in these theories a neutron electric dipole moment at the level of 10^{-24} cm is predicted, a result which has already been ruled out by experiment so it appears unlikely that the Higgs sector is the source of the *CP*-violating effects observed in neutral kaon decay.

A third proposed mechanism goes beyond the standard model and ascribes *CP*-violation to right-handed couplings contained in a model of the weak interactions based on the gauge group $SU(2)_L \times SU(2)_R \times U(1)$ (cf. section 10.4.1). In addition to the *CP*-violating phase which occurs in the standard model with six quarks, this model contains two new *CP*-violating phases with four quarks which can in principle result in a finite triple correlation coefficient in neutron β-decay (Herczeg 1983, Golowich and Valencia 1989). Additional possibilities arise in supersymmetric versions of the model (cf. section 10.4.3).

Finally one must consider the possibility that the observed *CP*-violation does not originate in the electro-weak sector at all, but is an example of strong *CP*-violation induced by instanton effects as parameterized by the vacuum angle θ in the *CP*-violating term L_θ in the QCD Lagrangian (cf. section 10.2.2) (Aoki and Hatsuda 1992). The neutron dipole length derived from this term has a value at the level of $(m_u\theta)/(m_N)^2$, where $m_u c^2 \approx 5.6$ MeV is the current mass of the *u* quark and $m_N c^2 \approx 1$ GeV is the nucleon mass. On the assumption that the dipole length is $< 10^{-25}$ cm this implies that $\theta < 3 \times 10^{-9}$ and this result is obviously consistent with $\theta = 0$ although how this comes about is not clear.

It was observed by Peccei and Quinn (1977) that, when the quark mass matrix is brought to real diagonal form by performing a U(1) chiral transformation on the fermion fields, the angle θ is simultaneously transformed according to the rule

$$\theta \to \theta' = \theta + \arg(\det(m)) \tag{10.48}$$

where m is the mass matrix. The possibility therefore arises of assigning the phase of the mass matrix determinant the value $-\theta$, in which case $\theta' = 0$ and the CP-violating term L_θ vanishes. One way of achieving this result is to have one of the quark masses vanish, in which case det(m) also vanishes and its phase can be assigned an arbitrary value. Since the u quark is the lightest quark, the first choice would be to set m_u equal to zero, but this solution is not supported by, for example, the observed n–p mass difference whose predicted value on this hypothesis would be an order of magnitude too large (Langacker 1979, Langacker and Pagels 1979, Dominguez 1978, 1979).

To account for the apparent vanishing of θ Peccei and Quinn advanced an alternative hypothesis valid in the case that at least one quark acquires its mass by Yukawa coupling to Higgs scalars in the electro-weak sector. In the absence of instanton effects the QCD Lagrangian in this case is assumed to admit an exact U(1) chiral symmetry such that the phases of the vacuum expectation values of Higgs fields, determined by minimizing the Higgs potential, can be fixed arbitrarily. When instanton effects are switched on the U(1) symmetry is broken, the Higgs potential becomes a function of θ and these phases assume definite values. It is then shown that, for a range of models at least, arg (det(m)) takes the value $-\theta$ as required and the strong CP-violation vanishes.

The difficulty with this scenario is that, when the gauge symmetry is spontaneously broken to give the quark a mass, the Peccei–Quinn U(1) symmetry is also broken spontaneously, thereby generating a neutral pseudo-scalar Goldstone boson, the axion, which acquires a small mass (1–100 MeV) through instanton effects (Weinberg 1978, Wilcsek 1978). However, experimental searches for the axion have been unsuccessful and this particular escape route from the strong CP-violation dilemma appears to be closed. One remaining possibility is that the energy of the true vacuum state is stationary with respect to θ and this result appears to be a feature of models based on dynamical symmetry breaking with fermion condensates rather than fundamental scalar fields (Eichten *et al* 1980, Lane 1981).

10.4.3 Baryon number non-conservation and neutron oscillations

10.4.3.1 Baryon number conservation

The observed stability of protons and neutrons bound in nuclei is conventionally expressed in terms of a law of baryon number conservation, which asserts that the difference between the numbers of baryons and anti-baryons in the Universe remains constant in time (Stueckelberg 1938, Wigner 1949). Baryon number conservation is therefore analogous to electric charge conservation, and both conservation laws may be associated with invariance under distinct U(1) global gauge transformations. The electromagnetic interaction is also invariant under local U(1) gauge transformations with the consequence that charged particles are coupled to a vector gauge field whose quanta, the photons, are observed to be massless to a precision $m_\gamma c^2 < 6 \times 10^{-25}$ GeV (Davis *et al* 1975). The local colour gauge symmetry $SU(3)_c$ is also believed to be exact and its quanta, the gluons, are assumed to be massless. There is, however, no direct observational basis for such a claim since the observed hadrons do not interact via the exchange of gauge bosons and their interactions have not yet been satisfactorily described in terms of the forces which govern their internal structure.

Unlike colour symmetry the broken symmetry associated with the flavour group $SU(3)_F$ of isospin and hypercharge appears to be purely global in character. Neither is their evidence for the existence of a massless gauge field associated with a local U(1) gauge symmetry for baryon number. Indeed, were such a field to exist, the coupling to it would have to be extremely weak to be consistent with the observed equality of inertial and gravitational mass (Lee and Yang 1955). Thus, if baryon number conservation is exact, it must be because of an unbroken global gauge symmetry which commutes with the local gauge symmetry of the theory, as indeed it does in the case of the standard model. If baryon number conservation is not exact, it could be because the global symmetry is explicitly broken by additional terms in the Lagrangian or because this conservation law stems from a spontaneously broken local gauge symmetry, or indeed from some combination of these effects.

As was first pointed out by Sakharov (1967), in the hot big bang model of the expanding Universe there is no macroscopic separation of matter from anti-matter and hence, in the framework of a theory in which baryon number conservation is exact, the observed matter–anti-matter asymmetry must be an intrinsic characteristic of the Universe. Conversely, the observed asymmetry may be understood in the context of a completely symmetric Universe provided there exist *CP*-violating and baryon number non-conserving interactions which are operative at a non-thermal equilibrium stage in the initial expansion. Among processes of particular significance in this respect are nucleon decay into leptons ($\Delta B = \Delta L = 1$) and neutron–anti-neutron (n–n̄) oscillations ($\Delta B = 2$, $\Delta L = 0$) (Gell-Mann and Pais 1955, Kuz'min 1970, Glashow 1979). Hydrogen–anti-hydrogen (H–H̄) oscillation ($\Delta B = \Delta L = 2$) is another baryon number violating process in the same general class (Feinberg *et al* 1978), although it conserves (B–L) which is not the case for n–n̄ oscillations. However, the two oscillation processes have otherwise many features in common and H–H̄ oscillations have been widely discussed in recent years (Arnellos and Marciano 1980, Mohapatra and Senjanovic 1982, Bilen'kii and Pontecorvo 1982, Sarker *et al* 1983, Parida 1983, 1983a, Deo and Maharana 1984, Nieves and Sharker 1984).

The phenomenon of particle–anti-particle oscillation was first considered for the case of the neutral kaon where the states K_0 and $\bar{K}_0$ are not mass eigenstates. Rather they are linear superpositions of mass eigenstates differing in strangeness by an amount $\Delta S = 2$. Since strangeness is not conserved in weak interactions the kaon state oscillates in time between the K_0 and $\bar{K}_0$ states with a period which, in the absence of external interactions is given by $\tau = \hbar/\delta mc^2$, where δm is the mass difference between the mass eigenstates (Gell-Mann and Pais 1955). Kaon oscillations have been studied experimentally in enormous detail, not only because of their intrinsic interest, but primarily because the mass eigenstates are not simultaneously CP eigenstates as was originally thought to be the case.

In the same way if baryon number violating $\Delta B = 2$ processes exist then the familiar n and n̄ states will not be mass eigenstates; ignoring small CP-violating amplitudes the mass eigenstates will be the 'Majorana' particles

$$n_1 = \sqrt{\tfrac{1}{2}}(\mathrm{n} + \bar{\mathrm{n}}) \qquad n_2 = \sqrt{\tfrac{1}{2}}(\mathrm{n} - \bar{\mathrm{n}}) \tag{10.49}$$

each having mean baryon number zero. Thus in principle (n–n̄) oscillations may occur even if the oscillation period is so great as to place the phenomenon beyond the range of observation (Bilen'kii and Pontecorvo 1982).

It has been pointed out above that baryon number B (and lepton number L) are exact global symmetries of the gauge group $\mathrm{SU(2)_c \times U_\gamma(1)}$ of the standard model and in this theoretical framework baryon number conservation is exact. Indeed in the limit of vanishing Cabibbo angle where the connection between quark families is broken, baryon number is separately conserved in each generation. Present evidence is that this actually happens for lepton number L assuming neutrinos are massless. We conclude therefore that, if the standard model represents a final description of the world, n–n̄ oscillations will be absolutely forbidden.

10.4.3.2 n–n̄ oscillations and grand unification

The difficulty with the standard model is that, in spite of its prodigious success, it contains far too many free parameters to be considered as the ultimate theory. It predicts neither the fermion masses nor the coupling constants and mixing angles and, most significantly, it provides no explanation for the observed quantization of electric charge. It does not explain why fermion generations repeat. These defects have lead to the search for a grand unified theory, the essential idea of which is that the gauge group $\mathrm{SU(3)_c \times SU(2)_L \times U_\gamma(1)}$ is embedded in a larger group whose symmetry restricts the allowed range of parameters which are free in the standard model (Langacker 1981, Goldhaber and Marciano 1986). Such a theory should be characterized by a single coupling constant to which the measured strong and electro-weak couplings will converge, and renormalization group analysis has shown that, for a range of grand unified theories, the weak, strong and electromagnetic interactions will be unified at a mass scale $m_x c^2 \geqslant 10^{15}$ GeV.

The extraordinarily small value of the ratio $m_w/m_x \simeq 10^{-13}$ for the mass scales of electro-weak and grand unification constitutes the so-called hierarchy problem for grand unified theories, for this feature is not 'natural' in the sense that it does not occur for a range of variation of the parameters of the theory. Rather its achievement requires an extremely delicate adjustment of these parameters; it turns out that the observability or otherwise of n–n̄ oscillations may hinge on a similar 'fine tuning' of parameters.

Since one cannot construct exact global symmetries whose generators distinguish between quark and lepton when these are associated in the same group representation as they are in grand unified theories, *B* and *L* non-conservation is a characteristic feature of such theories. Consequently the observation of *B*- and *L*-violating processes such as nucleon decay and matter–anti-matter oscillations would not only provide support for the theoretical notion of a grand unifying group, but would also help to identify the correct group from among several competing candidates and to unravel the pattern of spontaneous symmetry breaking. Specifically we would like to know whether there is a single grand unifying mass scale $m_x c^2 \geqslant 10^{15}$ GeV as in the 'minimal' SU(5) model (Georgi and Glashow 1974), or whether there are one or more stages of partial unification as in the left–right symmetric models. Indeed the experimental detection of n–n̄ oscillations would represent the clearest possible signature of intermediate mass scales. This is because the amplitude for n–n̄ oscillation must involve a six-quark operator with dimension $d=9$ in units of inverse mass, and must therefore have a mass dependence $\simeq m^{-5}$. Thus, if the grand unification mass m_x is the only mass scale entering the theory, n–n̄ oscillations, even if theoretically allowed, would in practice be unobservable (Kuo and Love 1980).

In the simplest version of SU(5) with a $\underline{10}$ and a $\underline{5}^*$ of fermions in each generation plus an adjoint representation and a $\underline{5}$ (and in some versions a $\underline{45}$) of Higgs bosons, there is a residual global symmetry conserving (*B–L*) thus permitting nucleon decay so far unobserved (Fiorini 1982, 1984, Park *et al* 1985, Hains 1986) and H–H̄ oscillations, but forbidding n–n̄ oscillations. It may, however, be possible to avoid this difficulty in a non-minimal extension of SU(5) by introducing a $\underline{15}$ of Higgs mesons which generate additional (*B–L*) violating amplitudes (Chang and Chang 1980). In this case the relevant amplitude is proportional to $m_x^{-4} m_{15}^{-1}$ and the observed stability of matter permits n–n̄ oscillations with $\tau^{n\bar{n}} \geqslant 10^7$ s provided $m_{15}c^2 \geqslant 10^5$ GeV. As we shall see below there is a serious difficulty with this and similar proposals (Kuz'min and Shaposhnikov 1983), which rely on the presence of low-mass Higgs bosons not explicitly required to break the symmetry.

The mass suppression of n–n̄ oscillations in SU(5) associated with the presence of a single grand unifying mass scale appears to be avoidable in models based on left–right symmetric groups $SU(2)_R \times SU(2)_L \times U(1)_{L+R}$

which are broken at mass scales intermediate between the electro-weak mass $m_w c^2 \simeq 10^2$ GeV and the grand unification mass $m_x c^2 \geqslant 10^{15}$ GeV (Mohapatra and Marshak 1980, Masiero and Mohapatra 1981). These models are most simply formulated in terms of the partial unification group $SU(2)_L \times SU(2)_R \times SU(4')$ where SU(4′) combines colour and (*B*–*L*) symmetry. This breaks down in successive stages to $SU_L(2) \times SU_R(2) \times U(1)_{B-L} \times SU(3)_c$ and ultimately to the standard model gauge group $SU(2)_L \times U(1)_y \times SU(3)_c$ at an intermediate mass scale m_{WR} where (*B*–*L*) symmetry and parity are simultaneously broken. In the minimal version of the theory observable n–n̄ oscillations can exist with $\tau_{n\bar{n}} \geqslant 10^7$ s assuming the presence of Higgs scalars Δ_{qq}, with the quantum numbers of the diquark, and $m_{\Delta qq} c^2 \geqslant 10^5$ GeV. However, this scheme encounters a new difficulty which arises because of the observed baryon–anti-baryon asymmetry. The problem is that, unless some linear combination $(B+aL)$ is conserved, with $a \neq 0$, $a \neq 1$, *B*-violating processes (such as n–n̄ oscillations) indicative of a left–right symmetry persisting down to mass scales $mc^2 \leqslant 10^{13}$ GeV will establish thermal equilibrium at temperatures $kT \geqslant mc^2$ and any existing baryon number excess will be wiped out (Weinberg 1980). Various means of avoiding the difficulty have been proposed, e.g. by introducing new *B*-violating processes at $kT \leqslant mc^2$ (Mohapatra and Pati 1981) or by postulating a scheme of partial unification which is not left–right symmetric (Fukugita *et al* 1982).

The lowest rank grand unifying group embedding parity restoration at intermediate mass scales, and thereby raising the possibility of observable n–n̄ oscillations, is SU(10) (Lüst *et al* 1982, Hazra *et al* 1983, Sokorac 1983). In theories based on this group the grand unification mass scale expands to 10^{18}–10^{19} GeV and left–right symmetry is broken at 10^8–10^9 GeV, a mass scale orders of magnitude greater than the mass $m_{\Delta qq} c^2 \geqslant 10^5$ GeV of the right-handed triplet of diquark Higgs bosons required to generate observable n–n̄ oscillations. This problem appears to arise in all schemes devised to produce observable n–n̄ oscillations, whether by non-minimal extensions of SU(5) (Chang and Chang 1980), in SU(6) (Fukugita *et al* 1982, Majumdar 1983) or in SO(10). The difficulty is that there is a restriction on the degree to which parameters in the underlying Lagrangian may be fine-tuned to produce any desired phenomenon, by the requirement to maintain the hierarchy of gauge boson masses which are acquired at various stages in the symmetry-breaking chain. This principle is expressed in the so-called extended survival hypotheses (Del Aguila and Ibanez 1981) which states in effect that only those components of the Higgs representations which are explicitly required to break the symmetry are not super heavy, i.e. the Higgs bosons acquire the maximum mass consistent with the postulated symmetry breaking pattern. Opinions are divided as to whether 'natural' mass scales, i.e. those consistent with this hypothesis, can give rise to observable n–n̄ oscillations in SO(10) (Parida 1983, Mohapatra and Senjanovic 1982).

One of the weaknesses of grand unified theories is that the fermions and bosons are organized in apparently unrelated representations of the unifying group and for this reason there has recently been great interest in supersymmetric versions of the theory (Bowick *et al* 1983, Ibanez 1982, Zwerner 1983, Mohapatra 1989). Supersymmetry unifies the fermion matter fields and the boson gauge fields and, since the observed natural world is not supersymmetric, the symmetry, if it exists, must be spontaneously broken, but at what mass scale is unknown. If it is broken close to the electro-weak mass scale this could be significant for n–n̄ oscillations, for in this respect supersymmetric theories are doubly advantageous. This is true first of all because, in place of the six-quark operators of dimension $d=9$ which are required in the conventional theory, one may have operators of dimension $d=7$ constructed from two quark fields and four scalar quark fields. Thus in supersymmetric SO(10) the diquark Higgs bosons required to generate n–n̄ oscillations with $\tau_{n\bar{n}} \geqslant 10^7$ s can have masses as high as 10^7 GeV (Kahara and Mohapatra 1983, Lüst 1983). Second, the extended survival hypothesis does not apply in these theories and the Higgs boson masses are constrained only by the boundary conditions on the renormalization group equations which fix the values of $\sin^2\theta_w$, and on the strong interaction parameters at the electro-weak mass scale (Aulakh and Mohapatra 1983, Majumdar *et al* 1984). Thus, although from the viewpoint of n–n̄ oscillations the supersymmetric models have not yet been properly explored, the prospects are encouraging. The general conditions under which n–n̄ oscillations with lifetimes in the range 10^6–10^{10} s are allowed in various forms of grand unified theory, with or without supersymmetry, have recently been subjected to a comprehensive review (Mohapatra 1989).

10.4.3.3 Detection of neutron oscillations

The neutron oscillation time $\tau_{n\bar{n}}$, while not being consistently defined in the literature, is most frequently taken to be equal to $\hbar/\delta mc^2$ where $\delta mc^2 = \langle \bar{n}| - \int L(\Delta B=2) d^3x |n\rangle$ is the matrix element describing n–n̄ mixing in the neutron mass matrix. The matrix elements of the six-quark operators contributing to the oscillation process have been evaluated in a number of models with results which are highly model dependent both as to sign and magnitude (Pasupathy 1982, Rao and Shrock 1982, Ozer 1982, Fajfer and Oakes 1983). The same Lagrangian which generates oscillations in free neutrons can, with the addition of a spectator quark line, give rise to oscillations in complex nuclei followed by annihilation and pion emission with an average energy of $\simeq 0.5$ GeV per pion. In these circumstances there may be additional matrix elements whose contribution to free neutron oscillation is negligible (Basecq and Wolfenstein 1983). Conversely the setting of a lower bound on $\tau_{n\bar{n}}$ in a complex nucleus does not imply that the same lower bound is valid for a free neutron (Kabir 1983).

The oscillation process for neutrons bound in nuclei has been analysed in detail by many authors (Kuz'min 1970, Mohapatra and Marshak 1980, Dover *et al* 1983, 1989) with the objective of setting an upper limit on δm derived from observations on nucleon stability (Giamati and Reines 1962, Learned *et al* 1979, Krishnaswami *et al* 1981, Cherry *et al* 1983, Jones *et al* 1984). The decay width of a nucleus $\Gamma_{n\bar{n}}(A,Z)$ due to neutron oscillation may be estimated from second-order perturbation theory (Chetyrkin *et al* 1980)

$$\Gamma_{n\bar{n}}(A,Z)=\frac{(A-Z)(\delta mc^2)^2.\Gamma}{\Delta\bar{M}^2c^4+(\Gamma/2)^2} \tag{10.50}$$

where $\bar{M}$ and $\Gamma \gg \Delta Mc^2=(\bar{M}-M(A,Z))c^2$ are the average mass and decay width of a quasi-nucleus containing an anti-neutron and $(A-Z-1)$ neutrons. If $\Gamma_m(A,Z)$ represents the experimental upper limit on the B-violating decay of the nucleus (A,Z) then $\Gamma_m(A,Z) \geqslant \Gamma_{n\bar{n}}(A,Z)/A$ and $\delta mc^2 \leqslant [\Gamma_m\Gamma A/4(A-Z)]^{\frac{1}{2}}$. Most estimates of Γ fix its value on the order of 0.2 ± 0.1 GeV, and a precise result for oxygen falls squarely within this range (Dover *et al* 1983). A direct search for n–n̄ oscillations in the oxygen nucleus using a 300-ton water Cerenkov detector in the Homestake gold-mine has given the result for the partial lifetime $T_0 \geqslant 1.4\times10^{30}$ years (Cherry *et al* 1983). This implies that, for neutrons bound in nuclei, $\tau_{n\bar{n}} \geqslant 2\times10^7$ s. Using more recent measurements of the two nucleon annihilation lifetime $T_0 \geqslant 2.4\times10^{31}$ years (Jones *et al* 1984), and $T_0 \geqslant 4.3\times10^{31}$ years (Takita *et al* 1985), this lower limit would be raised to $\tau_{n\bar{n}} \geqslant 7\times10^7$ s, and $\tau_{n\bar{n}} \geqslant 1.2\times10^8$ s, respectively.

Consider now the process of coherent generation of anti-neutrons in a beam of free neutrons. Since the neutron is created in a strong interaction process, e.g. fission, its initial state has baryon number $B=1$ whereas the anti-neutron has $B=-1$. Also, since neutron and anti-neutron have equal and opposite magnetic moments, and since it is impossible to eliminate magnetic fields entirely, it is necessary to take explicit account of the magnetic interaction. One must therefore solve for the time dependence of the two-component wavefunction

$$\psi=\begin{bmatrix}\psi_n\\ \psi_{\bar{n}}\end{bmatrix}$$

using the phenomenological Hamiltonian

$$H=\begin{bmatrix}mc^2-i\Gamma_\beta/2+\Delta E & \delta mc^2\\ \delta mc^2 & mc^2-i\Gamma_\beta/2-\Delta E\end{bmatrix} \tag{10.51}$$

where Γ_β represents the β-decay width and $\Delta E=\mu_n B$ is the magnetic energy of a neutron in a magnetic field B. the solution of the Schrödinger equation for the anti-neutron component $\psi_{\bar{n}}(t)$ is then

$$\psi_{\bar{n}}(t)=-i(\delta mc^2/\hbar\omega)\sin\omega t e^{-imc^2t/\hbar-\Gamma_\beta t/2\hbar}\psi_n(0) \tag{10.52}$$

where $\hbar\omega=[(\delta mc^2)^2+(\Delta E)^2]^{\frac{1}{2}}$. Thus the probability $P_{n\bar{n}}(t)$ that a neutron is converted into an anti-neutron in time t is given by

$$P_{n\bar{n}}(t)=(\delta mc^2/\hbar\omega)^2 \sin^2\omega t \qquad t\ll\hbar(\Gamma_\beta)^{-1}\simeq 10^3 \text{ s} \qquad (10.53)$$

Even supposing one were to create a magnetic environment with $B\leqslant 10^{-10}$ G, the corresponding magnetic energy $\Delta E\simeq 10^{-17}$ eV would still be some six orders of magnitude greater than the value δmc^2 that the nucleon stability evidence might lead one to expect. Thus in all realistic circumstances one may set $\hbar\omega\simeq\Delta E$. However, it is clearly unnecessary to establish ultra-low magnetic fields since we require only that $\omega t\ll 1$ in which case $P_{n\bar{n}}(t)\simeq(\delta mc^2 t/\hbar)^2$ and the anti-neutron population will grow quadratically in time independent of the precise value of the ambient field. Thus, assuming a growth time $t\leqslant 0.1$ s we require only that $B\leqslant 10^{-3}$ G, a condition which is relatively easy to establish using magnetic shields. When the condition $\omega t\ll 1$ fails then $P_{n\bar{n}}(t)\simeq(\delta mc^2/\Delta E)^2\leqslant 10^{-12}$ and the oscillation process is quenched. there is of course some experimental advantage in this situation since it provides a useful basis for a control experiment.

One may, however, envisage the performance of experiments involving long growth times and extended neutron flight paths where magnetic shielding would be impracticable and in these circumstances other avenues of approach must be explored. It has been shown, for example, that, whereas nothing is gained by applying spatially varying fields, in certain time-varying fields long-time solutions do exist describing quadratic growth with a coefficient equal to one-third of its value in the field-free case (Arndt *et al* 1981, Trower and Zovko 1982). A similar proposal is the suggestion to cancel the ambient field on average over a characteristic time interval by applying an additional intermittent field (Kabir *et al* 1984). Another idea which relies on a different principle is to 'dress' the neutrons in such a way as to cancel the magnetic interaction (cf. section 7.5.4).

The possibility of using bottled ultra-cold neutrons to achieve very long growth times for neutron oscillations has also been studied (Utsuro *et al* 1980, Chetyrkin *et al* 1980, Kazarnovski *et al* 1981, Marsh and McVoy 1983) but this idea is not as promising as it might seem because of the dephasing of n and n̄ components of the neutron wavefunction which takes place at each collision with the confining wall. This is due to the difference between the nuclear potentials for neutron and anti-neutron and leads to the regeneration of the baryon number eigenstates at each collision. Thus, if the mean time between collisions is τ_c, the mean number of collisions in the storage time t is $n_c=t/\tau_c$ and $P_{n\bar{n}}(t)\simeq n_c(\delta mc^2\tau_c/\hbar)^2=(\delta mc^2/\hbar)^2\,\tau_c t$. Thus the growth rate varies linearly rather than quadratically with time and the gain of $\approx 10^4$ in storage time achieved by using ultra-cold neutrons is insufficient to compensate for the reduction in intensity by factors of order 10^{-10}–10^{-11} thereby incurred. Similar problems of intensity and regeneration make it unlikely that the detection of oscillation-induced non-linear effects in neutron interferometry or spin-resonance experiments can seriously compete with the direct observation of anti-neutrons (Casella 1984).

Astrophysical systems provide one area in which the consequences of neutron oscillations might be manifested since it is well known, for example, that neutrino oscillations provide at least a possibility for solving the solar neutrino problem (Bahcall *et al* 1982). Existing measurements on the ratio of anti-protons to protons in the cosmic radiation imply that $\tau_{n\bar{n}} \geqslant 2.5 \times 10^3$ s although this is not a very sensitive bound because the neutron flux produced by spallation reactions of cosmic ray protons in the atmosphere is so low (Sawada *et al* 1981). The much higher fluxes of neutrons associated with solar flares could show the influence of neutron oscillations with $\tau_{n\bar{n}} \geqslant 10^5$ s which is the bound set by considering the n–n̄ annihilation rate which would photo-disintegrate the deuterium formed by neutron–proton fusion during the first stage of nucleosynthesis in the early Universe (Krishnan and Sivarem 1982) (cf. sections 1.2.1, 10.5.4).

The first direct search for anti-neutrons coherently generated in a beam of cold neutrons was carried out at the ILL Grenoble, and produced a lower limit $\tau_{n\bar{n}} > 10^6$ s (Fidecaro *et al* 1985), a conclusion subsequently confirmed by a second experiment at Pavia (Bressi *et al* 1989). This result has recently been improved by an order of magnitude in a second ILL experiment carried out using the apparatus illustrated in figure 10.6 (Baldo-Ceolin *et al* 1990). In this experiment the cold neutron beam had a wavelength spectrum peaking near 4 Å, and was transported to the main apparatus by a curved 60 m long guide tube which eliminated all fast neutrons and γ-rays from the beam.

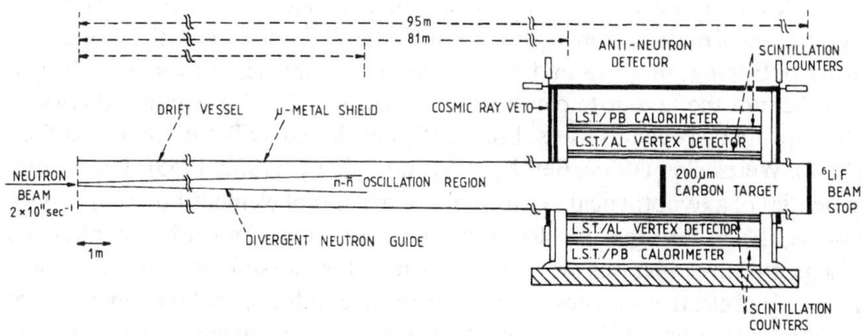

Figure 10.6. Apparatus used to search for neutron–anti-neutron oscillations in a 95 m drift tube in conditions of ultra-low magnetic field. Anti-neutrons created annihilate with neutrons in the carbon target releasing about five pions which are tracked in four quadrants of limited streamer tubes (LST). The γ-rays from π^0-decay are converted in 1.2 mm lead plates which also form the components of a calorimeter (Baldo-Ceolin *et al* 1990).

The apparatus itself consists of a 95 m long 1.2 m diameter drift tube of which about 75 m is enclosed in a μ-metal shield which reduces the internal magnetic field to <0.2 mG. Neutrons first enter a collimating guide and subsequently drift for about 0.1 s traversing a 200 μm carbon target where any anti-neutrons created during the drift phase are removed by annihilation. The transmitted neutrons are subsequently absorbed in a ^{6}Li beam stop. An n–n̄ annihilation event produces on average about five pions and these are recorded in a tracking device consisting of four quadrants each containing 10 planes of limited streamer tubes. Two layers of scintillation counters are placed before and after each plane to determine the pion direction by time of flight analysis, and after the second layer the planes are interspaced with lead plates a few millimetres thick. These provide some four radiation lengths for γ-rays produced in π° decay, and act not only to determine the direction of the π°, but also as a calorimeter to measure the total energy of the event. A veto system against external cosmic rays is obtained by siting scintillation counters around the pion detector, and shielded from it by 10 cm of lead to prevent the veto being activated by genuine events generated within the detector.

The system is about 45% efficient for detection of n–n̄ annihilation events and recorded an overall event rate of about 1.5 s^{-1}, mostly due to cosmic rays. However no genuine n–n̄ events were recorded leading to a lower limit $\tau_{n\bar{n}} > 10^7$ s.

10.5 The neutron in astrophysics and cosmology

10.5.1 Origin of the chemical elements

The idea that the power sources of sun and stars are derived from the conversion of mass into energy took hold in the 1920s following the pioneering work of Eddington, Jeans and others, with thermonuclear fusion of hydrogen into helium the favourite candidate (cf. section 1.2). The detailed theory of the p–p and C–N chains was subsequently developed by Bethe and Critchfield (1938), Weizsäcker (1938), Bethe (1939) and Salpeter (1952, 1955). The unique potential of a hypothetical neutron in the process of element building had, of course, been foreseen by Rutherford (1920) and, immediately after its discovery, investigations were initiated into the possible reaction products in a soup of electrons protons and neutrons in conditions of high temperature and pressure (Sterne 1933). The neutron also had a prominent place in various 'cold big bang' and 'continuous creation' model Universes, but these controversies are today mainly of historical interest (Tayler 1983).

The importance of neutron capture as an agent in the build up of the heavy elements (Gamow 1940) was underscored when it was discovered that the cosmic abundances of elements with $A \geqslant 65$ varied in approximate inverse ratio to their neutron capture cross-sections. This led Gamow and his associates (Alpher *et al* 1948) to propose the model of an initially hot Universe of photons and neutrons, whose decay and subsequent capture of protons to form deuterium would provide the seed material for the build up of heavy elements by successive neutron captures. Although this particular scenario fails because of the non-existence of stable or long-lived nuclear species at $A=5$ and $A=8$, the notion of a Universe created in a 'hot big bang' has survived. Since the discovery of the relict microwave background (Penzias and Wilson 1965) the 'hot big bang' has assumed an unassailable position in modern astrophysics as the 'standard model' for the origin of the Universe (Weinberg 1977).

10.5.2 Light element nucleosynthesis

Most of the information we have about stars has been derived from spectroscopic data and the identification of characteristic spectral lines in Fraunhofer spectra. A popular method for representing stellar observations is to plot the absolute visual magnitude M_V against the effective surface temperature as parameterized by the colour index $B-V$. Such a representation is known as a Hertzsprung–Russell diagram, and a simple plot of this type which summarizes the data on nearby stars is shown in figure 10.7. In determining cosmic abundances these data must be supplemented by chemical and isotopic analyses of meteorites which are usually classified according as they are predominantly metals, sulphides or silicates (Goldschmidt 1937). Existing tables of cosmic abundances derive in the main from the comprehensive tabulations of Suess and Urey (1956).

A smoothed-out cosmic abundance plot on a logarithmic scale is shown in figure 10.8. The rapid reduction in relative abundance with increasing A is a reflection of the corresponding growth in the height of the Coulomb barrier which inhibits fusion. The peak near ^{56}Fe is an indication of the maximal stability of iron. The double peaks at higher values of A may be correlated with magic numbers of neutrons at $N=50$, 82 and 126 (cf. section 4.2.1).

Although the relative abundances of hydrogen and helium do not vary much from star to star, the relative abundances of the heavier elements can decrease by an order of magnitude in old stars as compared with new stars, indicating that new stars are created in an unfolding evolutionary process from the debris of stars which have exploded in the past. These observations are consistent with current views which hold that light elements with $A \leqslant 65$ are formed by thermonuclear processing in conditions of hydrostatic equilibrium, while heavy elements are created by successive neutron captures at astrophysical sites, e.g. exploding stars, which favour the generation of high neutron fluxes (cf. section 10.5.7).

Under hydrostatic conditions the energy spectrum of each nuclear species is governed by a Maxwell–Boltzmann distribution at the ambient temperature.

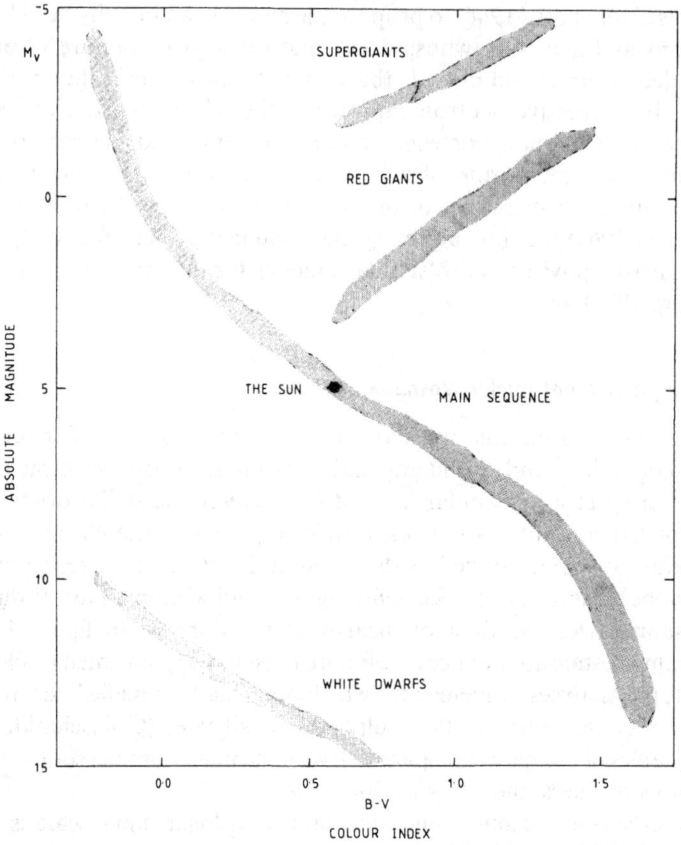

Figure 10.7. Schematic representation of the Hertzsprung–Russell diagram showing a plot of absolute visual magnitude M_V against colour index B–V, providing a measure of surface temperature for stellar populations in various stages of evolution. The quantities B and V represent the magnitudes measured in narrow wavelength ranges centred in the blue, and in the yellow, respectively. Since the magnitude is a linear function of the logarithm of the luminosity, the colour index B–V measures the ratio of the luminosities at two different wavelengths. Hence it provides an estimate of the temperature of the stellar surface assumed to radiate as a black body.

However, because charged particles must penetrate their mutual Coulomb barrier in order to interact, the Boltzmann factor $\exp[-E/kT]$ is compounded with the barrier tunnelling factor $\exp[-G(E)]$, which means that the operative cross-section is that appropriate to an energy, the so-called 'Gamow peak' E_g which may be as much as 5 to 10 times the mean thermal energy (cf. figure 5.13). An exception to this rule occurs when there is a resonance in the cross-section in which case the resonant energy is the relevant one, where the cross-section may be amplified by many orders of magnitude. In

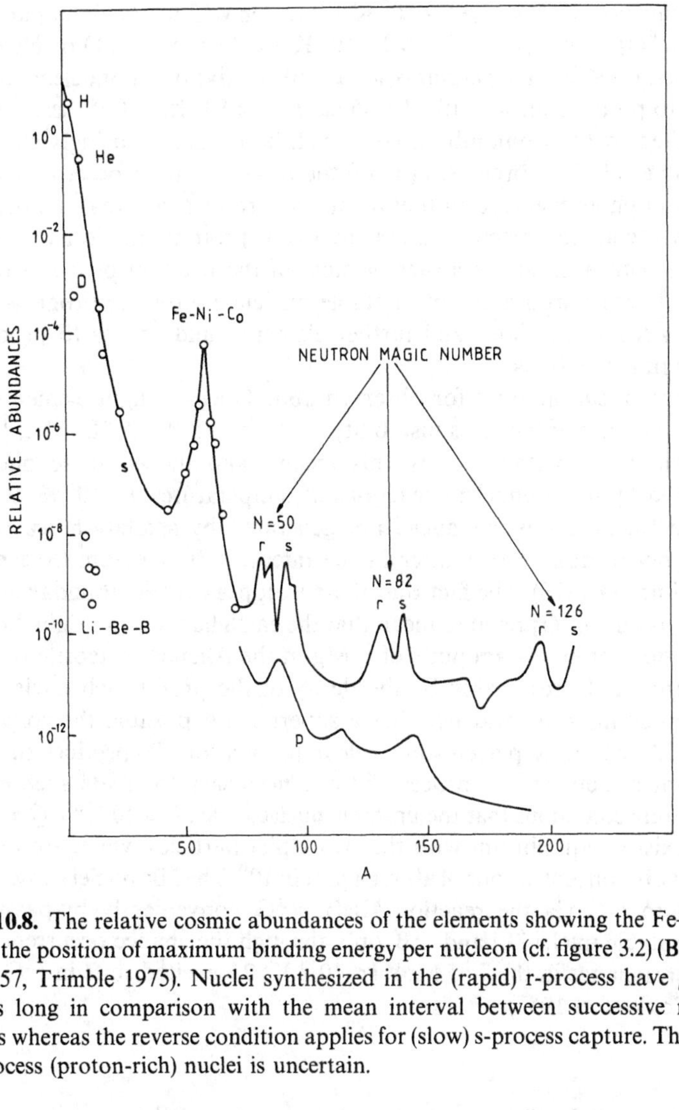

Figure 10.8. The relative cosmic abundances of the elements showing the Fe–Ni–Co peak at the position of maximum binding energy per nucleon (cf. figure 3.2) (Burbidge *et al* 1957, Trimble 1975). Nuclei synthesized in the (rapid) r-process have β-decay lifetimes long in comparison with the mean interval between successive neutron captures whereas the reverse condition applies for (slow) s-process capture. The origin of p-process (proton-rich) nuclei is uncertain.

first-generation stars such as the sun which are composed mainly of hydrogen, energy generation is mainly through the p–p cycle; in massive stars where high temperatures and densities may be achieved before hydrostatic equilibrium is established, the conversion of hydrogen into helium may be brought about through the agency of isotopes of carbon and nitrogen which act as catalysts in the sense that they are neither created nor destroyed in the process. This reaction chain is known as the C–N cycle.

In general the production of nuclei with $A \leqslant 65$ (with the exception of Li, Be and B) is viewed as taking place in four stages: (i) fusion of H into He in main sequence stars @ $T \geqslant 5\times10^7$ K, (ii) fusion of 3 (or 4) alpha particles to ^{12}C (or ^{16}O) in red giants @ $T \geqslant 2\times10^8$ K, (iii) fusion of C, O or Ne to form nuclei with $A \leqslant 28$ in supergiants @ $T \geqslant 10^9$ K, (iv) thermonuclear reactions in ^{28}Si to produce nuclei with $A \leqslant 65$ @ $T \geqslant 5\times10^9$ K. All of these reactions are limited by the Coulomb barrier which becomes increasingly effective at high values of Z. In processes (i)–(iii) the reaction rate proceeds faster than the rate of photodisintegration of the fusion products whereas in process (iv) there is a complex reaction chain involving protons, neutrons and alpha particles with a gradual transfer of most of the reaction products into the iron peak. Since iron is the most stable nucleus no further progress can be made by fusion reactions and further element building has to proceed by endothermic reactions.

Process (i) can account for observed abundances of light elements up to ^7Li but, because of the mass instability at $A=5$ and $A=8$, ^6Li, ^9Be, ^{10}B and ^{11}B cannot be synthesized by this means, and indeed these nuclei are consumed by thermonuclear reactions at temperatures $T \geqslant 10^6$ K. It is now believed that the missing nuclei are generated by spallation reactions of cosmic ray protons with matter in the intergalactic medium (Reeves *et al* 1970, Simpson 1973). The fact that these isotopes exist in abundance on the earth is usually interpreted to mean that the earth had a cold origin; however, observations of excess amounts of ^{26}Mg in the Allende meteorite (Lee *et al* 1976) produced presumably by the decay of the proton-rich nucleus ^{26}Al, has been adduced as evidence that a supernova explosion, the conjectured site of ^{26}Al synthesis, played some role in its creation. To produce the highly abundant nucleus ^{12}C in process (ii) it is necessary to create such extreme temperature conditions that the unstable nucleus ^8Be ($T_{\frac{1}{2}} \leqslant 10^{-16}$ s, $Q \simeq 94$ keV) can coexist in equilibrium with the two alpha particles which are its decay products, in concentrations of about a part in 10^9. The ^8Be nuclei subsequently convert to ^{12}C via the reaction $^8Be(\alpha,\gamma)^{12}C$, conveniently by-passing the missing stable nuclei ^6Li and ^{11}B, only through the chance occurrence of an s-wave resonance in the 7.65 MeV ($\simeq 10^{11}$ K) 2^+ excited state in ^{12}C (Hoyle 1954, Cook *et al* 1957).

10.5.3 Primordial helium synthesis and the number of light neutrino species

Although measured helium abundances vary from site to site, they typically fall within the range of 20–30% of the total mass, of which no more than a few per cent can be accounted for in terms of stellar thermonuclear processes. According to the hot big bang model of the expanding Universe, the helium is formed in the very early stages of expansion when there are many free neutrons present (Gamow 1940), through the sequence of electromagnetic and strong nuclear reactions $n+p \rightarrow {}^2H+\gamma$, ${}^2H+{}^2H \rightarrow {}^3He+n \rightarrow {}^3H+p$, ${}^3H+{}^2H \rightarrow {}^4He+n$. Since the variation of binding energy per nucleon over this mass range exhibits a strong maximum at ^{4}He (cf. table 3.1), it is the ultimate fate of most free neutrons to end up as constituents of this nucleus.

To trace the path from big bang to light element nucleosynthesis in the early Universe (Schramm 1977) one need only investigate developments below a temperature of about 10^{11} K, since above this temperature strong and electromagnetic interactions maintain the populations of the relevant particle species in a state of thermal equilibrium. Furthermore, the expansion time $t \simeq 10^{-2}$ s is much greater than the lifetime of heavy unstable particles and at this epoch only electrons, positrons, photons, nucleons and neutrinos remain.

The statistical balance by mass of neutrons and protons is maintained at its equilibrium value

$$X_n/X_p = \exp[-(m_n - m_p)c^2/kT] \tag{10.54}$$

through the action of the weak processes

$$\mathrm{n} + e^+ \leftrightarrows \mathrm{p} + \bar{\nu}_e \qquad \mathrm{p} + e^- \leftrightarrows \mathrm{n} + \nu_e \tag{10.55}$$

which at temperatures $> 10^{11}$ K ($t \simeq 10^{-2}$ s) proceed much faster than free neutron decay. However, the condition of thermal equilibrium ensures that these weak reaction rates can be computed using thermodynamics alone, once the characteristics of free neutron decay are known.

Since the characteristic weak scattering times increase as T^{-5} with falling temperature, in comparison with the expansion time which goes as T^{-2}, a point is reached at which these weak reactions can no longer maintain X_n/X_p at its equilibrium value. At this freeze-out temperature $T_f \approx 10^{10}$ K ($t \simeq 1$ s) the neutrinos decouple from matter and X_n/X_p decays with a time characteristic of free neutron decay. At $T \simeq 3 \times 10^9$ K ($t \approx 10$ s) the remaining $e^\pm$ pairs annihilate to photons, and thereafter matter and radiation stay in equilibrium until atoms form at $T \simeq 4 \times 10^3$ K ($t \simeq 0.5 \times 10^6$ years). At this point the universe switches over from radiation domination to matter domination and the photons decouple from the matter giving rise to the 2.9 K microwave background we observe today.

The onset of nucleosynthesis can be fixed quite precisely near $T \simeq 10^9$ K ($t \simeq 180$ s) when deuterium formed in the radiative capture of neutrons on protons remains stable against photodisintegration in the thermal radiation field. This is followed by a chain of strong interaction processes as summarized above, whose immediate effect is the conversion of virtually all free neutrons into helium. Knowing the relevant temperatures, the expansion rate of the Universe and the lifetime of the neutron, a simple calculation arrives at a value of about 25% for the ^{4}He/H mass ratio in very good accord with observation (Tayler 1968, 1979, 1980, 1981). Some further nuclear physics input leads to a 2×10^{-3}% abundance of deuterium by mass and lesser amounts of heavier elements up to ^{7}Li. The Coulomb barrier and the gaps at mass numbers 5 and 8 inhibit synthesis of heavier species through (n,γ), (p,γ) and (α,γ) reactions.

The conclusion then is that the helium abundance is determined by (a) n and p masses (b) the binding energy of the deuteron and (c) the ratio of the neutron lifetime to the characteristic expansion time.

$$\tau_e = \left(\frac{1}{V}\frac{dV}{dt}\right)^{-1} = \sqrt{(24\pi G\rho)} \qquad (10.56)$$

where G is the gravitational constant and ρ the density. Since helium is synthesized in the era of radiation domination where baryonic matter contributes $<10^{-3}\%$ of the energy density, ρ is determined by the temperature and the number of particle species which are relativistic at that temperature. In particular any significant increase in the measured value of the neutron lifetime leads to a corresponding increase in helium synthesis and conversely.

Suppose finally that there exist other families of massless, or almost massless, neutrinos, over and above the three known families. These particles would also contribute to the density and the universe would expand correspondly faster between freeze out and nucleosynthesis. Over this reduced period of time fewer neutrons would decay and more helium would be produced. On these grounds existing limits on helium abundances would appear to rule out more than one additional family of light two-component neutrinos, and indeed the same evidence would seem to point to the conclusion that known neutrino species cannot be four-component Dirac particles. Recent measurements of the decay width of the Z° intermediate vector boson have now settled the question of the number of light neutrino species (Adeva *et al* 1990, Decamp *et al* 1990, 1992, Akrawy *et al* 1991, Aarnio *et al* 1990, Abrams *et al* 1989) which is indeed limited to three in accord with the cosmological evidence (Steigman *et al* 1977, Yang *et al* 1984, Olive *et al* 1990, Walker *et al* 1991).

10.5.4 The solar neutrino problem

The proton–proton cycle of thermonuclear reactions is believed to be the predominant source of solar energy, the end-point of which is the fusion of four protons into a helium nucleus with the release of positrons, photons and neutrinos. The failure to observe more than about $(25\pm4)\%$ of the predicted capture rate of solar neutrinos in a ^{37}Cl target leading to an isobaric analogue state in ^{37}A constitutes 'the solar neutrino problem'. (Bahcall *et al* 1982, Bahcall and Ulrich 1988, Mikheev and Smirnov 1985, Bethe 1986).

In the first step of the chain two protons combine to form a deuteron

$$p + p \rightarrow {}^2H + e^+ + \nu_e \qquad (10.57)$$

generating neutrinos with energy $\leqslant 0.42$ MeV, insufficient to trigger the ^{37}Cl detector. This is a weak interaction whose rate determines the speed of the cycle. The p–*e*–p reaction,

$$p + e^- + p \rightarrow {}^2H + \nu_e \tag{10.58}$$

which occurs with a branching ratio of 0.25%, is an alternative to the p–p process, generating neutrinos of energy $\leqslant 1.44$ MeV which are detectable in ^{37}Cl.

The next step in the cycle is the fusion of hydrogen and deuterium to form ^{3}He

$$p + {}^2H \rightarrow {}^3He + \gamma \tag{10.59}$$

at which stage the process branches with fusion of $^3He + {}^3He$ to form ^{4}He (91%), or fusion of $^3He + {}^4He$ to form ^{7}Be (9%). Approximately 1% of the latter branch results in proton capture on ^{7}Be followed by the formation of ^{8}B whose β decay generates the 14.1 MeV neutrinos to which the ^{37}Cl detector primarily responds.

The predicted rate of neutrino counting based on the standard solar model is (7.9 ± 2.6) SNU, where 1 SNU $= 10^{-36}$ neutrino interactions per target atom per second (Bahcall and Ulrich 1988). Of the 2.6 SNU error on the predicted rate, approximately 0.3 SNU derives from the estimated error in the neutron lifetime. The role of the neutron lifetime in this analysis stems from the fact that the governing p–p reaction in (10.57) is nothing other than inverse neutron β decay with the spectator proton providing the energy, while the p–*e*–p reaction is the corresponding electron capture transition. However, since the two protons can react weakly only when they scatter in the singlet state because of the Pauli principle, and because the deuteron can exist only in the triplet state, this is a pure Gamow–Teller transition with a rate proportional to $g_A(0)^2$. Apart from meson exchange effects of order 1% this process can be calculated precisely knowing the temperature and the neutron lifetime (Bethe and Critchfield 1938, Salpeter 1952, 1952a, 1955, Bahcall and Wolf 1965, 1965a, Bahcall and May 1969, Bargholtz 1979).

The reason for the sensitivity of the neutrino rate to uncertainties in the value of the neutron lifetime is the following. If the neutron lifetime is reduced, the p–p and p–*e*–p reactions go faster in the same proportion. However, the solar luminosity is fixed so the increased rate of the weak interaction must be compensated by a reduction in temperature. It turns out that in these circumstances the relative number of ^{8}B nuclei is reduced with a corresponding reduction in the predicted rate of neutrino captures. However it now seems unlikely that the solar neutrino problem can be disposed of by this means, for recent results with new detectors have confirmed the original discrepancies. This is true both for the electron-neutrino scattering experiment at Kamiokande II, which is primarily sensitive to the ^{8}B neutrinos (Hirata *et al* 1988, 1990), and to the ^{7}Ga experiments, SAGE and GALLEX, which are also sensitive to the p–p neutrinos (Anselmann *et al* 1992, Abasov *et al* 1991). Indeed the current results strongly suggest that the solution to the problem is to be found in the phenomenon of neutrino oscillations (Bahcall and Bethe 1990, 1993, Hirata *et al* 1990a), as realized through the MSW mechanism (Mikheev and Smirnov 1986, Wolfenstein 1979).

10.5.5 Supernova core collapse

Type I supernovae are distinguished from type II supernovae in that the former reveal no hydrogen lines in their spectra. These are believed to be exploding white dwarfs with masses exceeding the Chandrasekhar stability limit of 1.4 solar masses (Chandrasekhar 1931, 1935, Landau 1932). Hydrogen is prominent in the spectra of type II supernovae which are exploding massive stars, of order 10–100 solar masses, which have rapidly evolved to the stage where their nuclear fuel is exhausted (Trimble 1982, 1983). When the core mass exceeds the Chandrasekhar limit, the pressure of relativistic degenerate electrons can no longer support it, and gravitational collapse sets in, releasing amounts of energy in excess of 10^{53} ergs. The time-scale is dictated by the weak interaction rates of leptons and nucleons, of which charged current processes dominate in the early stages of collapse, and neutral current processes take over towards the end. About 98% of the total energy released is carried off by neutrinos evenly spread over all three flavours, 1% is carried off as kinetic energy of the debris and a further 1% is emitted as gravitational radiation. No more than 10^{-2}% of the total is emitted in the form of visible or ultraviolet radiation.

The most recent supernova observed in our galaxy was 'Kepler's Star' which appeared in 1604. The 1987 type II supernova SN1987A exploded in the Large Magellanic Cloud, a nearby galaxy some 1.7×10^5 light years distant from the earth. It was remarkable in that the progenitor star could be precisely identified as the blue supergiant Sanduleak $-69°202$, rather than a red supergiant, with a mass estimated at 20 solar masses, and a radius of order 10^{12} m, an order of magnitude less than the radii of typically less dense red giants. Subsequent observations of the ultraviolet spectrum showed up the presence of the red giant envelope which had surrounded the massive star before it evolved into a blue giant. SN1987A was unique in that it was detected not only by its photon emission but also by its neutrino emission, the neutrinos being the first ever from an extra-galactic source to be detected on the earth. Altogether 19 neutrino events were recorded at the Kamiokande and IMB underground detectors (Hirata *et al* 1987, Bionata *et al* 1987). These detectors were primarily sensitive to electron anti-neutrinos which were observed through the classic reaction (Reines and Cowan 1953, 1953a, Cowan *et al* 1956, Bahcall *et al* 1987, 1987a)

$$\bar{\nu}_e+\mathrm{p}\rightarrow\mathrm{n}+e^+ \tag{10.60}$$

More than 90% of the light output from the supernova originated, not through dissipation of the shock energy, but from the decay of 6.1 day ^{56}Ni synthesized in the explosion. This decays to 77.1 day ^{56}Co which in turn decays to ^{56}Fe. The emitted γ-radiation, thermalized by Compton scattering and photoionization, provided the main power sources in the later stages of the supernova, whose light-curve decayed with the characteristic 77.1 day half-life.

The current model of a type II supernova explosion postulates a massive progenitor star composed of a large envelope of hydrogen enclosing successive layers of helium, carbon, oxygen and silicon (Arnett and Schramm 1973, Schramm and Arnett 1975). These surround an iron-nickel core of radius $\approx 2\times 10^3$ km, and mass in excess of 1.4 solar masses, which is supported by the pressure of a degenerate electron gas. When the core temperature reaches $\simeq 10^9$ K, photonuclear disintegration of the core nuclei sets in, and these are dissociated into nucleons and α particles. At a temperature $T\simeq 10^{10}$ K, and density $\rho\simeq 10^{10}$ g cm^{-3}, electron captures can occur

$$e^- + \mathrm{p} \to \mathrm{n} + \nu_e \tag{10.61}$$

and removal of the electron pressure causes the core to collapse. During this process, which lasts $<10^{-2}$ s, the core radius contracts to around 10–20 km and 10% of the available gravitational energy is released, mainly in the form of electron neutrinos. Since neutron β-decay is inhibited at a density $\rho\approx 10^8$ g cm^{-3}, and ceases for $\rho > 10^9$ g cm^{-3}, the core converts to a degenerate neutron gas; when the density reaches that of nuclear matter with $\rho\simeq 2\times 10^{14}$ g cm^{-3} the collapse stops and the remnant is formed. For core masses less than 2 solar masses the remnant is a neutron star; for more massive cores it is a black hole.

The initial neutronization phase described by (10.61) lasts a few milliseconds during which time a condition of hydrostatic equilibrium is established in the core which is composed mainly from n, p, e^-, together with ν_e and $\bar{\nu}_e$ which are trapped by diffusion through neutral current processes, e.g.

$$\begin{aligned} \nu_e(\bar{\nu}_e) + \mathrm{p} &\to \nu_e(\bar{\nu}_e) + \mathrm{p} \\ \nu_e(\bar{\nu}_e) + \mathrm{n} &\to \nu_e(\bar{\nu}_e) + \mathrm{n} \end{aligned} \tag{10.62}$$

During the next ≈ 10 s the process of neutronization is completed, the remnant cools to its final form, and the remaining energy is released. This release is evenly spread over all three neutrino flavours, generated in the main by a variety of neutral current processes e.g.

pair production $\quad e^+ + e^- \to \nu + \bar{\nu} \quad$ (10.63a)

neutrino bremsstrahlung $\quad \mathrm{n}+\mathrm{n}\to\mathrm{n}+\mathrm{n}+\nu+\bar{\nu} \quad$ (10.63b)

'plasmon' decay $\quad \gamma_{\mathrm{p}} \to \nu + \bar{\nu} \quad$ (10.63c)

where a 'plasmon' is a collective density excitation of a Fermi fluid. A significant contribution is also provided by the charged current processes

(Chiu and Salpeter 1964)

$$n+n \to n+p+e^- + \bar{\nu}_e \tag{10.64a}$$

$$n+p+e^- \to n+n+\nu_e \tag{10.64b}$$

which are in many ways analogous to the p–p and p–*e*–p reactions in the sun. The same processes also provide an important cooling mechanism for neutron stars (Bahcall and Wolf 1965).

The shock wave which causes the explosion, which is the visual signature of a supernova, is generated precisely at that point in time where nuclear matter density is reached in the core. This rebounds causing a hydrodynamic 'bounce' or pressure wave which spreads outwards carrying energy into the in-falling outer regions of the star. The outgoing shock wave neutronizes the material in its path through the reactions (10.61), the pressure defect being neutralized by the reverse reaction

$$\nu_e + n \to p + e^- \tag{10.65}$$

caused by neutrinos released from the shocked layers. The net result is that the shock wave propagates outwards without damping and the outer layers of the star are blown away.

In the normal case of a red giant supernova the shock takes several days to reach the surface of the star, although in the case of the blue star supernova SN1987A the measured delay was about 3 h, consistent with the theory. From the mean energy of 4.5 MeV of the detected $\bar{\nu}_e$ (assumed to constitute 1/6th of the total neutrino flux at the earth) the estimated energy release was 2×10^{53} ergs, again in good agreement with expectation. The data were consistent with a remnant of about 1.5 solar masses, i.e. a neutron star, but this prediction has yet to be confirmed by observation.

10.5.6 Neutron stars

The neutron star is a compact object as massive as the sun but with a radius of only about 10–20 km. It is unique in the history of astrophysics in being the only exotic stellar object whose existence had been predicted, and its properties elaborated in some detail, prior to its discovery. The black hole is an even more remarkable object whose existence is demanded by general relativistic cosmology but which has as yet to be identified unambiguously. The neutron star was first proposed by Baade and Zwicky (1934, 1934a, 1938) within a couple of years of the discovery of the neutron, when they suggested that a supernova signalled the final transition of a massive star into a neutron star, a high density cold degenerate incompressible neutron fluid, which they believed to be the most stable configuration of nuclear matter. Subsequently Tolman, Oppenheimer and Volkoff derived the equation of state for a neutron star by applying general relativistic hydrodynamical theory to a Fermi gas of free neutrons (Tolman 1939, Oppenheimer and Serber 1938, Oppenheimer and Volkoff 1939). In order to guarantee stability of the star against neutron β-decay sufficient protons and electrons must remain to inhibit this decay by filling the available momentum states as required by the Pauli principle. This requirement then sets an approximate lower limit to the mass of the star.

Today neutron stars are identified with the astronomical objects known as pulsars (Hewish *et al* 1968), of which more than 300 have been charted in the local galaxy. The first pulsar was discovered in the radio spectrum appearing as a double pulse with a stable period of 1.337 301 1 s. The very short period provided the principal clue to the identity of the source which could only be a spinning or pulsating object which was very small on celestial scales. It is now believed that the radiation is generated by charged particles accelerating in the strong magnetic flux ($\approx 10^{12}$ G) which is trapped when the progenitor star suffers gravitational collapse, and increases its central density by up to five orders of magnitude in a few milliseconds. This postulated association of pulsars and supernovae was confirmed with the discovery of a 0.033 s pulsar emitting at radio, optical and X-ray frequencies, from the centre of the Crab nebula, itself the embers of the great supernova or 'guest star' observed by the Chinese astronomers in AD 1054. The suggestion that a neutron star might be found within the crab nebula had been advanced even before pulsars were first identified (Cameron 1959), as indeed had the idea that neutron stars could provide the sources of X-ray from outside the solar system (Finzi 1964).

The precise structure of a neutron star cannot of course be determined from analysing its electromagnetic radiations alone, but requires a detailed theory involving an input from nuclear physics and elementary particle physics which is, to say the least, highly speculative. It is believed, however, that the neutron star has a metallic crust, extending perhaps 1 km into the core, which is itself composed mainly from a neutron superfluid (Migdal 1959) with small admixtures of order 1–2% of superconducting proton fluid and degenerate electrons required to maintain electrical neutrality. The conditions of β-stability and electrical neutrality jointly require that (Maxwell *et al* 1977)

$$\mu(\mathrm{n}) = \mu(\mathrm{p}) + \mu(e) = 2\mu(\mathrm{p}) \tag{10.66}$$

where $\mu(j)$ is the chemical potential of the jth particle species. Thus, at nuclear matter densities

$$\mu(\mathrm{n}) - \mu(\mathrm{p}) = \mu(\mathrm{p}) \simeq 100 \text{ MeV} \simeq 0.75 m_\pi c^2 \tag{10.67}$$

This result implies that, at rather higher densities, pions could exist in finite numbers in the ground state, generated by the strong interaction

$$\mathrm{n} \to \mathrm{p} + \pi^- \tag{10.68}$$

Since pions are bosons, at the absolute zero of temperature they would condense into the lowest energy state (Sawyer 1972, Scalapino 1972). A number of matter phases have been proposed for the deep interiors of neutron stars with $\rho > 3 \times 10^{14}\,\mathrm{g\,cm^{-3}}$, but most theories envisage some kind of negative pion condensate (Baym and Pethick 1975).

The weak interaction strengths of neutrons in neutron stars are not of course identical with those of free neutrons, even making due allowance for the operation of the Pauli principle. Although the CVC theorem requires that $g_V(0)$ remain constant, no such restriction applies to $g_A(0)$ which may indeed be 'quenched' (Blin-Stoyle 1975, Khanna *et al* 1978). It is well known for example that, in the β-decay of many mirror nuclei, $g_A(0)$ is reduced (Wilkinson 1970, 1971, 1973) and the phenomenon has been described in terms of a generalized Goldberger–Treiman relation applicable to nuclear matter (Oset and Rho 1979). In this proposal an analogy is drawn with the electrical polarizability of a dielectric medium to introduce the concept of an axial polarizability of nuclear matter, caused by the nucleon – Δ-isobar transition induced by the pion field. Magnetic fields are also important where they exist in strengths such that the spin energy of an electron associated with the Larmor procession is comparable with its rest mass, thereby disturbing the density of states for the emitted β particle. This happens in magnetic fields of order 10^{13} G which are believed to exist inside neutron stars (Gunn and Ostriker 1969). In such extreme conditions, not only the lifetime of the neutron, but also the electron spectrum and the various angular and polarization correlations are drastically altered (Ternov *et al* 1978, 1980, 1983).

10.5.7 Heavy element nucleosynthesis

Because of the absence of stable nuclei with mass numbers 5 and 8, and the relatively low density at the onset of big bang nucleosynthesis, nuclei more massive than ^{11}B cannot be synthesized in significant abundances in the early Universe. Although elements in the range up to the iron peak at $A \simeq 56$ are synthesized by charged particle reactions during the slow progress of stellar evolution through successive stages of thermonuclear burning from hydrogen to silicon, elements with $A > 65$ can be synthesized only through neutron capture. It is therefore necessary to identify suitable neutron sources.

At this point two time-scales must be recognized, namely τ_c the mean interval between neutron captures on a given nucleus, and τ_β the mean β-decay lifetime of nuclei in the reaction chain. Nuclei synthesized in chains for which $\tau_c \gg \tau_\beta$ are called (slow) s-process nuclei; if $\tau_c \ll \tau_\beta$ we have (rapid) r-process nuclei (Burbidge *et al* 1957, Cameron 1957). There also exist proton-rich stable nuclei, although with relatively low abundances, which cannot have been produced by neutron capture due to the absence of a stable seed nucleus. These are p-process nuclei; their origin is uncertain but is almost certainly connected in some way with supernovae (Audouze and Truran 1975).

All nuclei lying along the valley of β-stability in the Segré chart will have s-process contributions of the type

$$\binom{A}{Z} + \mathrm{n} \rightarrow \binom{A+1}{Z} + \gamma \rightarrow \binom{A+1}{Z+1} + e^- + \bar{\nu}_e \tag{10.69}$$

and nuclei on the neutron-rich side of the valley of stability will have r-process contributions. As an example consider the osmium isotopes (Trimble 1975); ^{186}Os is a pure s-process nucleus and ^{187}Os is almost pure s-process since ^{187}Re has a 5×10^{10} year half-life (cf. section 3.1). However ^{188}Os also has an r-process contribution through successive neutron captures in ^{186}Re (91 h) and ^{187}Re, followed by β-decay of ^{188}Re (17 h). On the other hand ^{184}Os (0.018%) has a p-contribution only, as indeed has ^{180}W (0.13%), since it is shielded from the neutron capture chain by the stable isobar ^{180}Ta (0.012%).

Studies of s-process nuclear abundances have led to the identification of the reaction ^{22}Ne(α, n) ^{25}Mg in the convective helium layers of thermally pulsed stars as the principal neutron source (Truran and Iben 1977). However, s-process nucleosynthesis takes place over such vast time-scales that the resulting nuclei are confined to that region of the nuclidic chart close to the stability line; thus the precise rates at which the weak decays proceed are unimportant. Conversely, to synthesize r-process neutron-rich nuclei, extremely intense short-lived neutron sources are required, lasting no more than perhaps a few seconds. Furthermore, to reach observed abundances with such a source requires the presence of substantial numbers of heavy ($A \simeq 60$) seed nuclei which can proceed to higher Z by β-decay.

For these reasons a site for synthesis of r-process nuclei has been postulated (Bruenn *et al* 1977) in the deep interior of a supernova near the mass-cut between the forming neutron star and the ejected matter. The mantle is assumed to expand adiabatically from conditions of peak temperature $T_0 \simeq 3 \times 10^{11}$ K and density $\rho_0 \simeq 10^{11}$ g cm^3, with time-dependent density determined according to an exponential law $\rho = \rho_0 e^{-t/\tau}$ where τ is a simple multiple of the free-fall time $(2\pi G\rho)^{-1/2}$. This model resembles the big bang in some respects except that the presence of a substantial number of species of seed nuclei, perhaps 20 or more, leads to a much more complex network of reactions.

It turns out then, as in the big bang, that the resulting abundances of r-process nuclei are determined essentially by the neutron–proton mass ratio X_n/X_p at the freeze out temperature $T_f \simeq 10^{10}$ K; this in turn is fixed by the rates of the elementary weak reactions

$$e^- + \mathrm{p} \rightarrow \mathrm{n} + \nu_e \qquad e^+ + \mathrm{n} \rightarrow \mathrm{p} + \nu_e \qquad \mathrm{n} \rightarrow \mathrm{p} + e^- + \bar{\nu}_e \tag{10.70}$$

Agreement with observed abundances can be obtained with freeze-out values of X_n/X_p in the range 4–8 but only for a very restricted range of values of T_0 and ρ_0 (Schramm and Barkat 1972).

There are several objections to this scenario not only in respect of its sensitivity to the initial conditions, but particularly because, at an assumed mass ejection rate of 0.2 solar masses per supernova, it could lead to over-production of r-process nuclei by factors $>10^3$. Explosive helium burning in massive stars initiated by the outgoing shock-wave in a supernova has been suggested as an alternative site (Klapdor *et al* 1981). In such a model, however, it is necessary to input the lifetime of each one of some thousands of nuclei participating in the reaction chain, from the β-stability line up to the neutron drip-line. In these circumstances the highly uncertain values of nuclear matrix elements become all important, and the simple dependence of the process on weak elementary reactions allied to neutron β-decay is lost.

References

Aarnio P *et al* (DELPHI Collaboration) 1990 *Phys. Lett.* **241B** 425, 449
Abasov A I *et al* 1991 *Phys. Rev. Lett.* **67** 3332
Abrahams K, Ratynski W, Stecher-Rasmussen F and Warming E 1966 *Nucl. Instr. Meth.* **45** 293
Abramowitz M and Stegun I A *Handbook of Mathematical Functions* (New York: Dover)
Abrams G S *et al* 1989 *Phys. Rev. Lett.* **63** 724, 2173
Adair R K 1952 *Phys. Rev.* **86** 160
— 1954 *Phys. Rev.* **94** 737
Adams J B 1966 *Proc. Phys. Soc. B* **89**, 189
Adelberger E G, Heckel B R, Stubbs B W and Rogers W F 1991 *Ann. Rev. Nucl. Part. Sci.* **41** 269
Ademollo M and Gatto R 1964 *Phys. Rev. Lett.* **13** 264
Adeva B *et al* 1991 *Z. Phys. C* **51** 179
Adkins G S, Nappi C R and Witten E 1983 *Nucl. Phys.* B **228** 552
Adler F T and Adler D B 1962 *Trans. Am. Nucl. Soc.* **5** 53
— 1964 *Trans. Am. Nucl. Soc.* **7** 86
Adler S L 1965 *Phys. Rev. Lett.* **14** 1051
— 1969 *Phys. Rev.* **177** 2426
Ageron P, Golub R, Hetzelt M, Mampe W, Pendlebury J M, Robson J and Smith K F 1977 *Proc. Int. Symp. on Neutron Inelastic Scattering* Vol 1 (Vienna: IAEA) 53
Ageron P, Mampe W, Bates J C and Pendlebury J M 1986 *Nucl. Instr. Meth.* A **249** 261
Aharonov Y and Ananden J 1987 *Phys. Rev. Lett.* **58** 1593
Aharonov Y and Susskind L 1967 *Phys. Rev.* **158** 1237
Airy GB 1835 *Trans. Camb. Phil. Soc.* **5** 283
Akheiser A and Pomeranchuk T 1948 *Sov. Phys.–JETP* **18** 475
Akrawy M Z et al 1991 *Z. Phys.* C **50** 373
Alaga G 1955 *Phys. Rev.* **100** 432
— 1957 *Nucl. Phys.* **4** 625
Alaga G, Alder K, Bohr A and Mottelson B R 1955 *Dan Mat. Fys. Medd.* **29** No 9
Alber D, Boebal O, Scharwz C, Diuve H, Hilscher D, Homeyer H, Jahnke U and Spellmeyer B 1989 *Z. Phys.* A **333** 319
Alder K, Stech B and Winther A 1957 *Phys. Rev.* **107** 728
Alefeld B, Christ J, Kukla D, Scherm R and Schmatz W 1965 *Berichte der Kernforschungsanlage*, Jülich Ju 1-194-NP
Aleksandrov D V, Belyatskii A F, Glukhov Yu A, Nikolskii E Yu, Novatskii B G, Ogloblin A A and Stepanov D N 1985 *JETP Lett.* **40** 909
Aleksandrov Yu A 1983 *Sov. J. Nucl. Phys.* **38** 1100
—1992 *Fundamental Properties of the Neutron* (Oxford: Clarendon)
Aleksandrov Yu A, Vrana M, Manrique G J, Machekhina T A and Sedlakova L N 1985 *Sov. Phys.–JETP* **62** 19
Aleksandrov Yu A *et al* 1986 *Sov. J. Nucl. Phys.* **44** 900
Alfimenkov V P, Borzakov S B, Thuan V V, Marseev Yu D, Pikelner L B, Khrykin A S and Sharapov E I 1983 *Nucl. Phys.* A **398** 93

Alfimenkov V P *et al* 1990 *Sov. Phys.–JETP Lett* **52** 373

Allasia D *et al* 1990 *Phys. Lett.* **249B** 366

Allen J S, Burman R L, Herrmansfeldt W B, Stahelin P and Braid T H 1959 *Phys. Rev.* **116** 134

Allet M, Hajdas W, Lang J, Lüscher M, Müller R, Naviliat-Cuncic O and Sromicki J 1992 *Phys. Rev. Lett.* **68** 572

Alpher R A, Bethe H A and Gamow G 1948 *Phys. Rev.* **73** 803

Als-Nielsen J, Bahnsen A and Brown W K 1967 *Nucl. Instr. Meth.* **50** 181

Als-Nielsen J 1984 *Physica B* **126** 145

— 1985 *Z. Phys.* B **61** 411

— 1986 *Physica* A **140** 376

Alsmiller R G, Santoro R T, Barish J and Gabriel T A 1975 *Nucl. Sci. Eng.* **57** 122

Altarelli G A and Ross G G 1988 *Phys. Lett.* **212B** 391

Altarev I S *et al* 1980 *Phys. Lett.* **80A** 419

— 1986 *JETP Lett.* **44** 460

— 1992 *Phys. Lett.* **276B** 242

Alvarez L and Bloch F 1940 *Phys. Rev.* **57** 111

Alvarez L W *et al* 1957 *Phys. Rev.* **105** 1127

Alvarez L W, Alvarez W, Asaro F and Michel H V 1980 *Science* **208** 4448

Amado R D and Brueckner K A 1959 *Phys. Rev.* **115** 778

Amaldi E and Fermi E 1936 *Phys. Rev.* **50** 899

Amaldi E, D'Agostino O, Fermi E, Pontecorvo B, Rasetti F and Segre E 1935 *Proc. R. Soc.* A **149** 522

Amaudruz P A *et al* 1991 *Phys. Rev. Lett.* **66** 2712

Amouyal A, Benoist P and Horowitz J 1957 *J. Nucl. Energy* **6** 79

Anders E, Sen Sarma R N and Kato P H 1956 *J. Chem. Phys.* **60** 707

Anderson H L, Fermi E, Long E A and Nagle D E 1952 *Phys. Rev.* **85** 956

Anderson H L, Johnson C S and Hincks E P 1963 *Phys. Rev.* **130** 2468

Ando M and Hosoya S 1972 *Phys. Rev. Lett.* **29** 281

— 1978 *J. Appl. Cryst.* **49** 6045

Andreev V N and Vladimirskii V V 1961 *Sov. Phys.–JETP* **41** 663

Anger H O 1958 *Rev. Sci. Instr.* **29** 27

Angerson W 1974 *Nucl. Phys.* B **69** 493

Anselman P *et al* 1992 *Phys. Lett.* **285B** 376

Aoki S and Hatsuda T 1992 *Phys. Rev.* D **45** 2427

Arima A and Iachello F 1974 *Phys. Lett.* **53B** 309

— 1975 *Phys. Rev. Lett.* **35** 1069

— 1976 *Ann. Phys., NY* **99** 253

— 1977 *Phys. Rev.* C **16** 2085

— 1978 *Ann. Phys., NY* **111** 201

— 1979 *Ann. Phys., NY* **123** 468

Arima A, Ohtsuka T, Iachello F and Talmi I 1977 *Phys. Lett.* **66B** 205

Armbruster P 1985 *Ann. Rev. Nucl. Part. Sci.* **35** 135

Arndt R A, Prasach V P and Riazzudin 1981 *Phys. Rev.* D **24** 1431

Arnellos L and Marciano W J 1982 *Phys. Rev. Lett.* **48** 1708

Arneodo A *et al* 1988 *Phys. Lett.* **211B** 493

— 1990 *Nucl. Phys.* B **333** 1

Arnett W D and Schramm D N 1973 *Astrophys. J.* **184** L47

Arnison *et al* 1983 *Phys. Lett.* **122B** 103
— 1983a *Phys. Lett.* **126B** 398
Arnold R G *et al* 1984 *Phys. Rev. Lett.* **52** 727
Arnold W R and Roberts A 1947 *Phys. Rev.* **71** 878
Ashman J *et al* 1988 *Phys. Lett.* **206B** 364
— 1989 *Nucl. Phys.* B **328** 1
Aston F 1919 *Nature* **104** 393
— 1920 *Nature* **105** 8, 547
— 1920a *Phil. Mag.* **39** 611
Atkinson R d'E and Houtermans F G 1929 *Z. Phys.* **54** 656
Attrep M and Kuroda P K 1968 *J. Inorg. Nucl. Chem.* **30** 699
Atwood D K, Horne M A, Shull C G and Arthur J 1984 *Phys. Rev. Lett.* **52** 1673
—1984a *Phys. Rev. Lett.* **53** 1300
Aubert J J *et al* 1974 *Phys. Rev. Lett.* **33** 1404
— 1983 *Phys. Lett.* **123B** 275
— 1985 *Nucl. Phys.* B **259** 189
— 1986 *Nucl. Phys.* B **272** 158
— 1987 *Nucl. Phys.* B **293** 740
Audouze J and Truran J 1975 *Astrophys. J.* **202** 204
Augustin J *et al* 1974 *Phys. Rev. Lett.* **33** 1406
Aulakh A H and Mohapatra R N 1983 *Phys. Rev.* D **28** 217
Austern N 1958 *Nucl. Phys.* **7** 195
Baade W and Zwicky F 1934 *Phys. Rev.* **45** 138
— 1934a *Proc. Natl. Acad. Sci.* **20** 254, 259
— 1938 *Astrophys. J.* **88** 411
Baade W, Burbidge G R, Hoyle F, Burbidge E M, Christy R F and Fowler W A 1956 *Publ. Astr. Soc. Pac.* **68** 296
Backenstoss G, Goebel K, Stadler B, Hegel U and Quitman D 1965 *Nucl. Phys.* **62** 449
Bacon G E 1951 *Proc. R. Soc.* A **209** 397
— 1966 *X-ray and Neutron Diffraction* (Oxford: Pergamon)
Bacon G E and Lowde R D 1948 *Acta. Cryst.* **1** 303
Badurek G, Rauch H and Zeilinger A 1988 Matter wave interferometry *Physica* B&C **151** 1
Bagnaia P *et al* (UA2 Collaboration) 1983 *Phys. Lett.* **129B** 130
Bahcall J N and Bethe H A 1990 *Phys. Rev. Lett.* **65** 2233
— 1993 *Phys. Rev.* D **47** 1298
Bahcall J N and May R M 1969 *Astrophys. J.* **155** 501
Bahcall J N and Wolf R A 1965 *Phys. Rev.* **140B** 1445, 1452
— 1965a *Astrophys. J.* **142** 1254
Bahcall J N, Huebner W F, Lubow S H, Parker P D and Ulrich R H 1982 *Rev. Mod. Phys.* **54** 767
Bahcall J N, Dar A and Piran T 1987 *Nature* **326** 135
Bahcall J N, Piran T and Press W H 1987a *Nature* **327** 682
Bahcall J N and Ulrich R K 1988 *Rev. Mod. Phys.* **60** 297
Baker B B and Copson E T 1950 *The Mathematical Theory of Huygens' Principle* 2nd edn (Oxford: Clarendon Press)
Baker M and Glashow S L 1962 *Phys. Rev.* **128** 2462
Baker R G and McNeill K G 1961 *Can. J. Phys.* **39** 1138

Balachandran A P, Nair V P, Rajeev S G and Stern A 1983 *Phys. Rev.* D **27** 1153
Baldo-Ceolin M *et al* 1990 *Phys. Lett.* **236B** 95
Baldwin G C and Klaiber G S 1947 *Phys. Rev.* **71** 3
— 1948 *Phys. Rev.* **73** 1156
Balyaev S T 1959 *Mat. Fys. Medd. Dan Vid. Selsk.* **31** No 1
Balysh A *et al* 1992 *Phys. Lett.* B **283** 32
Banner M *et al* 1983, *Phys. Lett.* **122B** 476
Barani M 1990 *Comments At. Mol. Phys.* **25** 9
Baranov P, Buinov G, Goden V, Kuznetzova V, Petrunken V, Talarenskaya L, Shirthenko S, Shtarkov L, Yurtchenko V and Yanulis Yu 1974 *Phys. Lett.* **52B** 122
Bardeen J, Cooper L N and Schrieffer J R 1957 *Phys. Rev.* **108** 1175
Bargholtz C 1979 *Astrophys. J. Lett.* **233** L161
Bargmann V 1949 *Rev. Mod. Phys.* **21** 488
Barish S J *et al* 1974 *Phys. Rev. Lett.* **33** 448
Barkan S, Bieber E, Burgy M T, Ketudat S, Krohn V F, Rice-Evans P and Ringo G R 1968 *Rev. Sci. Instr.* **39** 101
Barkas W H 1939 *Phys. Rev.* **55** 691
Barker F C, Brown B A, Jaus W and Rasche G 1992 *Nucl. Phys.* A **540** 501
Barkla C G 1911 *Phil. Mag.* **21** 648
Barkla C G and Martyn G H 1912 *Nature* **90** 435
Barnes V E *et al* 1964 *Phys. Rev. Lett.* **12** 204
Barschall H H 1952 *Phys. Rev.* **86** 431
— 1954 *Am. J. Phys.* **22** 517
Bartlett J H 1931 *Nature* **128** 408
— 1932 *Phys. Rev.* **41** 370
— 1936 *Phys. Rev.* **49** 102
Baruchel J and Schlenker M 1980 *Nukleonika* **25** 911
Baruchel J, Palmer S B and Schlenker M 1981 *J. Physique* **42** 1279
Baruchel J, Kuroda K, Liaud P, Michaelowicz A and Sillou D 1988 *J. Appl. Cryst.* **21** 28
Baruchel J, Sandonis T and Pearce A 1989 *Physica* B **156 & 157** 765
Basecq J and Wolfenstein L 1983 *Nucl. Phys.* B **224** 21
Bass L and Schrödinger E 1955 *Proc. R. Soc.* A **232** 1
Bass S D and Thomas A W 1993 *J. Phys. G.: Nucl. Phys.* **19** 925
Basov N G, Kriukov P G, Zakharov S D, Senitskii Yu Y and Tchekalc S V 1968 *IEEE J. Quantum Electron.* **4** 864
Bates J C 1983 *Nucl. Instr. Meth.* **216** 535
Batterman B W and Cole H 1964 *Rev. Mod. Phys.* **36** 681
Batty C J, Bonner B E, Kilvington A I, Tschalär C and Williams L E 1969 *Nucl. Instr. Meth.* **68** 273
Baudin G, Blair C, Hagemann R, Kremer M, Lucas M, Mervilat L, Molina R, Nief G, Prost-Marechal F, Regnaud F and Roth E 1972 *Cont. Rend.* D **275** 2291
Baum G *et al* 1983 *Phys. Rev. Lett.* **51** 1138
Baumann J, Gähler R, Klaus J and Mampe W 1988 *Phys. Rev.* D **37** 3107
Bauspiess W, Bonse U and Graeff W 1976 *J. Appl. Cryst.* **9** 68
Bauspiess W, Bonse U and Rauch H 1978 *Nucl. Instr. Meth.* **157** 495
Baym G and Pethick D 1975 *Ann. Rev. Nucl. Sci.* **25** 27
Bayman B F 1960 *Nucl. Phys.* **15** 33
Beck G 1928 *Z. Phys.* **47** 407

— 1928a *Z. Phys.* **50** 548
— 1930 *Z. Phys.* **61** 615
Becquerel H 1896 *Cont. Rend.* **122** 420, 501, 559, 689, 762, 1086
— 1896a *Cont. Rend.* **123** 855
— 1900 *Cont. Rend.* **130** 206, 372
Bée M 1988 *Quasi-Elastic Neutron Scattering* (Bristol: Adam Hilger)
Beg M A, Lee B W and Pais A 1964 *Phys. Rev. Lett.* **13** 517
Beg M A, Budny R V, Mohapatra R W and Sirlin A 1977 *Phys. Rev. Lett.* **38** 1252
Behrends R E and Sirlin A 1960 *Phys. Rev. Lett.* **4** 186
Bell J S 1964 *Physics* **1** 195
Bell J S and Jackiw R 1969 *Nuovo. Cimento* **60** 47
Bell R E and Elliott L G 1948 *Phys. Rev.* **74** 1552
Bender, Linke V and Rothe J 1968 *Z. Phys.* **212** 190
Bennett W H 1934 *Phys. Rev.* **45** 890
Benventi ACetal 1990 *Phys. Lett.* B **237** 592
Berard R W, Buskirk F R, Dally E B, Dye J N and Mruyama K 1973 *Phys. Lett.* **47B** 355
Berger H 1965 *Neutron Radiography* (Amsterdam: Elsevier)
— 1971 *Ann. Rev. Nucl. Sci.* **21** 335
Bergmann A A, Isakov A E, Popov Yu P and Shapiro F L 1957 *Zh. Exp. Teor. Fiz.* **33** 9
Bernard V, Kaiser N, Kambor J and Meissner U-G 1992 *Phys. Rev.* D **46** R2756
Bernard W and Callen H 1959 *Rev. Mod. Phys.* **31** 1017
Bernatowicz T, Brannon J, Brazzle R, Cowsik R, Hohenberg C and Podosek F 1992 *Phys. Rev. Lett.* **69** 2341
Bernstein H J 1967 *Phys. Rev Lett.* **18** 1102
— 1986 *Physica* B **137** 266
Bernstein J and Goldberger M L 1958 *Rev. Mod. Phys.* **30** 465
Berry M V 1984 *Proc. R. Soc.* A **392** 45
— 1987 *Nature* **326** 277
— 1987a *Proc. R. Soc.* A **414** 31
Berry M V and Mount K E 1972 *Rep. Prog. Phys.* **35** 315
Bessel F W 1832 *Pogg. Ann.* **25** 401
Beth R A 1936 *Phys. Rev.* **50** 115
Bethe H A 1935 *Phys. Rev.* **47** 633
— 1937 *Rev. Mod. Phys.* **9** 69
— 1938 *Phys. Rev.* **54** 436
— 1939 *Phys. Rev.* **55** 434
— 1940 *Phys. Rev.* **57** 1125
— 1949 *Phys. Rev.* **76** 38
— 1986 *Phys. Rev. Lett.* **56** 1305
Bethe H A and Peierls R 1935 *Proc. R. Soc.* A **148** 146
Bethe H A and Bacher R 1936 *Rev. Mod. Phys.* **8** 82
Bethe H A and Placzek G 1937 *Phys. Rev.* **51** 450
Bethe H A and Rose M E 1937 *Phys. Rev.* **51** 283
Bethe H A and Critchfield C L 1938 *Phys. Rev.* **54** 248
Bethe H A and Longmere C L 1950 *Phys. Rev.* **77** 647
Bethe H A and Goldstone J 1957 *Proc. R. Soc.* A **238** 551

Biedenharn L C and Rose M 1953 *Rev. Mod. Phys.* **25** 729
Bilen'kii S M and Hosek J 1982 *Phys. Rep.* **90** 73
Bilen'kii S M and Pontecorvo B M 1982 *Sov. J. Nucl. Phys.* **36** 99
Bilen'kii S M, Rynden R M, Smorodinskii Ya A and Ho-Tso-Hsiu 1960 *Sov. Phys.–JETP* **37** 1241
Bionata R M *et al* 1987 *Phys. Rev. Lett.* **58** 1494
Birgenau R J, Moncton D E and Zeilinger A (eds) *Physica* B&C **137**
Birr M, Heidemann A and Alefield B 1971 *Nucl. Instr. Meth.* **95** 435
Bitter T and Dubbers D 1987 *Phys. Rev. Lett.* **59** 251
Bjerge T and Westcott C H 1935 *Proc. R. Soc.* A **150** 709
Bjørken B J and Glashow S L 1964 *Phys. Lett.* **11** 255
Bjørken J D 1966 *Phys. Rev.* **148** 1467
— 1969 *Phys. Rev.* **179** 1547
Bjørken J D and Paschos E A 1969 *Phys. Rev.* **185** 1975
Blackett P M S 1925 *Proc. R. Soc.* A **107** 349
Blair J S 1954 *Phys. Rev.* **95** 1218
Blatt J M and Weisskopf V H 1952 *Theoretical Nuclear Physics* (New York: Wiley)
Blin-Stoyle R J 1953 *Proc. Phys. Soc.* A **66** 929, 1158
— 1956 *Rev. Mod. Phys.* **28** 75
— 1975 *Nucl. Phys.* 1 **254** 353
Blin-Stoyle R J and Perks M A 1954 *Proc. Phys. Soc.* A **67** 62
Blin-Stoyle R J and Freeman J M 1970 *Nucl. Phys.* A **150** 369
Bloch C 1954 *Phys. Rev.* **93** 1094
— 1930 *Z. Phys.* **61** 206
— 1932 *Z. Phys.* **74** 295
— 1936 *Phys. Rev.* **50** 259
— 1937 *Phys. Rev.* **51** 994
Bloch F and Møller C 1935 *Nature* **136** 912
Bloch F and Siegert A 1940 *Phys. Rev.* **57** 522
Bloch F, Hameresh M and Staub H 1943 *Phys. Rev.* **64** 47
Bloch F, Nicodemus D and Staub H 1948 *Phys. Rev.* **74** 1025
Block B and Feshbach H 1963 *Ann. Phys., NY* **23** 47
Blumberg S and Porter C E 1958 *Phys. Rev.* **110** 786
Bodek A 1979 *Phys. Rev.* D **20** 1471
Bodek A *et al* 1983 *Phys. Rev. Lett.* **50** 1431
— 1983a *Phys. Rev. Lett.* **51** 534
Bodu R, Bouzigues H, Morin N and Pfiffelmann J P 1972 *Cont. Rend* D **275** 1731
Bogoliubov N N 1958 *Nuovo Cimento* **7** 794
— 1958a *Sov. Phys.–JETP* **7** 41, 51
Bohigas O and Weidenmüller H A 1988 *Ann. Rev. Nucl. Part. Sci.* **38** 421
Bohm D and Bub J 1966 *Rev. Mod. Phys.* **38** 453
Bohr A 1952 *K. Danske Vidensk. Selsk., Mat.-fys. Medd.* **26** No 14
Bohr A 1956 *Proc. Int. Conf. on the Peaceful Uses of Atomic Energy (1955)* vol 2 (New York: United Nations) p 151
Bohr A and Mottelson B R 1952 *Physica* **18** 1066
— 1953 *Dan. Mat. Fys. Medd.* **27** No 16
— 1953a *Phys. Rev.* **90** 717
— 1955 *Dan. Mat. Fys. Medd.* **30** No 7

Bohr A, Mottelson B R and Pines D 1958 *Phys. Rev.* **110** 936
Bohr N 1913 *Phil. Mag.* **26** 1
— 1935 *Phys. Rev.* **48** 696
— 1936 *Nature* **137** 344
— 1939 *Nature* **143** 330
— 1939a *Phys. Rev.* **55** 418
Bohr N and Rosenfeld L 1933 *Det. Kgl. Dansk Vid. Selskab* **12** 8
Bohr N and Kalckar F 1937 *Kgl. Dansk. Vid. Selsk., Math. Fys. Med.* **14** 10
Bohr N and Wheeler J A 1939 *Phys. Rev.* **56** 426
Bohr N, Peierls R and Placzek G 1939 *Nature* **144** 200
Bollinger L M, Cote R E, Carpenter T R and Marion J P 1963 *Phys. Rev.* **132** 1640
Bondarenko L N, Kurguzov V V, Prokofiev Yu A, Rogov E Y and Spivak P E 1978 *Sov. Phys.–JETP Lett.* **28** 303
Bonse U 1979 *Neutron Interferometry* ed U Bonse and H Rauch (Oxford: Clarendon) p 3
Bonse U and Hart M 1965 *Z. Phys.* **188** 154
— 1965a *Appl. Phys. Lett.* **6** 155
— 1965b *Appl. Phys. Lett.* **7** 99
— 1966 *Z. Phys.* **190** 455
— 1966a *Z. Phys.* **194** 1
Bonse U and Wroblewski T 1983 *Phys. Rev. Lett.* **16** 1401
Bonse U and Rumpf A 1986 *Phys. Rev. Lett.* **56** 2441
Bopp P, Dubbers D, Hornig L, Klemt E, Last J, Schütze H, Freedman S J and Schärpf O 1986 *Phys. Rev. Lett.* **56** 919
Bopp P, Klemt E, Last J, Schütze H, Dubbers D and Freedman S J 1988 *Nucl. Instr. Meth.* A **267** 436
Born M 1926 *Z. Phys.* **37** 863
— 1926a *Z. Phys.* **38** 167, 803
Borrmann L 1941 *Phys. Z.* **42** 157
— 1950 *Z. Phys.* **127** 297
Bothe W 1930 *Z. Phys.* **63** 381
Bothe W and Becker H 1930 *Z. Phys.* **66** 289
Bothe W and Fränz H 1928 *Z. Phys.* **49** 1
Bouchiat M A and Bouchiat C C 1974 *Phys. Lett.* **48B** 111
Bourquin M *et al* 1983 *Z. Phys.* C **21** 27
Bourrely C, Leader E and Soffer J 1980 *Phys. Rep.* **59** 95
Bowick M J, Chase M K and Ramond P 1983 *Phys. Lett.* **128B** 185
Bowman C D, Auchampaugh G H and Fultz S L 1964 *Phys. Rev.* **13**, B676
Bowman J D, Garvey G T, Gould C R, Hayes A C and Johnston M B 1992 *Phys. Rev. Lett.* **68** 780
Bowman J D *et al* 1990 *Phys. Rev. Lett.* **65** 1192
Boyd G E and Larson Q V 1956 *J. Phys. Chem.* **60** 707
Bozoian M and Weber H J 1983 *Phys. Rev.* C **28** 811
Bracci L and Florentini G 1981 *Nucl. Phys.* A **364** 383
Bradner H and Isbell W M 1959 *Phys. Rev.* **114** 603
Bragg W H and Bragg W L 1913 *Proc. R. Soc.* A **88** 428
Bragg W L 1912 *Proc. Camb. Phil. Soc.* **17** 43
Brandenburg R A, Coon S A and Sauer P U 1978 *Nucl. Phys.* A **294** 305

Brandenburg R A, Chulik G S, Kim Y E, Klepacki D J, Machleidt R, Picklesheimer A and Thaler R M 1988 *Phys. Rev.* C **37** 781
Breidenbach M, Friedman J T, Kendall H W, Bloom E D, Coward D H, DeStaebler H, Drees J, Mo L W and Taylor R E 1969 *Phys. Rev. Lett.* **23** 935
Breit G 1937 *Phys. Rev.* **51** 248
— 1938 *Phys. Rev.* **53** 153
— 1947 *Phys. Rev.* **71** 215
Breit G and Condon E U 1936 *Phys. Rev.* **49** 904
Breit G and Feenberg E 1936 *Phys. Rev.* **50** 850
Breit G and Wigner E 1936 *Phys. Rev.* **49** 519, 642
Breit G, Condon E U and Present R D 1936 *Phys. Rev.* **50** 825
Breit G and Rustgi M L 1971 *Nucl. Phys.* A **161** 337
Bremmer H 1949 *Physica* **15** 593
— 1951 *Comm. Pure Appl. Math.* **4** 105
Brennan M H and Bernstein A M 1960 *Phys. Rev.* **120** 927
Bressi G *et al* 1989 *Z. Phys.* C **43** 175
Breunlich W H *et al* 1987 *Muon Catal. Fusion* **1** 67
Breunlich W H, Kammel P, Cohen J S and Leon M 1989 *Ann. Rev. Nucl. Part. Sci.* **39** 311
Brillouin L 1926 *Cont. Rend.* **183** 24
Brink D M and Satchler G R 1968 *Angular Momentum* (Oxford: Clarendon) p 59
Brix P and Kopfermann H 1949 *Z. Phys.* **126** 344
— 1958 *Rev. Mod. Phys.* **30** 567
Brockhouse B N and Hurst D J 1952 *Phys. Rev.* **88** 542
Brockhouse B N and Iyengar P K 1958 *Phys. Rev.* **111** 747
Brockhouse B N and Stewart A T 1955 *Phys. Rev.* **100** 756
— 1958 *Rev. Mod. Phys.* **30** 236
Brodsky S J, Ellis J and Karliner M 1988 *Phys. Lett.* **206B** 309
Brown G E and Myrer F 1983 *Phys. Lett.* **128B** 229
Brown G E and Rho M 1979 *Phys. Lett.* **82B** 177
Brown G E, De Domincis C T and Langer J S 1959 *Ann. Phys., NY* **6** 209
Brown G E, Rho M and Vento V 1980 *Phys. Lett.* **97B** 423
Brueckner K A 1952 *Phys. Rev.* **86** 106
— 1955 *Phys. Rev.* **97** 1353
— 1955a *Phys. Rev.* **100** 36
— 1958 *Phys. Rev.* **110** 597
— 1958a *Rev. Mod. Phys.* **30** 561
Brueckner K A and Goldman D T 1959 *Phys. Rev.* **116** 424
Brueckner K A and Jorna S 1974 *Rev. Mod. Phys.* **46** 325
Brueckner K A and Levinson C A 1955 *Phys. Rev.* **97** 1344
Bruenn S W, Arnett W D and Schramm D W 1977 *Astrophys. J.* **213** 213
Brugger R M 1965 *Thermal Neutron Scattering* (New York: Academic Press) p 53
Bruns H 1895 *Abh. Kgl. Sächs. Ges. Wiss. Math-Phys. Kl.* **21** 323
Budick B, Chen J and Lin H 1991 *Phys. Rev. Lett.* **67** 2630
Budnik A P and Rabatov N S 1973 *Phys. Lett.* **46B** 155
Bumiller F A, Buskirk F R and Stewart J W 1970 *Phys. Rev. Lett.* **25** 1774
Buras B 1963 *Nukleonika* **8** 259
Buras B and Leciejewicz J 1963 *Nukleonika* **8** 75

Burbidge E M, Burbidge G R, Fowler W A and Hoyle F 1957 *Rev. Mod. Phys.* **29** 547
Burbidge G R, Hoyle F, Burbidge E M, Christy R F and Fowler W A 1956 *Phys. Rev.* **103** 1145
Burgy M T, Krohn V E, Novey T B and Ringo G R 1958 *Phys. Rev.* **110** 1214
— 1960 *Phys. Rev.* **120** 1829
Busing W R and Levy H A 1957 *J. Chem. Phys.* **26** 563
Butler S T 1951 *Proc. R. Soc.* A **208** 559
Butterworth I, Egelstaff P A, London H and Webb F J 1957 *Phil. Mag.* **2** 915
Byrne J 1971 *J. Phys. B: At. Mol. Phys.* **4** 940
— 1978 *Fundamental Physics with Reactor Neutrons and Neutrinos* ed T von Egidy (Bristol: Institute of Physics) pp 28, 38
— 1978a *Nature* **275** 188
— 1979 *Neutron Interferometry* ed U Bonse and H Rauch (Oxford: Clarendon) p 273
Byrne J, Morse J, Smith K F, Shaikh F, Green K and Greene G L 1980 *Phys. Lett.* **92B** 274
Byrne J 1982 *Rep. Prog. Phys.* **45** 115
— 1984 *Nature* **310** 212
Byrne J and Farago P S 1965 *Proc. Phys. Soc.* **86** 801
Byrne J, Dawber P G, Spain J A, Dewey M S, Gilliam D M, Greene G L, Lamaze G P, Williams A P, Pauwels J, Eykens R, Van Gestel J, Lamberty A and Scott R D 1989 *Nucl. Instr. Meth.* A **284** 116
Byrne J, Dawber P G, Spain J A, Williams A P, Dewey M S, Gilliam D M, Greene G L, Lamaze G P, Scott R D, Pauwels J, Eykens R and Lamberty A 1990 *Phys. Rev. Lett.* **65** 289
Cabibbo N 1963 *Phys. Rev. Lett.* **10** 531
— 1964 *Phys. Lett.* **12** 137
Cabrera B 1982 *Phys. Rev. Lett.* **47** 1738
Cabrera B, Taber M, Gardner R, Bourg J 1983 *Phys. Rev. Lett.* **51** 1933
Calaprice F P and Holstein B R 1976 *Nucl. Phys.* A **273** 301
Callan C G and Gross D 1969 *Phys. Rev. Lett.* **22** 156
Callan C G and Treiman S B 1967 *Phys. Rev.* **162** 1494
Callan C G, Dashen R F and Gross D J 1976 *Phys. Lett.* **63B** 334
— 1977 *Phys. Lett.* **66B** 375
— 1978 *Phys. Lett.* **78B** 307
Callen H and Welton T R 1951 *Phys. Rev.* **83** 24
Cameron A G W 1957 *Publ. Astron. Soc. Pac.* **69** 201
— 1959 *Astrophys. J.* **130** 884
Carlitz R D, Collins J C and Mueller A H 1988 *Phys. Lett.* **214B** 229
Carlitz R D, Ellis S D and Savit R 1977 *Phys. Lett.* **68B** 443
Carnoy A S, Deutsch J, Gerard T A and Prieels R 1991 *Phys. Rev. Lett.* **65** 3249
Carnoy A S, Deutsch J and Holstein B R 1988 *Phys. Rev.* D **38** 1636
Carnoy A S, Deutsch J, Prieels R, Severijns N and Quin P A 1992 *J. Phys. G: Nucl. Phys.* **18** 823
Carruthers P and Nieto M 1968 *Rev. Mod. Phys.* **40** 411
Case K M 1960 *Ann. Phys., NY* **9** 1
Case K M and Zweifel P 1967 *Linear Transport Theory* (Reading, MA: Addison Wesley)

Case K M, de Hoffmann F and Placzek G D 1953 *Introduction to the Theory of Neutron Diffusion* (Washington, DC: US Government Printing Office)

Casella R C 1984 *Phys. Rev. Lett.* **53** 1023

Casimir H B G 1936 *Physica* **3** 936

Cassels J M 1951 *Proc. R. Soc.* A **208** 527

Cassen B and Condon E U 1936 *Phys. Rev.* **50** 846

Cavaignac J F, Vignon B and Wilson R 1977 *Phys. Lett.* **67B** 148

Chadwick J 1920 *Phil. Mag.* **40** 734

— 1932 *Nature* **129** 312

— 1932a *Proc. R. Soc.* **136** 692

— 1933 Bakerian Lecture *Proc. R. Soc.* A **142** 1

— 1933a *Br. J. Radiology* **6** 24

Chadwick J and Bieler E S 1927 *Phil. Mag.* **42** 923

Chadwick J and Goldhaber M 1934 *Nature* **134** 237

— 1935 *Proc. R. Soc.* A **151** 479

Chandrasekhar S 1935 *Mon. Not. R. Astron. Soc.* **95** 207

— 1931 *Astrophys. J.* **24** 81

— 1960 *Radiative Transfer* (New York: Dover) p 9

Chang L N and Chang N P 1980 *Phys. Lett.* **92B** 103

— 1980a *Phys. Rev. Lett.* **45** 1540

Cherry M L, Lande K, Lee C K, Steinberg R I and Cleveland B 1983 *Phys. Rev. Lett.* **50** 1354

Chertok B, Sheffield C, Lightbody J W, Penner S and Blum D 1973 *Phys. Rev.* C **8** 23

Chetyrkin K G, Zavarnosky M V, Kuz'min V A and Shaposhnikov M E 1980 *Phys. Lett.* **99B** 358

Chiao R Y and Wu Y-S 1986 *Phys. Rev. Lett.* **57** 933

Chinowski W 1977 *Ann. Rev. Nucl. Sci.* **27** 393

Chiu H-Y and Salpeter E E 1964 *Phys. Rev. Lett.* **12** 413

Chodos A and Thorn C B 1975 *Phys. Rev.* D **12** 2733

Chodos A, Jaffe R L, Johnson K and Thorn C B 1974 *Phys. Rev.* D **10** 2599

Chrien R E and Reich M 1967 *Nucl. Instr. Meth.* **53** 93

Christ J and Springer T 1962 *Nukleonik* **4** 23

Christen D K, Tasset F, Spooner S and Mook H A 1977 *Phys. Rev.* B **15** 4506

Christensen C J, Krohn V E and Ringo G R 1969 *Phys. Lett.* **28B** 411

— 1970 *Phys. Rev.* C **1** 1693

Christensen C J, Nielsen A, Bahnsen A, Brown W K and Rustad B M 1967 *Phys. Lett.* **26B** 11

— 1972 *Phys. Rev.* D **5** 1628

Christenson J H, Cronin J W, Fitch V L and Gurlay R 1964 *Phys. Rev. Lett.* **13** 138

Cimmino A., Opat G I, Klein A G, Kaiser H, Werner S A, Arif M and Clothier R 1989 *Phys. Rev Lett.* **63** 380

Clark M A and Robson J M 1960 *Can. J. Phys.* **38** 693

— 1961 *Can. J. Phys.* **39** 13

Close F E 1974 *Nucl. Phys.* B **80** 269

— 1979 *Rep. Prog. Phys.* **42** 1285

— 1990 *Nucl. Phys.* A **508** 413

Close F E, Robert R G and Ross G G 1983 *Phys. Lett.* **129B** 346

Code R F and Ramsey N F 1971 *Phys. Rev.* A **4** 1945

Cohen B L 1970 *Am. J. Phys.* **38** 766
Cohen B L and Handley T H 1953 *Phys. Rev.* **92** 101
Cohen E R 1962 *Nucl. Sci. Eng.* **2** 309
Cohen E R and Taylor B N 1987 *Rev. Mod. Phys.* **59** 112
Cohen J and Weber H J 1985 *Phys. Lett.* **165B** 229
Cohen M L 1984 *Am. J. Phys.* **52** 695
Cohen S and Swiatecki W J 1962 *Ann. Phys., NY* **19** 67
— 1963 *Ann. Phys., NY* **22** 406
Cohen-Tannoudji C and Haroche S 1969 *J. Physique* **30** 125, 153
Colella R, Overhauser A W and Werner S A 1975 *Phys. Rev. Lett.* **34** 1472
Condon E U 1936 *Phys. Rev.* **39** 459
Condon E U and Gurney R W 1929 *Phys. Rev.* **33** 127
Condon E U and Shortley G H 1935 *The Theory of Atomic Spectra* (Cambridge: Cambridge University Press) p 61
Cook C W, Fowler W A, Lauritsen C C and Lauritsen T 1957 *Phys. Rev.* **107** 508
Cooper L N 1956 *Phys. Rev.* **104** 1189
Copson E T 1935 *An Introduction to the Theory of Functions of a Complex Variable* (Oxford: Clarendon)
Corinaldesi E 1956 *Nucl. Phys.* **2** 420
Corngold M, Cohen V W and Ramsey N F 1956 *Phys. Rev.* **104** 283
Corngold N, Michael P and Woolman W 1963 *Nucl. Sci. Eng.* **15** 13
Coryell C D 1953 *Ann. Rev. Nucl. Sci.* **2** 305
Cottingham W N, Lacombe M, Loiseau B, Richard J M and Vinh-Mau R 1973 *Phys. Rev.* D **8** 800
Cowan C L, Reines F, Harrison F B, Kruse H W and McGuire A D 1956 *Science* **124** 103
Cox A E, Wynchank S A R and Collie C H 1965 *Nucl. Phys.* **74** 497
Crane H R, Lauritsen C C and Soltan A 1933 *Phys. Rev.* **44** 514 692
Creutz M 1980 *Phys. Rev. Lett.* **45** 313
Curie I 1931 *Cont. Rend.* **193** 1412
Curie I and Joliot F 1932 *Cont. Rend* **194** 273
— 1934 *Nature* **133** 201
— 1934a *Cont. Rend.* **198** 254, 408
Curie M S 1898 *Cont. Rend.* **126** 1101
D'Angelo N 1959 *Phys. Rev.* **114** 285
D'Auria J M, Carraz L C, Hansen P G, Jonson B, Mattson S, Ravn H L, Skarestad M and Westgaard L 1977 *Phys. Lett.* **66B** 233
Dabbs J W T, Harvey J A, Paya D and Horstmann H 1965 *Phys. Rev.* B **139** 756
Dabbs J W T, Roberts L D and Bernstein S 1955 *Report ORNL-CF-55-5-126* p 73
Daly R T and Holloway J H 1954 *Phys. Rev.* **96** 539
Dancoff S and Inglis D 1936 *Phys. Rev.* **50** 784
Danilov G S 1965 *Phys. Lett.* **18** 40
— 1971 *Phys. Lett.* **35B** 579
— 1972 *Sov. J. Nucl. Phys.* **14** 443
Danilyan G V 1980 *Sov. Phys. Usp.* **23** 323
Danner H R and Stiller H H 1961 *Physica* **27** 373
Danos M 1958 *Nucl. Phys.* **5** 3
Danos M and Fuller E G 1965 *Ann. Rev. Nucl. Sci.* **15** 29

Danos M and Greiner W 1964 *Phys. Lett.* **8** 113
— 1964a *Phys. Rev.* B **134** 283
— 1965 *Phys. Rev.* B **138** 876
Darwin C G 1914 *Phil. Mag.* **27** 315, 675
— 1922 *Phil. Mag.* **43** 800
— 1928 *Proc. R. Soc.* A **118** 654
Davis L A, Goldhaber A S and Nieto M M 1975 *Phys. Rev. Lett.* **35** 1402
Davisson C and Germer L H 1927 *Phys. Rev.* **30** 707
de Broglie L 1924 *Phil. Mag.* **47** 466
— 1925 *Ann. Phys., Paris* **3** 22
De Swart J and Marshak R E 1959 *Physica* **25** 1001
De Tar C E and Donoghue J F 1983 *Ann. Rev. Nucl. Part. Sci.* **33** 235
de Téramond G F and Gabioud B 1987 *Phys. Rev.* C **36** 691
Debye P 1912 *Ann. Phys., Lpz* **39** 789
Decamp D *et al* 1990 *Z. Phys.* C **48** 365
— 1992 *Z. Phys.* C **53** 1
Dee P I 1932 *Proc. R. Soc.* A **136** 727
DeGrand T, Jaffe R L, Johnson K and Kiskes J 1975 *Phys. Rev.* D **12** 2060
Del Aguila F and Ibanez L E 1981 *Nucl. Phys.* B **177** 60
Dennison D M 1927 *Proc. R. Soc.* A **115** 483
— 1940 *Phys. Rev.* **57** 457
Deo B B and Maharana K 1984 *Phys. Rev.* D **29** 1020
Deslattes R 1986 in NBS Special Publication 711 ed G L Greene (Washington: US Govt Printing Office) p 118
Diamond R M, Stephens F S and Swiatecki W J 1964 *Phys. Lett.* **11** 315
Dicke R H 1959 *Science* **129** 621
Dickinson W C, Lent E M and Bowman C D 1970 *Report UCRL 50848* p 1
Dirac P A M 1947 *Quantum Mechanics* 3rd edn (Oxford: Clarendon)
— 1928 *Proc. R. Soc.* A **117** 610
— 1928*a* *Proc. R. Soc.* A **118** 351
— 1930 *Proc. R. Soc.* A **126** 360
— 1931 *Proc. R. Soc.* A **133** 60
— 1948 *Phys. Rev.* **74** 817
Dobrozemski R, Kerschbaum E, Moraw G, Paul H, Stratowa C and Weinzierl P 1975 *Phys. Rev.* D **11** 510
Doi K, Minakawa N, Motohashi H and Masaki N 1971 *J. Appl. Cryst.* **4** 528
Dominguez C A 1978 *Phys. Rev. Lett.* **41** 605
— 1979 *Phys. Lett.* **86B** 171
— 1982 *Phys. Rev.* D **25** 1937
Donoghue J F and Johnson K 1980 *Phys. Rev.* D **21** 1975
Donoghue J F, Golowich E and Holstein B R 1975 *Phys. Rev* D **12** 2875
— 1986 *Phys. Rep.* **131** 319
Donoghue J F, Holstein B R and Klemt S W 1987 *Phys. Rev.* D **35** 934
Dorner B 1972 *Acta Cryst.* A **28** 319
Dover C B, Gal A and Richard M 1983 *Phys. Rev.* D **27** 1090
— 1989 *Nucl. Instr. Meth.* A **284** 13
Drabkin G M 1963 *Sov. Phys.–JETP* **16**, 781
— 1976 *Nucl. Instr. Meth.* **133** 453

Drabkin G M, Zabidarov E I, Kashman Ya A, Okorokov A I and Trunov V A 1965 *Sov. Phys.–JETP* **20** 1548

Dresden M 1987 *H A Kramers: Between Tradition and Revolution* (Heidelberg: Springer) p 454

Dress W B, Baird J K, Miller P D and Ramsey N F 1968 *Phys. Rev.* **170** 1200

Dress W B, Miller P D, Pendlebury J M, Perrin P and Ramsey N F 1977 *J. Phys.* D **15** 9

Dubbers D 1988 *Physica* B&C **151** 93

Dubbers D, El-Muzeini P, Kessler M and Last J 1989 *Nucl. Instr. Meth.* A **275** 294

Dubbers D, Mampe W and Döhner J 1990 *Europhys. Lett.* **11** 195

Duggal V P and Thaper C L 1962 *Rev. Sci. Instr.* **33** 49

Dumbrajs O, Koch R, Pelkuha H, Oades G C, Behrens H, De Swart J J and Kroll P 1983 *Nucl. Phys.* B **216** 277

Dunning J R, Pegram G B, Fink G A, Mitchell D P and Segre E 1935 *Phys. Rev.* **48** 704

Durbin R, Loar H and Steinburger J 1951 *Phys. Rev.* **83** 646

Eddington A S 1920 *Report of the 88th Meeting of the British Association for the Advancement of Science* (London: John Murray) p 34

Egelstaff P A 1951 *Nature* **168** 290

— 1954 *J. Nucl. Energy* **1** 57

— 1961 *Proc. Vienna Symposium on Inelastic Scattering of Neutrons in Solids and Liquids* (Vienna: IAEA) p 25

Egelstaff P A and Schofield P 1962 *Nucl. Sci. Eng.* **12** 260

Ehrenfest P 1992 *Nature* **109** 745

— 1927 *Z. Phys.* **45** 455

Ehrenfest P and Oppenheimer J R 1931 *Phys. Rev.* **37** 333

Eichten E, Lane K and Preskill J 1980 *Phys. Rev. Lett.* **45** 225

Einstein A 1905 *Ann. Phys., Lpz* **17** 132

— 1906 *Ann. Phys., Lpz* **19** 289, 371

— 1916 *Ann. Phys., Lpz* **49** 769

Einstein A, Podolsky B and Rosen N 1935 *Phys. Rev.* **47** 777

Ekstein H 1949 *Phys. Rev.* **76** 1328

— 1950 *Phys. Rev.* **78** 731

— 1951 *Phys. Rev.* **83** 721

— 1953 *Phys. Rev.* **89** 490

Elliott J P 1955 *Proc. R. Soc.* A **228** 424

— 1958 *Proc. R. Soc.* A **245** 128, 562

— 1985 *Rep. Prog. Phys.* **48** 171

Elliott J P and Dawber P G 1979 *Symmetry in Physics* (London: Macmillan)

Elliott J P and Skyrme T H R 1955 *Proc. R. Soc.* A **232** 561

Elliott R J and Lowde R D 1955 *Proc. R. Soc.* A **230** 46

Ellis C D and Wooster W A 1927 *Proc. R. Soc.* A **117** 109

Ellis J, Gaillard M K, Nanopoulos D W and Rudaz S 1981 *Phys. Lett.* **99B** 101

Elmroth T, Hagbert E, Hansen P G, Hardy J C, Jonson B, Ravn H C and Tidemand-Petersson P 1978 *Nucl. Phys.* A **304** 493

Elsasser W M 1925 *Naturwiss.* **13** 71

— 1934 *J. Phys. Rad.* **5** 389, 634

— 1936 *Cont. Rend.* **202** 1029

von Eötvös R, Pekar D and Fekete E 1922 *Ann. Phys., Lpz.* **68** 11

Ericson T 1960 *Nucl. Phys.* **17** 250

— 1960 *Phys. Rev. Lett.* **5** 430
— 1960a *Adv. Phys.* **9** 425
— 1963 *Ann. Phys., NY* **23** 390
Ericson T E O and Millar G A 1983 *Phys. Lett.* **132B** 32
Erikson R A 1953 *Phys. Rev.* **90** 779
Erozolimskii B G 1975 *Sov. Phys. Usp.* **18** 377
Erozolimskii B G and Mostovoi Yu A 1991 *Sov. J. Nucl. Phys.* **53** 260
Erozolimskii B G, Bondarenko L N, Mostovoi Yu A, Obinyakov B A, Fedunin V P and Frank A I 1971a *JETP Lett.* **13** 252
Erozolimskii B G, Bondarenko L N, Mostovoi Yu A, Obinyakov B A, Titov V A, Zakharova V P and Frank A I 1970a *Phys. Lett.* **33B** 351
— 1971 *Sov. J. Nucl. Phys.* **12** 178
Erozolimskii B G, Bondarenko L N, Mostovoi Yu A, Obinyakov B A, Zakharova V P and Titov V A 1969 *Sov. J. Nucl. Phys.* **8** 98
— 1968 *Phys. Lett.* **27B** 557
— 1970 *Sov. J. Nucl. Phys.* **11** 583
Erozolimskii B G, Frank A I, Mostovoi Yu A and Arzumanov S S 1976 *Sov. Phys.–JETP Lett.* **23** 663
Erozolimskii B G, Frank A I, Mostovoi Yu A, Arzumanov S S and Voitzik L R 1979 *Sov. J. Nucl. Phys.* **30** 356
Erozolimskii B G, Kuznetzov I A, Kuida I A, Mostovoi Yu A and Stepanenko I V 1990 *Sov. J. Nucl. Phys.* **52** 999
— 1991 *Phys. Lett.* **263B** 33
Erozolimskii B G, Mostovoi Yu A, Fedunin V P and Frank A I 1978 *Sov. J. Nucl. Phys.* **28** 48
Erozolimskii B G, Mostovoi Yu A, Fedunin V P, Frank A I and Khakhan O V 1974 *Sov. Phys.–JETP Lett.* **20** 345
Everett H 1957 *Rev. Mod. Phys.* **29** 454
Ewald P P 1916 *Ann. Phys., Lpz* **49** 1, 117
— 1917 *Ann. Phys., Lpz* **54** 519
Faessler A, Greiner W and Sheline R K 1965 *Nucl. Phys.* **62** 241
Faessler A and Plozajcjak M 1978 *Phys. Lett.* **76B** 1978
Fajfer F and Oakes R J 1983 *Phys. Lett.* **132B** 433
Falkoff D C and Uhlenbeck G E 1950 *Phys. Rev.* **79** 323
Fano U 1951 NBS Report No 1214 (Washington, DC: US Government Printing Office)
Feather N 1932 *Proc. R. Soc.* A **136** 709
— 1960 *Contemp. Phys.* **191** 257
Feenberg E 1932 *Phys. Rev.* **40** 40
— 1936 *Phys. Rev.* **49** 328
— 1947 *Rev. Mod. Phys.* **19** 239
— 1955 *Shell Theory of the Nucleus* (Princeton: Princeton University Press)
Feenberg E and Hammack K C 1949 *Phys. Rev.* **75** 1877
— 1951 *Phys. Rev.* **81** 285
Feinberg G 1957 *Phys. Rev.* **108** 878
Feinberg G, Goldhaber M and Steigman G 1978 *Phys. Rev.* D **18** 1301
Felcher G P, Lander G H and Brun T O 1971 *J. Physique Coll.* C **1** 575
Felcher G P 1981 *Phys. Rev.* B **24** 1595

Felcher G P, Kampwirth R D, Gray K E and Felici R 1984 *Phys. Rev. Lett.* **52** 1538
Felcher G P and Russell T P 1991 Methods of analysis and interpretation of neutron reflectivity data *Physica* B **173** 1
Feldman W C, Auchampaugh G F and Drake D M 1990 *Nucl. Instr. Meth.* A **287** 595
Fender B E, Hobbis L C W and Manning G 1980 *Phil. Trans. R. Soc.* B **290** 657
Fermi E 1930 *Z. Phys.* **60** 320
— 1933 *Ric. Sci.* **4** 491
— 1934 *Nuovo Cimento* **11** 1
— 1934a *Z. Phys.* **88** 161
— 1936 *Ric. Sci.* **7** 13
— 1947 *Science* **105** 27
— 1952 *Am. J. Phys.* **20** 536
Fermi E and Marshall L 1947 *Phys. Rev.* **71** 666
— 1947a *Phys. Rev.* **72** 1139
Fermi E and Zinn W H 1946 *Phys. Rev.* **70** 103
Fermi E, Amaldi E, D'Agostino O, Rasetti F and Segre F 1934a *Proc. R. Soc.* A **146** 483
Fermi E, Marshall J and Marshall L 1947 *Phys. Rev.* **72** 193
Fermi E, Rasetti F and D'Agostino O 1934 *Ric. Sci.* **5** 536
Fernbach S, Serber R and Taylor T B 1949 *Phys. Rev.* **75** 1352
Feshbach H 1974 *Rev. Mod. Phys.* **46** 1
Feshbach H and Schwinger J 1951 *Phys. Rev.* **84** 194
Feshbach H and Weisskopf V F 1949 *Phys. Rev.* **76** 1550
Feshbach H, Kerman A K and Koonin S E 1980 *Ann. Phys., NY* **125** 429
Feshbach H, Porter C E and Weisskopf V F 1954 *Phys. Rev.* **96** 448
Feynman R P 1948 *Rev. Mod. Phys.* **20** 367
Feynman R P, Vernon F L and Hellworth R W 1957 *J. Appl. Phys.* **28** 49
Feynman R P 1969 *Phys. Rev. Lett.* **23** 1415
Feynman R P and Gell-Mann M 1958 *Phys. Rev.* **109** 193
Fidecaro G, Fidecaro M, Lanceri L, Marchioro A, Mampe W, Baldo-Ceolin M, Mattioli F, Puglierin G, Batty C J, Green K, Prosper H B, Sharman P, Pendlebury J M and Smith K F 1985 *Phys. Lett.* **156B** 122
Fidecaro M, Finocchiaro G and Giancomelli G 1961 *Nuovo Cimento* **22** 657
Fierz M 1937 *Z. Phys.* **104** 553
Finkelstein K D, Shull C G and Zeilinger A 1986 *Physica* B **136** 131
Finzi A 1964 *Astrophys. J.* **139** 1398
Fiorini E 1982 *Phil. Trans. R. Soc.* A **304** 105
— 1984 *J. Physique Coll.* **C3** 151
Firk F W K 1964 *Nucl. Phys.* **52** 437
Firk F W K, Lynn J E and Moxon M C 1963 *Proc. Phys. Soc.* **82** 477
— 1963a *Nucl. Phys.* **41** 614
— 1963b *Nucl. Phys.* **44** 431
Fischbach E, Sudarsky D, Szafer A, Talmadge C and Aronson S H 1986 *Phys. Rev. Lett.* **56** 3
Fitch V L and Rainwater J 1953 *Phys. Rev.* **92** 789
Fizeau H 1862 *Ann. Chem. Phys.* **66** 429
Fleischer R L and Naeser C W 1972 *Nature* **240** 465
Fleischmann M and Pons S 1989 *J. Electroanal. Chem. Interfacial Electrochem.* **261** 301
Flerov G N, Oganessian Yu T S, Lobanov Yu V, Kuznetzov V I, Druin V A, Perelgen

V P, Gavrilov K A, Tretyakova S P and Plotko V M 1964 *Sov. J. At. Energy* **17** 1046
— 1964a *Phys. Lett.* **13** 73
Flerov G N, Pleve A A, Polikanov S M, Tretyakova S P, Martalogu N, Poenaru D, Sezin I, Vilcov I and Vilcov N 1967 *Nucl. Phys.* A **97** 444
Flerov G N, Polikanov S M, Karamyan A S, Pasyuk A S, Parfanovich D M, Tarantin N, Karnaukhov V A, Drun V A, Volkov V V, Semchinova A M, Oganessian Yu Ts, Khalisev V I, Khlebnikov G I, Nyasoedev B F and Gavrilov K A 1958 *Dokl. Akad. Nauk. SSR* **3** 546
— 1960 *JETP* **11** 61
Flerov G N, Korotkin Yu S, Mikheev V L, Miller M B, Polikanov S M and Stchegolev V A 1968 *Nucl. Phys.* A **106** 476
Foldy L L 1945 *Phys. Rev.* **67** 107
— 1951 *Phys. Rev.* **83** 688L
— 1952 *Phys. Rev.* **87** 688, 693
— 1953 *Phys. Rev.* **92** 178
— 1958 *Rev. Mod. Phys.* **30** 471
— 1966 *Preludes in Theoretical Physics* ed A de Shalit, H Feshbach and L Van Hove (Amsterdam: North-Holland) p 205
Ford K W and Bohm D 1950 *Phys. Rev.* **79** 745
Forte M, Heckel B R, Ramsey N F, Green K, Greene G L, Byrne J and Pendlebury J M 1980 *Phys. Rev. Lett.* **45** 2088
Forward R L 1963 Hughes Research Laboratories *Tech. Rep. No. 267* and *Proposal No 63M-3060/A2068* Part 2
Foudas C *et al* 1990 *Phys. Rev. Lett.* **64** 1207
Fowler W A, Delsasso L A and Lauritsen C C 1936 *Phys. Rev.* **49** 561
Francis N C and Watson K M 1953 *Phys. Rev.* **89** 328
— 1953a *Am. J. Phys.* **21** 659
Frank F C 1947 *Nature* **160** 525
Frankle C M *et al* 1991 *Phys. Rev. Lett.* **67** 554
— 1992 *Phys. Rev.* C **46** 778
Fränz H 1930 *Z. Phys.* **63** 370
Franzini P and Lee-Franzini J 1983 *Ann. Rev. Nucl. Sci.* **33** 1
Frauenfelder H, Babone R, von Goeler E, Levine N, Lewes H R, Peacock R N, Rossi A and de Pasquali G 1957 *Phys. Rev.* **106** 386
Fresnel A 1826 *Mémoires de l'Acad. Roy. des Sci.* **5** 339
Friedman J I and Kendall H W 1972 *Ann. Rev. Nucl. Sci.* **22** 203
Friedrich H and Heintz W 1978 *Z. Phys.* B **31** 423
Friedrich W, Knipping P and Laue M v 1913 *Ann. Phys., Lpz* **41** 971
Frisch O R 1939 *Nature* **143** 276
Fritzsch H and Minkowski P 1981 *Phys. Rep.* **73** 67
Fukugita M, Yanagida T and Yoshimura M 1982 *Phys. Lett.* **109B** 369
Funfer E 1937 *Naturwiss.* **25** 235
— 1938 *Z. Phys.* **111** 351
Furry W H 1936 *Phys. Rev.* **50** 784
Gábor D 1956 *Rev. Mod. Phys.* **28** 261
Gähler R and Golub R 1987 *Z. Phys.* B **65** 269
— 1988 *J. Physique* **49** 1295
Gähler R, Kalus J and Mampe W 1980 *J. Phys. E: Sci. Instrum.* **13** 546

— 1982 *Phys. Rev.* D **25** 2887
Gähler R, Klein A G and Zeilinger A 1981 *Phys. Rev.* A **23** 1611
Gales S, Hourani E, Hussonnois M, Shapiri J P, Stab L and Vergnes M 1984 *Phys. Rev. Lett.* **53** 759
Galster S, Klein H, Moritz J, Schmidt K H, Wegener D, Bleckwenn J 1971 *Nucl. Phys.* B **32** 221
Gammel J and Thaler R 1957 *Phys. Rev.* **107** 291
— 1957a *Phys. Rev.* **109** 1229
Gamow G 1928 *Z. Phys.* **51** 204
— 1938 *Phys. Rev.* **53** 595
— 1940 *Phys. Rev.* **70** 572
Gamow G and Teller E 1936 *Phys. Rev.* **49** 895
— 1938 *Phys. Rev.* **608**
Gans R 1915 *Ann. Phys., Lpz* **47** 709
Garcia A and Maya M 1978 *Phys. Rev.* D **17** 1376
Garcia A 1982 *Phys. Rev.* D **25** 1348
— 1983 *Phys. Rev.* D **28** 1655
Garg J B, Havens W W and Rainwater J 1964 *Phys. Rev.* B **136B** 177
Gari M and Schlitter J 1975 *Phys. Lett.* **59B** 118
Garrison J D 1964 *Ann. Phys., NY* **30** 269
Gartenhaus S 1955 *Phys. Rev.* **100** 900
Garwin R L, Lederman L M and Weinrich M 1957 *Phys. Rev.* **105** 1415
Gasser J and Leutwyler H 1982 *Phys. Rep.* **87** 77
Geiger H 1908 *Proc. R. Soc.* A **81** 174
— 1910 *Proc. R. Soc.* A **83** 492
Geiger H and Marsden E 1909 *Proc. R. Soc.* A **82** 495
— 1910 *Proc. R. Soc.* A **83** 492
— 1913 *Phil. Mag.* **25** 604
Geiger H and Nutall J M 1911 *Phil. Mag.* **22** 613
Gell-Mann M 1953 *Phys. Rev.* **92** 833
— 1958 *Phys. Rev.* **111** 362
— 1964 *Phys. Lett.* **8** 214
Gell-Mann M and Levy M 1960 *Nuovo Cimento* **16** 705
Gell-Mann M and Ne'eman Y 1964 *The Eightfold Way* (New York: W A Bengamin)
Gell-Mann M and Pais A 1955 *Phys. Rev.* **97** 1387
Gell-Mann M, Goldberger M L and Thirring W E 1954 *Phys. Rev.* **95** 1612
Georgi H and Glashow S L 1974 *Phys. Rev. Lett.* **32** 438
Gerlach W and Stern O 1921 *Z. Phys.* **8** 110
— 1922 *Z. Phys.* **9** 349, 353
Gerry C C 1984 *J. Phys. A: Math. Gen.* **17** L737
Gershstein S S and Ponomarev L I 1977 *Phys. Lett.* **72B** 80
Gershstein S S and Zel'dovich Yu B 1957 *Sov. Phys.–JETP* **2** 576
Gershstein S S, Popov Yu V, Ponomarev L I, Popov N P, Presnyakov L P and Somov L N 1981 *Sov. Phys.–JETP* **53** 872
Ghiorso A *et al* 1955 *Phys. Rev.* **99** 1048
Ghiorso A, Harvey B G, Choppin G R, Thompson S G and Seaborg G T 1955a *Phys. Rev.* **98** 1518

Ghiorso A, Nitschki J M, Alonso I R, Alonso C T, Nurmia M, Seaborg G T, Hulet E K and Lougheed P W 1974 *Phys. Rev. Lett.* **33** 1490
Ghiorso A, Nurmia M, Escola K and Escola P 1971 *Phys. Rev.* C **4** 1850
Ghiorso A, Nurmia M, Escola K, Harris J and Escola P 1970 *Phys. Rev. Lett.* **24** 1498
Ghiorso A, Sikkland T, Larsh A E and Latimer R M 1961 *Phys. Rev. Lett.* **6** 18
Ghiorso A, Sikkland T, Walton J R and Seaborg G T 1958 *Phys. Rev. Lett.* **1** 18
Ghiorso A, Thompson S G, Higgins G H, Harvey B G and Seaborg G T 1954 *Phys. Rev.* **95** 293
Ghiorso A, Thompson S G, Higgins G H, Seaborg G T, Studier M H, Fields P R, Fried S M, Diamond H, Mech J F, Pyle G L, Huizenga J R, Hirsch A, Manning W M, Browne C I and Ghoshal S N 1950 *Phys. Rev.* **80** 939
Giamati C C and Reines F 1962 *Phys. Rev.* **126** 2178
Gilman F J and Nir Y 1990 *Ann. Rev. Nucl. Part. Sci.* **40** 213
Gläser W, Rossat-Mignoud J, Schweizer J and Vetteer C (eds) 1989 Neutron scattering *Physica* **156&157** 1
Glashow S L 1961 *Nucl. Phys.* **22** 579
— 1979 *Neutrino 79* ed A Haatuft and C Jarlskog (Bergen: University of Bergen) p 518
Glashow S L, Iliopoulos J and Miani L 1970 *Phys. Rev.* D **2** 1285
Glasson J L 1921 *Phil. Mag.* **42** 596
Glasstone S and Edlund M C 1952 *The Elements of Nuclear Reactor Theory* (New York: Van Nostrand)
Glauber R J 1963 *Phys. Rev.* **131** 277
— 1963a *Phys. Rev.* **130** 2529
— 1963b *Phys. Rev. Lett.* **10** 84
Glendenning N K and Kramer G 1962 *Phys. Rev.* **126** 2159
Glück F and Tóth K 1990 *Phys. Rev.* D **41** 2160
Goldberger M L 1955 *Phys. Rev.* **97** 508
— 1955a *Phys. Rev.* **99** 979
Goldberger M L and Seitz F 1947 *Phys. Rev.* **17** 294
Goldberger M L and Treiman S B 1958 *Phys. Rev.* **110** 1178, 1478
Goldhaber M and Teller E 1948 *Phys. Rev.* **74** 1046
Goldhaber M and Marciano W J 1986 *Comments Nucl. Part. Phys.* **16** 23
Goldhaber M, Grodzins L and Sunyar A W 1958 *Phys. Rev.* **109** 1015
Goldschmidt V M 1937 *Skrifter Norske Videnskaps Akad. Oslo I. Mat: Naturv. Kl. No 4*
— 1938 *Geophysische Verteilungesetze der Elemente, Norske Videnskaps Akad., Oslo*
Goldstone J 1961 *Nuovo Cimento* **19** 154
Golikov V V, Luschikov V I and Shapiro F L 1972 *JINR Dubna Commun. P3-6556*
— 1973 *Sov. Phys.-JETP* **37** 41
Golowich E and Valencia D 1989 *Phys. Rev.* D **40** 112
Golub R and Gähler R 1987 *Phys. Lett.* **123A** 43
Golub R and Pendlebury J M 1979 *Rep. Prog. Phys.* **42** 439
— 1977 *Phys. Lett.* **62A** 337
Golub R and Lamoreaux S K 1992 *Phys. Rev. Lett.* **70** 517
Golub R, Richardson D and Lamoreaux S K 1991 *Ultra-Cold Neutrons* (Bristol: Adam Hilger)
Gomes L C, Walecka J D and Weisskopf V F 1958 *Ann. Phys., NY* **3** 241
Gorter C G 1936 *Physica* **3** 995

Goshal S N 1950 *Phys. Rev.* **80** 939
Goto E, Kolm H H and Ford K W 1963 *Phys. Rev.* **132** 387
Gottfried K 1967 *Phys. Rev. Lett.* **18** 1174
Graf A, Rauch H and Stern T 1979 *Atomkernenergie* **33** 298
Green A E S 1958 *Rev. Mod. Phys.* **30** 569
Green K and Thompson D 1990 *J. Phys. G: Nucl. Phys.* **16** L75
Greenberg O W 1964 *Phys. Rev. Lett.* **13** 598
Greenberg O W and Resnikoff M 1968 *Phys. Rev.* **163** 1844
Greenberger D M 1983 *Rev. Mod. Phys.* **55** 875
Greenberger D M and Overhauser A W 1979 *Rev. Mod. Phys.* **51** 43
Greenberger D M, Atwood D K, Arthur J, Shull C G and Schlenker M 1981 *Phys. Rev. Lett.* **47** 751
Greene G L, Kessler E G, Deslattes R D, Borner H 1986 *Phys. Rev. Lett.* **56** 819
Greene G L, Ramsey N F, Mampe W, Pendlebury J M, Smith K, Dress W D, Miller P D and Perrin P 1979 *Phys. Rev.* D **20** 2139
— 1977 *Phys. Lett.* **71B** 297
Greenwood R C and Chrien R E 1980 *Phys. Rev.* C **21** 498
Grenacs L 1985 *Ann. Rev. Nucl. Part. Sci.* **35** 455
Griffin J J and Rich M 1959 *Phys. Rev. Lett.* **3** 342
— 1960 *Phys. Rev.* **118** 850
Griffiths J H E 1939 *Proc. R. Soc.* A **170** 513
Grigor'ev V K, Grishin A P, Vladimirskii V V, Nikolaevskii E S and Zharkov D P 1967 *Sov. J. Nucl. Phys.* **6** 239
Groshev L V, Dvoretsky V N, Demidov A M, Luschikov V I, Nikolaev S A, Panin Y N, Pokotilovsky Y N, Strelkov A V and Shapiro F L 1973 *JINR Dubna. Commun. P3-7282*
Grover J R 1967 *Phys. Rev.* **157** 832
Guggenheimer K 1934 *J. Phys. Rad.* **5** 253
Guinier A 1937 *Cont. Rend.* **204** 1115
— 1939 *Ann. Phys., Paris* **12** 161
Gunn J and Ostriker J 1969 *Nature* **221** 454
Gurney R W and Condon E U 1928 *Nature* **122** 439
Gürsey F and Radicati L A 1964 *Phys. Rev. Lett.* **13** 173
Hafstad L R and Teller E 1936 *Phys. Rev.* **54** 681
Hagemann R, Devillers C, Lucas M, Lecomte T and Ruffenach J-C 1975 *The Oklo Phenomenon* ed P K Kuroda (Vienna: IAEA) 415
Hagen C R 1965 *Phys. ev.*B **140** 804
— 1990 *Phys. Rev. Lett.* **64** 2347
Hahn E L 1950 *Phys. Rev.* **80** 580
Hahn O and Strassmann F 1938 *Naturwiss.* **26** 755
— 1939 *Naturwiss.* **27** 11, 89, 163, 529
Hahn O, Meitner L and Strassmann F 1938 *Naturwiss.* **26** 475
Hains T 1986 *Proc. 23rd Int. Conf. on High-Energy Physics* (Singapore: World Scientific) p 1289
Halban H von and Preiswerk P 1936 *Cont. Rend.* **203** 73
Halpern J, Estermann I, Simpson O C and Stern O 1937 *Phys. Rev.* **52** 142
Halpern O 1964 *Phys. Rev.* **133** B579
Halpern O and Johnson M H 1937 *Phys. Rev.* **51** 992

— 1939 *Phys. Rev.* **55** 898
— 1937a *Phys. Rev.* **52** 52
Hamada T and Johnston I D 1962 *Nucl.Phys.* **34** 382
Hamer A N and Robbins E J 1960 *Geochem. Cosmogeochem. Acta* **19** 143
Hamermesh M 1949 *Phys. Rev.* **75** 1766
Hamermesh M and Eisner E 1950 *Phys. Rev.* **79** 888
Hammermesh M and Schwinger J 1947 *Phys. Rev.* **71** 678
Hamermesh M, Ringo G R and Wattenberg A 1952 *Phys. Rev.* **85** 483
Hamilton W A, Klein A G and Opat G I 1983 *Phys. Rev.* A **28** 3149
Hamilton W C 1957 *Acta Cryst.* **10** 629
— 1957a *Phys. Rev.* **110** 1050
— 1958 *Acta Cryst.* **11** 585
Hamilton W R 1828 *Trans. Royal Irish Acad.* **15** 69
— 1830 *Trans. Royal Irish Acad.* **16** 1
— 1831 *Trans. Royal Irish Acad.* **16** 193
— 1835 *Phil. Trans. R. Soc.* **125** 95
— 1837 *Trans. Royal Irish Acad.* **19** 1
Hammerschmied S, Rauch H, Clerc H and Kischko U 1981 *Z. Phys.* A **302** 323
Han M Y and Nambu Y 1965 *Phys. Rev.* **139** B1006
Haq R V, Pandey A and Bohigas O 1982 *Phys. Rev. Lett.* **48** 1086
Hardy J C, Towner I S, Koslowsky V T, Hagberg E and Schmeing H 1990 *Nucl. Phys.* A **509** 429
Harkins W D 1920 *Phys. Rev.* **15** 73
— 1921 *Am. Chem. Soc. J.* **43** 1038
— 1921a *Phil. Mag.* **42** 305
Haroche S 1970 *Phys. Rev. Lett.* **24** 861
Harris R A and Stodolsky L 1979 *J. Chem. Phys.* **70** 2789
Harris S, Muehlhause C and Thomas G 1950 *Phys. Rev.* **78** 632
— 1950a *Phys. Rev.* **79** 11
Harvey J A, Goldberg M D and Hughes D J 1951 *Phys. Rev.* **86** 1416
Hasert F J *et al* 1973 *Phys. Lett.* **46B** 138
Hättig H, Hunchen K and Waffler H 1970 *Phys. Rev. Lett.* **25** 941
Hauser W and Feshbach H 1952 *Phys. Rev.* **87** 366
Havens W W, Rabi I I and Rainwater L J 1949 *Phys. Rev.* **75** 1295
— 1951 *Phys. Rev.* **82** 345
Havens W W, Rainwater L J and Rabi I I 1947 *Phys. Rev.* **72** 634
Hawkesworth M R 1973 *Trans. Am. Nucl. Soc.* **17** 90
Haxel O, Jensen J D and Suess H E 1948 *Naturwiss.* **36** 376
— 1949 *Phys. Rev.* **75** 1766
— 1950 *Z. Phys.* **128** 295
Haymes R C 1959 *Phys. Rev.* **116** 1231
Hayter J B, Penfold J and Williams W C 1978 *J. Phys. E: Sci. Instrum.* **11** 454
Hayward E 1963 *Rev. Mod. Phys.* **35** 324
Hayward E and Fuller E G 1962 *Nucl. Phys.* **30** 613
— 1962a *Nucl. Phys.* **33** 431
Hayward E and Stovald T 1965 *Nucl. Phys.* **69** 241
Hazra C C, Parida M K and Sarker U 1983 *Phys. Lett.* **123B** 413
Heckel B, Forte M, Schärpf O, Green K, Greene G L, Ramsey N F, Byrne J and

Pendlebury J M 1984 *Phys. Rev.* C **29** 2389
Heckel B, Ramsey N F, Green K, Greene G L, Gähler R, Schärpf O, Forte M, Dress D, Miller P D, Golub R, Byrne J and Pendlebury J M 1982 *Phys. Lett.* **119B** 248
Hegerfeldt G C and Kraus K 1968 *Phys. Rev.* **170** 1185
Heisenberg W 1928 *Z. Phys.* **46** 619
— 1932 *Z. Phys.* **77** 1
— 1932a *Z. Phys.* **78** 156
— 1933 *Z. Phys.* **80** 587
— 1943 *Z. Phys.* **120** 513, 673
Heitler W 1954 *The Quantum Theory of Radiation* 3rd edn (Oxford: Clarendon)
Heitler W and Hu N 1947 *Nature* **159** 776
Heller L 1967 *Rev. Mod. Phys.* **39** 584
Henley E M, Hwang W-Y P and Kisslinger L S 1992 *Phys. Rev.* D **46** 431
Herb S *et al* 1977 *Phys. Rev. Lett.* **39** 252
Herczeg P 1983 *Phys. Rev.* D **28** 200
Herr W, Hintenberger H and Voshage H 1954 *Phys. Rev.* **95** 1691, 1960
Hess W N 1958 *Rev. Mod. Phys.* **30** 368
Hewat A W 1975 *Nucl. Instr. Meth.* **127** 361
Hewat A W and Bailey I 1976 *Nucl. Instr. Meth.* **137** 463
Hewish A, Bell S J, Pilkington J D H, Scott P F and Collins R A 1968 *Nature* **217** 709
Higgs P W 1964 *Phys. Lett.* **12** 132
— 1964a *Phys. Lett.* **13** 508
— 1966 *Phys. Rev.* **145** 1156
Hill D L and Wheeler J A 1953 *Phys. Rev.* **89** 1102
Hirata K S *et al* 1987 *Phys. Rev. Lett.* **58** 1490
— 1988 *Phys. Rev.* D **38** 448
— 1990 *Phys. Rev. Lett.* **65** 1297
— 1990a *Phys. Rev. Lett.* **65** 1301
Hirsch R L 1975 *Ann. Rev. Nucl. Sci.* **25** 79
Hodgson P E 1984 *Rep. Prog. Phys.* **47** 613
—1987 *Rep. Prog. Phys.* **50** 1172
Hofmann S, Reisdorf W, Munzenberg G, Hessberger F P, Schneider J R H and Armbruster P 1982 *Z. Phys.* A **305** 111
Hoffmann D C, Lawrence F O, Newherter J L and Rourke F M 1971 *Nature* **234** 132
Hofstadter R 1956 *Rev. Mod. Phys.* **28** 214
Hofstadter R, Busmiller F and Yearian M R 1958 *Rev. Mod. Phys.* **30** 482
Höhler G, Pietarinen E, Sabba-Slefanesan I, Borkowski F, Simon G G, Walther V H and Wendling R D 1976 *Nucl. Phys.* B **114** 505
Holden N E 1981 *Neutron Capture Cross Section Standards for BNL 325* 4th edn BNL-NCS-51388 UC-346
Holland B G and Pain L F 1971 *UKAEA Report NP 113* (Aldermaston: AWRE)
Holstein B R 1974 *Rev. Mod. Phys.* **46** 789
Holstein B R and Swift A R 1972 *Am. J. Phys.* **40** 829
Holstein B R and Treiman S B 1976 *Phys. Rev.* D **13** 3059
— 1977 *Phys. Rev.* D **16** 2396
— 1977a *Phys. Rev.* C **16** 753
Horie H and Arima A 1954 *Prog. Theor. Phys.* (*Kyoto*) **11** 509
— 1955 *Phys. Rev.* **99** 778

Horne M A 1986 *Physica* B **137** 260
Hoyle F 1954 *Astrophys J. Suppl.* **1** 121
— 1960 *Proc. R. Soc.* A 257 431
Hu N 1948 *Phys. Rev.* **74** 131
Huby R 1954 *Proc. Phys. Soc.* A **67** 1103
Hughes D J 153 *Ann. Rev. Nucl. Sci.* **3** 93
Hughes D J and Burgy M T 1949 *Phys. Rev.* **76** 1413
— 1951 *Phys. Rev.* **81** 498
Hughes D J, Burgy M T, Heller R B and Wallace J W 1949 *Phys. Rev.* **75** 565
Hughes D J, Harvey J A, Goldberg M D and Stafne M J 1953 *Phys. Rev.* **90** 497L
Hughes V W and Kute J 1983 *Ann. Rev. Nucl. Part. Sci.* **33** 611
Humblet J 1964 *Nucl. Phys.* **57** 386
Humblet J and Rosenfeld L 1961 *Nucl. Phys.* **26** 529
Hund F 1927 *Z. Phys.* **42** 93
Hurwitz H and Zweifel P F 1955 *J. Appl. Phys.* **26** 923
Huss T and Zupancic C 1953 *Kgl. Danske. Videnskab Selskab. Mat.-fys. Medd.* **28** No 1
Huygens C 1690 *Traité de la Lumière* (Leiden) Engl. transl. 1912 *Treatise on Light* (London: Macmillan)
Iachello F and Talmi I 1987 *Rev. Mod. Phys.* **59** 339
Ibáñez L E 1982 *Phys. Lett.* **114B** 243
Iliopoulos J 1980 *Contemp. Phys.* **21** 159
Inglis D R 1937 *Phys. Rev.* **51** 531
— 1954 *Phys. Rev.* **96** 1059
— 1953 *Rev. Mod. Phys.* **25** 390
— 1955 *Rev. Mod. Phys.* **27** 76
Isgur N and Karl G 1977 *Phys. Lett.* **72B** 109
— 1978 *Phys. Rev.* D **18** 4187
— 1979 *Phys. Rev.* D **19** 2653
— 1980 *Phys. Rev.* D **21** 3175
Ishikaa Y Wu and Saskawa S 1990 *Phys. Rev. Lett.* **64** 1875
Iwanenko D 1932 *Cont. Rend.* **195** 439
— 1932a *Nature* **129** 79
Jaccarino V, King J G, Setten R A and Stroke H H 1954 *Phys. Rev.* **94** 1798
Jackson J D, Treiman S B and Wyld H W 1957 *Phys. Rev.* **106** 517
— 1957a *Nucl. Phys.* **4** 206
Jacobi C G J 1837 *Crelle's J.* **17** 69
Jastrow R 1951 *Phys. Rev.* **81** 165
Jauch J M and Rohrlich F 1955 *The Theory of Photons and Electrons* (Reading, MA: Addison Wesley)
Jaynes E T and Cummings F W 1963 *Proc. IEEE 51* 89
Jeffries H 1923 *Proc. London Math. Soc.* **23** 428
Johns M W and Sargent B W 1954 *Can. J. Phys.* **32** 136
Johnson A, Ryde H and Sztarkur J 1971 *Phys. Lett.* **34B** 605
Johnson A, Ryde H and Hjorth S A 1972 *Nucl. Phys.* A **179** 753
Johnson J B 1928 *Phys. Rev.* **32** 97
Johnson K 1975 *Acta Phys. Pol.* B **6** 865
— 1978 *Phys. Lett.* **78B** 259
Johnson M H and Teller E 1955 *Phys. Rev.* **98** 783

Johnston C H, Pleasonton F and Carlson T A 1963 *Nucl. Phys.* **41** 167
Jones S E 1986 *Nature* **321** 127
— 1987 *Muon Cat. Fusion* **1** 21
Jones S E, Anderson A N, Caffrey A J, Walter J B, Watts K D, Bradbury J N, Gram P A W, Leon M, Maltrud H R and Paciotti M A 1983 *Phys. Rev. Lett.* **51** 1757
Jones S E, Palmer E P, Czirr J B, Decker D L, Jensen G L, Thorne J M and Taylor S F 1989 *Nature* **338** 737
Jones T J L and Williams W G 1980 *J. Phys.* E **13** 227
Jones T W *et al* (IMB Collaboration) 1984 *Phys. Rev. Lett.* **52** 720
Jost R 1947 *Helv. Phys. Acta* **20** 256
Kabir P K 1967 *Phys. Lett.* **24B** 601
— 1983 *Phys. Rev. Lett.* **51** 231
— 1983a *Nucl. Phys.* B **224** 21
Kabir P K, Karl G and Obryk E 1974 *Phys. Rev.* D **10** 1471
— 1975 *Can. J. Phys.* **53** 2661
Kabir P K, Nussinov S and Aharonov Y 1984 *Phys. Rev.* D **29** 1537
Kagan Y and Afazesev M 1966 *Sov. Phys.–JETP* **22** 377
Kahara S and Mohapatra R N 1983 *Phys. Lett.* **129B** 57
Kaina W, Soergel V, Thies H and Trost W 1977 *Phys. Lett.* **70B** 411
Kaiser H, Arif M and Werner S A 1986 *Physica* B **136** 134
Kaiser H, George E A and Werner S A 1984 *Phys. Rev.* A **29** 2276
Kaiser H, Rauch H, Badurek G, Bauspiess W and Bonse U 1979 *Z. Phys.* **291** 231
Kaiser H, Werner S A and George E 1983 *Phys. Rev. Lett.* **50** 560
Kalmann H 1948 *Research* **1** 254
Kapitsa P L 1975 *Sov. Phys.–JETP Lett.* **40** 9
— 1979 *Rev. Mod. Phys.* **51** 417
Kapitsa P L and Pitaevskii L P 1975 *Sov. Phys.–JETP* **40** 701
Kapur P L and Peierls R 1938 *Proc. R. Soc.* A **166** 277
Karim A, Arendt B H, Goyette R, Huang Y Y, Kleb R and Felcher G P 1991 *Physica* B **173** 17
Kato N 1961 *Acta Cryst.* **14** 526
— 1969 *Acta Cryst.* A **25** 119
Kato N and Lang A R 1959 *Acta Cryst.* **12** 787
Kavarnovskii M W, Kuz'min V A and Shaposhnikov M E 1981 *Sov. Phys.–JETP Lett.* **34** 47
Kearney P D, Klein A G, Opat G I and Gähler R 1980 *Nature* **287** 313
Keefe D 1982 *Ann. Rev. Nucl. Sci.* **32** 391
Keepin G R and Wimett T F 1958 *Nucleonics* **16** 86
Keilson J 1951 *Phys. Rev.* **82** 759L
Keller J B 1958 *Ann. Phys., NY* **4** 180
Kellogg J M B, Rabi I I, Ramsey N F and Zacharias J R 1939 *Phys. Rev.* **55** 318L
— 1939a *Phys. Rev.* **56** 728
— 1940 *Phys. Rev.* **57** 677
Kelvin Lord (W. Thomson) 1861 *Report of the 31st Meeting of the British Association for the Advancement of Science* (London: John Murray) Pt 2 p 27
Kemmer N 1938 *Proc. Camb. Phil. Soc.* **34** 354
Kenna B T and Kuroda P K 1964 *J. Inorg. Nucl. Chem.* **26** 493
Kerek A, Holm G B, Carle P and McDonald J 1972 *Nucl. Phys.* A **195** 159

Kerrmann P, Steinhauser K-A, Gähler R, Steyerl A and Mampe W 1985 *Phys. Rev. Lett.* **54** 1969
Khanna F C, Towner I S and Lee H C 1978 *Nucl. Phys.* A **305** 349
Khinchine A I 1934 *Math. Ann.* **109** 604
Kikuta S 1979 *Neutron Interferometry* ed U Bonse and H Rauch (Oxford: Clarendon) p 60
Kikuta S, Ishikawa I, Kohra K and Hishino S 1975 *J. Phys. Soc. Japan* **39** 471
Kikuta S, Kohra K, Minakawa N and Doi K 1971 *J. Phys. Soc. Japan* **31** 954
Kilvington A I, Golub R, Mampe W and Ageron P 1987 *Phys. Lett.* **125A** 416
Kirchhoff G 1883 *Ann. Phys., Lpz* **18** 663
Kirwan A J 1987 *Phys. Rev. Lett.* **58** 467
Kisslinger L S and Sorensen R A 1960 *Mat. Fys Medd. Dan. Vid Selsk* **32** No 9
Kladnik R and Kuscer I 1961 *Nucl. Sci. Eng.* **11** 116
Klapdor H V, Oda T, Metzinger J, Hillebrandt W and Thielemann F K 1981 *Z. Phys.* A **299** 213
Klein A G, Kearney P D, Opat G I, Cimmino A and Gähler R 1981 *Phys. Rev. Lett.* **46** 959
Klein A G and Opat G I 1975 *Phys. Rev.* D **11** 523
— 1976 *Phys. Rev. Lett.* **37** 238
Klein A G and Werner S A 1983 *Rep. Prog. Phys.* **46** 259
Klein A G, Kearney P D, Opat G I and Gähler R 1981 *Phys. Lett.* **83A** 71
Klein A G, Martin L J and Opat G I 1977 *Am. J. Phys.* **45** 295
Klein A G, Opat G I, Cimmino A, Zeilinger A, Treimer W and Gähler R 1981 *Phys. Rev. Lett.* **46** 1551
Klein O 1948 *Nature* **161** 897
Klemt E 1976 *Phys. Rev.* D **13** 3125
Klemt E, Bopp P, Hornig L, Last J, Freedman S J, Dubbers D and Schäerpf O 1988 *Z. Phys.* C **37** 179
Kline D B 1986 *Comments Nucl. Part. Phys.* **16** 131
Klinkenberg P F A 1952 *Rev. Mod. Phys.* **24** 63
Klippinger W E, Calaprice F P and Miller D 1978 *Bull. Am. Phys. Soc.* **23** 6903
Knowles J W 1956 *Acta Cryst.* **9** 61
Knyaz'kov V A, Kolomenskii E A, Lobashev V M, Nazarenko V A, Pirozhkov A N, Sobolev Yu V, Shablii A T and Shul'gina E V 1983 *Sov. Phys.–JETP Lett.* **38** 163
Kobayashi M and Maskawa T 1973 *Prog. Theor. Phys.* **49** 652
Koester L 1965 *Z. Phys.* **182** 328
— 1967 *Z. Phys.* **198** 187
— 1976 *Phys. Rev.* D **14** 907
Koester L and Nistler W 1975 *Z. Phys.* A **272** 189
Koester L and Steyerl A 1977 *Neutron Physics* (Berlin: Springer)
Koester L, Knopf K and Waschkowski W 1976a *Z. Phys.* A **277** 77
Koester L, Nistler W and Waschkowski W 1976 *Phys. Rev. Lett.* **36** 1021
Koester L, Waschkowski W and Klüver A 1986 *Physica* B **137** 282
Kofoed-Hansen O 1948 *Phys. Rev.* **74** 1785
Kofoed-Hansen O 1954 *K. Dansk Vidensk. Selsk Mat. Fys. Meddr.* **28** No 9
— 1958 *Rev. Mod. Phys.* **30** 449
Kohman T P, Mattauch J and Wapstra A H 1958 *Naturwiss.* **45** 174
— 1958a *Science* **127** 1431

Kolomenskii E A, Lobashev V M, Perozhkov A N, Smotritsky L M, Titov N A and Vesna V A 1981 *Phys. Lett.* **107B** 272
Konopinski E and Uhlenbeck G E 1935 *Phys. Rev.* **48** 1
Kopeliovich V B 1981 *Phys. Lett.* **103B** 157
Kossakowskii R, Grivot P, Liaud P, Schreckenbach K and Azuelas G 1989 *Nucl. Phys.* A **503** 473
Kostvintsev Yu Yu, Kushnir Yu A and Morozov V I 1976 *Sov. Phys.–JETP Lett.* **23** 118
Kostvintsev Yu Yu, Kushnir Yu A, Morozov V I and Terekhov G I 1980 *Sov. Phys.–JETP Lett.* **31** 236
Kostvintsev Yu Yu, Morozov V I and Terekhov G I 1982 *Sov. Phys.–JETP* **36** 514
— 1986 *Sov. Phys.–JETP Lett.* **44** 571
Kostvintsev Yu Yu, Morozov V I, Terekhov G I, Panin Yu N, Rogov E V and Forin A I 1987 *Proc. Int. Conf. on Neutron Physics, Kiev 1987* Vol 1, p 231 (Moscow: TsNII atominform)
Kotani T, Takebe H, Umezawa M and Yamaguchi Y 1952 *Prog. Theor. Phys.* **7** 469
— 1952a *Prog. Theor. Phys.* **8** 120
Kottler F 1923 *Ann. Phys., Lpz* **71** 457
— 1923a *Ann. Phys., Lpz* **72** 320
Kramers H A 1926 *Z. Phys.* **39** 828
— 1927 *Atti. Congresso Internazionale dei Fisici, Como* **2** 545
— 1934 *Physica* **1** 182
Kratz J V 1983 *Radiochem. Acta* **32** 35
Krishnaswamy M R, Menon M G K, Modal N K, Narasimham V S, Sreekantan N V, Ito N, Kawakami S, Hayashi Y and Miyake S 1982 *Phys. Lett.* **115B** 349
— 1981 *Phys. Lett.* **106B** 339
Krohn V E and Ringo G R 1960 *Phys. Rev.* **148** 1303
— 1973 *Phys. Rev.* D **8** 1305
— 1975 *Phys. Lett.* **55B** 175
Kronig R 1926 *J. Opt. Soc. Am.* **12** 547
— 1946 *Physica* **12** 543
Krueger H, Mineghetti D, Ringo G R and Winsberg L 1950 *Phys. Rev.* **80** 507
Krupchitsky P A 1987 *Fundamental Research with Polarized Slow Neutrons* (Berlin: Springer)
— 1989 *Nucl. Instr. Meth.* A **284** 71
Kruse U E and Ramsey N F 1951 *J. Math. Phys.* **39** 40
Krushnan V and Sivaram C 1982 *Astron. Space Sci.* **87** 205
Kruskal M and Schwarzchild 1954 *Proc. R. Soc.* A **223** 348
Kubo R 1957 *J. Phys. Soc. Japan* **12** 570
— 1966 *Rep. Prog. Phys.* **29** 255
Kügler K J, Moritz K, Paul W and Trinks U 1985 *Nucl. Instr. Meth.* A **228** 248
Kügler K J, Paul W and Trinks U 1978 *Phys. Lett.* **72B** 421
Kukulin V I, Troitski V E, Shirokov Yu M and Trubnikov S V 1972 *Phys. Lett.* **39B** 319
Kulidzhanov F G, Bradler J G and Kadeckova S 1987 *Sov. Phys. Solid State* **29** 228
Kumano S and Londergan J T 1991 *Phys. Rev.* D **44** 717
Kuo T K and Love S T 1980 *Phys. Rev. Lett.* **45** 93
Kuratsuji H 1987 *Phys. Lett.* **120A** 141
Kuroda P K 1956 *J. Chem. Phys.* **25** 1295

Kurz H and Rauch H 1969 *Z. Phys.* **220** 419
Kuscer I K and Corngold N 1965 *Phys. Rev.* **139** A981
Kusch P and Eck T G 1954 *Phys. Rev.* **94** 1799
Kuti J and Weisskopf V 1971 *Phys. Rev.* D **4** 3418
Kuz'min V A 1970 *Sov. Phys.–JETP Lett.* **12** 228
Kuz'min V A and Shaposhnikov M V 1983 *Phys. Lett.* **125B** 449
Kvardakov V V, Podurets K M, Chistvakov R R, Shil'shtein S Sh, Elyutin N O, Kulidzhanov F G, Bradler J and Kadeckova S 1987 *Sov. Phys. Solid State* **29** 228
Lamberty A and De Bièvre P 1991 *Int. J. of Mass. Spec. and Ion Processes* **108** 189
Lamberty A, Eykens R, Tagziria H, Pauwels J, De Bièvre P, Scott R D, Byrne J, Dawber P G, Gilliam D H and Greene G L 1990 *Nucl. Instr. Meth.* A **294** 393
Landau L D 1932 *Phys. Z. Sowjetunion* **1** 285
Landau L D and Lifshitz E M 1965 *Course of Theoretical Physics Vol 3 Quantum Mechanics* (London: Pergamon)
Landé A and Svenne J P 1969 *Nucl. Phys.* A **124** 241
Lander G H and Robinson R A (eds) 1986 Neutron Scattering *Physica* B&C **136**
Landkammer F J 1966 *Z. Phys.* **189** 113
Landsberg P T and Cole E A B 1967 *Physica* **37** 309
Lane A M and Pendlebury E D 1960 *Nucl. Phys.* **15** 39
Lane A M and Thomas R G 1958 *Rev. Mod. Phys.* **30** 257
Lane A M, Thomas R G and Wigner E P 1955 *Phys. Rev.* **97** 224
Lane K 1981 *Phys. Script.* **23** 1005
Lang A R 1958 *J. Appl. Phys.* **29** 597
— 1959 *Acta Cryst.* **12** 249
Lang D 1961 *Nucl. Phys.* **26** 434
Lang J M B and Le Couteur K J 1954 *Proc. Phys. Soc.* A **67** 586
Langacker P 1979 *Phys. Rev.* D **20** 2983
Langacker P and Pagels H 1979 *Phys. Rev.* D **19** 2070
Langacker P 1981 *Phys. Rep.* **72** 185
Langer L M and Price H C 1949 *Phys. Rev.* **75** 1109
Lassila K E, Hall M H, Ruppel H M, McDonald F A and Breit G 1962 *Phys. Rev.* **126** 881
Last J, Arnold M, Döhner J, Dubbers D and Freedman S J 1988 *Phys. Rev. Lett.* **60** 995
Lattes C M G, Occhialini G P S and Powell F1948 *Proc. Phys. Soc.* **61** 173
Laue M von 1912 *Sitzungsber. Math. Phys. Kl. Bayer Akad. Wiss.* **303**
— 1931 *Ergeb. exakt. Naturwiss.* **10** 133
Lawson J D 1957 *Proc. Phys. Soc.* B **70** 6
Lax M 1951 *Rev. Mod. Phys.* **23** 287
— 1952 *Phys. Rev.* **85** 621
— 1960 *Rev. Mod. Phys.* **32** 25
Le Caine J 1947 *J. Phys. Rad.* **72** 564
Leachman R B 1956 *Proc. Int. Conf. on the Peaceful Uses of Atomic Energy* (New York: United Nations) vol 2 p 193
Leader E and Anselmino M 1988 *Z. Phys.* C **41** 239
Learned J, Reines F and Soni A 1979 *Phys. Rev. Lett.* **43** 907
Lechner R E, Volino F, Dianoux A J, Hervet H and Stirling G C 1973 *Rep. 73L85* (Grenoble: ILL)
Lee I Y, Aleonard M M, Deleplanqu M A, El Masri Y and Newon J O 1977 *Phys.*

Rev. Lett. **38** 1454
Lee T, Papanastassiou D A, Wasserburg G T 1976 *Geophys. Res. Lett.* **3** 109
Lee T D and Yang C N 1955 *Phys. Rev.* **98** 1501
— 1956 *Phys. Rev.* **104** 254
Lee Y K, Mo L and Wu C S 1963 *Phys. Rev. Lett.* **10** 253
Leggett A J and Baym G 1989 *Phys. Rev. Lett.* **63** 191
— 1989a *Nature* **340** 45
Lekner J 1991 *Physica* B **173** 99
Lepore J V 1950 *Phys. Rev.* **79** 137
Leutwyler H and Roos M 1984 *Z. Phys.* C **25** 91
Levinger J S 1949 *Phys. Rev.* **76** 699
— 1951 *Phys. Rev.* **84** 43
Levinger J S and Bethe H A 1952 *Phys. Rev.* **85** 577
Levinson N 1949 *Kgl. Danske Vid. Selskab., Mat.-fys. Medd.* **25**
Li C W, Whaling W, Fowler W A and Lauritsen C C 1951 *Phys. Rev.* **83** 512
Lichtenberg A J and Liebermann M A 1983 *Regular and Stochastic Motion* (Heidelberg: Springer)
Lieder R M and Ryde H 1978 *Adv. Nucl. Phys.* **10** 1
Linke V 1969 *Nucl. Phys.* B **12** 669
Lipkin H J 1958 *Phys. Rev.* **110** 1395
— 1960 *Ann. Phys., NY* **9** 272
— 1978 *Phys. Rev. Lett.* **41** 1629
Lippmann B A 1950 *Phys. Rev.* **79** 481
Lippmann B A and Schwinger J 1950 *Phys. Rev.* **79** 469
Littlejohn B A 1986 *Phys. Rep.* **138** 193
Lloyd H 1833 *Trans. Royal Irish Acad.* **17** 45
Lobashev V M *et al* 1972 *Nucl. Phys.* A **197** 241
Lobashev V M, Porsev G D and Serebrov A P 1973 Konstantinova Institute of Nuclear Physics, Leningrad *Rept. No 37*
Lomon E L and Partovi M H 1973 *Phys. Rev.* D **8** 2307
Lone M A, Selander W N, Latoub J, Townes B W and Bartholomew G A 1982 Atomic Energy of Canada Ltd *Rept AECL-7839*
Louisell W 1964 *Radiation and Noise in Quantum Electronics* (New York: McGraw-Hill) p 212
Lovesey S W 1984 *Theory of Neutron Scattering* (Oxford: Clarendon)
Lovesey S W and Rimmer D E 1969 *Rep. Prog. Phys.* **32** 333
Lowde R D 1952 *Proc. Phys. Soc.* A **65** 857
— 1954 *Proc. R. Soc.* A **221** 206
Lüders G 1957 *Ann. Phys., NY* **2** 1
Luschikov V I, Pokotilovsky Y N, Strelkov A V and Shapiro F L 1969 *Sov. Phys.–JETP Lett.* **9** 23
Lüst D 1983 *Phys. Lett.* **125B** 295
Lüst D, Masiero A and Roncadelli M 1982 *Phys. Rev.* D **25** 3096
L'vov A I 1981 *Sov. J. Nucl. Phys.* **34** 597
Lynch H L, Meunier R and Ritson D M 1989 *Ann. Rev. Nucl. Part. Sci.* **39** 151
Lyttleton R A and Bondi H 1959 *Proc. R. Soc.* A **252** 313
— 1960 *Proc. R. Soc.* A **257** 442
Maier-Leibnitz H 1962 *Z. Angew. Phys.* **14** 738

Maier-Leibnitz H and Springer T 1962 *Z. Phys.* **167** 386
— 1963 *J. Nucl. Energy* **17** 217
Majkrzak C 1991 *Physica* B **173** 5
Majorana E 1932 *Nuovo Cimento* **9** 43
— 1933 *Z. Phys.* **82** 137
Majumdar P 1983 *Phys. Lett.* **121B** 25
Majumdar P, Raychaudhuri and Sarker U 1984 *Phys. Lett.* **137B** 181
Maliszewskii E 1960 *Nucl. Instr. Meth.* **9** 34
Mampe W, Ageron P, Bates C, Pendlebury J M and Steyerl A 1989 *Phys. Rev. Lett.* **63** 593
— 1989a *Nucl. Instr. Meth.* A **284** 111
Mampe W, Bondarenko L N, Morozov V I, Panin Yu N and Formin A I 1993 *JETP Lett.* **57** 82
Marciano W J 1991 *Ann. Rev. Nucl. Part. Sci.* **41** 469
Marciano W J and Sirlin A 1981 *Phys. Rev. Lett.* **46** 163
Marinsky J A, Glendenin L E and Coryell C D 1947 *J. Am. Chem. Soc.* **69** 2781
Markowitz P *et al* 1993 *Phys. Rev.* C **48** R5
Marra W C, Eisenberger E and Cho A Y 1979 *J. Appl. Phys.* **50** 6927
Marseguerra A M and Pauli G 1959 *Nucl. Instr. Meth.* **4** 140
Marsh S and McVoy K 1983 *Phys. Rev.* D **28** 2793
Marshak R E 1947 *Rev. Mod. Phys.* **19** 185
Marshak R E, Brooks H and Hurwitz H 1949 *Nucleonics* **5** 3, 59
Maruhn J and Greiner W 1972 *Z. Phys.* **251** 431
Mashoon B 1987 *Phys. Lett.* **122A** 299
— 1988 *Phys. Lett.* **126A** 393
Masiero A and Mohapatra R W 1981 *Phys. Lett.* **103B** 343
Masson O 1921 *Phil. Mag.* **41** 281
Masuda Y, Adachi T, Masaike A and Morimoto K 1989 *Nucl. Phys.* A **504** 269
Matfield R S 1971 *Atom* (Harwell: AERE) **174** 1
Mathews G J and Viola V E 1976 *Nature* **261** 382
Maurette M 1976 *Ann. Rev. Nucl. Sci.* **26** 319
Maxwell O, Brown G E, Campbell P K, Dashen R F and Manassah J T 1977 *Astrophys. J.* **216** 77
Mayer M G 1948 *Phys. Rev.* **74** 235
— 1949 *Phys. Rev.* **75** 1969
— 1950 *Phys. Rev.* **78** 16, 22
Mayer M G and Teller E 1949 *Phys. Rev.* **76** 1226
McArthur D A and Tollefsrud P B 1975 *Appl. Phys. Lett.* **26** 187
McEwan K A, Stirling W G, Taylor A D and Wilson C C (eds) 1992 Neutron Scattering *Physica* **180&181** Part A 1 and Part B 567
McFarlane W K *et al* 1985 *Phys. Rev.* D **32** 547
McKellar B H J 1975 *Nucl. Phys.* A **254** 349
McKinley W A and Feshbach H 1948 *Phys. Rev.* **74** 1759
McMaster W H 1961 *Rev. Mod. Phys.* **33** 8
McMillan E M and Abelson P H 1940 *Phys. Rev.* **57** 1185
McReynolds A W 1951 *Phys. Rev.* **83** 172, 233
Mecke R 1925 *Z. Phys.* **31** 709
Mehta M L and Gaudin M 1960 *Nucl. Phys.* **18** 420

Meitner L and Frisch O R 1939 *Nature* **143** 239, 471
Meitner L and Orthmann W O 1930 *Z. Phys.* **60** 143
Meldner H 1965 Private Communication to Myers and Swiatesci
— 1972 *Phys. Rev. Lett.* **28** 975
— 1972a *Phys. Rev. Lett.* **29** 78
Melkonian E, Rustad B M and Havens W W 1956 *Bull. Anm. Phys. Soc.* **1** 62
— 1959 *Phys. Rev.* **114** 1571
Mendeleev D I 1869 *J. prakt. Chem.* **106**
Merrill P W 1952 *Science* **115** 484
Merton T R 1919 *Proc. R. Soc.* A **96** 388
Merzbacher E, Crutchfield P W and Newson H W 1959 *Ann. Phys., NY* **8** 694
Messiah A 1964 *Quantum Mechanics* (Amsterdam: North-Holland)
Meyer L 1869 *Ann. Chem. Pharm.* **7** 354
Meyer-ter-Vehn J 1979 *Phys. Lett.* **84B** 10
Mezei F 1972 *Z. Phys.* **255** 146
— 1976 *Commun. Phys.* **1** 81
— 1978 *Fundamental Physics with Reactor Neutrons and Neutrinos* ed T von Egidy (Bristol: Institute of Physics) p 162
— 1980 (ed) *Neutron Spin Echo* (Heidelberg: Springer)
— 1983 *Physica* B **120** 51
Mezei F and Dagleish P A 1977 *Commun. Phys.* **2** 41
Mezei F and Murani A P 1979 *J. Magn. Magn. Mater.* **14** 211
Michel F C 1964 *Phys. Rev.* B **133** 329
— 1965 *Phys. Rev.* B **138** 408
Michelson A A 1881 *Amer. J. Sci.* **22** 120
— 1882 *Phil. Mag.* **13** 236
— 1891 *Phil. Mag.* **31** 338
— 1892 *Phil. Mag.* **34** 280
— 1920 *Astrophys. J.* **51** 257
Michelson A A and Pease F G 1921 *Astrophys. J.* **53** 249
Michelson A A, Gale H G and Pearson F 1925 *Astrophys. J.* **61** 140
Migdal A 1959 *Zh. Exp. Teor. Fiz.* **37** 249
Mikheev S P and Smirnov A Yu 1985 *Sov. J. Nucl. Phys.* **42** 913
Miller D A and Guinn V P 1976 *J. Radioanal. Chem.* **7** 107
Miller G A, Théberge S and Thomas A W 1980 *Phys. Rev.* D **22** 2838
— 1981 *Phys. Rev.* D **24** 216
Miller L G, Watanabe T and Junze J F 1971 *Trans. Am. Nucl. Soc.* **14** 517
Miller P D, Dress W B, Baird J K and Ramsey N F 1967 *Phys. Rev. Lett.* **19** 381
Miller W A and Wheeler J A 1983 *Proc. Int. Symp. on Foundations of Quantum Mechanics (Tokyo)* p 140
Mishra S R and Sciulli F 1989 *Ann. Rev. Nucl. Part. Sci.* **39** 259
Mitchell D P and Powers P N 1936 *Phys. Rev.* **50** 486
Mohapatra R N 1989 *Nucl. Instr. Meth.* A **284** 1
Mohapatra R N and Marshak R E 1980 *Phys. Rev. Lett.* **44** 1316
— 1980a *Phys. Lett.* **94B** 183
Mohapatra R N and Pati J C 1975 *Phys. Rev.* D **11** 566
— 1982 *Nucl. Phys.* B **177** 445
Mohapatra R N and Senjanovic G 1982 *Phys. Rev. Lett.* **49** 7

Mohapatra R N and Sidhu P P 1978 *Phys. Rev.* D **17** 1876
Møller C 1946 *Nature* **158** 403
— 1952 *The Theory of Relativity* (Oxford: Clarendon)
Møller P and Nilsson S G 1970 *Phys. Lett.* **31B** 283
Moon P B and Tillman J R 1935 *Nature* **135** 904
Moon R M 1986 *Physica* **137B** 19
Moon R M, Riste T and Koehler W C 1969C 1969 *Phys. Rev.* **181** 920
Moore G T 1970 *Am. J. Phys.* **38** 1177
Moravcsik M J and Noyes H P 1961 *Ann. Rev. Nucl. Sci.* **11** 95
Morinaga M and Gugelot P C 1963 *Nucl. Phys.* **46** 210
Morioka S and Ueda T 1978 *Prog. Theor. Phys.* **60** 299
Morozov V I 1989 *Nucl. Instr. Meth.* A **284** 108
Morpurgo G 1954 *Nuovo Cimento* **12** 60
Morse P M 1929 *Phys. Rev.* **34** 57
Morse P M and Rubenstein F J 1938 *Phys. Rev.* **54** 895
Moseley H G J 1913 *Phil. Mag.* **26** 1024
— 1914 *Phil. Mag.* **27** 703
Moses H E and Tuan S F 1959 *Nuovo Cimento* **13** 197
Mössbauer R L 1984 *J. Physique Coll.* C **3** 121
Moszkowski S A 1958 *Phys. Rev.* **110** 403
Mott N F 1929 *Proc. R. Soc.* A **124** 425
— 1932 *Proc. R. Soc.* A **135** 429
Mott N F and Massey H S W 1949 *Theory of Atomic Collisions* (Oxford: Clarendon)
Mottelson B R and Nilsson S G 1959 *Kgl. Danske. Videnskab. Selskab. Mat.-fys. Medd.* **1** No 8
Mottelson B R and Valatin J G 1960 *Phys. Lett.* **5** 511
Motz L and Schwinger J 1935 *Phys. Rev.* **48** 704
Mozer F S 1959 *Phys. Rev.* **116** 970
Mueller H 1943 OSRD Project OEMsr-576 *Rept. No 2*
Muehlhause C O 1963 *Report ANL-6797*
Münzenberg G 1988 *Rep. Prog. Phys.* **51** 57
Münzenberg G, Armbruster P, Folger H, Hessberger F P, Hofmann S, Keller J, Poppensieker K, Reisdorf W, Schmidt K H, Schött H J, Leino M E and Hingmann R 1984 *Z. Phys.* A **317** 235
Münzenberg G, Hofmann S, Hessberger F P, Reisdorf W, Schmidt K H, Schreider J H R, Armbruster P, Sahm C C and Thuma B 1981 *Z. Phys.* A **300** 107
Münzenberg G, Reisdorf W, Hofmann S, Agarwol Y K, Hessberger F P, Poppensieker K, Schneider J R H, Schneider W F W, Schmidt K H, Schött H J and Armbruster P 1984a *Z. Phys.* A **315** 145
Muskat E, Dubbers D and Schärpf O 1987 *Phys. Rev. Lett.* **58** 2047
Myers W D and Swiatesci W T 1966 *Nucl. Phys.* **81** 1
Myhrer F and Wroldsen J 1988 *Rev. Mod. Phys.* **60** 629
Nachtmann O 1968 *Z. Phys.* **215** 505
Nachtmann O and Pirner H J 1984 *Z. Phys.* C **21** 277
Nagaoka H 1904 *Phil. Mag.* **7** 445
Nakano T and Nishijima K 1953 *Prog. Theor. Phys.* **10** 581
Nambu Y 1960 *Phys. Rev. Lett.* **4** 380
Namiki M 1988 *Physica* B&C **151** 22

Nathans R, Shull C G, Shirane G and Andresen A 1959 *Phys. Chem. Solids* **10** 138
Naudet R 1975 *La Recherche* **6** 588
Naudet R, Filip A and Renson R 1975 *The Oklo Phenomenon* ed P K Kuroda (Vienna: IAEA) p 83
Nazarenko V A, Sayenko L F, Smotritskii L M and Yegorev A I 1972 *Nucl. Phys.* A **197** 241
Néel L 1932 *Ann. Phys., Paris* **18** 5
— 1936 *Cont. Rend.* **203** 304
— 1948 *Ann. Phys., Paris* **3** 137
Neddermeyer S H and Anderson C D 1937 *Phys. Rev.* **51** 884
Neher H V 1956 *Phys. Rev.* **103** 228
Nemenov L L 1980 *Sov. J. Nucl. Phys.* **31** 115
Nemenov L L and Ovchinnikov A A 1980 *Sov. J. Nucl. Phys.* **31** 659
Nesvizhevskii V V, Serebrov A P, Tai'daev R R, Kharitonov A G, Alfimenkov V P, Strelkov A V and Shvetsov V N 1992 *Sov. Phys.-JETP* **75** 405
Neuilly M, Bussac T, Fréjacques C, Nief G, Vendrys G and Yuon T 1972 *Cont. Rend.* **275D** 1847
Newlands J A R 1864 *Chem. News* **10** 94
Newport R J, Rainford B D and Cywinski R (eds) 1988 *Neutron Scattering at a Pulsed Source* (Bristol: Adam Hilger)
Newton I 1686 *Principia Mathematica Book III. The System of the World: Proposition VI* ed Florina Cajori (Berkeley: University of California)
Newton R G 1957 *Phys. Rev.* **105** 763
— 1960 *J. Math. Phys.* **1** 319
— 1977 *J. Math. Phys.* **18** 1348
Newton T D 1956 *Can. J. Phys.* **34** 804
Niemi A and Semenoff G 1985 *Phys. Rev. Lett.* **55** 927
— 1986 *Phys. Rev. Lett.* **56** 1019
— 1986a *Phys. Rep.* **135** 99
Nier A O, Booth E T, Dunning J R and Grosse A V 1940 *Phys. Rev.* **57** 546
Nieto M M and Simmons L M 1978 *Phys. Rev. Lett.* **41** 207
— 1979 *Phys. Rev.* **20** 1321, 1332, 1342
Nieves J F and Sharker O 1984 *Phys. Rev.* D **29** 1020
Nigam B P and Sundaresan M K 1958 *Phys. Rev.* **111** 284
Nilsson S G 1955 *Dan Mat. Fys. Medd.* **29** No 6
Nilsson S G and Prior O 1961 *K. Danske Videsk Selsk matt-fys Medd.* **32** No. 16
Nitschke J M, Leber R E, Nurmia M J and Ghiorso A 1979 *Nucl. Phys.* A **313** 236
Nix J R 1972 *Ann. Rev. Nucl. Sci.* **22** 65
Noble B 1950 *The Wiener-Hopf Technique* (Oxford: Pergamon)
Noddack I 1934 *Z. angew. Chemie* **37** 653
Noguera S and Desplanques B 1986 *Nucl. Phys.* A **457** 189
Nolan P J, Gifford D W and Twin P J 1985 *Nucl. Instr. Meth.* A **236** 95
Nolen J A and Schiffer J P 1969 *Phys. Lett.* **29B** 396
— 1969a *Ann. Rev. Nucl. Sci.* **19** 471
Nordheim L W 1949 *Phys. Rev.* **75** 1894
— 1950 *Phys. Rev.* **78** 294
Norman E B and Seamster A G 1979 *Phys. Rev. Lett.* **43** 1226
Noyes H P 1965 *Nucl. Phys.* **74** 508

— 1967 *Nucl. Phys.* A **95** 705
Nussenzweig H M 1959 *Nucl. Phys.* **11** 499
Nyakó B M 1984 *Phys. Rev. Lett.* **52** 507
Nyquist H 1928 *Phys. Rev.* **32** 371
Oganessian Yu Ts, Lobanov Yu V and Hussenoise M 1987a Press Release: Moscow News Weekly **34** p 10
Oganessian Yu Ts, Demin A G, Danilov N A, Flerov G N, Ivanov M P, Iljinov A S, Kolesnikov N N, Markov B N, Plotko V M and Tretyakova S P 1976 *Nucl. Phys.* A **273** 505
Oganessian Yu Ts, Demin A G, Hussenoise M, Tretyakova S P, Khartinov Yu P, Utyankov V K, Shrukowski I V, Constantinu U, Bruchertseifer H and Korotkin Yu S 1984 *Z. Phys.* A **319** 215
Oganessian Yu Ts, Hussenoise M, Demin A G, Kharitonov Yu P, Bruchertseifer H, Constantinescu O, Karolkin Yu S, Trelyakova S P and Utionkov V K 1984a *Radiochem. Acta* **37** 113
Oganessian Yu Ts, Lobanov Yu V and Hussenoise M 1987 JINR (Dubna) *Preprint* D7-87-392
Oganessian Yu Ts, Tretyakov Yu P, Iljenov A S, Demin A G, Pleve A A, Tretyakova S P, Plotko V M, Ivanov M P, Danilov N A, Korotkin Yu S and Flerov G N 1974 *Sov. Phys.–JETP Lett.* **20** 265
Okamoto K 1958 *Phys. Rev.* **110** 143
Okubo S 1962 *Prog. Theor. Phys.* **27** 949
Okubo S and Marshak R E 1958 *Ann. Phys., NY* **4** 166
Olive K, Schramm D N, Steigman G and Walker T 1990 *Phys. Lett.* **236B** 454
Oppenheimer J R and Phillips M 1935 *Phys. Rev.* **48** 500
Oppenheimer J R and Serber R 1938 *Phys. Rev.* **54** 540
Oppenheimer J R and Volkoff G M 1939 *Phys. Rev.* **55** 374
Ormand W E and Brown B A 1989 *Phys. Rev. Lett.* **62** 866
Ornstein L S and Uhlenbeck G E 1937 *Physica* **4** 478
Oseen C W 1915 *Ann. Phys., Lpz.* **48** 1
Oset E and Rho M 1979 *Phys. Rev. Lett.* **42** 47
Otsuka T, Arima A, Iachello F and Talmi I 1978 *Phys. Lett.* **76B** 139
Özer M 1982 *Phys. Rev.* D **26** 3159
Pagels H 1975 *Phys. Rep.* **16** 219
Pais A 1952 *Phys. Rev.* **86** 663
Palmer S B, Baruchel J, Drillat A, Patterson C and Fort D 1986 *J. Man. Magn. Mater.* **54–57** 1626
Panofky W K H 1968 *Proc. Int. Symp. on High Energy Physics, Vienna 1968* (Geneva: CERN)
Panofsky W K H, Aamodt R L and Hadley J 1951 *Phys. Rev.* **81** 565
Parida M K 1983 *Phys. Lett.* **126B** 220
— 1983a *Phys. Rev.* D **27** 2783
Park H S *et al* 1985 *Phys. Rev. Lett.* **54** 22
Parker E W 1970 *Astrophys. J.* **160** 383
Parker P D, Donovan P F, Kane J V and Mollenauer J F 1965 *Phys. Rev. Lett.* **14** 15
Parratt L G 1954 *Phys. Rev.* **95** 359
Pasupathy J 1982 *Phys. Lett.* **114B** 172
Patterson C, Palmer S B, Baruchel J and Ishikawa Y 1985 *Solid State Commun.* **55** 81

Paul W and Trinks U 1978 *La Récherche* **9** 1009
— 1978 *Fundamental Physics with Reactor Neutrons and Neutrinos* ed T von Egidy (Bristol: Institute of Physics) p 18
— 1978a *La Récherche* **9** 1008
Paul W, Anton F, Paul L, Paul S and Mampe W 1989 *Z. Phys.* C **45** 25
Pauli W 1930 *Collected Scientific Papers vol II*, ed V F Weisskopf and R Kronig (New York: Wiley) p 1316
— 1934 *Rapports 7me Conseil Solvay, Brussels, 1933* (Paris: Gauthier-Villars)
— 1936 *Ann. Inst. Henri Poincaré* **6** 104
— 1941 *Rev. Mod. Phys.* **13** 203
Pauwels J, Eykens R, Lamberty A, Van Gestel J, Tagzeria H, Scott R D, Byrne J, Dawber P G and Gilliam D M 1991 *Nucl. Instr. Meth.* A **303** 133
Pavlichenkov I M 1989 *Sov. J. Nucl. Phys.* **50** 189
Pawlicki G S and Smith E G 1952 *Phys. Rev.* **87** 221
Peaslee D C 1954 *Phys. Rev.* **95** 717
Peccei R D and Quinn H R 1977 *Phys. Rev. Lett.* **38** 1440
— 1977a *Phys. Rev.* D **16** 1791
Peebles P J E 1966 *Astrophys. J.* **146** 542
Peierls R E 1947 *Proc. Camb. Phil. Soc.* **44** 242
Peierls R E and Yoccoz J 1957 *Proc. Phys. Soc.* A **70** 381
Pendlebury J M and Smith K 1980 *Phil. Trans. R. Soc.* B **290** 617
Pendlebury J M, Smith K F, Golub R, Byrne J, McComb T J L, Sumner T J, Burnett S M, Taylor A R, Heckel B, Ramsey N F, Green K, Morse J, Kilvington A I, Baker C A, Clark S A, Mampe W, Ageron P and Miranda P C 1984 *Phys. Lett.* **136B** 327
Penfold J and Thomas R K 1990 *J. Phys. Condens. Matter* **2** 1369
Penfold J, Ward R C and Williams W G 1987 *J. Phys. E: Sci. Instrum.* **20** 1411
Penfold J 1991 *Physica* B **173** 1
Penning F M 1936 *Physica* **3** 873
Penzias A A and Wilson R E 1965 *Astrophys. J.* **142** 419
Perez A 1967 *Phys. Rev. Lett.* **18** 50
Perrier C and Segre E 1937 *Nature* **140** 193
— 1937a *J. Chem. Phys.* **5** 712
— 1939 *J. Chem. Phys.* **7** 155
Perrin F 1908, *Cont. Rend.* **146** 967
— 1908a *Cont. Rend.* **147** 475, 530
— 1932 *Cont. Rend.* **194** 1343
— 1932a *Cont. Rend.* **195** 236
Peter O 1946 *Z. Naturf.* **1** 577
Peterson S W and Levy H A 1951 *J. Chem. Phys.* **19** 1416
— 1952 *J. Chem. Phys.* **20** 704
Petrascheck D 1976 *Acta Phys. Austr.* **45** 217
Petrov Yu V 1980 *Nature* **285** 466
Petrzhak K and Flerov G N 1940 *Cont. Rend. Acad. Sci. USSR* **28** 500
— 1940a *Phys. Rev.* **58** 89L
Petukhov V A and Yakimenko M W 1963 *Nucl. Phys.* **49** 87
Pillay K K S and Thomas C C 1971 *J. Radioanal. Chem.* **7** 107
Placzek G 1946 *Phys. Rev.* **69** 423
— 1952 *Phys. Rev.* **86** 377

Placzek G and Bethe H A 1940 *Phys. Rev.* **57** 1075A
Placzek G and Van Hove L 1954 *Phys. Rev.* **93** 1207
Podurets K M, Somenkov V A, Chislyakov R R and Shil'stein S Sh 1989 *Physica* B **156&157** 694
Poincaré H 1892 *Théorie Mathematique de la Lumière* (Paris: George Carré) 187
Poisson S D 1819 *Mémoires de l'Acad. Roy. des Sci.* **3** 121
Polikhanov S M, Druin V A, Karnautkev V A, Mekheev V L, Pleve A A, Skobelev N B, Subbotin V G, Ter-Akopjan G M and Fomichev V A 1962 *Sov. Phys.–JETP* **15** 1016
Ponomarev L I 1990 *Contemp. Phys.* **31** 219
Porod G 1948 *Acta Phys. Austr.* **2** 255
— 1982 *Small Angle X-ray Scattering* ed O Glatter and O Kratky (London: Academic) p 17
Porter C E and Thomas R G 1956 *Phys. Rev.* **104** 483
Post E J 1967 *Rev. Mod. Phys.* **39** 475
Post R F 1956 *Rev. Mod. Phys.* **28** 338
— 1970 *Ann. Rev. Nucl. Sci.* **20** 509
Pound R V and Rebka G A 1960 *Phys. Rev. Lett.* **4** 337
Pound R V and Snider J L 1965 *Phys. Rev.* B **140** 788
Prange R E 1961 *Nucl. Phys.* **28** 369, 376
Prescott C Y *et al* 1978 *Phys. Lett.* **77B** 347
Prout W 1815 *Ann. Phil.* **6** 269
Puppi G 1948 *Nuovo Cimento* **5** 587
Purcell E M and Ramsey N F 1950 *Phys. Rev.* **78** 807
Purcell E M, Torrey H G and Pound R V 1946 *Phys. Rev.* **69** 37
— 1948 *Phys. Rev.* **73** 679
Quin P A 1993 *Nucl. Phys.* A **553** 319
Rabi I I, Millman S, Kusch P and Zacharias J R 1939 *Phys. Rev.* **55** 526
Rabi I I, Ramsey N F and Schwinger J 1954 *Rev. Mod. Phys.* **26** 167
Rabi I I, Zacharias J R, Millmann S and Kusch P 1938 *Phys. Rev.* **53** 318
Racah G 1943 *Phys. Rev.* **63** 367
Racah G and Talmi I 1952 *Physica* **18** 1097
Radcliffe J M 1971 *J. Phys.* A **4** 313
Radicati L A 1953 *Proc. Phys. Soc.* A **66** 139
— 1954 *Proc. Phys. Soc.* A **67** 39
Rainwater J 1950 *Phys. Rev.* **79** 432
Rambaugh L H and Locher G L 1936 *Phys. Rev.* **49** 855
Ramsey N F 1949 *Phys. Rev.* **76** 996
— 1950 *Phys. Rev.* **78** 695
— 1955 *Phys. Rev.* **100** 1191
— 1956 *Molecular Beams* (Oxford: Clarendon) (Paperback edn (1985))
— 1958 *Phys. Rev.* **109** 228
— 1982 *The Neutron and its Applications* ed P Schofield (Bristol: Institute of Physics) p 5
— 1990 *Ann. Rev. Nucl. Part. Sci.* **40** 1
Ramsey N F and Pound R V 1951 *Phys. Rev.* **81** 278
Ramsey N F and Silsbee H B 1951 *Phys. Rev.* **84** 506
Ramsey W 1895 *Nature* **51** 512
— 1907 *Nature* **76** 269

Rao S and Shrock R 1982 *Phys. Lett.* **116B** 238
Rarita W and Schwinger J 1941. *Phys. Rev.* **59** 436
Rasetti F 1929 *Phys. Rev.* **34** 367
— 1930 *Z. Phys.* **61** 598
Rauch H 1979 *Neutron Interferometry* ed U Bonse and H Rauch (Oxford: Clarendon) p 161
Rauch H, Seidl E, Bauspiess W and Bonse U 1978 *J. Appl. Phys.* **49** 2731
Rauch H, Treimer W and Bonse U 1974 *Phys. Lett.* **47A** 369
Rauch H, Zeilinger A, Badurek G, Wilfing A, Bauspiess W and Bonse U 1975 *Phys. Lett.* **54A** 425
Rayleigh Lord (Strutt J C) 1882 *Report of the 52nd Meeting of the British Association for the Advancement of Science* (London: John Murray) p 437
Rayleigh Lord (Strutt J C) and Ramsey W 1895 *Proc. R. Soc.* A **57** 187
Rayleigh Lord (Strutt J C) 1912 *Proc. R. Soc.* A **86** 207
— 1929 *The Theory of Sound Volume IIm* (1878) Ch 14 2nd edn (London: Macmillan)
Reeves H 1991 *Phys. Rep.* **201** 335
Reeves H, Fowler W A and Hoyle F 1970 *Nature* **226** 727
Regge T 1959 *Nuovo Cimento* **14** 951
— 1960 *Nuovo Cimento* **18** 947
Reines F and Cowan C L 1953 *Phys. Rev.* **90** 492
— 1953a *Phys. Rev.* **92** 830
Reusser D, Treichel M, Boehm F, Fisher P, Gabathuler K, Henrikson H E, Jörgens V, Mitchell L W, Nussbaum C and Vuilleumier J L 1992 *Phys. Rev.* D **45** 2548
Review of Particle Properties 1992 *Phys. Rev.* D **45** Part II 1
Reynolds H L, Biedenharn L C and Beard D B 1953 *Rept ORNL-1444*
Ribe F L 1975 *Rev. Mod. Phys.* **47** 7
Richardson D J, Kilvington A I, Green K and Lamoreaux S K 1988 *Phys. Rev. Lett.* **61** 2030
Ridley B W 1961 *Nucl. Phys.* **25** 483
Riehs P 1968 *Acta Phys. Austr.* **27** 225
Rietveld H M 1966 *Acta Cryst.* **20** 508
— 1967 *Acta Cryst.* **22** 151
— 1969 *J. Appl. Cryst.* **2** 65
Riska D O and Brown G E 1972 *Phys. Lett.* **38B** 193
Ritchie B 1979 *Phys. Rev.* A **20** 1915
Roberts R B, Meyer R C and Wang P 1939 *Phys. Rev.* **55** 510
Roberts T M 1991 *Physica* B **173** 157
Robinson A H and Barton J P 1972 *Trans. Am. Nucl. Soc.* **15** 140
Robson J M 1950 *Phys. Rev.* **77** 747
— 1950a *Phys. Rev.* **78** 311
— 1951 *Phys. Rev.* **83** 349
— 1958 *Can. J. Phys.* **36** 1450
— 1976 *Can. J. Phys.* **54** 1277
Rochester G D and Butler C C 1947 *Nature* **160** 855
Roll P G, Krotkev R and Dicke R H 1967 *Ann. Phys., NY* **26** 492
Roos M 1990 *Phys. Lett.* **246B** 179
Rose H J and Jones G A 1984 *Nature* **307** 245
Rose M E 1953 *Rept ORNL-1591*

Rosenbluth M N 1950 *Phys. Rev.* **79** 615
Rosenfeld L 1961 *Nucl. Phys.* **26** 594
Rosenthal J E and Breit G 1932 *Phys. Rev.* **41** 459
Rosenzweig N 1957 *Phys. Rev.* **108** 817
— 1958 *Phys. Rev. Lett.* **1** 101
Rosholt J N and Tatsumoto M 1971 *Proc. Second Lunar Sci. Conf.* (Cambridge, MA: MIT Press) vol 2 1577
Ross D A and Taylor J C 1973 *Nucl. Phys.* B **51** 116
Ross G G 1981 *Rep. Prog. Phys.* **44** 655
Rujula A de, Georgi H and Glashow S L 1975 *Phys. Rev.* D **12** 147
Rumbaugh L H and Locher G L 1936 *Phys. Rev.* **49** 855
Russel B 1946 *A History of Western Philosophy* (London: George Allen and Unwin)
Rustad B M and Ruby S L 1955 *Phys. Rev.* **97** 991
Rutherford E 1911 *Phil. Mag.* **21** 669
— 1919 *Phil. Mag.* **37** 538
— 1920 Bakerian Lecture, *Proc. R. Soc.* A **97** 374
— 1929 *Nature* **123** 313
Rutherford E and Geiger H 1908 *Proc. R. Soc.* A **81** 141, 162
Rutherford E and Royds T 1909 *Phil. Mag.* **17** 281
Ryder L H 1991 *Eur. J. Phys.* **12** 15
Sachs R G 1948 *Phys. Rev.* **74** 433
Sachs R G and Austern N 1951 *Phys. Rev.* **81** 705
Sagnac M G 1913 *Cont. Rend* **157** 708, 1410
Sakata S 1956 *Prog. Theor. Phys.* **16** 636
Sakharov A D 1967 *Sov. Phys.–JETP Lett.* **5** 24
Sakita B 1964 *Phys. Rev.* **136** B1756
Sakurai J J 1958 *Nuovo Cimento* **7** 649
Salam A 1968 *Elementary Particle Theory* ed N Svartholm (Stockholm: Almqvist and Wiksells) p 367
Salpeter E E 1951 *Phys. Rev.* **82** 60
— 1952 *Astrophys. J.* **115** 326
— 1952a *Phys. Rev.* **88** 547
— 1955 *Phys. Rev.* **97** 1237
Samosvat G S 1984 *J. Physique Coll.* **45** 51
Sarker U, Misra S P and Pakida M K 1983 *Phys. Lett.* **120B** 124
Sawada O, Fukugita M and Arafune J 1981 *Astrophys. J.* **248** 1162
Sawyer R F 1972 *Phys. Rev. Lett.* **29** 382
Scalapino D J 1972 *Phys. Rev. Lett.* **29** 386
Schärpf O 1989 *Physica* B **156 & 157** 631, 639
Schachinger L *et al* 1978 *Phys. Rev. Lett.* **41** 1348
Scharff-Goldhaber G 1957 *Nucleonics* **15** 122
Scharff-Goldhaber G and Weneser J 1955 *Phys. Rev.* **98** 212
Schelkunoff S A 1939 *Phys. Rev.* **56** 308
Schiff L I 1967 *Phys. Rev.* **160** 1257
Scheckenhofer K A and Steyerl A 1977 *Phys. Rev. Lett.* **39** 1310
Schlenker M and Baruchel J 1986 *Physica* **137B** 309
Schlenker M, Bauspiess W, Graeff W, Bonse U and Rauch H 1980 *J. Magn. Magn. Mater.* **15–18** 1507

Schmatz W, Springer T, Shelton J and Ibel K 197474 *J. Appl. Cryst.* **7** 96
Schmidt T 1939 *Z. Phys.* **106** 358
Schmiedmayer J, Rauch H and Riehs P 1988 *Phys. Rev. Lett.* **61** 1065
Schmiedmayer J, Riehs P, Harvey J A and Hill N W 1991 *Phys. Rev. Lett.* **66** 1015
Schneider C S 1973 *Rev. Sci. Instrum.* **44** 1594
— 1976 *Acta Cryst.* A **32** 375
Schneider C S and Shull C G 1971 *Phys. Rev.* B **3** 830
Schneider M B, Calaprice F P, Hallin A L, MacArthur D W and Schreiber D F 1983 *Phys. Rev. Lett.* **51** 1239
Schöberl F and Leeb H 1986 *Phys. Lett.* **166B** 355
Schoenborn B P, Casper D L D and Kammerer O F 1974 *J. Appl. Cryst.* **7** 508
Schofield P 1960 *Phys. Rev. Lett.* **4** 239
— 1961 *Proc. Vienna Symposium on Inelastic Scattering of Neutrons in Solids and Liquids* (Vienna: IAEA) p 39
Scholten O, Iachello F and Arima A 1978 *Ann. Phys., NY* **11** 325
Schopper H 1958 *Nucl. Instr. Meth.* **3** 158
— 1963 *Nucl. Instr. Meth.* **21** 338
Schori O, Gabroud B, Joseph C, Kerroud J P, Kuegger D, Tran M T, Truol P, Winkelmann E and Dahme W 1987 *Phys. Rev.* C **35** 2252
Schramm D N 1977 *Ann. Rev. Nucl. Sci.* **27** 37
Schramm D N and Arnett W D 1975 *Astrophys. J.* **198** 628
Schramm D N and Barkat Z 1972 *Astrophys. J.* **173** 195
Schröder U E 1972 *Acta Phys. Austr.* **36** 278
— 1980 *Nucl. Phys.* B **166** 103
— 1986 *Fortschr. Phys.* **24** 85
Schrödinger E 1926 *Ann. Phys., Lpz* **79** 1, 13, 45
— 1926a *Ann. Phys., Lpz* **80** 62
— 1926b *Ann. Phys. Lpz* **81** 102
— 1926c *Phys. Zeits* **27** 95
— 1926d *Naturwiss.* **28** 664
— 1928 *Collected Papers on Wave Mechanics* (Engl Transl) (London: Blackie)
— 1930 *Sitz. Preuss. Akad. Wiss. Berlin. Ber.* **24** 418
— 1935 *Naturwiss.* **23** 812
Schuler H and Schmidt T 1935 *Z. Phys.* **94** 47
Schütz G, Steyerl A and Mampe W 1980 *Phys. Rev. Lett.* **44** 1400
Schwartz C 1955 *Phys. Rev.* **97** 380
Schwinger J 1937 *Phys. Rev.* **51** 544
— 1937a *Phys. Rev.* **52** 1250
— 1946 *Phys. Rev.* **69** 681
— 1947 *Phys. Rev.* **72** 742
— 1948 *Phys. Rev.* **73** 407
— 1951 *Phys. Rev.* **82** 914
Schwinger J and Teller E 1937 *Phys. Rev.* **52** 286
Schwitters R F and Strauch K 1976 *Ann. Rev. Nucl. Sci.* **26** 89
Scott J M C 1954 *Phil. Mag.* **45** 751
Scott R D, Pauwels J, Eykens R, Byrne J, Dawber P G and Gilliam D M 1992 *Nucl. Instr. Meth.* A **314** 163
Seaborg G T and Loveland W 1987 *Contemp. Phys.* **28** 33

Seaborg G T and Perlman M L 1948 *J. Am. Chem. Soc.* **70** 1571
Seaborg G T, James R A and Ghiorso A 1949 *The Transuranium Elements: Research Papers* (New York: McGraw-Hill) Paper No 22.2 National Nuclear Energy Series Plutonium Project Reports Vol 14B Div IV
Seaborg G T, McMillan E M, Kennedy J W and Wahl A C, 1946 *Phys. Rev.* **69** 366
Sears V F 1977 *Acta Cryst.* A **33** 373
— 1978 *Can. J. Phys.* **56** 1261
— 1982 *Phys. Rep.* **82** 1
— 1982a *Phys. Rev.* D **25** 2023
— 1984 *Thermal Neutron Scattering Lengths and Cross Sections for Condensed Matter Research* Report AECL-8490
— 1989 *Neutron Optics* (Oxford: Clarendon)
Segre E 1952 *Phys. Rev.* **86** 21
Sehgal L M 1974 *Phys. Rev.* D **10** 1663
Seidl F G P 1954 *Report BNL-278*
Selove W 1952 *Rev. Sci. Instrum.* **23** 350
Semenoff G W and Sodano P 1986 *Phys. Rev. Lett.* **57** 1195
Senftle F E, Stieff L, Cuttitta F and Kuroda P K 1957 *Geochem. Cosmochem. Acta* **11** 189
Serber R 1947 *Phys. Rev.* **72** 1114
Shan R T 1971 *Nuovo Cimento* **5A** 591
Shapiro F L 1973 *JINR Dubna Report P7-7135*
Shapiro I S and Estulin I V 1957 *Sov. Phys.–JETP* **3** 626
Shekhter V M 1968 *Sov. J. Nucl. Phys.* **7** 756
Sherr R and Talmi I 1975 *Phys. Lett.* **56B** 212
Sherwood J E, Stephenson T E and Bernstein S 1954 *Phys. Rev.* **96** 1546
Shil'shtein S Sh, Somenkov V A and Kalanov M 1973 *Sov. Phys.–JETP* **36** 1170
Shirley J H 1963 *J. Appl. Phys.* **34** 783, 789
Shull C G 1951 *Phys. Rev.* **81** 626
— 1963 *Phys. Rev. Lett.* **10** 297
— 1968 *Phys. Rev. Lett.* **21** 1585
— 1969 *Phys. Rev.* **179** 752
Shull C G and Nathans R 1967 *Phys. Rev. Lett.* **19** 384
Shull C G and Smart J S 1949 *Phys. Rev.* **76** 1256
Shull C G, Atwood D K, Arthur J and Horne M A 1980 *Phys. Rev. Lett.* **44** 765
Shull C G, Billman K W and Wedgewood F A 1987 *Phys. Rev.* **153** 1415
Shull C G, Strauser W A and Woolan E O 1951 *Phys. Rev.* **83** 333
Shull C G, Woolan E O and Koehler W C 1951a *Phys. Rev.* **84** 912
Shull C G, Woolan E O and Strauser W A 1951b *Phys. Rev.* **81** 483
Siegert A F J 1937 *Phys. Rev.* **52** 787
— 1939 *Phys. Rev.* **56** 750
Signell P and Marshak R E 1957 *Phys. Rev.* **106** 832
— 1958 *Phys. Rev.* **109** 1229
Silverman M 1988 *Physica* B&C **151** 291
Simon G G, Schmitt Ch, Borkowski F and Walther V H 1980 *Nucl. Phys.* A **333** 381
Simpson J 1973 *Publ. Astron. Soc. Pac.* **85** 479
Simpson J A 1948 *Phys. Rev.* **73** 1389
— 1956 *Rev. Mod. Phys.* **28** 254

Sinclair R W and Brockhouse B N 1960 *Phys. Rev.* **120** 1638
Siratori K and Kita E 1980 *J. Phys. Soc. Japan* **48** 1443
Sirlin A 1967 *Phys. Rev.* **164** 1767
— 1974 *Phys. Rev. Lett.* **32** 966
— 1987 *Phys. Rev.* D **35** 2423
Skyrme T H R 1958 *Nucl. Phys.* **9** 615
— 1958a *Proc. R. Soc.* A **247** 260
— 1959 *Proc. R. Soc.* A **252** 236
— 1961 *Proc. R. Soc.* A **260** 127
Slaus I 1982 *Phys. Rev. Lett.* **48** 993
Smith J H, Purcell E M and Ramsey N F 1957 *Phys. Rev.* **108** 120
Smith K F, Crampin N, Pendlebury J M, Richardson D J, Shiers D, Green K, Kilvington A I, Moir J, Prosper H B, Thompson D, Ramsey N F, Heckel B R, Lamoreaux S K, Ageron P, Mampe W and Steyerl A 1990 *Phys. Lett.* **234B** 4
Smith P F 1989 *Ann. Rev. Nucl. Part. Sci.* **39** 73
Snell A H and Miller L C 1948 *Phys. Rev.* **74** 1217
Snell A H, Pleasanton F and McCord R V 1950 *Phys. Rev.* **78** 310
Soberman R K 1956 *Phys. Rev.* **102** 1399
Soddy F 1913 *Nature* **91** 57
Sokorac A 1983 *Phys. Rev.* D **28** 2329
Soller W 1924 *Phys. Rev.* **24** 158
Sommerfeld A 1965 *Optics* 3rd edn (Oxford: Pergamon)
Sommerfeld A and Runge J 1911 *Ann. Phys., Lpz.* **35** 289
Sonada H 1986 *Nucl. Phys.* B **266** 4
Sosnovski A N, Spivak P E, Prokofiev Yu A, Kutikov I E and Dobrinin Yu P 1959 *Nucl. Phys.* **10** 395
Specht H J 1974 *Rev. Mod. Phys.* **46** 773
Spitzer L 1952 *Astrophys. J.* **116** 299
— 1962 *The Physics of Fully Ionized Gases* 2nd edn (New York: Interscience)
Spivak P E 1988 *Sov. Phys.–JETP* **94** 1
Spivak P E, Sosnovsky A N, Prokofiev A Y and Sukolov V S 1956 *Proc. Int. Conf. on Peaceful Uses of Atomic Energy, Geneva, 1955* vol 2 (New York: United Nations) p 33
Spowart A R 1971 *J. Photogr. Sci.* **19** 1
Stahl R H and Ramsey N F 1954 *Phys. Rev.* **96** 1310
Staker W P 1950 *Phys. Rev.* **80** 52
Stamm M, Reiter G and Hüttenbach S 1989 *Physica* B **156 & 157** 564
Stamm M H, Hüttenbach S and Reiter G 1991 *Physica* B **173** 11
Stanford C P, Stephenson T E and Bernstein S 1954 *Phys. Rev.* **96** 983
Stanford G S 1961 *Rept ANL-6449*
Stapp H P, Ypsilantis T and Metropolis N 1957 *Phys. Rev.* **105** 302
Staudenmann J L, Werner S A, Colella R and Overhauser A W 1980 *Phys. Rev.* A **21** 1419
Steigman G, Schramm D N and Gunn J 1977 *Phys. Lett.* **66B** 202
Steinberg R I, Liaud P, Vignon B and Hughes V W 1974 *Phys. Rev. Lett.* **33** 41
Steiner D 1975 *Proc. Int. Conf. on Nuclear Cross Sections and Technology* ed R A Schrack and C D Bowman (NBS Special Publication)
Steiner F 1988 *Physica* B & C **151** 323
Steinsvoll O, Mahkrzak C F, Shirane G and Wickstead J 1984 *Phys. Rev.* B **30** 2377

Steinwedel H and Jensen J H D 1950 *Z. Naturf.* **59** 413
Stephens F S, Diamond R M and Perlman I 1959 *Phys. Rev. Lett.* **3** 435
Stephens F S and Simon R S 1972 *Nucl. Phys.* A **138** 257
Stern O 1921 *Z. Phys.* **7** 429
Sterne T E 1933 *Mon. Not. R. Astr. Soc.* **93** 736
Stevenson A F 1940 *Phys. Rev.* **58** 1061
Steyerl A 1969 *Phys. Lett.* **29B** 33
— 1972 *Z. Phys.* **252** 371
— 1972a *Z. Phys.* **254** 168
— 1972b *Nucl. Instr. Meth.* **101** 295
— 1975 *Nucl. Instr. Meth.* **125** 461
— 1989 *Physica* B **156 & 157** 528
Steyerl A, Malik S S and Iyengar L R 1991 *Physica* B **173** 47
Stodolsky L 1974 *Phys. Lett.* **50B** 353
— 1980 *Phys. Lett.* **96B** 127
Stokes G G 1852 *Trans. Camb. Phil. Soc.* **9** 399
— 1857 *Trans. Camb. Phil. Soc.* **10** 106
Stone R S and Slovacek R E 1959 *Nucl. Sci. Eng.* **6** 466
Stratowa C, Dobrozemsky R and Weinzierl P 1978 *Phys. Rev.* D **18** 3970
Stratton J A and Chiu L J 1939 *Phys. Rev.* **56** 99
Strutinsky V M 1966 *Sov. J. Nucl. Phys.* **3** 449
— 1967 *Nucl. Phys.* A **95** 420
Stueckelberg E C G 1938 *Helv. Phys. Acta* **11** 225, 299, 378
Sturm W J 1947 *Phys. Rev.* **71** 757
Sudarshan E C G and Marshak R E 1958 *Proc. of the Padua-Venice Conf. on Mesons and Recently discovered Particles* (Padua-Venice: Societa Italiana di Fisica)
Suess H E and Urey H E 1956 *Rev. Mod. Phys.* **28** 53
Summerfield G C 1964 *Ann. Phys., NY* **26** 72
Summhammer J, Badurek G, Rauch H and Kischko U 1982 *Phys. Lett.* **90A** 110
Summhammer J, Badurek G, Rauch H, Kischko U and Zeilinger A 1983 *Phys. Rev.* A **27** 2523
Sushkov O P and Flambaum V V 1982 *Sov. Phys. Usp.* **25** 1
Suter D, Mueller K T and Pines A 1988 *Phys. Rev. Lett.* **60** 1218
Sutherland W 1899 *Phil. Mag.* **47** 269
Swiatesci W T 1956 *Phys. Rev.* **101** 651
— 1956a *Phys. Rev.* **104** 993
— 1962 *Ann. Phys., NY* **19** 67
— 1963 *Ann. Phys., NY* **22** 406
Swinth K L 1974 *Brit. J. Non-destructive Testing* **16** 129
Symbalisty E M D and Schramm D N 1981 *Rep. Prog. Phys.* **44** 293
Szilard L 1935 *Nature* **136** 951
Szwed J 1983 *Phys. Lett.* **128B** 245
Szybisz L 1978 *Z. Phys.* **285** 223
Takita M *et al* 1986 *Phys. Rev.* D **34** 902
Tayler R J 1957 *Proc. Phys. Soc.* **708** 31
— 1968 *Nature* **217** 433
— 1979 *Nature* **288** 559
— 1980 *Rep. Prog. Phys.* **43** 253

— 1981 *Q. J. R. Astr. Soc.* **110** 1216
— 1983 *Q. J. R. Astr. Soc.* **24** 1
Teichman T and Wigner E P 1952 *Phys. Rev.* **87** 123
Teller E 1956 *Nucl. Sci. Eng.* **1** 313
Ternov I M, Rodionov V N, Lobanov A E and Dorofeev O F, 1983 *Sov. Phys.–JETP Lett.* **37** 342
Ternov I M, Rodionov V N, Zhulego V G and Studenikin A I 1978 *Sov. J. Nucl. Phys.* **28** 747
— 1980 *Ann. Phys., Lpz* **37** 406
Théberge S, Thomas A W and Miller G A 1980 *Phys. Rev.* D **22** 2838
— 1981 *Phys. Rev.* D **23** 2106
Thewles J T 1956 *Brit. J. Appl. Phys.* **7** 345
Thewles J T and Derbyshire R T B 1956 *Rep. M/TN37* (Harwell: AERE)
Thibault C, Klapisch R, Rigaud C, Poskanger A M, Prieels R, Lessard L and Reisdorf W 1975 *Phys. Rev.* C **12** 644
t'Hooft G 1971 *Nucl. Phys.* B **35** 167
— 1974 *Nucl. Phys.* B **79** 276
— 1976 *Phys. Rev. Lett.* **37** 8
Thomas A W, Théberge S and Miller G A 1981 *Phys. Rev.* D **24** 216
Thomas L H 1926 *Nature* **117** 514
— 1927 *Phil. Mag.* **3** 1
Thomas R G 1955 *Phys. Rev.* **97** 224
Thompson S G, Ghiorso A and Seaborg G T 1950 *Phys. Rev.* **80** 781, 79
Thompson W B 1957 *Proc. Phys. Soc.* B **70** 1
Thomson G P and Reid A 1927 *Nature* **119** 890
Thomson J J 1897 *Phil. Mag.* **44** 293
— 1904 *Phil. Mag.* **7** 237
— 1906 *Phil. Mag.* **11** 769
Thonemann P C 1956 *Nuclear Power* **1** 169
Thresher J J, Voss R G P and Wilson R W 1955 *Proc. R. Soc.* A **229** 492
Tiomno J and Wheeler J A 1949 *Rev. Mod. Phys.* **21** 144
Tolhoek H A and Cox J A M 1953 *Physica* **19** 101
Tolman R C 1939 *Phys. Rev.* **55** 364
— 1939a *Astrophys. J.* **90** 541, 568
Tomita A and Chiao R Y 1986 *Phys. Rev. Lett.* **57** 931
Towner I S 1992 *Nucl. Phys.* A **540** 478
Trautmann N and Hermann G 1967 *J. Radioanal. Chem.* **32** 53
Trebukhovskii Yu V, Vladimirskii V V, Grigor'ev V K and Ergakov V A 1959 *Sov. Phys.–JETP* **9** 931
Trimble V 1975 *Rev. Mod. Phys.* **47** 8877
— 1982 *Rev. Mod. Phys.* **54** 1183
— 1983 *Rev. Mod. Phys.* **55** 511
Trower W P and Zovko N 1982 *Phys. Rev.* D **25** 3088
Trubnikov S V 1981 *Sov. Phys.–JETP Lett.* **34** 134
Truran J W and Iben I 1977 *Astrophys. J.* **216** 797
Turchin V F 1965 *Slov Neutrons* (Jerusalem: Sivan Press)
Twin P J 1985 *Phys. Rev. Lett.* **55** 1380
Uffink J and Hilgevoord J 1988 *Physica* B&C **151** 309

Uhlenbeck G E and Goudsmidt G 1925 *Naturwiss.* **13** 953
— 1926 *Nature* **117** 264
Urey H W, Brickwedde F C and Murphy G M 1932 *Phys. Rev.* **40** 1
Utsuro M, Yoshiki H and Steyerl A 1980 *J. At. Ener. Soc. Japan* **22** 365
Van den Broek A 1913 *Phys. Z.* **14** 32
Van Hove L 1954 *Phys. Rev.* **95** 249
— 1958 *Physica* **24** 404
Van Kranendonk J and Van Vleck J H 1958 *Rev. Mod. Phys.* **30** 1
Vanhoy J R, Bilpuch E G, Shriner J F and Mitchell G E 1988 *Z. Phys.* A **331** 1
Verde M and Wick G C 1947 *Phys. Rev.* **71** 852
Verdet E 1869 *Lecons d'Optique Physique* (Paris: L'Imprimerie Imperial) vol 1 106
Vesman E A 1967 *Sov. Phys.–JETP Lett.* **5** 91
Vesna V A, Egorev A I, Kolomenskii E A, Lobashev V M, Perozhkov A N, Smotritsky L M and Titov N A 1981 *Nucl. Phys.* A **352** 181
Vesna V A, Kolomenskii E A, Lobashev V M, Nazarenko V A, Perozhkov A N, Smotritsky L M, Sobolev Yu V and Titov N A 1982 *Sov. Phys.–JETP Lett.* **36** 209
Villard P 1900 *Cont. Rend.* **130** 1178
Villars F 1947 *Phys. Rev.* **72** 256
Vinh Mau R 1983 *Proc. Int. Conf. on Nucl. Phys., Florence* (Bologna: Tipographica Compositare) p 61
Vinitskii S I, Ponomarev L I, Puzynin I V, Puznina T P, Somov L N and Faifman M P 1979 *Sov. Phys.–JETP* **47** 447
Vladimirskii V V 1961 *Sov. Phys.–JETP* **12** 740
Vladimirskii V V, Grigor'ev V K, Zharkov D P and Trebukhovskii Yu V 1961 *Bull. USSR, Acad. Sci. Phys. Ser. 1128*
Vogt E 1958 *Phys. Rev.* **112** 203
— 1960 *Phys. Rev.* **118** 724
Von Egidy T (ed) 1978 *Fundamental Physics with Reactor Neutrons and Neutrinos* (Bristol: Institute of Physics) p 162
Walker T P, Steigman G, Schramm D N, Olive K A and Kang H S 1991 *Astrophys. J.* **376** 51
Walt M and Barachall H H 1954 *Phys. Rev.* **93** 1062
Walter R L 1971 *Polarization Phenomena in Nuclear Reactions* ed H H Barschall and W Haeberli (Madison: University of Wisconsin Press) p 317
Wapstra A H and Nijgh G J 1955 *Physica* **21** 796
Waschkowski W and Koester L 1976 *Z. Naturf.* **319** 1115
Watson K M 1953 *Phys. Rev.* **89** 575
Webb F J 1963 *J. Nucl. Energy* **17** 187
Webster H C 1932 *Proc. R. Soc.* A **136** 428
Weigmann H W 1968 *Z. Phys.* **214** 7
Weinberg S 1958 *Phys. Rev.* **112** 1375
— 1967 *Phys. Rev. Lett.* **19** 1264
— 1975 *Phys. Rev.* D **11** 3583
— 1977 *The First Three Minutes* (London: Andre Deutsch)
— 1978 *Phys. Rev. Lett.* **40** 223
— 1980 *Phys. Rev.* D **22** 1464
Weiner R and Weise W 1985 *Phys. Lett.* **159B** 85
Weinfurter H and Badurek G 1990 *Phys. Rev. Lett.* **64** 1318

Weinfurter H, Badurek G and Rauch H 1989 *Physica* B **156&157** 528
Weinfurter H, Badurek G, Rauch H and Schwahn D 1988 *Z. Phys.* B **72** 195
Weinstock R 1944 *Phys. Rev.* **65** 1
Weiss P 1907 *J. Phys. Rad.* **4** 661
Weiss R J 1951 *Phys. Rev.* **83** 379
Weissberger W I 1965 *Phys. Rev. Lett.* **14** 1047
Weisskopf V F 1937 *Phys. Rev.* **52** 265
— 1951 *Science* **113** 101
Weisskopf V F and Ewing D H 1940 *Phys. Rev.* **57** 472
Weisskopf V H 1931 *Ann. Phys., Lpz* **9** 23
Weisskopf V H and Wigner E P 1930 *Z. Phys.* **63** 54
— 1930a *Z. Phys.* **65** 18
Weizsäcker C F v 1938 *Phys. Z.* **39** 633
— 1935 *Z. Phys.* **96** 431
— 1936 *Naturwiss* **24** 813
Wentzel G 1926 *Z. Phys.* **38** 518
Werner S A and Klein A G *Neutron Scattering* ed D L Price and K Sköld (New York: Academic Press) Ch. 4
Werner S A, Colella R, Overhauser A W and Eagen C F 1975 *Phys. Rev. Lett.* **35** 1053
West G B 1967 General Atomic Division, California Report GA-8310
Westcott C H 1955 *J. Nucl. Energy* **2** 59
— 1960 Effective Cross Section Values for Well-Moderated Thermal Neutron Spectra, AECL *Rept. No. 1101-CRRP-960*
Wheeler J A 1937 *Phys. Rev.* **52** 1083, 1107
— 1949 *Rev. Mod. Phys.* **21** 327
— 1955 *Niels Bohr and the Development of Physics* ed W Pauli (London: Pergamon) p 163
Wheeler J A 1978 *Mathematical Foundations of Quantum Mechanics* ed A R Marlow (New York: Academic) p 9
— 1989 *Ann. Rev. Nucl. Part. Sci.* **39** xiii
Wick G C, Wightman A S and Wigner E P 1952 *Phys.Rev.* **88** 101
Wien W 1898 *Berlin Phys. Gesell. Verh.* **17** 10
— 1898a *Wein Ann.* **65** 440
Wiener N 1930 *Acta Math. Stockh.* **55** 117
Wigner E P 1933 *Phys. Rev.* **43** 252
— 1933a *Z. Phys.* **83** 253
— 1937 *Phys. Rev.* **51** 106
— 1939 *Phys. Rev.* **56** 519
— 1946 *Phys. Rev.* **70** 606
— 1949 *Proc. Amer. Phil. Soc.* **93** 521
— 1957 *Proc. Conf. on Neutron Physics by Time-of-Flight. Gatlinburg* (1956) p 59 (Report ORNL-2309)
— 1958 *Ann. Math.* **67** 325
— 1963 *Am. J. Phys.* **31** 1
Wigner E P and Eisenbud L 1947 *Phys. Rev.* **72** 29
Wigner E P and Feenberg E 1942 *Rep. Phys. Soc. Prog. Phys.* **8** 274
Wilcsek F 1978 *Phys. Rev. Lett.* **40** 279
Wilets L and Jean M 1956 *Phys. Rev.* **102** 788

Wilkinson D H 1956 *Phil. Mag.* **1** 1031
— 1958 *Phys. Rev.* **109** 1603, 1610, 1614
— 1970 *Phys. Lett.* **31B** 447
— 1971 *Phys. Rev. Lett.* **27** 1018
— 1973 *Phys. Lett.* **48B** 169
— 1977 *Phys. Lett.* **66B** 105
— 1981 *Prog. Part. Nucl. Phys.* **6** 325
— 1982 *Nucl. Phys.* A **377** 474
— 1990 *Nucl. Phys.* A **511** 301
— 1991 *Nucl. Phys.* A **526** 131
— 1991 *Proc. Int. Conf. on Spin and Isospin in Nuclear Interactions, Telluride 1991* (New York: Plenum) p 361
Wilson K G 1974 *Phys. Rev.* D **10** 2445
Wing T and Fong P 1964 *Phys. Rev.* **136B** 923
Wolf E 1955 *Proc. R. Soc.* A **230** 246
Wolfenstein L 1951 *Phys. Rev.* **82** 690
— 1979 *Phys. Rev.* D **17** 2369
Wollan E O and Shull C G 1948 *Phys. Rev.* **73** 830
Woods R D and Saxon D S 1954 *Phys. Rev.* **95** 577
Wu C S 1964 *Rev. Mod. Phys.* **36** 618
Wu C S, Ambler E, Hayward R W, Hoppes D D and Hudson R 1957 *Phys. Rev.* **105** 1413
Wu C S, Lee Y K and Mo L 1977 *Phys. Rev. Lett.* **39** 72
Yamaguchi T, Tsushima K, Kohyama Y and Kubodera K 1989 *Nucl. Phys.* A **500** 429
Yang C N and Mills R 1954 *Phys. Rev.* **96** 191
Yang J, Turner M, Steigman G and Olive K 1984 *Astrophys. J.* **281** 493
Yennie D R, Ravenhall D G and Wilson R N 1954 *Phys. Rev.* **95** 500
Yoshiki H, Sakai K, Ogura M, Masuda Y, Nakajima T, Takayama T, Tanaka S and Yamaguchi A 1992 *Phys. Rev. Lett.* **68** 1323
Yoshimori A 1959 *J. Phys. Soc. Japan* **14** 807
Young T 1802 *Phil. Trans. R. Soc.* **12** 387
Yuan L C L 1949 *Phys. Rev.* **76** 1267
Yukawa H 1935 *Proc. Phys-Math. Soc. Japan* **17** 48
Zeilinger A 1981 *Nature* **294** 544
— 1986 *Physica* **137B** 235
Zeilinger A, Gähler R, Shull C G and Treimer W 1981 *Symp. on Neutron Scattering: Argonne National Laboratory AIP Conf. Proc.* No 89 p 93
Zeilinger A, Gähler R, Shull C G, Treimer W and Mampe W 1988 *Rev. Mod. Phys.* **60** 1067
Zeldes N 1956 *Nucl. Phys.* **2** 1
Zel'dovich Ya B 1954 *Dokl. Akad. Nauk SSSR* **95** 493
— 1959 *Sov. Phys.–JETP* **9** 1389
Zel'dovich Ya B and Sakharov A D 1957 *Zh. Exp. Teor. Fiz.* **32** 947
Zhu X *et al* 1992 *Phys. Rev.* C **46** 768
Ziebeck K R A and Brown P J 1980 *J. Phys. F: Met. Phys.* **10** 2015
Ziegler J F, Zabel J H, Cuomo J J, Bruzic V A, Gargill G S and O'Sullivan E J 1989 *Phys. Rev. Lett.* **62** 2929
Zinn W H 1947 *Phys. Rev.* **71** 752
Zwanziger D 1965 *Phys. Rev.* B **137** 647
Zweig G 1964 *CERN Rep. 8419TH* **401 & 402**
Zwerner F 1983 *Phys. Lett.* **132B** 103

Index

Physics

THEORETICAL NUCLEAR PHYSICS, John M. Blatt and Victor F. Weisskopf. An uncommonly clear and cogent investigation and correlation of key aspects of theoretical nuclear physics by leading experts: the nucleus, nuclear forces, nuclear spectroscopy, two-, three- and four-body problems, nuclear reactions, beta-decay and nuclear shell structure. 896pp. 5 3/8 x 8 1/2. 0-486-66827-4

QUANTUM THEORY, David Bohm. This advanced undergraduate-level text presents the quantum theory in terms of qualitative and imaginative concepts, followed by specific applications worked out in mathematical detail. 655pp. 5 3/8 x 8 1/2. 0-486-65969-0

ATOMIC PHYSICS AND HUMAN KNOWLEDGE, Niels Bohr. Articles and speeches by the Nobel Prize–winning physicist, dating from 1934 to 1958, offer philosophical explorations of the relevance of atomic physics to many areas of human endeavor. 1961 edition. 112pp. 5 3/8 x 8 1/2. 0-486-47928-5

COSMOLOGY, Hermann Bondi. A co-developer of the steady-state theory explores his conception of the expanding universe. This historic book was among the first to present cosmology as a separate branch of physics. 1961 edition. 192pp. 5 3/8 x 8 1/2. 0-486-47483-6

LECTURES ON QUANTUM MECHANICS, Paul A. M. Dirac. Four concise, brilliant lectures on mathematical methods in quantum mechanics from Nobel Prize-winning quantum pioneer build on idea of visualizing quantum theory through the use of classical mechanics. 96pp. 5 3/8 x 8 1/2. 0-486-41713-1

THE PRINCIPLE OF RELATIVITY, Albert Einstein and Frances A. Davis. Eleven papers that forged the general and special theories of relativity include seven papers by Einstein, two by Lorentz, and one each by Minkowski and Weyl. 1923 edition. 240pp. 5 3/8 x 8 1/2. 0-486-60081-5

PHYSICS OF WAVES, William C. Elmore and Mark A. Heald. Ideal as a classroom text or for individual study, this unique one-volume overview of classical wave theory covers wave phenomena of acoustics, optics, electromagnetic radiations, and more. 477pp. 5 3/8 x 8 1/2. 0-486-64926-1

THERMODYNAMICS, Enrico Fermi. In this classic of modern science, the Nobel Laureate presents a clear treatment of systems, the First and Second Laws of Thermodynamics, entropy, thermodynamic potentials, and much more. Calculus required. 160pp. 5 3/8 x 8 1/2. 0-486-60361-X

QUANTUM THEORY OF MANY-PARTICLE SYSTEMS, Alexander L. Fetter and John Dirk Walecka. Self-contained treatment of nonrelativistic many-particle systems discusses both formalism and applications in terms of ground-state (zero-temperature) formalism, finite-temperature formalism, canonical transformations, and applications to physical systems. 1971 edition. 640pp. 5 3/8 x 8 1/2. 0-486-42827-3

QUANTUM MECHANICS AND PATH INTEGRALS: Emended Edition, Richard P. Feynman and Albert R. Hibbs. Emended by Daniel F. Styer. The Nobel Prize–winning physicist presents unique insights into his theory and its applications. Feynman starts with fundamentals and advances to the perturbation method, quantum electrodynamics, and statistical mechanics. 1965 edition, emended in 2005. 384pp. 6 1/8 x 9 1/4. 0-486-47722-3

Physics

INTRODUCTION TO MODERN OPTICS, Grant R. Fowles. A complete basic undergraduate course in modern optics for students in physics, technology, and engineering. The first half deals with classical physical optics; the second, quantum nature of light. Solutions. 336pp. 5 3/8 x 8 1/2. 0-486-65957-7

THE QUANTUM THEORY OF RADIATION: Third Edition, W. Heitler. The first comprehensive treatment of quantum physics in any language, this classic introduction to basic theory remains highly recommended and widely used, both as a text and as a reference. 1954 edition. 464pp. 5 3/8 x 8 1/2. 0-486-64558-4

QUANTUM FIELD THEORY, Claude Itzykson and Jean-Bernard Zuber. This comprehensive text begins with the standard quantization of electrodynamics and perturbative renormalization, advancing to functional methods, relativistic bound states, broken symmetries, nonabelian gauge fields, and asymptotic behavior. 1980 edition. 752pp. 6 1/2 x 9 1/4. 0-486-44568-2

FOUNDATIONS OF POTENTIAL THERY, Oliver D. Kellogg. Introduction to fundamentals of potential functions covers the force of gravity, fields of force, potentials, harmonic functions, electric images and Green's function, sequences of harmonic functions, fundamental existence theorems, and much more. 400pp. 5 3/8 x 8 1/2. 0-486-60144-7

FUNDAMENTALS OF MATHEMATICAL PHYSICS, Edgar A. Kraut. Indispensable for students of modern physics, this text provides the necessary background in mathematics to study the concepts of electromagnetic theory and quantum mechanics. 1967 edition. 480pp. 6 1/2 x 9 1/4. 0-486-45809-1

GEOMETRY AND LIGHT: The Science of Invisibility, Ulf Leonhardt and Thomas Philbin. Suitable for advanced undergraduate and graduate students of engineering, physics, and mathematics and scientific researchers of all types, this is the first authoritative text on invisibility and the science behind it. More than 100 full-color illustrations, plus exercises with solutions. 2010 edition. 288pp. 7 x 9 1/4. 0-486-47693-6

QUANTUM MECHANICS: New Approaches to Selected Topics, Harry J. Lipkin. Acclaimed as "excellent" (*Nature*) and "very original and refreshing" (*Physics Today*), these studies examine the Mössbauer effect, many-body quantum mechanics, scattering theory, Feynman diagrams, and relativistic quantum mechanics. 1973 edition. 480pp. 5 3/8 x 8 1/2. 0-486-45893-8

THEORY OF HEAT, James Clerk Maxwell. This classic sets forth the fundamentals of thermodynamics and kinetic theory simply enough to be understood by beginners, yet with enough subtlety to appeal to more advanced readers, too. 352pp. 5 3/8 x 8 1/2. 0-486-41735-2

QUANTUM MECHANICS, Albert Messiah. Subjects include formalism and its interpretation, analysis of simple systems, symmetries and invariance, methods of approximation, elements of relativistic quantum mechanics, much more. "Strongly recommended." – *American Journal of Physics.* 1152pp. 5 3/8 x 8 1/2. 0-486-40924-4

RELATIVISTIC QUANTUM FIELDS, Charles Nash. This graduate-level text contains techniques for performing calculations in quantum field theory. It focuses chiefly on the dimensional method and the renormalization group methods. Additional topics include functional integration and differentiation. 1978 edition. 240pp. 5 3/8 x 8 1/2. 0-486-47752-5

Physics

MATHEMATICAL TOOLS FOR PHYSICS, James Nearing. Encouraging students' development of intuition, this original work begins with a review of basic mathematics and advances to infinite series, complex algebra, differential equations, Fourier series, and more. 2010 edition. 496pp. 6 1/8 x 9 1/4. 0-486-48212-X

TREATISE ON THERMODYNAMICS, Max Planck. Great classic, still one of the best introductions to thermodynamics. Fundamentals, first and second principles of thermodynamics, applications to special states of equilibrium, more. Numerous worked examples. 1917 edition. 297pp. 5 3/8 x 8. 0-486-66371-X

AN INTRODUCTION TO RELATIVISTIC QUANTUM FIELD THEORY, Silvan S. Schweber. Complete, systematic, and self-contained, this text introduces modern quantum field theory. "Combines thorough knowledge with a high degree of didactic ability and a delightful style." – *Mathematical Reviews.* 1961 edition. 928pp. 5 3/8 x 8 1/2. 0-486-44228-4

THE ELECTROMAGNETIC FIELD, Albert Shadowitz. Comprehensive undergraduate text covers basics of electric and magnetic fields, building up to electromagnetic theory. Related topics include relativity theory. Over 900 problems, some with solutions. 1975 edition. 768pp. 5 5/8 x 8 1/4. 0-486-65660-8

THE PRINCIPLES OF STATISTICAL MECHANICS, Richard C. Tolman. Definitive treatise offers a concise exposition of classical statistical mechanics and a thorough elucidation of quantum statistical mechanics, plus applications of statistical mechanics to thermodynamic behavior. 1930 edition. 704pp. 5 5/8 x 8 1/4. 0-486-63896-0

INTRODUCTION TO THE PHYSICS OF FLUIDS AND SOLIDS, James S. Trefil. This interesting, informative survey by a well-known science author ranges from classical physics and geophysical topics, from the rings of Saturn and the rotation of the galaxy to underground nuclear tests. 1975 edition. 320pp. 5 3/8 x 8 1/2. 0-486-47437-2

STATISTICAL PHYSICS, Gregory H. Wannier. Classic text combines thermodynamics, statistical mechanics, and kinetic theory in one unified presentation. Topics include equilibrium statistics of special systems, kinetic theory, transport coefficients, and fluctuations. Problems with solutions. 1966 edition. 532pp. 5 3/8 x 8 1/2. 0-486-65401-X

SPACE, TIME, MATTER, Hermann Weyl. Excellent introduction probes deeply into Euclidean space, Riemann's space, Einstein's general relativity, gravitational waves and energy, and laws of conservation. "A classic of physics." – *British Journal for Philosophy and Science.* 330pp. 5 3/8 x 8 1/2. 0-486-60267-2

RANDOM VIBRATIONS: Theory and Practice, Paul H. Wirsching, Thomas L. Paez and Keith Ortiz. Comprehensive text and reference covers topics in probability, statistics, and random processes, plus methods for analyzing and controlling random vibrations. Suitable for graduate students and mechanical, structural, and aerospace engineers. 1995 edition. 464pp. 5 3/8 x 8 1/2. 0-486-45015-5

PHYSICS OF SHOCK WAVES AND HIGH-TEMPERATURE HYDRO DYNAMIC PHENOMENA, Ya B. Zel'dovich and Yu P. Raizer. Physical, chemical processes in gases at high temperatures are focus of outstanding text, which combines material from gas dynamics, shock-wave theory, thermodynamics and statistical physics, other fields. 284 illustrations. 1966–1967 edition. 944pp. 6 1/8 x 9 1/4. 0-486-42002-7

Engineering

FUNDAMENTALS OF ASTRODYNAMICS, Roger R. Bate, Donald D. Mueller, and Jerry E. White. Teaching text developed by U.S. Air Force Academy develops the basic two-body and n-body equations of motion; orbit determination; classical orbital elements, coordinate transformations; differential correction; more. 1971 edition. 455pp. 5 3/8 x 8 1/2. 0-486-60061-0

INTRODUCTION TO CONTINUUM MECHANICS FOR ENGINEERS: Revised Edition, Ray M. Bowen. This self-contained text introduces classical continuum models within a modern framework. Its numerous exercises illustrate the governing principles, linearizations, and other approximations that constitute classical continuum models. 2007 edition. 320pp. 6 1/8 x 9 1/4. 0-486-47460-7

ENGINEERING MECHANICS FOR STRUCTURES, Louis L. Bucciarelli. This text explores the mechanics of solids and statics as well as the strength of materials and elasticity theory. Its many design exercises encourage creative initiative and systems thinking. 2009 edition. 320pp. 6 1/8 x 9 1/4. 0-486-46855-0

FEEDBACK CONTROL THEORY, John C. Doyle, Bruce A. Francis and Allen R. Tannenbaum. This excellent introduction to feedback control system design offers a theoretical approach that captures the essential issues and can be applied to a wide range of practical problems. 1992 edition. 224pp. 6 1/2 x 9 1/4. 0-486-46933-6

THE FORCES OF MATTER, Michael Faraday. These lectures by a famous inventor offer an easy-to-understand introduction to the interactions of the universe's physical forces. Six essays explore gravitation, cohesion, chemical affinity, heat, magnetism, and electricity. 1993 edition. 96pp. 5 3/8 x 8 1/2. 0-486-47482-8

DYNAMICS, Lawrence E. Goodman and William H. Warner. Beginning engineering text introduces calculus of vectors, particle motion, dynamics of particle systems and plane rigid bodies, technical applications in plane motions, and more. Exercises and answers in every chapter. 619pp. 5 3/8 x 8 1/2. 0-486-42006-X

ADAPTIVE FILTERING PREDICTION AND CONTROL, Graham C. Goodwin and Kwai Sang Sin. This unified survey focuses on linear discrete-time systems and explores natural extensions to nonlinear systems. It emphasizes discrete-time systems, summarizing theoretical and practical aspects of a large class of adaptive algorithms. 1984 edition. 560pp. 6 1/2 x 9 1/4. 0-486-46932-8

INDUCTANCE CALCULATIONS, Frederick W. Grover. This authoritative reference enables the design of virtually every type of inductor. It features a single simple formula for each type of inductor, together with tables containing essential numerical factors. 1946 edition. 304pp. 5 3/8 x 8 1/2. 0-486-47440-2

THERMODYNAMICS: Foundations and Applications, Elias P. Gyftopoulos and Gian Paolo Beretta. Designed by two MIT professors, this authoritative text discusses basic concepts and applications in detail, emphasizing generality, definitions, and logical consistency. More than 300 solved problems cover realistic energy systems and processes. 800pp. 6 1/8 x 9 1/4. 0-486-43932-1

THE FINITE ELEMENT METHOD: Linear Static and Dynamic Finite Element Analysis, Thomas J. R. Hughes. Text for students without in-depth mathematical training, this text includes a comprehensive presentation and analysis of algorithms of time-dependent phenomena plus beam, plate, and shell theories. Solution guide available upon request. 672pp. 6 1/2 x 9 1/4. 0-486-41181-8